Sex in Fungi

Molecular Determination and Evolutionary Implications

Sex in Fungi

Molecular Determination and Evolutionary Implications

Edited by

JOSEPH HEITMAN

Department of Molecular Genetics and Microbiology,
Duke University Medical Center, Durham, NC 27710

JAMES W. KRONSTAD

The Michael Smith Laboratories,
University of British Columbia, Vancouver, BC, V6T 2Z4, Canada

JOHN W. TAYLOR

Department of Plant and Microbial Biology,
University of California, Berkeley, Berkeley, CA 94720-3102

LORNA A. CASSELTON

Department of Plant Sciences,
University of Oxford, Oxford, OX1 3RB, United Kingdom

ASM
PRESS

Washington, D.C.

Photo credits:
Front cover: Young fruiting bodies of the homobasidiomycete *Coprinus cinereus* (*Coprinopsis cinerea*). *C. cinereus* has been used as a model fungus for studies of various aspects of sexual development because of its ease of cultivation and rapid development of the mushroom. (Photo by Kiyoshi Nakahori; courtesy Takashi Kamada)
Back cover:
[1] Sexual spores produced in four chains on a basidium of *Filobasidiella neoformans*, the sexual state of *Cryptococcus neoformans* (from R. A. Samson, J. A. Stalpers, and A. C. M. Weijman, *Antonie van Leeuwenhoek* **49**(4–5):447–456, 1983, with permission)
[2] Rosette of maturing asci of *Neurospora crassa* from an ascospore-color mutant *cys-3* × Normal (courtesy N. B. Raju)
[3] Scanning electron micrograph of a *Phycomyces blakesleeanus* zygospore, supported by the suspensor cells and decorated with spine-like appendages (©Walter Schröder)
[4] *Candida albicans* opaque **a/a** and α/α cells mating on skin (K. J. Daniels, S. Lachke, and D. R. Soll)

Ira Herskowitz: photo by Linda Russell, courtesy Anita Sil
John Raper, 1960s: courtesy Cardy Raper
Cardy Raper, August 2006: photo by Lorna Casselton

Copyright © 2007   ASM Press
American Society for Microbiology
1752 N Street, N.W.
Washington, DC 20036-2804

Library of Congress Cataloging-in-Publication Data

Sex in fungi : molecular determination and evolutionary implications / edited by Joseph Heitman ... [et al.].
    p. ; cm.
    Includes index.   1005453925
    ISBN-13: 978-1-55581-421-2
    ISBN-10: 1-55581-421-2
1. Fungi—Reproduction. I. Heitman, Joseph.
  [DNLM: 1. Fungi—physiology. 2. Evolution. 3. Fungi—genetics. 4. Genes, Mating Type, Fungal. 5. Reproduction, Asexual. QW 180 S5176 2007]

QK601.S49 2007
571.8′29—dc22

2007017650

All Rights Reserved
Printed in the United States of America

10 9 8 7 6 5 4 3 2 1

Address editorial correspondence to: ASM Press, 1752 N St., N.W., Washington, DC 20036-2904, U.S.A.

Send orders to: ASM Press, P.O. Box 605, Herndon, VA 20172, U.S.A.
Phone: 800-546-2416; 703-661-1593
Fax: 703-661-1501
Email: Books@asmusa.org
Online: estore.asm.org

Contents

Contributors   ix

Preface   xv

Dedication: Ira Herskowitz   xix

Appreciation: John and Cardy Raper   xxiii

I. GENERAL PRINCIPLES

1   The Evolution of *MAT*: the Ascomycetes   3
    GERALDINE BUTLER

2   Evolution of the Mating-Type Locus: the Basidiomycetes   19
    JAMES A. FRASER, YEN-PING HSUEH, KEISHA M. FINDLEY,
    AND JOSEPH HEITMAN

3   Mechanisms of Homothallism in Fungi and Transitions between
    Heterothallism and Homothallism   35
    XIAORONG LIN AND JOSEPH HEITMAN

4   Mating-Type Locus Control of Cell Identity   59
    BRYNNE C. STANTON AND CHRISTINA M. HULL

5   Rewiring Transcriptional Circuitry: Mating-Type Regulation in
    Saccharomyces cerevisiae and *Candida albicans* as a Model for
    Evolution   75
    ANNIE E. TSONG, BRIAN B. TUCH, AND ALEXANDER D. JOHNSON

II. ASCOMYCETES: FROM MODEL YEASTS TO PLANT AND HUMAN PATHOGENS

6   *Cochliobolus* and *Podospora*: Mechanisms of Sex Determination and the Evolution of Reproductive Lifestyle   93
    B. GILLIAN TURGEON AND ROBERT DEBUCHY

7   Sexual Reproduction and Significance of *MAT* in the Aspergilli   123
    PAUL S. DYER

8   The *mat* Genes of *Schizosaccharomyces pombe*: Expression, Homothallic Switch, and Silencing   143
    OLAF NIELSEN AND RICHARD EGEL

9   Decisions, Decisions: Donor Preference during Budding Yeast Mating-Type Switching   159
    JAMES E. HABER

10 MAT and Its Role in the Homothallic Ascomycete *Sordaria macrospora*   171
    STEFANIE PÖGGELER

11 Evolution of Silencing at the Mating-Type Loci in Hemiascomycetes   189
    LAURA N. RUSCHE AND MELEAH A. HICKMAN

12 The Evolutionary Implications of an Asexual Lifestyle Manifested by *Penicillium marneffei*   201
    MATTHEW C. FISHER

III. ASCOMYCETES: THE *CANDIDA MAT* LOCUS AND RELATED TOPICS

13 *MAT*, Mating, Switching, and Pathogenesis in *Candida albicans*, *Candida dubliniensis*, and *Candida glabrata*   215
    DAVID R. SOLL AND KARLA J. DANIELS

14 Evolution of *MAT* in the *Candida* Species Complex: Sex, Ploidy, and Complete Sexual Cycles in *C. lusitaniae*, *C. guilliermondii*, and *C. krusei*   235
    JENNIFER L. REEDY AND JOSEPH HEITMAN

15 Ascomycetes: the *Candida MAT* Locus: Comparing *MAT* in the Genomes of Hemiascomycetous Yeasts   247
    HÉLOÏSE MULLER, CHRISTOPHE HENNEQUIN, BERNARD DUJON, AND CÉCILE FAIRHEAD

IV. BASIDIOMYCETES: THE MUSHROOMS

16 Cloning the Mating-Type Genes of *Schizophyllum commune*: a Historical Perspective   267
    MARY M. STANKIS AND CHARLES A. SPECHT

17 The Origin of Multiple Mating Types in the Model Mushrooms *Coprinopsis cinerea* and *Schizophyllum commune*   283
    LORNA A. CASSELTON AND URSULA KÜES

18 Pheromones and Pheromone Receptors in *Schizophyllum commune* Mate Recognition: Retrospective of a Half-Century of Progress and a Look Ahead   301
THOMAS J. FOWLER AND LISA J. VAILLANCOURT

19 Analysis of Mating-Type Locus Organization and Synteny in Mushroom Fungi: Beyond Model Species   317
TIMOTHY Y. JAMES

20 Dikaryons, Diploids, and Evolution   333
JAMES B. ANDERSON AND LINDA M. KOHN

V.   BASIDIOMYCETES: PLANT AND ANIMAL PATHOGENIC YEASTS

21 History of the Mating Types in *Ustilago maydis*   351
FLORA BANUETT

22 Mating in the Smut Fungi: from *a* to *b* to the Downstream Cascades   377
REGINE KAHMANN AND JAN SCHIRAWSKI

23 Bipolar and Tetrapolar Mating Systems in the Ustilaginales   389
GUUS BAKKEREN AND JAMES W. KRONSTAD

VI.   ZYGOMYCETES, CHYTRIDIOMYCETES, AND OOMYCETES: THE FRONTIERS OF KNOWLEDGE

24 Sex in the Rest: Mysterious Mating in the Chytridiomycota and Zygomycota   407
ALEXANDER IDNURM, TIMOTHY Y. JAMES, AND RYTAS VILGALYS

25 How the Genome Is Organized in the Glomeromycota   419
TERESA E. PAWLOWSKA

26 Trisporic Acid and Mating in Zygomycetes   431
JOHANNES WÖSTEMEYER AND CHRISTINE SCHIMEK

27 Sexual Reproduction in Plant Pathogenic Oomycetes: Biology and Impact on Disease   445
HOWARD S. JUDELSON

VII.   THE IMPLICATIONS OF SEX

28 Origin, Evolution, and Extinction of Asexual Fungi: Experimental Tests Using *Cryptococcus neoformans*   461
JIANPING XU

29 Sex in Natural Populations of *Cryptococcus gattii*   477
DEE CARTER, NATHAN SAUL, LEONA CAMPBELL, TIEN BUI, AND MARK KROCKENBERGER

30 Why Bother with Sex? Answers from Experiments with Yeast and Other Organisms   489
MATTHEW R. GODDARD

31 Ploidy and the Sexual Yeast Genome in Theory, Nature, and Experiment   507
CLIFFORD ZEYL

32 Why Sex Is Good: on Fungi and Beyond   527
DUUR K. AANEN AND ROLF F. HOEKSTRA

Index   535

Contributors

DUUR K. AANEN
Laboratory of Genetics, Wageningen University and Research Center,
Arboretumlaan 4, 6703 BD Wageningen, The Netherlands

JAMES B. ANDERSON
Dept. of Ecology and Evolutionary Biology, University of Toronto, 3359
Mississauga Road North, Mississauga, Ontario L5L 1C6, Canada

GUUS BAKKEREN
Pacific Agri-Food Research Centre, Agriculture and Agri-Food Canada,
Summerland, BC, V0H 1Z0, Canada

FLORA BANUETT
Dept. of Biological Sciences, California State University, 1250 Bellflower Blvd.,
Long Beach, CA 90840

TIEN BUI
School of Molecular and Microbial Biosciences, University of Sydney, Sydney,
NSW 2006, Australia

GERALDINE BUTLER
UCD School of Biomolecular and Biomedical Research, Conway Institute,
University College Dublin, Belfield, Dublin 4, Ireland

LEONA CAMPBELL
School of Molecular and Microbial Biosciences, University of Sydney, Sydney,
NSW 2006, Australia [current address, Dept. of Biochemistry and Molecular
Biology, Saint Louis University School of Medicine, St. Louis, MO 63103]

DEE CARTER
School of Molecular and Microbial Biosciences, University of Sydney, Sydney,
NSW 2006, Australia

LORNA A. CASSELTON
Dept. of Plant Sciences, University of Oxford, Oxford, OX1 3RB, United Kingdom

KARLA J. DANIELS
Dept. of Biology, The University of Iowa, Iowa City, IA 52242

ROBERT DEBUCHY
CNRS, Institut de Génétique et Microbiologie, Bâtiment 400, UMR 8621, and Université Paris-Sud 11, Orsay, F-91405, France

BERNARD DUJON
Génétique Moléculaire des Levures (URA2171 CNRS and UFR927 Université P et M Curie), Institut Pasteur, 25 rue du Docteur Roux, F-75724 Paris Cedex 15, France

PAUL S. DYER
School of Biology, University of Nottingham, University Park, Nottingham NG7 2RD, United Kingdom

RICHARD EGEL
Institute of Molecular Biology and Physiology, University of Copenhagen, Copenhagen, Denmark

CECILE FAIRHEAD
Génétique Moléculaire des Levures (URA2171 CNRS and UFR927 Université P et M Curie), Institut Pasteur, 25 rue du Docteur Roux, F-75724 Paris Cedex 15, France

KEISHA M. FINDLEY
Dept. of Molecular Genetics and Microbiology, Duke University Medical Center, Durham, NC 27710

MATTHEW C. FISHER
Imperial College Faculty of Medicine, Dept. of Infectious Disease Epidemiology, St. Mary's Campus, Norfolk Place, London W2 1PG, United Kingdom

THOMAS J. FOWLER
Dept. of Biological Sciences, Southern Illinois University-Edwardsville, Edwardsville, IL 62026-1651

JAMES A. FRASER
School of Molecular and Microbial Sciences, University of Queensland, Brisbane, QLD 4072, Australia

MATTHEW R. GODDARD
School of Biological Sciences, University of Auckland, Private Bag 92019, Auckland 1142, New Zealand

JAMES E. HABER
Dept. of Biology, Brandeis University, MS029 Rosenstiel Center, Waltham, MA 02454-9110

JOSEPH HEITMAN
Dept. of Molecular Genetics and Microbiology, Duke University Medical Center, Durham, NC 27710

CHRISTOPHE HENNEQUIN
Faculté de Médecine P et M Curie, site St-Antoine, 27 rue Chaligny, F-75571 Paris Cedex 12, France

MELEAH A. HICKMAN
Institute for Genome Sciences and Policy, University Program in Genetics and
Genomics, Duke University, Durham, NC 27710

ROLF F. HOEKSTRA
Laboratory of Genetics, Wageningen University and Research Center,
Arboretumlaan 4, 6703 BD Wageningen, The Netherlands

YEN-PING HSUEH
Dept. of Molecular Genetics and Microbiology, Duke University Medical
Center, Durham, NC 27710

CHRISTINA M. HULL
Dept. of Biomolecular Chemistry and Dept. of Medical Microbiology &
Immunology, University of Wisconsin-Madison, School of Medicine and Public
Health, Madison, WI 53706

ALEXANDER IDNURM
Dept. of Molecular Genetics and Microbiology, Duke University Medical
Center, Durham, NC 27710

TIMOTHY Y. JAMES
Dept. of Evolutionary Biology, Uppsala University, 752 37 Uppsala, Sweden

ALEXANDER D. JOHNSON
Dept. of Biochemistry & Biophysics and Dept. of Microbiology & Immunology,
University of California San Francisco, San Francisco, CA 94143-2200

HOWARD S. JUDELSON
Dept. of Plant Pathology and Microbiology, University of California, Riverside,
CA 92521

REGINE KAHMANN
Max Planck Institute for Terrestrial Microbiology, 35043 Marburg, Germany

LINDA M. KOHN
Dept. of Ecology and Evolutionary Biology, University of Toronto, 3359
Mississauga Road North, Mississauga, Ontario L5L 1C6, Canada

MARK KROCKENBERGER
Faculty of Veterinary Science, University of Sydney, Sydney, NSW 2006,
Australia

JAMES W. KRONSTAD
The Michael Smith Laboratories, Dept. of Microbiology and Immunology,
University of British Columbia, Vancouver, BC, V6T 1Z4, Canada

URSULA KÜES
Institut für Forstbotanik, Georg-August-Universität Göttingen, Büsgenweg 2,
Göttingen D-37077, Germany

XIAORONG LIN
Dept. of Molecular Genetics and Microbiology, Duke University Medical
Center, Durham, NC 27710

HÉLOÏSE MULLER
Génétique Moléculaire des Levures (URA2171 CNRS and UFR927 Université
P et M Curie), Institut Pasteur, 25 rue du Docteur Roux, F-75724 Paris Cedex
15, France

OLAF NIELSEN
Institute of Molecular Biology and Physiology, University of Copenhagen, Copenhagen, Denmark

TERESA E. PAWLOWSKA
Dept. of Plant Pathology, Cornell University, 334 Plant Science Bldg., Ithaca, NY 14853-5904

STEFANIE PÖGGELER
Dept. of Genetics of Eukaryotic Microorganisms, Institute of Microbiology and Genetics, Georg-August University Göttingen, Grisebachstr. 8, 37077 Göttingen, Germany

JENNIFER L. REEDY
Dept. of Molecular Genetics and Microbiology, Duke University Medical Center, Durham, NC 27710

LAURA N. RUSCHE
Institute for Genome Sciences and Policy, Biochemistry Dept., Duke University, Durham, NC 27710

NATHAN SAUL
School of Molecular and Microbial Biosciences and Faculty of Veterinary Science, University of Sydney, Sydney, NSW 2006, Australia

CHRISTINE SCHIMEK
Dept. of General Microbiology and Microbial Genetics, Institute of Microbiology, Friedrich-Schiller-Universität Jena, Neugasse 24, 07743 Jena, Germany

JAN SCHIRAWSKI
Max Planck Institute for Terrestrial Microbiology, 35043 Marburg, Germany

DAVID R. SOLL
Dept. of Biology, The University of Iowa, Iowa City, IA 52242

CHARLES A. SPECHT
Dept. of Medicine, LRB-370D, Section of Infectious Diseases and Immunology, University of Massachusetts Medical School, 364 Plantation St., Worcester, MA 01605

MARY M. STANKIS
Dept. of Medicine, LRB-370D, Section of Infectious Diseases and Immunology, University of Massachusetts Medical School, 364 Plantation St., Worcester, MA 01605

BRYNNE C. STANTON
Dept. of Biomolecular Chemistry, University of Wisconsin-Madison, School of Medicine and Public Health, Madison, WI 53706

ANNIE E. TSONG
Dept. of Molecular and Cell Biology, University of California, Berkeley, Lawrence Berkeley National Labs, 1 Cyclotron Rd., Mailstop 84-355, Berkeley, CA 94720

BRIAN B. TUCH
Dept. of Biochemistry & Biophysics and Dept. of Microbiology & Immunology, University of California San Francisco, San Francisco, CA 94143-2200

B. GILLIAN TURGEON
Dept. of Plant Pathology, Cornell University, Ithaca, NY 14853

LISA J. VAILLANCOURT
Dept. of Plant Pathology, University of Kentucky, Lexington, KY 40546-0312

RYTAS VILGALYS
Dept. of Biology, Duke University, Durham, NC 27708

JOHANNES WÖSTEMEYER
Dept. of General Microbiology and Microbial Genetics, Institute of
Microbiology, Friedrich-Schiller-Universität Jena, Neugasse 24, 07743 Jena,
Germany

JIANPING XU
Dept. of Biology, McMaster University, 1280 Main St. West, Hamilton, Ontario,
L8S 4K1, Canada

CLIFFORD ZEYL
Dept. of Biology, Wake Forest University, P.O. Box 7325, Winston-Salem, NC
27109

Preface

Sexual reproduction is ubiquitous in nature, from organisms as diverse as fungi to plants and animals. As the engine that drives reassortment of genes to generate diversity, sex accelerates adaptation in the ever-changing environment and provides that more progeny avoid the relentless accumulation of deleterious mutations. In these ways, it plays a central role in the origin and success of species. As such, the molecular bases by which sexual identity and reproduction are defined and controlled have captured the interests of biologists for more than a century. These topics and interests have been pursued in a variety of organisms, with significant and wide-ranging contributions coming from explorations in the fungal kingdom. The insights that have come from investigating sexual reproduction in the major groups within the kingdom, including members of the Ascomycetes, Basidiomycetes, Chytridiomycetes, and Zygomycetes, are the subject of this book.

More than 40 years ago, John Raper published a thin text entitled *Genetics of Sexuality in Higher Fungi* (Ronald Press, New York, NY, 1966), which encapsulated much of what was known at that time on this topic in the basidiomycete fungi. While fascinating, the complex genetics of the system represented a puzzle and a challenge. How could it be that model mushroom species possessed literally thousands of mating types, or sexes, rather than the more pedestrian two sexes common in plants and animals and even many other fungi? The understanding came via molecular biology approaches whose advent and application were decades hence (1970s and 1980s). In parallel, advances in other fungal systems, notably the budding yeast *Saccharomyces cerevisiae*, provided further illuminating insights into the molecular details of cell-type specification, mating-type switching, pheromone perception and signaling, and cellular and nuclear fusion. This wealth of detailed molecular information on the wiring of a mating system provided a paradigm that guided research into the mechanisms of mating in all of the other fungi described in this book. With the advent of genomic approaches in the past 10 years, a window was opened on the entire genomes of many additional fungi, enabling kingdom-wide models of sex determination and sexual cycle evolution to be realized. The tremendous impact of comparative genomics on the analysis of mating is evident in many of the chapters in this book.

Here we have assembled chapters from a contingent of experts in the field to take stock of just how far knowledge of these fascinating biological systems and processes has progressed from 1966 to today. This includes chapters on the evolution and function of the mating-type locus, the specialized region of the genome that governs the establishment of cell type and orchestrates the sexual cycle in fungi. The species described in these chapters represent both the euascomycete and the hemiascomycete lineages of the prominent Ascomycetes phylum of fungi, as well as representatives of the Basidiomycetes, including the wood-rotting model fungus *Schizophyllum commune* that was the focus of John Raper's life work and his original text.

We have included representative chapters on both model and pathogenic fungi, given that many pathogenic fungi appear to have cryptic sexual cycles that may influence virulence or their evolution. Additionally, a section that looks forward to what we hope to learn in other fungal lineages is encapsulated in four chapters. Finally, the book concludes with a selection of chapters on the implications of sex, and studies of experimental evolution, to round out the discussions in a broader evolutionary context.

It is our hope and intent that the presentations throughout this book are not simply descriptions of the mating type loci, or a parts list of what fungi require for sexual reproduction. Rather, we hope this to be an exposé of the biological and molecular nature of sexual reproduction in an entire kingdom of life, one that is particularly amenable to genetic, molecular, and genomic analysis and that serves as a central paradigm to understand how sexual cycles function in, as well as drive, evolution. The biological principles that have emerged are profound and serve as general paradigms for how cell identity is established and maintained, how cells sense and respond to extracellular cues, the role of genetic rearrangements in generating changes in cell identity and fate, and how genomic regions governing sexual identity are organized and evolved.

We intend this new volume, *Sex in Fungi: Molecular Determination and Evolutionary Implications*, not only to encompass the current state of knowledge and to serve as a resource to guide the next several decades of study on these systems and organisms, but also to pay homage to those who made this effort possible. First, this text is dedicated in appreciation to John and Cardy Raper: to John for his insight in writing a text published in 1966 that is still cited to this day, and which foretold much of what was subsequently discovered on the molecular basis for transitions in mating behavior between out-crossing heterothallic systems and inbreeding homothallic organisms; and to Cardy, for carrying on with the molecular analysis of the *S. commune* mating type loci for several decades in her own laboratory at the University of Vermont after John. John and Cardy inspired a multitude of investigators, including many of the authors of the chapters in this book, and without them the field would certainly not be where it is today.

We also dedicate this text to Ira Herskowitz, who served as a champion for *Saccharomyces cerevisiae* as a premier model system and whose indefatigable efforts resulted in a molecular understanding of mating-type determination and mating-type switching. His elegant and powerful reductionist approaches, applying phage logic to a eukaryotic yeast, made possible the transition from complex genetics to textbook-clear models, establishing this yeast as a paradigm for all of biology. Again, Ira's intellectual leadership and his encouragement of other investigators, and of investigations with other fungi, helped inspire and drive the field to the current state documented in this book.

We are indebted to these three individuals for seeing sooner and farther than others, for sharing their vision with us, and for making this text both possible and worthwhile.

It has been our pleasure to serve as the co-editors, and also as authors, for this text and we invite you, the readers, to share your experiences with us. Moreover, we hope that this text inspires some of you to join us in this endeavor and to make it possible and necessary for the next text on this topic to be written.

In closing, we thank our families, our laboratory members, numerous readers, and our editor at ASM Press, Greg Payne, without whose efforts, forbearance, patience, and tolerance this book could not have been realized.

JOSEPH HEITMAN, Duke University
JAMES W. KRONSTAD, University of British Columbia
JOHN W. TAYLOR, University of California, Berkeley
LORNA A. CASSELTON, University of Oxford

Dedication:
Ira Herskowitz

It is a poignant and sobering realization that a book entitled *Sex in Fungi* will be published without a chapter from Ira Herskowitz. Although the book is poorer for lacking his chapter, we can celebrate Ira's many other contributions to this volume, contributions so deep and pervasive that they probably outweigh those of any other single scientist. In the numerous citations of his work, the many authors of this volume whose careers he helped guide, the influence of his writing and diagrammatic style, and finally, the scientific work itself, Ira's influence pervades this volume and continues to shape our research. Ira convinced us all that when he studied sex in fungi he was really studying fundamental problems in cell and molecular biology. Through his engaging personality, his accessible speaking style, and his many influential reviews and research articles, he provoked even the most narrowly focused biologist to think about fungal sex. Ira's work, and his influence on others, have thrust fungal mating into the pages of all major college textbooks in biology.

But it is the creative use of genetics as an exploratory tool that is Herskowitz's greatest legacy. Many scientists, from beginning students to accomplished professors, have tried to carry out "Ira-type" experiments—simple genetic studies designed to solve an outstanding biological problem in a new field. And some have succeeded, guided not only by the Herskowitz example, but often by the man himself. When it came to discussing a biological problem, Ira was generous with his time and seemingly tireless. Many biologists credit a discussion with him for pushing (or more often enticing) them to delve into new experimental realms or to develop a new way of conceptualizing old problems.

Ira's first papers were published in 1970, while a graduate student with Ethan Signer at MIT. He solved an important problem in gene regulation of bacteriophage λ: how a single regulatory protein turns on a whole battery of phage genes. Setting the style for much of his future work, Herskowitz attacked the problem using solely genetics, showing that the activator protein, called Q, must work at a single promoter, and the λ late genes must be transcribed as a very long operon. Thus began a lifelong affection for λ, with Ira publishing influential papers and reviews on the subject and even designing the poster that appeared in the second λ book from Cold Spring Harbor. (Based on the style of artwork, a colleague wrote that he was surprised to find that the poster didn't glow under black [UV] light). Although his lab formally stopped working on λ in the 1980s, Ira remained engrossed in the subject, continuing to teach and lecture on many aspects of it. Before his untimely death, he spoke of writing a small book tentatively titled *Two Proteins and a Decision*. It was to have told the story of how the bacteriophage λ cII and cIII gene products, synthesized when λ first enters a host cell, sense the cell's health and decide whether λ should replicate and lyse the cell or integrate into its genome. The title exemplifies Ira's presentation style: it reduces a complex biological problem to a single evocative phrase, rich in metaphor, with a bit of mystery and the promise of much to come.

After a short postdoctoral fellowship with David Botstein, which revealed unexpected evolutionary relationships among bacteriophage, Ira began his own laboratory at the University of Oregon in 1972. It was in Eugene that he began the work that would become most closely identified with him. Having taken the influential Yeast Genetics course at Cold Spring Harbor in 1971, Ira was one of a small group of scientists who realized that the powerful genetics in this simple eukaryote could be used to study important general problems in biology. Although this idea has now become a cliché, it was not so clear to most biologists working in the 1970s how much potential this approach held.

Ira and his colleagues in Eugene studied how the MAT locus determined the three yeast cell types—**a**, α, and **a**/α—and how some strains of yeast could switch their cell type. From these modest and seemingly specialized questions arose a great body of work with implications for nearly every branch of molecular and cell biology. Among other things, this early work showed that genes could be maintained in a "silent" state, that a small number of sequence-specific DNA binding proteins could control a large number of cellular characteristics, that programmed DNA rearrangements allowed cells to switch identities, and that yeast cells undergo regular asymmetric cell divisions. The work in these papers is beautifully described, the use of strong metaphors (the "wounding experiment," the "healing experiment," and the "cassette model") making the complex science understandable to all.

In 1981, Ira moved to the University of California-San Francisco (UCSF) and continued using yeast mating to explore and solve a variety of new problems in signal transduction, cell cycle control, polarized cell growth, meiosis, cell physiology, chromatin structure, gene expression, and RNA localization. Several years later, he turned his attentions to human genetics. The papers describing his first major effort in this new field appeared three weeks after his death.

Ira spent 22 years at UCSF, and it was here he supervised most of the 70-some graduate students and postdoctoral fellows who trained in his laboratory. His Genetics course, given to many generations of graduate students, reached its peak form at UCSF. Ira would enter the classroom with a pack of colored chalk and leave an hour later with every blackboard surface covered with boxes, arrows, and three-letter gene names. What happened in between is difficult to describe, but, as one colleague remarked, "it helped to have your seat belt fastened." They were wild rides indeed, rides whose memory continues to spark the imagination of those who witnessed them. Ira was also passionate about the free exchange of information and the value of collaborative efforts among scientists. He was instrumental in establishing these ideals at UCSF and elsewhere, largely through his own example. He convinced us all that a scientific community rich in ideas and ideals was worth fighting for, and he showed us all how to do it.

Ira's accomplishments, stunning by any standard, are based on a complex mixture of sensibilities. On the one hand, he worked for nearly 30 years on the mating behavior of a single-celled organism, steeped in the erudition and love of the organism that only a true specialist could muster. On the other hand, he was one of biology's great generalists, able to squeeze biological principle after biological principle from this tiny organism. And he could—and usually did—effortlessly relate his own work to pretty much any other subject in biology. Of course, the wide relevance of his work was no accident: Ira wisely chose his yeast projects to complement outstanding questions that often first arose in other areas of biology.

Ira Herskowitz died of pancreatic cancer on April 28, 2003, at the age of 56. He was one of the most influential geneticists of his generation, and one can only imagine how much thinner and less substantial this book—or indeed any account of biology in the past several decades—would have been without him.

Gerry Fink, Carol Gross, Karen Hopkin, Mary Maxon, Suzanne Noble, and Anita Sil provided valuable comments and editorial advice.

Appreciation: John and Cardy Raper

John Raper was born in 1911, the youngest of eight children who were brought up on a tobacco farm near Winston Salem in North Carolina. Their father was insistent that they all to go to university, and John in particular followed brother Ken to Chapel Hill. It was Ken who persuaded John to attend John Couch's legendary course on fungi. Couch was an inspired teacher, and John—or rather Red, as he preferred to be called—came under Couch's spell and fell in love with these wonderful organisms. Couch worked on water molds (bona fide fungi in those days) and was the first to show that these had sex by producing male and female organs. Red stayed on in Couch's laboratory after graduating and chose to study sex in *Achlya*. Red gained his Masters degree with Couch and then took *Achlya* to Harvard to continue on work that gained him his Ph.D. in Cap Western's laboratory. His work on *Achlya*, which continued until the 1950s, led him to discover the sex hormones (later shown to be steroids) that Couch had predicted to be the cause of sexual differentiation. He unraveled the remarkable sequence of events whereby the female sends out a signal to encourage any male in the vicinity to make his sex organs, and then how he signals back to induce her to make hers. This work established Red as an excellent scientist with a remarkable aptitude for ingenious experimentation.

After a postdoctoral period at Caltech, purifying the *Achlya* female hormone, and work as a radiobiologist on the Manhattan Project at Oak Ridge, Red was appointed to an Assistant Professorship at Chicago in 1946, where he met a promising young graduate student looking for an exciting research project. Red convinced

Carlene (Cardy) Allen not only that the sex life of *Achlya* would be exciting to study, but that he would be the best person to teach her how to be a critical and thoughtful scientist. She was not mistaken; he was a wonderful teacher who was never content to accept an experimental result unless he could approach it from a different angle and validate it. Cardy's influence on Red was equally important in this story. Cardy had taken Sewell Wright's courses on genetics and she persuaded Red to do the same. *Achlya*'s days were numbered, there was no chance of doing genetics with this organism, and the timing was right to find another sexual challenge. Fortuitously, Haig Papazian joined the laboratory in the late 1940s, and he convinced Red that Hans Kniep's work on sex in the mushroom *Schizophyllum commune* was worth revisiting. Here was an organism with which one could do genetics, it had multiple mating types, and it just had to be interesting! Red agreed, and so began the next phase of Red's research career, a genetic exploration of the complex mating type system of *S. commune*.

The Raper laboratory moved to Harvard in 1954 and established itself as the principal mecca for those interested in mushroom mating types. Cultures were collected from all over the world, and the genetic complexity of the *A* and *B* mating type loci of *Schizophyllum* was revealed by recombination studies and mutagenesis. This was a friendly and exciting place to work, and every visitor soon learned that the center for exchange of ideas was the tea room. Morning coffee, lunchtime sandwiches, and afternoon tea were a time for meeting and talking. Everyone was welcome to join Red in the tea room: the students, postdocs, technicians and assistants, and—most importantly for many years—the custodian of the glassware, whom Cardy still refers to as the resident psychiatrist because everyone turned to her in moments of personal crisis. The door was open and many visitors from other laboratories were welcome to join the debates. This was a perfect environment for promoting science and encouraging everyone to contribute. Of course, there were some rules, as was apparent to me when I spent the summer at Harvard in 1972. Red's chair was the one with arms and it was not a good idea to be found sitting in it! Of course you could have your say, but not for too long, or he would hum a tune to indicate that it was someone else's turn!

Red was convinced that the mating type genes of *Schizophyllum* were so interesting that they just had to be relevant to all eukaryotic organisms. How right he was! He died in 1974, before the molecular revolution that was to reveal the ubiquitous nature of the gene products that govern mating in mushrooms; the homeobox genes that regulate development in animals, flies, and many other organisms; and the peptide pheromones and their G-protein-coupled receptors that activate well-known intracellular signaling pathways. He was convinced that bipolar mushrooms were derived from tetrapolar forms, an idea that is substantiated by work described in this book. He had imaginative ideas about how the products of the mating type genes might regulate each other's activity, and though we now know that this is not at the protein level, it is true at the transcriptional level. He is remembered as a passionate scientist who taught his students to be rigorous in experimental design, but also to be independent and to have confidence in their own judgment and ability to do good science.

Red's passion for the mating type genes of *Schizophyllum* was shared by Cardy, and when the children were old enough, she returned to the laboratory and made her own contributions to the experimental work in progress. When Red died, it fell to Cardy to safeguard his legacy, to ensure that the culture collection was safe, that the science would continue, and that it would be handed on to future generations of scientists who would share their passion for this fungus. She succeeded in doing this, and as a result, she has become a highly respected (and much loved) scientist in her own right. She learned molecular biology techniques in Bill Timberlake's laboratory and took these to the University of Vermont at Burlington, where Red's last research student, Bob Ullrich, was working on a transformation system for *Schizophyllum*. Cardy funded her own laboratory and for some 20 years inspired young students and dedicated postdocs to work with her. Tom Fowler and Lisa Vaillancourt, in their chapter in this book, describe the wonderful molecular analysis of the *B* mating type genes of *Schizophyllum* that they carried out in Cardy's laboratory. Cardy continues to be as interested in the mushroom mating types as ever, and her enthusiasm and profound knowledge of *Schizophyllum,* derived from a research career spanning over 50 years, are formidable. Like Red, she can ask some penetrating questions, argue forcibly, and generate new and exciting ideas.

LORNA CASSELTON
August 2006

General Principles

I

Sex in Fungi: Molecular Determination
and Evolutionary Implications
Edited by Joseph Heitman et al.
© 2007 ASM Press, Washington, D.C.

Geraldine Butler

The Evolution of *MAT*:
The Ascomycetes

1

It is likely that fungi first appeared during the Precambrian period, approximately 1.5 billion years ago (32, 33). The Ascomycota is the largest fungal phylum, and diverged from the Basidiomycota from 968 to 1,208 million years ago (32, 33). The two major classes of Ascomycetes (Pezizomycotina and Saccharomycotina [Fig. 1.1]) contain some of the best-known and well-characterized yeast species. These range from filamentous yeasts that are pathogenic in animals (e.g., *Aspergillus fumigatus*) and plants (e.g., *Magnaporthe grisea*) to single-celled yeasts (e.g., *Saccharomyces cerevisiae*) important to biotechnology.

Mating in fungal species occurs through either heterothallic (self-sterile) or homothallic (self-fertile) mechanisms. In heterothallic species, mating takes place between two sexually compatible groups that are morphologically identical and are distinguished genetically by mating type. In homothallic species, each strain can mate with itself. Some fungi are pseudohomothallic—the strains are self-fertile, but the nuclei undergoing fusion are of different mating types (reviewed in references 13 and 42). Natural isolates of *S. cerevisiae* and related yeasts are an example of an unusual form of pseudohomothallism, as haploid strains can switch mating type, and nuclei of the opposite type then fuse.

Mating type in the Ascomycota is predominantly bipolar—determined by two possible DNA sequences at the mating-type locus (*MAT*). The molecular structure of *MAT* was first identified in *S. cerevisiae*, where it was shown that **a** and α isolates were distinguished by sequence information present at *MAT* (1). Analysis of the *MAT* loci of filamentous fungi led to the definition of an idiomorph, denoting sequences unrelated in structure but present at the same homologous locus (54). This term is now commonly used to describe alternative *MAT* structures in other fungi. The *MAT* idiomorphs encode transcription factors that regulate cell identity.

Our understanding of mating in ascomycetes has been greatly influenced by advances in genome sequencing, and to date (summer 2006) either complete or high-quality draft sequences are available or in progress for over 40 species (28). This chapter provides a summary of the structure and evolution of *MAT* in a range of species representing the major subphyla and sister groups (Fig. 1.1).

LOCATION OF THE *MAT* LOCUS IN ASCOMYCETES

Whereas the gene content and gene order at the *MAT* locus vary considerably across the ascomycetous yeasts, the position of *MAT* is surprisingly well conserved

Geraldine Butler, UCD School of Biomolecular and Biomedical Research, Conway Institute, University College Dublin, Belfield, Dublin 4, Ireland.

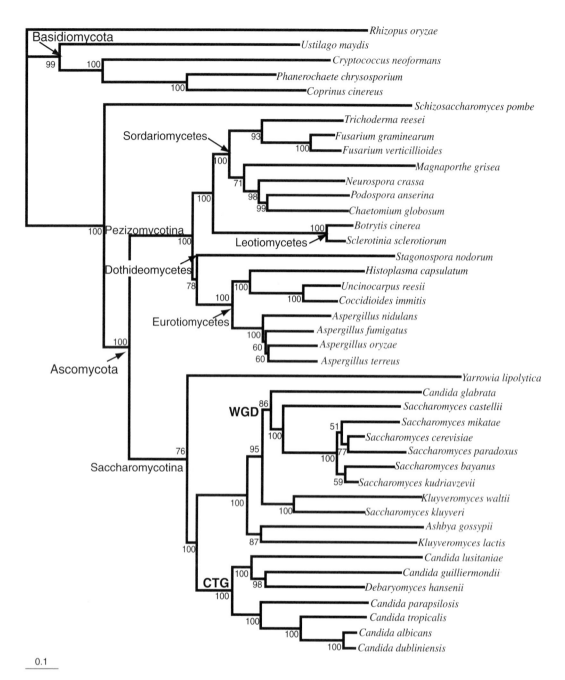

0.1

Figure 1.1 Phylogenetic relationships among the Ascomycota. A phylogenomic supertree was constructed using 4,805 gene families from fully sequenced genomes from 42 fungal species. The details will be published elsewhere (D. A. Fitzpatrick, unpublished data). *Rhizopus oryzae* was used as an outgroup. The divergence between the Basidiomycota and Ascomycota is shown, and the subphyla and classes discussed in the text are indicated. WGD; clade that has undergone a whole-genome duplication; CTG; clade in which CTG is translated as serine.

(Color Plate 1). In many of the Pezizomycotina (Color Plate 1), the *MAT* locus lies between homologs of *APN2* (encoding an abasic endonuclease/DNA lyase) and *SLA2* (encoding a protein that binds to cortical patch actin). This organization is seen in species from the classes Sordariomycetes (*Neurospora crassa* and *Gibberella zeae*), Eurotiomycetes (the soil fungi *Coccidioides immitis* and *Histoplasma capsulatum* [25]), and Lecanoromycetes (the lichen-forming ascomycete *Xanthoria polycarpa* [69]) (Color Plate 1). In the rice blast fungus *Magnaporthe grisea*, *APN2* is found on the left-hand side of *MAT*, but *SLA2* is at a different location in the genome. Within the genus *Aspergillus*, *A. nidulans* is homothallic and has a known sexual cycle, whereas *A. fumigatus* and *A. oryzae* are likely to be heterothallic, although they have not yet been shown to reproduce through sexual means (27, 62). The *A. nidulans* genome contains two unlinked genes associated with mating type, encoding proteins with either an HMG-domain or an alpha-box domain. *SLA2* is associated with the alpha-box gene, and homologs of *APN2* and *APC5* are located with the HMG gene (Color Plate 1) (27). Only one isolate of *A. oryzae* and one isolate of *A. fumigatus* have been sequenced at the genome level (27). In *A. oryzae* the alpha-box gene is flanked by *APN2* and *SLA2*, and in *A. fumigatus* the HMG gene is located beside orthologs of *APC5* and *APN2* (27). Sequencing of the *MAT* locus of other isolates of *A. fumigatus* revealed the existence of a second idiomorph (*MAT1-1*), in which the HMG gene is replaced with an alpha-box gene (62).

APN2 and SLA2 also surround the *MAT* locus in *Yarrowia lipolytica* and *Pichia angusta* (members of the subphylum Saccharomycotina, or the hemiascomycetes), and *SLA2* and its neighbor *SUI1* are found to the right of *MAT* in *Kluyveromyces lactis* and *Saccharomyces kluyveri* (Color Plate 1). The conservation across such a wide range of species suggests that this represents the ancestral configuration for all ascomycetes. However, among the group of yeasts that have undergone a whole-genome duplication event (WGD) (Color Plate 1) (40, 70, 83), the genes surrounding the *MAT* locus are very different. *SLA2* is no longer associated with the locus, and three different genes are found to the right of *MAT* in *Saccharomyces castellii*, *Candida glabrata*, and *S. cerevisiae* (Color Plate 1).

BUD5 is on the left side of all the post-WGD genomes shown. This region is found on chromosome III in *S. cerevisiae*, which is part of a duplicated region also represented on chromosome X (83). There are paralogs of some genes (*PHO87*/*YCR037C* and *PHO90*/*YJL198W*, *FEN1*/*YCR034W*, and *ELO1*/*YJL196C*) on both chromosomes. In *S. kluyveri* (which separated from the *Saccharomyces* lineage before the genome duplication event), the region to the left of the *MAT* locus contains several orthologs of genes from chromosome X in *S. cerevisiae*. *BUD5* is adjacent to *ECM25* (*YJL201W*), suggesting that there was a large deletion of at least 10 genes in the progenitor of the WGD clade. Both *K. lactis* and *Ashbya gossypii* (which also split from *Saccharomyces* before duplication) have orthologs of *S. cerevisiae* chromosome X genes adjacent to their *MAT* loci: *YJL207C* in *K. lactis* and *RCY1* (*YJL204C*) in *A.*

gossypii (9, 81). It is likely that both these genomes have undergone deletions, removing everything between *YJL207C* (or *YJL204C*) and *MAT*. It is interesting that *RCY1* (*YJL204C*) is the only *MAT* locus-linked gene shared with *Candida albicans*. One additional observation is that *DIC1* is immediately adjacent to the *MAT* locus in both *S. kluyveri* and *P. angusta*.

In summary, there has been obvious continuity in the location of the *MAT* locus in the ascomycetes, from lichen-forming yeasts to filamentous species and from plant to animal pathogens. This continuity is quite extraordinary when we consider that the functional *MAT* genes themselves have been replaced several times during evolution (discussed below). The association of *SLA2* with *MAT* in very distantly related species suggests that this gene may play some as yet unrecognized role in mating. The greatest variation in structure occurs among the *Saccharomyces* lineage, and also within the *Candida* clade. The variation in structure in these genomes may be associated with activity of the Ho endonuclease in the relatives of *Saccharomyces* and with changes in the structure of *MAT* in the *Candida* species.

EVOLUTION OF MATING-TYPE CASSETTES IN THE *SACCHAROMYCES* LINEAGE

Analysis of mating in *S. cerevisiae* provided a paradigm for our understanding of sexual pathways in ascomycetes, and indeed in many other fungi. The analysis of homothallic and heterothallic isolates led to the description of the mating loci, the discovery of the cassette-based system, and the identification and characterization of mating-type switching (31, 35, 36, 61). In *S. cerevisiae*, the mating type is determined by the information carried at the *MAT* locus on chromosome III (Fig. 1.2). The *MAT*α idiomorph encodes two proteins, α1 and α2, whereas the *MATa* idiomorph encodes a1. The α and a information is also found at two other loci near the telomeres of chromosome III—the *HML*α and *HMRa* cassettes. These loci are in regions of heterochromatin, and are silenced (not expressed). The sequences that determine mating type were delineated by comparing the *MATa* and *MATα* idiomorphs and the *HML* and *HMR* cassettes. The loci contain unique regions (Ya and Yα), surrounded by common sequences (W, X, Z1, and Z2). The promoters driving expression of α1, α2, and a1 lie within the Y region. *MAT*α and *HML*α are almost identical, whereas *HMRa* is considerably shorter than *MATa*. The α2 reading frame extends from Yα into the adjacent X region, which is shared with *MATa*. Because the copy of the 3′ end of α2 in the X region of *MATa* includes an ATG codon (corresponding to an internal methionine codon in α2), the resulting opening reading

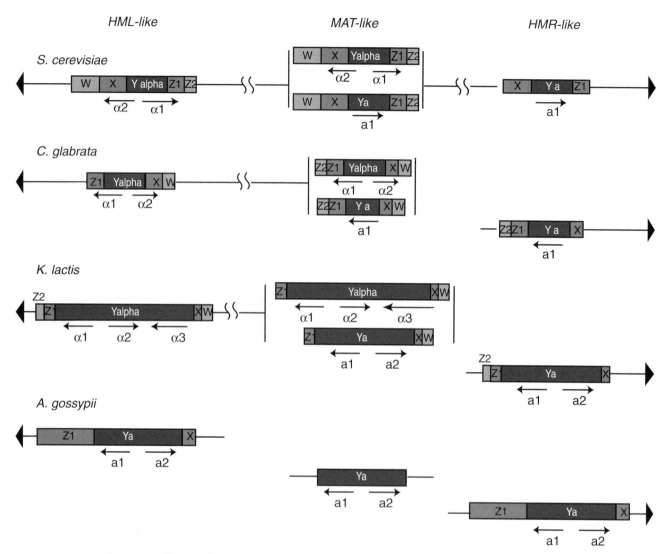

Figure 1.2 Organization of the *MAT* loci and silent cassettes in the *Saccharomyces* lineage. The figure is redrawn from reference 18. The two possible idiomorphs at the *MAT* locus are shown between horizontal lines for all species except *A. gossypii*, where only *MATa* was identified. The "Y" boxes encode **a** or alpha information and are specific to the idiomorphs. W, X, Y, and Z are shared sequence elements. It is not possible to determine the sizes of these regions in *A. gossypii*. Telomeres are indicated with arrowheads. The curly lines indicate when the cassettes are found on the same chromosome.

frame in *S. cerevisiae* is sometimes annotated as a putative gene ("*MATa2*"), but there is no evidence that it codes for a functional protein.

α1 (alpha-box motif), α2 (homeodomain motif), and **a**1 (homeodomain motif) proteins are transcriptional regulators that control the expression of pheromone and receptor genes and other factors involved in mating. In *MATα* haploid cells, α1 induces expression of α mating functions, and α2 represses **a** mating functions. In *MATa* haploid cells **a** mating functions are constitutively expressed. In diploid cells, the **a**1 and α2 proteins

combine to generate a repressor that switches off haploid-specific functions, including α1.

A haploid cell can switch mating type by replacing the information at the *MAT* locus with the sequence from either *HMLα* or *HMRa*. Switching occurs through unidirectional transposition, triggered by the action of the Ho endonuclease. Ho makes a staggered doublestrand break at a recognition sequence in the *MAT* locus that lies just right of the Y regions that are specific to the idiomorphs. The expression of *HO* is strongly regulated and occurs in G_1 phase in mother cells (i.e., cells that have divided by mitosis at least once)' (58). Transcription is prevented in diploid cells by binding of the a1/α2 repressor (53). In haploid cells, expression of *HO* occurs only in G$_1$ through the activity of a series of regulatory proteins targeted to the promoter (14, 57). Finally, transcription is prevented in daughter cells because of binding of the Ash1 repressor (5, 71).

A putative *MAT*-like locus (*MTL*) was first described in *C. glabrata* and its close relative *Kluyveromyces delphensis* by Wong et al. (84), and silent cassettes in *C. glabrata* were identified by Srikantha et al. (76). These were confirmed in the *C. glabrata* genome-sequencing project (20, 21). In addition, Butler et al. (9) sequenced a putative *HMRa* cassette from *K. delphensis*. The structure of the *C. glabrata* cassettes is very similar to that of *S. cerevisiae* cassettes, except that the *HML*-like and *MAT*-like loci are located on the same chromosome, and the *HMR* locus is elsewhere (21) (Fig. 1.2). The length of the common regions (W, X, Z1, and Z2) is also shorter. Mating-type switching has been described in a clinical isolate (7) and at a low level in environmental isolates (9). For two strains (CBS138 and RND13), switched products were identified by PCR and confirmed by sequencing (9). However, mating has never been observed, although the *C. glabrata* genome appears to contain all the requirements (9, 21). Only haploid cells have been identified, though both mating types exist in nature (7, 9, 76). In contrast, the *K. delphensis* strain CBS2170 is a *MATα/MATa* diploid (9). The inability of *C. glabrata* to mate may not be associated with the structure of the *MAT* locus. Fabre et al. (21) have shown that cells fail to respond to the presence of the mating pheromone α-factor and that there is a silencing defect, resulting in expression of both a and α genes. It is also possible that mating occurs at a low level in natural populations and remains to be discovered.

Prior to the advent of genome sequencing, the only other yeast in which similar silent and mating cassettes were identified was the sexual yeast *K. lactis* (2, 3, 34, 74). Switching occurs in a small proportion of cells in *K. lactis* and declines after several generations (34). Strains are usually haploid, and mating is inefficient. However, strains can spontaneously form diploids and it is possible to find spores of the opposite mating type in the asci (90). The structure of the mating-type loci was confirmed by sequencing the *MAT* locus from a *MATa* strain (9), and finally from the sequence of the entire genome (19). *K. lactis* has three mating cassettes, and similarly to *C. glabrata*, only the *HML*-like and *MAT*-like loci are found on the same chromosome (chromosome III) (Fig. 1.2) (21). *HMR* is located elsewhere, on

chromosome II. This is consistent with the early experiments of Herman and Roman (34), who showed that switching from *MATa* to *MTLα* occurs at a higher rate than the reciprocal switch. The *K. lactis* *MTLα* idiomorph encodes an additional gene of unknown function (α3), with little similarity to proteins in other fungal species. α3 is transcribed in an α-specific manner and is required for efficient mating, but it is expressed from both *MATα* and *HMLα* (2). Its biological role therefore remains to be elucidated.

A. gossypii is a filamentous yeast that has historically been used to generate large quantities of riboflavin (reviewed in reference 81). It is assumed to be an asexual haploid species—ascospores are generated, but they are uninucleate and haploid (81). The sequenced isolate contains three a-type cassettes, equivalent to *HML*, *MAT*, and *HMR* (18, 21, 81) (Fig. 1.2). Natural a,a,a and α,α,α isolates of *S. cerevisiae* are found at a low frequency (31), and at least one α,α,α isolate of *C. glabrata* has been described (76). It is therefore possible that other mating types of *A. gossypii* exist in nature and have yet to be identified. The overall structure of the cassettes in *A. gossypii* is similar to the cassette structures of the other yeasts in the *Saccharomyces* lineage, except that all three are found on different chromosomes. It is also impossible to determine the size and sequence of any X and Z1 regions present unless a *MATα* isolate is identified.

Silent mating cassettes cannot be identified in *C. albicans* (or related yeasts) or in *Y. lipolytica*. This suggests that they evolved in the *Saccharomyces* lineage following the split with the *Candida/Debaryomyces* clade (Fig. 1.1). Surprisingly, despite its phylogenetic position (Fig. 1.1) the genome of *S. kluyveri* also does not have any evidence of silent cassettes (12). It is possible that this species has lost the telomeric cassettes and become truly heterothallic.

Acquisition of the *HO* Endonuclease Gene
The *MAT* loci of *S. cerevisiae* (and its close relatives), *S. castellii*, *C. glabrata*, and *Zygosaccharomyces rouxii* contain recognition sites for the Ho endonuclease at the border between the Y and the Z regions, within the α1 coding sequence in Yα, and just after the a1 sequence in Ya (Color Plate 1 and Fig. 1.2) (9, 21). The consensus sequence CGCAAC was first described in *S. cerevisiae* and is completely conserved in *C. glabrata* *MTLa* and *K. delphensis* *MATa* and *MATα*. There is a minor alteration in *C. glabrata* *MTLα* (to CGCAGC), but this sequence is also recognized by the *S. cerevisiae* enzyme (60). These genomes also contain sequences encoding an Ho, i.e., homothallic endonuclease.

Ho belongs to a class of selfish mobile genetic elements, the homing endonuclease genes, or HEGs. HEGs typically encode a sequence-specific endonuclease that introduces a double-stranded break at its recognition site (reviewed in reference 8). The HEG open reading frame is usually found within its own recognition site, and chromosomes with alleles containing an HEG are therefore protected from cleavage. Cleavage elsewhere in the genome induces recombinational repair, and copying (or "homing") of the HEG at the new location. There are several families of HEG, named after motifs in the active site.

One HEG in *S. cerevisiae* is the *VDE* selfish genetic element that lies within the *VMA1/TFP1* gene, which encodes a vacuolar (H)-ATPase. Vma1 is translated as a single product, and the VDE element is excised in a protein-splicing reaction, leaving a functional ATPase. The VDE element contains a protein splicing motif and an endonuclease domain (with two LAGLIDADG motifs) that cuts at a 31-bp sequence within alleles of *VMA1* lacking VDE, resulting in copying of the VDE element (6, 29). VDE is widely distributed in the Saccharomycotina and probably moves by horizontal transfer (41).

The Ho endonuclease is most closely related to the VDE intein (39). It also encodes protein-splicing and endonuclease domains, but it contains an additional zinc finger domain at the C terminus required for mating-type switching (4, 63). Unlike other HEGs, Ho does not recognize other *HO* minus alleles but instead cuts at the sequence within the *MAT* loci. It has been proposed that Ho arose from a duplication of the *VMA1* intein that was "domesticated" and switched specificity (39). Keeling and Roger (39) also proposed that HO was integrated into an already existing inefficient mating switch system. This is supported by the analysis of Butler et al. (9) and Fabre et al. (21), who showed that silent mating cassettes are present in *K. lactis* and *A. gossypii*, but there is no evidence of the presence of Ho recognition sites within the *MAT* loci. Fabre et al. (21) have suggested that the remnant of an *HO* gene resides within the *K. lactis* genome. This sequence, however, is distantly related to the other *HO* genes and to the *VMA1* intein. The *K. lactis* sequence is most similar to a second HO-like sequence present in the *S. castellii* genome. Analyzing gene order also suggests that *HO* appeared between the *RIO2* and *SSB* genes in the immediate common ancestor of *S. cerevisiae*, *Z. rouxii*, *C. glabrata*, and *S. castellii* (9). It is therefore likely that the *K. lactis* gene and the second *S. castellii* gene evolved independently of *HO*, which was acquired late in the evolution of the hemiascomycetes (Fig. 1.3). The acquisition of *HO* also appears to correlate with rearrangements on the right-hand side of the *MAT* loci (Color Plate 1). This may be caused by an increased rate of chromosome rearrangement due to the activity of Ho.

EVOLUTION OF THE *MAT* LOCUS IN *CANDIDA* SPECIES

The term *Candida* was originally assigned to the so-called "imperfect" yeast species, those with no known sexual cycle. This ragbag definition covers a variety of species of diverse origins and provides little information regarding evolutionary relationships. For example, *C. glabrata* is more closely related to *S. cerevisiae* than it is to other *Candida* species, and *Debaryomyces hansenii* is a closer relative of *C. albicans* (Fig. 1.1). For the purposes of this discussion we define the *Candida* clade as the immediate relatives of *C. albicans*, including *C. dubliniensis*, *C. tropicalis*, *C. parapsilosis*, *C. lusitaniae*, *C. guilliermondii*, and *D. hansenii*. These share a relatively recent common ancestor (Fig. 1.1), and in all of these species the codon CUG is translated as serine rather than leucine (56, 77).

C. albicans was generally assumed to be completely asexual until analysis of the data from a genome-sequencing project led to the identification of a mating-type-like (*MTL*) locus (37, 78). Two idiomorphs were identified: *MTLa* and *MTLα*. Like *S. cerevisiae*, *MTLα* idiomorphs encode α1 and α2 proteins. *MTLa* contains genes encoding **a**1 and an additional HMG-box protein, **a**2 (Fig. 1.4). Both idiomorphs also contain alleles encoding three proteins, *PAP* [poly(A) polymerase], *OBP* (oxysterol-binding protein), and *PIK* (phosphatidylinositol kinase), with no known role in mating (Fig. 1.4) (37, 78). Unlike *S. cerevisiae*, haploid forms of *C. albicans* have never been identified, and isolates therefore contain two *MTL* idiomorphs. The vast majority are heterozygous **a**/α, and only a small percentage (3 to 7%) of natural isolates are homozygous for **a** or α (46, 49).

Mating between **a** and α cells was first shown using strains in which one *MTL* idiomorph had been deleted (38), or where loss and reduplication of one *MTL*-containing chromosome was induced by growing on sorbose (52). Mating appeared to be an extremely inefficient process, until an association was made with phenotypic switching. The switch between white and opaque forms was first shown to occur in certain *C. albicans* isolates by Slutsky et al. (75). The two phenotypes are easily distinguishable—opaque colonies are darker and flatter, and the cells are elongated with unusual pimple-like structures on the surface. In 2002, Miller and Johnson (55) demonstrated that only **a** or α

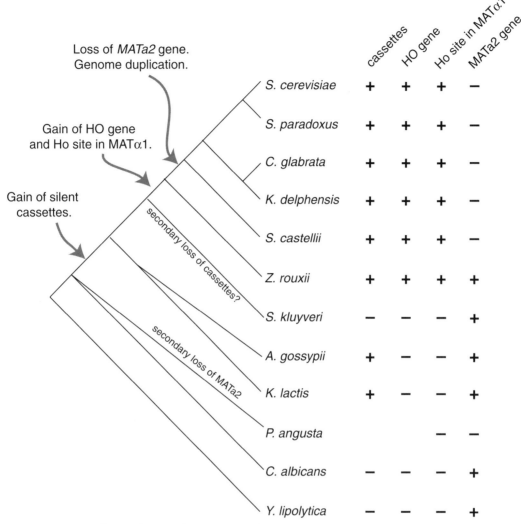

Figure 1.3 Evolution of MAT and HO in the Saccharomycotina. The phylogenetic tree is schematic and is based on reference 9 and Fig. 1.1. The figure is redrawn from reference 9, with additional information for *A. gossypii* from reference 15 and for *K. lactis* and *Y. lipolytica* from references 17 and 18. The *K. lactis* genome does contain a sequence with similarity to *HO*, but it is unlikely to be an ortholog. Blanks indicate missing data.

homozygotes undergo white-opaque switching and that switching is repressed by the a1/α2 heterodimer. They also showed that mating is most efficient in opaque cells.

The analysis of the structure of mating loci in *Candida* species is currently being assisted by the rapid growth in genome-sequencing projects. The sequence of *D. hansenii* was reported in 2005 as part of the hemiascomycete genome-sequencing program of Génolevures (66). The genome sequencing (at the time of writing) of *C. dubliniensis* is almost complete (http://www.sanger.ac.uk/Sequencing/Candida/dubliniensis/). A sequence survey of the *C. parapsilosis* genome was reported in 2005 (50),

and determination of the entire genome sequence is (at the time of writing) in progress by the Wellcome Trust Sanger Institute (http://www.sanger.ac.uk/Sequencing/Candida/parapsilosis/). In addition, the Fungal Genome Initiative of the Broad Institute of Harvard and MIT is sequencing the genomes of *C. tropicalis*, *C. lusitaniae*, *C. guilliermondii*, *Lodderomyces elongisporus*, and *C. albicans* WO-1 (http://www.broad.mit.edu/annotation/fgi/).

The *C. dubliniensis MAT* locus is very similar to that of *C. albicans*, both in structure and in function (Fig. 1.4). Mating occurs between **a** and α isolates and is dependent on white-opaque switching (65). However, the frequency of natural homozygous **a** or α isolates is up to

10 times higher than in *C. albicans* (33% versus 3 to 7%), and mating is promiscuous: *C. dubliniensis*/*C. albicans* hybrids can be formed. Both *C. tropicalis* and *C. parapsilosis* are somewhat less closely related to *C. albicans*, but the structures of their *MAT* loci are also similar (Fig. 1.4) (50). *C. tropicalis* and *C. parapsilosis* are both diploid, and the isolate of *C. tropicalis* sequenced by the Broad Institute is heterozygous at the *MAT* locus (50). However, white-opaque switching has never been identified in either species, and mating has not yet been observed. In addition, all isolates of *C. parapsilosis* tested to date are homozygous for *MTLa*. Whereas the overall structure of the *C. parapsilosis* *MTLa* locus resembles that of the other *Candida* species, *MTLa1* is a pseudogene. There is no evidence of an *MTLα* idiomorph in the sequenced isolate, or in any other isolate tested (50; G. Butler, unpublished data). There is also a remarkably low level of heterozygosity in *C. parapsilosis* isolates (26, 51), suggesting that little, if any, recombination is taking place. However, it remains possible that mating does occur and that *MTLα* isolates are present in the population at a very low rate, similar to the situation described for *MTLa* isolates of *Cryptococcus neoformans* serotype A (47).

C. albicans and *C. dubliniensis* have at best a cryptic sexual cycle, and meiosis has never been observed. However, *C. lusitaniae*, *C. guilliermondii*, and *D. hansenii* are apparently fully sexual species (21, 67, 82, 86). *Candida* (or *Clavispora*) *lusitaniae* and *Candida* (or *Pichia*) *guilliermondii* are haploid, and both mating types have been identified. The Broad Institute is currently sequencing an *MTLa* isolate of *C. guilliermondii* and an *MTLα* isolate of *C. lusitaniae*. Again, the overall organization of the *MTL* loci in these species is similar (Fig. 1.4). Most of the loci contain orthologs of the *PIK*, *PAP*, and *OBP* genes (Fig. 1.4). The only exception is *D. hansenii*, which has a single mating-type locus, containing both a-specific and α-specific genes (*MATa2*, *MATa1*, and *MATα1* [20, 21]). There are no adjacent orthologs of *PIK*, *PAP*, or *OBP*, although they are found at a different location on the same chromosome, in the same order as in the *MTLa* idiomorph in the other *Candida* species (Fig. 1.4).

The gene order immediately surrounding the *MTL* idiomorphs in *C. albicans*, *C. dubliniensis*, *C. parapsilosis*, and *C. tropicalis* is identical, although there is a break in synteny in *C. dubliniensis* upstream of the *STI1* gene (Fig. 1.4) (9). Genes upstream from *STI1* in the other species include three zinc finger proteins (*ZNC1*, *ZNC2* [*TAC1*], and *ZNC3*). Loss of heterozygosity at *TAC1* is associated with antifungal resistance in *C. albicans* (15). The same genes are found in inverse orienta-

tion on the left-hand side of the *C. guilliermondii* *MAT* locus. The genes on the left-hand side of the *C. lusitaniae* and *D. hansenii* *MAT* loci are orthologs of genes from the right-hand side of *MAT* in *C. guilliermondii*, but in inverse orientation. Interestingly, the gene order to the left of the *OBP*, *PAP*, and *PIK* gene region in *D. hansenii* is identical to the gene order on the left of *C. guilliermondii* *MAT*. The gene immediately to the right of *PIK* in *D. hansenii* is *CLN1*, associated with the *MAT* locus in other species, but missing at *MAT* in *D. hansenii* (Fig. 1.4). It is likely that a rearrangement between *MTLa* and *MTLα* idiomorphs occurred in the lineage leading to *D. hansenii*, leaving the *PAP*, *OBP*, *PIK*, and *CLN1* orthologs at one location, and joining the *MTLa* and *MTLα* genes at another. This probably resulted in homothallism. Similar fused *MAT* loci are also present in the genomes of the homothallic isolates *Pichia angusta* (Color Plate 1) (9), *Cochliobolus* (88), and *Gibberella* (87) (Fig. 1.5).

It is interesting that the *MAT* loci of almost all the *Candida* genomes sequenced to date contain orthologs of the *PIK*, *PAP*, and *OBP* genes. The origin and function of these genes are not clear. Homologs of *PAP1* and *OBP* (*OSH6*) are linked in the *S. cerevisiae* genome, but they are not associated with *MAT*. It is not yet known if the genes play any role in mating in the *Candida* species. It is notable, however, that the alleles at the **a** and α idiomorphs differ significantly (approximately 60% identity in *C. albicans*), suggesting that they diverged following their capture at the *MAT* locus.

Whereas the structures of the *MAT* loci are similar within the *Candida* clade, there are some differences. The *C. guilliermondii* *MTLa* idiomorph contains no identifiable ortholog of *MTLa1*, and the *C. lusitaniae* *MTLα* idiomorph and the *D. hansenii* fused *MAT* locus have no ortholog of *MTLα2* (Fig. 1.4). As *C. guilliermondii*, *C. lusitaniae*, and *D. hansenii* are all sexual species, it is clear that there must be significant differences in the regulation of the sexual cycle in comparison to *C. albicans*. Further analysis will be enabled by sequencing the *MTLα* idiomorph from *C. lusitaniae* and the *MTLa* idiomorph from *C. guilliermondii*.

Function of HMG Domain Proteins

All the species in the CTG clade (Fig. 1.4) code for an HMG domain DNA binding protein, *MTLa2*, either at the *MTLa* idiomorph or the *D. hansenii* fused *MAT* locus (Fig. 1.4). *MTLa2* was originally not identified in *C. albicans* (37), probably because it contains an intron. Orthologs of *MTLa2* are present in the genomes of many Saccharomycotina and Pezizomycotina (Color Plate 1). It is, however, missing in the *Saccharomyces*

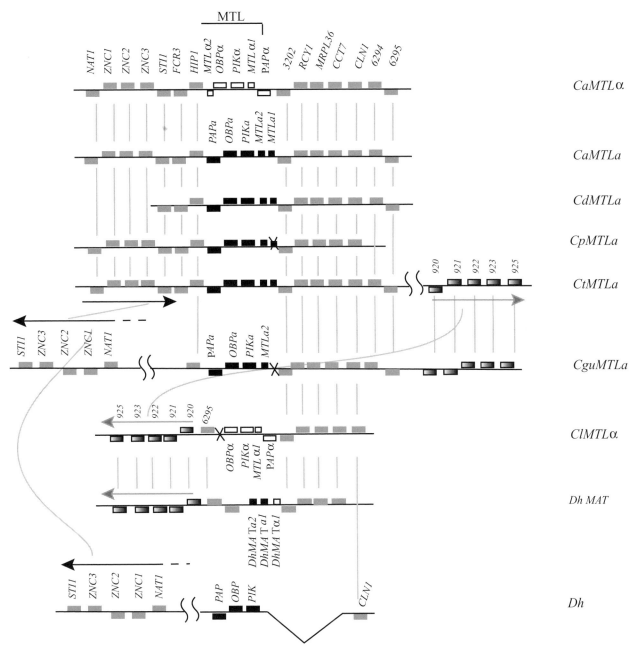

Figure 1.4 Organization of the *MTL* loci in the *Candida* clade. The order of the genes surrounding the *MTLα* and *MTLa* idiomorphs in *C. albicans* is shown in the top two lines. The gene names are taken from the Candida Genome Database (http://www.candidagenome.org), or gene number designations are from Assembly 19. The structures of *MTLα* idiomorphs from *C. dubliniensis* and *C. tropicalis* are similar to the *C. albicans MTLα* idiomorph and are not shown. The gene order for *D. hansenii* is taken from reference 16, and for the other species it is extrapolated from ongoing genome-sequencing projects at the Wellcome Trust Sanger Centre (http://www.sanger.ac.uk/Projects/Fungi/) and the Broad Institute (http://www.broad.mit.edu/annotation/fgi/). *MTLα*-specific genes are unfilled boxes and *MTLa*-specific genes are black. Orthologous genes are connected by gray lines; not all relationships are shown. Inversions in gene order are indicated by arrows. Wavy lines indicate regions of the chromosome where genes have been omitted. The "X" structures indicate the location of the *MTLa1* pseudogene in *C. parapsilosis* and the expected location of *MTLa1* in *C. guilliermondii* and of *MTLα2* in *C. lusitaniae*. The bottom line shows the gene order surrounding the *PAP*, *OBP*, and *PIK* region on chromosome E of *D. hansenii*, which is separate from the *MAT* locus. There is no gap between the *PIK* and *CLN1* orthologs. Abbreviations: Ca, *Candida albicans*; Cp, *Candida parapsilosis*; Cd, *Candida dubliniensis*; Cgu, *Candida guilliermondii*; Cl, *Candida lusitaniae*; Dh, *Debaryomyces hansenii*.

(a) Sordariomycetes

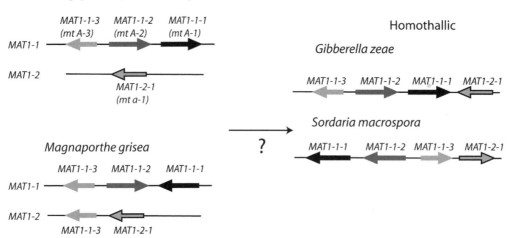

Heterothallic

Neurospora crassa, Podospora anserina
Fusarium oxysporum*, Gibberella fujikuroi

MAT1-1-3 (mt A-3)   MAT1-1-2 (mt A-2)   MAT1-1-1 (mt A-1)

MAT1-1

MAT1-2

MAT1-2-1 (mt a-1)

Homothallic

Gibberella zeae

MAT1-1-3   MAT1-1-2   MAT1-1-1   MAT1-2-1

Sordaria macrospora

?

MAT1-1-1   MAT1-1-2   MAT1-1-3   MAT1-2-1

Magnaporthe grisea

MAT1-1-3   MAT1-1-2   MAT1-1-1

MAT1-1

MAT1-2

MAT1-1-3   MAT1-2-1

(b) Dothideomycetes

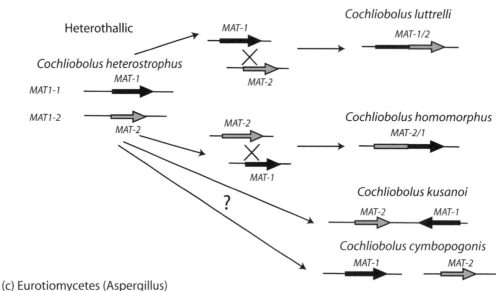

Homothallic

Heterothallic

Cochliobolus heterostrophus

MAT1-1   MAT-1

MAT1-2   MAT-2

MAT-1

MAT-2

Cochliobolus luttrelli

MAT-1/2

MAT-2

MAT-1

Cochliobolus homomorphus

MAT-2/1

?

Cochliobolus kusanoi

MAT-2   MAT-1

Cochliobolus cymbopogonis

MAT-1   MAT-2

(c) Eurotiomycetes (Aspergillus)

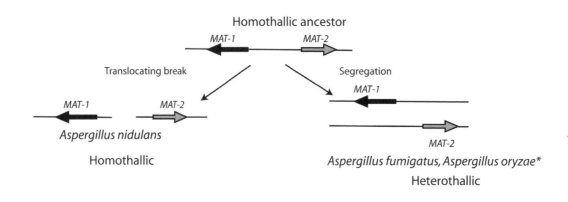

Homothallic ancestor

MAT-1   MAT-2

Translocating break

Segregation

MAT-1

MAT-1   MAT-2

MAT-2

Aspergillus nidulans

Homothallic

Aspergillus fumigatus, Aspergillus oryzae*

Heterothallic

Figure 1.5 Organization and evolution of the *MAT* loci in the Pezizomycotina. Alpha-box genes are shown in black, and HMG-domain genes are either gray with a black border (equivalent to *MTLa* in the Saccharomycotina) or solid light gray (*MAT1-1-3* orthologs). Orthologs of *MAT1-1-2* from the Sordariomycetes are shown as solid dark gray. (a) The gene names recommended in reference 67 are used for the Sordariomycetes with the equivalent names from *N. crassa* shown underneath. There is an ortholog of *MAT1-1-3* beside both idiomorphs in *M. grisea*, and this is indicated by its being drawn at both loci. (b) The structure and proposed evolution in *Cochliobolus* species are taken from reference 74. The fused genes in the homothallic species *C. luttrelli* and *C. homomorphus* most likely arose from crossing over between the alpha-box and HMG-domain genes in a heterothallic ancestor. Recombination between adjacent genes may have given rise to the locus in *C. kusanoi*. The origin of the unlinked *MAT* loci in *C. cymbopogonis* is not known. (c) Galagan et al. (27) suggest that both homothallic and putative heterothallic isolates of *Aspergillus* arose from a homothallic ancestor. In *A. fumigatus* and *A. oryzae* the alpha-box and HMG-domain genes are offset from each other, suggesting that they arose through gene loss. In *A. nidulans*, like *C. cymbopogonis*, the two genes are found in the same genome but are unlinked; *some species have not been shown to reproduce sexually, but the organization of the *MAT* locus suggests that they are heterothallic or have only recently lost the ability to mate.

---

lineage, including *S. cerevisiae*, *C. glabrata*, and *S. castellii* (Color Plate 1).

In 2003, Tsong et al. (78) demonstrated that in *C. albicans MTLa2* is required to activate expression of a-specific genes, a function that is constitutive in *S. cerevisiae*. The HMG protein also plays a role in mating in many of the filamentous yeasts (13, 42), suggesting that it was present in the ancestor of all the ascomycetes. In the *Saccharomyces* lineage, *MATa2* was lost, and **a** became the default cell state of haploids (Fig. 1.3). Curiously, this loss occurred during the same (relatively short) evolutionary interval as the whole-genome duplication. *P. angusta*, with its fused *MAT* locus, is the only other known member of the Saccharomycotina that does not have *MATa2* (Fig. 1.4) (9). It is possible that the gene was lost in this species, though it is also possible that it is present at another location not represented in the genome sequence survey. It has been suggested that *P. angusta* has a tetrapolar mating system (44), indicating the involvement of a second unlinked locus. Tetrapolar mating systems are likely to be rare in ascomycetes, but they have also been suggested in some species of *Glomerella* (11).

The *MAT* loci of most of the Saccharomycotina (including *Y. lipolytica*, at the base of the subphylum) contain four genes associated with mating (*MATa1*, *MATa2*, *MATα1*, and *MATα2*), suggesting that this is the ancestral structure (Color Plate 1). There are no orthologs of the homeodomain proteins *MATa1* and *MATα2* in equivalent positions in the filamentous yeasts, but similar proteins are present at the *MAT* loci in some basidiomycetes (48). It is therefore likely that these have been lost in the Pezizomycotina (Fig. 1.1 and Color Plate 1). In *S. cerevisiae*, the a1/α2 heterodimer represses haploid-specific transcription in diploids. In *C. albicans* and *C. dubliniensis*, a1/α2 were recruited as regulators of white-opaque switching. In other *Candida* species, however, there were subsequent losses of either *MATa1* or *MATα2*, suggesting that repression in diploids is either not necessary or achieved by different means and that the association of switching and mating is restricted to *C. albicans* and *C. dubliniensis* (and possibly also *C. tropicalis*).

MATING IN THE FILAMENTOUS ASCOMYCETES (PEZIZOMYCOTINA)

The mating-type locus has been described for a number of heterothallic and homothallic filamentous ascomycetes, including members of the classes Sordariomycetes (*Fusarium*, *Gibberella*, *Neurospora*, *Podospora*, and *Magnaporthe*), Leotiomycetes (*Pyrenopeziza* and *Rhynchosporium*), Dothideomycetes (*Cochliobolus*), and Eurotiomycetes (*Aspergillus*, *Coccidioides*, and *Histoplasma*) (16, 24, 25, 27, 30, 62, 72, 87, 88) (Color Plate 1 and Fig. 1.5). The nomenclature used varies widely from species to species and from the conventions applied to the Saccharomycotina, but the general structure is similar. The heterothallic isolates have a single *MAT* locus with two idiomorphs (Fig. 1.5). For the heterothallic Sordariomycetes, Turgeon and Yoder (79) have suggested that the *MAT* locus be designated *MAT1* and the two idiomorphs *MAT1-1* and *MAT1-2* and that the genes within the idiomorphs be indicated by the idiomorph symbol, followed by a hyphen and a number (e.g., *MAT1-1-1*). These conventions have been used in Fig. 1.5. The *MAT1-1* idiomorphs encode a protein with an alpha-box motif (*MAT1-1-1*) and are functionally equivalent to the *MATα* locus in the hemiascomycetes. *MAT1-1* contains two additional genes, *MAT1-1-2* and *MAT1-1-3* (HMG-domain). The *MAT1-2* idiomorph encodes an HMG-domain protein (*MAT1-2-1*), equivalent to the *MTLa2* gene in the *MATa* idiomorph in the Saccharomycotina. The *MAT1-1-1* (alpha-box) and *MAT1-2-1* (HMG domain) genes have been shown to be regulators of mating in several filamentous ascomycetes (13, 42). *MAT1-1-2* and *MAT1-1-3* are required at later stages in development.

In *Magnaporthe grisea* the *MAT1-1-1* gene is in the opposite orientation, relative to the other Sordariomycetes (Color Plate 1). The *MAT1-1-3* HMG-domain-like protein appears to lie outside the unique sequence and is present beside both *MAT1-1* and *MAT1-2* idiomorphs (GenBank accession no. AB080673 and AB080672; see the genome sequence at http://www.broad.mit.edu/annotation/fungi/magnaporthe/) (Fig. 1.5). It is not yet clear if *MAT1-1-3* has only recently become associated with *MAT-1* in the other species or if it was lost from *MAT1-1* only in *M. grisea*. There is some evidence that the gene plays a role in mating, because in *N. crassa*, the MAT1-1-3 protein (MT-A3) binds to the same DNA fragments in vitro as MAT1-2-1 (MT a-1) (23), and in *Podospora anserina*, the MAT1-1-3 ortholog (SMR2) interacts with the MAT1-1-1 alpha-box protein, FMR1 (89). However, a *MAT1-1-3* mutation in *N. crassa* can mate and produce asci (22). In addition, there is no *MAT1-1-3* associated with the *MAT1-1* locus in the Sordariomycetes *Cordyceps militaris* and *Cordyceps takaomontana* (85).

The *MAT* loci of other heterothallic Sordariomycetes are generally similar, with the notable exception of *Glomerella* species. Only *MAT1-2* (and not *MAT1-1*) idiomorphs have been identified from self-fertile and cross-fertile isolates of *Glomerella cingulata*, *G. lindemuthiana*, and *G. graminicola* (10, 11, 68, 80). Cisar and TeBeest (11) have suggested that mating in *G. cingulata* is regulated by multiple alleles at a single locus, whereas Vaillancourt et al. (80) hypothesized that two unlinked loci control mating in *G. graminicola*. It is clear that the mechanisms used in these species are complex and differ significantly from those used by other Sordariomycetes. It is difficult to draw any conclusions regarding the evolution of mating in these species without access to whole-genome data.

The *MAT1-2* loci in the Leotiomycetes *Pyrenopeziza brassicae* and *Rhynchosporium secalis* are similar to those of the Sordariomycetes, but there are differences at *MAT1-1* (24, 64, 72, 73). There is no *MAT1-1-2* ortholog, and although an HMG-domain protein has been identified at *MAT1-1*, it is not a close relative of the *MAT1-1-3* genes present in most of the Sordariomycetes. This is supported by an analysis of the draft genome sequences of two additional Leotiomycetes, *Sclerotinia sclerotiorum* and *Botrytis cinerea* (http://www.broad.mit.edu/). *S. sclerotiorum* is homothallic, and the presumptive *MAT* locus contains an alpha-box gene and an

HMG-domain gene between orthologs of *APN2* and *SLA2*. In *B. cinerea*, there is a single alpha-box gene, again between *APN2* and *SLA2*. Both species have an additional gene between *APN2* and the alpha-box protein (also present, but not annotated at the *MAT* loci of *P. brassicae* and *R. secalis*) with no significant similarity to proteins from other fungal species. The *P. brassicae MAT1-1* idiomorph also encodes a potential metallothionein-like protein (72). It is not known if either of these genes plays any role in mating. There is no *MAT1-1-2* ortholog present at *MAT* in either *S. sclerotiorum* or *B. cinerea*. The association of *MAT1-1-2* with the *MAT* locus may therefore have occurred only in the Sordariomycetes, after they diverged from the other species. The association of HMG-domain proteins with *MAT1-1* is interesting, as it may have occurred independently in both Leotiomycetes and Sordariomycetes. It is possible that these were acquired at *MAT1-1* by recombination with the *MAT1-2* idiomorph.

The structure of the *MAT* loci in heterothallic *Cochliobolus* and *Aspergillus* species is somewhat simpler, as each idiomorph contains only one gene, either *MAT-1* (alpha-box protein) or *MAT-2* (HMG domain) (Color Plate 1 and Fig. 1.5). The structure of a presumptive *MAT-1* in the genome sequence of the Dothideomycete *Stagonospora nodurum* resembles that of *Cochliobolus heterostrophus* (http://www.broad.mit.edu/). Mating has not been observed in *A. fumigatus* or *A. oryzae*, but two mating types have been identified for both, and it is assumed that they are heterothallic species (27, 62). The *MAT* loci of *H. capsulatum*, *Coccidioides posadasii*, and *C. immitis* have similar structures (25). An alpha-box domain protein (*MAT-1*) has been identified at the presumed *MAT* locus in *H. capsulatum*, and HMG-domain proteins (*MAT-2*) have been found in the two *Coccidioides* species. Mating has not been described in *Coccidioides*, but it is bipolar in *H. capsulatum* (43), suggesting that no other loci are required. The single gene locus may represent the ancestral structure for the Eurotiomycetes (*Aspergillus*, *Histoplasma*, and *Coccidioides*) and the Dothideomycetes (*Cochliobolus* and *Stagonospora*), as it has been found in members of both. However, Debuchy and Turgeon (17) have shown that the *MAT-1* gene in *Cochliobolus* species contains both α-box and HMG-box domains, suggesting that it may carry out the roles of both the *MAT1-1-1* and *MAT1-1-3* genes found in the Sordariomycetes.

There are a large number of homothallic species within the filamentous ascomycetes, and many are very closely related to heterothallic isolates (Fig. 1.5). The homothallic species contain genes from both *MAT* idiomorphs, and both are required to allow self-mating

(45). For example, the homothallic species *G. zeae* has *MAT1-1-1*, *MAT1-1-2*, *MAT1-1-3*, and *MAT1-2-1* at the same chromosomal location, whereas these are present in two alternative idiomorphs in the heterothallic species *Gibberella fujikuroi* (Fig. 1.5) (87). One of the most important unanswered questions is whether heterothallism or homothallism represents the ancestral state. Using a population genetics model, Nauta and Hoekstra (59) suggested that evolution from hetero- to homothallism is more likely than the reverse in the Sordariomycetes. The similarity in organization and sequence of the idiomorphs of *N. crassa*, *Podospora anserina*, *Fusarium oxysporum*, *Gibberella fujikuroi*, and *M. grisea* also suggests that they arose from a common heterothallic ancestor (Fig. 1.5) (87). Recombination between shared sequences outside the two idiomorphs could give rise to the single *MAT* locus seen in *G. zeae* and *Sordaria macrospora* (Fig. 1.5).

A detailed study of the structure of the *MAT* loci in *Cochliobolus* species provided clear evidence that heterothallism is an ancestral state (88). Five heterothallic isolates and two asexual species have the same organization, each with two alternative idiomorphs containing a single gene (*C. heterostrophus* is shown in Fig. 1.5). In the homothallic species, both genes are present in the same genome. In both *C. luttrelli* and *C. homomorphus*, the two *MAT* genes are fused. The *C. luttrelli* fusion most likely arose from a single crossover in an 8-bp sequence found towards the end of *MAT-1* and the beginning of *MAT-2* in *C. heterostrophus*. Similarly, the *MAT-2/1* fusion in *C. homomorphus* is derived from a crossover at a 9-bp sequence shared between the end of *MAT-1* and the beginning of *MAT-2*. In *C. kusanoi* and *C. cymbopogonis* the genes are not fused. They are linked in *C. kusanoi*, and the intervening region is derived from a β-glucosidase gene, found 3′ of both idiomorphs in *C. heterostrophus*.

In *Aspergillus* species, however, it is more likely that the ancestor was homothallic (27, 62). This is supported by the observation that within the idiomorphs of apparently heterothallic species *A. fumigatus* and *A. oryzae*, the alpha-box and HMG-domain genes are offset from each other. Secondly, the *MAT1-1* locus of *A. fumigatus* contains a 360-bp fragment of an HMG gene in addition to the alpha-box gene (62). The heterothallic structure could therefore have arisen from loss of the HMG domain protein (Fig. 1.5).

SUMMARY AND CONCLUSIONS

The phylum Ascomycota contains two of every three fungi currently known, including filamentous and non-filamentous, pathogenic and nonpathogenic, and sexual and asexual species. For many species (such as *A. fumi-*

gatus and *C. immitis*), although a sexual cycle has not been described, population studies suggest that recombination is occurring in the wild. Others (such as *C. parapsilosis*) may be truly asexual. This chapter describes the structure of the *MAT* locus in species from most of the major subphyla and classes within the ascomycetes. We have shown that the *MAT* loci probably evolved from a common structure at a common location but that dramatic changes in gene order and content have taken place. It is possible that in the ancestor of the ascomycetes, the *MAT* idiomorphs contained all of the determinants of mating, including two homeodomain proteins, one alpha-box protein and one HMG-domain protein. Components were then lost in specific lineages, such as the HMG-domain protein in the *Saccharomyces* lineage, and the a1/α2 homeodomain proteins in the Pezizomycotina. However, there is also a tendency to gain genes that may or may not be associated with mating; for example, there are HMG-domain genes present at *MAT1-1* in many of the Sordariomycetes, an undescribed gene in the Leotiomycetes, an α3 gene of unknown function at *MTLα* in *K. lactis*, and alleles of *PIK*, *PAP*, and *OPB* at both *MTL* idiomorphs in most of the *Candida* species. The ongoing genome-sequencing projects will provide an invaluable resource for further investigation, particularly to address the origin of these additional genes and the evolution of mating within the *Candida* clade.

References

1. Astell, C. R., L. Ahlstrom-Jonasson, M. Smith, K. Tatchell, K. A. Nasmyth, and B. D. Hall. 1981. The sequence of the DNAs coding for the mating-type loci of *Saccharomyces cerevisiae*. *Cell* **27:**15–23.

2. Astrom, S. U., A. Kegel, J. O. Sjostrand, and J. Rine. 2000. *Kluyveromyces lactis* Sir2p regulates cation sensitivity and maintains a specialized chromatin structure at the cryptic alpha-locus. *Genetics* **156:**81–91.

3. Astrom, S. U., and J. Rine. 1998. Theme and variation among silencing proteins in *Saccharomyces cerevisiae* and *Kluyveromyces lactis*. *Genetics* **148:**1021–1029.

4. Bakhrat, A., M. S. Jurica, B. L. Stoddard, and D. Raveh. 2004. Homology modeling and mutational analysis of Ho endonuclease of yeast. *Genetics* **166:**721–728.

5. Bobola, N., R. P. Jansen, T. H. Shin, and K. Nasmyth. 1996. Asymmetric accumulation of Ash1p in postanaphase nuclei depends on a myosin and restricts yeast mating-type switching to mother cells. *Cell* **84:**699–709.

6. Bremer, M. C., F. S. Gimble, J. Thorner, and C. L. Smith. 1992. VDE endonuclease cleaves *Saccharomyces cerevisiae* genomic DNA at a single site: physical mapping of the *VMA1* gene. *Nucleic Acids Res.* **20:**5484.

7. Brockert, P. J., S. A. Lachke, T. Srikantha, C. Pujol, R. Galask, and D. R. Soll. 2003. Phenotypic switching and

mating type switching of *Candida glabrata* at sites of colonization. *Infect. Immun.* **71:**7109–7118.

8. Burt, A., and V. Koufopanou. 2004. Homing endonuclease genes: the rise and fall and rise again of a selfish element. *Curr. Opin. Genet. Dev.* **14:**609–615.

9. Butler, G., C. Kenny, A. Fagan, C. Kurischko, C. Gaillardin, and K. H. Wolfe. 2004. Evolution of the *MAT* locus and its Ho endonuclease in yeast species. *Proc. Natl. Acad. Sci. USA.* **101:**1632–1637.

10. Chen, F., P. H. Goodwin, A. Khan, and T. Hsiang. 2002. Population structure and mating-type genes of *Colletotrichum graminicola* from *Agrostis palustris. Can. J. Microbiol.* **48:**427–436.

11. Cisar, C. R., and D. O. TeBeest. 1999. Mating system of the filamentous ascomycete, *Glomerella cingulata. Curr. Genet.* **35:**127–133.

12. Cliften, P., P. Sudarsanam, A. Desikan, L. Fulton, B. Fulton, J. Majors, R. Waterston, B. A. Cohen, and M. Johnston. 2003. Finding functional features in *Saccharomyces* genomes by phylogenetic footprinting. *Science* **301:**71–76.

13. Coppin, E., R. Debuchy, S. Arnaise, and M. Picard. 1997. Mating types and sexual development in filamentous ascomycetes. *Microbiol. Mol. Biol. Rev.* **61:**411–428.

14. Cosma, M. P., T. Tanaka, and K. Nasmyth. 1999. Ordered recruitment of transcription and chromatin remodeling factors to a cell cycle- and developmentally regulated promoter. *Cell* **97:**299–311.

15. Coste, A. T., V. Turner, F. Ischer, J. Morschhauser, A. Forche, A. Selmecki, J. Berman, J. Bille, and D. Sanglard. 2006. A mutation in Tac1p, a transcription factor regulating *CDR1* and *CDR2*, is coupled with loss of heterozygosity at Chromosome 5 to mediate antifungal resistance in *Candida albicans. Genetics* **172:**2139–2156.

16. Debuchy, R., and E. Coppin. 1992. The mating types of *Podospora anserina*: functional analysis and sequence of the fertilization domains. *Mol. Gen. Genet.* **233:**113–121.

17. Debuchy, R., and B. G. Turgeon. 2006. Mating-type structure, evolution and function in Euascomycetes, p. 293–324. *In* U. Kues and R. Fischer (ed.), *The Mycota*, vol. I. Springer, Heidelberg, Germany.

18. Dietrich, F. S., S. Voegeli, S. Brachat, A. Lerch, K. Gates, S. Steiner, C. Mohr, R. Pohlmann, P. Luedi, S. Choi, R. A. Wing, A. Flavier, T. D. Gaffney, and P. Philippsen. 2004. The *Ashbya gossypii* genome as a tool for mapping the ancient *Saccharomyces cerevisiae* genome. *Science* **304:**304–307.

19. Dujon, B. 2005. Hemiascomycetous yeasts at the forefront of comparative genomics. *Curr. Opin. Genet. Dev.* **15:**614–620.

20. Dujon, B., D. Sherman, G. Fischer, P. Durrens, S. Casaregola, I. Lafontaine, J. De Montigny, C. Marck, C. Neuveglise, E. Talla, N. Goffard, L. Frangeul, M. Aigle, V. Anthouard, A. Babour, V. Barbe, S. Barnay, S. Blanchin, J. M. Beckerich, E. Beyne, C. Bleykasten, A. Boisrame, J. Boyer, L. Cattolico, F. Confanioleri, A. De Daruvar, L. Despons, E. Fabre, C. Fairhead, H. Ferry-Dumazet, A. Groppi, F. Hantraye, C. Hennequin, N. Jauniaux, P. Joyet, R. Kachouri, A. Kerrest, R. Koszul, M. Lemaire, I. Lesur, L. Ma, H. Muller, J. M. Nicaud, M. Nikolski, S. Oztas, O. Ozier-Kalogeropoulos, S. Pellenz, S. Potier, G. F. Richard, M. L. Straub, A. Suleau, D. Swennen, F. Tekaia, M. Wesolowski-Louvel, E. Westhof, B. Wirth, M. Zeniou-Meyer, I. Zivanovic, M. Bolotin-Fukuhara, A. Thierry, C. Bouchier, B. Caudron, C. Scarpelli, C. Gaillardin, J. Weissenbach, P. Wincker, and J. L. Souciet. 2004. Genome evolution in yeasts. *Nature* **430:**35–44.

21. Fabre, E., H. Muller, P. Therizols, I. Lafontaine, B. Dujon, and C. Fairhead. 2005. Comparative genomics in hemiascomycete yeasts: evolution of sex, silencing, and subtelomeres. *Mol. Biol. Evol.* **22:**856–873.

22. Ferreira, A. V., Z. An, R. L. Metzenberg, and N. L. Glass. 1998. Characterization of *mat A-2, mat A-3* and *delta-matA* mating-type mutants of *Neurospora crassa. Genetics* **148:**1069–1079.

23. Ferreira, A. V., S. Saupe, and N. L. Glass. 1996. Transcriptional analysis of the mtA idiomorph of *Neurospora crassa* identifies two genes in addition to *mtA-1. Mol. Gen. Genet* **250:**767–774.

24. Foster, S. J., and B. D. Fitt. 2003. Isolation and characterisation of the mating-type (*MAT*) locus from *Rhynchosporium secalis. Curr. Genet.* **44:**277–286.

25. Fraser, J. A., and J. Heitman. 2006. Sex, *MAT*, and the evolution of fungal virulence, p. 13–33. *In* J. Heitman, S. G. Filler, J. E. Edwards, and A. P. Mitchell (ed.), *Molecular Principles of Fungal Pathogenesis*. ASM Press, Washington, DC.

26. Fundyga, R. E., R. J. Kuykendall, W. Lee-Yang, and T. J. Lott. 2004. Evidence for aneuploidy and recombination in the human commensal yeast *Candida parapsilosis. Infect. Genet. Evol.* **4:**37–43.

27. Galagan, J. E., S. E. Calvo, C. Cuomo, L. J. Ma, J. R. Wortman, S. Batzoglou, S. I. Lee, M. Basturkmen, C. C. Spevak, J. Clutterbuck, V. Kapitonov, J. Jurka, C. Scazzocchio, M. Farman, J. Butler, S. Purcell, S. Harris, G. H. Braus, O. Draht, S. Busch, C. D'Enfert, C. Bouchier, G. H. Goldman, D. Bell-Pedersen, S. Griffiths-Jones, J. H. Doonan, J. Yu, K. Vienken, A. Pain, M. Freitag, E. U. Selker, D. B. Archer, M. A. Penalva, B. R. Oakley, M. Momany, T. Tanaka, T. Kumagai, K. Asai, M. Machida, W. C. Nierman, D. W. Denning, M. Caddick, M. Hynes, M. Paoletti, R. Fischer, B. Miller, P. Dyer, M. S. Sachs, S. A. Osmani, and B. W. Birren. 2005. Sequencing of *Aspergillus nidulans* and comparative analysis with *A. fumigatus* and *A. oryzae. Nature* **438:**1105–1115.

28. Galagan, J. E., M. R. Henn, L. J. Ma, C. A. Cuomo, and B. Birren. 2005. Genomics of the fungal kingdom: insights into eukaryotic biology. *Genome Res.* **15:**1620–1631.

29. Gimble, F. S., and J. Thorner. 1992. Homing of a DNA endonuclease gene by meiotic gene conversion in *Saccharomyces cerevisiae. Nature* **357:**301–306.

30. Glass, N. L., S. J. Vollmer, C. Staben, J. Grotelueschen, R. L. Metzenberg, and C. Yanofsky. 1988. DNAs of the two mating-type alleles of *Neurospora crassa* are highly dissimilar. *Science* **241:**570–573.

31. Haber, J. E. 1998. Mating-type gene switching in *Saccharomyces cerevisiae. Annu. Rev. Genet.* **32:**561–599.

32. Heckman, D. S., D. M. Geiser, B. R. Eidell, R. L. Stauffer, N. L. Kardos, and S. B. Hedges. 2001. Molecular evidence for the early colonization of land by fungi and plants. *Science* 293:1129–1133.

33. Hedges, S. B., J. E. Blair, M. L. Venturi, and J. L. Shoe. 2004. A molecular timescale of eukaryote evolution and the rise of complex multicellular life. *BMC Evol. Biol.* 4:2.

34. Herman, A., and H. Roman. 1966. Allele specific determinants of homothallism in *Saccharomyces lactis*. *Genetics* 53:727–740.

35. Hicks, J., and J. N. Strathern. 1977. Interconversion of mating type in *S. cerevisiae* and the cassette model for gene transfer. *Brookhaven Symp. Biol.* 1977:233–242.

36. Hicks, J. B., J. N. Strathern, and I. Herskowitz. 1977. The cassette model of mating type interconversion, p. 457–462. *In* M. N. Hall and P. Linder (ed.), *DNA Insertion Elements, Plasmids and Episomes*. Cold Spring Harbor Laboratory Press, Plainview, NY.

37. Hull, C. M., and A. D. Johnson. 1999. Identification of a mating type–like locus in the asexual pathogenic yeast *Candida albicans*. *Science* 285:1271–1275.

38. Hull, C. M., R. M. Raisner, and A. D. Johnson. 2000. Evidence for mating of the "asexual" yeast *Candida albicans* in a mammalian host. *Science* 289:307–310.

39. Keeling, P. J., and A. J. Roger. 1995. The selfish pursuit of sex. *Nature* 375:283.

40. Keogh, R. S., C. Seoighe, and K. H. Wolfe. 1998. Evolution of gene order and chromosome number in *Saccharomyces*, *Kluyveromyces* and related fungi. *Yeast* 14:443–457.

41. Koufopanou, V., M. R. Goddard, and A. Burt. 2002. Adaptation for horizontal transfer in a homing endonuclease. *Mol. Biol. Evol.* 19:239–246.

42. Kronstad, J. W., and C. Staben. 1997. Mating type in filamentous fungi. *Annu. Rev. Genet.* 31:245–276.

43. Kwon-Chung, K. J. 1972. Sexual stage of *Histoplasma capsulatum*. *Science* 175:326.

44. Lahtchev, K. 2002. Basic genetics of *Hansenula polymorpha*, p. 8–20. *In* G. Gellissen (ed.), Hansenula polymorpha: *Biology and Applications*. Wiley-VCH, Weinheim, Germany.

45. Lee, J., T. Lee, Y. W. Lee, S. H. Yun, and B. G. Turgeon. 2003. Shifting fungal reproductive mode by manipulation of mating type genes: obligatory heterothallism of *Gibberella zeae*. *Mol. Microbiol.* 50:145–152.

46. Legrand, M., P. Lephart, A. Forche, F. M. Mueller, T. Walsh, P. T. Magee, and B. B. Magee. 2004. Homozygosity at the *MTL* locus in clinical strains of *Candida albicans*: karyotypic rearrangements and tetraploid formation. *Mol. Microbiol.* 52:1451–1462.

47. Lengeler, K. B., P. Wang, G. M. Cox, J. R. Perfect, and J. Heitman. 2000. Identification of the *MATa* mating-type locus of *Cryptococcus neoformans* reveals a serotype A *MATa* strain thought to have been extinct. *Proc. Natl. Acad. Sci. USA* 97:14455–14460.

48. Lin, X., C. M. Hull, and J. Heitman. 2005. Sexual reproduction between partners of the same mating type in *Cryptococcus neoformans*. *Nature* 434:1017–1021.

49. Lockhart, S. R., C. Pujol, K. J. Daniels, M. G. Miller, A. D. Johnson, M. A. Pfaller, and D. R. Soll. 2002. In *Candida albicans*, white-opaque switchers are homozygous for mating type. *Genetics* 162:737–745.

50. Logue, M. E., S. Wong, K. H. Wolfe, and G. Butler. 2005. A genome sequence survey shows that the pathogenic yeast *Candida parapsilosis* has a defective *MTLa1* allele at its mating type locus. *Eukaryot. Cell* 4:1009–1017.

51. Lott, T. J., R. J. Kuykendall, S. F. Welbel, A. Pramanik, and B. A. Lasker. 1993. Genomic heterogeneity in the yeast *Candida parapsilosis*. *Curr. Genet.* 23:463–467.

52. Magee, B. B., and P. T. Magee. 2000. Induction of mating in *Candida albicans* by construction of *MTLa* and *MTLalpha* strains. *Science* 289:310–313.

53. Mathias, J. R., S. E. Hanlon, R. A. O'Flanagan, A. M. Sengupta, and A. K. Vershon. 2004. Repression of the yeast *HO* gene by the MATalpha2 and MATa1 homeodomain proteins. *Nucleic Acids Res.* 32:6469–6478.

54. Metzenberg, R. L., and N. L. Glass. 1990. Mating type and mating strategies in *Neurospora*. *Bioessays* 12:53–59.

55. Miller, M. G., and A. D. Johnson. 2002. White-opaque switching in *Candida albicans* is controlled by mating-type locus homeodomain proteins and allows efficient mating. *Cell* 110:293–302.

56. Miranda, I., R. Silva, and M. A. Santos. 2006. Evolution of the genetic code in yeasts. *Yeast* 23:203–213.

57. Mitra, D., E. J. Parnell, J. W. Landon, Y. Yu, and D. J. Stillman. 2006. SWI/SNF binding to the *HO* promoter requires histone acetylation and stimulates TATA-binding protein recruitment. *Mol. Cell. Biol.* 26:4095–4110.

58. Nasmyth, K. 1993. Regulating the *HO* endonuclease in yeast. *Curr. Opin. Genet. Dev.* 3:286–294.

59. Nauta, M. J., and R. F. Hoekstra. 1992. Evolution of reproductive systems in filamentous ascomycetes. I. Evolution of mating types. *Heredity* 68:405–510.

60. Nickoloff, J. A., J. D. Singer, and F. Heffron. 1990. In vivo analysis of the *Saccharomyces cerevisiae* HO nuclease recognition site by site-directed mutagenesis. *Mol. Cell. Biol.* 10:1174–1179.

61. Oshima, Y. 1993. Homothallism, mating-type switching, and the controlling element model in *Saccharomyces cerevisiae*, p. 291–304. *In* M. N. Hall and P. Linder (ed.), *The Early Days of Yeast Genetics*. Cold Spring Harbor Laboratory Press, Plainview, NY.

62. Paoletti, M., C. Rydholm, E. U. Schwier, M. J. Anderson, G. Szakacs, F. Lutzoni, J. P. Debeaupuis, J. P. Latge, D. W. Denning, and P. S. Dyer. 2005. Evidence for sexuality in the opportunistic fungal pathogen *Aspergillus fumigatus*. *Curr. Biol.* 15:1242–1248.

63. Pietrokovski, S. 1994. Conserved sequence features of inteins (protein introns) and their use in identifying new inteins and related proteins. *Protein Sci.* 3:2340–2350.

64. Poggeler, S. 2001. Mating-type genes for classical strain improvements of ascomycetes. *Appl. Microbiol. Biotechnol.* 56:589–601.

65. Pujol, C., K. J. Daniels, S. R. Lockhart, T. Srikantha, J. B. Radke, J. Geiger, and D. R. Soll. 2004. The closely related species *Candida albicans* and *Candida dubliniensis* can mate. *Eukaryot. Cell* 3:1015–1027.

66. Richard, G. F., A. Kerrest, I. Lafontaine, and B. Dujon. 2005. Comparative genomics of hemiascomycete yeasts: genes involved in DNA replication, repair, and recombination. *Mol. Biol. Evol.* **22:**1011–1023.

67. Rodrigues de Miranda, L. 1979. *Clavispora*, a new yeast genus of the Saccharomycetales. *Antonie Leeuwenhoek* **45:**479–483.

68. Rodriguez-Guerra, R., M. T. Ramirez-Rueda, M. Cabral-Enciso, M. Garcia-Serrano, Z. Lira-Maldonado, R. G. Guevara-Gonzalez, M. Gonzalez-Chavira, and J. Simpson. 2005. Heterothallic mating observed between Mexican isolates of *Glomerella lindemuthiana*. *Mycologia* **97:** 793–803.

69. Scherrer, S., U. Zippler, and R. Honegger. 2005. Characterisation of the mating-type locus in the genus *Xanthoria* (lichen-forming ascomycetes, Lecanoromycetes). *Fungal Genet. Biol.* **42:**976–988.

70. Seoighe, C., and K. H. Wolfe. 1998. Extent of genomic rearrangement after genome duplication in yeast. *Proc. Natl. Acad. Sci. USA* **95:**4447–4452.

71. Sil, A., and I. Herskowitz. 1996. Identification of asymmetrically localized determinant, Ash1p, required for lineage-specific transcription of the yeast *HO* gene. *Cell* **84:**711–722.

72. Singh, G., and A. M. Ashby. 1998. Cloning of the mating type loci from *Pyrenopeziza brassicae* reveals the presence of a novel mating type gene within a discomycete *MAT 1-2* locus encoding a putative metallothionein-like protein. *Mol. Microbiol.* **30:**799–806.

73. Singh, G., P. S. Dyer, and A. M. Ashby. 1999. Intraspecific and inter-specific conservation of mating-type genes from the discomycete plant-pathogenic fungi *Pyrenopeziza brassicae* and *Tapesia yallundae*. *Curr. Genet.* **36:**290–300.

74. Sjostrand, J. O., A. Kegel, and S. U. Astrom. 2002. Functional diversity of silencers in budding yeasts. *Eukaryot. Cell* **1:**548–557.

75. Slutsky, B., M. Staebell, J. Anderson, L. Risen, M. Pfaller, and D. R. Soll. 1987. "White-opaque transition": a second high-frequency switching system in *Candida albicans*. *J. Bacteriol.* **169:**189–197.

76. Srikantha, T., S. A. Lachke, and D. R. Soll. 2003. Three mating type-like loci in *Candida glabrata*. *Eukaryot. Cell* **2:**328–340.

77. Sugita, T., and T. Nakase. 1999. Non-universal usage of the leucine CUG codon and the molecular phylogeny of the genus *Candida*. *Syst. Appl. Microbiol.* **22:**79–86.

78. Tsong, A. E., M. G. Miller, R. M. Raisner, and A. D. Johnson. 2003. Evolution of a combinatorial transcriptional circuit: a case study in yeasts. *Cell* **115:**389–399.

79. Turgeon, B. G., and O. C. Yoder. 2000. Proposed nomenclature for mating type genes of filamentous ascomycetes. *Fungal Genet. Biol.* **31:**1–5.

80. Vaillancourt, L. J., M. Du, J. Wang, J. Rollins, and R. Hanau. 2000. Genetic analysis of cross fertility between two self-sterile strains of *Glomerella graminicola*. *Mycologia* **92:**430–435.

81. Wendland, J., and A. Walther. 2005. *Ashbya gossypii*: a model for fungal developmental biology. *Nat. Rev. Microbiol.* **3:**421–429.

82. Wickerham, L. J., and K. A. Burton. 1954. A clarification of the relationship of *Candida guilliermondii* to other yeasts by a study of their mating types. *J. Bacteriol.* **68:**594–597.

83. Wolfe, K. H., and D. C. Shields. 1997. Molecular evidence for an ancient duplication of the entire yeast genome. *Nature* **387:**708–713.

84. Wong, S., M. A. Fares, W. Zimmermann, G. Butler, and K. H. Wolfe. 2003. Evidence from comparative genomics for a complete sexual cycle in the 'asexual' pathogenic yeast *Candida glabrata*. *Genome Biol.* **4:**R10.

85. Yokoyama, E., K. Yamagishi, and A. Hara. 2005. Heterothallism in *Cordyceps takaomontana*. *FEMS Microbiol. Lett.* **250:**145–150.

86. Young, L. Y., M. C. Lorenz, and J. Heitman. 2000. A *STE12* homolog is required for mating but dispensable for filamentation in *Candida lusitaniae*. *Genetics* **155:**17–29.

87. Yun, S. H., T. Arie, I. Kaneko, O. C. Yoder, and B. G. Turgeon. 2000. Molecular organization of mating type loci in heterothallic, homothallic, and asexual *Gibberella/Fusarium* species. *Fungal Genet. Biol.* **31:**7–20.

88. Yun, S. H., M. L. Berbee, O. C. Yoder, and B. G. Turgeon. 1999. Evolution of the fungal self-fertile reproductive life style from self-sterile ancestors. *Proc. Natl. Acad. Sci. USA* **96:**5592–5597.

89. Zickler, D., S. Arnaise, E. Coppin, R. Debuchy, and M. Picard. 1995. Altered mating-type identity in the fungus *Podospora anserina* leads to selfish nuclei, uniparental progeny, and haploid meiosis. *Genetics* **140:**493–503.

90. Zonneveld, B. J. M., and H. Y. Steensma. 2003. Mating, sporulation and tetrad analysis in *Kluyveromyces lactis*, p. 151–154. *In* K. Wolf, K. Breunig, and G. Barth (ed.), *Non-Conventional Yeasts in Genetics, Biochemistry and Biotechnology*. Springer, Berlin, Germany.

Sex in Fungi: Molecular Determination
and Evolutionary Implications
Edited by Joseph Heitman et al.
© 2007 ASM Press, Washington, D.C.

James A. Fraser
Yen-Ping Hsueh
Keisha M. Findley
Joseph Heitman

Evolution of the Mating-Type Locus:

2

The Basidiomycetes

Of all the fungi, production of sexual structures is most prominent in the members of the phylum *Basidiomycota*. We are all familiar with the sexually produced edible mushrooms and poisonous toadstools of the homobasidiomycetes that are synonymous with the concept of fungi (10). In the agricultural community, smuts are well known for their ability to infect important crop species such as maize, a process dependent on sex for the initial infection and subsequent formation of teliospore-filled tumors (16). And in the clinical setting, the deadly pathogen *Cryptococcus* is infamous for its ability to cause life-threatening meningitis, an infection that may begin with the inhalation of sexually produced basidiospores or desiccated yeast cells (32).

The sexual cycles of fungi have been well studied for several decades. Research has shown that although enormous variation exists between the sexual processes of different fungal species, there are two common, underlying paradigms: the bipolar and the tetrapolar mating-type systems (20). In the bipolar systems two mating types segregate at meiosis; a single locus orchestrates the sexual cycle, and for mating to occur, two cells that differ at this locus (most commonly referred to as **a** and α) must come together. While this system predominates in fungi of the phylum *Ascomycota*, such as in *Saccharomyces*

cerevisiae and *Neurospora crassa*, it is less common in the basidiomycete phylum, where it is the tetrapolar system that occurs most commonly. In the tetrapolar paradigm, progeny of four different mating types are produced at meiosis; two unlinked loci orchestrate the sexual cycle, and for mating to occur, cells that differ at *both* loci must come together.

The loci that orchestrate these processes have become a popular topic in contemporary microbiology, and ongoing studies in many laboratories are focused on elucidating the nature of these genomic sex-determining regions. Known as mating-type (*MAT*) loci, the genomes of bipolar ascomycetes usually contain only one of these small genomic features (typically spanning only a few kilobases or less) that differs in the one or two transcription factors it encodes (13, 15, 56). While exceptions exist, such as the *MAT* locus of *Candida albicans* and related species (31), which include three additional genes, the ascomycete loci exhibit similarities in the genes encoding transcription factors (homeodomain, alpha-domain, and HMG factors) that they contain. In contrast, the genomes of the basidiomycetes have evolved a tremendous level of diversity in the genomic regions involved in establishing cell identity, all of which are built on a common underlying theme.

James A. Fraser, School of Molecular and Microbial Sciences, University of Queensland, Brisbane, QLD 4072 Australia. **Yen-Ping Hsueh, Keisha M. Findley,** and **Joseph Heitman,** Department of Molecular Genetics and Microbiology, Duke University Medical Center, Durham, NC 27710.

REGULATION OF MATING IN THE *BASIDIOMYCOTA*

Approximately 50 to 65% of the homobasidiomycetes (mushroom fungi) have a tetrapolar mating system, meaning that they have two unlinked mating-type loci (39, 51). Because the basidiomycete tetrapolar mating-type system is present in members of at least two of the three major lineages—the smuts and mushrooms—it likely has an ancient origin (12).

How does the basidiomycete tetrapolar mating-type system function? Although there are many variations that have been characterized at the molecular level, conceptually these all involve derivatives of a common component set: one locus that encodes pheromones and pheromone receptors and another locus that encodes homeodomain transcription factors (20). For mating to

be successful, partners must differ at both loci. The shared role of these loci is to govern self versus nonself recognition to control the mating process, control that can be exerted during both the early and late stages of this developmental cascade. In a typical basidiomycete sexual cycle, cells undergo chemoattraction, fuse to form a dikaryon, and nuclear migration is orchestrated by a complex regulatory process involving clamp cell formation and fusion. The sexual cycle culminates with the formation of a basidium, where nuclear fusion and meiosis occur to produce haploid basidiospores (Fig. 2.1). This is a generic pathway—different species have modified this scheme to bypass certain steps. Some homobasidiomycetes (species that produce their basidia within a protective mushroom structure) do not employ a chemoattraction step and instead fuse with any avail-

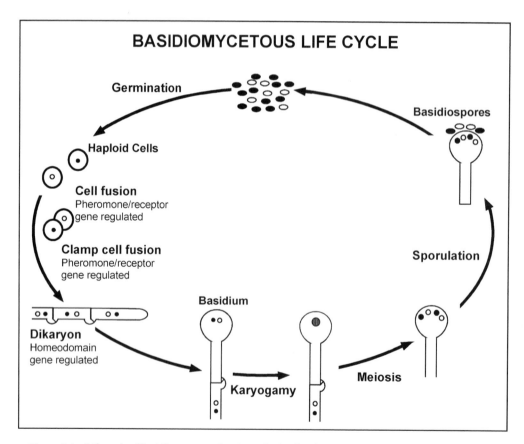

Figure 2.1   Life cycle of basidiomycetous fungi. Haploid cells of opposite mating type fuse, often via the action of the pheromone/receptor locus, to produce a filamentous dikaryon. The process of clamp cell formation and fusion (also under the control of the pheromone/receptor gene locus) functions to ensure the integrity of nuclear transmission in the dikaryon. The homeodomain locus is required for clamp cell formation and septation and to maintain the dikaryon once it has formed following nuclear division. The structure in which meiosis occurs, the basidium, is then formed, and the two nuclei fuse. Meiosis ensues, generating four nuclei, which ultimately give rise to basidiospores. These spores later germinate to produce individual haploid cells.

able partner of the same species (10). Some heterobasidiomycetes (species that *do not* produce their basidia within a protective mushroom structure) reportedly do not employ clamp cells in the mating process (16). Nevertheless, these underlying common themes are apparent in the sexual cycles of most basidiomycetes, and in all species studied thus far the *MAT* loci are intimately involved (14).

SENSING AND BEING SENSED: THE PHEROMONE/RECEPTOR LOCUS

In many basidiomycetes the first stage of mating usually involves the recognition that an appropriate mating partner is nearby, a process achieved through the export and sensing of pheromones (42). For strains of most species to be compatible they must produce different pheromones and receptors. This is accomplished when they encode different alleles of a *MAT* locus that contains two components of a pheromone production/sensing system: genes that encode pheromone-sensing class D members of the G protein-coupled receptor (GPCR) superfamily (28) and genes that encode small lipopeptide pheromones. Due to the way in which these loci were identified, the nomenclature is confusing; for example, in the heterobasidiomycete *Ustilago maydis* it is called the *a* locus, and in the homobasidiomycete *Coprinopsis cinerea* it is the *B* locus. For the sake of clarity, these will be referred to as the "pheromone/receptor locus" in this chapter.

The genes encoding pheromone produce an immature product that requires extensive posttranslational modification, with the immature peptide undergoing prenylation, proteolytic cleavage, and carboxy-terminal methylation to produce the mature, biologically active form. The interplay of the pheromone receptors and their lipid-modified ligands is well described by a comparison to the ascomycete *S. cerevisiae*, the best-studied system of this type (24). During the courtship stage of mating, *S. cerevisiae MATa* cells produce the small prenylated peptide a-factor, a pheromone that *MATα* cells detect via the G-protein coupled pheromone receptor Ste3. Conversely, *MATα* cells produce the unrelated peptide pheromone α factor, which *MATa* cells detect via the G-protein coupled pheromone receptor Ste2. Production of the mature pheromones is a complex pathway, and each is produced through independent pathways. The α-factor is produced via the action of multiple proteases liberating tandemly arrayed peptides from a larger precursor molecule that contains three or four copies of the pheromone that exit the cell through the canonical secretory pathway. a-Factor production is

even more complex, involving cleavage of the immature peptide via multiple proteases to give a single pheromone peptide that is farnesylated and carboxymethylated at the C terminus and exported via the Ste6 MDR-related pump. In response to these ligands interacting with the appropriate receptor, mating is initiated through a heterotrimeric G protein and downstream MAP kinase cascade. In the ascomycetes the genes encoding pheromones or pheromone receptors are scattered throughout the genome. Although *MATa* and *MATα* cells both encode all of the genes required to produce a factor, α factor, Ste2, and Ste3, their expression is regulated in response to the mating type of the cell such that only *MATα* cells produce α-factor and the Ste3 receptor, while only *MATa* cells produce a-factor and the Ste2 receptor (24).

How does this system compare to those present in the basidiomycetes? As described previously, with a few notable exceptions (addressed later in this chapter) the pheromone and pheromone receptor genes are usually contained in one of the two *MAT* loci. Furthermore, these genes always encode homologues of the a-factor/Ste3 *S. cerevisiae* paradigm; orthologues of the genes encoding α-factor or Ste2 have never been identified in this phylum, even in whole-genome analyses. In addition to exhibiting structural similarities to *S. cerevisiae* a-factor, the lipopeptide pheromones of the basidiomycetes appear to be processed and secreted by the same mechanisms that have been conserved since their divergence from the last common ancestor with the ascomycetes. This is highlighted by experiments showing that *S. cerevisiae* transformed with compatible *Schizophyllum commune* receptor/pheromone combinations are able to initiate the mating process, indicating that pheromone processing is conserved (17). And in *Cryptococcus neoformans*, candidate gene deletions of homologues of the *S. cerevisiae* pheromone processing and transporter pathway inhibit production of mature pheromone, abolishing the sexual cycle (B. Wickes, personal communication) (30).

The absence of a second independent pheromone production/sensing system therefore defines an important aspect of the basidiomycete archetype. Rather than a dual sensing system, the pheromone-based cell recognition system that regulates the fusion of mating cells (and some later stages of the mating process) is mediated by allelic variants of the same genes (14). Due to the limited information available from more basal lineages of fungi, it is currently impossible to determine if the Ste2/α-factor/Ste3/a-factor or the simpler Ste3/α-factor-only allelic systems are the ancestral form.

FORMING A PRODUCTIVE INTERACTION AFTER COURTSHIP: THE HOMEODOMAIN LOCUS

Following fusion of compatible strains, a second round of self/nonself recognition occurs, mediated by the second *MAT* locus. Again, most species must encode different alleles at this locus for strains to be compatible and enable the postfusion dikaryon to be maintained. Molecular analyses of this second type of locus revealed that these encode components similar to the *MAT* loci of the *Ascomycota*, which usually encode transcription factors of the HMG, α-box, or homeodomain types (13). However, as is elaborated below, in the *Basidiomycota* the transcription factors encoded are always homeodomain proteins. As for the pheromone/receptor locus, the nomenclature of this locus is confusing; in the heterobasidiomycete *U. maydis* it is called the *b* locus, and in the homobasidiomycete *C. cinerea* it is the *A* locus. Once again, for the sake of clarity these structures are referred to as "homeodomain loci" in this chapter.

The archetypal homeodomain *MAT* locus encodes a pair of proteins of differing homeodomain classes, one HD1 and one HD2, and these genes are typically divergently transcribed. The role of homeodomain proteins in regulating the fungal sexual cycle is well known. In *S. cerevisiae* the *a* locus encodes an HD2 gene (*a1*) while the α locus encodes an HD1 gene (α2), and upon mating these heterodimerize to form an active complex. In contrast, in the basidiomycetes where each locus encodes both an HD1 and an HD2 protein, different alleles of the homeodomain locus encode alternate versions of these factors. Upon fusion of compatible cells, the HD1 and HD2 proteins of the mating partners heterodimerize. However, dimerization is restricted to HD1 and HD2 proteins encoded by different alleles of the homeodomain locus. For example, the HD1 protein from one strain will be incompatible with the HD2 protein from the same strain but will heterodimerize with the HD2 protein from a mating strain that has a different allele of the homeodomain locus. This compatible mate recognition (the formation of a functional HD1/HD2 heterodimeric regulatory complex) will activate the homeodomain regulated pathway, which in most basidiomycetes is not defined but is necessary for the later stages of mating. The requirements for homeodomain heterodimerization therefore make this process an efficient self/nonself recognition system that ensures that active heterodimers are only formed in the dikaryon and do not arise in haploid cells.

Together, the pheromone/receptor and homeodomain loci create an efficient system in which self/nonself recognition occurs at multiple steps, both extra- and intracellularly. This elegant yet complex system lies at the core of all characterized basidiomycete mating systems; however, ongoing studies of these fascinating systems have uncovered numerous evolutionary permutations that provide for additional levels of sophistication.

TETRAPOLAR MATING-TYPE SYSTEMS: THE *USTILAGO MAYDIS* PARADIGM

U. maydis is a phytopathogen that infects maize plants, a process that occurs only when the fungus is in the dikaryotic filamentous phase produced by mating (7). Mating has therefore been intensively studied in this species, and it has subsequently arisen as a paradigmatic example of the tetrapolar sex-determining system. The earliest stage of this process, where two haploid cells identify each other and fuse, is mediated by pheromones produced by the biallelic pheromone/receptor (*a*) locus. Cells have one of two genotypes, either *a1* or *a2*, and strains must carry different alleles for mate recognition and cell-cell fusion to occur.

Sequence analysis of the pheromone/receptor locus of *U. maydis* has revealed that both alleles contain genes encoding homologues of the *S. cerevisiae* Ste3 pheromone receptor and the a-factor pheromone (8). The *a1* allele of the locus is the simplest form (~4.6 kb), and it carries a single pheromone gene (*mfa1*) and a single receptor gene (*pra1*), and therefore closely resembles the canonical basidiomycete pheromone/ receptor locus structure (Fig. 2.2). In contrast, the *a2* allele is larger (~8.5 kb) and contains a pheromone gene (*mfa2*), a receptor gene (*pra2*), and two additional genes (*lga2* and *rga2*). These additional genes, which are regulated by the sexual cycle and encode proteins that localize to the mitochondria, have been implicated in functional roles in mitochondrial morphology and virulence, likely via physical and functional interactions with the mitochondrial protein Mrb1 (9).

Following fusion, if the cells are heterozygous for the second, unlinked genomic sex-determining region, the homeodomain (*b*) locus, the dikaryon is maintained and cells switch to filamentous growth. Each *b* allele encodes two divergent homeodomain proteins, bE (HD1 class) and bW (HD2 class), with different alleles encoding alternate versions of these factors that can heterodimerize to govern the establishment of cell-type identity and completion of the sexual cycle (Fig. 2.2) (22, 37). In contrast to the biallelic pheromone/receptor locus, the homeodomain locus is multiallelic with at least 33 alternative versions identified, giving rise to dozens of distinct mating types. This enables any given isolate to mate with the vast majority of the population, while ensuring that active heterodimers are formed only in the dikaryon and do not arise in haploid cells.

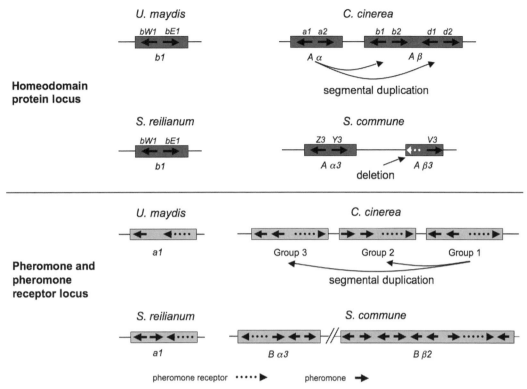

Figure 2.2 Schematic representation of *MAT* loci in the model tetrapolar basidiomycetes. The top panel depicts the homeodomain protein loci; arrows represent divergently transcribed homeodomain protein pairs. The bottom panel depicts the pheromone and pheromone receptor loci; the longer dotted arrows represent pheromone receptors, and the shorter solid arrows represent pheromone genes. Note that in both loci, segmental duplication events are evident in the evolution of *MAT*.

Because the biallelic pheromone/receptor and multiallelic homeodomain loci are unlinked and mating occurs only when the two differ, *U. maydis* has a tetrapolar mating system in which any given meiotic segregant can mate with only 25% of the progeny of a cross (*a1 b1* progeny mate with *a2 b2* but not with *a1 b1*, *a1 b2*, or *a2 b1*). However, in a random population the biallelic nature of the pheromone/receptor (*a*) locus enables fusion with 50% of the population, and the multiallelic homeodomain (*b*) locus enables dikaryon maintenance with ~97% of these individuals. The evolution of this tetrapolar system therefore restricts inbreeding and strongly drives an outbreeding lifestyle, increasing genetic diversity of the species, albeit not quite to the extent that would occur if both *MAT* loci were multiallelic.

AN ANCESTRAL FORM OF THE *U. MAYDIS* SYSTEM?

The smut *Sporisorium reilianum* is closely related to *U. maydis* and infects both maize and sorghum. Unlike its better-known relative, the role of the sexual cycle in in-

fection in *S. reilianum* is currently unknown, but the potential link between infection and mating has encouraged analysis of the mating-type system in this pathogen.

Studies at the genetic and molecular levels have revealed that like *U. maydis*, *S. reilianum* is also tetrapolar, with sexual compatibility determined by two unlinked loci. Subsequent analysis of the molecular structure of these loci revealed an organization highly similar to those of *U. maydis*: one locus encodes components of a pheromone/receptor system (the *a* locus), and the other encodes homeodomain transcription factors (the *b* locus) (Fig. 2.2) (54). Genetic and sequence analysis revealed that, like *U. maydis*, the *b* locus of *S. reilianum* encodes divergent homeodomain genes equivalent to *bE* (HD1) and *bW* (HD2) and that this locus is multiallelic (five alleles have been found).

However, these analyses have shown that the pheromone/pheromone receptor locus is quite different. In contrast to the biallelic locus of *U. maydis*, the *S. reilianum* pheromone/receptor locus is triallelic (8, 54). Each locus encodes a single pheromone receptor as well as *two* pheromone variants, each of which is recognized by one of the receptors encoded by the other alleles. For example, pheromone/receptor allele *a1* encodes two pheromones: one of these (Mfa1.2) is the ligand for the *a2* allele-encoded receptor. The other pheromone (Mfa1.3) is the ligand for the receptor encoded by the *a3* allele. In turn, alleles *a2* and *a3* both encode a pheromone (Mfa2.1 or Mfa3.1) that is the ligand of the *a1* receptor. An *a1* isolate can therefore detect and communicate with either an *a2* or an *a3* isolate. This system is organized such that each allele encodes the capacity to detect a pheromone from each of the others, in addition to reciprocating this communication by producing a pheromone detectable by the other alleles. Despite the difference apparent in this triallelic system, there are strong similarities with *U. maydis*. With the exception of the additional pheromone gene, the *a1* allele of *S. reilianum* is highly similar to the *a1* allele of *U. maydis*, just as the *S. reilianum a2* allele is homologous with the *U. maydis a2* locus, even down to the inclusion of the *lga2* and *rga2* genes. In contrast, the *S. reilianum a3* allele is unique.

The similarities between the homeodomain loci and differences between the pheromone/receptor loci raise the question of which mating-type system is the ancestral form. Although the simplicity of the *U. maydis* system would implicate it as the ancestral form, sequence analysis suggests otherwise. Work with *U. maydis* has identified a pheromone pseudogene in the *a2* allele, and comparison with the *S. reilianum a2* allele shows that an intact pheromone gene is present in this position (54, 58). This observation suggests that pseudogenization and gene deletion are two of the evolutionary forces affecting *MAT* locus structure in the *U. maydis* lineage and that *U. maydis* may be descended from an ancestral fungus that shared the more complex pheromone-encoding *MAT* locus with *S. reilianum*.

SEGMENTAL DUPLICATION AND THE EVOLUTION OF THOUSANDS OF MATING TYPES IN THE MUSHROOM *COPRINOPSIS CINEREA*

The ink cap mushroom *C. cinerea* is a saprophyte found in association with horse manure and one of the best-described model systems in the phylum *Basidiomycota*. *C. cinerea* completes its entire life cycle within 2 weeks in the laboratory, enabling genetic and molecular analyses to be performed with this genetically tractable multicellular fungus. The earliest genetic studies in this system revealed an enormous repertoire of tetrapolar mating types, estimated to be more than 12,000. This incredible diversity ensures that essentially any two individuals in the wild can mate and renders the outcrossing breeding system of *C. cinerea* extremely efficient.

Unlike the tetrapolar systems of *U. maydis* and *S. reilianum*, the *MAT* loci of *C. cinerea* do not control chemoattraction of dissimilar isolates as control of cell-cell fusion by pheromones has been dispensed with. Each encounter of two different strains results in promiscuous cell-cell fusion, and two nuclei of different mating types are not required until the postfusion steps in the sexual cycle. Thus, the mating-type locus genes control only the later aspects of mating, including dikaryon maintenance and fruiting body formation (14, 35). Unlike the *U. maydis* monokaryotic yeast form, *C. cinerea* grows as a monokaryotic (uninucleate) mycelium. This monokaryotic mycelium fuses freely with any other *C. cinerea* mycelium encountered, without a requirement for self or nonself recognition. However, if the fused monokaryons are dissimilar at both *MAT* loci, they establish a stable dikaryotic form in which two haploid parental nuclei are paired in each hyphal compartment. These nuclei remain unfused throughout the growth of dikaryons until the mushroom fruiting body forms, and it is there that nuclear fusion and meiosis occur.

The structures of the two *MAT* loci of *C. cinerea* show many similarities to their counterparts in *U. maydis*, with one locus encoding homeodomain genes (*A*) and the other encoding pheromones and pheromone receptors (*B*) (Fig. 2.2). However, both are multiallelic; 167 *A* and 79 *B* alleles are estimated to be present in nature. The structures of these alleles are also very different. In the homeodomain locus, the common theme of divergently transcribed HD1- and HD2-encoding homeodomain genes is present, but unlike in *U. maydis*, where a single HD1/HD2 pair is present, in *C. cinerea* the complexity of the locus is increased and tandemly arrayed copies of this motif are present. Susumi Ohno first suggested the power of gene duplication as an evolutionary tool over 30 years ago (48), and the *C. cinerea MAT* loci are excellent examples of this evolutionary strategy enabling the production of thousands of mating types, creating a highly selective outbreeding sexual life cycle that reduces the incidence of sibling mating within a given population. In the homeodomain (*A*) locus, two linked subloci (α and β) exist; in the archetypal form, α encodes one HD1/HD2 pair of homeodomain proteins, and β lies 7 kb downstream and encodes three HD1/HD2 pairs (14). This is a theoretical structure, derived from the sequencing and analysis of five different *A* alleles. In reality, gene deletion has also been at work during the evolution of the homeodomain locus. This has resulted in each allele encoding only four or five of the homeodomain genes expected to have existed in the recent ancestor, which likely encoded eight. Consequently, functional heterodimers can only form between HD1 and HD2 proteins derived from different alleles, enabling stable dikaryon formation.

The evolutionary theme of segmental duplication is also apparent in the pheromone/receptor (*B*) locus. *B* consists of three tandemly arrayed subloci referred to as Group 1, Group 2, and Group 3, each of which encodes a pheromone receptor and (usually) two pheromone genes; in this respect, each sublocus resembles the pheromone/receptor locus of *S. reilianum* (Fig. 2.2). To date, 13 different *B* loci have been sequenced, revealing a patchwork assembly of common pheromone/receptor cassettes. From 13 alleles, two Group 1, five Group 2, and seven Group 3 cassettes were identified. Together, this gives rise to 70 possible combinations, approaching the predicted number of permutations that exist in nature. However, the mechanism by which these can be shuffled has not yet been determined, as each cassette abuts the next without a common intervening region in which recombination could occur.

Strains only need to differ in one of the β subloci to be compatible; for example, even if the Group 1 and Group 2 subloci are identical, variation in Group 3 is sufficient to enable mating. These observations contribute to the model that pheromones encoded within each group only activate receptors in the same group, avoiding self-activation as would occur should cross talk occur between subloci as seen for the homeodomain locus. Phylogenetic analyses have confirmed that each sublocus has arisen via segmental duplication events, showing that this is a common evolutionary theme in the evolution of both the homeodomain and the pheromone/receptor loci in this species. Again, gene loss and gain, likely via unequal crossing over, are also implicated as some subloci encode only one pheromone gene and others encode three.

EVOLVING BEYOND THE PARADIGM—THE PROMISCUOUS MUSHROOM *SCHIZOPHYLLUM COMMUNE*

As in *C. cinerea*, the control of cell-cell fusion by pheromones has been dispensed with in *S. commune*, a closely related split gill mushroom that is a decay fungus growing on plant material (e.g., wood) that is already dead, and whose white appearance leads to the name white rot. Each encounter of two strains results in promiscuous cell-cell fusion, and postfusion steps in the sexual cycle require two nuclei of different mating types (38). More than 15,000 different mating types are esti-

mated to occur in this species (51). First characterized by the classic studies of John Raper and colleagues, *S. commune* is one of the oldest systems used in studies to elucidate the genetics of mating in basidiomycete fungi, and one of the most complex (51). The *S. commune* tetrapolar loci are multiallelic with an estimated 288 homeodomain (*A*) locus and 81 pheromone/receptor (*B*) locus alleles yielding over 20,000 potential permutations, a number of mating types greater than reported for the ink cap mushroom. The similarities between these two mushroom systems are further highlighted by the fact that *S. commune* also encodes subloci at both the homeodomain and pheromone/receptor loci, in agreement with Raper's model that multiallelism is a unique system that has arisen only once in fungi (51). The homeodomain (*A*) locus specificities include 9 *Aα* and 32 *Aβ* alleles, and the pheromone/receptor locus has 9 *Bα* and 9 *Bβ* alleles.

While multiallelism is a common theme shared between these two mushroom fungi, the structure of the loci is not. The homeodomain locus of *S. commune* is very different; while *Aα* and *Aβ* alleles exist, they are not closely linked. Instead, they lie 5 centimorgans (an estimated ~650 kb) apart and are readily recombined in a genetic cross. The structures of the subloci themselves are also different (51, 55). While in most isolates the homeodomain locus contains two divergently transcribed HD1/HD2-encoding genes as in *C. cinerea*, in at least one case just the HD2 gene is present (3) (Fig. 2.2). The importance of this becomes apparent when the *Aβ* locus is considered. Although only a single locus has been sequenced to date, analysis shows that it barely resembles a traditional homeodomain locus at all. Rather than a divergently transcribed HD1/HD2 gene pair, there is a single gene present that shares similarity only with the homeodomain-encoding motif of the *Aα* HD2 gene—the rest of the gene is unique, with no similar sequences present in available databases (55). Given that the *Aα* and *Aβ* loci are functionally redundant, with deletion of both the divergently transcribed HD1/HD2-encoding genes of the *Aα* locus not affecting the ability to complete the sexual cycle (52), these observations suggest that the traditional homeodomain locus (*Aα*) could potentially be in the process of being lost, to be replaced with a new, novel locus (*Aβ*) that has adopted this function (55). Further sequencing of the numerous *Aβ* alleles in nature could potentially shed light on how this unusual system has evolved.

The arrangement of the pheromone/receptor (*B*) locus also differs from its *C. cinerea* counterpart. Rather than three subloci, there are two (Fig. 2.2). Three variants of the first of these (the *Bα* locus) have been sequenced, and each encodes a single receptor and two or three pheromones, similar to *C. cinerea*. The second sublocus (the *Bβ* locus) resides 8 kb downstream, and only one has been sequenced. Remarkably, in addition to a single pheromone receptor gene, this sublocus encodes eight different pheromones. And unlike *C. cinerea*, these loci can undergo cross talk, with some α pheromones capable of activating the β pheromone receptor, and vice versa. Also, some pheromones and pheromone receptors are functionally redundant, further complicating the interpretation of potential allele combinations possible without causing self-activation.

This therefore raises a quandary—how can these loci freely recombine without generating self-fertile isolates? Some insight into this dilemma derives from the observation that despite a theoretical 81 combinations of the *Bα* and *Bβ* subloci, only 59 have ever been found in nature or isolated in the laboratory. There is evidence that the combinations achievable in this system may be regulated at the level of recombination. The interlocus 8-kb region is littered with tandemly arrayed repeats and GC-rich regions. In mapping studies, the distance between these loci ranges from 0.3 to 8 centimorgans, depending on which subloci are involved, and this is greater than would be expected based on physical distance. While the mechanism driving this recombination process is not yet understood, and sequencing of additional loci is required, this potentially represents an important evolutionary pathway capable of promoting gene shuffling while restricting the formation of self-activating combinations at the pheromone/receptor locus. Alternatively, self-activating combinations may be selected against in nature or lead to speciation.

THE EVOLUTION OF MULTIALLELIC BIPOLAR BASIDIOMYCETES

Early work indicated that although the tetrapolar mating-type system predominates in the mushroom species, almost 25% are bipolar (51). There are other bipolar mushroom species distributed across several different genera, including *Marasmius*, *Sistotrema*, and *Collybia* (47, 57). As noted by Raper, this wide distribution of the bipolar species implies that the tetrapolar mating system could be a primitive state from which the bipolar mating system evolved independently in different clades based on simple genetic changes (51).

Raper proposed three hypotheses on the origin of the bipolar mating systems. First, self-compatible mutations could occur in either of the tetrapolar loci that lead to bipolar mating behavior, such as those observed in mutants identified independently in both nature and the laboratory setting (40, 50, 51). Second, a chromosomal translocation event could take place that results in a fusion of the pheromone/receptor and homeodomain loci into a nonrecombining region that creates bipolarity. Finally, Raper proposed that the regulatory function of one factor could be gradually assumed by the other. From the viewpoint of population genetics, the main difference between a tetrapolar and a bipolar mating system is that the degree of inbreeding between the siblings is more restricted in tetrapolar species (25% compared to 50% in bipolar species). Therefore, the repeated emergence of bipolar systems in the basidiomycetes indicates a trend to evolve towards inbreeding.

There is now molecular evidence to support some of these models. The mushrooms *Coprinellus disseminatus* and *Pholiota nameko* both support Raper's theory that bipolar systems could evolve by self-compatible mutations occurring in either of the tetrapolar loci. Although both pheromone/receptor and homeodomain loci have been identified in these species, genetic analyses have shown that the pheromone/receptor locus does not confer *MAT* specificity and is not linked to mating type (1, 34). In contrast, the homeodomain locus (again encoding an HD1/HD2 divergently transcribed pair) is still a multiallelic mating-type determinant.

Characterization of the complete *MAT* loci from *C. disseminatus* has shown a remarkable similarity to the other multiallelic mushrooms. The multiallelic homeodomain locus (*A*, conferring an estimated 123 mating types) is composed of two pairs of divergently transcribed HD1/HD2-encoding genes, while the pheromone locus is more diffuse, with a cluster of three pheromone genes and two pheromone receptor genes disrupted by the inclusion of four other genes not involved in pheromone sensing or production (34). In both of these species, for two strains to mate they must differ only at the homeodomain locus, and the pheromone/receptor locus segregates independently of mating type in genetic crosses. These mushrooms therefore represent the end of the *MAT* locus complexity spectrum opposite to *C. cinerea* and *S. commune*: they have abandoned the pheromone system as a mating-type determinant (either because the receptor is constitutively activated or the pheromone and receptor are self-compatible), creating a system more analogous to the bipolar ascomycetes where inbreeding is supported.

THE EVOLUTION OF THE BIALLELIC BIPOLAR BASIDIOMYCETES *USTILAGO HORDEI* AND *MICROBOTRYUM VIOLACEUM*

The mechanism employed by *C. disseminatus* and *P. nameko* has given rise to unusual systems where these mushrooms are still multiallelic, even though they are bipolar (1, 34). This is therefore very different from the bipolar system seen in the ascomycetes. Unlike these mushrooms, there are a number of examples where the bipolar systems that have evolved resemble even more closely the ascomycete lineage; that is, a biallelic bipolar system.

Several direct examples have been identified that support Raper's second model for the formation of a bipolar system, where a chromosomal translocation event has taken place to result in a close linkage of *A* and *B* loci, which ultimately led to the fusion of the *A* and *B* factors into a nonrecombining region, creating bipolarity. For example, in the pathogen of small-grain cereals *Ustilago hordei*, a species closely related to *U. maydis* and *S. reilianum*, the two small tetrapolar *MAT* loci are now linked, forming a single larger locus that spans ~500,000 bp, more than one-sixth of the ~2.8-Mb-chromosome on which it resides (6, 41). Recombination is suppressed across the entire region between the homeodomain (*b*) and pheromone/receptor (*a*) loci at the distal ends of the *MAT* locus, which is rich in transposons and repetitive elements in addition to numerous genes not involved in the sexual cycle (2, 4). The *U. hordei* mating-type locus is no longer multiallelic, resulting in only two different mating-type alleles, *MAT-1* (*a1 b1*, separated by 500 kb) and *MAT-2* (*a2 b2*, separated by 430 kb), to give rise to a bipolar system. It is striking that heterologous expression studies have revealed that the *U. hordei b* genes are functional in *U. maydis*, further illustrating the close evolutionary link between the two fungi (5).

A similar biallelic, bipolar scenario has been observed in the more distantly related smut *Microbotryum violaceum*, a basidiomycetous fungus that causes anther smut disease of flowering plants. Again, mating is controlled by two mating-type alleles, *a1* and *a2*. In this case the mating-type identity is conferred by large sex chromosomes (2.9 and 3.5 Mb) that exhibit clear heteromorphism and are also rich in repetitive elements (25–27). While in this case the mating-type-specific regions of the genome have not yet been sequenced, this system is likely to share features with that present in *U. hordei*.

CRYPTOCOCCUS NEOFORMANS, A COMPLEX AND UNUSUAL MATING-TYPE SYSTEM

An even more complex example of the tetrapolar-bipolar transition is *C. neoformans*, a ubiquitous human fungal pathogen with a bipolar mating-type system involving haploid **a** and α cells. Sequencing of the *Cryptococcus MAT* locus revealed that it is unusually large

(>100 kb), contains ~25 genes, and is composed almost entirely of divergent alleles of a common gene set (19, 43). In addition to including homeodomain and pheromone/receptor genes, the *C. neoformans MAT* alleles also include multiple additional genes involved in the sexual cycle, including several genes involved in the pheromone-sensing mitogen-activated protein kinase cascade, as well as genes involved in meiosis and sporulation. In contrast, in the other basidiomycetes whose genomes are now complete (*U. maydis*, *C. cinerea*, and *Phanerochaete chrysosporium*), these genes are scattered throughout the genome (36, 46).

Subsequent studies in the sibling species *Cryptococcus gattii* have revealed that this species also contains an equivalently large mating-type locus arranged in a similar fashion (19). In *C. neoformans*, *MAT* encompasses ~6% of the chromosome upon which it resides (44) and therefore represents a proportionately larger sex-specific region than even the "sex chromosomes" present in some animals and plants. The unusual structure of the *C. neoformans MAT* locus has been characterized in **a** and α isolates of a *Cryptococcus* species cluster diverging over 40 million years. While the size and content of *MAT* are conserved, gene positions within *MAT* are highly rearranged. Using techniques similar to those applied to studies of animal and plant sex chromosomes, including analysis of gene order, phylogeny, and synonymous substitution rates, the evolutionary steps that fashioned the extant bipolar *MAT* locus system from an ancestral tetrapolar system were deduced (Fig. 2.3) (19). First, *MAT* expanded via the acquisition of sex-related genes into two unlinked clusters on different autosomes. Next, these independent regions fused to form a single larger sex-determining genomic structure in one mating type. The opposite mating type was then converted to an equivalent structure via gene conversion between the linked and unlinked sex-determining regions. These steps are highlighted by the presence of four evolutionary gene strata similar to those seen in the sex chromosomes of mammals, birds, and plants. Finally, *MAT* has been subject to ongoing intra- and interallelic gene conversion and inversions that suppress recombination, giving rise to the extant structures observed today (19).

The evolution of *MAT* in *Cryptococcus* therefore shares remarkable parallels with the human Y chromosome (19). Like animal and plant sex chromosomes (and the *MAT* loci of *U. hordei* and *M. violaceum*), *MAT* is rich in repetitive elements, with transposons and other repeats representing ~25% of the structure, roughly five times higher than in other genomic regions. The accumulation of transposable elements has therefore emerged as a common theme in those basidiomycetes that have

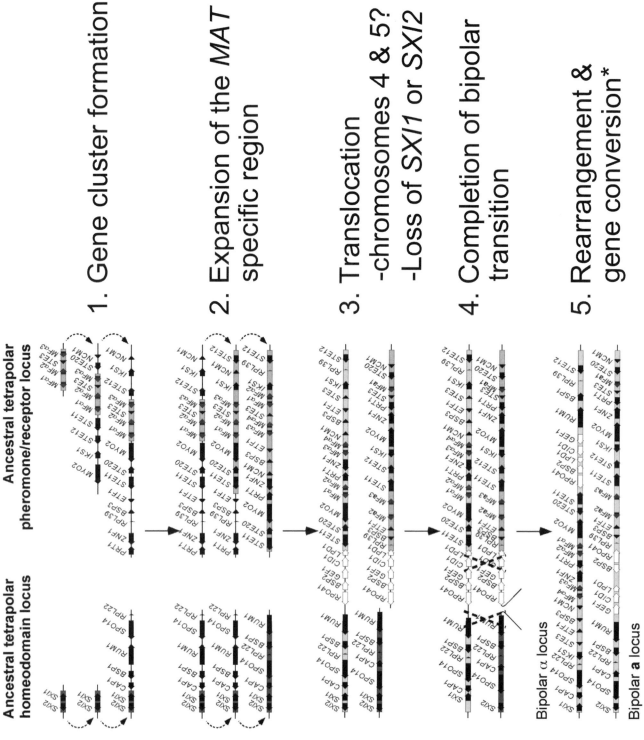

1. Gene cluster formation

2. Expansion of the *MAT* specific region

3. Translocation
-chromosomes 4 & 5?
-Loss of *SXI1* or *SXI2*

4. Completion of bipolar transition

5. Rearrangement & gene conversion*

Figure 2.3 Model for the evolution of *MAT* in *C. neoformans*. *MAT* has evolved via four main steps, beginning with gene acquisition into two unlinked sex-determining regions, forming two discrete gene clusters involved in mating. Next, these fused into a single structure via a chromosomal translocation in one mating type. The translocation resulted in a tripolar intermediate state that was converted to a bipolar system via gene conversion between the linked and unlinked sex-determining regions. Lastly, *MAT* has been subjected to ongoing intra- and interallelic gene conversion and inversions that suppress recombination.

evolved from a tetrapolar to a bipolar system. Beyond these features common to most sex chromosomes, *MAT* has further similarity to the human Y chromosome. Analysis of the phylogeny of each gene of *MAT* identified ancient boundaries of the expanding sex-determining regions where parts of the *RPO41* and *LPD1* genes are *MAT* specific, the rest is not—similar to the amelogenin locus that spans an ancient pseudoautosomal boundary in primates (33). In short, the *Cryptococcus MAT* locus bears many of the hallmarks of the evolutionary events responsible for the formation of the human Y chromosome, and all within a structure a fraction of the size.

Almost 40 years ago, Ohno first proposed how animal sex chromosomes might have been forged (49). In his model, the original master sex-determining gene was autosomal and only later captured onto the incipient sex chromosome in conjunction with other genes with sex-related functions. Building on this more discrete genomic foundation, the evolution of distinct sex chromosomes was then proposed to have occurred via suppression of recombination, by mechanisms including inversions, leading to the divergence of large genomic tracts and the emergence of heteromorphic sex chromosomes. Gene capture and suppression of recombination therefore punctuate sex chromosome evolution since differentiation of the nascent sex chromosomes is only accelerated when recombination is decreased.

Beyond the complexities of the *Cryptococcus* system that it shares with sex chromosomes of animals and the *U. hordei* and *M. violaceum* bipolar fungi, it has evolved some unique features of its own. In contrast to other basidiomycetes, where the pheromone genes are different from each other when multiple genes are present, in *Cryptococcus* a single pheromone gene has undergone triplication or quadruplication with each usually yielding an identical product. As with duplicated genes involved in fertility on the human Y chromosome (53), the copies of the *Cryptococcus* pheromone genes are arranged in inverted repeats that are maintained by intra-allelic gene conversion, enabling gene repair in the absence of a paired chromosome for recombinational repair. In the *a* and *α* loci only a single pheromone receptor is present, and it functions to recognize the pheromone from the opposite cell type.

Another unique feature of the *Cryptococcus* system is apparent when considering the homeodomain-encoding genes. Unlike the other basidiomycetes, each of the *Cryptococcus MAT* alleles contains only a single homeodomain gene; the *α* locus contains an HD1 gene (*SXI1α*), while the *a* locus contains an HD2 gene (*SXI2a*). A functional heterodimer can still therefore form following fusion of opposite cell types, but this effectively abolishes the restrictions normally placed on HD1/HD2 evolution; that is, in a haploid cell self-self interactions do not occur as both an HD1 and HD2 homeodomain protein will never be present. Also, that only one HD1/HD2 pair is present rather than two or multiple possible compatible pairs may serve to further restrict outbreeding potential if not all *SXI1α* alleles are compatible with all *SXI2a* alleles in the population.

RECOMBINATION AND THE EVOLUTION OF *MAT* LOCI

A recurrent topic in the evolution of these different mating-type themes is recombination, a process that is synonymous with the sexual cycle. In *S. commune*, regulation of recombination between different pheromone/receptor subloci is hypothesized to prevent the formation of self-activating combinations. Recent studies in *C. neoformans* have also implicated recombination in *MAT* locus dynamics.

A dramatic expansion of the genetic map was noticed in the regions that are adjacent to the *MAT* locus in *Cryptococcus*, which implies the existence of recombination hot spots in the regions flanking *MAT*. Through genetic analysis of strains bearing dominant selectable markers inserted at defined physical distances from the right and left borders of *MAT*, recombination was found to be elevated 10- to 50-fold compared to the genome-wide average (29, 45). Furthermore, crossovers were observed to occur on both sides of *MAT* more frequently than would be predicted based on the product of the individual recombination rates, illustrating an example of negative interference. This results in an increased rate of crossover events on both sides of *MAT* and, as a result, the *MAT* locus can be replaced onto the homologous chromosome as a unit—an effective switch of *MAT*. Fine mapping of the 3′ recombination hot spot indicates that it is located ~1 to ~5 kb away from the right *MAT* border (29).

Global mapping of recombination hot spots in *S. cerevisiae* revealed a correlation between hot spot activity and a high G+C sequence content. Interestingly, sequence analysis of the G+C base-pair composition for *Cryptococcus MAT* and flanking regions also revealed that the estimated hot spot positions are in close proximity to G+C peaks, suggesting that a high G+C content might contribute to these recombination hot spots. Indeed, when the partial G+C-rich sequence was deleted, recombination frequency was reduced to close to the level of the genome-wide average. Moving part of the G+C-rich sequence to another location on the *MAT* chromosome did not increase recombination, suggesting that either the hot spot activity is context dependent and could involve higher-order chromosome structure or that there are additional flanking sequences involved yet to be defined. In addition to the G+C peaks that flank the *MAT* locus, a minor G+C peak was identified within *MAT* and lies between two highly conserved divergently transcribed genes (*RPO41* and *BSP2*). This G+C content peak was found in all three *Cryptococcus* varieties and both mating types.

We propose that these recombination hot spots likely contributed to several of the hypothesized steps in the evolution of the *MAT* locus gene cluster. First, we hypothesize that the ancestral tetrapolar *MAT* loci were flanked by the G+C-rich recombination hot spots. These *MAT*-linked recombination hot spots may have facilitated capture of genes into both mating-type alleles by enabling rapid assimilation of the flanking regions in linkage to both the *a* and *α* alleles before these flanking genes were captured by inversions into the *MAT* locus. Second, the hot spots might have mediated the fusion/translocation event between the two ancestral tetrapolar loci, and the collapse of the tripolar intermediate to a bipolar one. The G+C-rich sequence that lies inside the *MAT* locus might serve as a hot spot to drive local gene conversions between the *a* and *α* alleles, which is evident by the fact that the *a* and *α* alleles of *RPO41* and *BSP2* share over 99% sequence identity at the nucleotide level and are therefore species-specific but not mating-type-specific genes, despite their residence within the confines of the borders of the *MAT* locus (19).

Several examples of fungal gene clusters have been identified and studied in detail; many are responsible for the production of secondary metabolites or involved in biosynthetic pathways. These include the gene clusters for aflatoxin, sterigmatocystin, and sirodesmin production and the *DAL* gene cluster for allantoin utilization (11, 21, 59). How these clusters were assembled during evolution is not well understood. One common feature of these clusters is that they exhibit reduced recombination, which is also an important feature of the *MAT* gene cluster of *Cryptococcus*, although mechanisms that suppress recombination could be different. Interestingly, based on previous studies (59), although recombination is suppressed within the *DAL* gene cluster in *S. cerevisiae*, the flanking regions of the cluster exhibit increased meiotic recombination. Thus, we propose that studies on the evolution of *C. neoformans MAT* gene cluster formation may have broader general implications for how gene clusters evolve.

IMPLICATIONS FOR TRANSITIONS BETWEEN HETEROTHALLIC AND HOMOTHALLIC SEXUAL CYCLES

Studies on the molecular events that have transpired during the evolution of *MAT* in the phylum *Basidiomycota* also promise to provide insights into the transitions that have occurred between heterothallic outcrossing modes of sexual reproduction and homothallic inbreeding cycles. The fact that independent mating-type determinants have arisen that are unlinked and lie on different chromosomes in the genome reveals an underlying plasticity in the ways in which mating type is genetically determined. In the ascomycete lineage, the genes encoding the pheromones and pheromone receptors are under the transcriptional control of the *MAT* locus but are not components of *MAT* itself. By contrast, in many basidiomycetes, the genes encoding pheromones and pheromone receptors have become an integral part of one of the two *MAT* loci in those fungi with a tetrapolar system.

How might this have arisen? We hypothesize that in an ancestral fungus either or both *MAT* loci might have been organized in a self-compatible fashion. By this we mean that the locus encoding the pheromones and pheromone receptors encoded a pheromone that acted as an activating ligand for the pheromone receptor encoded by the linked gene. This would then serve to promote an autocrine signaling loop in which the producing cell would respond to its own pheromone, producing a situation in which ligand-receptor compatibility could evolve. Examples of mutations that lead to pheromone-receptor self-activation, or constitutive pheromone receptor activity, are known in *S. commune* and *C. cinerea*. For example, in *C. cinerea*, a mutation (R96H) in the second intracellular loop of the Rcb36 pheromone receptor results in activation of the receptor by a pheromone ligand that is incompatible with the wild-type receptor (50). In another case, an F67W mutation in the Phb3.26 pheromone enables activation of a receptor that is incompatible with the wild-type pheromone (50). In *S. commune*, similar pheromone and receptor mutations that alter compatibility have been described (18). Mutations that constitutively activate pheromone receptor signaling, or novel chimeric receptor genes, also confer self-compatibility (23, 50). Thus, mutations can readily alter an incompatible receptor-pheromone gene pair into a compatible pair.

We propose that transitions between self-activating and self-incompatibility may have occurred not only by mutation but also by recombination. If one imagines that two alleles of the pheromone and pheromone receptor locus that are both self-compatible exist in the population, a recombination event between the pheromone and receptor genes would produce two recombinant loci that are no longer self-compatible but instead are now intercompatible (Fig. 2.4). This could effect a transition from a bipolar system in which the pheromone receptor locus was not linked to mating type into a tetrapolar system in which it is now a mating-type determinant. Transitions in the opposite direction would lead to the production of isolates capable of responding to their own pheromone, which would be the first step in a transition from heterothallic to homothallic reproduction. The recent studies by James et al. on *C. disseminatus* (34) revealed that the *MAT* locus encoding pheromones and pheromone receptors is no longer linked to mating type in this species. The hypothesis that the pheromone receptor might be constitutively activated was advanced. Alternatively, this may represent an example in which the pheromone and receptor are self-compatible, and these models remain to be tested by genetic and molecular approaches.

Transitions between heterothallic and homothallic reproduction might also be effected by similar events that could transpire at the *MAT* locus encoding the homeodomain cell-type identity determinants (Fig. 2.4). A shared feature of *MAT* organization in *U. maydis*, *S. commune*, and *C. cinerea* is that two homeodomain genes are present, one HD1 and one HD2, which are divergently transcribed. These are not self-compatible, whereas the gene products from two different alleles are intercompatible for these heterothallic fungi. Examples are known in which fusion events have produced activated HD1-HD2 fusions in *C. cinerea* (40), and thus, it is clear that mutations can lead to self-activating alleles, at least at the *A* locus for *C. cinerea*. Similar to the hypotheses advanced above for the pheromone receptor locus, recombination events could occur between the two linked homeodomain genes of two different *MAT* alleles (for example, in *U. maydis*, recombination between a *bE1 bW1* allele and a *bE2 bW2* allele would

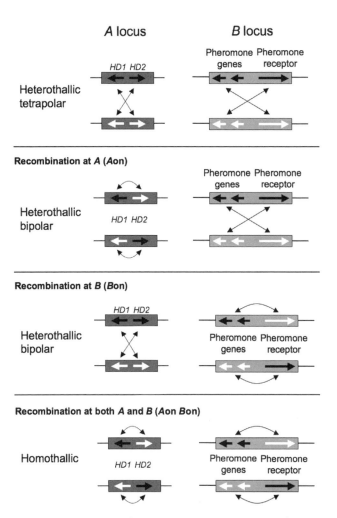

Figure 2.4 *MAT* recombination may trigger transitions between bipolar and tetrapolar, and homothallic-heterothallic mating systems. Recombination at either of the *MAT* loci in a tetrapolar basidiomycete may lead to self-activation of one of the loci, which transforms the mating system to bipolar. Similarly, recombination events at both *MAT* loci would convert a tetrapolar heterothallic basidiomycete to a homothallic life cycle.

produce *bE1 bW2* and *bE2 bW1*) to give rise to alleles encoding two homeodomain proteins that are self-compatible. By these transitions, a species that is heterothallic might undergo events at one, the other, or both *MAT* alleles, ultimately resulting in an isolate in which both alleles are self-compatible, similar to the *Aon Bon* isolates that have been described for several species. This could then effect transitions from bipolar to tetrapolar mating behavior, and the reciprocal, and also promote transitions between heterothallic and homothallic sexual cycles, and the converse. Our view is that sequencing additional *MAT* loci, and characterizing in detail the sexual cycles of additional basidiomycete species, including more detailed studies of *C. disseminatus* and the homothallic species *Filobasidiella depauperata*, will further illuminate the genetic and developmental plasticity that underlies sexual reproduction and cell type identity determination in the basidiomycetes. These analyses can be also extended to other fungal species and phyla.

References

1. Aimi, T., R. Yoshida, M. Ishikawa, D. Bao, and Y. Kitamoto. 2005. Identification and linkage mapping of the genes for the putative homeodomain protein (*hox1*) and the putative pheromone receptor protein homologue (*rcb1*) in a bipolar basidiomycete, *Pholiota nameko*. *Curr. Genet.* 48:184–194.

2. Anderson, C. M., D. A. Willits, P. J. Kosted, E. J. Ford, A. D. Martinez-Espinoza, and J. E. Sherwood. 1999. Molecular analysis of the pheromone and pheromone receptor genes of *Ustilago hordei*. *Gene* 240:89–97.

3. Asada, Y., C. Yue, J. Wu, G. P. Shen, C. P. Novotny, and R. C. Ullrich. 1997. *Schizophyllum commune* A alpha mating-type proteins, Y and Z, form complexes in all combinations in vitro. *Genetics* 147:117–123.

4. Bakkeren, G., G. Jiang, R. L. Warren, Y. Butterfield, H. Shin, R. Chiu, R. Linning, J. Schein, N. Lee, G. Hu, D. M. Kupfer, Y. Tang, B. A. Roe, S. Jones, M. Marra, and J. W. Kronstad. 2006. Mating factor linkage and genome evolution in basidiomycetous pathogens of cereals. *Fungal Genet. Biol.* 43:655–666.

5. Bakkeren, G., and J. W. Kronstad. 1993. Conservation of the *b* mating-type gene complex among bipolar and tetrapolar smut fungi. *Plant Cell* 5:123–136.

6. Bakkeren, G., and J. W. Kronstad. 1994. Linkage of mating-type loci distinguishes bipolar from tetrapolar mating in basidiomycetous smut fungi. *Proc. Natl. Acad. Sci. USA* 91:7085–7089.

7. Banuett, F. 1992. *Ustilago maydis*, the delightful blight. *Trends Genet.* 8:174–180.

8. Bolker, M., M. Urban, and R. Kahmann. 1992. The *a* mating type locus of U. *maydis* specifies cell signaling components. *Cell* 68:441–450.

9. Bortfeld, M., K. Auffarth, R. Kahmann, and C. W. Basse. 2004. The *Ustilago maydis a2* mating-type locus genes *lga2* and *rga2* compromise pathogenicity in the absence of the mitochondrial p32 family protein Mrb1. *Plant Cell* 16:2233–2248.

10. Brown, A. J., and L. A. Casselton. 2001. Mating in mushrooms: increasing the chances but prolonging the affair. *Trends Genet.* 17:393–400.

11. Brown, D. W., J. H. Yu, H. S. Kelkar, M. Fernandes, T. C. Nesbitt, N. P. Keller, T. H. Adams, and T. J. Leonard. 1996. Twenty-five coregulated transcripts define a sterigmatocystin gene cluster in *Aspergillus nidulans*. *Proc. Natl. Acad. Sci. USA* 93:1418–1422.

12. Burnett, J. H. 1975. *Mycogenetics*. John Wiley & Sons, London, United Kingdom.

13. Butler, G., C. Kenny, A. Fagan, C. Kurischko, C. Gaillardin, and K. H. Wolfe. 2004. Evolution of the *MAT* locus and its Ho endonuclease in yeast species. *Proc. Natl. Acad. Sci. USA* 101:1632–1637.

14. Casselton, L. A., and N. S. Olesnicky. 1998. Molecular genetics of mating recognition in basidiomycete fungi. *Microbiol. Mol. Biol. Rev.* 62:55–70.

15. Dooijes, D., M. van de Wetering, L. Knippels, and H. Clevers. 1993. The *Schizosaccharomyces pombe* mating-type gene *mat-Mc* encodes a sequence-specific DNA-binding high mobility group box protein. *J. Biol. Chem.* 268:24813–24817.

16. Feldbrugge, M., J. Kamper, G. Steinberg, and R. Kahmann. 2004. Regulation of mating and pathogenic development in *Ustilago maydis*. *Curr. Opin. Microbiol.* 7:666–672.

17. Fowler, T. J., S. M. DeSimone, M. F. Mitton, J. Kurjan, and C. A. Raper. 1999. Multiple sex pheromones and receptors of a mushroom-producing fungus elicit mating in yeast. *Mol. Biol. Cell* 10:2559–2572.

18. Fowler, T. J., M. F. Mitton, L. J. Vaillancourt, and C. A. Raper. 2001. Changes in mate recognition through alterations of pheromones and receptors in the multisexual mushroom fungus *Schizophyllum commune*. *Genetics* 158:1491–1503.

19. Fraser, J. A., S. Diezmann, R. L. Subaran, A. Allen, K. B. Lengeler, F. S. Dietrich, and J. Heitman. 2004. Convergent evolution of chromosomal sex-determining regions in the animal and fungal kingdoms. *PLoS Biol.* 2:2243–2255.

20. Fraser, J. A., and J. Heitman. 2003. Fungal mating-type loci. *Curr. Biol.* 13:R792–R795.

21. Gardiner, D. M., A. J. Cozijnsen, L. M. Wilson, M. S. Pedras, and B. J. Howlett. 2004. The sirodesmin biosynthetic gene cluster of the plant pathogenic fungus *Leptosphaeria maculans*. *Mol. Microbiol.* 53:1307–1318.

22. Gillissen, B., J. Bergemann, C. Sandmann, B. Schroeer, M. Bolker, and R. Kahmann. 1992. A two-component regulatory system for self/non-self recognition in *Ustilago maydis*. *Cell* 68:647–657.

23. Gola, S., J. Hegner, and E. Kothe. 2000. Chimeric pheromone receptors in the basidiomycete *Schizophyllum commune*. *Fungal Genet. Biol.* 30:191–196.

24. Herskowitz, I., J. Rine, and J. N. Strathern. 1992. *Mating Type Determination and Mating-Type Interconversion in* Saccharomyces cerevisiae, vol. 2. Cold Spring Harbor Laboratory Press, Cold Spring Harbor, NY.

25. Hood, M. E. 2002. Dimorphic mating-type chromosomes in the fungus *Microbotryum violaceum*. *Genetics* 160:457–461.

26. Hood, M. E. 2005. Repetitive DNA in the automictic fungus *Microbotryum violaceum*. *Genetica* 124:1–10.

27. Hood, M. E., J. Antonovics, and B. Koskella. 2004. Shared forces of sex chromosome evolution in haploid-mating and diploid-mating organisms: *Microbotryum violaceum* and other model organisms. *Genetics* 168:141–146.

28. Horn, F., E. Bettler, L. Oliveira, F. Campagne, F. E. Cohen, and G. Vriend. 2003. GPCRDB information system

for G protein-coupled receptors. *Nucleic Acids Res.* 31: 294–297.

29. Hsueh, Y. P., A. Idnurm, and J. Heitman. 2006. Recombination hot spots flank the *Cryptococcus* mating-type locus: implications for the evolution of a fungal sex chromosome. *PLoS Genet.* 2:e184.

30. Hsueh, Y. P., and W. C. Shen. 2005. A homolog of Ste6, the a-factor transporter in *Saccharomyces cerevisiae*, is required for mating but not for monokaryotic fruiting in *Cryptococcus neoformans*. *Eukaryot. Cell* 4:147–155.

31. Hull, C. M., and A. D. Johnson. 1999. Identification of a mating type-like locus in the asexual pathogenic yeast *Candida albicans*. *Science* 285:1271–1275.

32. Idnurm, A., Y. S. Bahn, K. Nielsen, X. Lin, J. A. Fraser, and J. Heitman. 2005. Deciphering the model pathogenic fungus *Cryptococcus neoformans*. *Nat. Rev. Microbiol.* 3:753–764.

33. Iwase, M., Y. Satta, Y. Hirai, H. Hirai, H. Imai, and N. Takahata. 2003. The amelogenin loci span an ancient pseudoautosomal boundary in diverse mammalian species. *Proc. Natl. Acad. Sci. USA* 100:5258–5263.

34. James, T. Y., P. Srivilai, U. Kues, and R. Vilgalys. 2006. Evolution of the bipolar mating system of the mushroom *Coprinellus disseminatus* from its tetrapolar ancestors involves loss of mating-type-specific pheromone receptor function. *Genetics* 172:1877–1891.

35. Kamada, T. 2002. Molecular genetics of sexual development in the mushroom *Coprinus cinereus*. *Bioessays* 24: 449–459.

36. Kamper, J., R. Kahmann, M. Bolker, L. J. Ma, T. Brefort, B. J. Saville, F. Banuett, J. W. Kronstad, S. E. Gold, O. Muller, M. H. Perlin, H. A. Wosten, R. de Vries, J. Ruiz-Herrera, C. G. Reynaga-Pena, K. Snetselaar, M. McCann, J. Perez-Martin, M. Feldbrugge, C. W. Basse, G. Steinberg, J. I. Ibeas, W. Holloman, P. Guzman, M. Farman, J. E. Stajich, R. Sentandreu, J. M. Gonzalez-Prieto, J. C. Kennell, L. Molina, J. Schirawski, A. Mendoza-Mendoza, D. Greilinger, K. Munch, N. Rossel, M. Scherer, M. Vranes, O. Ladendorf, V. Vincon, U. Fuchs, B. Sandrock, S. Meng, E. C. Ho, M. J. Cahill, K. J. Boyce, J. Klose, S. J. Klosterman, H. J. Deelstra, L. Ortiz-Castellanos, W. Li, P. Sanchez-Alonso, P. H. Schreier, I. Hauser-Hahn, M. Vaupel, E. Koopmann, G. Friedrich, H. Voss, T. Schluter, J. Margolis, D. Platt, C. Swimmer, A. Gnirke, F. Chen, V. Vysotskaia, G. Mannhaupt, U. Guldener, M. Munsterkotter, D. Haase, M. Oesterheld, H. W. Mewes, E. W. Mauceli, D. DeCaprio, C. M. Wade, J. Butler, S. Young, D. B. Jaffe, S. Calvo, C. Nusbaum, J. Galagan, and B. W. Birren. 2006. Insights from the genome of the biotrophic fungal plant pathogen *Ustilago maydis*. *Nature* 444:97–101.

37. Kamper, J., M. Reichmann, T. Romeis, M. Bolker, and R. Kahmann. 1995. Multiallelic recognition: nonself-dependent dimerization of the bE and bW homeodomain proteins in *Ustilago maydis*. *Cell* 81:73–83.

38. Kothe, E. 1999. Mating types and pheromone recognition in the homobasidiomycete *Schizophyllum commune*. *Fungal Genet. Biol.* 27:146–152.

39. Kothe, E. 1996. Tetrapolar fungal mating types: sexes by the thousands. *FEMS Microbiol. Rev.* 18:65–87.

40. Kues, U., B. Gottgens, R. Stratmann, W. V. Richardson, S. F. O'Shea, and L. A. Casselton. 1994. A chimeric homeodomain protein causes self-compatibility and constitutive sexual development in the mushroom *Coprinus cinereus*. *EMBO J.* 13:4054–4059.

41. Lee, N., G. Bakkeren, K. Wong, J. E. Sherwood, and J. W. Kronstad. 1999. The mating-type and pathogenicity locus of the fungus *Ustilago hordei* spans a 500-kb region. *Proc. Natl. Acad. Sci. USA* 96:15026–15031.

42. Lengeler, K. B., R. C. Davidson, C. D'Souza, T. Harashima, W. C. Shen, P. Wang, X. Pan, M. Waugh, and J. Heitman. 2000. Signal transduction cascades regulating fungal development and virulence. *Microbiol. Mol. Biol. Rev.* 64:746–785.

43. Lengeler, K. B., D. S. Fox, J. A. Fraser, A. Allen, K. Forrester, F. S. Dietrich, and J. Heitman. 2002. Mating-type locus of *Cryptococcus neoformans*: a step in the evolution of sex chromosomes. *Eukaryot. Cell* 1:704–718.

44. Loftus, B. J., E. Fung, P. Roncaglia, D. Rowley, P. Amedeo, D. Bruno, J. Vamathevan, M. Miranda, I. J. Anderson, J. A. Fraser, J. E. Allen, I. E. Bosdet, M. R. Brent, R. Chiu, T. L. Doering, M. J. Donlin, C. A. D'Souza, D. S. Fox, V. Grinberg, J. Fu, M. Fukushima, B. J. Haas, J. C. Huang, G. Janbon, S. J. Jones, H. L. Koo, M. I. Krzywinski, J. K. Kwon-Chung, K. B. Lengeler, R. Maiti, M. A. Marra, R. E. Marra, C. A. Mathewson, T. G. Mitchell, M. Pertea, F. R. Riggs, S. L. Salzberg, J. E. Schein, A. Shvartsbeyn, H. Shin, M. Shumway, C. A. Specht, B. B. Suh, A. Tenney, T. R. Utterback, B. L. Wickes, J. R. Wortman, N. H. Wye, J. W. Kronstad, J. K. Lodge, J. Heitman, R. W. Davis, C. M. Fraser, and R. W. Hyman. 2005. The genome of the basidiomycetous yeast and human pathogen *Cryptococcus neoformans*. *Science* 307:1321–1324.

45. Marra, R. E., J. C. Huang, E. Fung, K. Nielsen, J. Heitman, R. Vilgalys, and T. G. Mitchell. 2004. A genetic linkage map of *Cryptococcus neoformans* variety *neoformans* serotype D (*Filobasidiella neoformans*). *Genetics* 167:619–631.

46. Martinez, D., L. F. Larrondo, N. Putnam, M. D. Gelpke, K. Huang, J. Chapman, K. G. Helfenbein, P. Ramaiya, J. C. Detter, F. Larimer, P. M. Coutinho, B. Henrissat, R. Berka, D. Cullen, and D. Rokhsar. 2004. Genome sequence of the lignocellulose degrading fungus *Phanerochaete chrysosporium* strain RP78. *Nat. Biotechnol.* 22:695–700.

47. Murphy, J., and O. J. Miller. 1997. Diversity and local distribution of mating alleles in *Marasmiellus praeacutus* and *Collybia subnuda* (Basidiomycetes, Agaricales). *Can. J. Bot.* 75:8–17.

48. Ohno, S. 1970. *Evolution by Gene Duplication.* Allen and Unwin, London, United Kingdom.

49. Ohno, S. 1967. *Sex Chromosomes and Sex-Linked Genes.* Springer-Verlag, New York, NY.

50. Olesnicky, N. S., A. J. Brown, Y. Honda, S. L. Dyos, S. J. Dowell, and L. A. Casselton. 2000. Self-compatible *B* mutants in coprinus with altered pheromone-receptor specificities. *Genetics* 156:1025–1033.

51. Raper, J. 1966. *Genetics of Sexuality in Higher Fungi.* The Ronald Press, New York, NY.

52. **Robertson, C. I., K. A. Bartholomew, C. P. Novotny, and R. C. Ullrich.** 1996. Deletion of the *Schizophyllum commune* Aα locus: the roles of Aα *Y* and *Z* mating-type genes. *Genetics* **144:**1437–1444.

53. **Rozen, S., H. Skaletsky, J. D. Marszalek, P. J. Minx, H. S. Cordum, R. H. Waterston, R. K. Wilson, and D. C. Page.** 2003. Abundant gene conversion between arms of palindromes in human and ape Y chromosomes. *Nature* **423:**873–876.

54. **Schirawski, J., B. Heinze, M. Wagenknecht, and R. Kahmann.** 2005. Mating type loci of *Sporisorium reilianum*: novel pattern with three *a* and multiple *b* specificities. *Eukaryot. Cell* **4:**1317–1327.

55. **Shen, G. P., D. C. Park, R. C. Ullrich, and C. P. Novotny.** 1996. Cloning and characterization of a *Schizophyllum* gene with A beta 6 mating-type activity. *Curr. Genet.* **29:**136–142.

56. **Staben, C., and C. Yanofsky.** 1990. *Neurospora crassa a* mating-type region. *Proc. Natl. Acad. Sci. USA* **87:**4917–4921.

57. **Ullrich, R., and J. Raper.** 1974. Number and distribution of bipolar incompatibility factors in *Sistotrema brinkmannii. Am. Nat.* **108:**507–518.

58. **Urban, M., R. Kahmann, and M. Bolker.** 1996. The biallelic *a* mating type locus of *Ustilago maydis*: remnants of an additional pheromone gene indicate evolution from a multiallelic ancestor. *Mol. Gen. Genet.* **250:**414–420.

59. **Wong, S., and K. H. Wolfe.** 2005. Birth of a metabolic gene cluster in yeast by adaptive gene relocation. *Nat. Genet.* **37:**777–782.

*Sex in Fungi: Molecular Determination
and Evolutionary Implications*
Edited by Joseph Heitman et al.
© 2007 ASM Press, Washington, D.C.

Xiaorong Lin
Joseph Heitman

3

Mechanisms of Homothallism in Fungi and Transitions between Heterothallism and Homothallism

The sexual cycles of fungi serve as paradigms to understand cell type control and specification, how cells produce and respond to extracellular cues involving pheromones, and how transitions in mating behaviors impact evolutionary specialization to unique niches in nature (57, 59, 60, 76, 82). Sexual systems are classified into heterothallism or homothallism in fungi. Homothallic fungi can fertilize themselves and produce sexual products in a culture derived from a single spore or cell (self-fertile), while heterothallic fungi require another compatible individual for sexual reproduction to occur (self-sterile).

The process of sexual development is similar in both homothallic and heterothallic fungi, and in ascomycetes and basidiomycetes. It typically involves a cell fusion event that creates a dikaryotic state (either transient or prolonged) and then a nuclear fusion event that produces a diploid state, which is followed by meiosis to generate recombinant progeny (139–143) (Fig. 3.1). What distinguishes heterothallic from homothallic species is the initiation event for the sexual cycle. Heterothallic fungi require a compatible partner for mating, whereas homothallic fungi do not and are able to self-mate. Compared to heterothallism, homothallism is in many cases less well understood. This chapter focuses on ex-

amples of homothallic fungi and how their sexuality is governed.

Several mechanisms of homothallism have been defined genetically and molecularly and involve (i) two nuclei with compatible mating-type information packaged in a single spore, (ii) mating-type switching, (iii) fused mating-type loci, (iv) the presence of two unlinked mating-type loci in a single genome, and (v) same-sex mating or monokaryotic/homokaryotic fruiting. Remarkably, the last form of homothallism occurs with only one mating type present and culturing isolates of these homothallic species under appropriate conditions leads to a complete sexual cycle in the absence of an isolate of opposite mating type. Other fungi (such as *Ashbya gossypii* and *Eremothecium coryli*) may exhibit homothallic sexual cycles in which the underlying molecular nature remains to be established. In some examples, this may involve novel mechanisms in which ploidy changes occur in the absence of cell (or even nuclear) fusion, and mechanisms other than heterozygosity for the mating-type locus enable completion of the sexual cycle. Known mutations that enable these types of ploidy and *MAT*-independent meiotic/sporulation cycles in *Saccharomyces cerevisiae* could be involved in these atypical homothallic fungi. We consider paradigmatic examples of

Xiaorong Lin and Joseph Heitman, Departments of Molecular Genetics and Microbiology, Pharmacology and Cancer Biology, and Medicine, Duke University Medical Center, Durham, NC 27710.

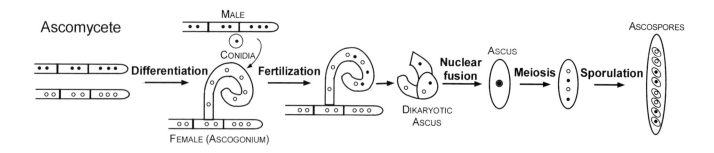

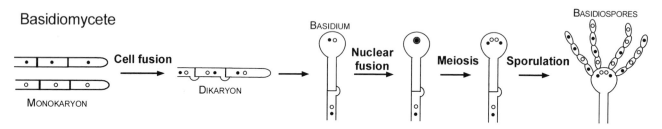

Figure 3.1 Heterothallic sexual reproduction in ascomycetes (upper row) and basidiomycetes (bottom row) shares several conserved features. A cell fusion event creates a dikaryotic state, and then, nuclear fusion occurs followed by meiosis and sporulation. The small circles represent nuclei. The solid and open circles indicate nuclei from different individuals.

each form of homothallism and then survey the fungal kingdom for examples and modifications on these themes, many of which remain to be elucidated in genetic and molecular detail.

In the fungal kingdom, homothallism is common in Zygomycetes and Ascomycetes and relatively less common in the Basidiomycetes. Conversions between heterothallic and homothallic sexual cycles are common evolutionary transitions in fungi, and whether homothallism evolved from heterothallism or vice versa is a controversial topic, and there is evidence supporting both hypotheses (145). Because it is not uncommon to find both patterns of sexuality in the same genus, and sometimes even within the same species, conversion between different sexual strategies may vary in different genera. The transition between heterothallic and homothallic sexual cycles is in essence a question of the balance between outcrossing and inbreeding, and thus, these transitions are likely a response to environmental selective pressures that favor one or the other breeding strategy. Transitions between cross- and self-fertilization are common in plants and represent an analogous transition between outbreeding and inbreeding that mirror the heterothallic-homothallic transitions in fungi (154). Given the plasticity of the mechanisms by which fungi can transit between these two distinct modes of sexual reproduction, their further analysis promises to teach us much about how fungal mating type and identity are defined, how the mating-type locus controls sexual reproduction, and the con-

sequences of these transitions on population structure and evolutionary success. It is our contention that further elucidation and understanding of prominent transitions in the mode of sexual reproduction will expand and deepen our understanding of the more general roles of sex and of the evolutionary pressures to which it is subject and that lead to changes in the relative level of inbreeding versus outcrossing.

While it may be convenient to define fungi as either homothallic or heterothallic, there are well-defined species that encompass both types of reproductive modes as described later in this chapter. One well-established example is *S. cerevisiae*, in which the original two isolates characterized by Lindegren and Winge were distinguished by a mutation in a single gene, *HO*, with Ho$^+$ strains exhibiting homothallic mating behavior and ho$^-$ strains exhibiting heterothallic mating behavior (61, 62, 109). While the majority of natural isolates of *S. cerevisiae* are diploid, Ho$^+$, and homothallic, up to 20% of the population is ho$^-$ (of 22 strains analyzed in one study, 17 were *HO/HO* **a**/α, 3 were *ho/ho* **a**/α, and 2 were *ho/HO* **a**/α) (109), indicating that even within a well-defined species both types of mating strategy are apparent (77, 118). These findings are echoed and expanded by the recent appreciation that the heterothallic fungus *Cryptococcus neoformans* with an established sexual cycle occurring between **a** and α isolates is also capable of homothallic mating behavior involving same-sex mating of α isolates (102). Thus, not only can both mating strategies occur within a given species, but also individual isolates can be endowed with the capacity to choose between a heterothallic outcrossing mode or a homothallic inbreeding mode of reproduction. These recent discoveries bring homothallic and heterothallic sexual reproduction to the forefront of current thoughts on the role of the *MAT* locus and the pheromone response pathway in sexual reproduction and illustrate that a broader view of the mechanisms of cell identity, meiosis, and genome reduction will be required if we are to understand more completely the diversity of sexual strategies adopted by fungi.

Finally, it is our hope that by illuminating these examples in the fungal kingdom this will stimulate thought, research, and insight into studies on sexual reproduction in other kingdoms of life that might have adopted similar unusual approaches to the generation of recombinant offspring better fit to meet environmental challenges.

MATING-TYPE SWITCHING IN YEASTS: PARADIGMATIC MODEL SYSTEMS FOR THE MOLECULAR BASIS OF HOMOTHALLISM

Saccharomyces cerevisiae

S. cerevisiae is a premier model system for understanding transitions in modes of sexual reproduction (159). The genetic basis for homothallism in *S. cerevisiae* was

defined by classic studies of Roman and colleagues resolving the long-standing debate that raged between Winge and Lindegren as to whether the species was heterothallic or homothallic. As is often the case in such scientific disputes, both turned out to be correct, depending on which isolate was under examination! The further genetic and molecular basis for homothallism was elucidated by classic studies of Oshima and Herskowitz and colleagues, who developed the cassette model for mating-type switching (61).

The cloning of the active *MAT* locus and of the silent *MAT* cassettes *HML* and *HMR* (63, 120) enabled the molecular basis of switching to be defined as a gene conversion event elicited by a DNA break near *MAT* provoked by the action of the Ho endonuclease (Fig. 3.2) (161). The silent copies of *MAT* that reside at *HML* and *HMR* are kept silent by virtue of a modified chromatin structure invoked by the action of the Sir protein complex (149) (see the chapter by Rusche and Hickman [chapter 11]). This altered chromatin structure serves to maintain these *MAT* copies as transcriptionally silenced, which is necessary for haploid strains to exhibit a single defined mating type and be competent for mating. In strains in which the Sir proteins have been inactivated by mutation, both alleles of *MAT* are expressed and the haploid cell then behaves as if it were an a/α diploid.

Silencing of the *HML* and *HMR* cassettes also prevents the Ho endonuclease from cleaving at these regions

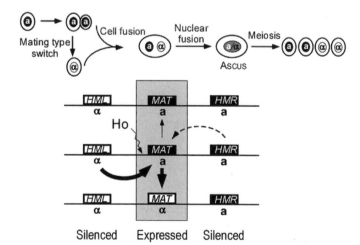

Figure 3.2 Mating-type switching in *Saccharomyces cerevisiae*. The mother cell (depicted as **a**) undergoes mating-type switching to α and then mates with the daughter cell (**a**) to create a heterozygous diploid (**a**/α). This diploid cell can either amplify by mitosis through budding (not depicted) under rich media conditions or, in response to nitrogen limitation and nonfermentable carbon source, undergo meiosis to produce four meiotic ascospores with two **a** cells and two α cells. Solid and open circles indicate nuclei of **a** and α mating type. Only the shaded locus (*MAT*) is expressed. The thickness of the arrow represents the efficiency of the switching event. Arrows indicate the direction of the switching event.

of the genome and thereby ensures that only a single double-stranded break will occur near the active *MAT* cassette, generating then a unidirectional mode of switching, from *HML* or *HMR* to *MAT* (81). There is a further bias in that the *a* allele at *MAT* more often recombines with the α allele that is most commonly resident at *HML* and the α allele at *MAT* more often recombines with the *a* allele that is most commonly resident at *HMR*, increasing the efficacy of productive mating-type switching (Fig. 3.2) (8, 68, 163, 184). The molecular basis for this interesting example of directional gene conversion is discussed in considerable detail in the chapter by Haber (chapter 9).

How did this homothallic *MAT* switching system arise? Earlier studies on mating-type switching in *Kluyveromyces lactis* indicated that a low level of mitotic gene conversion can occur to effect mating-type switching in the absence of the Ho endonuclease (58). Comparative genomic studies reveal that the three *MAT* cassettes appeared in a common progenitor to the hemiascomycete lineage some 100 million years ago (9). The appearance of three *MAT* cassettes must have involved gene duplication events in which the active *MAT* cassette was copied to two other genomic locations. Although these three cassettes are often resident on the same chromosome and this could have been the ancestral arrangement, a common chromosomal location is not obligatory for mating-type switching, as this type of arrangement has been broken in some species (see the chapter by Muller et al. [chapter 15]), including *Candida glabrata* and *A. gossypii*, and mating-type switching can still occur in *S. cerevisiae* isolates in which *HML* or *HMR* has been relocated to another chromosome, albeit at lower efficiency in some examples (53).

How then did mating-type silencing and high-efficiency mating-type switching evolve in *S. cerevisiae*? *HML* and *HMR* reside near the telomeres of chromosome III, and genes positioned near telomeres can be subject to a process known as telomeric silencing, which involves several of the same components involved in mating-type silencing. Telomeric silencing of introduced, integrated transgenes is somewhat less efficient than mating-type silencing (~99% versus ~99.99%), and thus, an initial subtelomeric location for the silent mating-type cassettes may have been the first evolutionary step towards a more efficient homothallic mating-type-switching system. Later evolutionary events may have moved the silent *MAT* cassettes to their current, more chromocentric position (~12 and ~22 kb from the telomeres), necessitating the evolution of additional mechanisms by which heterochromatin forms at these genomic locales to ensure that only one *MAT* cassette is

actively expressed (see the chapter by Rusche and Hickman [chapter 11]).

With respect to the Ho endonuclease that is required for high-efficiency mating-type switching, comparative genomics revealed that it was acquired in a common progenitor of the hemiascomycetous yeasts, including *S. cerevisiae* and related species that undergo high-efficiency mating-type switching. The current evolutionary model is that the *HO* gene was captured from a mobile element and then conscripted to serve a specific role in mating-type switching (9). Interestingly, mating-type switching is restricted to cells that have divided at least once (mother cells), and this is thought to give rise to a pattern of cell divisions and cell types that maximizes the potential for each cell to mate and reenter the diploid state (160). In essence, this promotes inbreeding and a rapid return to the diploid state, which is the preferred ploidy for *S. cerevisiae* and related yeasts.

The natural *S. cerevisiae* population consists of both Ho$^+$ and ho$^-$ isolates that are homothallic and heterothallic, and spore products of an Ho$^+$ strain can readily mate with haploid ho$^-$ isolates (109). Thus, both homothallism and heterothallism can occur within a given species, depending on conditions and isolates.

Schizosaccharomyces pombe

The *S. cerevisiae* mating-type-switching system serves as a Rosetta stone for understanding how homothallic cycles can evolve. Yet equally striking is that a very similar system evolved apparently independently in the fission yeast *S. pombe* (see the chapter by Nielsen and Egel [chapter 8]). The two systems share similar features, including (i) the presence of three mating-type cassettes resident on the same chromosome, (ii) a DNA lesion provoking an efficient unidirectional switching event, and (iii) the fact that only one cassette is expressed and the others are transcriptionally silenced. However, the structure, sequence, and even the nature of the *MAT*-encoded gene products of the two yeasts are completely different. The enzymes and nature of the lesion that incites mating-type switching also differ between *S. pombe* and *S. cerevisiae*. Instead of a DNA double-strand break provoked by the Ho endonuclease in *S. cerevisiae*, some unusual type of replication-induced break involving a nick and possibly the retention or processing of ribonucleotides from the primer of an Okazaki fragment appear to be involved in *S. pombe* (1, 24, 78, 177, 178) (Fig. 3.3). The precise molecular nature by which chromatin is silenced and the enzymes and proteins involved also differ in considerable detail between these two homothallic yeast species.

That two highly efficient mating-type-switching systems evolved completely independently as the basis of

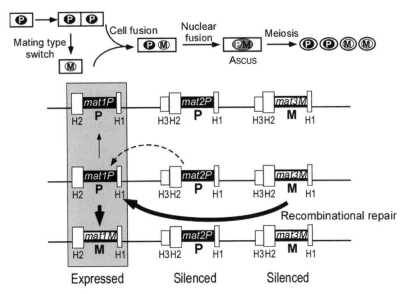

Figure 3.3 Mating-type switching in *Schizosaccharomyces pombe*. The mother cell (P) undergoes a mating-type switch to M and then mates with the daughter cell (P) to create a heterozygous diploid cell (P/M), which undergoes meiosis immediately to produce four meiotic ascospores with two P cells and two M cells. Solid and open small circles indicate nuclei of P and M mating type. H1 and H2 are conserved flanking sequences for all three cassettes. H3 is a conserved element of the *mat2P* and *mat3M* silent cassettes. These conserved sequences are involved in regulation of mating-type expression and switching (for details, see the chapter by Nielsen and Egel [chapter 8]). Only the shaded locus (*mat1*) is expressed. The thickness of the arrow represents the efficiency of the switching event. Arrows indicate the direction of the switching event.

homothallism in two ascomycetous yeasts diverged by ~250 million years of evolution suggests that a common evolutionary pressure gave rise to this poignant example of convergent evolution. This pressure likely involves an advantage to inbreeding (selfing), as closely related siblings can undergo mating-type switching and mating. Selection is not exclusively to return to the diploid state, because for *S. pombe* most natural isolates are haploid and the diploid state is transient.

SECONDARY HOMOTHALLISM

Mating-type switching can be either bidirectional or unidirectional, and it is one form of secondary homothallism. Mating-type switching has been described in both of the model yeasts *S. cerevisiae* and *S. pombe* as discussed earlier, and in filamentous ascomycetes such as *Chromocrea spinulosa*, *Sclerotinia trifoliorum*, *Botryotinia fuckeliana*, *Cryphonectria parasitica*, *Ceratocystis coerulescens*, and *Glomerella cingulata* (19, 20, 56, 104, 106, 130), and also in the basidiomycete fungus *Agrocybe aegerita* (discussed later in the section on Basidiomycetes in this chapter). It is remarkable that an organism can switch from one mating type or even sex to another, and we know of clear examples in amphib-

ians and fishes in which environmental conditions, such as temperature or the density and sex of other individuals, influence the development of sex (5, 132, 133, 151). But in fungi, mating type is governed genetically by the mating-type locus, which is not known to be influenced by just environmental conditions.

The other form of secondary homothallism (also known as pseudohomothallism) is accomplished by packaging nuclei of compatible mating types into the same spore. After germination, the single ascospore can then give rise to a self-fertile heterokaryotic mycelium with nuclei of compatible mating types. Secondary homothallism of this form has been well documented in *Neurospora tetrasperma*, *Podospora anserina*, and *Gelasinospora tetrasperma* (139) (Fig. 3.4). Both forms of secondary homothallism are essentially functional heterothallism. We turn now to a consideration of sexual cycles in ascomycetes, which is necessary to understand the mechanisms of homothallism.

ASCOMYCETES

In some ascomycetes, such as *Neurospora crassa* and related species, sexual differentiation is independent of sexuality because even in heterothallic species, strains of

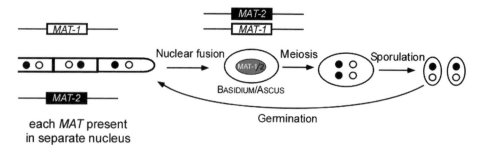

Figure 3.4   One form of pseudohomothallism involves packaging two nuclei of compatible mating type into a single spore. The dikaryon contains two nuclei of different but compatible mating types (for example, *MAT-1* and *MAT-2*). After nuclear fusion in the basidium/ascus, meiosis occurs and four nuclei are produced. Two nuclei of each mating type are packaged into one spore, which gives rise to a dikaryon competent for sexual reproduction after germination.

each mating type can produce both receptor (female) and donor (male) elements. However, a compatible interaction that initiates the development of fruiting bodies and produces ascospores in heterothallic fungi involves two strains of opposite mating types that contribute a compatible pair of elements. During the complex process of sexual reproduction, mating-type products regulate the initial attraction for cell fusion (fertilization) and coordinated migration of nuclei of opposite mating types into the ascogenous hyphae (20, 26, 155).

In ascomycetes, homothallic and heterothallic species can be commonly found in the same genus. For example, *Neurospora*, *Sordaria*, and *Aspergillus*, three genetically well-known filamentous ascomycete genera, include both homothallic and heterothallic species. Mating-type sequences have also been found in many homothallic and asexual ascomycetes, suggesting that *MAT* genes may have a role in regulating sexual reproduction even in homothallic fungi, and asexual species may arise from sexual species or may even still have cryptic sex (127). In addition to the mating-type genes, genes involved in sexual signaling between compatible heterothallic isolates, such as those encoding pheromone and pheromone receptors, are also present in homothallic fungi (29). Although much is understood about the genetic control and signaling pathways that operate during mating in heterothallic fungi, less is known about the functions of these genes and how they are regulated in homothallic species. Recent studies indicate that similar signaling processes may be required for sexual development in homothallic and in heterothallic fungi (54, 136). For example, the G protein-coupled receptor GprD negatively regulates sexual development in the homothallic fungus *Aspergillus nidulans* and deletion of this gene causes uncontrolled sexual development and stunted vegetative hyphal growth (54). In addition, increased expression of *MAT* genes and activa-

tion of the pheromone-responsive mitogen-activated protein (MAP)-kinase signaling pathway were observed during sexual reproduction in this homothallic fungus (P. S. Dyer, M. Paoletti, F. A. Seymour, and D. B. Archer, presented at the 8th European Conference on Fungal Genetics, Vienna, Austria, 2006).

The *MAT* locus of heterothallic ascomycetes contains completely different nonallelic DNA sequences termed idiomorphs that establish opposite mating types (21, 112, 169), and these idiomorphs are embedded within conserved flanking sequences. The bipolar mating system (with a single *MAT* locus) is the most common mechanism by which sexuality is orchestrated in heterothallic ascomycetes. The basic *MAT* gene structure has the *MAT-1* idiomorph encoding a protein with an α domain motif and the *MAT-2* idiomorph encoding a DNA-binding protein with a high-mobility group (HMG) motif (20). In some species the *MAT-1* locus harbors more than one gene. For example, the *S. cerevisiae* *MATα* locus contains the *MATα1* (α domain) and *MATα2* (homeodomain) genes. The *S. cerevisiae* *MATa* locus contains the single *MATa1* (homeodomain) gene (*MATa2* is likely to be nonfunctional). Transcription of cell-type-specific genes depends on the interaction of mating-type proteins (60, 64). Characterization of the functions of *MAT* target genes will be necessary to understand the mechanisms of cellular and nuclear recognition during mating that are elicited by the interaction of mating-type genes.

Because the *MAT* genes appear to evolve faster than other regions of the genome, the sequence and structure of *MAT* and its genes vary among different species (135, 168). In contrast to the fast-evolving sequence of the *MAT* genes, their function is largely conserved among different species regardless of the pattern of sexuality (20). For example, *mtA-1* from *Neurospora africana* can function like *mtA-1* from *N. crassa* in promoting

N. crassa mating (48). *MAT* genes from the heterothallic fungi *P. anserina* and *N. crassa* are functional in promoting fertilization in the heterothallic fungus *Cochliobolus heterostrophus*, and vice versa, even though the DNA sequences of the *MAT* genes are not highly conserved between *N. crassa/P. anserina* and *C. heterostrophus* (3). Similarly, introduction of mating-type genes from the homothallic species *Sordaria macrospora* into the heterothallic species *P. anserina* induces fruiting-body development without a partner of the opposite mating type (138). Finally, expression of *MAT* genes from the homothallic species *Cochliobolus luttrellii* in the heterothallic species *C. heterostrophus MAT*Δ strain confers self-fertility (188). These results indicate that mating-type genes can be functionally interchangeable with respect to fertilization, not only between different heterothallic fungi but also between homothallic and heterothallic species.

Although the functions of *MAT* genes may be conserved with respect to fertilization (cellular recognition) and manipulation of *MAT* genes among species can change the patterns of sexuality between homothallism and heterothallism, the ability to undergo meiosis and the generation of viable progeny appear to be species specific. For example, transgenic *N. crassa* expressing *mtA-1* from *N. africana* produced mating structures, but no perithecia were formed (48). *P. anserina* transformants with the *S. macrospora MAT* genes could not produce ascospores in the fruiting bodies (138), and a transgenic *C. heterostrophus MAT*Δ strain harboring the *C. luttrellii MAT* locus exhibits reduced ascospore production (188) (see the chapter by Turgeon and Debuchy [chapter 6]). Finally, transforming a *MAT1/2* fusion allele into *Aspergillus fumigatus*, a fungus with no known sexual cycle, induced hyphal aggregates, and

overexpression of *MAT1/2* genes even induced the formation of cleistothecium, but no ascospores were formed (P. S. Dyer, presented at the 16th Congress of the International Society for Human and Animal Mycology, Paris, France, 2006). Thus, how and to what extent mating-type genes are involved in the sexual reproduction process beyond cell recognition and how they contribute to the evolution of sexual patterns remain to be elucidated.

Primary Homothallism

Two *MAT* Idiomorphs Coexisting in the Same Genome

In heterothallic fungi, the *MAT* genes are involved in cellular and nuclear recognition to establish a biparental cellular state with nuclei of opposite mating types in the same cytoplasm. Most primary homothallic ascomycete species examined so far contain a full complement of *MAT* genes carried by the two *MAT* idiomorphs of closely aligned heterothallic species (6, 48, 187, 188). In homothallic fungi that retain both *MAT-1* and *MAT-2* idiomorphs, the mating-type structure can have two types of arrangements: *MAT-1* and *MAT-2* are fused together or closely linked; or alternatively, the two idiomorphs are unlinked and located in different regions of the genome (Fig. 3.5).

Two explanations have been proposed for the existence of both *MAT-1* and *MAT-2* in homothallic species. One explanation is that *MAT* genes are required for postfertilization events as demonstrated by genetic and mutational analyses (6, 50, 51, 128). Interaction of the products encoded by the two mating-type idiomorphs in a single genome of the homothallic species mimics the interaction that occurs in heterothallic fungi after the

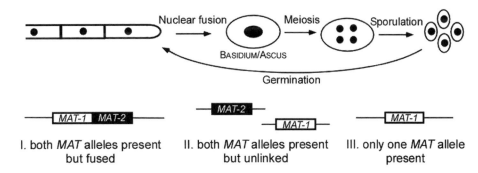

Figure 3.5 Three molecular mechanisms of primary homothallism. The homokaryotic hyphae (a monokaryon in this figure) can undergo nuclear fusion, meiosis, and sporulation in the absence of a partner. The *MAT* structure in these homothallic fungi can be as follows: two fused idiomorphs are present in a single genome (I); two idiomorphs are present in a single genome but located at different genomic regions (II); and only *MAT-1* is present, and *MAT-2* is absent (III).

cellular recognition step that leads to cell fusion and can initiate the subsequent processes of meiosis and spore generation. The other explanation is that even though each nucleus contains both idiomorphs, different nuclei could functionally act as different mating types depending on differential transcription of either one or the other of the *MAT* idiomorphs. In this case, primary homothallism would be conceptually similar to the type of secondary homothallism in which two nuclei with compatible mating types are packaged into one spore.

Closely Linked or Fused MAT *Loci*

Cochliobolus. The genus *Cochliobolus* includes both heterothallic and homothallic species (168). In contrast to the highly conserved *MAT* structural organization of all heterothallic *Cochliobolus* species, the organization of *MAT* in homothallic species differs in that all homothallic *Cochliobolus* species have both *MAT* alleles present in a single genome, either linked to each other or fused (188). This variation in homothallic *MAT* structure indicates that homothallism in this genus is likely derived from a heterothallic ancestor. One of the mechanisms proposed for this conversion of reproduction mode is a rare recombination event between small identity islands (as small as 8 nucleotides) shared between the *MAT* loci. Recombination is normally otherwise suppressed within *MAT* because the sequences of the two idiomorphs are significantly diverged (188). The structural analyses of *MAT* sequences from homothallic and heterothallic *Cochliobolus* species support the hypothesis that heterothallism is the ancestral state in this genus (6, 20, 47, 121, 130, 188) (see the chapter by Turgeon and Debuchy [chapter 6]).

Stemphylium. Like the closely related homothallic *Cochliobolus* species, most homothallic *Stemphylium* species contain both *MAT1-1* and *MAT1-2* genes in the same genome (168). However, in contrast to the varied structures of *MAT* found in homothallic *Cochliobolus* species, self-fertile *Stemphylium* isolates with fused *MAT1-1* and *MAT1-2* genes all share the same *MAT* structure: *MAT1-1* was inverted and joined to a forward-oriented *MAT1-2* region. A single fusion event followed by lateral transfer across lineages is hypothesized to have been responsible for the evolution of selfing *Stemphylium* species from outcrossing ancestors in this genus (74). Again, heterothallism is proposed to be the ancestral state in this genus.

Sordaria. The genus *Sordaria* is closely related to *Neurospora* and *Podospora*. It encompasses homothallic species such as *S. fimicola*, *S. humana*, *S. papyricola*,

and *S. macrospora*. The well-characterized species *S. macrospora* has fused *MAT-1* and *MAT-2* sequences resident at its *MAT* locus, and all four open reading frames (ORFs) of *MAT* are transcriptionally expressed (138). Although *S. macrospora* is homothallic, it has a dikaryotic phase in its life cycle, similar to heterothallic filamentous ascomycetes. The two nuclei in the apical cell of the crozier fuse, undergo karyogamy and meiosis, and immediately sporulate. Thus, nuclear fusion occurs in the ascus (35). Similar nuclear pairing and fusion events also occur in *Gibberella zeae* during perithecial development (166), and dikaryotic ascogenous hyphae are commonly observed in filamentous ascomycete species. Mutations of mating-type genes, such as *smta-1*, which encodes an HMG protein, render *G. zeae* self-sterile, although female structures such as ascogonia and protoperithecia still form normally and this self-sterile mutant can still outcross with other strains (137) (see the chapter by Pöggeler [chapter 10]).

Unlinked MAT *Loci*

Most homothallic ascomycete species have *MAT-1* and *MAT-2* fused together or closely linked (187, 188). The single *MAT* locus in these homothallic fungi is normally located at the same physical location as the *MAT* locus of related heterothallic species. Homothallic fungi with two genetically unlinked *MAT* idiomorphs are uncommon. The two unlinked idiomorphs can be located on the same chromosome, or even on two different chromosomes (Fig. 3.5). This type of *MAT* organization in homothallic euascomycetes is unusual, with *A. nidulans*, *Cochliobolus cymbopogonis*, and *Neosartorya fischeri* as the only reported examples (42, 150, 188). Whether these structural differences in the configuration of the mating-type loci have any biological significance is unknown.

Aspergillus. *N. fischeri* is a homothallic species closely related to the asexual species *A. fumigatus*, whose sexual cycle has not as yet been described (41, 43, 147, 176). Recent studies from Rydholm et al. showed that two *MAT* idiomorphs are present in the *N. fischeri* genome at distinct unlinked genomic locations. The sequences flanking the *MAT-1* locus harbor the same genes in syntenic orientation as those flanking the *MAT-2* locus. The copies flanking *MAT-2* are pseudogenes, while the copies flanking *MAT-1* appear to be intact (150) (Fig. 3.6). Because the region upstream of *MAT-2* codes for numerous transposases, it was hypothesized that the *MAT-2* region results from a segmental break mediated by transposons that generated the current homothallic species *N. fischeri* from a heterothallic ancestor (150) (Fig. 3.6).

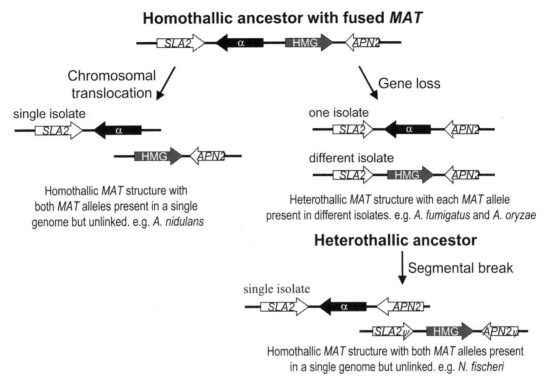

Figure 3.6 *Aspergillus MAT* structures and their evolution. Arrangement of the *MAT* idiomorphs and their flanking genes for *A. nidulans*, *N. fischeri*, *A. fumigatus*, and *A. oryzae* are shown. Homologous coding regions are color coded and indicated in the corresponding boxes. Arrows indicate the direction of expression. Hypotheses regarding the transition between homothallism and heterothallism in this genus are depicted. See the text and the chapters by Butler (chapter 1) and Dyer (chapter 7) for details.

In contrast to most other *Aspergillus* species, the sexual cycle of *A. nidulans* is well characterized. This homothallic model fungus contains *MAT-1* and *MAT-2* idiomorphs located on different chromosomes (chromosomes 6 and 3, respectively) (42). While other *Aspergillus* species, such as *A. fumigatus* and *A. oryzae*, have only one idiomorph present in a single isolate and both idiomorphs share conserved flanking sequences on both sides of the *MAT* locus, *A. nidulans* has only one of the two flanking loci adjacent to the *MAT-1* and *MAT-2* idiomorphs (42) (Fig. 3.6).

The *A. nidulans MAT* arrangement is also different from the homothallic species *N. fischeri*, where both flanking loci are present at the two unlinked idiomorphs. This unique arrangement of the *A. nidulans MAT* locus led to the proposal that homothallism is the ancestral state and that a transition from homothallism to heterothallism could have occurred in *A. oryzae* and *A. fumigatus* by gene loss, although neither species has a defined sexual cycle (42) (see the chapter by Geraldine Butler [chapter 1]). Alternatively, homothallism in *Aspergillus* could have arisen from a heterothallic ancestor by a chromosomal translocation that moved *MAT-1*

and *MAT-2* from an allelic configuration to two unlinked genomic regions, and both are therefore now present (Fig. 3.6). This hypothesis is based on the fact that the two idiomorphs present in the two characterized homothallic species, *N. fischeri* and *A. nidulans*, are unlinked and have different *MAT* structures, and the majority of asexual species in this genus show a similar "heterothallic" *MAT* locus (see the chapter by Dyer [chapter 7]). Sexual reproduction in *A. nidulans* is correlated with significantly increased expression of *MAT* genes and representative genes of the pheromone-responsive MAP-kinase signaling pathway. However, in contrast to other heterothallic species, expression of this MAP-kinase pathway appeared to be independent of *MAT* gene control (Dyer et al., 8th Eur. Conf. Fungal Genet., 2006).

Only One *MAT* Idiomorph Present in the Genome

As described above, most homothallic ascomycete species contain both *MAT-1* and *MAT-2* idiomorphs in the genome and these are either linked or unlinked (6, 48, 187, 188). Some homothallic species contain only one of the two idiomorphs, such as the *Neurospora* species *N. africana*, *N. lineolata*, *N. galapagosensis*, and *N. dodgei* (47–49). In these cases, a *MAT-1* homologue is invariably present while a *MAT-2* homologue is absent (6, 20, 134) (Fig. 3.5). This preference of *MAT-1* over *MAT-2* in homothallic fungi that retain only one *MAT* idiomorph suggests that, at least in these species, regulation of meiosis does not require the products of *MAT-2* (112). It is possible that the *MAT-2*-encoded HMG protein product may not be necessary for sexual reproduction, or its function could be supplanted by one of the *MAT-1* products, as some *MAT-1* products also encode HMG proteins (20). These are reasonable explanations based on the observation that in the heterothallic species *Sordaria brevicollis*, strains of both mating types can produce perithecia in unmated cultures, but only *mtA* (*MAT-1*) strains can generate ascospores (148). The homothallic species *Neospora terricola* is an example that might support both hypotheses. *N. terricola* contains the *mta-1* (HMG), *mtA-1* (α domain), and *mtA-2* genes but is missing the *mtA-3* (HMG) gene of its heterothallic relatives (6). This loss of the *mtA-3* (HMG) gene could be due to a dual function of the *mta-1* (HMG) gene, or *mtA-3* (HMG) gene function is not required for meiosis in this fungus. Phylogenetic analyses of sequences of *MAT* and glyceraldehyde-3-phosphate dehydrogenase within *Neurospora* indicate that the homothallic strains with both idiomorphs present in a single genome are strictly separated from those homothallic species with only *MAT-1* (135). This suggests that distinct events gave rise to the two different *MAT* locus patterns associated with homothallism in this genus (135), and that the homothallic species with only one idiomorph could have evolved directly from heterothallic ancestors and independently from those homothallic species with both *MAT* idiomorphs.

Ashbya gossypii

Similar to its close relative *S. cerevisiae*, the filamentous cotton pathogen *A. gossypii* contains three *MAT* loci (28). However, unlike the model yeast, all three loci encode *MATa* information and each copy resides on a different chromosome (36) (see the chapter by Muller et al. [chapter 15]). Surprisingly, this fungus can produce eight spore asci from a single isolate cultured under laboratory conditions (A. Gladfelter, personal communication). Because *A. gossypii* contains multiple nuclei within a single hyphal compartment, nuclear fusion could occur prior to meiosis and sporulation. Genomic analysis reveals no homologue of the Ho endonuclease, which mediates mating-type switching in yeasts. However, the *A. gossypii* genome does have pheromone genes encoding both mating-type pheromones MFα and

MF**a** (36), which could potentially promote karyogamy (nuclear fusion). It is not clear whether sporulation in *A. gossypii* represents a true homothallic sexual cycle or a form of asexual reproduction. Interestingly, there is no apparent homologue of *RME1*, a meiotic inhibitor, in the *A. gossypii* genome. It is possible that meiosis could occur in strains with only mating type **a**, similar to *S. cerevisiae rme1/rme1* mutants, where meiosis and sporulation can occur in **a**/**a** or α/α strains. Meiosis is normally restricted to **a**/α cells in *S. cerevisiae* due to repression of Rme1 by the **a**1/α2 complex (113). Morphological studies of the sporulation process at the cellular level and genetic studies of recombination, such as tests of roles of Spo11 or Dmc1 homologues, will allow these questions to be addressed.

Eremothecium coryli (*Nematospora coryli*), a close relative of *A. gossypii* and also a plant pathogen, can produce pseudohyphae, true hyphae, and yeast cells of various forms (27). Many strains have been observed to produce asci with eight spores from the yeast form under laboratory conditions or in plants, with no apparent conjugation (27, 40). Whether there is ploidy shift during this process, either through endoreplication or cell-cell and/or nuclear-nuclear fusion, is not clear (84). The *MAT* structure of this fungus is not known, and there is no genome sequence available, and thus, further analyses are needed to define the mechanism of homothallism in this organism.

Pneumocystis carinii

Pneumocystis is a general name for a group of unicellular pathogens that cause pneumonia in animals. It has been defined as a fungus with a small genome (23). Because it is an obligate pathogen and cannot be cultured in vitro, the basic biological processes of its life cycle are poorly understood. However, based on cytological studies, the life cycle of *P. carinii* may include both asexual and sexual cycles. The trophic form undergoes asexual replication by binary fission (165), and the cystic form (ascus) contains intracystic bodies (ascospores) with two, four, or eight nuclei in which synaptonemal complexes have been reported, which could be indicative of sexual reproduction (22, 108, 185). Only one mating-type-like region is known in this fungus, and there is no evidence of mating-type switching (156), suggesting that *P. carinii* could be homothallic or that a second mating-type remains to be discovered.

Absence of *MAT*

The gene products encoded by *MAT* play established roles in cellular and nuclear recognition during the early steps of sexual reproduction. However, *MAT* gene products may not be invariably required for meiosis and spore generation. For example, *Podospora anserina* mating-type gene mutants can still undergo meiosis and sporulation regardless of the nuclear content in the ascus, whether it is two nuclei of opposite mating types, two identical nuclei, or a single nucleus (189).

It is conceivable that some fungi could have sexual cycles without *MAT* sequences in their genome. The ascomycete yeast *Lodderomyces elongisporus* is closely related to pathogenic *Candida* species such as *C. parapsilosis*, *C. tropicalis*, and *C. albicans* (75). Recent genome sequencing (8.7× coverage and high-quality assembly) reveals that *MAT* sequences are not present in this fungus (http://www.broad.harvard.edu/annotation/genome/lodderomyces_elongisporus/Home.html). Because *L. elongisporus* forms ascus-like structures from a single yeast cell cultured on mating-inductive media, this fungus has been classified as homothallic (4, 86). However, the homothallic sexual cycle needs to be confirmed as no conjugation has been observed and the ascus-like structure could be derived from yeast cells. Determination of nuclear ploidy during the process of ascospore formation will help to clarify the nature of the process. If sexual products are indeed produced, then the single cell normally formed in the ascus-like structure could be the product prior to meiosis or the only viable spore remaining postmeiosis. Studies to determine how sexual reproduction is controlled in species with no genetically definable mating types, if this occurs, will be of interest to establish the fundamental requirements for sexual reproduction.

Homothallic Fungi with Heterothallic Life Cycles

Most heterothallic ascomycete fungi with only one *MAT* idiomorph present in the genome undergo obligate outcrossing. By outcrossing, genetic information can be exchanged and populations with diverse genotypes can be generated. However, in situations where geographic barriers preclude any encounter between strains of different mating types, or mating-type distribution is distorted, undergoing obligate outcrossing limits the ability to mate and produce sexual spores, which are easily dispersed, dormant, and resistant to desiccation and radiation (131).

Homothallic fungi can overcome this barrier because they are self-fertile. For example, homothallic *Neurospora* species tend to occur at the extremes of the ecological range and they colonize at least as successfully as their heterothallic counterparts (6). In addition, homothallic fungi reap the benefit of DNA repair mechanisms associated with meiosis to maintain their genomic integrity. However, because homothallic fungi are self-fertile, meiotic recombination during sexual reproduction in homothallic species may not always generate genetically distinct progeny, and this may promote homogeneity in the population as indicated by very low levels of DNA polymorphisms in *Neurospora* homothallic species across different continents (47).

Self-fertility does not preclude homothallic species from exchanging genetic information with different strains. Most homothallic fungi can still outcross with genetically and phenotypically distinct strains (35, 65). Cell fusion between two homothallic individuals could occur through hyphal anastomosis, and nuclear content can be interchanged (35, 37). For example, both *S. fimicola* and *S. macrospora* are self-fertile, and a culture derived from a single homokaryotic ascospore is capable of completing the life cycle, yet they may also be outcrossed. In the homothallic species *A. nidulans*, when genetically different individuals are present, it has been reported to preferentially undergo outcrossing compared to selfing (65). This kind of relative heterothallism does not appear to be related to mating type (65).

Primary Heterothallic Fungi with Homothallic Life Cycles

Sordaria brevicollis

Most heterothallic fungi undergo obligate outcrossing. However, a few heterothallic fungi have also been demonstrated to be able to self. For example, homothallism has been demonstrated in *S. brevicollis*, which was considered to be an exclusively heterothallic filamentous ascomycete (148). *S. brevicollis* has two mating types. Isolates of either mating type can develop female protoperithecia structures, of which a small proportion can develop into perithecia in the absence of the opposite mating type. However, only *mtA* (*MAT-1*) strains can form ascospores within the perithecia formed in an unmated culture.

Homothallism in *S. brevicollis* is a true sexual process and is not attributable to a mitotic event as has been observed in *Podospora arizonensis* (103). Heterokaryon analysis during homothallic development reveals that each ascus contains recombinant progeny from a diploid nucleus that arose from fusion between two haploid individuals marked with different spore colors. Mating type switching is not responsible because all of the spores produced during this process are of the *mtA* mating type (148). The unique ability of *MAT-1* strains to reproduce sexually is reminiscent of the homothallic fungi with only one *MAT* idiomorph present in their genome, most of which contain the *MAT-1* allele (47, 48, 74). However, under laboratory conditions, this type of homothallic fruiting is inefficient in the primarily heterothallic species *S. brevicollis* compared to crossing with a partner of the opposite mating type (148). The relative inefficiency of homothallic sporulation under laboratory conditions may indicate that unknown environmental conditions stimulate more robust development via this route or that a mutation bypasses mating-type regulation to allow inefficient sexual reproduction, as is known to be the case with *rme1/rme1* mutants of *S. cerevisiae* **a**/**a** or α/α strains.

BASIDIOMYCETES

During sexual development, both heterothallic ascomycetes and basidiomycetes produce the dikaryotic state (Fig. 3.1). However, basidiomycetes are quite different in that the dikaryotic state is long-lived and the dikaryotic mycelia can grow indefinitely. The nuclei in the dikaryotic hyphae of basidiomycetes divide and migrate coordinately, and the formation and fusion of clamp connections ensure that both nuclei are faithfully transmitted as hyphae extend. Nuclear fusion and meiosis occur in the basidia (Fig. 3.1). Unlike ascomycetes, in which heterothallism is generally bipolar and controlled by a nonallelic single mating-type locus, heterothallism in basidiomycetes is often tetrapolar and controlled by two structurally complex unlinked loci which are typically multiallelic. Moreover, each locus can harbor subloci (16, 83). As a result, thousands of mating types can be present in a single basidiomycete species (see the chapters by Fraser et al., Casselton and Kües, and Fowler and Vaillancourt [chapters 2, 17, and 18]).

Secondary Homothallism

Similar to pseudohomothallism in ascomycetes, mating-type switching is one form of secondary homothallism in basidiomycetes. The other form of secondary homothallism involves packaging two nuclei possessing complementary mating types into a single basidiospore that gives rise to a heterokaryotic mycelium competent to complete sexual reproduction upon germination.

Agrocybe aegerita

A. aegerita is a tetrapolar basidiomycete with two unlinked factors, *A* and *B* (111). Three types of homokaryotic fruiting bodies have been described in *A. aegerita*: barren basidia, bisporic basidia, and tetrasporic basidia. Isolates that produce either barren basidia or bisporic basidia develop mycelia without clamp connections, and thus, fruiting in these isolates is considered typical homokaryotic fruiting as described in *A. aegerita* (33,

110), *Pholiota nameko* (2), *Polyporus ciliatus* (158), and *Schizophyllum commune* (34). Isolates that produce tetrasporic basidia give rise to mycelium bearing clamp connections and mating-type segregates in the spores produced (92). This type of homokaryotic fruiting results from dikaryotization because of the emergence of switched mating types at both the *A* and *B* loci (92). The spores with changed mating types at both the *A* and *B* loci were always compatible with the types expressed by the original fruiting homokaryon, and this mating-type switching is directional and stable (92). Using auxotrophic markers unlinked to the mating types to track strain identity, the change of mating type in the progeny was confirmed to be caused by mating-type switching and not by contamination. It is hypothesized that there are potentially expressible *A* and *B* factors stored as silent copies in the *A. aegerita* genome. Mechanisms involving silent copies transposed into expressed loci are proposed for mating-type switching in *A. aegerita*, similar to the cassette systems of *S. cerevisiae* and *S. pombe*, and the genomic sequence for this species will allow this model to be rigorously tested.

Agaricus bisporus var. bisporus

A. bisporus, the white button mushroom, has a bipolar mating system (186). *A. bisporus* var. *bisporus* forms two spores on one basidium, and each spore contains two postmeiotic nuclei with compatible mating-type alleles. These spores germinate to produce fertile heterokaryotic mycelium and thus are pseudohomothallic (80). *A. bisporus* var. *burnettii* produces four uninucleate spores per basidium. This tetrasporic variety is self-sterile and thus heterothallic (79). That intervarietal matings between these two varieties are completely fertile suggests that they indeed belong to the same species (18). Because this species can be pseudohomothallic or heterothallic, it is considered amphithallic (30, 144).

It has been shown that a single locus linked to the *MAT* locus, *BSN*, is responsible for the number of spores produced per basidium (72, 73, 186). Strains with the *Bsn-t/t* genotype produce tetrasporic basidia and are heterothallic. Strains with the *Bsn-b/b* genotype form bisporic basidia and are pseudohomothallic. Strains with a heterozygous *Bsn-t/b* genotype are normally heterothallic, although some are self-fertile due to variable penetrance of the dominant *Bsn-t* allele (11, 72, 73). This example indicates that loci outside, or linked to, the mating-type locus can influence the mode of sexual reproduction. It would be of interest to determine the molecular nature of the *BSN* locus, which may encode a protein functioning in the nuclear segregation pathway. Understanding the molecular basis of *Bsn* may enable manipulation of the reproductive mode of this commercially important mushroom. Amphithallism is not uncommon in basidiomycetes, and 9% of agarics are considered to be amphithallic (11, 13). Whether a mechanism similar to that in *A. bisporus* is responsible for amphithallism in other fungi is not yet known.

Microbotryum violaceum

The anther-smut fungus *M. violaceum* (*Ustilago violacea*) is an obligate basidiomycete plant parasite. It has a bipolar mating system and size-dimorphic sex chromosomes (66, 67). *M. violaceum* produces diploid teliospores in the place of pollen in infected plants and renders the infected plant sterile (123). Insects disperse the diploid teliospores from diseased to healthy plants, and the teliospores then undergo meiosis and give rise to four haploid cells. These meiotic haploid cells amplify locally by budding on the plant surface, and two cells of opposite mating type can readily conjugate to produce new infectious dikaryons, which can enter the host tissue and grow endophytically. Because early conjugation among the immediate progeny of a single meiosis event maintains deleterious recessive alleles linked to mating type (intratetrad selfing) (45, 164), intratetrad mating is frequent. This represents a form of selfing/pseudohomothallism (45).

Primary Homothallism

Sistotrema brinkmannii

Primary homothallic fungi represent a minority of basidiomycetes (83, 94). The wood-rotting species *S. brinkmannii* is an aggregate of biological species with different patterns of sexualities including homothallism, bipolar heterothallism, and tetrapolar heterothallism. Homothallic and bipolar heterothallic strains within the *S. brinkmannii* species complex can successfully hybridize and produce progeny, of which all of the heterothallic progeny share the same mating type as the bipolar progenitor. Thus, homothallism and bipolar heterothallism in *S. brinkmannii* are not fundamentally different and these subspecies are related and have not been sexually isolated to form independent species (94, 170). It is speculated that isolates exhibiting primary homothallism may arise through mutation or deletion of the entire bipolar incompatibility locus (171).

A. bisporus var. eurotetrasporus

As discussed earlier, *A. bisporus* contains varieties with both pseudohomothallic and heterothallic lifestyles. A third type of life cycle, primary homothallism, is found in some rare European tetrasporic strains and leads to

production of uninucleate haploid spores (11, 12, 14). These single-spore progeny are self-fertile through homokaryotic fruiting (181). The name *A. bisporus* var. *eurotetrasporus* has been assigned to this new variety because these isolates differ from *A. bisporus* var. *bisporus* in having tetrasporic basidia and from *A. bisporus* var. *burnettii* by virtue of their longer spores (10). Morphological and genetic comparisons of different homothallic isolates indicate that the homothallic variety is highly homogeneous. The new variety *A. bisporus* var. *eurotetrasporus* can interbreed with *A. bisporus* var. *bisporus*, suggesting that differences in sexual reproduction style constitute little impediment to sexual compatibility between varieties (10). This is reminiscent of interfertility between isolates of different sexual styles in *S. brinkmannii* and between variety *burnettii* and variety *bisporus* in *A. bisporus*. Homokaryotic fruiting has also been described in *Agaricus bitorquis* (107), and may be more widespread in the genus *Agaricus* than currently appreciated. The genetic basis for this example of homothallism is not known.

Heterothallic Fungi with Homothallic Life Cycles

Cryptococcus neoformans

Most homothallic fungi maintain the ability to outcross. In contrast, few heterothallic fungi have been demonstrated to be able to self. The basidiomycete *C. neoformans*, the most common fungal agent causing meningitis, has a bipolar mating system and an unusually large mating-type locus (39, 95, 117). The traditional mating process of this heterothallic fungus involves haploid cells of α and **a** mating type (87) and produces an equal proportion of **a** and α basidiospores in four chains (71, 87, 88) (Fig. 3.7). Although this sexual cycle of *C. neoformans* has been known to occur under laboratory conditions for more than 3 decades, the sharply skewed mating-type distribution in the *Cryptococcus* population in favor of the α mating type (>99%) represents a paradox as to how sexual reproduction might occur in this essentially unisexual population (90).

Haploid *C. neoformans* α strains can also undergo a transition from yeast to filamentous growth and sporulation through monokaryotic fruiting (32, 91, 183). Fruiting occurs under conditions similar to those that promote traditional mating between two strains of opposite **a** and α mating types: nitrogen starvation, desiccation, darkness, and the presence of mating pheromone (71, 89, 102, 180). Monokaryotic fruiting was considered to be mitotic and asexual, but recently it was shown to be a modified form of sexual reproduction

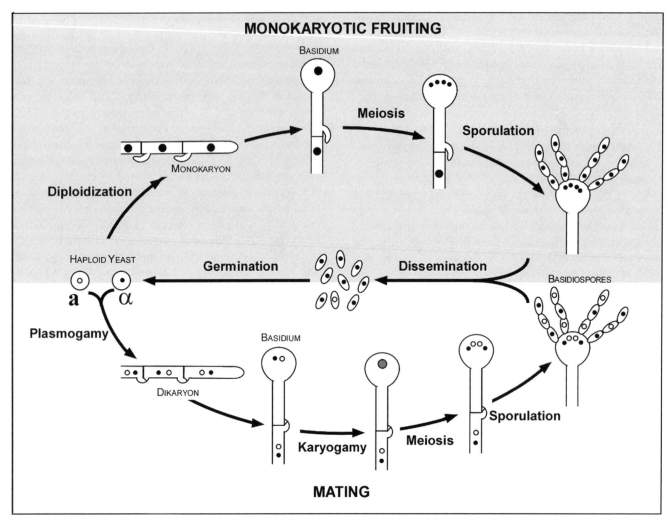

Figure 3.7   The sexual cycle of *Cryptococcus neoformans*. During mating (the heterothallic life cycle), **a** and α yeast cells undergo cell-cell fusion and produce dikaryotic hyphae. At the stage of basidium development, the two parental nuclei fuse and undergo meiosis to produce four meiotic products that form chains of basidiospores by repeated mitosis and budding. During monokaryotic fruiting (the homothallic life cycle), cells of one mating type, e.g., α cells, become diploid α/α cells, either by endoreplication or by cell fusion followed by nuclear fusion between two α cells. At the stage of basidium development, meiosis occurs and haploid basidiospores of one mating type are produced in four chains.

that involves cells of the same mating type, most commonly α (102). During monokaryotic fruiting, haploid nuclei diploidize, either before filamentation, during filamentation, or in the basidia immediately preceding sporulation. Diploidization can occur through either mating between cells of the same mating type or endoreplication (Fig. 3.7). How this process is genetically controlled and the environmental cues that stimulate it remain to be defined. Meiosis has been demonstrated to occur during the monokaryotic fruiting process, as complete genome reduction was observed and recombinant

progeny were obtained from a heterozygous α/α diploid isolate created by the fusion of two genetically distinct α strains.

Recombination during fruiting occurs at a frequency similar to that observed during the traditional a-α sexual cycle, consistent with a meiotic process (102, 105). Moreover, deletion of two meiosis-specific genes, *DMC1* (encoding a recombinase) and *SPO11* (encoding a key enzyme that initiates meiotic recombination by introducing double-strand breaks), severely impairs monokaryotic fruiting. In contrast to the four long basidiospore chains formed in the wild type, only stunted spore chains were observed in the *dmc1Δ* or *spo11Δ* mutants and some basidia formed only two chains (dyads); viability of *dmc1Δ* spores is also drastically reduced (102; X. Lin and J. Heitman, unpublished results).

The observation that some **a** isolates can also initiate monokaryotic fruiting independent of the *MAT* locus raised questions regarding the role of mating type in the differences in fruiting behavior observed between α and **a** isolates (167). A recent study applied quantitative trait loci mapping to an inbred population and demonstrated that the *MAT*α allele enhances hyphal growth during monokaryotic fruiting as a quantitative trait locus (101). These observations may explain why monokaryotic fruiting is most commonly observed in α isolates and why α isolates predominate in clinical and environmental populations. However, compared to mating between **a** and α cells, selfing through monokaryotic fruiting is less efficient under laboratory conditions, indicating that *C. neoformans* is a heterothallic fungus with the ability to self in the absence of the opposite mating type. It is not clear whether *C. neoformans* has evolved this alternative mechanism for individual strains, mostly α, to produce spores as a response to the distorted distribution of mating types in the environment or whether the enhanced ability of α cells to undergo fruiting provided the means for this disparity to have arisen.

Manipulation of the mating-type genes, for example, transforming the α cell identity gene *SXI1*α into **a** strains or the **a** cell identity gene *SXI2*a into α strains, produces strains that mimic the diploid or dikaryotic state and enable more robust monokaryotic fruiting in response to temperature and environmental cues (69, 70). Yet, deletion of the *SXI1*α gene does not impair fruiting of the α isolates tested, indicating that additional regulatory elements, possibly those downstream of Sxi1α/Sxi2a, are involved. The contributions of the remaining individual genes in the *MAT* locus to fruiting behavior remain to be established. The *Cryptococcus MAT* locus encodes products (>20) of diverse functions (39). Ste20 (a PAK kinase) (124), mating pheromone (101, 153), and Ste12 (a transcription factor) (182) are encoded by the *MAT* locus and are involved in hyphal growth during fruiting. Genes outside the *MAT* locus may also contribute to the fruiting process, either due to their involvement in the pheromone-sensing pathway or by virtue of key functional roles in the general processes of morphogenesis, genome duplication, and meiosis.

A closely related species, *Filobasidiella depauperata*, a mycoparasite (44), lacks a yeast-like form and grows exclusively filamentously. This fungus produces four chains of basidiospores from a single spore culture and thus is considered homothallic (52, 87). The hyphae are monokaryotic and lack clamp connections (89). Synaptonemal complexes were observed during sporulation, consistent with a meiotic nature to this reproduction mode (89). This apparent homothallism could be similar to the monokaryotic fruiting observed in *C. neoformans* or the homokaryotic fruiting observed in *Agaricus* species, or it could be similar to the homothallism in *A. nidulans*, where both mating-type sequences are present in a single genome. Sequence information about the *F. depauperata MAT* locus will help to define the nature of homothallism in this species.

Coprinus cinereus (*Coprinus macrorhizus*)

C. cinereus is a tetrapolar heterothallic homobasidiomycete and a model fungus commonly used to study developmental processes. Two types of monokaryotic fruiting have been described (172). First, *fis*⁺ strains produce monokaryotic fruiting bodies under inductive conditions, such as growth on media containing cell-free extracts from fruiting bodies of *C. macrorhizus* or *Lentinus edodes* (AMP and cyclic AMP have been shown to be the main active ingredients [174]). Second, *fis*ᶜ strains constitutively produce monokaryotic fruiting bodies.

During the monokaryotic fruiting process in *fis*ᶜ strains, young basidia contain only one haploid nucleus, which divides mitotically into two daughter nuclei, and these then fuse to form a diploid nucleus. The fused nucleus then divides, and four basidiospores are subsequently formed on a basidium. Because synaptonemal complexes and typical meiotic chromosome behaviors were observed, this indicates that the monokaryotic fruiting process can be a meiotic sexual process (114). Basidium development was further confirmed by fluorescence microscopy, revealing a shift in the relative DNA content in basidia in the monokaryotic basidiocarp of the *fis*ᶜ strain, consistent with the meiotic nature of the fruiting process (125). In these studies, the meiotic products originated from a single nucleus, and it is not known if the monokaryotic strain can fuse with another strain of similar mating type to exchange DNA and produce genetically distinct progeny.

Although basidiomycetes normally form fruiting bodies in the dikaryotic state, it has been known for a long time that monokaryons can also form fruiting bodies under certain conditions, such as nutritional depletion (55, 116, 179, 183). Mutations outside the mating-type locus can also enable fruiting, including *su-A* (25), *fis*ᶜ (114, 125, 172, 174), and *pcc1* (119). Mutations in the mating-type locus can also overcome self-incompatibility, such as the self-fertile *Amut Bmut* strains (162). The finding that homothallic fruiting can be a sexual process involving karyogamy and meiosis is of considerable interest from the viewpoint of the molecular mechanisms governing homothallism and the evolution of this sexual pattern. For example, loss of selfing inhibition in homokaryotic fruiting isolates may occur during the evolution of homothallism. The ability to undergo homokaryotic fruiting provides heterothallic fungi with both the opportunity to interbreed when compatible partners are available and the ability to produce sexual spores in the absence of a compatible partner (15). Although the complex mating-type loci in basidiomycetes play important roles in establishing sexual identity and orchestrating sexual reproduction (17, 38, 85, 126), among homobasidiomycetes, two individuals can fuse irrespective of mating type through hyphal anastomosis. The observation of homokaryotic fruiting suggests that mating-type genes, while essential in establishing and maintaining compatible heterokaryotic status, may not be essential for meiosis and fruiting morphogenesis (99).

Homokaryotic fruiting may be widespread in fungi (31, 158) and has been demonstrated to occur in monokaryotic hyphae in the absence of a dikaryotic phase in some heterothallic basidiomycetes. For example, fruiting bodies have been found on the monokaryotic mycelia of *Peniophora ludoviciana* (7), *Schizophyllum commune* (146), *Sistotrema brinkmannii* (94), *Cryptococcus neoformans* (183), and *Fomes cajanderi* (122), although the basidia of these heterothallic species normally form on dikaryotic mycelia during traditional mating with compatible mating partners (83).

Chemical substances and extracts of fruiting bodies of the same or unrelated species have been shown to be able to induce monokaryotic fruiting in some higher basidiomycetes (96, 115, 172, 175), and cell cycle regulation and signal transduction have been implicated in regulating fruiting body development (173). However, because there is considerable variability in natural populations (98), the genetic basis for homokaryotic fruiting in basidiomycetes may differ between species (97, 100), and the nature of fruiting, as a meiotic or mitotic process, in most of these fungi is unknown. In conclusion, the full significance and impact of homokaryotic fruiting are not yet clear and future studies will be necessary to address the molecular mechanisms of the process and its evolutionary impact.

CONVERSIONS BETWEEN HETEROTHALLISM AND HOMOTHALLISM

Homothallic species may originate from heterothallic predecessors, and the homothallic lifestyle may have a selective advantage under certain ecological pressures

(129). This hypothesis is consistent with the repeated occurrence of homothallism within numerous genera and the fact that many heterothallic fungi achieve homothallism in the form of pseudohomothallism by mating-type switching or packaging two compatible nuclei into one spore. This hypothesis is also supported by the prevalence of homothallism in filamentous ascomycetes and of homokaryotic fruiting in heterothallic basidiomycetes. The observation that homothallism and heterothallism are often found among related species and that one species can even comprise isolates with different patterns of sexuality (such as *S. cerevisiae*, *A. bisporus*, and *S. brinkmannii*) suggests that the two modes of sexual reproduction are fundamentally similar from a molecular perspective. This is again supported by the relative ease with which the sexual reproductive mode can be altered in many systems by manipulation of the *MAT* loci.

The conversion from heterothallism to homothallism can usually be achieved by expressing genes from two compatible mating-type loci in one individual (188). This type of mating-type manipulation can reconfigure sexual mode not only in the same species but also across species. For example, when a heterothallic *N. crassa* strain carries both mating types, it is able to undergo self-mating (46, 157). *C. heterostrophus* carrying the *C. luttrellii MAT* locus can form fertile reproductive structures via selfing. A conversion from homothallism to obligate heterothallism can be achieved through deleting part of the fused *MAT* locus. For example, the homothallic species *G. zeae* has a *MAT* locus containing closely linked counterparts of the *MAT* genes found in heterothallic species in this group (187). Deletion of *MAT-1* or *MAT-2* from a wild-type strain renders the strain self-sterile, yet able to cross with wild-type isolates and with each other (93).

Although conversions between heterothallism and homothallism can be achieved through manipulations of the *MAT* locus, the ability for sexual reproduction to proceed successfully is not exclusively controlled by the *MAT* locus. For example, the asexual ascomycete fungus *Bipolaris sacchari* is a close relative of the heterothallic species *C. heterostrophus* and one isolate tested contains a *MAT-2* gene (isolates of the other mating type have also been found; see the chapter by Turgeon and Debuchy [chapter 6]). This *MAT-2* is functional when assayed in a *C. heterostrophus MATΔ* strain based on the fact that the transgenic *C. heterostrophus* is fertile when crossed to *C. heterostrophus* tester strains (152). However, transforming *C. heterostrophus MAT* genes (*MAT-1* or *MAT-2*) with all possible combinations into one *B. sacchari* isolate does not confer upon the resulting transgenic *B. sacchari* the ability to initiate sexual reproduction, even though these *MAT* genes are fully expressed. This indicates that elements other than *MAT* are essential for sexual reproduction (152) (see the chapter by Turgeon and Debuchy [chapter 6]). That elements in addition to *MAT* govern sexual pattern also derives from experiments in which introduction of the homothallic species *N. africana MtA-1* locus into the heterothallic species *N. crassa* did not confer homothallism. This suggests that differences other than *MAT* between *N. africana* and *N. crassa* are important in defining sexual behavior (48).

Homothallic individuals contain all of the necessary genetic information for full sexual expression. With respect to the mating-type locus, homothallic fungi generally contain all of the mating information in either a single genome in the cases of primary homothallism and mating-type switching or multiple genomes within a single cell in the case of pseudohomothallism where nuclei with compatible mating types coexist in a single spore. Heterothallic species may have evolved from a homothallic ancestor by separating the mating-type idiomorphs. Correspondingly, homothallic species may have arisen from heterothallic predecessors, either by acquiring both idiomorphs or by evolving a bypass mechanism in cases where there is only a single *MAT* allele or even possibly no *MAT* locus at all. That such transitions between homothallism and heterothallism have commonly occurred throughout the fungal kingdom during evolution, and in both directions, highlights the plasticity of sexual determination and reproduction. These transitions have occurred in response to ongoing selective pressures either to restrict outbreeding and promote inbreeding, or to engage in more prominent outbreeding as a means to diversify the population. Future studies on *MAT* and the homothallic-heterothallic dichotomy promise to further the understanding of how sexual reproduction transitions enable speciation, adaptation, and evolution.

References

1. **Arcangioli, B., and R. de Lahondes.** 2000. Fission yeast switches mating type by a replication-recombination coupled process. *EMBO J.* **19:**1389–1396.

2. **Arita, I.** 1978. *Pholiota nameko,* p. 475–496. *In* S. T. Chang and W. A. Hayes (ed.), *The Biology and Cultivation of Edible Mushrooms.* Academic Press, New York, NY.

3. **Arnaise, S., D. Zickler, and N. L. Glass.** 1993. Heterologous expression of mating-type genes in filamentous fungi. *Proc. Natl. Acad. Sci. USA* **90:**6616–6620.

4. **Barnett, J. A., R. W. Payne, and D. Yarrow.** 2000. *Yeasts: Characteristics and Identification,* 3rd ed. Cambridge University Press, Cambridge, United Kingdom.

5. **Baroiller, J. F., and H. D'Cotta.** 2001. Environment and sex determination in farmed fish. *Comp. Biochem. Physiol. C Toxicol. Pharmacol.* **130:**399–409.

6. **Beatty, N. P., M. L. Smith, and N. L. Glass.** 1994. Molecular characterization of mating-type loci in selected homothallic species of *Neurospora, Gelasinospora* and *Anixiella. Mycol. Res.* **98:**1309-1316.

7. **Biggs, R.** 1938. Cultural studies in the *Thelephoraceae* and related fungi. *Mycologia* **30:**64–78.

8. **Bressan, D. A., J. Vazquez, and J. E. Haber.** 2004. Mating-type-dependent constraints on the mobility of the left arm of yeast chromosome III. *J. Cell Biol.* **164:**361–371.

9. **Butler, G., C. Kenny, A. Fagan, C. Kurischko, C. Gaillardin, and K. H. Wolfe.** 2004. Evolution of the MAT locus and its Ho endonuclease in yeast species. *Proc. Natl. Acad. Sci. USA* **101:**1632–1637.

10. **Callac, P., I. J. de Haut, M. Imbernon, J. Guinberteau, C. Desmerger, and I. Theochari.** 2003. A novel homothallic variety of *Agaricus bisporus* comprises rare tetrasporic isolates from Europe. *Mycologia* **95:**222–231.

11. **Callac, P., S. Hocquart, M. Imbernon, C. Desmerger, and J. M. Olivier.** 1998. Bsn-t alleles from French field strains of *Agaricus bisporus. Appl. Environ. Microbiol.* **64:** 2105–2110.

12. **Callac, P., M. Imbernon, J. Guinberteau, L. Pirobe, S. Granit, and J. M. Olivier.** 2000. Discovery of a wild Mediterranean population of *Agaricus bisporus,* and its usefulness for breeding work. *Mushroom Sci.* **15:**245–252.

13. **Callac, P., C. Spataro, A. Caille, and M. Imbernon.** 2006. Evidence for outcrossing via the Buller phenomenon in a substrate simultaneously inoculated with spores and mycelium of *Agaricus bisporus. Appl. Environ. Microbiol.* **72:**2366–2372.

14. **Callac, P., I. Theochari, and R. W. Kerrigan.** 2002. The germplasm of *Agaricus bisporus:* main results after ten years of collecting in France, in Greece, and in North America. *Acta Hortic.* **579:**49–55.

15. **Calvo-Bado, L., R. Noble, M. Challen, A. Dobrovin-Pennington, and T. Elliott.** 2000. Sexuality and genetic identity in the *Agaricus* section *Arvenses. Appl. Environ. Microbiol.* **66:**728–734.

16. **Casselton, L. A., and U. Kues.** 1994. Mating-type genes in homobasidiomycetes, p. 307–321. *In* J. W. Wessels and F. Meinhardt (ed.), *The Mycota,* vol. 1. Springer-Verlag, Berlin, Germany.

17. **Casselton, L. A., and N. S. Olesnicky.** 1998. Molecular genetics of mating recognition in basidiomycete fungi. *Microbiol. Mol. Biol. Rev.* **62:**55–70.

18. **Challen, M. P., R. W. Kerrigan, and P. Callac.** 2003. A phylogenetic reconstruction and emendation of *Agaricus* section *Duploannulatae. Mycologia* **95:**61–73.

19. **Cisar, C. R., and D. O. TeBeest.** 1999. Mating system of the filamentous ascomycete, *Glomerella cingulata. Curr. Genet.* **35:**127–133.

20. **Coppin, E., R. Debuchy, S. Arnaise, and M. Picard.** 1997. Mating types and sexual development in filamentous ascomycetes. *Microbiol. Mol. Biol. Rev.* **61:**411–428.

21. **Cozijnsen, A. J., and B. J. Howlett.** 2003. Characterisation of the mating-type locus of the plant pathogenic ascomycete *Leptosphaeria maculans. Curr. Genet.* **43:**351–357.

22. **Cushion, M. T.** 1998. Taxonomy, genetic organization, and life cycle of *Pneumocystis carinii*. *Semin. Respir. Infect.* **13:**304–312.

23. **Cushion, M. T., and A. G. Smulian.** 2001. The *Pneumocystis* genome project: update and issues. *J. Eukaryot. Microbiol.* **Suppl:**182S–183S.

24. **Dalgaard, J. Z., and A. J. Klar.** 1999. Orientation of DNA replication establishes mating-type switching pattern in *S. pombe*. *Nature* **400:**181–184.

25. **Day, P. R.** 1963. Mutations of the A mating type factor in *Coprinus lagopus*. *Genet. Res. Camb.* **4:**55–64.

26. **Debuchy, R., and B. G. Turgeon.** 2006. Mating-type structure, evolution, and function in euascomycetes, p. 293–323. *In* U. Kües and R. Fischer (ed.), *Growth, Differentiation and Sexuality*, vol. I. Springer-Verlag, Berlin, Germany.

27. **de Hoog, G. S., C. P. Kurtzman, H. J. Phaff, and M. W. Miller.** 2000. *Eremothecium borzi* emend. Kurtzman, p. 201–208. *In* C. P. Kurtzman and J. W. Fell (ed.), *The Yeasts: a Taxonomic Study*. Elsevier, Amsterdam, The Netherlands.

28. **Dietrich, F. S., S. Voegeli, S. Brachat, A. Lerch, K. Gates, S. Steiner, C. Mohr, R. Pohlmann, P. Luedi, S. Choi, R. A. Wing, A. Flavier, T. D. Gaffney, and P. Philippsen.** 2004. The *Ashbya gossypii* genome as a tool for mapping the ancient *Saccharomyces cerevisiae* genome. *Science* **304:** 304–307.

29. **Dyer, P. S., M. Paoletti, and D. B. Archer.** 2003. Genomics reveals sexual secrets of *Aspergillus*. *Microbiology* **149:** 2301–2303.

30. **Eileen Kennedy, M., and J. H. Burnett.** 1956. Amphithallism in fungi. *Nature* **177:**882–883.

31. **Elliott, T. J.** 1985. Developmental genetics—from spore to sporephore, p. 451–465. *In* D. Moore, L. A. Casselton, D. A. Wood, and J. C. Frankland (ed.), *Developmental Biology of Higher Fungi*. Cambridge University Press, Cambridge, United Kingdom.

32. **Erke, K. H.** 1976. Light microscopy of basidia, basidiospores, and nuclei in spores and hyphae of *Filobasidiella neoformans* (*Cryptococcus neoformans*). *J. Bacteriol.* **128:**445–455.

33. **Esser, K., and F. Meinhardt.** 1977. A common genetic control of dikaryotic and monokaryotic fruiting in the basidiomycete *Agrocybe aegerita*. *Mol. Gen. Genet.* **155:** 113–115.

34. **Esser, K., F. Saleh, and F. Meinhardt.** 1979. Genetics of fruit body production in higher basidiomycetes. 2. Monokaryotic and dikaryotic fruiting in *Schizophyllum commune*. *Curr. Genet.* **1:**85–88.

35. **Esser, K., and J. Straub.** 1958. Genetic studies on *Sordaria macrospora* Auersw., compensation and induction in gene-dependent developmental defects. *Z. Vererbungsl.* **89:**729–746.

36. **Fabre, E., H. Muller, P. Therizols, I. Lafontaine, B. Dujon, and C. Fairhead.** 2005. Comparative genomics in hemiascomycete yeasts: evolution of sex, silencing, and subtelomeres. *Mol. Biol. Evol.* **22:**856–873.

37. **Fields, W. G.** 1970. An introduction to the genus *Sordaria*. *Neurospora Newsl.* **16:**14–17.

38. **Fowler, T. J., S. M. DeSimone, M. F. Mitton, J. Kurjan, and C. A. Raper.** 1999. Multiple sex pheromones and receptors of a mushroom-producing fungus elicit mating in yeast. *Mol. Biol. Cell* **10:**2559–2572.

39. **Fraser, J. A., S. Diezmann, R. L. Subaran, A. Allen, K. B. Lengeler, F. S. Dietrich, and J. Heitman.** 2004. Convergent evolution of chromosomal sex-determining regions in the animal and fungal kingdoms. *PLoS Biol.* **2:**e384.

40. **Frazer, H. L.** 1944. Observations on the method of transmission of internal boll disease of cotton by the cotton stainer-bug. *Ann. Appl. Biol.* **21:**271–290.

41. **Frisvad, J. C., and R. A. Samson.** 1990. Chemotaxonomy and morphology of *Aspergillus fumigatus* and related taxa, p. 201–208. *In* R. A. Samson and J. I. Pitt (ed.), *Modern Concepts in Penicillium and Aspergillus Classification*. Plenum Press, New York, NY.

42. **Galagan, J. E., S. E. Calvo, C. Cuomo, L. J. Ma, J. R. Wortman, S. Batzoglou, S. I. Lee, M. Basturkmen, C. C. Spevak, J. Clutterbuck, V. Kapitonov, J. Jurka, C. Scazzocchio, M. Farman, J. Butler, S. Purcell, S. Harris, G. H. Braus, O. Draht, S. Busch, C. D'Enfert, C. Bouchier, G. H. Goldman, D. Bell-Pedersen, S. Griffiths-Jones, J. H. Doonan, J. Yu, K. Vienken, A. Pain, M. Freitag, E. U. Selker, D. B. Archer, M. A. Penalva, B. R. Oakley, M. Momany, T. Tanaka, T. Kumagai, K. Asai, M. Machida, W. C. Nierman, D. W. Denning, M. Caddick, M. Hynes, M. Paoletti, R. Fischer, B. Miller, P. Dyer, M. S. Sachs, S. A. Osmani, and B. W. Birren.** 2005. Sequencing of *Aspergillus nidulans* and comparative analysis with *A. fumigatus* and *A. oryzae*. *Nature* **438:**1105–1115.

43. **Geiser, D. M., W. E. Timberlake, and M. L. Arnold.** 1996. Loss of meiosis in *Aspergillus*. *Mol. Biol. Evol.* **13:** 809–817.

44. **Ginns, J., and D. W. Malloch.** 2003. *Filobasidiella depauperata* (Tremellales): haustorial branches and parasitism of *Verticillium lecani*. *Mycol. Prog.* **2:**137–140.

45. **Giraud, T.** 2004. Patterns of within population dispersal and mating of the fungus *Microbotryum violaceum* parasitising the plant *Silene latifolia*. *Heredity* **93:**559–565.

46. **Glass, N. L., J. Grotelueschen, and R. L. Metzenberg.** 1990. *Neurospora crassa* A mating-type region. *Proc. Natl. Acad. Sci. USA* **87:**4912–4916.

47. **Glass, N. L., R. L. Metzenberg, and N. B. Raju.** 1990. Homothallic Sordariaceae from nature: the absence of strains containing only the a mating-type sequence. *Exp. Mycol.* **14:**274–289.

48. **Glass, N. L., and M. L. Smith.** 1994. Structure and function of a mating-type gene from the homothallic species *Neurospora africana*. *Mol. Gen. Genet.* **244:**401–409.

49. **Glass, N. L., S. J. Vollmer, C. Staben, J. Grotelueschen, R. L. Metzenberg, and C. Yanofsky.** 1988. DNAs of the two mating-type alleles of *Neurospora crassa* are highly dissimilar. *Science* **241:**570–573.

50. **Griffiths, A. J. F.** 1982. Null mutants of *A* and *a* mating type alleles of *Neurospora crassa*. *Can. J. Genet. Cytol.* **24:**167–176.

51. **Griffiths, A. J. F., and A. M. Delange.** 1978. Mutations of the a mating type in *Neurospora crassa*. *Genetics* **88:** 239–254.

52. Gueho, E., L. Improvisi, R. Christen, and G. S. de Hoog. 1993. Phylogenetic relationships of *Cryptococcus neoformans* and some related basidiomycetous yeasts determined from partial large subunit rRNA sequences. *Antonie Leeuwenhoek* **63**:175–189.

53. Haber, J. E., L. Rowe, and D. T. Rogers. 1981. Transposition of yeast mating-type genes from two translocations of the left arm of chromosome III. *Mol. Cell. Biol.* **1**:1106–1119.

54. Han, K. H., J. A. Seo, and J. H. Yu. 2004. A putative G protein-coupled receptor negatively controls sexual development in *Aspergillus nidulans*. *Mol. Microbiol.* **51**:1333–1345.

55. Hanna, W. F. 1928. Sexual stability in monosporous mycelia of *Coprinus lagopus*. *Ann. Bot.* **42**:379–388.

56. Harrington, T. C., and D. L. McNew. 1997. Self-fertility and uni-directional mating-type switching in *Ceratocystis coerulescens*, a filamentous ascomycete. *Curr. Genet.* **32**:52–59.

57. Heitman, J. 2006. Sexual reproduction and the evolution of microbial pathogens. *Curr. Biol.* **16**:R711–R725.

58. Herman, A., and H. Roman. 1966. Allele specific determinants of homothallism in *Saccharomyces lactis*. *Genetics* **53**:727–740.

59. Herskowitz, I. 1988. Life cycle of the budding yeast *Saccharomyces cerevisiae*. *Microbiol. Rev.* **52**:536–553.

60. Herskowitz, I. 1989. A regulatory hierarchy for cell specialization in yeast. *Nature* **342**:749–757.

61. Herskowitz, I., and Y. Oshima. 1981. Control of cell type in *Saccharomyces cerevisiae*: mating type and mating-type interconversion, p. 181–209. *In* J. N. Strathern, E. W. Jones, and J. R. Broach (ed.), *The Molecular Biology of the Yeast* Saccharomyces, *Life Cycle and Inheritance*. Cold Spring Harbor Laboratory, Cold Spring Harbor, NY.

62. Herskowitz, I., J. Rine, and J. Strathern. 1991. Mating-type determination and mating-type interconversion in *Saccharomyces cerevisiae*, p. 583–656. *In* J. R. Broach, J. R. Pringle, and E. W. Jones (ed.), *The Molecular and Cellular Biology of the Yeast* Saccharomyces. Cold Spring Harbor Laboratory Press, Cold Spring Harbor, NY.

63. Hicks, J., J. N. Strathern, and A. J. Klar. 1979. Transposable mating type genes in *Saccharomyces cerevisiae*. *Nature* **282**:478–483.

64. Hiscock, S. J., and U. Kües. 1999. Cellular and molecular mechanisms of sexual incompatibility in plants and fungi. *Int. Rev. Cytol.* **193**:165–295.

65. Hoffmann, B., S. E. Eckert, S. Krappmann, and G. H. Braus. 2001. Sexual diploids of *Aspergillus nidulans* do not form by random fusion of nuclei in the heterokaryon. *Genetics* **157**:141–147.

66. Hood, M. E. 2002. Dimorphic mating-type chromosomes in the fungus *Microbotryum violaceum*. *Genetics* **160**:457–461.

67. Hood, M. E., J. Antonovics, and B. Koskella. 2004. Shared forces of sex chromosome evolution in haploid-mating and diploid-mating organisms: *Microbotryum violaceum* and other model organisms. *Genetics* **168**:141–146.

68. Houston, P. L., and J. R. Broach. 2006. The dynamics of homologous pairing during mating type interconversion in budding yeast. *PLoS Genet.* **2**:e98.

69. Hull, C. M., M. J. Boily, and J. Heitman. 2005. Sex-specific homeodomain proteins Sxi1alpha and Sxi2a coordinately regulate sexual development in *Cryptococcus neoformans*. *Eukaryot. Cell* **4**:526–535.

70. Hull, C. M., R. C. Davidson, and J. Heitman. 2002. Cell identity and sexual development in *Cryptococcus neoformans* are controlled by the mating-type-specific homeodomain protein Sxi1alpha. *Genes Dev.* **16**:3046–3060.

71. Hull, C. M., and J. Heitman. 2002. Genetics of *Cryptococcus neoformans*. *Annu. Rev. Genet.* **36**:557–615.

72. Imbernon, M., P. Callac, P. Gasqui, R. W. Kerrigan, and J. Velcko. 1996. *BSN*, the primary determinant of basidial spore number and reproductive mode in *Agaricus bisporus*, maps to chromosome I. *Mycologia* **88**:749–761.

73. Imbernon, M., P. Callac, S. Granit, and L. Pirobe. 1995. Allelic polymorphism at the mating type locus in *Agaricus bisporus* var. *burnettii*, and confirmation of the dominance of its tetrasporic trait. *Mushroom Sci.* **14**:11–19.

74. Inderbitzin, P., J. Harkness, B. G. Turgeon, and M. L. Berbee. 2005. Lateral transfer of mating system in *Stemphylium*. *Proc. Natl. Acad. Sci. USA* **102**:11390–11395.

75. James, S. A., M. D. Collins, and I. N. Roberts. 1994. The genetic relationship of *Lodderomyces elongisporus* to other ascomycete yeast species as revealed by small-subunit rRNA gene sequences. *Lett. Appl. Microbiol.* **19**:308–311.

76. Johnson, A. D. 1995. Molecular mechanisms of cell-type determination in budding yeast. *Curr. Opin. Genet. Dev.* **5**:552–558.

77. Johnston, J. R., C. Baccari, and R. K. Mortimer. 2000. Genotypic characterization of strains of commercial wine yeasts by tetrad analysis. *Res. Microbiol.* **151**:583–590.

78. Kaykov, A., and B. Arcangioli. 2004. A programmed strand-specific and modified nick in *S. pombe* constitutes a novel type of chromosomal imprint. *Curr. Biol.* **14**:1924–1928.

79. Kerrigan, R. W., M. Imbernon, P. Callac, C. Billette, and J. M. Olivier. 1994. The heterothallic life cycle of *Agaricus bisporus* var. *burnettii*, and the inheritance of its tetrasporic trait. *Exp. Mycol.* **18**:193–210.

80. Kerrigan, R. W., J. C. Royer, L. M. Baller, Y. Kohli, P. A. Horgen, and J. B. Anderson. 1993. Meiotic behavior and linkage relationships in the secondarily homothallic fungus *Agaricus bisporus*. *Genetics* **133**:225–236.

81. Klar, A. J., J. B. Hicks, and J. N. Strathern. 1982. Directionality of yeast mating-type interconversion. *Cell* **28**:551–561.

82. Kohn, L. M. 2005. Mechanisms of fungal speciation. *Annu. Rev. Phytopathol.* **43**:279–308.

83. Koltin, Y., J. Stamberg, and P. A. Lemke. 1972. Genetic structure and evolution of the incompatibility factors in higher fungi. *Bacteriol. Rev.* **36**:156–171.

84. Koopsman, A. 1977. A cytological study of *Nematospora coryli* Pegl. *Genetica* **47**:187–195.

85. Kües, U., and Y. Liu. 2000. Fruiting body production in Basidiomycetes. *Appl. Microbiol. Biotechnol.* **54**:141–152.

86. **Kurtzman, C. P.** 2000. *Lodderomyces* van der Walt, p. 254–255. *In* C. P. Kurtzman and J. W. Fell (ed.), *The Yeasts: a Taxonomic Study*, 4th ed. Elsevier, Amsterdam, The Netherlands.

87. **Kwon-Chung, K. J.** 1975. A new genus, *Filobasidiella*, the perfect state of *Cryptococcus neoformans*. *Mycologia* **67:**1197–1200.

88. **Kwon-Chung, K. J.** 1976. Morphogenesis of *Filobasidiella neoformans*, the sexual state of *Cryptococcus neoformans*. *Mycologia* **68:**821–833.

89. **Kwon-Chung, K. J.** 1977. Heterothallism vs. self-fertile isolates of *Filobasidiella neoforms* (*Cryptococcus neoformans*). Proc. 4th International Conference on Mycoses. **356:**204–213. Pan American Health Organization, Washington, DC.

90. **Kwon-Chung, K. J., and J. E. Bennett.** 1978. Distribution of alpha and **a** mating types of *Cryptococcus neoformans* among natural and clinical isolates. *Am. J. Epidemiol.* **108:**337–340.

91. **Kwon-Chung, K. J., Y. C. Chang, R. Bauer, E. C. Swann, J. W. Taylor, and R. Goel.** 1995. The characteristics that differentiate *Filobasidiella depauperata* from *Filobasidiella neoformans*. *Stud. Mycol.* **38:**67–79.

92. **Labarere, J., and T. Noel.** 1992. Mating type switching in the tetrapolar basidiomycete *Agrocybe aegerita*. *Genetics* **131:**307–319.

93. **Lee, J., T. Lee, Y. W. Lee, S. H. Yun, and B. G. Turgeon.** 2003. Shifting fungal reproductive mode by manipulation of mating type genes: obligatory heterothallism of *Gibberella zeae*. *Mol. Microbiol.* **50:**145–152.

94. **Lemke, P. A.** 1969. A reevaluation of homothallism, heterothallism and the species concept in *Sistotrema brinkmanni*. *Mycologia* **60:**57–76.

95. **Lengeler, K. B., D. S. Fox, J. A. Fraser, A. Allen, K. Forrester, F. S. Dietrich, and J. Heitman.** 2002. Mating-type locus of *Cryptococcus neoformans*: a step in the evolution of sex chromosomes. *Eukaryot. Cell* **1:**704–718.

96. **Leonard, T. J., and S. Dick.** 1968. Chemical induction of haploid fruiting bodies in *Schizophyllum commune*. *Proc. Natl. Acad. Sci. USA* **59:**745–751.

97. **Leslie, J. F., and T. J. Leonard.** 1979. Monokaryotic fruiting in *Schizophyllum commune*: genetic control of the response to mechanical injury. *Mol. Gen. Genet.* **175:**5–12.

98. **Leslie, J. F., and T. J. Leonard.** 1980. Monokaryotic fruiting in *Schizophyllum commune*: survey of a population from Wisconsin. *Am. Midl. Nat.* **103:**367–374.

99. **Leslie, J. F., and T. J. Leonard.** 1984. Nuclear control of monokaryotic fruiting in *Schizophyllum commune*. *Mycologia* **76:**760–763.

100. **Leslie, J. F., and T. J. Leonard.** 1979. Three independent genetic systems that control initiation of a fungal fruiting body. *Mol. Gen. Genet.* **175:**257–260.

101. **Lin, X., J. C. Huang, T. G. Mitchell, and J. Heitman.** 2006. Virulence attributes and hyphal growth of *Cryptococcus neoformans* are quantitative traits and the MATα allele enhances filamentation. *PLoS Genet.* **2:**e187, 1–14.

102. **Lin, X., C. M. Hull, and J. Heitman.** 2005. Sexual reproduction between partners of the same mating type in *Cryptococcus neoformans*. *Nature* **434:**1017–1021.

103. **Mainwaring, H. R., and I. M. Wilson.** 1968. The life cycle and cytology of an apomictic *Podospora*. *Trans. Br. Mycol. Soc.* **51:**663–677.

104. **Marra, R. E., P. Cortesi, M. Bissegger, and M. G. Milgroom.** 2004. Mixed mating in natural populations of the chestnut blight fungus, *Cryphonectria parasitica*. *Heredity* **93:**189–195.

105. **Marra, R. E., J. C. Huang, E. Fung, K. Nielsen, J. Heitman, R. Vilgalys, and T. G. Mitchell.** 2004. A genetic linkage map of *Cryptococcus neoformans* variety *neoformans* serotype D (*Filobasidiella neoformans*). *Genetics* **167:**619–631.

106. **Marra, R. E., and M. G. Milgroom.** 2001. The mating system of the fungus *Cryphonectria parasitica*: selfing and self-incompatibility. *Heredity* **86:**134–143.

107. **Martinez-Carrera, D., J. F. Smith, M. P. Challen, T. J. Elliott, and C. F. Thurston.** 1995. Homokaryotic fruiting in *Agaricus bitorquis*: a new approach. *Mushroom Sci.* **14:**29–36.

108. **Matsumoto, Y., and Y. Yoshida.** 1984. Sporogony in *Pneumocystis carinii*: synaptonemal complexes and meiotic nuclear divisions observed in precysts. *J. Protozool.* **31:**420–428.

109. **McCusker, J. H.** 2006. *Saccharomyces cerevisiae*: an emerging and model pathogenic fungus, p. 245–259. *In* J. Heitman, S. G. Filler, J. E. Edwards, and A. P. Mitchell (ed.), *Molecular Principles of Fungal Pathogenesis*. ASM Press, Washington, DC.

110. **Meinhardt, F., and K. Esser.** 1981. Genetic studies of the basidiomycete *Agrocybe aegerita*. 2. Genetic control of fruit body formation and its practical implications. *Theor. Appl. Genet.* **60:**265–268.

111. **Meinhardt, F., and J. F. Leslie.** 1982. Mating types of *Agrocybe aegerita*. *Curr. Genet.* **5:**65–68.

112. **Metzenberg, R. L., and N. L. Glass.** 1990. Mating type and mating strategies in *Neurospora*. *Bioessays* **12:**53–59.

113. **Mitchell, A. P., and I. Herskowitz.** 1986. Activation of meiosis and sporulation by repression of the RME1 product in yeast. *Nature* **319:**738–742.

114. **Miyake, H., K. Tanaka, and T. Ishikawa.** 1980. Basidiospore formation in monokaryotic fruiting bodies of a mutant strain of *Coprinus macrorhizus*. *Arch. Microbiol.* **126:**207–211.

115. **Mizushina, Y., L. Hanashima, T. Yamaguchi, M. Takemura, F. Sugawara, M. Saneyoshi, A. Matsukage, S. Yoshida, and K. Sakaguchi.** 1998. A mushroom fruiting body-inducing substance inhibits activities of replicative DNA polymerases. *Biochem. Biophys. Res. Commun.* **249:**17–22.

116. **Moore, D.** 1998. *Fungal Morphogenesis*. Cambridge University Press, Cambridge, United Kingdom.

117. **Moore, T. D., and J. C. Edman.** 1993. The alpha-mating type locus of *Cryptococcus neoformans* contains a peptide pheromone gene. *Mol. Cell. Biol.* **13:**1962–1970.

118. **Mortimer, R. K., P. Romano, G. Suzzi, and M. Polsinelli.** 1994. Genome renewal: a new phenomenon revealed from a genetic study of 43 strains of *Saccharomyces cerevisiae* derived from natural fermentation of grape musts. *Yeast* **10:**1543–1552.

119. Murata, Y., M. Fujii, M. E. Zolan, and T. Kamada. 1998. Molecular analysis of *pcc1*, a gene that leads to A-regulated sexual morphogenesis in *Coprinus cinereus*. *Genetics* **149**:1753–1761.

120. Nasmyth, K. A., and K. Tatchell. 1980. The structure of transposable yeast mating type loci. *Cell* **19**:753–764.

121. Nauta, M. J., and R. F. Hoekstra. 1992. Evolution of reproductive systems in filamentous ascomycetes. II. Evolution of hermaphroditism and other reproductive strategies. *Heredity* **68**:537–546.

122. Neuhauser, K. S., and R. L. Gilbertson. 1971. Some aspects of bipolar heterothallism in *Fomes cajanderi*. *Mycologia* **63**:722–735.

123. Ngugi, H. K., and H. Scherm. 2006. Mimicry in plant-parasitic fungi. *FEMS Microbiol. Lett.* **257**:171–176.

124. Nichols, C. B., J. A. Fraser, and J. Heitman. 2004. PAK kinases Ste20 and Pak1 govern cell polarity at different stages of mating in *Cryptococcus neoformans*. *Mol. Biol. Cell* **15**:4476–4489.

125. Oishi, K., I. Uno, and T. Ishikawa. 1982. Timing of DNA replication during the meiotic process in monokaryotic basidiocarps of *Coprinus macrorhizus*. *Arch. Microbiol.* **132**:372–374.

126. O'Shea, S. F., P. T. Chaure, J. R. Halsall, N. S. Olesnicky, A. Leibbrandt, I. F. Connerton, and L. A. Casselton. 1998. A large pheromone and receptor gene complex determines multiple B mating-type specificities in *Coprinus cinereus*. *Genetics* **148**:1081–1090.

127. Paoletti, M., C. Rydholm, E. U. Schwier, M. J. Anderson, G. Szakacs, F. Lutzoni, J. P. Debeaupuis, J. P. Latge, D. W. Denning, and P. S. Dyer. 2005. Evidence for sexuality in the opportunistic fungal pathogen *Aspergillus fumigatus*. *Curr. Biol.* **15**:1242–1248.

128. Perkins, D. D. 1984. Advantages of using the inactive-mating-type *a* m1 strain as a helper component in heterokaryons. *Fungal Genet. Newsl.* **31**:41–42.

129. Perkins, D. D. 1991. In praise of diversity, p. 3–26. *In* J. W. Bennett and L. L. Lasure (ed.), *More Gene Manipulations in Fungi*. Academic Press, San Diego, CA.

130. Perkins, D. D. 1987. Mating-type switching in filamentous ascomycetes. *Genetics* **115**:215–216.

131. Perkins, D. D., and B. C. Turner. 1988. *Neurospora* from natural populations: toward the population biology of a haploid eukaryote. *Exp. Mycol.* **12**:91–131.

132. Pieau, C., and M. Dorizzi. 2004. Oestrogens and temperature-dependent sex determination in reptiles: all is in the gonads. *J. Endocrinol.* **181**:367–377.

133. Pieau, C., M. Dorizzi, and N. Richard-Mercier. 1999. Temperature-dependent sex determination and gonadal differentiation in reptiles. *Cell. Mol. Life Sci.* **55**:887–900.

134. Pöggeler, S. 2001. Mating-type genes for classical strain improvements of ascomycetes. *Appl. Microbiol. Biotechnol.* **56**:589–601.

135. Pöggeler, S. 1999. Phylogenetic relationships between mating-type sequences from homothallic and heterothallic ascomycetes. *Curr. Genet.* **36**:222–231.

136. Pöggeler, S. 2000. Two pheromone precursor genes are transcriptionally expressed in the homothallic ascomycete *Sordaria macrospora*. *Curr. Genet.* **37**:403–411.

137. Pöggeler, S., M. Nowrousian, C. Ringelberg, J. J. Loros, J. C. Dunlap, and U. Kück. 2006. Microarray and real-time PCR analyses reveal mating-type-dependent gene expression in a homothallic fungus. *Mol. Genet. Genomics* **275**:492–503.

138. Pöggeler, S., S. Risch, U. Kück, and H. D. Osiewacz. 1997. Mating-type genes from the homothallic fungus *Sordaria macrospora* are functionally expressed in a heterothallic ascomycete. *Genetics* **147**:567–580.

139. Raju, N. B. 1992. Functional heterothallism resulting from homokaryotic conidia and ascospores in *Neurospora tetrasperma*. *Mycol. Res.* **96**:103–116.

140. Raju, N. B. 1992. Genetic control of the sexual cycle in *Neurospora*. *Mycol. Res.* **96**:241-262.

141. Raju, N. B. 1980. Meiosis and ascospore genesis in *Neurospora*. *Eur. J. Cell Biol.* **23**:208–223.

142. Raju, N. B. 1978. Meiosis nuclear behaviour and ascospore formation in five homothallic species of *Neurospora*. *Can. J. Bot.* **56**:754–763.

143. Raju, N. B., and D. D. Perkins. 1994. Diverse programs of ascus development in pseudohomothallic species of *Neurospora*, *Gelasinospora*, and *Podospora*. *Dev. Genet.* **15**:104–118.

144. Raper, C. A., J. R. Raper, and R. E. Miller. 1972. Genetic analysis of the life cycle of *Agaricus bisporus*. *Mycologia* **64**:1088-1117.

145. Raper, J. R. 1966. *Genetics of Sexuality in Higher Fungi*. Ronald Press Co., New York, NY.

146. Raper, J. R., and G. S. Krongelb. 1958. Genetic and environmental aspects of fruiting in *Schizophyllum commune*. *Mycologia* **50**:707–740.

147. Raper, K. B., and D. I. Fennell. 1965. *The Genus Aspergillus*. Williams & Wilkins, Baltimore, MD.

148. Robertson, S. J., D. J. Bond, and N. D. Read. 1998. Homothallism and heterothallism in *Sordaria brevicollis*. *Mycol. Res.* **102**:1215–1223.

149. Rusche, L. N., A. L. Kirchmaier, and J. Rine. 2003. The establishment, inheritance, and function of silenced chromatin in *Saccharomyces cerevisiae*. *Annu. Rev. Biochem.* **72**:481–516.

150. Rydholm, C., P. S. Dyer, and F. Lutzoni. 2007. DNA sequence characterization and molecular evolution of *MAT1* and MAT2 mating-type loci of the self-compatible ascomycete mold *Neosartorya fischeri*. *Eukaryot. Cell* **23** [Epub ahead of print].

151. Sarre, S. D., A. Georges, and A. Quinn. 2004. The ends of a continuum: genetic and temperature-dependent sex determination in reptiles. *Bioessays* **26**:639–645.

152. Sharon, A., K. Yamaguchi, S. Christiansen, B. A. Horwitz, O. C. Yoder, and B. G. Turgeon. 1996. An asexual fungus has the potential for sexual development. *Mol. Gen. Genet.* **251**:60–68.

153. Shen, W. C., R. C. Davidson, G. M. Cox, and J. Heitman. 2002. Pheromones stimulate mating and differentiation via paracrine and autocrine signaling in *Cryptococcus neoformans*. *Eukaryot. Cell* **1**:366–377.

154. Shimizu, K. K., J. M. Cork, A. L. Caicedo, C. A. Mays, R. C. Moore, K. M. Olsen, S. Ruzsa, G. Coop, C. D. Bustamante, P. Awadalla, and M. D. Purugganan. 2004. Darwinian selection on a selfing locus. *Science* **306**:2081–2084.

155. Shiu, P. K., and N. L. Glass. 2000. Cell and nuclear recognition mechanisms mediated by mating type in filamentous ascomycetes. *Curr. Opin. Microbiol.* **3:**183–188.

156. Smulian, A. G., T. Sesterhenn, R. Tanaka, and M. T. Cushion. 2001. The *ste3* pheromone receptor gene of *Pneumocystis carinii* is surrounded by a cluster of signal transduction genes. *Genetics* **157:**991–1002.

157. Staben, C., and C. Yanofsky. 1990. *Neurospora crassa* a mating-type region. *Proc. Natl. Acad. Sci. USA* **87:**4917–4921.

158. Stahl, U., and K. Esser. 1976. Genetics of fruit body production in higher basidiomycetes. I. Monokaryotic fruiting and its correlation with dikaryotic fruiting in *Polyporus ciliatus*. *Mol. Gen. Genet.* **148:**183–197.

159. Strathern, J., J. Hicks, and I. Herskowitz. 1981. Control of cell type in yeast by the mating type locus. The alpha 1-alpha 2 hypothesis. *J. Mol. Biol.* **147:**357–372.

160. Strathern, J. N., and I. Herskowitz. 1979. Asymmetry and directionality in production of new cell types during clonal growth: the switching pattern of homothallic yeast. *Cell* **17:**371–381.

161. Strathern, J. N., A. J. Klar, J. B. Hicks, J. A. Abraham, J. M. Ivy, K. A. Nasmyth, and C. McGill. 1982. Homothallic switching of yeast mating type cassettes is initiated by a double-stranded cut in the MAT locus. *Cell* **31:**183–192.

162. Swamy, S., I. Uno, and T. Ishikawa. 1984. Morphogenetic effects of mutations at the A and B incompatibility factors of *Coprinus cinereus*. *J. Gen. Microbiol.* **130:**3219–3224.

163. Szeto, L., M. K. Fafalios, H. Zhong, A. K. Vershon, and J. R. Broach. 1997. Alpha2p controls donor preference during mating type interconversion in yeast by inactivating a recombinational enhancer of chromosome III. *Genes Dev.* **11:**1899–1911.

164. Tellier, A., L. M. Villareal, and T. Giraud. 2005. Maintenance of sex-linked deleterious alleles by selfing and group selection in metapopulations of the phytopathogenic fungus *Microbotryum violaceum*. *Am. Nat.* **165:**577–589.

165. Thomas, C. F., Jr., and A. H. Limper. 2004. *Pneumocystis pneumonia*. *N. Engl. J. Med.* **350:**2487–2498.

166. Trail, F., and R. Common. 2000. Perithecial development by *Gibberella zeae*: a light microscopy study. *Mycologia* **92:**130–138.

167. Tscharke, R. L., M. Lazera, Y. C. Chang, B. L. Wickes, and K. J. Kwon-Chung. 2003. Haploid fruiting in *Cryptococcus neoformans* is not mating type alpha-specific. *Fungal Genet. Biol.* **39:**230–237.

168. Turgeon, B. G. 1998. Application of mating type gene technology to problems in fungal biology. *Annu. Rev. Phytopathol.* **36:**115–137.

169. Turgeon, B. G., H. Bohlmann, L. M. Ciuffetti, S. K. Christiansen, G. Yang, W. Schafer, and O. C. Yoder. 1993. Cloning and analysis of the mating type genes from *Cochliobolus heterostrophus*. *Mol. Gen. Genet.* **238:**270–284.

170. Ullrich, R. C. 1973. Sexuality, incompatibility, and intersterility in the biology of the *Sistotrema brinkmannii* aggregate. *Mycologia* **65:**1234–1249.

171. Ullrich, R. C., and J. R. Raper. 1975. Primary homothallism—relation to heterothallism in the regulation of sexual morphogenesis in *Sistotremai*. *Genetics* **80:**311–321.

172. Uno, I., and T. Ishikawa. 1971. Chemical and genetical control of induction of monokaryotic fruiting bodies in *Coprinus macrorhizus*. *Mol. Gen. Genet.* **113:**229–239.

173. Uno, I., and T. Ishikawa. 1973. Metabolism of adenosine 3′,5′-cyclic monophosphate and induction of fruiting bodies in *Coprinus macrorhizus*. *J. Bacteriol.* **113:**1249–1255.

174. Uno, I., and T. Ishikawa. 1973. Purification and identification of the fruiting-inducing substances in *Coprinus macrorhizus*. *J. Bacteriol.* **113:**1240–1248.

175. Urayama, T. 1969. Stimulative effect of extracts from fruit bodies of *Agaricus bisporus* and some other hymenomycetes on primordia formation in *Marasmius* sp. *Trans. Mycol. Soc. Jpn.* **10:**73–78.

176. Varga, J., and B. Toth. 2003. Genetic variability and reproductive mode of *Aspergillus fumigatus*. *Infect. Genet. Evol.* **3:**3–17.

177. Vengrova, S., and J. Z. Dalgaard. 2004. RNase-sensitive DNA modification(s) initiates *S. pombe* mating-type switching. *Genes Dev.* **18:**794–804.

178. Vengrova, S., and J. Z. Dalgaard. 2006. The wild-type *Schizosaccharomyces pombe mat1* imprint consists of two ribonucleotides. *EMBO Rep.* **7:**59–65.

179. Verrinder-Gibbins, A. M., and B. C. Lu. 1984. Induction of normal fruiting on originally monokaryotic cultures of *Coprinus cinereus*. *Trans. Br. Mycol. Soc.* **82:**331–335.

180. Wang, P., J. R. Perfect, and J. Heitman. 2000. The G-protein beta subunit Gpb1 is required for mating and haploid fruiting in *Cryptococcus neoformans*. *Mol. Cell. Biol.* **20:**352–362.

181. Wessels, J. G. 1993. Fruiting in the higher fungi. *Adv. Microb. Physiol.* **34:**147–202.

182. Wickes, B. L., U. Edman, and J. C. Edman. 1997. The *Cryptococcus neoformans STE12alpha* gene: a putative *Saccharomyces cerevisiae* STE12 homologue that is mating type specific. *Mol. Microbiol.* **26:**951–960.

183. Wickes, B. L., M. E. Mayorga, U. Edman, and J. C. Edman. 1996. Dimorphism and haploid fruiting in *Cryptococcus neoformans*: association with the alpha-mating-type. *Proc. Natl. Acad. Sci. USA* **93:**7327–7331.

184. Wu, C., K. Weiss, C. Yang, M. A. Harris, B. K. Tye, C. S. Newlon, R. T. Simpson, and J. E. Haber. 1998. Mcm1 regulates donor preference controlled by the recombination enhancer in *Saccharomyces* mating-type switching. *Genes Dev.* **12:**1726–1737.

185. Wyder, M. A., E. M. Rasch, and E. S. Kaneshiro. 1998. Quantitation of absolute *Pneumocystis carinii* nuclear DNA content. Trophic and cystic forms isolated from infected rat lungs are haploid organisms. *J. Eukaryot. Microbiol.* **45:**233–239.

186. Xu, J., R. W. Kerrigan, P. A. Horgen, and J. B. Anderson. 1993. Localization of the mating type gene in *Agaricus bisporus*. *Appl. Environ. Microbiol.* **59:**3044–3049.

187. Yun, S. H., T. Arie, I. Kaneko, O. C. Yoder, and B. G. Turgeon. 2000. Molecular organization of mating type loci in heterothallic, homothallic, and asexual *Gibberella/Fusarium* species. *Fungal Genet. Biol.* **31:**7–20.

188. **Yun, S. H., M. L. Berbee, O. C. Yoder, and B. G. Turgeon.** 1999. Evolution of the fungal self-fertile reproductive life style from self-sterile ancestors. *Proc. Natl. Acad. Sci. USA* **96:**5592–5597.

189. **Zickler, D., S. Arnaise, E. Coppin, R. Debuchy, and M. Picard.** 1995. Altered mating-type identity in the fungus *Podospora anserina* leads to selfish nuclei, uniparental progeny, and haploid meiosis. *Genetics* **140:**493–503.

*Sex in Fungi: Molecular Determination
and Evolutionary Implications*
Edited by Joseph Heitman et al.
© 2007 ASM Press, Washington, D.C.

Brynne C. Stanton
Christina M. Hull

4

Mating-Type Locus Control of Cell Identity

DISTINCT CELL IDENTITIES ARE ESSENTIAL FOR PROPER SEXUAL DEVELOPMENT

How cells respond to and interact with their environments is a critical aspect of life for virtually all cells. Cellular responses from microbes to mammalian cells are dictated in large part by developmental programs that provide cells with distinct properties and behaviors. In this way cells are imbued with an "identity" that allows them to carry out designated functions (1). For example, neurons respond to nerve growth factors because they express receptors specific for nerve growth factors, whereas hepatocytes do not respond to such factors. Hepatocytes, on the other hand, take up lipid vesicles that other cells cannot. These innate abilities are essential for proper development and survival of an organism and often result from differential expression of a common set of genes. That is, the DNA in each cell is identical, and the differences among cells are determined through differences in the expression levels of genes in the genome. Each cell type contains the same information, but the resulting cell types are distinct.

Cell identity can also be established through the expression of differentially encoded genes. In these cell types the genomes are not identical to one another, and genes that reside in one cell type do not reside in the other, thus creating distinct cell types. This strategy has been utilized with great success in the determination of distinct sexes, a critical requirement for sexual reproduction. Although the advantages and disadvantages of sex have been debated (10), it is clear that sex has been maintained as a common form of eukaryotic reproduction, and it certainly contributes to genetic diversity within a population (61, 81). From yeast to humans, the drive to reproduce sexually has driven the evolution of mechanisms to maintain distinct sexes. In multicellular organisms "male" and "female" designate the opposite types required to mate and develop progeny. In human cells, male and female are designated by the X and Y chromosomes in which male-determining factors are encoded by the Y chromosome (70). The expression of sex-specific factors has also been adopted by many microbes, including fungi, and the resulting sexes are known as "mating types" (20).

Mating type information in fungi resides in a specialized region of the genome known as the mating type or *MAT* locus (25, 26). Akin to sex chromosomes in larger eukaryotes, the *MAT* locus encompasses a region of the genome in which different information resides in the same genomic position. Because the genes in one mating

Brynne C. Stanton, Department of Biomolecular Chemistry, University of Wisconsin—Madison, School of Medicine and Public Health, Madison, WI 53706.   **Christina M. Hull,** Departments of Biomolecular Chemistry, and Medical Microbiology & Immunology, University of Wisconsin—Madison, School of Medicine and Public Health, Madison, WI 53706.

type are distinct from those in another mating type, a region of the genome results in which homologous chromosomes contain identical DNA sequences, except for the *MAT* locus region. This critical region is required for cells to distinguish self from nonself prior to mating, as well as to specify new cell types after mating. These distinctions are essential for maintaining fungal life cycles, and improper cell identity specification can lead to dead-end reproductive events. Cell identity is therefore critical for survival.

In this chapter we address how *MAT* loci from budding ascomycetes to filamentous ascomycetes to basidiomycetes function to specify cell identity. From simple systems with only two, stable mating types to complex ones with thousands of mating types, we see how the *MAT* locus exerts its control to maintain proper cell identities and facilitate effective sexual reproduction. We begin with an introduction to the cell identity determination paradigm established in the budding yeast *Saccharomyces cerevisiae*. From this foundation we compare and contrast other systems in related ascomycetes and filamentous ascomycetes. Finally, we explore how cell identity determination affects sexual reproduction in basidiomycetes. Through these examples, we see how cell identity determination lays the foundation for effective sexual reproduction and survival across diverse fungi.

THE *SACCHAROMYCES CEREVISIAE* PARADIGM

Nowhere have the mechanisms of cell identity determination been studied more thoroughly than in the budding yeast *S. cerevisiae*. Using this system as a foundation, we compare and contrast cell identity control systems of other fungi in an effort to understand the diverse mechanisms of cell identity determination adopted by fungi.

In *S. cerevisiae*, a single-celled ascomycete, growth can take place either clonally through haploid cell budding, or sexually through mating followed by meiosis. *S. cerevisiae* cells can exist in three cell types (**a**, α, and **a**/α), and the identity of each cell type is specified by the actions of transcription factors encoded at the *MAT* locus (Fig. 4.1A) (26). In **a** cells, the *MAT* locus encodes the homeodomain transcription factor **a**1. In α cells, the *MAT* locus encodes the α-domain transcription factor α1 and the homeodomain transcription factor α2. It is the actions of α1 and α2 in haploid cells that control haploid cell identity. In α cells, α1 is responsible for activating α-specific genes, and α2 is responsible for repressing **a**-specific genes. This combination of transcriptional control specifies the α cell type. In **a** cells, **a**1 has no known

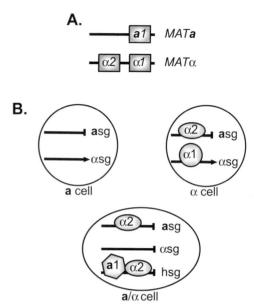

Figure 4.1 *MAT* control of cell identity in *S. cerevisiae*. (A) In *S. cerevisiae*, *MATa* in **a** cells encodes the homeodomain transcription factor **a**1. *MATα* in α cells encodes the α-domain protein α1 and the homeodomain transcription factor α2. In diploid cells, both *MATa* and *MATα* alleles are present. (B) In **a** cells, the **a**1 protein is present but has no known function. The **a**-specific genes (**a**sg) are constitutively expressed, whereas the α-specific genes (αsg) are not expressed. In α cells, α2 represses **a**-specific genes while α1 activates α-specific genes to specify the α cell type. In an **a**/α diploid, α2 maintains repression of **a**-specific genes, but it also heterodimerizes with **a**1 to form a transcriptional regulatory complex that represses haploid-specific genes (including α1). This repression establishes the diploid cell type, allowing continued sexual development.

functions, and thus, the **a** cell represents a "default" state (Fig. 4.1B). As a result of these two distinct transcriptional states, **a** and α cells exhibit specific mating behaviors. **a** cells produce **a** pheromone and a receptor for α pheromone, while α cells produce α pheromone and a receptor for **a** pheromone. When **a** and α cells encounter one another, they have the capacity to respond to a mating partner and carry out the mating process. Fusion of the **a** and α cell types results in the formation of a third cell type, the **a**/α diploid. This cell type is characterized by the presence of both **a**1 and α2 in the same cell, which form a transcriptional regulatory complex that represses the expression of haploid-specific genes. Through the direct repression of ~30 genes in the genome, **a**1-α2 specifies the diploid cell type with properties distinct from either haploid cell type and thus acts as a crucial regulator of cell type and development (21, 22). This transcriptional state makes the cells incapable of mating but competent to undergo meiosis and sporulation in response to appropriate environmental conditions (26, 33).

If the transcriptional states of the cells are perturbed, the results can be disastrous. For example, in haploid α cells in the absence of α1, the cells repress a-specific genes (through the action of α2), but they cannot activate α-specific genes. Their identities are unclear, and they are sterile. This is an undesirable state for a haploid cell attempting to mate with a partner. By the same token, misexpression in diploids can be equally disastrous. For example, if a1 does not function in a diploid, the cells express only α information, and they adopt the cell identity of an α haploid. In this case, the diploid cell will mate with haploid a cells and create triploid cells that do not make viable progeny during meiosis and sporulation. This is also a dead-end fate for the cell.

An additional type of cell identity crisis can occur because S. cerevisiae and its close relatives have the ability to switch mating types by altering the information expressed at MAT. This is accomplished by maintaining silent cassettes that contain a copy of a information (HMRa) or a copy of α information (HMLα) that can be recombined into the MAT locus (26). As long as the cassettes are silenced, only the information at MAT is expressed. However, defects in silencing can lead to expression of the silent cassettes and thus simultaneous expression of both a and α information. The result is a diploid expression pattern (a/α) in a haploid (a or α) cell. The consequence is a haploid cell that can neither mate nor undergo meiosis. Because of the dire consequences of defects in silencing the silent cassettes, the cell goes to great lengths to maintain an efficient silencing system with an infrequent failure rate.

In summary, studies of cell identity determination in S. cerevisiae have emphasized the importance of proper cell identity for survival and established the basic mechanisms used to govern this essential process. From this foundation, we can compare the diverse and interesting mechanisms adopted by other fungi to control cell identity and sexual development.

MAT CONTROL IN OTHER BUDDING ASCOMYCETES

MAT control of cell identity in ascomycetes is controlled across the board by different combinations of MAT-encoded transcription factors. Generally, three kinds of transcription factors are encoded by MAT: homeodomain-containing proteins like a1 and α2, α-domain proteins like α1, and high-mobility-group (HMG) domain-containing proteins. The homeodomain regulators found in MAT loci fall into two distinct classes: those containing the typical homeodomain structure,

known as HD1 (represented by proteins such as a1), and those containing an atypical homeodomain structure, known as HD2 (represented by proteins such as α2). The heterodimeric regulatory complexes formed by MAT homeodomain proteins generally consist of one protein from each class (e.g., a1-α2). This regulatory architecture has been conserved in many ascomycetes as well as in basidiomycetes, and additional examples are discussed later.

The HMG domain-containing proteins are distinct because they have been lost from the S. cerevisiae lineage (11). As the MAT loci of other budding ascomycetes have been identified and characterized, it has become clear that S. cerevisiae and its close relatives have a modified MAT locus that does not contain the HMG-domain transcriptional regulator common to many ascomycete MAT loci. These factors are essential for MAT control in ascomycetes like Yarrowia lipolytica (42) and Schizosaccharomyces pombe (37) as well as filamentous ascomycetes that are addressed later.

Schizosaccharomyces pombe

For most ascomycetes, the precise mechanisms by which these transcriptional regulators specify cell identity is not known; however, MAT control in S. pombe has been explored in more detail. Like S. cerevisiae, this fission yeast has two mating types and a mating-type switching system (37). The haploid cell types are known as plus (P) and minus (M), and the diploid state is P/M. Plus and minus cells sense and respond to one another through the secretion and detection of pheromones. When two opposing haploid cell types encounter one another, conjugation tube formation occurs and is followed by cell fusion to form a P/M diploid that immediately undergoes meiosis.

The MAT locus responsible for cell type determination in S. pombe is mat1, and each mat1 allele houses two genes (Fig. 4.2A). In P cells mat1-P contains $mat1-P_c$ and $mat1-P_m$, an α-domain protein and homeodomain protein, respectively. In M cells mat1-M contains $mat1-M_m$, which encodes a protein with no similarity to known domains, and $mat1-M_c$, which encodes an HMG-domain transcription factor that does not have an S. cerevisiae MAT counterpart (37). These apparent transcription factors control mate detection and cell fusion as well as the initiation of meiosis, and all are essential for sexual development (17, 79).

In haploid cells, $mat1-P_c$ and $mat1-M_c$ are expressed at constitutively low levels and are required to specify P or M identity (Fig. 4.2B). In response to a mating partner or nitrogen limitation, these genes are induced and their products upregulate the expression of the P and M

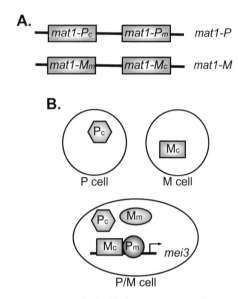

Figure 4.2 *MAT* control of cell identity in *S. pombe*. (A) The *MAT* locus of *S. pombe* contains two alleles that specify the P or M cell type. Haploid P cells contain the *mat1-P* allele that encodes mat1-P$_c$ (homeodomain) and mat1-P$_m$ (α domain) proteins. The *mat1-M* allele is contained within M cells and encodes the mat1-M$_c$ (HMG domain) and mat1-M$_m$ (unknown domain) proteins. All of the *MAT* proteins are required for sexual development. (B) In haploid cells, P$_c$ or M$_c$ is expressed constitutively to specify the P or M cell identity, respectively. In diploid P/M cells, all four *MAT* proteins are expressed. M$_c$ and P$_m$ are thought to bind to the *mei3* promoter to activate its expression (the product of which is a direct inducer of meiosis).

pheromones and their receptors. Establishment of the pheromone signal under nitrogen-limiting conditions activates the expression of *mat1-P$_m$* and *mat1-M$_m$*. Cells can then fuse to create a P/M diploid that continues to express pheromones and pheromone receptors as well as all four of the *mat1* proteins (79). The P/M diploid then usually undergoes meiosis immediately because of the actions of the P$_m$ and M$_c$ proteins. After cell fusion, these proteins are proposed to bind directly to the promoter of the *mei3* gene and activate its expression (Fig. 4.2B) (74). Although the other *mat1* proteins, P$_c$ and M$_m$, are required for the activation of the *mei3* gene, the specific roles of these proteins after cell fusion are unknown. Activation of *mei3* is sufficient to induce meiosis and sporulation. This coordinate regulation of *mei3* by *mat1*-encoded proteins ensures that meiosis is tightly regulated and occurs in the P/M diploid only after P and M cell fusion. In contrast to *S. cerevisiae*, where repression by *MAT* components is a key factor in regulation of sexual development, *S. pombe mat1* components are required to activate genes to specify the diploid state and initiate meiosis (74, 77).

The specified "identity" in this case is that of a cell competent to undergo meiosis and sporulation rather than a distinct growth phase (such as diploid). *MAT* control in this situation is responsible for a rapid progression to the spore and subsequent restoration of the haploid cell types P and M. *mat1* is essential for maintaining mating types with the capacity to mate and sporulate when severe conditions, such as low nutrient availability, arise. This process is also dependent on the production of pheromones and receptors, a situation very different from that in *S. cerevisiae*, where pheromone production is repressed in diploid cells and meiosis is induced only under nutrient-limiting conditions. Although there is a general understanding of how *mat1* components control cell identity in *S. pombe*, much remains to be discovered. The mechanisms by which *mat1* proteins function are largely unknown, and *mei3* is the only identified target of P$_m$ and M$_c$. Understanding precisely how *mat1* in *S. pombe* controls the fate of both haploids and diploids during sexual development remains to be explored and promises to reveal novel mechanisms of cell identity control.

Human Fungal Pathogens

An ascomycete in which cell identity determination has been studied in detail is the human fungal pathogen *Candida albicans* (52). Because of efforts to explore the relationships between cell identity and pathogenesis in *C. albicans*, a great deal is known about cell identity determination. In *C. albicans* cell identity is controlled by a *MAT* locus, but the locus components as well as cell identity regulation are quite different from those of *S. cerevisiae* and other fungi. *C. albicans* is an obligate diploid long thought to have lost its sexual cycle; however, the discovery of a mating-type-like (*MTL*) locus revealed the presence of both **a** and α information, as would be expected for an ascomycete diploid (31). As shown in Fig. 4.3A, *MTL***a** contains an **a**1 and an **a**2 (HMG-domain protein), and *MTL*α contains an α1 and α2, but each allele of the *MTL* locus also contains genes that encode homologs of a poly(A) polymerase (*PAP***a** and *PAP*α), a phosphatidyl inositol-3 kinase (*PIK***a** and *PIK*α), and an oxysterol binding protein (*OBP***a** and *OBP*α). The significance of *PAP*, *PIK*, and *OBP* homologs in the *MTL* locus is not known, but the discovery of transcription factor homologs in the *MTL* locus allowed the creation of strains with haploid cell identities and mating behaviors (32, 48).

Through careful analysis of strains harboring deletions in predicted transcription factors, the roles of **a**1, **a**2, α1, and α2 have been delineated. Interestingly, the regulatory circuit differs significantly from that of *S. cerevisiae* in several respects. First, *C. albicans* encodes

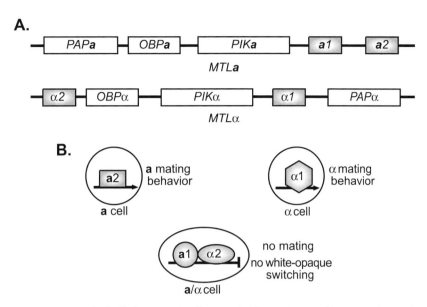

Figure 4.3 *MAT* control of cell identity in *C. albicans*. (A) The mating-type-like (*MTL*) locus of *C. albicans* has two alleles, each housing transcription factors and three other kinds of genes. *MTLa* contains the homeodomain transcription factor **a**1 and the HMG-domain protein **a**2. *MTLα* encodes the α-domain protein α1 and the homeodomain transcription factor α2. In addition to these proteins, both *MTLa* and *MTLα* encode diverged alleles of a poly(A) polymerase (*PAP*), an oxysterol binding protein (*OBP*), and a phosphatidyl inositol kinase (*PIK*). The roles of *PAP*, *OBP*, and *PIK* in cell type determination are unknown. (B) Cells that contain only **a** information (**a**/**a** or **a**/Δ at the *MTL* locus) mate as **a** cells because the putative transcription factor **a**2 activates genes involved in establishing the **a**-type mating. Cells that contain only α information (α/α or α/Δ at the *MTL* locus) mate as α cells because the predicted transcription factor α1 activates genes involved in α-type mating. Cells containing both **a** and α information (**a**/α at the *MTL* locus) do not mate with other cells because **a**1 and α2 work together to repress genes that are necessary for mating and white-opaque switching. Cells that are unable to switch from white to opaque cannot undergo mating, and **a**1 and α2 repress this switch.

the HMG-domain protein **a**2. This protein is conserved among the *MAT* loci of other ascomycetes but appears to have been lost in *S. cerevisiae* (11). While the **a**1 protein has no apparent regulatory role of its own (as in *S. cerevisiae*), **a**2 is responsible for establishing the "**a**" cell type in cells that contain only *MTLa* information, allowing them to exhibit **a** mating behavior. In cells containing *MTLα*, α1 establishes the "α" cell type allowing them to mate as α cells (similar to its role in *S. cerevisiae*), but the α2 protein has no regulatory role in α cells. In cells containing both *MTLa* and *MTLα* (most isolates of *C. albicans*), **a**1 and α2 act as they do in *S. cerevisiae* diploid cells. That is, **a**1 and α2 act in concert to regulate transcription and establish the **a**/α state (Fig. 4.3B). In these ways, the transcriptional regulators encoded by the *MTL* locus establish cell type and facilitate mating; however, unlike in most fungi, distinct mating types in *C. albicans* are not the only requirement for efficient mating. The cells must also be of the proper morphological type.

Cells of *C. albicans* can grow as either "white" or "opaque" cells, and they can switch reversibly between these growth forms (66); however, it is only the opaque forms of **a** and α cells that can mate with one another (51). Furthermore, the ability to switch between these forms is controlled by **a**1 and α2. Just as **a**1-α2 activity specifies the diploid state, this activity also restricts switching from white to opaque. The *C. albicans MTL* locus controls both mating type and cell type, inextricably linking these processes at the level of transcription.

A detailed expression analysis determined the transcriptional profiles of cells with targeted deletions of the *MTL* regulator components in every possible combination in both the white and opaque phases (where applicable) (71). This extensive analysis allowed the identification of genes involved in mating, those involved in white-opaque switching, and those involved in both. It was found that **a**1-α2 in *C. albicans* represses only seven genes directly, many of which are homologs of the haploid-specific genes in *S. cerevisiae*. What was striking is that in cells lacking either **a**1 or α2 (i.e., capable of switching from white to opaque) the expression patterns of several hundred genes were altered, and the expression of **a**-specific and α-specific genes occurs only in the opaque phase. The a1-α2 heterodimer regulates white-to-opaque switching, and this switching is essential for establishing mating type (Fig. 4.3B). The system to establish mating types in other fungi has been altered in *C. albicans* to make the specification of mating type dependent on morphological type.

It appears that close relatives of *C. albicans* share the same regulatory fate. A similar *MTL* locus has been identified in *Candida dubliniensis*, and **a** and α strains of *C. dubliniensis* (as well as those of *C. albicans* and *C. dubliniensis*) can mate with one another (59). Although it remains to be determined what the significance of mating might be in the life cycle of these *Candida* species, the discovery of mating types and mechanisms of control is a rich area in which to explore the relationships between cell identity and pathogenesis.

Although much is known about cell identity determination in *C. albicans*, little is known about the mechanisms of mating-type determination in other human fungal pathogens. Mating-type genes have been identified in fungi such as *Candida parapsilosis* (47), *Candida glabrata* (68, 78), and *Aspergillus fumigatus* (54, 75), but their significance in terms of cell identity remains to be determined. A phylogenetic outlier among the human fungal pathogens is *Cryptococcus neoformans*, and cell identity determination in this organism is described in a later section on basidiomycete fungi.

MAT CONTROL IN FILAMENTOUS ASCOMYCETES

In addition to the budding yeasts, the *MAT* loci of many filamentous ascomycetes have been studied, including *Neurospora crassa*, *Podospora anserina*, and numerous plant pathogens such as *Cochliobolus heterostrophus*. During sexual development in these fungi and their relatives, male and female structures from mycelia of opposite mating types fuse with one another. A nucleus from the male structure enters the female hyphal cell to stimulate the formation of a fruiting body. At this point, the nuclei do not fuse but instead undergo repeated mitotic divisions resulting in a fruiting body that contains plurinucleate cells composed of nuclei from both mating types. These nuclei are subsequently packaged into cells that contain one nucleus from each mating type. After these dikaryotic cells form, karyogamy occurs followed by meiosis and spore production (15).

There are two key recognition steps in this process that are under the control of the *MAT* locus. First, male and female structures must detect and fuse exclusively with their counterpart of the opposite mating type.

MAT components are required to establish haploid cell identity to distinguish mating types. Second, when two nuclei are subsequently packaged together into a single cell, it is critical that one (and only one) nucleus of each mating type is included in the dikaryotic cell. *MAT* genes are essential for specifying nuclear identity within a common cytoplasm. These events are controlled by a single *MAT* locus that contains predicted transcription factors (38).

Although the organization of the *MAT* locus may differ somewhat among the filamentous ascomycetes, the basic components and the roles they play in specifying cell and nuclear identity are constant: cell identity determination in filamentous ascomycetes requires an HMG domain protein and an α-domain protein (38). All *MAT* loci from filamentous ascomycetes identified to date contain some combination of genes that encode these classes of proteins.

Cochliobolus heterostrophus

The most rudimentary of *MAT* loci is found in the corn blight fungus *C. heterostrophus* and its close relatives. In these fungi the *MAT* locus of each mating type (designated *MAT1-1* and *MAT1-2*) contains a single gene (Fig. 4.4A). *MAT1-1* houses *MAT1-1-1*, which encodes

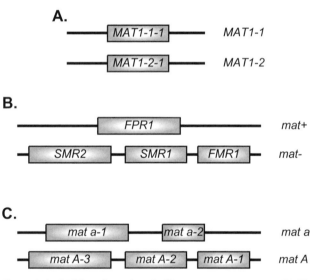

Figure 4.4 *MAT* configurations in filamentous ascomycetes. (A) The mating-type (*MAT*) locus of *C. heterostrophus* has two alleles: *MAT1-1* encodes the α-domain protein MAT1-1-1, and *MAT1-2* encodes the HMG-domain protein MAT1-2-1. (B) The *mat+* allele of *P. anserina* encodes the HMG-domain protein FPR1. The *mat−* allele encodes the HMG-domain protein SMR2, a protein of unknown class, SMR1, and the α-domain protein FMR1. (C) The *mat a* allele of *N. crassa* encodes the HMG-domain protein MAT a-1 and a protein of unknown class, MAT a-2. The *mat A* allele encodes the HMG-domain protein MAT A-3, a protein of unknown class, MAT A-2, and the α-domain protein MAT A-1.

an α-domain protein, and *MAT1-2* houses *MAT1-2-1*, which encodes an HMG-domain protein (72). These proteins are necessary for sexual development because they must both be expressed for mating to occur (45). They are also sufficient; the ectopic expression of *MAT1-1-1* in a *MAT1-2* strain (or vice versa) drives sexual development (72). This most basic of *MAT* configurations reveals the power that a simple transcriptional control circuit can exert on the relatively complex life cycle of a eukaryote. Variations on this basic regulatory scheme can be seen throughout the filamentous ascomycetes. Ultimately, however, there are no known direct targets of the *MAT*-encoded regulators, and the mechanisms by which they act remain largely unknown.

Neurospora crassa and Podospora anserina

Other filamentous ascomycetes, such as *N. crassa* and *P. anserina*, exhibit a more complex version of the *MAT* locus; however, the classes of proteins they contain and their ultimate functions are very similar to those of *C. heterostrophus*. In *P. anserina*, *N. crassa*, and related fungi, one *MAT* allele houses a single gene, which encodes an HMG-domain protein. The other allele houses three genes, which encode an HMG-domain protein, a protein of unknown class, and an α-domain protein (15, 16) (Fig. 4.4B and C). In *P. anserina*, it is the constitutive expression of the solo HMG-domain protein, FPR1, in *mat+* cells that confers *mat+* behavior. In the *mat−* locus, which houses *SMR2*, *SMR1*, and *FMR1*, it is the expression of the α-domain protein (FMR1) that specifies *mat−* behavior (Fig. 4.4B) (14). This expression pattern allows the sexual structures on mycelia of opposite mating types to fuse with one another and initiate sexual development (Fig. 4.5A).

MAT proteins must also function later in development to specify nuclear identities for proper nuclear pairing (Fig. 4.5B). In *mat+* nuclei, nuclear identity is determined by the HMG-domain protein FPR1. The two proteins responsible for communicating *mat−* nuclear identity are FMR1 (α-domain protein) and SMR2 (HMG-domain protein) (14). These proteins have been shown to interact in a yeast two-hybrid assay, and both are required for proper *mat−* behavior (2). Once the nuclei of distinct mating types recognize one another and pair in a new cell, the protein SMR1 (unknown class) is required for continued development. Mutants in SMR1 are sterile and cannot progress beyond dikaryon formation, and data suggest that SMR1 allows cells to escape from a developmental arrest that occurs during internuclear recognition (3).

In *N. crassa*, *mat* proteins have many functions similar, but not necessarily identical, to those of their *P. anserina* counterparts. There are two mating types that are speci-

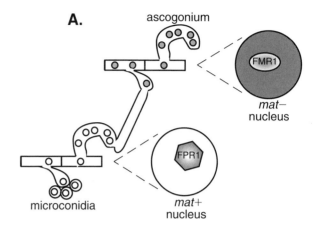

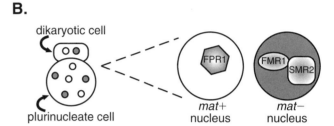

Figure 4.5 *MAT* control of cell and nuclear identity in *P. anserina*. (A) Fertilization occurs between opposite mating types when a male structure fuses with a female structure and the male nucleus is transferred. In these cells, mating type (either *mat+* or *mat−*) is specified by the expression of mating-type-specific proteins: *FMR1* is expressed in *mat+* cells, and *FPR1* is expressed in *mat−* cells. (B) After fertilization and repeated mitotic divisions, a plurinucleate cell results. The nuclei from this cell must be paired and sequestered to a dikaryotic cell that contains one nucleus of each mating type. The *mat+* nuclear identity is specified by the FPR1 protein, and the *mat−* nuclear identity is specified by the FMR1 and SMR2 proteins.

fied by the functions of genes housed at *mat a* and *mat A* (23). Genes in *mat a* encode MAT a-1 (HMG domain) and MAT a-2 (unknown class). Genes in *mat A* encode MAT A-3 (HMG domain), MAT A-2 (unknown class), and MAT A-1 (α domain) (57, 58). The proteins MAT a-1 and MAT A-1 act like their *P. anserina* counterparts and specify haploid mating types; however, MAT A-3 and MAT A-2 (counterparts to SMR2 and SMR1) are not required for nuclear recognition and subsequent sexual development as they are in *P. anserina* (18). These factors enhance the efficiency of spore formation but are not essential for the process to occur.

The *mat*-encoded proteins of *N. crassa* are, however, required for an additional process that is specific to filamentous ascomycetes. When *N. crassa* and its relatives grow vegetatively, hyphal fusion can occur between haploid mycelia. If the cells undergoing fusion have different alleles at specific loci in the genome (*het* loci), then the resulting heterokaryon will be rapidly destroyed to prevent incompatible strains from growing together (63). In *N. crassa* the *mat* locus also functions as a *het* locus to determine whether the cells are vegetatively compatible. In this process, like sexual development, the products of *mat* specify the outcome of cell fusion. However, unlike sexual development, during vegetative growth both alleles of *mat* cannot coexist, and cells rely on the MAT A-1 and MAT a-1 proteins to convey this information to the growing heterokaryon (27).

Although these proteins transmit the cell identity signal in both sexual development and vegetative incompatibility, the regions of the proteins responsible for conveying the signals are distinct. During sexual development, the HMG domain of MAT a-1 is indispensable, but this domain is not required to ascertain vegetative incompatibility. However, the acidic C-terminal domain that is expendable for sexual development is essential for determining vegetative incompatibility (56). Based on this information, the domains in MAT a-1 appear to be compartmentalized and responsible for different stages of the life cycle, underscoring the diversity of functions that can be controlled by fungal mating-type loci.

CONTROL OF CELL IDENTITY IN BASIDIOMYCETES

Control of cell identity by the *MAT* locus in basidiomycetes expands the simple transcription factor-encoding locus seen in ascomycetes to a more complex structure (13). In most basidiomycetes, there are two *MAT* loci: one encodes homeodomain transcription factors, and the other encodes cell type-specific pheromones and pheromone receptors. In most cases these *MAT* loci are not linked, resulting in a tetrapolar mating system. Studies with several basidiomycetes have revealed interesting permutations of the tetrapolar *MAT* structure, and we explore below how such changes have influenced cell type.

Ustilago maydis

The classic system in which basidiomycete *MAT* structure and control have been characterized is that of the corn smut *U. maydis* (7). *U. maydis* has a tetrapolar mating system in which there are four mating types that result from the presence of two distinct *MAT* loci. One *MAT* locus (*b*) encodes the homeodomain DNA binding proteins bE and bW, and the second, unlinked locus (*a*) encodes mating-type-specific pheromones and pheromone receptors (Fig. 4.6A) (13, 38). In the haploid state, cell type is distinguished by the expression of different pheromones and pheromone receptors (*mfa1/pra1* and *mfa2/pra2*) (Fig. 4.6B). This mechanism readily specifies

which cells are compatible for fusion (6). When haploid cells sense one another, they initiate mating and the upregulation of gene expression at the *b* locus in preparation for fusion (73). Once haploid cells fuse with one another, the components of the *b* locus (bE and bW) interact to direct the formation of a new cell type, the dikaryon. In this cell type, the haploid nuclei do not fuse with one another but are instead maintained as individual nuclei until later in development. This dikaryotic cell is specified when a heterodimeric complex of the homeodomain proteins bE and bW (akin to a1-α2) forms to create a novel transcriptional regulatory complex (Fig. 4.6B). Because every cell contains both bE and bW, development into a dikaryon is limited to those cells that contain a bE-bW complex composed of proteins from distinct alleles of bE and bW (e.g., bE1-bW2 or bE2-bW1) (7). There are over 25 documented alleles of the b proteins, presenting the opportunity for hundreds of allelic combinations (13). For bE and bW to interact, they must have compatible interaction domains in the amino terminus of each protein (independent of the homeodomain DNA binding region) (35). Once a compatible regulatory pair is formed, the bE-bW complex binds to specific promoter sequences and activates transcription. Using differential expression techniques and microarray analysis, over 200 genes have been identified as *b* regulated; however, thus far, only three have been identified as direct targets (34, 64). The promoters of these directly regulated genes (*dik6*, *polX*, and *lga2*) contain a *bbs* (*b* binding sequence) to which bE-bW binds (9). This regulation not only establishes the dikaryon but also leads to subsequent sexual development in planta.

Coprinopsis cinerea

C. cinerea is a basidiomycete mushroom in which studies have identified the classic basidiomycete *MAT* locus architecture where two unlinked *MAT* loci (*A* and *B*) govern cell identity (41). The *A* locus contains genes encoding regulatory DNA binding proteins, and the *B* locus houses pheromone and pheromone receptor genes (Fig. 4.7A) (4, 50). Unlike the situation found in many of its basidiomycete relatives (and other fungi), however, the initial fusion events during *C. cinerea* sexual development are not mediated by pheromones and pheromone receptors (13). In fact, cells are not restricted from fusing with any potential partner, and in this freely fusing population all fusion events are allowed. It is only after cell fusion that the specification of identity takes place (41). After hyphal fusion, the DNA-binding proteins housed in the *A* locus interact to assess the compatibility of the fusants.

Each individual *A* locus contains a variable number of genes that are subdivided into three divergently tran-

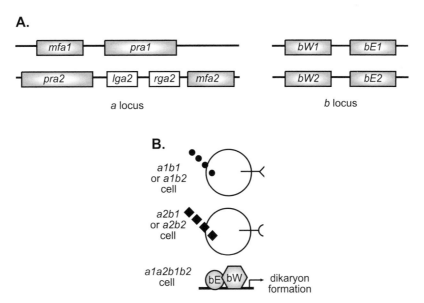

Figure 4.6 *MAT* control of cell identity in *U. maydis*. (A) The *a1 MAT* allele houses the cell-type-specific pheromone gene *mfa1* and pheromone receptor gene *pra1*. *a2* contains their counterparts, *mfa2* and *pra2*, as well as *lga2* and *rga2*. The products of *lga2* and *rga2* do not belong to any known class of proteins and have no obvious role in sexual development but have been found to be involved in mitochondrial function (8). The *b* locus encodes two homeodomain transcription factors, bE and bW, with *bE1* and *bW1* housed in the *b1* allele and *bE2* and *bW2* housed in the *b2* allele. (B) Haploid cell types (large, open circles) are specified by the expression of the pheromone *mfa1* (●) and pheromone receptor *pra1* (forked receptor on the cell surface) or the pheromone *mfa2* (◆) and pheromone receptor *pra2* (semicircular receptor on cell surface). Pheromones are sensed by their corresponding receptors on the opposite cell type. These signals activate cell fusion when two cells of opposite mating types encounter one another. After cell fusion, the proteins encoded by the *b* locus (bE and bW) interact with one another in specific combinations to specify the dikaryotic state and prepare cells for further sexual development.

scribed gene pairs (*a*, *b*, and *d*) (55). The three gene pairs are further grouped into two subloci, α (includes the *a* pair) and β (includes the *b* and *d* pairs) (Fig. 4.7A). The genes housed within the *a*, *b*, or *d* pairs encode two classes of homeodomain-containing proteins, HD1 and HD2, with each pair generally including one HD1 and one HD2 protein (39).

The fate of fused hyphae is determined by the interactions of the homeodomain proteins after cell fusion: if different alleles of an HD1 and HD2 protein from the same gene pair interact, this complex will form a heterodimeric transcriptional regulator that stimulates dikaryotic growth (40). For example, if the HD1 and HD2 proteins from a single gene pair can interact with one another, then sexual development will progress. If, however, the alleles of HD1 and HD2 are all the same in all of the gene pairs, then this results in a dead-end fusion event, and the cells do not progress through sexual development.

The HD1-HD2 interactions are mediated by variable regions of the homeodomain proteins (amino terminal to the homeodomain itself) that determine compatibility

among the **A** locus proteins (5). Similarly positioned regions within the *U. maydis* proteins bE and bW have also been shown to perform these functions (80). These regions are very diverged among proteins, and this dissimilarity allows not only for the assessment of compatibility but also for diverse functions to be provided by each protein within the heterodimeric complex.

To assess the contributions of HD1 and HD2 proteins in sexual development, a series of experiments were carried out to evaluate the importance of nuclear localization and DNA binding to the function of the HD1-HD2 complex. First, it was found that transferring the predicted nuclear localization sequences from HD1 to HD2 allowed HD2 to translocate into the nucleus, suggesting that one role of HD1 is to shuttle the HD1-HD2 complex to the nucleus where it can then regulate transcription (67).

Second, the function of the HD1-HD2 complex is dependent on the presence of the HD2 homeodomain but not the HD1 homeodomain, suggesting that the primary function of HD2 is to bind DNA and regulate transcription (67). These findings lead to a model in

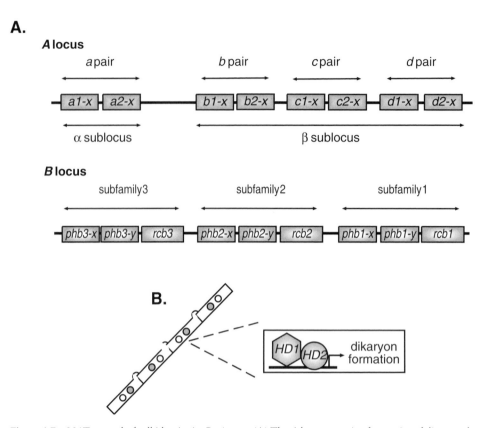

Figure 4.7 *MAT* control of cell identity in *C. cinerea.* (A) The *A* locus contains four pairs of divergently transcribed genes that each encode two homeodomain transcription factors, HD1 and HD2. The *a* pair is contained within the α sublocus, while the β sublocus contains the *b* and *d* pairs. (These pairs are not always complete and sometimes encode only one of the two genes.) Lettering within the gene name refers to the gene pair from which it originated, while the number indicates the type of homeodomain it represents (*1* for HD1 and *2* for HD2). The *B* locus contains three subfamilies of genes. Each subfamily has two genes encoding pheromones (*phb*) and one gene encoding a pheromone receptor (*rcb*). The pheromone products of one subfamily are proposed to interact with the pheromone receptor of an allelic subfamily. (B) The products of both the *A* and *B* mating-type loci act to initiate and maintain the dikaryotic state. The filamentous dikaryon contains one nucleus from each mating type in each cell and characteristic clamp connections (domed structures between adjacent filament cells). The products of the *B* locus (pheromones and pheromone receptors) control nuclear migration and clamp cell fusion, while the products of the *A* locus (HD1 and HD2 homeodomain proteins) control the expression of genes required to drive dikaryon formation, specifically to coordinate nuclear division and initiate clamp cell formation.

which the primary role of the HD1 is to target the HD2 to the nucleus where the HD2 can regulate gene expression. These studies expand on the mechanism by which homeodomain proteins (now known to be modular in the activities they provide) interact to regulate their target genes. Studies of the HD1 and HD2 proteins of *C. cinerea* have advanced the view of homeodomain protein regulation in basidiomycetes and have refined our current understanding of the molecular mechanisms by which homeodomain proteins control cell identity.

Once the dikaryotic state has been established by the homeodomain proteins from the *A* locus, the dikaryon continues to develop in a *B* locus-dependent manner (53).

Like the *A* locus, multiple subfamilies exist within the *B* locus (Fig. 4.7B). The genes housed within each subfamily encode one pheromone receptor and two pheromones (24, 62). After cell fusion, pheromones from one subfamily can bind to a different pheromone receptor allele from the same subfamily. The binding of pheromone to a compatible receptor signals initial nuclear migration and clamp cell fusion within the dikaryon (60, 69). Unlike the *A* locus, however, relatively little is known about the mechanisms of pheromone-pheromone receptor functions in *C. cinerea* sexual development. Future studies will likely reveal novel mechanisms of pheromone-mediated signaling in basidiomycetes.

Cryptococcus neoformans

An unusual basidiomycete, *C. neoformans* is a human fungal pathogen that causes meningoencephalitis, primarily in immunocompromised individuals (12). It is also a haploid, budding yeast with two stable mating types, **a** and α, that mate with one another and undergo sexual development. The factors that control this process are encoded at the *MAT* locus, but in this case the locus architecture is quite different from that of other fungi. Even though *C. neoformans* is a basidiomycete, it contains only a single *MAT* locus that encodes components homologous to those found in the unlinked loci of other basidiomycetes (i.e., homeodomain proteins, pheromones, and pheromone receptors) (Fig. 4.8A). The *C. neoformans MAT* locus is much larger than most (over 100 kb) and also encodes many genes never seen before in *MAT* loci (19, 36, 44). Alleles of each of the genes in the locus are present in both *MATa* and *MATα* with two exceptions. The Sex Inducer genes, *SXI1α* and *SXI2a*, are distinct from one another and, unlike the other genes in the locus, do not represent divergent alleles of a common ancestor gene (*SXI1α* is located in *MATα* only, and *SXI2a* is located in *MATa* only). The *SXI* genes both encode predicted homeodomain transcription factors similar to a1-α2 and bE-bW (28).

Sexual development in *C. neoformans* takes place when haploid **a** and α cells encounter one another under nutrient-specific conditions and fuse (30). In this case, as in *U. maydis*, the transcriptional regulators appear to exert their effects after cell fusion, suggesting that haploid cell identity is controlled by another mechanism. The most likely possibility is that mating-type-specific pheromones and pheromone receptors specify the haploid mating types; MFa pheromone is produced by **a** cells, MFα is produced by α cells, and cell type-specific receptors sense the pheromone of a mating partner (65). Unlike in *U. maydis*, however, where *MAT* encodes distinct pheromones and receptors, the predicted pheromones (MFa and MFα) and pheromone receptors (STE3a and

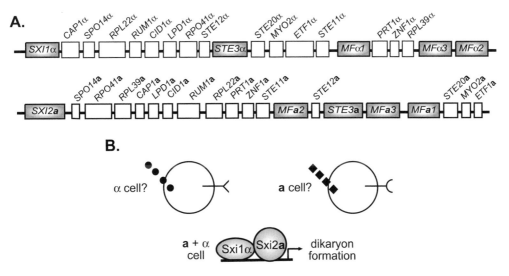

Figure 4.8 *MAT* control of cell identity in *C. neoformans*. (A) The mating-type (*MAT*) locus of *C. neoformans* contains pheromones, pheromone receptors, and homeodomain transcription factors as well as many genes with no apparent role in cell type determination. A schematic of the *MAT* locus for *C. neoformans* var. *neoformans* is shown. *MATa* encodes three copies of the MFa pheromone, the putative pheromone receptor Ste3a, and the homeodomain transcription factor Sxi2a. *MATα* encodes three copies of the MFα pheromone, the putative pheromone receptor Ste3α, and the homeodomain transcription factor Sxi1α. The remaining genes have related but diverged alleles in both *MATa* and *MATα*. Note: schematic diagram is not to scale. (**B**) The determinants of haploid cell identity are unknown; however, it is likely that pheromones and pheromone receptors play a role. The expression of *MFa* could specify the **a** mating type, and the expression of *MFα* could specify the α mating type. Pheromones would be sensed by surface receptors of opposite mating types to activate cell fusion when two cells of opposite mating types are in close proximity to one another. After cell fusion has taken place, the homeodomain transcription factors Sxi1α and Sxi2a are predicted to interact with one another to activate the expression of genes involved in the specification of the dikaryotic state. The dikaryon is then capable of undergoing further sexual development.

STE3α) in *C. neoformans* are only moderately diverged from one another, similar to the other 15 genes in *MAT* that have alleles in both *MATa* and *MATα*. There are two obvious possibilities for specifying haploid cell identity: either the pheromone and pheromone receptor alleles in each mating type have diverged from one another sufficiently to confer pheromone-receptor binding specificity, or there are other factors or mechanisms at play in determining haploid cell types. The mechanism by which haploid cell identity is established in *C. neoformans* remains to be elucidated. Once haploid cells have fused with one another, the homeodomain proteins Sxi1α and Sxi2**a** are transcriptionally upregulated and initiate formation of the dikaryon (Fig. 4.8B) (28).

It is not yet known precisely how Sxi1α and Sxi2**a** act at the molecular level, but it has been shown that they interact with one another in a two-hybrid assay and that the ectopic expression of either protein in a haploid of the opposite mating type initiates diploid sexual development. That is, expression of *SXI1α* in **a** cells or expression of *SXI2a* in α cells results in the induction of complete sexual development including meiosis and sporulation in those haploid cells. In all cases, both Sxi1α and Sxi2**a** must be present for sexual development to occur (28, 29). These findings lead to a model in which Sxi1α and Sxi2**a** interact with one another and bind DNA as a complex to regulate genes responsible for controlling the dikaryotic state and subsequent sexual development (as has been seen in other fungi) (Fig. 4.8B). Although it remains to be determined how Sxi1α and Sxi2**a** exert their effects at the molecular level, the sequencing of several *C. neoformans* genomes and the development of whole genome microarrays will facilitate the elucidation of the molecular pathways that are required for establishing both haploid and dikaryon cell identities crucial for the maintenance of the sexual cycle.

Another way in which cell identity is important in *C. neoformans* is in its ecology. There has been a long-standing observation that *C. neoformans* α strains vastly outnumber **a** strains in both clinical and environmental isolates (43). There is no clear explanation for why natural populations would be skewed toward a single mating type, but it may be related to the finding that α cells have the capacity to essentially ignore the constraints of cell identity and engage in sexual development in the absence of an **a** mating partner. This process, called monokaryotic fruiting, occurs under extreme nutrient limitation and desiccation and results in the formation of filaments, basidia, and spores that resemble sexual structures (76). It has been discovered that fruiting involves the fusion of two haploid α cells or nuclei in a manner that is promoted by mating factors

(46). The fused α/α diploid cells then initiate a filamentation and sporulation pathway similar to that of sexual development. Furthermore, through restriction fragment length polymorphism analysis using a meiotic map (46, 49), it was shown that spores formed during fruiting are recombinant and appear to be the products of meiosis. Thus, α cells have the ability to carry out a sexual development process in the absence of a mating partner of opposite mating type. In addition, the cell type-specific factor Sxi1α, required for classical sexual development, is not necessary for monokaryotic fruiting (29). It is as yet unclear what roles **a**-α sexual development and α-α monokaryotic fruiting play in the natural life cycle of the organism, and it remains to be determined how haploid cell identity is controlled and when it is important. What is clear is that *C. neoformans* has the potential to develop in the absence of an opposite mating type, and this ability may explain the prevalence of α strains in the environment and suggest a mechanism by which α strains can have increased fitness and survival rates over their **a** counterparts in the environment.

SUMMARY

Establishing and maintaining distinct cell types are key for cell survival and reproduction, and there is no exception for fungi. Diverse fungi from *S. cerevisiae* to *C. neoformans* are under pressure to devise strategies to maintain their sexual cycles, but they must also respond rapidly to changes in their environments. As we continue to explore the basis of cell identity in the context of mating type and sexual development, we are finding links to other kinds of cell identities and the varied responses that accompany them. Inevitably, we will continue to discover new strategies that fungi have undertaken to develop and integrate cell identity information in an effort to survive.

References

1. Alberts, B., A. Johnson, J. Lewis, M. Raff, K. Roberts, and P. Walter. 2002. *Molecular Biology of the Cell.* Garland Science, Oxford, United Kingdom.

2. Arnaise, S., R. Debuchy, and M. Picard. 1997. What is a bona fide mating-type gene? Internuclear complementation of *mat* mutants in *Podospora anserina*. *Mol. Gen. Genet.* **256**:169–178.

3. Arnaise, S., D. Zickler, S. Le Bilcot, C. Poisier, and R. Debuchy. 2001. Mutations in mating-type genes of the heterothallic fungus *Podospora anserina* lead to self-fertility. *Genetics* **159**:545–556.

4. Bakkeren, G., and J. W. Kronstad. 1993. Conservation of the *b* mating-type gene complex among bipolar and tetrapolar smut fungi. *Plant Cell* **5**:123–136.

5. Banham, A. H., R. N. Asante-Owusu, B. Gottgens, S. Thompson, C. S. Kingsnorth, E. Mellor, and L. A. Casselton. 1995. An N-terminal dimerization domain permits homeodomain proteins to choose compatible partners and initiate sexual development in the mushroom *Coprinus cinereus*. *Plant Cell* 7:773–783.

6. Banuett, F. 1995. Genetics of *Ustilago maydis*, a fungal pathogen that induces tumors in maize. *Annu. Rev. Genet.* 29:179–208.

7. Banuett, F., and I. Herskowitz. 1994. Morphological transitions in the life cycle of *Ustilago maydis* and their genetic control by the *a* and *b* loci. *Exp. Mycol.* 18:247–266.

8. Bortfeld, M., K. Auffarth, R. Kahmann, and C. W. Basse. 2004. The *Ustilago maydis a2* mating-type locus genes *lga2* and *rga2* compromise pathogenicity in the absence of the mitochondrial p32 family protein Mrb1. *Plant Cell* 16:2233–2248.

9. Brachmann, A., G. Weinzierl, J. Kamper, and R. Kahmann. 2001. Identification of genes in the bW/bE regulatory cascade in *Ustilago maydis*. *Mol. Microbiol.* 42:1047–1063.

10. Burt, A. 2000. Perspective: sex, recombination, and the efficacy of selection—was Weismann right? *Evolution Int. J. Org. Evolution* 54:337–351.

11. Butler, G., C. Kenny, A. Fagan, C. Kurischko, C. Gaillardin, and K. H. Wolfe. 2004. Evolution of the *MAT* locus and its HO endonuclease in yeast species. *Proc. Natl. Acad. Sci. USA* 101:1632–1637.

12. Casadevall, A., and J. R. Perfect. 1998. Cryptococcus neoformans. ASM Press, Washington, DC.

13. Casselton, L. A., and N. S. Olesnicky. 1998. Molecular genetics of mating recognition in basidiomycete fungi. *Microbiol. Mol. Biol. Rev.* 62:55–70.

14. Coppin, E., and R. Debuchy. 2000. Co-expression of the mating-type genes involved in internuclear recognition is lethal in *Podospora anserina*. *Genetics* 155:657–669.

15. Coppin, E., R. Debuchy, S. Arnaise, and M. Picard. 1997. Mating types and sexual development in filamentous ascomycetes. *Microbiol. Mol. Biol. Rev.* 61:411–428.

16. Debuchy, R., and E. Coppin. 1992. The mating types of *Podospora anserina*: functional analysis and sequence of the fertilization domains. *Mol. Gen. Genet.* 233:113–121.

17. Egel, R., O. Nielsen, and D. Weilguny. 1990. Sexual differentiation in fission yeast. *Trends Genet.* 6:369–373.

18. Ferreira, A. V., Z. An, R. L. Metzenberg, and N. L. Glass. 1998. Characterization of mat A-2, mat A-3 and ΔmatA mating-type mutants of *Neurospora crassa*. *Genetics* 148:1069–1079.

19. Fraser, J. A., S. Diezmann, R. L. Subaran, A. Allen, K. B. Lengeler, F. S. Dietrich, and J. Heitman. 2004. Convergent evolution of chromosomal sex-determining regions in the animal and fungal kingdoms. *PLoS Biol.* 2:e384.

20. Fraser, J. A., and J. Heitman. 2003. Fungal mating-type loci. *Curr. Biol.* 13:R792–R795.

21. Galgoczy, D. J., A. Cassidy-Stone, M. Llinas, S. M. O'Rourke, I. Herskowitz, J. L. DeRisi, and A. D. Johnson. 2004. Genomic dissection of the cell-type-specifica-tion circuit in *Saccharomyces cerevisiae*. *Proc. Natl. Acad. Sci. USA* 101:18069–18074.

22. Galitski, T., A. J. Saldanha, C. A. Styles, E. S. Lander, and G. R. Fink. 1999. Ploidy regulation of gene expression. *Science* 285:251–254.

23. Glass, N. L., J. Grotelueschen, and R. L. Metzenberg. 1990. *Neurospora crassa* A mating-type region. *Proc. Natl. Acad. Sci. USA* 87:4912–4916.

24. Halsall, J. R., M. J. Milner, and L. A. Casselton. 2000. Three subfamilies of pheromone and receptor genes generate multiple B mating specificities in the mushroom *Coprinus cinereus*. *Genetics* 154:1115–1123.

25. Herskowitz, I. 1985. A master regulatory locus that determines cell specialization in yeast. *Harvey Lect.* 81:67–92.

26. Herskowitz, I., J. Rine, and J. Strathern. 1992. Mating-type determination and mating-type interconversion in *Saccharomyces cerevisiae*, p. 583–656. *In* E. W. Jones, J. R. Pringle, and J. R. Broach (ed.), *The Molecular and Cellular Biology of the Yeast Saccharomyces*, vol. 2. *Gene Expression*. Cold Spring Harbor Press. Cold Spring Harbor, NY.

27. Hiscock, S. J., and U. Kues. 1999. Cellular and molecular mechanisms of sexual incompatibility in plants and fungi. *Int. Rev. Cytol.* 193:165–295.

28. Hull, C. M., M.-J. Boily, and J. Heitman. 2004. The cell type-specific homeodomain proteins Sxi1α and Sxi2a coordinately regulate sexual development in *Cryptococcus neoformans*. *Eukaryot. Cell* 4:526–535.

29. Hull, C. M., R. C. Davidson, and J. Heitman. 2002. Cell identity and sexual development in *Cryptococcus neoformans* are controlled by the mating-type-specific homeodomain protein Sxi1α. *Genes Dev.* 16:3046–3060.

30. Hull, C. M., and J. Heitman. 2002. Genetics of *Cryptococcus neoformans*. *Annu. Rev. Genet.* 36:557–615.

31. Hull, C. M., and A. D. Johnson. 1999. Identification of a mating type-like locus in the asexual pathogenic yeast *Candida albicans*. *Science* 285:1271–1275.

32. Hull, C. M., R. M. Raisner, and A. D. Johnson. 2000. Evidence for mating of the "asexual" yeast *Candida albicans* in a mammalian host. *Science* 289:307–310.

33. Johnson, A. D. 1995. Molecular mechanisms of cell-type determination in budding yeast. *Curr. Opin. Genet. Dev.* 5:552–558.

34. Kahmann, R., and J. Kamper. 2004. *Ustilago maydis*: how its biology relates to pathogenic development. *New Phytol.* 164:31–42.

35. Kamper, J., M. Reichmann, T. Romeis, M. Bolker, and R. Kahmann. 1995. Multiallelic recognition: nonself-dependent dimerization of the bE and bW homeodomain proteins in *Ustilago maydis*. *Cell* 81:73–83.

36. Karos, M., Y. C. Chang, C. M. McClelland, D. L. Clarke, J. Fu, B. L. Wickes, and K. J. Kwon-Chung. 2000. Mapping of the *Cryptococcus neoformans MATα* locus: presence of mating type-specific mitogen-activated protein kinase cascade homologs. *J. Bacteriol.* 182:6222–6227.

37. Kelly, M., J. Burke, M. Smith, A. Klar, and D. Beach. 1988. Four mating-type genes control sexual differentiation in the fission yeast. *EMBO J.* 7:1537–1547.

38. Kronstad, J. W., and C. Staben. 1997. Mating type in filamentous fungi. *Annu. Rev. Genet.* 31:245–276.

39. Kues, U., R. N. Asante-Owusu, E. S. Mutasa, A. M. Tymon, E. H. Pardo, S. F. O'Shea, B. Gottgens, and L. A. Casselton. 1994. Two classes of homeodomain proteins specify the multiple a mating types of the mushroom *Coprinus cinereus*. *Plant Cell* 6:1467–1475.

40. Kues, U., A. M. Tymon, W. V. Richardson, G. May, P. T. Gieser, and L. A. Casselton. 1994. A mating-type factors of *Coprinus cinereus* have variable numbers of specificity genes encoding two classes of homeodomain proteins. *Mol. Gen. Genet.* 245:45–52.

41. Kües, U. 2000. Life history and developmental processes in the basidiomycete *Coprinus cinereus*. *Microbiol. Mol. Biol. Rev.* 64:316–353.

42. Kurischko, C., M. B. Schilhabel, I. Kunze, and E. Franzl. 1999. The MATA locus of the dimorphic yeast *Yarrowia lipolytica* consists of two divergently oriented genes. *Mol. Gen. Genet.* 262:180–188.

43. Kwon-Chung, K. J., and J. E. Bennett. 1978. Distribution of α and a mating types of *Cryptococcus neoformans* among natural and clinical isolates. *Am. J. Epidemiol.* 108:337–340.

44. Lengeler, K. B., D. S. Fox, J. A. Fraser, A. Allen, K. Forrester, F. S. Dietrich, and J. Heitman. 2002. Mating-type locus of *Cryptococcus neoformans*: a step in the evolution of sex chromosomes. *Eukaryot. Cell* 1:704–718.

45. Leubner-Metzger, G., B. A. Horwitz, O. C. Yoder, and B. G. Turgeon. 1997. Transcripts at the mating type locus of *Cochliobolus heterostrophus*. *Mol. Gen. Genet.* 256: 661–673.

46. Lin, X., C. M. Hull, and J. Heitman. 2005. Sexual reproduction between partners of the same mating type in *Cryptococcus neoformans*. *Nature* 434:1017–1021.

47. Logue, M. E., S. Wong, K. H. Wolfe, and G. Butler. 2005. A genome sequence survey shows that the pathogenic yeast *Candida parapsilosis* has a defective *MTLa1* allele at its mating type locus. *Eukaryot. Cell* 4:1009–1017.

48. Magee, B. B., and P. T. Magee. 2000. Induction of mating in *Candida albicans* by construction of *MTLa* and *MTLα* strains. *Science* 289:310–313.

49. Marra, R. E., J. C. Huang, E. Fung, K. Nielsen, J. Heitman, R. Vilgalys, and T. G. Mitchell. 2004. A genetic linkage map of *Cryptococcus neoformans* variety *neoformans* serotype D (*Filobasidiella neoformans*). *Genetics* 167:619–631.

50. May, G., L. L. Chevanton, and P. J. Pukkila. 1991. Molecular analysis of the *Coprinus cinereus* mating type A factor demonstrates an unexpectedly complex structure. *Genetics* 128:528–538.

51. Miller, M. G., and A. D. Johnson. 2002. White-opaque switching in *Candida albicans* is controlled by mating-type locus homeodomain proteins and allows efficient mating. *Cell* 110:293–302.

52. Odds, F. C. 1988. *Candida and Candidosis: a Review and Bibliography*, 2nd ed. Bailliere Tindall, London, United Kingdom.

53. O'Shea, S. F., P. T. Chaure, J. R. Halsall, N. S. Olesnicky, A. Leibbrandt, I. F. Connerton, and L. A. Casselton. 1998. A large pheromone and receptor gene complex determines multiple B mating type specificities in *Coprinus cinereus*. *Genetics* 148:1081–1090.

54. Paoletti, M., C. Rydholm, E. U. Schwier, M. J. Anderson, G. Szakacs, F. Lutzoni, J. P. Debeaupuis, J. P. Latge, D. W. Denning, and P. S. Dyer. 2005. Evidence for sexuality in the opportunistic fungal pathogen *Aspergillus fumigatus*. *Curr. Biol.* 15:1242–1248.

55. Pardo, E. H., S. F. O'Shea, and L. A. Casselton. 1996. Multiple versions of the A mating type locus of *Coprinus cinereus* are generated by three paralogous pairs of multiallelic homeobox genes. *Genetics* 144:87–94.

56. Philley, M. L., and C. Staben. 1994. Functional analyses of the *Neurospora crassa* MT a-1 mating type polypeptide. *Genetics* 137:715–722.

57. Poggeler, S. 2001. Mating-type genes for classical strain improvements of ascomycetes. *Appl. Microbiol. Biotechnol.* 56:589–601.

58. Poggeler, S., and U. Kuck. 2000. Comparative analysis of the mating-type loci from *Neurospora crassa* and *Sordaria macrospora*: identification of novel transcribed ORFs. *Mol. Gen. Genet.* 263:292–301.

59. Pujol, C., K. J. Daniels, S. R. Lockhart, T. Srikantha, J. B. Radke, J. Geiger, and D. R. Soll. 2004. The closely related species *Candida albicans* and *Candida dubliniensis* can mate. *Eukaryot. Cell* 3:1015–1027.

60. Raper, J. R. 1966. *Genetics of Sexuality in Higher Fungi*. Ronald Press Co., New York, NY.

61. Rice, W. R., and A. K. Chippindale. 2001. Sexual recombination and the power of natural selection. *Science* 294:555–559.

62. Riquelme, M., M. P. Challen, L. A. Casselton, and A. J. Brown. 2005. The origin of multiple B mating specificities in *Coprinus cinereus*. *Genetics* 170:1105–1119.

63. Saupe, S. J., C. Clave, and J. Begueret. 2000. Vegetative incompatibility in filamentous fungi: *Podospora* and *Neurospora* provide some clues. *Curr. Opin. Microbiol.* 3: 608–612.

64. Scherer, M., K. Heimel, V. Starke, and J. Kamper. 2006. The Clp1 protein is required for clamp formation and pathogenic development of *Ustilago maydis*. *Plant Cell* 18:2388–2401.

65. Shen, W. C., R. C. Davidson, G. M. Cox, and J. Heitman. 2002. Pheromones stimulate mating and differentiation via paracrine and autocrine signaling in *Cryptococcus neoformans*. *Eukaryot. Cell* 1:366–377.

66. Slutsky, B., M. Staebell, J. Anderson, L. Risen, M. Pfaller, and D. R. Soll. 1987. "White-opaque transition": a second high-frequency switching system in *Candida albicans*. *J. Bacteriol.* 169:189–197.

67. Spit, A., R. H. Hyland, E. J. C. Mellor, and L. A. Casselton. 1998. A role for heterodimerization in nuclear localization of a homeodomain protein. *Proc. Natl. Acad. Sci. USA* 95:6228–6233.

68. Srikantha, T., S. A. Lachke, and D. R. Soll. 2003. Three mating type-like loci in *Candida glabrata*. *Eukaryot. Cell* 2:328–340.

69. Swiezynski, K. M., and P. R. Day. 1960. Migration of nuclei in *Coprinus lagopus*. *Genet. Res.* 1:129–139.

70. Tilmann, C., and B. Capel. 2002. Cellular and molecular pathways regulating mammalian sex determination. *Recent Prog. Horm. Res.* 57:1–18.

71. Tsong, A. E., M. G. Miller, R. M. Raisner, and A. D. Johnson. 2003. Evolution of a combinatorial transcriptional circuit: a case study in yeasts. *Cell* 115:389–399.

72. Turgeon, B. G., H. Bohlmann, L. M. Ciuffetti, S. K. Christiansen, G. Yang, W. Schafer, and O. C. Yoder. 1993. Cloning and analysis of the mating type genes from *Cochliobolus heterostrophus*. *Mol. Gen. Genet.* 238:270–284.

73. Urban, M., R. Kahmann, and M. Bolker. 1996. Identification of the pheromone response element in *Ustilago maydis*. *Mol. Gen. Genet.* 251:31–37.

74. Van Heeckeren, W. J., D. R. Dorris, and K. Struhl. 1998. The mating-type proteins of fission yeast induce meiosis by directly activating *mei3* transcription. *Mol. Cell. Biol.* 18:7317–7326.

75. Varga, J. 2003. Mating type gene homologues in *Aspergillus fumigatus*. *Microbiology* 149:816–819.

76. Wickes, B. L., M. E. Mayorga, U. Edman, and J. C. Edman. 1996. Dimorphism and haploid fruiting in *Cryptococcus neoformans*: association with the α-mating type. *Proc. Natl. Acad. Sci. USA* 93:7327–7331.

77. Willer, M., L. Hoffmann, U. Styrkarsdottir, R. Egel, J. Davey, and O. Nielsen. 1995. Two-step activation of meiosis by the mat1 locus in *Schizosaccharomyces pombe*. *Mol. Cell. Biol.* 15:4964–4970.

78. Wong, S., M. A. Fares, W. Zimmermann, G. Butler, and K. H. Wolfe. 2003. Evidence from comparative genomics for a complete sexual cycle in the 'asexual' pathogenic yeast *Candida glabrata*. *Genome Biol.* 4:R10.

79. Yamamoto, M. 1996. The molecular control mechanisms of meiosis in fission yeast. *Trends Biochem. Sci.* 21:18–22.

80. Yee, A. R., and J. W. Kronstad. 1993. Construction of chimeric alleles with altered specificity at the *b* incompatibility locus of *Ustilago maydis*. *Proc. Natl. Acad. Sci. USA* 90:664–668.

81. Zeyl, C., and G. Bell. 1997. The advantage of sex in evolving yeast populations. *Nature* 388:465–468.

Sex in Fungi: Molecular Determination
and Evolutionary Implications
Edited by Joseph Heitman et al.
© 2007 ASM Press, Washington, D.C.

Annie E. Tsong
Brian B. Tuch
Alexander D. Johnson

5

Rewiring Transcriptional Circuitry:

Mating-Type Regulation in *Saccharomyces cerevisiae* and *Candida albicans* as a Model for Evolution

Life on earth is astonishing in its variety of form, function, and habitat. Over one million species have been identified, and it is estimated that as many as 10 to 20 million more have yet to be discovered (14). Surprisingly, genome sequencing has revealed not only that the diversity of life-forms is achieved with a small repertoire of protein families, but also that differences in a life-form's complexity are not directly proportional to the number of genes harbored in its genome. In response to this apparent disconnect, researchers have proposed that organismal diversity often arises from changes in gene regulation—that is, changes in the control of when and where genes are expressed—rather than from changes in the genes themselves (9, 10, 15). Conversely, it has recently become evident that phenotypes often remain constant, even as underlying transcriptional circuits diverge (40, 60).

In principle, transcriptional regulatory circuits may change through a variety of mechanisms (Fig. 5.1). For example, an individual gene can come under the control of an existing transcriptional regulatory circuit through changes in the *cis* elements within its promoter. Conversely, a *trans* factor may evolve such that its binding specificity or activity has been changed, resulting in altered gene expression for whole groups of genes. In both cases, small changes that affect protein-DNA interactions have the potential to effect major downstream consequences in the development or life cycle of an organism.

In recent years, vivid examples of transcriptional circuit evolution have been uncovered in systems ranging from butterfly wing pigmentation to yeast ribosomal modules (6, 60). However, complexity and technical limitations in the favored systems have generally limited analysis to a subset of the *cis* or *trans* elements involved in a given change.

The mating-type determination circuit of the budding yeast *Saccharomyces cerevisiae* has long served as a model system for the general problem of how eukaryotic cells maintain differentiated cell types. Due in part to the variety of genetic, genomic, and biochemical tools available in *S. cerevisiae* and in part to the vast body of research on its sexual cycle, the *S. cerevisiae* mating-type regulatory circuit has recently emerged as a system for studies in molecular evolution. In this chapter, we discuss insights into transcriptional evolution that have arisen from detailed comparisons of the mating circuitry

Annie E. Tsong, Department of Molecular and Cell Biology, University of California, Berkeley, Lawrence Berkeley National Labs, 1 Cyclotron Rd., Mailstop 84-355, Berkeley, CA 94720. Brian B. Tuch and Alexander D. Johnson, Department of Biochemistry & Biophysics, Department of Microbiology & Immunology, University of California, San Francisco, CA 94143-2200.

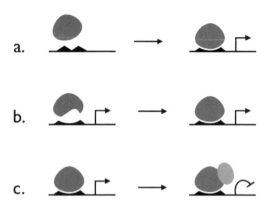

Figure 5.1 Mechanisms underlying the evolution of transcriptional regulation. In principle, transcriptional regulatory circuits may change through a variety of mechanisms. For instance, an individual gene can come under the control of an existing transcriptional regulatory circuit through changes in the *cis* elements within its promoter (a); a *trans* factor may evolve such that its binding specificity has changed (b); or a *trans* factor may evolve such that its activity has changed—for instance, by gaining an interaction with another protein (c).

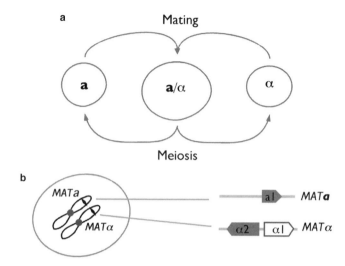

Figure 5.2 *S. cerevisiae* has three distinguishable mating types. (a) *S. cerevisiae* forms three mating types: **a**, α, and **a**/α. **a** and α cells can mate with each other to form an **a**/α cell. **a**/α cells cannot mate, but they can undergo meiosis, regenerating **a** and α cells. (b) Mating type is determined by a short segment of DNA called the *MAT*, or mating type locus. Cells that express only *MATa* or *MATα* are **a** or α cells, respectively, while cells that express both alleles are **a**/α. The *MAT* loci encode sequence-specific DNA binding proteins. a1 and α2 are homeodomain proteins, and α1 is an α-domain protein.

in *S. cerevisiae* and the related ascomycete *Candida albicans*. We first review major concepts in mating-type regulation in *S. cerevisiae* which have been established over the past several decades. We then review more recent work describing mating-type determination in *C. albicans*, a pathogenic yeast that last shared a common ancestor with *S. cerevisiae* some 200 to 800 million years ago. Close comparison of the two circuits reveals multiple changes in transcriptional regulation, which fall into distinct mechanistic classes, and also provides multiple examples of both phenotypic variance and phenotypic conservation accompanying transcriptional regulatory evolution.

SACCHAROMYCES CEREVISIAE MATING-TYPE DETERMINATION

Each of the Three Cell Types Has a Different Transcriptional Program

S. cerevisiae forms three easily distinguishable cell types (22, 23, 31, 56). Two cell types, called **a** and α, can mate with each other to form a third cell type, **a**/α. In most laboratory strains, **a** and α cells are haploid, whereas **a**/α cells are diploid. An **a**/α cell cannot mate; however, it can undergo meiosis and sporulation to regenerate **a** and α haploid cells (Fig. 5.2a).

a, α, and **a**/α cells each follow distinct transcriptional regulatory programs. **a** and α cells are specialized for mating with each other: **a** cells express the **a**-specific genes (asgs), which include a mating pheromone for communicating with α cells (**a**-factor) and a pheromone receptor which detects pheromone signals from α cells (STE2). Likewise, α cells express the α-specific genes

(αsgs), which include the mating pheromone α-factor and the **a**-factor pheromone receptor (STE3).

In addition to the asgs and αsgs, **a** and α cells also express a shared group of genes called the haploid-specific genes (hsgs). Several of the hsgs have mating functions specifically needed by both **a** and α cells; for example, some encode signal transduction pathway components which communicate the pheromone signal to the cell nucleus, while others direct the downstream mating process of cell-cell fusion. Other hsgs include genes involved in the regulation of meiosis (e.g., the repressor of meiosis, RME1), bud site determination (e.g., AXL1), and DNA repair pathway choice (e.g., NEJ1). In a broad sense, the hsgs cover aspects of biology that differ between haploids and diploids.

The Three Different Transcription Programs Are Put into Motion by DNA Binding Proteins Encoded at the *MAT* Locus

The distinct transcriptional programs expressed by **a** cells, α cells, and **a**/α cells are entirely determined by a short segment of DNA called the *MAT* locus (Fig. 5.2b). The *MAT* locus exists in two alleles, *MATa* or *MATα*, each of which encodes unique sequence-specific DNA binding proteins. Cells that express only *MATa* or *MATα* are **a** or α cells, respectively, while cells that express both alleles are **a**/α. *MATa* encodes the homeodomain protein a1, whereas *MATα* encodes the α-domain protein α1 and the homeodomain protein α2. The means by which these three sequence-specific DNA binding proteins define the three mating types have been meticulously studied over the past several decades and are briefly summarized below.

asg Expression

In *S. cerevisiae* **a** cells, asgs are constitutively activated by the MADS box transcription factor Mcm1 in con-junction with the transcription factor Ste12 (28). Mcm1 is expressed in all three cell types, and Ste12 is expressed at high levels in **a** and α cells and at low levels in **a**/α cells (13, 34). To restrict expression of the asgs to **a** cells, asgs are repressed when Mcm1 binds to asg promoters in conjunction with α2, a situation that arises only in α and **a**/α cells (Fig. 5.3a).

The interaction between α2 and Mcm1 and the way these two proteins bind to a specific DNA sequence in

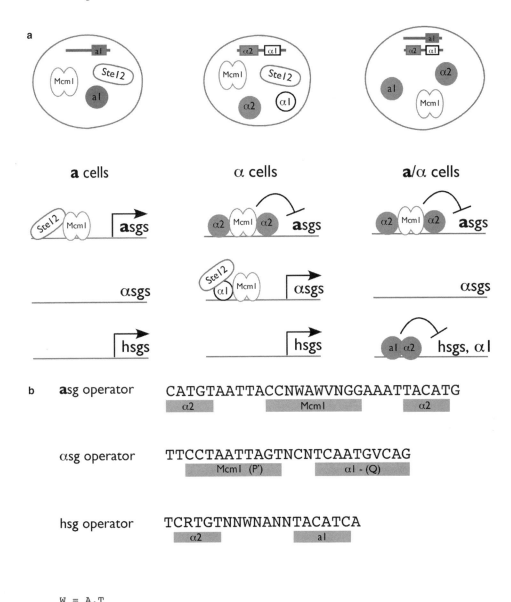

Figure 5.3 The three cell types of *S. cerevisiae* follow unique transcriptional programs. (a) **a**, α, and **a**/α cells each follow distinct transcriptional regulatory programs, directed by the sequence-specific DNA binding proteins encoded by *MATa* and *MATα*. These proteins control mating type by regulating three groups of genes: the **a**-specific genes (asgs), the α-specific genes (αsgs), and the haploid-specific genes (hsgs). The unique combination of site-specific DNA binding proteins in each cell type results in differential expression of the asgs, αsgs, and hsgs. (b) Different *cis* elements ("operators"), bound by the *MAT*-encoded DNA binding proteins, are required for the regulation of the asgs, αsgs, and hsgs.

GENERAPRNCPES

the asg promoter have been extensively characterized in both structural and mutational studies (44, 53, 59). Mcm1 and α2 cooperatively bind as a heterotetramer to a DNA sequence called the asg operator (Fig. 5.3b), a highly defined DNA sequence consisting of an Mcm1 dimer binding site flanked by two α2 binding sites (34). On its own, α2 binds DNA with low affinity through its homeodomain. However, an interaction with Mcm1 locks it into a configuration optimized for recognition of the asg operator (52). The interaction between Mcm1 and α2 has been shown through both crystallographic and mutational analyses to be dependent on a critical nine-residue "linker" region in α2 (63). Each of the residues in this linker region has been shown to contact Mcm1 in a crystal structure of the α2-Mcm1 complex bound to an asg operator (59); furthermore, the effects of mutating each of these residues have been quantified in vitro and in vivo (44).

αsg Expression

In α cells, the α1 protein activates αsgs by cooperatively binding with Mcm1 to a site upstream of αsgs called the QP′ element, which consists of an Mcm1 dimer binding site flanked by a single presumptive α1 binding site (Fig. 5.3a and b) (2, 55). As in the case of asg transcription, Ste12 is also required for full αsg transcription (28). An important distinction in the activation of the asgs and αsgs by Mcm1 lies in their requirement for a MAT-encoded cofactor. Whereas Mcm1 requires its binding partner α1 to activate transcription from the QP′ element of αsgs, Mcm1 activates transcription asg operators without any input from MAT. This difference in cofactor requirements appears to be due to differences in the AT content flanking the Mcm1 site (1). The AT nucleotides flanking Mcm1 sites in the asg operators are required for function; consistent with this, the AT content surrounding Mcm1 sites at αsgs is substantially lower.

hsg Expression

In a/α cells, a1 from the MATa locus and α2 from the MATα locus form a heterodimer which recognizes the hsg operator through its two homeodomains and represses the hsgs (Fig. 5.3a and b) (16, 17, 37). The coexpression of a1 and α2 in a/α cells but not a or α cells ensures that haploid a and α cells express the hsgs but diploid a/α cells do not. Like the α2-Mcm1 interaction, the a1-α2 interaction and the interaction of the heterodimer with DNA have been characterized in detail using both structural and mutational approaches (30, 37, 43). Notably, the regions of α2 required for interaction with a1 are entirely separate from those required for interaction with Mcm1, suggesting that these two activities evolved independently (17).

Mating-Type Switching and Silent Cassettes

Although any individual haploid S. cerevisiae cell expresses only MATa or MATα, all cells carry transcriptionally silenced copies of both loci (24). These copies, called HMR and HML, reside near the telomeres of the chromosome which carries expressed MAT locus (Fig. 5.4). a and α cells participate in a highly coordinated genetic rearrangement whereby either HMR or HML is copied into the expressed MAT locus. This process, called "mating-type switching," results in the switching of a haploid cell to the opposite mating type (58). The mechanics by which HML and HMR are transcriptionally silenced are reviewed in reference 41, and the mechanics by which these silent loci recombine with the expressed MAT locus are discussed in detail in chapter 13 of this volume, as well as reviewed in reference 20. One consequence of mating-type switching is that even when isolated from larger S. cerevisiae populations, individual haploid a and α cells can produce mating partners to regenerate a/α diploids, a form better able to survive certain environmental and genetic hardships (39).

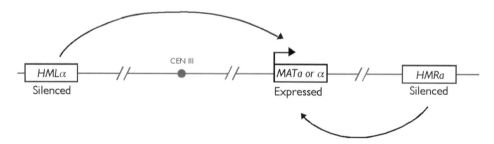

Figure 5.4 *S. cerevisiae* carries silent copies of both *MATa* and *MATα*. Individual haploid *S. cerevisiae* cells express only *MATa* or *MATα*, but they carry transcriptionally silenced copies of both loci, called *HMR* and *HML*. Information from either *HMR* or *HML* can be copied into the expressed *MAT* locus in a highly regulated process of gene conversion, resulting in the switching of mating types.

MATING-TYPE REGULATION IN *C. ALBICANS*

Overview of Mating in *C. albicans*

In contrast to its relative *S. cerevisiae*, whose benign biological properties have benefited humanity for thousands of years in the brewing, baking, and wine-making industries, *C. albicans* is an opportunistic pathogen which causes life-threatening infections in immunocompromised humans. *C. albicans* and *S. cerevisiae* last shared a common ancestor some 200 to 800 million years ago, and their genomes are approximately as diverged as humans are from fish (8, 12, 21). The *C. albicans* and *S. cerevisiae* genomes each carry approximately 6,000 genes, two-thirds of which are clearly related through orthology (5).

Many aspects of *C. albicans* biology permit efficient colonization of mammalian hosts, including its ability to grow in both a budding yeast form and a true hyphal form consisting of multicellular filaments. *C. albicans* has also evolved to respond rapidly to environmental conditions that arise in its host, such as the presence of serum, changes in pH, and changes in temperature (for reviews, see references 8 and 46). Thus, despite their common origin, *C. albicans* and *S. cerevisiae* have diverged significantly in their environmental specialization.

Unlike *S. cerevisiae*, whose mating-type determination circuit has served as an archetypal fungal mating system for decades, *C. albicans* was classified as an asexual, obligate diploid until only recently. However, in 1999 a locus with similarities to the mating-type locus of *S. cerevisiae* was described (Fig. 5.5) (26). As in *S. cerevisiae*, the *MAT* locus of *C. albicans* exists as one of two versions, **a** or α, each of which occupies an allelic position on homologous chromosomes. Each allele encodes sequence-specific DNA binding proteins with homology to those encoded at the *S. cerevisiae MAT* locus; moreover, the directions of transcription, location of introns, and overall configuration of the a1, α1, and α2 genes within the locus are conserved (Fig. 5.5). Most *C. albicans* strains, including the common laboratory strain SC5314, have the **a**/α configuration, explaining in part why mating had not been observed despite decades of investigation.

In 2000, mating of *C. albicans* in the lab was reported using two different approaches to construct mating-competent strains. One approach involved the construction of **a** and α strains by deleting either the *MATa*-encoded transcription factors or the *MATα*-encoded transcription factors from the **a**/α starting strain (27). The other approach involved inducing the loss and rereplication of the chromosome that carries the *MAT* locus (chromosome 5), resulting in strains that carry either two copies of *MATa* or two copies of *MATα* (42). Both experiments demonstrated that **a** cells can mate specifically with α cells (and vice versa), producing nonmating **a**/α tetraploid progeny that carry the genomes of both parents.

The overall configurations of the *S. cerevisiae* and *C. albicans MAT* loci are similar, but there are numerous distinguishing features (compare Fig. 5.2b and 5.5). First, the *C. albicans MATa* locus encodes an additional sequence-specific DNA binding protein called a2; this gene is absent from *S. cerevisiae* (7, 61). Second, unlike the *S. cerevisiae MAT* locus, which exclusively encodes transcriptional regulators, the *C. albicans MAT* locus

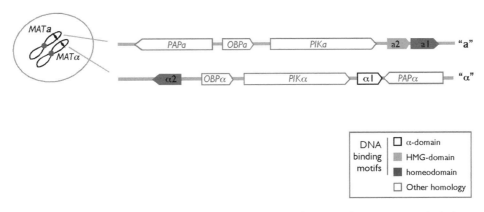

Figure 5.5 The *C. albicans MAT* locus. The *C. albicans MAT* locus encodes 10 genes, 4 of which are sequence-specific DNA binding proteins (see key). *C. albicans* a1, α1, and α2 have orthologs in the *S. cerevisiae MAT* locus and are conserved in their direction of transcription, location of introns, and overall configuration within the locus. The *C. albicans MATa* locus also encodes several additional genes. These include the HMG domain protein a2, as well as **a** and α alleles of an oxysterol binding protein (*OBP*), a phosphatidyl inositol kinase (*PIK*), and a poly(A) polymerase (*PAP*).

also harbors six additional genes whose **a** and α versions vary significantly at the protein level: an oxysterol binding protein (*OBPa* and α), a phosphatidyl inositol kinase (*PIKa* and α), and poly(A) polymerase (*PAPa* and α) (26). Whether the **a** and α alleles of these genes have differential functions related to mating type remains to be determined. Finally, no equivalents of the *HML* or *HMR* silent loci are present in *C. albicans*; hence, *C. albicans* cells cannot switch to the opposite mating type.

The Role of White-Opaque Switching in *C. albicans* Mating

Curiously, mating between **a** and α cells was infrequent—in quantitative mating assays in vitro, it was shown that only 1 in 1 million cells mated over the course of several days on standard laboratory media. Under identical conditions, by contrast, *S. cerevisiae* reaches nearly 100% mating efficiency in much less time. The discovery of a surprising relationship between mating type and a phenomenon called white-opaque switching helped to explain *C. albicans*' mysteriously low mating efficiency (Fig. 5.6 and 5.7) (45).

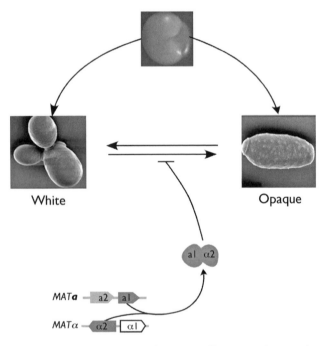

Figure 5.6 White-opaque switching in *C. albicans* is under *MAT* locus control. (Top) White colony with an opaque sector. (Middle) Scanning electron micrographs of white and opaque cells. White cells are rounder, while opaque cells are elongated and show distinct cell membrane properties. (Bottom) The switch from white to opaque is blocked by the a1-α2 protein complex; as a result, cells expressing both *MATa* and *MATα* cannot switch. Images courtesy of Mathew Miller, University of California, San Francisco.

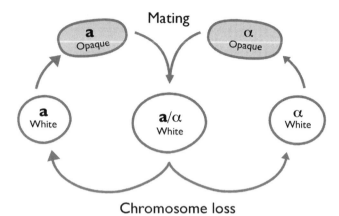

Figure 5.7 The white-opaque transition is required for mating in *C. albicans*. In order to mate, *C. albicans* **a** and α cells must first switch to the mating-competent opaque state.

White-opaque switching is an epistable, heritable switch between two vastly different cellular forms, first described in the clinical isolate WO-1 (51). Because switching occurs only once every several thousand generations, colonies consisting of predominantly white or opaque cells can be isolated: white cells form dome-shaped colonies on solid agar, whereas opaque cells form flatter, matte colonies. At the cellular level, white cells are round, whereas opaque cells are elongated with altered cell surface morphology. The cells show vastly different pathogenic profiles, suggesting that the white-opaque switch plays a role in the ability of *C. albicans* to colonize mammalian hosts (reviewed in reference 54).

Two key observations have linked mating-type regulation and the white-opaque switch (45). First, the white-opaque switch is repressed by the coordinated activity of a1 and α2, encoded at the *MATa* and *MATα* loci (Fig. 5.6). Second, **a** and α cells in the opaque state mate with each other more than a million times more efficiently than their white cell counterparts, at a frequency of up to 40%. The discovery that a1-α2 represses the white-opaque switch suggested why only a subset of clinical isolates are able to switch; the vast majority of clinical isolates, which cannot switch, were shown to be **a**/α, whereas the 3 to 7% of clinical isolates capable of switching were shown to be naturally **a**/**a** or α/α (38). Additionally, the discovery that opaque cells are mating competent (whereas genetically identical white cells are not) explained why mating observed in the laboratory was so inefficient in early experiments. Initial mating experiments were carried out using white cells, and the low frequency of mating observed was likely due to a small fraction of spontaneously arising opaque cells.

In summary, whereas *S. cerevisiae* **a** and α cells are competent to mate without further differentiation, *C. albicans* **a** and α cells must first undergo a large-scale transition to an epistable, heritable, mating-competent form prior to mating (Fig. 5.7; compare to Fig. 5.2a). This transition is inhibited by a1-α2, a condition that prevents a/α cells from mating.

The *C. albicans* Mating-Type Regulatory Circuit

Mating-type determination in *C. albicans* requires four transcriptional regulators encoded at the mating-type locus: a1 and a2 from the *MATa* locus, and α1 and α2 from the *MATα* locus (Fig. 5.8) (61). a2 positively regulates a-type mating (the ability to mate with α cells) by activating the asgs, whereas α1 positively regulates α-type mating (the ability to mate with a cells) by activating the αsgs. On their own, a1 and α2 have no known activities; together, however, they cooperate to repress several genes, as well as the transition from the white to the mating-competent opaque state.

As in *S. cerevisiae*, the *C. albicans* asgs and αsgs encode functions specific for a-type and α-type mating, respectively. The asgs include α-factor receptor, as well as several enzymes required for the processing of a-factor (the gene for a-factor itself has yet to be described), while the αsgs include a-factor receptor and α factor.

Several differences distinguish the regulation of the asgs and αsgs in *S. cerevisiae* and *C. albicans* (Fig. 5.9). In *S. cerevisiae*, asgs and αsg are differentially expressed in a and α cells, regardless of environmental conditions. In *C. albicans*, however, neither the asgs nor the αsgs are expressed differentially in white a and α cells. Detectable differential expression of the αsgs occurs only after the switch to the opaque state, whereupon the αsgs are specifically upregulated several hundred-fold in α cells. Differential expression of the asgs requires an additional step: a cells must both undergo the transition to the opaque state, as well as exposure to α-factor, before asgs are detectably upregulated in a cells over α cells (Fig. 5.9) (4, 61).

Although the requirements for asg and αsg expression have not been fully investigated, we now know that expression of asgs requires a2, as well as a highly specified asg operator of 26 bp which contains binding sites for a2 and Mcm1 (Fig. 5.10) (62).

The a1-α2 complex of *C. albicans* controls several classes of genes. Combined, nine genes have been shown to be directly repressed by a1-α2 by both transcriptional assays and chromatin immunoprecipitation (57, 61). Since a1-α2 is expressed only in a/α cells, these genes are upregulated in white a and α cells relative to a/α cells. The binding site for a1-α2 in *C. albicans* is similar to that of *S. cerevisiae* (26).

Because a1-α2 controls the white-opaque switch, it is also a formal regulator of the almost 450 genes that are differentially regulated between the white and opaque states (36, 61). In both a and α opaque cells, approximately 250 genes are upregulated and 200 genes are downregulated. The genes upregulated in the opaque state include a handful of genes required for mating, explaining why opaque cells are competent to mate. However, the

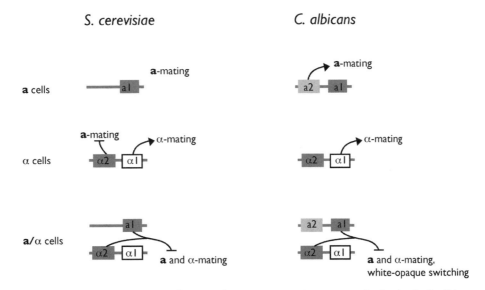

Figure 5.8 Comparison of the *C. albicans* and *S. cerevisiae* mating-type genetic circuits. In *C. albicans*, a2 from the *MATa* locus activates a-type mating, while α1 from the *MATα* locus activates α-type mating. In an a/α cells, which carry both *MATa* and *MATα*, a1 and α2 work together to repress the white-opaque switch, which precedes mating. This circuit differs from that of *S. cerevisiae*, where a-type mating requires no input from the *MAT* locus and is instead repressed by α2 in α and a/α cells.

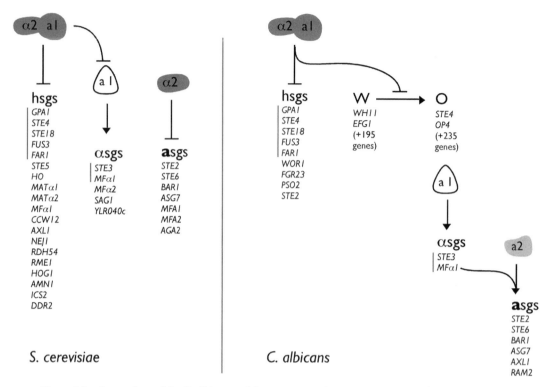

Figure 5.9 Comparison of the *C. albicans* and *S. cerevisiae* mating-type transcriptional circuits. In *C. albicans* white-phase cells, a1-α2 represses nine genes directly, several of which are orthologous to hsgs of *S. cerevisiae* (orthologous genes regulated by orthologous proteins are marked with a gray line). a1-α2 also represses the switch from the white phase to the mating-competent opaque phase. If either component of a1-α2 is absent (for instance, in **a** or α cells), the cell may switch to the opaque state under appropriate environmental conditions. This switch is accompanied by the up- or downregulation of nearly 450 genes. αsgs are activated by α1, but only if the cells are in the opaque phase. asgs are activated by a2, but only if the cells are in the opaque phase and have been exposed to α pheromone. Note two major differences between the *S. cerevisiae* and *C. albicans* mating circuits. First, in contrast to *S. cerevisiae*, relief of a1-α2 repression is necessary but not sufficient for mating competence in *C. albicans*, as *C. albicans* must further undergo the white-opaque transition prior to mating. Second, a2 is required for activation of the asgs in *C. albicans*. In *S. cerevisiae*, asgs are on by default and are repressed in α and a/α cells by α2. Hence, α2 has two roles in *S. cerevisiae*: to repress the hsgs with a1, and to repress the asgs. Despite these changes, the overall output of the circuit is identical in *S. cerevisiae* and *C. albicans*: **a** cells mate with α cells to form nonmating a/α cells.

vast majority of differentially regulated genes are not obviously related to mating competence and instead are involved in diverse aspects of biology such as metabolic specialization, cell adhesion, and stress response (36). Many of the differentially regulated genes lack clear orthologs in *S. cerevisiae*; it is possible that the functions of these genes are related to *C. albicans'* requirement for a warm-blooded host.

A Master Regulator of White-Opaque Switching

Among the genes directly repressed by a1-α2 in *C. albicans*, one has been identified by several groups as the principal regulator of white-opaque switching (25, 57, 66). A deletion of this gene, *WOR1*, prevents **a** and α

cells from switching to the opaque form; conversely, a pulse of ectopically expressed *WOR1* is sufficient to convert white **a** and α cells to the opaque form.

Both a1-α2 and *WOR1* bind directly to the *WOR1* promoter, suggesting a simple mechanism by which a1-α2 and *WOR1* regulate white-opaque switching (Fig. 5.11): in a/α cells, *WOR1* is repressed by a1-α2. In **a** and α cells, *WOR1* is expressed at low levels; however, on occasion the level of Wor1 protein accumulates to high enough concentrations to occupy and activate transcription from its own promoter, resulting in a positive-feedback loop and differentiation into the opaque state. This feedback loop can then be perpetuated through multiple cell divisions. This model explains why the opaque state is blocked in a/α cells, why

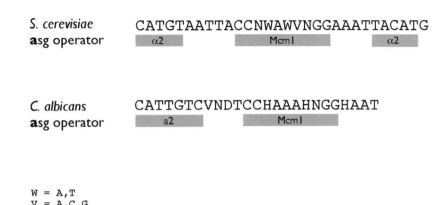

Figure 5.10   asg activation in *C. albicans* requires the asg operator, which comprises an a2 binding site and an Mcm1 binding site. The *S. cerevisiae* asg operator is shown for comparison. The *C. albicans* a2 recognition site and the *S. cerevisiae* α2 recognition site differ by a single base pair deletion; note also that the AT content flanking the Mcm1 site is higher in *S. cerevisiae* than in *C. albicans*.

opaque cells arise infrequently in **a** or α cells, and why once they switch from white to opaque, the switch is stable for many generations. It is also consistent with the observation that *WOR1* expression is upregulated fourfold in **a** or α white cells over **a**/α white cells and is induced an additional 40-fold in **a** or α opaque cells (61). While the simple *WOR1* positive-feedback loop can account for many of the properties of white-opaque switching, it is likely that the stability of the two states requires additional mechanisms.

SIMILARITIES AND DIFFERENCES BETWEEN *S. CEREVISIAE* AND *C. ALBICANS* MATING-TYPE CIRCUITS

S. cerevisiae and *C. albicans* share several important overall similarities in their mating-type regulatory circuits. First, **a** cells can mate with α cells, forming non-

mating **a**/α cells. Second, mating type is determined by orthologous site-specific DNA binding proteins encoded at the *MAT* locus. Many specific aspects of the mating-type circuits have also been preserved through evolution: for instance, orthologous genes are regulated by orthologous transcription factors in many cases (e.g., *STE3* is regulated by α1 in both species). However, the abundance of differences in the mating-type regulatory circuits has provided a rich opportunity to observe the effects of evolution since the two yeasts last shared a common ancestor. In the following section, we discuss multiple categories of differences in the transcriptional circuits of *C. albicans* and *S. cerevisiae* and how they may have arisen. Much of this discussion is speculative; however, the last category we discuss, the replacement of positive with negative regulation of the asgs, has been substantially explored through experimental and bioinformatic means (62).

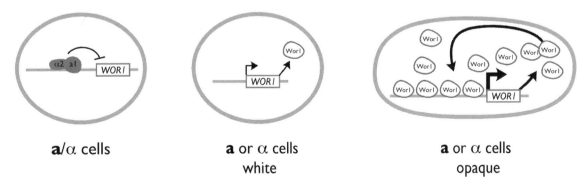

Figure 5.11   *WOR1* controls the white-opaque switch. a1-α2 represses a master regulator of white-opaque switching, *WOR1*. In **a** and α cells, which lack a1-α2, *WOR1* is expressed at low levels. On occasion, the concentration of Wor1 protein surpasses a threshold and activates transcription from its own promoter, resulting in a positive-feedback loop and differentiation into the opaque state.

Changes in the Regulation of Individual Gene Targets

Many of the differences in transcriptional circuitry between *C. albicans* and *S. cerevisiae* involve instances in which an individual gene is under mating-type control in one organism, but not the other. For instance, of the 19 genes definitively shown to be both bound and transcriptionally regulated by a1-α2 in *S. cerevisiae,* only 5 overlap with the 9 genes bound and regulated by a1-α2 in *C. albicans* (Fig. 5.9). Differences in regulation of single genes could arise in two ways: either the gene was mating-type regulated in the common ancestor, with regulation being lost in one lineage by mutation of the *cis* regulatory element, or the gene was not mating-type regulated in the common ancestor, and a binding site for a mating-type transcriptional regulator was gained in one lineage but not the other.

In principle, these two scenarios can be distinguished by determining the presence or absence of an a1-α2 regulatory site in orthologous gene promoters in species that diverged from *S. cerevisiae* prior to *C. albicans,* such as *Yarrowia lipolytica.* In practice, this approach has been hampered by a lack of clear orthology in many of the genes that are differentially regulated by a1-α2 in *S. cerevisiae* and *C. albicans,* limited knowledge of a1-α2 sequence recognition in other species, and a high rate of a1-α2 binding site gain and loss (B. B. Tuch, unpublished results). Additional analysis of mating circuitry in other yeasts, in combination with sequencing of more yeast genomes, may clarify the pathways leading to differences in a1-α2 regulation in *S. cerevisiae* and *C. albicans.*

While we do not yet know the regulatory evolutionary history of a1-α2 controlled genes, it is clear that several differences in a1-α2 regulation reflect specific differences in *S. cerevisiae* and *C. albicans* biology. For instance, several of the genes repressed by a1-α2 in *S. cerevisiae* but not *C. albicans* are involved in haploid-specific biological functions—processes such as a DNA repair process optimized for the haploid genome (*NEJ1*) and bud site selection optimized for the self-mating which results from haploid mating-type switching (*AXL1*). Consistent with the lack of a1-α2 regulation of these processes in *C. albicans,* haploid-diploid differences are irrelevant in *C. albicans,* which undergoes a diploid-tetraploid mating cycle (3, 4).

Changes in the Requirement for an External Signal in Expression of the asgs

As described above, asgs are constitutively activated in *S. cerevisiae* **a** cells and are further upregulated upon pheromone exposure. In *C. albicans,* asgs are detectably expressed only upon pheromone exposure. The primary difference in asg expression between *S. cerevisiae* and *C. albicans* thus appears to be in the basal levels of transcription.

This difference could be effected by changing the level or activity of a transcriptional activator involved in asg expression, such as *STE12.* In *S. cerevisiae,* the transcriptional regulator *STE12* is involved in both the basal asg expression and the pheromone-induced upregulation of asgs. Consistent with this, *STE12* in *S. cerevisiae* is expressed at higher levels in **a** and α cells than in **a**/α cells and is further upregulated in **a** and α cells upon pheromone exposure. The *C. albicans STE12* ortholog, on the other hand, is expressed equally in all three cell types, regardless of whether they are in the white or opaque phases; its levels increase only in opaque cells upon pheromone exposure (4, 61). Interestingly, the Ste12 DNA recognition site is highly overrepresented in asg promoters in *S. cerevisiae,* but not in orthologous promoters of *C. albicans* (Tuch, unpublished). The gain of additional Ste12 sites in *S. cerevisiae* asg promoters, coupled with increased relative levels of Ste12 protein in **a** cells over **a**/α cells, easily accounts for the higher basal asg expression level. Whether these are the only differences in the circuitry relevant to this change in asg regulation remains to be determined.

Incorporation of the White-Opaque Switch into Mating-Type Control

The incorporation of the white-opaque switch into the *C. albicans* mating circuitry is undoubtedly the most phenotypically dramatic of the differences between *S. cerevisiae* mating and *C. albicans* mating. The white-opaque switch is accompanied by massive changes in cell shape, cell wall, and cell membrane architecture and involves differential regulation of nearly 450 genes, most of which fall into distinct functional groups. What biological purpose is served by the integration of mating and white-opaque switching? It has been proposed that specific steps in mating, such as cell-cell signaling and membrane fusion, are impossible in certain mammalian microenvironments; because the opaque state is stable only in specific environments, the requirement for white-opaque switching in mating likely restricts *C. albicans* mating to those environments most amenable to mating (3, 18, 32). Alternatively, it is possible that mating is advantageous only in the environments which permit the white-opaque switch.

Careful dissection of the white and opaque transcriptional networks may well reveal how the interdependence between the white-opaque switch and mating evolved. For instance, the gain of an a1-α2 binding site upstream of *WOR1* would be sufficient to bring the entire white-opaque transcriptional program under mating-type regulation. Once integrated into the a1-α2 regulatory circuit, a gene required for mating, such as *STE4*, could fall under *WOR1* regulation and lose its original a1-α2 site; this would preserve the overall circuit logic, since *WOR1* is itself under a1-α2 regulation. However, this change would have a major consequence: mating would now be dependent on *WOR1*. While transcriptional network structure has been studied in theoretical terms (for an example, see reference 33) and in general terms (29), its roles in evolution remain largely uncharacterized. The white-opaque switch has the potential to serve as a powerful model for understanding how permutations of transcriptional networks give rise to major developmental programs.

Replacement of a Positive Branch of Mating-Type Regulation in One Organism with a Negative Branch in Another

In *S. cerevisiae*, the asgs are on by default. To restrict their expression to a cells, asg expression is repressed by α2, encoded at *MAT*α. In *C. albicans*, however, asgs are off by default. As described above, their activation requires the *MAT*α-encoded HMG domain protein a2, as well as the switch to the opaque state and pheromone stimulation. Mapping of *MAT* loci and functional data from diverse ascomycetes onto a phylogenetic tree shows that activation of asgs by an a2 ortholog represents the ancestral mode of asg regulation (Fig. 5.12). a2 was recently lost in the *S. cerevisiae* lineage; because only *S. cerevisiae* and its close relatives regulate asgs through repression by α2, α2-mediated repression of the asgs must have evolved in recent evolutionary time (7, 61).

In order for asgs to transition from positive regulation by a2 to negative regulation by α2, at least two steps were required: (i) their expression had to become independent of the activator a2, and (ii) their expression had to come under negative control by α2. The comprehensive characterization of asg regulation in *S. cerevisiae* over the past several decades, in combination with phylogenetic analyses, suggests a series of subtle changes in both *cis*- and *trans*-regulatory elements which account for both steps.

Whereas asg activation in *C. albicans* requires both Mcm1 and a2, asg activation in *S. cerevisiae* requires only Mcm1. As previously mentioned, a high AT content surrounding the Mcm1 binding site in *S. cerevisiae* asgs is required for Mcm1 to bind DNA without a cofactor; thus, independence from the activator a2 may have been the result of a simple increase in the AT content surrounding the ancestral Mcm1 binding site (1). Indeed, Mcm1 binding sites within the *C. albicans* asg operators show a lower surrounding AT content than those within *S. cerevisiae* asg operators (Fig. 5.10) (62).

Evolution of dependence on α2 required both the acquisition of α2 binding sites in the asg operator, and the acquisition of the α2-Mcm1 interaction domain. Acquisition of the α2 binding sites was certainly accelerated by the similarities of the a2 and α2 recognition sites: the a2 site converts to an α2 site with a single base-pair deletion (Fig. 5.10) (62). Evolution of the nine-amino-acid region required for α2-Mcm1 interaction, possibly less trivial, is still eminently plausible. Several factors likely facilitated evolution of the interaction. First, the prior presence of Mcm1 at asg operators expanded the promoter interaction surfaces available to α2 to include both protein and DNA. Second, the linker region, nine amino acids of which contact Mcm1, is disordered in free α2 protein; its evolution may thus have been accelerated, being less subject to structural constraints (48). Third, only a fraction of the interactions between α2 and Mcm1 are required to establish cooperative binding (44); once a weak cooperative interaction was established, strengthening the interaction would be straightforward to select.

Examination of asg regulation across multiple yeast species reveals the order in which these *cis* and *trans* element changes occurred: asg regulation in the *Kluyveromyces lactis* lineage (including the yeasts *K. lactis*, *Kluyveromyces waltii*, *Eremothecium gossypii*, and *Saccharomyces kluyveri*) carries hallmarks of both positive regulation by a2 and negative regulation by α2 (62). Taking into account the phylogenetic relationship of *S. cerevisiae* to *K. lactis* and *C. albicans*, this observation indicates that the α2-mediated repression of the asgs evolved prior to the loss of a2 (Fig. 5.13).

Implications of asg Regulon Evolution

A detailed understanding of how asgs evolved from positive to negative regulation has provided multiple general insights into the process of transcriptional evolution. First, this example demonstrates that major rewiring of a circuit can be achieved with the coordination of remarkably small changes to *cis* and *trans* elements. Changes as minor as increases in AT content surrounding a binding site, a single base-pair deletion from an existing binding site, and a small change in the amino acid sequence of a sequence-specific DNA binding protein, allowing it to interact with an additional protein, have resulted in conversion of a regulon from being positively to being negatively regulated. Second, the temporal coordination of changes in *cis* and *trans* has permitted a dramatically altered circuit to evolve,

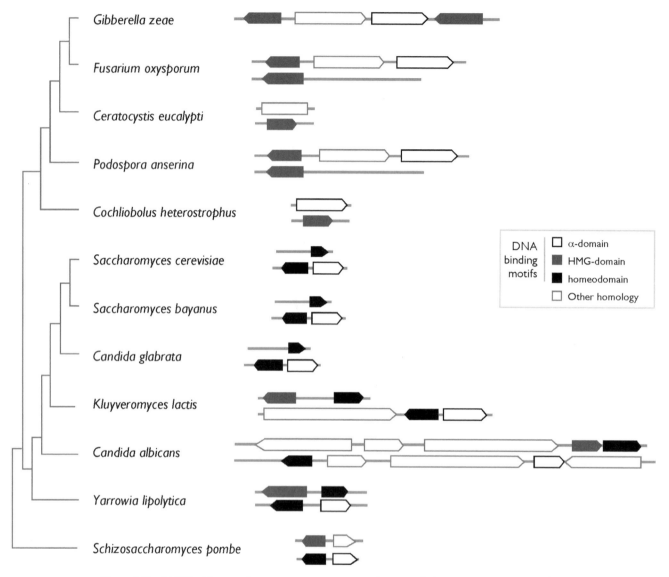

Figure 5.12 *MAT* loci from various ascomycetes are mapped onto a phylogenetic tree. An HMG box ortholog of *C. albicans* a2 is present in the *MAT* loci of extant species spanning up to 1 billion years of evolution; moreover, this ortholog is required for a-type mating in many species, including *K. lactis*, *C. albicans*, *Y. lipolytica*, *C. heterostrophus*, *G. zeae*, *P. anserina*, and *S. pombe*. The lineage carrying *S. cerevisiae* lost this regulator after its divergence from *K. lactis*.

apparently without relinquishing appropriate regulation of the asgs during the transition. Third, this example underscores the role of phenotype as the interface between genotype and natural selection: natural selection does not seem to "care" about molecular details, as long as the appropriate phenotypic output is achieved.

In addition to providing a molecular-level snapshot of transcriptional evolution, understanding how the asg regulon evolved has also resolved a long-standing mystery in *S. cerevisiae* biology. "Although one might have imagined that **a** cells would contain activator and repressor proteins analogous to those of α cells," ob-

served Ira Herskowitz, "they contain neither." Why is the *S. cerevisiae* mating-type circuit so lopsided? By comparing the *S. cerevisiae* mating-type circuit to that of *C. albicans* and other yeasts, we now know that the ancestor of *S. cerevisiae* did in fact carry an activator of **a**-type behavior analogous to α1 in α cells and that it was lost during evolution; we also know that the asg repression activity of α2 is a recent evolutionary innovation that compensates for the loss of the activator. The *C. albicans* mating-type circuitry, on the other hand, is precisely what "one might have imagined": each of the two mating-type alleles contributes one positive regulator

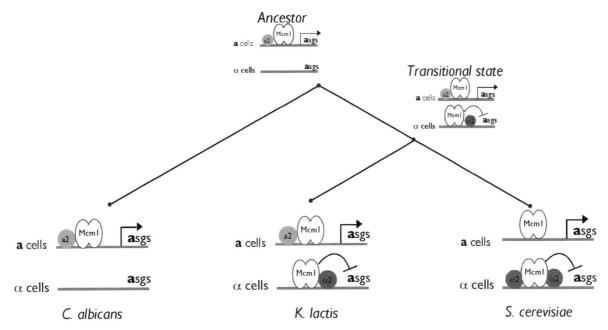

Figure 5.13 The asgs were positively regulated by a2 in the common ancestor of *C. albicans* and *S. cerevisiae* but are negatively regulated by α2 in *S. cerevisiae*. Yeasts in the *K. lactis* lineage, including *K. lactis*, *E. gossypii*, *S. kluyveri*, and *K. waltii*, show hallmarks of both positive and negative regulation of the asgs, as evaluated by bioinformatic analysis of both asg operators and α2 and Mcm1 protein sequence. Given the phylogenetic relationships of *S. cerevisiae*, *K. lactis*, and *C. albicans*, this suggests that α2-mediated repression of the asgs evolved prior to the loss of a2.

of its respective mating type, and each allele contributes one half of a heterodimeric repressor such that when both alleles are present in the same cell, mating is shut off.

Intriguingly, it has been shown that positive and negative transcriptional regulations result in differing systemic properties, even when the two modes encode identical logical output (49, 64). Studies with multiple phyla have also shown that sex determination is less evolutionarily conserved, on a whole, than other developmental programs (11, 19, 35, 47, 65), potentially because changes in sex determination pathways closely track with reproductive isolation and speciation (50). Determining whether the changes in asg regulation were due to the vagaries of neutral evolution or were the result of subtle environmental or sexual selection will certainly prove to be a challenging avenue of research.

CONCLUSIONS

Mating-type regulation in *S. cerevisiae* has long served as a model system to understand principles of gene regulation, cell-to-cell signaling, polarized growth, cell fusion, and many other aspects of cell and molecular biology. In this chapter we have highlighted a new use of this model system, namely in understanding how transcriptional circuits evolve over evolutionary timescales.

Close comparison of mating-type regulation in *S. cerevisiae* and *C. albicans* has revealed an abundance of examples in transcriptional evolution. At one extreme, strength of selection has ensured that even reversing the means by which the entire asg regulon is regulated has not perturbed the overall output of the mating-type regulatory circuit. At the other extreme, several major changes in biology appear to hinge on the mechanistically straightforward acquisition of mating-type regulatory sites in front of one or a few genes. Combined, these examples speak to both the plasticity of gene regulation in evolution and the inextricable link between phenotype and environmental selection.

References

1. **Acton, T. B., H. Zhong, and A. K. Vershon.** 1997. DNA-binding specificity of Mcm1: operator mutations that alter DNA-bending and transcriptional activities by a MADS box protein. *Mol. Cell. Biol.* **17**:1881–1889.

2. **Bender, A., and G. F. Sprague, Jr.** 1987. MAT alpha 1 protein, a yeast transcription activator, binds synergistically with a second protein to a set of cell-type-specific genes. *Cell* **50**:681–691.

3. **Bennett, R. J., and A. D. Johnson.** 2005. Mating in Candida albicans and the search for a sexual cycle. *Annu. Rev. Microbiol.* **59**:233–255.

4. Bennett, R. J., M. A. Uhl, M. G. Miller, and A. D. Johnson. 2003. Identification and characterization of a *Candida albicans* mating pheromone. *Mol. Cell. Biol.* **23:** 8189–8201.

5. Braun, B. R., M. van Het Hoog, C. d'Enfert, M. Martchenko, J. Dungan, A. Kuo, D. O. Inglis, M. A. Uhl, H. Hogues, M. Berriman, M. Lorenz, A. Levitin, U. Oberholzer, C. Bachewich, D. Harcus, A. Marcil, D. Dignard, T. Iouk, R. Zito, L. Frangeul, F. Tekaia, K. Rutherford, E. Wang, C. A. Munro, S. Bates, N. A. Gow, L. L. Hoyer, G. Kohler, J. Morschhauser, G. Newport, S. Znaidi, M. Raymond, B. Turcotte, G. Sherlock, M. Costanzo, J. Ihmels, J. Berman, D. Sanglard, N. Agabian, A. P. Mitchell, A. D. Johnson, M. Whiteway, and A. Nantel. 2005. A human-curated annotation of the Candida albicans genome. *PLoS Genet.* **1:** 36–57.

6. Brunetti, C. R., J. E. Selegue, A. Monteiro, V. French, P. M. Brakefield, and S. B. Carroll. 2001. The generation and diversification of butterfly eyespot color patterns. *Curr. Biol.* **11:** 1578–1585.

7. Butler, G., C. Kenny, A. Fagan, C. Kurischko, C. Gaillardin, and K. H. Wolfe. 2004. Evolution of the MAT locus and its Ho endonuclease in yeast species. *Proc. Natl. Acad. Sci. USA* **101:** 1632–1637.

8. Calderone, R. A. 2002. *Candida and Candidiasis.* ASM Press, Washington, DC.

9. Carroll, S. B., J. K. Grenier, and S. D. Weatherbee. 2001. *From DNA to Diversity.* Blackwell Science, Inc., Malden, MA.

10. Davidson, E. H. 2001. *Genomic Regulatory Systems.* Academic Press, San Diego, CA.

11. de Bono, M., and J. Hodgkin. 1996. Evolution of sex determination in caenorhabditis: unusually high divergence of tra-1 and its functional consequences. *Genetics* **144:** 587–595.

12. Dujon, B., D. Sherman, G. Fischer, P. Durrens, S. Casaregola, I. Lafontaine, J. De Montigny, C. Marck, C. Neuveglise, E. Talla, N. Goffard, L. Frangeul, M. Aigle, V. Anthouard, A. Babour, V. Barbe, S. Barnay, S. Blanchin, J. M. Beckerich, E. Beyne, C. Bleykasten, A. Boisrame, J. Boyer, L. Cattolico, F. Confanioleri, A. De Daruvar, L. Despons, E. Fabre, C. Fairhead, H. Ferry-Dumazet, A. Groppi, F. Hantraye, C. Hennequin, N. Jauniaux, P. Joyet, R. Kachouri, A. Kerrest, R. Koszul, M. Lemaire, I. Lesur, L. Ma, H. Muller, J. M. Nicaud, M. Nikolski, S. Oztas, O. Ozier-Kalogeropoulos, S. Pellenz, S. Potier, G. F. Richard, M. L. Straub, A. Suleau, D. Swennen, F. Tekaia, M. Wesolowski-Louvel, E. Westhof, B. Wirth, M. Zeniou-Meyer, I. Zivanovic, M. Bolotin-Fukuhara, A. Thierry, C. Bouchier, B. Caudron, C. Scarpelli, C. Gaillardin, J. Weissenbach, P. Wincker, and J. L. Souciet. 2004. Genome evolution in yeasts. *Nature* **430:** 35–44.

13. Fields, S., and I. Herskowitz. 1987. Regulation by the yeast mating-type locus of STE12, a gene required for cell-type-specific expression. *Mol. Cell. Biol.* **7:** 3818–3821.

14. Futuyma, D. J. 1997. *Evolutionary Biology.* Sinauer, Sunderland, MA.

15. Gerhart, J., and M. Kirschner. 1997. *Cells, Embryos, and Evolution.* Blackwell Science, Inc., Malden, MA.

16. Goutte, C., and A. D. Johnson. 1994. Recognition of a DNA operator by a dimer composed of two different homeodomain proteins. *EMBO J.* **13:** 1434–1442.

17. Goutte, C., and A. D. Johnson. 1993. Yeast a1 and alpha 2 homeodomain proteins form a DNA-binding activity with properties distinct from those of either protein. *J. Mol. Biol.* **233:** 359–371.

18. Gow, N. A. 2002. Candida albicans switches mates. *Mol. Cell* **10:** 217–218.

19. Haag, E. S., S. Wang, and J. Kimble. 2002. Rapid coevolution of the nematode sex-determining genes fem-3 and tra-2. *Curr. Biol.* **12:** 2035–2041.

20. Haber, J. E. 1998. Mating-type gene switching in Saccharomyces cerevisiae. *Annu. Rev. Genet.* **32:** 561–599.

21. Hedges, S. B. 2002. The origin and evolution of model organisms. *Nat. Rev. Genet.* **3:** 838–849.

22. Herskowitz, I. 1988. Life cycle of the budding yeast Saccharomyces cerevisiae. *Microbiol. Rev.* **52:** 536–553.

23. Herskowitz, I., J. Rine, and J. Strathern. 1992. *Mating-Type Determination and Mating-Type Interconversion in Saccharomyces cerevisiae,* vol. II. Cold Spring Harbor Laboratory Press, Cold Spring Harbor, NY.

24. Hicks, J., J. N. Strathern, and A. J. Klar. 1979. Transposable mating type genes in Saccharomyces cerevisiae. *Nature* **282:** 478–483.

25. Huang, G., H. Wang, S. Chou, X. Nie, J. Chen, and H. Liu. 2006. Bistable expression of WOR1, a master regulator of white-opaque switching in Candida albicans. *Proc. Natl. Acad. Sci. USA* **103:** 12813–12818.

26. Hull, C. M., and A. D. Johnson. 1999. Identification of a mating type-like locus in the asexual pathogenic yeast Candida albicans. *Science* **285:** 1271–1275.

27. Hull, C. M., R. M. Raisner, and A. D. Johnson. 2000. Evidence for mating of the "asexual" yeast Candida albicans in a mammalian host. *Science* **289:** 307–310.

28. Hwang-Shum, J. J., D. C. Hagen, E. E. Jarvis, C. A. Westby, and G. F. Sprague, Jr. 1991. Relative contributions of MCM1 and STE12 to transcriptional activation of a- and alpha-specific genes from Saccharomyces cerevisiae. *Mol. Gen. Genet.* **227:** 197–204.

29. Ihmels, J., S. Bergmann, and N. Barkai. 2004. Defining transcription modules using large-scale gene expression data. *Bioinformatics* **20:** 1993–2003.

30. Jin, Y., J. Mead, T. Li, C. Wolberger, and A. K. Vershon. 1995. Altered DNA recognition and bending by insertions in the alpha 2 tail of the yeast a1/alpha 2 homeodomain heterodimer. *Science* **270:** 290–293.

31. Johnson, A. D. 1995. Molecular mechanisms of cell-type determination in budding yeast. *Curr. Opin. Genet. Dev.* **5:** 552–558.

32. Johnson, A. D. 2003. The biology of mating in Candida albicans. *Nat. Rev. Microbiol.* **1:** 106–116.

33. Kashtan, N., and U. Alon. 2005. Spontaneous evolution of modularity and network motifs. *Proc. Natl. Acad. Sci. USA* **102:** 13773–13778.

34. Keleher, C. A., C. Goutte, and A. D. Johnson. 1988. The yeast cell-type-specific repressor alpha 2 acts cooperatively with a non-cell-type-specific protein. *Cell* **53:** 927–936.

35. **Kulathinal, R. J., L. Skwarek, R. A. Morton, and R. S. Singh.** 2003. Rapid evolution of the sex-determining gene, transformer: structural diversity and rate heterogeneity among sibling species of Drosophila. *Mol. Biol. Evol.* **20:**441–452.

36. **Lan, C. Y., G. Newport, L. A. Murillo, T. Jones, S. Scherer, R. W. Davis, and N. Agabian.** 2002. Metabolic specialization associated with phenotypic switching in Candida albicans. *Proc. Natl. Acad. Sci. USA* **99:**14907–14912.

37. **Li, T., M. R. Stark, A. D. Johnson, and C. Wolberger.** 1995. Crystal structure of the MATa1/MAT alpha 2 homeodomain heterodimer bound to DNA. *Science* **270:**262–269.

38. **Lockhart, S. R., C. Pujol, K. J. Daniels, M. G. Miller, A. D. Johnson, M. A. Pfaller, and D. R. Soll.** 2002. In Candida albicans, white-opaque switchers are homozygous for mating type. *Genetics* **162:**737–745.

39. **Luchnik, A. N., V. M. Glaser, and S. V. Shestakov.** 1977. Repair of DNA double-strand breaks requires two homologous DNA duplexes. *Mol. Biol. Rep.* **3:**437–442.

40. **Ludwig, M. Z., C. Bergman, N. H. Patel, and M. Kreitman.** 2000. Evidence for stabilizing selection in a eukaryotic enhancer element. *Nature* **403:**564–567.

41. **Lustig, A. J.** 1998. Mechanisms of silencing in Saccharomyces cerevisiae. *Curr. Opin. Genet. Dev.* **8:**233–239.

42. **Magee, B. B., and P. T. Magee.** 2000. Induction of mating in Candida albicans by construction of MTLa and MTLalpha strains. *Science* **289:**310–313.

43. **Mak, A., and A. D. Johnson.** 1993. The carboxy-terminal tail of the homeo domain protein alpha 2 is required for function with a second homeo domain protein. *Genes Dev.* **7:**1862–1870.

44. **Mead, J., H. Zhong, T. B. Acton, and A. K. Vershon.** 1996. The yeast alpha2 and Mcm1 proteins interact through a region similar to a motif found in homeodomain proteins of higher eukaryotes. *Mol. Cell. Biol.* **16:**2135–2143.

45. **Miller, M. G., and A. D. Johnson.** 2002. White-opaque switching in Candida albicans is controlled by mating-type locus homeodomain proteins and allows efficient mating. *Cell* **110:**293–302.

46. **Odds, F. C.** 1988. *Candida and Candidosis*, 2nd ed. W. B. Saunders Company, St. Louis, MO.

47. **O'Neil, M. T., and J. M. Belote.** 1992. Interspecific comparison of the transformer gene of Drosophila reveals an unusually high degree of evolutionary divergence. *Genetics* **131:**113–128.

48. **Phillips, C. L., M. R. Stark, A. D. Johnson, and F. W. Dahlquist.** 1994. Heterodimerization of the yeast homeodomain transcriptional regulators alpha 2 and a1 induces an interfacial helix in alpha 2. *Biochemistry* **33:**9294–9302.

49. **Savageau, M. A.** 1983. Regulation of differentiated cell-specific functions. *Proc. Natl. Acad. Sci. USA* **80:**1411–1415.

50. **Singh, R. S., and R. J. Kulathinal.** 2000. Sex gene pool evolution and speciation: a new paradigm. *Genes Genet. Syst.* **75:**119–130.

51. **Slutsky, B., M. Staebell, J. Anderson, L. Risen, M. Pfaller, and D. R. Soll.** 1987. "White-opaque transition": a second high-frequency switching system in Candida albicans. *J. Bacteriol.* **169:**189–197.

52. **Smith, D. L., and A. D. Johnson.** 1992. A molecular mechanism for combinatorial control in yeast: MCM1 protein sets the spacing and orientation of the homeodomains of an alpha 2 dimer. *Cell* **68:**133–142.

53. **Smith, D. L., and A. D. Johnson.** 1994. Operator-constitutive mutations in a DNA sequence recognized by a yeast homeodomain. *EMBO J.* **13:**2378–2387.

54. **Soll, D. R.** 1992. High-frequency switching in Candida albicans. *Clin. Microbiol. Rev.* **5:**183–203.

55. **Sprague, G. F., Jr., R. Jensen, and I. Herskowitz.** 1983. Control of yeast cell type by the mating type locus: positive regulation of the alpha-specific STE3 gene by the MAT alpha 1 product. *Cell* **32:**409–415.

56. **Sprague, G. F., Jr., and J. Thorner.** 1992. Pheromone response and signal transduction during the mating process of *Saccharomyces cerevisiae*, p. 657–744. *In* E. W. Jones, J. R. Pringle, and J. R. Broach (ed.), *The Molecular and Cellular Biology of the Yeast.* Cold Spring Harbor Laboratory Press, Cold Spring Harbor, NY.

57. **Srikantha, T., A. R. Borneman, K. J. Daniels, C. Pujol, W. Wu, M. R. Seringhaus, M. Gerstein, S. Yi, M. Snyder, and D. R. Soll.** 2006. TOS9 regulates white-opaque switching in *Candida albicans. Eukaryot. Cell* **5:**1674–1687.

58. **Strathern, J. N., and I. Herskowitz.** 1979. Asymmetry and directionality in production of new cell types during clonal growth: the switching pattern of homothallic yeast. *Cell* **17:**371–381.

59. **Tan, S., and T. J. Richmond.** 1998. Crystal structure of the yeast MATalpha2/MCM1/DNA ternary complex. *Nature* **391:**660–666.

60. **Tanay, A., A. Regev, and R. Shamir.** 2005. Conservation and evolvability in regulatory networks: the evolution of ribosomal regulation in yeast. *Proc. Natl. Acad. Sci. USA* **102:**7203–7208.

61. **Tsong, A. E., M. G. Miller, R. M. Raisner, and A. D. Johnson.** 2003. Evolution of a combinatorial transcriptional circuit: a case study in yeasts. *Cell* **115:**389–399.

62. **Tsong, A. E., B. B. Tuch, H. Li, and A. D. Johnson.** 2006. Evolution of alternative transcriptional circuits with identical logic. *Nature* **443:**415–420.

63. **Vershon, A. K., and A. D. Johnson.** 1993. A short, disordered protein region mediates interactions between the homeodomain of the yeast alpha 2 protein and the MCM1 protein. *Cell* **72:**105–112.

64. **Wall, M. E., W. S. Hlavacek, and M. A. Savageau.** 2004. Design of gene circuits: lessons from bacteria. *Nat. Rev. Genet.* **5:**34–42.

65. **Whitfield, L. S., R. Lovell-Badge, and P. N. Goodfellow.** 1993. Rapid sequence evolution of the mammalian sex-determining gene SRY. *Nature* **364:**713–715.

66. **Zordan, R. E., D. J. Galgoczy, and A. D. Johnson.** 2006. Epigenetic properties of white-opaque switching in Candida albicans are based on a self-sustaining transcriptional feedback loop. *Proc. Natl. Acad. Sci. USA* **103:**12807–12812.

Ascomycetes:
From Model Yeasts
to Plant and
Human Pathogens

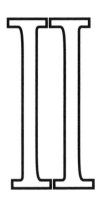

*Sex in Fungi: Molecular Determination
and Evolutionary Implications*
Edited by Joseph Heitman et al.
© 2007 ASM Press, Washington, D.C.

B. Gillian Turgeon
Robert Debuchy

6

Cochliobolus and *Podospora*: Mechanisms of Sex Determination and the Evolution of Reproductive Lifestyle

Cellular specialization leading to sexual reproduction is controlled in fungi by mating-type (*MAT*) genes. There are two basic "fruitful" fungal sexual reproductive styles: the first requires partners of opposite mating type, whereas the second is accomplished by a single individual without participation of a partner. Fungi that require a genetically distinct partner to complete the sexual process are called "self-incompatible" (heterothallic); species able to undergo sexual reproduction without a partner are "self-compatible" (homothallic) (17). In addition, noncanonical mating strategies, which do not follow the genetic "rules" associated with the best-known mating behaviors, have been reported (64, 67). Our article addresses the structural, functional, and evolutionary issues that correspond to each mating lifestyle. We focus on two filamentous ascomycetes: *Cochliobolus heterostrophus* (representing the Dothideomycetes), which has been used extensively as a model to address questions of mating-type structure and evolution of reproductive lifestyle, and *Podospora anserina* (representing the Sordariomycetes), which is the most rigorously studied filamentous ascomycete, in terms of structure and function of *MAT* proteins. This

chapter builds on a chapter by the same authors that appeared recently in *The Mycota* (30). Certain parts of that article are necessary as background here; however, our intention here is to capture the essence of that report, update information where appropriate, and consider models by which reproductive style evolves and mating-type proteins cooperate to control the complex sexual developmental pathway.

Molecular comparisons of alternate sequences at *MAT* show that they are nonallelic. The term "idiomorph" has been applied to designate alternate forms of a locus that occupy the same chromosomal position but are "detectably similar neither in DNA sequence nor in the proteins for which they code, and possibly are not even of common origin" (59). By definition, one idiomorph is, by structure and by function, completely unlike the other, and because of this the two are not expected to recombine with each other. While this view of *MAT* architecture is essentially correct, the issue of whether the extant dissimilar *MAT* genes have separate evolutionary histories or do, in fact, share a common origin, is far from settled. Furthermore, as is described below, recombination events between pockets of homology in these regions are

---

B. Gillian Turgeon, Department of Plant Pathology, Cornell University, Ithaca, NY 14853. **Robert Debuchy,** CNRS, Institut de Génétique et Microbiologie, Batiment 400, UMR 8621, and Université Paris-Sud 11, Orsay, F-91405, France.

likely the mechanism by which fungi switch from self-incompatible to self-compatible lifestyles or vice versa. In the self-incompatible ascomycetes, the size of the idiomorphs varies (1 to 6 kb) (30); however, all are flanked by sequences that tend to be highly similar for strains of opposite mating type. The mechanisms by which these loci evolved and the factors contributing to their maintenance as unique addresses in otherwise similar genomes are subject to much speculation.

Although mating-type systems in fungi were described at the beginning of the last century (12, 31), molecular elucidation of *MAT* loci was not possible until much later. As described elsewhere in this volume, the first *MAT* gene cloned was from the budding yeast *Saccharomyces cerevisiae* (44); subsequent analyses revealed that *S. cerevisiae MAT* is a master regulatory locus and that mating type is determined by the activity of transcription regulators encoded by *MAT* (42, 43, 55). The cue to mate sets in motion a signal transduction pathway that culminates in the activation of the target structural genes needed for sexual reproduction (genes for pheromones, receptors, karyogamy, meiosis, etc.). In yeast, each haploid cell has a complete array of these genes, which are differentially expressed to determine cell type.

Additional ascomycete *MAT* genes have been isolated and characterized in detail from several yeasts, including *Schizosaccharomyces pombe* (49), and from many filamentous fungi (30). Comparative sequence analyses of these genes have revealed that all encode putative transcriptional regulators; conserved DNA-binding motifs

have been identified in the putative proteins encoded by the *MAT* genes of all of these species. One of the DNA binding proteins (Mc) of *S. pombe* has a conserved region known as an HMG motif (74, 76, 87) that is found in members of the high-mobility group (HMG) of DNA-binding proteins. These include proteins involved in determining cell fate, such as the human SRY protein, a sex-determining factor (50, 73). A protein with a conserved HMG motif is encoded by one of the two *MAT* idiomorphs of all filamentous ascomycetes examined to date and provides the basis for *MAT* locus nomenclature (80) and for a procedure to clone *MAT* genes from a wide range of fungi (1). The alternate idiomorph encodes a protein that contains a motif called the α-box, found in the transcription factor known as α1 encoded by one of the *S. cerevisiae MAT* idiomorphs. In the Dothideomycetes, this appears to be the only *MAT1-1*-encoded protein. In the Sordariomycetes, it is one of three proteins encoded at *MAT1-1*. As with the yeasts, each haploid strain has a complete array of all genes needed for sexual reproduction and it is the differential expression of these genes that determines cell type. Thus, yeast and filamentous fungal *MAT* loci are similar in terms of overall genomic organization and they all encode functionally similar proteins that are regulators of transcription.

Sexual reproduction in filamentous ascomycetes is a complex developmental process requiring self/nonself recognition between cells and between nuclei (Fig. 6.1). If conditions are favorable, the sexual cycle begins with

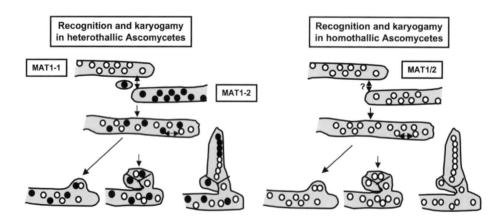

Figure 6.1   Cell-cell recognition and nucleus-nucleus recognition are necessary for mating. In the case of heterothallic matings (left), self-nonself recognition is required between mating partners. The male partner is a microconidium, macroconidium, or hyphal cell. A second recognition step occurs between nuclei; unlike nuclei must recognize each other and pair. For homothallic strains (right), the cell-cell recognition step is likely not required: it is known, for example, that homothallic *Sordaria macrospora* and *Aspergillus nidulans* do not require external cells for fertilization. Current hypotheses suggest that homothallic species must have a mechanism that alters some nuclei so they are different from parental nuclei and that these now unlike nuclei can then pair to form the transient diploid. The process by which this is achieved is unknown at present (but see "Mating-Type Structure and Function in *Podospora anserina*"). The process of ascospore formation, once the diploid is formed, has been described in detail (66, 93).

differentiation of female tissues called ascogonia, from which protoperithecia develop. In some fungi, such as *P. anserina*, *MAT* genes are not required for this step (26). Cell fusion occurs, following cell-cell recognition between the strain acting as the female and a strain of opposite mating type, acting as the male. The female partner is a specialized coiled hypha called the trichogyne that issues from the protoperithecium; the male participant is a microconidium, macroconidium, or hyphal cell. Fertilization is achieved when a nucleus of the male cell enters the trichogyne. Formation of fruiting bodies starts once fertilization is complete; the sexual spores (ascospores) are formed in asci as the fruiting body develops. Nuclei of opposite mating type do not fuse immediately after fertilization; instead, they divide and remain separate in the female cytoplasm until the fruiting body has been erected. The second phase of recognition occurs when nuclei of opposite mating type discriminate self from nonself and pairs of nuclei of opposite mating type enter specialized ascogenous hyphae called croziers. There, two nuclei of opposite mating type undergo an additional mitotic division and then donate one nucleus of each type, usually to the penultimate cell in the crozier where karyogamy occurs. The fused nucleus is the only diploid cell in the entire life cycle and is diploid only briefly before meiosis occurs. The ascus develops from this cell. Commonly, in filamentous ascomycetes, a mitotic division follows meiosis and the mature ascus contains eight ascospores, twice the number of ascomyceteous yeasts.

In self-incompatible ascomycetes, the *MAT* genes are required for recognition of strains of opposite mating type, as well as nuclei of opposite mating type. Pheromones and receptors are involved in the former. What role do the *MAT* genes play in self-compatible species? One would suppose that cell-cell recognition, mediated by pheromones and receptors, is unnecessary but what about the nuclear recognition step? What mechanism do like nuclei of self-compatible species use to form a diploid? How is this different from the self-incompatible situation, where only unlike nuclei fuse? Cytologically, the program of ascus development appears identical in heterothallic and homothallic *Neurospora* spp., i.e., two nuclei are partitioned in the penultimate cell of the crozier (66). Similar developmental events in the crozier, mirroring those described by Raju, have been described for *Gibberella zeae* (75). The recognition mechanism remains unknown at present.

MAT nomenclature is not uniform (77, 80). It has been proposed, as new *MAT* loci are characterized at the molecular level, that the formal name for the *MAT* locus should be *MAT1*, the two idiomorphs would be *MAT1-1* and *MAT1-2*, and the genes housed at *MAT1-1* and *MAT1-2* would be *MAT1-1-1*, *MAT1-1-2*, etc., and *MAT1-2-1*, *MAT1-2-2*, etc., as needed (80). Fungi for which there is rich mating-type literature, such as *P. anserina* or *Neurospora crassa*, are generally referred to by original terminology. Thus, the *MAT1-1* and *MAT1-2* idiomorphs are *mat−* and *mat+*, respectively, for *P. anserina* and *mat A* and *mat a*, respectively, for *N. crassa*; the *MAT1-1-1*, *MAT1-1-2*, and *MAT1-1-3* genes are *FMR1*, *SMR1*, and *SMR2*, respectively, for *P. anserina* and *mat A-1*, *mat A-2*, and *mat A-3*, respectively, for *N. crassa*; and the *MAT1-2-1* genes are *FPR1* and *mat a-1* for *P. anserina* and *N. crassa*, respectively.

MATING-TYPE STRUCTURE AND FUNCTION IN *COCHLIOBOLUS* SPECIES

Members of the ascomycete genus *Cochliobolus* (anamorph *Bipolaris* or *Curvularia*) have diverse reproductive strategies, i.e., some are self-incompatible and some are self-compatible, while others have no known sexual state and thus are termed "asexual" although they group robustly, in phylogenetic treatments, with *Cochliobolus* spp. that do have sexual states. Thus, the evolution of reproductive strategy can be examined in closely related species within the genus. Additionally, the *Cochliobolus* spp. (more than 30 species) are favorable subjects for analyses of sexual development mechanisms and evolution of mating strategy because they are easily grown in culture (88), have an efficient sexual stage readily produced in the laboratory (52) (Fig. 6.2), and are tractable from the molecular perspective, i.e., they are readily transformed and targeted gene knockout is highly efficient (19, 79, 86, 91). In addition, the homothallic *Cochliobolus* spp. analyzed so far (one exception) carry both *MAT* genes tightly linked (within a kilobase of each other) on the same chromosome, an arrangement that facilitates molecular manipulation. Since gene replacement via transformation is extremely efficient in these fungi, targeted mutation and shuffling of *MAT* among different genetic backgrounds can be achieved at will. *C. heterostrophus* has been developed as a model for the genus; *mat* deletion strains have proven to be very useful to address questions of *MAT* function.

Structure of the Mating-Type Locus in Self-Incompatible *Cochliobolus* spp.

C. heterostrophus

The *C. heterostrophus MAT* genes were the first Dothideomycete *MAT* genes identified (72). The genes were identified using a strategy whose underlying prediction

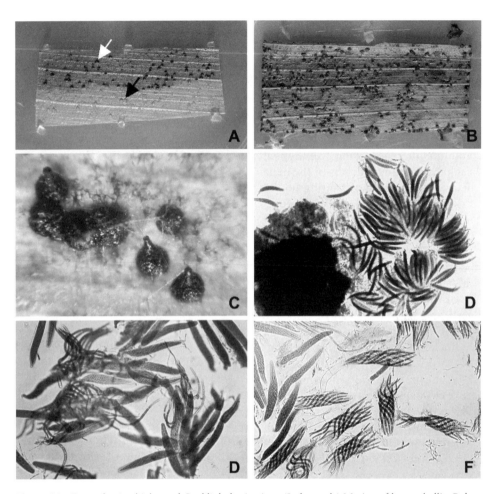

Figure 6.2   Reproductive biology of *Cochliobolus* in vitro. (Left panels) Mating of heterothallic *C. heterostrophus*; (right panels) selfing of homothallic *C. luttrellii*. (Top row) Mating plates; agar plugs bearing mycelium of each mating type (left) or the same mating type (right) are placed on opposite sides of a senescent corn leaf substrate. Fruiting bodies (pseudothecia) form in 4 to 7 days; color reflects the strain that acted as the female in the case of heterothallic matings. In block A, an albino and a pigmented strain were mated. (Middle row) Left, close-up of pseudothecia (note beaks); right, crushed pseudothecium and contents (asci). (Bottom row) Asci and ascospores. *Cochliobolus* ascospores are multicellular, multinucleate, and filamentous. Note asci from selfs (right) usually contain complete tetrads.

was that the presence of both mating-type genes in a single haploid strain would render the transformed strain self-compatible. For this experiment, a mating-type 2 strain was transformed with a cosmid library of DNA of a mating-type 1 strain and transformants were screened for the ability to form pseudothecia when selfed. A transformant was identified which formed abundant pseudothecia in a self-mating assay and, in addition, could outcross to tester strains of both mating types. Recovery of the transforming cosmid and sequencing of the portion of the cosmid responsible for inducing pseudothecium formation revealed a single opening reading frame (ORF) with homology to the *S. cerevisiae* MAT α1 protein. The *C. heterostrophus* mating-type 2 counter-

part was identified by probing a mating-type 2 cosmid library with a restriction fragment carrying DNA from the mating-type 1 flanks.

The *MAT1-1* idiomorph is 1,297 bp while *MAT1-2* is 1,171 bp and the start codons of *MAT1-1-1* and *MAT1-2-1* are 47 and 36 bp into the *MAT1-1* and *MAT1-2* idiomorphs, respectively. The stop codons are 44 and 45 bp before the end of the *MAT1-1* and *MAT1-2* idiomorphs, respectively. Upstream of both *MAT* genes in the common flanking DNA is an ORF (termed *ORF1*), with homology to an *S. cerevisiae* ORF (accession number U22383) encoding a protein of unknown function. Downstream of both *MAT* genes in the common flanking DNA is a gene (*BGL1*) encoding β-glucosidase (85).

A gene encoding a GTPase-activating protein lies upstream of *ORF1*. Despite the consistent association with *MAT*, deletion of *ORF1* has no effect on *C. heterostrophus* mating or on any other detectable phenotype and neither does the GTPase-activating gene, when deleted (85).

Both *MAT1-1-1* and *MAT1-2-1* genes have the same orientation with respect to their chromosome. For *C. heterostrophus*, as noted above, each ORF occupies almost all of the idiomorph sequence (Fig. 6.3); in both cases, transcription starts and stops in the common flanking regions (54). Furthermore, there is evidence for alternative splicing and multiple *MAT* transcripts (54). Curiously, however, it has been demonstrated that untranslated leader sequences at the 5′ end of the *MAT* ORF corresponding to the 5′ ends of the multiple transcripts are not required for *MAT* function (85). A portion (160 bp) of the untranslated 3′ end region, however, is absolutely required for fertility and must be present in both mating partners (85). Wirsel et al. (85) speculated that this region must be intact in order for proper localization of *MAT* transcripts and that proper localization precedes nuclear recognition and ensuing meiosis.

Other *Cochliobolus* spp.

Sequencing of the complete idiomorphs of four additional self-incompatible *Cochliobolus* species (*C. ellisii*, *C. carbonum*, *C. victoriae*, and *C. intermedius*) confirmed the one-idiomorph/one-gene structure found in *C. heterostrophus* (91). Sequence conservation is >90% among closely related *Cochliobolus MAT* genes, and the *C. carbonum* or *C. victoriae MAT* genes, for example, can be substituted for *C. heterostrophus MAT* genes, as they confer the same mating abilities as do the *C. heterostrophus MAT* genes (Table 6.1) (22). Note that a cross between a transgenic *C. heterostrophus* ΔMAT{CcarbMAT1-1-1} strain and a *C. carbonum* strain is not productive, indicating that introduction of the *C. carbonum MAT* gene does not eliminate the species barrier (Table 6.1).

A screen for mating type of a collection of 44 *C. victoriae* strains from diverse geographical origins revealed that the entire collection was *MAT1-2* (22). Since it was clear that a *C. heterostrophus mat* deletion strain could be transformed with either *MAT1-1* or *MAT1-2* and be fully functional, an attempt was made to create a mating pair of *C. victoriae* strains. One parent was a wild-type *C. victoriae MAT1-2* strain, and

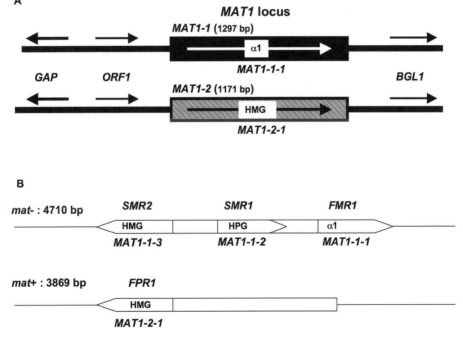

Figure 6.3 Organization of the *C. heterostrophus* and *P. anserina MAT* loci. Lines represent sequences common to *MAT1-1* and *MAT1-2* or *mat−* and *mat+*. Boxes represent *mat*-specific sequences. (Top) *C. heterostrophus*. Arrows indicate ORFs and direction of transcription. *GAP*, GTPase-activating protein; *ORF1*, conserved ORF of unknown function; *BGL1*, β-glucosidase. (Bottom) *P. anserina*. Arrowed boxes represent mating-type genes. *P. anserina* nomenclature for mating-type genes is shown above the corresponding box. Standard nomenclature for mating-type genes is shown under the corresponding box. Protein domains encoded by mating-type genes are indicated in the corresponding box.

Table 6.1   Function of heterologous *Cochliobolus MAT* genes[a]

MAT1-1 parent	MAT1-2 parent	Pseudothecia	Ascospores
Chet WT	Chet WT	+	+
ChetΔMAT	Chet WT	−	−
Chet WT	ChetΔMAT	−	−
ChetΔMAT{CcarbMAT1-1-1}	Chet WT	+	+
Chet WT	ChetΔMAT{CvictMAT1-2-1}	+	+
Ccarb WT	Ccarb WT	+	+
ChetΔMAT{CcarbMAT1-1-1}	Ccarb WT	−	−
CcarbΔMAT{ChetMAT1-1-1}	Ccarb WT	+	+
Ccarb WT	CcarbΔMAT{ChetMAT1-2-1}	+	+

[a]ΔMAT, deletion of the MAT idiomorph, i.e., ChetΔMAT is a *C. heterostrophus* strain in which the *MAT1-2* idiomorph has been deleted; this strain can act as a mating-type 1 or 2 strain, depending on which *MAT* transgene is incorporated by transformation; { }, addition of a *MAT* transgene; +, formation of pseudothecia or ascospores; −, no formation of pseudothecia or ascospores.

the other was a transgenic *C. victoriae mat1-2* deletion strain carrying either *C. carbonum* or *C. heterostrophus MAT1-1-1* (*CvictΔmat{CcarbMAT1-1-1}* or *CvictΔmat{ChetMAT1-1-1}*). Crosses of the *C. victoriae* transgenic strain to wild-type *C. carbonum MAT1-2* were successful; however, no cross of an engineered *C. victoriae* strain to a wild-type *C. victoriae* strain produced pseudothecia (22). It was subsequently determined that all of the *C. victoriae* strains examined were female sterile; thus, a cross of any two *C. victoriae* strains would be expected to be sterile. The success of the *C. carbonum* × *C. victoriae* cross confirmed old reports (61) that interspecific crosses can occur between some isolates of *C. victoriae* and *C. carbonum*. In this study (22), *MAT1-1* and *MAT1-2* segregated 1:1; however, attempts to cross progeny issuing from a *C. carbonum* × *C. victoriae* cross were unsuccessful, and none of the progeny could be backcrossed (22).

Other Dothideomycetes spp.

Other self-incompatible Dothideomycetes for which there are *MAT* molecular data include *Mycosphaerella graminicola* (84), *Leptosphaeria maculans* (27), *Phaeosphaeria nodorum* (8), *Didymella rabiei* (7), *Pyrenophora teres* (68), *Leptosphaeria biglobosa* "canadensis" group (83), *Ascochyta lentis* (20), and *Crivellia papaveracea*, an isolate from opium poppy, which groups with *Alternaria* spp. (46). For *C. papaveracea*, the mycelia from 14 single ascospores from a field-collected fruiting body were found to carry either a *MAT1-1-1* or *MAT1-2-1* gene and thus have a heterothallic architecture. *MAT* sequences obtained from two of these *C. papaveracea* isolates of opposite mating type revealed that the 5′ ends of the mating-type genes are downstream of *ORF1*, as in the closely related heterothallic *C. heterostrophus* (91). Attempts to

persuade these isolates to mate in a heterothallic manner in the laboratory were unsuccessful. In all cases, *MAT1-1* and *MAT1-2* idiomorphs consist of only one ORF corresponding to genes encoding the α1 and HMG proteins.

The actual size of the Dothideomycete idiomorphs varies with species (between ~1,100 [*C. heterostrophus*] and ~4,500 [*P. nodorum*]); nevertheless, the overall characteristics described for the *C. heterostrophus* ORFs are conserved. The type of gene on the *MAT* flanks and the relative positions of these genes vary (Table 6.2). For most of these taxa, the closest gene on the 3′ flank has not been reported. For *C. heterostrophus* this is the *BGL1* gene, which is a full 4 kb from the 3′ end of the *MAT* gene. Most Dothideomycete taxa carry *ORF1* on the 5′ flank. An exception is *M. graminicola*, which carries a DNA lyase, also present flanking *MAT* in most of the Pyrenomycetes for which there are genome data (30). It may be significant that in phylogenetic analyses, *M. graminicola* "bounces" around when *MAT* is used as a character for analysis. For example, Bennett et al. (8) showed a tree built with α-box sequences that placed *M. graminicola MAT1-1-1* with the Discomycete *Pyrenopeziza brassicae*, while a tree built with the HMG box placed *M. graminicola MAT1-2-1* on a branch of its own. Goodwin et al. (37) performed similar analyses and also found that α-box sequences grouped *M. graminicola* (and *Septoria passerinii*) with *P. brassicae* and with a Hypocreales cluster. The next most closely related group included the Sordariomycetes *Neurospora* and *Podospora*. As in the Bennett et al. (8) analysis, phylogenetic treatment of the HMG-box sequences grouped *M. graminicola* and *S. passerinii* on a branch separate from any other. It is unclear, at this point, whether these *MAT* genes have a Dothideomycete or Sordariomycete history. Internal transcribed spacer (ITS) analyses (9, 37)

Table 6.2 Genes adjacent to *MAT* in Dothideomycetes

Organism	5′ of *MAT*	3′ of *MAT*	Comments
C. heterostrophus	*ORF1*	*BGL1*	*ORF1* to *MAT* = 1,079 bp
A. alternata	*ORF1*	?[a]	*ORF1* to *MAT* = 4 bp
P. nodorum	*ORF1*	?	3′ end of *ORF1* extends into idiomorph
L. maculans	*ORF1*	DNA lyase	
A. rabiei	*ORF1*	?	*ORF1* to *MAT1-2* = 51 bp; *ORF1* overlaps *MAT1-1* idiomorph by 16 bp
A. lentis	*ORF1*	DNA lyase	3′ end of *ORF1* extends into idiomorph
S. passerinii	?	?	
M. graminicola	DNA lyase	?	

[a] ?, unknown.

suggest that *Mycosphaerella* groups with Dothideomycetes.

Structure of the Mating-Type Locus in Self-Compatible Dothideomycetes spp.

Cochliobolus spp.

Yun et al. (91) compared extant *MAT* sequences from self-incompatible and self-compatible *Cochliobolus* species. Strikingly, the structural organization of *MAT* loci of all self-incompatible species examined (*C. heterostrophus, C. carbonum, C. victoriae, C. ellisii, C. sativus,* and *C. intermedius*) was the same; each strain carries either *MAT1-1* or *MAT1-2* (Fig. 6.3). In stark contrast, the structural organization of each *MAT* locus is unique in all self-compatible *Cochliobolus* species, although all four species carry both *MAT* genes in one genome. In *C. luttrellii* and *C. homomorphus*, the genes are fused into a single ORF; the gene order in *C. luttrellii* (5′*MAT1-1-1::MAT1-2-13′*) is reversed in *C. homomorphus* (5′*MAT1-2-1::MAT1-1-13′*). In the remaining two cases, the genes are not fused. In *C. kusanoi*, the organization is 5′*MAT1-2-13′--3′MAT1-1-15′*, and part of the sequence between the genes is similar to a portion of the β-glucosidase gene normally found 3′ of both *MAT* genes in self-incompatible *C. heterostrophus* (Fig. 6.3). To the 5′ of *MAT1-2-1* is a perfect inverted repeat of a 561-bp region containing 123 bp of the 5′ end of the *MAT1-1-1* ORF (which lacks the α1 encoding region but contains the HMG encoding sequence) fused to the 5′ end of *MAT1-2-1* and 145 bp of a different fragment of the β-glucosidase gene, separated from each other by 293 bp. *C. cymbopogonis* carries both homologs of the self-incompatible idiomorphs, but these are not closely linked. It is not known if they reside on the same or different chromosomes. Thus, the *MAT* genes have close physical association in three *Cochliobolus* self-compatible species but not in the fourth.

ORF1, linked to the *MAT* genes in three of four homothallic species and in all *Cochliobolus* heterothallic species examined to date, is consistently found on the *MAT* 5′ flank and is more conserved than the *MAT* genes themselves; in *C. cymbopogonis*, there are two copies of *ORF1* and each is linked to a different *MAT* gene. *C. kusanoi* is an exception; gel blots and PCR amplifications indicate that *ORF1* exists in the genome but suggest that it is not closely linked to either *MAT-1* or *MAT-2*.

Comparisons among *MAT* Regions of Self-Incompatible and Self-Compatible *Cochliobolus* Species

Detailed comparative analyses of *MAT* sequences and their flanking regions (92) showed that the level of *MAT* gene similarity is not related to reproductive lifestyle; *MAT* and *ORF1* are highly conserved among closely related species and less conserved among distantly related species. Identities at the nucleotide and amino acid levels between *MAT* gene sequences and proteins are shown in Table 6.3. Previous molecular and phylogenetic analyses showed that species within the genus *Cochliobolus* fall into two large groups (11, 77, 91). Heterothallic *C. heterostrophus* is a member of one group, which contains all known *Cochliobolus* pathogens of cereals, while heterothallic *C. ellisii* represents the other. Homothallic *C. luttrellii*, closely related to *C. heterostrophus*, falls in the first group. Its close relationship to *C. heterostrophus* makes these two taxa an ideal pair for lifestyle comparison. Homothallic *C. homomorphus* is also in the first group but does not have a known close heterothallic relative. Homothallic *C. cymbopogonis* and *C. kusanoi* are in the second group, but neither is the closest relative of heterothallic *C. ellisii*.

Pairwise percent identity of *MAT* sequences reflects the relative positions of these species in a phylogenetic tree,

Table 6.3 Percent nucleotide and amino acid identities of *Cochliobolus MAT1-1-1* and *MAT1-2-1* genes and encoded proteins[a]

Organism	MAT-1-1-1/MAT1-2-1					
	C. heterostrophus	*C. ellisii*	*C. luttrellii*	*C. homomorphus*	*C. kusanoi*	*C. cymbopogonis*
C. heterostrophus		**58.9/65.1**	**90.1/92.1**	**70.6/77.7**	**68.3/63.6**	**56.1/63.6**
C. ellisii	56.2/64.8		**64.2/64.3**	**65.4/69.1**	**63.0/68.1**	**69.1/68.7**
C. luttrellii	92.2/92.2	63.9/62.0		**74.7/78.4**	**70.2/64.4**	**60.8/56.9**
C. homomorphus	70.9/79.3	54.9/66.0	78.6/79.6		**69.7/70.5**	**46.7/47.7**
C. kusanoi	68.6/79.5	57.1/64.6	73.9/72.5	71.8/74.4		**57.2/59.7**
C. cymbopogonis	49.0/55.9	65.6/66.0	66.1/52.8	48.8/58.4	48.0/58.0	

[a]Percent nucleotide values for *MAT-1-1-1/MAT1-2-1* are given in **boldface**; percent amino acid values for *MAT-1-1-1/MAT1-2-1* are underscored.

constructed using ribosomal DNA (ITS) and glyceraldehyde-3-phosphate dehydrogenase (*GPD*) sequences (11, 90), suggesting vertical, not lateral, inheritance of *MAT* (however, see "*Stemphylium* spp." below). As noted above, *MAT* genes from species within a group are more similar to each other than are *MAT* genes from species of similar mating style. For example, percent nucleotide identity between *MAT* genes in the two self-incompatible *Cochliobolus* species (*C. heterostrophus* and *C. ellisii*), each representing one of the large *Cochliobolus* groupings, is only 58.9 for *MAT1-1-1* and 65.1 for *MAT1-2-1*. In contrast, percent nucleotide identity between the *MAT* genes of self-incompatible *C. heterostrophus* and self-compatible *C. luttrellii* is 90.1 for *MAT1-1-1* and 92.1 for *MAT1-2-1*. Similarly, the species with *MAT* genes showing the highest similarity to those of self-incompatible *C. ellisii* is self-compatible *C. cymbopogonis*. Previous phylogenetic analyses strongly support the hypothesis that each homothallic *Cochliobolus* species arose independently from a different heterothallic ancestor (91).

The MAT1-1-1 and MAT1-2-1 signature sequences and intron positions within these domains are highly conserved among MAT proteins of all species, both self-incompatible and self-compatible (92). The amino-terminal end of the MAT1-1-1 proteins is highly conserved among members within each large *Cochliobolus* group, while for MAT1-2-1, the amino-terminal end is conserved in all. The carboxy-terminal portions of the MAT proteins are not conserved in the terminal 3' half of MAT1-1-1, or third of MAT1-2-1, except in proteins from very closely related species. However, the extreme carboxy-terminal end is conserved among MAT1-1-1 proteins. *C. ellisii* MAT1-1-1, however, has an insert of 10 amino acids not found in the other MAT1-1-1 proteins, including that of *C. cymbopogonis*.

ORF1 sequence conservation ranges from 62.5% identity for two unrelated homothallic species, *C. kusanoi* and *C. cymbopogonis* (*MAT1-1-1* and *MAT1-2-1* are 57.2 and 59.7%, respectively), to 94.8% for the closely related species *C. heterostrophus* and *C. luttrellii*

(92). Interestingly, the two copies of *ORF1* in *C. cymbopogonis* are not identical (93.3% identity), suggesting that the duplication of this gene in the genome did not occur recently. The high level of conservation among *ORF1* sequences has proven to be useful as a basis for design of PCR primers aimed at cloning *ORF1*-linked *MAT* genes from fungi whose *MAT* genes themselves are too divergent for cloning by usual procedures (91).

Similarity among noncoding DNAs flanking *MAT* varies with the relatedness of the species (92). Within species, percent identity in the 5' flank from the 3' end of *ORF1* to the 5' end of the *MAT* ORF is 96.6% for *C. heterostrophus MAT1-1* and *MAT1-2* and 97.0% for *C. ellisii MAT1-1* and *MAT1-2*. Interspecies percent identity in the same region for two closely related species, heterothallic *C. heterostrophus* and homothallic *C. luttrellii*, is 79.4% for *MAT1-1-1* and 81.2% for *MAT1-2-2*. For *C. ellisii* and *C. cymbopogonis* it is 63.2 and 69.4%. Percent identity in the same region when any other pair of species is compared ranges from 10.5 to 51.1%. Within this region is a short stretch (about 190 bp, corresponding to base-pair positions 4280 to 4475 of the *C. heterostrophus MAT-1* locus, accession no. AF029913) that is highly conserved (80.0 to 85.7% identity) among all species. Interestingly, the 5' flanking regions of the two *MAT* genes of *C. cymbopogonis* are only 77.6% identical. There is 100% identity for the 5' flanks of *C. kusanoi*, which is a reflection of the presence of an exact repeat of the *MAT1-1-1 5'* region fused to *MAT1-2-1* (92).

In Table 6.4, the region between the 3' end of *ORF1* and the 5' end of the *MAT* gene is compared among species. The size of this region is remarkably constant (1,097 to 1,117 bp), even for homothallic *C. cymbopogonis*, which has two copies of this region and no apparent linkage between the two *MAT* genes. The distance is somewhat less for *C. homomorphus* than for the others, likely because the gene order is reversed. There is no meaningful comparison to be made for *C. kusanoi*, since *ORF1* appears unlinked to *MAT*.

Table 6.4 Intergenic distance is conserved in the *Cochliobolus* species[a]

C. heterostrophus		C. ellisii		C. luttrellii	C. homomorphus	C. cymbopogonis	
MAT1-1	MAT1-2	MAT1-1	MAT1-2	MAT1/2	MAT2/1	MAT1-1	MAT1-2
1,110	1,117	1,115	1,115	1,097	883	1,105	1,114

[a] Values give the distance (in base pairs) between the 3′ end of *ORF1* and the 5′ end of the *MAT* ORF.

DNA 3′ of *MAT* (the 3′ flank) is more divergent than that of the 5′ flank; only *C. luttrellii* shares significant identity (78.0 to 82.0%) with the *C. heterostrophus MAT-1* or *MAT-2* flanks; 3′ flanks of the other three homothallic species range from 13.1 to 49.0% identity to either of the two heterothallics (92). Furthermore, there is no conserved region (not even a short stretch as on the 5′ flanks) among the 3′ flanks of all species, except between those of *C. cymbopogonis* and *C. ellisii*, where several islands of similarity (70 to 95%) are found. Note again that the 3′ flanking regions of the two *MAT* genes of *C. cymbopogonis* are nonidentical (only 62.9% identity).

MAT Structure in Other Dothideomycete Self-Compatible Species

Characterization of *MAT* loci from self-compatible Dothideomycete species outside the genus *Cochliobolus* has been reported for *Mycosphaerella zeae-maydis* (*Didymella zeae-maydis*) (89), the sexual state of certain *Stemphylium* spp. (45), and for certain *Crivellia* sp. (46). *D. zeae-maydis MAT* has complete versions of the *C. heterostrophus MAT* counterparts tandemly arranged, head to tail, separated by ~1 kb of noncoding sequence.

Self-compatible isolates of *Stemphylium* spp. collected from field ascomata have two types of *MAT* locus configuration (45). In the first type, the two *MAT* genes are linked, although *MAT1-1-1* and its flanks are inverted and connected to *MAT1-2-1* oriented as for *Cochliobolus MAT1-2-1*, ~200 bp upstream of the *MAT1-2-1* gene, and between it and *ORF1* (Fig. 6.4). The second type of self-compatible isolates carry only *MAT1-1-1*, a situation analogous to certain *Neurospora* self-compatible isolates (e.g., *N. africana*) in which only the *mat A-1* gene is detected (35). Self-compatible isolates collected from field ascomata were demonstrated to undergo self-mating in the laboratory.

The observation that *MAT1-2-1* was not detected in some *Stemphylium* self-compatible isolates suggests that the HMG regulatory domain might be dispensable for sexual cycle control. Another possibility may be proposed, based on the finding that each *MAT1-1-1* gene of *Cochliobolus* spp. appears to encode a protein with both an HMG and an α1 domain (S.-W. Lu and B. G. Turgeon, unpublished data) (see "Conserved Motifs in MAT1-1-1 and MAT1-2-1" below); the regulatory functions of the lost *MAT1-2-1* gene may be taken on by the remaining *MAT1-1-1* gene. Evolution of the target genes of MAT1-2-1 would then be necessary to bring them under the control of the HMG domain of MAT1-1-1. Supporting this hypothesis is the fact that *Stemphylium* MAT proteins have a structure similar to those of *Cochliobolus* spp. and thus the MAT1-1-1 protein indeed carries a HMG domain.

In a separate study by Inderbitzin et al. (46), a fungal isolate (a *Crivellia* species with a *Brachycladium papaveris* asexual state) from the opium poppy, originally thought to be a *Pleospora* species, was found to group with *Alternaria* spp., a sister genus to *Cochliobolus*, when subjected to rigorous phylogenetic analyses. Regarding *MAT*, each single-conidium isolate was found to have an incomplete *MAT1-2-1* gene (missing 43 bp of the 3′ end) fused to the 5′ end of the *MAT1-1* idiomorph, upstream of the *MAT1-1-1* gene, and is inferred to be homothallic (Fig. 6.4).

Structure of the Mating-Type Locus in Asexual Dothideomycetes

Asexual Relatives of Cochliobolus

The first report documenting that fungi that have never been demonstrated to reproduce sexually have *MAT* genes can be traced to 1996, when Sharon et al. (70) reported *MAT* homologs in *Bipolaris sacchari*, a pathogen of sugarcane and an asexual species related to *Cochliobolus* (70). Complete sequencing of the *MAT1-2-1* gene from one isolate revealed the putative translation product to be 100% identical to *C. heterostrophus* MAT1-2-1. Additional asexual *Bipolaris* spp. shown to harbor *MAT* genes (screened for *MAT1-2-1*) include *B. australis*, *B. clavata*, *B. curvispora*, *B. cylindrica*, *B. indica*, *B. sorghicola*, *B. urochloae*, and *B. zeae* (1). The initial investigation of mating-type distribution and phylogenetic relationships of asexual *B. sacchari* was refined using several molecular characters and more isolates (C. Schoch, G. Saenz, and B. G. Turgeon, unpublished data). Combined data from both *MAT* genes, plus the 5′ *MAT* flank, the 3′ *MAT* flank, and several additional sequences, confirmed that *B. sacchari* isolates exist in the

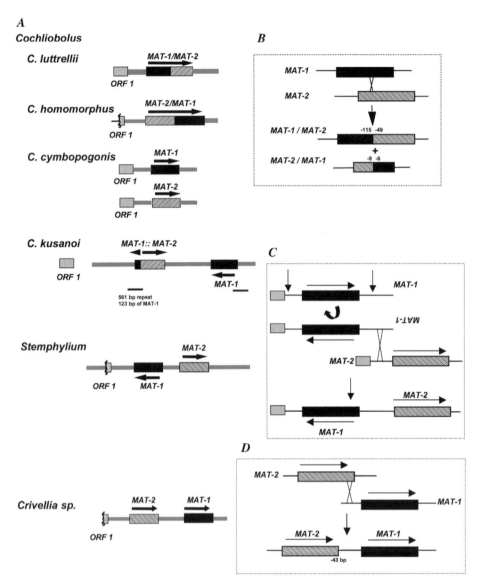

Figure 6.4 Organization of *MAT* loci in homothallic *Cochliobolus*, *Stemphylium*, and *Crivellia* species. (A) *MAT* organization. (B through D) Proposed recombination events responsible for converting heterothallic ancestors to extant homothallic species, as described in the text ("The Direction of *MAT* Evolution") and references 45, 46, and 91.

field as both mating types. All data sets indicated that the *B. sacchari* isolates fall into two well-supported groups, with representatives of both mating types in each group. Another asexual species, *Alternaria alternata*, was found also to contain *MAT1-1* or *MAT1-2* idiomorphs, which are structurally similar to those of *C. heterostrophus* (2).

The cloned *MAT* genes of *B. sacchari* and *A. alternata* were tested in sexual *C. heterostrophus* and shown to induce the formation of fertile pseudothecia (Table 6.5) (2, 64). This test demonstrated that not only do asexual *Bipolaris* and *Alternaria* possess *MAT* genes

with the same structure as *MAT* from sexual *Cochliobolus* but also the *MAT* gene from the asexual species is a functional ortholog of *MAT* in *Cochliobolus*. A quick test for function is the test that was used initially to identify the *C. heterostrophus MAT* genes (see "*C. heterostrophus*" above); haploid strains carrying both *MAT* genes form pseudothecia in self-mating assays. Expression of *C. heterostrophus MAT1-1-1* in a *B. sacchari MAT1-2* strain, expression of both *C. heterostrophus MAT* genes in *B. sacchari*, and expression of *C. heterostrophus MAT1-1-1* and *MAT1-2-1* genes in separate *B. sacchari* strains (and mating of the strains) failed in all cases to

Table 6.5 Functional analysis of *MAT* genes from asexual species[a]

MAT1-1 parent	MAT1-2 parent	Pseudothecia	Ascospores
Chet WT	Chet WT	+	+
Chet WT{Chet MAT1-2-1} self		+[b]	−
Bsacc WT	Bsacc WT	−	−
Chet WT	Chet ΔMAT{BsaccMAT1-2-1}[c]	+	+[d]
—	Bsacc WT{Chet MAT1-1-1} self	−	−
—	Bsacc WT{Chet MAT1-2-1;Chet MAT1-1-1} self	−	−
Bsacc WT{Chet MAT1-1-1}	Bsacc WT{Chet MAT1-2-1}	−	−
Chet WT	Chet ΔMAT{AaltMAT1-2-1}	+[e]	(+)[f]
Chet ΔMAT{AaltMAT1-1-1}	Chet WT	+[e]	(+)[f]
Chet ΔMAT{AaltMAT1-1-1}	ChetΔMAT{AaltMAT1-2-1}	+	−

[a]Chet, *C. heterostrophus*; Bsacc, *B. sacchari*; Aalt, *A. alternata*; WT, wild type; +, formation of pseudothecia or ascospores; −, no formation.
[b]Pseudothecia were indistinguishable from those formed in a Chet WT cross; occasional asci and ascospores were seen.
[c]The opposite cross, e.g., *MAT1-1* parent, Chet ΔMAT{Bsacc MAT1-1-1} × *MAT1-2* parent, Chet WT, has not been done.
[d]Pseudothecia were indistinguishable from those formed in a Chet WT cross; fertility was WT.
[e]Pseudothecia were indistinguishable from those formed in a Chet WT cross but fewer in number.
[f]Only a few asci and ascospores were formed.

induce the formation of pseudothecia. Northern blot analyses failed to detect any *B. sacchari MAT1-2-1* transcript in the same *B. sacchari* isolate tested, although the heterologous *C. heterostrophus MAT1-1-1* and *MAT1-2-1* genes were transcribed in this isolate (70). Taken together, these results suggest that the sexual defect cannot be attributed to a defect in transcription of the endogenous *B. sacchari MAT1-2-1* gene. In contrast, reverse transcription-PCR analyses of the *A. alternata MAT* genes indicated that both are transcribed. These data support the argument that asexual Dothideomycetes arise from sexual progenitors and indicate that the cause of asexuality may be defects in any gene in the mating-type pathway, such as the target genes of the *MAT* regulatory factors.

MAT in Other "Asexual" Dothideomycetes
Since these studies, researchers in the laboratory of Mary Berbee (University of British Columbia) have examined field collections of additional Dothideomycete taxa thought to be primarily asexual, including more *Alternaria* spp. (e.g., *A. brassicae*, *A. brassicicola*, and *A. tenuissima*) and *Stemphylium* spp. (45). Goodwin et al. (37) reported on mating-type idiomorphs of the barley pathogen *Septoria passerinii*. In all cases, it has been demonstrated that these taxa carry *MAT* genes.

For *Alternaria* spp. Berbee et al. (10) screened many *A. tenuissima*, *A. brassicicola*, and *A. brassicae* isolates for *MAT*. Each species had both *MAT* genes, in separate isolates, suggesting that each is potentially, or was, self-incompatible. Phylogenetic analyses grouped all three *Alternaria* species with a fourth *Alternaria*, *A. gaisen*, and all four grouped with *Cochliobolus*. The ITS sequences from isolates of opposite mating type were identical, suggesting that there was a history of sexual recombination during the evolution of each species. If the species had not undergone sexual reproduction, differences would be expected in the ITS sequences, as each mating type would have a separate history. This type of analysis can indicate only that, at some point in the history of the species, ancestors were sexual. It does not resolve whether extant strains carrying opposite-mating-type genes are sexual.

For *Stemphylium* spp., 106 geographically diverse isolates were typed for *MAT* and found to carry both separate and linked *MAT* genes, indicating that some strains are potentially heterothallic while others are homothallic (see "Other Dothideomycetes spp." above) (45). Where separate, *MAT* loci have either a *MAT1-1-1* or *MAT1-2-1* gene in the same orientation as in heterothallic *Cochliobolus* strains, between conserved upstream *ORF1* and downstream *BGL1* genes.

In no case, to our knowledge, has it been proven experimentally that a pair of "asexual" strains carrying opposite-mating-type genes has been crossed successfully. This highlights, once again, the fact that a mutation in any one of hundreds of genes involved in sexual development may render a particular fungus asexual. It is also possible that conditions conducive to mating have not been found.

EVOLUTION OF MATING TYPE IN THE DOTHIDEOMYCETES
How dissimilar regions (idiomorphs) of homologous chromosomes or dimorphic sex chromosomes evolved is largely unknown. One hypothesis for the *MAT*

idiomorphs is that the small pockets of identity in the otherwise unlike *MAT* idiomorphs reflect a common ancestry (26). Conceivably, these pockets are remnants of a series of mutagenic events in a single ancestral gene(s). The mutations, coupled with recombination suppression, might have led to the highly divergent extant *MAT* genes that now encode different products (21, 78). Furthermore, the recent finding that several conserved motifs, some with known function, are apparent when MAT1-1-1 and MAT1-2-1 proteins are compared (30; Lu and Turgeon, unpublished) supports this hypothesis (see "Conserved Motifs in MAT1-1-1 and MAT1-2-1" below). A similar scenario has been proposed for the evolution of the Y chromosome. Indeed, acquisition of sex-determining genes and recombination isolation are characteristics shared between fungal mating-type loci and mammalian X and Y chromosomes; it has been suggested that these elements may be early steps linking the evolution of sex chromosomes in diverse organisms (34, 51, 82).

A somewhat less daunting question and one that is approachable experimentally, using tools of phylogenetics and molecular analysis of function, is the question of how different reproductive lifestyles, i.e., self- or non-self-compatibility, evolve. Molecular and phylogenetic data have been used to determine that the direction of evolution of reproductive lifestyle is likely from heterothallism to homothallism in *Cochliobolus* (91). The discovery that *Cochliobolus* homothallic species appear to arise from heterothallic ones led to a consideration of whether the evolutionary clock could be advanced by transfer of homothallic *MAT* genes into a heterothallic *MAT*-null strain and whether it could be turned back by transfer of heterothallic *MAT* genes into a homothallic *MAT*-null strain. As is described in "Are *MAT* Genes from Self-Incompatible and Self-Compatible *Cochliobolus* Species Interchangeable?" below, both are possible. Thus, it appears that the *MAT* gene alone is required to convert a single species from one reproductive tactic to the other.

The Direction of *MAT* Evolution: Phylogenetic and Structural Analyses of *MAT* Genes in Selected Dothideomycetes Species

MAT structure has been examined in many species of *Cochliobolus* (91) and *Stemphylium* (45) from diverse geographical locations and in isolates of *Crivellia* (46), and this information, in addition to phylogenetic treatments, supports the hypothesis that the direction is likely from self-incompatible to self-compatible in all three cases. Furthermore, comparison of *MAT* sequences from strains differing in reproductive lifestyle has revealed mechanisms underlying changes in reproductive mode.

Cochliobolus

As noted in "*Cochliobolus* spp." above, Yun et al. (91) compared extant *MAT* sequences from self-incompatible and self-compatible species. The structural organization of *MAT* loci of all self-incompatible species examined was the same, while each *MAT* locus from self-compatible species was different. The fused *MAT* genes in self-compatible species provided clues to the genetic mechanism mediating the change from one lifestyle to the other. The *C. luttrellii MAT1-1-1* gene is missing 345 nucleotides from the 3′ end and 147 nucleotides from the 5′ end of the *MAT1-2-1* gene, compared to the *C. heterostrophus* self-incompatible homologs. The deletions are consistent with the hypothesis that a crossover event occurred within the dissimilar ancestral genes (*C. heterostrophus* is used as a surrogate ancestor here) at positions corresponding to the fusion junction. Indeed, inspection of the *C. heterostrophus* genes reveals 8 bp of sequence identity precisely at the proposed crossover site, which would explain this arrangement. A single crossover within this region would yield two chimeric products, one of which is identical to the fused *MAT* gene actually found in *C. luttrellii* (Fig. 6.4), which contains both an HMG and an α-box domain. The other hypothetical fused product would lack both the HMG and the α-box domains. A similar scenario can be proposed for *C. homomorphus*; in this case the fused gene is missing 27 nucleotides from the 3′ end of *MAT1-2-1* and 21 nucleotides from the 5′ end of *MAT1-1-1* compared with the *C. heterostrophus MAT* genes. Examination of the *C. heterostrophus MAT* sequences (again acting as a surrogate heterothallic ancestor, in this case, more distantly related) at positions corresponding to the *C. homomorphus* fusion junction reveals 9 bp of identity (with one mismatch) and thus a putative recombination point. Thus, Yun et al. (91) proposed that the mechanism of conversion from self-incompatibility to self-compatibility is likely a recombination event between small islands of identity in the otherwise dissimilar *MAT* sequences.

Phylogenetic analyses support a convergent origin for self-compatibility (91). None of the self-compatible species cluster together in any parsimony or maximum likelihood tree. Thus, phylogenetic evidence clearly supports independent evolution of self-compatibility in the four *Cochliobolus* species and underpins the structural evidence.

Stemphylium spp.

To investigate the direction of reproductive lifestyle in *Stemphylium*, Inderbitzin et al. (45) constructed a phylogeny using sequence data from four loci unrelated to mating type, as well as the *MAT* genes. As noted in "*MAT* Structure in Other Dothideomycete Self-Compatible

Species" above, two types of *MAT* loci were detected in self-compatible isolates; in the first case, the *MAT* genes are linked, while in the second case only *MAT1-1-1* is found. Phylogenetic and comparative structural analyses led to the conclusion that linked *MAT* regions are derived from ancestral separate *MAT* regions, as suggested by the basal position of isolates with separate *MAT* regions in phylogenetic analyses. Furthermore, there was 100% support from all analyses for the monophyly of *MAT1-1-1* from fused regions when it was used as a character for phylogenetic analyses and for the monophyly of *MAT1-2-1* from fused regions when it was used as a character for phylogenetic analyses, while organismal phylogenies built with genes other than mating-type genes support polyphyly of the isolates with fused *MAT* regions. Thus, in contrast to the *Cochliobolus* history, where self-fertility appears to have originated in different species by independent mating-type gene fusions or rearrangements, Inderbitzin et al. (45) found evidence for a single fusion, present in 76 self-fertile isolates, that appears to have been laterally transferred across lineages. The self-compatible isolates, which harbor only *MAT1-1-1*, may represent a lineage that evolved selfing ability independently, with unknown genetic changes responsible for its evolution.

A recombination event in which a region of DNA encompassing an ancestral *MAT1-1-1* gene and its flanking regions was inverted, such that a short sequence of identity was created, and fused to *MAT1-2-1* (Fig. 6.4) (45) is proposed as the mechanism which fused *MAT* regions from separate *MAT* genes of self-incompatible progenitors. If the sequence of a *MAT1-1-1* locus in an incompatible strain is aligned to the same region in a self-compatible strain with both *MAT* regions, it is clear that the flanking sequence carrying *ORF1* is identical, until ~300 bp 3′ of *ORF1*, at which point the sequence loses identity but, in fact, is the reverse complement of the *MAT1-1-1* region for about 1,740 bp. At this point, the sequence returns to identity with the 5′ flank of a *MAT1-2-1* gene, in the native orientation.

Thus, homothallism in sexual species with a *Stemphylium* asexual state appears to have evolved independently from self-incompatible ancestors, by fusion of *MAT1-1* and *MAT1-2*, by lateral transfer of this fused region across lineages, and in certain cases, where *MAT1-1* is present only, by unknown means.

Crivellia

As in *Cochliobolus*, isolates of outcrossing *C. papaveracea* (which groups phylogenetically with *A. alternata*) had one or the other *MAT* idiomorph, but not both, whereas all strains of selfing *Crivellia* had both *MAT* genes. Inderbitzin et al. (46) proposed that a crossover

between *MAT* genes in outcrossing ancestors might have resulted in the fused *MAT* loci found in the homothallic species (Fig. 6.4), as suggested by Yun et al. (91) for the fused *MAT* genes in self-compatible *Cochliobolus* spp. Unlike the fused *C. homomorphus* or *C. luttrellii MAT* genes, the *MAT* genes in the self-compatible *Crivellia* isolate are not fused into a single ORF; the *MAT1-1-1* gene, as well as most of the 5′ idiomorph flanking sequence, is fused to the 3′ end of *MAT1-2-1*, 43 bp before the stop codon. A 4-bp motif, present at the fusion junction, is present at the corresponding regions of the *MAT1-1* noncoding idiomorph region in *C. papaveracea* and in the *MAT1-2-1* gene of *A. alternata* (which was used as a surrogate *C. papaveracea MAT1-2* strain) was proposed as the crossover site.

Are *MAT* Genes from Self-Incompatible and Self-Compatible *Cochliobolus* Species Interchangeable?

Conversion of a Self-Incompatible to a Self-Compatible Strain

To determine whether *MAT* genes alone can control reproductive style, a sterile *C. heterostrophus MAT* deletion strain was transformed with a construct carrying the fused *C. luttrellii MAT1-1-1::MAT1-2-1* gene (91) (Fig. 6.5A). Transformants were selfed and crossed to albino *C. heterostrophus MAT1-1* and *MAT1-2* tester strains. Abundant pseudothecia formed when the transformants were selfed or crossed, most of which were fertile (1 to 10% of wild-type ascospore production). Pseudothecia and progeny from selfs were pigmented, whereas approximately one-half of the pseudothecia and one-half of the progeny from crosses were albino, indicating that self-incompatible *C. heterostrophus* expressing a *MAT* from a self-compatible species can both self and outcross. Thus, the *C. luttrellii MAT1-1-1::MAT1-2-1* gene alone conferred selfing ability to normally self-incompatible *C. heterostrophus* without impairing its ability to cross (91). A similar experiment was done with the *C. homomorphus MAT1-2-1::MAT1-1-1* gene. In this case, transformants were able to make pseudothecia, but these were barren. Since the only difference in the *C. luttrellii* and *C. homomorphus* experiments is the *MAT* gene itself (otherwise the genetic background of *C. heterostrophus* is constant), we can conclude that differences in these *MAT* sequences determine fertility (Fig. 6.5B).

Conversion of a Self-Compatible to a Self-Incompatible Species

In a reverse of the above experiment, the *C. luttrellii MAT1-1-1::MAT1-2-1* gene was deleted (Lu and Turgeon, unpublished), rendering the strain completely sterile. This

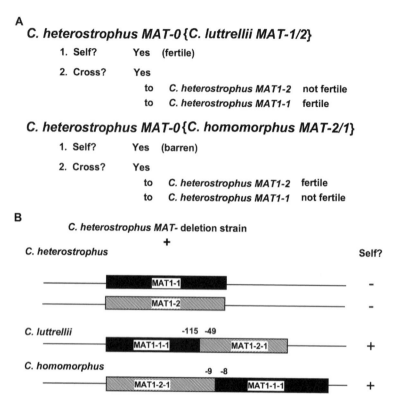

Figure 6.5 Experiments to convert heterothallic *C. heterostrophus* to homothallism. (A) A *mat*-deletion strain of *C. heterostrophus* carrying the *MAT1-1/2* gene from homothallic *C. luttrellii* can self and is fertile. It can also outcross to *C. heterostrophus* strains of either mating type. Only a cross to a *C. heterostrophus* mating type 1 strain is fertile, however, perhaps reflecting the requirement that both *MAT* 3′ untranslated regions be present for fertility (see "*C. heterostrophus*") (85). In contrast, a *mat*− deletion strain of *C. heterostrophus* carrying the *MAT1-2/1* gene from homothallic *C. homomorphus* can self but is not fertile. It can also outcross to *C. heterostrophus* strains of either mating type. Only a cross to a *C. heterostrophus* mating type 2 strain is fertile, however, perhaps for reasons described above. (B) *MAT* determines reproductive style. A *mat*− deletion strain of *C. heterostrophus* is sterile. If either *C. heterostrophus MAT1-1* or *MAT1-2* is introduced at the native *MAT* site, transgenic strains are sterile but can cross to a strain of opposite mating type. If, however, the *C. luttrellii MAT1-1/2* or *C. homomorphus MAT1-2/1* genes are introduced, the strains are able to self, as described for panel A. Since the genetic background of the original strain is held constant, it is the *MAT* gene alone that determines mating style.

strain was then transformed, separately, with the *MAT1-1* or *MAT1-2* idiomorphs from self-incompatible *C. heterostrophus*, yielding a *MAT*-deleted *C. luttrellii* strain carrying *C. heterostrophus MAT1-1-1* and a *MAT*-deleted *C. luttrellii* strain carrying *C. heterostrophus MAT1-2-1*. Otherwise the genetic background of the two transgenic strains is identical. When these strains were crossed, it was found that (i) a *C. luttrellii* transgenic strain carrying *Chet-MAT1-1-1* and a *C. luttrellii* transgenic strain carrying *ChetMAT1-2-1* can mate in a self-incompatible manner, and the fertility of the cross is similar to that of a wild-type *C. luttrellii* self; (ii) a *C. luttrellii* transgenic strain carrying *ChetMAT1-1-1* can mate with the parental wild-type *C. luttrellii MAT1-1;MAT1-2* strain, indicating that the latter

is able to outcross, a result which was expected but has not been demonstrated previously; (iii) each transgenic strain of *C. luttrellii* is also able to self, although all pseudothecia produced are smaller than those of the wild type and fertility is low (the number of asci was about 5 to 10% of the number of wild-type asci, and full tetrads were found). No recombinants were found in ascospores isolated from these tiny pseudothecia, demonstrating that all sexual structures originated from a self. Asci that are produced when any of these self-compatible strains self always contain full tetrads (Turgeon and Lu, unpublished).

These data support the argument that in *Cochliobolus* spp., and perhaps other ascomycetes also, the primary determinant of reproductive mode is *MAT* itself and that a self-incompatible strain can be made self-compatible or a self-compatible strain can be made self-incompatible by exchange of *MAT* genes. This bolsters the argument that a change in reproductive lifestyle is initiated by a recombination event. The ability to self, observed in transgenic *C. luttrellii* strains generated in this study, also suggests that both MAT1-1-1 and MAT1-2-1 proteins of *Cochliobolus* spp. carry a set of equivalent transcription regulatory activities capable of promoting sexual development alone, in a suitable genetic background.

Conserved Motifs in MAT1-1-1 and MAT1-2-1

Although the Dothideomycete *MAT* locus has a nonallelic architecture, consisting of two alternative forms with dissimilar nucleotide sequences, each harboring a single gene encoding a unique protein that determines mating type, in fact, the MAT proteins of *Cochliobolus* and relatives are partially conserved. Close inspection has revealed that the MAT1-1-1 protein has not only an α-box domain, but also a novel HMG box domain which is fused to and located downstream of the α-box domain (30; Lu and Turgeon, unpublished) (Fig. 6.6). Interestingly, the *S. cerevisiae MATa* idiomorph (the *MAT1-1* counterpart) carries two genes; one corresponds to the α-box protein (MAT1-1-1) of filamentous ascomycetes, and the other is a protein with a yeast-specific homeodomain, which has been shown to be both structurally and functionally related to the HMG box domain (15). In addition to the conserved HMG motif (motif 1, Fig. 6.6) in both MAT1-1-1 and MAT1-2-1, there are at least two other conserved motifs. A cautionary note: the functional significance of these domains awaits experimental assays. Possibly related to this is the observation that transfer of either *C. heterostrophus MAT1-1* or *C. heterostrophus MAT1-2* into a *mat*-deleted strain of *C. luttrellii* induces formation of small pseudothecia (see above), sug-

gesting that these proteins have both MAT1-1 and MAT1-2 functions.

Clues to Reproductive Lifestyle Conversion from Fungi with Nonstandard Mechanisms

There are several examples in the literature of fungi that have unusual mating mechanisms (Table 6.6). We note these examples because insight into mechanisms contributing to their unusual reproductive behavior may contribute to our understanding of mating control in "canonical" self-compatible fungi. The Turgeon laboratory has done some molecular work with *Chromocrea spinulosa/Hypocrea spinulosa*, while Raju and Perkins have generated beautiful genetic and cytological data for *Coniochaeta* (67). Strains of *C. spinulosa* were sent to the Turgeon group by the late John Fincham; to our knowledge, mating-type switching in this fungus had not been worked on since Mathieson (56) worked with Fincham. The *C. spinulosa* sexual reproduction phenomenon is as follows (Fig. 6.7). Two types of strain are known. One is self-compatible and the other is self-incompatible. The latter mates only with the self-compatible strains, and when they mate, a 1:1 ratio of self-incompatible to self-compatible progeny is observed. The former, when mated, also yields progeny which segregate 1:1, self-incompatible to self-compatible.

This phenomenon is similar to that reported for self-compatible *Coniochaeta tetraspora* (67); however, in this case, although eight uninucleate ascospores are delineated in the ascus, only one-half are alive at maturity. If these are allowed to self again, meiotic progeny again are 50% self-compatible (and alive, of course) and 50% dead. Raju and Perkins (67) speculate that ascospore death is developmentally regulated and likely occurs before karyogamy, but not before perithecial initial formation, since vegetative spores show normal viability and ability to self. The ascospore death phenotype segregates as a single Mendelian element; "second-division

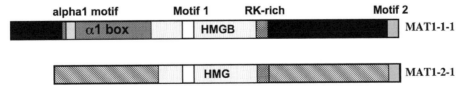

Figure 6.6 Diagrammatic representation of the MAT1-1-1 and MAT1-2-1 proteins, showing conserved motifs. These motifs are found in all Dothideomycete taxa examined including *Cochliobolus* (*Bipolaris*), *Pleospora* (*Stemphylium*), *Phaeosphaeria* (*Stagonospora*), *Leptosphaeria maculans*, and *Alternaria alternata*. A signature motif within the HMG box (stippled rectangle) of MAT1-2-1 is also found in the MAT1-1-1 protein (Motif 1, white box). A motif found in the α1 box (light gray stippled) is found at the C-terminal end of both MAT proteins (Motif 2, gray box). A third common stretch is RK rich (checked box) (Lu and Turgeon, unpublished). To date, functional analyses of these motifs have not been done.

Table 6.6 Fungi that produce both self-fertile and self-sterile progeny

Species	Mating lifestyle	Progeny	Reference(s)
Chromocrea spinulosa (= *Hypocrea spinulosa*)	Self-fertile (large spore)	50% self-sterile/small spore, 50% self-fertile/large spore	56
Sclerotinia trifolium	Self-fertile (large spore)	50% self-sterile/small spore, 50% self-fertile/large spore	81
Ceratocystis coerulescens	Self-fertile	50% self-sterile, 50% self-fertile	40
Coniochaeta tetraspora	Self-fertile	50% self-fertile, 50% dead	67
			16
Botryotinia fuckeliana	Self-sterile	Most asci	32
	Self-fertile	6% asci are 25, 50% self-fertile	
Cryphonectria parasitica	Self-fertile	50% MAT-1, 50% MAT-2; cryptic idiomorphs	57, 58

segregation frequencies indicate that the factor is located in a chromosome arm at a fixed distance from the centromere. Thus, it appears each and every ascus must originate from a diploid nucleus that is heterozygous for the factor" (67). This means that the factor is not inherited vertically and that reprogramming occurs during each new selfing event. We speculate that the same type of segregation event happens in *Chromocrea*, although in this case the meiotic products are 50% self-compatible and 50% self-incompatible.

Preliminary molecular characterization of the mating-type loci has been undertaken for both types of *C. spinulosa* strain (Yun and Turgeon, unpublished) (Fig. 6.8A). Self-incompatible strains are all *MAT1-1* and have the three-gene structure typical of Pyrenomycetes. Self-compatible strains have *MAT1-1* plus a second version of *MAT1-1* linked to a *MAT1-2-1* gene. These findings are the result of PCR amplification of *MAT*, using DNA from vegetative (nonmating) cultures of each type

of strain, which suggests, but does not prove, that there are two versions of *MAT* in the self-compatible strain. The architecture of the version containing both *MAT* genes with an exact repeat hints at an intramolecular mechanism, whereby the *MAT1-2-1* gene might be eliminated, resulting in *MAT1-1* only (Fig. 6.8B). Further work, however, is needed to determine whether the developmentally controlled event is a genetic or an epigenetic phenomenon in which one of the two mating-type genes is silenced. Raju and Perkins (67) state it best: "If altered ascospore size or ascospore viability is a pleiotropic effect of one of the mating-type idiomorphs, and if the effect is expressed only in spores in which the opposite mating-type idiomorph is inactive or absent, this might account both for ascospore dimorphism in *Sclerotinia* and *Chromocrea* and for ascospore death in *Coniochaeta*."

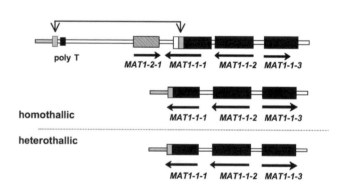

Figure 6.8 Organization of the *C. spinulosa* mating-type loci in homothallic and heterothallic strains. Heterothallic strains all carry the three-gene complement typical of Pyrenomycete *MAT1-1* strains. Homothallic strains carry two types of *MAT* locus, as described in "Clues to Reproductive Lifestyle Conversion from Fungi with Nonstandard Mechanisms." Arrows on the top line indicates repeated sequences in the version of the *MAT* locus that carries both *MAT1-1* and *MAT1-2*. One hypothesis is that these repeated sequences may promote an intramolecular recombination that would eliminate *MAT1-2*, leaving all three *MAT1-1* genes.

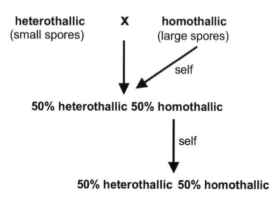

Figure 6.7 *Chromocrea spinulosa* mating scheme. Selfs and crosses both yield progeny that are 50% homothallic and 50% heterothallic. The latter are all *MAT1-1*.

MATING-TYPE STRUCTURE AND FUNCTION IN *P. ANSERINA*

Mating-Type Structure in *P. anserina*

P. anserina is a self-incompatible sordariomycete which contains two allelic idiomorphs denominated *mat−* and *mat+* (65) and corresponding to *MAT1-1* and *MAT1-2* in the standard terminology (80). The *mat−* idiomorph contains three genes, called *FMR1* (*MAT1-1-1*), *SMR1* (*MAT1-1-2*), and *SMR2* (*MAT1-1-3*) (28, 29). The *mat+* idiomorph contains only one gene, *FPR1* (*MAT1-2-1*) (29). Figure 6.3 summarizes the structures of the idiomorphs and the names of the genes in the standard terminology. *FPR1* and *SMR2* encode transcription regulatory proteins of the HMG family, while *FMR1* encodes a transcription factor of the MATα1 family. SMR1 belongs to the HPG family, which is defined by the presence of three invariant amino acids, histidine, proline, and glycine, within a conserved region encompassing 25 residues (30). The molecular function of SMR1 is not yet established. The biological role of *SMR1* and of the three other *P. anserina* mating-type genes has been determined by genetic analyses after crossing different mating-type gene mutants with compatible tester strains. The genetic analyses comprised qualitative and quantitative examinations of progeny and they were associated with cytological observations of the development inside the fruiting body. *P. anserina* is easily amenable to genetic and cytological analyses: the sexual cycle from fertilization to ascospore projection is performed in 4 days, and each step of the perithecium development has been described in detail (94).

Mating-Type Genes Control Fertilization and Postfertilization Developmental Steps during Fruiting-Body Formation

Deletion of the *MAT* locus was the first step towards the identification of the mating-type genes and the determination of their functions. Strains with a deletion of the *mat+* idiomorph (*PaΔmat*) still produce microconidia and protoperithecia, indicating that mating-type genes are not involved in the developmental switch from vegetative to sexual reproduction, nor in the differentiation of the prefertilization male and female structures. Transformation of the *PaΔmat* strain with a *mat−* DNA fragment encompassing *FMR1* restores the ability to fertilize a *mat+* strain, demonstrating that this gene is the activator for the mat− functions required for fertilization (28). However, *FMR1* alone is not sufficient to restore a complete sexual cycle upon crossing to a *mat+* tester strain. The completion of the sexual cycle requires a DNA fragment containing *SMR1* and *SMR2*, indicat-

ing that these genes are required after fertilization. In *N. crassa*, the homologs of *SMR1* (*mat A-2*) and *SMR2* (*mat A-3*) are also required after fertilization, but surprisingly they appear to be redundant genes (33). In *P. anserina*, *SMR1* and *SMR2* mutants display different phenotypes, proving that these genes have different functions (3). The resident *SMR1* gene has been disrupted by the insertion of a selectable marker at a site localized after the region encoding the HPG domain, resulting in the *smr1-r* strain. A cross between *smr1-r* and a *mat+* tester strain produces a great number of fertilized perithecia, but their development is arrested at an early stage after fertilization and no progeny are produced. The disruption of the resident *SMR2* gene after a region encoding 24 residues results in a strain called *smr2-r*. This strain displays a fertilization efficiency similar to that of a wild-type *mat−* strain, but the production of progeny is strongly reduced. The number of asci is about 0.4% of the number of asci recovered in a wild-type cross. Moreover, more than 99% of these asci contain only genetic markers from the *smr2-r* parent, suggesting that an *smr2-r* nucleus passes through the entire sexual cycle and yields progeny without any fusion and genetic exchange with the nucleus from the *mat+* partner strain. These asci have been called uniparental asci. Their frequency in a wild-type cross is less than 10^{-6} (S. Arnaise, unpublished results), and their presence in progeny constitutes a hallmark of mutations in mating-type genes, except *SMR1*. Further investigations with *FMR1* allowed us to identify a mutation leading to the formation of uniparental progeny. A disruption of the resident *FMR1* yields a truncated gene encoding 228 residues and containing the complete α1 domain (3). The α1 domain is supposed to be the master regulatory domain for mat− fertilization functions, based on the functions of MAT α1 in *S. cerevisiae* (reviewed in reference 43). As expected, the corresponding strain, called *fmr1-r*, is able to fertilize a *mat+* tester strain as efficiently as a wild-type *mat−* strain, but the progeny represent 1% of the progeny of a wild-type cross and these contain 50% *fmr1-r* uniparental asci. These features indicate that *FMR1* is required for postfertilization developmental events. Interestingly, it has been established by yeast two-hybrid assays that the FMR1 region between residues 211 and 248 is necessary and sufficient for an interaction with SMR2. The region of SMR2 that is required for the interaction with FMR1 has been localized between residues 1 and 121, upstream of the HMG domain. Moreover, mutations which affect the postfertilization functions of *SMR2* have been found to also affect the FMR1/SMR2 interaction in the yeast two-hybrid system: *smr2^{I26T}* and *smr2^{L39P}* reduce the interaction between SMR2 and

Table 6.7 Phenotype of mating-type gene mutations in *P. anserina*[a]

Mating-type idiomorph	Mating-type gene mutation	Fertilization phenotype[b]	Progeny upon selfing	Progeny upon mating with a tester strain of:	
				Same mating type	Opposite mating type
mat−	WT	mat−, self-incompatible			Biparental
mat−	*fmr1-r*[c]	mat−, weakly self-compatible	Yes	Uniparental *fmr1-r*	Biparental and uniparental *fmr1-r* asci
mat−	*smr1-r*[c] *smr1*[W193A][d] *smr1*[A347D][d]	All strains are mat− and self-incompatible			All strains yield barren perithecia
mat−	*smr2-r*[c] *smr2*[I26T][e] *smr2*[L39P][e] *smr2*[g398a][e] *smr2*[R163T][e] *smr2*[Y171H][e] *smr2*[S176F][e] *smr2*[W199*][e] *smr2*[E202A][e]	All strains are mat− and weakly self-compatible	All strains give a progeny	Uniparental *smr2*[m] ND ND ND ND ND ND ND	All strains yield biparental and uniparental *smr2*[m] asci
mat+	WT	mat+, self-incompatible			Biparental
mat+	*fpr1-r*[f] *fpr1*[T66I][e] *fpr1*[M112I][e] *fpr1*[K208E][e]	All strains are mat+ and weakly self-compatible	No Yes Yes Yes	No No ND ND	All strains yield biparental and uniparental *fpr1*[m] asci
Pa∆*mat*	Complete deletion of the *mat+* idiomorph[g]	Very weakly mat+, very weakly mat−	No	ND	∆*mat* uniparental asci in crosses with *mat+* and *mat−* tester strain

[a]Abbreviations: WT, wild type; ND, not determined.
[b]When used as male on *mat+* and *mat−* tester strains.
[c]Reverse-genetic mutation (3).
[d]Reverse-genetic mutations (E. Coppin, X. Robellet, S. Arnaise, K. Bouhouche, D. Zickler, and R. Debuchy, unpublished data).
[e]Selected as suppressor of *smr1-r* (4).
[f]Reverse-genetic mutation (4).
[g]Reverse-genetic mutation (23).

FMR1, while *smr2*[S176F] increases this interaction (Table 6.7 lists mutant phenotypes) (E. Coppin and R. Debuchy, unpublished results). These observations suggest that these two *mat−* proteins interact to control the same postfertilization event.

Because *FPR1* is the sole gene present in the *mat+* idiomorph, this gene was supposed to be the activator for the mat− functions required for fertilization. In agreement with this assumption, introduction of *FPR1* in a *mat+*-deleted strain restores mat+ fertilization ability. However, the function of FPR1 is not limited to fertilization. The resident *FPR1* has been disrupted with a selectable marker, leaving a truncated gene which encodes a

protein with 282 residues containing the complete HMG domain (4). The corresponding strain, called *fpr1-r*, is as proficient as a *mat+* wild-type strain for fertilization and produces quantitatively normal progeny. A close examination of the progeny reveals that they contain 0.4% of *fpr1-r* uniparental asci, indicating that *FPR1* is required for a developmental event after fertilization.

Taken together, these results indicate that the mating-type genes are involved in fertilization and postfertilization development. Mutant alleles of *FPR1*, *FMR1*, and *SMR2* were found to confer a characteristic and common feature to the corresponding mutant strains: they produce uniparental asci upon crossing with a tester strain of opposite mating type. This feature suggests that these three genes are likely to be involved in a similar pathway after fertilization. In contrast, mutations in *SMR1* lead to barren perithecia and never result in uniparental ascus, indicating that its function is different from that of *FMR1*, *SMR2*, and *FPR1*.

Control of Fertilization by Mating-Type Genes: a Dual Action of Activation and Repression

Further analyses of mating-type mutants provided evidence that the genes involved in fertilization have a dual action of activation and repression (4). This dual action was revealed by the association of mutations that increase the number of microconidia with mating-type genes, resulting in an increased detection of fertilization events. In this genetic context, we observed that the *fmr1-r* strain is able to self-fertilize and to fertilize a *mat−* tester strain, although with a much weaker efficiency than fertilization of a *mat+* tester strain (Table 6.7). This self-compatible behavior indicates that the *fmr1-r* strain produces male cells which express the *mat+* mating type at a weak level. The expression of *mat+* fertilization functions in an *fmr1-r* strain suggests that the wild-type *FMR1* gene represses these functions in *mat−* sexual organs and that the *fmr1-r* mutant has lost, at least partly, this repressive action. This rationale leads to the conclusion that *FMR1* has an activating action on mat− fertilization functions and a repressive action on mat+ fertilization functions in *mat−* sexual organs. *SMR2* was not required for the expression of mat− fertilization functions, but surprisingly, the disruption and several other mutations of *SMR2* relieved the expression of mat+ fertilization functions (Table 6.7), indicating that the wild-type *SMR2* represses the mat+ functions required for fertilization in *mat−* sexual organs. A similar rationale based on *FPR1* mutations (Table 6.7) leads to the conclusion that the wild-type *FPR1* gene could have a repression activity on the mat− fertilization functions in *mat+* sexual organs. The model for the control of fertilization by mating-type genes is summarized in Figure 6.9.

Intercellular recognition leading to fertilization requires a pheromone/receptor system that allows female organs to recognize male cells of opposite mating type. This hormonal mechanism in *P. anserina* was described by George Bistis, who observed that the trichogyne produced by female organs grew towards male cells of opposite mating type (G. Bistis, personal communication). This observation suggests that the male cells produce a pheromone which is recognized by receptors from the trichogyne. *P. anserina* contains two pheromone genes, *MFM* and *MFP*, which are specifically transcribed in *mat−* and *mat+* strains, respectively (24). Quantitative reverse transcription experiments have been performed to measure the transcription of *MFM* in a wild-type *mat+* strain and an *fpr1*[T66I] mutant (Table 6.7), to investigate the molecular basis of self-fertility. The transcription of *MFM* in the *fpr1*[T66I] mutant strain is as low as in the *mat+* wild-type strain (F. Bidard, E. Coppin, and R. Debuchy, unpublished results). This result indicates that self-fertility of the *fpr1*[T66I] strain does not result from a transcriptional increase of *MFM* but is likely to result from the derepression of the genes required for the posttranscriptional expression steps of *MFM*, as predicted by Coppin et al. (24).

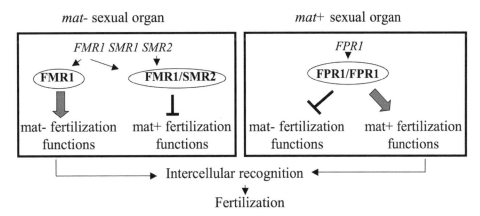

Figure 6.9 Control of fertilization by mating-type genes in *P. anserina*. Gene actions are symbolized as follows: arrows connote positive regulation, and lines ending in bars connote repression. Proteins are displayed in association as established by yeast two-hybrid assays (Coppin and Debuchy, unpublished).

Postfertilization Function of *FMR1*, *SMR2*, and *FPR1*

FMR1, *SMR2*, and *FPR1* Are Required for Nucleus Migration into Ascogenous Hyphae

The models for the biological functions of *FMR1*, *SMR2*, and *FPR1* during postfertilization development are based on the specific occurrence of uniparental asci and on the cytological observations of developmental events associated with the formation of the uniparental progeny (94). The genetic markers of the uniparental asci come exclusively from the parent containing the mutant mating type, indicating that the developmental alteration leading to uniparental progeny takes place before karyogamy and meiosis. In a wild-type strain, fertilization is followed by the migration of the male nucleus down the trichogyne toward the body of the female organ. The male nucleus undergoes a series of mitotic divisions in cells containing female nuclei. These mitotic divisions lead to the formation of plurinucleate cells containing about 10 to 20 nuclei of different mating types (93). Pairs of nuclei are then cellularized in specialized dikaryotic hyphae. These hyphae, called ascogenous hyphae, maintain the dikaryotic state until the formation of a crozier, which consists of one basal and one lateral cell, each of them uninucleate, and one upper dikaryotic cell (Fig. 6.1). Nuclei of the dikaryotic cell fuse, and meiosis ensues immediately while the cell is elongating to form the ascus. After a postmeiotic mitosis, ascospores are delineated inside the ascus and eventually projected outside the fruiting body. Genetic markers from both parents are retrieved in the ascospores from the same ascus, indicating that male and female nuclei, namely, *mat+* and *mat−* nuclei, have contributed to the dikaryotic state of the ascogenous hypha. In crosses involving any one of the *FMR1*, *SMR2*, or *FPR1* mutants and a wild-type partner strain, croziers contain mainly one haploid nucleus (94). The upper cell either forms an ascus or remains as an uninucleate cell, while the basal cell generally divides a second time. All perithecia become filled with several rows of uninucleate cells. Taken together, these results suggest that one type of nucleus enters the ascogenous hyphae in a mutant × wild-type cross. This nucleus is likely to be the nucleus carrying the mutant mating-type gene, although there is as yet no straightforward evidence for this assumption, except the observation of uniparental asci. Viable ascospores are formed after a haploid meiosis during which random segregation of chromosomes results by chance in even distribution.

These cytological observations indicate that the defect of *FMR1*, *SMR2*, and *FPR1* mutants is the migration of the defective nucleus instead of *mat+*/*mat−* pairs into the ascogenous hypha. This defect can be interpreted according to two different models: the internuclear recognition model and the random segregation model.

The Internuclear Recognition Model

The internuclear recognition hypothesis proposes that segregation of nuclei from the plurinucleate cell to the ascogenous hypha requires a recognition step between nuclei of opposite mating types inside the plurinucleate cell. This recognition step would ensure that one *mat+* and one *mat−* nucleus cellularize to form the ascogenous hypha. Internuclear recognition has been supposed as early as 1956 by Haig P. Papazian, who described the male nucleus migrating down the trichogyne and passing on its way the trichogyne's own nucleus or nuclei and proposed: "For them to be thus distinguished they must have a different phenotype. This nuclear phenotype may be controlled in two ways: it may be determined by the mating-type genes, ..." (63). The model for the control of internuclear recognition is based on the model for the control of fertilization, as intercellular recognition and internuclear recognition share many common features. The nuclear phenotype hypothesized by H. P. Papazian has been called nuclear identity (94). Nuclei with different nuclear identities recognize each other and participate in the formation of the ascogenous hyphae, very much like sexual organs with different mating types recognize each other. In this model, uniparental progeny has been interpreted as corresponding to self-fertilization (4). *FMR1* and *SMR2* are the activator of the *mat−* function required for determining the *mat−* nuclear identity, while *FPR1* is the activator of *mat+* functions required for determining the mat+ nuclear identity. *FMR1* and *SMR2* are the repressor of the mat+ functions required for internuclear recognition in *mat−* nuclei. Conversely, *FPR1* is also a repressor of the *mat−* functions required for internuclear recognition in *mat+* nuclei. Mutations leading to the formation of uniparental progeny are supposed to relieve the repressor activity of the mating-type regulatory proteins. Therefore, nuclei carrying mutant mating-type genes express both mat+ and mat− functions required for internuclear recognition and trigger self-recognition. These selfish nuclei can engage themselves in the developmental events followed by wild-type pairs of nuclei and yield eventually uniparental asci. The model for the internuclear recognition is summarized in Fig. 6.10.

The internuclear recognition model implies that FMR1 and SMR2, which determine the mat− nuclear identity, return specifically to the nucleus that encodes

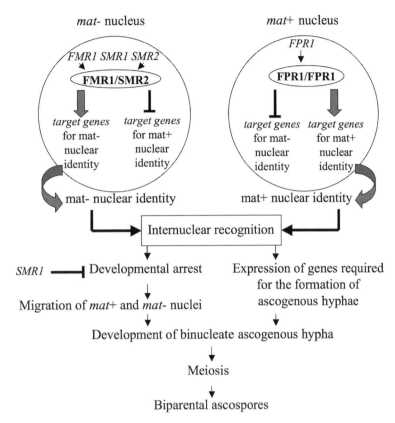

Figure 6.10 Internuclear recognition model. Gene actions are symbolized as in Fig. 6.9. The *mat+* and *mat−* nuclei recognize each other inside plurinucleate cells based on their nuclear identity. The mat− nuclear identity is determined by the FMR1/SMR2 heterodimer that returns specifically to *mat−* nuclei. The mat+ nuclear identity is determined by the FPR1 homodimer that returns specifically to *mat+* nuclei. Internuclear recognition results from the superimposition of the nuclear identity of *mat−* and *mat+* nuclei, which triggers nuclear migration and expression of the genes required for ascogenous hyphae formation. The developmental arrest is overcome by the action of *SMR1*. Uniparental *mat−* asci result from the loss of the repressive action of the FMR1/SMR2 heterodimer, which allows basal expression of the target genes for mat+ nuclear identity. Expression of mat− and mat+ nuclear identity in the same nucleus triggers self-recognition, migration of the *mat* mutant nucleus and development of uninucleate ascogenous hypha. A similar rationale is applied to *mat+* mutations that yield a uniparental *mat+* progeny (see "The Internuclear Recognition Model" for more details).

them, namely, the *mat−* nucleus. These transcription factors should not diffuse to adjacent *mat+* nuclei. If they did, the mat− nuclear identity would not be limited to *mat−* nuclei and recognition would be impossible. Conversely, FPR1, which determines the mat+ nuclear identity, is expected to return to the *mat+* nucleus and should not diffuse to adjacent *mat−* nuclei. This property has been termed nucleus-restricted expression and has been initially observed in *N. crassa* by Burton and Metzenberg for factors involved in sulfur metabolism (14). Nucleus-restricted expression of *FMR1*, *SMR1*, and *SMR2* has been tested according to the following rationale. Each *fmr1-r*, *smr1-r*, and *smr2-r* strain (Table 6.7) has been crossed with a *mat+* strain that contains a wild-type copy of the gene which is disrupted in its partner (3). Restoration of a wild-type progeny would indicate that the wild-type product can complement the defect of the disrupted mating-type gene by diffusing from one nucleus to another during internuclear recognition. This internuclear complementation would provide evidence that the ectopic gene expression is not nucleus restricted. If the progeny displays the typical features of the progeny obtained from a cross between a mutant strain and a wild-type strain, this implies that no internuclear complementation occurs during internuclear recognition. This result would suggest that the product of the ectopic copy has a nucleus-restricted expression. Although several lines of evidence indicate that an ectopic copy of

SMR2 is expressed in mat+ nuclei, this ectopic copy does not complement the smr2-r mutation in the mat− nucleus. In contrast to SMR2, SMR1 can complement the smr1-r disruption, demonstrating that its action is not restricted to the nucleus containing the SMR1 gene. SMR1 is not involved in internuclear recognition, and its expression was not expected to be nucleus restricted. FMR1 was found to complement the defect of fmr1-r, indicating that FMR1 can diffuse from one nucleus to another in the absence of SMR2 (S. Arnaise, unpublished results). However, the FMR1/SMR2 interaction evidenced in the two-hybrid assays suggests that SMR2 may prevent the diffusion of FMR1 to mat+ adjacent nuclei when FMR1 is in the wild-type mat− context. A copy of SMR2 in the mat+ nucleus containing FMR1 reduces the FMR1 internuclear complementation, as expected from the SMR2 effect on FMR1 internuclear diffusion. Nucleus-restricted expression of FPR1 has not yet been established.

The internuclear recognition model implies that the products of the target genes involved in nuclear identity also have a nucleus-restricted expression. These products should return to the surface or remain in the vicinity of the nucleus in which the corresponding genes are transcribed, defining thus the nuclear identity. Superimposition of different nuclear identities from mat+ and mat− nuclei is supposed to trigger nucleus migration and the expression of the genes necessary for ascogenous hyphae formation. FMR1, SMR2, and FPR1 mutations leading to uniparental progeny are supposed to result in the expression of mat− and mat+ nuclear identity in the same nucleus, thus triggering migration of this nucleus and formation of uninucleate ascogenous hyphae.

Internuclear recognition may take place also in Homobasidiomycetes, after hyphal fusion between compatible strains. The nuclei from the fused cells become the "donors" and migrate through the cells of the "recipient" compatible strain towards hyphal tips (reviewed in reference 18). Once two compatible nuclei are present within the hyphal tip cell, they form a dikaryotic hypha that can be considered, at least morphologically, to be equivalent to the euascomycete ascogenous hypha. However, the dikaryotic stage is rapidly followed by karyogamy and meiosis in Euascomycetes, while it can be maintained indefinitely in Homobasidiomycetes, until appropriate environmental conditions induce formation of the fruiting body. Nuclear migration in Homobasidiomycetes requires that the nuclei of the donor be distinguished from those of the recipient strain, which do not migrate. The nuclear migration and its associated putative internuclear recognition are con-trolled by a pheromone/receptor system (reviewed in reference 18). However, investigation of the phero-mone functions in P. anserina indicates that their role is limited to intercellular recognition during fertilization (24). This result suggests that in spite of seemingly sim-ilar events, development in Euascomycetes and Ho-mobasidiomycetes is controlled by different molecular systems.

The Random Segregation Model

The formation of mat+/mat− dikaryotic ascogenous hypha can be explained by an alternative model that re-lies on random segregation of pairs of nuclei from the plurinucleate cell into the ascogenous hypha. Migration of a compatible mat+/mat− couple would result in an ascogenous hypha, while migration of incompatible pairs (mat+/mat+ or mat−/ mat−) would either abort or yield paraphyses or simply be impossible. The ran-dom segregation model requires that heterokaryotic mat+/mat− pairs have specific properties leading to the development of the ascogenous hyphae. This model has been proposed by Metzenberg and Glass in the N. crassa context: "It is possible, but unproven, that this diploid cell (ascogenous hypha cell) contains a novel regulatory product specified by cooperation of the A and a idiomorphs analogous to that in a/α diploids of S. cerevisiae (41)" (59). This new heterodimer would be responsible for the gene expression required for ascoge-nous hypha formation. This model has prompted sev-eral teams to search for a possible interaction between the products of opposite mating types in different or-ganisms. In N. crassa, two-hybrid experiments estab-lished the ability of MAT a-1 (homolog of FPR1) to interact with MAT A-1 (homolog of FMR1) (5). How-ever, mutations that interfere with this interaction corre-late with mutations that eliminate vegetative incompati-bility, but not mating, suggesting that this interaction is not essential for the sexual cycle. An interaction be-tween SMT A-1 (homolog of FMR1) and SMT a-1 (ho-molog of FPR1) has also been established in Sordaria macrospora, but no mutation provides information of the relevance of this interaction to the sexual cycle (47). In P. anserina, all possible pairings of FMR1, SMR2, and FPR1 have been tested in the yeast two-hybrid system. SMR1 has not been included in these assays because fusions of SMR1 and GAL4 are not expressed in yeast for an unknown reason. No interaction be-tween FPR1 and any other mat− specific product has been detected (E. Coppin and R. Debuchy, unpublished results). Therefore, there is no evidence of any interac-tion between opposite mating-type products that would be essential for the sexual cycle. However, it must be emphasized that the yeast two-hybrid system failed to detect many established interactions, notably the yeast a1 and α2 interaction, which yields a heterodimer that controls sexual cycle in yeast diploid cells (reviewed in reference 48). The random segregation model can account for self-fertilization and uniparental progeny if one assumes that the FMR1/SMR2 heterodimer and FPR1 repress the target genes that are activated by the putative regulatory product specified by cooperation of the *mat+* and *mat−* idiomorphs (Fig. 6.11). Mutations in *FMR1*, *SMR2*, and *FPR1* would allow the expression of these target genes that should be required for the formation of ascogenous hyphae. Consequently, nuclei with mutant *FPR1*, *FMR1*, and *SMR2* could trigger the formation of uninucleate ascogenous hyphae and uniparental progeny.

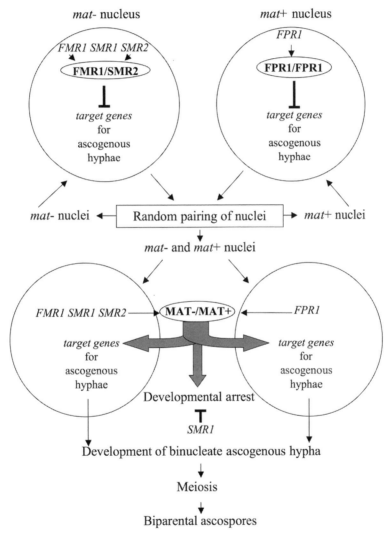

Figure 6.11 Random segregation model. Gene actions are symbolized as in Fig. 6.9. During their mitoses in the plurinucleate cells, the FMR1/SMR2 heterodimer and FPR1 homodimer repress the expression of target genes involved in ascogenous hyphae formation. Random pairing of nuclei yield *mat−/mat−* and *mat+/mat+* pairs which do not undergo further development, while *mat+/mat−* pairs produce a regulatory product specified by cooperation of the *mat+* and *mat−* idiomorphs. This product is symbolized by MAT+/MAT−. It activates the expression of the genes required for ascogenous hyphae formation. Uniparental *mat−* asci result from the loss of repressive action of the FMR1/SMR2 on the target genes for ascogenous hyphae. Expression of these genes allows the mutant nucleus to trigger the development of ascogenous hyphae and to yield a *mat−* uniparental progeny. A similar rationale is applied to *mat+* mutations that yield uniparental *mat+* progeny (see "The Random Segregation Model" for more details).

Postfertilization Functions of Mating-Type Genes in Euascomycete Fungi

Few data are available about the postfertilization functions of mating-type genes in the main self-incompatible model systems, *C. heterostrophus* and *N. crassa*. In *C. heterostrophus*, each idiomorph contains a single gene (78). A sequence encoding the 3′ untranslated region of idiomorph messengers is necessary during postfertilization development, but its function remains as yet unknown (85). Further investigations of the mating-type functions would require mutational analysis of the mating-type genes. *N. crassa* has a mating-type structure identical to that of *P. anserina* (reviewed in reference 30). Many mutations in the mating-type genes have been analyzed but none of them leads to the formation of uniparental asci. This difference has led to the conclusion that postfertilization functions of mating-type genes are very different in *N. crassa* and *P. anserina*. However, it must be noted that the *N. crassa* mating-type mutations are all different from those in *P. anserina* mating-type genes. Therefore, phenotypic differences between mutations in a mating-type gene in *N. crassa* and its homolog in *P. anserina* may be attributed to specific properties of different alleles instead of interpreted as an indication of a different function of the considered mating-type gene. A striking observation is that independent screening in *P. anserina* for mutations which allow uniparental ascus formation yields identical alleles (E. Coppin, unpublished data), indicating that these mutations must obey some yet unknown but specific constraints. It is therefore possible that the *N. crassa* mating-type mutations tested until now do not obey these constraints. A conclusive experiment would be to introduce in a given *N. crassa* mating-type gene a mutation that induces the formation of uniparental asci in *P. anserina*. The search for uniparental progeny in *N. crassa* is an important issue because it would allow a unified view of mating-type functions in Sordariomycetes. A first hint for the possibility of uniparental progeny in *N. crassa* was provided by Edward Barry, who reported the occurrence of a unique uniparental ascus of *a* mating type (6). Unfortunately, no further sequencing of the *mat a-1* gene has been done to detect a possible mutation leading to a weak frequency of uniparental asci, as is the case for instance with the *fpr1*^{K208E} mutation in *P. anserina* (4).

Self-compatible species raise a challenging question about the role of mating-type genes in relation to the nuclear content of ascogenous hyphae. In self-compatible species, ascogenous hyphae contain two genetically identical nuclei. Either the ascogenous hyphae can accommodate not less and not more than two nuclei for reasons of space, or the two nuclei are epigenetically different. Shifting from the homothallic reproductive mode to the heterothallic reproductive mode in *G. zeae* has helped to solve this question (53). *G. zeae* carries linked counterparts of the *P. anserina* idiomorphs, symbolized as *MAT1-1;MAT1-2* (90). Deletion of either *MAT1-1* or *MAT1-2* mating-type sequences resulted in two strains that have a self-incompatible mating-type structure and behavior. In crosses of the *mat1-1;MAT1-2* strain (deleted for *mat1-1*) with a wild-type strain, perithecia contain *mat1-1;MAT1-2* nuclei and nuclei of the self-compatible parent. Lee et al. (53) analyzed the progeny to establish the proportion of the two possible nuclear recognition events: internuclear recognition of two wild-type nuclei and internuclear recognition between a wild-type nucleus and a *mat1-1;MAT1-2* nucleus. Surprisingly, they found only asci contributed by both parents, suggesting that wild-type and *mat1-1;MAT1-2* nuclei recognize each other preferentially. Similar results were obtained in crosses of *MAT1-1;mat1-2* strain (deleted for *mat1-2*) with a wild-type strain. The absence of uniparental wild-type progeny leads to the conclusion that space occupation is not the reason for dikaryotic ascogenous hyphae. The presence of progeny coming exclusively from both parents strongly suggests that the wild-type nucleus, although genetically *MAT1-1* and *MAT1-2*, is committed to be functionally either *MAT1-1* or *MAT1-2*, as proposed by Metzenberg and Glass to explain self-compatible mating in *Neurospora* sp. (59). According to this hypothesis, Lee et al. observed a progeny only when a functionally *MAT1-1* nucleus was paired with a *mat1-1;MAT1-2* nucleus or when a functionally *MAT1-2* nucleus was paired with a *MAT1-1;mat1-2* nucleus. The *mat1-1;MAT1-2* nuclei (or *MAT1-1;mat1-2* nuclei) do not give any uniparental progeny, indicating that the deletion of one mating type leads to a defect in the formation of ascogenous hyphae. In contrast, a cross between *mat1-1;MAT1-2* and *MAT1-1;mat1-2* is fertile, indicating that the two compatible nuclei bring all information necessary for the sexual cycle. However, these data provide no proof supporting either the internuclear recognition model or the random segregation model.

The Function of *SMR1*

SMR1 belongs to the HPG family, which is defined by the presence of three invariant amino acids, histidine, proline, and glycine, within a conserved region encompassing 25 residues. The HPG family is present in the MAT1-1 idiomorph of all Sordariomycetes known to date (30). Genome-wide searches for SMR1/MAT1-1-2 homologs in fungi outside the Sordariomycete taxon failed to yield any protein with a significant similarity, suggesting that the HPG family is specific to the sordariomycete group. Based on the high isoelectric point of the conserved domain of SMR1, Debuchy et al. have proposed that it defines a new DNA-binding domain (28). However, subsequent investigations of SMR1 function do not support this hypothesis. Prediction of SMR1 subcellular localization with PSORTII (60) failed to detect any nuclear localization signal and predicts a cytoplasmic localization. Green fluorescent protein tagging of SMR1 does not affect the function of SMR1 and displays a cytosolic localization, in agreement with the in silico prediction (E. Coppin, X. Robelet, S. Arnaise, K. Bouhouche, D. Zickler, and R. Debuchy, unpublished results). Alanine scanning or replacement of alanine by aspartic acid reveals only two residues that are essential for the SMR1 function: tryptophan 193 and alanine 347 (Table 6.7). Surprisingly, mutation of the three invariant residues of the conserved region to alanine does not affect the function of SMR1. This may support the idea that the conserved region is involved in protein-protein interaction, since these regions are considered to be more tolerant to mutations than regions involved in DNA binding. If the molecular function of SMR1 remains unknown, its biological functions are better understood. Crosses involving smr1-r, or the mat− strain with the smr1^{W193A} mutation, and wild-type mat+ strain result in a complete arrest of the development of the fruiting body at a stage preceding the formation of the ascogenous hyphae (Coppin et al., unpublished). Internuclear complementation of smr1-r disruption indicates that SMR1 product can diffuse from one nucleus to another (3). SMR1 can indeed be localized in mat+ or in mat− nuclei, or in both nuclei, without any deleterious effect on the progeny, unlike FMR1, SMR2, or FPR1. However, its biological function is closely connected to the function of these three genes. In order to assess the effect of the vegetative expression of FMR1, SMR2, and FPR1, Coppin and Debuchy observed that their expression in the same nucleus inside ascospores resulted in a lethal phenotype (25). This lethal phenotype can be suppressed by the expression of SMR1. The vegetative lethal phenotype suggests that FMR1, SMR2, and FPR1 expression during fruiting-body development triggers a developmental arrest. This developmental arrest is likely to correspond to the complete development arrest observed in crosses involving smr1-r or smr1^{W193A}. According to this model, the biological function of SMR1 would be to overcome the developmental arrest associated with FPR1, FMR1, and SMR2 expression. Uninucleate ascogenous hyphae can differentiate even in the absence of SMR1, which is required for an as yet unknown reason specifically for the formation of mat+/mat− ascogenous hyphae.

Future Prospects

The analysis of evolution of reproductive lifestyles in Dothideomycetes has led to a conclusive statement regarding the direction of evolution, but the origin and evolution of mating types in Euascomycetes are still unknown. However, a hint for future direction of research appears in this review. It is the observation that in Dothideomycetes, the MAT1-1-1 gene encodes an α1 and also, tentatively, an HMG box (Lu and Turgeon, unpublished) while in P. anserina, the α1 and HMG proteins encoded by FMR1 and SMR2 interact physically (Coppin and Debuchy, unpublished). The physical and functional linkage between the α1 and HMG domains encoded by the MAT1-1 idiomorph is a common feature to these distantly related fungi, suggesting that the ancestral mating-type structure in Euascomycetes might have consisted of α1- and HMG-encoding genes. Whether the ancestral mating-type structure contains only one gene (as in Dothideomycetes) or two genes (as in Sordariomycetes) should be solved by extending structural analyses to other groups. The MAT1-2 idiomorph may have diverged from the ancestral MAT1-1 allele by loss of the α1-encoding region.

The major challenge in the field of mating-type gene function is the identification of the target genes of the mating-type regulatory proteins, notably those required for internuclear recognition or ascogenous hyphae development, according to the considered model. Many genetic approaches have been used unsuccessfully in P. anserina to identify these target genes. It seems that genetic approaches have reached their limits in this system and that further progress will require approaches more suitable to the analysis of complex gene wiring. Microarray analyses of carefully selected mating-type mutants may allow significant progress in this area and eventually allow an understanding of how two nuclei of determined mating type are selected in a syncytium in which several nuclei with different mating types are present. As emphasized in "Postfertilization Functions of Mating-Type Genes in Euascomycete Fungi" above, it is also necessary to determine if the mating-type genes control the same developmental steps through identical pathways in different organisms, a question that is not yet resolved in close relatives such as P. anserina and N. crassa. Preliminary reports on the functions of mating-type genes in Aspergillus nidulans suggest that their functions are very different from those inferred in P. anserina (62). In contrast, the finding of an HMG-encoding region in MAT1-1-1 in C. heterostrophus opens the way for a unification with Sordariomycetes. It would be of high interest to test if the HMG domain of the *MAT1-1-1* gene has any implication in fertilization. Sordariomycetes results would suggest that the HMG domain of *C. heterostrophus* MAT1-1-1 is not involved in the MAT1-1 fertilization functions. Importantly, these questions show that the analyses of evolution of mating types based on their structure could not be separated from analyses of their function. An evo-devo analysis is necessary to understand the origin and evolution of the mating-type system.

Several developmental events take place during the sexual cycle, and future analyses of mating-type functions will determine their connection with these developmental events and possibly extend their functions beyond what is known today. Meiosis may require mating-type genes in *N. crassa*, as suggested by Debuchy and Turgeon (30) based on the effect of meiotic silencing by unpaired DNA (MSUD) on ectopic copies of *mat A* and *mat a* idiomorphs (30). MSUD is a process that silences unpaired genes and all homologous copies during meiosis in *N. crassa* (72). Idiomorphs at the *mat* locus are immune to MSUD, although they are unpaired during meiosis. By contrast, ectopic copies of *mat a* or *mat A* idiomorph trigger MSUD and result in barren perithecia (36, 71). As MSUD silences genes specifically during meiosis, this effect reveals that mating-type information is required during meiosis, or shortly after this step. The immunity of the *N. crassa* mating-type locus to MSUD raised the question of the mechanism of this immunity. Is the mating-type locus protected from MSUD by a subnuclear structure such as the nucleolus, which protects the ribosomal DNA repeats from another silencing mechanism called repeat-induced point mutation (RIP)? This silencing mechanism is homology dependent and introduces mutations in duplicated sequences (69). Does this hypothetical subnuclear structure protect a duplication in the idiomorph against RIP? RIP has been demonstrated in *P. anserina* (13, 38, 39); comparison of RIP efficiency on targets inserted in various genomic loci, including *MAT*, should indicate whether this locus is immune to RIP. The temporal relation between RIP and the main steps controlled by mating-type genes after fertilization opens another research area.

B.G.T. acknowledges the commendable work of Shun-Wen Lu, Sung-Hwan Yun, and Patrik Inderbitizin, former lab members, all of whom generously provided data prior to publication. B.G.T. also thanks P. Inderbitzin for critical reading of the manuscript. R.D. is very much indebted to S. Arnaise, V. Berteaux-Lecellier, and E. Coppin for critical reading of the section about mating-type functions in P. anserina and for their suggestions.

References

1. **Arie, T., S. K. Christiansen, O. C. Yoder, and B. G. Turgeon.** 1997. Efficient cloning of ascomycete mating type genes by PCR amplification of the conserved MAT HMG box. *Fungal Genet. Biol.* **21:**118–130.

2. **Arie, T., I. Kaneko, T. Yoshida, M. Noguchi, Y. Nomura, and I. Yamaguchi.** 2000. Mating-type genes from asexual phytopathogenic ascomycetes *Fusarium oxysporum* and *Alternaria alternata. Mol. Plant-Microbe Interact.* **13:**1330–1339.

3. **Arnaise, S., R. Debuchy, and M. Picard.** 1997. What is a bona fide mating-type gene? Internuclear complementation of mat mutants in *Podospora anserina. Mol. Gen. Genet.* **256:**169–178.

4. **Arnaise, S., D. Zickler, S. Le Bilcot, C. Poisier, and R. Debuchy.** 2001. Mutations in mating-type genes of the heterothallic fungus *Podospora anserina* lead to self-fertility. *Genetics* **159:**545–556.

5. **Badgett, T., and C. Staben.** 1999. Interaction between and transactivation by mating type polypeptides of *Neurospora crassa. Fungal Genet. Newsl. Suppl.* **46:**73.

6. **Barry, E.** 1996. Uniparental progeny in *Neurospora crassa. Fungal Genet. Newsl.* **43:**14.

7. **Barve, M. P., T. Arie, S. S. Salimath, F. J. Muehlbauer, and T. L. Peever.** 2003. Cloning and characterization of the mating type (MAT) locus from *Ascochyta rabiei* (teleomorph: *Didymella rabiei*) and a *MAT* phylogeny of legume-associated *Ascochyta* spp. *Fungal Genet. Biol.* **39:**151–167.

8. **Bennett, R. S., S. H. Yun, T. Y. Lee, B. G. Turgeon, E. Arseniuk, B. M. Cunfer, and G. C. Bergstrom.** 2003. Identity and conservation of mating type genes in geographically diverse isolates of *Phaeosphaeria nodorum. Fungal Genet. Biol.* **40:**25–37.

9. **Berbee, M. L.** 2001. The phylogeny of plant and animal pathogens in the Ascomycota. *Physiol. Mol. Plant Pathol.* **59:**165–187.

10. **Berbee, M. L., B. P. Payne, G. J. Zhang, R. G. Roberts, and B. G. Turgeon.** 2003. Shared ITS DNA substitutions in isolates of opposite mating type reveal a recombining history for three presumed asexual species in the filamentous ascomycete genus *Alternaria. Mycol. Res.* **107:**169–182.

11. **Berbee, M. L., M. Pirseyedi, and S. Hubbard.** 1999. *Cochliobolus* phylogenetics and the origin of known, highly virulent pathogens, inferred from ITS and glyceraldehyde-3-phosphate dehydrogenase gene sequences. *Mycologia* **91:**964–977.

12. **Blakeslee, A. F.** 1904. Sexual reproduction in the Mucorineae. *Proc. Natl. Acad. Sci. USA* **40:**205–319.

13. **Bouhouche, K., D. Zickler, R. Debuchy, and S. Arnaise.** 2004. Altering a gene involved in nuclear distribution increases the repeat-induced point mutation process in the fungus *Podospora anserina. Genetics* **167:**151–159.

14. **Burton, E. G., and R. L. Metzenberg.** 1972. Novel mutation causing derepression of several enzymes of sulfur metabolism in *Neurospora crassa. J. Bacteriol.* **109:**140–151.

15. Butler, G., C. Kenny, A. Fagan, C. Kurischko, C. Gaillardin, and K. H. Wolfe. 2004. Evolution of the *MAT* locus and its Ho endonuclease in yeast species. *Proc. Natl. Acad. Sci. USA* **101:**1632–1637.

16. Cain, R. F. 1961. Studies of soil fungi. III. New species of *Coniochaeta*, *Chaetomidium*, and *Thielavia*. *Can. J. Bot.* **39:**1231–1239.

17. Carlile, M. J. 1987. Genetic exchange and gene flow: their promotion and prevention, p. 203–214. *In* A. D. M. Rayner, C. M. Brasier, and D. Moore (ed.), *Evolutionary Biology of the Fungi*. Cambridge University Press, New York, NY.

18. Casselton, L. A., and N. S. Olesnicky. 1998. Molecular genetics of mating recognition in basidiomycete fungi. *Microbiol. Mol. Biol. Rev.* **62:**55–70.

19. Catlett, N., B.-N. Lee, O. Yoder, and B. Turgeon. 2003. Split-marker recombination for efficient targeted deletion of fungal genes. *Fungal Genet. Newsl.* **50:**9–11.

20. Cherif, M., M. I. Chilvers, H. Akamatsu, T. L. Peever, and W. J. Kaiser. 2006. Cloning of the mating type locus from *Ascochyta lentis* (teleomorph : *Didymella lentis*) and development of a multiplex PCR mating assay for Ascochyta species. *Curr. Genet.* **50:**203–215.

21. Christiansen, S. K., A. Sharon, O. C. Yoder, and B. G. Turgeon. 1993. Functional conservation of mating type genes and interspecific mating, p. 63. *In Proceedings of the 17th Fungal Genetics Conference*, Asilomar, CA.

22. Christiansen, S. K., S. Wirsel, O. C. Yoder, and B. G. Turgeon. 1998. The two *Cochliobolus* mating type genes are conserved among species but one of them is missing in C. *victoriae*. *Mycol. Res.* **102:**919–929.

23. Coppin, E., S. Arnaise, V. Contamine, and M. Picard. 1993. Deletion of the mating-type sequences in *Podospora anserina* abolishes mating without affecting vegetative functions and sexual differentiation. *Mol. Gen. Genet.* **241:**409–414.

24. Coppin, E., C. de Renty, and R. Debuchy. 2005. The function of the coding sequences for the putative pheromone precursors in *Podospora anserina* is restricted to fertilization. *Eukaryot. Cell* **4:**407–420.

25. Coppin, E., and R. Debuchy. 2000. Co-expression of the mating-type genes involved in internuclear recognition is lethal in *Podospora anserina*. *Genetics* **155:**657–669.

26. Coppin, E., R. Debuchy, S. Arnaise, and M. Picard. 1997. Mating types and sexual development in filamentous ascomycetes. *Microbiol. Mol. Biol. Rev.* **61:**411–428.

27. Cozijnsen, A. J., and B. J. Howlett. 2003. Characterisation of the mating-type locus of the plant pathogenic ascomycete *Leptosphaeria maculans*. *Curr. Genet.* **43:**351–357.

28. Debuchy, R., S. Arnaise, and G. Lecellier. 1993. The *mat−* allele of *Podospora anserina* contains three regulatory genes required for the development of fertilized female organs. *Mol. Gen. Genet.* **241:**667–673.

29. Debuchy, R., and E. Coppin. 1992. The mating types of *Podospora anserina*—functional analysis and sequence of the fertilization domains. *Mol. Gen. Genet.* **233:**113–121.

30. Debuchy, R., and B. G. Turgeon. 2006. Mating-type structure, evolution and function in Euascomycetes,

p. 293–324. *In* U. Kües and R. Fischer (ed.), *The Mycota*, vol. 1. Springer-Verlag, Berlin, Germany.

31. Edgerton, C. W. 1914. Plus and minus strains in the genus *Glomerella*. *Am. J. Bot.* **1:**244–254.

32. Faretra, F., E. Antonacci, and S. Pollastro. 1988. Sexual behaviour and mating system of *Botryotinia fuckeliana*, teleomorph of *Botrytis cinerea*. *J. Gen. Microbiol.* **134:** 2543–2550.

33. Ferreira, A. V., Z. An, R. L. Metzenberg, and N. L. Glass. 1998. Characterization of mat A-2, mat A-3 and delta-matA mating-type mutants of *Neurospora crassa*. *Genetics* **148:**1069–1079.

34. Fraser, J. A., and J. Heitman. 2004. Evolution of fungal sex chromosomes. *Mol. Microbiol.* **51:**299–306.

35. Glass, N. L., R. L. Metzenberg, and N. B. Raju. 1990. Homothallic Sordariaceae from nature: the absence of strains containing only the *a* mating type sequence. *Exp. Mycol.* **14:**274–289.

36. Glass, N. L., S. J. Vollmer, C. Staben, J. Grotelueschen, R. L. Metzenberg, and C. Yanofsky. 1988. DNAs of the two mating-type alleles of *Neurospora crassa* are highly dissimilar. *Science* **241:**570–573.

37. Goodwin, S. B., C. Waalwijk, G. H. Kema, J. R. Cavaletto, and G. Zhang. 2003. Cloning and analysis of the mating-type idiomorphs from the barley pathogen *Septoria passerinii*. *Mol. Genet. Genomics* **269:**1–12.

38. Graia, F., O. Lespinet, B. Rimbault, M. Dequard-Chablat, E. Coppin, and M. Picard. 2001. Genome quality control: RIP (repeat-induced point mutation) comes to *Podospora*. *Mol. Microbiol.* **40:**586–595.

39. Hamann, A., F. Feller, and H. D. Osiewacz. 2000. The degenerate DNA transposon Pat and repeat-induced point mutation (RIP) in *Podospora anserina*. *Mol. Gen. Genet.* **263:**1061–1069.

40. Harrington, T. C., and D. L. McNew. 1997. Self-fertility and uni-directional mating-type switching in *Ceratocystis coerulescens*, a filamentous ascomycete. *Curr. Genet.* **32:** 52–59.

41. Herskowitz, I. 1988. The life cycle of the budding yeast *Saccharomyces cerevisiae*. *Microbiol. Rev.* **52:**536–553.

42. Herskowitz, I. 1987. A master regulatory locus that determines cell specialization in yeast. *Harvey Lect.* **81:**67–92.

43. Herskowitz, I. 1989. A regulatory hierarchy for cell specialization in yeast. *Nature* **342:**749–757.

44. Hicks, J., J. N. Strathern, and A. J. S. Klar. 1979. Transposable mating type genes in *Saccharomyces cerevisiae*. *Nature* **282:**478–483.

45. Inderbitzin, P., J. Harkness, B. G. Turgeon, and M. L. Berbee. 2005. Lateral transfer of mating system in *Stemphylium*. *Proc. Natl. Acad. Sci. USA* **102:**11390–11395.

46. Inderbitzin, P., R. A. Shoemaker, N. R. O'Neill, B. G. Turgeon, and M. L. Berbee. Systematics and mating systems of two fungal pathogens of opium poppy: the heterothallic *Crivellia papaveracea* with a *Brachycladium penicillatum* asexual state and a homothallic species with a *B. papaveris* asexual state. *Can. J. Bot.* **84:**1304–1326.

47. Jacobsen, S., M. Wittig, and S. Poggeler. 2002. Interaction between mating-type proteins from the homothallic fungus *Sordaria macrospora*. *Curr. Genet.* **41:**150–158.

48. Johnson, A. D. 1995. Molecular mechanisms of cell-type determination in budding yeast. *Curr. Opin. Genet. Dev.* **5**:552–558.

49. Kelly, M., J. Burke, M. Smith, A. Klar, and D. Beach. 1988. Four mating-type genes control sexual differentiation in the fission yeast. *EMBO J.* **7**:1537–1548.

50. Koopman, P. 1995. The molecular biology of *SRY* and its role in sex determination in mammals. *Reprod. Fertil. Dev.* **7**:713–722.

51. Lahn, B. T., and D. C. Page. 1999. Four evolutionary strata on the human X chromosome. *Science* **286**:964–967.

52. Leach, J., B. R. Lang, and O. C. Yoder. 1982. Methods for selection of mutants and *in vitro* culture of *Cochliobolus heterostrophus*. *J. Gen. Microbiol.* **128**:1719–1729.

53. Lee, J., T. Lee, Y. W. Lee, S. H. Yun, and B. G. Turgeon. 2003. Shifting fungal reproductive mode by manipulation of mating type genes: obligatory heterothallism of *Gibberella zeae*. *Mol. Microbiol.* **50**:145–152.

54. Leubner-Metzger, G., B. A. Horwitz, O. C. Yoder, and B. G. Turgeon. 1997. Transcripts at the mating type locus of *Cochliobolus heterostrophus*. *Mol. Gen. Genet.* **256**:661–673.

55. Marsh, L., A. M. Neiman, and I. Herskowitz. 1991. Signal transduction during pheromone response in yeast. *Annu. Rev. Cell Biol.* **7**:699–728.

56. Mathieson, M. J. 1952. Ascospore dimorphism and mating type in *Chromocrea spinulosa* (Fuckel) Petch n. comb. *Ann. Bot.* **26**:449–466.

57. McGuire, I. C., R. E. Marra, and M. G. Milgroom. 2004. Mating-type heterokaryosis and selfing in *Cryphonectria parasitica*. *Fungal Genet. Biol.* **41**:521–533.

58. McGuire, J. C., J. E. Davis, M. L. Double, W. L. MacDonald, J. T. Rauscher, S. McCawley, and M. G. Milgroom. 2005. Heterokaryon formation and parasexual recombination between vegetatively incompatible lineages in a population of the chestnut blight fungus, *Cryphonectria parasitica*. *Mol. Ecol.* **14**:3657–3669.

59. Metzenberg, R. L., and N. L. Glass. 1990. Mating type and mating strategies in *Neurospora*. *Bioessays* **12**:53–59.

60. Nakai, K., and P. Horton. 1999. PSORT: a program for detecting sorting signals in proteins and predicting their subcellular localization. *Trends Biochem. Sci.* **24**:34–36.

61. Nelson, R., and T. Hebert. 1960. The inheritance of pathogenicity and mating type crosses of *Helminthosporium carbonum* and *Helminthosporium victoriae*. *Phytopathology* **50**:649.

62. Orbach, M. J., and B. G. Turgeon. 2006. The XXIII Fungal Genetics Conference, March 15–20, 2005, Asilomar Conference Grounds, Pacific Grove, California. *Fungal Genet. Biol.* **43**:669–678.

63. Papazian, H. P. 1956. Sex and cytoplasm in the fungi. *Trans. N. Y. Acad. Sci.* **18**:388–397.

64. Perkins, D. 1987. Mating-type switching in filamentous Ascomycetes. *Genetics* **115**:215–216.

65. Picard, M., R. Debuchy, and E. Coppin. 1991. Cloning the mating types of the heterothallic fungus *Podospora anserina*—developmental features of haploid transformants carrying both mating types. *Genetics* **128**:539–547.

66. Raju, N. B. 1980. Meiosis and ascospore genesis in *Neurospora*. *Eur. J. Cell Biol.* **23**:208–223.

67. Raju, N. B., and D. D. Perkins. 2000. Programmed ascospore death in the homothallic ascomycete *Coniochaeta tetraspora*. *Fungal Genet. Biol.* **30**:213–221.

68. Rau, D., F. J. Maier, R. Papa, A. H. Brown, V. Balmas, E. Saba, W. Schaefer, and G. Attene. 2005. Isolation and characterization of the mating-type locus of the barley pathogen *Pyrenophora teres* and frequencies of mating-type idiomorphs within and among fungal populations collected from barley landraces. *Genome* **48**:855–869.

69. Selker, E. U. 2002. Repeat-induced gene silencing in fungi. *Adv. Genet.* **46**:439–450.

70. Sharon, A., K. Yamaguchi, S. Christiansen, B. A. Horwitz, O. C. Yoder, and B. G. Turgeon. 1996. An asexual fungus has the potential for sexual development. *Mol. Gen. Genet.* **251**:60–68.

71. Shiu, P. K., and R. L. Metzenberg. 2002. Meiotic silencing by unpaired DNA: properties, regulation and suppression. *Genetics* **161**:1483–1495.

72. Shiu, P. K., N. B. Raju, D. Zickler, and R. L. Metzenberg. 2001. Meiotic silencing by unpaired DNA. *Cell* **107**:905–916.

73. Sinclair, A. H., P. Berta, M. S. Palmer, J. R. Hawkins, B. L. Griffiths, M. J. Smith, J. W. Foster, A. M. Frischauf, R. Lovell-Badge, and P. N. Goodfellow. 1990. A gene from the human sex-determining region encodes a protein with homology to a conserved DNA binding motif. *Nature* **346**:240–244.

74. Sugimoto, A., Y. Iino, T. Maeda, Y. Watanabe, and M. Yamamoto. 1991. *Schizosaccharomyces pombe ste11+* encodes a transcription factor with an HMG motif that is a critical regulator of sexual development. *Genes Dev.* **5**:1990–1999.

75. Trail, F., and R. Common. 2000. Perithecial development by *Gibberella zeae*: a light microscopy study. *Mycologia* **92**:130–138.

76. Travis, A., A. Amsterdam, C. Belanger, and R. Grosschedl. 1991. *LEF-1*, a gene encoding a lymphoid-specific protein, with an HMG domain, regulates T-cell receptor α enhancer function. *Genes Dev.* **5**:880–894.

77. Turgeon, B. G. 1998. Application of mating type gene technology to problems in fungal biology. *Annu. Rev. Phytopathol.* **36**:115–137.

78. Turgeon, B. G., H. Bohlmann, L. M. Ciuffetti, S. K. Christiansen, G. Yang, W. Schafer, and O. C. Yoder. 1993. Cloning and analysis of the mating type genes from *Cochliobolus heterostrophus*. *Mol. Gen. Genet.* **238**:270–284.

79. Turgeon, B. G., R. C. Garber, and O. C. Yoder. 1987. Development of a fungal transformation system based on selection of sequences with promoter activity. *Mol. Cell. Biol.* **7**:3297–3305.

80. Turgeon, B. G., and O. C. Yoder. 2000. Proposed nomenclature for mating type genes of filamentous ascomycetes. *Fungal Genet. Biol.* **31**:1–5.

81. Uhm, J. Y., and H. Fujii. 1983. Ascospore dimorphism in *Sclerotinia trifoliorum* and cultural characters of strains from different-sized spores. *Phytopathology* **73**:565–569.

82. Uyenoyama, M. K. 2005. Evolution under tight linkage to mating type. *New Phytol.* **165:**63–70.

83. Voigt, K., A. J. Cozijnsen, R. Kroymann, S. Poggeler, and B. J. Howlett. 2005. Phylogenetic relationships between members of the crucifer pathogenic *Leptosphaeria maculans* species complex as shown by mating type (*MAT1-2*), actin, and beta-tubulin sequences. *Mol. Phylogenet. Evol.* **37:**541–557.

84. Waalwijk, C., O. Mendes, E. C. Verstappen, M. A. de Waard, and G. H. Kema. 2002. Isolation and characterization of the mating-type idiomorphs from the wheat septoria leaf blotch fungus *Mycosphaerella graminicola*. *Fungal Genet. Biol.* **35:**277–286.

85. Wirsel, S., B. Horwitz, K. Yamaguchi, O. C. Yoder, and B. G. Turgeon. 1998. Single mating type-specific genes and their 3′ UTRs control mating and fertility in *Cochliobolus heterostrophus*. *Mol. Gen. Genet.* **259:**272–281.

86. Wirsel, S., B. G. Turgeon, and O. C. Yoder. 1996. Deletion of the *Cochliobolus heterostrophus* mating type (*MAT*) locus promotes function of *MAT* transgenes. *Curr. Genet.* **29:**241–249.

87. Wright, J. M., and G. H. Dixon. 1988. Induction by torsional stress of an altered DNA conformation 5′ upstream of the gene for a high mobility group protein from trout and specific to flanking sequences by the gene product HMG-T. *Biochemistry* **27:**576–581.

88. Yoder, O. C. 1988. *Cochliobolus heterostrophus*, cause of Southern Corn Leaf Blight, p. 93–112. *In* G. S. Sidhu (ed.), *Genetics of Plant Pathogenic Fungi*, vol. 6. Academic Press, San Diego, CA.

89. Yun, S. H. 1998. *Molecular Genetics and Manipulation of Pathogenicity and Mating Determinants in* Cochliobolus heterostrophus *and* Mycosphaerella zeae-maydis. Ph.D. thesis. Cornell University, Ithaca, NY.

90. Yun, S. H., T. Arie, I. Kaneko, O. C. Yoder, and B. G. Turgeon. 2000. Molecular organization of mating type loci in heterothallic, homothallic, and asexual *Gibberella/Fusarium* species. *Fungal Genet. Biol.* **31:**7–20.

91. Yun, S. H., M. L. Berbee, O. C. Yoder, and B. G. Turgeon. 1999. Evolution of the fungal self-fertile reproductive life style from self-sterile ancestors. *Proc. Natl. Acad. Sci. USA* **96:**5592–5597.

92. Yun, S. H., and B. G. Turgeon. 1999. Molecular comparison of mating-type loci and adjacent chromosomal regions from self-fertile and self-sterile *Cochliobolus* species. *Plant Pathol. J.* **15:**131–136.

93. Zickler, D. 1973. *La méiose et les mitoses au cours du cycle de quelques ascomycàetes.* Thèse d'Etat. Université Paris-Sud, Orsay, France.

94. Zickler, D., S. Arnaise, E. Coppin, R. Debuchy, and M. Picard. 1995. Altered mating-type identity in the fungus *Podospora anserina* leads to selfish nuclei, uniparental progeny, and haploid meiosis. *Genetics* **140:**493–503.

Sex in Fungi: Molecular Determination
and Evolutionary Implications
Edited by Joseph Heitman et al.
© 2007 ASM Press, Washington, D.C.

Paul S. Dyer

7

Sexual Reproduction and Significance of *MAT* in the Aspergilli

The genus *Aspergillus* comprises over 180 species that are united by the presence of the "aspergillum," which consists of a specialized, enlarged conidiophore bearing phialides and characteristic radiating asexual conidia (55, 61). The aspergilli have been one of the most studied fungal groups, as its members include a series of species of key importance in the food, biotechnology, and medical sectors, in addition to the model organism *A. nidulans*, which has long been used for laboratory physiological and genetic studies (20, 57). They also have a worldwide distribution and are one of the most ubiquitous and abundant of all groups of fungi (3, 59). It has more recently been recognized that the aspergilli provide an ideal opportunity to investigate the genetic basis of reproductive mode in fungi because the genus includes both known asexual and sexual species—with the latter exhibiting either homothallic (self-fertilizing) or heterothallic (obligate outcrossing) breeding systems. Most of these species are believed to have arisen from a common ancestor, as the aspergilli form an essentially monophyletic group (although the confidence level is comparatively low and there are exceptions) (71). Hence, it is of particular interest to elucidate the molecular basis for changes in reproductive state.

TAXONOMY OF THE ASPERGILLI AND THEIR PERFECT STATES

Anamorphic *Aspergillus* species are associated with eight holomorphic genera, which span the historic subgenera of *Aspergillus* (21, 61). Only three of these, *Emericella*, *Eurotium*, and *Neosartorya*, include more than three species (55) (Table 7.1). Of the 184 species of *Aspergillus*, sexual states are known for ca. 70 species (55). Of these, the vast majority exhibit homothallic breeding systems, with only four heterothallic species described: one from the genus *Emericella* (*E. heterothallica*) (41) and three from the genus *Neosartorya* (*N. fennelliae*, *N. spathulata*, and *N. udagawae*) (31, 74, 76). However, it is important to note that some homothallic aspergilli retain the ability to outcross, even with isolates of different vegetative compatibility groups (13), and thus, they are not restricted to self-fertilization. One study involving *Emericella nidulans* indicated that, at least for this species, propagation is primarily by clonal

Paul S. Dyer, School of Biology, University of Nottingham, University Park, Nottingham NG7 2RD, United Kingdom.

Table 7.1 Ascomycetous teleomorphs of *Aspergillus* subgenera

Teleomorph	No. of species[a]	Representative species and related asexual species[b]
Chaetosartorya	3	*C. cremea, C. chrysella, A. wentii*
Emericella	28	*E. nidulans, A. sydowii, A. versicolor*
Eurotium	19	*E. repens, E. rubrum, A. restrictus*
Fennellia	3	*F. nivea, F. flavipes, A. terreus*
Hemicarpenteles	2	*H. acanthosporus, A.clavatus*
Neosartorya	12	*N. fischeri, N. fennelliae, A. fumigatus*
Petromyces	2	*P. alliaceus, A. flavus, A. oryzae, A. niger[c]*
Sclerocleista	2	*S. ornata, A. sparsus*

[a]Data compiled from Pitt et al. (55).

[b]Asexual species of economic, ecological, or medical importance. Relatedness based on phylogenetic analysis of large subunit ribosomal RNA gene sequence (54).

[c]Relatedness has relatively low support (54, 71).

means but recombination events are frequent enough to disrupt the stable maintenance of clonal genotypes (22).

Strictly speaking, in cases where a teleomorph has been identified, then the species holomorph should be known by this name, but this ICBN (International Code of Botanical Nomenclature) convention has rarely been followed by *Aspergillus* researchers as witnessed by the continued use of "*Aspergillus nidulans*" rather than "*Emericella nidulans*" in most published works. The teleomorph name is used where relevant in this chapter. Phylogenetic relationships between the various teleomorphic genera have been described by both Peterson (54) and Tamura et al. (71).

MORPHOLOGICAL ASPECTS OF SEXUAL DEVELOPMENT

Ascomata development in the family *Trichocomaceae*, which includes the aspergilli, has been described by Benjamin (2) and Malloch and Cain (46). It is particularly intriguing to note the diversity of morphological form shown by the various teleomorph states, given their supposed common origin. Almost all of the sexual aspergilli form asci within enclosed maternal tissues (generally termed cleistothecia), but the extent of tissue development varies considerably. *Neosartorya* species develop asci within loose pseudoparenchymatous wall tissue several layers thick, while *Eurotium* species typically form ascomata with a distinct peridium wall composed of a single layer of large, irregular cells free of adjoining tissues. In contrast, *Emericella* species develop ascomata with walls of compacted, interwoven hyphae several layers thick surrounded by masses of thick-walled "Hülle" cells, and the ascomata of *Petromyces* species develop within compact, hardened masses of hyphal stromata (46, 59). It has been speculated that the ancestral type of ascomata may have been characterized by a

well-developed pseudoparenchymatous wall enclosing the asci, borne within a loose hyphal stroma. There has then been gradual simplification and reduction of the peridium, together with either specialization or loss of stroma tissues during evolution (46). At the most extreme, some aspergilli (e.g., *Eurotium athecius* [formerly classified as *Edyuillia athecia*]) produce naked asci that develop from ascogonial coils while the adjacent mycelium remains undifferentiated (54, 59). Most aspergilli exhibit species-specific ornamentation of the ascospore wall, with some exquisite forms evident, although how this deposition is determined remains a mystery.

Particular attention has been focused on the morphology of sexual development in *E. nidulans* as reviewed in depth by Champe et al. (10) and also described by Zonneveld (84) and Braus et al. (5). In addition, a detailed ultrastructural study of cleistothecium development has been reported (68). Briefly, sexual development requires growth on a surface with exposure to air and is time dependent following asexual sporulation. The sexual cycle is favored by incubation in the dark and sealing of plates (after an initial period of ca. 24 h of asexual growth, unsealed) to limit air exchange. The first stage in differentiation is the appearance of cleistothecial initials, being coiled lumps of cells ca. 6 μm in diameter, which then undergo further coiling and expand to around 10 μm in diameter. Glassy clusters of globular Hülle cells simultaneously develop that cluster around and envelop the developing cleistothecia. Of particular relevance to this chapter is the repeated failure to detect the presence of trichogynes (chemotactic hyphae involved in pheromone response [12]) or differentiated ascogonia and antheridia in this homothallic species. There is then rapid development and thickening of the peridium as hyphae become flattened and fuse to form a dense, interwoven layer with spaces between the hyphae filled with a poorly characterized noncellular, electron-dense substance termed "cleistin." Once the cleistothecia reach ca. 100 μm in diameter, a centrum containing ascogenous hyphae and crozier tips becomes visible. It has been suggested that Hülle cells may act as "nurse" cells to the developing cleistothecia, for instance, supplying laccase to enable production of the cleistin matrix, which might be a polymeric product of phenolic material (10), or being involved in energy release though α-(1,3)-glucan breakdown (80). Indeed, it is the nest-like aggregation of hyphae around the developing cleistothecia that led to the naming of the species (from the Latin *nidulans*, meaning nest former) (5). Finally, mature cleistothecia may reach 170 to 200 μm in diameter and contain thousands of asci (Fig. 7.1). The nest hyphae later degenerate and detach from the mature cleistothecial envelope. Fertility, in terms of cleistothecial abundance, varies from strain to strain (13), but standard wild types may produce about 3,000 cleis-

tothecia per cm^2 on suitable media (10). Octad analysis has demonstrated that cleistothecia are not necessarily the result of a single fertilization event but may arise from two or more fertilizations or the incorporation of fertilization coils found close together (29).

GENETIC AND PHYSIOLOGICAL DETERMINANTS OF SEXUAL DEVELOPMENT

There is considerable interest in determining the genetic and physiological processes that govern sexual development in fungi (see elsewhere in this volume), and *E. nidulans* has received much attention in this respect as a model filamentous ascomycete. Research efforts have been facilitated over recent years by the release of genome sequence data and the availability of tools for functional genomic experimentation (1). A series of genes have been

Figure 7.1 Massed cleistothecia of *E. nidulans* (diameter, 150 to 200 μm) visible as dark spheres surrounded by lighter, smaller, glistening Hülle cells. Inset shows higher magnification; note resemblance to certain jeweled eggs produced by Carl Fabergé in the late 19th and early 20th centuries.

identified as being involved in various aspects of sexual development in *E. nidulans* as follows (Fig. 7.2).

Perception of Environmental Signals

Sexual sporulation is favored, and asexual sporulation reduced, by incubation in the dark. Conversely, the onset of sexual development is delayed or repressed, and asexual conidiation favored, by exposure to red light during early stages of growth (49). Sexual reproduction is also repressed by adverse environmental conditions such as high salt or low nutrient concentration. This is likely to link to the ecology of *E. nidulans*, which is a soil organism. Thus, asexual sporulation and consequent wind dispersal of conidia would be induced by light at the soil surface, whereas the sexual cycle leading to production of environmentally resistant ascospores (57) might be advantageous within the soil and therefore be triggered by darkness. The following genes have been identified as being involved with perception of such environmental signals in *E. nidulans*.

FphA

A gene for a red-light-sensing phytochrome termed FphA has been identified, which is thought to be involved with repression of sexual development in the light because this block was significantly reduced, al-though not totally abolished, in Δ*fphA* gene deletion mutants (4). Green fluorescent protein (GFP)-tagging of the protein indicated localization to the cytoplasm.

VeA

FphA was suggested to act upstream of another key environmental sensor, the so-called "velvet" *veA* gene. In contrast to FphA, the VeA gene product activates sexual development and is a negative regulator of asexual development in the dark (the effect is inhibited by light). Overexpression of *veA* resulted in sexual morphogenesis under high salt and liquid conditions normally unfavorable for sex, whereas gene deletion resulted in the total inability to undergo sexual development even under conditions favorable for sex (35). It has been suggested that among other activities VeA interacts with FluG, a likely extracellular development signal for asexual conidiation (5). Interestingly VeA localization seems to be light dependent. There is evidence that VeA is found abundantly in the cytoplasm during light exposure but is located mainly in the nucleus in the dark, i.e., apparently a nuclear-cytoplasmic shuttling occurs, dependent on illumination (69; A. M. Calvo, personal communication; S. Krappmann, O. Valerius, and G. H. Braus, unpublished data). Furthermore, a VelB (velvet-like) protein interacting with VeA has been discovered. Deletion of *velB* results in the same phenotype as *veA* deletion, i.e., total

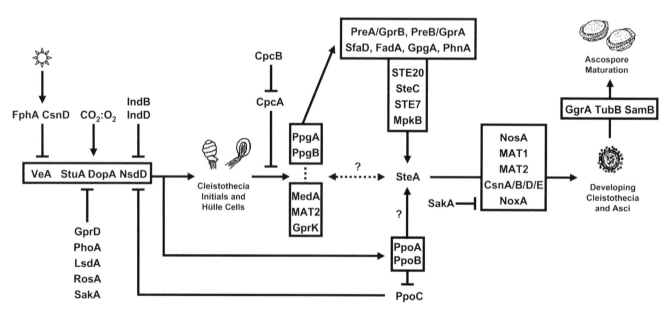

Figure 7.2 Hypothetical genetic model showing pathways involved with control of cleistothecial development in *E. nidulans*. See the text for details of particular genes encoding regulatory factors. Gene products are enclosed in the same box when the particular order of gene activity (or nature of interaction) is unknown or has not yet been experimentally verified. Cleistothecial initials refer to the very earliest stage of cleistothecial development.

lack of sexual development. VelB localization is dependent on VeA, indicating the possible existence of a velvet protein complex which regulates cleistothecial formation by transmitting external stimuli to the nucleus (Krappmann et al., unpublished).

It is noted that a partial mutation in *veA*, resulting in a truncated protein lacking the first 37 amino acids (due to a single nucleotide difference in the initiation codon), is present in many laboratory stock cultures. This *veA1* mutation results in a lack of red-light sensitivity, leading to increased asexual sporulation and reduced sexual development in the dark (35, 49). The mutation also largely eliminates the formation of aerial hyphae during asexual growth, whereas these are produced by the wild-type *veA* (giving the latter a velvety appearance), and blocks the formation of cleistothecia at 42°C and can thus be considered a thermosensitive mutation (49). Suppressors of the *veA1* mutation have been described (10). Significantly, VeA1 was found predominantly in the cytoplasm independent of illumination—most likely due to nonfunctionality of a nuclear localization signal in the truncated protein (69). Due to the fact that *veA1* strains are so widespread and often contain many useful phenotypic markers, much genetic manipulation work is by necessity performed with strains of a *veA1* background. It is cautioned that this may introduce complication(s) with respect to interpreting effects on sexual reproduction and fertility levels on genes involved with sexual reproduction in *E. nidulans*; for example, the reduction in red-light sensitivity seen in Δ*fphA* mutants was observed only in a *veA* wild-type background (4). Meanwhile, a mutant screen has been reported which identified a series of isolates able to undergo sexual development in the light. The genetic basis has yet to be determined for most, but one *silA* mutant showed disruption of a zinc-finger motif gene (36).

LsdA, PhoA, CpcA, and CpcB

Other genes involved in environmental sensing include *lsdA*, which inhibits sexual development under high salt concentrations (44); *phoA*, a cyclin-dependent kinase that suppresses sexual development under low-phosphorus conditions (8); and the cross-pathway-control genes *cpcA* and *cpcB*, which are involved in sensing of amino acid levels (30). CpcA suppresses development at the microcleistothecium stage under limiting amino acid levels, whereas CpcB represses *cpcA* expression in the presence of amino acids, thereby allowing sexual development. Consistent with the latter findings, it is noted that mutant strains deficient in certain amino acid biosynthetic pathways (e.g., *argB*, *trpB*, and *hisB*) are unable to undergo the sexual cycle (10).

Signal Transduction Pathways

In common with other fungi, *E. nidulans* contains a series of signal transduction pathways that allow response to signals from the external environment (20). For relevance to the topic of sexual reproduction, discussion is now focused first on pathways involved with pheromone response and then on other signaling factors that may have an impact on sexual development.

Pheromone Production and a MAP Kinase Signal Transduction Pathway

Heterothallic ascomycetes are able to detect partners of the opposite mating type by the presence of mating-type-specific diffusible peptide pheromones. Binding of a hydrophilic α-type pheromone, or a hydrophobic a-type pheromone, to a cognate receptor triggers the dissociation of a coupled heterotrimeric G protein, resulting in signal transmission through a mitogen-activated protein (MAP) kinase cascade ultimately targeting a homeodomain transcription factor (19). The result is twofold: cell cycle arrest and activation of mating-specific genes (19, 45). This signaling process has been best characterized in *Saccharomyces cerevisiae*, but there is accumulating evidence that components similar to those in yeasts are highly conserved across heterothallic fungal taxa and are responsible for chemotactic growth in filamentous fungi (9, 12, 45, 56; see also the results of K. Borkovich and colleagues on *Neurospora crassa* described in the work of M. Paoletti, F. A. Seymour, M. J. C. Alcocer, K. Navgeet, A. M. Calvo, D. B. Archer, and P. S. Dyer [submitted for publication]).

Given the role of pheromone production and signaling in partner detection, it is therefore perhaps surprising that similar elements appear to be involved in sexual reproduction of the self-fertile *E. nidulans*. Experimental methods were initially used to identify putative genes associated with a pheromone-signaling pathway, and more recently genome data have been used extensively to identify signaling components. Sequential stages of the putative pheromone-signaling pathway are described below.

Pheromone Precursors PpgA and PpgB

Fungal pheromones are produced from precursor products which undergo enzymatic processing to yield active yeast α- (hydrophilic) or a- (hydrophobic) factor-like pheromones. BLAST searching of the newly released *E. nidulans* genome revealed the presence of a putative pheromone precursor gene (*ppgA*) encoding a PPGA precursor product that was predicted to yield two shorter α-factor-like pheromones PPGA(An1) and PPGA(An2) following enzymatic processing (18). PPGA(An1) and

PPGA(An2) exhibit slight divergence in amino acid sequence. A series of genes involved with the processing and efflux of both α- and a-factor pheromones were also identified (18, 20). No precursor gene for a hydrophobic a-factor-like pheromone could be detected. However, such genes are very short (likely <100 bp) and highly divergent so a gene(s) may have been present but missed due to low homology. A detailed bioinformatic screening of the *Aspergillus niger* genome has revealed various candidate a-factor-like *ppgB* pheromone precursor genes, and it is hoped that information arising may help identify a homologue from *E. nidulans* (P. S. Dyer and H. Roubos, unpublished data).

Pheromone Receptors PreA(GprB) and PreB(GprA)

BLAST searching of the newly released *E. nidulans* genome also revealed the presence of two genes encoding putative G-protein-coupled pheromone receptors. These genes were termed *preA* and *preB* (according to convention used elsewhere [56]) and exhibited similarity to *S. cerevisiae STE3* a-receptor and *STE2* α-receptor family genes, respectively; i.e., PreB would be the predicted receptor for *ppgA*-encoded pheromones, whereas PreA would be the receptor for any *ppgB*-encoded pheromone(s). Experimental characterization of *preA* and *preB* was then undertaken by Seo et al. (65) who cotermed the genes *gprB* (= *preA*) and *gprA* (= *preB*) to be consistent with their other studies (28). Deletion of the individual receptor genes had no appreciable effect on vegetative growth rate, asexual sporulation, or the development of Hülle cells. In contrast, there was a marked effect on sexual fertility, with only a few, small cleistothecia produced, which had significantly reduced numbers of ascospores, although these remained viable. Meiotic crosses were then used to generate double Δ*gprA*(*preB*)Δ*gprB*(*preA*) deletion strains, which were again found to produce Hülle cells but were now completely sterile, being unable to produce cleistothecia. These results indicated that GprA(PreB) and GprB(PreA) are required for self-fertilization, although all the various deletion mutants remained able to outcross in forced heterokaryons (65). Upregulation of *nsdD* (a transcription factor required for sex, described in "NsdD" below) in Δ*gprA*(*preB*) or Δ*gprB*(*preA*) single-deletion mutants only partially restored the developmental defect, while upregulation in the double Δ*gprA* (*preB*)Δ*gprB*(*preA*) deletion mutant caused production of large cleistothecia that were immature and defective in ascospore production. Collectively, these results suggested an early pivotal role in initiation of cleistothecial development for NsdD, whereas GprA/B(PreB/A) signaling

might be required for later stages of morphogenesis, karyogamy, meiosis, and ascosporogenesis. It was also noted that levels of *nsdD* mRNA in GPCR deletion mutants after sexual induction were lower than in the wild type, suggesting that GprA/B(PreB/PreA) are required to maintain elevated levels of *nsdD* expression (though no effect on *steA* transcript abundance was seen).

G-Protein Subunits FadA, SfaD, and GpgA

G proteins are heterotrimeric units composed of alpha (α), beta (β), and gamma (γ) subunits. Two G-protein subunits, an α-subunit encoded by a gene termed *fadA* (mnemonic for fluffy autolytic dominant; related to the class of "fluffy" mutants exhibiting undifferentiated aerial hyphae and sterility) and a β-subunit encoded by *sfaD*, were first shown to be required for cleistothecial development, although Hülle cells continued to be formed in both Δ*fadA* and Δ*sfaD* deletion strains under conditions favorable for sex (60). Indeed, the Δ*sfaD* strain exhibited a 22-fold increase in Hülle cell formation (a double Δ*fadA*Δ*sfaD* strain showed a 13-fold increase). In a later study, Seo et al. (66) used genomic data to identify a putative G-protein γ-subunit encoded by a gene termed *gpgA*. Deletion of the gene resulted in restricted vegetative growth and reduced asexual sporulation together with the failure to form cleistothecia and severely reduced ability to outcross. However, as with the Δ*sfaD* G-protein β-subunit deletion (60), the resulting strain showed significantly higher production of Hülle cells than did the wild type. These observations lead to the conclusion that the G-protein subunits SfaD, GpgA, and most likely FadA are all required for normal sexual fruiting-body formation. It was also noted that SfaD and GpgA were apparently the only G-protein β- and γ-subunits present in the *E. nidulans* genome (whereas at least three α-subunits were present). This raised the question of how signaling specificity could be achieved given the presence of multiple G-protein-coupled receptors (GPCRs), and it was speculated that this might be explained by the presence of varying levels of GPCRs during the life cycle allowing a differential response; for example, under conditions favorable for sex, levels of the pheromone receptors GprA(PreA) and GprB(PreA) might increase considerably, triggering a subsequent signaling cascade leading to cleistothecial development rather than any other response (66).

Most recently a phosducin-like protein, PhnA, has been identified which is required for correct Gβγ-mediated signaling (67). Deletion of *phnA* resulted in a phenotype almost identical to that of *sfaD*, i.e., reduced biomass, altered asexual sporulation, and a block in fruiting-body formation (although enhanced Hülle cell formation was noted). However, there was no clear impact on *nsdD* expression. Other proteins from this family are essential positive regulators of Gβγ signaling via their roles as molecular chaperones for Gβγ assembly (67).

Members of a MAP Kinase Cascade—STE20, SteC, STE7, MpkB, SteA, and STE50

A combination of experimental and genomic approaches has identified an entire MAP kinase signaling cascade linked to pheromone sensing in *E. nidulans*. The first element to be identified was *steA*, which encodes a homeodomain protein related to *S. cereviseae* Ste12p (the end target of the yeast pheromone-signaling MAP kinase cascade). Deletion of *steA* resulted in sterility, with the lack of cleistothecia or ascogenous hyphae, although Hülle cells were formed (75). SteA contains both N-terminal homeodomain and C-terminal zinc-finger domains, and it has been suggested that the zinc-finger domain may interact act with the VeA protein to regulate sexual fertility (47). Next, a homologue of the *S. cerevisiae* MAP kinase kinase kinase (MAPKKK) *STE11* was cloned and characterized. The gene *steC* was found to be required for initiation of cleistothecium development, although Hülle cells were again formed throughout the colony (79). Further extensive BLAST searches later identified the "missing" elements of the pathway—namely, homologues of the *S. cerevisiae* MAPKKKK *STE20*, MAPKK *STE7*, and MAPK *FUS3* gene (later termed *mpkB*)—in the *E. nidulans* genome (20). The whole pathway was then the subject of a postgenomic study by Paoletti et al. (submitted). A semiquantitative reverse transcription (RT)-PCR approach was used to monitor the expression of key genes with possible involvement in a pheromone response pathway during sexual development of *E. nidulans*. Expression of these genes was barely detectable from unsealed plates, indicating low-level expression under conditions favoring vegetative growth and asexual conidiation. In contrast, strong expression of all representative genes of the pheromone response pathway was detected in extracts from sealed plates, in which sexual development was occurring. This included markedly increased expression of the putative α-pheromone precursor (*ppgA*), both a- and α-factor-like pheromone receptors [*preA*(*gprB*) and *preB*(*gprA*), respectively], the subunits of a heterotrimeric G protein (*fadA*, *sfaD*, and *gpgA*), three kinases of the MAP kinase cascade (*steC*, the *STE7* equivalent from *S. cerevisiae*, and *mpkB*), and a protein kinase regulator (*STE50* equivalent), as well as the target homeodomain transcription factor (*steA*). Expression of the actin control gene remained constant. Paoletti et al. (submitted) also deleted the intermediate *mpkB* MAP kinase gene (equivalent of *S. cerevisiae*

FUS3) as part of the study. This was found to result in total sterility, i.e., the lack of both Hülle cells and cleistothecia.

The data described are of significance because they indicate that pheromone signaling, normally associated with sex in outcrossing heterothallic species, is also required for correct sexual development of the self-fertile *A. nidulans* (Paoletti et al., submitted). However, the precise role of pheromone signaling remains unknown and the cellular and hyphal location of gene and pheromone expression has yet to be determined. Thus, it is unclear whether pheromones are involved in cell-to-cell fusion (recall that there is no evidence of trichogyne formation in *E. nidulans*) or some other role such as nuclear recognition (14). It may be speculated that suitable environmental conditions may trigger pheromone production and release in *A. nidulans*, leading to activation and reinforcement of the signaling cascade through increased gene transcription and resultant sexual development. It is also conceivable that observed expression changes may be partly attributable to hyphal anastomosis in the developing cleistothecia (51), and it is cautioned that deletion of certain of the MAP kinase elements can have pleiotropic effects. Pheromone precursor and receptor genes have also been shown to be required for normal sexual development in the homothallic *Sordaria macrospora* (see chapter 10 by S. Pöggeler).

Additional Signal Transduction Pathways

In addition to the a- and α-factor GPCRs, Han et al. (28) identified a further seven GPCR-encoding genes. There was no clearly distinguishable phenotype arising from disruption of three of these genes (*gprC*, *gprE*, and *gprG*). In contrast, deletion of a fourth gene, *gprD*, resulted in restricted vegetative hyphal growth and uncontrolled activation of sexual development leading to the formation of small colonies densely packed with cleistothecia. Thus, it was concluded that the GprD-mediated signaling cascade negatively regulates sexual development and is required for proper proliferation of *E. nidulans*. Consistent with this, *nsdD* transcripts were very elevated in a Δ*gprD* deletion strain. However, the binding ligand(s) and downstream activators were not identified, although it was speculated that a protein kinase C signal pathway was involved (28). Analysis of combined *gprA*(*preB*), *gprB*(*preA*), and *gprD* mutants indicated that GprD functions upstream of GprA/B(PreB/PreA) (65). In ongoing work, a total of 16 GPCRs have now been identified and examinations of mRNA transcript levels during the life cycle of *E. nidulans* have been made. This has led to the detection of a further receptor "GprK" that may be necessary for sexual fruiting body formation. Deletion of *gprK* resulted in a phenotype similar to that caused by deletion of *sfaD* or *phnA*, i.e., highly enhanced accumulation of Hülle cell aggregates but a block in further cleistothecial development. Transcripts of *gprK* mRNA were not detectable in conidia or ascospores but were highly expressed during vegetative growth and sexual development (J. H. Yu, personal communication).

Finally, other signal transduction pathways are able to influence sexual development in *E. nidulans*. Kawasaki et al. (34) identified a gene, *sakA*, with homology to the MAP kinase *HOG1* from *S. cerevisiae*, which is the target of a signaling pathway involved with detection of environmental osmotic and oxidative stress. It was found that a Δ*sakA* mutant displayed precocious sexual development and produced at least twofold-increased numbers of cleistothecia compared to the wild type under conditions promoting the sexual cycle. However, this effect was dependent on *steA* with a Δ*sakA*Δ*steA* double deletion being unable to form cleistothecia (although large numbers of Hülle cells were produced). This led to the conclusion that SakA may be a repressor of *steA*-dependent sexual development under environmentally unfavorable conditions, reminiscent of cross talk between the Hog1 pathway and pheromone response in yeast.

Role of Transcription Factors/Regulatory Proteins

A diverse range of regulatory proteins with probable transcription factor activity have also been shown to be involved with sexual development in *E. nidulans*, although the precise nature of interaction of these factors has yet to be determined. These are listed below.

StuA and MedA

The developmental modifiers StuA (which contains a conserved APSES domain found in several yeast proteins, with characteristics of helix-turn-helix transcription factors) and MedA (no clearly conserved domain) are best known for their role in the control of asexual conidiation (11). Mutation of *stuA* results in stunted conidiophores lacking metulae and phialides (11, 16, 82), while *medA* mutant strains exhibit delayed differentiation of phialides and conidia, resulting in the proliferation of branching chains of sterigmata (6, 11). However, these proteins also regulate sexual development. *medA* mutants produce only unorganized masses of Hülle cells and no cleistothecia or ascogenous tissues (6, 11), whereas mutation in *stuA* results in total sterility with the failure to form cleistothecia, ascopores, or Hülle cells (11, 82). It has been suggested that StuA and MedA help to establish correct spatiotemporal gradi-

ents of signaling molecules to allow appropriate developmental responses and that StuA may function as an activator of the sexual cycle or may block expression of a repressor of sexual development (6, 16).

DopA

Sexual morphogenesis has been shown to require the activity of the protein DopA, a member of a leucine zipper-like protein family. Sexual reproduction was completely abolished in a Δ*dopA* deletion strain, which also showed greatly reduced asexual conidiation (53). It was suggested that DopA acts upstream of the transcription factor SteA (described above).

NsdD

As introduced earlier, a putative GATA-type zinc-finger transcription factor NsdD was identified from a screen of UV-induced *nsd* (mnemonic for never in sexual development) sterile mutants that is essential for sex. A Δ*nsdD* deletion strain exhibited retarded hyphal growth, premature asexual conidiation, and the failure to produce cleistothecia or Hülle cells, even under conditions favorable for sex (although a few fertile cleistothecia were formed from a point inoculum) (27). In contrast, overexpression of *nsdD* led to a dramatic increase in numbers of cleistothecia and morphogenesis under conditions normally repressing sex (e.g., cleistothecia formed on high-salt media, and Hülle cell formation occurred in submerged culture). NsdD was concluded to act as a positive activator of sex, rather than a repressor of asexual development. Multicopy expression of *nsdD* was also able to overcome the reduced fertility effect of the *veA1* mutation, suggesting that *nsdD* acted downstream of *veA* or was in an overlapping pathway. A yeast two-hybrid screen was later used to search for proteins interacting with NsdD, and at least two Interactor of NsdD proteins, "IndB" and "IndD," were identified (42). These exhibited possible VeA-dependent expression (repressed in *veA* background, but highly induced in Δ*veA* mutant), suggesting a possible role in inhibition of NsdD function possibly due to binding to the zinc-finger region, thereby preventing DNA binding and transcription by NsdD. Meanwhile, analysis of several allelic mutants of *nsdD* producing truncated polypeptides suggested that expression of *nsdD* is subject to negative autoregulation and repressed by high-salt conditions (26).

NosA and RosA

A C6 zinc-finger protein-encoding gene, *pro1*, has been described from *S. macrospora*, which is required for sexual development beyond the protoperithecial stage (see chapter 10 by S. Pöggeler). Two related genes have since been identified from *E. nidulans*, but with very contrasting roles in fruiting-body formation. The first *nosA* (mnemonic for number of sexual spores) has a role similar to that of *pro1*, being involved with activation of fruiting-body development (77). A Δ*nosA* deletion strain exhibited wild-type asexual growth and developed possible primordia (in a *veA* background). However, for the most part further development was blocked, with hardly any Hülle cells produced—although very occasionally small cleistothecia were formed that contained ascospores. In contrast, a second gene, *rosA* (for repressor of sexual development), was shown to inhibit sexual development under certain culture conditions (78). A Δ*rosA* deletion strain showed little difference with respect to the wild type in timing of sexual development under conditions favorable for growth (a slight enhancement in numbers of cleistothecia was seen in a *veA*, but not *veA1*, background). However, the Δ*rosA* strain was also able to undergo sexual development under low-glucose and high-osmolarity conditions, which repressed sex in the wild type, and Hülle cells were produced under submerged culture. NosA and RosA shared 43% amino acid identity, and 44 and 38% homology to Pro1, respectively (77). Interestingly, NosA and RosA may be paralogues because *Aspergillus* species appear to be the only Ascomycetes taxa with two proteins with similarity to Pro1: screening of other ascomycete genomes reveals only single orthologues. Vienken et al. (77, 78) undertook further work to try and assess the relationship(s) between NosA/RosA and other transcription factors. A strain was constructed with constitutive expression of *nsdD* in a Δ*nosA* background. Only primordia were formed, suggesting that NosA most likely acts downstream of NsdD, or in a parallel pathway. Further evidence that NosA acts at a later stage in development came from the finding that overexpression of *nosA* had no obvious developmental phenotype (77). This was consistent with the possibility that VeA, NsD, and RosA act early in development to control genes inducing sexual primordia (Fig. 7.2), whereas NosA was involved with maturation of already existing primordia, i.e., NosA was unable to trigger de novo development of primordia. Meanwhile, transcript levels of *nsdD*, *veA*, and *stuA* were increased in a Δ*rosA* deletion strain, and preliminary experimentation suggested that RosA represses, directly or indirectly, NosA function (77).

COP9—CsnD and CsnE

A further regulator of early sexual differentiation is the COP9 signalosome. This is a conserved multiprotein complex found throughout the eukaryota, where it is known to have various roles in developmental processes.

Two components have so far been characterized from *E. nidulans*, encoded by the genes *csnD* and *csnE* (7). These appear to have at least two distinct roles in sexual morphogenesis. First, deletion of *csnD* abolished the repression of light on sexual development; i.e., sexual development was initiated independent of the light signal. This indicated that the COP9 signalosome is essential for light-dependent signaling and suppression of sex in the light (possibly by a negative posttranscriptional effect on VeA expression?). In contrast, deletion of *csnD* secondly resulted in a block in development after the formation of cleistothecial primordia, although Hülle cells were formed as normal. Other effects on vegetative growth and secondary metabolism were also evident. A similar phenotype was seen in Δ*csnE* deletion strains. Thus, the COP9 elements appear to both inhibit sexual development in the light and at the same time be required for maturation of cleistothecia. These independent functions are supported by the observation that overexpression of VeA in a Δ*csnD* background failed to overcome the block in development at the primordial stage (7). Work is now under way to characterize other subunits of the signalosome, with genes *csnA* through *H* encoding a total of eight subunits having been identified. Deletion of *csnA* and *csnB* has been found to result in the same block in sexual development at the stage of primordia as seen in Δ*csnD* and Δ*csnE* strains (Hülle cells were formed), providing evidence that a correctly assembled signalosome is required to link protein turnover to fungal sexual development (S. Busch, E. U. Schwier, K. Nahlik, O. Bayram, O. Draht, K. Helmstaedt, S. Krappmann, W. N. Lipscomb, O. Valerius, and G. H. Braus, unpublished data).

MAT1 and MAT2

Most recently, mating-type (*MAT*) genes have been shown to be involved in sexual development of *E. nidulans*. As described elsewhere in this volume (see particularly chapter 6 by B. G. Turgeon and R. Debuchy), *MAT* genes are best known as master loci controlling sexual development in heterothallic fungi, in which they are responsible for conferring mating type. By convention, *MAT1-1* (abbreviated to *MAT-1*) ascomycete isolates contain a conserved alpha (α)-domain protein-encoding *MAT* gene, whereas *MAT1-2* (abbreviated to *MAT-2*) isolates contain a conserved high-mobility group (HMG)-domain protein-encoding *MAT* gene (15). The presence of these DNA binding domains is consistent with the role of *MAT* proteins as transcriptional activators (9). Gene disruption studies have confirmed that *MAT* genes are required for normal sexual development in heterothallic species (15). Given their initial identification through their role in outcrossing, it is significant that there is now accumulating evidence that *MAT* genes also play critical roles in the development of homothallic fungi. In the case of *E. nidulans*, experimental means were first used to identify an HMG *MAT2* family mating-type gene from this species (17a; Paoletti et al., submitted). Genomic analysis was subsequently used to identify an α-domain *MAT1* family mating-type gene (18). Interestingly, these genes were found to be unlinked, unlike the situation in most other homothallic fungi characterized to date (see chapter 6 by Turgeon and Debuchy), and were therefore termed *MAT2* and *MAT1*, respectively, to recognize the location at different loci within the genome, following proposed nomenclature for *MAT* genes (15; Paoletti et al., submitted). This observation has implications for the evolution of breeding systems within the aspergilli (see "*MAT* and the Evolution of Homothallism and Heterothallism" below). Homologues of the *MAT1-1-2* and *MAT1-1-3* genes found in some euascomycete taxa (15) were not found at the *E. nidulans MAT* loci.

MAT1 and *MAT2* were then subjected to experimental characterization as described by Paoletti et al. (submitted). First, RT-PCR studies demonstrated that both *MAT* genes were expressed at a low level during vegetative growth and asexual conidiation but showed marked increases in expression following sealing of plates and concomitant sexual development. Second, overexpression of both *MAT1* and *MAT2* genes together resulted in a suppression of vegetative growth and stimulation of sexual differentiation (leading to formation of mature cleistothecia) under conditions unfavorable for sex. This effect was not observed in strains overexpressing either *MAT1* or *MAT2* alone, indicating that balanced *MAT* gene expression is required for correct sexual development. Third, gene disruption resulted in various blocks in sexual development. Δ*MAT2* deletion strains formed numerous Hülle cells and limited numbers of cleistothecia, but the latter were sterile and smaller in size than those formed by the parental control. In parallel with this observation, inhibition of MAT2 activity by antisense RNA manipulation (expression of an inverted *MAT2* copy within the same genome) resulted in much delayed sexual development. Meanwhile, disruption of *MAT1* also resulted in a block in the formation of ascospores. The gene disruption results were similar to the findings of a preliminary report by Miller et al. (48), who also indicated that MedA might influence *MAT* expression. Finally, it was noted that overexpression of the *MAT* genes did not result in increased expression of either the pheromone precursor gene *ppgA* or the pheromone receptor genes *preA*(*gprB*) and *preB*(*gprB*). This was surprising given that a key

role of *MAT* genes in heterothallic species is to regulate expression of pheromone signaling, with pheromone expression under the control of *MAT* genes (12, 45). The apparent lack of induction by *MAT* genes alone led to the suggestion that other factors may also control pheromone production in *E. nidulans*, representing a possible adaptation for homothallism (Paoletti et al., submitted). Thus, the pheromone-signaling pathway may be activated under environmental conditions favorable for sex leading to a pheromone response signal similar to that in heterothallic species, but without the need for a compatible mate. This might in turn stimulate *MAT* expression, and it was noted that a *Schizosaccharomyces pombe* Ste11 (thought to be a functional equivalent of *A. nidulans* SteA) binding site was present in the *MAT1* and *MAT2* promoter regions.

It has been suggested that *MAT* gene expression might be involved in recognition of compatible MAT1 and MAT2 nuclei in heterothallic species (14). It may therefore be speculated that *MAT* gene expression might also contribute to nuclear identity within the ascogenous hyphae of homothallic species, and a future experimental goal is to tag nuclei to see if there is differential expression of *MAT1* and *MAT2* genes in nuclei within the ascogenous hyphae. Indeed, this might be one factor involved in the phenomenon of "relative heterothallism" seen in *E. nidulans* (29, 57). This refers to the fact that, although self-fertile, *E. nidulans* retains the ability to outcross and in certain of these heterokaryons there is preferential fusion of nuclei from different partners (29); i.e., there must be some form of nuclear identity and recognition process occurring, to which *MAT* genes may contribute (Paoletti et al., submitted). In a recent commentary, Scazzochio (62) further speculated that one reason that the aspergilli may be able to form stable vegetative diploids (an unusual feature among ascomycetes, most species being committed to sex following diploid formation) is that expression of both mating-type genes would be in the same state (possibly off) and therefore would avoid progression to meiosis. To support this theory, it was noted that such diploids can form fruiting bodies, but these have poor fertility and failure of meiosis consistent with aberrant *MAT* gene expression.

Thus, to summarize, *MAT* gene expression in *E. nidulans* may be primarily required for initial morphogenesis and later stages of sexual development (e.g., nuclear identity) rather than principal control of pheromone signaling, representing an advanced evolutionary stage in homothallism (for comparison, see data for *S. macrospora* in chapter 10 by S. Pöggeler).

Physiological Aspects and Specific Developmental Events

A number of genes have been shown to be involved with specific physiological processes or developmental events in cleistothecial development of *E. nidulans* (in addition to those already described above) as follows.

Carbohydrate Metabolism

Zonneveld (84) reported that morphogenesis is accompanied by the appearance of α-glucanase activity that hydrolyzes α-(1,3)-glucan in the hyphal wall. It was suggested that this release of glucose provides a carbon and energy source for the formation of cleistothecia. A gene, *mutA*, encoding an α-1,3 glucanase (mutanase) was subsequently identified by Wei et al. (80) and shown to be expressed mainly in the Hülle cells. Surprisingly, a Δ*mutA* deletion strain exhibited wild-type levels of fertility, suggesting that additional carbon sources are utilized for sexual development. Wei et al. (81) later identified a gene, *hxtA*, encoding a putative high-affinity hexose transporter, which was repressed under high-glucose conditions but showed high levels of expression in carbon-starved vegetative hyphae and during sexual development. An HxtA::GFP fusion protein showed particular localization to the ascogenous hyphae. It was proposed that HxtA is involved with sugar metabolism and uptake during sex but that there is redundancy in the sugar uptake system because a Δ*hxtA* deletion strain showed no inhibition of sexual development.

Lipid Metabolism and Psi Hormonal Signaling

E. nidulans provides one of the best-characterized examples of the role of lipid molecules acting as signaling compounds influencing hyphal morphogenesis in a filamentous ascomycete. In the 1990s Champe and coworkers undertook extensive biochemical studies which led to the identification of a family of hydroxylated fatty acid compounds (derived mainly from linoleic acid) that greatly influenced sporulation in *E. nidulans*. Addition of certain members resulted in precocious sexual induction; hence, they were termed "psi" factors (psiα from linoleic and psiβ from oleic acid moieties). The most commonly encountered members, psiAα, psiBα, and psiCα, were interconvertible yet demonstrated opposing biological activities: psiAα induced asexual sporulation, whereas psiBα and psiCα induced sexual sporulation (10). Curiously, *veA1* mutants were found to be highly refractory to the sexual-inducing activity of psiC.

Later, Tsitsigiannis et al. (73) drew upon the fortuitous identification of a linoleate dioxygenase from *Gaeumannomyces graminis* to isolate by genome analysis a gene, *ppoA*, encoding a lipid body protein from *E. nidulans* that was involved in psiBα production. Deletion of *ppoA* resulted in a sixfold reduction in sexual spore production but a fourfold increase in asexual sporulation. Overexpression of *ppoA* led to increased sexual sporulation relative to asexual sporulation. GFP tagging revealed localization of PpoA to asexual metulae, Hülle cells, and cleistothecial initials. It was therefore concluded that PpoA is required for balancing anamorph/teleomorph development through effecting oxylipin production. There was also evidence that *ppoA* interacts with (is regulated by?) *VeA*, virtually no expression being detected in a Δ*veA* strain, and is regulated by the COP9 signalosome, with misscheduled and upregulated expression in *csnD* mutant strains (these were originally identified as psi-overproducing aconidial *aco* mutant strains).

Two further oxylipin biosynthetic genes, *ppoB* and *ppoC*, were subsequently identified (72). Deletion of *ppoB* resulted in lowered production of the oleic acid derivative psiBβ and a decreased level of sexual sporulation. Gene *ppoB* appeared to antagonistically mediate expression of *ppoA* and *ppoC* (downregulation of *ppoA* but upregulation of *ppoC* in a Δ*ppoB* strain), suggesting a regulatory loop among the three genes. Meanwhile, a triple Δ*ppoA*Δ*ppoB*Δ*ppoC* mutant was constructed. Given the supposed role of psi factors in sex, the latter surprisingly exhibited misscheduled and increased activation of sexual development with production of Hülle cells and even cleistothecia in liquid culture. This was correlated with greatly increased *veA* expression in the triple mutant (though not in the Δ*ppoB* single mutant). This observation was not easily explainable and showed that other parameters regulate sporulation independent of the *ppo* genes—but it was nevertheless argued that the Ppo proteins and oxylipins have an important role in spatial and temporal signaling cascades that determine the ratio of asexual to sexual differentiation (Fig. 7.2). The target sites/receptors for the psi factors remain tantalizingly unknown. One exciting observation from ongoing work is that a plant oxygenase can partially remediate the sexual development phenotype of *ppo* deletion when expressed in deletion strains. This strengthens the hypothesis that oxylipins may be cross-kingdom signals whose importance has hitherto not been recognized (M. Brodhagen, D. I. Tsitsigiannis, and N. Keller, personal communication).

Oxidation State

It has been proposed that changes in the cellular oxidation state might be important triggers of hyphal differentiation (43, 64). A gene, *noxA*, has been identified from *E. nidulans* that encodes an NADPH oxidase able to produce reactive oxygen species (43). Deletion of this gene had no obvious impact on asexual growth but resulted in a block in sexual development beyond the formation of cleistothecial initials (though Hülle cells were formed). Expression of *noxA* was independent of the transcription factors SteA and StuA but suppressed by SakA. NoxA was localized in Hülle cells and cleistothecial initials, and it was proposed that reactive oxygen species might lead to a hyperoxidant state and have roles in triggering cell proliferation, apoptotic death, and cell wall cross-linking and polymerization. A related catalase-peroxidase (laccase II)-encoding gene, *cpeA*, has been identified which showed greatly increased expression during early sexual development and was localized mainly within Hülle cells (63, 64). However, deletion of *cpeA* did not have any adverse impact on asexual or sexual spore production. Expression appeared to be induced by StuA, though not by NsdD or VeA. It was speculated that CpeA provided some form of protection against oxidative damage, e.g., the Hülle cells might provide protection for the dikaryotic mycelium.

Late Stages of Sexual Development— Ascospore Production

Certain genes have been detected with specific roles leading to ascospore formation in *E. nidulans*. A gene, *tubB*, encoding an alpha-tubulin protein, has been shown to be required specifically for ascosporogenesis. A Δ*tubB* deletion strain exhibited normal asexual growth and sexual development up to the point of ascus formation, but thereafter only a single nuclear mass was observed, indicating a block in karyogamy and/or a block in meiosis following zygote formation (37). A mutagen-sensitive *nuv* strain appeared to be blocked at meiosis due to impaired recombination ability (50). Similarly, gene *ggrA* (related to fungal F-box proteins, which are substrate receptors for ubiquitin ligases) is required for development of mature ascospores. A Δ*ggrA* deletion strain was shown to form Hülle cells and cleistothecia containing asci but was then blocked at meiosis and ascopore formation (39). Ascospores were apparently formed in strains containing a mutation in the zinc-finger *samB* gene (associated with onset of polarized growth and nuclear positioning) but were not viable with spore lysis occurring. The sad end result was merely red liquid inside the cleistothecia (40). Meanwhile 20 mutants blocked at various stages in ascospore development (e.g., no crozier tips, arrest at various stages of meiotic prophase or metaphase, and nondelineation of ascospores) have been identified from a mu-

tant screen (70). Although the genetic basis of these mutations has yet to be determined, it was estimated that some 50 to 100 genes were involved specifically in ascosporogenesis. Finally, the role of a putative protein kinase, ImeB, is currently under investigation. The homologue *IME2* in *S. cerevisiae* is expressed only during meiosis and is essential for sporulation. Initial results have yielded the unexpected finding that ImeB may be a negative regulator of sexual development in *E. nidulans*, as a targeted deletion of *imeB* resulted in the formation of enhanced numbers of sexual structures (F. Sari, O. Bayram, S. Irniger, and G. H. Braus, unpublished data).

Sexual Morphogenesis—Putting It All Together

As described above (in the subsections of "Genetic and Physiological Determinants of Sexual Development"), over 40 genes have been described, all of which are associated with sexual development in *E. nidulans*. Thus, this species provides one of the best models for understanding how sexual morphogenesis is controlled in a filamentous ascomycete because a framework of genes is already known. In addition, a large number of *acl* (acleistothecial), *bsd* (block in sexual development), and *asd* (abnormal in sexual development) mutant strains have been isolated that show deviations from normal sexual development (together with *nsd* mutants from complementation groups other than *nsdD*), although the precise genetic basis of mutation is unknown (27, 83). Conversely, *dcl* (dense cleistothecia) strains with precocious and greatly increased production (10-fold or higher) of cleistothecia have been described (63, 83). Furthermore, the availability of new genomic resources (1) will provide an invaluable tool to allow analysis of gene expression in mutants blocked in various stages of sexual development and allow identification of further genes involved with sexual development, e.g., the use of microarrays to compare gene expression under sexual versus asexual growth conditions. An expressed sequence tag screen for transcripts preferentially expressed during early sexual morphogenesis has already been performed, and at least 17 transcripts with increased abundance were identified. These included a ribosomal protein-encoding gene (33). Three other housekeeping genes were subsequently found to also exhibit increased expression during the sexual cycle, but this was thought to arise from increased levels of protein synthesis during sex rather than these genes being stage-specific transcripts (32).

Although many genes are known to be involved with sexual reproduction, we are still some way from understanding how overall gene expression and development are integrated. Most studies to date have focused on the effects of disruption of a specific gene and have clearly provided valuable information. Some studies have also attempted to assess gene expression within the context of other regulatory factors or within whole signaling pathways. This is gradually making it possible to build up a picture of the overall sexual machinery involved with control of the sexual cycle in *E. nidulans*. A hypothetical genetic model showing control pathways is presented in Fig. 7.2. This scheme draws on many of the observations described above, and in particular the findings of Seo et al. (65), Vienken and Fischer (77), and Paoletti et al. (submitted).

A critical observation to support the scheme shown in Fig. 7.2 concerns the presence or absence of Hülle cells in the various gene disruption strains. It is clear that Hülle cells can be formed independent of the development of mature (ascospore containing) cleistothecia, e.g., as seen in *steA*, *steC*, *gpgA*, and various other deletion strains. Hülle cells are also formed by three species of "asexual" aspergilli (*A. raperi*, *A. puniceus*, and *A. silvaticus*) and single mating types of *E. heterothallicus*, which fail to form fertile cleistothecia (10). However, Hülle cells were not formed in the *stuA* mutant or in the *veA* or *nsdD* deletion strains. One interpretation of this observation is that VeA, NsdD, and StuA are critical for initial activation of the sexual developmental process, which encompasses both formation of cleistothecial initials (defined as the very earliest stage in cleistothecial formation, i.e., initial coiling of hyphae into ascogonial coils or similar in *E. nidulans* before wall formation) and Hülle cell development. However, beyond this point Hülle cell development may proceed with some degree of independence from further true sexual development (formation of ascogenous hyphae and maturation of cleistothecia). The latter sexual processes would be likely to include pheromone signaling, as confirmed by the fact that disruption of most of the elements of the pheromone response pathway results in loss of fertility, but not loss of Hülle cell development—indeed, these are produced in abundance by *sfaD* and *gpgA* deletion strains. Latter-stage genes would also include *nosA*, *MAT1*, *MAT2*, and the *csn* elements, which are required for fertility. Note also the role of the oxylipin-encoding genes *ppoA*, *ppoB*, and *ppoC* acting as a secondary control circuit regulating the whole developmental cascade. It is nevertheless cautioned that this model remains speculative. Thus, relations between certain key elements remain unclear, e.g., the interactions of VeA, StuA, DopA, and NsdD with each other. Also the extent of formation of initials (the very earliest stages of cleistothecium development) is not addressed in many reports.

One other particularly intriguing observation is that while deletion of some genes results in a supposed total loss of fertility (e.g., *medA*, *stuA*, *sfaD*, and *steA*), in other instances [e.g., *nsdD*, *gprA(preB)*, *gprB(preA)*, and *nosA*] some fertile cleistothecia may still be formed (though these may be smaller than the normal wild type) even though development is almost entirely blocked. This indicates that the genetic pathways controlling sex exhibit some "leakiness," allowing bypass of certain stages, possibly an adaptation to ensure ascospore production despite mutational damage. This leakiness is evident in strains manipulated to overexpress genes positively inducing sex in an otherwise sterile background; for example, increased dosage of *nsdD* partially rescued the developmental defect of GPCR mutants (65). There is also at least one example of this phenomenon where the genes involved have not yet been characterized. Here, mutation in a single dominant gene, *yB*, results in mutants severely deficient in both laccase I and laccase II, which are almost totally acleistothecial but produce abundant masses of Hülle cells and very occasional cleistothecia. This defect could be overcome by propagation on a copper-supplemented medium, suggesting that the *yB* mutation was due to a defect in copper transport or storage. Linked suppressors were also identified (10).

ASEXUALITY AND *MAT* IN THE ASPERGILLI

The genus *Aspergillus* includes many species of major economic or medical importance: *A. oryzae*, used in the production of Asian foodstuffs such as soy sauce, miso, and sake; *A. niger*, used for commercial production of citric acids and the topic of much biotechnological interest; *A. flavus*, a significant food spoilage organism and source of carcinogenic aflatoxins; and *A. fumigatus*, which can cause life-threatening infections in immunocompromised hosts and allergenic reactions in otherwise healthy individuals. As a result there is long-standing interest in strain improvement of *Aspergillus* species and also in determination of the genetic basis of traits of interest such as metabolite production or pathogenicity factors. However, such work has been impeded by the lack of a known perfect (sexual) state in all of these key species, which has precluded the use of the sexual cycle for studies of classical genetics. Indeed, the prevalence of asexuality within the *Aspergillus* genus as a whole is surprising given the supposed evolutionary benefits of sex (see chapter 32 by D. K. Aanen and R. F. Hoekstra and reference 17 for further discussion of this topic). Current research is therefore focusing on trying to determine the genetic reasons for asexuality within the aspergilli, given that all members of this group are thought to have arisen from sexual ancestors (25). Various approaches have been taken to address the question of asexuality as follows.

Genome Screening

Genome sequence data for *A. fumigatus* and *A. oryzae* are now publicly available (1, 20). BLAST searches were therefore undertaken to determine if there were mutation(s) in any key gene required for sexual reproduction, which might explain the lack of sexuality. However, screening revealed that both species contained apparently functional (i.e., encoding full-length protein and lacking any mutation resulting in frameshift or internal stop codon) copies of at least 215 genes implicated in the sexual cycle, including genes involved with mating processes, pheromone response, meiosis, and fruiting-body development (20). It remains possible that there may be mutations in other so-far-unknown genes required for sex, but at present genome screening has failed to explain the reason for asexuality in *A. fumigatus* and *A. oryzae*. A similar genome screen has been undertaken with *A. niger*. This again revealed that almost all sex-related genes appeared to encode functional proteins, although at least one possible mutation was detected (53a). Finally, genome data for the asexual species *Aspergillus clavatus*, a significant allergen and producer of the toxin patulin, have just become available. Intriguingly, a preliminary screen reveals the presence of an α-domain *MAT1* family mating-type gene, but correct translation of this gene would rely on editing of an aberrant intron (ending in TAGG rather than consensus TAG) (GenBank entry AAKD03000020, from 42770 to 41599), suggesting a possible mutation in this *MAT* gene that may be a factor conferring asexuality.

Presence and Distribution of *MAT* Genes

In heterothallic species it is necessary to have isolates of compatible *MAT1* and *MAT2* genotypes for sex to occur. Thus, one possible reason for asexuality might be that the asexual species have arisen from a narrow genetic base (e.g., one or a few closely related individuals) all of the same mating type, which have subsequently become reproductively isolated. Alternatively, genetic drift may have occurred during evolution, leading to the loss of one mating type. In either case, one reason why the asexual species would not be able to undergo sexual reproduction would be the lack (or scarcity) of isolates of the opposite mating type. With the release of genome sequence data it has now become possible to devise PCR-based methods to perform the necessary mating-

type screens to determine whether there is such a bias in mating-type distribution.

Such a screen was first performed for *A. fumigatus* (52). BLAST analysis of genome sequence data had revealed the presence of an HMG *MAT2* family mating-type gene at a *MAT* locus bordered by genes characteristically found flanking *MAT* genes in other euascomycete species (20). Paoletti et al. (52) used a degenerate PCR approach to identify the presence of a complementary α-domain *MAT1* family mating-type gene in certain laboratory isolates and demonstrated that the *MAT* locus of *A. fumigatus* has an idiomorph structure characteristic of heterothallic euascomycete fungi (15). A multiplex-PCR test to determine mating type was then designed and used to screen 290 worldwide isolates of *A. fumigatus*. Results revealed the presence of *MAT1-1* and *MAT1-2* genotypes in similar proportions (43 and 57%, respectively). Further screening of *A. fumigatus* has more recently been undertaken with populations collected from aerial samples from Dublin, Ireland, and results again indicated a 1:1 ratio of *MAT1-1:MAT1-2* genotypes (C. M. O'Gorman, H. T. Fuller, and P. S. Dyer, unpublished data). Thus, overall results for *A. fumigatus* indicate that isolates of both mating types are present in nature, failing to provide a reason for asexuality.

Similar *MAT* screens are currently being made with other "asexual" aspergilli. This has led to the identification of both *MAT1-1* and *MAT1-2* genotypes of *A. oryzae* and to the discovery that the genes are found at a single *MAT* locus that again exhibits an idiomorph organization (20, 52; M. Paoletti, N. Yamamoto, D. B. Archer, K. Kitamoto, and P. S. Dyer, unpublished data). Both *MAT1-1* and *MAT1-2* genotypes have also been detected in *A. flavus*, *A. sojae*, and *A. parasiticus* (M. Paoletti and P. S. Dyer, unpublished data). Furthermore, screening of data from an ongoing *A. terreus* genome-sequencing project reveals the presence of a putative α-domain *MAT1* family mating-type gene (locus ATEG_08812.1 as part of GenBank accession NW_001471196) at the same *MAT* locus as seen in *A. fumigatus* and *A. oryzae*, bounded by flanking *SLA2* and *DNA lyase* genes (Fig. 7.3) (data from Broad Institute website: http://www.broad.mit.edu/annotation/genome/aspergillus_terreus/Blast.html). (It is cautioned that screening of genome databases may reveal sequences with homology to HMG family proteins but that the *MAT2* family HMG-encoding genes have particular sequence motifs that distinguish them from other HMG-type genes.)

The one exception so far concerns screening of members of the *A. niger* "black aspergilli" group. Screening of over 150 isolates has revealed a very strong bias towards isolates containing an α-domain *MAT1* family

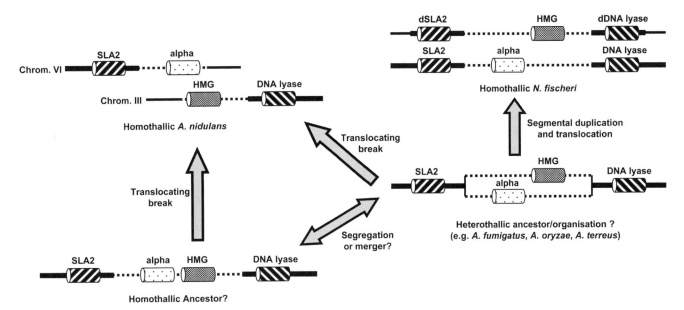

Figure 7.3 Organization of *MAT* loci within the aspergilli showing possible evolution of breeding systems from either a homothallic or heterothallic ancestor (not to scale). Adapted in part from reference 20. Textured blocks indicate mating-type genes (*MAT1* α-domain or *MAT2* HMG family) or flanking genes (*SLA2* or *DNA lyase*). Dotted lines indicate idiomorph region; heavy bold lines indicate conserved sequence flanking the idiomorph region; the suffix 'd' indicates disabled pseudogene. Note that the illustration does not show all genes present in the flanking regions (e.g., an *APC* gene is also present, but syntenic order varies according to species).

mating-type gene, with very few *MAT2* HMG-type isolates detected (J. Varga, S. Kocsubé, K. Pál, F. Debets, G. Eyres, S. E. Baker, R. A. Samson, and P. S. Dyer, unpublished data).

Expression of Sex-Related Genes

Another possible cause of asexuality might be lack of expression of key genes required for sexual reproduction, encompassing a variety of genes shown in Fig. 7.2. Relatively little work has been undertaken in this area regarding the aspergilli, but some observations are available. Paoletti et al. (52) used RT-PCR to monitor expression of pheromone precursor (*ppgA*), pheromone receptor (*preA* and *preB*), and mating-type (*MAT-1* and *MAT-2*) genes from *A. fumigatus* during growth on media, under conditions known to induce sexual reproduction in *E. nidulans*. And indeed, expression of all these key genes was observed. This was the first time that pheromone precursor and receptor gene expression had been observed in a supposedly asexual species, although *MAT* gene expression had already been observed in other asexual species (52). Similar observations have been made with *A. oryzae*, with expression of both pheromone-related and *MAT* genes seen in axenic culture, albeit at levels lower than those observed in *A. fu-*

migatus (M. Paoletti, E. U. Schwier, D. B. Archer, K. Kitamoto, and P. S. Dyer, unpublished data). In contrast, it has not yet been possible to detect *MAT* gene expression in *A. niger* (53a). It is noted that overexpression of the *veA* locus in *A. fumigatus* did not result in any obvious phenotype (38).

Population Biology Approaches

If a species is truly asexual, then it may be predicted that populations would show increased clonality due to a lack of sexual recombination. However, accumulating data suggest that this is not the case for some of the asexual aspergilli. Geiser et al. (24) found evidence of recombination among group I type *A. flavus* isolates from Australia. In a similar fashion, population genetic studies of *A. fumigatus* from a set of 106 global isolates yielded evidence of recombination within the samples (52). Recombination was also detected in a smaller sample of *A. fumigatus* "fumigatus" isolates from five continents (58). The latter results are consistent with earlier studies of genetic variation within *A. fumigatus*, which provide evidence for recombination (reviewed in reference 17).

Thus, in summary, the genetic cause(s) for asexuality has yet to be identified for most of the asexual aspergilli that have been subjected to study. To the contrary, the evidence instead indicates that many of these species may have only recently lost sexuality OR might even retain some potential for sexual reproduction, of possible great significance. It has been suggested that isolates with a spectrum of sexual fertility may be present in nature (17). Studies are currently in progress to try to induce the sexual cycle of *A. fumigatus* and *A. oryzae*. One important factor is the need to identify suitable in vitro conditions in order to induce any sexual cycle, and some challenging of assumptions and thinking outside the box may be required. For example, the standard temperature for inducing the sexual cycle of *E. nidulans* is 37°C as originally utilized by Pontecorvo (57). However, a project was recently undertaken in which five independent wild-type strains were incubated at a range of temperatures from 25 to 37°C under conditions promoting the sexual cycle. It was found that the optimum temperature for production of cleistothecia was in fact 32°C, with an average of a 1.9-fold increase in numbers compared to those obtained at 37°C. This pattern was present in all isolates and was repeated when isolates were grown on seven different commonly used growth media (data from S. Tranter and P. S. Dyer). Thus, a range of test incubation conditions will be required. Also some biotic interaction may be required to induce the sexual cycle, noting that *A. fumigatus* is particularly common in compost heaps, presumed to be rich in microbial diversity. In addition, certain micronutrients such as manganese might be required for fruiting-body formation (5, 84). The one true example of ancient asexuality may prove to be among the *A. niger* black aspergilli group. This would be consistent with phylogenetic analyses which indicate that the *A. niger* group is not closely related to any *Aspergillus* species with a known sexual state (54, 71).

MAT AND THE EVOLUTION OF HOMOTHALLISM AND HETEROTHALLISM

One final topic of interest concerns the evolution of breeding systems within the aspergilli. A long-standing question in fungal biology is whether homothallism or heterothallism is the ancestral state of sexual reproduction. Convincing evidence has been presented elsewhere that heterothallism is the ancestral state in *Cochliobolus* and *Stemphylium* species (see chapter 6 by Turgeon and Debuchy). However, the prevalence of homothallism in the aspergilli (over 65 homothallic versus only 4 known heterothallic species) provides evidence that homothallism might be the ancestral state in this genus. Phylogenetic analyses provide further support for heterothallism being a derived character. Both Geiser et al. (23) and Varga et al. (76) found that asexual, homothallic, and

heterothallic taxa were scattered throughout different clades of the section *Fumigati* and, together with tree-length data, interpreted this to indicate that there had been multiple independent losses of the *Neosartorya* sexual state and that heterothallism was possibly derived from homothallism through loss of self-compatibility.

The availability of genome sequence has now provided a new tool with which to address this question. It might be argued that if homothallism were the ancestral state, then it would be likely that most homothallic aspergilli would share a similar *MAT* locus organization (recalling that *MAT* genes are key determinants of sexual breeding systems [15]), whereas some variation might be present in *MAT* locus organization of the heterothallic taxa. Conversely, if heterothallism was the ancestral state, then the heterothallic taxa might share a common *MAT* locus organization similar to that seen in other heterothallic ascomycetes (15), with some variation instead present in the *MAT* organization of homothallic taxa. Unfortunately, so far there are only limited data available, this coming from the homothallic *E. nidulans* and *Neosartorya fischeri* and the asexual *A. oryzae*, *A. fumigatus*, and *A. terreus*. Information is not yet available for any heterothallic species, although investigations with *E. heterothallicus* are under way (F. A. Seymour and P. S. Dyer, unpublished data).

Given these limitations, some observations can nevertheless be made. All three asexual species exhibit a single *MAT* locus with similar organization and synteny to known heterothallic euascomycetes, as shown by the presence of conserved genes (*SLA2* and *APN1 DNA lyase*) flanking the locus (15) (Fig. 7.3). Indeed, *A. fumigatus* and *A. oryzae* display conserved synteny over 1.7 Mb on either side of the *MAT* locus (20). *A. fumigatus* and *A. oryzae* also have *MAT1-1* and *MAT1-2* idiomorphs at this locus and, as discussed above, may have the potential for crossing in a heterothallic manner. In contrast, *E. nidulans* shows an unusual *MAT* organization compared to other homothallic fungi, with the α-domain *MAT1* and HMG *MAT2* family genes found on different chromosomes (18; Paoletti et al., submitted). The α-domain *MAT1* and HMG *MAT2* gene loci of *E. nidulans* have been subjected to extensive analysis (20). It was found that one flank of the HMG *MAT2* locus is syntenic with 409 kb downstream of the *A. fumigatus*/*A. oryzae MAT* loci (the other flank showing no synteny), whereas the α-domain *MAT1* locus is syntenic with 34 kb upstream of the *A. fumigatus*/*A. oryzae MAT* loci (the other flank again showing no synteny) (Fig. 7.3). Most recently, *MAT* loci from the homothallic *N. fischeri* (a close taxonomic relative of *A. fumigatus*) have been examined using both experimental and genomic approaches (60a). This has revealed an unexpected and novel organization of *MAT* loci (Fig. 7.3). Both α-domain *MAT1* and HMG *MAT2* mating-type genes are present within the same genome. But unlike *E. nidulans*, an uninterrupted *MAT* locus resembling that of *A. fumigatus* and *A. oryzae* was present, at which the *N. fischeri* α-domain *MAT1* gene was located. The *N. fischeri* HMG *MAT2* gene was found elsewhere in the genome, flanked by nonfunctional pseudogenes resembling genes bordering the *MAT1* locus.

So what are the implications of these data for the evolution of homothallism and heterothallism in the aspergilli? Galagan et al. (20) interpreted the data to indicate that there was most likely an ancestral homothallic ancestor for the aspergilli, which contained both α-domain *MAT1* and HMG *MAT2* genes at the same locus. A translocating break then occurred to give the *MAT* organization seen in extant *E. nidulans* (Fig. 7.3). Meanwhile some form of segregation event resulted in division of the α-domain and HMG genes giving rise to the present "heterothallic" organization of *A. oryzae* and *A. fumigatus* (Fig. 7.3). This interpretation is consistent with data for these species, the prevalence of homothallism in the aspergilli, and also the discovery of 360 bp of *MAT-2* sequence within the flanking sequence common to both *MAT1-1* and *MAT1-2* idiomorphs of *A. fumigatus* (suggesting that the α-domain and HMG-domain mating-type genes were once united) (52). However, there are some difficulties with this hypothesis. It predicts a homothallic ancestor with α-domain and HMG genes present together, and yet in neither homothallic *E. nidulans* nor *N. fischeri* has this *MAT* organization been seen. Rydholm et al. (60a) suggested that self-fertility in *N. fischeri* arose due to segmental duplication and translocation of the *MAT* chromosomal region rather than the major translocating break seen in *E. nidulans*. Instead it is possible that a heterothallic state, as represented by *A. oryzae* and *A. fumigatus*, is ancestral. Indeed, it is conceivable that the majority of the asexual species in the *Aspergillus* genus may exhibit a heterothallic *MAT* locus organization, in this case making heterothallism the most common state and therefore the most likely ancestral state (ca.110 of 180 species!). Overall it must be concluded that the jury is still out as we await *MAT* sequence data from an increased number of species of *Aspergillus* from a broader phylogenetic sampling.

CONCLUSIONS—SEXUAL REPRODUCTION AND *MAT* IN THE ASPERGILLI

The aspergilli represent a fascinating group of fungi in which to study the molecular determination and evolutionary implications of sexual reproduction. They ex-

hibit a remarkable fluidity in reproductive lifestyle, with asexual, heterothallic, and homothallic taxa found to be intermixed. This ability to evolve different reproductive strategies might be one factor contributing to the undoubted success of the aspergilli, which are abundant in ecosystems worldwide (3). It is intriguing to note that transposase genes have been found upstream of *MAT* loci in certain *Aspergillus* species, which might help to explain the particular ability of the aspergilli to evolve different breeding systems (60a). A long-term goal is now to manipulate the life cycles of medically, economically, and scientifically important species of *Aspergillus* for the benefit of all.

Many thanks are extended to Gerhard Braus, Ana Calvo, Nancy Keller, Claudio Scazzocchio, and Jaehyuk Yu for sharing unpublished information. The Institute for Genomic Research (TIGR) and Broad Institute (United States) are also thanked for access to data from the A. clavatus *and* A. terreus *genome-sequencing projects, respectively. Funding to support research is provided by the Biotechnology and Biological Sciences Research Council, United Kingdom.*

References

1. Archer, D. B., and P. S. Dyer. 2004. From genomics to post-genomics in *Aspergillus*. *Curr. Opin. Microbiol.* **7:** 499–504.

2. Benjamin, C. R. 1955. Ascocarps of *Aspergillus* and *Penicillium*. *Mycologia* **47:**669–687.

3. Bennett, J. W., and M. A. Klich. 1992. Aspergillus: *Biology and Industrial Applications* (preface). Butterworth-Heinemann, Boston, MA.

4. Blumenstein, A., K. Vienken, R. Tasler, J. Purschwitz, D. Veith, N. Frankenberg-Dinkel, and R. Fischer. 2005. The *Aspergillus nidulans* phytochrome FphA represses sexual development in red light. *Curr. Biol.* **15:**1833–1838.

5. Braus, G. H., S. Krappman, and S. E. Eckert. 2002. Sexual development in ascomycetes. Fruit body formation of *Aspergillus nidulans*, p. 215–244. *In* H. D. Osiewacz (ed.), *Molecular Biology of Fungal Development*. Marcel Dekker, New York, NY.

6. Busby, T. M., K. Y. Miller, and B. L. Miller. 1996. Suppression and enhancement of the *Aspergillus nidulans medusa* mutation by altered dosage of the *bristle* and *stunted* genes. *Genetics* **143:**155–163.

7. Busch, S., S. E. Eckert, S. Krappmann, and G. H. Braus. 2003. The COP9 signalosome is an essential regulator of development in the filamentous fungus *Aspergillus nidulans*. *Mol. Microbiol.* **49:**717–730.

8. Bussink, H. J., and S. A. Osmani. 1998. A cyclin-dependent kinase family member (PHOA) is required to link developmental fate to environmental conditions in *Aspergillus nidulans*. *EMBO J.* **17:**3990–4003.

9. Casselton, L. A. 2002. Mate recognition in fungi. *Heredity* **88:**142–147.

10. Champe, S. P., D. L. Nagle, and L. N. Yager. 1994. Sexual sporulation, p. 429–454. *In* S. D. Martinelli and J. R.

Kinghorn (ed.), Aspergillus: *50 Years on.* Elsevier, Amsterdam, The Netherlands.

11. Clutterbuck, A. J. 1969. A mutational analysis of conidial development in *Aspergillus nidulans. Genetics* **63**:317–327.

12. Coppin, E., C. de Renty, and R. Debuchy. 2005. The function of the coding sequences for the putative pheromone precursor in *Podospora anserina* is restricted to fertilization. *Eukaryot. Cell* **4**:407–420.

13. Croft, J. H., and J. L. Jinks. 1977. Aspects of the population genetics of *Aspergillus nidulans,* p. 349–360. *In* J. E. Smith and J. A. Pateman, *Genetics and Physiology of As-pergillus.* Academic Press, London, United Kingdom.

14. Debuchy, R. 1999. Internuclear recognition: a possible connection between Euascomycetes and homobasidiomycetes. *Fungal. Genet. Biol.* **27**:218–223.

15. Debuchy, R., and B. G. Turgeon. 2006. Mating-type structure, evolution, and function in euascomycetes, p. 293–323. *In* U. Kües and R. Fischer (ed.), *The Mycota I: Growth, Differentiation and Sexuality.* Springer-Verlag, Berlin, Germany.

16. Dutton, J. R., S. John, and B. L. Miller. 1997. StuAp is a sequence-specific transcription factor that regulates developmental complexity in *Aspergillus nidulans. EMBO J.* **16**:5710–5721.

17. Dyer, P. S., and M. Paoletti. 2005. Reproduction in *Aspergillus fumigatus:* sexuality in a supposedly asexual species? *Med. Mycol.* **43**(Suppl. 1):S7–S14.

17a. Dyer, P. S., M. Paoletti, M. J. Alcocer, and D. B. Archer. 2003. Identification of a mating-type gene in the homothallic fungus *Aspergillus nidulans. Fungal Genet. Newsl.* **50**(Suppl.):145.

18. Dyer, P. S., M. Paoletti, and D. B. Archer. 2003. Genomics reveals sexual secrets of *Aspergillus. Microbiology* **149**: 2301–2303.

19. Elion, E. A. 2000. Pheromone response, mating and cell biology. *Curr. Opin. Microbiol.* **3**:573–581.

20. Galagan, J., S. E. Calvo, C. Cuomo, L.-J. Ma, J. Wortman, S. Batzoglou, S.-I. Lee, M. Brudno, B. Meray, C. C. Spevak, J. Clutterbuck, V. Kapitonov, J. Jurka, C. Scazzocchio, M. Farman, J. Butler, S. Purcell, S. Harris, G. H. Braus, O. Draht, S. Busch, C. D'Enfert, C. Bouchier, G. H. Goldman, D. Bell-Pedersen, S. Griffiths-Jones, J. H. Doonan, J. Yu, K. Vienken, A. Pain, M. Freitag, E. U. Selker, D. B. Archer, M. A. Penalva, B. R. Oakley, M. Momany, T. Tanaka, T. Kumagai, K. Asai, M. Machida, W. C. Nierman, D. W. Denning, M. Caddick, M. Hynes, M. Paoletti, R. Fischer, B. Miller, P. Dyer, M. S. Sachs, S. A. Osmani, and B. W. Birren. 2005. Sequencing and comparative analysis of *Aspergillus nidulans. Nature* **438**: 1105–1115.

21. Gams, W., M. Christensen, A. H. Onions, J. I. Pitt, and R. A. Samson. 1985. Infrageneric taxa of *Aspergillus,* p. 55–62. *In* R. A. Samson and J. I. Pitt (ed.), *Advances in* Penicillium *and* Aspergillus *Systematics.* Plenum Press, New York, NY.

22. Geiser, D. M., M. L. Arnold, and W. E. Timberlake. 1994. Sexual origins of British *Aspergillus nidulans* isolates. *Proc. Natl. Acad. Sci. USA* **91**:2349–2352.

23. Geiser, D. M., J. C. Frisvad, and J. W. Taylor. 1998. Evolutionary relationships in *Aspergillus* section *Fumigati* inferred from partial β-tubulin and hydrophobin DNA sequences. *Mycologia* **90**:831–845.

24. Geiser, D. M., J. I. Pitt, and J. W. Taylor. 1998. Cryptic speciation and recombination in the aflatoxin-producing fungus *Aspergillus flavus. Proc. Natl. Acad. Sci. USA* **95**: 388–393.

25. Geiser, D. M., W. E. Timberlake, and M. L. Arnold. 1996. Loss of meiosis in *Aspergillus. Mol. Biol. Evol.* **13**: 809–817.

26. Han, K. H., K. Y. Han, M. S. Kim, D. B. Lee, J. H. Kim, S. K. Chae, K. S. Chae, and D. M. Han. 2003. Regulation of nsdD expression in *Aspergillus nidulans. J. Microbiol.* **41**:259–261.

27. Han, K. H., K. Y. Han, J. H. Yu, K. S. Chae, K. Y. Jahng, and D. M. Han. 2001. The *nsdD* gene encodes a putative GATA-type transcription factor necessary for sexual development of *Aspergillus nidulans. Mol. Microbiol.* **41**: 299–309.

28. Han, K. H., J. A. Seo, and J. H. Yu. 2004. A putative G protein-coupled receptor negatively controls sexual reproduction in *Aspergillus nidulans. Mol. Microbiol.* **51**: 1333–1345.

29. Hoffmann, B., S. E. Eckert, S. Krappman, and G. H. Braus. 2001. Sexual diploids of *Aspergillus nidulans* do not form by random fusion of nuclei in the heterokaryon. *Genetics* **157**:141–147.

30. Hoffmann, B., C. Wanke, S. A. Kirsten, K. Paglia, and G. H. Braus. 2000. c-Jun and RACK1 homologues regulate a control point for sexual development in *Aspergillus nidulans. Mol. Microbiol.* **37**:28–41.

31. Horie, Y., M. Miyaji, K. Nishimura, M. F. Franco, and K. I. R. Coelho. 1995. New and interesting species of *Neosartorya* from Brazilian soil. *Mycoscience* **36**:199–204.

32. Jeong, H. Y., G. B. Cho, K. Y. Han, J. Kim, D. M. Han, K. Y. Jahng, and K. S. Chae. 2001. Differential expression of house-keeping genes of *Aspergillus nidulans* during sexual development. *Gene* **262**:215–219.

33. Jeong, H. Y., D. M. Han, K. Y. Jahng, and K. S. Chae. 2000. The rpl16a gene for ribosomal protein L16A identified from expressed sequence tags is differentially expressed during sexual development of *Aspergillus nidulans. Fungal Genet. Biol.* **31**:69–78.

34. Kawasaki, L., O. Sánchez, K. Shiozaki, and J. Aguirre. 2002. SakA MAP kinase is involved in stress signal transduction, sexual development and spore viability in *Aspergillus nidulans. Mol. Microbiol.* **45**:1153–1163.

35. Kim, H. S., K. Y. Han, K. J. Kim, D. M. Han, K. Y. Jahng, and K. S. Chae. 2002. The *veA* gene activates sexual development in *Aspergillus nidulans. Fungal Genet. Biol.* **37**:72–80.

36. Kim, J. H., M. S. Kim, Y. H. Cheon, K. S. Chae, and D. M. Han. 2002. Isolation and characterisation of mutants that can sexually develop in the presence of visible light. *Fungal Genet. Newsl.* **50**(Suppl.):68.

37. Kirk, K. E., and N. R. Morris. 1991. The tubB alpha-tubulin gene is essential for sexual devlopement in *Aspergillus nidulans. Genes Dev.* **5**:2014–2023.

38. Krappmann, S., O. Bayram, and G. H. Braus. 2005. Deletion and exchange of the *Aspergillus fumigatus veA* locus via a novel recyclable marker module. *Eukaryot. Cell* **4:** 1298–1307.

39. Krappmann, S., N. Jung, B. Medic, R. A. Prade, and G. H. Braus. 2006. The *Aspergillus nidulans* F-box protein GrrA links SCF activity to meiosis. *Mol. Microbiol.* **61:** 76–88.

40. Kuger, M., and R. Fischer. 1998. Integrity of a Zn finger-like domain in SamB is crucial for morphogenesis in ascomycetous fungi. *EMBO J.* **17:** 204–214.

41. Kwon, K. J., and K. B. Raper. 1967. Sexuality and cultural characteristics of *Aspergillus heterothallicus. Am. J. Bot.* **54:** 36–48.

42. Kwon, N. J., D. M. Han, and K. S. Chae. 2003. veA-dependent expression of IndB and indD encoding proteins that interact with NSDD, a GATA-type transcription factor required for sexual development in *Aspergillus nidulans. Fungal Genet. Newsl.* **50(Suppl.):** 70.

43. Lara-Ortíz, T., H. Rosas-Riveros, and J. Aguirre. 2003. Reactive oxygen species generated by microbial NADPH oxidase NoxA regulate sexual development in *Aspergillus nidulans. Mol. Microbiol.* **50:** 1241–1255.

44. Lee, D. W., S. Kim, S. J. Kim, D. M. Han, K. Y. Jahng, and K. S. Chae. 2001. The *lsdA* gene is necessary for sexual development inhibition by a salt in *Aspergillus nidulans. Curr. Genet.* **39:** 237–243.

45. Lengeler, K. B., R. C. Davidson, C. D'Souza, T. Harashima, W. C. Shen, G. P. Wang, X. Pan, M. Waugh, and J. Heitman. 2000. Signal transduction cascades regulating fungal development and virulence. *Microbiol. Mol. Biol. Rev.* **64:** 746–785.

46. Malloch, D., and R. F. Cain. 1972. The trichomataceae: ascomycetes with *Aspergillus, Paecilomyces,* and *Penicillium* imperfect stages. *Can. J. Bot.* **50:** 2613–2628.

47. Miller, B. L., and E. J. Telfer. 2003. The conserved zinc finger domain of *Aspergillus nidulans* SteAp regulates the frequency of cleistothecial initial formation in a veA1 background. *Fungal Genet. Newsl.* **50(Suppl.):** 90.

48. Miller, K. Y., A. Nowell, and B. L. Miller. 2005. Differential regulation of fruitbody development and meiosis by the unlinked *Aspergillus nidulans* mating type loci. *Fungal Genet. Newsl.* **52(Suppl.):** 184.

49. Mooney, J. L., and L. N. Yager. 1990. Light is required for conidiation in *Aspergillus nidulans. Genes Dev.* **4:** 1473–1482.

50. Osman, F., B. Tomsett, and P. Strike. 1993. The isolation of mutagen-sensitive *nuv* mutants of *Aspergillus nidulans* and their effects on mitotic recombination. *Genetics* **134:** 445–454.

51. Pandey, A., M. G. Roca, N. D. Read, and N. L. Glass. 2004. Role of a mitogen-activated protein kinase pathway during conidial germination and hyphal fusion in *Neurospora crassa. Eukaryot. Cell* **3:** 348–358.

52. Paoletti, M., C. Rydholm, E. U. Schwier, M. J. Anderson, G. Szakacs, F. Lutzoni, J. P. Debeaupuis, J. P. Latgé, D. W. Denning, and P. S. Dyer. 2005. Evidence for sexuality in the opportunistic fungal pathogen *Aspergillus fumigatus. Curr. Biol.* **15:** 1242–1248.

53. Pascon, R. C., and B. L. Miller. 2000. Morphogenesis in *Aspergillus nidulans* requires Dopey (DopA), a member of a novel family of leucine zipper-like proteins conserved from yeast to humans. *Mol. Microbiol.* **36:** 1250–1264.

53a. Pel, H. J., J. H. de Winde, D. B. Archer, P. S. Dyer, G. Hofmann, P. J. Schaap, G. Turner, R. P. de Vries, R. Albany, K. Alberman, et al. 2007. Genome sequencing and analysis of the versatile cell factory *Aspergillus niger* CBS 513.88. *Nature Biotech.* **25:** 221–231.

54. Peterson, S. W. 2000. Phylogenetic relationships in *Aspergillus* based on rDNA sequence amalysis, p. 323–355. *In* R. A. Samson and J. I. Pitt (ed.), *Integration of Modern Taxonomic Methods for* Penicillium *and* Aspergillus *Classification.* Harwood Academic Publishers, Amsterdam, The Netherlands.

55. Pitt, J. H., R. A. Samson, and J. C. Frisvad. 2000. List of accepted species and their synonyms in the family *Trichocomaceae,* p. 9–72. *In* R. A. Samson and J. I. Pitt (ed.), *Integration of Modern Taxonomic Methods for* Penicillium *and* Aspergillus *Classification.* Harwood Academic Publishers, Amsterdam, The Netherlands.

56. Pöggeler, S., and U. Kück. 2001. Identification of transcriptionally expressed pheromone receptor genes in filamentous ascomycetes. *Gene* **280:** 9–17.

57. Pontecorvo, G. 1953. The genetics of *Aspergillus nidulans. Adv. Genet.* **5:** 142–238.

58. Pringle, A., D. M. Baker, J. L. Platt, J. P. Wares, J. P. Latge, and J. W. Taylor. 2005. Cryptic speciation in the cosmopolitan and clonal human pathogenic fungus *Aspergillus fumigatus. Evolution* **59:** 1886–1899.

59. Raper, K. B., and D. I. Fennell. 1965. *The Genus* Aspergillus. The Williams & Wilkins Company, Baltimore, MD.

60. Rosén, S., J. H. Yu, and T. H. Adams. 1999. The *Aspergillus nidulans sfaD* gene encodes a G protein β subunit that is required for normal growth and repression of sporulation. *EMBO J.* **18:** 5592–5600.

60a. Rydholm, C., P. S. Dyer, and F. Lutzoni. 2007. DNA sequence characterization and molecular evolution of *Mat1* and *MAT2* mating-type loci of the self-compatible ascomycete mold *Neosartorya fischeri. Eukaryot. Cell,* in press.

61. Samson, R. A. 1992. Current taxonomic schemes of the genus *Aspergillus* and its teleomorphs, p. 355–390. *In* J. W. Bennett and M. A. Klich (ed.), Aspergillus: *Biology and Industrial Applications.* Butterworth-Heinemann, Boston, MA.

62. Scazzocchio, C. 2006. *Aspergillus* genomes: secret sex and the secrets of sex. *Trends Genet.* **22:** 521–525.

63. Scherer, M., and R. Fischer. 1998. Purification and characterization of laccase II of *Aspergillus nidulans. Arch. Microbiol.* **170:** 78–84.

64. Scherer, M., H. Wei, R. Liese, and R. Fischer. 2002. *Aspergillus nidulans* catalase-peroxidase gene (*cpeA*) is transcriptionally induced during sexual development through the transcription factor StuA. *Eukaryot. Cell* **1:** 725–735.

65. Seo, J. A., K. H. Han, and J. H. Yu. 2004. The *gprA* and *gprB* genes encode putative G protein-coupled receptors

required for self-fertilization in *Aspergillus nidulans*. *Mol. Microbiol.* **53**:1611–1623.

66. Seo, J. A., K. H. Han, and J. H. Yu. 2005. Multiple roles of a heterotrimeric G-protein γ-subunit in governing growth and development of *Aspergillus nidulans*. *Genetics* **171**:81–89.

67. Seo, J. A., and J. H. Yu. 2006. The phosducin-like protein PhnA is required for Gβγ-mediated signaling for vegetative growth, developmental control, and toxin biosynthesis in *Aspergillus nidulans*. *Eukaryot. Cell* **5**:400–410.

68. Sohn, K. T., and K. S. Yoon. 2002. Ultrastructural study on the cleistothecial development in *Aspergillus nidulans*. *Mycobiology* **30**:117–127.

69. Stinnett, S. M., E. A. Espeso, L. Cobeño, L. Araújo-Bazán, and A. M. Calvo. 2007. *Aspergillus nidulans* VeA subcellular localization is dependent on the importin α carrier and on light. *Mol. Microbiol.* **63**:242–255.

70. Swart, K., D. V. Van Heemst, M. Slakhorst, F. Debets, and C. Heyting. 2001. Isolation and characterization of sexual sporulation mutants of *Aspergillus nidulans*. *Fungal Genet. Biol.* **33**:25–35.

71. Tamura, M., K. Kawahara, and J. Sugiyama. 2000. Molecular phylogeny of *Aspergillus* and associated teleomorphs in the Trichocomaceae (Eurotiales), p. 357–372. *In* R. A. Samson and J. I. Pitt (ed.), *Integration of Modern Taxonomic Methods for* Penicillium *and* Aspergillus *Classification*. Harwood Academic Publishers, Amsterdam, The Netherlands.

72. Tsitsigiannis, D. I., T. M. Kowieski, R. Zarnowski, and N. P. Keller. 2005. Three putative oxylipin biosynthetic genes integrate sexual and asexual development in *Aspergillus nidulans*. *Microbiology* **151**:1809–1821.

73. Tsitsigiannis, D. I., R. Zarnowski, and N. P. Keller. 2004. The lipid body protein, PpoA, coordinates sexual and asexual sporulation in *Aspergillus nidulans*. *J. Biol. Chem.* **279**:11344–11353.

74. Udagawa, S., and M. Takada. 1985. Contribution to our knowledge of *Aspergillus* teleomorphs: some taxonomic problems, p. 429–435. *In* R. A. Samson and J. I. Pitt (ed.), *Advances in* Penicillium *and* Aspergillus *Systematics*. Plenum Press, New York, NY.

75. Vallim, M. A., K. Y. Miller, and B. L. Miller. 2000. *Aspergillus* SteA (sterile12–like) is a homeodomain-

C$_2$/H$_2$-Zn^{+2} finger transcription factor required for sexual reproduction. *Mol. Microbiol.* **36**:290–301.

76. Varga, J., Z. Vida, B. Tóth, F. Debets, and Y. Horie. 2000. Phylogenetic analysis of newly described *Neosartorya* species. *Antonie Leeuwenhoek* **77**:235–239.

77. Vienken, K., and R. Fischer. 2006. The Zn(II)$_2$Cys$_6$ putative transcription factor NosA controls fruiting body formation in *Aspergillus nidulans*. *Mol. Microbiol.* **61**:544–554.

78. Vienken, K., M. Scherer, and R. Fischer. 2005. The Zn(II)$_2$Cys$_6$ putative *Aspergillus nidulans* transcription factor repressor of sexual development inhibits sexual development under low-carbon conditions and in submersed culture. *Genetics* **169**:619–630.

79. Wei, H., N. Requena, and R. Fischer. 2003. The MAPKK kinase SteC regulates conidiophore morphology and is essential for heterokaryon formation and sexual development in the homothallic fungus *Aspergillus nidulans*. *Mol. Microbiol.* **47**:1577–1588.

80. Wei, H., K. M. Scherer, A. Singh, R. Liese, and R. Fischer. 2001. *Aspergillus nidulans* α-1,3 glucanase (mutanase), *mutA*, is expressed during sexual development and mobilizes mutan. *Fungal Genet. Biol.* **34**:217–227.

81. Wei, H., K. Vienken, R. Weber, S. Bunting, N. Requena, and R. Fischer. 2004. A putative high affinity hexose transporter, *hxtA*, of *Aspergillus nidulans* is induced in vegetative hyphae upon starvation and in ascogenous hyphae during cleistothecium development. *Fungal Genet. Biol.* **41**:148–156.

82. Wu, J., and B. L. Miller. 1997. *Aspergillus* asexual reproduction and sexual reproduction are differentially affected by transcriptional and translational mechanisms regulating *stunted* gene expression. *Mol. Cell. Biol.* **17**:6191–6201.

83. Zonneveld, B. J. M. 1974. α-1,3 glucan synthesis correlated with α-1,3 glucanase synthesis, conidiation and fructification in morphological mutants of *Aspergillus nidulans*. *J. Gen. Microbiol.* **81**:445–451.

84. Zonneveld, B. J. M. 1977. Biochemistry and ultrastructure of sexual development in *Aspergillus nidulans*, p. 59–80. *In* J. E. Smith and J. A. Pateman, *Genetics and Physiology of* Aspergillus. Academic Press, London, United Kingdom.

*Sex in Fungi: Molecular Determination
and Evolutionary Implications*
Edited by Joseph Heitman et al.
© 2007 ASM Press, Washington, D.C.

Olaf Nielsen
Richard Egel

8

The *mat* Genes of *Schizosaccharomyces pombe*: Expression, Homothallic Switch, and Silencing

LIFE CYCLE—OVERVIEW

Schizosaccharomyces pombe (fission yeast) is essentially a haploid organism. The cylindrical cells grow at the poles and divide symmetrically at a central septum. The resulting sister cells are fully equivalent by morphological and physiological criteria, and the mother-daughter asymmetry observed in budding yeasts does not apply to fission yeasts. Mating activities occur only in preparation for meiosis and sporulation. The sexual differentiation pathway is activated at the end of the vegetative growth phase when the supply of nutrients declines—about one cell division cycle before starved cells become fully stationary. Haploid cells of opposite mating types, M (Minus) and P (Plus), can then conjugate to form diploid zygotes. Usually the zygotes will proceed into meiosis and sporulation without delay, producing asci with four haploid spores. The ascus wall dissolves eventually by self-digestion to liberate the dormant spores, which allow spreading and survival under conditions unfavorable for growth. Once the supply of nutrients improves, the spores will germinate into haploid vegetative cells, thus completing the life cycle (Fig. 8.1). Only a few uncommitted zygotes can ever be diverted to diploid growth—requiring selective techniques to detect diploid colonies from such events. Diploid cells heterozygous for mating type can sporulate without conjugation.

Since zygote formation is an integral part of the sexual differentiation process, it is tightly coregulated with entry into meiosis, and the entire pathway is activated only when the cells detect a decline in the supply of nutrients. The main function of the M and P mating types is to establish a complementary recognition system by which the two cell types can identify each other, by means of secreted pheromones and corresponding membrane-embedded receptors, as well as complementary agglutinins. Hence, M and P cells differ only in the expression of a few genes, which are required for the synthesis or processing of the cell-type-specific components. These type-specific genes are under control of the *mat1* locus, which carries different mating-type genes in M and P cells. When the cellular mating type is switched, the M- or P-specifying DNA cassette at *mat1* is replaced by the alternative version, which is copied from one of two unexpressed storage loci.

In this chapter we review our current understanding of the fission yeast mating-type system. For a more detailed account of the sexual differentiation pathway we refer to recent in-depth reviews on this topic (59, 69, 89, 90).

Olaf Nielsen and Richard Egel, Institute of Molecular Biology and Physiology, University of Copenhagen, Copenhagen, Denmark.

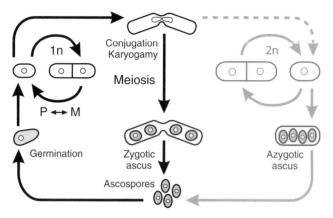

Figure 8.1 The life cycle of *S. pombe*. Vegetative cells are predominantly haploid (1n). Rarely occurring diploid growth of zygotes (2n) can be selected for in the laboratory. Homothallic switch of mating type (P ↔ M) occurs during vegetative growth.

ORGANIZATION OF THE *mat* REGION

A high degree of self-fertility is observed in Leupold's *h*[90] strain, showing about 90% sporulation in clonal cultures (54)—herein considered the wild-type strain. The mating-type region is a cluster of three ~1-kb DNA cassettes situated on the right arm of chromosome II (Fig. 8.2). The centromere-proximal cassette, *mat1*, which constitutes the expressed locus, exists in two different versions: *mat1-M* (found in M cells) and *mat1-P* (in P cells). The genes expressed from these allelic alternatives activate transcription of, respectively, M- and P-specific genes elsewhere in the genome. Two silenced storage cassettes, *mat2-P* and *mat3-M*, are situated 17 and 29 kb, respectively, centromere-distal to *mat1*. Their P or M information is utilized as the template for DNA copy transposition during mating-type switching.

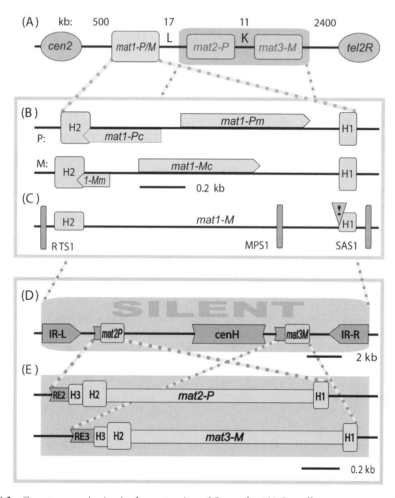

Figure 8.2 Cassette organization in the *mat* region of *S. pombe*. (A) Overall arrangement on the right arm of chromosome 2 (not to scale); (B) open reading frames at *mat1-P* and *mat1-M*; (C) *cis*-acting elements at *mat1-M* (!, site of imprint); (D) overall organization of the silent *mat2/3* domain; (E) local organization of *mat2-P* and *mat3-M*.

All three mating-type cassettes are flanked by identical pairs of "homology boxes," H1 (59 bp) and H2 (135 bp); a third one (H3, 57 bp) is found next to H2 of *mat2-P* and *mat3-M* exclusively (43). These matching sequences are crucial for the alignment of donor and acceptor loci in the initial and final stages of mating-type switching (see below).

The entire storage region containing the *mat2-P* and *mat3-M* donor cassettes is wrapped up in an inactive, heterochromatin-like structure. Two inverted repeats (IR-L and IR-R, 2.1 kb in length) serve as flanking borders to prevent spreading of the silencing effect to adjacent genes (64, 78). The silenced interval between *mat2-P* and *mat3-M* (K region) carries a 4.3-kb *cenH* segment homologous to centromeres (28) but is devoid of ordinary genes. The *cenH* segment acts as a global organizer of heterochromatin establishment. In addition, locally acting silencing signals RE2 and RE3 are placed outside the H3 box of *mat2* and *mat3*, respectively. The interval between *mat1* and *mat-2-P* (L region) contains several other genes, of which *let1* is essential; selection for its presence is thought to suppress deletions, if recombination between homologous sequences during mating-type switching goes awry. In fact, circular DNA representing such excised deletions can be observed in certain strains (11).

HETEROTHALLIC STRAINS

Facilitated by the repetitive arrangement of homology boxes at three closely spaced subloci, the *mat* region of *S. pombe* is prone to genetic rearrangements, which can yield various types of heterothallic derivatives. Two of the most frequently occurring strains, h^{+N} and h^{-S}, were already isolated at the start of fission yeast genetics from a heterogeneous stock collection culture, together with h^{90} (54). These heterothallic types are still widely used in mutant strain collections. Both can spontaneously arise from the h^{90} configuration, albeit by entirely different mechanisms (Fig. 8.3). Notably, however, no single-mutant strain equivalent to Ho endonuclease deficiency in the budding yeast *Saccharomyces cerevisiae* has ever been detected in *S. pombe*, and no homolog of the *HO* gene is present in the *S. pombe* genome.

The "stable" h^{-S} strain is purely M by mating type. Its genotype can be written as *mat1M--L--mat2:3M*. Due to a clean loop-out deletion between *mat2* and *mat3*, it has permanently lost the silent *mat-P* cassette, together with the entire K region between *mat2* and *mat3*. In turn, it carries a silent fusion cassette, *mat2:3-M*, flanked by *mat2-* and *mat3*-related sequences to the left and to the right, respectively. The complex configuration

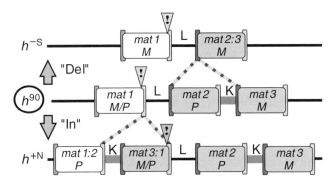

Figure 8.3 Heterothallic strains by rearrangement (spacer regions not to scale). The wild-type configuration of the homothallic h^{90} strain (middle lane) can lead to h^{-S} by deletion (Del), or h^{+N} by insertion (In), between homology boxes of different *mat* cassettes. The h^{+N} insertion results from aberrant switching events (see Fig. 8.5C). Silent cassettes are shaded; imprintable sites to initiate mating-type switches are indicated (!).

of "normal" h^{+N}, on the other hand, requires a more circuitous explanation. Instead of having a simple *P* cassette at *mat1*, this strain carries a long insertion, generating fusion cassette *mat1:2-P*, the entire K region from between *mat2* and *mat3*, and another fusion cassette *mat3:1-(M* or *P)*. Hereafter, the entire L--*mat2P*--K--*mat3M* region is the same as for h^{90}. Notably, potential sites to initiate mating-type switching are retained in both h^{-S} and h^{+N}. Yet, h^{-S} can switch only to the same mating type (M), and h^{+N} can switch only at a silenced cassette. Secondary rearrangements can regenerate the h^{90} configuration from h^{+N} or give rise to yet other heterothallic strains. Altogether, about 10 different configurations of *mat* cassettes have been described in *S. pombe* (reviewed in reference 20).

TRANSCRIPTIONAL REGULATION

The activation of mating-type genes requires profound changes in the transcriptional program in prestationary cells. The HMG-box protein Ste11 is a key transcription factor in the mating and meiosis pathway (reviewed in references 88 and 89). At its target genes, Ste11 activates transcription by binding to TTTCTTTGTT motifs present in these promoter regions. Starvation induces differentiation by causing accumulation of Ste11 and stimulation of Ste11-dependent transcription. Initially, *ste11* induction is mediated by other transcription factors in response to decreasing levels of intracellular cyclic AMP (cAMP), as well as other signals in the environmental stress response (reviewed in reference 80). In addition the *ste11* gene is autoregulated by the Ste11 protein itself. As a further safeguard, the Ste11 protein is modulated by phosphorylation during growth, but no longer in starved cells. By all these means starvation appears to trigger a positive-feedback loop to prepare the cells for sexual reproduction.

In particular, Ste11 controls expression from the DNA cassette inserted at *mat1* and hence is indirectly required for expression of all M- and P-specific genes. In addition, Ste11 is directly involved in the activation of M-specific genes (see below). Finally, most genes required for mating in both cell types (e.g., components downstream of the receptors in the pheromone signal transduction pathway) are under direct control of Ste11.

Each of the *mat1-M* and *mat1-P* cassettes encodes two divergently transcribed genes (Fig. 8.2B) (43); one of these controls the cellular mating type (*mat1-Mc* in M-cells and *mat1-Pc* in P cells), the other one is required for entry into meiosis (*mat1-Mm* in M cells and *mat1-Pm* in P cells). The six known M-specific genes (Table 8.1) are activated by the expression of *mat1-Mc*. Like Ste11, the Mat1-Mc protein contains a DNA-binding HMG-box, but it cannot activate transcription on its own. It functions by recruiting Ste11 to a weaker version of the Ste11 binding site (CTTTGTT) present in the promoter regions of M-specific genes (47). The mechanism by which the Mat1-Pc protein activates three known P-specific genes (Table 8.1) is less well understood. The *map1* gene, which encodes a putative transcription factor of the MADS-box family, is also required for P-specific gene activation (61, 86); presumably, Map1/Mat1-Pc heterodimers bind to and activate these genes, similar to the activation of a-specific genes in budding yeast.

PHEROMONE COMMUNICATION

Both pheromones are secreted peptides. P-factor is a 23-amino-acid unmodified peptide (37), made by proteolytic cleavage while a larger precursor protein is transported along the conventional secretory pathway. M-factor is a 9-amino-acid peptide, whose C-terminal cysteine is carboxymethylated and farnesylated (17)—modifications commonly found in fungal pheromones. Mature M-factor is transported out of the cell by a specific ATP-dependent ABC transporter, Mam1 (12).

The pheromones bind to G-protein-coupled receptors present at the surface of the opposite cell types, and this activates an intracellular mitogen-activated protein kinase (MAPK) signal transduction pathway, which prepares the partner cells for mating (reviewed in reference 60). The consequences of pheromone stimulation are G_1 arrest of the cell cycle, transcriptional induction of target genes, and morphological elongation towards the pheromone source. The G_1 arrest is accompanied by downregulation of the cyclin-dependent kinase complexes that activate S phase. This is mediated by a combination of increased cyclin proteolysis and induction of the cyclin-dependent kinase inhibitor protein Rum1. Arresting the cell cycle at the same stage in both cell types is, of course, essential to synchronize both parental nuclei before karyogamy. In a bulk culture, the pheromone-induced G_1 arrest occurs about one cell cycle before all the cells in a nonresponding culture stop dividing for lack of nutrients.

Both pheromone receptors, Mam2 for P-factor and Map3 for M-factor, contain characteristic seven-transmembrane domains (7TM), conveying the signal across the plasma membrane. At the inside, receptor activation leads to GDP-GTP exchange at the receptor-coupled

Table 8.1 Cell-type-specific genes, controlled by the *mat* gene products

Gene	Biochemical function	Pathway	Reference(s)
M-specific genes, controlled by *mat1-Mc* (HMG-box transcriptional cofactor)			
mam1	ABC transporter	Export of M-factor	12
mam2	7TM receptor	Response to P-factor	46
mam3	M agglutinin	Sexual cell agglutination	—[a]
mfm1-mfm3	M-factor precursor	M mating pheromone	17, 47
sxa2	Serine carboxypeptidase	Degradation of P-factor	53
P-specific genes, controlled by *mat1-Pc* (transcriptional cofactor)			
map2	P-factor precursor	P mating pheromone	37
map3	7TM receptor	Response to M-factor	74
map4	P agglutinin	Sexual cell agglutination	—[a]
SPAC1565.03	Sequence orphan	?	—[b]
P/M-specific genes, controlled by *mat1-Pm* + *mat1-Mm* (homeobox transcription factor + transcriptional cofactor)			
mei3	Inactivation of Pat1 protein kinase	Derepression of meiosis	55, 82, 85

[a]Briefly cited as unpublished data in reference 90.
[b]O. Nielsen, unpublished data.

Gpa1 G-protein, which is common to both mating types. In slight variation of the standard scheme, Gpa1 of *S. pombe* operates as a monomeric Gα subunit, without a corresponding Gβγ dimer (reviewed in reference 34). While the direct target of the activated Gpa1 is not yet known, activation of another G-protein, Ras1, is also required to induce the pheromone response. Downstream of Gpa1 and Ras1 the signal is conveyed by protein kinases. Byr2, Byr1, and Spk1 form a distinctive MAPKKK-MAPKK-MAPK activation cascade (related to mitogen-activated protein kinases). Notably, the Byr2 MAPKK can directly bind to Ras1-GTP, to be activated by the pheromone signal. Apart from this sexual function, Ras1 also contributes to the morphogenesis of the cylindrical shape in vegetative cells. The vegetative and sexual subfunctions of Ras1 appear to be activated by two different guanine nucleotide exchange factors, Efc25 and Ste6, respectively (65). In particular, the mating-specific *ste6* gene is hardly expressed at all in vegetative cells but is induced by nitrogen starvation and, further yet, by the pheromone response itself (36, 57).

Pheromone signaling induces transcription of a large number of genes required for mating and meiosis. Many of the pheromone-responsive genes are under Ste11 control, and indeed Ste11 has been demonstrated to be a direct target for the pheromone-signaling pathway (48). Hence, Ste11 activity is stimulated both by nitrogen starvation and by pheromone sensing. In fact, genetic evidence suggests that the nitrogen-starvation signal itself is partly conveyed by the pheromone-signaling pathway. It is interesting that both the pheromone response and the environmental stress response are mediated by different versions of MAPK activation models, as represented by Byr2-Byr1-Spk1 and Wak1/Win1-Wis1-Sty1, respectively. Both these signal transduction cascades converge on *ste11* expression.

The nutritional state is being monitored by additional 7TM-protein-coupled receptor components. Most work has been done on the sensing of glucose by Git3-Gpa2 (reviewed in reference 34), which only plays a minor role for sexual responsiveness in fission yeast. However, the recent characterization of Stm1-Gpa2 as a presumptive nitrogen sensor (14) has opened an important new lead to a fuller understanding. Both Git3 and Stm1 are 7TM proteins which interact with the Gα subunit of the trimeric G-protein Gpa2$_α$-Git5$_β$-Git11$_γ$. Upon activation of Git3 (or Stm1, too, presumably), Gpa2-GTP is liberated from the Gβγ dimer, thus activating Cyr1/Git2 adenylate cyclase as its downstream effector. Conversely, a decline in the intracellular concentration of cAMP is necessary for the induction of sexual activities. Due to the absence of cAMP in *cyr1*⁻ mutant cells, the pheromone-inducible G$_1$ arrest can occur even without nutritional depletion (reviewed in reference 60).

Expression of M- and P-specific genes is slightly asymmetric. While M cells produce significant levels of pheromone and agglutinin on their own, P cells do so only after being stimulated by M-factor pheromone (37). The elongation of cells towards the mating partner helps to establish cell-to-cell contact prior to mating. This process involves profound alterations in the cytoskeleton. Elongation ceases along the cellular axis; instead, tip growth is redirected towards the pheromone gradient. Again, the Ras1 protein plays an important role in these morphological changes (10). The actual merging of cell walls and membranes requires further reorganization of the cytoskeleton. A mating-specific, pheromone-induced formin protein (Fus1) is assumed to interact with common profilin (Cdc3) and tropomysin (Cdc8) to redirect still-unknown lytic enzymes to the fusion point (66).

Likewise induced by the pheromone response, cytoplasmic microtubules rearrange in bundles that agitate the nuclei. At first, these lateral movements position both nuclei with juxtaposed spindle pole bodies in the neck of the zygote, shortly followed by karyogamy. Thereafter, characteristic "horsetail" nuclear movements continue for the most part of meiotic prophase, when the telomeres—instead of centromeres—are attached to the spindle pole body (reviewed in reference 32). These movements are thought to bring the three pairs of homologous chromosomes into approximate alignment for crossing over, as a dynamic substitute for synaptonemal complexes, which are missing in *S. pombe*.

Furthermore, the meiosis-specific genes *mat1-Mm* and *mat1-Pm* are both subject to pheromone induction, linking meiosis directly to the pheromone response (2, 85). They are also designated *Mi* and *Pi* for being inducible. The corresponding Mat1-Mm and Mat1-Pm transcription factors activate the *mei3* gene as a common target (82, 85). Hence, *mei3* can be expressed only after zygote formation, where it releases the meiotic program from the restrictive bonds imposed in haploid cells. During the normal cell cycle, as well as in stationary haploid cells, expression of meiotic genes is strictly prevented. The direct inducer of meiosis, Mei2, is effectively inactivated by the Pat1 protein kinase. Therefore, Pat1 activity needs to be eliminated before meiosis can be started. This is in fact accomplished by the Mei3 protein, which strongly binds to Pat1 as an unproductive pseudosubstrate (55).

HOMOTHALLIC SWITCH

The homothallic wild-type h^{90} strain of *S. pombe* is self-fertile, allowing abundant spore formation in clonal cultures started from a single spore. Upon nitrogen starvation, about 90% of cells mate and sporulate, mostly by forming zygotes between sister cells (54). This indicates a high potential of interconversion between the alternative mating types P and M. Closer inspection of the switching pattern revealed a cyclic course through four distinguishable cell types, brought about by predominantly asymmetric cell divisions (22, 49, 58). Thereby, cells of either mating type are differentiated as unswitchable and switchable subtypes. Each mitotic cell division from a given subtype tends to give rise to different daughter cells: only one of these is preserving the subtype of the mother cell, whereas its sister converts to the next stage around the cyclic ratchet (Fig. 8.4). This model provides for switching frequencies of up to 50% per cell division on average, or 25% per newly emerging cell. The actual frequencies observed tend to be a little lower—e.g., due to unproductive switching to the same mating type or loss of the switchable state by DNA repair without a switch (see below).

The complexity of this switching pattern reflects a variety of different mechanisms that contribute to the molecular details of mating-type interconversion in *S. pombe*. Evidently, the competence of switching is determined a full cell cycle ahead of the switching event itself to be effective. In general terms, *mat1* DNA is merely "imprinted" during replication in one cell cycle, while the replacement of the resident cassette is delayed until

S phase in the subsequent cycle. Hence, the modifying imprint has to remain intact throughout the intervening stages. Correspondingly, mutations that alter the efficiency of mating-type switching can affect the various aspects of the complex switching cycle to different extents. In particular, mutations in 10 *swi* genes were allocated to three functional classes—in turn affecting imprinting at *mat1*, initiation of copy choice recombination at the H1 boxes, and resolution of cassette transposition at the H2 box boundaries (23, 30).

CREATING THE IMPRINT AT *mat1* AND MAINTAINING IT THROUGH THE CELL CYCLE

While the actual mating type of a cell is determined by the genetic information inside the expressible *mat1* cassette (P or M), the competence of switching is caused by a site- and strand-specific modification of DNA at *mat1* (reviewed in references 7 and 21). As for the biochemical nature of this imprint, interpretations have varied from double-strand (ds) break, to single-strand (ss) nick or an alkaline- and RNase-sensitive modification. The earlier notion of a ds break has been ascribed to a shearing artifact during extraction (3, 15), but the final solution of ss-specific alternatives is still pending.

The sensitizing event is affecting one or two nucleotides, which may be altered to ribonucleotides in the backbone or by yet-to-be-characterized modification at the corresponding base residues (15, 83, 84). On the other hand, the primary imprints appear to be very labile during extraction so that they are often converted to dephosphorylated nicks (3, 41). No enzyme directly implicated in the creation of the imprint has been identified as yet.

Although the structural details are still unknown, the imprinting event takes place in close association with DNA replication at *mat1*. To this effect, the corresponding replication fork must invariably approach the imprintable site at the H1 box from the centromere-distal direction on the standard map (Fig. 8.2C). Potentially interfering replication from the proximal side is effectively barred by a unidirectional terminator signal, RTS1, some 700 bp outside the H2 box at *mat1* (16).

There are other *cis*-acting sequence elements around the *mat1* H1 box that influence the efficiency of imprinting and/or mating-type switching. These are scattered over some 500 to 600 bp, which is remarkably long for a specific recognition sequence. Outside H1, spanning 130 to 160 bp away from the imprint, a switch-activating sequence (SAS1) is absolutely required; it provides a specific binding site for the Sap1 protein (5), although the mechanism of Sap1 participation is still unclear.

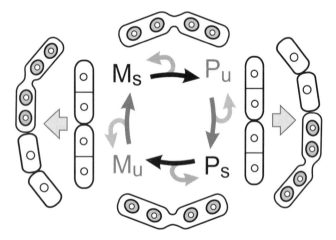

Figure 8.4 The cycle of homothallic mating-type switching in *S. pombe*, passing four stages of asymmetric cell division ($P_u \rightarrow P_u + P_s$; $P_s \rightarrow P_s + M_u$; $M_u \rightarrow M_u + M_s$; $M_s \rightarrow M_s + P_u$). When switchable cells divide (P_s or M_s), one of their respective daughters will have switched mating type. On depleted medium, this allows conjugation and ascospore formation with the unswitched sister cell. If unswitchable cells divide (P_u or M_u), both daughter cells retain the same mating type and are not mutually inhibited by mating pheromones, even though one of these becomes imprinted. This leads to a single switch in the subsequent quadruplet of cousin cells.

Inside *mat1-M*, about 340 bp away from the imprint, a directional *mat1* pausing site (MPS1) is temporarily stalling the imprint-conducive replication fork (83). Between SAS1 and MPS1, the imprint is created anew in the specific strand that corresponds to the lagging strand at the replication fork. Notably, a predominant initiation site for lagging-strand synthesis in this area has been mapped close to the MPS1. Two previously identified switching factors, Swi1 and Swi3, are part of a recently characterized "fork protection complex" (62, 63), which is implicated in the stabilization of stalled replication forks in general.

In each mating type, the nucleotides affected by the imprint are 1 or 2 thymines, starting a contiguous run of 7 Ts at *mat1-M* or 10 Ts at *mat1-P*. However, the actual sequence close to the imprint can be changed substantially without compromising its ability to be imprinted (42). Hence, the overall structure of the region appears to be more important than the local sequence. There are indications that maintaining the imprint during the cell cycle requires active protection to be shielded from surveillance by damage repair systems, perhaps by the specific stabilization of nucleosomes in this region. This shielding can be overcome by ectopic insertion of a derepressible *nmt1* promoter distal of the H1 box (35). When transcription is induced into *mat1*, the imprint is erased by transcription-coupled repair mechanisms. This newly developed inducible switching system will become very useful in the future to reveal further details of the switching mechanism in *S. pombe*.

Moreover, the imprint's capacity to be repaired becomes vitally important if both donor cassettes for mating-type switching have been deleted (50). Such strains are readily viable in *S. pombe*—very much in contrast to the analogous situation in *S. cerevisiae*, where a ds cut at the switchable *MAT* locus, as caused by HO endonuclease, is lethal in the absence of appropriate donor cassettes (51). Another characteristic difference concerns the switching pattern in diploid cells. While all switching activities cease in a/α diploid cells of *S. cerevisiae*, due to the shutdown of *HO* transcription, mating-type switching continues in diploid cells of *S. pombe*. Moreover, the two homologous *mat*-bearing chromosomes continue to follow their individual switching cycles, quite independently of one another (19).

COPY TRANSPOSITION FROM *mat2/3* TO *mat1* AND DIRECTIONAL BIAS OF DONOR CHOICE

After the imprint has been created during lagging-strand synthesis, it is located in the template for leading-strand synthesis in the S phase of the next cell cycle. This causes the incoming replication fork to be arrested precisely at the imprint (42). Although the stalled fork can be repaired in situ (mainly in the absence of donor cassettes, see above), the wild-type configuration in a homothallic strain is resolved by a switch of *mat1* cassette on the imprinted strand (Fig. 8.5). The lagging-strand template is not imprinted itself; it likely supports reinitiation of the new lagging strand at the MPS1 pausing site, ahead of the stalled and collapsing replication fork; and this newly made segment of the lagging strand is modified by a new imprint (not shown in Fig 8.5). After completion of the lagging strand throughout the conserved *mat1* cassette, this branch of the fork contributes to a yet-unswitched daughter cell, which in turn can switch at the next division, according to Fig. 8.4.

The 3′ end of the arrested leading strand (Fig. 8.5A), however, results in mating-type switching of the other daughter cell (6) (reviewed in reference 21). This end is temporarily liberated from the collapsing fork. By means of the recombinational repair machinery, it can invade the matching sequence at the H1 box of either *mat2-P* or *mat3-M* (Fig. 8.5B, i). The actual choice between these potential donor cassettes for mating-type switching is biased by the residing *mat1-M* or *mat1-P* cassette (see below). At its new target, the 3′ end serves as a primer for break-induced repair synthesis along the silent donor cassette. This does not involve a lasting attachment with its template but leads to an extended ss product. Both strands of the donor cassette reassociate after a short "displacement loop"—remaining genetically unaffected by the switching process. The neosynthesized strand is usually terminated at the opposite border of the cassette (Fig. 8.5B, ii). Potential foldback structures at the H2/H3 boundary are thought to guide the processing of the liberated 3′ end (68) (reviewed in reference 21), which then reassociates with H2 of the *mat1* cassette to be exchanged (Fig. 8.5B, iii). This neosynthesized strand is rendered ds by acting as a template itself, in the same round of replication (4). The aborted leading-strand template inside the *mat1* cassette, together with the original imprint that halted the fork, is degraded in this process.

Presumably, h^{+N} cells arise from h^{90} during aberrant mating-type switching, if molecular intermediates are not properly resolved (21, 23). If switching of *mat1-P* to M fails to be terminated at the H2/H3 boundary of *mat3-M*, repair synthesis continues throughout the entire K region, only to be resolved at the H2/H3 boundary of *mat2-P* (Fig. 8.5C). The leftmost *mat1:2-P* cassette is expressible but cannot switch. The *mat3:1* cassette, however, is still switching but cannot be expressed. Note that the local RE3 silencer from *mat3-M*

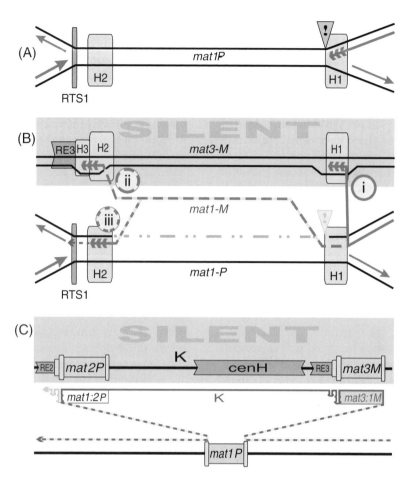

Figure 8.5 Mating-type switching at the imprinted *mat1-P* cassette. (A) A replication fork approaching from the left is halted outside *mat1* at RTS1. The leading strand approaching from the right is halted at the imprint (!). (B) Facilitated by various recombinational repair factors, the leading 3' end is liberated from H1 at *mat1*, swapping templates with H1 at *mat3-M* (i). Repair synthesis pursues throughout *mat3-M*, to be terminated at the H2/H3 boundary (ii). The processed 3' end reenters *mat1* at H2 (iii), to be joined with the arrested lagging strand at RTS1. This strand is duplicated immediately thereafter (not shown), to complete the newly formed *mat1-M* cassette. The intact strand of the resident *mat1-P* cassette is duplicated as well and imprinted anew (not shown), whereas the previously imprinted strand is degraded. (C) If switching of *mat1-P* to M fails to be terminated at the H2/H3 boundary of *mat3-M*, repair synthesis continues throughout the entire K region, only to be resolved at the H2/H3 boundary of *mat2-P*. This aberrant switching event results in the heterothallic h^{+N} configuration.

is retained in the long insertion, whereas RE2 from *mat2-P* is not. Accordingly, this strain is predominantly P by mating type. Due to the presence of silent M cassettes, h^{+N} is not completely stable but can give rise to secondary rearrangements.

The waxing and waning of switch-related conversion intermediates during S phase has been followed by specific PCR analysis (6). It took about 30 min from their first appearance to their resolution, which extends into late S to early G_2. On the other hand, in contrast to most other organisms in which heterochromatin is replicated rather late, the silent *mat2/3* region of *S. pombe* is

replicated in early S phase (44). This indicates that newly replicated donor cassettes are used as templates for the switching process. At this nascent stage, when the heterochromatin structure has not yet been fully reestablished, the donor cassettes appear to be most readily accessible, in contrast to the remaining span of the cell cycle when they are refractory to recombination events.

The choice of donor cassette is biased in such a way that the mating type of switchable cells, in fact, is switched with about 90% efficiency. Hence, cells must sense the temporary state of *mat1* (M or P) and conduct strand invasion to the opposite donor cassette accordingly. As to the mechanism of donor choice, the yield of effective mating-type switches has been diminished to about 10% by swapping the internal parts of both donor cassettes into *mat2-M* and *mat3-P*, as carried by the artificial h^{09} strain (76). This indicates that *cis*-active cues for donor choice must lie outside *mat2* and *mat3* cassettes, acting independently of the respective coding capacity inside.

More recently, the biased guidance of strand invasion has been attributed to a *mat*-specific protein complex, Swi5-Swi2, assumed to be equivalent to the Swi5-Sfr1 complex required for the homology search in general recombination (1). In P cells, localization of the Swi5-Swi2 complex is restricted to a recombination enhancer outside the H1 box at *mat3*. In M cells, however, it spreads across the entire extent of heterochromatin around *mat2* and *mat3* (39). Accordingly, preferential switching in P cells is limited to *mat3-M*. In M cells, on the other hand, the closer-positioned *mat2-P* cassette is used preferentially by proximity, even though both *mat2* and *mat3* should be available as targets. It is not yet clear whether the redistribution of Swi5-Swi2 in M cells is due to the mere presence of the *mat1-M* cassette, or rather to its differential expression at the appropriate time. In the latter case, the sensitive phase should be at G_1, since newly switched nuclei, during the long G_2 phase of the *S. pombe* cell cycle, carry both *mat1-M* or *mat1-P* cassettes on their respective sister chromatids.

The differential binding of Swi5-Swi2 is strictly dependent upon the covering of heterochromatin by Swi6 (1), which plays a central role in silencing (see below). All mutants known to affect the regular deposition of Swi6 do, in fact, compromise the biased directionality of mating-type switching, and newly isolated directionality mutants also affect the localization of Swi6 (79). Hence, the different aspects of directionality bias and silencing of donor cassettes are highly interrelated.

SILENCING OF STORAGE CASSETTES *mat2* AND *mat3*

The donor cassettes are embedded in a 20-kb stretch of heterochromatin, which is devoid of coding genes other than the internal parts of *mat2-M* and *mat3-P* themselves. The entire region differs from ordinary chromatin in various ways. Transcription of embedded genes by RNA polymerase II is severely restricted, meiotic crossovers are virtually absent, and targeted insertions by gene displacement procedures are more difficult to make than elsewhere in the genome. The heterochromatin in this area is organized by multiple pathways of nucleo-

some modification, starting from several scattered nucleation centers and followed by autocatalytic extension. Outward, it is confined by flanking boundaries to shield adjacent areas of coding genes against harmful spreading effects (reviewed in references 7 and 29).

The degree of silencing in the wild-type configuration, as well as relieving these restrictions in a mutant background, can be assessed by the occurrence of aberrant asci from haploid meiosis. This indicates that both mating types are being expressed in the same haploid cell, which is a biologically hazardous condition. More quantitatively, silencing can be assayed by expression of ectopically inserted marker genes at specified positions. Both approaches have been used extensively for mutant screening at the colony level. In complementary analyses, the distribution of different proteins at various subregions of heterochromatin is conveniently assayed by chromatin immunoprecipitation, or chromatin affinity adsorption of tagged proteins, followed by quantitative PCR of included DNA and hybridization to a panel of sequence probes to cover the range of heterochromatin.

In addition to the silenced *mat2/3* region, and similar to many other organisms, heterochromatin in fission yeast is present at 20- to 70-kb arrays of pericentromeric repeats, as well as lesser blocks at subtelomeric positions. Hence, molecular dissection of the well-defined, yet nonessential, *S. pombe mat2/3* region has been advanced as a tractable model system for heterochromatin organization in general. A full coverage of this specialized field is beyond the scope of this review, and only certain highlights are given here in summary.

Notably, the 4.3-kb *cenH* segment between *mat2* and *mat3* shares 96% sequence similarity to pericentromeric repeats (28). It is essential, yet not sufficient, for full silencing capacity; other elements are required in addition. Close to the silent cassettes, within 0.5 kb at the centromere-proximal side in both cases, repressive elements RE2 at *mat2* and RE3 at *mat3* were detected by marker gene derepression screening among appropriate deletions (25, 77).

As mentioned above, inverted repeats of 2.1 kb (IR-L and IR-R) mark the boundaries of the silent region (64, 78). They are identical in sequence but have no striking resemblance elsewhere in the entire genome. Their boundary effect cannot be duplicated by indiscriminate insertion of other inverted repeats of similar length, suggesting that they may be the target of specific factors. Among several mutants relieving the boundary effect, *epe1* has been characterized further (9), belonging to a conserved family of histone demethylases (81).

The striking similarity of 100% between IR-L and IR-R is very unusual indeed for sequences of noncoding

DNA. Equivalent repeats are present in another strain related to *S. pombe*, tentatively termed *Schizosaccharomyces kambucha* (71). Although these repeats and other intergenic sequences differ between these variant strains by 2%, IR-L and IR-R of *S. kambucha* are 100% identical with one another as well. Presumably, therefore, sequence equalization is actively maintained by mutual genetic interaction between the inverted repeats. How this is brought about, apparently without concomitant inversion of the intervening region, remains to be investigated.

The main route to heterochromatin involves specific deacetylation of histones and methylation at the same lysine residues instead (reviewed in references 24 and 29). Trimethylated lysine 9 on histone 3, as catalyzed by the SU(VAR)3-9/Clr4 methyltransferase, provides the platform for binding of chromodomain proteins, such as Swi6 and Chp2 in *S. pombe*. Among numerous histone deacetylases, Clr3 and Sir2 are especially important for *mat2/3* silencing.

Heterochromatin formation can be started in various independent ways. A widely conserved mechanism is based on small RNA interference molecules, which in this case are derived from, and limited to, the *cenH* element between *mat2* and *mat3* (31). Nucleation from the cassette-associated RE2 and RE3 elements, on the other hand, are under different control. As for RE3, two binding sites for the ATF/CREB family proteins Atf1 and Pcr1 are critical for nucleation proficiency (38). Also, the Clr3 histone deacetylase is essential at RE3, but not at *cenH* (87). Yet-to-be-characterized proteins Esp1, Esp2, and Esp3 are specifically involved in silencing at RE2 and RE3 (75). Moreover, the Sir2 histone deacetylase is highly important for Swi6 deposition at telomeres and throughout the silent *mat2/3* region, whereas overall Swi6 loading in the absence of Sir2 is little affected at centromeres (27).

EVOLUTIONARY REMARKS

For all they have in common, the fission yeast *S. pombe* and the budding yeast *S. cerevisiae* clearly share a common ancestor within ascomycetous fungi. Yet further back in evolution they also share a common ancestor with animals, including humans and other mammals. Hence, both these yeasts are used extensively as genetically tractable model systems for basic mechanisms of eukaryotic cell biology in general. As to particular aspects, however, residual similarity of either yeast with mammalian cells can vary greatly, so that their relevance as a suitable model has to be judged anew from case to case.

It is most reasonable to assume that the common ancestor of ascomycetes and basidiomycetes already was a filamentous fungus, well capable of hyphal growth (56). In particular, the capacity to form dikaryons in both groups and the similarity of forming clamp connections in basidiomycetes with forming croziers at the tip of ascogenous hyphae in ascomycetes have often been considered homologous ancestral traits. According to this view, all yeast-like fungi represent lineages of simplification by regression. In various respects the successful and diverse group of budding yeasts has moved away from the common ancestor considerably faster than the narrow group of fission yeasts (72, 73). Thereby, *S. pombe* has more faithfully retained certain basic characteristics, such as cell cycle regulation, cytokinesis, intron frequency, spliceosome machinery, and heterochromatin organization. The characteristic differences between *S. cerevisiae* and *S. pombe* indicate that these yeasts represent separate lines of independent origin. Thus, the propagation of single cells throughout the entire life cycle is assumed to have originated several times from ancestral ascomycetous fungi. In fact, yet other yeast forms have arisen independently among basidiomycetous fungi (52).

Conceivably, the budding and fission modes of cell division in yeasts have been derived from different modes of dispersal as a perpetuated minicycle of conidiation—fission yeast cells and budding yeasts resembling arthrospores and microconidia, respectively. As the principal role of microconidia in other fungi is to serve as a mobile partner in fertilization, it may not merely be fortuitous that newly shed buds of *S. cerevisiae* in the haploid phase are ready for fertilization whenever a potential mating partner is around. More similar to macroconidia, *S. pombe* cells keep on dividing vigorously irrespective of mating type or ploidy, only engaging in sexual reproduction upon nutritional deprivation.

The main events in sexual reproduction consist of syngamy (cytoplasmic fusion), karyogamy (nuclear fusion), and meiosis. In *S. pombe*, these processes principally occur in succession, without intervening nuclear divisions at heterokaryotic or diploid stages. This direct coupling is controlled by the mating-type genes at *mat1-M* and *mat1-P*. Similarly to many other fungi, the functional mating-type locus accommodates two alternative stretches of nonhomologous DNA, sometimes referred to as idiomorphs. Two of the *mat1*-encoded transcription factors induce the cell-type-specific mating pheromones and corresponding receptors. The other two combine to derepress meiosis in the diploid zygote. Successful pheromone response in turn induces various activities required for cell wall fusion, nuclear movements, and karyogamy, as well as the meiosis-directing subfunctions of the mating-type locus itself. Likewise resembling other fungi, the pheromones consist of processed peptides, one of which is prenylated and carboxymethylated. This processing has much in common with hormone processing in vertebrates as well. The main components of signal transduction, too, are widely conserved from yeasts to animals. Even one of the *mat*-encoded transcription factors, Mat1-Mc, shows significant similarity to the mammalian Sry factor, involved in male sex determination (70).

While many regulatory components have been conserved by evolution, their mutual interactions can vary considerably; for example, the mechanisms of G-protein coupling to 7TM-receptors in budding and fission yeasts—to monitor the nutritional state or the availability of potential mates—are strikingly different in many details (reviewed in reference 34). In each of these cases, a specific Gα component is charged with ATP upon receptor activation, but the downstream signaling takes various different routes. In the pheromone response of *S. cerevisiae*, the liberated Gβγ dimer does most of the downward signaling, whereas in *S. pombe*, with no corresponding Gβγ dimer being present, the Gα-ATP component itself must play an active role. Also, the main role of the Ras-like G-protein of *S. cerevisiae* is to activate adenylate cyclase in the glucose-cAMP pathway, whereas its counterpart in *S. pombe* directly affects the MAP kinase cascade in the pheromone response.

The critical inducer of meiosis in *S. pombe* is the RNA-binding Mei2 protein, but the mechanism of Mei2 action is still unclear. Mei2-like proteins are present in certain other fungi and in plants (33, 40), yet not in *S. cerevisiae*. The Pat1 protein kinase, which is counteracting Mei2 in vegetative cells, is represented by close homologs throughout ascomycetes and basidiomycetes; *S. cerevisiae* has even two of them. In fungi other than *S. pombe*, however, their biological function remains to be determined. Overall, a cascade of various transcription factors direct the ordered expression of a complex program, the step-by-step transcriptome for meiosis (57), which includes a conserved core of related proteins in comparison to budding yeasts (13, 67). About 1,000 genes are differentially expressed during the entire course from nitrogen starvation through pheromone response, meiosis, and sporulation. Among these, 75 genes have orthologs that are induced during meiosis in two different strains of *S. cerevisiae*. These define a core meiotic transcriptome, which is enriched for proteins involved in cell cycle regulation, recombination, and chromosome cohesion (57).

Differential splicing during meiosis, in a Mei2-dependent manner, has been demonstrated in *S. pombe* for *mes1* mRNA, which is necessary for meiosis II specifically (45). Other meiosis-related transcripts are likewise controlled at the level of mRNA splicing (8). Even though *S. cerevisiae* has rather few introns in vegetatively important genes, there is substantial meiosis-specific splicing in the meiotic transcriptome as well (18, 26).

The coupling of meiosis with preceding karyogamy is commonly observed in most fungi. This is likely a primitive trait going back to primordially haploid protists. The additional coupling of karyogamy with conjugation may be primitive as well; alternatively, it may represent a secondary simplification, since filamentous ascomycetes, considered ancestral to fission yeasts, are highly polymorphic in this respect. Nevertheless, the present coupling of syngamy with meiosis and sporulation in fission yeasts appears to be analogous to primordially haploid protists where meiosis-mediated sexual cycles first developed, likely in close association with the differentiation of a dormant stage in the life cycle. Only certain consequences of the homothallic mating-type switch seem to distract from such a general view, calling for a special explanation (see below).

The hallmark of meiosis is programmed recombination at a formidable scale. Just two parental sets of chromosomes can lead to vastly different progeny, diverging exponentially with the number of heteroallelic differences. If this indeed is the major driving force for the maintenance of meiotic sex, it is somewhat surprising that different yeasts have developed intricate systems of mating-type switching twice independently. The frequent selfing allowed in homothallic lineages can lead only to isogenic progeny, so that meiotic recombination runs idle for want of genetic heterogeneity. What other evolutionary driving force might have counteracted this apparent futility of homothallic meiosis in the repeated establishment of homothallism in yeasts? Ensuring the formation of dormant spores appears to be that additional factor.

The tight coupling to ascospore formation puts meiosis in a critical position as to long-term survival in a dormant state. This is especially crucial in yeasts, which generally lack the ability of mold-like fungi to produce resistant vegetative spores (conidia), in addition to sexually generated ascospores. Even though stationary yeast cells are more resistant than dividing cells, they are no match to ascospores when it comes to desiccation, freezing, or being eaten by a snail or fruit fly. Yeasts are adapted to exploit sugar-rich droplets produced by plants. Hence, if a suitable but isolated microenvironment happens to be colonized by a single yeast spore, the resulting yeast clone will briefly flourish until the limited resources are depleted. Chances are rather low, however, that these cells may ever reach another suitable substrate if no ascospores can be formed at all for lack of potential mating partners. The capability of switching mating type, therefore, has high survival value under such conditions, and genetic rearrangements to this effect are readily selected for.

It is remarkable indeed that budding and fission yeasts have reached very similar results in mobilizing silently stored information—albeit by rather different means, as to the molecular components controlling all the details. The recurrent interconversion of expressed mating-type cassettes from silenced storage sites has become a powerful asset in a mixed survival strategy. At the scale of populations, the recombinational benefit of meiosis can still prevail by interbreeding wherever a microenvironment is simultaneously seeded by spores of different descent. In addition, the total number of ascospores available throughout many disconnected and ephemeral microenvironments is effectively multiplied by the ability of isolated homothallic clones to sporulate well on their own.

Doing research in yeast genetics is both fun and earnest. Seriously enough, the commodity of yeast cells in experimental work provides a strong incentive for exploring model similarities to higher organisms, so as to learn more about fundamental aspects in our own cells as well. Still, the more casual pursuit of curiosity-driven research has revealed a lot of molecular detail on how yeasts are running their own affairs in private. Tempering with genomic DNA deliberately for a specific purpose is not a trivial matter. As fission yeast does just this in shuffling its *mat* genes during homothallic switching, it has incited scientific interest to find out how this is accomplished.

We thank Søren Kjaerulff and Michi Egel-Mitani for critical comments to the manuscript.

References

1. Akamatsu, Y., D. Dziadkowiec, M. Ikeguchi, and H. Shinagawa. 2003. Two different Swi5-containing protein complexes are involved in mating-type switching and recombination repair in fission yeast. *Proc. Natl. Acad. Sci. USA* **100:**15770–15775.

2. Aono, T., H. Yanai, F. Miki, J. Davey, and C. Shimoda. 1994. Mating pheromone-induced expression of the mat1-Pm gene of *Schizosaccharomyces pombe*: identification of signalling components and characterization of upstream controlling elements. *Yeast* **10:**757–770.

3. Arcangioli, B. 1998. A site- and strand-specific DNA break confers asymmetric switching potential in fission yeast. *EMBO J.* **17:**4503–4510.

4. Arcangioli, B. 2000. Fate of mat1 DNA strands during mating-type switching in fission yeast. *EMBO Rep.* **1:**145–150.

5. Arcangioli, B., and A. J. Klar. 1991. A novel switch-activating site (SAS1) and its cognate binding factor (SAP1) required for efficient mat1 switching in *Schizosaccharomyces pombe*. *EMBO J.* **10:**3025–3032.

6. Arcangioli, B., and R. de Lahondes. 2000. Fission yeast switches mating type by a replication-recombination coupled process. *EMBO J.* **19:**1389–1396.

7. Arcangioli, B., and G. Thon. 2004. Mating-type cassettes: structure, switching and silencing, p. 129–147. *In* R. Egel (ed.), *Molecular Biology of* Schizosaccharomyces pombe. Springer-Verlag, Berlin, Germany.

8. Averbeck, N., S. Sunder, N. Sample, J. A. Wise, and J. Leatherwood. 2005. Negative control contributes to an extensive program of meiotic splicing in fission yeast. *Mol. Cell* **18:**491–498.

9. Ayoub, N., K. Noma, S. Isaac, T. Kahan, S. I. Grewal, and A. Cohen. 2003. A novel *jmjC* domain protein modulates heterochromatization in fission yeast. *Mol. Cell. Biol.* **23:**4356–4370.

10. Bauman, P., Q. C. Cheng, and C. F. Albright. 1998. The Byr2 kinase translocates to the plasma membrane in a Ras1-dependent manner. *Biochem. Biophys. Res. Commun.* **244:**468–474.

11. Beach, D. H., and A. J. Klar. 1984. Rearrangements of the transposable mating-type cassettes of fission yeast. *EMBO J.* **3:**603–610.

12. Christensen, P. U., J. Davey, and O. Nielsen. 1997. The *Schizosaccharomyces pombe mam1* gene encodes an ABC transporter mediating secretion of M-factor. *Mol. Gen. Genet.* **255:**226–236.

13. Chu, S., J. DeRisi, M. Eisen, J. Mulholland, D. Botstein, P. O. Brown, and I. Herskowitz. 1998. The transcriptional program of sporulation in budding yeast. *Science* **282:**699–705.

14. Chung, K. S., M. Won, S. B. Lee, Y. J. Jang, K. L. Hoe, D. U. Kim, J. W. Lee, K. W. Kim, and H. S. Yoo. 2001. Isolation of a novel gene from *Schizosaccharomyces pombe*: stm1$^+$ encoding a seven-transmembrane loop protein that may couple with the heterotrimeric Galpha 2 protein, Gpa2. *J. Biol. Chem.* **276:**40190–40201.

15. Dalgaard, J. Z., and A. J. Klar. 1999. Orientation of DNA replication establishes mating-type switching pattern in S. pombe. *Nature* **400:**181–184.

16. Dalgaard, J. Z., and A. J. Klar. 2001. A DNA replication-arrest site RTS1 regulates imprinting by determining the direction of replication at mat1 in *S. pombe*. *Genes Dev.* **15:**2060–2068.

17. Davey, J. 1992. Mating pheromones of the fission yeast *Schizosaccharomyces pombe*: purification and structural characterization of M-factor and isolation and analysis of two genes encoding the pheromone. *EMBO J.* **11:**951–960.

18. Davis, C. A., L. Grate, M. Spingola, and M. Ares, Jr. 2000. Test of intron predictions reveals novel splice sites, alternatively spliced mRNAs and new introns in meiotically regulated genes of yeast. *Nucleic Acids Res.* **28:**1700–1706.

19. Egel, R. 1984. The pedigree pattern of mating-type switching in *Schizosaccharomyces pombe*. *Curr. Genet.* **8:**205–210.

20. Egel, R. 1989. Mating-type genes, meiosis and sporulation, p. 31–73. *In* A. Nasim, P. Young, B. F. Johnson (ed.), *Molecular Biology of the Fission Yeast.* Academic Press, San Diego, CA.

21. Egel, R. 2005. Fission yeast mating-type switching: programmed damage and repair. *DNA Repair* 4:525–536.

22. Egel, R., and B. Eie. 1987. Cell lineage asymmetry in *Schizosaccharomyces pombe*: unilateral transmission of a high-frequency state for mating-type switching in diploid pedigrees. *Curr. Genet.* 12:429–433.

23. Egel, R., D. H. Beach, and A. J. Klar. 1984. Genes required for initiation and resolution steps of mating-type switching in fission yeast. *Proc. Natl. Acad. Sci. USA* 81:3481–3485.

24. Ekwall, K. 2004.The roles of histone modifications and small RNA in centromere function. *Chromosome Res.* 12:535–542.

25. Ekwall, K., O. Nielsen, and T. Ruusala. 1991. Repression of a mating type cassette in the fission yeast by four DNA elements. *Yeast* 7:745–755.

26. Engebrecht, J .A., K. Voelkel-Meiman, and G. S. Roeder. 1991. Meiosis-specific RNA splicing in yeast. *Cell* 66:1257–1268.

27. Freeman-Cook, L. L., E. B. Gomez, E. J. Spedale, J. Marlett, S. L. Forsburg, L. Pillus, and P. Laurenson. 2005. Conserved locus-specific silencing functions of *Schizosaccharomyces pombe sir2⁺. Genetics* 169:1243–1260.

28. Grewal, S. I., and A. J. Klar. 1997. A recombinationally repressed region between *mat2* and *mat3* loci shares homology to centromeric repeats and regulates directionality of mating-type switching in fission yeast. *Genetics* 146:1221–1238.

29. Grewal, S. I., and J. C. Rice. 2004. Regulation of heterochromatin by histone methylation and small RNAs. *Curr. Opin. Cell Biol.* 16:230–238.

30. Gutz, H., and H. Schmidt. 1985. Switching genes in *Schizosaccharomyces pombe. Curr. Genet.* 9:325–331.

31. Hall, I. M., G. D. Shankaranarayana, K. Noma, N. Ayoub, A. Cohen, and S. I. Grewal. 2002. Establishment and maintenance of a heterochromatin domain. *Science* 297:2232–2237.

32. Hiraoka, Y., and Y. Chikashige. 2004. Telomere organization and nuclear movements, p. 191–205. *In* R. Egel (ed.), *Molecular Biology of* Schizosaccharomyces pombe. Springer-Verlag, Berlin, Germany.

33. Hirayama, T., C. Ishida, T. Kuromori, S. Obata, C. Shimoda, M. Yamamoto, K. Shinozaki, and C. Ohto. 1997. Functional cloning of a cDNA encoding Mei2-like protein from *Arabidopsis thaliana* using a fission yeast pheromone receptor deficient mutant. *FEBS Lett.* 413:16–20.

34. Hoffman, C. S. 2005. Except in every detail: comparing and contrasting G-protein signaling in *Saccharomyces cerevisiae* and *Schizosaccharomyces pombe. Eukaryot. Cell* 4:495–503.

35. Holmes, A. M., A. Kaykov, and B. Arcangioli. 2005. Molecular and cellular dissection of mating-type switching steps in *Schizosaccharomyces pombe. Mol. Cell. Biol.* 25:303–311.

36. Hughes, D. A., N. Yabana, and M. Yamamoto. 1994. Transcriptional regulation of a Ras nucleotide exchange factor gene by extracellular signals in fission yeast. *J. Cell Sci.* 107:3635–3642.

37. Imai, Y., and M. Yamamoto. 1994. The fission yeast mating pheromone P-factor: its molecular structure, gene structure, and ability to induce gene expression and G$_1$ arrest in the mating partner. *Genes Dev.* 8:328–338.

38. Jia, S., K. Noma, and S. I. Grewal. 2004. RNAi-independent heterochromatin nucleation by the stress-activated ATF/CREB family proteins. *Science* 304:1971–1976.

39. Jia, S., T. Yamada, and S. I. Grewal. 2004. Heterochromatin regulates cell type-specific long-range chromatin interactions essential for directed recombination. *Cell* 119:469–480.

40. Kaur, J., J. Sebastian, and I. Siddiqi. 2006. The *Arabidopsis mei2*-like genes play a role in meiosis and vegetative growth in *Arabidopsis. Plant Cell* 18:545–559.

41. Kaykov, A., and B. Arcangioli. 2004. A programmed strand-specific and modified nick in *S. pombe* constitutes a novel type of chromosomal imprint. *Curr. Biol.* 14:1924–1928.

42. Kaykov, A., A. M. Holmes, and B. Arcangioli. 2004. Formation, maintenance and consequences of the imprint at the mating-type locus in fission yeast. *EMBO J.* 23:930–938.

43. Kelly, M., J. Burke, M. Smith, A. Klar, and D. Beach. 1988. Four mating-type genes control sexual differentiation in the fission yeast. *EMBO J.* 7:1537–1547.

44. Kim, S. M., D. D. Dubey, and J. A. Huberman. 2003. Early-replicating heterochromatin. *Genes Dev.* 17:330–335.

45. Kishida, M., T. Nagai, Y. Nakaseko, and C. Shimoda. 1994. Meiosis-dependent mRNA splicing of the fission yeast *Schizosaccharomyces pombe mes1⁺* gene. *Curr. Genet.* 25:497–503.

46. Kitamura, K., and C. Shimoda. 1991. The *Schizosaccharomyces pombe mam2* gene encodes a putative pheromone receptor which has a significant homology with the *Saccharomyces cerevisiae* Ste2 protein. *EMBO J.* 10:3743–3751.

47. Kjaerulff, S., D. Dooijes, H. Clevers, and O. Nielsen. 1997. Cell differentiation by interaction of two HMG-box proteins: Mat1-Mc activates M cell-specific genes in *S. pombe* by recruiting the ubiquitous transcription factor Ste11 to weak binding sites. *EMBO J.* 16:4021–4033.

48. Kjaerulff, S., I. Lautrup-Larsen, S. Truelsen, M. Pedersen, and O. Nielsen. 2005. Constitutive activation of the fission yeast pheromone-responsive pathway induces ectopic meiosis and reveals Ste11 as a mitogen-activated protein kinase target. *Mol. Cell. Biol.* 25:2045–2059.

49. Klar, A. J. 1990. The developmental fate of fission yeast cells is determined by the pattern of inheritance of parental and grandparental DNA strands. *EMBO J.* 9:1407–1415.

50. Klar, A. J., and L. M. Miglio. 1986. Initiation of meiotic recombination by double-strand DNA breaks in *S. pombe. Cell* 46:725–731.

51. Klar, A. J., J. N. Strathern, and J. A. Abraham. 1984. Involvement of double-strand chromosomal breaks for mating-type switching in *Saccharomyces cerevisiae*. *Cold Spring Harb. Symp. Quant. Biol.* **49**:77–88.

52. Kurtzman, C. P. 1994. Molecular taxonomy of the yeasts. *Yeast* **10**:1727–1740.

53. Ladds, G., E. M. Rasmussen, T. Young, O. Nielsen, and J. Davey. 1996. The *sxa2*-dependent inactivation of the P-factor mating pheromone in the fission yeast *Schizosaccharomyces pombe*. *Mol. Microbiol.* **20**:35–42.

54. Leupold, U. 1950. Die Vererbung von Homothallie und Heterothallie bei *Schizosaccharomyces pombe*. *C. R. Trav. Lab. Carlsberg Ser. Physiol.* **24**:381–480.

55. Li, P., and M. McLeod. 1996. Molecular mimicry in development: identification of *ste11*⁺ as a substrate and *mei3*⁺ as a pseudosubstrate inhibitor of *ran1*⁺ kinase. *Cell* **87**:869–880.

56. Liu, Y. J., and B. D. Hall. 2004. Body plan evolution of ascomycetes, as inferred from an RNA polymerase II phylogeny. *Proc. Natl. Acad. Sci. USA* **101**:4507–4512.

57. Mata, J., R. Lyne, G. Burns, and J. Bahler. 2002. The transcriptional program of meiosis and sporulation in fission yeast. *Nat. Genet.* **32**:143–147.

58. Miyata, H., and M. Miyata. 1981. Mode of conjugation in homothallic cells of *Schizosaccharomyces pombe*. *J. Gen. Appl. Microbiol.* **27**:365–369.

59. Nielsen, O. 2004. Mating-type control and differentiation, p. 281–296. *In* R. Egel (ed.), *Molecular Biology of* Schizosaccharomyces pombe. Springer-Verlag, Berlin, Germany.

60. Nielsen, O., and J. Davey. 1995. Pheromone communication in the fission yeast *Schizosaccharomyces pombe*. *Semin. Cell Biol.* **6**:95–104.

61. Nielsen, O., T. Friis, and S. Kjaerulff. 1996. The *Schizosaccharomyces pombe map1* gene encodes an SRF/MCM1-related protein required for P-cell specific gene expression. *Mol. Gen. Genet.* **253**:387–392.

62. Noguchi, E., C. Noguchi, L. L. Du, and P. Russell. 2003. Swi1 prevents replication fork collapse and controls checkpoint kinase Cds1. *Mol. Cell. Biol.* **23**:7861–7874.

63. Noguchi, E., C. Noguchi, W. H. McDonald, J. R. Yates III, and P. Russell. 2004. Swi1 and Swi3 are components of a replication fork protection complex in fission yeast. *Mol. Cell. Biol.* **24**:8342–8355.

64. Noma, K., C. D. Allis, and S.I. Grewal. 2001. Transitions in distinct histone H3 methylation patterns at the heterochromatin domain boundaries. *Science* **293**:1150–1155.

65. Papadaki, P., V. Pizon, B. Onken, and E. C. Chang. 2002. Two Ras pathways in fission yeast are differentially regulated by two Ras guanine nucleotide exchange factors. *Mol. Cell. Biol.* **22**:4598–4606.

66. Petersen, J., O. Nielsen, R. Egel, and I. M. Hagan. 1998. FH3, a domain found in formins, targets the fission yeast formin Fus1 to the projection tip during conjugation. *J. Cell Biol.* **141**:1217–1228.

67. Primig, M., R. M. Williams, E. A. Winzeler, G. G. Tevzadze, A. R. Conway, S. Y. Hwang, R. W. Davis, and R. E. Esposito. 2000. The core meiotic transcriptome in budding yeasts. *Nat. Genet.* **26**:415–423.

68. Rudolph, C., C. Kunz, S. Parisi, E. Lehmann, E. Hartsuiker, B. Fartmann, W. Kramer, J. Kohli, and O. Fleck. 1999. The *msh2* gene of *Schizosaccharomyces pombe* is involved in mismatch repair, mating-type switching, and meiotic chromosome organization. *Mol. Cell. Biol.* **19**: 241–250.

69. Shimoda, C., and T. Nakamura. 2004. Control of late meiosis and ascospore formation, p. 311–327. *In* R. Egel (ed.), *Molecular Biology of* Schizosaccharomyces pombe. Springer-Verlag, Berlin, Germany.

70. Sinclair, A. H., P. Berta, M. S. Palmer, J. R. Hawkins, B. L. Griffiths, M. J. Smith, J. W. Foster, A. M. Frischauf, R. Lovell-Badge, and P. N. Goodfellow. 1990. A gene from the human sex-determining region encodes a protein with homology to a conserved DNA-binding motif. *Nature* **346**:240–244.

71. Singh, G., and A. J. Klar. 2002. The 2.1-kb inverted repeat DNA sequences flank the *mat2,3* silent region in two species of *Schizosaccharomyces* and are involved in epigenetic silencing in *Schizosaccharomyces pombe*. *Genetics* **162**:591–602.

72. Sipiczki, M. 2000. Where does fission yeast sit on the tree of life? *Genome Biol.* **1**(2):REVIEWS1011. PMID: 11178233

73. Sipiczki, M. 2004. Fission yeast phylogenesis and evolution, p. 431–443. *In* R. Egel (ed.), *Molecular Biology of* Schizosaccharomyces pombe. Springer-Verlag, Berlin, Germany.

74. Tanaka, K., J. Davey, Y. Imai, and M. Yamamoto. 1993. *Schizosaccharomyces pombe map3*⁺ encodes the putative M-factor receptor. *Mol. Cell. Biol.* **13**:80–88.

75. Thon, G., and T. Friis. 1997. Epigenetic inheritance of transcriptional silencing and switching competence in fission yeast. *Genetics* **145**:685–696.

76. Thon, G., and A. J. Klar. 1993. Directionality of fission yeast mating-type interconversion is controlled by the location of the donor loci. *Genetics* **134**:1045–1054.

77. Thon, G., K. P. Bjerling, and I. S. Nielsen. 1999. Localization and properties of a silencing element near the *mat3-M* mating-type cassette of *Schizosaccharomyces pombe*. *Genetics* **151**:945–963.

78. Thon, G., P. Bjerling, C. M. Bunner, and J. Verhein-Hansen. 2002. Expression-state boundaries in the mating-type region of fission yeast. *Genetics* **161**:611–622.

79. Thon, G., K. R. Hansen, S. P. Altes, D. Sidhu, G. Singh, J. Verhein-Hansen, M. J. Bonaduce, and A. J. Klar. 2005. The Clr7 and Clr8 directionality factors and the Pcu4 cullin mediate heterochromatin formation in the fission yeast *Schizosaccharomyces pombe*. *Genetics* **171**:1583–1595.

80. Toone, W. M, and N. Jones. 2004. Stress responses in *S. pombe*, p. 57–72. *In* R. Egel (ed.), *Molecular Biology of* Schizosaccharomyces pombe. Springer-Verlag, Berlin, Germany.

81. Tsukada, Y., J. Fang, H. Erdjument-Bromage, M. E. Warren, C. H. Borchers, P. Tempst, and Y. Zhang. 2006. Histone demethylation by a family of JmjC domain-containing proteins. *Nature* **439**:811–816.

82. Van Heeckeren, W. J., D. R. Dorris, and K. Struhl. 1998. The mating-type proteins of fission yeast induce meiosis

by directly activating *mei3* transcription. *Mol. Cell. Biol.* **18:**7317–7326.

83. **Vengrova, S., and J. Z. Dalgaard.** 2004. RNase-sensitive DNA modification(s) initiates *S. pombe* mating-type switching. *Genes Dev.* **18:**794–804.

84. **Vengrova, S., and J. Z. Dalgaard.** 2006. The wild-type *Schizosaccharomyces pombe mat1* imprint consists of two ribonucleotides. *EMBO Rep.* **7:**59–65.

85. **Willer, M., L. Hoffmann, U. Styrkarsdottir, R. Egel, J. Davey, and O. Nielsen.** 1995. Two-step activation of meiosis by the *mat1* locus in *Schizosaccharomyces pombe. Mol. Cell. Biol.* **15:**4964–4970.

86. **Yabana, N., and M. Yamamoto.** 1996. *Schizosaccharomyces pombe map1*⁺ encodes a MADS-box-family protein required for cell-type-specific gene expression. *Mol. Cell. Biol.* **16:**3420–3428.

87. **Yamada, T., W. Fischle, T. Sugiyama, C. D. Allis, and S. I. Grewal.** 2005. The nucleation and maintenance of heterochromatin by a histone deacetylase in fission yeast. *Mol. Cell* **20:**173–185.

88. **Yamamoto, M.** 1996. Regulation of meiosis in fission yeast. *Cell Struct. Funct.* **21:**431–436.

89. **Yamamoto, M.** 2004. Initiation of meiosis, p. 297–309. *In* R. Egel (ed.), *Molecular Biology of* Schizosaccharomyces pombe. Springer-Verlag, Berlin, Germany.

90. **Yamamoto, M., Y. Imai, and Y. Watanabe.** 1997. Mating and sporulation in *Schizosaccharomyces pombe*, p. 1037–1106. *In* J. R. Pringle, J. R. Broach, and E. W. Jones (ed.), *The Molecular and Cellular Biology of the Yeast* Saccharomyces, vol. 3. Cold Spring Harbor Laboratory Press, Cold Spring Harbor, NY.

Color Plates

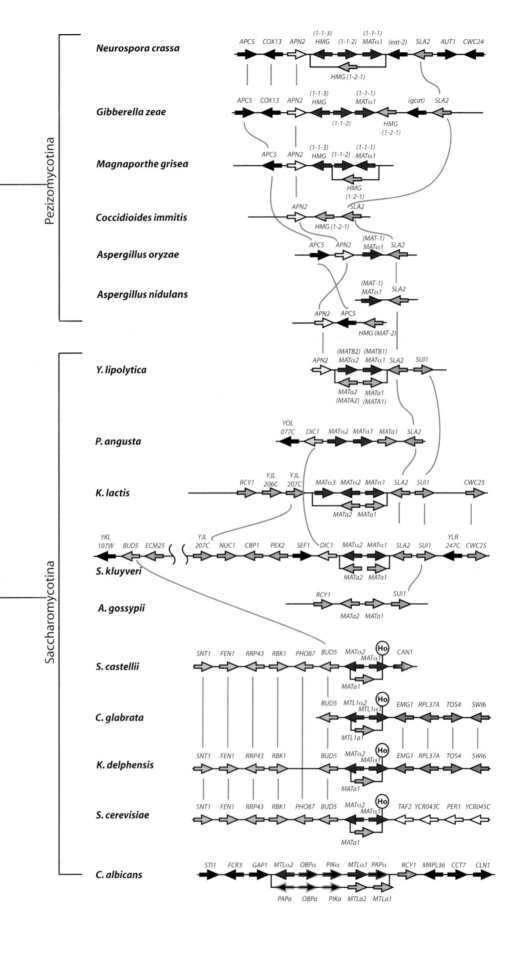

Color Plate 1 (chapter 1) Comparative organization around the *MAT* locus in the Ascomycota. The main horizontal line shows the organization of the *MAT* locus in homothallic species, or the α idiomorph (where known) in heterothallic species. The organization of the **a** idiomorph is represented by the offset box below the α idiomorph. The nomenclature suggested in reference 67 is used for the Pezizomycotina (e.g., 1-1-1 represents *MAT1-1-1*). Orthologous genes are connected by gray lines. Conserved groups of genes are indicated by color: red, α idiomorph; green, **a** idiomorph; yellow, *APN2*; purple, *SLA2* and homologs of *S. cerevisiae* XIV; orange, homologs of *S. cerevisiae* chromosome X; blue, homologs of *S. cerevisiae* chromosome III (*YCR033W-YCR038W*); white, homologs of *S. cerevisiae* chromosome III (*YCR042C-YCR045C*); gray, homologs of *S. cerevisiae* chromosome XI (*YLR186W-YLR182W*); gray gradient, *CAN1* (*YEL063C*). The position of an Ho endonuclease site in *MATα1* is marked where present. The figure was redrawn from Butler et al. (9), with additional information for *G. zeae* (73), *M. grisea* (http://www.broad.mit.edu/annotation/fgi/ and GenBank accession numbers AB080673 and AB080672), *C. immitis* (http://www.broad.mit.edu/annotation/fgi/), *Aspergillus* (21), *Y. lipolytica* (17), and *A. gossypii* (15).

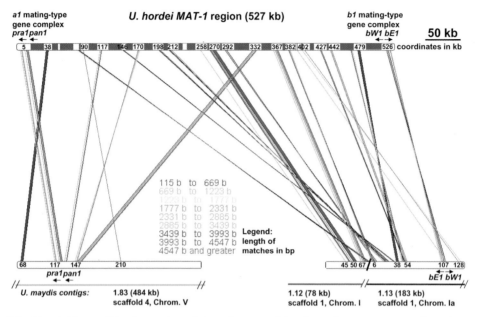

Color Plate 2 (chapter 23)   Comparison of the regions containing the mating-type gene complexes from *U. hordei* and *U. maydis*. The top box represents the 527-kb *U. hordei MAT-1* region (refer to EMBL accession no. *AM118080*). The bottom boxes represent the *U. maydis* contigs harboring the homologs (compare with Fig. 23.2). The overview illustrates the gross rearrangements present in these areas and illustrates the conservation of synteny in local regions, as well as the accumulation of repetitive sequences (blue boxes) in *U. hordei MAT-1*. See the legend to Fig. 23.2 for explanation of gene names. Note that the genes are not drawn to scale and the figure indicates the approximate sizes of the blocks of genes to display their arrangements and orientations relative to the repeats found in the *MAT-1* locus. Numbers refer to coordinates in kilobases. See reference 8 for more details. Reprinted from *Fungal Genetics & Biology* (reference 8, Fig. 2) © 2006, with permission from Elsevier.

Sex in Fungi: Molecular Determination
and Evolutionary Implications
Edited by Joseph Heitman et al.
© 2007 ASM Press, Washington, D.C.

James E. Haber

9

Decisions, Decisions:
Donor Preference during Budding Yeast Mating-Type Switching

Many fungi have homothallic life cycles in which a haploid spore can give rise to cells of the opposite mating type, leading to conjugation and the formation of diploids able to go through meiosis. Mating-type switching in the budding yeast *Saccharomyces cerevisiae* has been particularly well studied, both in terms of the homologous recombination mechanism that leads to the switch itself and the remarkable donor preference system that ensures efficient exchange of mating types. Sexual identity is controlled by the mating-type (*MAT*) locus (reviewed in references 12, 16, and 52). In *MATα* cells, two key proteins are encoded: Matα1, a positive regulator of a set of α-specific genes, and Matα2, a corepressor of both a-specific genes and, in diploids, of a more general set of haploid-specific genes (44). *MATa* encodes Mata1, which is not required for expressing a-specific genes but is a corepressor with Matα2, so that *MATa*/*MATα* diploids turn off haploid mating functions and are capable of initiating meiosis. The difference between *MATa* and *MATα* is the presence of 650 and 750 bp, respectively, of "Y" sequences that encode most of the *MAT* open reading frames (ORFs), but *MATa* and *MATα* share additional identical sequences that include extensions of the coding regions (X and Z1).

Beyond the coding regions, *MAT* is defined in terms of sequences that it shares with two other loci, *HML* and *HMR*, which are located near the two ends of the same chromosome (Fig. 9.1). *HML*, which normally carries Yα sequences and is designated *HMLα*, shares regions W, X, Z1, and Z2, whereas *HMRa* shares only X and Z1. *HML* and *HMR* serve as donors in the recombination/replacement process of *MAT* switching; they differ from *MAT* in that they are not expressed, even though they carry all of the promoter/enhancer sequences needed for expression. Flanking *HML* and *HMR* are *cis*-acting silencer sequences, designated E and I, that recruit the Sir2 histone deacetylase and the Sir3 and Sir4 proteins. These proteins establish a region of heterochromatin characterized by a highly ordered array of nucleosomes and the absence of transcription (28, 40, 60). The silencer sequences around *HML* and *HMR* are different, and indeed the degree of silencing of the two loci is not equivalent. For example, mutations that prevent acetylation of four lysines in histone H4 lead to the complete unsilencing of *HMLα* but have much less effect on the silencing of *HMRa* (38). Moreover, when a temperature-sensitive Sir3 protein is inactivated, the kinetics of unsilencing *HML* are significantly

James E. Haber, Brandeis University, MS029 Rosenstiel Center, Waltham, MA 02454-9110.

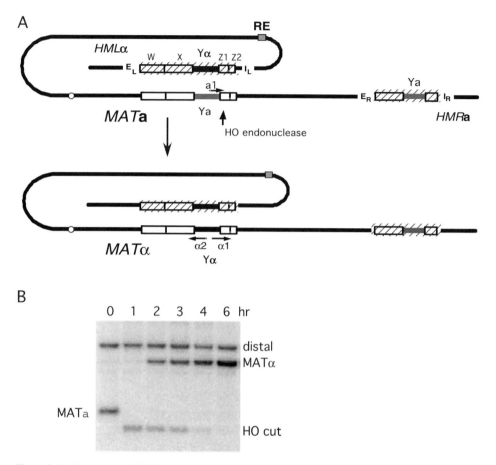

Figure 9.1 Mating-type (*MAT*) gene switching in *S. cerevisiae*. (A) HO endonuclease cleavage of *MAT*a promotes homologous recombination with one of two donor loci, in this case *HML*α, to repair the locus and to replace the Ya sequences with Yα sequences. Ya and Yα encode regulators of a-, α-, and haploid-specific genes to establish the cell's mating type. *MAT* shares homology with both *HML* and *HMR* in the X and Z1 regions and shares additional homology with *HML* in W and Z2. Both *HML* and *HMR* are maintained in an unexpressed and heterochromatic state (shown as hatched lines) by Sir2-mediated gene silencing that is organized through the E and I silencer sequences. The preferential use of *HML* by *MAT*a cells and the selection of *HMR* by *MAT*α cells is controlled by a small *cis*-acting RE. (B) Physical monitoring of *MAT*a switching to *MAT*α. The *GAL::HO* gene was induced for 1 h, and DNA was extracted at intervals for Southern blot analysis. The StyI restriction endonuclease cleaves in Ya but not in Yα, so that there are different restriction fragments for *MAT*a and *MAT*α probed with a *MAT*-distal probe that also hybridizes with a common, more distal segment. In this experiment *HMR* was deleted (Southern blot courtesy of Neal Sugawara).

faster than for *HMR* (A. Walther and J. E. Haber, unpublished data).

The mechanism of *MAT* switching emerged from the insightful pioneering work of Takano and Oshima (34, 50); they first imagined that *MAT* switching involved the transposition of "controlling elements" from the *HML* and *HMR* loci to *MAT*, where they would activate expression of already-present mating-type-specific genes. Important insights came from the often-unpublished work of Hawthorne (15) and culminated in the insightful "cas-

sette" model of Hicks, Strathern, and Herskowitz (17). This model suggested that *MAT* switching involved the replacement of mating-type-specific gene sequences at the *MAT* locus by mating-type-specific ORF sequences copied from either *HML* or *HMR*, where they lacked promoters, but would be expressed only when present at *MAT*. There were many important contributors to the evolution of this idea to our present understanding, where there are silenced, but complete mating-type sequences at the donors. Readers are encouraged to read earlier, historical reviews about the characterization of *MAT*-encoded regulatory genes and the silencing of *HML* and *HMR* (10, 11, 45).

MAT switching is catalyzed by the expression of the *HO* gene encoding HO endonuclease, which makes a 4-bp 3' overhanging double-stranded break (DSB) a few base pairs from the Y/Z1 border, in Z1 (23, 32). The HO recognition site is large ($\geq$24 bp for efficient cleavage) and somewhat degenerate, as both the Ya-Z1 and Yα-Z1 regions can be cut efficiently. The difference in the recognition of the cleavage sites in *MAT***a** and *MAT*α is illustrated in the alleles in the sequenced S288c strain. In that strain *MAT***a** actually carries an A-to-T base-pair variant in position Z_{11} that greatly reduces HO cleavage compared to most *MAT***a** alleles (41). This "stuck" mutation actually is present in *HMR***a** (i.e., *HMR***a**-*stk*), and when *MAT*α is induced to switch, the cells become *MAT***a**-*stk*. When *MAT***a**-*stk* switches using *HML*α, many cells switch to *MAT*α, replacing the Z_{11} allele back to A, but some cells retain the Z_{11}T variant and become *MAT*α-*stk*, which proves to be almost completely resistant to HO cleavage and also Matα1$^-$ and sterile. So the portion of the cleavage site in Ya or Yα influences the efficiency of cleavage.

The *HO* gene is under elaborate cell cycle and developmental control so that it is only expressed in "mother" cells at the late G_1 phase of the cell cycle (27, 30, 31). Among the unusual features of this regulation is the asymmetric localization of the mRNA encoding the Ash1 repressor of *HO* transcription to daughter cells (8, 27) so that *HO* is transcribed only in mother cells.

When *HO* is placed under the control of a galactose-inducible promoter (21), *MAT* switching can be initiated synchronously at any point in the cell cycle (7). There is no significant difference in the kinetics of switching whether *HO* is induced in asynchronous cells or in G_2-arrested cells or when a composite regulatory region is used that restricts *HO* expression to the normal G_1 phase, but only in the presence of galactose. The HO protein is very rapidly degraded, so it is possible to deliver a "pulse" of HO to cells and follow the subsequent repair of the DSB. Because the switching event is so synchronous, it is possible to follow these molecular events of *MAT* switching in great detail. Figure 9.1B shows a Southern blot in which one can observe the kinetics of replacing *MAT***a** by *MAT*α, a process that takes about 1 h. HO-induced recombination involving other DNA sequences can be studied by inserting a cloned HO cleavage site into other regions (33, 43). The DSB can be repaired in two different ways. About 10% of the time the complementary 4-bp 3' single-stranded overhanging ends can anneal and religate using the end-joining proteins common to all

eukaryotes: the Ku70/Ku80 proteins, DNA ligase 4, Xrcc4 (known as Lif1), and XLF-Cernunnos (known as Nej1) (53). But most DSBs are repaired by a gene conversion event leading to the replacement of one *MAT* allele by another.

MAT SWITCHING BY GENE CONVERSION

MAT switching occurs by a mechanism of DSB repair known as synthesis-dependent strand annealing (SDSA) (Fig. 9.2). There are a number of distinct, slow steps in *MAT* switching that have been identified by a combination of Southern blot and PCR analyses, as well as by chromatin immunoprecipitation techniques, that monitor the recruitment of recombination proteins to the sequences undergoing recombination/repair (Fig. 9.2). This process has been extensively reviewed (12, 13), and here we can be satisfied with a brief description. The first key step is the 5' to 3' resection of DSB ends to leave long (and remarkably stable) 3'-ended single-stranded DNA (ssDNA) that first binds the ssDNA binding RPA proteins and then assembles a Rad51 nucleoprotein filament that is the active complex that can locate homologous sequences in order to repair the DSB by homologous recombination (56). Rad51 loading requires Rad52 and is assisted by the Rad55-Rad57 proteins (47). Resection requires Cdk1 kinase activity, and thus, neither resection nor *MAT* switching occurs in G_1-arrested cells (2, 19). This is not a contradiction with the assertion above that normal *MAT* switching is naturally a G_1 event, as it occurs after the "start" point when Swi4/Swi6 activates transcription of *HO* and other G_1-specific genes. *MAT* switching occurs efficiently in cells blocked after "start" but before initiating S phase by inhibiting Cdc7 kinase (19).

Once Rad51 binds to ssDNA at *MAT*, there is a search for homology that culminates in strand invasion and the formation of a D-loop. Strand invasion can apparently occur without the Rad54 protein, but this step is not entirely normal, as later steps fail to occur in this mutant (47, 62). It now appears that the Snf2-Snf5 chromatin remodeling enzymes are needed for strand invasion (4). After strand invasion, the PCNA clamp protein is required for the initiation of new DNA synthesis (primer extension) (57). The great majority of repair events occur by an SDSA mechanism in which the newly extended strand is displaced and can anneal with the other end of the DSB, allowing all the newly synthesized DNA to be located at the recipient locus (20). Before the second strand can be synthesized, however, the single-stranded Ya region must be clipped off by a complex using the Rad1-Rad10 nucleotide excision repair endonuclease and the Msh2-Msh3 mismatch repair proteins

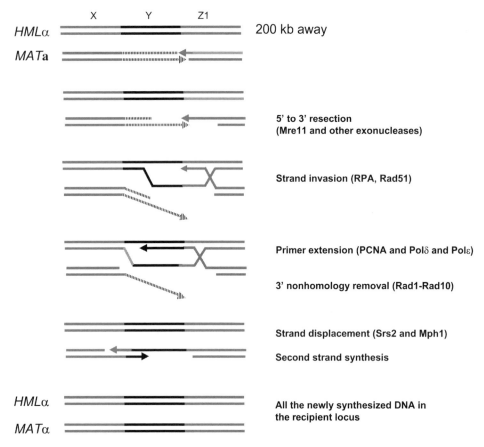

Figure 9.2 Molecular events during *MAT*a switching. Following HO cleavage, the ends of the DSB are resected by 5′ to 3′ exonucleases, leaving 3′-ended ssDNA that first recruits the ssDNA binding complex, RPA, and then the Rad51 recombinase. The Rad51 nucleoprotein filament engages in a search for homologous sequences that will serve as a template for DSB repair. Rad51 promotes strand exchange and the formation of a D-loop in the Z region. PCNA-dependent DNA synthesis copies the Yα sequences into the adjacent W region. Unlike normal semiconservative DNA replication where the newly synthesized strand remains base paired with its template, the new strand is displaced by 3′ to 5′ helicases, allowing it to base-pair with complementary single-stranded W sequences on the opposite side of the DSB. Removal of the nonhomologous 3′-ended segment by Rad1-Rad10 endonuclease, assisted by the mismatch repair proteins, Msh2-Msh3, allows the initiation of the DNA synthesis of the second strand. All newly synthesized DNA ends up in the recipient, leaving the donor unaltered.

(35, 46). Analysis of various temperature-sensitive mutations in essential DNA replication genes has shown that the short patch of new DNA synthesis in *MAT* switching does not require the Mcm helicase proteins or Polα-primase (57) or the Cdc7 kinase (19). How the late steps in recombination occur is not understood in detail.

MICROSCOPIC ANALYSIS OF *MAT* SWITCHING

The search for homology and subsequent steps in recombination can also be studied microscopically, using fluorescently tagged chromosomal regions near *MAT*

and adjacent to *HML*. A revolution in chromosome biology occurred with the development of fluorescently tagging chromosomal regions using arrays of bacterial LacO or TetO operators to which the cognate LacI-green fluorescent protein (GFP) or TetR-GFP fluorescent fusion repressor proteins can bind. Examination of cells collected and fixed at intervals during *MAT* switching indicated that *HML* and *HMR* were usually not juxtaposed prior to the induction of the DSB; however, after HO induction the spots near *HML* and *MAT* frequently merged and remained associated for about 1 h, consistent with other measurements of the duration of *MAT* switching (3). About 40% of the cells showed such pairing at the maximum, in a population where about 60% of cells actually completed *MAT* switching. The lower proportion of cells with paired spots than of those that completed switching could be explained if the process of recombination was not entirely synchronous so that some cells completed repair earlier than others; alternatively it is possible that the association between *MAT* and its donor is not entirely continuous.

Recently Houston and Broach (18) used live-cell imaging to study the association of *HML* with an HO-cut *MAT* locus, each marked with adjacent GFP-tagged arrays. Consistent with our observation of pairing of GFP-tagged *HML* and *MAT* after HO induction, they saw such HO-induced pairing, where two spots are overlapping, but they reported that the associations are very short-lived (≤6 min) and occur repeatedly. Between these periods of apparent synapsis the two spots move as far apart as they did before HO induction. The authors suggest that the periods of association represent short periods of nonprocessive DNA synthesis in which the copying of about 700 bp of new DNA synthesis would require several cycles of pairing and elongation. These results challenge several notions about what has been learned from examining *MAT* switching in bulk populations.

The idea that there might be a series of short replication steps is not inconsistent with the finding that there is about a 30-min delay from the initiation of the first primer extension after strand invasion to the completion of switching (61). However, monitoring of *MAT* switching of synchronized cells in solution by Southern blotting, PCR, or chromatin immunoprecipitation has previously been interpreted to indicate that synapsis results in a stable association; but this switching could be explained if a steady state of pairings and unpairings were achieved so that most of the images in fixed cells would appear to have synapsed chromosome regions. It is possible that the dissociations of the two spots reflect nonprocessive DNA synthesis, but it is hard to understand why then the spots go completely apart and then, in an unknown way, find the right location again (or must there be a completely new homology search?). In SDSA mechanisms it is assumed that eventually the newly synthesized strand is dissociated from its template, but here it seems that the entire replication machinery would be dissociated. Other studies of HO-induced recombination suggest that there is no significant replication slippage when gene conversions must copy repeated sequences (although there are frequent out-of-register annealings when the newly synthesized strands complete SDSA) (36, 37). It would be of great interest to examine synapsis in cells that were able to accomplish normal strand invasion but not carry out new DNA synthesis, for example, in cells with a temperature-sensitive PCNA protein.

Unfortunately the Houston and Broach (18) experiments stopped before there was evidence that switching was complete; if switching was complete, the *HML* and *MAT* spots should have come apart and resumed their unpaired behavior, as seen before HO induction. One explanation for the findings of Houston and Broach is that they fluorescently tagged the left side of the HO-cut *MAT* locus and the right side of *HML*, whereas the measurements made on fixed cells used *MAT* and *HML* both marked on the same (right) side (3). During *MAT* switching, the left side of *MAT*, with 700 bp of nonhomologous sequences, cannot efficiently initiate recombination but may wait until resection exposes the W and X regions that can then anneal with the strand that was synthesized after strand invasion in the homologous Z regions. The on-and-off associations of the left-side mark at *MAT* could be explained by the fact that this end of the DSB is not obliged to invade or stay associated with the donor. This question needs to be resolved, but in any case these live cell measurements will provide the basis for describing the way that cells search for homologous sequences and how DNA repair synthesis occurs.

GENETIC CONTROL OF DONOR PREFERENCE

There is a another, fascinating aspect of *MAT* switching—donor preference—that enables *MAT*a cells to recombine with *HML*α about 90% of the time and with *HMR*a only about 10%. Similarly, *MAT*α cells recombine 90% of the time with *HMR*a. With this arrangement of alleles, donor preference ensures that most switches will produce cells of the opposite mating type. However, in nature there are strains in which *HML* carries Ya and not Yα and *HMR* carries Yα, but the rules of *MAT* switching and donor utilization are unchanged (14, 22, 34, 50, 64). Donor preference is also not changed if *HML*α is deleted and replaced with *HMR*α (58).

Donor choice is enacted through a *cis*-acting "recombination enhancer" (RE) located about 17 kb from *HML* on the left arm of chromosome III (64). To facilitate the analysis of donor preference, we replaced *HMR*a by *HMR*α-*B* (differing from normal α sequences by a single base-pair change that creates a BamHI restriction site). This makes it possible to examine on a Southern blot what proportion of cells that did switch from *MAT*a used *HML*α or used *HMR*α-*B* (Fig. 9.3). How RE was found involved making a large number of chromosomal truncations and internal deletions; during this work we found that *HML*α could be moved to

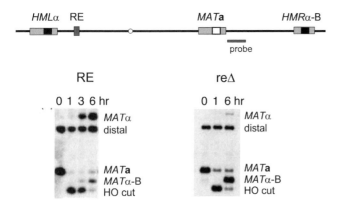

Figure 9.3 Kinetics and outcomes of *MAT*a switching influenced by RE. Following HO cleavage, *MAT*a can recombine with either *HML*α or with *HMR*α-B (BamHI). A galactose-inducible *HO* gene is turned on for 1 h, sufficient to cleave *MAT*a in nearly all cells, as shown in the Southern blots below. Cells are returned to glucose medium to prevent further HO cleavage, and the appearance of DSB repair products can be seen at 3 and 6 h. In cells with an intact RE, 90% of the switches are to *MAT*α (i.e., using *HML*α) and a few switch to *MAT*α-B (using *HMR*α-B). Fewer than 10% of the cells repair the DSB by nonhomologous end joining, restoring *MAT*a. When RE is deleted (reΔ), nearly all the gene conversions use *HMR*α-B as the donor. In the Southern blots, DNA is digested with the StyI that does not cleave in Yα and with BamHI, allowing all three outcomes to be distinguished. An additional StyI fragment, distal to *MAT*, also hybridizes to the probe. Southern blots provided by Eric Coïc.

other locations along the left arm of chromosome III and its preferential usage in *MAT*a cells was maintained (65). Surprisingly, when RE is deleted, *MAT*a cells do not exhibit 50:50 usage of *HML*α; rather, without RE *HML* usage falls to about 10%, similar to what is seen in *MAT*α cells (64).

RE lies within a 2.5-kb region lacking any ORFs between the *KAR4* and *SPB1* genes. Identification of the important sequences in this region came from the comparison of DNA sequences of the same, conserved region in other *Saccharomyces* species (63). Analogous RE sequences from *S. bayanus* and from *S. carlsbergensis* work in place of the *S. cerevisiae* RE, and an alignment of their sequences revealed several well-conserved regions, unimaginatively named A, B, C, D, and E (Fig. 9.4). The most striking feature of these sequences is that both D and E have up to 15 iterations of the motif TTT(A/G), a feature not found in any other part of the genome. Within this repeated region are multiple consensus sequences for the binding of the forkhead family of transcription factors (see below). There are several "sterile" (i.e., noncoding) transcripts in the region. The entire RE region can be replaced with much smaller segments that have full or nearly full activity. A 275-bp region embracing A, B, C, D, and E and deleted for all other sequences in the intergenic region has been well studied (48, 63, 64).

The key to how RE is regulated came from the discovery that region C contains a Matα2-Mcm1 repressor binding site that turns off a-specific genes elsewhere in the genome (49, 63); but here there are no obvious genes. Indeed there is a second Matα2-Mcm1 binding site at the other end of the region. RE is conserved in the sensu stricto *Saccharomyces* spp., but it is absent in more distant relatives that still have an HO gene, including *S. servazzi*, *S. castellii*, and more distantly related fungi in which *KAR4* and *SPB1* are still linked (68) (see also the Yeast Gene Order Browser [http://wolfe.gen.tcd.ie/ygob/]).

The idea that RE is an activator of the use of *HML* and that this function is repressed in *MAT*α received direct support from the demonstration that Matα2-Mcm1 binds

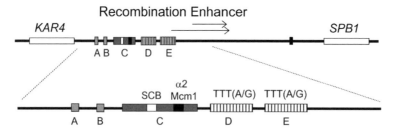

Figure 9.4 DNA sequence elements in the RE. RE is located in a 2.5-kb noncoding region. A minimum enhancer of about 750 bp contains five elements (A, B, C, D, and E) conserved among sensu stricto *Saccharomyces* spp. Regions D and E contain up to 15 repeats of the 4-bp sequence TTT(A/G). Regions A, D, and E contain binding sites for Fkh1, while the SCB sequence in element C (white box) binds Swi4 and Swi6. Region C contains an α2-Mcm1 repressor binding site that prevents other proteins from binding to RE in *MAT*α cells. There is a second α2-Mcm1 binding site on the right side of the element. Several sterile transcripts are found in RE; their role is unknown. RE function can be obtained when the RE region is deleted and replaced by four copies of the A sequence or four or five copies of D or E.

to RE and that this binding establishes a region of strongly positioned nucleosomes across RE (63). But this chromatin change does not extend into *KAR4* or *SPB1*, and there is in fact no mating-type-dependent change in gene expression of any genes along the left arm of chromosome III, according to various microarray experiments (see Saccharomyces Genome Database, Expression connection [http://db.yeastgenome.org/cgi-bin/expression/expressionConnection.pl]). Donor preference in cells lacking Matα1 (which is required to activate expression of *trans*-acting α-specific genes) has no effect on donor preference (i.e., *HMR* is the main donor), but deleting or mutating *MATα2* causes the preferential use of *HML* (51, 66). Mutations that eliminate Matα2-Mcm1 binding at the RE also activate *HML* in *MATα* cells (63).

In *MATa* cells, RE is bound by several proteins. As noted above, there are multiple binding sites for forkhead proteins in regions D and E, but there is also a single site in region A as well. In fact, four copies of the 22-bp region A (4×A) are sufficient to confer most, but not all, of *HML* preferential usage when all the rest of RE is deleted (48). Chromatin immunoprecipitation confirmed that both Fkh1 and Fkh2 can bind to RE in *MATa* cells but are excluded from binding in *MATα*. As expected, when RE is replaced by 4×A, *HML* donor preference is nearly as high in *MATα* as in *MATa*, as there now is no repression of Fkh1 binding; the finding that the use of *HML* is 55% in *MATα* versus 65% in *MATa* suggests that there is some other *trans*-acting mating-type-dependent regulation of donor preference that has not been identified. Genetic experiments demonstrated that deleting *FKH1*, but not *FKH2*, had a profound effect on donor preference (48). In the 4×A construct, *fkh1Δ* eliminated donor preference completely; in the intact RE, *fkh1Δ* reduced donor preference from 85% to about 35% (where an RE deletion would be 10%).

In Fkh1-regulated genes such as *CLB2* or *SWI5*, Fkh1 binding depends on the binding of the Mcm1 protein, which also binds to RE at the Matα2-Mcm1 operator region (1, 25, 29, 63); however, the finding that Fkh1 works in the 4×A construct argues that Mcm1 may be required to open up the chromatin structure of RE to allow other transcription factors to bind but when binding is unimpeded Mcm1 is dispensable. Indeed, when the Mcm1 binding site is altered by a 2-bp substitution, the RE in a *MATa* cell is again covered with highly positioned nucleosomes, but now in the absence of Matα2! How nucleosomes are positioned in this case is not yet understood. The role of Mcm1 in facilitating Fkh1 binding can also be substituted by inserting a different promoter region derived from the copper-inducible *CUP1* gene (9). This inserted region was

effective even under conditions where the promoter is not active, in the absence of copper, but the promoter may still bind the copper-inducible activator protein. The fact that there was still significant residual use of *HML* in *fkh1Δ* strains suggested that there might be another important protein binding to RE. An inspection of the evolutionary conservation of region C in sensu stricto strains revealed another short conserved region, a cell cycle regulation sequence known as SCB to which Swi4 and Swi6 bind (6). Deleting SCB or eliminating Swi4 or Swi6 each had the same effects: in the intact RE, donor preference was reduced to about 35% and a double mutant such as *scbΔ fkh1Δ* resembles RE*Δ*. These data suggest that there are no other key *trans*-acting proteins required for RE activation, but the recent discovery of a third Fkh homolog, Hcm1 (39), suggests another factor to be evaluated. Another protein that contributes to *MATa* donor preference is Chl1, a putative DNA helicase implicated in the establishment of sister chromatid cohesion (26, 59). Chl1 appears to act in the Swi4/6 pathway (6). Interestingly, none of the other cohesion-defective or chromosome instability mutants that are either epistatic with or synthetically lethal with *chl1Δ*, including deletions of *KAR3*, *CTF18*, *BIM1*, *YDR332W*, *NUP170*, or *CTF4*, affect donor preference (6). Donor preference is also not perturbed by other proteins that are reported to bind in the RE region, Mbp1, Ace2, Ste12, or Ndd1 (6). The fact that *HML* usage is higher in *MATa* than *MATα* in situations such as the replacement of RE by 4×A might suggest that there is another a-specific gene that increases *HML* usage or an α-specific gene that assists *HMR* usage. No candidate has emerged from surveys of mating-type regulated genes.

The binding of both Fkh1 and Swi4/6 to RE has been evaluated by chromatin immunoprecipitation at different times in the cell cycle (6). Not surprisingly, Swi4/6 bind strongly in G_1-arrested cells but not in S-phase-arrested or G_2-arrested cells; this is the same behavior as that seen at other Swi4/6-regulated genes. In contrast, whereas Fkh1 could be detected at *CLB2* or *SWI5* promoter regions at all times in the cell cycle, even though transcription is seen only in G_2 cells (1, 69), RE binds Fkh1 only in G_2-arrested cells. Perhaps there is a facilitating posttranslational modification. These results suggest that RE may act by maintaining some chromosome state that is "refreshed" at each phase of the cell cycle by a different activator. In this regard, the recent discovery that another Fkh homolog, Hcm1, binds to its gene targets preferentially in S phase opens the possibility that there are separate activators of RE in G_1, S, and G_2. Hcm1 has yet to be tested for its role in donor preference. The question then emerges: if one arrests cells in G_2, is Fkh1 the only necessary activator of donor preference? In fact, in both G_1- and G_2-arrested cells, *HML* donor preference is reduced compared to that in cycling cells (from 85 to 90% down to 50%, remembering that deleting RE has 10% usage). But *fkh1Δ* does not eliminate all preference in G_2-arrested cells, and *scbΔ* does not prevent all *HML* activation in G_1 arrest; hence, it seems that there are epigenetically "remembered" contributions of these factors in phases of the cell cycle when they are not bound to RE.

Thus, there are two transcription activators that work at RE, in spite of the lack of a nearby ORF. There are several *MATa*-specific noncoding transcripts detected in the intact region, centromere proximal to region E (9, 49), but in many of the constructs discussed above all of the sequences in the region have been replaced (for example with 4×A) without affecting donor preference, so the role of these transcripts is unknown. The 5′ end of the longer transcript maps within region E. Nevertheless it is important to know more about such transcripts. Ercan et al. (9) reported that these transcripts are not dependent on Fkh1. The transcripts are cell cycle regulated, with a peak occurring in late G_1 and early S, which would be consistent with the time that Swi4/6 are bound to RE; the SCB dependence of these transcripts needs to be examined. It would seem that these transcripts are not required for the preferential use of *HML*, as the donor preference mechanism is largely intact in G_2-arrested cells where there is little or no transcription.

To pursue this point further, the entire RE has been replaced with four LexA binding sites. The LexA-VP16 fusion protein is a very potent transcription activator in yeast transcription assays, yet it does not activate RE; in contrast, expressing LexA-Fkh1 activates *HML* usage (E. Coïc and J. E. Haber, unpublished data). It has not been established if LexA-Swi4 or Swi6 will also function as isolated entities.

One other pair of proteins that may play a role in donor preference are the Ku proteins. Deletion of Ku80 appears to reduce *HML* usage in *MATa* cells, with no effect in *MATα* cells (42). The reductions are not as strong as deleting the transcription factors discussed above; moreover, an equivalent deletion of Ku70 had little effect (42). Ku proteins have been implicated in the tethering of silenced sequences at the nuclear periphery, and the effect of deleting them may be to change the localization of *HML* or *HMR*. Whether there are other proteins that bind to RE has not been thoroughly investigated. Indeed Ku proteins are associated with *HML* and *HMR* prior to *MAT* switching. A report that Ku

proteins bind to RE in *MATa* cells prior to switching (42) was not verified in a survey of Ku binding sites along chromosome III (M. Marvin and E. J. Louis, personal communication).

Finally, an alteration in *MAT* switching was reported in cells lacking the DNA helicase Srs2 or another helicase Sgs1 (18). Sgs1 has previously been shown to affect strand annealing of mismatched DNA strands and in controlling the proportion of gene conversion events that are accompanied by crossover. Here *sgs1Δ* greatly slowed down the synapsis of HO-cut *MAT* with *HML*, although donor preference was not altered. The Srs2 helicase has previously been shown to play an important role in channeling DSBs into an SDSA pathway and also to be required for resumption of cell cycle progression after a DSB activates the DNA damage checkpoint (24, 54, 55). Houston and Broach now report that *srs2Δ* delays cell division after *MAT* switching but does not prevent cell cycle progression; moreover, *srs2Δ* increased *HML* usage in *MATα* cells but did not alter *MATa* donor selection. It is not evident what role the helicase would play in *MATα*'s normally preferred use of *HMR*, though it is worth noting that *HMR* is inherently more silent than *HML*, as mentioned above. Hence, it is possible that Srs2 plays some role in strand invasion in the more silenced sequence.

HOW DOES RE WORK?

RE appears to regulate some aspect of the accessibility of *HML*, whether at its normal location or at other sites on the left arm of chromosome III, so that when RE is either deleted or repressed in *MATα*, *HML* usage is very low. Indeed, in a *MATα* cell that lacks *HMR*, its ability to use *HML* is markedly reduced, so that about one-third of cells die after HO cleavage, even though there is a DNA damage checkpoint that arrests cells in G_2/M for a long time to ensure that a DSB can be repaired (66). This result suggests that *HML* either cannot be found by the Rad51 nucleoprotein filament or that the silent locus has become hypersilenced in some way that prevents strand invasion.

But there is substantial evidence that argues against the idea that *HML* has been turned into a lump of concrete that requires RE for access. First, RE is able to affect recombination that does not involve *HML*, *MAT*, or HO endonuclease. If the entire *HML* region and its adjacent silencers are replaced by a *leu2* gene with a mutation, the rate with which this allele recombines with a different *leu2* allele located either at *MAT* or on another chromosome is ≥10 times higher in *MATa* cells than in *MATα* cells (64). Thus, the effect of RE is not dependent on the silenced state of *HML*. Indeed, in the *leu2* assay deleting Sir proteins needed for silencing has no effect. Second, normal *MAT* switching donor preference is not altered when *HML* is unsilenced (necessitating the introduction of a mutation to prevent HO cutting) (Coïc and Haber, unpublished). Third, if *MAT* is deleted and moved to a location about 40 kb to the right of RE and on the same chromosome arm, the use of *HML* in *MAT*α cells becomes 90% instead of 10% (5). This suggests that *HML* and the left arm of chromosome III are arranged in a folded or tethered structure that precludes *HML* from interacting with *MAT* on the right arm of chromosome III but does not prevent productive recombination when the locus is in the same domain. It should be noted that if *MAT* is moved to chromosome V, the use of *HML* as a donor is still under the control of RE (67).

Taken together the observed rules of donor preference favor the idea that RE governs the architecture of the left arm of chromosome III and does so by using the transcriptional activation domains of Swi4/6 and Fkh1 as long-distance regulators that interact with folding or tethering sites within the donor sequences as well as other sites along the left arm of chromosome III. Thus, RE is a kind of locus control region that affects recombination but not transcription along the left arm of chromosome III. Microarray experiments have confirmed that RE does not change the expression level of any genes along the left arm of chromosome III.

RE acts over a distance of about 50 kb. If RE is deleted at 29 kb from the left telomere and inserted at 74 kb, it still activates, though less strongly, *HML* at its normal position and strongly activates *HML* at 41, 72, and 91 kb (5). The centromere appears to be a partial barrier to activity as shown by inserting RE at 116 kb to the right of *CEN3* (113 kb). *HML* use at 72 kb, across the centromere, was only 32%, but when *CEN3* was deleted and inserted at 117 kb so that RE and its donor were on the same chromosome arm, usage increased to 56%.

An indication that some of the RE-responsive sequences lie in or very close to the silent loci came from inserting a second copy of RE near *HMR* (5). Now, with RE near both donors, the use of *HML* was reduced from 90% to about 50%. A molecular dissection of the silent locus needs to be done to identify such an element. Indeed, the least understood aspect of the donor preference system is the basis by which the left arm is excluded when RE is inactive or repressed. The experiments showing RE's effect on *leu2* recombination also demonstrated that there must be more than one "sequestering sequence" on the left arm because in those experiments *HML* was completely deleted. One way to find the interacting targets of RE would be to use chromatin immunoprecipitation on CHIP experiments to identify other sites on chromosome III where either Fkh1 or Swi4-Swi6 were located in *MAT*a cells but not in *MAT*α cells; however, so far a survey of published experiments has not yielded any candidate sites.

Another approach to identify specific RE-responsive sequences has been to delete all the relevant elements from chromosome III and insert them in roughly similar locations on the next-largest yeast chromosome, chromosome V (5). With *HMR*α on the left arm and *MAT*a and *HMR*α-B on the right arm and no RE, the use of the left arm donor was about 40%, suggesting that there was no strong donor preference (the two donors were nearly equidistant from *MAT*). But when RE was inserted about 15 kb proximal to the left donor, its use increased to 93%. These data suggest that RE can activate the nearby *HMR* locus and again suggest that one RE-responsive element may reside within the silent locus itself. Whether RE inserted proximal to the right donor would have the same effect has not yet been tested.

HMR in its normal location does seem to have some unusual properties. One way to assess donor preference is in competition with another recombination process. If *MAT* is situated between a pair of 1-kb flanking repeated sequences (created for example when a *MAT-URA3*-pBR322 plasmid is inserted at *ura3* on chromosome V), then a DSB can be repaired either by ectopic gene conversion using *HML* or *HMR* or by single-strand annealing between flanking *URA3* genes, leading to deletion of *MAT* and pBR322 sequences (67). *HML* usage is strongly controlled by RE, but even in *MAT*α cells, *HMR* usage in an ectopic recombination event is quite limited, as if *HMR* is unable to interact with the HO-cut *MAT* on chromosome V. This constraint is not attributable to a nearby tRNA gene that has been suggested to be tethered to the nucleolus along with other tRNA genes (5), and at present there is no indication how *HMR* might be discouraged from interchromosomal interactions.

MICROSCOPIC ANALYSIS OF DONOR PREFERENCE

By placing LacO/LacI-GFP or TetO/TetR-CFP marks near *HML*, *MAT*, and *HMR*, Bressan et al. (3) showed that the relative locations of the three loci in G$_1$ cells (when switching normally is initiated) were significantly different in *MAT*a and *MAT*α, but the surprise was that in *MAT*a the three sites were equidistant whereas in *MAT*α, *HMR* and *MAT* were closer together and *HML* somehow moved further away. If *HML* and *HMR* are the same distance from *MATa*, on average, then donor preference cannot simply be explained by initial proximity; instead it would seem that *HML* must be much more mobile, moving both closer to *MAT* and further away (to maintain the same average distance). An indication that the left arm of chromosome III is indeed more mobile was provided in studies of the live-cell motion of a LacI-GFP cluster near *HML* in diploids where both homologs had the same GFP tag. In diploids that are a-mating (*MATa/matΔ*), the two *HML*-adjacent spots much more frequently come close together than was seen in α-mating (*matΔ/MATα*) cells (3). Measurements now need to be made between tags at *HML* and *MAT* in haploids to relate the differences in motion in the diploid to the more relevant haploid situation.

FUTURE PROSPECTS

RE represents one of the most fascinating and best-defined *cis*-acting loci that control aspects of chromosome positioning, folding, or tethering. The availability of new cytological resources as well as a wide variety of molecular and genetic tools opens the way to a detailed understanding of the ways chromosomes are arranged and constrained in the nucleus. Similarly, the use of microscopic as well as chromatin immunoprecipitation techniques makes it possible to expand in vivo biochemical approaches to define in great detail how the ends of a DSB locate and effect strand exchange with distant donor sequences. It will be amusing to look back in 5 years at how naïve we were.

References

1. Althoefer, H., A. Schleiffer, K. Wassmann, A. Nordheim, and G. Ammerer. 1995. Mcm1 is required to coordinate G2-specific transcription in *Saccharomyces cerevisiae*. *Mol. Cell. Biol.* 15:5917–5928.

2. Aylon, Y., and M. Kupiec. 2005. Cell cycle-dependent regulation of double-strand break repair: a role for the CDK. *Cell Cycle* 4:259–261.

3. Bressan, D. A., J. Vazquez, and J. E. Haber. 2004. Mating type-dependent constraints on the mobility of the left arm of yeast chromosome III. *J. Cell Biol.* 164:361–371.

4. Chai, B., J. Huang, B. R. Cairns, and B. C. Laurent. 2005. Distinct roles for the RSC and Swi/Snf ATP-dependent chromatin remodelers in DNA double-strand break repair. *Genes Dev.* 19:1656–1661.

5. Coic, E., G. F. Richard, and J. E. Haber. 2006. Saccharomyces cerevisiae donor preference during mating-type switching is dependent on chromosome architecture and organization. *Genetics* 173:1197–1206.

6. Coic, E., K. Sun, C. Wu, and J. E. Haber. 2006. Cell cycle-dependent regulation of *Saccharomyces cerevisiae* donor preference during mating-type switching by SBF (Swi4/Swi6) and Fkh1. *Mol. Cell. Biol.* 26:5470–5480.

7. Connolly, B., C. I. White, and J. E. Haber. 1988. Physical monitoring of mating type switching in *Saccharomyces cerevisiae*. *Mol. Cell. Biol.* 8:2342–2349.

8. Cosma, M. P. 2004. Daughter-specific repression of Saccharomyces cerevisiae HO: Ash1 is the commander. *EMBO Rep.* 5:953–957.

9. Ercan, S., J. C. Reese, J. L. Workman, and R. T. Simpson. 2005. Yeast recombination enhancer is stimulated by transcription activation. *Mol. Cell. Biol.* 25:7976–7987.

10. Haber, J. E. 1998. Mating-type gene switching in *Saccharomyces cerevisiae*. *Annu. Rev. Genet.* 32:561–599.

11. Haber, J. E. 1992. Mating-type gene switching in *Saccharomyces cerevisiae*. *Trends Genet.* 8:446–452.

12. Haber, J. E. 2002. Switching of *Saccharomyces cerevisiae* mating-type genes, p. 927–952. *In* R. C. N. Craig, M. Gellert, and A. Lambowitz (ed.), *Mobile DNA II*. ASM Press, Washington, DC.

13. Haber, J. E. 2006. Transpositions and translocations induced by site-specific double-strand breaks in budding yeast. *DNA Repair* 5:998–1009.

14. Harashima, S., and Y. Oshima. 1980. Functional equivalence and co-dominance of homothallic genes HM alpha/hm alpha and HMa/hma in Saccharomyces yeasts. *Genetics* 95:819–831.

15. Hawthorne, D. C. 1963. A deletion in yeast and its bearing on the structure of the mating type locus. *Genetics* 48:1727–1729.

16. Herskowitz, I. 1989. A regulatory hierarchy for cell specialization in yeast. *Nature* 342:749–757.

17. Hicks, J., J. Strathern, and I. Herskowitz. 1977. The cassette model of mating-type interconversion, p. 457–462. *In* A. Bukhari, J. Shapiro, and S. Adhya (ed.), *DNA Insertion Elements, Plasmids, and Episomes*. Cold Spring Harbor Press, Cold Spring Harbor, NY.

18. Houston, P. L., and J. R. Broach. 2006. The dynamics of homologous pairing during mating type interconversion in budding yeast. *PLoS Genet.* 2:e98.

19. Ira, G., A. Pellicioli, A. Balijja, X. Wang, S. Fiorani, W. Carotenuto, G. Liberi, D. Bressan, L. Wan, N. M. Hollingsworth, J. E. Haber, and M. Foiani. 2004. DNA end resection, homologous recombination and DNA damage checkpoint activation require CDK1. *Nature* 431:1011–1017.

20. Ira, G., D. Satory, and J. E. Haber. 2006. Conservative inheritance of newly synthesized DNA in double-strand break-induced gene conversion. *Mol. Cell. Biol.* 26:9424–9429.

21. Jensen, R. E., and I. Herskowitz. 1984. Directionality and regulation of cassette substitution in yeast. *Cold Spring Harb. Symp. Quant. Biol.* 49:97–104.

22. Klar, A. J., J. B. Hicks, and J. N. Strathern. 1982. Directionality of yeast mating-type interconversion. *Cell* 28:551–561.

23. Kostriken, R., and F. Heffron. 1984. The product of the HO gene is a nuclease: purification and characterization

of the enzyme. *Cold Spring Harb. Symp. Quant. Biol.* **49:**89–96.

24. Krejci, L., S. Van Komen, Y. Li, J. Villemain, M. S. Reddy, H. Klein, T. Ellenberger, and P. Sung. 2003. DNA helicase Srs2 disrupts the Rad51 presynaptic filament. *Nature* **423:**305–309.

25. Kumar, R., D. M. Reynolds, A. Shevchenko, A. Shevchenko, S. D. Goldstone, and S. Dalton. 2000. Forkhead transcription factors, Fkh1p and Fkh2p, collaborate with Mcm1p to control transcription required for M-phase. *Curr. Biol.* **10:**896–906.

26. Liras, P., J. McCusker, S. Mascioli, and J. E. Haber. 1978. Characterization of a mutation in yeast causing nonrandom chromosome loss during mitosis. *Genetics* **88:**651–671.

27. Long, R. M., R. H. Singer, X. Meng, I. Gonzalez, K. Nasmyth, and R. P. Jansen. 1997. Mating type switching in yeast controlled by asymmetric localization of ASH1 mRNA. *Science* **277:**383–387.

28. Loo, S., and J. Rine. 1995. Silencing and heritable domains of gene expression. *Annu. Rev. Cell. Dev. Biol.* **11:**519–548.

29. Maher, M., F. Cong, D. Kindelberger, K. Nasmyth, and S. Dalton. 1995. Cell cycle-regulated transcription of the CLB2 gene is dependent on Mcm1 and a ternary complex factor. *Mol. Cell. Biol.* **15:**3129–3137.

30. Nasmyth, K. 1993. Regulating the HO endonuclease in yeast. *Curr. Opin. Genet. Dev.* **3:**286–294.

31. Nasmyth, K. 1987. The determination of mother cell-specific mating type switching in yeast by a specific regulator of *HO* transcription. *EMBO J.* **6:**243–248.

32. Nickoloff, J. A., E. Y. Chen, and F. Heffron. 1986. A 24-base-pair DNA sequence from the *MAT* locus stimulates intergenic recombination in yeast. *Proc. Natl. Acad. Sci. USA* **83:**7831–7835.

33. Nickoloff, J. A., J. D. Singer, M. F. Hoekstra, and F. Heffron. 1989. Double-strand breaks stimulate alternative mechanisms of recombination repair. *J. Mol. Biol.* **207:**527–541.

34. Oshima, Y., and I. Takano. 1971. Mating types in Saccharomyces: their convertibility and homothallism. *Genetics* **67:**327–335.

35. Pâques, F., and J. E. Haber. 1997. Two pathways for removal of nonhomologous DNA ends during double-strand break repair in *Saccharomyces cerevisiae*. *Mol. Cell. Biol.* **17:**6765–6771.

36. Pâques, F., W. Y. Leung, and J. E. Haber. 1998. Expansions and contractions in a tandem repeat induced by double-strand break repair. *Mol. Cell. Biol.* **18:**2045–2054.

37. Pâques, F., G.-F. Richard, and J. E. Haber. 2001. Expansions and contractions in 36-bp minisatellite by gene conversion in yeast. *Genetics* **158:**155–166.

38. Park, E. C., and J. W. Szostak. 1990. Point mutations in the yeast histone H4 gene prevent silencing of the silent mating type locus HML. *Mol. Cell. Biol.* **10:**4932–4934.

39. Pramila, T., W. Wu, S. Miles, W. S. Noble, and L. L. Breeden. 2006. The forkhead transcription factor Hcm1 regulates chromosome segregation genes and fills the S-phase gap in the transcriptional circuitry of the cell cycle. *Genes Dev.* **20:**2266–2278.

40. Ravindra, A., K. Weiss, and R. T. Simpson. 1999. High-resolution structural analysis of chromatin at specific loci: *Saccharomyces cerevisiae* silent mating-type locus *HMRa*. *Mol. Cell. Biol.* **19:**7944–7950.

41. Ray, B. L., C. I. White, and J. E. Haber. 1991. Heteroduplex formation and mismatch repair of the "stuck" mutation during mating-type switching in *Saccharomyces cerevisiae*. *Mol. Cell. Biol.* **11:**5372–5380.

42. Ruan, C., J. L. Workman, and R. T. Simpson. 2005. The DNA repair protein yKu80 regulates the function of recombination enhancer during yeast mating type switching. *Mol. Cell. Biol.* **25:**8476–8485.

43. Rudin, N., and J. E. Haber. 1988. Efficient repair of HO-induced chromosomal breaks in *Saccharomyces cerevisiae* by recombination between flanking homologous sequences. *Mol. Cell. Biol.* **8:**3918–3928.

44. Strathern, J., J. Hicks, and I. Herskowitz. 1981. Control of cell type in yeast by the mating type locus. The alpha 1-alpha 2 hypothesis. *J. Mol. Biol.* **147:**357–372.

45. Strathern, J. N. 1988. Control and execution of mating type switching in *Saccharomyces cerevisiae*, p. 445–464. *In* R. Kucherlapati and G. R. Smith (ed.), *Genetic Recombination*. American Society for Microbiology, Washington, DC.

46. Sugawara, N., F. Paques, M. Colaiacovo, and J. E. Haber. 1997. Role of *Saccharomyces cerevisiae* Msh2 and Msh3 repair proteins in double-strand break-induced recombination. *Proc. Natl. Acad. Sci. USA* **94:**9214–9219.

47. Sugawara, N., X. Wang, and J. E. Haber. 2003. In vivo roles of Rad52, Rad54, and Rad55 proteins in Rad51-mediated recombination. *Mol. Cell* **12:**209–219.

48. Sun, K., E. Coic, Z. Zhou, P. Durrens, and J. E. Haber. 2002. Saccharomyces forkhead protein Fkh1 regulates donor preference during mating-type switching through the recombination enhancer. *Genes Dev.* **16:**2085–2096.

49. Szeto, L., M. K. Fafalios, H. Zhong, A. K. Vershon, and J. R. Broach. 1997. Alpha2p controls donor preference during mating type interconversion in yeast by inactivating a recombinational enhancer of chromosome III. *Genes Dev.* **11:**1899–1911.

50. Takano, I., and Y. Oshima. 1970. Mutational nature of an allele-specific conversion of the mating type by the homothallic gene HO alpha in Saccharomyces. *Genetics* **65:**421–427.

51. Tanaka, K., T. Oshima, H. Araki, S. Harashima, and Y. Oshima. 1984. Mating type control in *Saccharomyces cerevisiae*: a frameshift mutation at the common DNA sequence, X, of the HML alpha locus. *Mol. Cell. Biol.* **4:**203–211.

52. Tsong, A. E., B. B. Tuch, H. Li, and A. D. Johnson. 2006. Evolution of alternative transcriptional circuits with identical logic. *Nature* **443:**415–420.

53. Valencia, M., M. Bentele, M. B. Vaze, G. Herrmann, E. Kraus, S. E. Lee, P. Schar, and J. E. Haber. 2001. NEJ1 controls non-homologous end joining in Saccharomyces cerevisiae. *Nature* **414:**666–669.

54. Vaze, M., A. Pellicioli, S. Lee, G. Ira, G. Liberi, A. Arbel-Eden, M. Foiani, and J. Haber. 2002. Recovery from checkpoint-mediated arrest after repair of a double-strand break requires srs2 helicase. *Mol. Cell* **10:**373.

55. Veaute, X., J. Jeusset, C. Soustelle, S. C. Kowalczykowski, E. Le Cam, and F. Fabre. 2003. The Srs2 helicase prevents recombination by disrupting Rad51 nucleoprotein filaments. *Nature* **423:**309–312.

56. Wang, X., and J. E. Haber. 2004. Role of Saccharomyces single-stranded DNA-binding protein RPA in the strand invasion step of double-strand break repair. *PLoS Biol.* **2:**104–111.

57. Wang, X., G. Ira, J. A. Tercero, A. M. Holmes, J. F. Diffley, and J. E. Haber. 2004. Role of DNA replication proteins in double-strand break-induced recombination in *Saccharomyces cerevisiae*. *Mol. Cell. Biol.* **24:**6891–6899.

58. Weiler, K. S., and J. R. Broach. 1992. Donor locus selection during Saccharomyces cerevisiae mating type interconversion responds to distant regulatory signals. *Genetics* **132:**929–942.

59. Weiler, K. S., L. Szeto, and J. R. Broach. 1995. Mutations affecting donor preference during mating type interconversion in *Saccharomyces cerevisiae*. *Genetics* **139:**1495–1510.

60. Weiss, K., and R. T. Simpson. 1998. High-resolution structural analysis of chromatin at specific loci: *Saccharomyces cerevisiae* silent mating type locus HMLa. *Mol. Cell. Biol.* **18:**5392–5403.

61. White, C. I., and J. E. Haber. 1990. Intermediates of recombination during mating type switching in *Saccharomyces cerevisiae*. *EMBO J.* **9:**663–673.

62. Wolner, B., S. van Komen, P. Sung, and C. L. Peterson. 2003. Recruitment of the recombinational repair machinery to a DNA double-strand break in yeast. *Mol. Cell* **12:**221–232.

63. Wu, C., K. Weiss, C. Yang, M. A. Harris, B. K. Tye, C. S. Newlon, R. T. Simpson, and J. E. Haber. 1998. Mcm1 regulates donor preference controlled by the recombination enhancer in *Saccharomyces* mating-type switching. *Genes Dev.* **12:**1726–1737.

64. Wu, X., and J. E. Haber. 1996. A 700 bp cis-acting region controls mating-type dependent recombination along the entire left arm of yeast chromosome III. *Cell* **87:**277–285.

65. Wu, X., and J. E. Haber. 1995. *MATa* donor preference in yeast mating-type switching: activation of a large chromosomal region for recombination. *Genes Dev.* **9:**1922–1932.

66. Wu, X., J. K. Moore, and J. E. Haber. 1996. Mechanism of *MAT* alpha donor preference during mating-type switching of *Saccharomyces cerevisiae*. *Mol. Cell. Biol.* **16:**657–668.

67. Wu, X., C. Wu, and J. E. Haber. 1997. Rules of donor preference in *Saccharomyces* mating-type gene switching revealed by a competition assay involving two types of recombination. *Genetics* **147:**399–407.

68. Zhou, Z., K. Sun, E. A. Lipstein, and J. E. Haber. 2001. A Saccharomyces servazzii clone homologous to Saccharomyces cerevisiae chromosome III spanning KAR4, ARS 304 and SPB1 lacks the recombination enhancer but contains an unknown ORF. *Yeast* **18:**789–795.

69. Zhu, G., P. T. Spellman, T. Volpe, P. O. Brown, D. Botstein, T. N. Davis, and B. Futcher. 2000. Two yeast forkhead genes regulate the cell cycle and pseudohyphal growth. *Nature* **406:**90–94.

Sex in Fungi: Molecular Determination and Evolutionary Implications
Edited by Joseph Heitman et al.
© 2007 ASM Press, Washington, D.C.

Stefanie Pöggeler

10

MAT and Its Role in the Homothallic Ascomycete *Sordaria macrospora*

INTRODUCTION

Sordaria macrospora belongs to the family *Sordariaceae* within the ascomycetes. Phylogenetic studies have revealed that species of the genus *Sordaria* are closely related to the genus *Neurospora* (12, 77) which includes the fungal model organism *Neurospora crassa* (18, 32, 74). The *Sordariaceae* family is presently composed of 7 to 10 genera characterized by dark, usually ostiolate, fruiting bodies and unitunicate asci (26, 50). The genetic breeding mechanism of species in the ascomycete family *Sordariaceae* can be either heterothallic (self-incompatible), homothallic (compatible), or pseudohomothallic (pseudocompatible). In heterothallic species, like *N. crassa*, sexual reproduction occurs only between morphologically identical partners that are distinguished by their mating type. In *N. crassa*, these are termed *MATA* and *MATa*. The alternative versions of the mating-type locus on homologous chromosomes of the mating partners are called idiomorphs because they are completely dissimilar in the genes they carry (66). In homothallic species, like *S. macrospora*, a mycelium derived from a uninucleate ascospore is self-fertile and able to perform all steps of meiosis. Pseudohomothallic members of the *Sordariaceae*, e.g., *Neurospora tetrasperma* and *Podospora anserina*, develop four-spored asci, in which most ascospores carry two nuclei, one of each mating type. A typical binucleate ascospore gives rise to a self-fertile mycelium, but only because it is a heterokaryon with nuclei of opposite mating types. A few asci contain five ascospores, of which three are binucleate and two are small uninucleate. The latter produce homokaryons of opposite mating types.

Recent reviews on mating-type systems in filamentous ascomycetes have been published by Shiu and Glass (91), Souza et al. (93), and Debuchy and Turgeon (22). Mating-type genes have been identified not only in heterothallic and pseudohomothallic but also in homothallic members of the *Sordariaceae*. This review provides a comprehensive and up-to-date overview of the mating-type genes and their roles in the homothallic ascomycete *S. macrospora*.

LIFE CYCLE OF *SORDARIA MACROSPORA*

The ubiquitous genus *Sordaria* comprises heterothallic and homothallic species (57). The homothallic *S. macrospora* is the most studied species in the genus. Starting in the 1950s, *S. macrospora* was used in genetic studies (28, 31, 41), and it has been subsequently

Stefanie Pöggeler, Department of Genetics of Eukaryotic Microorganisms, Institute of Microbiology and Genetics, Georg-August University Göttingen, Grisebachstr. 8, 37077 Göttingen, Germany.

proved to be an excellent model system for investigating not only meiotic pairing and recombination (105, 106, 108) but also fruiting-body development (28, 60, 70, 81, 83). The natural habitat of *S. macrospora* is the dung of herbivorous animals. Thus, *S. macrospora* belongs to the group of coprophilous fungi that are of ecological importance in recycling nutrients from animal feces. In the laboratory, the life cycle of *S. macrospora* can be completed within 7 days. As illustrated in Fig. 10.1, a culture derived from a single ascospore is capable of completing the life cycle. After 2 days, sexual development starts from the vegetative mycelium with the formation of female gametangia, called ascogonia. These are enwrapped by sterile hyphae to make closed spherical fruiting-body precursors, the protoperithecia. In contrast to its heterothallic relative *N. crassa*, *S. macrospora* does not form any asexual spore such as macroconidia or microconidia. Moreover, its protoperithecia do not produce any trichogynes. However, similar to heterothallic filamentous ascomycetes, the homothallic *S. macrospora* has a dikaryophase in its life cycle. During this phase, pairs of nuclei synchronously and repeatedly divide inside the ascogonium, thereby producing a great number of nuclei. The nuclei then migrate in pairs to the developing dikaryotic ascogenous hyphae emerging from the ascogonium. The ascogenous hypha tips develop a U-shaped hook (crozier) containing two nuclei. These nuclei divide mitotically in synchrony, and subsequently, two septa appear, dividing the hook cell into three sections: a lateral and a basal cell with one nucleus each, and an apical cell with two nuclei (Fig. 10.1). The two nuclei in the apical cell then fuse. Immediately after karyogamy within the young ascus, the diploid nucleus undergoes meiosis, which is followed by a postmeiotic mitosis, resulting in the formation of eight nuclei. Each of the eight nuclei is incorporated into its own ascospore (27, 28). Fruiting bodies of *S. macrospora*, called perithecia, harbor between 50 and 150 asci. After 7 days of development, mature black ascospores are forcibly discharged through an apical pore, the ostiolum, at the neck of the fruiting body. The self-fertilization modus of *S. macrospora* is termed autogamous fertilization. This means that a pairwise fusion of nuclei present within the ascus-initial occurs, without cell fusion having taken place beforehand (28).

Even though single-spore cultures of *S. macrospora* are self-fertile, they may be crossed (28, 31). For example, when two *S. macrospora* strains are paired in a petri dish, each mycelium gives rise to perithecia and homokaryotic asci. Nevertheless, in regions where the two mycelia meet, nuclei are interchanged through hyphal anastomosis and thereby produce perithecia containing heterozygous asci. However, it is difficult to distinguish between self-fertile and crossed hybrid perithecia in crosses of wild-type strains. To circumvent this problem, suitable markers can be employed to detect such crossed hybrid perithecia. Suitable markers used in formal genetic studies are either ascospore color mutants or sterile mutant strains (28, 41) (Fig. 10.1). In crosses with sterile strains, complementation of the genetic defects results in formation of fertile hybrid perithecia only in the contact zone of two mutant mycelia. In *S. macrospora* not all such pairings are successful. In some cases, fertility substantially decreased when strains of different origin were mated. In these cases, hybrid perithecia with a reduced number of ascospores are formed, most of which are inviable. Pulsed-field gel electrophoresis has shown that *S. macrospora* strains from different culture collections possess chromosomes of different sizes and that these polymorphic karyotypes contribute to reducing fertility in forced crosses between two strains of different origin. Therefore, the intraspecific chromosome length polymorphism might have consequences on the speciation process of a homothallic fungus capable of sexual but not of asexual spore formation (82).

STRUCTURE OF THE *S. MACROSPORA* MATING-TYPE LOCUS

To better understand the phenomenon of homothallism in mycelial ascomycetes, homothallic members of the *Sordariaceae* have been probed with *MatA* and *Mata* sequences from *N. crassa*. These analyses revealed that sequences hybridizing to the *N. crassa* mating-type idiomorphs are conserved in both heterothallic and homothallic *Sordariaceae* species (34). To date, two groups of homothallic species have been distinguished: the first group contains only sequences similar to the *N. crassa* *MATA* idiomorph (A-type), and the second group contains sequences similar to both the *MATA* and *MATa* idiomorphs (A/a-type) (34). The *Neurospora* species *N. africana*, *N. dodgei*, *N. galapagosensis*, and *N. lineolata* are A-type homothallic species, while *N. pannonica*, *N. terricola*, and *S. macrospora* belong to the A/a-type of homothallic *Sordariaceae* (6, 34, 86).

Cloning and sequencing of the entire mating-type locus of *S. macrospora* revealed that sequences homologous to the *MATA* and *MATa* idiomorphs of *N. crassa* are directly linked in the mating-type locus of *S. macrospora*. Figure 10.2 presents a comparative genetic map of the mating type loci from *S. macrospora* and *N. crassa*.

The *MATA* idiomorph of *N. crassa* contains three genes: *mat A-1*, *mat A-2*, and *mat A-3* (29, 30, 33, 36).

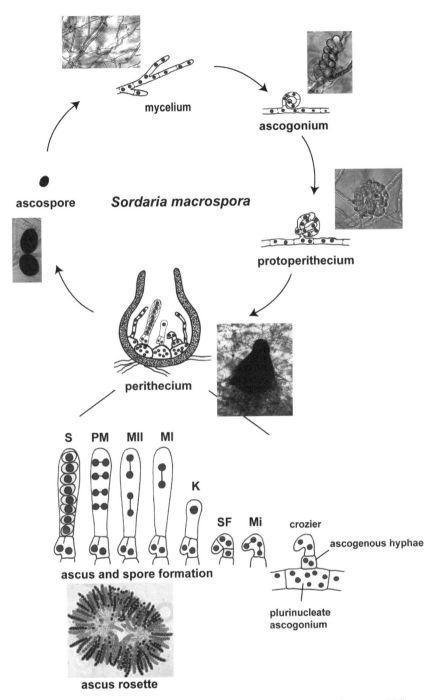

Figure 10.1 Life cycle of *S. macrospora*. Mi, synchronous mitosis; SF, septum formation; K, karyogamy; MI, first meiotic division; MII, second meiotic division; PM, postmeiotic mitosis; S, spore formation.

The MAT A-1-polypeptide, which is the major regulator of sexual development in *MatA* strains, accommodates a DNA-binding motif that shows similarity to the *Saccharomyces cerevisiae* MATα1p mating-type protein (89). The *N. crassa* gene *mat A-2* encodes a protein without a characteristic DNA-binding motif, but with a conserved region with three invariant histidine, proline, and glycine residues, which is therefore called the HPG domain (22). The *mat A-3* gene encodes a protein with a high-mobility group (HMG) domain (30). The HMG

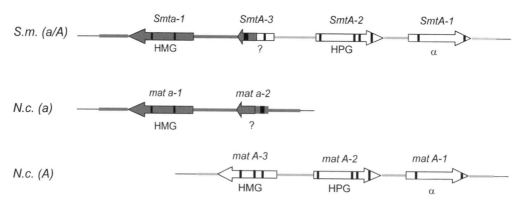

Figure 10.2 The mating-type locus of *S. macrospora*. Map of the mating type (*MAT*) locus of *S. macrospora* showing exons, introns, and intergenic regions compared to the *MATa* (*a*) and *MATA* (*A*) locus of *N. crassa*. The direction of transcription of each of the four *MAT* genes is indicated by an arrow. *MATa*-specific sequences are in gray, *MATA*-specific sequences are in white and light gray, respectively. Introns are indicated as black bars. HMG, high-mobility group domain; HPG, conserved histidine, proline, and glycine domain; α, α domain; ?, no conserved domain.

motif is a DNA-binding motif found in several transcription factors (37). The *a*-idiomorph of *N. crassa* contains two genes, *mat a-1* and *mat a-2*. The *mat a-1* gene encodes an HMG-domain protein that is the major regulator of mating in *a* strains, whereas *mat a-2* codes for a protein of unknown function (79, 94). In the mating-type locus of *S. macrospora*, four different open reading frames (ORFs) were identified, *Smt a-1*, *Smt A-3*, *Smt A-2*, and *SmtA-1* (79). Proteins encoded by *Smta-1*, *SmtA-1*, and *SmtA-2* show strong sequence similarities with the corresponding *N. crassa* mating-type ORFs and with mating-type proteins encoded by other members of the *Sordariaceae*. However, the *SmtA-3* gene has a chimeric character and has sequence similarity to the *mat A-3* and the *mat a-2* ORF of *N. crassa*. The *S. macrospora SmtA-3* lacks the region encoding the HMG domain encoded by the *mat A-3* gene of *N. crassa*. As in *S. macrospora*, the mating-type locus of the homothallic *Gibberella zeae* carries four genes structurally identical to the *MAT1-1* and *MAT1-2* mating-type genes of its heterothallic relative *Gibberella fujikuroi*. In contrast to *S. macrospora*, a fused mating-type gene was not identified in the *G. zeae* MAT locus and the *MAT1-1-3* gene of *G. zeae* encodes a protein with an HMG domain (103).

PHYLOGENETIC ANALYSIS OF MATING-TYPE SEQUENCES FROM *SORDARIA* AND *NEUROSPORA*

To elucidate how reproductive modes might have evolved in different members of the *Sordariaceae*, phylogenetic relationships of *MATA*- and *MATa*-specific

mating-type genes from heterothallic and homothallic members of the genera *Sordaria* and *Neurospora* have been examined. The *MAT* genes of members of the *Sordariaceae* appear to evolve more quickly than other regions of the genome, a phenomenon also observed for members of the genus *Cochliobolus*. However, mating-type genes are highly conserved within species, thus making them useful for phylogenetic analysis of closely related species (97). Phylogenetic trees based on *gpd* sequences of the conserved glyceraldehyde-3-phosphate dehydrogenase gene and on gene fragments of *mat a-1* and *mat A-1* homologous have shown that *Neurospora* and *Sordaria* are monophyletic units. Moreover, they have also revealed a strict separation of heterothallic and homothallic species within both genera and a separation of homothallic strains of *A/a*-type and *A*-type within *Neurospora* (77). The phylogenetic analyses suggest that changes in the reproductive strategy may represent a single event in each genus (77). This finding is in contrast to the proposed polyphyletic evolutionary origin of homothallism in the genus *Cochliobolus* (104) but in agreement with findings in the *Gibberella/Fusarium graminearum* clade and in the genus *Stemphylium* (43, 71). Similar to *Cochliobolus*, *Stemphylium*, and *Fusarium*, homothallism in *A/a*-type *Sordaria* and *Neurospora* species appears to be derived from heterothallic ancestors. The molecular cause of the fusion of *MAT* regions in *Sordariaceae* is so far unexplained, but most probably, a recombination event led to the close linkage of *MATA*- and *MATa*-specific sequences on one chromosome in the homothallic *A/a*-type *Sordariaceae*. Under distinct environmental conditions, *MATA* strains of the heterothallic *Sordaria brevicollis* were observed to produce perithecia and ascospores (89). Therefore, *A*-type homothallic strains in the genus *Neurospora* may have evolved directly from *MATA* strains of heterothallic ancestors.

TRANSCRIPTIONAL EXPRESSION OF THE FOUR *S. MACROSPORA* MATING-TYPE GENES

Sequence analysis of the cDNA from *Smta-1*, *SmtA-3*, *SmtA-2*, and *SmtA-1* revealed that introns in each gene are spliced, indicating that all of the mating-type genes identified in the mating-type locus of *S. macrospora* are transcriptionally active. Surprisingly, reverse transcription (RT)-PCR experiments demonstrated cotranscription of the *SmtA-3* gene and the *MATa*-specific *Smta-1* gene and optional splicing of two introns within the *SmtA-3* gene (79). Due to the optional splicing of the two introns in *SmtA-3*, three different classes of the dicistronic *SmtA-3/Smta-1* have been defined (Fig. 10.3). The smallest class of transcripts lacks four precisely spliced introns, two in *SmtA-3* and two in *Smta-1*. The second class of transcripts still contains the second intron of the *SmtA-3* gene, while the third class of transcripts retains both *SmtA-3* introns. Therefore, depending on the pattern of splicing of the introns, the encoded SMTA-3 polypeptides have a length of either 124 amino acids (aa) (splicing of both introns, SMTA-3-3), 116 aa (splicing of the first intron only, SMTA-3-2), or 52 aa (no splicing of introns, SMTA-3-1). A high degree of iden-

tity to the *N. crassa* MATA-3 protein is found within the first 91 N-terminal amino acids of SMTA-3-3 and SMTA-3-2 (89%) and the 52 aa of SMTA-3-1 (81%). The 35 and 43 C-terminal amino acids of the 116-residue SMTA-3-2 protein and the 124-residue SMTA-3-3 protein, respectively, show a weak similarity to the MATa-2 protein of *N. crassa* (79). Similarly, inefficient or optional splicing of introns has been noted for transcripts of the *N. crassa mat A-1* gene as well as for two introns in the 5' untranslated sequences of mating-type genes from *Cochliobolus heterostrophus* (55, 90).

The *SmtA-3* introns may be spliced in a stage-specific and/or tissue-specific manner. In other words, different proteins at different developmental stages and in different tissues would be produced. However, further experiments are required before determining whether *SmtA-3* and/or *Smta-1* gene expression is under developmental or tissue-specific control.

The unusual gene structure as well as gene cotranscription in *S. macrospora* may imply a specific function. In the heterothallic *N. crassa*, homologues of the cotranscribed *S. macrospora* genes *Smta-1* and *SmtA-3* are never located within the same nucleus. As has been shown for *P. anserina*, one of the functions of mating-type genes is to establish nuclear identity, thereby allowing recognition between nuclei of different mating types within the syncytial ascogonium after fertilization. Internuclear recognition seems to be a prerequisite for the pairwise migration of nuclei of opposite mating types into the ascogenous hyphae (3, 107). Moreover, nuclei of homothallic ascomycetes have been suggested to be functionally heterothallic in order to circumvent the recognition problem (17). Hence, it may be conceivable that homothallic ascomycetes have evolved a mechanism that allows alternate expression of either mating type to guarantee internuclear recognition before pairwise migration into ascogenous hyphae. Thus, cotranscription of *SmtA-3* and *Smta-1* might entail differential expression of the downstream *Smta-1* gene. The *P. anserina* FPR1, a homologue of the *S. macrospora Smta-1* gene, was shown to be involved in establishing nuclear identity. It might be that the downstream *Smta-1* gene is expressed only as a dicistronic messenger together with the upstream *SmtA-3* gene, whereas the *SmtA-3* gene might be also expressed as a monocistronic mRNA. Such a mechanism would lead to the expression of alternate mating types in the mycelium of *S. macrospora*.

To date, whether *SmtA-3* encodes a protein essential for sexual reproduction or whether it serves as a regulatory mini-ORF for the downstream *Smta-1* gene remains still to be elucidated. Finally, it cannot be excluded that cotranscription has no special function in the mating-type locus of *S. macrospora*.

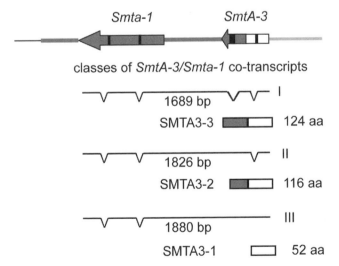

Figure 10.3 Cotranscription of *SmtA-3* and *Smta-1* and optional splicing of introns. Schematic representation of the *S. macrospora* genes *Smta-1* and *SmtA-3* and different classes of cotranscripts and encoded SMTA-3 proteins.

MATING-TYPE GENES FROM *S. MACROSPORA* INDUCE PERITHECIAL DEVELOPMENT IN A HETEROTHALLIC FUNGUS

Sequence conservation and transcriptional expression of the mating-type genes in the homothallic *S. macrospora* suggest that the mating-type genes encode functional proteins. In heterothallic filamentous ascomycetes, at least two important steps in sexual reproduction are regulated by the mating-type-encoded transcription factors: (i) the initial fertilization event mediated by chemoattraction between reproductive structures of two compatible partners; and (ii) before karyogamy, the paired migration of nuclei of opposite mating types into the ascogenous hyphae (17, 22, 91).

To investigate the functional conservation of mating-type genes of *S. macrospora*, cosmid clones containing the entire mating-type locus from *S. macrospora* were transformed into both *P. anserina MAT−* and *MAT+* strains. Similar to *N. crassa MATA* strains, *P. anserina MAT−* strains contain three genes (*FMR1* = *N. crassa mat A-1*; *FMR2* = *N. crassa mat A-2*; and *SMR3* = *N. crassa mat A-3*) at the mating-type locus, while the *P. anserina MAT+* locus contains only one gene (*FPR1*) encoding an HMG-domain protein which is a homologue of the *N. crassa mat a-1* gene (20, 21, 76). After introduction of the *S. macrospora* mating-type information, 60% of the *P. anserina MAT−* transformants and 37% of the *MAT+* transformants were capable of inducing fruiting-body development without crossing with a mating partner of the opposite mating type. As seen in Fig. 10.4, unfertilized protoperithecia of *MAT−* and *MAT+* strains are small in comparison with fertile perithecia produced in a *MAT−* ×

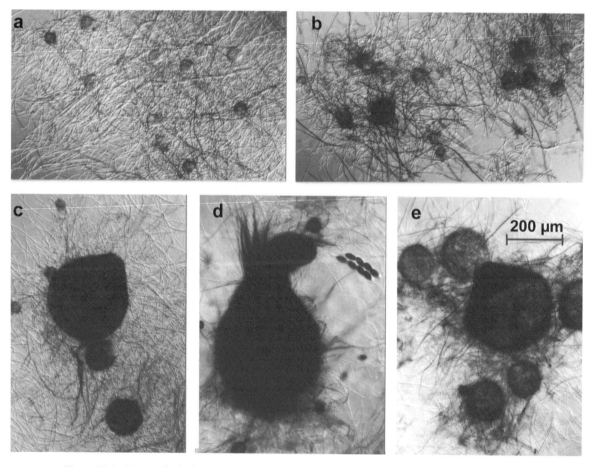

Figure 10.4 Fruiting-body development in *P. anserina MAT−* transformants and *MAT+* transformants carrying the entire mating-type locus of *S. macrospora* compared to *MAT−* and *MAT+* and to the fertilized *P. anserina* wild-type strain. *P. anserina MAT−* (a) and *MAT+* (b) strains form only protoperithecia; fertilization of a *MAT−* strain results in the formation of perithecia (d); and *MAT−* (c) and *MAT+* (e) transformants carrying the entire mating-type locus of *S. macrospora* produce fruiting bodies.

MAT+ cross. Like fertilized *MAT*− and *MAT*+ strains, transformants carrying the *MAT* locus of *S. macrospora* develop large flask-shaped perithecia from the ball-shaped protoperithecia. However, fruiting bodies of the transformants do not form tufts of hairs near the ostiolum, which is a characteristic feature of perithecia from *P. anserina*. The most striking difference between the fruiting bodies from transformants and *P. anserina* wild-type strains is that spore formation is never observed in fruiting bodies of transformants. Instead of asci with four ascospores, perithecia of transformants contain only a gelatinous mass and have no structures such as hook cells, croziers, asci, or spores. Similar results were obtained when *MATA*-specific genes *SmtA-1*, *SmtA-2*, and *SmtA-3* were separately transformed into a *MAT*+ strain and the *Smta-1* gene was separately transformed into a *MAT*−strain of *P. anserina* (85). Thus, the transformation studies reveal that *S. macrospora* mating-type genes are capable of inducing perithecial development in *P. anserina*. Similarly, *N. crassa* mating-type genes can provide fertilization functions in *P. anserina* and reciprocal introduction of *P. anserina* genes confers mating activity in *N. crassa* (2). In these experiments, functional conservation of vegetative incompatibility and postfertilization function have not been demonstrated in *N. crassa* and *P. anserina* (2). In addition, Glass and Smith (35) showed that introduction of an ectopic *mat A-1* gene of the homothallic *N. africana* in either a sterile A^{m64} or a^{m1} strain of *N. crassa* enables the transformants to mate as a *MATA* strain and to develop perithecia. However, examination of the contents of the transformants' perithecia revealed no indication of asci and ascospores. These results indicate that mating-type products are interchangeable with respect to the fertilization not only between heterothallic but even between homothallic and heterothallic members of the *Sordariaceae*. The absence of ascospore formation can be explained by the interference of the resident mating-type locus with the transgene. Crossing and selfing experiments with *C. heterostrophus*, *N. crassa*, and *P. anserina* transgenic strains have demonstrated this interference. Although plenty of barren fruiting bodies are formed by transformants carrying the resident mating-type locus and an ectopic copy of the opposite mating-type locus, ascospore production has been shown to be extremely low (15, 36, 76, 98). In contrast, no interference is found when a transgene is expressed in a strain not containing a resident mating-type locus (13, 15, 76, 101). While ectopic mating-type copies confer fertility in delta (Δ) *MAT* strains of *P. anserina* and *C. heterostrophus*, in *N. crassa* the mating-type sequence must even reside at the mating-type locus to gain complete functionality (13).

To test whether an interference of the resident mating-type locus with the ectopic *S. macrospora* mating-type genes might cause the failure of ascospore production of *P. anserina MAT*− and *MAT*+ transformants, the entire *MAT* locus of *S. macrospora* was introduced into a Δ*MAT* strain of *P. anserina* (15). Interestingly, *P. anserina* Δ*MAT* transformants carrying an ectopic copy of the mating locus of *S. macrospora* develop fertile perithecia containing rosettes of asci (S. Pöggeler, U. Kück, and H. D. Osiewacz, unpublished results) (Fig. 10.5). However, in most cases, ascus rosettes of the perithecia formed comprise only 3 to 20 asci and the frequency of asci with less than four ascospores is increased (Fig. 10.5f and g). Similarly, the mating-type locus of the homothallic *Cochliobolus luttrellii* was shown to confer self-fertility on a Δ*MAT* strain of the heterothallic *C. heterostrophus*. Like the results obtained with the Δ*MAT* strain of *P. anserina*, ascospore production in the transgenic self-fertile *C. heterostrophus* is only 1 to 10% of wild-type ascospore production (104). Thus, expression of the mating-type genes from the homothallic *S. macrospora* confers self-fertility to the heterothallic *P. anserina* when it does not interfere with a resident mating-type locus.

A Δ*SMTA-1* MUTANT IS INCAPABLE OF FORMING FERTILE FRUITING BODIES

Results from the heterologous expression of the *S. macrospora* mating-type genes in *P. anserina* suggest that the *S. macrospora* mating-type genes are functional and most probably are involved in fruiting-body development and ascosporogenesis. To further determine the role of the HMG-domain transcription factor SMTa-1 during vegetative and sexual development of the homothallic ascomycete *S. macrospora*, a Δ*Smta-1* mutant strain was generated (84). No irregularities in vegetative growth and mycelial morphology were observed when the mutant strain was compared to wild-type *S. macrospora* (84). Although the *S. macrospora* Δ*Smta-1* mutant forms ascogonia and protoperithecia, it is unable to perform the transition to mature fruiting bodies (Fig. 10.1). When grown on fructification medium, Δ*Smta-1* can produce 240 protoperithecia/cm^2, which is similar to the wild-type strain production (around 230 protoperithecia/cm^2). However, fruiting bodies and ascospores that are formed in the wild-type strain (39 perithecia/cm^2) were never observed in the Δ*Smta-1* mutant even after extended incubation time. Thus, the Δ*Smta-1* mutant is completely sterile and resembles phenotypically previously described *S. macrospora* *pro1*, *pro11*, and *pro22* mutants which also do not produce any perithecia. The *pro1*, *pro11*, and *pro22* genes encode a zinc finger transcription factor, a WD40 repeat

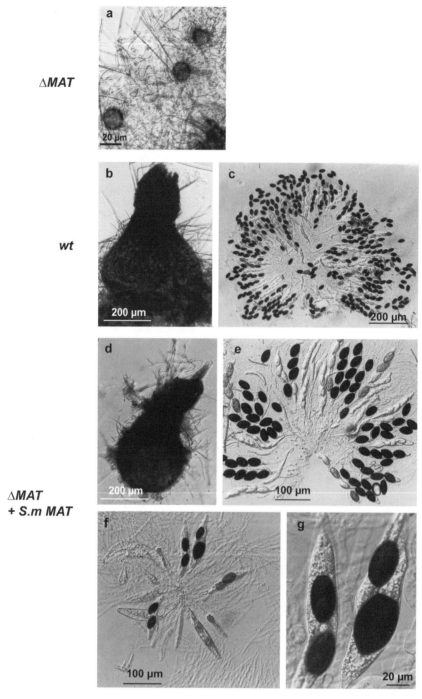

Figure 10.5 Fruiting-body and ascospore development in *P. anserina* Δ*MAT* transformants carrying the entire mating-type locus of *S. macrospora* (*S.m MAT*) compared to the *P. anserina* Δ*MAT* strain (Δ*MAT*) and the fertilized *P. anserina MAT−* wild-type strain (*wt*). The *P. anserina* Δ*MAT* strain forms only protoperithecia (a); fertilization of a *MAT−* strain results in the formation of perithecia (b) with four-spored asci (c); self-fertile Δ*MAT* transformants carrying the entire mating-type locus of *S. macrospora* produce perithecia (d) containing either rosettes with a reduced number of asci (e and f). Some asci harbor only two ascospores (g).

protein, and a putative membrane protein, respectively (60, 69, 81).

Due to the presence of *A*-type homothallic species in the genus *Neurospora* (34, 35) and *MATA* strains of the heterothallic *Sordaria brevicollis* that were shown to produce fruiting bodies with ascospores under certain cultural conditions (89), it was not clear whether *MATa*-specific genes are essential for sexual reproduction in homothallic members of the family *Sordariaceae*. Deletion of *Smta-1* encoding a putative HMG-domain transcription factor converts the self-fertile *S. macrospora* to a self-sterile fungus, no longer able to produce fruiting bodies and ascospores. This finding demonstrates that, at least in the homothallic *S. macrospora*, the *Smta-1* gene is required for fruiting-body development and sexual reproduction. Similarly, deletion of the HMG-domain protein encoding *MAT1-2* gene of the homothallic *Nectriaceae* member *G. zeae* resulted in self-sterile strains. In contrast to the *S. macrospora* Δ*Smta-1* mutant, one-half of the Δ*MAT1-2* strains of *G. zeae* produce perithecium-like structures; however, these structures do not contain any ascospores (24). In *G. zeae*, the *MAT1-2* deletion strain is still able to outcross with a homothallic tester strain (53). Likewise, the *S. macrospora* Δ*Smta-1* strain has been shown to display a 1:1 segregation of the *hph* marker gene when outcrossed with a spore color mutant, thus indicating that the Δ*Smta-1* strain retains the ability to sexually reproduce (84).

MATING-TYPE-DEPENDENT GENE EXPRESSION IN *S. MACROSPORA*

In the yeast *S. cerevisiae*, the unique proteins encoded by the two different mating-type loci directly act as transcriptional regulators on mating-type-specific expression of pheromone and pheromone receptor genes (5, 40). Therefore, as in the yeast *S. cerevisiae*, the expression of pheromone genes and pheromone receptor genes in filamentous ascomycetes might be directly controlled by the transcription factors encoded by the mating-type genes (19).

Pheromone and Pheromone Receptor Genes

Interestingly, genes encoding two different pheromone precursors and two pheromone receptors have been found not only in the heterothallic *N. crassa* but also in the homothallic *S. macrospora* (8, 78, 80). The pheromone precursor gene *ppg1* is predicted to encode an α-factor-like peptide pheromone, and the *ppg2* gene is predicted to encode an **a**-factor-like lipopeptide pheromone (78). The deduced gene products of *pre1* and *pre2* show structural similarities to the yeast **a**-factor receptor Ste3p and α-factor receptor Ste2p, respectively (80).

Male and female fertility of heterothallic mycelial ascomycetes depends on interactions of pheromones with their specific receptors. When pheromone genes were deleted, male spermatia were no longer able to fertilize the female partner, proving that pheromones are crucial for the fertility of male spermatia (16, 49, 99). In *P. anserina*, the function of pheromones is restricted to fertilization, while the *N. crassa* lipopeptide-pheromone gene *mfa-1* has also been shown to be involved in female sexual development, ascospore production, and vegetative growth of both mating types (16, 49). Similarly, deletion of the *N. crassa pre-1* pheromone receptor gene does not affect vegetative growth or male fertility. However, protoperithecia from Δ*pre-1 MATA* mutants are female sterile because their trichogynes are unable to recognize and fuse with *MATa* spermatia (48).

For *S. macrospora*, it has recently been demonstrated that disruption of the pheromone precursor *ppg1* gene, encoding the α-factor-like peptide pheromone, prevents production of the peptide pheromone but does not affect vegetative growth or fruiting-body or ascospore development (61). Similarly, no effect on vegetative growth or fruiting-body and ascospore development was observed in the single pheromone mutant Δ*ppg2* and the receptor mutant Δ*pre1* and Δ*pre2* strains, respectively. However, double-knockout strains lacking any compatible pheromone/receptor pair (Δ*pre2*/Δ*ppg2*, Δ*pre1*/Δ*ppg1*) and the double-pheromone mutant strain (Δ*ppg1*/Δ*ppg2*) have a drastically reduced number of perithecia and sexual spores, whereas deletion of both receptor genes (Δ*pre1*/Δ*pre2*) completely eliminates fruiting-body and ascospore formation. Taken together, these results suggest that pheromones and pheromone receptors are required for optimal sexual reproduction of the homothallic *S. macrospora*. Moreover, by heterologously expressing the pheromone/receptor pair *ppg1*/*pre2* in *S. cerevisiae* the receptor PRE2 was shown to facilitate all aspects of the *S. cerevisiae* pheromone response in *MATa* cells lacking the Ste2p receptor when activated by the *S. macrospora* peptide pheromone. Therefore, one may conclude that the receptors encoded by *pre2* and *pre1* may also function as G-protein-coupled receptors in *S. macrospora* (61).

In the heterothallic *N. crassa*, *MATA* or *MATa* mating-type-specific transcription factors are present in *MATA* or *MATa* strains, respectively. In *N. crassa*, pheromone-encoding genes are predominantly expressed in a mating-type-specific manner (8). As mentioned in "Transcriptional Expression of the Four *S. macrospora* Mating-Type Genes" above, both types of mating-type regulators are expressed in the same individual of the homothallic ascomycete *S. macrospora* (85). It is thus not surprising that both *S. macrospora* pheromone precursor genes and pheromone receptor genes are transcriptionally expressed during sexual development in the same individual (78, 80). RT-PCR analysis of the Δ*Smta-1* mutant that lacks the *MATa*-specific HMG transcription factor SMTa-1 reveals a 500-fold reduction of lipopeptide *ppg2* gene expression, thus indicating that SMTa-1 has a direct or indirect impact on the activation of *ppg2* gene expression. Similarly, in the heterothallic *N. crassa*, the *a*-specific lipopeptide pheromone gene *mfa-1* is not expressed in a *MATa^{m1}* mutant strain that has a frameshift mutation in the *mat a-1* coding region (8, 94). Previously, the *N. crassa* MATa-1 polypeptide produced in *Escherichia coli* was shown to bind to specific DNA sequences having core motif CTTTG or CTTCG, and it was shown that this DNA binding is due to the HMG domain (75). The CTTTG motif corresponds to the pheromone response element recognized by the PRF1 HMG transcription factor in the basidiomycete *Ustilago maydis* (39). The *S. macrospora* SMTa-1 protein shows 86.0% identity with the corresponding MATa-1 protein of *N. crassa*, and the 80-aa HMG domain of both proteins displays 96.0% identity. This finding strongly suggests that the cellular functions and the specific DNA-binding recognition sites for both proteins are highly similar, possibly identical. Furthermore, analysis of the 5′ noncoding region of the *S. macrospora ppg2* gene revealed the presence of the above-mentioned HMG-domain protein binding sequences: two CTTTG and four CTTCG sequences. These sequences, similar to the *N. crassa* binding sites, might thus be binding sites for the SMTa-1 protein, which might then directly activate transcription of *ppg2*. However, the action of SMTa1 on *ppg2* transcription might also be an indirect activation of a transcriptional activator that again might bind the 5′ noncoding sequence of *ppg2*.

In contrast, no clear effect on the expression of the complementary pheromone gene *ppg1* has been observed in the *S. macrospora* Δ*Smta-1* strain. Similarly, the deletion of *Smta-1* had no clear effect on the transcriptional expression of the pheromone receptor genes *pre1* and *pre2*. This result does not come as a surprise, since transcription of pheromone receptor genes has previously been shown to be mating-type independent in the heterothallic *N. crassa* (80). However, an RT-PCR approach provides information only on the transcriptional expression of *ppg1*, *pre1*, and *pre2*; thus, a posttranscriptional effect of SMTa-1 on the expression of *ppg1*, *pre1*, and *pre2* cannot be ruled out. For example, in *P. anserina*, the orthologue of the *S. macrospora* SMTa-1, FPR1, was shown to not only activate the expression of the *MAT−*-specific pheromone gene *mfm* (= *ppg1* in *S. macrospora*) but also repress the expression of the

complementary *MAT+*-specific pheromone gene *mfp* (= *ppg2* in *S. macrospora*) posttranscriptionally (16).

An *S. macrospora* Δ*ppg2* pheromone knockout mutant was observed to still be able to form fruiting bodies and ascospores (62). Therefore, the 500-fold downregulation of *ppg2* gene expression cannot be the cause for sterility of the Δ*Smta-1* mutant.

SMTa-1 Affects the Expression of Several Genes Including a Common Set of Developmental Genes Deregulated in Sterile *pro* Mutants

The high percentage of sequence identity between *S. macrospora* and *N. crassa* allows cross-species microarray hybridizations using *N. crassa* cDNA microarrays for hybridization with targets prepared from *S. macrospora* to be performed (69, 70). Using this cross-species microarray technology, 74 and 33 genes that are at least twofold up- and down-regulated, respectively, in the *S. macrospora* Δ*Smta-1* mutant strain were identified. Of these 107 genes, 80 *N. crassa*-derived genes have homologues with known or putative function and were sorted into 10 putative functional categories (84). The largest category is "metabolism," into which 34 genes have been assigned. However, a number of genes that are involved in differentiation processes in other fungi or have homology to signal transduction components have been also identified. By means of cDNA subtraction and microarray analysis, Lee et al. (54) recently identified 171 downregulated genes in a *mat1-2*-deleted strain of the homothallic *G. zeae*. The *MAT1-2* gene of *G. zeae* encodes a homologue of the *S. macrospora Smta-1* gene. Similar to the microarray data obtained with the *S. macrospora* Δ*Smta-1* mutant strain, most genes that are downregulated in the *G. zeae mat1-2* mutant strain have been assigned to the metabolism category, but several genes encoding proteins involved in signal transduction and sexual development have also been identified (54).

The analysis of three other independent mutant strains affected in fruiting-body formation in *S. macrospora* (*pro1*, *pro11*, and *pro22*) showed that only 10 genes are deregulated in a way similar to that of the Δ*Smta-1* mutant strain. Thus, this set of deregulated genes seems to be fruiting body development-specific rather than mating-type-specific regulated. Among the genes down-regulated in all mutant strains examined are, for example, two genes encoding putative transcription factors. A putative bZIP transcription factor (NCU08055.2) and a putative forkhead domain transcription factor (NCU06173.2) show a four- to fivefold decrease in mRNA level in the Δ*Smta-1* mutant strain and are also down-regulated in the *S. macrospora pro* mutants. NCU08055.2 is a putative homologue of IDI-4, a bZIP transcription factor inducing autophagy and cell death in *P. anserina* when overexpressed (23). The second transcription factor gene down-regulated in the Δ*Smta-1* mutant and the *pro* mutants, NCU06173.2, is a homologue of the *Schizosaccharomyces pombe* forkhead transcription factor Sep1p. The *S. pombe* Sep1p regulates mitotic gene transcription and is involved in cell separation at the end of mitosis (11, 87).

Of the 107 genes that are regulated differentially in the Δ*Smta-1* mutant strain, 56 and 24 genes are down- and up-regulated, respectively, only in the Δ*Smta-1* mutant and are not affected in the *pro* mutants. Thus, the regulation of these genes seems to be *Smta-1* specific rather than fruiting-body specific. These data indicate that *Smta-1* is involved in the regulation of pathways distinct from those regulated by the *pro* genes. Of the 80 genes regulated differentially in Δ*Smta-1* alone, several encode components of signal transduction pathways. The gene homologous to NCU00041.2 encodes the γ-subunit GNG-1 of a heterotrimeric G-protein. In the Δ*Smta-1* mutant, the *gng-1* gene is up-regulated. Recently, it was demonstrated that the deletion of *gng-1* negatively influences levels of Gα proteins (GNA-1, GNA-2, and GNA-3) in plasma membrane fractions isolated from various tissues of *N. crassa* and that this leads to a significant reduction in the amount of intracellular cyclic AMP (51). Thus, up-regulation of *gng-1* in Δ*Smta-1* might lead to an up-regulation of genes encoding Gα subunits and adenylate cyclases. Interestingly, the microarray study has demonstrated that the transcript level of adenylate cyclase gene *cr-1* (NCU08377.2) is indeed elevated in the Δ*Smta-1* mutant strain. In *N. crassa*, adenylate cyclase together with the Gα subunit GNA-1 is known to contribute to proper fruiting-body and ascospore development (44). Similarly, cylic AMP signaling might be involved in sexual development of *S. macrospora*.

Another component of a signal transduction pathway exhibiting a >5-fold decreased mRNA level in the Δ*Smta-1* mutant strain is the homologue to NCU02131.2. The *N. crassa* gene encodes a putative homologue of the *S. pombe* guanine nucleotide exchange factor Rgfp3. In *S. pombe*, Rgfp3 regulates the activity of the small GTPase Rho1p which is involved in β-1,3-glucan biosynthesis and cell integrity during septation (95). In this context, it is worth noting that in the Δ*Smta-1* mutant strain, the NCU04431.2 homologue encoding a putative endo-1,3-β-glucanase is one of the most strongly down-regulated genes. In addition to the endo-1,3-β-glucanase gene, the expression of a further six genes involved in cell wall biosynthesis is deregulated in Δ*Smta-1* (84).

The genes affected by SMTA-1 are numerous and include the pheromone gene *ppg2* and several genes for components of signaling cascades. Deletion of *Smta-1* results in sterility of *S. macrospora*, and thus, similar to other sterile *S. macrospora* mutants, genes up-regulated during fruiting-body development are down-regulated in the Δ*Smta-1* mutant. Taken together, *S. macrospora* SMTA-1 seems to control sexual reproduction by regulating a variety of essential cellular processes. To elucidate the molecular mechanism underlying the complex regulatory network controlled by the mating-type proteins, the challenge will be to identify direct targets not only of SMTA-1 but also of all other mating-type-encoded transcription factors.

INTERACTION NETWORK OF *S. MACROSPORA* MATING-TYPE PROTEINS

Transcription of cell-type-specific genes in the ascomycete yeast *S. cerevisiae* has been shown to rely on the interaction mating-type proteins (40, 42). In *S. cerevisiae*, each mating-type locus carries two genes, the *MATα* locus *MATα1* and *MATα2* and the *MATa* locus *MATa1* and *MATa2*. The Matα1p protein is known to be a transcriptional activator of α-specific genes and like the SMTA-1 protein of *S. macrospora* carries the α-domain as a DNA-binding motif. The gene product of *MATα2* is a homeodomain protein and acts as a negative regulator of a-specific genes. In the *MATa* locus, *MATa1* is the only gene to encode a functional protein, Mata1p, which is also a homeodomain transcription factor. However, unlike the Matα2p homeodomain protein of α cells, Mata1p does not play a role in determining the a cell type. Rather, a-specific genes are expressed because they are not repressed by Matα2p, and α-specific genes are not expressed because there is no Matα1p activator present. Mata1p does, however, have a role in diploid cells, where in conjunction with Matα2p it represses transcription of haploid-specific genes (46).

Thus, heterodimerization provides a way of regulating transcription factor function by joining different functional domains. Consequently, a similar mechanism might exist not only in heterothallic but also in homothallic filamentous ascomycetes; in these fungi each mating-type idiomorph might encode one-half of a heteromeric transcription factor that controls the expression of genes required for the development of dikaryotic ascogenous hyphae. This model may explain why ascogenous hyphae of filamentous ascomycetes contain two nuclei of opposite mating types and never nuclei of only one mating type in heterothallic fungi (17).

Interactions between Mating-Type Proteins from *S. macrospora*

A two-hybrid approach in conjunction with protein cross-linking analysis demonstrated that in *S. macrospora* SMTA-1 and SMTa-1, proteins homologous to polypeptides encoded by opposite mating partners of the heterothallic *N. crassa*, are able to form a heterodimer (45) (Fig. 10.6). Similarly, two-hybrid studies of *N. crassa* mating-type genes established that MAT A-1 can interact with MAT a-1 (4). In contrast to these results, two-hybrid analysis performed with the mating-type proteins of *P. anserina* failed to detect any interaction between proteins encoded by opposite-mating-type loci. Nevertheless, an interaction between *MAT*--specific transcription factors FMR1 and SMR2, which are homologues of the *N. crassa* MAT A-1 and MAT A-3, respectively, has been reported (17, 22). Mutagenesis of *P. anserina* mating-type genes gave rise to a model in which mating-type proteins control recognition between nuclei of opposite mating types within the multinucleate ascogonium instead of controlling the development of ascogenous hyphae (107). In this model, the interacting proteins FMR1 and SMR2 establish nuclear identity by mutual interaction after fertilization (1, 3). Because an interaction between SMTA-1 and SMTA-3 was not detected in *S. macrospora* or between MAT A-1 and MAT A-3 in *N. crassa* (4), a different mode of action may exist for mating-type proteins of *P. anserina*. In addition to the heterodimerization of SMTA-1 and SMTa-1, two-hybrid analyses revealed that SMTA-1 can also homodimerize.

An MCM1 Homologue Interacts with SMTA-1 and Is Required for Fruiting-Body Development in *S. macrospora*

To carry out their roles as transcription factors in the budding yeast *S. cerevisiae*, each of the mating-type proteins Matα1p and Matα2p works together with the minichromosome maintenance protein 1 (Mcm1p) (7, 47). The budding yeast Mcm1p is an essential sequence-specific homodimeric DNA-binding protein and a member of the MADS-box transcription factor family. The MADS-box is a highly conserved sequence motif characteristic of this family of transcriptional regulators. This motif was identified after sequence comparison of Mcm1p, AGAMOUS, DEFICIENS, and SRF (serum response factor), and the name MADS was derived from the first letters of these four "founders" (92). MADS-box proteins interact with diverse sequence-specific transcription factors to repress or activate different sets of genes (65). In addition to its function in cell-type-specific gene expression, the MADS-box protein Mcm1p from *S. cerevisiae* has a well-defined role in the control of genes that determine general metabolism (64), minichromosome maintenance (73), and the regulation of the cell cycle (59, 63).

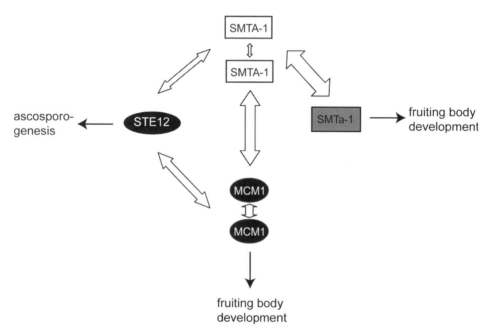

Figure 10.6 Interaction network of mating-type proteins and the transcription factors STE12 and MCM1. Interactions validated by two-hybrid analysis and by means of Far Western and protein cross-linking experiments, respectively (45, 67). Proteins are indicated by ovals and protein interactions with arrows. The thickness of the arrows reflects the strength of the interaction.

To elucidate whether additional transcription factors are involved in the sexual development of the homothallic ascomycete *S. macrospora*, the *S. macrospora mcm1* gene encoding a putative homologue of the *S. cerevisiae* Mcm1p protein was cloned and functionally analyzed (67). Deletion of the *mcm1* gene led to a pleiotropic phenotype, including reduced biomass, increased hyphal branching, and reduced hyphal compartment length during vegetative growth as well as sexual sterility. The Δ*mcm1* mutant is capable of producing only protoperithecia but is unable to form either ascospores or perithecia. A two-hybrid analysis has demonstrated that MCM1 may physically interact with itself and with the α-domain mating-type protein SMTA-1. Only the first N-terminal 155 aa of MCM1 containing the MADS-box domain are sufficient for homodimerization and interaction with SMTA-1 (Fig. 10.6). Similarly, the conserved MADS-box domain of the *S. cerevisiae* Mcm1p was shown to be sufficient for dimerization and DNA binding as well as for interaction with different cofactors (9, 14).

In *S. cerevisiae*, Mcm1p activates α-specific genes, e.g., α-specific pheromone and pheromone receptor genes, together with the α-domain transcription factor Matα1p, and represses **a**-specific genes (e.g., *MATa*-specific pheromone and receptor genes) together with homeodomain transcription factor Matα2p (7, 40). The *S. macrospora* MCM1 is capable of interacting only with SMTA-1; no positive interactions between MCM1 and other proteins encoded by the *S. macrospora* mating-type locus have been detected to date (67). This result does not come as a surprise, since unlike *S. cerevisiae* the mating-type locus of *S. macrospora* does not encode a Matα2p-like homeodomain transcription factor. As mentioned in "Pheromone and Pheromone Receptor Genes" above, in *S. macrospora*, the α-domain protein SMTA-1 interacts with the mating-type encoded HMG-domain protein SMTa-1. Thus, the HMG-domain protein SMTa-1 may be recruited via SMTA-1 into a complex which contains, among others, the MCM1 protein. Interestingly, with respect to fruiting-body and ascospore development, the phenotype of a Δ*Smta-1* mutant resembles that of the Δ*mcm1* mutant (see "Mating-Type Genes from *S. macrospora* Induce Perithecial Development in a Heterothallic Fungus" above). Similar to Δ*mcm1*, a Δ*Smta-1* mutant was shown to be sterile and to produce only protoperithecia (84).

A STE12 Homologue of *S. macrospora* Interacts with SMTA-1 and the MADS-Box Protein MCM1 and Is Required for Ascosporogenesis

The pleiotropic phenotype of the *S. macrospora* Δ*mcm1* mutant suggests that the *S. macrospora* MCM1 protein

might be involved in a wide range of functions via interaction with diverse transcriptional regulators. From a yeast two-hybrid screen, among several other proteins, a putative homologue of the *S. cerevisiae* homeodomain protein Ste12p was identified as an MCM1 interaction partner (68). The protein is termed STE12 in *S. macrospora*. In *S. cerevisiae*, Ste12p is known to be a transcriptional factor active downstream of Fus3p/Kss1 MAP kinases. It plays a central role in a signaling pathway from the plasma membrane to the nucleus in a cell-type-specific manner. In haploid yeast cells, Ste12p is required for the response to mating pheromone produced by the opposite mating type and for invasive growth in response to limited nutrients. In diploid yeast cells, Ste12p regulates pseudohyphal development in response to nitrogen starvation (38, 58, 88). Transcriptional regulation of different classes of genes is thereby triggered through interactions with different transcriptional regulators. Ste12p homodimerizes or binds either Mcm1p or Matα1p to regulate pheromone-responsive genes or cell-type-specific genes, respectively, and binds with Tec1p to activate genes required for filamentous growth (10, 25, 58, 102).

Two-hybrid and biochemical studies showed that the N-terminus of the *S. macrospora* STE12 associates with MCM1. In addition, the *S. macrospora* STE12 is able to interact with the mating-type protein SMTA-1 (Fig. 10.6). These results are in agreement with data from *S. cerevisiae* (46). In *MATα* cells of the budding yeast, Ste12p activates the expression of the α-specific *STE3* pheromone receptor gene and α-factor precursor genes in a complex with Mcm1p and the mating-type protein Matα1p, the homologue of the *S. macrospora* SMTA-1 protein (46, 102). This indicates a role for STE12 in a regulatory network involved in controlling of mating and sexual development. Homologues of the *S. cerevisiae* transcription factor Ste12p have also been characterized in several filamentous ascomycetes. While in *Aspergillus nidulans* and *N. crassa* Ste12p homologues are required for sexual development (56, 100), Ste12p homologues of the plant pathogens *Colletotrichum lagenarium* and *Magnaporthe grisea* are involved in pathogenicity (72, 96). Analysis of an *S. macrospora* Δ*ste12* knockout mutant demonstrated that STE12 is not needed for fruiting-body formation and vegetative growth but is involved in ascosporogenesis (68). The *S. macrospora* Δ*ste12* mutant is able to form protoperithecia and perithecia, but the latter contain a drastically reduced number of asci with predominantly inviable ascospores. Particularly, cell walls of asci and ascospores appear to be fragile. In contrast, Δpp-1 strains of *N. crassa* lacking the *ste12* homologue are female sterile and show a complete absence of protoperithecia, but similar to *S. macrospora* cause ascospore lethality when used as male partner (56). In the homothallic *A. nidulans*, Δ*steA* strains are sterile and do not differentiate fruiting bodies, whereas the *MST12* of the *M. grisea* gene was shown to be dispensable for mating and sexual reproduction but essential for plant infection (72, 100).

As previously mentioned, despite being self-fertile, *S. macrospora* carries two pheromone precursor genes and two pheromone receptor genes. These genes encode two pheromone receptor pairs (PPG1/PRE2 and PPG2/PRE1), which upon interaction are supposed to trigger pheromone-induced responses (61, 78, 80). Moreover, pheromones and pheromone receptors are required for optimal sexual reproduction in *S. macrospora* (62). The phenotypes of double-knockout strains lacking any compatible pheromone receptor pairs (Δ*pre1*/Δ*ppg1*, Δ*pre2*/Δ*ppg2*) display a drastically reduced number of perithecia and sexual spores and thus resemble the phenotype of the *S. macrospora* Δ*st12* mutant strain. These common phenotypes suggest that STE12 might be a pheromone response regulator. However, unlike the Δ*ste12* mutant strain, fragile ascus and ascospore cell walls have never been observed in Δ*pre1*/Δ*ppg1* and Δ*pre2*/Δ*ppg2* mutant strains. This implies that STE12, in addition to its role in pheromone signal transduction, might be involved in cell wall integrity of asci and ascospores. Interestingly, the *S. cerevisiae* Ste12p by interaction with Mcm1p is known to play a major role in maintaining cell wall integrity as it can be targeted to the promoters of a set of genes involved in cell wall biology (52).

SUMMARY

S. macrospora is a homothallic ascomycete that is closely related to the heterothallic *N. crassa*. Analyses of the mating-type locus of *S. macrospora* revealed that this locus contains sequences homologous to both the *MATa* and *MATA* idiomorphs of *N. crassa* (85). The mating-type locus of *S. macrospora* contains four different ORFs: *Smt a-1*, *Smt A-3*, *Smt A-2*, and *SmtA-1*, all of which are transcribed (79). The proteins encoded by *Smta-1* and *SmtA-1* contain domains typical for eukaryotic transcription factors. When expressed in haploid strains of both mating types of the related heterothallic *P. anserina*, the mating-type locus of *S. macrospora* can induce fruiting-body formation (85). Deletion of the mating-type gene *Smta-1* encoding an HMG protein results in sterility of *S. macrospora*. A Δ*Smta-1* mutant strain is morphologically wild type during vegetative growth, but it is unable to produce perithecia or ascospores. Thus, mating-type genes of the homothallic

S. macrospora are functional genes that are involved in sexual reproduction. At least the mating-type proteins SMTa-1 and SMTA-1 seem to act as transcriptional regulators and to control sexual reproduction by regulating a variety of genes involved in different cellular processes. Furthermore, regulatory interactions between mating-type-gene-encoded proteins and other transcription factors like the MADS-box protein MCM1 and the homeodomain protein STE12 may represent a mechanism to control fertilization, fruiting-body development, and ascosporogenesis in *S. macrospora* (Fig. 10.6). However, it is also apparent that to unravel the function of mating-type proteins and to better understand the complexity of sexual development in *S. macrospora* requires further studies to identify upstream components regulating mating-type proteins and downstream targets whose expression is dependent on the mating-type-gene-encoded proteins. Due to its self-fertility, *S. macrospora* provides an ideal model to investigate the function of mating-type genes and mating-type regulated process. For example, recessive mutations that affect postfertilization perithecial development will remain undetected in heterothallic species until the mutant allele is available in both mating types, thus allowing homozygous crosses. In contrast, in the homothallic *S. macrospora*, recessive mutations can directly be tested for defects in fruiting-body development. Moreover, *S. macrospora* produces only meiotically derived ascospores, while asexual spores, such as conidia, are absent. Thus, there is no interference between two different developmental programs in *S. macrospora*, making it, for example, easier to analyze differentially expressed genes involved in fruiting-body development. Furthermore, a comparative study of mating-type-dependent processes in homothallic and heterothallic fungi will enable us to increase our knowledge on how homothallic and heterothallic reproductive mechanisms have evolved in fungi.

I thank N. Nolting, U. Kück, and H. D. Osiewacz for sharing unpublished data and E. Coppin for providing the P. anserina *ΔMAT strain. I am grateful to R. Willmott for English-language editing of the manuscript. The work was funded by the Deutsche Forschungsgemeinschaft (SFB480 and PO523/3-1) and the Ruhr-University of Bochum.*

References

1. **Arnaise, S., R. Debuchy, and M. Picard.** 1997. What is a bona fide mating-type gene? Internuclear complementation of mat mutants in *Podospora anserina*. *Mol. Gen. Genet.* **256:**169–178.
2. **Arnaise, S., D. Zickler, and N. L. Glass.** 1993. Heterologous expression of mating-type genes in filamentous fungi. *Proc. Natl. Acad. Sci. USA* **90:**6616–6620.

3. **Arnaise, S., D. Zickler, S. Le Bilcot, C. Poisier, and R. Debuchy.** 2001. Mutations in mating-type genes of the heterothallic fungus *Podospora anserina* lead to self-fertility. *Genetics* **159:**545–556.

4. **Badgett, T. C., and C. Staben.** 1999. Interaction between and transactivation by the mating type polypeptides of *Neurospora crassa. Fungal Genet. News. Suppl.* **46:**127.

5. **Bardwell, L.** 2005. A walk-through of the yeast mating pheromone response pathway. *Peptides* **26:**339–350.

6. **Beatty, N. P., M. L. Smith, and N. L. Glass.** 1994. Molecular characterization of mating-type loci in selected homothallic species of Neurospora, Gelasinospora and Anixiella. *Mycol. Res.* **98:**1309–1316.

7. **Bender, A., and G. F. Sprague, Jr.** 1987. MAT alpha 1 protein, a yeast transcription activator, binds synergistically with a second protein to a set of cell-type-specific genes. *Cell* **50:**681–691.

8. **Bobrowicz, P., R. Pawlak, A. Correa, D. Bell-Pedersen, and D. J. Ebbole.** 2002. The *Neurospora crassa* pheromone precursor genes are regulated by the mating type locus and circadian clock. *Mol. Microbiol.* **45:**795–804.

9. **Bruhn, L., J. J. Hwang-Sum, and G. F. Sprague.** 1992. The N-terminal 96 residues of MCM1, a regulator of cell type-specific genes in *Saccharomyces cerevisiae*, are sufficient for DNA binding, transcription activation, and interaction with α 1. *Mol. Cell. Biol.* **12:**3563–3572.

10. **Bruhn, L., and G. F. J. Sprague.** 1994. MCM1 point mutants deficient in expression of alpha-specific genes: residues important for interaction with alpha 1. *Mol. Cell. Biol.* **14:**2534–2544.

11. **Buck, V., S. S. Ng, A. B. Ruiz-Garcia, K. Papadopoulou, S. Bhatti, J. M. Samuel, M. Anderson, J. B. A. Millar, and C. J. McInerny.** 2004. Fkh2p and Sep1p regulate mitotic gene transcription in fission yeast. *J. Cell Sci.* **117:**5623–5632.

12. **Cai, L., R. Jeewon, and K. D. Hyde.** 2006. Phylogenetic investigations of Sordariaceae based on multiple gene sequences and morphology. *Mycol. Res.* **110:**137–150.

13. **Chang, S., and C. Staben.** 1994. Directed replacement of mt A by mt a-1 effects a mating type switch in *Neurospora crassa. Genetics* **138:**75–81.

14. **Christ, C., and B. K. Tye.** 1991. Functional domains of the yeast transcription/replication factor MCM1. *Genes Dev.* **5:**751–763.

15. **Coppin, E., S. Arnaise, V. Contamine, and M. Picard.** 1993. Deletion of the mating-type sequences in *Podospora anserina* abolishes mating without affecting vegetative functions and sexual differentiation. *Mol. Gen. Genet.* **241:**409–414.

16. **Coppin, E., C. de Renty, and R. Debuchy.** 2005. The function of the coding sequences for the putative pheromone precursors in *Podospora anserina* is restricted to fertilization. *Eukaryot. Cell* **4:**407–420.

17. **Coppin, E., R. Debuchy, S. Arnaise, and M. Picard.** 1997. Mating types and sexual development in filamentous ascomycetes. *Microbiol. Mol. Biol. Rev.* **61:**411–428.

18. **Davis, R., and D. D. Perkins.** 2002. *Neurospora:* a model of model microbes. *Nat. Rev. Genet.* **3:**7–13.

19. **Debuchy, R.** 1999. Internuclear recognition: a possible connection between euascomycetes and homobasidiomycetes. *Fungal Genet. Biol.* **27:**218–223.

20. **Debuchy, R., S. Arnaise, and G. Lecellier.** 1993. The mat-allele of *Podospora anserina* contains three regulatory genes required for the development of fertilized female organs. *Mol. Gen. Genet.* **241:**667–673.

21. **Debuchy, R., and E. Coppin.** 1992. The mating types of *Podospora anserina:* functional analysis and sequence of the fertilization domains. *Mol. Gen. Genet.* **233:**113–121.

22. **Debuchy, R., and B. G. Turgeon.** 2006. Mating-type structure, evolution, and function in euascomycetes, p. 293–323. *In* U. Kües and R. Fischer (ed.), *Growth, Differentiation and Sexuality*, vol. I. Springer-Verlag, Berlin, Germany.

23. **Dementhon, K., S. J. Saupe, and C. Clave.** 2004. Characterization of IDI-4, a bZIP transcription factor inducing autophagy and cell death in the fungus *Podospora anserina. Mol. Microbiol.* **53:**1625–1640.

24. **Desjardins, A. E., D. W. Brown, S.-H. Yun, R. H. Proctor, T. Lee, R. D. Plattner, S.-W. Lu, and B. G. Turgeon.** 2004. Deletion and complementation of the mating type (MAT) locus of the wheat head blight pathogen *Gibberella zeae. Appl. Environ. Microbiol.* **70:**2437–2444.

25. **Dolan, J. W., C. Kirkman, and S. Fields.** 1989. The yeast STE12 protein binds to the DNA sequence mediating pheromone induction. *Proc. Natl. Acad. Sci. USA* **86:**5703–5707.

26. **Eriksson, O. E.** 2004. Outline of Ascomyta—2006. *Myconet* **12:**1–82.

27. **Esser, K.** 1982. *Cryptogams—Cyanobacteria, Algae, Fungi, Lichens.* Cambridge University Press, London, United Kingdom.

28. **Esser, K., and J. Straub.** 1958. Genetische Untersuchungen an *Sordaria macrospora* Auersw.: Kompensation und Induktion bei genbedingten Entwicklungsdefekten. *Z. Vererbungsl.* **89:**729–746.

29. **Ferreira, A. V., Z. An, R. L. Metzenberg, and N. L. Glass.** 1998. Characterization of *mat A-2, mat A-3* and delta-matA mating-type mutants of *Neurospora crassa. Genetics* **148:**1069–1079.

30. **Ferreira, A. V., S. Saupe, and N. L. Glass.** 1996. Transcriptional analysis of the mtA idiomorph of *Neurospora crassa* identifies two genes in addition to *mtA-1. Mol. Gen. Genet.* **250:**767–774.

31. **Fields, W. G.** 1970. An introduction to the genus Sordaria. *Neurospora Newsl.* **16:**14–17.

32. **Galagan, J. E., S. E. Calvo, K. A. Borkovich, E. U. Selker, N. D. Read, D. Jaffe, W. FitzHugh, L. J. Ma, S. Smirnov, S. Purcell, B. Rehman, T. Elkins, R. Engels, S. Wang, C. B. Nielsen, J. Butler, M. Endrizzi, D. Qui, P. Ianakiev, D. Bell-Pedersen, M. A. Nelson, M. Werner-Washburne, C. P. Selitrennikoff, J. A. Kinsey, E. L. Braun, A. Zelter, U. Schulte, G. O. Kothe, G. Jedd, W. Mewes, C. Staben, E. Marcotte, D. Greenberg, A. Roy, K. Foley, J. Naylor, N. Stange-Thomann, R. Barrett, S. Gnerre, M. Kamal, M. Kamvysselis, E. Mauceli, C. Bielke, S. Rudd, D. Frishman, S. Krystofova, C. Rasmussen, R. L. Metzenberg, D. D. Perkins,

S. Kroken, C. Cogoni, G. Macino, D. Catcheside, W. Li, R. J. Pratt, S. A. Osmani, C. P. DeSouza, L. Glass, M. J. Orbach, J. A. Berglund, R. Voelker, O. Yarden, M. Plamann, S. Seiler, J. Dunlap, A. Radford, R. Aramayo, D. O. Natvig, L. A. Alex, G. Mannhaupt, D. J. Ebbole, M. Freitag, I. Paulsen, M. S. Sachs, E. S. Lander, C. Nusbaum, and B. Birren. 2003. The genome sequence of the filamentous fungus *Neurospora crassa*. *Nature* **422:**859–868.

33. Glass, N. L., J. Grotelueschen, and R. L. Metzenberg. 1990. *Neurospora crassa* A mating-type region. *Proc. Natl. Acad. Sci. USA* **87:**4912–4916.

34. Glass, N. L., R. L. Metzenberg, and N. B. Raju. 1990. Homothallic Sordariaceae from nature: the absence of strains containing only a mating type sequences. *Exp. Mycol.* **14:**274–289.

35. Glass, N. L., and M. L. Smith. 1994. Structure and function of a mating-type gene from the homothallic species *Neurospora africana*. *Mol. Gen. Genet.* **244:**401–409.

36. Glass, N. L., S. J. Vollmer, C. Staben, J. Grotelueschen, R. L. Metzenberg, and C. Yanofsky. 1988. DNAs of the two mating-type alleles of *Neurospora crassa* are highly dissimilar. *Science* **241:**570–573.

37. Grosschedl, R., K. Giese, and J. Pagel. 1994. HMG domain proteins: architectural elements in the assembly of nucleoprotein structures. *Trends Genet.* **10:**94–100.

38. Gustin, M. C., J. Albertyn, M. Alexander, and K. Davenport. 1998. MAP kinase pathways in the yeast *Saccharomyces cerevisiae*. *Microbiol. Mol. Biol. Rev.* **62:**1264–1300.

39. Hartmann, H. A., R. Kahmann, and M. Bölker. 1996. The pheromone response factor coordinates filamentous growth and pathogenicity in *Ustilago maydis*. *EMBO J.* **15:**1632–1641.

40. Herskowitz, I. 1989. A regulatory hierarchy for cell specialization in yeast. *Nature* **342:**749–757.

41. Heslot, H. 1958. Contribution à l'étude cytogénétique des Sordariacées. *Rev. Cytol. Biol. Végétal.* **19**(Suppl. 2):1–255.

42. Hiscock, S. J., and U. Kües. 1999. Cellular and molecular mechanisms of sexual incompatibility in plants and fungi. *Int. Rev. Cytol.* **193:**165–295.

43. Inderbitzin, P., J. Harkness, B. G. Turgeon, and M. L. Berbee. 2005. Lateral transfer of mating system in Stemphylium. *Proc. Natl. Acad. Sci. USA* **102:**11390–11395.

44. Ivey, F. D., A. M. Kays, and K. A. Borkovich. 2002. Shared and independent roles for a Gαi protein and adenylyl cyclase in regulating development and Stress responses in *Neurospora crassa*. *Eukaryot. Cell* **1:**634–642.

45. Jacobsen, S., M. Wittig, and S. Pöggeler. 2002. Interaction between mating-type proteins from the homothallic fungus *Sordaria macrospora*. *Curr. Genet.* **41:**150-158.

46. Johnson, A. D. 1995. Molecular mechanisms of cell-type determination in budding yeast. *Curr. Opin. Genet. Dev.* **5:**552–558.

47. Keleher, C. A., C. Goutte, and A. D. Johnson. 1988. The yeast cell-type-specific repressor alpha 2 acts cooperatively with a non-cell-type-specific protein. *Cell* **53:**927–936.

48. Kim, H., and K. A. Borkovich. 2004. A pheromone receptor gene, *pre1*, is essential for mating type-specific directional growth and fusion of trichogynes and female fertility in *Neurospora crassa*. *Mol. Microbiol.* **52:**1781–1798.

49. Kim, H., R. L. Metzenberg, and M. A. Nelson. 2002. Multiple functions of *mfa-1*, a putative pheromone precursor gene of *Neurospora crassa*. *Eukaryot. Cell* **1:**987–999.

50. Kirk, P., P. Cannon, J. C. David, and J. Stalpers. 2001. *Ainsworth & Bisby's Dictionary of the Fungi*, 9th ed. CABI International, Wallingford, United Kingdom.

51. Krystofova, S., and K. A. Borkovich. 2005. The heterotrimeric G-protein subunits GNG-1 and GNB-1 form a Gβγ dimer required for normal female fertility, asexual development, and Gα protein levels in *Neurospora crassa*. *Eukaryot. Cell* **4:**365–378.

52. Kuo, M., E. Nadeau, and E. Grayhack. 1997. Multiple phosphorylated forms of the *Saccharomyces cerevisiae* Mcm1 protein include an isoform induced in response to high salt concentrations. *Mol. Cell. Biol.* **17:**819–832.

53. Lee, J., T. Lee, Y. W. Lee, S. H. Yun, and B. G. Turgeon. 2003. Shifting fungal reproductive mode by manipulation of mating type genes: obligatory heterothallism of *Gibberella zeae*. *Mol. Microbiol.* **50:**145–152.

54. Lee, S. H., S. Lee, D. Choi, Y. W. Lee, and S. H. Yun. 2006. Identification of the down-regulated genes in a mat1-2-deleted strain of *Gibberella zeae*, using cDNA subtraction and microarray analysis. *Fungal Genet. Biol.* **43:**295–310.

55. Leubner-Metzger, G., B. A. Horwitz, O. C. Yoder, and B. G. Turgeon. 1997. Transcripts at the mating type locus of *Cochliobolus heterostrophus*. *Mol. Gen. Genet.* **256:**661–673.

56. Li, D., P. Bobrowicz, H. H. Wilkinson, and D. J. Ebbole. 2005. A mitogen-activated protein kinase pathway essential for mating and contributing to vegetative growth in *Neurospora crassa*. *Genetics* **170:**1091–1104.

57. Lundquist, N. 1972. Nordic Sordariaceae s.lat. *Symb. Bot. Ups.* **20:**1–314.

58. Madhani, H. D., and G. R. Fink. 1997. Combinatorial control required for the specificity of yeast MAPK signaling. *Science* **275:**1314–1317.

59. Maher, M., F. Cong, D. Kindelberger, K. Nasmyth, and S. Dalton. 1995. Cell cycle-regulated transcription of the CLB2 gene is dependent on Mcm1 and a ternary complex factor. *Mol. Cell. Biol.* **15:**3129–3137.

60. Masloff, S., S. Pöggeler, and U. Kück. 1999. The *pro1*⁺ gene from *Sordaria macrospora* encodes a C_6 zinc finger transcription factor required for fruiting body development. *Genetics* **152:**191–199.

61. Mayrhofer, S., and S. Pöggeler. 2005. Functional characterization of an a-factor-like *Sordaria macrospora* peptide pheromone and analysis of its interaction with its cognate receptor in *Saccharomyces cerevisiae*. *Eukaryot. Cell* **4:**661–672.

62. Mayrhofer, S., J. M. Weber, and S. Pöggeler. 2006. Pheromones and pheromone receptors are required for proper sexual development in the homothallic ascomycete *Sordaria macrospora*. *Genetics* **172:**1521–1533.

63. McInerny, C. J., J. F. Partridge, G. E. Mikesell, D. P. Creemer, and L. L. Breeden. 1997. A novel Mcm1-dependent element in the SWI4, CLN3, CDC6, and CDC47 promoters activates M/G$_1$-specific transcription. *Genes Dev.* **11**:1277–1288.

64. Messenguy, F., and E. Dubois. 1993. Genetic evidence for a role for MCM1 in the regulation of arginine metabolism in *Saccharomyces cerevisiae*. *Mol. Cell. Biol.* **13**:2586–2592.

65. Messenguy, F., and E. Dubois. 2003. Role of MADS box proteins and their cofactors in combinatorial control of gene expression and cell development. *Gene* **316**:1–21.

66. Metzenberg, R. L., and N. L. Glass. 1990. Mating type and mating strategies in Neurospora. *Bioessays* **12**:53–59.

67. Nolting, N., and S. Pöggeler. 2006. A MADS-box protein interacts with a mating-type protein and is required for fruiting body development in the homothallic ascomycete *Sordaria macrospora*. *Eukaryot. Cell* **5**:1043–1056.

68. Nolting, N., and S. Pöggeler. 2006. A STE12 homologue of the homothallic ascomycete *Sordaria macrospora* interacts with the MADS box protein MCM1 and is required for ascosporogenesis. *Mol. Microbiol.* **62**:853–868.

69. Nowrousian, M., C. Ringelberg, J. C. Dunlap, J. J. Loros, and U. Kück. 2005. Cross-species microarray hybridization to identify developmentally regulated genes in the filamentous fungus *Sordaria macrospora*. *Mol. Gen. Genomics* **273**:137–149.

70. Nowrousian, M., C. Würtz, S. Pöggeler, and U. Kück. 2004. Comparative sequence analysis of *Sordaria macrospora* and *Neurospora crassa* as a means to improve genome annotation. *Fungal Genet. Biol.* **41**:285–292.

71. O'Donnell, K., T. J. Ward, D. M. Geiser, H. Corby Kistler, and T. Aoki. 2004. Genealogical concordance between the mating type locus and seven other nuclear genes supports formal recognition of nine phylogenetically distinct species within the *Fusarium graminearum* clade. *Fungal Genet. Biol.* **41**:600–623.

72. Park, G., C. Xue, L. Zheng, S. Lam, and J. R. Xu. 2002. MST12 regulates infectious growth but not appressorium formation in the rice blast fungus *Magnaporthe grisea*. *Mol. Plant-Microbe Interact.* **15**:183–192.

73. Passmore, S., R. Elble, and B. K. Tye. 1989. A protein involved in minichromosome maintenance in yeast binds a transcriptional enhancer conserved in eukaryotes. *Genes Dev.* **3**:921–935.

74. Perkins, D. D., and R. H. Davis. 2000. Neurospora at the millennium. *Fungal Genet. Biol.* **31**:153–167.

75. Philley, M. L., and C. Staben. 1994. Functional analyses of the *Neurospora crassa* MT a-1 mating type polypeptide. *Genetics* **137**:715–722.

76. Picard, M., R. Debuchy, and E. Coppin. 1991. Cloning the mating types of the heterothallic fungus *Podospora anserina*: developmental features of haploid transformants carrying both mating types. *Genetics* **128**:539–547.

77. Pöggeler, S. 1999. Phylogenetic relationships between mating-type sequences from homothallic and heterothallic ascomycetes. *Curr. Genet.* **36**:222–231.

78. Pöggeler, S. 2000. Two pheromone precursor genes are transcriptionally expressed in the homothallic ascomycete *Sordaria macrospora*. *Curr. Genet.* **37**:403–411.

79. Pöggeler, S., and U. Kück. 2000. Comparative analysis of the mating-type loci from *Neurospora crassa* and *Sordaria macrospora*: identification of novel transcribed ORFs. *Mol. Gen. Genet.* **263**:292–301.

80. Pöggeler, S., and U. Kück. 2001. Identification of transcriptionally expressed pheromone receptor genes in filamentous ascomycetes. *Gene* **280**:9–17.

81. Pöggeler, S., and U. Kück. 2004. A WD40 repeat protein regulates fungal cell differentiation and can functionally be replaced by the mammalian homologue striatin. *Eukaryot. Cell* **3**:232–240.

82. Pöggeler, S., S. Masloff, S. Jacobsen, and U. Kück. 2000. Karyotype polymorphism correlates with intraspecific infertility in the homothallic ascomycete *Sordaria macrospora*. *J. Evol. Biol.* **13**:281–289.

83. Pöggeler, S., M. Nowrousian, and U. Kück. 2006. Fruiting-body development in ascomycetes, p. 326–355. *In* U. Kües and R. Fischer (ed.), *Growth, Differentiation and Sexuality*, 2nd ed., vol. I. Springer-Verlag, Berlin, Germany.

84. Pöggeler, S., M. Nowrousian, C. Ringelberg, J. Loros, J. Dunlap, and U. Kück. 2006. Microarray and real time PCR analyses reveal mating type-dependent gene expression in a homothallic fungus. *Mol. Gen. Genomics* **275**:492–503.

85. Pöggeler, S., S. Risch, U. Kück, and H. D. Osiewacz. 1997. Mating-type genes from the homothallic fungus *Sordaria macrospora* are functionally expressed in a heterothallic ascomycete. *Genetics* **147**:567–580.

86. Randall, T. A., and R. L. Metzenberg. 1995. Species-specific and mating-type specific DNA regions adjacent to mating-type idiomorphs in the genus Neurospora. *Genetics* **141**:119–136.

87. Ribar, B., A. Grallert, E. Olah, and Z. Szallasi. 1999. Deletion of the sep1(+) forkhead transcription factor homologue is not lethal but causes hyphal growth in *Schizosaccharomyces pombe*. *Biochem. Biophys. Res. Commun.* **263**:465–474.

88. Roberts, C. J., B. Nelson, M. J. Marton, R. Stoughton, M. R. Meyer, H. A. Bennett, Y. D. He, H. Dai, W. L. Walker, T. R. Hughes, M. Tyers, C. Boone, and S. H. Friend. 2000. Signaling and circuitry of multiple MAPK pathways revealed by a matrix of global gene expression profiles. *Science* **287**:873–880.

89. Robertson, S. J., D. J. Bond, and N. D. Read. 1998. Homothallism and heterothallism in *Sordaria brevicollis*. *Mycol. Res.* **102**:1215–1223.

90. Saupe, S., L. Stenberg, K. T. Shiu, A. J. Griffiths, and N. L. Glass. 1996. The molecular nature of mutations in the mt A-1 gene of the *Neurospora crassa* A idiomorph and their relation to mating-type function. *Mol. Gen. Genet.* **250**:115–122.

91. **Shiu, P. K., and N. L. Glass.** 2000. Cell and nuclear recognition mechanisms mediated by mating type in filamentous ascomycetes. *Curr. Opin. Microbiol.* 3:183–188.

92. **Shore, P., and A. D. Sharroks.** 1995. The MADS-box family of transcription factors. *Eur. J. Biochem.* 229:1–13.

93. **Souza, C. A., C. C. Silva, and A. V. Ferreira.** 2003. Sex in fungi: lessons of gene regulation. *Genet. Mol. Res.* 2:136–147.

94. **Staben, C., and C. Yanofsky.** 1990. *Neurospora crassa a* mating-type region. *Proc. Natl. Acad. Sci. USA* 87:4917–4921.

95. **Tajadura, V., B. Garcia, I. Garcia, P. Garcia, and Y. Sanchez.** 2004. *Schizosaccharomyces pombe* Rgf3p is a specific Rho1 GEF that regulates cell wall beta-glucan biosynthesis through the GTPase Rho1p. *J. Cell Sci.* 117:6163–6174.

96. **Tsuji, G., S. Fujii, S. Tsuge, T. Shiraishi, and Y. Kubo.** 2003. The *Colletotrichum lagenarium Ste12*-like gene CST1 is essential for appressorium penetration. *Mol. Plant-Microbe Interact.* 16:315–325.

97. **Turgeon, B. G.** 1998. Application of mating-type gene technology to problems in fungal biology. *Annu. Rev. Phytopathol.* 36:115–137.

98. **Turgeon, B. G., H. Bohlmann, L. M. Ciuffetti, S. K. Christiansen, G. Yang, W. Schäfer, and O. C. Yoder.** 1993. Cloning and analysis of the mating type genes from *Cochliobolus heterostrophus. Mol. Gen. Genet.* 238:270–284.

99. **Turina, M., A. Prodi, and N. K. Alfen.** 2003. Role of the Mf1-1 pheromone precursor gene of the filamentous ascomycete *Cryphonectria parasitica. Fungal Genet. Biol.* 40:242–251.

100. **Vallim, M. A., K. A. Miller, and B. L. Miller.** 2000. Aspergillus SteA (sterile12-like) is a homeodomain-C2/H2-Zn+2 finger transcription factor required for sexual reproduction. *Mol. Microbiol.* 36:290–301.

101. **Wirsel, S., B. G. Turgeon, and O. C. Yoder.** 1996. Deletion of the *Cochliobolus heterostrophus* mating-type (MAT) locus promotes the function of MAT transgenes. *Curr. Genet.* 29:241–249.

102. **Yuan, Y. O., I. L. Stroke, and S. Fields.** 1993. Coupling of cell identity to signal response in yeast: interaction between the alpha 1 and STE12 proteins. *Genes Dev.* 7:1584–1597.

103. **Yun, S. H., T. Arie, I. Kaneko, O. C. Yoder, and B. G. Turgeon.** 2000. Molecular organization of mating type loci in heterothallic, homothallic, and asexual Gibberella/Fusarium species. *Fungal Genet. Biol.* 31:7–20.

104. **Yun, S.-H., M. L. Berbee, O. C. Yoder, and B. G. Turgeon.** 1999. Evolution of the fungal self-fertile reproductive life style from self-sterile ancestors. *Proc. Natl. Acad. Sci. USA* 96:5592–5597.

105. **Zickler, D.** 1977. Development of the synaptonemal complex and the "recombination nodules" during meiotic prophase in the seven bivalents of the fungus *Sordaria macrospora* Auersw. *Chromosoma* 61:289–316.

106. **Zickler, D.** 2006. From early homologue recognition to synaptonemal complex formation. *Chromosoma* 115:158–174.

107. **Zickler, D., S. Arnaise, E. Coppin, R. Debuchy, and M. Picard.** 1995. Altered mating-type identity in the fungus Podospora anserina leads to selfish nuclei, uniparental progeny, and haploid meiosis. *Genetics* 140:493–503.

108. **Zickler, D., P. Moreau, A. D. Huynh, and A. M. Slezec.** 1992. Correlation between pairing initiation sites, recombination nodules and meiotic recombination in *Sordaria macrospora. Genetics* 132:135–148.

*Sex in Fungi: Molecular Determination
and Evolutionary Implications*
Edited by Joseph Heitman et al.
© 2007 ASM Press, Washington, D.C.

Laura N. Rusche
Meleah A. Hickman

11

Evolution of Silencing at the Mating-Type Loci in Hemiascomycetes

INTRODUCTION

As outlined elsewhere in this book, some fungi, including the model yeasts *Saccharomyces cerevisiae* and *Schizosaccharomyces pombe*, have the ability to switch mating types as often as once per generation. In these organisms, the mating type of a haploid cell is generally determined by a "cassette," encoding master transcriptional regulators, expressed at the active *MAT* locus. In many cases, there are two types of cassettes, termed **a** and α, which specify two distinct mating types. Switching is achieved by a gene conversion event, in which a new cassette of the opposite mating type replaces the old cassette at the active *MAT* locus. These replacement cassettes are derived from the cryptic, or silent, mating-type loci, which are not expressed but instead serve as repositories for mating-type genes. It is important to emphasize that these cassettes include the promoters as well as the coding portions of the genes. Consequently, the same promoter will simultaneously be expressed at the active locus and be repressed at the silent locus. Hence, a special regional silencing mechanism, in the form of a distinct chromatin structure, is required to maintain the transcriptional repression of the silent mating-type loci. This silencing is crucial to maintaining cell identity, and

in the absence of silencing a haploid cell expresses both **a** and α cassettes and consequently cannot mate. Silencing of mating-type loci has been well characterized in both *S. cerevisiae* and *S. pombe*, and although the end results are the same (silent mating-type loci are repressed), the mechanisms by which this end is reached are quite different. In this chapter, we explore how a unique silencing mechanism evolved in the hemiascomycete class of fungi, which includes the model yeast *S. cerevisiae*.

A Paradigm: Silencing in *S. cerevisiae*

The silencing of the cryptic mating-type loci, *HMR* and *HML*, in *S. cerevisiae* (reviewed in reference 60) is mediated by *cis*-acting silencers, known as *E* and *I*, which flank the two loci. Each of the four silencers has a binding site for the origin recognition complex (ORC), as well as binding sites for either Rap1p or Abf1p or both (Fig. 11.1). These three silencer binding proteins recruit the silencing proteins, Sir1 through Sir4, to the silencers. The assembly of Sir proteins into silenced chromatin at the *HM* loci involves two phases, nucleation and spreading (37, 47, 61). First, the four Sir proteins assemble at the silencer via interactions with the silencer

Laura N. Rusche, Institute for Genome Sciences and Policy, Biochemistry Department, Duke University, Durham, NC 27710. **Meleah A. Hickman,** Institute for Genome Sciences and Policy, University Program in Genetics and Genomics, Duke University, Durham, NC 27710.

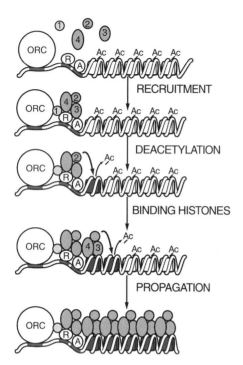

Figure 11.1 Model for Sir-mediated silencing. 1 through 4, Sir1p through Sir4p; ORC, origin recognition complex; R, Rap1p; A, Abf1p; Ac, acetyl group.

binding proteins. Then, Sir2p, Sir3p, and Sir4p spread along the chromosome via interactions with histones. Unlike the other three Sir proteins, Sir1p does not spread. Instead, its presence stabilizes the assembly of Sir proteins at the silencer. Sir2p is an NAD^+-dependent deacetylase, and its activity is required for the spreading of all three Sir proteins (37, 61). Sir3p and Sir4p bind preferentially to deacetylated tails of histones H3 and H4 (9, 34). These observations inspire a "sequential deacetylation" model for the propagation of silenced chromatin (61). In this model, Sir2p deacetylates the neighboring nucleosome, creating new high-affinity binding sites for Sir3p and Sir4p, which in turn recruit additional Sir2p to the newly deacetylated nucleosome.

Sir-mediated silencing also occurs at telomeres in *S. cerevisiae* (30). The degenerate telomeric repeat sequence, TG_{1-3}, found at the ends of the chromosomes, generates binding sites for Rap1p, leading to the association of multiple Rap1p molecules with the telomeres (46). Although single Rap1p binding sites not associated with ORC do not recruit Sir proteins, the presence of tandem Rap1p binding sites is thought to recruit the Sir proteins. The association of Sir proteins with telomeres is also stabilized by the presence of the Ku complex, which binds DNA ends and also interacts with

Sir4p (50, 59). Sir1p is not thought to act at the telomeres. Once Sir2p, Sir3p, and Sir4p are recruited to the telomeres, they spread into the subtelomeric regions via the same mechanism by which they spread at the silent mating-type loci. Reporter genes inserted into the Sir-containing chromatin at telomeres are subject to silencing. However, there are no endogenous subtelomeric genes known to be silenced by Sir proteins in *S. cerevisiae*. Hence, telomeric Sir-containing chromatin probably serves a structural role.

A Different Paradigm: Silencing in *S. pombe*

Another well-studied model yeast, *S. pombe*, also possesses three mating-type loci, of which one is expressed. However, silencing of the two unexpressed mating-type loci, *mat2* and *mat3*, involves a different set of proteins than those employed in *S. cerevisiae* (reviewed in reference 31). In *S. pombe*, a 20-kb region encompassing both *mat2* and *mat3* is subject to silencing, and as in *S. cerevisiae*, a discrete DNA sequence acts as a silencer to recruit silencing proteins to this region (39, 43). However, the silencer binding proteins, atf1 and pcr1, are not related to the silencer binding proteins in *S. cerevisiae*. Also in contrast to *S. cerevisiae*, a second sequence recruits silencing proteins via a mechanism that involves small, noncoding complementary RNAs (55). This sequence, known as *cenH*, is homologous to the outer centromeric repeat sequences, which also recruit silencing proteins. Recruitment of silencing factors through *cenH* involves a protein-RNA complex known as RITS (RNA-induced transcriptional silencing) that is related to the RISC posttranscriptional silencing complex (55, 75). RITS contains small RNA derived from the centromeric repeat sequences as well as argonaute (ago1), the key protein in RISC complexes. Presumably, complementarity between small RNAs and the *cenH* sequence targets RITS to this genomic location. However, the details of the association between RITS and chromatin are still being investigated.

Two important silencing factors in *S. pombe* are a methyltransferase, clr4, which specifically methylates lysine 9 of histone H3, and a chromodomain protein, swi6, which binds preferentially to H3 tails that are methylated on lysine 9. These silencing proteins are thought to spread by a sequential modification mechanism akin to that proposed for *S. cerevisiae* Sir proteins. Specifically, clr4 is recruited through the silencer and *cenH* sequences and targeted to methylate lysine 9 of histone H3 in adjacent nucleosomes. These methylated nucleosomes are then bound by swi6, which in turn recruits additional clr4. In addition, clr4 and swi6 form silenced chromatin at the telomeres and centromeres by a similar mechanism, and the pericentromeric heterochromatin is important for proper centromere function (23).

Most of the factors known to contribute to silencing in *S. pombe* are well conserved among eukaryotes. The homolog of swi6 is HP1 (heterochromatin protein 1) in *Drosophila* and mammals, and clr4 is homologous to SuVar3-9. The RNA interference (RNAi) proteins that act in silencing, including argonaute (ago1) and dicer (dcr1), are also well conserved. Together, these proteins form heterochromatic structures in *Drosophila* and mammalian cells, particularly in the pericentromeric regions (reviewed in reference 14). A key point to note is that none of these heterochromatin proteins (swi6, clr4, ago1, or dcr1) have identifiable homologs in *S. cerevisiae*. Additionally, three of the four Sir proteins (1, 3, and 4) do not have identifiable homologs beyond the close relatives of *S. cerevisiae*. Thus, "classical" heterochromatin, defined by HP1 and methylation of H3 on lysine 9, does not exist in *S. cerevisiae*. Instead, an alternate type of silenced chromatin involving the Sir proteins has evolved. This chapter reviews what is known about silencing in species related to *S. cerevisiae* and explores the origins and development of the unique Sir-mediated silencing mechanism.

The Hemiascomycetes

S. cerevisiae belongs to the hemiascomycete class of fungi, which consists primarily of unicellular yeasts.

Due to their relatively small genome sizes and relationship to *S. cerevisiae*, a favorite organism for genetic and genomic studies, many of these yeasts have been sequenced, providing an exceptional opportunity for comparative genomics (20). The phylogenetic relationships of the species discussed in this chapter are illustrated in Fig. 11.2. Comparisons of average protein sequence identity suggest that the diversity among the hemiascomycetes is slightly greater than that among the chordates (21). For comparison, *S. pombe* belongs to the archiascomycete class, which is thought to have diverged from the hemiascomycetes around 1 billion years ago (35). Currently, there are no other sequenced genomes among the archiascomycetes.

One important event in the evolution of *S. cerevisiae* and its relatives was a whole-genome duplication, which occurred approximately 100 million years ago, after the divergence of *Saccharomyces* and *Kluyveromyces* species (17, 42, 76, 77). Subsequent to this whole-genome duplication, most of the duplicated genes returned to a single-copy status, although roughly 10% of *Saccharomyces* genes are duplicate pairs derived from the whole-genome duplication. As is discussed below, the whole-genome duplication played an important role in the development of Sir-mediated silencing.

For clarity, in the remainder of this chapter, gene and protein names are preceded by the initials of the genus and species of the host organism.

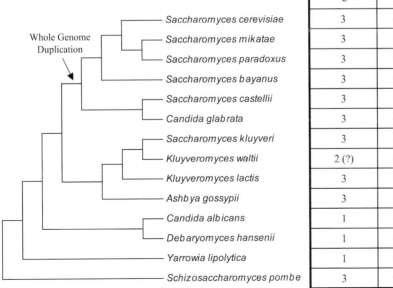

Mating Loci	Sir2	Sir4	Sir3	Sir1	
3	+	+	+	+	*Saccharomyces sensu stricto* species
3	+	+	+	+	
3	+	+	+	+	
3	+	+	+	+	
3	+	+	+	-	
3	+	+	+	-	
3	+	+	-	-	
2 (?)	+	+	-	-	
3	+	+	-	-	
3	+	+	-	-	
1	+	-	-	-	
1	+	-	-	-	
1	+	-	-	-	
3	+	-	-	-	

Figure 11.2 Relationship of species discussed. The phylogenetic tree is based on the 25S ribosomal DNA sequences, using Mega3.1 software, under the Maximum Parsimony model (500 replicates).

COMPARATIVE GENOMICS OF SILENCING

Which Paradigm?

To understand how silencing mechanisms have evolved in the hemiascomycetes, one of the first issues to address is when classical heterochromatin was lost. Analyses of sequenced hemiascomycete genomes fail to detect homologs of the proteins involved in classical heterochromatin formation, even though homologs are easily detected in nonhemiascomycete fungal genomes. Furthermore, the RNAi machinery is also missing from this node in the phylogeny (53). It is not clear what led to the loss of these silencing proteins, but their absence may have presented an evolutionary opportunity for the development of an alternative silencing mechanism, namely Sir-mediated silencing. In the next section, we explore the evolution of the Sir proteins and then examine the changing nature of the genomic regions most likely to be silenced: silent mating-type cassettes, telomeres, and centromeres. Our analysis has been facilitated by a recent survey of silencing proteins in seven hemiascomycete species (24).

Sir Proteins

The deacetylase Sir2p is the most widespread and well-conserved of the Sir proteins. In fact, unlike the other Sir proteins, which are restricted to the hemiascomycetes, Sir2p has homologs among all domains of life, including eubacteria and archea (reviewed in reference 4). Furthermore, the genomes of many species encode multiple Sir2 family members. For example, in *S. cerevisiae* there are five Sir2 deacetylases (*SIR2* and *HST1* through *HST4*) (5). Therefore, when examining the evolving role of Sir2p in silencing, one must take care to identify the correct ortholog of Sir2p. In the case of the hemiascomycetes, this identification is made more certain by using synteny as well as homology. Orthologs of Sir2p have been identified in all examined hemiascomycetes (24); however, the silencing potential of these orthologs has been examined only in a few species. In *Kluyveromyces lactis*, KlSir2p contributes to silencing of mating-type loci (12), and in *Candida glabrata* CgSir2p has been shown to be important in subtelomeric silencing (19). Additionally, the Sir2 orthologs from *Candida albicans* and *K. lactis* have both been shown to partially suppress the mating defect due to the lack of silencing in a *sir2Δ* strain of *S. cerevisiae* (12, 57). Thus, in the cases that have been examined, the silencing function of Sir2p is conserved. Beyond the hemiascomycetes, SpSir2p contributes to classical silencing at centromeres, telomeres, and mating-type loci in *S. pombe*, where it deacetylates H3-K9, thereby promoting methylation of this lysine and the association of SpSwi6 (26, 65). In *Drosophila*, the Sir2 homolog most similar to ScSir2p participates in repression mediated by polycomb group proteins (27). In addition, mutations in DmSir2 affect position effect variegation mediated by HP1 (1, 54). Therefore, the Sir2 deacetylase most likely had an ancient role in silencing.

To participate in silencing, Sir2p, which could theoretically deacetylate any nucleosome in the genome, must be recruited to the correct site through an adaptor protein. In *S. cerevisiae*, ScSir4p plays this role. Does Sir4p recruit Sir2p to chromatin in other species? Homologs of Sir4p are found among some of the hemiascomycetes. However, unlike the homologs of Sir2p, the sequence conservation among the homologs of Sir4p is low, and the identification of these homologs is often based solely on conserved synteny. The species most distant from *S. cerevisiae* in which Sir4p has been identified are *Kluyveromyces* and *Ashbya* species. Sir4p could not be identified in *Debaryomyces hansenii* based on synteny or homology (24). However, given the poor sequence conservation, it is hard to be certain that Sir4p does not exist in *D. hansenii* or other, more distant, species. Whether these homologs of Sir4p interact functionally with Sir2p has been investigated only in two species. In *C. glabrata*, CgSir4p participates in telomeric silencing (38). The *K. lactis SIR4* gene was identified by its ability to complement a *sir4Δ* mutation in *S. cerevisiae* (3). Additionally, the deletion of *KlSIR4* partially disrupts silencing at *HML* (3). In the future it will be important to extend these studies to other species to determine whether Sir4p always recruits Sir2p to chromatin or whether Sir2p sometimes has other partners. In addition, it will be interesting to identify proteins that associate with Sir2p in species apparently lacking Sir4p.

It is interesting that Sir2p, which has an enzymatic function, has been well conserved over the course of evolution, whereas its adaptor, Sir4p, has experienced dramatic changes in its sequence. This difference in rates of evolution is likely due to the more stringent structural requirements of an enzyme compared to an architectural protein. The one conserved feature of Sir4p is a C-terminal coiled-coil domain (3, 24), which is essential for silencing in *S. cerevisiae* (11, 52). Another interesting observation regarding Sir4p is that there are two, tandemly repeated *SIR4* genes in *Ashbya gossypii* and *K. lactis* (8). In both cases, the duplicated genes are substantially diverged from one another. Studies aimed at investigating the functions of these duplicated genes will be interesting.

Sir3p and Sir1p are even more restricted in their distribution across species (24). Thus, the minimal silencing mechanism requires Sir2p and Sir4p, but not Sir3p or Sir1p. Sir3p is a paralog of the DNA replication protein Orc1p and arose in the whole-genome duplication. ScSir3p is required for silencing, yet a distinct Sir3 protein is clearly not needed for silencing in preduplication species. Sir1p arose after Sir3p, appearing right at the emergence of the *Saccharomyces* sensu stricto clade (24). ScSir1p is not essential for silencing, and therefore, its absence in some species poses less of a puzzle than the absence of Sir3p, at least regarding the mechanism of silencing. However, a mystery does surround the origin of the *SIR1* gene and its neighbors. In *Saccharomyces* species, *SIR1* is located near a telomere, and all of the genes from *SIR1* to the end of the chromosome are absent from the sequenced preduplication species.

Where Do Silencing Proteins Act?

All of the hemiascomycetes contain *MAT* alleles that determine mating type. However, only some possess additional, silenced copies of mating-type cassettes, which allow for switching of mating type (homothallism). *Yarrowia lipolytica* and *C. albicans* contain only a single mating-type locus. *D. hansenii* also has a single mating-type locus, although this locus contains both **a** genes and α genes. It is not known if these genes are alternately expressed, which would seem to be necessary for haploid identity and mating. *Kluyveromyces waltii* has only two identified mating-type loci on one contig, but this genome is not fully assembled and another locus may exist. *A. gossypii*, *K. lactis*, *C. glabrata*, and *S. cerevisiae* all have three mating-type loci. Interestingly, in all four of these species, two of the three mating-type loci are located near telomeres (24) and, in those cases in which the expression status is known, these are the two silenced loci. Thus, it is probable (and in some cases shown) that Sir proteins form silenced chromatin at the telomere-proximal mating-type loci in these species.

Clearly one function of the Sir proteins is to silence mating-type loci. However, in species with a single mating-type locus, Sir2p may participate in the silencing of other loci. In addition, all three mating-type loci of *A. gossypii* contain **a**-type cassettes, and hence, silencing of two of these cassettes does not seem likely to serve a biological function. One reasonable hypothesis is that Sir2p contributes to the silencing of other loci, and two candidate regions at which Sir2p might act are telomeres and centromeres, both of which are silenced by classical heterochromatin in *S. pombe*. In fact, a role at telomeres is highly likely since Sir proteins are already known to form a special structure in the subtelomeric regions of *S. cerevisiae* and *C. glabrata* (10, 30, 38). In addition, in *K. lactis*, a *URA3* reporter gene displays variegated expression when placed at the junction between the telomere repeat sequences and the subtelomeric region (32). The role of the *K. lactis* Sir proteins in telomeric silencing has not been examined, although the deletion of *KlSIR4* does result in longer telomeres (3), suggesting an involvement in chromatin structure at the telomere. In *C. albicans*, telomeric silencing has not been reported. However, the deletion of *CaSIR2*, in conjunction with other, unknown mutation(s), results in an increased frequency of phenotypic switching (reference 57; A. D. Johnson, personal communication). Perhaps this increased phenotypic switching is due to variable expression of subtelomeric genes, which affect the properties of the cell surface. It is also notable that the silent mating-type cassettes are located near telomeres, consistent with the chromosome ends being within the "sphere of influence" of the Sir proteins. These observations suggest a scenario in which silencing at the telomeres mediated by Sir2p and perhaps Sir4p preceded and enabled the development of telomere-proximal silent mating-type loci.

Another genomic feature often associated with heterochromatin is the centromere. Most eukaryotes, including *S. pombe*, have centromeres composed of repetitive sequences, and some of these repetitive sequences are incorporated into pericentromeric heterochromatin, which is required for proper centromere function. However, many of the hemiascomycetes have unusual "point" centromeres, in which a relatively short sequence (<0.5 kb) specifies the centromere (20). In *S. cerevisiae*, it is thought that one or at most two nucleosomes at the centromere contain a histone H3 variant, and no special pericentric chromatin has been observed to flank this variant nucleosome. Furthermore, the silencing proteins ScSir2p, ScSir3p, and ScSir4p are not associated with centromeres (66). (ScSir1p is associated with centromeres but is not thought to generate silenced chromatin at the centromeres [66].) One possibility is that the development of a centromeric structure that does not require heterochromatin for stability occurred in parallel with the loss of classical heterochromatin and RNAi.

Not all hemiascomycetes have point centromeres, and it is possible that Sir2-containing silenced chromatin contributes to the stability of these larger centromeres. For example, *C. albicans* centromeres do not contain conserved sequence elements (63) as are observed in the point centromeres of *S. cerevisiae*. However, *C. albicans* centromeres are also not composed of direct or complex repetitive elements, as found at the centromeres of other eukaryotes. (Two of the eight *C. albicans* centromeres are flanked by inverted repeats [63, 64].) What is the nature of the chromatin at these centromeres, and is pericentromeric heterochromatin required for proper function? If so, does CaSir2p participate in this heterochromatin? Interestingly, plasmids containing centromere sequences are not mitotically stable in *C. albicans* (63), perhaps suggesting that a special chromatin structure that does not form efficiently is required for proper segregation.

EXPERIMENTAL INVESTIGATIONS OF SILENCING

Candida glabrata

The only postduplication species other than *S. cerevisiae* in which silencing has been examined experimentally is *C. glabrata*. *C. glabrata* is the second most common cause of yeast infection in humans, after *C. albicans* (25, 58). Although both of these pathogenic species bear the name "*Candida,*" they are phylogenetically distant (Fig. 11.2). *C. albicans*, which clusters with most other *Candida* species, is a diploid, preduplication species, with a single mating-type locus. In contrast, *C. glabrata* is more closely related to *S. cerevisiae* and is a haploid, postduplication species with three mating-type loci (6, 70). The genome of *C. glabrata* has been completely sequenced (22).

Silencing in *C. glabrata* has been studied primarily at the telomeres. As in *S. cerevisiae*, CgSir2p, CgSir3p, CgSir4p, and CgRap1p are all required for telomeric silencing (16, 19, 38), and by analogy these proteins are likely to function at the mating-type loci. The *CgHMLα* locus is, indeed, silenced, whereas the *CgHMRa* locus is not effectively silenced (H. Muller and C. Fairhead, personal communication). Thus, *MAT*α strains express both types of mating-type genes, leading to a phenotypically diploid cell. An extremely interesting observation is that telomeric silencing is regulatable (19). In contrast, in *S. cerevisiae*, silencing at both the telomeres and mating-type loci occurs constitutively under examined conditions. The regulation of silencing in *C. glabrata* is made possible by an auxotrophy for nicotinic acid, the precursor of NAD$^+$. It is thought that in nicotinic acid-poor environments, the cellular level of NAD$^+$ drops, reducing the function of the NAD$^+$-dependent deacetylase CgSir2p and hence silencing (19). The ability to regulate silencing in this way may facilitate the infection of the urinary tract, which is poor in nicotinic acid. It is thought that subtelomeric silencing is reduced in the urinary tract, leading to the induction of several subtelomeric *EPA* genes, which encode adhesion proteins found in the cell wall. The expression of these genes enables the cells to adhere to epithelial cells and colonize the urinary tract (41).

C. glabrata reproduces mitotically and is not known to have a sexual stage in its life cycle. Nevertheless, examination of the sequenced genome has revealed that in addition to possessing three mating-type loci (70), *C. glabrata* also has genes involved in meiosis and producing and responding to mating pheromones (78). These authors have argued that it is highly unlikely that *C. glabrata* would have retained all these genes if they had no function. Furthermore, it is unlikely that so many genes, whose only known function in *S. cerevisiae* is in the sexual cycle, would have gained new functions in *C. glabrata*. Therefore, the authors conclude that *C. glabrata* is likely to have an undiscovered sexual cycle. Consistent with this conclusion, two clinical isolates of *C. glabrata* were found to arrest a *MAT*α strain of *S. cerevisiae* in G$_0$, suggesting that these isolates secrete a pheromone similar to a-factor (J. Piskur, personal communication). Furthermore, although *C. glabrata* has a predominantly clonal population structure, indicative of mitotic growth, evidence of recombination, consistent with a rare sexual cycle, was observed (18). Perhaps mating occurs only on rare occasions when the *CgHMRa* locus is actually silenced.

It is interesting to speculate that the adaptation to a predominantly asexual life cycle has enabled Sir-mediated silencing to become regulatable. If mating were an essential part of the life cycle, the silent mating loci would need to remain silenced at all times, and hence, derepressing silencing by perturbing the function of CgSir2p would not be feasible.

Kluyveromyces lactis

The only other hemiascomycete yeast in which silencing has been examined experimentally is the preduplication species *K. lactis*. *K. lactis* is a budding yeast that was originally isolated from milk-derived products, although it grows on a wide range of carbon sources. Interest in cultivating *K. lactis* for biotechnology led to the development of its genetics, and its phylogenetic position as a preduplication species makes *K. lactis* an important alternative model system to *S. cerevisiae*. The genome of *K. lactis* has been completely sequenced (22).

K. lactis strains are heterothallic, with a and α mating types. Mating does occur, and the resulting diploids sporulate spontaneously (36). Several of the elements of the *S. cerevisiae* pheromone response pathway have been identified in *K. lactis*, and in the latter organism their role in regulation of the mating response differs from that seen in *S. cerevisiae* (reviewed in reference 15). Silent mating-type alleles in *K. lactis* were first identified in 1966 by Herman and Roman (36), who showed that there are two independent, unlinked loci and that mating-type switching occurred at a very low frequency. The mechanism for switching is still unknown, as the endonuclease that triggers switching in *S. cerevisiae* has not been identified in *K. lactis* (7).

As outlined above, *K. lactis* contains homologs of *SIR2* and *SIR4* (3, 12), and deletion of either *KlSIR2* or *KlSIR4* leads to increased expression of α1 and α3 mRNA at *HML* (2, 3). In addition, both KlSir2p and KlSir4p can partially complement the mating defects of *sir2Δ* or *sir4Δ S. cerevisiae* strains (3, 12), illustrating their conserved function. Curiously, deletions of *KlSIR2* and *KlSIR4* have much less profound effects on mating efficiency than do *sir* deletions in *S. cerevisiae*, and mating ability varies substantially from strain to strain (3, 12). In the case of *sir4Δ*, which showed only a slight reduction in mating, a second tandemly duplicated copy of *SIR4* had not been identified when this study was conducted, and it is possible that deletion of both Sir4 alleles would result in a greater decrease in mating. In any event, mating efficiency is determined by many factors in addition to silencing of the mating-type loci, and mating may not occur for reasons other than incomplete silencing of the mating-type loci.

In contrast to Sir2p and Sir4p, no distinct homolog of Sir3p has been identified in *K. lactis*. As described above, in *S. cerevisiae* Sir3p and Orc1p are paralogs (42), and the nearest equivalent of *ScSIR3* in *K. lactis* is *KlORC1*. The potential role of KlOrc1p in silencing the *HM* loci has not been investigated to date but presents two hypotheses regarding Sir silencing: either KlOrc1p acts in place of Sir3p or KlSir2p and KlSir4p can achieve silencing without Sir3p. Sir1p is also missing from *K. lactis*.

Examination of the structure of the *HM* loci in *K. lactis* has revealed several intriguing differences compared to *S. cerevisiae*. First, *HMLα* in *K. lactis* contains three genes, α1, α2, and α3, whereas there are only two α genes in *S. cerevisiae* (2). The α3 protein does not have obvious homology to other proteins, and its function is not known. Another interesting observation is that the silencers and silencer-binding proteins differ significantly between these two species (67). Three DNA elements, named A-, B-, and C-boxes, have been identified as important for silencer function at *HMLα* in *K. lactis*. *HMRa* also contains A- and B-boxes. None of these DNA elements correspond to ORC or Rap1 binding sites, which are required for silencer function in *S. cerevisiae*. The B-box binds the essential protein KlReb1p, which is required for silencing and mating

(67). KlReb1 contains two Myb domains, which are related to a helix-loop-helix motif. Other Myb-containing proteins from *S. cerevisiae* (ScRap1p) and *S. pombe* (SpTaz1p) have been shown to be important in telomeric silencing (33, 40, 45). The A-box contains an exact consensus binding site for ScUme6p, but binding of KlUme6p to the silencer has not yet been tested. The C-box contains a sequence with weak similarity to the ScAbf1 binding site, but rigorous testing of this claim or its implications has not been conducted. Thus, the silencers in *K. lactis* and *S. cerevisiae* are remarkably divergent. Future experiments aimed at identifying silencers in other species will reveal whether silencers have evolved rapidly and, if so, whether this rapid evolution has implications in the development of species identity.

The divergence of the *K. lactis* and *S. cerevisiae* silencers is also consistent with the rapid evolution of Sir4p, which apparently interacts with different silencer binding proteins in these two species. However, KlSir4p may also possess the ability to interact with KlRap1p, as silencing of a telomeric reporter gene in *K. lactis* is reduced by the mutation of KlRap1p (32). Another possibility is that one of the two KlSir4 proteins acts at silencers through KlReb1p, and the other acts at telomeres through KlRap1p.

IMPACT OF WHOLE-GENOME DUPLICATION ON SILENCING PROTEINS

Two of the four Sir proteins, Sir2p and Sir3p, have paralogs that arose in the whole-genome duplication. Thus, it is likely that the genome duplication enabled these proteins to become more specialized for silencing.

Silencing in Preduplication Species—Why Is Sir3p Not Needed?

Sir3p, which is found only in postduplication hemiascomycete species, appears to be derived from its paralog, Orc1p, which is highly conserved and widely distributed throughout the eukaryotes. Orc1p is the largest subunit of the origin recognition complex, which binds to origins of DNA replication. Comparison of Orc1p and Sir3p sequences from multiple species reveals that the N termini are highly similar. In contrast, the C terminus of Sir3p has experienced accelerated evolution. It is therefore likely that the N termini of Orc1p and Sir3p have a conserved function, whereas the C terminus of Sir3p has evolved a new function. One question is whether the preduplication Orc1p had a silencing function, perhaps mediated by the N terminus, or whether the silencing function arose after the duplication event. A related question is how Sir-mediated silencing occurs in preduplication species, such as *K. lactis*, that lack a distinct Sir3p. Does KlOrc1p fill the role of Sir3p, or do KlSir2p and KlSir4p achieve silencing without the function provided by Sir3p in postduplication species? Consistent with the former hypothesis, the single *ORC1/SIR3* gene from another preduplication species, *S. kluyveri*, can partially complement both *orc1Δ* and *sir3Δ* mutations in *S. cerevisiae* (74). In addition, ScOrc1p itself functions in silencing in *S. cerevisiae*. The ScORC complex binds to all four silencers, and its only known role is to recruit ScSir1p (73), which in turn recruits the other Sir proteins. However, if Orc1p participates in silencing in preduplication species, its role must be different there from what it is in *S. cerevisiae* for two reasons. First, ScSir1p, the binding partner of ScOrc1p, does not exist in examined preduplication species (24). Second, at least in *K. lactis*, ORC is not one of the silencer binding proteins (67).

The other possibility, that in some species silencing could occur independently of Sir3p, is consistent with observations in *S. cerevisiae* that suggest a distinct role for ScSir3p compared with ScSir2p and ScSir4p. For example, in coimmunoprecipitation studies, ScSir2p and ScSir4p are closely associated, whereas ScSir3p is unassociated under many conditions (28, 51). Additionally, overexpression of ScSir3p, but not ScSir2p or ScSir4p, increases the extent of telomeric silenced chromatin, and under these conditions ScSir3p appears to spread farther along the chromosome than ScSir2p and ScSir4p do (71). Finally, the deletion of the deacetylase ScRpd3p partially suppresses the mating defect of *sir3Δ* but not *sir2Δ* or *sir4Δ* strains (69). The mechanism by which this suppression occurs is unknown, but its occurrence does suggest that under some circumstances ScSir2p and ScSir4p can achieve silencing without ScSir3p.

Additional Functions of Sir2p?

An interesting feature of the deacetylase Sir2p is that it appears to have much broader effects on chromatin structure than would be expected if it simply silenced a few genes in the mating-type loci and near the telomeres. In *S. cerevisiae*, ScSir2p modulates the structure of chromatin in the ribosomal DNA repeat to reduce recombination (29, 68), and it has been implicated in other functions such as modulation of DNA replication (56). In other species, Sir2p has similarly broad effects on DNA metabolism. For example, the deletion of *KlSIR2* leads to sensitivity to ethidium bromide and other DNA-damaging agents (12). One likely mechanism by which Sir2p has such broad effects on chromatin

is its recruitment into multiple complexes that target different substrates. Thus, the evolution of different functions for Sir2p in various hemiascomycete species could result from other chromatin-associated proteins gaining or losing the ability to interact with Sir2p.

An interesting example of the evolution of Sir2p is the emergence of a new paralog, Hst1p (named for Homolog of Sir Two), which arose in the whole-genome duplication. ScHst1p is found in a complex with the DNA-binding repressor protein ScSum1p, and this complex contributes to repression of midsporulation genes (49, 62, 79). In contrast to the Sir complex, the Sum1 complex acts locally at promoters and does not generate an extended repressive chromatin structure (48). Thus, in *S. cerevisiae*, these two paralogous deacetylases participate in repressive complexes with different properties. This raises the question of the ancestral function of Sir2p prior to the whole-genome duplication. It is possible that the ancestral protein possessed both Sir2-like and Hst1-like functions. Alternatively, it may have possessed the properties of only one of these two deacetylases and evolved a new function after the duplication. Clearly, KlSir2p has properties similar to ScSir2p in that it represses mating-type loci (2). Whether it also acts with KlSum1p has not been investigated. Also unknown is whether the orthologs of ScHst1p in other postduplication species perform similar functions.

EVOLUTION OF A NEW TYPE OF SILENCED CHROMATIN

Clearly, the Sir proteins have been evolving for millions of years and now constitute a complex, finely tuned silencing machinery. But how does a new type of silenced chromatin first arise? An interesting case study suggests one possible mechanism. In *S. cerevisiae*, the ScSum1 repressor protein, which interacts with the deacetylase ScHst1p, as described above, can be converted from a nonspreading repressor protein into a spreading silencing protein by a single amino acid change. This neomorphic mutation, known as *ScSUM1-1*, was originally identified as a suppressor of a *sir2* mutation (44). The mutation results in the substitution of isoleucine for threonine 988 (13) and causes ScSum1p to relocalize from the genes that it normally represses to the silencers at the mating-type loci (62, 72). Once at the silencers, mutant ScSum1-1p spreads along the chromosome, and this spreading requires the deacetylase activity of ScHst1p (48). Thus, an alternate type of silenced chromatin, composed of ScSum1p and its partners, forms in the place of the absent Sir proteins. Clearly, the wild-type ScSum1p must possess almost all the properties necessary for spreading and simply needs a small adjustment—the single amino acid mutation—to spread. This example highlights the mechanistic similarities between repressive complexes that do and do not spread, and it is easy to imagine that similar types of mutations have arisen during the course of evolution, giving rise to novel silencing proteins. Might an ancestor of Sir4p have been a promoter-specific repressor as well?

CONCLUSION

An examination of the distribution of silencing proteins in the hemiascomycete class of fungi, coupled with experimental studies of silencing in a few species, suggests the following speculative model for the evolution of Sir-mediated silencing. First, the components of classical heterochromatin and RNAi were lost early in the hemiascomycete lineage. This loss necessitated adaptations, such as the development of alternative chromatin structures at centromeres to maintain stable chromosome segregation. The absence of classical heterochromatin also provided an opportunity for the evolution of an alternative silencing mechanism. At the core of this alternative silencing mechanism was the conserved histone deacetylase Sir2p, which participates in the formation of silenced chromatin in a variety of fungal and nonfungal species. At first, Sir2p and its adaptor Sir4p may have contributed to the formation of subtelomeric chromatin and perhaps pericentromeric heterochromatin as well. Later, silent mating-type cassettes developed at telomere-proximal locations, benefiting from the preexisting silenced domain at the ends of chromosomes. More recently Sir3p and finally Sir1p arose, enhancing the silencing mechanism. This vision is clearly *Saccharomyces*-centric, and no doubt in unexamined hemiascomycete species other proteins unrelated to Sir3p or Sir1p participate with Sir2p and Sir4p in silencing. We look forward to their discovery.

We thank Greg Crawford, Patrick Lynch, and Kristin Scott for comments on the manuscript and members of the Rusche lab for helpful discussions. Figure 1, which is a modified version of a published figure (61) created by Ann Kirchmaier, is used with the permission of Ann Kirchmaier and the American Society for Cell Biology. Research in the Rusche lab is supported by a grant from the National Institutes of Health (GM073991).

References

1. **Astrom, S. U., T. W. Cline, and J. Rine.** 2003. The *Drosophila melanogaster* sir2+ gene is nonessential and has only minor effects on position-effect variation. *Genetics* **163:**931–937.

2. **Astrom, S. U., A. Kegel, J. O. Sjostrand, and J. Rine.** 2000. *Kluyveromyces lactis* Sir2p regulates cation sensitivity and maintains a specialized chromatin structure at the cryptic alpha-locus. *Genetics* **156:**81–91.

3. **Astrom, S. U., and J. Rine.** 1998. Theme and variation among silencing proteins in *Saccharomyces cerevisiae* and *Kluyveromyces lactis*. *Genetics* **148:**1021–1029.

4. **Blander, G., and L. Guarente.** 2004. The Sir2 family of protein deacetylases. *Annu. Rev. Biochem.* **73:**417–435.

5. **Brachmann, C. B., J. M. Sherman, S. E. Devine, E. E. Cameron, L. Pillus, and J. D. Boeke.** 1995. The *SIR2* gene family, conserved from bacteria to humans, functions in silencing, cell cycle progression, and chromosome stability. *Genes Dev.* **9:**2888–2902.

6. **Brockert, P. J., S. A. Lachke, T. Srikantha, C. Pujol, R. Galask, and D. R. Soll.** 2003. Phenotypic switching and mating type switching of *Candida glabrata* at sites of colonization. *Infect. Immun.* **71:**7109–7118.

7. **Butler, G., C. Kenny, A. Fagan, C. Kurischko, C. Gaillardin, and K. H. Wolfe.** 2004. Evolution of the *MAT* locus and its Ho endonuclease in yeast species. *Proc. Natl. Acad. Sci. USA* **101:**1632–1637.

8. **Byrne, K. P., and K. H. Wolfe.** 2005. The Yeast Gene Order Browser: combining curated homology and syntenic context reveals gene fate in polyploid species. *Genome Res.* **15:**1456–1461.

9. **Carmen, A. A., L. Milne, and M. Grunstein.** 2002. Acetylation of the yeast histone H4 N terminus regulates its binding to heterochromatin protein SIR3. *J. Biol. Chem.* **277:**4778–4781.

10. **Castano, I., S. J. Pan, M. Zupancic, C. Hennequin, B. Dujon, and B. P. Cormack.** 2005. Telomere length control and transcriptional regulation of subtelomeric adhesins in *Candida glabrata*. *Mol. Microbiol.* **55:**1246–1258.

11. **Chang, J. F., B. E. Hall, J. C. Tanny, D. Moazed, D. Filman, and T. Ellenberger.** 2003. Structure of the coiled-coil dimerization motif of Sir4 and its interaction with Sir3. *Structure* **11:**637–649.

12. **Chen, X. J., and G. D. Clark-Walker.** 1994. sir2 mutants of *Kluyveromyces lactis* are hypersensitive to DNA-targeting drugs. *Mol. Cell. Biol.* **14:**4501–4508.

13. **Chi, M. H., and D. Shore.** 1996. *SUM1-1*, a dominant suppressor of *SIR* mutations in *Saccharomyces cerevisiae*, increases transcriptional silencing at telomeres and *HM* mating-type loci and decreases chromosome stability. *Mol. Cell. Biol.* **16:**4281–4294.

14. **Choo, K. H.** 2001. Domain organization at the centromere and neocentromere. *Dev. Cell* **1:**165–177.

15. **Coria, R., L. Kawasaki, F. Torres-Quiroz, L. Ongay-Larios, E. Sanchez-Paredes, N. Velazquez-Zavala, R. Navarro-Olmos, M. Rodriguez-Gonzalez, R. Aguilar-Corachan, and G. Coello.** 2006. The pheromone response pathway of *Kluyveromyces lactis*. *FEMS Yeast Res.* **6:**336–344.

16. **De Las Penas, A., S. J. Pan, I. Castano, J. Alder, R. Cregg, and B. P. Cormack.** 2003. Virulence-related surface glycoproteins in the yeast pathogen *Candida glabrata* are encoded in subtelomeric clusters and subject to RAP1- and SIR-dependent transcriptional silencing. *Genes Dev.* **17:**2245–2258.

17. Dietrich, F. S., S. Voegeli, S. Brachat, A. Lerch, K. Gates, S. Steiner, C. Mohr, R. Pohlmann, P. Luedi, S. Choi, R. A. Wing, A. Flavier, T. D. Gaffney, and P. Philippsen. 2004. The *Ashbya gossypii* genome as a tool for mapping the ancient *Saccharomyces cerevisiae* genome. *Science* 304: 304–307.

18. Dodgson, A. R., C. Pujol, M. A. Pfaller, D. W. Denning, and D. R. Soll. 2005. Evidence for recombination in *Candida glabrata. Fungal Genet. Biol.* 42:233–243.

19. Domergue, R., I. Castano, A. De Las Penas, M. Zupancic, V. Lockatell, J. R. Hebel, D. Johnson, and B. P. Cormack. 2005. Nicotinic acid limitation regulates silencing of *Candida* adhesins during UTI. *Science* 308:866–870.

20. Dujon, B. 2005. Hemiascomycetous yeasts at the forefront of comparative genomics. *Curr. Opin. Genet. Dev.* 15:614-620.

21. Dujon, B. 2006. Yeasts illustrate the molecular mechanisms of eukaryotic genome evolution. *Trends Genet.* 22:375–387.

22. Dujon, B., D. Sherman, G. Fischer, P. Durrens, S. Casaregola, I. Lafontaine, J. De Montigny, C. Marck, C. Neuveglise, E. Talla, N. Goffard, L. Frangeul, M. Aigle, V. Anthouard, A. Babour, V. Barbe, S. Barnay, S. Blanchin, J. M. Beckerich, E. Beyne, C. Bleykasten, A. Boisrame, J. Boyer, L. Cattolico, F. Confanioleri, A. De Daruvar, L. Despons, E. Fabre, C. Fairhead, H. Ferry-Dumazet, A. Groppi, F. Hantraye, C. Hennequin, N. Jauniaux, P. Joyet, R. Kachouri, A. Kerrest, R. Koszul, M. Lemaire, I. Lesur, L. Ma, H. Muller, J. M. Nicaud, M. Nikolski, S. Oztas, O. Ozier-Kalogeropoulos, S. Pellenz, S. Potier, G. F. Richard, M. L. Straub, A. Suleau, D. Swennen, F. Tekaia, M. Wesolowski-Louvel, E. Westhof, B. Wirth, M. Zeniou-Meyer, I. Zivanovic, M. Bolotin-Fukuhara, A. Thierry, C. Bouchier, B. Caudron, C. Scarpelli, C. Gaillardin, J. Weissenbach, P. Wincker, and J. L. Souciet. 2004. Genome evolution in yeasts. *Nature* 430:35–44.

23. Ekwall, K., J. P. Javerzat, A. Lorentz, H. Schmidt, G. Cranston, and R. Allshire. 1995. The chromodomain protein Swi6: a key component at fission yeast centromeres. *Science* 269:1429–1431.

24. Fabre, E., H. Muller, P. Therizols, I. Lafontaine, B. Dujon, and C. Fairhead. 2005. Comparative genomics in hemiascomycete yeasts: evolution of sex, silencing, and subtelomeres. *Mol. Biol. Evol.* 22:856–873.

25. Fidel, P. L., Jr., J. A. Vazquez, and J. D. Sobel. 1999. *Candida glabrata*: review of epidemiology, pathogenesis, and clinical disease with comparison to *C. albicans. Clin. Microbiol. Rev.* 12:80–96.

26. Freeman-Cook, L. L., E. B. Gomez, E. J. Spedale, J. Marlett, S. L. Forsburg, L. Pillus, and P. Laurenson. 2005. Conserved locus-specific silencing functions of *Schizosaccharomyces pombe sir2+. Genetics* 169:1243–1260.

27. Furuyama, T., R. Banerjee, T. R. Breen, and P. J. Harte. 2004. SIR2 is required for polycomb silencing and is associated with an E(Z) histone methyltransferase complex. *Curr. Biol.* 14:1812–1821.

28. Ghidelli, S., D. Donze, N. Dhillon, and R. T. Kamakaka. 2001. Sir2p exists in two nucleosome-binding complexes with distinct deacetylase activities. *EMBO J.* 20:4522–4535.

29. Gottlieb, S., and R. E. Esposito. 1989. A new role for a yeast transcriptional silencer gene, *SIR2*, in regulation of recombination in ribosomal DNA. *Cell* 56:771–776.

30. Gottschling, D. E., O. M. Aparicio, B. L. Billington, and V. A. Zakian. 1990. Position effect at *S. cerevisiae* telomeres: reversible repression of Pol II transcription. *Cell* 63:751–762.

31. Grewal, S. I., and J. C. Rice. 2004. Regulation of heterochromatin by histone methylation and small RNAs. *Curr. Opin. Cell Biol.* 16:230–238.

32. Gurevich, R., S. Smolikov, H. Maddar, and A. Krauskopf. 2003. Mutant telomeres inhibit transcriptional silencing at native telomeres of the yeast *Kluyveromyces lactis. Mol. Genet. Genomics* 268:729–738.

33. Hansen, K. R., P. T. Ibarra, and G. Thon. 2006. Evolutionary-conserved telomere-linked helicase genes of fission yeast are repressed by silencing factors, RNAi components and the telomere-binding protein Taz1. *Nucleic Acids Res.* 34:78–88.

34. Hecht, A., T. Laroche, S. Strahl-Bolsinger, S. M. Gasser, and M. Grunstein. 1995. Histone H3 and H4 N-termini interact with SIR3 and SIR4 proteins: a molecular model for the formation of heterochromatin in yeast. *Cell* 80: 583–592.

35. Hedges, S. B. 2002. The origin and evolution of model organisms. *Nat. Rev. Genet.* 3:838–849.

36. Herman, A., and H. Roman. 1966. Allele specific determinants of homothallism in *Saccharomyces lactis. Genetics* 53:727–740.

37. Hoppe, G. J., J. C. Tanny, A. D. Rudner, S. A. Gerber, S. Danaie, S. P. Gygi, and D. Moazed. 2002. Steps in assembly of silent chromatin in yeast: Sir3-independent binding of a Sir2/Sir4 complex to silencers and role for Sir2-dependent deacetylation. *Mol. Cell. Biol.* 22:4167–4180.

38. Iraqui, I., S. Garcia-Sanchez, S. Aubert, F. Dromer, J. M. Ghigo, C. d'Enfert, and G. Janbon. 2005. The Yak1p kinase controls expression of adhesins and biofilm formation in *Candida glabrata* in a Sir4p-dependent pathway. *Mol. Microbiol.* 55:1259–1271.

39. Jia, S., K. Noma, and S. I. Grewal. 2004. RNAi-independent heterochromatin nucleation by the stress-activated ATF/CREB family proteins. *Science* 304:1971–1976.

40. Kanoh, J., M. Sadaie, T. Urano, and F. Ishikawa. 2005. Telomere binding protein Taz1 establishes Swi6 heterochromatin independently of RNAi at telomeres. *Curr. Biol.* 15:1808–1819.

41. Kaur, R., R. Domergue, M. L. Zupancic, and B. P. Cormack. 2005. A yeast by any other name: *Candida glabrata* and its interaction with the host. *Curr. Opin. Microbiol.* 8:378–384.

42. Kellis, M., B. W. Birren, and E. S. Lander. 2004. Proof and evolutionary analysis of ancient genome duplication in the yeast *Saccharomyces cerevisiae. Nature* 428:617–624.

43. Kim, H. S., E. S. Choi, J. A. Shin, Y. K. Jang, and S. D. Park. 2004. Regulation of Swi6/HP1-dependent heterochromatin assembly by cooperation of components of the mitogen-activated protein kinase pathway and a histone deacetylase clr6. *J. Biol. Chem.* 279:42850–42859.

44. Klar, A. J., S. N. Kakar, J. M. Ivy, J. B. Hicks, G. P. Livi, and L. M. Miglio. 1985. *SUM1*, an apparent positive regulator of the cryptic mating-type loci in *Saccharomyces cerevisiae*. *Genetics* **111:**745–758.

45. Kyrion, G., K. Liu, C. Liu, and A. J. Lustig. 1993. RAP1 and telomere structure regulate telomere position effects in *Saccharomyces cerevisiae*. *Genes Dev.* **7:**1146–1159.

46. Longtine, M. S., N. M. Wilson, M. E. Petracek, and J. Berman. 1989. A yeast telomere binding activity binds to two related telomere sequence motifs and is indistinguishable from RAP1. *Curr. Genet.* **16:**225–239.

47. Luo, K., M. A. Vega-Palas, and M. Grunstein. 2002. Rap1-Sir4 binding independent of other Sir, yKu, or histone interactions initiates the assembly of telomeric heterochromatin in yeast. *Genes Dev.* **16:**1528–1539.

48. Lynch, P. J., H. B. Fraser, E. Sevastopoulos, J. Rine, and L. N. Rusche. 2005. Sum1p, the origin recognition complex, and the spreading of a promoter-specific repressor in *Saccharomyces cerevisiae*. *Mol. Cell. Biol.* **25:**5920–5932.

49. McCord, R., M. Pierce, J. Xie, S. Wonkatal, C. Mickel, and A. K. Vershon. 2003. Rfm1, a novel tethering factor required to recruit the Hst1 histone deacetylase for repression of middle sporulation genes. *Mol. Cell. Biol.* **23:**2009–2016.

50. Mishra, K., and D. Shore. 1999. Yeast Ku protein plays a direct role in telomeric silencing and counteracts inhibition by rif proteins. *Curr. Biol.* **9:**1123–1126.

51. Moazed, D., A. Kistler, A. Axelrod, J. Rine, and A. D. Johnson. 1997. Silent information regulator protein complexes in *Saccharomyces cerevisiae*: a SIR2/SIR4 complex and evidence for a regulatory domain in SIR4 that inhibits its interaction with SIR3. *Proc. Natl. Acad. Sci. USA* **94:**2186–2191.

52. Murphy, G. A., E. J. Spedale, S. T. Powell, L. Pillus, S. C. Schultz, and L. Chen. 2003. The Sir4 C-terminal coiled coil is required for telomeric and mating type silencing in *Saccharomyces cerevisiae*. *J. Mol. Biol.* **334:**769–780.

53. Nakayashiki, H. 2005. RNA silencing in fungi: mechanisms and applications. *FEBS Lett.* **579:**5950–5957.

54. Newman, B. L., J. R. Lundblad, Y. Chen, and S. M. Smolik. 2002. A *Drosophila* homologue of Sir2 modifies position-effect variegation but does not affect life span. *Genetics* **162:**1675–1685.

55. Noma, K., T. Sugiyama, H. Cam, A. Verdel, M. Zofall, S. Jia, D. Moazed, and S. I. Grewal. 2004. RITS acts in cis to promote RNA interference-mediated transcriptional and post-transcriptional silencing. *Nat. Genet.* **36:**1174–1180.

56. Pappas, D. L., Jr., R. Frisch, and M. Weinreich. 2004. The NAD(+)-dependent Sir2p histone deacetylase is a negative regulator of chromosomal DNA replication. *Genes Dev.* **18:**769–781.

57. Perez-Martin, J., J. A. Uria, and A. D. Johnson. 1999. Phenotypic switching in *Candida albicans* is controlled by a *SIR2* gene. *EMBO J.* **18:**2580–2592.

58. Pfaller, M. A., R. N. Jones, S. A. Messer, M. B. Edmond, and R. P. Wenzel. 1998. National surveillance of nosocomial blood stream infection due to species of *Candida* other than *Candida albicans*: frequency of occurrence and antifungal susceptibility in the SCOPE Program. *Diagn. Microbiol. Infect. Dis.* **30:**121–129.

59. Roy, R., B. Meier, A. D. McAinsh, H. M. Feldmann, and S. P. Jackson. 2004. Separation-of-function mutants of yeast Ku80 reveal a Yku80p-Sir4p interaction involved in telomeric silencing. *J. Biol. Chem.* **279:**86–94.

60. Rusche, L. N., A. L. Kirchmaier, and J. Rine. 2003. The establishment, inheritance, and function of silenced chromatin in *Saccharomyces cerevisiae*. *Annu. Rev. Biochem.* **72:**481–516.

61. Rusche, L. N., A. L. Kirchmaier, and J. Rine. 2002. Ordered nucleation and spreading of silenced chromatin in *Saccharomyces cerevisiae*. *Mol. Biol. Cell* **13:**2207–2222.

62. Rusche, L. N., and J. Rine. 2001. Conversion of a gene-specific repressor to a regional silencer. *Genes Dev.* **15:**955–967.

63. Sanyal, K., M. Baum, and J. Carbon. 2004. Centromeric DNA sequences in the pathogenic yeast *Candida albicans* are all different and unique. *Proc. Natl. Acad. Sci. USA* **101:**11374–11379.

64. Selmecki, A., A. Forche, and J. Berman. 2006. Aneuploidy and isochromosome formation in drug-resistant *Candida albicans*. *Science* **313:**367–370.

65. Shankaranarayana, G. D., M. R. Motamedi, D. Moazed, and S. I. Grewal. 2003. Sir2 regulates histone H3 lysine 9 methylation and heterochromatin assembly in fission yeast. *Curr. Biol.* **13:**1240–1246.

66. Sharp, J. A., D. C. Krawitz, K. A. Gardner, C. A. Fox, and P. D. Kaufman. 2003. The budding yeast silencing protein Sir1 is a functional component of centromeric chromatin. *Genes Dev.* **17:**2356–2361.

67. Sjostrand, J. O., A. Kegel, and S. U. Astrom. 2002. Functional diversity of silencers in budding yeasts. *Eukaryot. Cell* **1:**548–557.

68. Smith, J. S., and J. D. Boeke. 1997. An unusual form of transcriptional silencing in yeast ribosomal DNA. *Genes Dev.* **11:**241–254.

69. Smith, J. S., E. Caputo, and J. D. Boeke. 1999. A genetic screen for ribosomal DNA silencing defects identifies multiple DNA replication and chromatin-modulating factors. *Mol. Cell. Biol.* **19:**3184–3197.

70. Srikantha, T., S. A. Lachke, and D. R. Soll. 2003. Three mating type-like loci in *Candida glabrata*. *Eukaryot. Cell* **2:**328–340.

71. Strahl-Bolsinger, S., A. Hecht, K. Luo, and M. Grunstein. 1997. SIR2 and SIR4 interactions differ in core and extended telomeric heterochromatin in yeast. *Genes Dev.* **11:**83–93.

72. Sutton, A., R. C. Heller, J. Landry, J. S. Choy, A. Sirko, and R. Sternglanz. 2001. A novel form of transcriptional silencing by Sum1-1 requires Hst1 and the origin recognition complex. *Mol. Cell. Biol.* **21:**3514–3522.

73. Triolo, T., and R. Sternglanz. 1996. Role of interactions between the origin recognition complex and SIR1 in transcriptional silencing. *Nature* **381:**251–253.

74. van Hoof, A. 2005. Conserved functions of yeast genes support the duplication, degeneration and complementation model for gene duplication. *Genetics* **171:**1455–1461.

75. Verdel, A., S. Jia, S. Gerber, T. Sugiyama, S. Gygi, S. I. Grewal, and D. Moazed. 2004. RNAi-mediated targeting of heterochromatin by the RITS complex. *Science* **303**: 672–676.

76. Wolfe, K. H., and D. C. Shields. 1997. Molecular evidence for an ancient duplication of the entire yeast genome. *Nature* **387**:708–713.

77. Wong, S., G. Butler, and K. H. Wolfe. 2002. Gene order evolution and paleopolyploidy in hemiascomycete yeasts. *Proc. Natl. Acad. Sci. USA* **99**:9272–9277.

78. Wong, S., M. A. Fares, W. Zimmermann, G. Butler, and K. H. Wolfe. 2003. Evidence from comparative genomics for a complete sexual cycle in the 'asexual' pathogenic yeast *Candida glabrata*. *Genome Biol.* **4**:R10.

79. Xie, J., M. Pierce, V. Gailus-Durner, M. Wagner, E. Winter, and A. K. Vershon. 1999. Sum1 and Hst1 repress middle sporulation-specific gene expression during mitosis in *Saccharomyces cerevisiae*. *EMBO J.* **18**:6448–6454.

*Sex in Fungi: Molecular Determination
and Evolutionary Implications*
Edited by Joseph Heitman et al.
© 2007 ASM Press, Washington, D.C.

Matthew C. Fisher

12

The Evolutionary Implications of an Asexual Lifestyle Manifested by *Penicillium marneffei*

Throughout the eukaryotes, sexual reproduction is an almost universal phenomenon. However, within the Kingdom Fungi this relationship is not so clear-cut. Fungi exhibit a spectrum of reproductive modes and life cycles; reviews of the better-known species have shown that sexual reproduction is often facultative, can be rare, and in over one-half of the known Ascomycota is unknown (53). Over the last decade, research has shown that many of these asexual mitosporic taxa undergo cryptic recombination via unknown, or unobserved, mechanisms and this work has shown that wholly asexual fungi are a rarity (5, 53, 55). On the other hand, there is also convincing evidence that asexual species derive from sexual ancestors and that species occur that exhibit overwhelmingly clonal populations, as well as relatively long-lived clonal lineages (1, 7, 21). Here, we argue that a truly comprehensive understanding of the evolution of sexuality in fungi needs to also account for the occurrence of their antithesis, the mitosporic asexual species. There is a need to understand, and model, the evolutionary forces that create and maintain asexuality, and to determine the eventual fate of these taxa relative to their sexual conspecifics. Recent work on the spatial population genetics of one such mitosporic species, the biverticilliate mycosis agent *Penicillium marneffei*, is shedding much-needed light on the processes and consequences of asexuality in fungi.

RELATIVE COSTS OF RECOMBINING AND ASEXUAL LIFE CYCLES IN FUNGI

The accumulation of annotated fungal genome projects has shown that almost all species so far examined contain the genes that are necessary for mating processes, pathway signaling, and meiosis. This work has proved that mitosporic taxa are a derived state, a conclusion that had previously been reached by examining the phylogenetic relationships between mitosporic and meiosporic species (53). However, the evolutionary forces governing this relationship are far less clearly understood. The paradox of asexuality holds that, while sexual species are in the majority, this occurs in the face of weighty costs. Sex is energetically expensive due to the need for specialized reproductive structures and the production of males, known as the "twofold cost to sex" (46). Further, sex disrupts favorable gene combinations, leading to recombinational

Matthew C. Fisher, Imperial College Faculty of Medicine, Department of Infectious Disease Epidemiology, St Mary's Campus, Norfolk Place, London, W2 1PG, United Kingdom.

load (4). These costs to sex allow a trait with a short-term fitness advantage to be favored by natural selection, and in direct competition, the sexual "tortoise" tends to lose to the asexual "hare."

Arguments accounting for the comparative rarity in nature of truly asexual taxa focus on the negative consequences of asexuality. It is clear that meiotic recombination is necessary to preserve the long-term physical integrity of a genome; witness the evolutionary degradation of the Y chromosome as a consequence of being sheltered from recombination (10). Further, convincing theories have been developed showing that the mean fitness of asexual populations will decrease due to increases in genetic load as a consequence of the accumulation of mutations with negative fitness correlations. Broadly speaking, these are long-term costs that occur in asexuals due to a reduction in the efficiency of natural selection, and there is now increasing experimental evidence backing this concept (26). While it is unclear as to the rates at which such long-term costs accrue within natural asexual populations, and therefore how sexual taxa avoid being overwhelmed by the asexual "hares," it is now clear that an understanding of the process of natural selection is going to provide the key to understanding the relative rarity of wholly asexual life cycles in fungi.

THE ROLE OF CHANCE AND NATURAL SELECTION IN THE EVOLUTION OF ASEXUALITY

Evolution is a stochastic process and is most often conceptualized by the classic Wright-Fisher model (13). This model idealizes species as a group of individuals, of population size N, that draw equally, and synchronously, from an infinite pool of gametes, in this manner creating the next generation, $N + 1$. No species fits this idealized world, and separate sexes, overlapping generations and spatial structure, violate the implicit assumptions of the model. The sum of deviations from the Wright-Fisher model is determined by the genetic "effective" population size (N_e). N_e is a very important population genetic parameter, as it quantifies the probability, and time to fixation, of genetic mutations within nonideal natural populations such that the probability of fixation of a mutation in a haploid species is equal to its initial frequency, $1/N_e$. Where s is the selective coefficient associated with a specific mutation, then the theoretical limit at which genetic drift balances natural selection is estimated as $N_e s < 1$ (44). A consequence of this relationship is that within populations with large N_e, natural selection is at its most efficient and deleterious mutations tend to be purged while beneficial mutations

are incorporated into genomes. The converse is true, however, for populations with small N_e. Here, genetic drift has the powerful tendency to greatly reduce the efficiency of natural selection, allowing the fixation of mutations with negative fitness effects.

Consider now the mutation pressures to which the loci which control mating are exposed. Successful fungal mating is controlled by large, complex loci of up to 500 kb (37) and comprising two or more idiomorphs, each containing many open reading frames (35). A consequence of the large size of the sequence space that codes for mating is that this locus presents a large mutational target. Therefore, these mating-associated loci experience a greater rate of conversion to nonfunctional copies than do smaller housekeeping genes. In fungi where sexual reproduction is facultative, loss-of-function mutations in mating-type loci (also termed "Female-sterile strains" by Leslie and Klein [38]) are, generally speaking, selectively neutral. As a consequence, in populations of large N_e, fixation of such loss-of-function mutations will be rare. However, in populations with a small N_e there is a greatly increased probability that fixation of nonfunctional mating-type loci will occur, and the sexual viability of the population will therefore be lost through the process of genetic drift. In heterothallic species, such drift-type processes will also operate on the frequencies of the mating types themselves. This is because bouts of sexual reproduction tend to restore balanced mating-type frequencies. However, if sexual reproduction is facultative and rare, and N_e is small, then there is a probability that disappearance of the mating types themselves will occur by simple drift alone, rendering a population wholly asexual (38). These arguments show that, in fungi, the evolution of asexuality is theoretically possible under neutral scenarios and by chance alone and may occur even when the mating-type loci under consideration are under positive or balancing selection. In this particular case, the fixation of a mating-type locus under positive selection will occur when genetic drift overwhelms natural selection. Therefore, this evolutionary model will be particularly relevant in fungi for which N_e tends to be low, and we argue here that an in-depth description of this parameter is essential to an understanding of how variation in rates of recombination leads to the evolution of asexual fungal species.

ESTIMATING THE EFFECTIVE POPULATION SIZES OF FUNGI

Until now, it has been difficult to determine the effective population sizes of fungi in order to ascertain the potential strength of natural selection. This is because a description of the quantity N_e relies on describing the amounts, and partitioning, of nucleotide sequence variation within natural populations. However, the molecular revolution has allowed the development of rapid and relatively inexpensive techniques for surveying genomes for molecular diversity in order to ascertain the degree to which populations approximate an idealized panmixia. At mutation-drift equilibrium, the magnitude of silent nucleotide variation within a haploid population is equivalent to $2N_eu = \pi$, where u is the per-nucleotide mutation rate and π is an estimate of the level of synonymous (silent) variation within a species. Therefore, if we have an estimate for the background mutation rate and an observation of the levels of neutral variation in genomes, π, then we can estimate N_e. A recent survey of the literature by Lynch (42) has determined levels of genetic variation in 17 fungal species; estimates of molecular diversity range from 0.00695 to 0.26457, with a mean of 0.0402. Kasuga et al. (32) have shown that fungal synonymous nucleotide substitution rates u vary from 0.9×10^{-9} to $16.7 \times$

10^{-9}, and here we take a value (also used in reference 42) of 1.6×10^{-9} substitutions per nucleotide per generation. Using this value we have derived estimates of N_e (Table 12.1), these equating to an overall estimation of an average N_e for fungi of $\sim 10^7$.

However, an interesting observation here is found by comparing N_e values of mitosporic fungi that lack an observed sexual stage of their life cycle with those of recombining, meiosporic species. N_e for mitosporic taxa is $\sim 2^6$ compared to $\sim 1.5^7$ for meiosporic taxa, a difference of almost 1 magnitude, and this is highly significant (t test; $P = 0.003$). Therefore, there appears to be a relationship between effective population size and reproductive mode in fungi, and low effective population sizes appear to be correlated, at least to a first approximation, with asexual modes of recombination. Here is evidence that the phenomenon of asexuality in fungi is associated with a population genetic parameter that describes the rate at which natural selection is expected to operate on a population, and which has the potential to

Table 12.1 Estimates of noncoding variation in fungi and derived effective population sizes[a]

Species	Reproductive mode[b]	Estimate of π	N_e	Reference
Candida albicans	Mitosporic	0.00695	2.2×10^6	52
Candida glabrata	Mitosporic	0.01417	4.4×10^6	12
Penicillium marneffei	Mitosporic	0.00815	2.5×10^6	Fisher, unpublished
Coccidioides posadasii	Mitosporic	0.00297	9.3×10^5	34
Coccidioides immitis	Mitosporic	0.00125	3.9×10^5	34
Auxarthron zuffianum	Meiosporic	0.26457	8.3×10^7	33
Cryptococcus neoformans	Meiosporic	0.03189	1.0×10^7	51
C. neoformans	Meiosporic	0.03022	9.4×10^6	24
C. neoformans	Meiosporic	0.01934	6.0×10^6	40
Heterobasidion annosum	Meiosporic	0.10043	3.1×10^7	29
H. annosum	Meiosporic	0.03807	1.2×10^7	29
Histoplasma capsulatum	Meiosporic	0.05105	1.6×10^7	31
Mycosphaerella graminicola	Meiosporic	0.03492	1.1×10^7	2
Neurospora crassa	Meiosporic	0.01899	5.9×10^6	11
N. crassa	Meiosporic	0.053	1.7×10^7	60
Neurospora discreta	Meiosporic	0.02613	8.2×10^6	11
Neurospora intermedia	Meiosporic	0.01612	5.0×10^6	11
Neurospora sitophila	Meiosporic	0.00203	6.3×10^5	11
Saccharomyces cerevisiae	Meiosporic	0.00802	2.5×10^6	14
S. cerevisiae	Meiosporic	0.06722	2.1×10^7	49
S. cerevisiae	Meiosporic	0.0442	1.4×10^7	36
S. cerevisiae	Meiosporic	0.0518	1.6×10^7	47
S. cerevisiae	Meiosporic	0.0478	1.5×10^7	59
S. cerevisiae	Meiosporic	0.0456	1.4×10^7	59
Saccharomyces paradoxus	Meiosporic	0.00388	1.2×10^6	30
Uncinocarpus reesii	Meiosporic	0.05646	1.8×10^7	33

[a]Calculated using an estimated mutation rate of 1.6×10^{-9}. Modified from reference 42.
[b]Species are classed as meiosporic (sexual) if there are known meiotic phases of the life cycle or as mitosporic (asexual) if meiosis is unrecorded. Species undergoing cryptic recombination, i.e., *Coccidioides* sp. and *Candida albicans*, are conservatively classed as mitosporic (asexual) because although recombination has been characterized by population genetic analyses (53), it has not been determined whether this is ancestral, rare, or ongoing.

drive the evolution of asexuality purely through genetic-drift-type mechanisms.

IS THERE A LINK BETWEEN SMALL EFFECTIVE POPULATION SIZE, ASEXUALITY, AND ENDEMISM IN FUNGI?

A significant problem with the above calculations is that the effect of population subdivision has not been factored into the estimates of N_e. Numerous data on the metapopulation structure of fungi have recently accumulated, and increasingly it has been shown that fungi show the full continuum of population genetic structures ranging from global panmixia (e.g., *Aspergillus fumigatus* [48]) to pronounced biogeographic population genetic structure (e.g., *Histoplasma capsulatum*, *P. marneffei*, *Coccidioides* sp., and *Blastomyces dermatitidis*). Theoretical work has shown that spatial subdivision acts to decrease N_e through the concerted effects of demographic, environmental, and genetic stochasticity (28, 58). Using this result, we can make the prediction that fungi with pronounced metapopulation structures and geographically endemic distributions are associated with smaller effective population sizes than are seen in cosmopolitan species. If this is the case, then following from the arguments detailed above, *there will be a link between* N_e, *fertility, and geographical distributions in fungi*. In order to test these theories, there is a need to measure population genetic parameters and mating systems in naturally occurring species that cover the whole spectrum of population genetic structures, from panmixia (random) to highly structured (heterogeneous). There have been rapid advances in our description of the population genetics of fungi (53, 54), and research is reaching the stage where broad syntheses are becoming possible. However, there is a need to uncover species that exhibit the extremes of metapopulation structure, as it is within these species that the effect of life history strategies on the genetics of their populations will be seen with greatest clarity. To this end, we have been deciphering the genetical structure of one of the most enigmatic fungi on earth, the human pathogenic fungus *P. marneffei*.

P. MARNEFFEI, ITS BIOLOGY, AND ITS MATING SYSTEM

P. marneffei is a fungus of the family *Trichocomaceae* that has emerged since 1990 as a significant agent of human mycosis. The fungus causes a disease, penicilliosis marneffei, that occurs in immunosuppressed patients. As a consequence, the disease has emerged in concert with the spread of human immunodeficiency virus in southeast Asia, where it is classified as an AIDS-related indicator disease (39). The organism is highly endemic within a relatively narrow region of tropical southeast Asia, extending from Assam in northeast India across Myanmar, Thailand, Laos, and southern China (Guangxi province and Hong Kong), and reaching its known easterly limits on the island of Taiwan. *P. marneffei* is a relatively recent addition to the lexicon of known fungi of southeast Asia, having been discovered in 1956 as an agent of infection of bamboo rats (6), where it was easily identified due to it secreting in culture a characteristic, and highly visible, red diffusible pigment. Within the bamboo rat species *Rhizomys* and *Cannomys*, the infection occurs with a high prevalence, and molecular data have shown that humans and bamboo rats share genetically identical strains, raising the possibility that bamboo rats are a zoonotic reservoir for the pathogen (19, 21). *P. marneffei* has a simple haploid life cycle whereby filamentous soil mycelia give rise to aerially borne conidia (8), and the species is morphologically asexual. A phylogenetic analysis of the species demonstrated that *P. marneffei* is closely related to species of the *Penicillium* subgenus *Biverticillium*, including those with sexual *Talaromyces* teleomorphs (41), and that *T. stipitatus* is the sister species to *P. marneffei* (S. de Hoog, unpublished observations). While the mating types of other *Trichocomaceae* are known, thus far the mating system of *Penicillium* remains unclear. However, the recent undertaking of the genome sequences for *P. marneffei* and *T. stipitatus* in 2006 will enable idiomorphs to be identified, enabling the development of markers with which to identify whether the species are homothallic or, as has been determined for the closely related species *Aspergillus fumigatus* (48), are heterothallic.

MULTILOCUS GENOTYPING OF P. MARNEFFEI

To learn the basic features of a population, i.e., its geographic extent, effective population size, reproductive mode, and evolutionary history, requires data on genetic variation from a representative sample of individuals. Modern molecular methods, specifically multilocus genotyping, provide a high-throughput and powerful method for characterizing genetic variation. Multilocus genotyping works by targeting specific loci in order to index the amount of genetic variation residing between genomes within species. Multilocus genotyping of *P. marneffei* relies on two closely allied methodologies. The first, multilocus sequence typing (MLST), works by indexing the sequence variation from ca. 500 nucleotides of each of 5 to 10 genes to characterize genetic diversity (43). Although developed for bacteria, MLST has become a popular method for typing fungi because it directly samples the polymorphism present within nucleotide sequences, and each new study can use, and add to, all previously obtained data through the MLST websites (http://www.mlst.net). Successful fungal MLST schemes have been developed for *Candida albicans* (52), *Coccidioides* sp. (34), *H. capsulatum* (31), and *Cryptococcus* sp. (25, 40), proving the utility of the method for typing agents of mycoses. The second method, multilocus microsatellite typing (MLMT), is used when more highly variable loci are needed in species where genetic variation is low or when closely related isolates need to be discriminated. MLMT schemes work by amplifying loci that contain tandem repeats of di-, tri-, or tetranucleotide motifs and then scoring the resulting length polymorphisms by using a DNA sequencer. MLMT schemes are managed in the same manner as MLST schemes via a website located at http://www.multilocus.net, and the utility of MLMT schemes in mycology has been demonstrated by their ability to uncover global patterns of dispersal in the fungi *Coccidioides immitis*, *Coccidioides posadasii* (22, 23), and *P. marneffei* (21).

An MLMT scheme has been developed for *P. marneffei* based on the amplification of 21 microsatellite-containing loci within three multiplexed PCRs (19, 20). Following these amplifications, size polymorphisms are scored and, for each locus, alleles are then determined. Therefore, each isolate is assigned a multilocus microsatellite type (MT) "barcode" of 21 digits that is subsequently curated and then uploaded to the *P. marneffei* MLMT website, held at http://pmarneffei.multilocus.net/. Thus far, the MLMT database for *P. marneffei* contains the barcodes for around 200 isolates of the fungus, collected from humans and bamboo rats across southeast Asia. Within this data set, there are 108 unique MTs, each of these MTs being composed of between 1 and 10 isolates that are genetically identical at all 21 microsatellite loci. Within Thailand, 83 MTs were recovered from an analysis of human and bamboo rat isolates, and of these, five are shared between humans and bamboo rats (21). These data have shown that there is no evidence for fungal lineages that are specific to bamboo rats and therefore that there is a high likelihood that all lineages of *P. marneffei* are equally infectious to both bamboo rats and humans. While, on the face of it, this finding appears to support the hypothesis that bamboo rats are zoonotic reservoirs for human infections of the fungus, the alternative hypothesis that humans and bamboo rats are infected from a common environmental source is equally likely. So far, there is little epidemiological evidence for linkage between infected humans and bamboo rats (56). However, there has also been rather little progress on finding potential environmental reservoirs of the fungus. *P. marneffei* grows slowly in unsterilized soil (57; M. C. Fisher, unpublished observations) and has ever been recovered only once from soil, and this was at the site of a bamboo rat burrow (9). Therefore, the complete life cycle and environmental reservoirs of *P. marneffei* are still not known, and much work is still needed in order to determine the natural history of the fungus.

SPATIAL COMPONENTS OF GENETIC DIVERSITY IN *P. MARNEFFEI* AND THEIR EFFECT ON N_e

P. marneffei belongs to a group of fungi that are known as "endemic mycoses," which, as detailed above, exhibit highly constrained spatial ranges. In other endemic mycoses, such as those caused by *C. immitis* and *C. posadasii*, there is a strong biogeographic signal, with certain genotypic clusters being strongly associated with certain geographic areas (22). The generality of this relationship has been determined by assessing the degree of association between genetic and geographic distance in *P. marneffei* by comparing the spatial origin of each bamboo rat or clinical isolate with their associated MT. Principal-components analyses reveal that the first principal component (PC1) describes 18.8% of the total geographic variation and that this diversity has a strong spatial component, falling into two clusters that broadly correspond to isolates from the East of southeast Asia (called the Eastern clade) and those from the West (called the Western clade), as depicted in Fig. 12.1. Within each of these major clades, there are a series of smaller, nested clusters that correspond to different geographical regions, namely, India, North/Central Thailand, Eastern Thailand, Vietnam, Southern China, and Taiwan (Fig. 12.2). On a broad scale, therefore, *P. marneffei* exhibits a highly heterogeneous spatial population genetic structure, with different genotypes and genotypic clusters being found in different geographical areas.

This broad biogeographical pattern was investigated at smaller scales, by calculating the "clone distance" that is covered by each of the observed *P. marneffei* MTs. This distance was measured as the geodesic line between the most distant representatives of each MT. Analysis of this data set using a geographical information system showed that 22% of *P. marneffei* MTs were recovered from two or more sites in Thailand and that this equated to an average radial dispersal distance of 28.6 km (21). Thus, it appears that long-distance dispersal of *P. marneffei* in nature

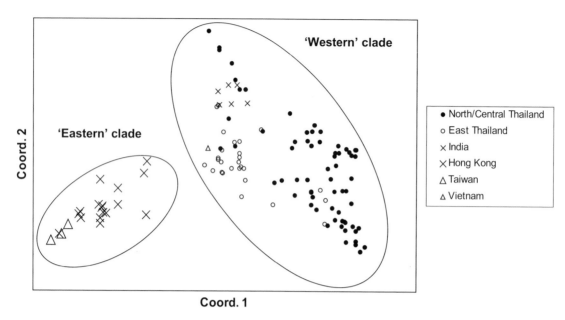

Figure 12.1 Principal-components analysis of the *P. marneffei* MTs.

is relatively rare. This approach was then refined by correlating genetic distances against geographic distances in order to investigate whether clonal dispersal was random or could be expressed as a function of distance. This analysis was performed by creating two matrices; the first is a pairwise geographic-distance matrix, and the second is a pairwise matrix of genetic distances, calculated as $1 -$ (the total number of shared alleles at all loci/n), where n is the number of loci compared; in this dataset, n equals 21. The autocorrelation coefficient r is then calculated from these two matrices and compared against a null model of a single randomly mating population with no spatial structure. If isolates are more closely related than is expected under a model of random mating across the whole population, then r will be positive (50). As a consequence of this relationship, assuming spatial structure, then r will exhibit a negative correlation with distance. Investigations of the relatedness coefficients within pools of isolates from two populations, humans in Thailand and bamboo rats in Thailand, showed that r is high at the local level but decays rapidly with increasing distance (21). The distance at which $r = 0$, and becomes nonsignificant is, 367 km for the human data set and 109 km for the bamboo rats. The difference in r between isolates from humans and rats is most likely due to the increased range over which humans move, relative to the rats. The observed rapid decay of r alongside increasing geographic distances for both host species shows that, within Thailand at least, populations of *P. marneffei* are local and effective long-

distance dispersal is negligible. Therefore, at both local and larger scales, *P. marneffei* exhibits extensive spatial structure and strains are most likely to be related to those within a circle of ~150 km in radius.

In order to calculate the global effective population size of *P. marneffei*, we developed an MLST scheme for *P. marneffei* (Fisher, unpublished) and assessed the levels of genetic diversity that are found in five genes, *GasA*, *GasC*, *StuA*, *ICL*, and *AbaA*. This sequencing project found levels of diversity corresponding to 0.815% nucleotide divergence for noncoding sites, which corresponds to an estimated value for N_e of 2.5×10^6. This value, as detailed above, is broadly comparable to that seen for other mitosporic fungal species (Table 12.1). What happens, however, when we take into account the effect that spatial structure has on the global effective population size of *P. marneffei*? Here, we restructured the data set into two populatons (Western and Eastern clades) and then four populations (India, Thailand, Hong Kong, and Taiwan) and recalibrated N_e to take into account the levels of genetic diversity found within each population. These analyses show that population structure has a pronounced effect on N_e, with the mean value equating to fewer than 500,000 individuals for the total four-population data set, shown in Fig. 12.3. Here, it is clear that population structure has an inverse relationship with the effective population size that is seen in *P. marneffei*. A consequence of this finding is that the stochastic role of genetic drift is underestimated unless the true genetic structure of natural populations

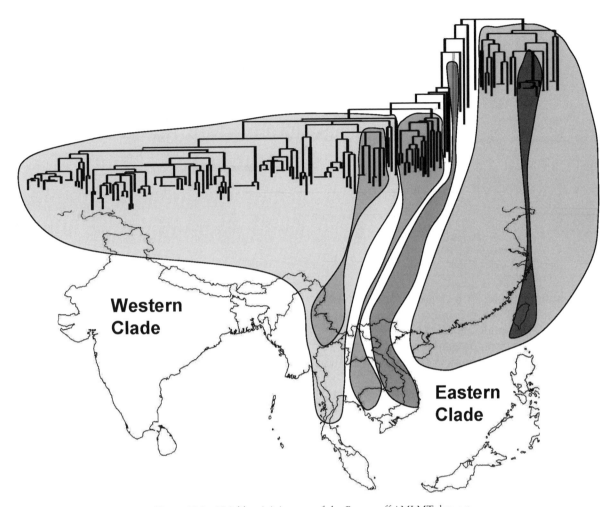

Figure 12.2 Neighbor-joining tree of the *P. marneffei* MLMT data set.

is taken into account. These arguments are important to consider for the following reason: much recent work has assumed that microbial population sizes are huge and global in their range (18, 42). However, if, as we have found, some microbial species have spatially structured populations, then assumptions on the rate of fixation of mutations, genome evolution, and speciation may need to be reassessed for some species. Moreover, the theoretical impact of population structure on the evolution of asexuality needs to be taken into account, as I argue below.

THE PREDOMINANT MODE OF REPRODUCTION IN *P. MARNEFFEI* IS ASEXUAL

As detailed in Table 12.1, there is a relationship between reproductive mode and N_e in fungi. In order to demonstrate the reproductive mode in *P. marneffei*, we took a number of population genetic approaches. Our first approach used the program eBURST (15) to search for groups of closely related MTs and then to determine the pattern of descent within these eBURST groups. Under a model of simple asexual reproduction, founding genotypes increase in frequency and, over time, diversify into a group of closely related genotypes, known as a clonal complex. These groups are identified by eBURST as links between genotypes that differ by only 1 of 21 MLMT genotypes, known as single locus variants. Recombination has the effect of randomly shuffling alleles into new genetic backgrounds, with the result that clonal complexes are broken up (16). Therefore, species that undergo little recombination will be represented within eBURST diagrams as highly connected networks, while the connectivity within recombining species will be decreased. A simulated example of this is shown in Fig. 12.4 where the effects of increasing

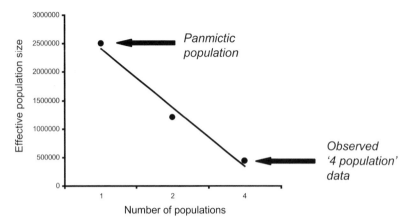

Figure 12.3   The relationship between effective population size and population structure inferred from the *P. marneffei* MLST data set.

the recombination rate are to increase the number of BURST groups from 139 to 238 and to more than double the number of singletons (unrelated isolates) within the data set. Analysis of the *P. marneffei* MLMT data set from Thailand showed that, of 83 MTs, there exist 13 clonal complexes that contain 43 MTs (21). These data show that over 50% of MTs are contained within clonal complexes, suggesting that there is a significant asexual component to the data set. A rough estimate of the relative contributions of mutation and recombination to the generation of diversity in this data set can be found by identifying the changes between MTs that incorporate unique alleles and are therefore most likely due to mutation, compared to the changes that utilize preex-

isting alleles and are most likely due to recombination. This calculation indicates that the ratio of mutation to recombination within the *P. marneffei* genome is 5:1, a figure that is close to that observed for many highly clonal bacteria and surprisingly high for a fungus (21).

The extreme population structure of *P. marneffei* has the effect of artificially "imposing" a clonal population structure on the MLMT data set, by decreasing the probability that isolates will be able to swap genes across geographical distances. In order to control for this effect of population substructure, the global data set was restructured into populations comprising isolates that fell within a circle of radius 150 km, corresponding to our best estimate for a local population of

A. Recombination = 0   **B.** Recombination = x11

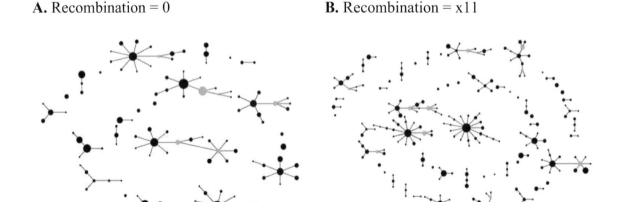

Figure 12.4   eBURST groupings of two simulated populations evolving under a Wright/Fisher model, with *n* = 1,000 and identical mutation rates. For population A, recombination = 0, and for population B, the recombination rate is twice the background mutation rate. Simulations and diagrams were produced by Katy Turner.

P. marneffei. Within these populations, under a model of complete clonal reproduction, all alleles will be co-inherited as a single block and diversification will be solely through the effects of mutation. In this case, the coinheritance of loci will lead to the accumulation of linkage disequilibrium within the data set. The levels of linkage disequilibrium between loci within multilocus data sets can be estimated through the use of a convenient summary statistic, known as the index of association (45). This measure determines the correlations between alleles at different loci by comparing the variation in genetic distance between individuals relative to those observed within an artificially recombined data set. A modified statistic, $\bar{r}_d$, standardizes the index of association by correcting for the inflating effect of including extra loci in the analysis (P.-M. Agapow and A. Burt [ed.], *Multilocus v1.2.2.* [http://www.bio.ic.ac.uk/evolve/software/multilocus/]). This analysis showed that $\bar{r}_d$ is highly significant for *P. marneffei*, and our finding is replicated within three well-defined populations where isolates have been recovered from small geographic areas (Table 12.2). This result confirms that loci are in linkage disequilibria, a result that is most likely due to extensive clonal reproduction.

Taken together, the findings that (i) identical MTs are recovered from separate hosts and separate species, (ii) well-defined clonal complexes occur, (iii) the estimated ratio of mutation to recombination is 5:1, and (iv) multilocus linkage disequilibria are high show that the major mode of reproduction in *P. marneffei* is asexual. The occurrence of recombination within these populations cannot be ruled out; however, the balance of evidence shows that both clades of *P. marneffei* are evolving in an overwhelmingly clonal manner.

THE IMPLICATIONS OF A CLONAL LIFESTYLE ON THE EVOLUTIONARY TRAJECTORY OF *P. MARNEFFEI*

In summary, we have put forward an argument suggesting that in fungi where sexual reproduction is facultative loss-of-function mutations in mating-type loci will,

generally speaking, be selectively neutral. We then argue that while this is of little consequence in species with cosmopolitan distributions and large N_e, species that are spatially genetically structured will be exposed to the effects of stochastic genetic drift and will accrue mating-type biases and nonfunctional mating-type loci as a consequence of a breakdown in the strength of natural selection. As a result, asexuality will inevitably build up in species that exhibit (i) structured populations and (ii) facultative sexual modes of reproduction.

Low frequencies of recombination lead to a decrease in the mean fitness of a population. This is because recombination increases the rate of adaptation of a population by increasing the variance in phenotypic characters between parents and their offspring. This component of variance is known as "additive genetic variance" (4). The effect of asexual reproduction in decreasing additive genetic variance, and thus decreasing the rate of adaptation of the population (17), will tend to lead to populations that are evolutionarily static and less well able to respond to changes in the environment (26). A corollary of this theory is that a particular clonal lineage is expected to be less fit within a range of different environments as a consequence of reduced additive variance. In such a scenario, strong disruptive selection may be expected to lead to the evolution of locally adapted lineages that will tend to outcompete any incoming immigrants, thus imposing a limit to dispersal, and even in species that have evolved an effective dispersal stage (such as conidia in *P. marneffei*). The environment is heterogeneous, at almost any scale one chooses, and it has been shown that the different clonal lineages of *P. marneffei* are associated with different environments within southeast Asia (21). This is evidence, although not proof, that there are associations between genetic and ecological diversity within *P. marneffei*. In this case, dispersal in *P. marneffei* to new geographical areas and envionments may be limited by the prior occurrence of locally adapted lineages. Our finding that long-distance dispersal is rare may be construed as evidence that natural selection *is* in fact occurring within populations of *P. marneffei*, leading to the evolution of locally adapted lineages. Indeed, if

Table 12.2 Multilocus linkage disequilibria for three populations of *P. marneffei* in the Eastern and Western clades

Species	Population (*n*)	$\bar{r}_d{}^a$	*P*
P. marneffei (Chiang Mai, Western clade)	95	0.071	0.001
P. marneffei (Ubon Ratchathani, Western clade)	21	0.029	0.004
P. marneffei (Hong Kong, Eastern clade)	13	0.094	0.040

$^a\bar{r}_d$ ranges from 1 to 0, and its significance is tested via Monte-Carlo permutations following Agapow and Burt, Multilocus v1.2.2. program (see the text).

selective coefficients are high, then adaptation will occur even within populations with small N_e. However, metapopulation models have also shown that even under assumptions of strict neutrality, founder effects determine which lineages are excluded (27), showing that local adaptation need not be invoked to explain the heterogeneous population structure that is seen with *P. marneffei*.

Under the above scenarios, spatial metapopulations linked to an accumulation of loss-of-sex mutations will lead to an overall decrease in N_e, which will therefore reinforce the effect of genetic drift. Such a system can, theoretically at least, lead to positive-feedback loops that will increase the rate at which asexual reproduction occurs within a population and may lead to mating competence becoming irrevocably lost. Our work shows that *P. marneffei* is one of the most asexual fungi yet found, and we argue that *P. marneffei*'s extreme geographic endemism and fragmented population structure are directly linked to its asexuality. It has been argued that such group level selective disadvantages to asexual species will lead to increases in their extinction rates, and our observations on the relative rarities of truly asexual fungi would tend to confirm these hypotheses. The generality of our observations and arguments relies on the collection of population data for a range of fungal species that vary in their reproductive modes, as advocated by Barraclough et al. (3), and this approach will enable comparative analyses that assess (i) how many asexual fungi exist, (ii) whether N_e scales with recombination rate, (iii) whether mating competence scales with N_e, and (iv) whether geographic range correlates with N_e and recombination rate. Rapid current advances in genomic and genotyping technologies provide the tools by which these questions can be answered. The answers themselves will provide a key insight into how fungal populations evolve and what processes determine the eventual fate of asexual taxa relative to their sexual conspecifics.

References

1. **Anderson, J. B., and L. M. Kohn.** 1995. Clonality in soil-borne, plant-pathogenic fungi. *Annu. Rev. Phytopathol.* 33:369–391.
2. **Banke, S., A. Peschon, and B. A. McDonald.** 2004. Phylogenetic analysis of globally distributed Mycosphaerella graminicola populations based on three DNA sequence loci. *Fungal Genet. Biol.* 41:226–238.
3. **Barraclough, T. G., C. W. Birky, Jr., and A. Burt.** 2003. Diversification in sexual and asexual organisms. *Evolution Int. J. Org. Evolution* 57:2166–2172.
4. **Barton, N. H., and B. Charlesworth.** 1998. Why sex and recombination? *Science* 281:1986–1990.
5. **Burt, A., D. A. Carter, G. L. Koenig, T. J. White, and J. W. Taylor.** 1996. Molecular markers reveal cryptic sex in the human pathogen Coccidioides immitis. *Proc. Natl. Acad. Sci. USA* 93:770–773.
6. **Capponi, M., P. Sureau, and G. Segretain.** 1956. Penicillose de Rhizomys sinensis. *Bull. Soc. Pathol. Exot.* 49:418–421.
7. **Carbone, I., J. B. Anderson, and L. M. Kohn.** 1999. Patterns of descent in clonal lineages and their multilocus fingerprints are resolved with combined gene genealogies. *Evolution* 53:11–21.
8. **Chariyalertsak, S., T. Sirisanthana, K. Supparatpinyo, J. Praparattanapan, and K. E. Nelson.** 1997. Case-control study of risk factors for Penicillium marneffei infection in human immunodeficiency virus-infected patients in northern Thailand. *Clin. Infect. Dis.* 24:1080–1086.
9. **Chariyalertsak, S., P. Vanittanakom, K. E. Nelson, T. Sirisanthana, and N. Vanittanakom.** 1996. Rhizomys sumatrensis and Cannomys badius, new natural animal hosts of Penicillium marneffei. *J. Med. Vet. Mycol.* 34:105–110.
10. **Charlesworth, B.** 1996. The evolution of chromosomal sex determination and dosage compensation. *Curr. Biol.* 6:149–162.
11. **Dettman, J. R., D. J. Jacobson, and J. W. Taylor.** 2003. A multilocus genealogical approach to phylogenetic species recognition in the model eukaryote Neurospora. *Evolution* 57:2703–2720.
12. **Dodgson, A. R., C. Pujol, D. W. Denning, D. R. Soll, and A. J. Fox.** 2003. Multilocus sequence typing of Candida glabrata reveals geographically enriched clades. *J. Clin. Microbiol.* 41:5709–5717.
13. **Ewens, W. J.** 1979. *Mathematical Population Genetics.* Springer, Berlin, Germany.
14. **Fay, J. C., H. L. McCullough, P. D. Sniegowski, and M. B. Eisen.** 2004. Population genetic variation in gene expression is associated with phenotypic variation in Saccharomyces cerevisiae. *Genome Biol.* 5:R26.
15. **Feil, E. J., B. C. Li, D. M. Aanensen, W. P. Hanage, and B. G. Spratt.** 2004. eBURST: inferring patterns of evolutionary descent among clusters of related bacterial genotypes from multilocus sequence typing data. *J. Bacteriol.* 186:1518–1530.
16. **Feil, E. J., and B. G. Spratt.** 2001. Recombination and the population structures of bacterial pathogens. *Annu. Rev. Microbiol.* 55:561–590.
17. **Felsenstein, J.** 1974. Evolutionary advantage of recombination. *Genetics* 78:737–756.
18. **Finlay, B. J.** 2002. Global dispersal of free-living microbial eukaryote species. *Science* 296:1061–1063.
19. **Fisher, M. C., D. Aanensen, S. de Hoog, and N. Vanittanakom.** 2004. Multilocus microsatellite typing system for Penicillium marneffei reveals spatially structured populations. *J. Clin. Microbiol.* 42:5065–5069.
20. **Fisher, M. C., G. S. de Hoog, and N. Vannittanakom.** 2004. A highly discriminatory multilocus microsatellite typing system (MLMT) for Penicillium marneffei. *Mol. Ecol. Notes* 4:515–518.
21. **Fisher, M. C., W. P. Hanage, S. de Hoog, E. Johnson, M. D. Smith, N. J. White, and N. Vanittanakom.** 2005.

Low effective dispersal of asexual genotypes in heterogeneous landscapes by the endemic pathogen *Penicillium marneffei. PLoS Pathog.* 2:159–165.

22. **Fisher, M. C., G. L. Koenig, T. J. White, G. San-Blas, R. Negroni, I. G. Alvarez, B. Wanke, and J. W. Taylor.** 2001. Biogeographic range expansion into South America by *Coccidioides immitis* mirrors New World patterns of human migration. *Proc. Natl. Acad. Sci. USA* 98:4558–4562.

23. **Fisher, M. C., B. Rannala, V. Chaturvedi, and J. W. Taylor.** 2002. Disease surveillance in recombining pathogens: multilocus genotypes identify sources of human *Coccidioides* infections. *Proc. Natl. Acad. Sci. USA* 99:9067–9071.

24. **Franzot, S. P., J. S. Hamdan, B. P. Currie, and A. Casadevall.** 1997. Molecular epidemiology of *Cryptococcus neoformans* in Brazil and the United States: evidence for both local genetic differences and a global clonal population structure. *J. Clin. Microbiol.* 35:2243–2251.

25. **Fraser, J. A., S. S. Giles, E. C. Wenink, S. G. Geunes-Boyer, J. R. Wright, S. Diezmann, A. Allen, J. E. Stajich, F. S. Dietrich, J. R. Perfect, and J. Heitman.** 2005. Same-sex mating and the origin of the Vancouver Island *Cryptococcus gattii* outbreak. *Nature* 437:1360–1364.

26. **Goddard, M. R., H. C. Godfray, and A. Burt.** 2005. Sex increases the efficacy of natural selection in experimental yeast populations. *Nature* 434:636–640.

27. **Gourbiere, S., and F. Gourbiere.** 2002. Competition between unit-restricted fungi: a metapopulation model. *J. Theor. Biol.* 217:351–368.

28. **Higgins, K., and M. Lynch.** 2001. Metapopulation extinction caused by mutation accumulation. *Proc. Natl. Acad. Sci. USA* 98:2928–2933.

29. **Johannesson, H., and J. Stenlid.** 2003. Molecular markers reveal genetic isolation and phylogeography of the S and F intersterility groups of the wood-decay fungus *Heterobasidion annosum. Mol. Phylogenet. Evol.* 29:94–101.

30. **Johnson, L. J., V. Koufopanou, M. R. Goddard, R. Hetherington, S. M. Schafer, and A. Burt.** 2004. Population genetics of the wild yeast *Saccharomyces paradoxus. Genetics* 166:43–52.

31. **Kasuga, T., T. J. White, G. Koenig, J. McEwen, A. Restrepo, E. Castaneda, C. Da Silva Lacaz, E. M. Heins-Vaccari, R. S. De Freitas, R. M. Zancope-Oliveira, Z. Qin, R. Negroni, D. A. Carter, Y. Mikami, M. Tamura, M. L. Taylor, G. F. Miller, N. Poonwan, and J. W. Taylor.** 2003. Phylogeography of the fungal pathogen *Histoplasma capsulatum. Mol. Ecol.* 12:3383–3401.

32. **Kasuga, T., T. J. White, and J. W. Taylor.** 2002. Estimation of nucleotide substitution rates in Eurotiomycete fungi. *Mol. Biol. Evol.* 19:2318–2324.

33. **Koufopanou, V., A. Burt, T. Szaro, and J. W. Taylor.** 2001. Gene genealogies, cryptic species, and molecular evolution in the human pathogen *Coccidioides immitis* and relatives (Ascomycota, Onygenales). *Mol. Biol. Evol.* 18:1246–1258.

34. **Koufopanou, V., A. Burt, and J. W. Taylor.** 1997. Concordance of gene genealogies reveals reproductive isolation in the pathogenic fungus *Coccidioides immitis. Proc. Natl. Acad. Sci. USA* 94:5478–5482.

35. **Kronstad, J. W., and C. Staben.** 1997. Mating type in filamentous fungi. *Annu. Rev. Genet.* 31:245–276.

36. **Laize, V., F. Tacnet, P. Ripoche, and S. Hohmann.** 2000. Polymorphism of Saccharomyces cerevisiae aquaporins. *Yeast* 16:897–903.

37. **Lee, N., G. Bakkeren, K. Wong, J. E. Sherwood, and J. W. Kronstad.** 1999. The mating-type and pathogenicity locus of the fungus Ustilago hordei spans a 500-kb region. *Proc. Natl. Acad. Sci. USA* 96:15026–15031.

38. **Leslie, J. F., and K. K. Klein.** 1996. Female fertility and mating type effects on effective population size and evolution in filamentous fungi. *Genetics* 144:557–567.

39. **Li, P. C., M. C. Tsui, and K. F. Ma.** 1992. *Penicillium marneffei*: indicator disease for AIDS in South East Asia. *AIDS* 6:240–241.

40. **Litvintseva, A. P., R. Thakur, R. Vilgalys, and T. G. Mitchell.** 2005. Multilocus sequence typing reveals three genetic subpopulations of *Cryptococcus neoformans* var. grubii (serotype A), including a unique population in Botswana. *Genetics* 172:2223–2238.

41. **Lobuglio, K. F., J. I. Pitt, and J. W. Taylor.** 1993. Phylogenetic analysis of two ribosomal DNA regions indicates multiple independent losses of a sexual Talaromyces state among asexual Penicillium species in subgenus Biverticillium. *Mycologia* 85: 592–604.

42. **Lynch, M.** 2006. The origins of eukaryotic gene structure. *Mol. Biol. Evol.* 23:450–468.

43. **Maiden, M. C., J. A. Bygraves, E. Feil, G. Morelli, J. E. Russell, R. Urwin, Q. Zhang, J. Zhou, K. Zurth, D. A. Caugant, I. M. Feavers, M. Achtman, and B. G. Spratt.** 1998. Multilocus sequence typing: a portable approach to the identification of clones within populations of pathogenic microorganisms. *Proc. Natl. Acad. Sci. USA* 95:3140–3145.

44. **Maruyama, T., and M. Kimura.** 1980. Genetic variability and effective population size when local extinction and recolonization of subpopulations are frequent. *Proc. Natl. Acad. Sci. USA* 77:6710–6714.

45. **Maynard-Smith, J., N. H. Smith, M. O'Rourke, and B. G. Spratt.** 1993. How clonal are bacteria? *Proc. Natl. Acad. Sci. USA* 90:4384–4388.

46. **Maynard Smith, J.** 1978. *The Evolution of Sex.* Cambridge Univ. Press, Cambridge, United Kingdom.

47. **Mosse, M. O., P. Linder, J. Lazowska, and P. P. Slonimski.** 1993. A comprehensive compilation of 1001 nucleotide sequences coding for proteins from the yeast Saccharomyces cerevisiae (= ListA2). *Curr. Genet.* 23:66–91.

48. **Paoletti, M., C. Rydholm, E. U. Schwier, M. J. Anderson, G. Szakacs, F. Lutzoni, J. P. Debeaupuis, J. P. Latge, D. W. Denning, and P. S. Dyer.** 2005. Evidence for sexuality in the opportunistic fungal pathogen Aspergillus fumigatus. *Curr. Biol.* 15:1242–1248.

49. **Sakai, A., T. Chibazakura, Y. Shimizu, and F. Hishinuma.** 1992. Molecular analysis of POP2 gene, a gene required for glucose-derepression of gene expression in Saccharomyces cerevisiae. *Nucleic Acids Res.* 20:6227–6233.

50. **Smouse, P. E., and R. Peakall.** 1999. Spatial autocorrelation analysis of individual multiallele and multilocus genetic structure. *Heredity* 82(Pt. 5):561–573.

51. **Sugita, T., R. Ikeda, and T. Shinoda.** 2001. Diversity among strains of Cryptococcus neoformans var. gattii as revealed by a sequence analysis of multiple genes and a chemotype analysis of capsular polysaccharide. *Microbiol. Immunol.* **45:**757–768.

52. **Tavanti, A., N. A. Gow, S. Senesi, M. C. Maiden, and F. C. Odds.** 2003. Optimization and validation of multilocus sequence typing for *Candida albicans. J. Clin. Microbiol.* **41:**3765–3776.

53. **Taylor, J., D. Jacobson, and M. Fisher.** 1999. The evolution of asexual fungi: reproduction, speciation and classification. *Annu. Rev. Phytopathol.* **37:**197–246.

54. **Taylor, J. W., and M. C. Fisher.** 2003. Fungal multilocus sequence typing—it's not just for bacteria. *Curr. Opin. Microbiol.* **6:**351–356.

55. **Taylor, J. W., D. J. Jacobson, S. Kroken, T. Kasuga, D. M. Geiser, D. S. Hibbett, and M. C. Fisher.** 2001. The phylogenetic species concept in fungi. *Fungal Genet. Biol.* **31:** 21–32.

56. **Vanittanakom, N., C. R. Cooper, Jr., M. C. Fisher, and T. Sirisanthana.** 2006. Penicillium marneffei infection and recent advances in the epidemiology and molecular biology aspects. *Clin. Microbiol. Rev.* **19:**95–110.

57. **Vanittanakom, N., M. Mekaprateep, P. Sriburee, P. Vanittanakom, and P. Khanjanasthiti.** 1995. Efficiency of the flotation method in the isolation of Penicillium marneffei from seeded soil. *J. Med. Vet. Mycol.* **33:**271–273.

58. **Whitlock, M. C., and N. H. Barton.** 1997. The effective size of a subdivided population. *Genetics* **146:**427–441.

59. **Winzeler, E. A., D. R. Richards, A. R. Conway, A. L. Goldstein, S. Kalman, M. J. McCullough, J. H. McCusker, D. A. Stevens, L. Wodicka, D. J. Lockhart, and R. W. Davis.** 1998. Direct allelic variation scanning of the yeast genome. *Science* **281:**1194–1197.

60. **Yeadon, P. J., and D. E. A. Catcheside.** 1999. Polymorphism around cog extends into adjacent structural genes. *Curr. Genet.* **35:**631–637.

Ascomycetes: the Candida MAT Locus and Related Topics

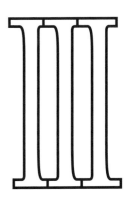

*Sex in Fungi: Molecular Determination
and Evolutionary Implications*
Edited by Joseph Heitman et al.
© 2007 ASM Press, Washington, D.C.

David R. Soll
Karla J. Daniels

13

MAT, Mating, Switching, and Pathogenesis in *Candida albicans*, *Candida dubliniensis*, and *Candida glabrata*

It is presumed that pathogenic yeasts, like other microorganisms, mate in order to generate rare variants that have a competitive edge in an altered environment. If a yeast is haploid in nature, mating generates phenotypic variation through changes in the combination of the alleles of different genes. If a yeast is diploid in nature, phenotypic variation is generated not only through such combinations but also through changes in the combination of alleles at the same locus. The basic assumption is that whether frequent or rare, sex is essential for the long-term survival of the species. It is, therefore, assumed that the primary role of mating is recombination and in turn that sexual systems are maintained during evolution, primarily because of their essential role in adaptation. But in the face of this logic, one is confronted with the overwhelming results of population studies of the modes of propagation. These studies reveal time after time that the population structure of many pathogenic yeasts, including *Candida albicans* and *Candida glabrata*, is primarily clonal (62). In the majority of these population studies, there have been strong indications of low levels of recombination, but it is not at all clear when these events occurred or how frequently they continue to occur in natural popula-

tions. It is, therefore, reasonable to wonder whether *C. albicans* or *C. glabrata* could survive without recombination resulting from mating, or why their mating systems are maintained, if recombination is in fact rare or incidental. Indeed, there is every reason to believe that *Candida parapsilosis*, the fourth most prevalent yeast pathogen, has actually begun to lose its mating system (50).

Two general explanations can be considered for the maintenance of mating systems in the pathogenic yeasts. First, it may be that even though mating is rare, it plays a long-term role in species maintenance and, therefore, is continuously selected for. We may in fact have no realistic perception of the time frame in which selection functions to maintain mating systems or of whether selection is continuous or sporadic. Alternatively, it may be that the mating process plays an additional and more immediate role than simply providing rare recombinants in natural populations. If, for instance, mating played a direct and immediate role in pathogenesis, then the selective pressure to maintain it would be continuous and immediate. In chapter 5, A. E. Tsong, B. B. Tuch, and A. D. Johnson have reviewed the details of the mating-type-like (*MTL*) locus in *C. albicans* and discussed its evolution in the hemiascomycetes. In this

David R. Soll and Karla J. Daniels, Department of Biological Sciences, The University of Iowa, Iowa City, IA 52242.

chapter, this discussion continues, focusing on the relationships between mating, switching, and pathogenesis in *C. albicans* and the comparative aspects of the mating systems in *C. albicans*, *C. dubliniensis*, *C. parapsilosis*, and *C. glabrata*. In particular, mating is considered within the context of the pathogenic lifestyles of these organisms.

POPULATION STRUCTURE INDICATES RARE RECOMBINATION BETWEEN STRAINS OF *C. ALBICANS*

Until 1999, it was generally assumed that *C. albicans*, the most pervasive yeast pathogen, was asexual. This assumption was based primarily on negative data, the absence of an *MTL* locus and the absence of credible observations of mating-like fusion. Studies of population structure, however, consistently suggested that genetic exchange occurred frequently enough to have a slightly disruptive effect on patterns of clonal propagation (24, 62). Pujol et al. (65), employing multilocus enzyme electrophoresis at 13 polymorphic loci, provided the first evidence for departures from random segregation at single loci and recombination between loci in a study of 55 *C. albicans* isolates. They concluded that the most parsimonious hypothesis for their results was that *C. albicans* had a predominantly clonal population structure. Since the Pujol et al. (65) study, more than 10 critical studies of the population structure of *C. albicans* have been performed, employing a variety of independent genetic fingerprinting methods (62). All of these studies supported the conclusion that *C. albicans* had a clonal mode of reproduction, but in every report in which a significant number of loci were tested against Hardy-Weinberg expectations for departure from segregation, roughly one-third of the loci exhibited genetic frequencies compatible with panmixia. Anderson et al. (3) found that recombination occurred in mitochondrial DNA and identified 3 of 48 strains with phylogenetically incompatible nuclear and mitochondrial DNA polymorphisms. Forty-five of the 48 strains had congruent mitochondrial and nuclear DNA polymorphisms. These results again led to the conclusion that the large majority of strains propagated clonally but that rare recombinatorial events did occur. Pujol et al. (66) further demonstrated that independent genetic fingerprinting methods separated a collection of unrelated strains into the same genetically related groups (clades). The description of clades continues to develop (10, 80). Strains from the different clades were found to inhabit the same geographical niches side by side, suggesting that genetic exchange between strains of different clades was possible, but rare. The combined results on population studies through the 1990s did not significantly contradict the notion, based on the absence of a mating-type locus, that *C. albicans* was asexual (62).

DISCOVERY OF THE *MTL* LOCUS IN *C. ALBICANS* AND CHANGED VIEWS

The identification of a locus in *C. albicans* containing genes homologous to the mating-type genes (*MAT* genes) of *Saccharomyces cerevisiae* (31) changed our perception of the sexuality of *C. albicans*. Rather than continue to emphasize the general clonality of populations, interest turned to the low level of recombination. The possibility was entertained that *C. albicans* **a** and α strains mated in the host. But when the unique features of the *C. albicans* mating system began to emerge, it became evident that mating in this pathogen might not be so straightforward.

In contrast to the cassette system in *S. cerevisiae*, which is made up of three mating-type loci, two silent (*HML* and *HMR*) and one expressed (*MAT*) (Fig. 13.1A) (25, 27), *C. albicans*, which apparently is an obligate diploid, contained one *MTL* locus on chromosome 5 (Fig. 13.1B) (31). The original strain that Hull and Johnson (31) analyzed, strain CAI8, was an auxotrophic derivative of strain SC5313, isolated from a patient with disseminated candidiasis in 1984. The *MTL* copy on one homolog of chromosome 5 in strain CAI8 harbored genes *MTLα1* and *MTLα2*, which are homologous to *S. cerevisiae MATα1* and *MATα2*, and the *MTL* locus on the other chromosome 5 homolog harbored gene *MTLa1*, which was homologous to *S. cerevisiae MATa1*, and gene *MTLa2*, which had no *S. cerevisiae* homolog (31, 81). The configuration of the *MTL* locus in *C. albicans* suggested a different scenario for achieving mating competency from that elucidated in *S. cerevisiae* (Fig. 13.1C). Because *C. albicans* was an obligate diploid, it would have to become **a**/**a** or α/α to mate, and do so by *MTL* homozygosis (Fig. 13.1D), which would contrast markedly with **a**/α *S. cerevisiae*, which undergoes meiosis to haploid **a** or α to achieve mating competency. Furthermore, if **a**/**a** and α/α cells fused in the mating process, the *C. albicans* zygote would be tetraploid (Fig. 13.1D), in contrast to the diploid zygote that results from *S. cerevisiae* mating (Fig. 13.1C). And as we shall see, that was not where the unique aspects of the mating process of *C. albicans* ended.

THE UNIQUE DEPENDENCY OF MATING ON SWITCHING IN *C. ALBICANS*

Soon after identification of the *MTL* locus in *C. albicans*, Hull et al. (32) used complementation of auxotrophic markers to detect cell type-dependent fusion

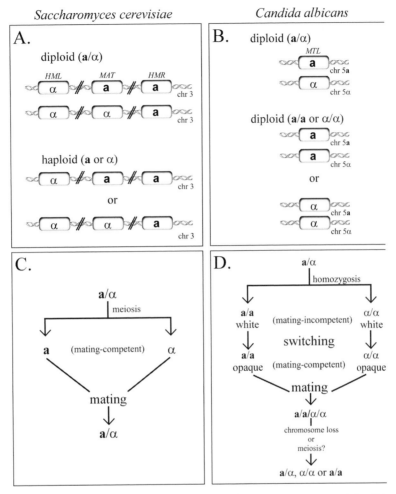

Figure 13.1 A comparison of the mating systems of *Saccharomyces cerevisiae* and *Candida albicans*. (A and B) The respective configurations of mating loci. (C and D) The respective steps in the mating programs and acquisition of mating competence. In panel D, the question mark next to meiosis indicates that the process has not been demonstrated.

between auxotrophic **a** and α strains in a mouse model, and Magee and Magee (51) used a similar selection strategy to identify fusion in vitro. In both studies, however, the frequency of fusion seemed low. Miller and Johnson (55) then made two extraordinary observations that explained why the frequencies of mating in the original two studies were low, and which revealed the unique role switching played in mating. They found that whereas the **a**/α laboratory strain CAI8 was incapable of undergoing white-opaque switching, both *MTL***a** and *MTL*α deletion strains, which were genetically α/− and **a**/−, respectively, were enabled for white-opaque switching. They demonstrated that the **a**1-α2 repressor complex not only repressed mating in **a**/α cells, but also repressed white-opaque switching. They further demonstrated that when **a**/− and α/− cells were mixed in the opaque phase, they

fused at frequencies many orders of magnitude higher than **a**/− and α/− cells mixed in the white phase (55). Their results indicated, therefore, that in order to switch from white to opaque, an **a**/α cell had to undergo *MTL*-homozygosis to **a**/**a** and α/α, and in order to mate, *MTL*-homozygous cells then had to switch from white to opaque (Fig. 13.1D). Lockhart et al. (44) then demonstrated that a majority of natural strains were **a**/α and that these strains also had to undergo *MTL*-homozygosis to either **a**/**a** or α/α to switch from white to opaque, thus generalizing the observations of Miller and Johnson (55). Lockhart et al. (45) further demonstrated, using microscopically identified cell-cell fusion as an assay, that for a number of unrelated natural strains from different clades, **a**/**a** and α/α cells fused only when expressing the opaque phenotype (Fig. 13.2).

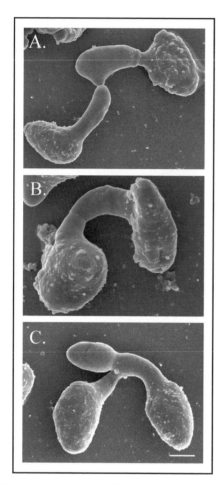

Figure 13.2 Scanning electron micrographs of select stages in the mating process of *Candida albicans*. (A) Contact of conjugation tubes; (B) fusion of conjugation tubes; (C) daughter bud formation from conjugation bridge. Scale bar, 1 μm.

SO WHAT IS THE WHITE-OPAQUE TRANSITION?

The discovery that *C. albicans* had uniquely inserted into its mating process a complex phenotypic transition from white to opaque raised a fundamental question: why? *S. cerevisiae* underwent a superficially similar mating process but did not have to undergo the white-opaque transition, or anything resembling it, to mate. Could the inclusion of the white-opaque transition in the *C. albicans* mating system have something to do with its pathogenic lifestyle, or more specifically, with the intimate relationship *C. albicans* had evolved with its animal host? To explore this issue, we must first consider the unique features of the white-opaque transition (Fig. 13.3A).

In 1987, Slutsky et al. (70) first identified this transition in a systemic isolate (WO-1) from a bone marrow transplant patient at the University of Iowa Hospitals

and Clinics. The patient subsequently died of this infection. Slutsky et al. (70) found that when cells from a white colony were clonally plated, $\sim 10^{-3}$ formed opaque colonies, and when opaque cells from an opaque colony were clonally plated, $\sim 10^{-3}$ formed white colonies. We now know that WO-1 is α/α and, therefore, undergoes white-opaque switching (Fig. 13.3C). In their original description, Slutsky et al. (70) reported that opaque cells were elongate or bean-shaped and twice as large as white cells but still contained one nucleus and approximately the same amount of DNA. They also demonstrated that switching was reversible and that both extreme cold (25°C) and heat (>35°C) induced opaque cells to switch to white.

In the same year, Anderson and Soll (1) identified by scanning electron microscopy (SEM) a unique morphological feature of the opaque state, the cell wall "pimple" (Fig. 13.3E). They also demonstrated the presence of an opaque-specific antigen distributed in a punctuate fashion on the wall. Employing transmission electron microscopy (TEM) and indirect immunogold labeling, Anderson et al. (2) demonstrated that channels traversed the pimples, from their distal surface to the plasma membrane (Fig. 13.3F), and that the opaque-specific 14.5-kDa antigen localized at the pimple apex (Fig. 13.3G).

In the early 1990s, several genes were identified that were expressed only in the opaque phase (e.g., *OP4* and *SAP1*) or the white phase (e.g., *WH11*) (30, 57, 58, 75, 84). Through the 1990s and early 2000s, the list of opaque- and white-specific or enriched genes continued to accumulate (71, 72), culminating in a microarray analysis of opaque- and white-regulated genes by Lan et al. (42), which identified over 300 genes regulated by the white-opaque transition, approximately 6% of the gene repertoire of *C. albicans*. The unique and complex nature of the transition was further revealed in studies of promoter regulation (40, 46, 76, 77). Functional analyses of the promoters of phase-specific genes revealed coordinate regulation by different upstream regulatory sequences, indicating that different *trans*-acting factors were responsible. Hence, the switch from one phase to the other did not activate genes in the second phase through a single *trans*-acting factor, but rather through multiple factors, leading to complex models of gene regulation (71, 72). Recently, a putative master switch gene, referred to as *TOS9* or *WOR1*, was identified that plays a key role in the switch event (29, 74, 89). Deletion of this gene blocks switching, locking cells in the white phase. Misexpression in the white phase causes mass conversion to opaque, and misexpression of the gene in opaque cells inhibits temperature-induced mass conversion to white.

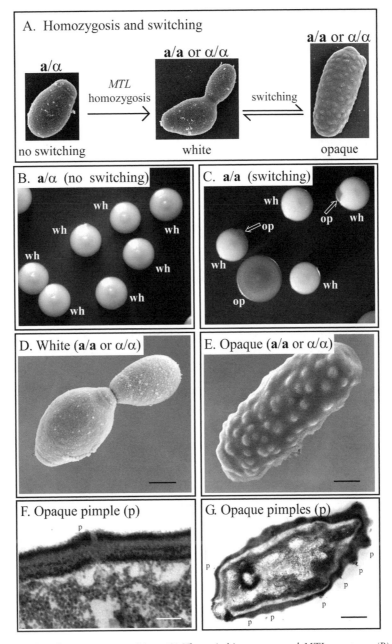

Figure 13.3 The white-opaque transition. (A) The switching system and *MTL* genotype; (B) a/α cells don't switch; (C) a/a (or α/α) cells switch (op, opaque; wh, white); (D) SEM of white cell; (E) SEM of opaque cell; (F) TEM of pimple channel; (G) TEM of pimples around cell surface. Scale bars, 1 μm in panels D, E, and G and 100 nm in panel F.

Since temperature-induced mass conversion from opaque to white depends first on a decrease in the *TOS9 (WOR1)* protein and then on a round of DNA replication, a two-step model has been proposed involving levels of the gene product above (opaque) and below (white) a threshold and a change in chromatin state at an as-yet-unidentified locus (*GENE-X*) (74).

THE TEMPERATURE PARADOX

With the preceding review of white-opaque switching as background, we can now return to the more salient question related to the present discussion. Why has *C. albicans* invested so much in inserting the white-opaque transition into the mating process, when *S. cerevisiae* mates readily in the absence of such a transition?

Switching becomes even more paradoxical when one considers that the opaque phenotype is unstable at physiological temperature. If opaque cells are transferred from 25 to ≥37°C, they mass convert to white (67, 70, 71, 75). Upon increasing the temperature from 25 to ≥37°C, cells retain their opaque identity for 4 to 6 h and then mass convert semisynchronously to the opaque phenotype concurrently with the second cell doubling, which depends on a second round of DNA replication (71, 75). The kinetics of commitment in the population follow a pattern consistent with an imprinting model in which a temperature-induced change in chromatin state occurs at the site of the switch event, from a dominant opaque to recessive white state (71). The shift in temperature from 25 to ≥37°C results in an immediate cessation of opaque phase-specific gene transcription, including *TOS9* expression (74), and the return of these cells to 25°C prior to the point of phenotypic commitment blocks the switch and reactivates these genes within 1 h (58, 74). A return to 25°C after the point of commitment to the white phenotype, however, does not reactivate opaque phase-specific genes.

At face value, the sensitivity of opaque phase cells to physiological temperature seemed inconsistent with mating in the host. The life history of *C. albicans* appears to have become intimately tied to its host. *C. albicans* lives as a commensal in healthy hosts, ready to become pathogenic in response to one or more predisposing changes in host physiology. One would therefore expect *C. albicans* to mate in the host, but this seemed unlikely given the temperature sensitivity of the opaque phenotype. Observations made in the late 1990s on the differences between white and opaque in the colonization of skin appeared to provide a possible answer. While white cells of strain WO-1 (α/α) were highly virulent in the mouse model for systemic infection, opaque cells were not (36). This seemed consistent with the sensitivity of opaque cells to physiological temperature. Indeed, when mice that had been injected with opaque cells expired or were sacrificed, the majority of yeast cells in the kidneys had switched to white. But the opposite occurred in the mouse model for cutaneous (skin) infection. While opaque cells readily colonized the skin of newborn mice (Fig. 13.4B), white cells did not (Fig. 13.4A) (37). The temperature of mouse and human skin is approximately 32°C, a temperature that readily supports the opaque phenotype. Lachke et al. (39) therefore hypothesized that skin might represent a body niche in which mating could occur. Experiments in which opaque **a/a** and opaque α/α cells were mixed on the skin of newborn mice supported the hypothesis (39). In the opaque phase, mixtures of **a/a** and α/α cells colonized skin and fused (Fig.

13.4C). The fusion frequencies were high, and on some skin patches involved more than 50% of cells. While this seemed to provide an explanation for why opaque cells were mating competent but sensitive to physiological temperature, there were still two caveats. First, while skin supported shmooing, conjugation tube growth, and fusion, mating on skin rarely culminated in the formation of a daughter bud. Perhaps mating was initiated on skin but completed elsewhere. Perhaps fusants would have to be transported from skin to the mouth or genitalia in order to complete mating. This possibility would incorporate host behavior (e.g., transport by hand from skin to mouth) in the *C. albicans* mating program. The second caveat was that skin represented a minor host niche for *C. albicans* commensalism and pathogenesis. Why, then, would fusion be limited to skin in the evolution of the host-pathogen relationship of *C. albicans*? It seemed likely that more remained to this story.

FUSION AND GENE REGULATION: SOME POSSIBLE CLUES

Because one might assume that *C. albicans* mates within the context of host-pathogen interactions, were there any unique features of the system which could provide clues to explain why the white-opaque transition was inserted as an essential step in the process? Comparisons with the *S. cerevisiae* mating system provided a possible clue. At the cellular level, the *C. albicans* mating process appeared highly similar to that of *S. cerevisiae* (8, 45). α-Pheromone stimulated **a**/− or **a/a** cells, and **a**-pheromone stimulated α/− or α/α cells, to shmoo. Although the responding opaque cells of *C. albicans* were morphologically quite distinct from the respective **a** and α cells of *S. cerevisiae*, pheromone induced similar polarization and evagination (Fig. 13.2A), which represent the initial steps in conjugation tube formation. The tubes of opposite mating types of both species fused end to end to form a zygote in which the mating cells were connected by a conjugation bridge (Fig. 13.2B). In both species, the nuclei of the parent cells migrated into the bridge and fused. A daughter bud grew from the bridge (Fig. 13.2C). The fused nucleus then divided, and one daughter nucleus entered the daughter cell. For *S. cerevisiae*, the signal transduction pathway initiated by pheromone-receptor interactions has been carefully described (4, 69), as well as the genes activated downstream (68).

In 2002, two studies (15, 52) provided evidence that a similar pathway regulated the mating response in *C. albicans*. Deletion of *C. albicans* homologs of genes along this pathway blocked or affected mating. Furthermore, several transcription-profiling studies revealed

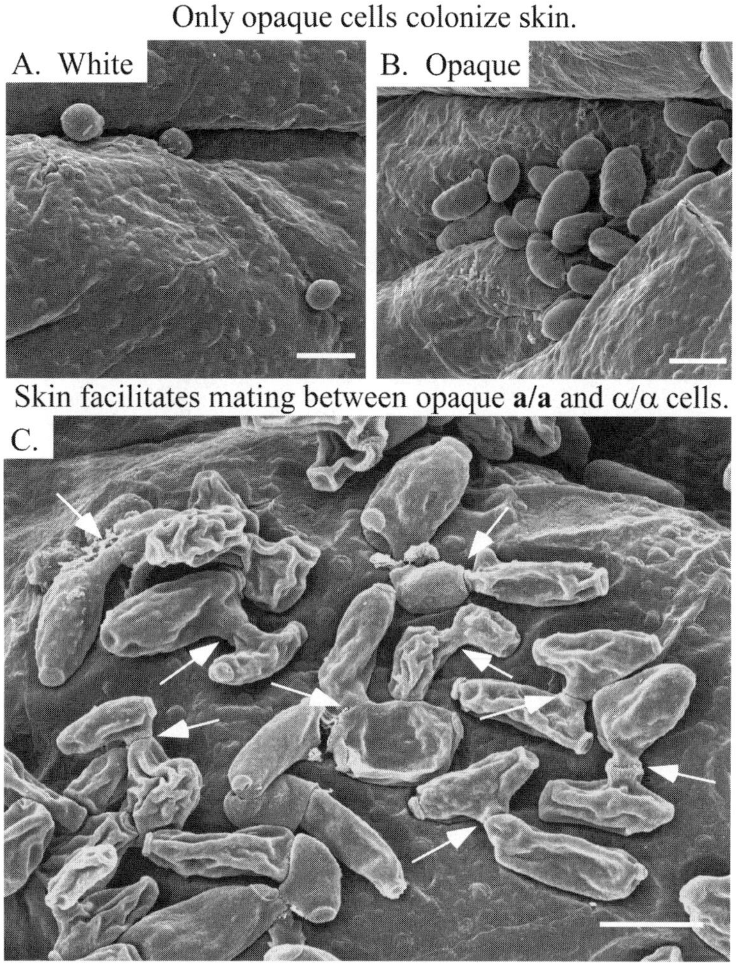

Only opaque cells colonize skin.

A. White   B. Opaque

Skin facilitates mating between opaque **a/a** and α/α cells.

C.

Figure 13.4 Skin facilitates opaque-cell colonization and mating. (A) White cells poorly colonize skin. (B) Opaque cells densely colonize skin. (C) Opaque **a/a** and α/α cells mate on skin at high frequency. See Kvaal et al. (37) and Lachke et al. (39) for details. Scale bars, 5 μm.

that many of the genes involved in the signal transduction pathway, in the maturation, transport, and release of pheromone, and in adaptation to the pheromone signal, were up-regulated in pheromone-treated opaque cells (7, 49, 88). Of 15 mating-associated genes up-regulated by α-pheromone in *S. cerevisiae* **a** cells, 14 were also up-regulated by α-pheromone in *C. albicans* **a/a** cells. Several mating-associated genes that had not been observed to be up-regulated by pheromone in *S. cerevisiae* were found to be up-regulated in *C. albicans*. What may, however, prove to be more interesting is the regulation of two groups of genes in *C. albicans* without parallels in *S. cerevisiae*. First, pheromone was found to up-regulate 13 genes that were either up-regulated or associated with filamentation and to down-regulate 1 gene that was down-regulated during filamentation (7, 88). Ten of the 14

genes had no known homolog in *S. cerevisiae*. Why would the mating factors have such an extensive effect on filamentation genes? The answer might be that filamentation also plays a role in *C. albicans* mating. Consistent with this notion, it was observed in the first studies of the cell biology of *C. albicans* mating that the conjugation tubes could grow remarkably long (7, 18, 45). Zhao et al. (88) demonstrated that under similar conditions of α-pheromone induction, *S. cerevisiae* cells formed shmoos with very short conjugation tubes (Fig. 13.5A), while *C. albicans* formed long conjugation tubes that were reminiscent of hyphae (Fig. 13.5B). Why would *C. albicans* require long conjugation tubes in the mating process? This issue is returned to in a later section of this chapter.

The second group of genes unique to *C. albicans* mating are opaque specific (45, 88). *OP4*, *SAP1*, and

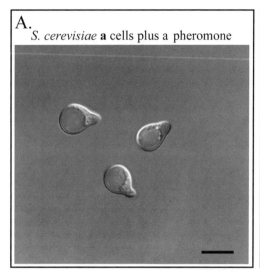
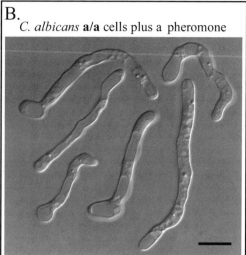

Figure 13.5 α-Pheromone treatment causes shmooing and blunt conjugation tube formation in *S. cerevisiae* **a** cells (A). It causes shmooing and long conjugation tube formation in *C. albicans* (B). Scale bars, 5 μm.

SAP3 are differentially expressed in opaque cells. These genes are down-regulated in opaque **a/a** cells treated with α-pheromone. *WHII* is differentially expressed in white cells (75). It is not up-regulated in pheromone-treated opaque cells (45, 88). The latter observation indicates that selective down-regulation of opaque-specific genes does not represent a pheromone-induced switch to white. These results are intriguing because they suggest that to become mating competent, an **a/a** or **α/α** cell must switch to opaque, but part of the subsequent response of such mating-competent cells to pheromone is the down-regulation of a number of opaque-specific genes.

BUT WHY SWITCH? AN UNEXPECTED RESULT PROVIDES A CLUE

As the complexity of the mating system of *C. albicans* continued to unfold, the inclusion of the complex white-opaque switch made less sense. Why were *S. cerevisiae* cells immediately mating competent upon meiosis from **a/α** to **a** or **α**, while *C. albicans*, upon *MTL* homozygosis, had to turn on and off large numbers of phase-specific or enriched genes and then change every aspect of cellular architecture in the white-to-opaque transition to be mating competent? Why had the *C. albicans* system not evolved so that cells were immediately mating competent upon becoming *MTL* homozygous? And if the opaque phenotype was required for mating for some as-yet-unidentified reason, why didn't the system evolve so that cells were immediately opaque after *MTL*

homozygosis? In the same line of questioning, what role did the *MTL*-homozygous white phase play? Was it simply a transition phenotype to opaque and mating?

Possible answers to the preceding questions emerged from an unexpected result. In their analysis of α-pheromone-induced gene expression in opaque **a/a** cells, Lockhart et al. (49) performed a presumptive negative control, namely, α-pheromone treatment of white **a/a** cells. Their expectation that this was a negative control was based on the observation that white cells neither released pheromone nor shmooed nor mated when mixed with opaque cells of opposite mating types (45, 55). Surprisingly, Lockhart et al. (49) observed that α-pheromone selectively up-regulated expression of *CAG1*, *STE4*, and *STE2* in white cells to approximately the same induced levels as in opaque cells. α-Pheromone did not, however, up-regulate *FIG1* and *KAR4* in white cells, as it did in opaque cells. These results, although reproducible, were met with skepticism. The experiment was, therefore, repeated and expanded by Daniels et al. (19), who added to the list of genes up-regulated by α-pheromone in white cells *CEK2*, *CEK1*, and *SST2*. Daniels et al. (19) also verified that α-pheromone had no detectable effect on the cell cycle of white **a/a** cells, while blocking opaque **a/a** cells in G_1, and caused no polarization or conjugation tube formation (shmooing) in white **a/a** cells, as it did in opaque **a/a** cells (7, 45, 49, 55). Daniels et al. (19) also demonstrated that white cells, like opaque cells, possessed pheromone receptors, although the distribution of these receptors and the redistribution of receptors resulting from receptor occupancy differed from that of opaque cells. Why would white cells, which are mating incompetent, possess pheromone receptors and exhibit a partial response to pheromone? And why would opaque cells signal white cells through the release of pheromone? Could it be that this unexpected result provided a key to understanding why the white-opaque transition was included as an essential step in mating and to elucidating the role the white phenotype played in the basic biology of *C. albicans*?

OPAQUE CELLS SIGNAL WHITE CELLS TO FORM BIOFILMS

Daniels et al. (19) speculated on a possible role for opaque-cell signaling of white cells, from the perspective that it must serve some purpose in mating. They knew that in nature 10% of established strains were *MTL* homozygous and, hence, could switch and mate. They hypothesized that on occasion those strains would overlap in a host niche, expressing the dominant white phenotype at 37°C (Fig. 13.6A). White cells would spontaneously switch on rare occasions to opaque (Fig. 13.6A). If this event hap-

pened in overlapping **a/a** and α/α opaque-cell populations within signaling distance, cells would be blocked in G₁ and therefore unable to switch back to white as a result of high temperature (Fig. 13.6B). The **a/a** opaque cell, through release of a-pheromone, would then induce polarization and shmooing in the α/α opaque cell, and the α/α opaque cell, through release of α-pheromone, would induce polarization and shmooing in the **a/a** opaque cell (Fig. 13.6C). Opposing gradients of the a-pheromone and α-pheromone would be generated to direct conjugation tubes of the α/α and **a/a** cells, respectively, to migrate towards each other, a process referred to as "chemotropism" (Fig. 13.6C). Contact would result in end-to-end fusion (Fig. 13.6D). Because the switch events are rare, the **a/a** and α/α conjugation tubes would on average be separated by relatively long distances, which would explain why pheromone upregulates filamentation-associated genes for the generation of very long conjugation tubes. Within such a scenario, the opposing pheromone gradients would have to extend over long distances and would have to be stable for extended periods of time. Such standing soluble gradients would be prone to disruption by vibrations and to dissipation

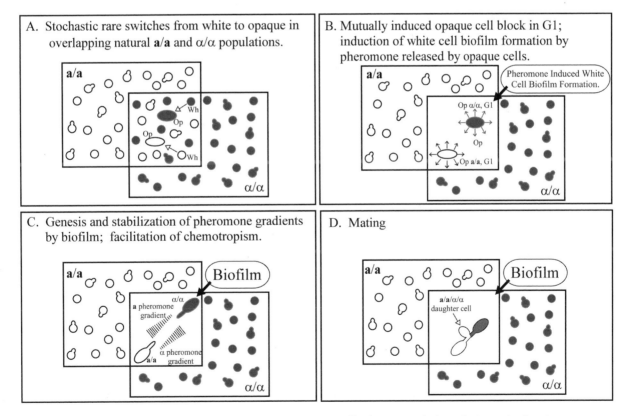

Figure 13.6 A hypothesis that switching is involved in biofilm formation. The hypothesis was developed by Daniels et al. (19) that in overlapping **a/a** and α/α populations, rare **a/a** and α/α opaque cells signal majority white cells of opposite mating types to form a biofilm that facilitates mating by stabilizing pheromone gradients.

through diffusion. Daniels et al. (19) considered that in separating molecules by electrophoresis, problems of diffusion and mechanical disruption were solved by separating molecules in polyacrylamide or agarose gels. By analogy, Daniels et al. (19) hypothesized that opaque cells signaled majority white cells through the release of pheromone to form a white-cell biofilm that stabilized the opposing pheromone gradients over extended distances and for extended time periods to allow chemotropism to proceed without interruption (Fig. 13.6C). Because biofilms play such a fundamental role in both bacterial and fungal pathogenesis, this hypothesis tied together for the first time switching, mating, and pathogenesis. It also provided a possible explanation for why so rare an event as mating was maintained during the rapid evolution of a pathogen.

To test their hypothesis, Daniels et al. (19) first assessed whether pheromone selectively induced cohesiveness in suspension between white, but not opaque, cells and whether it induced the formation of a monolayer of cells adhering to a substratum, the first step in biofilm formation (26, 59). In both cases they found that it did. In Fig. 13.7A, it is clear that α-pheromone induces white, but not opaque, cells to form an incipient biofilm on a plastic surface (19). Second, Daniels et al. (19) assessed whether a minority of opaque cells enhanced biofilm maturation in a majority of opaque cells. They found that 1% opaque cells doubled the thickness of a majority white-cell biofilm (19). Finally, they tested whether a white-cell biofilm promoted chemotropism between minority opaque a/a and opaque α/α cells. It did. In 85% of cases in which an opaque a/a and an opaque α/α cell were within 25 cell diameters in a protected three-dimensional white-cell biofilm, the respective conjugation tubes were oriented towards each other (Fig. 13.7B). In an unprotected two-dimensional white-cell biofilm, however, only 15% were correctly oriented in the monolayer at the biofilm edge; the majority were randomly oriented (Fig. 13.7D). While the results of all of the experiments performed by Daniels et al. (19) supported different aspects of their hypothesis, none proved that this process occurred in vivo or that it played a role in the general pathogenesis of *C. albicans* in the human host. They did, however, provide the first indication that switching and mating might play a role in pathogenesis.

MAINTENANCE OF THE C. *ALBICANS* MATING SYSTEM: CHROMOSOME 5 HETEROZYGOSITY AND VIRULENCE

In nature, roughly 10% of *C. albicans* strains are *MTL* homozygous (a/a or α/α) while the majority is *MTL* heterozygous (a/α) (43, 44). Lockhart et al. (44) first demonstrated that natural a/α strains underwent spontaneous *MTL* homozygosis in vitro, in some strains at extremely high frequency. Wu et al. (86) subsequently demonstrated that the great majority of these strains did so by the loss of one chromosome 5 homolog followed by duplication of the retained homolog, resulting in the condition of "uniparental disomy." Growth on minimal medium containing sorbose induces chromosome loss (34) and can lead to *MTL* homozygosis (6, 33, 51). Hence, one may assume that there is in nature a robust avenue of *MTL* homozygosis leading to a/a and α/α strains.

But is there as robust an avenue from the *MTL*-homozygous state to a/α? As noted, studies of population structure indicate that propagation of *C. albicans* is primarily clonal, suggesting in turn that mating as an avenue for *MTL* heterozygosis is not robust. Hence, one may wonder why a/α strains, not *MTL*-homozygous strains, predominate in nature, and in turn how *C. albicans* maintains its mating system. *S. cerevisiae* has evolved a unique solution to this problem. It retains a and α information in silent loci and an independent expression locus that is either a or α. The expression locus dictates mating type. Hence, a haploid a cell retains α information and an α cell retains a information. The expression locus can switch from a to α or vice versa by recombination between a DNA copy of the silent locus of opposite mating type (i.e., gene conversion) and the expression locus. When *C. albicans* becomes a/a or α/α, it loses alternative mating-type information. The only way back is by mating.

In considering this conundrum, Lockhart et al. (48) set forth the hypothesis that the predominance of a/α strains in nature may be accomplished through competition. Natural a/α strains may be more competitive in colonizing host niches than their a/a or α/α offspring. Therefore, if spontaneous a/a or α/α offspring did not mate, they would be diluted out of the population by their more competitive parent strains. To test this hypothesis, Lockhart et al. (48) first compared virulence in the murine model for systemic infection. They generated survival curves for mice injected with either the parent a/α strain or a spontaneous *MTL*-homozygous offspring of that strain. Second, they performed competition experiments between cells of the parent a/α strain and cells of the spontaneous *MTL*-homozygous offspring, by coinjection, and analyzed the contribution of competing strains in kidney colonization. For the unrelated natural a/α strains P37037, P37039, and P5063, spontaneous *MTL*-homozygous offspring cells proved far less virulent than parent a/α cells in single-strain injection experiments (Fig. 13.8A, B, and C). Furthermore, in competition experiments, parent strains represented the majority genotype in the infected kidneys. Lockhart et al. (48) tentatively concluded from these results that

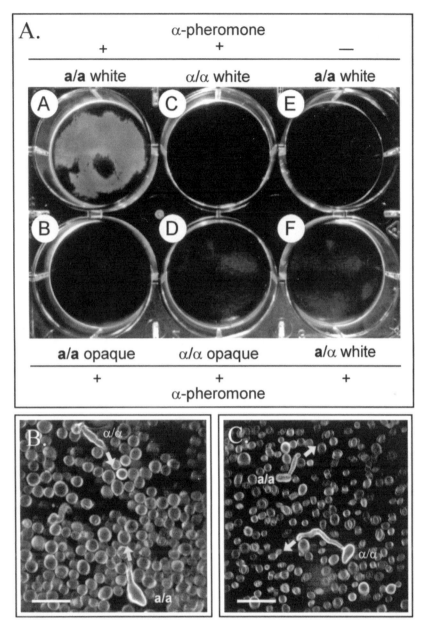

Figure 13.7 Opaque cells induce white cells through pheromone to form a biofilm that facilitates mating. Proof for some of the steps in the hypothesis of Daniels et al. (19). (A) α-Pheromone induces **a/a** white cells, but not **a/a** opaque cells, to form a biofilm on a plastic surface of a multiwell plate. + and −, presence and absence, respectively, of chemically synthesized 13-mer α-pheromone. (B) Facilitation of chemotropism between a rare **a/a** and α/α opaque cell in a three-dimensional biofilm. Arrows denote direction of conjugation tube growth. (C) Lack of chemotropism in the two-dimensional layer of cells at the edge of a three-dimensional biofilm. Scale bars, 5 μm.

the mating type genotype may regulate virulence and found that the addition of *MTLα2* to a spontaneously generated **a/a** offspring strain increased competitiveness in mixed infections with the parent **a/a** strain, supporting this conclusion.

However, Wu et al. (86), who analyzed polymorphisms (i.e., heterozygosities) along chromosome 5,

found that of nine spontaneous *MTL*-homozygous offspring, several of which had been compared with parent **a/α** strains in the Lockhart et al. (48) study, eight were generated by loss of one chromosome 5 homolog followed by duplication of the remaining homolog, while only one was generated by mitotic recombination. None were generated by gene conversion. Hence, the former

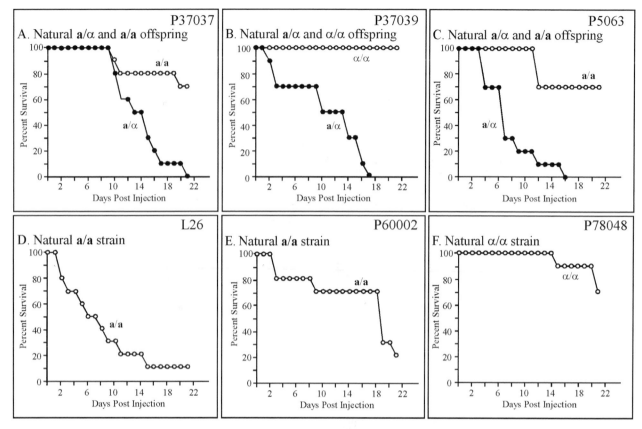

Figure 13.8 While the great majority of natural **a**/α strains of *C. albicans* are virulent, the great majority of spontaneous *MTL*-homozygous offspring are relatively avirulent, and natural *MTL*-homozygous strains run the gamut of virulence. Virulence was measured in the mouse model for systemic infection of natural **a**/α cells (A through C), *MTL*-homozygous offspring of natural **a**/α cells (A through C), and natural *MTL*-homozygous strains (D through F). Strain names are in the upper right-hand corner of each panel. Each panel contains data for host survival over time (i.e., the percentage of injected mice still alive without signs of extreme morbidity). See Lockhart et al. (48) and Wu et al. (87).

eight *MTL*-homozygous strains had undergone homozygosis for all genes along chromosome 5, while the latter single strain had undergone homozygosis for approximately 40% of the genes on chromosome 5 (86). The loss of polymorphisms along chromosome 5 resulting from uniparental disomy is diagrammed in Fig. 13.9A, B, and C. The discovery by Wu et al. (86) led to an alternative explanation for the loss of virulence and competitiveness among spontaneous *MTL*-homozygous offspring, namely, that the loss of heterozygosity for genes along chromosome 5 other than those harbored at the *MTL* locus was the reason for the loss of virulence.

Wu et al. (87) subsequently deleted *MTLa1* or *MTLα2* in five natural **a**/α strains freshly collected from patients and then compared virulence between the **a**/α strains, spontaneous **a**/**a** or α/α offspring derived from these strains, and *MTLa1* or *MTLα2* deletion derivatives of these natural strains. They found that for **a**/α

strains exhibiting normal levels of virulence in the mouse model for systemic infection, the virulence of the deletion derivatives was more comparable to that of the parent **a**/α strains than to that of the relatively avirulent spontaneous *MTL*-homozygous offspring (87). The parent **a**/α and *MTLa1* or *MTLα2* deletion derivatives shared heterozygosities at non-*MTL* genes along chromosome 5. These results, therefore, supported the conclusion that the heterozygosity of one or more genes along chromosome 5 other than those harbored at the *MTL* locus was the major factor imparting virulence in **a**/α strains.

The results of Wu et al. (87) should not be construed as challenging the hypothesis formulated by Lockhart et al. (48) that natural **a**/α strains predominate in nature because they are more competitive than their spontaneous **a**/**a** and α/α offspring in colonizing the host. By losing the heterozygosity of genes other than those at

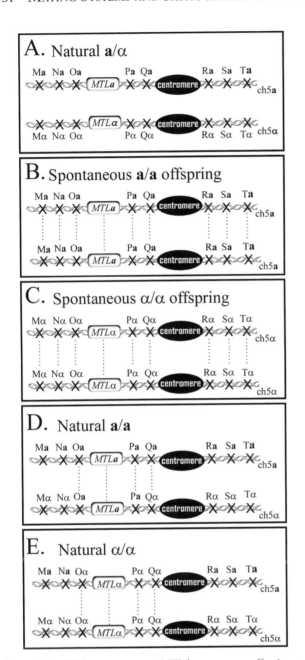

Figure 13.9 While spontaneous *MTL*-homozygous offspring are generated from natural **a**/α strains by loss of one homolog of chromosome 5 followed by duplication of the retained homolog, resulting in uniparental disomy, natural *MTL*-homozygous strains arise by mitotic recombination. Heterozygosities are presented along chromosome 5 for natural **a**/α strains (A), spontaneous **a**/**a** offspring of natural **a**/α strains (B), spontaneous α/α offspring of natural **a**/α strains (C), natural **a**/**a** strains (D), and natural α/α strains (E). Vertical dotted lines refer to homozygosity. Details from Wu et al. (87).

demonstrated that tetraploids generated by mating were less virulent than their parent strains, suggesting that the fusion products of the mating process can regain virulence only through a return to the diploid state.

One caveat, however, still clouded this story. If spontaneous **a**/**a** and α/α offspring lost competitiveness, then what was the explanation for the success of natural **a**/**a** and α/α strains, which make up approximately 10% of all *C. albicans* strains in nature (48, 80)? If the hypothesis of Lockhart et al. (48) were correct, then strains collected from patients that were *MTL* homozygous, and hence established in nature, should be heterozygous for the other genes along chromosome 5 that impart virulence. To test this prediction, Wu et al. (87) set out to compare the heterozygosity of genes along chromosome 5 between natural *MTL*-homozygous strains and natural *MTL*-heterozygous strains. But first they compared the relative virulence of natural **a**/**a** and α/α strains to that of natural **a**/α strains by single-strain injection experiments in the mouse model for systemic infection. Surprisingly, they found that natural *MTL*-heterozygous strains were on average not as virulent as natural *MTL*-homozygous strains. The virulence of natural *MTL*-homozygous strains in the mouse model ran the gamut from highly virulent (Fig. 13.8D), like *MTL*-heterozygous strains (Fig. 13.8A, B, and C), to relatively avirulent (Fig. 13.8F), like spontaneous *MTL*-homozygous offspring (Fig. 13.8A, B, and C). While 90% of 10 natural **a**/**a** strains killed or caused extreme morbidity in 90 to 100% of mice within 22 days, only 30% of 10 natural *MTL*-homozygous strains did so in that time period. Of seven spontaneous *MTL*-homozygous offspring, 0% did so in that period. Again, if the hypothesis of Lockhart et al. (48) is correct, natural *MTL*-homozygotes should also be less heterozygous on average for genes along chromosome 5 other than those at the *MTL* locus. This was exactly what Wu et al. (87) found. Sites along the first two-thirds of chromosome 5 on the side of the SfiI restriction site that harbored the *MTL* locus and the putative centromere were on average less heterozygous in natural *MTL*-homozygous strains than in natural *MTL*-heterozygous strains, while sites along chromosome 5 on the other side exhibited the same levels of heterozygosity (87). Sites bordering the *MTL* locus exhibited the highest levels of homozygosity in *MTL*-homozygous strains. These results were consistent with the hypothesis that (i) the preponderance of *MTL*-heterozygous strains in nature and maintenance of the mating system are both the result of increased virulence and competitiveness, and (ii) the heterozygosity of chromosome 5 genes other than those harbored at the *MTL* locus is responsible for both increased virulence and competitiveness in the

the *MTL* locus along chromosome 5 through uniparental disomy, spontaneous *MTL* homozygotes become less competitive and for that reason do not accumulate in nature. Ibrahim et al. (33) further colonization of the host. The results of Wu et al. (87) also demonstrated that while the majority of *MTL*-homozygous strains spontaneously generated in natural a/α populations result from the loss of one chromosome 5 homolog followed by duplication of the retained homolog, the majority of *MTL*-homozygous strains established in nature result from apparently complex recombinative events along two-thirds of chromosome 5. There was no clear-cut indication of gene conversion.

WHAT QUESTIONS CONCERNING *C. ALBICANS* MATING MUST STILL BE ANSWERED?

Since *C. albicans* is the most pervasive fungal pathogen colonizing humans, interest by the scientific community in any aspect of its basic biology will wane if there is no connection to pathogenesis. Interest in mating, therefore, might not have been sustained if a link to pathogenesis had not been discovered. That link was strongly suggested by the observations of Daniels et al. (19) that opaque cells signaled white cells to form biofilms to facilitate mating. That link, however, is far from securely established, and several additional questions concerning mating, switching, and pathogenesis remain unanswered. At the time this chapter was written, it was still not known if tetraploid fusants, the product of mating, returned to the diploid state by meiosis. Although orthologs to a number of genes involved in meiosis in *S. cerevisiae* had been identified in the *C. albicans* genome (*IME2*, *IME1*, *RIMIO1*, *MEK1*, and *AMA1*) (82), meiosis itself had not been observed. Orthologs of some key meiotic genes in *S. cerevisiae* were not found in *C. albicans*, but the possibility existed that their functions could have been replaced by other genes. Bennett and Johnson (6) had demonstrated that when tetraploid fusion products, identified by complementation of markers on chromosomes 1, 3, and 5, were incubated at high temperature in presporulation medium formulated for *S. cerevisiae*, cells lost random chromosomes with time. While this nonmeiotic and relatively haphazard mechanism could be the route back to the diploid state, one must wonder if meiosis, a more decisive mechanism, may not occur in nature, especially in light of the observation by Ibrahim et al. (33) that tetraploids are less virulent in the mouse model for systemic infection. Conceivably a decisive meiotic event would more readily reestablish competitiveness. Where might meiosis occur? Could the process have been missed because of the in vitro conditions employed to search for it? Could it occur in biofilms? In vivo strategies may, therefore, be necessary to identify meiosis.

And where and how does mating take place in the human host? Does temperature sensitivity restrict at least the first steps of mating to skin, or does it play a role as an escape mechanism? Could it be that if a spontaneous switch to opaque occurs in an a/a population within a host and it is not accompanied within signaling distance by a switch to opaque by a cell in an overlapping α/α population, temperature drives the cell back to the white phase? Hence, stability of opaque cells in the host is achieved only through a pheromone-induced G_1 block. Without it, temperature causes reversion to white. Methods must, therefore, be established to identify mating in vivo. Is mating restricted to biofilms induced in white populations by rare opaque cells? More interestingly, could mating play a broader role in biofilm formation in majority a/α populations? Could a progression of events in a/α populations be at the basis of biofilm formation that includes (i) *MTL* homozygosis to one or both mating types, (ii) spontaneous switching, (iii) opaque induction of white-cell biofilm formation in cells that have undergone *MTL* homozygosis to the opposite mating type, and finally, (iv) mating?

SWITCHING AND MATING IN *C. DUBLINIENSIS*

C. dubliniensis represents a species highly related to, but distinct, from *C. albicans* (16, 79). It was first identified as atypical *C. albicans* strains in the early 1990s (11, 54, 64) but soon achieved species status based on both genetic and phenotypic characteristics. Consistent genetic differences were found in DNA fingerprinting patterns with complex probes, randomly amplified polymorphic DNA patterns, multilocus enzyme electrophoresis patterns, ribosomal DNA sequences, hybridization patterns with microsatellite DNA sequences and a *C. dubliniensis* species-specific mid-repeat sequence dispersed throughout its genome. A genomic comparison of genes in *C. albicans* and *C. dubliniensis*, using a *C. albicans* microarray, revealed a high degree of similarity (*C. dubliniensis* genomic DNA hybridized to 96% of *C. albicans* microarray sequences), but a significant degree of sequence divergence (56). *C. dubliniensis*, like *C. albicans*, formed chlamydospores, underwent 3153A-like switching, and formed true hyphae. *C. dubliniensis*, however, exhibited both phenotypic and genotypic hypervariability, the latter most notable in electrophoretic karyotypes (35, 63). The general conclusion was, therefore, that *C. dubliniensis* had separated from *C. albicans* quite recently in evolutionary time and was phenotypically and genotypically less stable than *C. albicans*.

Pujol et al. (63) first demonstrated that like *C. albicans*, *C. dubliniensis* underwent white-opaque switching and cell type-dependent mating. They found that natural strains of *C. dubliniensis* contained an *MTL*α locus organized like that of *C. albicans*, and data from the *C. dubliniensis* genome-sequencing project (http://www.sanger.ac.uk/Projects/Candida/dubliniensis/) revealed similarity of the *MTLa* locus. Natural strains included a/α, a/a, and α/α mating locus genotypes, as for *C. albicans*, but Pujol et al. (63) found that the proportion of natural *MTL*-homozygous strains was 33%, a much higher proportion than that of *C. albicans* (4 to 10%) (33, 44, 80), suggesting either a high rate of *MTL* homozygosis (63) or less of an a/α competitive advantage (48, 87). Almost all natural *C. dubliniensis MTL*-homozygous strains were members of the Group I clade. *C. dubliniensis* switched from white to opaque but in several strains rapidly switched back to white at high frequency at colony edges, a phenomenon not observed with *C. albicans* (63). As was the case with *C. albicans*, the appearance of opaque sectors in colonies of a/α strains was due to *MTL* homozygosis. Opaque sectors contained elongate opaque cells with surface pimples, as in *C. albicans*, but unlike *C. albicans*, these opaque cells sometimes grew to extraordinary lengths (63), again a phenomenon not observed in *C. albicans*. As was the case with *C. albicans*, mixing *C. dubliniensis* a/a and α/α cells resulted in shmooing and the formation of long conjugation tubes (63). These tubes fused and daughter buds formed on conjugation bridges. *C. dubliniensis* a/a cells also shmooed in response to chemically synthesized *C. albicans* α-pheromone (63). *C. dubliniensis*, like *C. albicans*, readily fused on the skin of newborn mice.

C. ALBICANS AND C. DUBLINIENSIS MATE IN VITRO

Given that *C. albicans* and *C. dubliniensis* are genetically related and that mating in each species requires a white-to-opaque switch, Pujol et al. (63) tested whether opaque *C. albicans* and *C. dubliniensis* cells could mate. By vitally staining *C. albicans* opaque a/a cells green with fluorescein isothiocyanate-concanavalin A and *C. dubliniensis* opaque a/a cells red with rhodamine-concanavalin A, they demonstrated mating between the two species (Fig. 13.10A and B) and the formation of a daughter cell from the conjugation bridge (Fig. 13.10C, D, E, and F). Hoechst 3342 staining of nuclei demonstrated interspecies karyogamy (Fig. 13.10D).

Maintenance of these two closely related species in the same geographical locales suggests that if mating between them did occur in nature, the offspring might not be competitive. Pujol et al. (63) found that the in vitro frequency of interspecies mating was higher than that of intraspecies mating. If this was the case in nature, one would expect a continuum between the two species, which

C.a. - C.d. fusion and nuclear migration

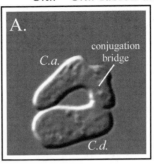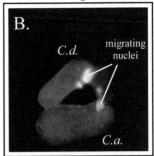

Nuclear fusion, daughter cell formation

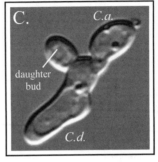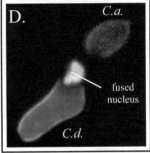

Nuclear division

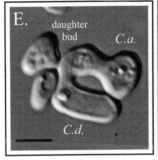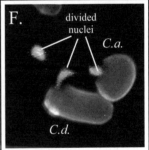

Figure 13.10 *Candida albicans* (*C.a.*) and *Candida dubliniensis* (*C.d.*) can mate in vitro. (A and B) Fusion of conjugation tubes and nuclear migration to the conjugation bridge. (C and D) Nuclear fusion, daughter bud formation. (E and F) Nuclear division with one daughter nucleus migrating into daughter cell. *C. albicans* and *C. dubliniensis* cells were distinguished by vital dye. See Pujol et al. (63) for details. Scale bar, 2 μm.

does not appear to be the case. These results suggest that barriers exist in nature, which are not manifested in vitro to interfere with interspecies mating.

CANDIDA PARAPSILOSIS: POSSIBLE LOSS OF A MATING SYSTEM

Candida parapsilosis may be best known for causing clinical outbreaks, but it has emerged in recent years as one of the four most common agents of yeast bloodstream infections (22, 61). Based upon a number of different genetic measurements, *C. parapsilosis* has proven to be more closely related to *C. albicans* than *S. cerevisiae* in the phylogenetic tree for the hemiascomycetes. As is the case for *C. albicans*, *C. parapsilosis* is probably diploid (50, 83). Although switching systems have been described in *C. parapsilosis*, there have been no reports of the white-opaque transition. It is not clear, however, if it has been properly searched for in this species. To identify the mating system of *C. parapsilosis*, Logue et al. (50) sequenced both ends of 8,000 clones of random genomic sequences with average insert sizes of 5 kb. The average sequence identity for deduced proteins (N = 3,898) between *C. parapsilosis* and *C. albicans* was 59%. The identity for *C. dubliniensis* and *C. albicans* was 82% (N = 5,511). Logue et al. (50) identified 19 genes that functioned in the mating process, 3 were α-specific, 11 were *a*-specific, and 9 were of general function. They found that *C. parapsilosis* contained an *MTL* locus with an *a* copy containing the same genes found in the *MTL* locus of *C. albicans*, *PAPa*, *OBPa*, and *PIKa*, as well as *MTLa2* and *MTLa1*, in that order. However, the *C. parapsilosis* *MTLa1* homolog contained four stop codons, and it is therefore a pseudogene. They found no evidence for a functional *MTLa1* gene positioned elsewhere in the genome. They found the *MTLa1* pseudogene in seven additional *C. parapsilosis* isolates. Because the gene product of *MTLa1* interacts with that of *MTLα2* to form a repressor complex that blocks mating in both *S. cerevisiae* and *C. albicans* and in addition blocks switching in the latter, the absence of a viable *MTLa1p* brings into question the existence of either a repressed mating or white-opaque switching system. Surprisingly, *MTLα* sequences, while present in the *C. parapsilosis* genome, were not at the same *MTL* locus harboring *MTLa* genes. Logue et al. (50) concluded that *C. parapsilosis* either had a defective mating system and hence did not mate or contained a system quite different from that of *C. albicans*. But if *C. parapsilosis* is not using its mating system, why would it retain the majority of mating-associated genes? Could it be that it has evolved a divergent mating system, or could it be that it recently possessed a mating system, but that the system is rapidly disintegrating? Or could it be that the aberrant mating system and associated genes now play a role in pathogenesis, and that is what preserves them in their present configuration?

C. GLABRATA, A C. ALBICANS MIMIC WITH S. CEREVISIAE FEATURES

Based upon the alignment of coding regions of the rRNAs and the proportion of shared genes, *C. glabrata* has been found to be far more highly related to *S. cere-*visiae than to *C. albicans* (5, 13, 14, 20, 85). Even so, it has emerged as the second most pervasive pathogenic *Candida* species, causing many of the diseases associated with *C. albicans* (60, 61). It lives as a commensal in the oral cavity, like *C. albicans*, and opportunistically proliferates and invades tissue in response to a compromising condition in the host (60). *C. glabrata*, like *C. albicans*, is a cause of yeast vaginitis and life-threatening systemic infections. It preferentially inhabits the oral cavities of the aged (47), which is worrisome, given that it is on average far less susceptible than *C. albicans* to antifungal drugs (9, 23, 28, 53). Like *C. albicans*, *C. glabrata* multiplies as a budding yeast or in a pseudohyphal growth form (17, 38). While it is capable of forming elongate tubes in colonies, these tubes do not compartmentalize (38). These tubes are morphologically similar to *C. albicans* conjugation tubes, not hyphae, undergoing apical reversion to the budding growth form like the former. *C. glabrata* is also capable of switching, but the two switching systems so far identified do not include the white-opaque transition (38, 41). All *C. glabrata* strains tested have been found to switch between a number of phenotypes that can be distinguished by graded colony coloration, either on agar containing $CuSO_4$ or phloxine B. This switching system is referred to as "core switching." Cells can also switch between a smooth and irregular wrinkled colony phenotype. This system is referred to as "irregular wrinkle switching."

In phylogenetic progressions of the hemiascomycetes that incorporate events related to mating, *C. glabrata* and *S. cerevisiae* are positioned at the end, while *C. albicans* branches from the base, far closer to the ancestral organism (8, 13, 73). It was, therefore, no surprise to find that the mating-type genes of *C. glabrata* were organized similarly to those of *S. cerevisiae*. In 2003, Wong et al. (85) sequenced approximately 3,000 plasmids containing random genomic inserts of 7 to 15 kb. By comparing these sequences to those in the *S. cerevisiae* genome, they identified orthologs involved in mating and meiosis, including genes contained in the *MATα* locus, *MATα1* and *MATα2*. Other mating-related genes included *MFα2*, *STE13*, *STE6*, *FIG4*, *FAR1*, *CDC24*, *SGV1*, *AKR1*, *FUS3*, *KAR5*, *LSG1*, *OPY2*, and *SAG1*. In the same year, Srikantha et al. (78) cloned the three mating-type-like loci *MTL1*, *MTL2*, and *MTL3*, using degenerative primers for the conserved carboxy termini of *S. cerevisiae MATα2*. They found that, as was the case in *S. cerevisiae*, the *MTLa* locus contained *MTLa1* and *MTLa2* oriented divergently and the *MTLα* locus contained *MTLα1* and *MTLα2* oriented divergently. Of 38 unrelated natural strains of *C. glabrata* in which the *MTL* genotype consisted of the three *MTL* loci, the majority contained two copies of *MTLa* and one of *MTLα*, or one copy of *MTLa* and two copies of *MTLα*, suggesting that as is the case in *S. cerevisiae*, *C. glabrata* contained a cassette system with two silent loci and one expressed locus. Northern analysis suggested that only one of the three loci, *MTL1*, was the expression locus and dictated mating type (78).

This configuration suggested that mating-type switching between *a* and *α* could occur, and the identification of the *HO* endonuclease gene strengthened this suggestion (13). The Ho endonuclease cuts at sites flanking the *MATa* and *MATα* loci and is necessary for gene conversion at the expressed *MAT* locus, in the process of mating-type switching in *S. cerevisiae*. Two studies then demonstrated mating-type switching in *C. glabrata*, one in vitro (13) and one in vivo (12). Butler et al. (13), using a PCR strategy, demonstrated a switch from *a* to *α* and from *α* to *a* at the *MTL1* site in two unrelated strains. They verified that cleavage occurred at an Ho site. Brockert et al. (12) analyzed the *MTL* genotype of clones from a population of *C. glabrata* infecting the vaginal canal of a patient and found that one clone from the population had changed genotype at the *MTL1* locus. They verified that the clones evolved from one strain by DNA fingerprinting with the complex DNA probe Cg6.

The presence of a functional cassette system for mating-type switching seems to suggest that mating must occur in *C. glabrata* populations. Indeed, as is the case in *C. albicans*, multilocus sequence typing of a collection of 165 *C. glabrata* isolates, using six loci for analysis, provided strong evidence for clonal propagation, but 14 examples of phylogenetic incompatibility suggested that recombination took place, although these results did not prove that recombination through mating is still occurring (21). But in the face of these suggestive observations, neither Butler et al. (13) nor Srikantha et al. (78) nor Brockert et al. (12) reported that they had observed a mating event. One might argue that mating occurs only rarely, only between select strains of opposite mating types, and only under very specific environmental conditions. Therefore, no one may have performed a rigorous enough search. There is, however, an additional possibility. What if mating does not occur in *C. glabrata*, but the mating system, including mating-type switching, plays a role in pathogenesis, and for that reason is preserved evolutionarily? Given the recent revelations that switching and mating may be intimately involved in pathogenesis in *C. albicans* (19), this possibility must also be entertained for *C. glabrata*.

CONCLUDING REMARKS

We have argued that although *C. albicans* and, probably, *C. glabrata* possess the genes which one might assume are sufficient for a functional mating system, it has not been established that mating occurs with any regularity. Indeed, studies of population structure in both organisms have revealed predominantly clonal modes of propagation. Because of the discovery that opaque cells signal white cells to form biofilms, we have suggested that mating systems in the infectious *Candida* species may have been subverted into functioning in host colonization and pathogenesis. Furthermore, we suggest that the selective pressure to maintain these systems and associated genes may persist not only through recombination but also through their role in virulence.

The work performed in our laboratory was funded by NIH grants AI2392 and DE014219. We are indebted to T. Srikantha, W. Wu, S. Lockhart, and C. Pujol in our lab for suggestions and to J. Collins for assembling the manuscript.

References

1. **Anderson, J. M., and D. R. Soll.** 1987. Unique phenotype of opaque cells in the white-opaque transition of *Candida albicans*. *J. Bacteriol.* **169:**5579–5588.

2. **Anderson, J., R. Mihalik, and D. R. Soll.** 1990. Ultrastructure and antigenicity of the unique cell wall pimple of the *Candida* opaque phenotype. *J. Bacteriol.* **172:**224–235.

3. **Anderson, J. B., C. Wickens, M. Khan, L. E. Cowen, N. Federspiel, T. Jones, and L. M. Kohn.** 2001. Infrequent genetic exchange and recombination in the mitochondrial genome of *Candida albicans*. *J. Bacteriol.* **183:**865–872.

4. **Bardwell, L.** 2005. A walk-through of the yeast mating pheromone response pathway. *Peptides* **26:**339–350.

5. **Barns, S. M., D. J. Lane, M. L. Sogin, C. Bibeau, and W. G. Weisburg.** 1991. Evolutionary relationships among pathogenic *Candida* species and relatives. *J. Bacteriol.* **173:** 2250–2255.

6. **Bennett, R. J., and A. D. Johnson.** 2003. Completion of a parasexual cycle in *Candida albicans* by induced chromosome loss in tetraploid strains. *EMBO J.* **22:**2505–2515.

7. **Bennett, R. J., M. A. Uhl, M. G. Miller, and A. D. Johnson.** 2003. Identification and characterization of a *Candida albicans* mating pheromone. *Mol. Cell. Biol.* **23:** 8189–8201.

8. **Bennett, R. J., M. G. Miller, P. R. Chua, M. E. Maxon, and A. D. Johnson.** 2005. Nuclear fusion occurs during mating in *Candida albicans* and is dependent on the *KAR3* gene. *Mol. Microbiol.* **55:**1046–1059.

9. **Blaschke-Hellmessen, R.** 1996. Fluconazole and itraconazole susceptibility testing with clinical yeast isolates and algae of the genus *Prototheca* by means of the Etest. *Mycoses* **39**(Suppl. 2):39–43.

10. **Blignaut, E., C. Pujol, S. Lockhart, S. Joly, and D. R. Soll.** 2002. Ca3 fingerprinting of *Candida albicans* isolates from human immunodeficiency virus-positive individuals reveals a new clade in South Africa. *J. Clin. Microbiol.* **40:**826–836.

11. **Boerlin, P., F. Boerlin-Petzold, C. Durussel, M. Ado, J.-L. Pagani, J.-P. Chave, and J. Bille.** 1995. Cluster of oral atypical *Candida albicans* isolates in a group of human

immunodeficiency virus-positive drug users. *J. Clin. Microbiol.* **33:**1129–1135.

12. **Brockert, P. J., S. A. Lachke, T. Srikantha, C. Pujol, R. Galask, and D. R. Soll.** 2003. Phenotypic switching and mating type switching of *Candida glabrata* at sites of colonization. *Infect. Immun.* **12:**7109–7118.

13. **Butler, G., C. Kenny, A. Fagan, C. Kurischko, C. Gaillardin, and K. H. Wolfe.** 2004. Evolution of the *MAT* locus and its Ho endonuclease in yeast species. *Proc. Natl. Acad. Sci. USA* **101:**1632–1637.

14. **Cai, J., I. N. Roberts, and M. D. Collins.** 1996. Phylogenetic relationships among members of the ascomycetous yeast genera *Brettanomyces, Debaryomyces, Dekkera,* and *Kluyveromyces* deduced by small-subunit rRNA gene sequences. *Int. J. Syst. Bacteriol.* **46:**542–549.

15. **Chen, J., J. Chen, S. Lane, and H. Liu.** 2002. A conserved mitogen-activated protein kinase pathway is required for mating in *Candida albicans. Mol. Microbiol.* **46:**1335–1344.

16. **Coleman, D. C., D. J. Sullivan, D. E. Bennett, G. P. Moran, H. J. Barry, and D. B. Shanley.** 1997. Candidiasis: the emergence of a novel species *Candida dubliniensis. AIDS* **11:**557–567.

17. **Csank, C., and K. Haynes.** 2000. *Candida glabrata* displays pseudohyphal growth. *FEMS Microbiol. Lett.* **189:**115–120.

18. **Daniels, K. J., S. R. Lockhart, P. Sundstrum, and D. R. Soll.** 2003. During *Candida albicans* mating, the adhesin Hwp1 and the first daughter bud localize to the a/a portion of the conjugation bridge. *Mol. Biol. Cell* **14:**4920–4930.

19. **Daniels, K. J., T. Srikantha, S. R. Lockhart, C. Pujol, and D. R. Soll.** 2006. Opaque cells signal white cells to form biofilms in *Candida albicans. EMBO J.* **25:**2240–2252.

20. **Diezmann, S., C. J. Cox, G. Schonian, R. J. Vilgalys, and T. G. Mitchell.** 2004. Phylogeny and evolution of medical species of *Candida* and related taxa: a multigenic analysis. *J. Clin. Microbiol.* **42:**5624–5635.

21. **Dodgson, A. R., C. Pujol, D. W. Denning, D. R. Soll, and A. J. Fox.** 2003. Multilocus sequence typing of *Candida glabrata* reveals geographically enriched clades. *J. Clin. Microbiol.* **41:**5709–5717.

22. **Edmond, M. B., S. E. Wallace, D. K. McClish, M. A. Pfaller, R. N. Jones, and R. P. Wenzel.** 1999. Nosocomial bloodstream infections in United States hospitals: a three-year analysis. *Clin. Infect. Dis.* **29:**239–244.

23. **Fortun, J., A. Lopez-San Roman, J. J. Velasco, A. Sanchez-Sousa, E. de Vicente, J. Nuno, C. Quereda, R. Barcena, G. Monge, A. Candela, A. Honrubia, and A. Guerrero.** 1997. Selection of *Candida glabrata* strains with reduced susceptibility to azoles in four liver transplant patients with invasive candidiasis. *Eur. J. Clin. Microbiol. Infect. Dis.* **16:**314–318.

24. **Graser, Y., M. Volovsek, J. Arrington, G. Schonian, W. Presber, T. G. Mitchell, and R. Vilgalys.** 1996. Molecular markers reveal that population structure of the human pathogen *Candida albicans* exhibits both clonality and recombination. *Proc. Natl. Acad. Sci. USA* **93:**12473–12477.

25. **Haber, J. E.** 1998. Mating-type gene switching in *Saccharomyces cerevisiae. Annu. Rev. Genet.* **32:**561–599.

26. **Hawser, S. P., and L. J. Douglas.** 1994. Biofilm formation by *Candida* species on the surface of catheter materials in vitro. *Infect. Immun.* **62:**915–921.

27. **Herskowitz, I., J. Rine, and J. N. Strathern.** 1992. Mating-type determination and mating-type interconversion in *Saccharomyces cerevisiae,* p. 583–656. *In* E. W. Jones, J. R. Pringle, and J. R. Broach (ed.), *The Molecular and Cellular Biology of the Yeast* Saccharomyces. Cold Spring Harbor Laboratory Press, Cold Spring Harbor, NY.

28. **Hitchcock, C. A., G. W. Pye, P. F. Troke, E. M. Johnson, and D. W. Warnock.** 1993. Fluconazole resistance in *Candida glabrata. Antimicrob. Agents. Chemother.* **37:**1962–1965.

29. **Huang, G., H. Wang, S. Chou, X. Nie, J. Chen, and H. Liu.** 2006. Bistable expression of *WOR1,* a master regulator of white-opaque switching in *Candida albicans. Proc. Natl. Acad. Sci. USA* **103:**12813–12818.

30. **Hube, B., M. Monod, D. Schofield, A. Brown, and N. Gow.** 1994. Expression of seven members of the gene family encoding aspartyl proteinases in *Candida albicans. Mol. Microbiol.* **14:**87–99.

31. **Hull, C. M., and A. D. Johnson.** 1999. Identification of a mating type-like locus in the asexual pathogenic yeast *Candida albicans. Science* **285:**1271–1275.

32. **Hull, C. M., R. M. Raisner, and A. D. Johnson.** 2000. Evidence for mating of the "asexual" yeast *Candida albicans* in a mammalian host. *Science* **289:**307–310.

33. **Ibrahim, A. S., B. B. Magee, D. C. Sheppard, M. Yang, S. Kauffman, J. Becker, J. E. Edwards, Jr., and P. T. Magee.** 2005. Effects of ploidy and mating type on virulence of *Candida albicans. Infect. Immun.* **73:**7366–7374.

34. **Janbon, G., F. Sherman, and E. Rustcheko.** 1999. Appearance and properties of L-sorbose-utilizing mutants of *Candida albicans* obtained on a selective plate. *Genetics* **153:**653–664.

35. **Joly, S., C. Pujol, and D. R. Soll.** 2002. Microevolutionary changes and chromosomal translocations are more frequent at RPS loci in *Candida dubliniensis* than in *Candida albicans. Infect. Genet. Evol.* **2:**19–37.

36. **Kvaal, C. A., T. Srikantha, and D. R. Soll.** 1997. Misexpression of the white phase-specific gene *WH11* in the opaque phase of *Candida albicans* affects switching and virulence. *Infect. Immun.* **65:**4468–4475.

37. **Kvaal, C., S. A. Lachke, T. Srikantha, K. Daniels, J. McCoy, and D. R. Soll.** 1999. Misexpression of the opaque phase-specific gene *PEP1 (SAP1)* in the white phase of *Candida albicans* confers increased virulence in a mouse model of cutaneous infection. *Infect. Immun.* **67:**6652–6662.

38. **Lachke, S. A., S. Joly, K. Daniels, and D. R. Soll.** 2002. Phenotypic switching and filamentation in *Candida glabrata. Microbiology* **148:**2661–2674.

39. **Lachke, S. A., S. R. Lockhart, K. J. Daniels, and D. R. Soll.** 2003. Skin facilitates *Candida albicans* mating. *Infect. Immun.* **71:**4970–4976.

40. **Lachke, S. A., T. Srikantha, and D. R. Soll.** 2003. The regulation of *EFG1* in white-opaque switching in *Can-*

dida albicans involves overlapping promoters. *Mol. Microbiol.* 48:523–536.

41. Lachke, S. A., T. Srikantha, L. Tsai, K. Daniels, and D. R. Soll. 2000. Phenotypic switching in *Candida glabrata* involves phase-specific regulation of the metallotheionein gene *MT-II* and the newly discovered hemolysin gene *HLP. Infect. Immun.* 68:884–895.

42. Lan, C. Y., G. Newport, L. A. Murillo, T. Jones, S. Scherer, R. W. Davis, and N. Agabian. 2002. Metabolic specialization associated with phenotypic switching in *Candida albicans. Proc. Natl. Acad. Sci. USA* 99:14907–14912.

43. Legrand, M., P. Lephart, A. Forsche, F.-M. C. Mueller, T. Walsh, P. T. Magee, and B. B. Magee. 2004. Homozygosity at the *MTL* locus in clinical strains of *Candida albicans*: karyotypic rearrangements and tetraploid formation. *Mol. Microbiol.* 52:1451–1462.

44. Lockhart, S. R., C. Pujol, K. Daniels, M. Miller, A. Johnson, and D. R. Soll. 2002. In *Candida albicans*, white-opaque switchers are homozygous for mating type. *Genetics* 162:737–745.

45. Lockhart, S. R., K. J. Daniels, R. Zhao, D. Wessels, and D. R. Soll. 2003. Cell biology of mating in *Candida albicans. Eukaryot. Cell* 2:49–61.

46. Lockhart, S. R., M. Nguyen, T. Srikantha, and D. R. Soll. 1998. A MADS box protein consensus binding site is necessary and sufficient for activation of the opaque-phase specific gene *OP4* of *Candida albicans. J. Bacteriol.* 180:6607–6616.

47. Lockhart, S. R., S. Joly, K. Vargas, J. Swails-Wenger, L. Enger, and D. R. Soll. 1999. Natural defenses against *Candida* colonization breakdown in the oral cavities of the elderly. *J. Dent. Res.* 78:857–868.

48. Lockhart, S. R., W. Wu, J. B. Radke, and D. R. Soll. 2005. Increased virulence and competitive advantage of a/α over a/a or α/α offspring conserves the mating system of *Candida albicans. Genetics* 169:1883–1890.

49. Lockhart, S. R., R. Zhao, K. J. Daniels, and D. R. Soll. 2003. α-Pheromone-induced "shmooing" and gene regulation require white-opaque switching during *Candida albicans* mating. *Eukaryot. Cell* 2:847–855.

50. Logue, M. E., S. Wong, K. H. Wolfe, and G. Butler. 2005. A genome sequence survey shows that the pathogenic yeast *Candida parapsilosis* has a defective *MTLa1* allele at its mating type locus. *Eukaryot. Cell* 4:1009–1017.

51. Magee, B. B., and P. T. Magee. 2000. Induction of mating in *Candida albicans* by construction of MTLa and MTLalpha strains. *Science* 289:310–313.

52. Magee, B. B., M. Legrand, A. M. Alarco, M. Raymond, and P. T. Magee. 2002. Many of the genes required for mating in *Saccharomyces cerevisiae* are also required for mating in *Candida albicans. Mol. Microbiol.* 46:1345–1351.

53. Marichal, P., H. Vanden Bossche, F. C. Odds, G. Nobels, D. W. Warnock, V. Timmerman, C. Van Broeckhoven, S. Fay, and P. Mose-Larsen. 1997. Molecular biological characterization of an azole-resistant *Candida glabrata* isolate. *Antimicrob. Agents Chemother.* 41:2229–2237.

54. McCullough, M., B. Ross, and P. Reade. 1995. Characterization of genetically distinct subgroup of *Candida albicans* strains isolated from oral cavities of patients infected with human immunodeficiency virus. *J. Clin. Microbiol.* 33:696–700.

55. Miller, M. G., and A. D. Johnson. 2002. White-opaque switching in *Candida albicans* is controlled by mating-type locus homeodomain proteins and allows efficient mating. *Cell* 110:293–302.

56. Moran, G., C. Stokes, S. Thewes, B. Hube, D. C. Coleman, and D. Sullivan. 2004. Comparative genomics using *Candida albicans* DNA microarrays reveals absence and divergence of virulence-associated genes in *Candida dubliniensis. Microbiology* 150:3363–3382.

57. Morrow, B., T. Srikantha, and D. R. Soll. 1992. Transcription of the gene for a pepsinogen, *PEP1*, is regulated by white-opaque switching in *Candida albicans. Mol. Cell. Biol.* 12:2997–3005.

58. Morrow, B., T. Srikantha, J. Anderson, and D. R. Soll. 1993. Coordinate regulation of two opaque-specific genes during white-opaque switching in *Candida albicans. Infect. Immun.* 61:1823–1828.

59. Mukherjee, P. K., G. Zhou, R. Munyon, and M. A. Ghannoum. 2005. *Candida* biofilm: a well-designed protected environment. *Med. Mycol.* 43:191–208.

60. Odds, F. C. 1988. Candida *and Candidosis*, 2nd ed. Bailliere Tindall, London, England.

61. Pfaller, M. A., D. J. Diekema, R. N. Jones, H. S. Sader, A. C. Fluit, R. J. Hollis, and S. A. Messer. 2001. International surveillance of bloodstream infections due to *Candida* species: frequency and occurrence and in vitro susceptibilities to fluconazole, ravuconazole, and viriconazole of isolates collected from 1997 through 1999 in the SENTRY antimicrobial surveillance program. *J. Clin. Microbiol.* 39:3254–3259.

62. Pujol, C., A. Dodgson, and D. R. Soll. 2005. Population genetics of ascomycetes pathogenic to humans and animals, p. 149–188. *In* J.-P. Xu (ed.), *Evolutionary Genetics of Fungi.* Horizon Scientific Press, Norfolk, United Kingdom.

63. Pujol, C., K. J. Daniels, T. Srikantha, S. R. Lockhart, J. Geiger, and D. R. Soll. 2004. The two closely related species *Candida albicans* and *Candida dubliniensis* can mate. *Eukaryot. Cell* 3: 1015–1027.

64. Pujol, C., F. Renaud, M. Mallie, T. de Meeus, and J. M. Bastide. 1997. Atypical strains of *Candida albicans* recovered from AIDS patients. *J. Med. Vet. Mycol.* 35:115–121.

65. Pujol, C., J. Reynes, F. Renaud, M. Raymond, M. Tibayrenc, F. J. Ayala, F. Janbon, M. Mallie, and J. M. Bastide. 1993. The yeast *Candida albicans* has a clonal mode of reproduction in a population of infected human immunodeficiency virus-positive patients. *Proc. Natl. Acad. Sci. USA* 90:9456–9459.

66. Pujol, C., S. Joly, S. R. Lockhart, S. Noel, M. Tibayrenc, and D. R. Soll. 1997. Parity among the randomly amplified polymorphic DNA method, multilocus enzyme electrophoresis, and Southern blot hybridization with the moderately repetitive DNA probe Ca3 for fingerprinting *Candida albicans. J. Clin. Microbiol.* 35:2348–2358.

67. Rikkerink, E. H., B. B. Magee, and P. T. Magee. 1988. Opaque-white phenotype transition: a programmed mor-

phological transition in *Candida albicans. J. Bacteriol.* 170:895–899.

68. **Roberts, C. J., B. Nelson, M. J. Matron, R. Stoughton, M. R. Meyer, H. A. Bennett, Y. D. He, H. Dai, W. L. Walker, T. R. Hughes, M. Tyers, C. Boone, and S. H. Friend.** 2000. Signaling and circuitry of multiple MAPK pathways revealed by a matrix of global gene expression profiles. *Science* 287:873–880.

69. **Schwartz, M. A., and H. D. Madhani.** 2004. Principles of MAP kinase signaling specificity in *Saccharomyces cerevisiae. Annu. Rev. Genet.* 38:725–748.

70. **Slutsky, B., M. Staebell, J. Anderson, L. Risen, M. Pfaller, and D. R. Soll.** 1987. "White-opaque transition": a second high-frequency switching system in *Candida albicans. J. Bacteriol.* 169:189–197.

71. **Soll, D. R.** 2003. *Candida albicans,* p. 165–201. *In* A. Craig and A. Scherf (ed.), *Antigenic Variation.* Academic Press, London, United Kingdom.

72. **Soll, D. R.** 2004. Mating-type locus homozygosis, phenotypic switching and mating: a unique sequence of dependencies in *Candida albicans. Bioessays* 26:10–20.

73. **Soll, D. R.** The evolution of a mating system uniquely dependent upon switching and pathogenesis in *Candida albicans. In* F. Basquero, C. Nombela, G. H. Cassell, and J. A. Gutierrez (ed.), *Introduction to the Evolutionary Biology of Bacterial and Fungal Pathogens,* in press.

74. **Srikantha, T., A. R. Borneman, K. J. Daniels, C. Pujol, W. Wu, M. R. Seringhaus, M. Gerstein, S. Yi, M. Snyder, and D. R. Soll.** *TOS9* regulates white-opaque switching in *Candida albicans. Eukaryot. Cell* 5:1674–1687.

75. **Srikantha, T., and D. R. Soll.** 1993. A white-specific gene in the white-opaque switching system of *Candida albicans. Gene* 131:53–60.

76. **Srikantha, T., A. Chandrasekhar, and D. R. Soll.** 1995. Functional analysis of the promoter of the phase-specific *WH11* gene of *Candida albicans. Mol. Cell. Biol.* 15: 1797–1805.

77. **Srikantha, T., L. Tsai, and D. R. Soll.** 1997. The *WH11* gene of *Candida albicans* is regulated in two distinct developmental programs through the same transcription activation sequences. *J. Bacteriol.* 179:3837–3844.

78. **Srikantha, T., S. A. Lachke, and D. R. Soll.** 2003. Three mating type-like loci in *Candida glabrata. Eukaryot. Cell* 2:328–340.

79. **Sullivan, D. J., K. Haynes, J. Bille, P. Boerlin, L. Rodero, S. Lloyd, M. Henman, and D. Coleman.** 1997. Widespread geographic distribution of oral *Candida dubliniensis* strains in human immunodeficiency virus-infected individuals. *J. Clin. Microbiol.* 35:960–964.

80. **Tavanti, A., A. D. Davidson, M. J. Fordyce, N. A. R. Gow, M. C. J. Maiden, and F. C. Odds.** 2005. Population structure and properties of *Candida albicans* as determined by multilocus sequence typing. *J. Clin. Microbiol.* 43:5601–5613.

81. **Tsong, A. E., M. G. Miller, R. M. Raisner, and A. D. Johnson.** 2003. Evolution of a combinatorial transcriptional circuit: a case study in yeasts. *Cell* 115:389–399.

82. **Tzung, K.-W., R. M. Williams, S. Scherer, N. Federspiel, T. Jones, N. Hansen, V. Bivolarevic, L. Huizar, C. Komp, R. Surzycki, R. Tamse, R. W. Davis, and N. Agabian.** 2001. Genomic evidence for a complete sexual cycle in *Candida albicans. Proc. Natl. Acad. Sci. USA* 98:3249–3253.

83. **Whelan, W. L., and K. J. Kwon-Chung.** 1988. Auxotrophic heterozygosities and the ploidy of *Candida parapsilosis* and *Candida krusei. J. Med. Vet. Mycol.* 26:163–171.

84. **White, T. C., H. Miyasaki, and N. Agabian.** 1993. Three distinct secreted aspartyl proteinases in *Candida albicans. J. Bacteriol.* 175:6126–6135.

85. **Wong, S., M. A. Fares, W. Zimmermann, G. Butler, and K. H. Wolfe.** 2003. Evidence from comparative genomics for a complete sexual cycle in the 'asexual' pathogenic yeast *Candida glabrata. Genome Biol.* 4:R10.

86. **Wu, W., C. Pujol, S. R. Lockhart, and D. R. Soll.** 2005. Mechanisms of mating type homozygosis in *C. albicans. Genetics* 169:1311–1327.

87. **Wu, W., S. R. Lockhart, C. Pujol, T. Srikantha, and D. R. Soll.** 2007. Heterozygosity of genes on the sex chromosome regulates *Candida albicans* virulence. *Mol. Microbiol.,* in press.

88. **Zhao, R., K. J. Daniels, S. R. Lockhart, K. M. Yeater, L. L. Hoyer, and D. R. Soll.** 2005. Unique aspects of gene expression during *Candida albicans* mating and possible G$_1$ dependency. *Eukaryot. Cell* 4:1175–1190.

89. **Zordan, R. E., D. J. Galgoczy, and A. D. Johnson.** 2006. Epigenetic properties of white-opaque switching in *Candida albicans* are based on a self-sustaining transcriptional feedback loop. *Proc. Natl. Acad. Sci. USA* 103: 12807–12812.

*Sex in Fungi: Molecular Determination
and Evolutionary Implications*
Edited by Joseph Heitman et al.
© 2007 ASM Press, Washington, D.C.

Jennifer L. Reedy
Joseph Heitman

14

Evolution of *MAT* in the *Candida* Species Complex:
Sex, Ploidy, and Complete Sexual Cycles in *C. lusitaniae*, *C. guilliermondii*, and *C. krusei*

The genus *Candida* affords an excellent opportunity to study the evolution of sexual reproduction, as this genus contains closely related asexual and sexual species. Historically, *Candida* described a group of yeasts that lacked sexual cycles and ascospore production, propagated through asexual budding, and formed pseudo- and/or true hyphae. However, some of the 151 *Candida* species were subsequently found to represent the anamorphic form of a teleomorphic species and thus are capable of sexual reproduction and sporulation alone or when mated with cells of a compatible mating type. Many of the species within this clade are human pathogens, and thus, understanding the life cycle of these organisms and developing robust genetic systems for the study of these species are of considerable import. The best-studied species to date are the human pathogens *Candida albicans* and *Candida glabrata*. However, neither of these species is as yet known to possess a complete meiotic sexual cycle.

Candida lusitaniae, *Candida guilliermondii*, and *Candida krusei* are three members of this genus that are exciting prospects for further study. First, all three possess complete sexual cycles. Second, they are all opportunistic human pathogens. And third, the genomes of *C. lusitaniae* and *C. guilliermondii* have been sequenced and *C. krusei* has been advanced for genome sequencing, which will facilitate robust comparative genomic studies. These organisms are attractive candidates for the development of classical genetic systems for the study of mating, meiosis, sporulation, and virulence in the *Candida* species complex. Furthermore, comparisons between the sexual and asexual *Candida* species will help elucidate the mechanisms and evolutionary timing of the loss or modification of sexual reproduction within this complex genus and provide insights into the evolution of signaling cascades with dual roles in mating and virulence.

EVOLUTIONARY RELATIONSHIPS OF THE *CANDIDA* COMPLEX SPECIES

Because species belonging to the genus *Candida* were classified originally quite broadly (based primarily upon their vegetative growth as budding yeasts, ability to produce pseudohyphae or true hyphae, and lack of a sexual cycle),

Jennifer L. Reedy, Duke University Medical Center, 320 CARL Bldg, Research Drive, Box 3546, Durham, NC 27710. Joseph Heitman, Duke University Medical Center, 322 CARL Bldg, Research Drive, Box 3546, Durham, NC 27710.

the resulting genus is polyphyletic. Multiple phylogenies have been constructed containing either some or all of the medically relevant *Candida* species. Most published phylogenies are based on single genes or regions such as the large or small ribosomal subunits (5, 9, 31, 43), topo-isomerase II (24), cytochrome *b* (80), or the actin 1 gene (10). The most robust analysis containing all of the pathogenic *Candida* species, including the sexual species *C. lusitaniae*, *C. krusei*, and *C. guilliermondii*, employed six nuclear genes (four protein-encoding genes; *EF2*, *ACT1*,

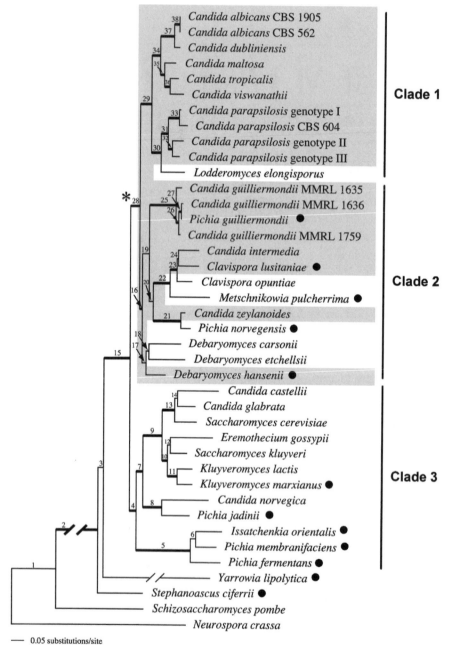

Figure 14.1 Phylogeny of *Candida* anamorphic and teleomorphic species. Species are designated by the teleomorph or anamorph designation or both. An asterisk indicates the proposed point of CUG codon capture and the expansion of the *MTL* locus to include *PAP1*, *OBP1*, and *PIK1* genes. Species with gray shading are those that encode CTG as serine rather than leucine. Black dots indicate teleomorphic species with a *Candida* anamorph. Node numbers are indicated above each branch. Phylogeny is adapted from Diezmann et al. (11).

RPB1, and *RPB2*; and the 18S and the 26S ribosomal DNA genes) to construct a phylogeny (11) (Fig. 14.1). These analyses have shown that *C. guilliermondii* and *C. lusitaniae* are more closely related to *C. albicans*, whereas *C. krusei* is more closely related to *C. glabrata* and *Saccharomyces cerevisiae*. In addition to these medically relevant *Candida* species, there are other known *Candida* species that possess sexual cycles (Table 14.1) (4, 10, 19, 29, 32, 33).

Another interesting character of the *Candida* genus is the evolution of CUG codon reassignment. This nonuniversal codon usage was first recognized in the asexual species *Candida cylindracea* (25, 79). Subsequently, many *Candida* species were shown to translate CUG codons as serine rather than leucine. The tRNA responsible for this transition arose from a serine tRNA approximately 270 million years ago, prior to the divergence of the *Candida* and *Saccharomyces* lineages (47, 60, 61, 65, 67, 70). The "ambiguous intermediate theory" suggests that the common ancestor could decode CUG as either serine or leucine, and that codon reassignment occurred only after divergence, approximately 170 million years ago (47, 62, 68). This theory is supported by evidence that some *Candida* species are still able to decode CUG as either serine or leucine, such as *Candida zeylanoides* (66). Both *C. guilliermondii* and *C. lusitaniae*, like *C. albicans*, decode CUG as serine; however, *C. krusei* translates this codon as leucine similarly to *C. glabrata* and *S. cerevisiae* (55, 83).

CANDIDA LUSITANIAE (TELEOMORPH CLAVISPORA LUSITANIAE)

C. lusitaniae was first described in 1970 by van Uden and do Carmo-Sousa, who isolated the organism from the gastrointestinal tract of animals (71). Of the three species discussed in this chapter, the sexual life cycle has been the best studied for this species. The teleomorphic form of *Candida lusitaniae*, *Clavispora lusitaniae*, was subsequently described by Rodrigues de Miranda in 1979 after noting mating between previously collected strains and a newly isolated sample from citrus peel juice (30). *C. lusitaniae* can be isolated from human and animal samples, as well as from environmental sources such as decaying trees and fruit (4, 33). In the clinic, *C. lusitaniae* accounts for ~1% of candidemia but is notable for its propensity to develop resistance to amphotericin B (1, 8, 48, 50, 81).

C. lusitaniae is an experimentally tractable haploid organism. A first draft of the genome sequence completed by the Broad Institute suggests a genome size of 16 Mb arranged into nine supercontigs (*Candida lusitaniae* sequencing project, Broad Institute of Harvard and MIT [http://www.broad.mit.edu]). Contour-clamped homogeneous electric field gel analyses of multiple isolates observed six to eight distinguishable chromosomal bands (26, 73). The haploid nature of the genome makes *C. lusitaniae* a more attractive species for manipulation than the diploid *C. albicans*. Techniques for transformation and gene disruption by homologous recombination have already been established for *C. lusitaniae* (82, 83).

Table 14.1 *Candida* species with known teleomorphs

Anamorph	Teleomorph	Sexual type
Candida bimundalis	*Pichia americana*	Heterothallic
Candida ciferrii	*Stephanoascus ciferrii*	Heterothallic
Candida edax	*Stephanoascus smithiae*	Heterothallic
Candida famata	*Debaryomyces hansenii*	Homothallic
Candida guilliermondii var. *guilliermondii*	*Pichia guilliermondii*	Heterothallic
Candida guilliermondii var. *membranaefaciens*	*Pichia ohmeri*	Heterothallic
Candida kefyr	*Kluyveromyces marxianus*	Homothallic
Candida krusei	*Issatchenkia orientalis/Pichia kudriavzevii*	Heterothallic
Candida lambica	*Pichia fermentans*	Homothallic
Candida lipolytica	*Yarrowia lipolytica*	Heterothallic
Candida lusitaniae	*Clavispora lusitaniae*	Heterothallic
Candida norvegensis	*Pichia norvegensis*	Homothallic
Candida pelliculosa	*Pichia anomala*	Heterothallic
Candida pintolopesii	*Kazachstania pintolopesii*	Homothallic
Candida pulcherrima	*Metschnikowia pulcherrima*	Homothallic
Candida sorbosa	*Issatchenkia occidentalis*	Heterothallic
Candida utilis	*Pichia jadinii*	Homothallic
Candida valida	*Pichia membranifaciens*	Mixed

Mating between strains of *C. lusitaniae* is consistent with a heterothallic species containing a single mating-type-determining locus. Historically, the two mating types in this biallelic system were designated h+ and h−, but they were subsequently referred to as **a** and α, respectively (30). Original studies suggested that the ratio of α to **a** cells in the environment was skewed 6:1 (16). However, a larger study of 76 clinical isolates from 60 patients demonstrated an equal distribution of mating types and also revealed no correlation between mating type and severity of disease or site of isolation (14, 16). In this larger study, all isolates of *C. lusitaniae* tested were capable of mating (57).

Identification of *C. lusitaniae* in the clinical setting is usually accomplished based on carbon assimilation profiles, most notably the ability to assimilate L-rhamnose and to ferment cellobiose (51, 56). However, assimilation profiles cannot reliably type all strains. Due to the robustness of mating, it was proposed that ability to mate with tester strains could be used as a technique for positively identifying *C. lusitaniae* isolates (14, 53). As a proof of principle, five strains that were indistinguishable as either *C. lusitaniae* or *Candida pulcherrima* by standard clinical carbon assimilation profiles were positively identified as *C. lusitaniae* based on mating preference (53). However, other studies have suggested that the efficiency of sporulation can vary among isolates (16, 34).

Although *C. lusitaniae* is regarded as heterothallic, there is a single report of a strain capable of producing spores in the absence of a mating partner either through isogamous conjugation or bud-parent cell conjugation (34). Sequencing of a portion of the large ribosomal subunit demonstrated that this strain was similar to other isolates of *C. lusitaniae* (34); however, no further studies have been conducted with this strain. One possibility is that this isolate could be a diploid, but there have been no recorded accounts of diploids being isolated either in nature or in the laboratory as by-products of mating events. Presumably, diploids formed by mating rapidly undergo meiosis and thus only haploid cells have been isolated from matings and in the environment. Interestingly, sequence analysis of the D1/D2 domain of the large-subunit ribosomal DNA revealed the presence of significant polymorphisms within an interbreeding population of *C. lusitaniae* (34). There was no association between these polymorphisms and mating type, suggesting that there is recombination and sexual reproduction within the natural population.

Multiple types of media are capable of inducing mating including dilute potato dextrose agar, 1% malt extract media, sodium acetate, yeast carbon base, V8, and SLAD (14, 83). Effective matings are supported by low concentrations of ammonium and require a solid support, as increasing the concentration of ammonium sulfate or incubating cells in liquid media blocked mating (14, 83). Additionally, the efficiency of conjugation and ascospore formation is optimal at temperatures between 18 and 28°C (14). Within 24 h of coincubating cells of opposite mating types, conjugating cells can be observed (Fig. 14.2). Scanning and transmission electron microscopy revealed that the conjugation tube bridging mating **a** and α cells averages 1 μm in length (14).

At the point of cell fusion, the conjugation tube contains a central septal perforation through which nuclear transfer presumably occurs. The nucleus of one parent traverses the conjugation tube and undergoes karyogamy and meiosis inside the mating partner cell (Fig. 14.2). The parent that donates a nucleus is referred to as the "head cell" and remains grossly unchanged throughout the process of meiosis and ascospore formation. The other parent, the nucleus acceptor, becomes the ascus (Fig. 14.2). Labeling studies suggest that the nuclear transfer is highly polarized with one parental strain in a mating serving primarily as the nucleus donor (14). Based on the small set of strains tested there was no definitive linkage between the mating type of the parent and whether the cell served as a nuclear donor or acceptor (14). However, a more robust analysis using multiple labeling techniques is necessary to establish this point rigorously. After 48 h of incubation one to four clavate, echinulate ascospores are formed, sometimes containing a small oil droplet (14, 16, 57). Ascospores are readily released upon maturation and often agglutinate; empty asci and free ascospores can be readily observed at 72 h (14, 57).

An initial study to elucidate the signaling pathways controlling mating in *C. lusitaniae* focused on homologs of the *S. cerevisiae* mitogen-activated protein (MAP) kinase pheromone response cascade. In *S. cerevisiae* the MAP kinase pheromone response cascade regulates both mating and filamentation (pseudohyphae) (39). Deletion of the Ste12 transcription factor results in loss of mating ability (17) and a filamentation defect in *S. cerevisiae* (39). The *C. albicans* homolog of Ste12 is Cph1, the deletion of which reduced pseudohyphal filamentation in response to certain environmental stimuli (38) and also blocked mating between homozygous **a** and α strains (44). On rich media *C. lusitaniae* propagates via budding, but it can form pseudohyphae under conditions of nutrient deprivation on V8, filament, or SLAD media (30, 57). Interestingly, the *C. lusitaniae* Ste12 homolog, Cls12, is required for mating but dispensable for filamentation under the conditions studied (83). The divergent functions of this conserved transcription factor demonstrate how studying both sexual

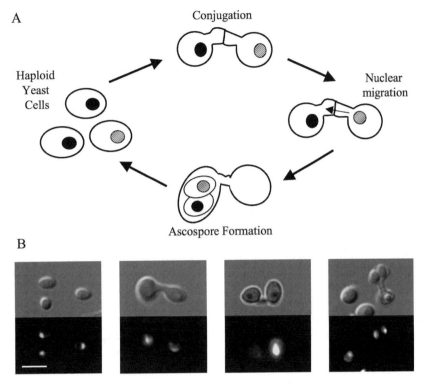

Figure 14.2 Life cycle of *C. lusitaniae*. (A) Diagrammatic representation of the *C. lusitaniae* life cycle. (B) Differential interference contrast images of each stage of the *C. lusitaniae* life cycle. Cells of opposite mating types were incubated on dilute potato dextrose agar media, stained with DAPI (4′,6′-diamidino-2-phenylindole), and photographed at 24 to 72 h after coincubation. Scale bar is 5μm. Haploid yeast cells of opposite mating types undergo conjugation. Following conjugation one parental nuclei traverses the conjugation tube to enter the cell of the mating partner. Asci formed contain 1 to 2 clavate ascospores per ascus.

and asexual species within the genus *Candida* could provide interesting insights into the evolution of signal transduction cascades involved not only in mating and filamentation but also in other pathways important for pathogenesis.

CANDIDA GUILLIERMONDII (TELEOMORPH PICHIA GUILLIERMONDII)

Candida guilliermondii is a haploid heterothallic yeast whose teleomorphic form is *Pichia guilliermondii*. The anamorph has been cultured from a variety of ecological niches, including human clinical specimens, insects, fruit, and decaying matter (4, 33). Depending upon the geographical location, *C. guilliermondii* has been identified as the causative agent in 2 to 10% of all candidal bloodstream infections (27, 58). Therefore, most studies of this fungus have been oriented toward topics such as clinical epidemiology and antifungal susceptibility, and relatively little work has been done to characterize the

sexual cycle and the pathways involved in sexual reproduction, meiosis, and sporulation. Vegetative yeast cells are ellipsoidal or ovoid, reproduce via budding, and can form pseudohyphae, but not true hyphae (4).

Early studies employing pulsed-field gel electrophoresis suggest that, unlike *C. albicans*, *C. guilliermondii* is a haploid yeast with a genome size of approximately 12 Mb (12). The number of chromosomal bands observed has ranged from six to eight (2, 12, 45). The Broad Institute recently released a 12X first draft of the *C. guilliermondii* genome organized into nine supercontigs which confirms the estimated genome size (*C. guilliermondii* sequencing project; see Broad Institute URL above).

The sexual cycle of *C. guilliermondii* was first identified in 1952 by Wickerham and Burton, after recognizing that some yeasts previously classified to non-ascospore-forming genera actually represented the anamorphic form of a sexual species (77). The teleomorph species *P. guilliermondii* was formally described in 1966 (76) and subsequent DNA complementarity tests confirmed the teleomorph-anamorph relationship (28). The initial matings were carried out on malt extract sporulation media at 25°C for 6 to 14 days. Positive mating mixtures were identified by the presence of conjugating cells and ascospore formation. The rate of sporulation is relatively low; the most efficiently sporulating pair produced only 4% ascospores (76). In an attempt to increase ascospore formation, germinated ascospores were backcrossed with the parental mating-competent strains with the aim of producing a strain with a higher mating efficiency; however, this was unsuccessful (77). The asci are formed from two conjugated cells and contain one or two hat-shaped spores. Upon maturation the asci rupture, releasing the ascospores, which become swollen and refractile (76, 77).

Recently, some cryptic species were identified within the *C. guilliermondii* clade (2, 3, 59, 72). Notably *Candida fermentati* and *Candida carpophila*, although phenotypically indistinguishable from *C. guilliermondii*, were shown to be genetically different on the basis of DNA reassociation and electrophoretic karyotyping and thus were described as separate species (72). Additionally, it was noted that strains of *C. fermentati* were capable of ascospore formation when mixed together, and this teleomorphic form was designated *Pichia caribbica* (72). Thus, one possibility for the low rate of ascosporulation described previously within the *C. guilliermondii* clade could be the presence of cryptic species within this group that are capable of cell-cell fusion during mating but impaired in meiosis and spore viability due to genetic divergence.

Although sexual reproduction has been reported in the laboratory with low efficiency, it is unclear what role sexual reproduction plays in the environment. Multilocus sequence typing analysis of 32 strains of *C. guilliermondii* isolated from Ontario (Canada), China, and the Philippines suggests that the population is primarily clonal (36). However, due to the small sample size and the limited molecular variation among the isolates, sexual reproduction among natural populations could not be excluded. To further address this question, a larger population and the identification of more markers for typing of strains will be required.

CANDIDA KRUSEI (TELEOMORPH ISSATCHENKIA ORIENTALIS/PICHIA KUDRIAVZEVII)

Candida krusei is the anamorphic form of *Issatchenkia orientalis* (*I. orientalis* will be renamed *Pichia kudriavzevii* based upon a multigene analysis that demonstrates grouping of the *Issatchenkia* species within the *Pichia* clade [C. Kurtzman, personal communication]). Isolates can be obtained from a wide variety of environmental sources including fruit juice, tea, and decaying organic material as well as clinical samples (4, 30, 49). Opportunistic infections caused by *C. krusei* are associated with prior treatment with fluconazole to which *C. krusei* is highly resistant (15, 54, 74).

Unlike *C. guilliermondii* and *C. lusitaniae*, *C. krusei* is likely diploid (22a). Some studies suggest that the genome is approximately 11 Mb and contains three to six chromosomes (13, 18, 78). However, using pulsed-field gel electrophoresis, Doi et al. estimated an average of eight chromosomes and a genome of 20 Mb (this study also provided estimates of the chromosome number and genome size for *C. albicans*, *C. guilliermondii*, and *C. parapsilosis* that were similar to those determined by genome sequencing) (12). Additional support for the diploid nature of the *C. krusei* genome is the demonstration of heterozygosity at the orotidine-5′-phosphate decarboxylase locus (75). *C. krusei* has currently been proposed as a candidate for genome sequencing at the Sanger Center (Geraldine Butler, personal communication).

There is limited description of the sexual cycle of *C. krusei*. Ascospore formation was originally reported by Kudryavtsev, who observed the formation of a single persistent spheroidal spore per ascus. Spore formation is infrequent, and up to two spores per asci have been observed (30).

THE EVOLUTION AND STRUCTURE OF THE MATING-TYPE LOCUS IN THE CANDIDA SPECIES COMPLEX

The best-studied species in the *Candida* genus are *C. albicans* and *C. glabrata*. *C. glabrata* is more closely related to *S. cerevisiae* than to *C. albicans*, which is reflected in the organization of its mating-type loci. The structure of the mating-type-like (*MTL*) locus of *C. glabrata* is similar to that in *S. cerevisiae* in that *C. glabrata* contains three *MTL* loci, containing either **a**- or α-specific information (64). Mating has never been observed in *C. glabrata*. *C. albicans* was also thought to lack a sexual cycle; however, in 1999 the *MTL* locus of *C. albicans* was identified (21). The majority of isolates were found to be *MTL* heterozygotes possessing both **a** and α information, rendering them incapable of mating. Strains engineered to possess only **a** or α information and naturally occurring **a**/**a** or α/α isolates (~3 to 10% of clinical samples) are capable of mating (22, 37, 41, 46). These strains can switch from the white to the opaque (mating-efficient) cell type and can subsequently undergo shmooing and cell fusion (35, 37, 40, 41, 52, 63). The product of mating is a tetraploid **a**/**a**/α/α yeast cell that can be induced to undergo a parasexual cycle of chromosome loss in vitro to return to a diploid state (6).

The *MTL* locus alleles of *C. albicans* contain genes encoding homologs of the transcription factors **a**1, α1, and α2, the primary regulators of cell type in *S. cerevisiae*. The *MTL***a** allele of *C. albicans* also contains the HMG-box transcriptional regulator, **a**2, which is not present in *S. cerevisiae* (7, 21, 23). Additionally, both *MTL* loci contain three extra genes encoding poly(A) polymerase (*PAP1*), phosphatidylinositol kinase (*PIK1*), and an oxysterol binding protein (*OBP1*) (Fig. 14.3). The **a** and α alleles of these genes share approximately 60% identity (21, 23). The function of the transcriptional regulators also differs between *S. cerevisiae* and *C. albicans*. In haploid *S. cerevisiae* an **a**-type cell is the default cell type, whereas specification as an α cell requires α1 to turn on α-specific genes, and α2 to repress **a**-specific genes. After mating, formation of an **a**1-α2 dimer represses haploid-specific genes and promotes meiosis. The transcriptional networks have been rewired in *C. albicans*, such that **a**2 activates **a**-specific genes and α1 activates α-specific genes, and the **a**1-α2 dimer represses mating by inhibiting the white-to-opaque transition (23, 52). Thus, *C. albicans* differs from *S. cerevisiae* in several crucial ways including the structure of the mating-type loci and control of both haploid and diploid specific gene expression.

Recently, the *MTL***a** locus of *C. parapsilosis*, an asexual species closely related to *C. albicans*, was sequenced (42). The structure of the *C. parapsilosis* *MTL***a** locus is similar to the *C. albicans* *MTL***a** locus, with conserved syntenic relationships among the genes contained in the locus (Fig. 14.3). However, sequence analysis revealed that the **a**1 gene is a pseudogene containing four stop codons (42). Although hybridization studies suggested that *C. parapsilosis* might contain α information as well, the locus could not be identified (42). Interestingly, preliminary analyses of the *MTL*α locus of *C. lusitaniae* and the *MTL***a** locus of the *C. guilliermondii* strain sequenced by the Broad Institute reveal that these *MTL* loci also contain the same conserved set of genes as *C. albicans* (see the Broad Institute genome sequencing website above) (Fig. 14.3). However, similarly to *C. parapsilosis*, the **a**1 gene of *C. guilliermondii* appears to be defective, and interestingly *C. lusitaniae* appears to be missing α2 (J. Reedy, J. Heitman, and G. Butler, unpublished data).

These findings have interesting implications regarding the regulation of the sexual life cycle in these species. Cell identity, as an **a** or α cell, could be controlled by **a**2 or α1 as in *C. albicans*. However, since either **a**1 or α2 appears to be defective or absent, it is

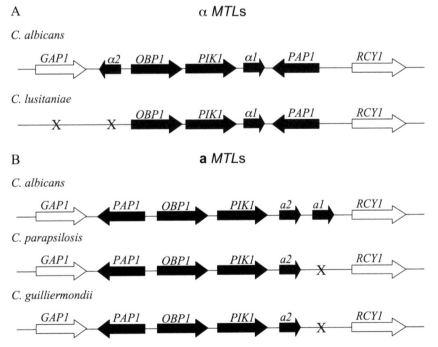

Figure 14.3 Mating-type-like loci of *Candida* species. (A) Depiction of the *MTL*α loci of *C. albicans* and *C. lusitaniae*. (B) *MTL***a** loci of *C. albicans*, *C. parapsilosis*, and *C. guilliermondii*. The genes contained within the *MTL* loci are represented by black arrows. The genes flanking the *MTL* loci are depicted in white. The absence of a gene in a particular *MTL* locus is denoted by an X.

unclear how the transcriptional roles ascribed to this heterodimer in other yeasts, primarily repression of haploid-specific genes and phenotypic switching, or activation of meiosis, are regulated in *C. lusitaniae* and *C. guilliermondii*. We hypothesize that since *C. lusitaniae* and *C. guilliermondii* are not known to occur as stable diploids and mating is rapidly followed by only a transient diploid phase from which meiosis and sporulation immediately ensue, the ancestral roles of the a1/α2 heterodimer in the diploid state may have been lost. How *a* genes would be repressed in α cells in the absence of a functional α2 gene remains an interesting open question in how cell fate is molecularly specified.

CONCLUSIONS

The majority of work with *Candida* has focused on the human pathogens *C. albicans* and *C. glabrata*. *C. albicans* is the most prevalent of the *Candida* species isolated from patient specimens, and *C. glabrata* has increased in incidence but more importantly can become highly resistant to azole antifungals, rendering treatment more difficult. Although neither *C. albicans* nor *C. glabrata* is known to have a complete meiotic sexual cycle, several of the *Candida* species that infect humans are sexually competent, mate, undergo meiosis, and sporulate. These include *C. lusitaniae* (of interest clinically due to a propensity to develop resistance to amphotericin B), *C. krusei* (which has high natural resistance to azole antifungals), and *C. guilliermondii*. Understanding the life cycle of these *Candida* species is important in developing a complete understanding of these pathogenic fungi. Since *C. albicans* is the most common clinically isolated species, the majority of effort has been focused on understanding this yeast. However, the lack of a complete meiotic sexual cycle has limited the use of classical genetics in *C. albicans*. In addition, heterologous expression experiments are complicated by the alternative genetic code of *C. albicans* whereby the CUG codon encodes serine rather than leucine. Thus, developing sexual models in species that are more closely related to *C. albicans* may provide interesting information regarding the evolution of the signaling pathways that govern sexual reproduction and also provide useful systems for heterologous expression studies and for comparison with *C. albicans*.

Investigations of the signal transduction pathways involved in mating and potentially meiosis and sporulation in *C. albicans* have relied heavily upon the established paradigms in *S. cerevisiae*. This reliance upon *S. cerevisiae* was natural, as it is the most closely related ascomycete to *C. albicans* in which the pathways regulating sexual development have been extensively studied. Thus, to determine whether *C. albicans* can undergo meiosis (or altered forms of meiosis) and to identify the changes that potentially rendered this fungus defective in sporulation, attempts to find homologs of the crucial signaling components from *S. cerevisiae* in *C. albicans* were made (69). These studies have demonstrated that not all of the meiosis and sporulation genes are present in *C. albicans*, but the implications of this are still unknown. A potentially more robust comparison will be between *C. albicans* and other closely related *Candida* species capable of sexual reproduction, including meiosis and sporulation. If these species also lack these key meiotic components and yet undergo meiosis, we will need to broaden our view of which machinery is necessary for the process. The study of the *MTL* locus of *C. parapsilosis*, taken together with the preliminary data regarding the *MTL* loci of *C. lusitaniae* and *C. guilliermondii*, has already highlighted the potential for intriguing studies of the differentiation of signal transduction cascades and cell identity circuitry within these species.

One pervasive question is whether sexual or clonal/asexual reproduction is the ancestral state in the *Candida* complex, a question which comparative genomics between the sexual and asexual *Candida* species could help elucidate. Interestingly, there appears to be an inverse relationship between the ability to form ascospores and success as a commensal/pathogen. Species with incomplete sexual cycles, such as *C. albicans*, *C. glabrata*, and *C. parapsilosis*, account for the majority of candidal infections, suggesting that meiosis may be disadvantageous. A similar phenomenon occurs among dermatophytic fungi which cause cutaneous infections, where those species with complete sexual cycles are less successful pathogens (J. Kwon-Chung, personal communication). That sexual *Candida* species exist argues that loss or restriction of sexual reproduction is the evolved state, possibly due to the energy expenditure required to undergo meiosis or to limit genetic exchange in a pathogen highly evolved to its host niche (20). Thus, continued study of the *Candida* species complex will provide insight on interesting evolutionary questions regarding the evolution of signal transduction pathways, sexual reproduction, commensalism, and pathogenesis.

We thank the members of the Heitman lab for stimulating discussions and helpful suggestions. We also thank C. Kurtzman, J. Kwon-Chung, J. Kronstad, and J. McCusker for comments on the chapter.

References

1. **Ahearn, D. G., and M. S. McGlohn.** 1984. In vitro susceptibilities of sucrose-negative *Candida tropicalis*, *Candida lusitaniae*, and *Candida norvegensis* to amphotericin B, 5-fluorocytosine, miconazole, and ketoconazole. *J. Clin. Microbiol.* 19:412–416.

2. **Bai, F. Y.** 1996. Separation of *Candida fermentati comb. nov.* from *Candida guilliermondii* by DNA base composition and electrophoretic karyotyping. *Syst. Appl. Microbiol.* 19:178–181.

3. **Bai, F. Y., H. Y. Liang, and J. H. Jia.** 2000. Taxonomic relationships among the taxa in the *Candida guilliermondii* complex, as revealed by comparative electrophoretic karyotyping. *Int. J. Syst. Evol. Microbiol.* 50(Pt. 1):417–422.

4. **Barnett, J. A., R. W. Payne, and D. Yarrow.** 2000. *Yeasts: Characteristics and Identification*, 3rd ed. Cambridge University Press, Cambridge, United Kingdom.

5. **Barns, S. M., D. J. Lane, M. L. Sogin, C. Bibeau, and W. G. Weisburg.** 1991. Evolutionary relationships among pathogenic *Candida* species and relatives. *J. Bacteriol.* 173:2250–2255.

6. **Bennett, R. J., and A. D. Johnson.** 2003. Completion of a parasexual cycle in *Candida albicans* by induced chromosome loss in tetraploid strains. *EMBO J.* 22:2505–2515.

7. **Bennett, R. J., and A. D. Johnson.** 2005. Mating in *Candida albicans* and the search for a sexual cycle. *Annu. Rev. Microbiol.* 59:233–255.

8. **Blinkhorn, R. J., D. Adelstein, and P. J. Spagnuolo.** 1989. Emergence of a new opportunistic pathogen, *Candida lusitaniae*. *J. Clin. Microbiol.* 27:236–240.

9. **Cai, J., I. N. Roberts, and M. D. Collins.** 1996. Phylogenetic relationships among members of the ascomycetous yeast genera *Brettanomyces*, *Debaryomyces*, *Dekkera*, and *Kluyveromyces* deduced by small-subunit rRNA gene sequences. *Int. J. Syst. Bacteriol.* 46:542–549.

10. **Daniel, H. M., T. C. Sorrell, and S. A. Meyer.** 2001. Partial sequence analysis of the actin gene and its potential for studying the phylogeny of *Candida* species and their teleomorphs. *Int. J. Syst. Evol. Microbiol.* 51:1593–1606.

11. **Diezmann, S., C. J. Cox, G. Schonian, R. J. Vilgalys, and T. G. Mitchell.** 2004. Phylogeny and evolution of medical species of *Candida* and related taxa: a multigenic analysis. *J. Clin. Microbiol.* 42:5624–5635.

12. **Doi, M., M. Homma, A. Chindamporn, and K. Tanaka.** 1992. Estimation of chromosome number and size by pulsed-field gel electrophoresis (PFGE) in medically important *Candida* species. *J. Gen. Microbiol.* 138:2243–2251.

13. **Essayag, S. M., G. G. Baily, D. W. Denning, and J. P. Burnie.** 1996. Karyotyping of fluconazole-resistant yeasts with phenotype reported as *Candida krusei* or *Candida inconspicua*. *Int. J. Syst. Bacteriol.* 46:35–40.

14. **François, F., T. Noël, R. Pépin, A. Brulfert, C. Chastin, A. Favel, and J. Villard.** 2001. Alternative identification test relying upon sexual reproductive abilities of *Candida lusitaniae* strains isolated from hospitalized patients. *J. Clin. Microbiol.* 39:3906–3914.

15. **Fukuoka, T., D. A. Johnston, C. A. Winslow, M. J. de Groot, C. Burt, C. A. Hitchcock, and S. G. Filler.** 2003. Genetic basis for differential activities of fluconazole and voriconazole against *Candida krusei*. *Antimicrob. Agents Chemother.* 47:1213–1219.

16. **Gargeya, I. B., W. R. Pruitt, R. B. Simmons, S. A. Meyer, and D. G. Ahearn.** 1990. Occurrence of *Clavispora lusitaniae*, the teleomorph of *Candida lusitaniae*, among clinical isolates. *J. Clin. Microbiol.* 28:2224–2227.

17. **Hartwell, L. H.** 1980. Mutants of *Saccharomyces cerevisiae* unresponsive to cell division control by polypeptide mating hormone. *J. Cell Biol.* 85:811–822.

18. **Hayford, A. E., and M. Jakobsen.** 1999. Characterization of *Candida krusei* strains from spontaneously fermented maize dough by profiles of assimilation, chromosome profile, polymerase chain reaction and restriction endonuclease analysis. *J. Appl. Microbiol.* 87:29–40.

19. **Hazen, K. C.** 1995. New and emerging yeast pathogens. *Clin. Microbiol. Rev.* 8:462–478.

20. **Heitman, J.** 2006. Sexual reproduction and the evolution of microbial pathogens. *Curr. Biol.* 16:R711-R725.

21. **Hull, C. M., and A. D. Johnson.** 1999. Identification of a mating type-like locus in the asexual pathogenic yeast *Candida albicans*. *Science* 285:1271–1275.

22a. **Jacobsen, M. D., N. A. Gow, M. C. Maiden, D. J. Shaw, and F. C. Odds.** 2007. Strain typing and determination of population structure of *Candida krusei* by multilocus sequence typing. *J. Clin. Microbiol.* 45:317–323.

22. **Hull, C. M., R. M. Raisner, and A. D. Johnson.** 2000. Evidence for mating of the "asexual" yeast *Candida albicans* in a mammalian host. *Science* 289:307–310.

23. **Johnson, A.** 2003. The biology of mating in *Candida albicans*. *Nat. Rev. Microbiol.* 1:106–116.

24. **Kato, M., M. Ozeki, A. Kikuchi, and T. Kanbe.** 2001. Phylogenetic relationship and mode of evolution of yeast DNA topoisomerase II gene in the pathogenic *Candida* species. *Gene* 272:275–281.

25. **Kawaguchi, Y., H. Honda, J. Taniguchi-Morimura, and S. Iwasaki.** 1989. The codon CUG is read as serine in an asporogenic yeast *Candida cylindracea*. *Nature* 341:164–166.

26. **King, D., J. Rhine-Chalberg, M. A. Pfaller, S. A. Moser, and W. G. Merz.** 1995. Comparison of four DNA-based methods for strain delineation of *Candida lusitaniae*. *J. Clin. Microbiol.* 33:1467–1470.

27. **Krcmery, V., and A. J. Barnes.** 2002. Non-*albicans Candida* spp. causing fungaemia: pathogenicity and antifungal resistance. *J. Hosp. Infect.* 50:243–260.

28. **Kurtzman, C. P.** 1992. DNA relatedness among phenotypically similar species of *Pichia*. *Mycologia* 84:72–76.

29. **Kurtzman, C. P.** 1994. Molecular taxonomy of the yeasts. *Yeast* 10:1727–1740.

30. **Kurtzman, C. P., and J. W. Fell (ed.).** 1998. *The Yeasts, a Taxonomic Study*, 4th ed. Elsevier, Amsterdam, The Netherlands.

31. **Kurtzman, C. P., and C. J. Robnett.** 1998. Identification and phylogeny of ascomycetous yeasts from analysis of nuclear large subunit (26S) ribosomal DNA partial sequences. *Antonie Leeuwenhoek* 73:331–371.

32. **Kurtzman, C. P., C. J. Robnett, J. M. Ward, C. Brayton, P. Gorelick, and T. J. Walsh.** 2005. Multigene phylogenetic analysis of pathogenic candida species in the

Kazachstania (*Arxiozyma*) *telluris* complex and description of their ascosporic states as *Kazachstania bovina* sp. nov., *K. heterogenica* sp. nov., *K. pintolopesii* sp. nov., and *K. slooffiae* sp. nov. *J. Clin. Microbiol.* **43:**101–111.

33. Kwon-Chung, K. J., and J. E. Bennett. 1992. *Medical Mycology*. Lea & Febiger, Philadelphia, PA.

34. Lachance, M. A., H. M. Daniel, W. Meyer, G. S. Prasad, S. P. Gautam, and K. Boundy-Mills. 2003. The D1/D2 domain of the large-subunit rDNA of the yeast species *Clavispora lusitaniae* is unusually polymorphic. *FEMS Yeast Res.* **4:**253–258.

35. Lachke, S. A., S. R. Lockhart, K. J. Daniels, and D. R. Soll. 2003. Skin facilitates *Candida albicans* mating. *Infect. Immun.* **71:**4970–4976.

36. Lan, L., and J. Xu. 2006. Multiple gene genealogical analyses suggest divergence and recent clonal dispersal in the opportunistic human pathogen *Candida guilliermondii*. *Microbiology* **152:**1539–1549.

37. Legrand, M., P. Lephart, A. Forche, F. M. Mueller, T. Walsh, P. T. Magee, and B. B. Magee. 2004. Homozygosity at the *MTL* locus in clinical strains of *Candida albicans*: karyotypic rearrangements and tetraploid formation. *Mol. Microbiol.* **52:**1451–1462.

38. Liu, H., J. Kohler, and G. R. Fink. 1994. Suppression of hyphal formation in *Candida albicans* by mutation of a STE12 homolog. *Science* **266:**1723–1726.

39. Liu, H., C. A. Styles, and G. R. Fink. 1993. Elements of the yeast pheromone response pathway required for filamentous growth of diploids. *Science* **262:**1741–1744.

40. Lockhart, S. R., K. J. Daniels, R. Zhao, D. Wessels, and D. R. Soll. 2003. Cell biology of mating in *Candida albicans*. *Eukaryot. Cell* **2:**49–61.

41. Lockhart, S. R., C. Pujol, K. J. Daniels, M. G. Miller, A. D. Johnson, M. A. Pfaller, and D. R. Soll. 2002. In *Candida albicans*, white-opaque switchers are homozygous for mating type. *Genetics* **162:**737–745.

42. Logue, M. E., S. Wong, K. H. Wolfe, and G. Butler. 2005. A genome sequence survey shows that the pathogenic yeast *Candida parapsilosis* has a defective *MTLa1* allele at its mating type locus. *Eukaryot. Cell* **4:**1009–1017.

43. Lott, T. J., R. J. Kuykendall, and E. Reiss. 1993. Nucleotide sequence analysis of the 5.8S rDNA and adjacent ITS2 region of *Candida albicans* and related species. *Yeast* **9:**1199–1206.

44. Magee, B. B., M. Legrand, A. M. Alarco, M. Raymond, and P. T. Magee. 2002. Many of the genes required for mating in *Saccharomyces cerevisiae* are also required for mating in *Candida albicans*. *Mol. Microbiol.* **46:**1345–1351.

45. Magee, B. B., and P. T. Magee. 1987. Electrophoretic karyotypes and chromosome numbers in *Candida* species. *J. Gen. Microbiol.* **133:**425–430.

46. Magee, B. B., and P. T. Magee. 2000. Induction of mating in *Candida albicans* by construction of *MTL*a and *MTL* alpha strains. *Science* **289:**310–313.

47. Massey, S. E., G. Moura, P. Beltrão, J. R. Garey, M. F. Tuite, and M. A. S. Santos. 2003. Comparative evolutionary genomics unveils the molecular mechanism of re-

assignment of the CTG codon in *Candida* spp. *Genome Res.* **13:**544–557.

48. Merz, W. G. 1984. *Candida lusitaniae*: frequency of recovery, colonization, infection, and amphotericin B resistance. *J. Clin. Microbiol.* **20:**1194–1195.

49. Merz, W. G., J. E. Karp, D. Schron, and R. Saral. 1985. Increased incidence of fungemia caused by *Candida krusei*. *J. Clin. Microbiol.* **24:**581–584.

50. Merz, W. G., U. Khazan, M. A. Jabra-Rizk, L. Wu, G. J. Osterhout, and P. F. Lehmann. 1992. Strain delineation and epidemiology of *Candida* (*Clavispora*) *lusitaniae*. *J. Clin. Microbiol.* **30:**449–454.

51. Michel-Nguyen, A., A. Favel, C. Chastin, M. Selva, and P. Regli. 2000. Comparative evaluation of a commercial system for identification of *Candida lusitaniae*. *Eur. J. Clin. Microbiol. Infect. Dis.* **19:**393–395.

52. Miller, M. G., and A. D. Johnson. 2002. White-opaque switching in *Candida albicans* is controlled by mating-type locus homeodomain proteins and allows efficient mating. *Cell* **110:**293-302.

53. Noël, T., A. Favel, A. Michel-Nguyen, A. Goumar, K. Fallague, C. Chastin, F. Leclerc, and J. Villard. 2005. Differentiation between atypical isolates of *Candida lusitaniae* and *Candida pulcherrima* by determination of mating type. *J. Clin. Microbiol.* **43:**1430–1432.

54. Orozco, A. S., L. M. Higginbotham, C. A. Hitchcock, T. Parkinson, D. Falconer, A. S. Ibrahim, M. A. Ghannoum, and S. G. Filler. 1998. Mechanism of fluconazole resistance in *Candida krusei*. *Antimicrob. Agents Chemother.* **42:**2645–2649.

55. Pesole, G., M. Lotti, L. Alberghina, and C. Saccone. 1995. Evolutionary origin of nonuniversal CUGSer codon in some *Candida* species as inferred from a molecular phylogeny. *Genetics* **141:**903–907.

56. Ramani, R., S. Gromadzki, D. H. Pincus, I. F. Salkin, and V. Chaturvedi. 1998. Efficacy of API 20C and ID 32C systems for identification of common and rare clinical yeast isolates. *J. Clin. Microbiol.* **36:**3396–3398.

57. Rodrigues de Miranda, L. 1979. *Clavispora*, a new yeast genus of the *Saccharomycetales*. *Antonie Leeuwenhoek* **45:**479–483.

58. Sandven, P. 2000. Epidemiology of candidemia. *Rev. Iberoam. Micol.* **17:**73–81.

59. San Millán, R. M., L. Wu, I. F. Salkin, and P. F. Lehmann. 1997. Clinical isolates of *Candida guilliermondii* include *Candida fermentati*. *Int. J. Syst. Bacteriol.* **47:**385–393.

60. Santos, M. A., G. Keith, and M. F. Tuite. 1993. Nonstandard translational events in *Candida albicans* mediated by an unusual seryl-tRNA with a 5'-CAG-3' (leucine) anticodon. *EMBO J.* **12:**607–616.

61. Santos, M. A., and M. F. Tuite. 1995. The CUG codon is decoded in vivo as serine and not leucine in *Candida albicans*. *Nucleic Acids Res.* **23:**1481–1485.

62. Santos, M. A. S., C. Cheeseman, V. Costa, P. Moradas-Ferreira, and M. F. Tuite. 1999. Selective advantages created by codon ambiguity allowed for the evolution of an alternative genetic code in *Candida* spp. *Mol. Microbiol.* **31:**937–947.

63. Soll, D. R., S. R. Lockhart, and R. Zhao. 2003. Relationship between switching and mating in *Candida albicans*. *Eukaryot. Cell* **2:**390–397.

64. Srikantha, T., S. A. Lachke, and D. R. Soll. 2003. Three mating type-like loci in *Candida glabrata*. *Eukaryot. Cell* **2:**328–340.

65. Sugita, T., and T. Nakase. 1999. Non-universal usage of the leucine CUG codon and the molecular phylogeny of the genus *Candida*. *Syst. Appl. Microbiol.* **22:**79–85.

66. Suzuki, T., T. Ueda, and K. Watanabe. 1997. The 'polysemous' codon—a codon with multiple amino acid assignment caused by dual specificity of tRNA identity. *EMBO J.* **16:**1122–1134.

67. Suzuki, T., T. Ueda, T. Yokogawa, K. Nishikawa, and K. Watanabe. 1994. Characterization of serine and leucine tRNAs in an asporogenic yeast *Candida cylindracea* and evolutionary implications of genes for tRNA(Ser)CAG responsible for translation of a non-universal genetic code. *Nucleic Acids Res.* **22:**115–123.

68. Tuite, M. F., and M. A. S. Santos. 1996. Codon reassignment in *Candida* species: an evolutionary conundrum. *Biochimie* **78:**993–999.

69. Tzung, K. W., R. M. Williams, S. Scherer, N. Federspiel, T. Jones, N. Hansen, V. Bivolarevic, L. Huizar, C. Komp, R. Surzycki, R. Tamse, R. W. Davis, and N. Agabian. 2001. Genomic evidence for a complete sexual cycle in *Candida albicans*. *Proc. Natl. Acad. Sci. USA* **98:**3249–3253.

70. Ueda, T., T. Suzuki, T. Yokogawa, K. Nishikawa, and K. Watanabe. 1994. Unique structure of new serine tRNAs responsible for decoding leucine codon CUG in various *Candida* species and their putative ancestral tRNA genes. *Biochimie* **76:**1217–1222.

71. van Uden, N., and H. Buckley. 1970. *Candida Berkhout*, p. 893–1087. *In* J. Lodder (ed.), *The Yeasts: A Taxonomic Study*. North Holland, Amsterdam, The Netherlands.

72. Vaughan-Martini, A., C. P. Kurtzman, S. A. Meyer, and E. B. O'Neill. 2005. Two new species in the *Pichia guilliermondii* clade: *Pichia caribbica* sp. nov., the ascosporic state of *Candida fermentati*, and *Candida carpophlia* comb. nov. *FEMS Yeast Res.* **5:**463–469.

73. Vazquez, J. A., A. Beckley, S. Donabedian, J. D. Sobel, and M. J. Zervos. 1993. Comparison of restriction enzyme analysis versus pulsed-field gradient gel electrophoresis as a typing system for *Torulopsis glabrata* and *Candida* species other than *C. albicans*. *J. Clin. Microbiol.* **31:**2021–2030.

74. Vos, M. C., H. P. Endtz, D. Horst-Kreft, J. Doorduijn, E. Lugtenburg, H. A. Verbrugh, B. Löwenberg, S. de Marie, C. van Pelt, and A. van Belkum. 2006. *Candida krusei* transmission among hematology patients resolved by adapted antifungal prophylaxis and infection control measures. *J. Clin. Microbiol.* **44:**1111–1114.

75. Whelan, W. L., and K. J. Kwon-Chung. 1988. Auxotrophic heterozygosities and the ploidy of *Candida parapsilosis* and *Candida krusei*. *J. Med. Vet. Mycol.* **26:**163–171.

76. Wickerham, L. J. 1966. Validation of the species *Pichia guilliermondii*. *J. Bacteriol.* **92:**1269.

77. Wickerham, L. J., and K. A. Burton. 1954. A clarification of the relationship of *Candida guilliermondii* to other yeasts by a study of their mating types. *J. Bacteriol.* **68:**594–597.

78. Wickes, B. L., J. B. Hicks, W. G. Merz, and K. J. Kwon-Chung. 1992. The molecular analysis of synonymy among medically important yeasts within the genus *Candida*. *J. Gen. Microbiol.* **138:**901–907.

79. Yokogawa, T., T. Suzuki, T. Ueda, M. Mori, T. Ohama, Y. Kuchino, S. Yoshinari, I. Motoki, K. Nishikawa, S. Osawa, et al. 1992. Serine tRNA complementary to the nonuniversal serine codon CUG in *Candida cylindracea*: evolutionary implications. *Proc. Natl. Acad. Sci. USA* **89:**7408–7411.

80. Yokoyama, K., S. K. Biswas, M. Miyaji, and K. Nishimura. 2000. Identification and phylogenetic relationship of the most common pathogenic *Candida* species inferred from mitochondrial cytochrome *b* gene sequences. *J. Clin. Microbiol.* **38:**4503–4510.

81. Yoon, S. A., J. A. Vazquez, P. E. Steffan, J. D. Sobel, and R. A. Akins. 1999. High-frequency, in vitro reversible switching of *Candida lusitaniae* clinical isolates from amphotericin B susceptibility to resistance. *Antimicrob. Agents Chemother.* **43:**836–845.

82. Young, L. Y., C. M. Hull, and J. Heitman. 2003. Disruption of ergosterol biosynthesis confers resistance to amphotericin B in *Candida lusitaniae*. *Antimicrob. Agents Chemother.* **47:**2717–2724.

83. Young, L. Y., M. C. Lorenz, and J. Heitman. 2000. A *STE12* homolog is required for mating but dispensable for filamentation in *Candida lusitaniae*. *Genetics* **155:**17–29.

*Sex in Fungi: Molecular Determination
and Evolutionary Implications*
Edited by Joseph Heitman et al.
© 2007 ASM Press, Washington, D.C.

Héloïse Muller
Christophe Hennequin
Bernard Dujon
Cécile Fairhead

Ascomycetes: the *Candida MAT* Locus:

15

Comparing *MAT* in the Genomes of Hemiascomycetous Yeasts

In *Saccharomyces cerevisiae*, sexual reproduction has been studied for decades and remains an active field of genetic, biochemical, and structural studies in the actors of the signal transduction cascade that governs mating (9, 22, 64). This encompasses pheromone receptors, regulators of G-proteins and mitogen-activated protein kinases, and transcription factors, many of which were first identified in this model yeast (15, 28, 55). The basic steps of the life cycle of *S. cerevisiae* grown in the laboratory are the following; it is haplodiplontic, i.e., it can grow vegetatively as both haploid or diploid cells. In response to pheromone signaling, two haploid cells of opposite mating types (a and α) can fuse to form a diploid, which can then undergo meiosis to give rise to four haploid cells (two a and two α) (Fig. 15.1) (27). The mating type is determined in haploid cells by the presence of one of the two possible alleles, or idiomorphs, at the *MAT* locus on chromosome III: *MATa* or *MATα*. Each idiomorph contains specific transcription factors that regulate three categories of genes: *asg* (a-specific genes), *αsg* (α-specific genes), and *hsg* (haploid-specific genes) (Fig. 15.2). These cell-type-specific genes are involved in

pheromone expression and sensing, transduction cascade, and inhibition of meiosis (29).

In a cells, transcription factors from *MATa*, a1 and a2, are expressed, but are not required for the expression of *asg* (induced by the non-cell-type-specific Mcm1p transcription factor) and *hsg*. In α cells, transcription factors from *MATα*, α1 and α2, are expressed; α1 inducing *asg* and α2 repressing *asg*. In diploid cells, both idiotypes, *MATa* and *MATα*, are present in the cell and are expressed. The proteins a1 and α2 form a heterodimer that represses expression of α1 and *hsg*; thus, the three categories of genes are turned off (29) (Fig. 15.2).

In addition to the *MAT* locus, the *S. cerevisiae* genome contains two silent loci, *HML* and *HMR*, containing, respectively, the α and a information. This organization allows mating-type switching by gene conversion of the *MAT* locus with one of the two silent loci. This phenomenon is initiated by a double-strand break made by the Ho endonuclease at the *MAT* locus and occurs at every generation in mother cells, leading to mixed populations of both a and α cells (29).

Héloïse Muller, Bernard Dujon, and Cécile Fairhead, Génétique Moléculaire des Levures (URA2171 CNRS and UFR927 University P.M. Curie), Institut Pasteur, 25 rue du Docteur Roux, F-75724 Paris Cedex 15, France. **Christophe Hennequin,** Faculté de médecine P et M Curie, site St-Antoine, 27 rue Chaligny F-75571 Paris Cedex 12, France.

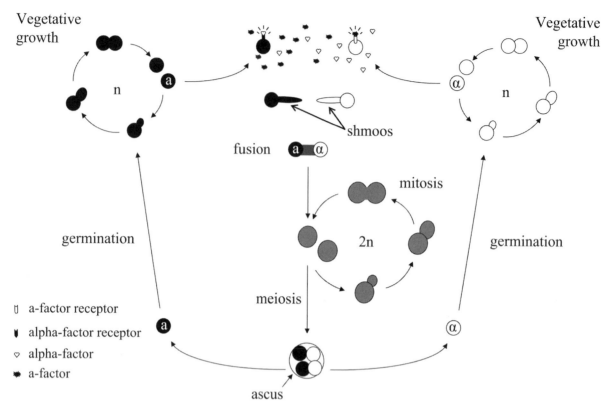

Figure 15.1 Sexual cycle of an *ho* mutant of *S. cerevisiae*. Haploid α cells are shown in white, **a** cells in black, and diploids in gray. Mating-type switching in wild-type *HO* cells is not illustrated, for the sake of simplicity. See the text for details.

S. cerevisiae serves as a model for many physiological processes, including mating, and was also the first eukaryote to be completely sequenced (in 1996) (24). Since then, many other hemiascomycetes have been partially or completely sequenced. Low-coverage sequencing of 13 hemiascomycete species (*Saccharomyces bayanus, S. exiguus, S. servazzii, S. kluyveri, Zygosaccharomyces rouxii, Kluyveromyces lactis, K. marxianus, K. thermotolerans, Debaromyces hansenii, Pichia angusta, P. sorbitophila, Candida tropicalis,* and *Yarrowia lipolytica*) was accomplished in the Génolevures I program, in 2000 (57). In 2003, two papers reported sequencing of six new species of the *Saccharomyces* complex (*S. mikatae, S. kudriavzevii, S. bayanus, S. castellii, S. kluyveri,* and *S. paradoxus*) (10, 37), and in 2004, high-coverage sequences of hemiascomycete genomes were published: four in the Génolevures II program (*Candida glabrata, K. lactis, D. hansenii,* and *Y. lipolytica*) (16), *Ashbya gossypii* by Dietrich et al. (13), *Kluyveromyces waltii* by Kellis et al. (36), and *Candida albicans* by Jones et al. (34). The Génolevures consortium is currently assembling the sequences of *Z. rouxii* and *K. thermotolerans*. Even though

some data were already available about the presence and organization of *MAT* loci in certain species, such as *Y. lipolytica, K. lactis, C. glabrata,* and *C. albicans* (1, 31, 38, 39, 58, 65), the availability of new sequences allows comparative genomics of mating-related genes on a larger scale than previously possible.

We review here the data on nine species that have high sequence coverage, with reference to *S. cerevisiae*, as shown in Fig. 15.3. Furthermore, data from species where partial sequencing covered the *MAT* and other related loci are included when appropriate. The large variety of lifestyles in this monophyletic group illustrates the plasticity of reproduction and ploidy in the evolution of species (Table 15.1); some yeasts have characterized haploid and diploid phases, others are found in a single state of ploidy, accompanied, for some species at least, with apparent absence or rarity of meiosis and cell mating. Some species are described as homothallic (self-fertile) like *S. cerevisiae*, others are described as heterothallic (self-sterile), and as noted above, others have no described sexual cycle. These physiological differences must be

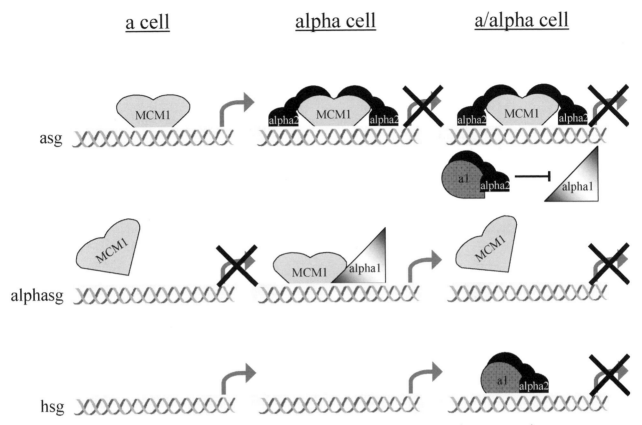

Figure 15.2 Regulation of mating type in *S. cerevisiae*. The drawing represents the promoters of *asg*, α*sg*, and *hsg* in different cell types, indicated at top (see the text for details), with the double helix representing the DNA, the arrow representing transcription, and transcription factors represented by various shapes with their names, shown as fixed to the DNA or not. In diploid cells, the *a1*/α*2* complex is shown as inhibiting α*1*, while in reality it is repressed at the transcription level.

the reflection of different genome contents, which we examine for several features, including primary genes for mating-type determination, their organization into cassettes, genes needed for silencing, and the gene encoding the Ho endonuclease.

SPECIES WITH SEVERAL *MAT*-LIKE LOCI

Multiple Blast hits in some species, using *S. cerevisiae MAT* genes as query, indicate that there are several copies of these genes. Additional genes are in fact part of additional loci, similar to the silent *HML* and *HMR* loci in *S. cerevisiae*. The species closest to *S. cerevisiae* that have several sexual loci represent the majority of species examined and are discussed first. These include *Z. rouxii* and *K. thermotolerans*, whose genomes are in the finishing stage, but where assembled contigs approach actual chromosome sizes and are considered to be chromosomes here (Génolevures consortium, personal communication).

Identification of Mating-Type-Determining Genes

The *MAT* locus in *S. cerevisiae* contains the mating-type-determining genes that encode three proteins with known functions: the homeodomain proteins a1 and α2 and the "α-domain" protein α1. All are transcription factors that regulate genes involved in the primary steps of the sexual cycle—pheromone production, pheromone sensing, signal transduction, meiosis, and mating-type switching (29, 33). This organization is apparent in all hemiascomycetous genomes, even in species that have reduced mating. We describe homologs of these genes encoded by putative *MAT*-like cassettes in the genomes of species containing multiple cassettes (see corresponding paragraph for description of cassettes). Indeed, homologs of the *a1*, α*1*, and α*2* genes are found by Blast searches in high-coverage genomes (Table 15.2) (18, 31; this work) but also in some partially sequenced genomes (*S. castellii*, *Kluyveromyces delphensis*, and *S. kluyveri*) (7). Additional genes such as *a2* and α*3* are

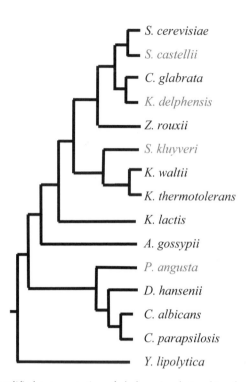

Figure 15.3 Simplified representation of phylogenic relationship of all species analyzed. Data from B. Dujon (15a). Species names in black are from high-coverage genomes, and names in gray are from partial genomes. In the case of *Ashbya gossypii*, we have kept the name of the genus used by the sequencing group, instead of *Eremothecium*.

members aligned here. The *a1* gene from *S. cerevisiae* contains two introns that are conserved in *C. glabrata*, the most closely related species among the yeasts with completely assembled genomes (18). *Z. rouxii a1* also contains an intron, but it is not in the same location as those of *S. cerevisiae* (C. Neuvéglise, personal communication, and this work), implying that the introns are of different origins. The conserved domain of α2 also corresponds to a homeodomain (residue numbers 132 to 189 in *S. cerevisiae*) recognized by Pfam (http://pfam .wustl.edu/ hmmsearch.shtml). We found no initiator codon in the sequence homologous to α2 from *K. thermotolerans*. This could be because of the presence of an as-yet-undetected exon containing the initiator codon, or it could indicate that this gene is not functional. The alignment of α1 proteins shows higher sequence conservation in this protein than the previous two, even though the alignment is subdivided into several segments.

In addition to the known transcription factor genes, the *MATa* locus from *S. cerevisiae* contains the *a2* gene. The *a2* gene corresponds to the 3′-terminal part of the α2 gene, the part that is encoded by the *X* box (see below), which is not exchanged during mating-type switching. Since it has no known function, it is possible that the *a2* gene in *S. cerevisiae* is not a true gene but a fortuitous open reading frame. The *MATa* loci of other species also contain two genes: *a1*, which is conserved as explained above, and a second gene often named *a2* when identified, as in *C. glabrata* (7). Following this, we have also named new genes in *MATa a2*. *a2* genes can be classified into two distinct groups. The first group of *a2* genes includes the dubious one from *S. cerevisiae* and one from *C. glabrata* that share 44% similarity (Fig. 15.4D). Since the *a2* gene in *C. glabrata* has no start codon, it reinforces the notion that this gene may be nonfunctional.

considered separately. Figure 15.4A, B, and C shows a multiple alignment of a1, α1, and α2, respectively, from completely assembled genomes. The conserved domain of a1 corresponds to the homeodomain that spans from residue numbers 71 to 127 in *S. cerevisiae*; the rest of the protein does not show conservation throughout all

Table 15.1 Characteristics of species with high-coverage sequences of genomes[a]

Species	Habitat	Life cycle	Mating	Multiple MAT loci	HO gene
S. cerevisiae	Grape must	Haplontic or diplontic	Homothallic	Yes	Yes
C. glabrata	Humans	Haplontic	None described	Yes	Yes
Z. rouxii	Sugary foods, fruit	Haplontic	Homothallic	Yes	Yes
K. waltii[b]	Probably insects	Haplontic	Homothallic?	Yes	No
K. thermotolerans[c]	Fruit	Diplontic?	Sporulation described	Yes	Relic
K. lactis	Dairy products	Haplontic	Heterothallic	Yes	Relic
A. gossypii	Cotton plant	Haplontic	Sporulation described	Yes	No
D. hansenii	Salted foods	Haplontic	Homothallic	No	No
C. albicans	Humans	Diploid	Parasexual cycle	No	No
Y. lipolytica	Oily foods	Haplontic or diplontic	Heterothallic	No	No

[a]All data are from reference 40.
[b]Homothallism is supposed because of mating reported in haploid cultures.
[c]Sporulation is described in strains that may be diploid, so that homo- or heterothallism cannot be deduced.

Table 15.2 Systematic nomenclature, or coordinates on contigs, when appropriate, of homologs of genes from *MAT* cassettes, in species with multiple cassettes

Species	Loci	a1	a2	α1	α2	α3
S. cerevisiae	*MAT*			YCR040w	YCR039c	NA[a]
	HML/HMR	YCR097w	YCR096c	YCL066w	YCL067c	NA
C. glabrata	*MAT*			CAGL0B01243g	CAGL0B01265g	NA
	HML/HMR	CAGL0E00341g	CAGL0E00319g	CAGL0B00242g	CAGL0B00264g	NA
Z. rouxii	*MAT/HML/HMR*	3580	3584	"6731, 6492"	"6732, 6493"	NA
K. waltii	*MAT/HML/HMR*	12991[b]	ctg_123: 21506.. 22153[c]	12992	12995	NA
K. thermotolerans	*MAT*	16735	16734			NA
	HML/HMR	17133	17132	17123	17122	na
K. lactis	*MAT*	KLLA0C03135g	KLLA0C03157g			
	HML/HMR	KLLA0B14553g	KLLA0B14575g	KLLA0C00352g	KLLA0C00374g	KLLA0C00396g
A. gossypii	*MAT*	AFR643c	AFR643w-A			NA
	HML/HMR	ADL394c, AER456w	ADL393w, AER455c			NA

[a]NA, not applicable.
[b]Two copies of gene *a1* are found in the genome (Fig. 15.5), corresponding to a single protein file (12991).
[c]*a2* not annotated, coordinates are given.

This *a2* sequence could be a pseudogene from an ancestor common to the *cerevisiae-glabrata* species, pointing to a possible difference in the sexual reproduction of the ancestor (21), or it could be the result of a gene conversion event during *MAT* conversion that would have included a part of the α allele. Perhaps this is simply a "filler" sequence, kept because gene conversion events between **a** and α cassettes are more efficient if cassettes are above a certain size or of similar sizes. This would explain the conservation of this apparent pseudogene in modern species.

The second group includes all other species (Fig. 15.4E). These genes form a true group because they share a conserved box, the HMG domain, and some of them have a known function in more-distant species (see below).

Finally, one must note the presence of an additional gene in the *MAT*α cassette of *K. lactis*, α3. This gene has no homolog in any of the genomes examined here. In cells, the two copies of α3 from *HML*- and *MAT*-like cassettes are expressed, but this gene's function is unclear, even though the double-deletion mutant is unable to mate (1).

Description of *MAT*-Like Loci and the Silenced "Cassettes"
Seven species with high-coverage genome sequences contain multiple *MAT*-like cassettes, as shown in Fig. 15.5 (1, 7, 18, 58; this work).

In *S. cerevisiae*, the three cassettes are on chromosome III, the *MAT* locus being in a central location on the right arm of the chromosome, while *HML* and *HMR* are in subtelomeric locations, respectively, on the left and right arm of the chromosome. Subtelomeres in yeasts are defined as gene-poor regions at the ends of chromosomes, encompassing several tens of kilobases, from the telomeric sequence to the first essential gene on the chromosome (19, 45). These subtelomeres contain many genes from duplicated families and have particular chromatin structures that can give rise to complex transcriptional regulation such as silencing (18) (see corresponding paragraph). This is the case of the silent *HML* and *HMR* loci themselves. In *S. cerevisiae*, *HML* typically bears the α-type information and *HMR* bears the **a**-type information, although *HMLa* and *HMR*α also occur. Loci from new genomes were assigned to be *MAT/HML/HMR*-like according to their location on chromosomes (central or subtelomeric) and to the presence of the **a** or α information in them. Apart from *S. cerevisiae*, experimental evidence that the extra copies are silenced is available only for *K. lactis* where genes in *HML*- and *HMR*-like loci are silenced, except for α3 (1). In *S. cerevisiae*, the three cassettes can be subdivided into boxes according to sequence identity. In addition to mating-type-specific sequences, named Ya and Yα, *MAT*, *HML*, and *HMR* share identical sequences on each side of the Y box, named X and Z1 boxes (see Table 15.2 for box sizes). *MAT* and *HML* share larger identical sequences that define a W box and a Z2 box (Fig. 15.4 and

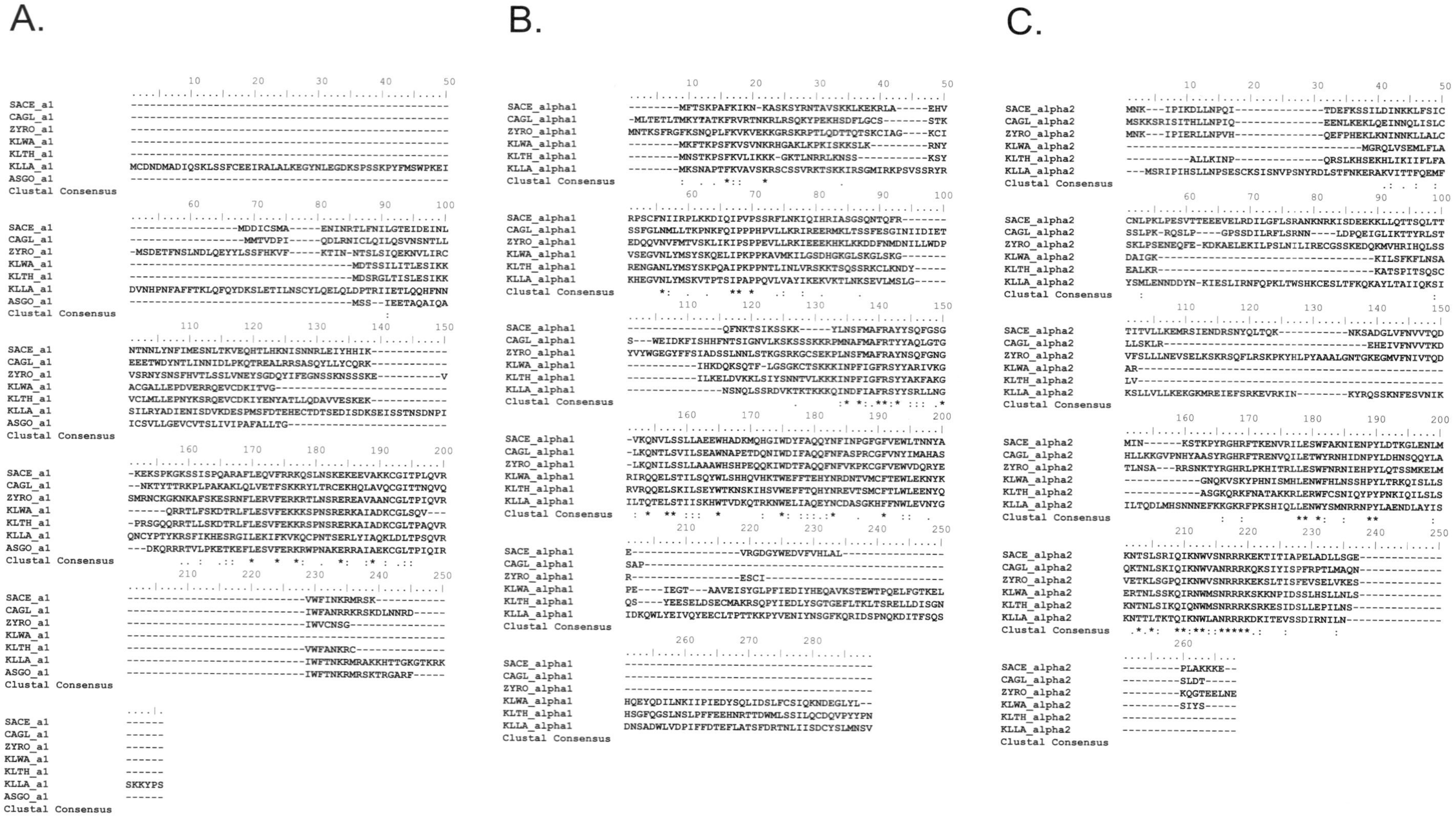

A.

B.

C.

Figure 15.4 Multiple alignments of genes from *MAT*-like loci. Alignments were done using Clustal W (61), with sequences translated from genomic data. Species are indicated thus: SACE, *S. cerevisiae*; CAGL, *C. glabrata*; ZYRO, *Z. rouxii*; KLWA, *K. waltii*; KLTH, *K. thermotolerans*; KLLA, *K. lactis*; ASGO, *A. gossypii*. (A) Alignment of a1 proteins; (B) alignment of α1 proteins; (C) alignment of α2 proteins; (D) alignment of dubious proteins a2; (E) alignment of HMG proteins a2 (see the text for details).

E.

```
                    10        20        30        40        50
           ....|....|....|....|....|....|....|....|....|....|
ZYRO_a2    MDFASECVLPKVVVKTESHKDFLFKNKKKPQSNSLQFKFVG---QKVN
KLWA_a2    ----MESEAWTQGYECQVPRKCQNSREIKKGIQK--HNTFPSI
KLTH_a2    -MILSELNKVENNNWTEGYAECHLP-KVANQIKKKFGTQKTLGKDRFPCI
KLLA_a2    ----MANSLRRUTFFKLUTTEDEDTIPKILQP------NNN
ASGO_a2    ----------MTRTINLQLPKRTST---------------YSSN
Clustal Consensus

                    60        70        80        90        100
           ....|....|....|....|....|....|....|....|....|....|
ZYRO_a2    ETNRVAEINLSSEVNKGSSKETNQIITRPRNCFIIMRSIFHNVIV-R--
KLWA_a2    GREQAHLLNYQFISDHKTKQNRSPSSKPKNKFILMRSSLHGVIQ-RLQF
KLTH_a2    AQRKCFKQICHFVTSGKTDKIKQS-GLKSRNKFILMRGLLHCMVQ-QLRN
KLLA_a2    SVAFFNLKRTGKAFTNDTIKENTKKYTRPRNQFVLMRTLFNRRVNNHILQ
ASGO_a2    FLKPAGCPKYEFVEGSKKP------SRPRNKFIIMRTIFH----------
Clustal Consensus                              ::   * *:**  ::

                   110       120       130       140       150
           ....|....|....|....|....|....|....|....|....|....|
ZYRO_a2    ..SLQKYEISSLQHVSAITSQLWGKNDGIFQLYFELLSQFEEHWHLNIYP
KLWA_a2    FSRAESLKESKVELVSKIASKLWRINKGAFQTYHELLAQFDE--YKKSFNK
KLTH_a2    CSDPFNANRAKVEEVSKIASAIWRTNKGVFQSYHEMLAVFED--SKSTNI
KLLA_a2    YYNKSKLEKKMFTLIISKITSELMNESSPDLKSYFSLLATLEKNWHKYTHY
ASGO_a2    ....NSSSKIVSAIWKHSPDQFQKYFQLLAEFEQNWHKHNHS
Clustal Consensus     * *.:*   * :*  * :.  *    *. :  : ::

                   160       170       180       190       200
           ....|....|....|....|....|....|....|....|....|....|
ZYRO_a2    EYRYHKVNKISRQLENKLVYQNMLDRMRYFTASSLIANLEDILILPPAHP
KLWA_a2    PTNKLSLKKLTEAIGRECIPS---------------------
KLTH_a2    TISKSTLKKLTEAIGRECIPS---------------------
KLLA_a2    CSWDRNSAQTLSMEPIELSQVR---------------------
ASGO_a2    PAALTDAEAFRVIARSLHPQ---------------------
Clustal Consensus

                   210       220       230       240       250
           ....|....|....|....|....|....|....|....|....|....|
ZYRO_a2    PPSSSTYTYPALRRPAAAAAAKAQAQAQAQAANAQAANPAAPAPPAT
KLWA_a2    ------------------TYSYETLTNGAHTTAIE---------
KLTH_a2    ------------------TYSYETLTNGADTTAME---------
KLLA_a2    ------PRLISSLTVGAGSSVSTYTLRELLLVRKIKSRKKRKTTLTQDSSP
ASGO_a2    ------------------PRVIKRRQQ---------
Clustal Consensus

                   260       270       280       290       300
           ....|....|....|....|....|....|....|....|....|....|
ZYRO_a2    NAERNAKTKPAPGGKVSKQRIHPKKKPSPPPRLRRPASAVQQIACQTFKS
KLWA_a2    -----SKKTYKAFCGRFSLNEATGRVRKKKPIQTF---------A
KLTH_a2    -----SKKTYKAFCGRFSLKEATGRVRKKKPTQTF---------
KLLA_a2    TTKFKYKFKKQPKTKMNSNLSKLRFKSKQPPTPPEENSN----VFKKRYT
ASGO_a2    -----KKKVKMLCGRFSR---------
Clustal Consensus                            *

                   310
           ....|....|.
ZYRO_a2    STWLNVEDLFIPNQ
KLWA_a2    LRSANIEDVFLI--
KLTH_a2    LRSAKIEDIFLI--
KLLA_a2    SENRIIEDLFLM--
ASGO_a2    -----IEDVFSSL-
Clustal Consensus      :**:.*
```

D.

```
                    10        20        30        40        50
           ....|....|....|....|....|....|....|....|....|....|
SACE_a2    ----------MRSIEND---------------------YRGHRFT
CAGL_a2    QEENLKEKLQEINNQLISLCSSLPKRQSLPGPSSDILRFLSRNNLDPQEI
Clustal Consensus                         ::.*:*:

                    60        70        80        90        100
           ....|....|....|....|....|....|....|....|....|....|
SACE_a2    --RSNVQLTQ--KNKSADGLVFNVVTQDMINKSTKP-----YRGHRFT
CAGL_a2    GLIKTTYRLSTLLSKLREHEIVFNVVTKDHLLKKGVPNHYAASYRGHRFT
Clustal Consensus   ::*.: .:   :      ******:*  *  *  *******

                   110       120       130       140       150
           ....|....|....|....|....|....|....|....|....|....|
SACE_a2    KENVRILESWFAKNIENPYLDTKGLENLMKNTSLSRIQIKNWVSNRRRKE
CAGL_a2    RENVQILETWTRNHIDNPYLDHNSQQYLAQKTNLSKIQIKNWVANRRRKQ
Clustal Consensus  ***.***:*  :*:.:******.*  : * *  *.******.*****

                   160       170
           ....|....|....|....|....
SACE_a2    KITIIAPELADILSGEPLAKKKE
CAGL_a2    -----ICKEKGPHCSF------
Clustal Consensus      *. *   *
```

Figure 15.4 *(Continued)*.

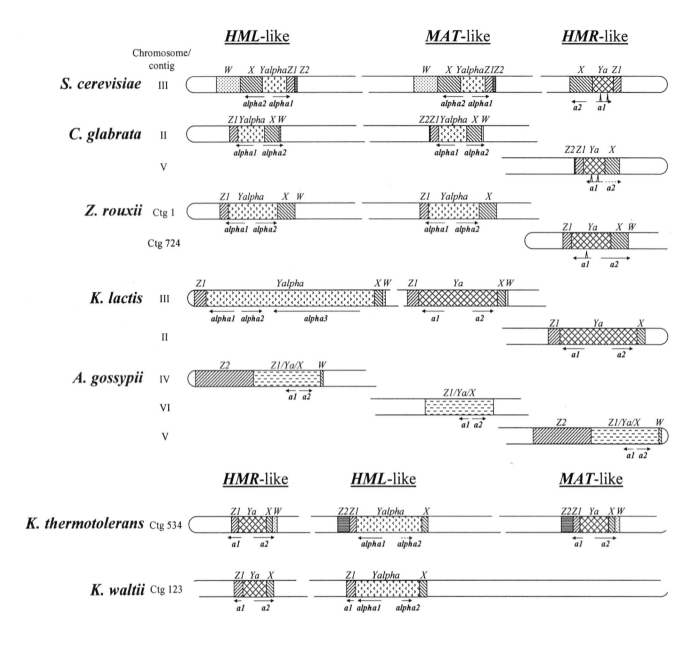

Figure 15.5 *MAT*-, *HMR*-, and *HML*-like loci in species with multiple cassettes. Species names are indicated on the left, close to chromosome or contig number. Attribution of type, *HML*-, *HMR*-, and *MAT*-like, to cassettes is shown above chromosomes. Chromosomes are drawn as double lines inside which cassette "boxes" are drawn (see the text for details); rounded ends represent telomeres.

Table 15.2). The promoter sequences of the *a1*, *a2*, *α1*, and *α2* genes are all in the Y boxes, and thus the *a1* and *a2* genes or the *α1* and *α2* genes are transcriptionally divergent: from start to end, the *a1* and *α1* genes are directed towards the Z1 box and the *a2* and *α2* genes are directed towards the X box (29). In the other species, we have identified boxes of identical and mating-type-specific sequences within *MAT*-, *HML*- and *HMR*-like loci in the same way as in *S. cerevisiae*, as shown in Fig.

15.5 and Table 15.3. Mating-type-specific regions are named *Ya* or *Yα* according to homology of gene content with *S. cerevisiae*; sequences shared by the three loci are named X and Z1; X is next to *a2* or *α2*, and Z1 is next to *a1* or *α1*; sequences shared by two of the three loci are named W, which lies next to the first X box, and Z2, which lies adjacent to the Z1 box.

As seen in Fig. 15.5, *C. glabrata*, *Z. rouxii*, and *K. lactis* have the same global organization of the loci with

Table 15.3 Sequence identity boxes in *MAT*-like cassettes[a]

Species	Sequences[b]							
	W		X (size)	Ya (size)	Yα (size)	Z1 (size)	Z2	
	Size	Loci					Size	Loci
S. cerevisiae	730	HML, MAT	704	642	747	239	88	MAT, HML
C. glabrata	18	HML, MAT	451	673	859	258	7	MAT, HMR
Z. rouxii	4	HML, HMR	537	1,233	1,549	260		
K. waltii			232	757	1,986	277		
K. thermotolerans	151	HMR, MAT	191	898	2,022	210	368	MAT, HML
K. lactis	82	HML, MAT	250	2,393	5,240	359		
A. gossypii	89	HML, HMR		2,091[c]			1,795	HML, HMR

[a]Loci are indicated only for the W and Z2 sequences, not for sequences present once at *MAT* (Y box) or present at all loci (X and Z1 boxes).
[b]Sizes are in base pairs.
[c]2,091 is in fact the size of Z1/Ya/X, see the text and Fig. 15.5.

two on the same chromosome (the *HML*- and *MAT*-like cassettes) and a third in a subtelomeric region of another chromosome. The size of cassettes in *C. glabrata* is smaller than in *S. cerevisiae*, especially the X and W boxes (Table 15.3). The sequenced strain of *C. glabrata* is *MATα*. Even though it has no known sexual cycle, both **a**- and α-type information can be found among natural strains (58). In *Z. rouxii*, the *MAT*-like locus on contig 1 from the sequenced strain encodes α-type information and shares X, Yα, and Z1 boxes with the *HML*-like locus. The *HML*-like and *HMR*-like loci share a further four identical nucleotides in a small W box (Table 15.3). *K. lactis* has longer cassettes than *S. cerevisiae*. The sequenced strain is *MATa*, and the three *MAT*-, *HML*-, and *HMR*-like loci share the Z1 and X boxes that were previously named R and L, respectively, by Astrom et al. (1).

K. thermotolerans is the only species apart from the *Saccharomyces* complex to have the three cassettes on the same contig (this work). They have been tentatively named *MAT*-, *HML*-, and *HMR*-like according to their relative position on the chromosome as described above. In an intriguing way, the *HMR*- and *HML*-like cassettes are in fact in the same subtelomere at approximately 24,000 and 32,000 bp, respectively, from the end. The third cassette is positioned centrally, at coordinates 284,800 bp on a 1,629,000-bp-long chromosome. This configuration is reminiscent of a very distant yeast, *Schizosaccharomyces pombe*. This is the first described case of such organization in a hemiascomycete, i.e., where the three cassettes are on the same chromosome, but the *MAT*-like cassette is not central to the two others (Fig. 15.5).

We have found only two cassettes on the same contig of the genome of *K. waltii* (18). One has a Ya box, and the other has a Yα box. We characterized an X box and

a Z box. The absence of a third cassette prohibited the search for W and additional Z boxes (Table 15.2). Without a complete assembly of this genome, we cannot determine the true number of cassettes, nor whether they are central or subtelomeric (see also "Conservation of Synteny around Cassettes" below). The small distance, approximately 8,000 bp, between the two cassettes resembles the organization in *K. thermotolerans*, so that perhaps the two cassettes are the silent ones and the two related species share this nonconventional organization of *MAT* loci.

The three cassettes found in *A. gossypii* all contain type **a** information and are on three different chromosomes (18). Two of the three cassettes are subtelomeric and have been arbitrarily named *HML*- and *HMR*-like. X and Z1 boxes shared by the three cassettes are likely to exist, but without any Yα in the genome, the limits between Z1, Y, and X could not be determined.

In *S. cerevisiae*, the *MAT* genes are entirely included in the cassettes, and as a consequence, the copies carried by *MAT* and one of the silent loci are identical. In other species, some STOP codons are external to the cassettes and this implies that the 3′ end of these genes differs between the copies. This is the case in *C. glabrata* for α2 and the dubious a2 gene, in *Z. rouxii* for a2, and in *K. thermotolerans* for a1. These alternate forms of a gene would not modify the putative regulation of mating type if transcriptional regulation in these species is the same as in *S. cerevisiae*, where only the copy from the *MAT*-like locus is expressed.

As can be seen in Table 15.3, considerable variation is observed in both size and location of the boxes in the cassettes. Amid the sequences that are always present, X, Y, and Z1, the Z1 sequences are the most homogeneous in size. The Y boxes are always the largest, and Ya is always smaller than Yα. This reflects the fact that the α1 gene is always larger than the *a1* gene, and also, in the case of *K. lactis*, the fact that Yα carries an additional third gene. The other two boxes, *W* and *Z2*, show variation in presence or absence and also in their location at *MAT*-, *HML*-, and/or *HMR*-like loci. Some of them are of dubious short length, such as *Z2* in *C. glabrata* or *W* in *Z. rouxii*, but we have included these nonetheless, because even short sequence identity may be of importance in recombinational mechanisms of putative cassette conversion events.

Conservation of Synteny around Cassettes

Butler et al. (7) examined conservation of synteny around the central *MAT*-like loci in several species. This defines roughly two groups: (i) species closely related to *S. cerevisiae* which share large synteny blocks with *S. cerevisiae*, in which the *BUD5* gene is always present; and (ii) more-distant species which share between themselves, excluding *S. cerevisiae*, a small synteny block consisting essentially of the *SLA2* and *SUI1* genes. *Z. rouxii* and *K. thermotolerans* share with *K. lactis* a synteny block near the *MAT* genes larger than *SLA2* and *SUI1* (this work). In *A. gossypii*, the *SUI1* gene is included in the cassette next to the *a1* gene (on the opposite side from *a2*) and so is present at all three loci, but the rest of the syntenic block is present only at the *MAT*-like locus. In the case of *K. waltii*, we searched for homologs of *SLA2* and *SUI1* but could not find homologs of these genes close to any of the two cassettes described in Fig. 15.5. Nonetheless, the presence of an *SLA2* homolog on a small isolated contig may point to a putative unassembled *MAT*-like locus.

Genome context was also examined at the *HML*- and *HMR*-like loci (this work). Globally, syntenic conservation around *MAT*- and *HML*-like loci confirms our primary identification of cassettes. A syntenic block is found near *HML*-like loci of all species studied here, except *A. gossypii*. The chromosomal region of more than 60 kb, spanning from *YCL025c* to *YCL064c*, is conserved near *HML* between *S. cerevisiae* and *C. glabrata* (using the gene nomenclature from *S. cerevisiae*, where genes are numbered from centromere to telomere). A part of the block, from *YCL050c* to *YCL064c* (roughly 22 kb in *S. cerevisiae*), is found in *Z. rouxii*, *K. waltii*, *K. thermotolerans*, and *K. lactis*. In all genomes except in *S. cerevisiae*, the syntenic fragment is on the α2-side (or *X*-side) of the *HML*-like cassettes, whereas it is on the α1-side (*Z*-side) in *S. cerevisiae*, where a specific inversion of the cassettes must have taken place, as is also visible from the orientation of the two cassettes relative to the telomere.

No conserved blocks were found near *HMR*-like loci, except between *S. cerevisiae* and *C. glabrata*, its closest relative, where a 44-kb region of synteny spans from genes *YCR079w* to *YCR095c*, even though the *HMR*-like locus in *C. glabrata* is on a chromosome different from that of the other two loci. Like *HML*, the *HMR* locus of *S. cerevisiae* is inverted compared to that of *C. glabrata*, because the syntenic genes are on the *X* side of the *HMR* cassette of *S. cerevisiae* and on the *Z* side of the *HMR*-like cassette of *C. glabrata*.

Silencing of the Additional Cassettes

Heterochromatin formation in hemiascomycetous yeasts is essential for the silencing of *HMR*- and *HML*-like loci to ensure mating-type determination. In its absence, **a**- and α-type information is expressed in all cells and they are phenotypically diploid. Conservation of proteins involved in silencing in *S. cerevisiae* was studied, but it was determined that a number of silencing factors have appeared in branches close to *S. cerevisiae* and are therefore not found by Blast search in distant species (18). The most important silencing factors in *S. cerevisiae* are the DNA binding factors Orc1p, Rap1p, and Abf1p, which bind directly to DNA sequences known as silencers, and the Sir proteins needed for establishment and maintenance of silencing: Sir1p, Sir2p, Sir3p, and Sir4p. Rap1p and Abf1p are not found in species distant from *S. cerevisiae*, contrary to Orc1p, which is present from *S. cerevisiae* to *Y. lipolytica*. Conservation of Orc1p may result from its belonging to the ORC complex that is required for initiation of DNA replication, on which a strong selective pressure is operated. Sir2p and other proteins that perform posttranscriptional modification of histones H3 and H4 are essential for creating a heterochromatin-like structure and are well conserved among hemiascomycetes, but this is not the case for the rest of the Sir proteins. Homologs of the *SIR1* gene were not detected in the genomes of any yeast except for a putative highly diverged homolog in *S. castellii* (5). This leaves open the question of whether this gene appeared de novo in *Saccharomyces* or whether it has diverged rapidly in all genomes. The *SIR3* gene in *S. cerevisiae* results from the duplication of *ORC1* in the whole-genome duplication (13, 36) that occurred in the branch containing *Saccharomyces* and *C. glabrata*, where it acquired a novel function. In species where a *SIR3* homolog is absent, it is possible that Orc1p itself assumes the silencing function. A weakly conserved *SIR4* homolog is found in *C. glabrata* and not in any other species. Functional homologs of Sir4p that do not share any sequence similarity have been found in *K. lactis* (2). For *K. waltii* and *A. gossypii*, synteny data and observation of a conserved coiled-coil domain at the carboxy-terminal part of the encoded protein have allowed identification of putative homologs of the *SIR4* gene (18).

The global mechanism of formation of heterochromatin must be conserved, and as we have seen, some functional homologs of key proteins are found in distant yeasts, even though the precise correspondence to all proteins present in *S. cerevisiae* has yet to be determined.

The Ho Endonuclease

In *S. cerevisiae*, mating-type switching is driven by the Ho endonuclease. Ho initiates a double-strand break at the *MAT* locus, which is repaired by gene conversion using one of the two silent cassettes *HML* and *HMR* as a template (29). Expression of the *HO* gene is highly regulated; it is expressed only in haploid "experienced" mother cells (cells that have budded at least once) at the end of G$_1$ phase, and the protein is rapidly degraded by the ubiquitin-26 S proteasome system (3, 29). The *HO* gene is related in sequence to the major class of inteins, selfish mobile genetic elements that can insert in frame into host genes, and is thought to derive from the "domestication" of the fungal *VMA1* protein-splicing intein element (3, 30). Two domains essential to Vma1p can still be recognized in Ho: first and perhaps surprisingly, the protein-splicing domain which autoexcises the intein from the host protein, and second, a homing endonuclease domain, with characteristic LAGLIDADG motifs, which mediates copying of the intein gene into an orthologous unoccupied intein integration point (23). Ho contains an additional zinc finger domain at its C terminus (51). In the mating-type-switching process, the Ho endonuclease cuts its site at the *MAT* locus; however, this does not result in the insertion of a copy of the *HO* gene at the cut site, but instead in the insertion of a copy of one of the silent loci.

HO homologs were searched for in hemiascomycete genomes, and it has been reported that none of the species without three cassettes have an *HO* gene (7, 18). On the other hand, a functional *HO* gene is not systematically found in species with three cassettes. The *HO* gene is indeed present in all *Saccharomyces* sensu stricto species, as well as in related species such as *C. glabrata*, *K. delphensis*, *S. castellii*, and *Z. rouxii* (7). Figure 15.6

Figure 15.6 Multiple alignments of Ho proteins. The *HO* translation from *S. cerevisiae* is the wild-type one (54), with the positions of residues that are different in the mutant underlined. Alignments were done using Clustal W, with sequences translated from genomic data. From C terminus to N terminus, boxed sequences correspond, in order, to the first LAGLIDADG motif, the first NLS motif, the second LAGLIDADG, and the second NLS.

```
                          10         20         30         40         50
                 ....|....|....|....|....|....|....|....|....|....|
S._cerevisiae_HO  MLSENTTILM ANGEIKDIAN VTANSYVMCA DGSAARVINV TQGYQKIYNI
S._castellii_HO   MIEEGTRIIM ADGQIKDIAD VTVNSYVMCE DGSSSRVTSV SKDVQTVYNV
C._glabrata_HO    MFEKGTFILM ADGHLEDISA IKSNSYVMCE DGTPGRVAYT TKAKQTIYEI
K._delphensis_HO  MFDINTTVVM ANGEIKKVCE LTVNSLVMCD DGTPARITHI SSAHHTTYEI
Z._rouxii_HO      MLEEGTKLLM ANGQIKDVGK LDVGEMVCMA DGSSAKVTSV ARDVQTTYQI
Clustal Consensus *:. .* ::* *:*.::.:   .. *** **:.::   :. *:

                          60         70         80         90        100
                 ....|....|....|....|....|....|....|....|....|....|
S._cerevisiae_HO  QQKTKHRAFE GEPGRLDPRR RTVYQRLALQ CTAGHKLSVR VPTKPLLEKS
S._castellii_HO   TQRTRHRAYE GEPGRIDPLR RQIYQRLEIN CTATHRLNLR TLTKPTLENS
C._glabrata_HO    VQKTKHRANE GEPGRLDPRR RTVYNRLGFN CSATHKLVLK TPSIPTLENN
K._delphensis_HO  YQKTKHRANE GEPGRLDPRR KTVYQRLGFN CTGSHLIPLR VPAIACLENS
Z._rouxii_HO      LQKTKHRANE GEAAEKDPLR REIHHRLGFQ CSVAHELALR TSMKPSVENC
Clustal Consensus *:*:*** * **... ** *  :::** :: *:    :. .  :*:

                         110        120        130        140        150
                 ....|....|....|....|....|....|....|....|....|....|
S._cerevisiae_HO  GRNATKYKVR WRNLQQCQTL DGRIIIPKN HHKTFPMTVE GEFAAKRFIE
S._castellii_HO   FK-RNHYVVK WKRMHNVTTI DGRVISIPKI HHKDFLMTPE GEVAAKLFLT
C._glabrata_HO    PR-RPNLTVK WRCLEEILTT DGRAITVPKN HHKNFPKTKE GQLQAQNFMR
K._delphensis_HO  TI-KPNLTVR WRCLEDVVTN DGRLISIPKN HHKNFPKTQE GFLLADNFIK
Z._rouxii_HO      FK-RNHFKVC WKNLEDTLTL DGRIIKIPKT HHKDFPMTPE GQLAAKGFLD
Clustal Consensus   :   * *::.* *** .:* * ** ***   * *. *:

                         160        170        180        190        200
                 ....|....|....|....|....|....|....|....|....|....|
S._cerevisiae_HO  EMERSKGEYF NFDIEVRDLD YLDAQLRISS CIRFGPVLTG NGVLSKFLTG
S._castellii_HO   QLQEEFGPQF NYNLELRDID YLSTQICQTT MLHYTPLLTG NGILSEFLTG
C._glabrata_HO    DKRLIWGPDI DYEIQVRDLE YLDASMRVTS TLKCNPVFSG NGILSNFLSG
K._delphensis_HO  ERSAITGDYI DYLIEVRDLD YLDTSMRVTS ALKCSPILCG NGILSKFLSG
Z._rouxii_HO      EKENSTGRFA EYNVQVRDLD ILEAQVRVNS FLRFNPLLEG NGVLSEFLTG
Clustal Consensus  :   * :*.:: **:. *:.: . :  *::. :* **:**.**:*

                         210        220        230        240        250
                 ....|....|....|....|....|....|....|....|....|....|
S._cerevisiae_HO  RSDLVTPAVK SMAWWLGLWL GDGTTKEPEI SVDSLDPKLM ESLRENAKIW
S._castellii_HO   QKHLITPYTL SMAWLLGLWL GDGTTKKPEI SVDSIDTHLM HSLIQLGSLW
C._glabrata_HO    QRHLITPAIV SMAWLLGLWI GDGTTKEPEI TVDSVDEKLM ESLTVLGRYW
K._delphensis_HO  KPHLITPSIT AMAWLLGLWL GDGTTKEPEI TMDSLDAPLM NSLIVLGRKW
Z._rouxii_HO      QKGLNSPAVL TMAWLLGLWL GDGTTKEPEI SVDSHDTGLM EGLIERGKIW
Clustal Consensus  :   * :**:****: .*****.*** ::** *  **  .*  *

                         260        270        280        290        300
                 ....|....|....|....|....|....|....|....|....|....|
S._cerevisiae_HO  GLYLTVCDDH VPLRAKHVRL HYGDGPDENR KTRNLRKNNP FWKAVTILKF
S._castellii_HO   GLDAVYKECP VPLRAKHVRL YYG---KGSP KERKFRKDNI FWNILLDLRF
C._glabrata_HO    GLYPTYKDEK VPLRAKHVRL YYGKGPEERR KTRNLRKNNP FWNTIQNLGI
K._delphensis_HO  GIYPTYKDEK VPLRAKHVRL YYGNEPEEKR KTRNLRKNNP FWNTVLALGF
Z._rouxii_HO      GLYPEYKDEQ IPLRAKHVKL FYGSECDGHR RNRHLRKNNP FWNCVVNLKF
Clustal Consensus *:   . :*******.* .** .:    :**:**** **  *  : *

                         310        320        330        340        350
                 ....|....|....|....|....|....|....|....|....|....|
S._cerevisiae_HO  KRDLDGEKQI PEFMYGEHIE VREAFLAGLI DSDGYVVKKG EGPESYKIAI
S._castellii_HO   KREDDGVKQI PEFMWTEDIE IREALLAGLI DSDGYVMKTT PNSDIFIVSI
C._glabrata_HO    KREIDGEKQV PEFMWHEDIE IREAFLAGLI DSDGYVVKKK ENPDVYKVSI
K._delphensis_HO  KKENSGEKYV PEFMWTEDIE IREAFLAGLI DSDGYVSKRK DNPDVYKVSI
Z._rouxii_HO      KRELDGEKQI PSFMWTEDLE IREAFLAGLI DSDGYVSKRK NPLDSFKVSI
Clustal Consensus *:: .* * : .**: *:.* :***:****** ******.*   . : :: :*

                         360        370        380        390        400
                 ....|....|....|....|....|....|....|....|....|....|
S._cerevisiae_HO  QTVYSSIMDG IVHISRSLGM SATVTTRSAR EEIIEGRKVQ CQFTYDCNVA
S._castellii_HO   PTIYPSIMEG IVNIARSLGM TATVTTKSAK QNVIENRQIQ CKFAYECTIS
C._glabrata_HO    QTIYPSVMNA IVHIARSLGI SATVTTRSAR NEVIEGRRVQ CRFTYDCNIS
K._delphensis_HO  QTIYPCIMDA VVHIARSLGI SATVTTRSAR PEVIEGRLVN CQFTYDCNIS
Z._rouxii_HO      QTVYPSIMGG IVHITRSLGM PVTVTTRSAK TATIVGRTVS CHFTYDCHLA
Clustal Consensus *:*.:.* : * :.*****: .*****:**: .::. * .: *:*:*:*

                         410        420        430        440        450
                 ....|....|....|....|....|....|....|....|....|....|
S._cerevisiae_HO  GGTTLQNVLS YCRSGHKTRE VPPIIKREPV YFSFTDDFQG ESTVVGLTIE
S._castellii_HO   GKTSLQNVLS YCRSGLKHRE PPEEVIRNPV YFGFNIEPSS QKNTVGLAIE
C._glabrata_HO    GSTPLQNVLS YCRSGHKRRK APEVVIRDPQ YVGFIDRKLR EAEVYGIHLD
K._delphensis_HO  GSSALQNVLS YCRSGHKRRK APESVIREPQ FFGFIDKKIS EADVKGIHFE
Z._rouxii_HO      GRTPMQKVLS YCRSGHVKVKT EPEYVERSPI YFGFNEEKRG SNNVVGVTTN
Clustal Consensus * ::.:*.*** ***** .  .* : . *. * .:.*.*:*:*

                         460        470        480        490        500
                 ....|....|....|....|....|....|....|....|....|....|
S._cerevisiae_HO  GHKNFLLGNK IEVKSCRGCC VGEQHKISQK KNLKHCVACP RKGIKYFYKD
S._castellii_HO   DNKQILLENK ICIPVCTKQC EKEQPKLTVT KNLKQCIAEP RKGVKYFYRD
C._glabrata_HO    QPRNILLGNK IVVYSCTETC ETDAVHITKI KNLKHCVSCP RTGVRYFYRD
K._delphensis_HO  DERNILLGNK VVAHCCKQEC LTEGNRLKDT KKLKYCVSCP RSGVRYFYRD
Z._rouxii_HO      SDKRILLDNK IVIHACGDHC KAEQPKLTTT RCLKYCIACP RKGVRYFYRD
Clustal Consensus  ::.*** **  *   * .. : .  *::** *  : *:.:** * .* ::**

                         510        520        530        540        550
                 ....|....|....|....|....|....|....|....|....|....|
S._cerevisiae_HO  WSGKNRVCAR CYGRYKFSGH HCINCKYVPE AREVKKAKDK GEKLGITPEG
S._castellii_HO   WTGDHRVCAR CHGRYKFSGH RCLNCKYVPE AREVRKAVAK GPKAIVSPDG
C._glabrata_HO    WTGKSHVCGR CYGRYKFSGY RCLHCKYIPE AREIKKARMR GEETRLFEQS
K._delphensis_HO  WTGKNRVCGR CYGRYKFSGY RCVNCMVYPE AREVKKAKMK GEELGIDPTG
Z._rouxii_HO      WSGRHLICGR CYGRYKFSGY RCLHCQYVPE SREIKRAKLR GEELGTSPDG
Clustal Consensus *:*. : *.* *.***:**.  * :* *:** :**::* .:.*:   . . .

                         560        570        580        590
                 ....|....|....|....|....|....|....|....|....
S._cerevisiae_HO  LPVKGPECIK CGGILQFDAV RGPHKSCGNN AGARIC---- ---------
S._castellii_HO   IVVRGLECIR CSGILVFDEI RGPKINASP  NPIMAI---- ---------
C._glabrata_HO    Y-ITGLVCSR CEGILTFDEI RGP--AKKQA VQZ------- ---------
K._delphensis_HO  IPVRGICCPR CSGILNFDEI RGPSSAHRRA MIN------- ---------
Z._rouxii_HO      TTVSGLICGK CNGILKFDEI RGPRKVTTTT DISSDIPASN ILSDISVTV
Clustal Consensus  : * : ** .* *** ** : ***
```

shows the high conservation of these Ho proteins. In addition to intein and endonuclease domains containing the two LAGLIDADG motifs, two nuclear localization signal (NLS) regions that target the protein to the nucleus (3) are also conserved (boxed in Fig. 15.6). The inactive Ho protein of standard laboratory strains of *S. cerevisiae* has four mutations at positions 189, 223, 405, and 475, underlined in Fig. 15.6 (54). The major mutation is at the end of the first LAGLIDADG motif, where a glycine residue is replaced by a serine residue. All sequences aligned here show the wild-type glycine residue at this position. Other important residues include three lysines, involved in site recognition, at positions 99, 308, and 417 (4). Lys99 is replaced by an asparagine in all genes examined, but Lys308 and Lys417 are conserved in all species.

In the genomes of more-distant species with multiple cassettes, *K. lactis* and *K. thermotolerans*, only relics of *HO* are found (41), in which the LAGLIDADG motifs are conserved (18). *K. waltii* and *A. gossypii* do not have a recognizable *HO* homolog in their genomes. It is possible that the gene or a pseudogene is present in *K. waltii* but not contained within the sequenced portion of the genome.

In *S. cerevisiae*, the Ho cut site is present in all three cassettes at the border between the *Y* and *Z1* boxes, but the endonuclease cuts only at *MAT*. We have examined this sequence in all species where such boxes were defined, and a partially degenerate site can be identified only in *C. glabrata* (18; this work). Indeed, there is some experimental evidence for mating-type switching in *C. glabrata* (6, 7). The absence of sites is surprising in the case of the other species where multiple cassettes and an apparently intact *HO* gene are found.

SPECIES WITH A UNIQUE *MAT* LOCUS

Hemiascomycete species more distantly related to *S. cerevisiae* usually have a single *MAT*-like locus; these loci are organized in various fashions depending on the species, but nonetheless, all encode distant homologs of some of the genes housed at *MAT* in *S. cerevisiae*. The additional *a2* genes described in these species are all of the HMG type (see above) and are experimentally characterized as bona fide genes in some species, even if they have different functions: *a2* represses conjugation in *Y. lipolytica* diploid cells (39), but positively regulates *asg* in *C. albicans* cells (62).

In *S. kluyveri*, given its location on the phylogenetic tree, it is surprising not to find additional cassettes, but indeed, there are no published reports of mating-type switching in this species (2). Its unique *MAT*-like locus encodes either type **a** (*a1* and *a2*) or α (α*1* and α*2*) information (7).

Y. lipolytica has a *MAT* locus that bears A or B information (38). It must be noted that *MATA1* has no similarity to the *a1* gene in other species (39). The *MATB* locus is homologous to *MAT*α in *S. cerevisiae*, with *MATB1* and *MATB2* genes corresponding to α*1* and α*2*.

P. angusta and *D. hansenii* both have a single *MAT*-like locus that is a mosaic of *MATa* and *MAT*α genes (*a1*, α*1*, and α*2* in *P. angusta* and *a1*, *a2*, and α*1* in *D. hansenii*). Both species are reported to be homothallic (40, 42). It is possible that they are true homothallic species with only one mating type, i.e., any cell can mate with any other. Nonetheless, it is possible that opposite mating types exist, allowing **a** type and α type to be differentially expressed within the unique *MAT*-like locus, with mating-type switching the result of a change in transcriptional regulation inside the *MAT*-like locus. Only experimental analysis of gene expression and cellular mating will determine the situation in both *P. angusta* and *D. hansenii*.

C. albicans has a more complex *MAT*-like locus (or mating-type-like [MTL]) because it contains not only homologs of the genes encoding transcription factors but also three other genes not implicated in regulation of mating type (31, 62). Thus, *MTLa* carries *a1*, *a2* (encoding a protein with an HMG domain), *PAPa* [poly(A) polymerase], *OBPa* (oxysterol-binding protein-like protein), and *PIKa* (phosphatidylinositol kinase); and *MTL*α carries α*1*, α*2*, *PAP*α, *OBP*α, and *PIK*α. These additional genes are conserved in hemiascomycetes but are not linked to each other or to the *MAT* locus. Their presence in the *MAT* locus of *C. albicans* is intriguing, especially as their location is not identical in both idiomorphs. Since the genome is diploid, the two idiomorphs are present in the sequenced strain and the specific DNA sequence is 8,742 bp long in *MATa* and 8,861 bp long in *MAT*α (31). The *a1* and *a2* genes are in the same orientation, and both contain two introns (7, 62). The second intron of the *a1* gene is conserved in *S. cerevisiae* (31).

In the diploid *C. parapsilosis*, the *MATa* locus is homologous to the one of *C. albicans*, but the *MTLa1* open reading frame is a pseudogene: it has four internal STOP codons, and its predicted introns are not spliced (44). It is not known whether a *MAT*α locus exists; if it does, it is not in the same chromosomal context as the *MATa* locus, based on hybridization and PCR data (44).

DISCUSSION

The hemiascomycetous yeasts represent a monophyletic group with small genomes, but the range of evolutionary distances within it is comparable to those in the whole phylum of chordates, and we have seen that even though sexual cycles rely on conserved mechanisms, species seem to have evolved special features presumably adapted to their lifestyle and environment.

The species of hemiascomycetes phylogenetically closer to *S. cerevisiae* contain several cassettes. It is interesting to recall that the two model yeasts *S. cerevisiae* and *S. pombe* (an archaeascomycete) are genetically set to be heterothallic (i.e., cells can mate only if they have different mating types) but are in reality homothallic as a consequence of complex mechanisms of mating-type switching. Since these two species are so distant phylogenetically and since underlying mechanisms of switching are different in spite of an overall similarity (there is no double-strand break created by an endonuclease in *S. pombe*; nonetheless, switching occurs by replacement of an expressed cassette by a copy of one of two silent cassettes), it is assumed that they have acquired their cassettes and mating-type-switching mechanisms independently in evolution (17). The configuration of cassettes in *S. pombe* differs from that of *S. cerevisiae*: the three mating-type loci are closely spaced, and the expressed locus is not central to the other two. The present study shows that species of the hemiascomycete phylum can have an organization of mating-type loci more similar to that seen in *S. pombe* than to that of *S. cerevisiae*. Another point concerns the orientation of cassettes: even though inversion of cassettes can be observed between species, within each species, when two or more cassettes are on the same chromosome, they are always in a direct-repeat orientation (WXYZ or ZYXW) relative to each other. This may be the consequence of gene conversion or may be necessary for gene conversion to occur. In *S. cerevisiae*, mating-type switching is rarely accompanied by crossover, which leads to chromosomal deletions because of the direct orientation of cassettes. Fusions of *MAT* with *HMRa* (Hawthorne's deletion) and of *MAT* with *HMLα* (alpha-ring chromosome) are known to be viable (29, 59). In an inverted orientation, crossovers would lead to inversion of the chromosomal fragment between *MAT* and the silent cassette. Although it is difficult to imagine, inversion of chromosomal segments may be more of a disadvantage than a deletion, and this would be enough to maintain the selection pressure on direct orientation of the cassettes. Perhaps these rare crossover events play no role, and only mechanisms of gene conversion govern the orientation of cassettes. In the three species with *HML*- and *MAT*-like cassettes on one chromosome and an *HMR*-like cassette on another, two have the same orientation of cassettes (i.e., the *HML*- and *HMR*-like cassettes are in inverted orientation relative to the telomere), but *Z. rouxii* seems to have *HML*- and *MAT*-like cassettes as in *C. glabrata*, and the *HMR*-like cassette as

in *S. cerevisiae*, as if *HML*- and *HMR*-like cassettes were inverted relative to each other.

We have also seen that sequence identity boxes within cassettes are variable in size and location, and it is known that in *S. cerevisiae*, donor preference (i.e., the choice of silent cassette with opposite information for replacement of the *MAT* cassette) does not depend on the sequence of cassettes but on the chromosomal context (25). Nonetheless, cassette conversion extends beyond the Y box into the other boxes. Subtle differences in conversion mechanisms in the different species could therefore explain the differences observed in cassette sequences. The question of how cassettes remain nearly 100% identical in species without demonstrated mating-type switching remains unanswered. Rare events of mitotic gene conversion may be sufficient in species with no functional Ho endonuclease to maintain the sequence identity.

All of these variations, in both cassette configuration and sequence composition between species along with conservation between cassettes inside a species, could point to an independent origin of cassettes among hemiascomycetes. Nonetheless, there are conserved blocks of synteny around *MAT*- and *HML*-like loci in all species, which imply one of two scenarios: either the three cassettes arose once in an ancestor to these species, and chromosomal rearrangements are the cause of differences, or only the *MAT* cassette was present in an ancestor to these species, and the extra cassettes arose several times by duplication of segments of various lengths, but with a strong selective pressure on the localization of the *HML*-like cassette. This intriguing possibility needs further investigation and will benefit from other genomic sequences in the hemiascomycetes.

All hemiascomycete species share homologous transcription factors expressed from the *MAT* loci. Since transcription factors are DNA-binding proteins that bind to specific DNA sequences, it is possible to define consensus binding sites in *S. cerevisiae*. Examination of promoters of homologs of the *asg*, *αsg*, and *hsg* in all species studied did not reveal any conserved binding sites. This is in accordance with the report of Kellis et al. (37), who identified conserved consensus sequences in promoters by cross-species comparisons, but none concerning the mating-type-specific genes.

Pathogenicity and apparent asexuality are often linked in fungi, as in *Aspergillus fumigatus*, *Pneumocystis carinii*, and *Cryptococcus neoformans*, and this is no different in hemiascomycetes, among which the two human pathogens *C. albicans* and *C. glabrata* have reduced or no apparent sexuality through different primary causes. *C. glabrata* is exclusively known as a haploid, but genome survey confirms the presence of homologs of all genes needed for the complete sexual cycle in *S. cerevisiae*: the *MAT*-like locus as well as *HML*- and *HMR*-like loci; the proteins needed for silencing these loci, with the exception of Sir1p, which may not be essential, as exemplified by the silencing of surface adhesion genes in its absence in *C. glabrata* (12); proteins of the transduction cascade; pheromones and proteins needed for their maturation; and the Ho endonuclease (7, 18, 58, 65). The two different mating types, **a** and α, are found in *C. glabrata* isolates, but population studies point to clonal reproduction with little, but some, evidence of sexual recombination (14) (C. Hennequin, unpublished data). The apparent absence of mating in this species still needs further investigation. In *C. albicans*, mating genes have also been found, but it is isolated only as a diploid. Nonetheless, parasexuality was demonstrated with the fusion of two diploids (32, 47). Conservation of mating pathways is again demonstrated by cross-species mating between *C. albicans* and the closely related species *C. dubliniensis* (52). This parasexuality resembles the sexuality of *S. cerevisiae* as it was shown to involve the same pathways and proteins (46, 67) but is much more complex because of the presence of supplementary steps of phenotypic switching from white to opaque form that occurs only in homozygote cells for the *MAT*-like locus and is necessary for cell fusion (48, 56). It has been proposed that this additional layer of regulation, compared to *S. cerevisiae*, may allow a closer control of mating in mammalian hosts and illustrates the role of rewiring of transcriptional circuitry in evolution (62). Finally, the status of the phytopathogen *A. gossypii* is uncertain. We have seen that the sequenced strain, the only strain characterized to date, is haploid and that the three cassettes contain **a**-type information, thus prohibiting the appearance of an α cell in the population. Nonetheless, sporulation in culture is reported (40; P. Philippsen, personal communication). This could be the result of either monokaryotic fruiting in this filamentous hemiascomycete, involving diploidization and meiotic recombination, as has been reported for *C. neoformans* (43), or packaging of mitotic nuclei (P. Philippsen, personal communication).

Among the hemiascomycetes presented here, different evolutionary paths leading to homothallism originated from a heterothallic ancestor to hemiascomycetes, even though it is believed that this heterothallic ancestor must itself be descended from a single-sexed homothallic organism (21). The configuration of the hypothesized heterothallic ancestor could have resembled that of the contemporary *Y. lipolytica*. One path to homothallism consists of mixing **a** and α loci in a diploid cell to obtain a mosaic of transcription factors in a homothallic yeast as is the case for *D. hansenii* and *P. angusta* (42). This situation can still give rise to two different hypotheses for underlying mechanisms: first, all cells can fuse with all others, a mechanism proposed for the distant ascomycete *Cochliobolus* (66); second, the mosaic *MAT* locus is subject to differential transcriptional control and secondary homothallism occurs by mating-type switching through regulatory change, a mechanism that was proposed for some filamentous ascomycetes by Coppin et al. (11). This mechanism is also suggested in the appearance of two different mating types from a single-sexed ancestor (21). The second path to homothallism consists of a complex system of silent copies of the two idiomorphs of the *MAT* locus that can be interchanged with the *MAT* locus itself, via poorly efficient mitotic gene conversion. Then, high-efficiency switching was acquired with the Ho endonuclease, giving rise to the type of homothallism of the model yeast *S. cerevisiae*, which appeared in the ancestor to both *Saccharomyces* and *Kluyveromyces*. *HO* gene sequences in genomes indicate that it was acquired at the same time as the appearance of the silent cassettes (or quickly thereafter) at the divergence of *Candida* and *Kluyveromyces* species. The preferred scenario for the origin of Ho is that the intein Vma1p was "hijacked" by the cell in order to increase the frequency of mating-type-switching (7). A "reverse" scenario was proposed by Keeling and Roger (35), in which it is the selfish intein element which takes advantage of the mating-type-switching system for its own propagation, by enabling mating in clonal subpopulations. Although an attractive hypothesis for *Saccharomyces*, it cannot be generalized: in the case of the highly efficient mating-type-switching system of *S. pombe*, for the moment, no endonuclease or other DNA modifying enzyme has been shown to be involved.

Finally, understanding the advantages of the evolutionary strategy of reduced sexuality or of homothallic or heterothallic sex in the species studied remains an open field for future research. Indeed, the advantages of reduced sexuality seem to be partially explained in pathogenic species by the dependency of the pathogen to its host. Infrequent sexuality would favor stability of genomes and thus continued adaptation to the host by avoiding meiotic recombination. In fact, nonmating is also observed in nonpathogenic symbionts, as in all the Glomeromycota (50). Still, apparently asexual fungi seem to resort to mating or parasexuality some of the time, or under certain ill-defined conditions. Also, the apparent logic of genome stability linked to reduced sexuality is complicated by the recent observation that the pathogenic nonmaters *C. albicans* and *C. glabrata* undergo high rates of chromosomal rearrangements as measured by Fischer et al. (20). A more classic example, in *C. glabrata*, concerns recombination between the duplicated *EPA* genes in subtelomeres, combined with complex transcriptional regulation in these chromosomal regions (8, 12). The *FLO* genes of *S. cerevisiae* undergo similar dynamics, especially in brewing strains with highly specialized lifestyles (26, 60, 63). These genome rearrangement mechanisms are actively involved in variation of genome content in order to adapt to changes in the host environment and probably ensure the minimal level of genome plasticity required for species to survive in the absence of frequent sexual recombination.

As for fully sexual species, the advantages of homothallism, which possibly facilitates inbreeding and thus homozygosis (49) versus heterothallism, are not obvious, especially because the succession of changes between the two states in the evolutionary history of the hemiascomycetes commonly occurs. Natural populations of *S. cerevisiae* can contain both heterothallic and homothallic isolates (49). Unknown contingencies in the lifestyles, added to particular evolutionary strategies, determine the present state of reduced sexuality, homothallism, or heterothallism in species. Such contingencies and strategies may include different repertoires of recombination genes for meiosis or for *MAT* switching available to the cell (53), adaptation of lifestyle to the environment, such as obligate polyploidy, or haploidy, and other genomic and external features that determine the way genomes evolve. Further analyses of the natural habitat, lifestyle, and even ploidy of these species, and of the more recent adaptations of their genomes to the conditions under which we now study them, will help us to decipher the evolutionary strategies followed by each species. This is where future research should take us, because there must lie the conditions that have shaped the major characteristics of the genomes we are examining.

We thank the Génolevures consortium and the Génoscope for communicating sequences of Z. rouxii and K. thermotolerans prior to publication. We thank Cécile Neuvéglise for help in identifying the intron in the a1 gene of Z. rouxii and P. Philippsen for sharing unpublished data. This work was supported in part by ACI no. MIC0314 and the GDR 2354 CNRS. B.D. is a member of the Institut Universitaire de France. H.M. is a recipient of a doctoral fellowship of the Ministère de la Recherche through the University Paris 6.

References

1. **Astrom, S. U., A. Kegel, J. O. Sjostrand, and J. Rine.** 2000. *Kluyveromyces lactis* Sir2p regulates cation sensitivity and maintains a specialized chromatin structure at the cryptic α-locus. *Genetics* 156:81–91.

2. **Astrom, S. U., and J. Rine.** 1998. Theme and variation among silencing proteins in *Saccharomyces cerevisiae* and *Kluyveromyces lactis*. *Genetics* 148:1021–1029.

3. **Bakhrat, A., K. Baranes, O. Krichevsky, I. Rom, G. Schlenstedt, S. Pietrokovski, and D. Raveh.** 2006. Nuclear import of ho endonuclease utilizes two nuclear localization signals and four importins of the ribosomal import system. *J. Biol. Chem.* 281:12218–12226.

4. **Bakhrat, A., M. S. Jurica, B. L. Stoddard, and D. Raveh.** 2004. Homology modeling and mutational analysis of Ho endonuclease of yeast. *Genetics* 166:721–728.

5. **Bose, M. E., K. H. McConnell, K. A. Gardner-Aukema, U. Muller, M. Weinreich, J. L. Keck, and C. A. Fox.** 2004. The origin recognition complex and Sir4 protein recruit Sir1p to yeast silent chromatin through independent interactions requiring a common Sir1p domain. *Mol. Cell. Biol.* 24:774–786.

6. **Brockert, P. J., S. A. Lachke, T. Srikantha, C. Pujol, R. Galask, and D. R. Soll.** 2003. Phenotypic switching and mating type switching of *Candida glabrata* at sites of colonization. *Infect. Immun.* 71:7109–7118.

7. **Butler, G., C. Kenny, A. Fagan, C. Kurischko, C. Gaillardin, and K. H. Wolfe.** 2004. Evolution of the MAT locus and its Ho endonuclease in yeast species. *Proc. Natl. Acad. Sci. USA* 101:1632–1637.

8. **Castano, I., S. J. Pan, M. Zupancic, C. Hennequin, B. Dujon, and B. P. Cormack.** 2005. Telomere length control and transcriptional regulation of subtelomeric adhesins in *Candida glabrata*. *Mol. Microbiol.* 55:1246–1258.

9. **Chasse, S. A., P. Flanary, S. C. Parnell, N. Hao, J. Y. Cha, D. P. Siderovski, and H. G. Dohlman.** 2006. Genome-scale analysis reveals Sst2 as the principal regulator of mating pheromone signaling in the yeast *Saccharomyces cerevisiae*. *Eukaryot. Cell* 5:330–346.

10. **Cliften, P., P. Sudarsanam, A. Desikan, L. Fulton, B. Fulton, J. Majors, R. Waterston, B. A. Cohen, and M. Johnston.** 2003. Finding functional features in *Saccharomyces* genomes by phylogenetic footprinting. *Science* 301:71–76.

11. **Coppin, E., R. Debuchy, S. Arnaise, and M. Picard.** 1997. Mating types and sexual development in filamentous ascomycetes. *Microbiol. Mol. Biol. Rev.* 61:411–428.

12. **De Las Penas, A., S. J. Pan, I. Castano, J. Alder, R. Cregg, and B. P. Cormack.** 2003. Virulence-related surface glycoproteins in the yeast pathogen *Candida glabrata* are encoded in subtelomeric clusters and subject to RAP1- and SIR-dependent transcriptional silencing. *Genes Dev.* 17:2245–2258.

13. **Dietrich, F. S., S. Voegeli, S. Brachat, A. Lerch, K. Gates, S. Steiner, C. Mohr, R. Pohlmann, P. Luedi, S. Choi, R. A. Wing, A. Flavier, T. D. Gaffney, and P. Philippsen.** 2004. The *Ashbya gossypii* genome as a tool for mapping the ancient *Saccharomyces cerevisiae* genome. *Science* 304:304–307.

14. **Dodgson, A. R., C. Pujol, M. A. Pfaller, D. W. Denning, and D. R. Soll.** 2005. Evidence for recombination in *Candida glabrata*. *Fungal Genet. Biol.* 42:233–243.

15. **Dohlman, H. G.** 2002. G proteins and pheromone signaling. *Annu. Rev. Physiol.* 64:129–152.

15a. Dujon, B. 2006. Yeasts illustrate the molecular mechanisms of eukaryotic genome evolution. *Trends Genet.* 22:375–387.

16. Dujon, B., D. Sherman, G. Fischer, P. Durrens, S. Casaregola, I. Lafontaine, J. De Montigny, C. Marck, C. Neuveglise, E. Talla, N. Goffard, L. Frangeul, M. Aigle, V. Anthouard, A. Babour, V. Barbe, S. Barnay, S. Blanchin, J. M. Beckerich, E. Beyne, C. Bleykasten, A. Boisrame, J. Boyer, L. Cattolico, F. Confanioleri, A. De Daruvar, L. Despons, E. Fabre, C. Fairhead, H. Ferry-Dumazet, A. Groppi, F. Hantraye, C. Hennequin, N. Jauniaux, P. Joyet, R. Kachouri, A. Kerrest, R. Koszul, M. Lemaire, I. Lesur, L. Ma, H. Muller, J. M. Nicaud, M. Nikolski, S. Oztas, O. Ozier-Kalogeropoulos, S. Pellenz, S. Potier, G. F. Richard, M. L. Straub, A. Suleau, D. Swennen, F. Tekaia, M. Wesolowski-Louvel, E. Westhof, B. Wirth, M. Zeniou-Meyer, I. Zivanovic, M. Bolotin-Fukuhara, A. Thierry, C. Bouchier, B. Caudron, C. Scarpelli, C. Gaillardin, J. Weissenbach, P. Wincker, and J. L. Souciet. 2004. Genome evolution in yeasts. *Nature* 430:35–44.

17. Egel, R. 2005. Fission yeast mating-type switching: programmed damage and repair. *DNA Repair* (Amsterdam) 4:525–536.

18. Fabre, E., H. Muller, P. Therizols, I. Lafontaine, B. Dujon, and C. Fairhead. 2005. Comparative genomics in hemiascomycete yeasts: evolution of sex, silencing, and subtelomeres. *Mol. Biol. Evol.* 22:856–873.

19. Fairhead, C., and B. Dujon. 2006. Structure of *Kluyveromyces lactis* subtelomeres: duplications and gene content. *FEMS Yeast Res.* 6:428–441.

20. Fischer, G., E. P. Rocha, F. Brunet, M. Vergassola, and B. Dujon. 2006. Highly variable rates of genome rearrangements between hemiascomycetous yeast lineages. *PLoS Genet.* 2:e32.

21. Fraser, J. A., and J. Heitman. 2004. Evolution of fungal sex chromosomes. *Mol. Microbiol.* 51:299–306.

22. Gehret, A., A. Bajaj, F. Naider, and M. E. Dumont. 2006. Oligomerization of the yeast alpha-factor receptor: implications for dominant negative effects of mutant receptors. *J. Biol. Chem.* 281:20698–20714.

23. Gimble, F. S., and J. Thorner. 1992. Homing of a DNA endonuclease gene by meiotic gene conversion in *Saccharomyces cerevisiae*. *Nature* 357:301–306.

24. Goffeau, A., B. G. Barrell, H. Bussey, R. W. Davis, B. Dujon, H. Feldmann, F. Galibert, J. D. Hoheisel, C. Jacq, M. Johnston, E. J. Louis, H. W. Mewes, Y. Murakami, P. Philippsen, H. Tettelin, and S. G. Oliver. 1996. Life with 6000 genes. *Science* 274:546, 563–567.

25. Haber, J. E. 1998. Mating-type gene switching in *Saccharomyces cerevisiae*. *Annu. Rev. Genet.* 32:561–599.

26. Halme, A., S. Bumgarner, C. Styles, and G. R. Fink. 2004. Genetic and epigenetic regulation of the FLO gene family generates cell-surface variation in yeast. *Cell* 116:405–415.

27. Herskowitz, I. 1988. Life cycle of the budding yeast *Saccharomyces cerevisiae*. *Microbiol. Rev.* 52:536–553.

28. Herskowitz, I. 1995. MAP kinase pathways in yeast: for mating and more. *Cell* 80:187–197.

29. Herskowitz, I., J. Rine, and J. N. Strathern. 1992. Mating-type dertermination and mating-type interconversion

in *Saccharomyces cerevisiae*, p. 583–656. *In* E. W. Jones, J. R. Pringle, and J. R. Broach (ed.), *The Molecular and Cellular Biology of the Yeast* Saccharomyces, vol. II. *Gene Expression.* Cold Spring Harbor Laboratory Press, Cold Spring Harbor, NY.

30. Hirata, R., Y. Ohsumk, A. Nakano, H. Kawasaki, K. Suzuki, and Y. Anraku. 1990. Molecular structure of a gene, VMA1, encoding the catalytic subunit of H(+)-translocating adenosine triphosphatase from vacuolar membranes of *Saccharomyces cerevisiae*. *J. Biol. Chem.* 265:6726–6733.

31. Hull, C. M., and A. D. Johnson. 1999. Identification of a mating type-like locus in the asexual pathogenic yeast *Candida albicans*. *Science* 285:1271–1275.

32. Hull, C. M., R. M. Raisner, and A. D. Johnson. 2000. Evidence for mating of the "asexual" yeast *Candida albicans* in a mammalian host. *Science* 289:307–310.

33. Johnson, A. D. 1995. Molecular mechanisms of cell-type determination in budding yeast. *Curr. Opin. Genet. Dev.* 5:552–558.

34. Jones, T., N. A. Federspiel, H. Chibana, J. Dungan, S. Kalman, B. B. Magee, G. Newport, Y. R. Thorstenson, N. Agabian, P. T. Magee, R. W. Davis, and S. Scherer. 2004. The diploid genome sequence of *Candida albicans*. *Proc. Natl. Acad. Sci. USA* 101:7329–7334.

35. Keeling, P. J., and A. J. Roger. 1995. The selfish pursuit of sex. *Nature* 375:283.

36. Kellis, M., B. W. Birren, and E. S. Lander. 2004. Proof and evolutionary analysis of ancient genome duplication in the yeast *Saccharomyces cerevisiae*. *Nature* 428:617–624.

37. Kellis, M., N. Patterson, M. Endrizzi, B. Birren, and E. S. Lander. 2003. Sequencing and comparison of yeast species to identify genes and regulatory elements. *Nature* 423:241–254.

38. Kurischko, C., P. Fournier, M. Chasles, H. Weber, and C. Gaillardin. 1992. Cloning of the mating-type gene MATA of the yeast *Yarrowia lipolytica*. *Mol. Gen. Genet.* 232:423–426.

39. Kurischko, C., M. B. Schilhabel, I. Kunze, and E. Franzl. 1999. The MATA locus of the dimorphic yeast *Yarrowia lipolytica* consists of two divergently oriented genes. *Mol. Gen. Genet.* 262:180–188.

40. Kurtzman, C. P., and J. W. Fell. 2000. *The Yeasts: a Taxonomic Study.* Elsevier, Amsterdam, The Netherlands.

41. Lafontaine, I., G. Fischer, E. Talla, and B. Dujon. 2004. Gene relics in the genome of the yeast *Saccharomyces cerevisiae*. *Gene* 335:1–17.

42. Lahtchev, K. L., V. D. Semenova, I. I. Tolstorukov, I. van der Klei, and M. Veenhuis. 2002. Isolation and properties of genetically defined strains of the methylotrophic yeast *Hansenula polymorpha* CBS4732. *Arch. Microbiol.* 177:150–158.

43. Lin, X., C. M. Hull, and J. Heitman. 2005. Sexual reproduction between partners of the same mating type in *Cryptococcus neoformans*. *Nature* 434:1017–1021.

44. Logue, M. E., S. Wong, K. H. Wolfe, and G. Butler. 2005. A genome sequence survey shows that the pathogenic yeast *Candida parapsilosis* has a defective *MTLa1* allele at its mating type locus. *Eukaryot. Cell* 4:1009–1017.

45. Louis, E. J. 1995. The chromosome ends of *Saccharomyces cerevisiae*. *Yeast* **11**:1553–1573.

46. Magee, B. B., M. Legrand, A. M. Alarco, M. Raymond, and P. T. Magee. 2002. Many of the genes required for mating in *Saccharomyces cerevisiae* are also required for mating in *Candida albicans*. *Mol. Microbiol.* **46**:1345–1351.

47. Magee, B. B., and P. T. Magee. 2000. Induction of mating in *Candida albicans* by construction of MTLa and MTLα strains. *Science* **289**:310–313.

48. Miller, M. G., and A. D. Johnson. 2002. White-opaque switching in Candida albicans is controlled by mating-type locus homeodomain proteins and allows efficient mating. *Cell* **110**:293–302.

49. Mortimer, R. K. 2000. Evolution and variation of the yeast (*Saccharomyces*) genome. *Genome Res.* **10**:403–409.

50. Pawlowska, T. E. 2005. Genetic processes in arbuscular mycorrhizal fungi. *FEMS Microbiol. Lett.* **251**:185–192.

51. Pietrokovski, S. 1994. Conserved sequence features of inteins (protein introns) and their use in identifying new inteins and related proteins. *Protein Sci.* **3**:2340–2350.

52. Pujol, C., K. J. Daniels, S. R. Lockhart, T. Srikantha, J. B. Radke, J. Geiger, and D. R. Soll. 2004. The closely related species *Candida albicans* and *Candida dubliniensis* can mate. *Eukaryot. Cell* **3**:1015–1027.

53. Richard, G. F., A. Kerrest, I. Lafontaine, and B. Dujon. 2005. Comparative genomics of hemiascomycete yeasts: genes involved in DNA replication, repair, and recombination. *Mol. Biol. Evol.* **22**:1011–1023.

54. Russell, D. W., R. Jensen, M. J. Zoller, J. Burke, B. Errede, M. Smith, and I. Herskowitz. 1986. Structure of the *Saccharomyces cerevisiae* HO gene and analysis of its upstream regulatory region. *Mol. Cell. Biol.* **6**:4281–4294.

55. Schwartz, M. A., and H. D. Madhani. 2004. Principles of MAP kinase signaling specificity in *Saccharomyces cerevisiae*. *Annu. Rev. Genet.* **38**:725–748.

56. Soll, D. R., S. R. Lockhart, and R. Zhao. 2003. Relationship between switching and mating in *Candida albicans*. *Eukaryot. Cell* **2**:390–397.

57. Souciet, J., M. Aigle, F. Artiguenave, G. Blandin, M. Bolotin-Fukuhara, E. Bon, P. Brottier, S. Casaregola, J. de Montigny, B. Dujon, P. Durrens, C. Gaillardin, A. Lepingle, B. Llorente, A. Malpertuy, C. Neuveglise, O. Ozier-Kalogeropoulos, S. Potier, W. Saurin, F. Tekaia, C. Toffano-Nioche, M. Wesolowski-Louvel, P. Wincker, and J. Weissenbach. 2000. Genomic exploration of the hemiascomycetous yeasts. 1. A set of yeast species for molecular evolution studies. *FEBS Lett.* **487**:3–12.

58. Srikantha, T., S. A. Lachke, and D. R. Soll. 2003. Three mating type-like loci in *Candida glabrata*. *Eukaryot. Cell* **2**:328–340.

59. Strathern, J. N., E. Spatola, C. McGill, and J. B. Hicks. 1980. Structure and organization of transposable mating type cassettes in *Saccharomyces* yeasts. *Proc. Natl. Acad. Sci. USA* **77**:2839–2843.

60. Teunissen, A. W., and H. Y. Steensma. 1995. Review: the dominant flocculation genes of *Saccharomyces cerevisiae* constitute a new subtelomeric gene family. *Yeast* **11**:1001–1013.

61. Thompson, J. D., D. G. Higgins, and T. J. Gibson. 1994. CLUSTAL W: improving the sensitivity of progressive multiple sequence alignment through sequence weighting, position-specific gap penalties and weight matrix choice. *Nucleic Acids Res.* **22**:4673–4680.

62. Tsong, A. E., M. G. Miller, R. M. Raisner, and A. D. Johnson. 2003. Evolution of a combinatorial transcriptional circuit: a case study in yeasts. *Cell* **115**:389–399.

63. Verstrepen, K. J., A. Jansen, F. Lewitter, and G. R. Fink. 2005. Intragenic tandem repeats generate functional variability. *Nat. Genet.* **37**:986–990.

64. Wang, Y., and H. G. Dohlman. 2004. Pheromone signaling mechanisms in yeast: a prototypical sex machine. *Science* **306**:1508–1509.

65. Wong, S., M. A. Fares, W. Zimmermann, G. Butler, and K. H. Wolfe. 2003. Evidence from comparative genomics for a complete sexual cycle in the 'asexual' pathogenic yeast *Candida glabrata*. *Genome Biol.* **4**:R10.

66. Yun, S. H., M. L. Berbee, O. C. Yoder, and B. G. Turgeon. 1999. Evolution of the fungal self-fertile reproductive life style from self-sterile ancestors. *Proc. Natl. Acad. Sci. USA* **96**:5592–5597.

67. Zhao, R., K. J. Daniels, S. R. Lockhart, K. M. Yeater, L. L. Hoyer, and D. R. Soll. 2005. Unique aspects of gene expression during *Candida albicans* mating and possible G$_1$ dependency. *Eukaryot. Cell* **4**:1175–1190.

Basidiomycetes: the Mushrooms

*Sex in Fungi: Molecular Determination
and Evolutionary Implications*
Edited by Joseph Heitman et al.
© 2007 ASM Press, Washington, D.C.

Mary M. Stankis
Charles A. Specht

16

Cloning the Mating-Type Genes of *Schizophyllum commune*: A Historical Perspective

The intriguing nature of the mating interactions of the higher fungi has long piqued the curiosity of mycologists and geneticists alike. Forty years ago, John Raper's book, *Genetics of Sexuality in Higher Fungi* (39), summarized the scientific observations on this topic and, more importantly, presented the extensive classical genetic work done in his laboratory, mostly utilizing the easily manipulated homobasidiomycete *Schizophyllum commune*. Although other mushroom fungi, notably *Coprinopsis cinerea* (formerly *Coprinus cinereus*), were candidates for the study of mating interactions, *S. commune* was John Raper's choice of a model system for the analysis of the genes that control mating interactions between individuals and initiate sexual development. *S. commune* is easily collected in nature, where the fan-shaped, grey fruiting body is found on fallen hardwoods; it is of worldwide distribution; and it is amenable to culture in the laboratory where all phases of the life cycle can be completed within 10 to 14 days on defined media, unlike most fungal pathogens and many mycorrhizal fungi. Sporulation and spore germination are rapid and reliable. As a heterothallic species that is haploid during much of its life cycle, *S. commune* is highly suited to classical genetic studies.

Genetics of Sexuality in Higher Fungi became a definitive reference on fungal genetics. Without sacrificing scientific content, the conversational style of the little grayish-brown book seemed reminiscent of an earlier time, when scientific writing took a more elegant and understandable form. To young investigators studying mating-type genetics in the 1970s and 1980s at the University of Vermont (UVM), it had the distinction of being the most well-worn reference we used. Although we never had the pleasure of meeting John Raper, we appreciated his wry humor; quotations from *Genetics of Sexuality in Higher Fungi* are inserted occasionally in this chapter.

We ourselves were fortunate to be part of the genealogy of *Schizophyllum* geneticists: we began our work on mating type in *S. commune* at UVM under the tutelage of Robert Ullrich, a student of John Raper's from Harvard University and an authority on mating in *Schizophyllum* as well as the fungal forest pathogens *Sistotrema* and *Armillaria*; Carlene "Cardy" Raper, John Raper's wife and colleague and keeper of all the minutiae of classical genetic information as well as a pioneer of molecular techniques for this organism; and Charles Novotny, recruited

Mary M. Stankis and Charles A. Specht, University of Massachusetts Medical School, Department of Medicine, LRB-370D, Section of Infectious Diseases and Immunology, 364 Plantation St., Worcester, MA 01605.

from bacterial molecular genetics and one of the earliest researchers at UVM to utilize recombinant DNA techniques. In these three laboratories, the molecular genetic techniques were developed in studies of *Schizophyllum* that eventually provided some of the answers to the ancient mysteries of mating. Though the principal investigators have since retired, the work done in their laboratories not only has been seminal to the ongoing investigation of mating-type interactions in *Schizophyllum* but also has served as a starting point for molecular work with many other fungi as well.

The specific goals of this chapter are to describe the life cycle of *S. commune* and to ascribe function to the principal genetic determinants, the mating-type genes, that govern it, and to present the strategies that were devised for the successful isolation of these genes. In only occasionally discussing these results in the light of what is known about mating-type genes in other basidiomycete fungi and yeasts, we are deferring to the more significant overviews proffered in this book by other authors.

THE LIFE CYCLE

. . . [S]exuality in the higher fungi is no more mystifying than elsewhere; the facts, examined in proper sequence, are simple enough, but it does seem that there are quite a few of them.

J. Raper (39)

S. commune exhibits a high degree of differentiation as the organism progresses from a haploid mycelium to a dikaryotic, sporulating fruiting body, a process that is well-defined morphologically, physiologically, biochemically, and genetically. John Raper's *Genetics of Sexuality in Higher Fungi* (39) as well as book chapters by Carlene Raper (34, 35) present and review the work discussed in this section.

Several distinct stages may be delineated as *S. commune* advances through its life cycle (Fig. 16.1A). Initially, the germination of a haploid spore gives rise to a homokaryotic mycelium comprised of uninucleate, haploid cells. Each individual homokaryon has genes at two different mating-type loci, historically designated the *A* and *B* factors, that decide its mating type. In the world-wide population of *S. commune*, a large number of *A* and *B* mating specificities are conferred by the extensive series of alleles of the genes residing at these loci. When two homokaryons have different alleles of the *A* and *B* genes (shown in Fig. 16.1A as *Ai Bi* and *Aj Bj*), they are compatible. Heterokaryotic cells, which are the predecessors of the fertile dikaryon, are formed when two uninucleate cells from each compatible homokaryon fuse. Any two homokaryons may join to form heterokaryotic cells; however, only infertile heterokaryons arise from interactions between homokaryons with identical *A* and/or *B* genes.

The conversion from mated homokaryons to the fertile dikaryon ensues with the reciprocal migration of nuclei from each mate into and throughout the hyphae of the other (shown for one mate in Fig. 16.1B). The migrating nuclei ultimately reach the growing hyphal tips, where they pair, but do not fuse, with the resident nucleus. The cells of the dikaryon are thus binucleate, with one nucleus derived from each mate. In each successive cell division, the two nuclei divide synchronously, a process called conjugate division (Fig. 16.1C). Conjugate division sustains a perfect 1:1 ratio of both parental nuclear types in each cell of the dikaryon. This genetically balanced dikaryon that is found in Basidiomycetes contrasts with most other groups of fungi where heterokaryosis is found: there, the ratio of component nuclear types is variable (34). The dikaryon occurs in Ascomycetes, too, but it is a transient stage just prior to karyogamy and meiosis and is confined to the ascus; it cannot be propagated vegetatively. In contrast, the Basidiomycete dikaryon, in which the two haploid genomes can interact although they are maintained in separate nuclei, can be propagated indefinitely. The dikaryotic mycelium, while unique, may be considered

Figure 16.1 (A) Life cycle of *S. commune* in diagrammatic representation. See the text for details. (B) Nuclear migration (shown for one mate). In a compatible interaction, a reciprocal exchange of nuclei takes place: in the fusion cell, the nucleus of each mate divides, and the daughter nuclei migrate into the cells of the other mating type. The donor nuclei rapidly divide, as they migrate throughout the existing mycelium of the other mate. Eventually they reach the apical cells. The nuclei move at a speed exceeding the rate of radial growth of the mycelia, i.e., at velocities of 1 to 6 mm per h, depending on the temperature (30). At these velocities, a migrating nucleus can traverse one *S. commune* cell, which is typically 100 μm in length, in 1 min. Microtubules and microfibrils are associated with the migrating nuclei (42). Nuclear migration requires the rapid dissolution of the septa between cells, a process that has been correlated with the production of a specific hydrolytic enzyme, R-glucanase (61). Nuclear migration is regulated by the *B* mating-type genes. (C) Conjugate nuclear division. The migrating nuclei arrive at the growing hyphal tips, where they pair, but do not fuse, with the resident nucleus. A mechanism for segregating paired nuclei, common to many Basidiomycetes, ensures that each newly formed cell has two nuclei, one from each parent. This is accomplished through the formation of a unique structure called a hook cell or clamp connection. Prior to nuclear division, a short branch arises on the side of the apical cell, into which one of the nuclei moves. Both nuclei then

A. Life cycle of *Schizophyllum commune*

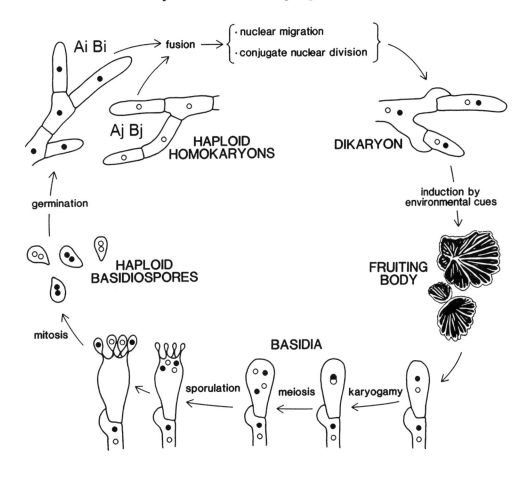

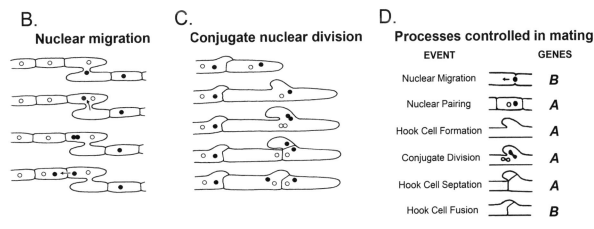

B.
Nuclear migration

C.
Conjugate nuclear division

D.
Processes controlled in mating

EVENT		GENES
Nuclear Migration		*B*
Nuclear Pairing		*A*
Hook Cell Formation		*A*
Conjugate Division		*A*
Hook Cell Septation		*A*
Hook Cell Fusion		*B*

divide in synchrony and new cell walls form perpendicular to the spindles to generate three cells. One of each of the daughter nuclei is retained in the new apical cell, one in the hook cell and one in the subterminal cell. The hook or clamp cell then fuses with the subterminal cell, providing a bridge for the previously entrapped daughter nucleus to move into the subapical cell, where it pairs with its partner. This restores the dikaryotic condition to the penultimate cell. The presence of fused hook cells (clamp connections) can be used to identify dikaryotic mycelia. The formation of the hook cell and synchronized nuclear division are regulated by the *A* genes, and the fusion of the hook cell with the subterminal cell is regulated by the *B* genes. (D) Processes controlled by the *A* and *B* mating-type genes of *S. commune*. Adapted from Stankis et al. (54).

genetically and physiologically equal to the diploid phase of other organisms.

The *S. commune* dikaryon has the potential for unlimited vegetative growth, but in response to appropriate environmental cues, including a brief exposure to light (45; reviewed in reference 60), formation of the gilled fruiting bodies occurs (Fig. 16.1A). Fruiting-body development progresses through several stages (24, 31, 41). The mature mushroom is grey, fan shaped, and 1 to 4 cm wide and has radiating, spore-bearing gills. A layer of spore-producing cells, the basidia, blanket the gills and the lower surface of the fruiting body. In the basidia, karyogamy occurs between the pair of compatible nuclei, and this transiently diploid nucleus undergoes immediate meiotic division. The four haploid meiotic nuclei each migrate into one of four spores atop the basidium and subsequently undergo a mitotic division. The spores are discharged from the basidium.

GENETIC DETERMINANTS OF THE LIFE CYCLE: *A* AND *B* MATING-TYPE LOCI

Unraveling the tangled skein of interrelations between genetic factors, mating patterns, and morphogenetic sequences is certainly worthy of our best efforts, but this has not prevented our appreciation of many of the details as well-nigh ludicrous by any standards other than those imposed by the evolutionary history of the higher fungi.

J. Raper (39)

The stages of the life cycle in *S. commune* are governed by the *A* and *B* mating-type genes that regulate mating beween individual homokaryons. As we have seen, compatible matings effect the formation of the dikaryon and culminate in fruiting, meiosis, and sporulation. Two homokaryons must have different alleles of the *A* genes and different alleles of the *B* genes in order to initiate the entire sequence of events that leads to the formation of the dikaryon. Only part of the developmental sequence is expressed in heterokaryons that are alike at either *A* or *B*. To illustrate, when *A* mating-types genes are the same and *B* genes are different, nuclear migration into and throughout the mate occurs, but further maturation is checked. Development is arrested in a continuous cycle of septal dissolution and nuclear movement between cells, resulting in submerged mycelium with little aerial hyphae, known as a "flat" phenotype (see Fowler and Vaillancourt, chapter 18, this volume.) In contrast, when *B* mating-type genes of two mates are identical, nuclear migration does not ensue, indicating that the migration of nuclei and concomitant dissolution of septa are

controlled by the *B* genes (Fig. 16.1D). When the *A* genes of two homokaryons are different and the *B* genes are the same, a heterokaryon can be formed in which there is nuclear pairing, conjugate nuclear division, the development of unfused hook cells (referred to as clamp cells in many other basidiomycetes), and hook cell septation. These processes, then, are regulated by genes at the *A* locus (Fig. 16.1D). The fusion of the hook cell to the penultimate cell is effected by the action of the *B* genes. The dikaryon is therefore the consequence of a developmental sequence that integrates two different, but coordinating, pathways: one regulated by the *A* mating-type locus and the other by the *B* mating-type locus.

Copious classical genetic observations by Kniep in the 1920s (39) on the segregation of the *A* and *B* genes revealed that they are located on different chromosomes. More recently it has been shown by DNA-DNA hybridization experiments, in which chromosomes separated by pulsed-field gel electrophoresis were probed with labeled DNA of isolated *A* and *B* genes, as well as other genes known to reside in the *A* and *B* linkage groups, that the *A* locus resides on the largest chromosome and the *B* locus is on the smallest chromosome (3, 17).

THE *A* AND *B* FACTORS ARE COMPLEX

Classical genetic analysis further revealed that the *A* and *B* mating-type genes in *S. commune* are each comprised of two linked subloci, α and β, which are separable by recombination. The cardinal rule of the mating game, then, is this: two homokaryons are fully compatible if they possess different alleles of the genes at *A*α and/or *A*β, and different alleles at *B*α and/or *B*β. The genes at α and β are functionally redundant within the *A* or *B* mating-type locus on the physiological level, because the interaction of either two different α products or two different β products triggers the relevant *A* or *B* developmental sequence.

In *S. commune* and other Basidiomycetes, the genetic variability ensured by mating is further enhanced by the multiallelic mating-type system. Classical genetics indicated that each of the four mating-type subloci has an extensive number of alleles. A statistical analysis of alleles identified in the worldwide population of *S. commune* indicates that there are 9 alleles of *A*α, 32 alleles of *A*β, 9 alleles of *B*α, and 9 alleles of *B*β (20, 40, 53). The specific mating type of a homokaryotic strain, then, is designated by its individual combination of alleles at *A*α and *A*β and at *B*α and *B*β; for example, *Schizophyllum* strain 4-40 bears the *A*α4 *A*β6 *B*α1 *B*β1 alleles.

For the *A* locus alone, with 9 alleles of *A*α and 32 alleles of *A*β, there are 288 possible different mating-type specificities, and when these are combined with all possible *B* specificities, the total number of predicted mating types is over 20,000!

> The title that appears on this book is something of a second choice, a compromise dictated by the practical necessity of a clear rapport between title and content, and it is with some regret that the title "Sexes by the Thousands" has been replaced.
>
> J. Raper (39)

Because it forbids self-compatibility and restricts sib-compatibility, a multiallelic system increases outbreeding potential within a population and thus reassortment and recombination of genetic traits. The functional redundancy of the two subloci within each mating-type locus and the large number of alleles at each further ensure this. Globally, there is a random distribution of the different alleles of the *A* and *B* genes.

MUTATIONS THAT MAP TO THE MATING-TYPE LOCI

Extraordinarily exact and regular in its supervision of mycelial interactions, the mating-type system of *S. commune* is not influenced by external factors such as nutrition and environmental conditions. It may, however, be disrupted by mutation, both in its intrinsic loci and in loci which map elsewhere in the genome and are thought to be targeted by the mating-type gene products (21, 37, 38). Primary mutations map to the *A*β, *B*α, and *B*β loci and were the consequence of attempts to mutate one allele of a given locus into an allele of different specificity, although no such mutations resulted. The basis of selection was the newly acquired ability to mate with strains carrying the original, unmutated version of the mating-type locus. Despite exhaustive efforts to mutate one allele into another, all of the numerous primary mutations isolated resulted only in the loss of self-recognition and self-incompatibility by the affected locus. All the primary mutations exert their effect constitutively and are referred to as the *con* mutations. *Con* mutations occur at a frequency of about 5×10^{-8} and can be induced by various mutagenic agents (21, 38, 39). The effect of a *con* mutation in a mating-type locus of a homokaryon simulates the effect of two different wild-type alleles of the same locus in a heterokaryon. For example, a homokaryon with a *con* mutation in *A*β (no *con* mutations were ever isolated in *A*α) exhibits the *A*-regulated series of events: nuclear pairing in each cell of the mycelium, hook cell formation, conjugate division, and

hook cell septation. Homokaryons carrying a *con* mutation in *B*α or *B*β display continual nuclear migration and disruption of septa. A homokaryon in which a *con* mutation in *A*β is combined with a *con* mutation in *B*α or *B*β closely mimics a true dikaryon and will fruit and sporulate.

The original strain collection of John Raper and his colleagues as well as many of its derivatives, containing all the predicted *B* mating-type alleles, and most of the *A* mating-type alleles, is currently available through the Fungal Genetics Stock Center (accession numbers 9098 through 9350). The collection includes many auxotrophs, mutants with modified sexual development, and specific mutants of the mating-type genes themselves (36). Without them, neither the isolation of the mating-type loci nor our current understanding of it would have been possible.

ISOLATION OF THE A MATING-TYPE GENES

In the 1960s and 1970s, many different models were proposed for the mechanism of action of the mating genes in Basidiomycetes (for a discussion of these, see reference 33). On a molecular level, any explanation must answer two questions, the first posed by multiallelism: *how is self distinguished from a large number of nonself entities?* The second question, when one observes the regulation of nuclear migration and septal dissolution by the *B* mating-type genes, and the control of conjugate nuclear division and hook-cell formation by the *A* genes, is the following: once self/nonself recognition between the alleles has occurred, *how do their products act as "master switches" to regulate the genes downstream in the developmental process?* In the 1980s, the concept that different gene products from two different individuals could interact to initiate a developmental program was unfolding. In the ascomycete yeast *Saccharomyces cerevisiae*, the regulatory proteins al and α2, encoded by *MAT* genes from different mates, were shown to form a heterodimer that blocks the transcription of haploid-specific genes, allowing only diploid cell functions to be expressed after mating (10, 15, 16). The challenge in studying mating type in Basidiomycetes would be that the multiallelic mating genes generate a large number of different proteins that can or cannot successfully interact, depending on self/nonself recognition, to regulate the developmental cascade.

> All later work in this area is the direct outgrowth of these earlier investigations, and present endeavor owes much of its success to the meticulous observations and keen perception of those who compiled this early history.
>
> J. Raper (39)

As we have seen, the many years of inquiry into the self/nonself recognition system of *S. commune* by John and Carlene Raper and their colleagues set the stage for the isolation of the mating-type genes and an investigation of how they regulate sexual development on a molecular level. By the late 1970s and early 1980s the next step was the development of a molecular toolbox in *Schizophyllum* to enable study on a molecular level (reviewed in reference 35): the Ullrich and Novotny laboratories took up the gauntlet at the University of Vermont, where Cardy Raper also established a laboratory in 1984.

The isolation of a gene from a DNA library commonly entails the complementation of a specific mutation through the importation of the wild-type gene into the mutant organism via genetic transformation. Some selective process, for example, the restoration of prototrophy in the case of the isolation of a nutritional gene, becomes the basis for recognition of a successfully transformed cell. The next step in the isolation of the mating-type genes was the development of an efficient transformation system in *S. commune*. To this end, the wild-type *TRP1* gene of *Schizophyllum* (encoding indole-3-glycerolphosphate synthetase) was isolated by complementation of a *trpC* mutant of *Escherichia coli* (29) and was used to transform protoplasts of a *trp1* mutant *S. commune* strain (28) to prototrophy (Trp+). (It may be noted that *S. commune* genetic notation follows the standard rules governing yeast nomenclature.) *S. commune* was the first filamentous basidiomycete for which a transformation system was developed to enable the study of genetic determinants on a molecular level.

The efficiency of the transformation system was sufficiently high, using protoplasts made from basidiospores, to allow the isolation of nutritional genes, such as *URA1*, from plasmid clone banks through direct complementation of the relevant *Schizophyllum* auxotrophic mutation (12). One of these genes was *PAB1* (encoding synthesis of para-aminobenzoic acid), a mere 0.3 centimorgans (cM) from *Aα* (12, 39). This made it feasible to isolate the *Aα* mating-type gene by "walking" the chromosome from *PAB1*, but how would one know when one got there? A selection process was conceived based on the following rationale. In *S. commune*, integration of transforming DNA is predominantly a nonhomologous event, and a typical transformant is merodiploid for the gene of interest. Because, in a normal mating, the intracellular presence of the product of two different *Aα* alleles activates the *A*-regulated series of developmental events, the introduction by transformation of *Aα* DNA into the genome of a homokaryotic recipient carrying a different *Aα* allele should trigger the

same cascade. Thus, transformants at each step of the walk would be screened for the induction of *A*-regulated events that could be easily observed by microscope, such as the production of unfused hook cells (Fig. 16.2A).

Genomic DNA was isolated from a strain carrying the *Aα4* allele to make the library used in the chromosomal walk from *PAB1*. The library was constructed in a cosmid vector in which the *S. commune TRP1* gene would be the selectable marker in transformation (28). The recipient of the cosmid DNA at each step of the walk was a strain with the markers *Aα1 pab1 ade5 Aβ6 trp1* (*ADE5* encodes phosphoribosylaminoimidazole synthetase). Transformants selected on the basis of tryptophan prototrophy (Trp+) were then observed microscopically for hook cell formation. On the third step of the walk, about 50 kb from *PAB1*, cosmids with approximately 9 kb of sequence in common were found that conferred *Aα* activity in transformants: the Trp+ transformants exhibited hook cells characteristic of *A*-activated development. In mating tests, they formed dikaryons with either *Aα1 Aβ6* or *Aα4 Aβ6* strains with compatible *B* genes and subsequently fruited (13, 57). Thus, the addition of *Aα4* DNA to the genome of the recipient *Aα1* strain successfully activated the cascade of *A*-regulated events. Analyses of the transformants showed that the *Aα4* transforming DNA integrated unlinked to the resident *Aα1* locus and revealed that the *Aα* DNA can operate in *trans* (13, 57). This argued that the *Aα* DNA possessed its own promoter sequences and that it encoded some diffusable product. In addition, the *Aα4* sequence had no effect when used to transform strains carrying *Aα4* as a resident allele; *A*-regulated activity was detected only when the *Aα4* sequence was used to transform strains in which the *Aα* allele was different. Sequences from the cosmids carrying *Aα4* activity were then used as probes to recover the *Aα* alleles from gene libraries made from strains carrying *Aα1* and *Aα3* alleles (55), and eventually the *Aα5* allele (64) and *Aα6* allele (47).

DNA-DNA hybridization experiments revealed that *Aα* DNA is encoded in a single copy in the *S. commune* genome, i.e., there are no silent cassettes of multiple *Aα* alleles like those observed in yeasts (4, 56). This technique also provided the first approximation of the degree of relatedness between the alleles. A 1.2-kb *Aα4*-specific DNA fragment was hybridized to two DNA blots: one blot carried DNA from nine different strains, each encoding one of the nine *Aα* alleles from the worldwide population of *S. commune*; and a second blot had DNA from several different *Aα4* strains collected from geographically diverse areas.

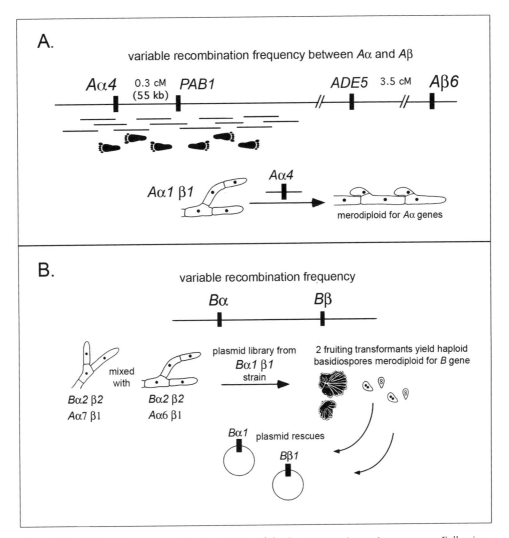

Figure 16.2 (A) Selection strategy for the isolation of the *S. commune* *A*α mating-type gene. Following the isolation of the linked *PAB1* gene (12), a chromosomal "walk" to *A*α4 was accomplished by DNA-DNA hybridization of overlapping clones from a cosmid library (13). Hypothesizing that the addition of *A*α4 DNA to the genome of the recipient *A*α1 strain, to create transformants merodiploid for *A*α, would activate *A*-regulated developmental events, transformants at each step of the walk were screened by light microscopy for unfused hook cells. (B) Selection strategy for the isolation of the *S. commune* *B*α and *B*β mating-type genes. A mixture of protoplasts prepared from two strains having compatible *A* mating types (*A*α7 *A*β1 and *A*α6 *A*β1) and the same *B* mating type (*B*α2 *B*β2) were transformed with a genomic library constructed with DNA from a *B*α1 *B*β1 strain. Full sexual development, i.e., the formation of fruiting bodies, could occur if a transformant, merodiploid for either the *B*α or *B*β mating-type genes, mated with an *A*-compatible regenerate, or fused with one during protoplast manipulation. Two fruiting transformants yielded haploid basidiospores which revealed the *B*-activated phenotype (49). The transforming *B*α1 and *B*β1 DNA was recovered using plasmid rescue techniques (12).

The *A* factor becomes "Curiouser and Curiouser."

J. Raper (39)

Astonishingly, the *A*α4 probe bound only to DNA from *A*α4 strains and not to DNA from strains carrying any of the other eight alleles. An *A*α1 DNA probe dis-closed a parallel pattern: it hybridized only to the DNA from those strains that carried the *A*α1 allele (13, 52, 57). Strains that carried the same *A*α allele as the hybridizing DNA but that were collected from different geographical areas displayed restriction polymorphisms within the locus.

Further restriction mapping and DNA-DNA hybridization analyses revealed that the DNA fragments that trigger $A\alpha$ development are allelic in that they are embedded in the same physical region of the genome, with well-conserved, homologous DNA showing few restriction-site polymorphisms surrounding the $A\alpha$ locus. The mating-type activity, however, is encoded within a heterologous subregion consisting of approximately 5 to 8 kb (for a graphic depiction of the degree of DNA relatedness of the $A\alpha1$, $A\alpha3$, and $A\alpha4$ mating-type regions, see reference 52). Such a degree of heterogeneity may act as a block to normal recombination during meiosis, accounting for the fact that intragenic recombinants of the mating-type subloci have never been found in *S. commune* (or any basidiomycete).

AN OVERVIEW OF THE $A\alpha$ LOCUS

A total of 7,600 bp of $A\alpha1$, 13,190 bp of $A\alpha3$, 11,372 bp of $A\alpha4$, 7,005 bp of $A\alpha5$, and 10,115 bp of $A\alpha6$ DNA has been sequenced (13, 18, 27, 47, 55, 64). These sequences encode the $A\alpha$ mating activity as defined in transformation and mating experiments as well as flanking regions (Fig. 16.3A). Sequence data begin in a region about 50 kb from *PAB1*.

On the basis of classical genetic analysis, each $A\alpha$ was previously thought of as a single gene and $A\alpha$-regulated development was thought to proceed from the interaction of different alleles of that gene. However, sequence analysis showed that the heterogeneous DNA of most $A\alpha$ loci, as exemplified by $A\alpha3$, $A\alpha4$, $A\alpha5$, and $A\alpha6$, encodes two divergently transcribed genes, designated Z and Y, that are involved in mating (Fig. 16.3A). There are probably nine distinct sets in nature corresponding to the nine known $A\alpha$ mating-type specificities in the worldwide population of *S. commune*. $A\alpha3$ strains contain Z3 and Y3, $A\alpha4$ strains carry Z4 and Y4, etc., with the one known exception to this pattern being $A\alpha1$, which contains only Y1. This absence of a Z gene in $A\alpha1$ strains is probably the result of a natural deletion (52). Transformation experiments (51) utilizing this characteristic of $A\alpha1$, that is, its absence of a Z gene, first showed that because Y genes from other specificities do not activate an $A\alpha1$ strain, it is the product of Z from one mating partner that interacts with the product of Y from the other partner to form the functional regulatory complex that activates $A\alpha$ developmental events (Fig. 16.3C). This interaction of Y and Z proteins has been verified genetically via the use of an $A\alpha$ deletion strain (43), through use of the yeast two-hybrid system (26), and through direct binding assays (1). No $A\alpha$ mating types are known that contain only a

Z gene, and while Southern analysis deems it unlikely (52), this will remain in question until all the $A\alpha$ alleles are isolated.

That a heteromultimeric protein complex activates development in other filamentous basidiomycete fungi, notably *Coprinopsis cinerea* (23) and *Ustilago maydis* (14), was shown at about the same time. In *S. commune*, the initiation of $A\alpha$-regulated development is impossible in homokaryons because the Z and Y proteins encoded by a particular $A\alpha$ locus, such as Z4 and Y4, do not interact in a manner that initiates development in an $A\alpha4$ homokaryon. In addition, the likelihood that recombination between Y and Z from different alleles would occur during meiosis, creating recombinants (for example, *Z3 Y4)* that would activate the A series of developmental events, is very low, because the DNA of these genes is so dissimilar. Such homokaryons might also be compromised in nature, because they would undergo a wasteful process of partial development.

A comparison of the DNA sequences of the Y or Z alleles revealed some interesting features. Each intron within the Y or Z gene is positionally conserved among their alleles. Within the exons of Y or Z, DNA identity varies sharply: for example, a comparison of Y3 with Y4 using a 50-bp window shows that some regions are highly conserved (>85% identity) whereas others are very dissimilar (only 20% identical). In fact, initial DNA sequence comparisons of both the Y and Z mating-type genes, along with the DNA-DNA hybridization studies mentioned above, indicated such low base pair identity between alleles (averaging about 50%) that the allelic relationship of the Y and Z alleles was called into question. Only when the deduced protein sequences were aligned did their relatedness as alternate forms of the same genes become clear.

THE Y AND Z GENE PRODUCTS

The defining feature of both the predicted Y and Z proteins (Fig. 16.3B), which is shared with the corresponding mating-type proteins of other basidiomycetes, as well as the proteins encoded by *MATa1* and *MATα2* of *Saccharomyces cerevisiae*, is a homeodomain motif (HD), which identifies them as transcriptional regulators of developmental genes via their DNA binding properties (46). HD proteins have been identified in organisms ranging from fungi to humans, and their structure, both alone and in complex with target DNA sequences, shows remarkable conservation. In fungi, the homeodomain sequences of mating-type proteins fall into two distinct classes, designated HD1 and HD2 (22). HD1 proteins include the Z proteins of *S. commune*, the bE proteins of

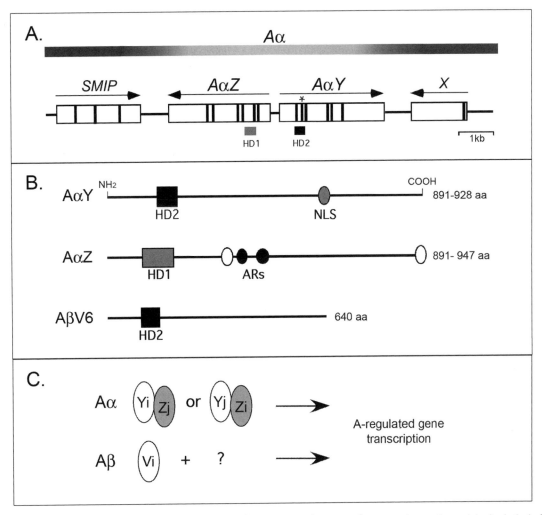

Figure 16.3 (A) A generalized map of the *Aα* mating-type region of *S. commune*. Sequences demonstrating mating activity in *Aα1*, *Aα3*, *Aα4*, *Aα5*, and *Aα6* specificities average about 50% similarity, although they are embedded in a region of DNA common to all strains examined, as denoted by the shaded line. The transition from common sequence, encoding *SMIP* (*S. commune* mitochondrial intermediate peptidase), at the left end, to the heterogeneous *Aα* region is gradual in all *Aα* mating types, except *Aα1*, where the transition (not shown) is extremely abrupt due to a natural deletion of the *Aα1* Z gene. The right boundary of the common sequence is approximately 7 to 8.5 kb from that of the left, except in *Aα1*, where heterogeneous DNA extends only 4.5 kb (52). The *Aα* Y and Z mating-type genes are a dyad of divergently transcribed homeodomain (HD1 and HD2) genes whose products regulate sexual development. Gene *X* has no apparent function in mating. The coding region of each gene is boxed, and introns conserved among the alleles of each gene are shown as vertical black lines. The starred intron in HD2 is inserted between the codons for the highly conserved amino acid residues W and F. The direction of transcription is depicted by an arrow. GenBank accession numbers of DNA sequences used to generate map: *Aα1*, U13942 and M97179; *Aα3*, L43072, U13943, and M97180; *Aα4*, U13944 and M97181; *Aα5*, U22049; *Aα6*, AF274566. (B) A generalized map of the *Aα* Y and Z proteins and the *Aβ* V6 protein. Each protein is depicted as a solid line with NH₂- and COOH-termini as denoted for *Aα* Y. The Y proteins all contain an HD2 homeodomain, a predicted coiled coil that overlaps a 28-residue bipartite nuclear localization sequence (NLS) that is embedded within a basic region (6, 44), and a serine-rich region (not shown). These characteristics have been observed for each Y protein studied to date, and the position of each feature is approximately the same in each Y allele. The Z proteins also contain features bearing on their presumptive function as transcriptional regulators, which reside in the same relative position in each allele: an atypical homeodomain sequence (HD1), two highly acidic 30-amino-acid regions rich in glutamate and aspartate (ARs), and two predicted coiled-coil regions (white ovals). The C-terminal regions of Z which are predicted to form a coiled coil display extremely high identity among the alleles (47, 55, 64). The interacton of coiled coils is the mechanism by which both homo- and heterodimerizations of regulatory transcription factors occur in many organisms. Discussions and graphic presentations of the Y and Z regions which are involved in mating interactions, specificity, and binding are available (1, 64, 66). The V6 protein, encoded by *Aβ*, displays an HD2 homeodomain motif. (C) The active regulatory complex encoded by *Aα* and *Aβ*. Activation of development follows mating via the interaction of Y and Z proteins from compatible mates. An active complex is formed from the HD1 protein from one mate in combination with the HD2 protein from the other mate, that is, Yi in combination with Zj or Yj in combination with Zi. The heteromultimer is postulated to be a transcription factor that directs development down a new pathway by binding upstream of specific target genes (51). The V protein, encoded at the *Aβ* locus, may be the HD2 partner of an HD1-HD2 pair, similar to the composition of the active *Aα* products.

U. maydis, the HD1 a, b, and d proteins of *C. cinerea*, the Sxi1α protein of *Cryptococcus neoformans* and the MATα2 protein of *Saccharomyces cerevisiae*. The HD2 proteins include the Y proteins of *S. commune*, the bW proteins of *U. maydis*, the HD2 a, b, and d proteins of *C. cinerea*, the Sxi2a protein of *C. neoformans* and the MATa1 protein of *S. cerevisiae*. In all these fungi, development is activated through the interaction of an HD1 and an HD2 protein, i.e., Z + Y, bE + bW, a1 + a2, Sxi1α + Sxi2a, and MATα2 + MATa1.

An alignment of the HD2 amino acid sequences of the Y proteins with those of the HD2 *U. maydis* bW mating-type proteins, and the HD2 A proteins of *C. cinerea* reveals their evolutionary relationship (9). This is additionally supported by the fact that they all have an intron inserted into the same position in the DNA sequences encoding their homeodomains (9), between the codons for the highly conserved amino acid residues W and F. Of the 60 amino acids that constitute the HD2 homeodomain of the *S. commune* Y1-Y6 proteins, there is approximately 90% identity.

Mutational analysis of the *S. commune* HD2 homeodomain by Luo et al. (25) demonstrated that the function of Y1 and Y3 proteins is sensitive to substitution and deletion within this homeodomain. The ability of the *Y* gene to activate *A*-regulated events in a transformant of a different *A*α specificity varied according to the particular mutation, but in general, loss of activity resulted from those mutations in the third helix of the HD2 domain that changed or deleted residues important to the current paradigm for HD-DNA complex formation. In every HD-DNA complex whose structure has been determined, the third helix, the "recognition helix," makes contact with the major groove of the target DNA and provides the majority of the base-specific and sugar phosphate backbone contacts (63). However, an extensive mutation in the flexible "N-terminal arm" extending from the N terminus of the first helix had no effect on the function of Y. In most homeodomains, this region wraps around the target DNA to make specific contacts in the minor groove and is thought to contribute some degree of specificity to the binding of the HD to target DNA (62).

The HD1 homeodomain is said to be atypical. In the Z proteins, there are three extra amino acids found between helices 1 and 2 as in the 63-amino-acid three-amino-acid loop extension (TALE) homeobox, found in a subgroup of homeobox genes (5). A second insertion of 10 to 35 amino acids lies between helices 2 and 3 and accounts for the unique length of each Z homeodomain as well as increasing their differences in sequence. (See

Burglin [7] for an alignment of the HD1 domain of the *S. commune* Z proteins with the TALE homeodomains of MATα2 of *S. cerevisiae*, MAT1-Pi of *S. pombe*, *C. cinerea* Aβ HD1 proteins, and the *U. maydis* bE proteins; as in the *S. commune* Z proteins, there are additional sequences between helices 2 and 3 of the *U. maydis* bE protein.) Because large deletions within the HD1 homeodomain appeared to have no effect on the ability of Z3 and Z4 proteins to activate development, the HD1 domain was not considered to be essential (25). However, some of the N-terminal 17 amino acids of the HD1 are essential for activity of the Z5 protein (64), and as the authors point out, none of the deletions of Z3 and Z4 involved the N-terminal 17 amino acids of the HD1 domain, so this small region may be essential. Overall, however, one can conclude that the HD1 domain is less critical to DNA binding than the HD2 domain. Similar experimental data have been obtained for the nonessential role of the HD1 homeodomain in heterodimer function in *C. cinerea* (2) and *S. cerevisiae* (19).

PROMOTER REGIONS

The mating genes are similar to other *S. commune* genes (and many fungal genes in general) in that there are no obvious promoter sequences such as a TATA box or CAAT box. We originally designated promoter regions (55) based on the likelihood that essential promoter elements would map 5′ of the genes and that deletion of the promoter region would inactivate the *Y* or *Z* gene in transformation experiments. Based on this, *Y* and *Z* appear to have promoter regions of approximately 100 bp each that lie central to the two divergently transcribed genes. Results from transformations using *Z4* and *Y4* show that the essential promoter elements do not overlap (51). Comparison of the putative promoter regions of *Z3*, *Z4*, and *Z5* reveals one sequence that appears to be conserved (GGGNNGGAANT) −10 to −12 bp upstream (64). Its deletion in *Z5* abolished activity.

DO *Y* AND *Z* FIT THE CURRENT PARADIGM FOR MATING INTERACTIONS?

Since there are so many facets of the system that are imperfectly known, it would appear temerariously presumptive to add at this time yet another inadequate—if hopefully, not inaccurate as well—model for the action of the incompatibility factors in the higher fungi. Two extenuations, however, may make this venture less hazardous than were similar attempts in the past: (a) the precedents available from the elucidation of the details of regulatory mechanisms at the subcellular level and (b) results obtained during the past few years regarding the genetics of the incompatibility system, particularly the disruptions caused by mutative changes in the genetic components of the system.

J. Raper (39)

The mating genes *MATa1* and *MATa2* of *S. cerevisiae* are probably the best-known example in the fungal kingdom of a dyad consisting of one typical and one TALE homeodomain gene. As described above, other examples are the *U. maydis bE* and *bW* mating genes, the *a*, *b*, and *d* gene pairs of the *A* locus of *C. cinerea*, and as we have seen, the *Y* and *Z* genes of *S. commune*, indicating that this pattern of two linked, divergently transcribed mating genes is evolutionarily conserved (see James, chapter 19, this volume). In *Encephalitozoon cuniculi*, a closely linked pair of homeobox genes has been found, one a typical and one a TALE homeobox gene, suggesting that the Microsporidia also contain a mating-type locus (8). While their function is unknown, it does show that a locus with a typical homeobox gene and a TALE homeobox gene is an ancestral feature in the fungal lineage (8).

The ability of these homeodomain proteins to interact is the common theme in regulation of development in *S. cerevisiae* and in Basidiomycetes: sexual compatibility occurs in the fusion cell where an active regulatory complex is formed from HD1 proteins from one mate and HD2 proteins from the other mate. The heteromultimer is postulated to be a transcription factor that directs development down a new pathway by binding upstream of specific target genes involved in development, although, as in the case of *S. cerevisiae*, the interaction may require additional proteins not involved in self/nonself recognition. Such combinatorial control of gene expression in eukaryotes, of course, allows a relatively modest number of proteins to regulate many different pathways (62). In chapter 22 of this volume, Kahmann and Schiwarski describe the actual DNA target sites of the *U. maydis* bE+bW heterodimer and identify some of the target genes, giving us insights into the cascade of sexual development that is initiated by mating. In a fungal fusion cell which recognizes the mate as *self*, an active form of the transcription factor will not result, and thus sexual development does not proceed.

FLANKING GENES AND BOUNDARIES

The *Aα* locus is flanked by two genes that have no function in the mating process. *SMIP*, located to the left of the *Z* mating gene (Fig. 16.3A) and oriented in the opposite direction, encodes *S. commune* mitochondrial in-

termediate peptidase, an enzyme involved in processing nuclear-encoded precursor proteins that are destined for the mitochondrial matrix or inner membrane (18, 65). *MIP* is a requisite for mitochondrial function in *S. cerevisiae*, and expression of *SMIP* cDNA in a yeast *mip1* deletion mutant rescues the respiratory-deficient phenotype, evincing conservation of function of this enzyme in fungi. The proteins predicted by *S. commune*, yeast, and rat *MIP* sequences are 54 to 57% similar over 700 amino acids (18).

Both sequence and transformation data suggest that the left boundary of the *Aα* mating-type locus, then, is 3′ of *SMIP* (Fig. 16.3A). This gene is now known to be conserved at the *A* mating-type locus boundary of other hymenomycete species and has been used to isolate the corresponding *MAT* locus from several of these (see chapter 19 by James, this volume).

Transformation experiments with DNA fragments extending ~12 kb immediately downstream of *Y3* indicated no genes present in this region that, like *Z* and *Y*, activated the *A* pathway of development when introduced into a recipient strain with a different *Aα* specificity (55). Northern analysis, however, enticingly displayed a >1.7-kb transcript encoded immediately downstream of *Y1*, *Y3*, and *Y4* (65). Were *Y* and *Z* the only genes in the *Aα* locus? Sequencing located a gene (27) in *Aα1*, *Aα3*, and *Aα4* DNA, oriented in the direction opposite to *Y*, and this was designated *X* (Fig. 16.3A).

Despite conservation of this gene, *X*-disrupted mutants exhibited no discernible phenotype under laboratory conditions. In addition, matings in which one or both mates carried a deletion of *X* showed no abnormalities in sexual development: the mutation does not obstruct the *A*- or *B*-regulated events of dikaryosis, fruiting, sporulation, or the viability of progeny (27). Thus, based on both the sequence data and functional analysis of *X*, the right limit of the *Aα* mating-type locus falls between *Y* and *X* (Fig. 16.3A).

CLONING *Aβ*

Aα and *Aβ* genes are functionally redundant in that they regulate a phenotypically identical developmental pathway. To determine the nature of the *Aβ* locus, and to explore the evolutionary relationships between *Aα* and *Aβ* genes, it was necessary to clone the latter (48). *Aα* and *Aβ*, though residing on the same chromosome, are some 1 to 19 cM apart (39). The *Aα4* mating genes, *Y4* and *Z4*, were isolated by walking the chromosome from *PAB1*, and subsequent *Aα* loci (*Aα1*, *Aα3*, *Aα5*, and *Aα6*) were cloned using common sequences that flanked the heterogeneous region. However, the alleles of Aα do not cross-hybridize to each other or, more importantly, to the Aβ locus, so Aβ could not be cloned in that way. In addition, the closest known marker, ADE5, was estimated to be more than 500 kb away, based on recombination frequencies (39) and the walk from PAB1 to Aα: in the S. commune Aα region, 0.6 cM corresponds to about 100 kb of DNA (13). How, then, was Aβ cloned? By brute force.

An Aα4 Aβ1 Bα1 Bβ6 trp1 strain was cotransformed with an Aα4 Aβ6 plasmid library and a TRP1 plasmid. Successful transformation and integration of Aβ6 in the genome would yield a homokaryon with genes encoding two different Aβ specificities. A total of 10,777 Trp+ transformants were obtained and mated against an Aα4 Aβ1 strain with a compatible B mating type. Three Trp+ transformants were identified as being capable of mating with the Aβ1 strain. The transforming DNA was recovered via plasmid rescue techniques, and subsequent transformations and genetic analyses proved it to contain Aβ6 DNA (48). A single large open reading frame, 1,920 bp long, was deduced from cDNA and genomic DNA sequences and named V6 (GenBank accession no. U17434). The AβV gene encodes a protein belonging to the HD2 class of homeodomain mating-type proteins like the Y genes of the Aα sublocus. However, V is smaller than the Y and Z proteins and lacks other motifs which support the hypothesis that Y and Z are transcriptional activators (Fig. 16.3B). The most immediate question, though, about Aβ is whether or not V is the only gene in the locus: from what we now know of the structure of the corresponding A locus of C. cinerea, one would expect the V gene to be the HD2 partner of an HD1-HD2 gene pair, similar to the Y-Z pair in the Aα sublocus (Fig. 16.3B; see also chapter 17, Casselton and Kües, this volume). DNA-DNA and DNA-RNA hybridizations with of V6 with Aβ1, Aβ5, Aβ20, and Aβ23 strains showed no cross-hybridization, suggesting that, similar to Aα, the degree of relatedness of the alleles of Aβ is not detectable using these techniques. Further elucidation of the basic differences in the structure of the Aα and Aβ components may bear on the mystery of why there are so many more mating types of Aβ than Aα (32 versus 9) and why the A con mutations that activate the A pathway in homokaryons all map to Aβ.

> Unraveling the tangled skein of interrelations between genetic factors, mating patterns, and morphogenetic sequences . . . during the past fifteen years might well be likened to a fast-paced and complex detective novel—the apprehension of each culprit

has led, usually by the most intriguing indirection, to yet another corpse.

                                                                                                    J. Raper (39)

ISOLATION OF THE B GENES

Information from several previous studies was incorporated to create a successful strategy to isolate active fragments from the Bα and Bβ loci of S. commune. As we had seen, after walking from PAB1, experiments using Aα DNA to transform a recipient of a different Aα specificity showed activation of the developmental events under control of the A genes: in particular, hook cell formation, which could easily be screened with the light microscope (13, 51). Thus, one might predict that DNA encoding B activity could be directly isolated from homokaryotic recipients that were selected for the activation of B-regulated events through the addition of Bα or Bβ DNA of a different specificity to form a merodiploid. But this experiment, which parallels that in which the first Aα gene was isolated, presents special difficulties when attempting to isolate a B gene. Following selection with a nutritional marker, transformants would have to be screened visually and microscopically for the "flat," B-activated phenotype: little aerial hyphae, formation of irregular, submerged mycelium, and continuously migrating nuclei. The probability that transformants still wild-type for a single B specificity would quickly overgrow any "flat" transformant was very high. An alternative strategy of using a homokaryotic recipient with a constitutive mutation in Aβ to overcome the necessity of screening for "flat" transformants had already been employed with no success: none of several thousand Schizophyllum Trp+ transformants, genetically transformed with a cosmid library of genomic DNA having a selectable TRP1 marker, displayed B-activated development, which in combination with the A con mutation would be expected to phenotypically behave as a dikaryon (M. Stankis, unpublished results.)

In the early 1960s, Parag (32) had mutagenized a heterokaryon carrying different A genes and identical B genes and selected mutants that would form fruiting bodies, thus indicating that the B-regulated series of developmental events had been activated successfully as a prelude to fruiting and sporulation. Among the progeny of the mutated heterokaryon were homokaryons carrying con mutations mapping to the B subloci. The new strategy for isolation of a B gene took advantage of the formation of a similar common-B heterokaryon (49). A mixture of protoplasts was prepared from two trp1 strains with different A mating types and the same B mating type (*Bα2 Bβ2*) and was cotransformed with plasmid DNA containing *TRP1* (p*TRP1*) as the selectable marker as well as a genomic library constructed with DNA from a *Bα1 Bβ1* strain (Fig. 16.2B). Full sexual development, i.e., the formation of fruiting bodies, could occur if a transformant expressed either *Bα* or *Bβ* mating-type DNA derived from the genomic library and either mated with an *A*-compatible regenerate or fused with one during protoplast manipulation.

This plasmid library, made from *Schizophyllum* strain 4-40 DNA, and represented by 28,000 clones, had been used to isolate five genes, including the *PAB1*, *ADE5*, and *TRP1* genes (12, 50), and it seemed reasonable that it would include DNA from the *Bα1* and *Bβ1* loci. The vector for the plasmid library, however, did not have a selectable marker for transformation of *S. commune*; therefore, cotransformation with *TRP1* was necessary, but might be used to advantage. In order to increase the odds of finding the nonselectable gene, the ratio of library DNA to p*TRP1* DNA was increased to approximately seven to one, so that the total number of Trp+ transformants would underrepresent the number of transformants that integrated genomic DNA from the plasmid library. Multiple integration events might also occur in transformants (12) and contribute to finding *B* gene transformants. Cotransformation studies were used to estimate the amounts of DNA and the numbers of protoplasts needed to screen the library (50).

Among 1,450 Trp+ transformants, 12 areas showed possible initiation of a fruiting body after 2 weeks of incubation. These areas were subcultured, and three of them developed normal fruiting structures and produced spores. Spores were germinated from two of the three fruiting transformants, and two distinct groups of progeny were classified. Approximately one half displayed the normal morphology of a homokaryon and carried the genetic markers found in the strains used to make protoplasts. Significantly, the other half had the morphology typical of *B*-activated development: their growth was submerged, and microscopy showed numerous lateral branches, the "flat" phenotype. Like *A* mating-type transformants, these progeny exhibited a *B* specificity compatible with that of the recipient, *Bα2 Bβ2*, and thus, the transformants were considered to have integrated either *Bα1* or *Bβ1* DNA from the genomic library. Further genetic analysis revealed that the transforming *B* DNAs had integrated ectopically. That the products could act in *trans* precluded the possibility that a mutagenic event at the recipient *B* locus yielded the *B*-activated phenotype (49).

Several plasmids from each transformant were recovered following rescue in *E. coli* and were individually tested for their ability to activate *B*-regulated development. In contrast to the primary screen, where two strains were used and activation of *B*-regulated development yielded dikaryotic hyphae capable of forming fruiting bodies, the transformation of a homokaryotic strain by itself with individual plasmids yielded the anticipated "flat" phenotype, activated by two compatible *B* genes.

In order to determine the identity of the *B* genes that were cloned, i.e., *Bα1* or *Bβ1*, two *Schizophyllum* strains were individually transformed with each plasmid. *B*-regulated development would not be activated if the strain and the transforming DNA encoded the same *Bα* or *Bβ* mating-type specificity, that is, transformants would not display a *B*-activated "flat" phenotype. The results of these transformations indicated that one plasmid encoded an active portion of the *Bα1* locus and the other plasmid encoded an active region of the *Bβ1* locus. Plasmid pBtr2 yielded transformants with the *B*-activated phenotype in a strain of *Bα2 β1* specificity but not in a strain of *Bα1 β6*, therefore encoding *Bα1* mating-type specificity. Conversely, and fortuitously, plasmid pBtr4 yielded transformants with the *B*-activated phenotype when the recipient strain was of *Bα1 Bβ6* specificity, but not in a strain carrying *Bα2 Bβ1*, thus encoding the other possible locus, *Bβ1* (49).

Asgeirsdottir et al. (3) demonstrated that DNA probes from the *Bα1* and *Bβ1* regions both hybridized to a single chromosome (1.6 Mbp), the smallest chromosome found in *Schizophyllum* strain 4-40, consistent with the tight linkage of the *Bα* and *Bβ* loci (11, 20).

Probing with DNA fragments from the plasmid containing *Bα1* activity led Wendland et al. (59) to the selection of a 8.2-kb genomic DNA fragment that activated *B*-regulated development in each of the eight other *Bα* specificities. Vaillancourt et al. (58) similarly showed that *Bβ1* activity was completely encoded in an 8.5-kb fragment. We now know from the work of these authors that both *Bα1* and *Bβ1* are complex loci, containing several genes encoding cell signaling components, pheromones, and pheromone receptors. Thus, as in the ascomycete yeasts and nonfilamentous Basidiomycetes, pheromone signaling plays a crucial role in mate recognition, but unlike these other fungal species, the genes that encode these signaling molecules are multiallelic and contribute to the generation of the large numbers of mating specificities that we see in filamentous Basidiomycetes. A detailed description of the structure and organization of the *B* locus of *S. commune* and an account of the highly specific interactions demanded by this multiallelic pheromone-receptor system, as well as the basis of many of the *B con* mutants isolated by earlier researchers, are given by Fowler and Vaillancourt in chapter 18, this volume.

In summary, the multiallelic mating-type systems of the Basidiomycetes provide us with a fascinating setting in which to study the protein-protein interactions involved in both self/nonself recognition and transcriptional regulation of developmental genes. We can only conclude with John Raper's own words (39) on the topic:

For such studies, some preliminary work has already been done, and there is available a modest amount of relevant background information.

References

1. Asada, Y., C. Yue, J. Wu, G.-P. Shen, C. Novotny, and R. C. Ullrich. 1997. *Schizophyllum commune* Aα mating-type proteins, Y and Z, form complexes in all combinations *in vitro*. *Genetics* 147:117–123.

2. Asante-Owusu, R. N., A. H. Banham, H. U. Böhnert, E. J. C. Mellor, and L. A. Casselton. 1996. Heterodimerization between two classes of homeodomain proteins in the mushroom *Coprinus cinereus* brings together potential DNA-binding and activation domains. *Gene* 172:25–31.

3. Asgeirsdottir, S. A., F. H. J. Schuren, and J. G. H. Wessels. 1994. Assignment of genes to pulse-field separated chromosomes of *Schizophyllum commune*. *Mycol. Res.* 98:689–693.

4. Beach, D. H. 1983. Cell type switching by DNA transposition in fission yeast. *Nature* 305:682–683.

5. Bertolino, E., B. Reimund, D. Wildt-Perinic, and R. G. Clerc. 1995. A novel homeobox protein which recognizes a TGT core and functionally interferes with a retinoid-responsive motif. *J. Biol. Chem.* 270:31178–31188.

6. Boulikas, T. 1994. Putative nuclear localization signals (NLS) in protein transcription factors. *J. Cell. Biochem.* 55:32–58.

7. Burglin, T. R. 1997. Analysis of TALE superclass homeobox genes (MEIS, PBC, KNOX, Iroquios, TGIF) reveals a novel domain conserved between plants and animals. *Nucleic Acids Res.* 25:4173–4180.

8. Burglin, T. R. 2003. The homeobox genes of *Encephalitozoon cuniculi* (Microsporidia) reveal a putative mating-type locus. *Dev. Gene Evol.* 213:50–52.

9. Burglin, T. R. 2005. Homeodomain proteins, p. 179–222. *In* R. A. Meyers (ed.), *Encyclopedia of Molecular Cell Biology and Molecular Medicine*. Wiley-VCH Verlag GmbH & Co., Weinheim, Germany.

10. Dranginis, A. M. 1990. Binding of yeast a1 and alpha 2 as a heterodimer to the operator DNA of a haploid-specific gene. *Nature* (London) 347:682–685.

11. Fowler, T. J., M. F. Mitton, E. I. Rees, and C. A. Raper. 2004. Crossing the boundary between the Bα and Bβ mating-type loci in *Schizophyllum commune*. *Fungal Genet. Biol.* 41:89–101.

12. Froeliger, E. H., A. Munoz-Rivas, C. A. Specht, R. C. Ullrich, and C. P. Novotny. 1987. The isolation of specific genes from the basidiomycete *Schizophyllum commune*. *Curr. Genet.* 12:547–554.

13. Giasson, L., C. A. Specht, C. Milgrim, C. P. Novotny, and R. C. Ullrich. 1989. Cloning and comparison of Aα mating-type alleles of the Basidiomycete *Schizophyllum commune*. *Mol. Gen. Genet.* 218:72–77.

14. Gillissen, B., J. Bergemann, C. Sandmann, B. Schoee, and M. Bolker, and R. Kahmann. 1992. A two-component regulatory system for self/non-self recognition in *Ustilago maydis*. *Cell* 68:647–657.

15. Goutte, C., and A. D. Johnson. 1988. a1 protein alters the DNA binding specificity of alpha 2 repressor. *Cell* 52:875–882.

16. Herskowitz, I. 1989. A regulatory hierarchy for cell specialization in yeast. *Nature* (London) 342:749–757.

17. Horton, J. S., and C. A. Raper. 1991. Pulsed-field gel electrophoretic analysis of *Schizophyllum commune* chromosomal DNA. *Curr. Genet.* 19:77–80.

18. Isaya, G., W. R. Sakati, R. A. Rollins, G.-P. Shen, L. C. Hanson, R. C. Ullrich, and C. P. Novotny. 1995. Mammalian mitochondrial intermediate peptidase: structure/function analysis of a new homologue from *Schizophyllum commune* and relationship to thimet oligopeptidases. *Genomics* 28:450–461.

19. Johnson, A. D. 1995. Molecular mechanisms of cell-type determination in budding yeast. *Curr. Opin. Genet. Dev.* 5:552–558.

20. Koltin, Y., J. R. Raper, and G. Simchen. 1967. Genetic structure of the incompatibility factors of *Schizophyllum commune*: the *B* factor. *Proc. Natl. Acad. Sci. USA* 57:55–63.

21. Koltin, Y., J. Stamberg, N. Bawnik, R. Tamarkin, and R. Werczberger. 1979. Mutational analysis of natural alleles in and affecting the *B* incompatibility factor of *Schizophyllum*. *Genetics* 93:383–391.

22. Kües, U., and L. Casselton. 1992. Homeodomains and regulation of sexual development in basidiomycetes. *Trends Genet.* 8:154–155.

23. Kües, U., W. V. J. Richardson, A. M. Tymon, E. S. Mutasa, B. Gottgens, S. Gaubatz, A. Gregoriades, and L. A. Casselton. 1992. The combination of dissimilar alleles of the Aα and Aβ gene complexes, whose proteins contain homeodomain motifs, determines sexual development in the mushroom *Coprinus cinereus*. *Genes Dev.* 6:568–577.

24. Leonard, T. J., and S. Dick. 1968. Chemical induction of haploid fruiting in *Schizophyllum commune*. *Proc. Natl. Acad. Sci. USA* 59:745–751.

25. Luo, Y., R. C. Ullrich, and C. P. Novotny. 1994. Only one of the paired *Schizophyllum commune* Aα mating-type, putative homeobox genes encodes a homeodomain essential for Aα-regulated development. *Mol. Gen. Genet.* 244:318–324.

26. Magae, Y., C. Novotny, and R. Ullrich. 1995. Interaction of the Aα Y and Z mating-type homeodomain proteins of *Schizophyllum commune* detected by the two-hybrid system. *Biochem. Biophys. Res. Commun.* 211:1071–1076.

27. Marion, A. L., K. A. Bartholomew, J. Wu, H. Yang, C. P. Novotny, and R. C. Ullrich. 1995. The Aα mating-type locus of *Schizophyllum commune*: structure and function of gene *X*. *Curr. Genet.* **29:**143–149.

28. Munoz-Rivas, A. M., C. A. Specht, B. J. Drummond, E. Froeliger, C. P. Novotny, and R. C. Ullrich. 1986. Transformation of the Basidiomycete *Schizophyllum commune*. *Mol. Gen. Genet.* **205:**103–106.

29. Munoz-Rivas, A. M., C. A. Specht, C. P. Novotny, and R. C. Ullrich. 1986. Isolation of the DNA sequence coding indole-3-glycerol phosphate synthetase and phosphoribosylanthranilate isomerase of *Schizophyllum commune*. *Curr. Genet.* **10:**909–913.

30. Niederpruem, D. J. 1980. Direct studies of dikaryotization in *Schizophyllum commune*. 1. Live intercellular nuclear migration patterns. *Arch. Microbiol.* **128:**162–171.

31. Palmer, G. E., and J. S. Horton. 2006. Mushrooms by magic: making connections between signal transduction and fruiting body development in the basidiomycete fungus *Schizophyllum commune*. *FEMS Microbiol. Lett.* **262:**1–8.

32. Parag, Y. 1962. Mutations in the B incompatibility factor of *Schizophyllum commune*. *Proc. Natl. Acad. Sci. USA* **48:**743–750.

33. Raper, C. A. 1978. Control of development by the incompatibility system in basidiomycetes, p. 3–29. *In* M. N. Schwalb and P. G. Miles (ed.), *Genetics and Morphogenesis in the Basidiomycetes*. Academic Press, New York, NY.

34. Raper, C. A. 1983. Controls for development and differentiation in the dikaryon in Basidiomycetes, p. 195–238. *In* J. Bennett and A. Ciegler (ed.), *Secondary Metabolism and Differentiation in Fungi*. Marcel Dekker, New York, NY.

35. Raper, C. A. 1988. *Schizophyllum commune*, a model for genetic studies of the Basidiomycotina, p. 511–522. *In* G. S. Sidhu (ed.), *Genetics of Pathogenic Fungi*, vol. 6 of D. S. Ingrains and P. H. Williams (ed.), *Advances in Plant Pathology*. Academic Press, New York, NY.

36. Raper, C. A. 2004. Why study Schizophyllum? *Fungal Genetics Newsl.* **51:**30–36.

37. Raper, C.A., and J. R. Raper. 1966. Mutations modifying sexual morphogenesis in *Schizophyllum*. *Genetics* **54:**1151–1168.

38. Raper, C.A., and J. R. Raper. 1973. Mutational analysis of a regulatory gene for morphogenesis in *Schizophyllum*. *Proc. Natl. Acad. Sci. USA* **70:**1427–1431.

39. Raper, J. R. 1966. *Genetics of Sexuality in Higher Fungi*. Ronald Press, New York, NY.

40. Raper, J. R., M. G. Baxter, and A. H. Ellingboe. 1960. The genetic structure of the incompatibility factors of *Schizophyllum commune*: the A factor. *Proc. Natl. Acad. Sci. USA* **44:**889–900.

41. Raudaskoski, M., and R. Vauras. 1982. Scanning electron microscope study of fruit body differentiation in *Schizophyllum commune*. *Trans. Br. Mycol. Soc.* **78:**475–481.

42. Raudaskoski, M., V. Salo, and S. S. Niini. 1988. Structure and function of the cytoskeleton in filamentous fungi. *Karstenia* **28:**49–60.

43. Robertson, C. I., K. A. Bartholomew, C. P. Novotny, and R. C. Ullrich. 1996. Deletion of the *Schizophyllum commune* Aα locus: the roles of Aα *Y* and *Z* mating-type genes. *Genetics* **144:**1437–1444.

44. Robertson, C. I., A. M. Kende, K. Toenjes, C. P. Novotny, and R. C. Ullrich. 2002. Evidence for interaction of *Schizophyllum commune* Y mating-type proteins in vivo. *Genetics* **160:**1461–1467.

45. Ruiter, M. H., J. H. Sietsma, and J. G. H. Wessels. 1988. Expression of dikaryon-specific mRNAs of *Schizophyllum commune* in relation to incompatibility genes, light, and fruiting. *Exp. Mycol.* **12:**60–69.

46. Scott, M. P., J. W. Tamkun, and G. W. Hartzell III. 1989. The structure and function of the homeodomain. *Biochim. Biophys. Acta* **989:**25–48.

47. Shen, G.-P., Y. Chen, D. Song, Z. Peng, C. P. Novotny, and R. C. Ullrich. 2001. The Aα6 locus: its relation to mating-type regulation of sexual development in *Schizophyllum commune*. *Curr. Genet.* **39:**340–345.

48. Shen, G.-P., D.-C. Park, R. C. Ullrich, and C. P. Novotny. 1996. Cloning and characterization of a *Schizophyllum* gene with Aβ6 mating-type activity. *Curr. Genet.* **29:**136–142.

49. Specht, C. A. 1995. Isolation of the Bα and Bβ mating-types loci of *Schizophyllum commune*. *Curr. Genet.* **28:**374–379.

50. Specht, C. A., A. Munoz-Rivas, C. P. Novotny, and R. C. Ullrich. 1988. Transformation of *Schizophyllum commune*, an analysis of parameters for improving transformation frequencies. *Exp. Mycol.* **12:**357–366.

51. Specht, C. A., M. M. Stankis, L. Giasson, C. P. Novotny, and R. C. Ullrich. 1992. Functional analysis of the homeodomain-related proteins of the Aα locus of *Schizophyllum commune*. *Proc. Natl. Acad. Sci. USA* **89:**7174–7178.

52. Specht, C. A., M. M. Stankis, C. P. Novotny, and R. C. Ullrich. 1994. Mapping the heterogeneous DNA region that determines the nine Aα mating-type specificities of *Schizophyllum commune*. *Genetics* **137:**709–714.

53. Stamberg, J., and Y. Koltin. 1972. The organization of the incompatibility factors in higher fungi: the effects of structure and symmetry on breeding. *Heredity* **30:**15–26.

54. Stankis, M. M., C. A. Specht, and L. Giasson. 1990. Sexual incompatibility in *Schizophyllum commune*: from classical genetics to a molecular view. *Semin. Dev. Biol.* **1:**195–206.

55. Stankis, M. M., C. A. Specht, H. Yang, L. Giasson, R. C. Ullrich, and C. P. Novotny. 1992. The Aα mating locus of *Schizophyllum commune* encodes two dissimilar multiallelic homeodomain proteins. *Proc. Natl. Acad. Sci. USA* **89:**7169–7173.

56. Strathern, J. N., E. Spatola, C. McGill, and J. B. Hicks. 1980. Structure and organization of transposable mating-type cassettes in *Saccharomyces* yeast. *Proc. Natl. Acad. Sci. USA* **77:**2839–2843.

57. Ullrich, R. C., L. Giasson, C. A. Specht, M. M. Stankis, and C. P. Novotny. 1990. The Aα multiallelic mating-type genes of *Schizophyllum commune*, p. 271–288. *In* E. H. Davidson, J. V. Ruderman, and J. W. Posakony

(ed.), *Developmental Biology*. Wiley-Liss, New York, NY.

58. **Vaillancourt, L. J., M. Raudaskoski, C. A. Specht, and C. A. Raper.** 1997. Multiple genes encoding pheromones and a pheromone receptor define the Bβ1 mating-type specificity in *Schizophyllum commune*. *Genetics* **146:** 541–551.

59. **Wendland, J., L. J. Vaillancourt, J. Hegner, K. B. Lengeler, K. J. Laddison, C. A. Specht, C. A. Raper, and E. Kothe.** 1995. The mating-type locus Bβ1 of *Schizophyllum commune* contains a pheromone receptor gene and putative pheromone genes. *EMBO J.* **14:**5271–5278.

60. **Wessels, J. G. H.** 1992. Gene expression during fruiting in *Schizophyllum commune*. *Mycol. Res.* **96:**609–620.

61. **Wessels, J. G. H., and J. R. Marchant.** 1974. Enzymatic degradation in hyphal wall preparations from a monokaryon and a dikaryon of *Schizophyllum commune*. *J. Gen. Microbiol.* **83:**359–368.

62. **Wolberger, C.** 1999. Multiprotein-DNA complexes in transcriptional regulation. *Annu. Rev. Biophys. Biomol. Struct.* **28:**29–56.

63. **Wolberger, C., A. K. Vershon, B. Liu, A. D. Johnson, and C. O. Pabo.** 1991. Crystal structure of a MAT alpha 2 homeodomain-operator complex suggests a general model for homeodomain-DNA interactions. *Cell* **67:**517–528.

64. **Wu, J., R. C. Ullrich, and C. P. Novotny.** 1996. Regions in the Z5 mating gene of *Schizophyllum commune* involved in Y-Z binding and recognition. *Mol. Gen. Genet.* **252:**739–745.

65. **Yang, H., G.-P. Shen, D. C. Park, C. P. Novotny, and R. C. Ullrich.** 1995. The Aα mating-type transcripts of *Schizophyllum commune*. *Exp. Mycol.* **19:**16–25.

66. **Yue, C., M. Osier, C. P. Novotny, and R. C. Ullrich.** 1997. The specificity determinant of the Y mating-type proteins of *Schizophyllum commune* is also essential for Y-Z protein binding. *Genetics* **145:**253–260.

Sex in Fungi: Molecular Determination
and Evolutionary Implications
Edited by Joseph Heitman et al.
© 2007 ASM Press, Washington, D.C.

Lorna A. Casselton
Ursula Kües

17

The Origin of Multiple Mating Types in the Model Mushrooms *Coprinopsis cinerea* and *Schizophyllum commune*

Historically, *Coprinopsis cinerea* (*Coprinus lagopus* sensu *Buller*, *C. macrorhizus*) and *Schizophyllum commune* were the first hymenomycete species shown to be heterothallic, and it is not surprising that studies on mating type have, until recently, focused on these two species. The Portuguese mycologist Mathilde Bensaude, working in France in 1918 (10), and the German mycologist Hans Kniep, who was publishing his observations between 1915 and 1918 (55), were both fascinated by the complex nuclear behavior that occurs during cell division in the fertile dikaryotic mycelium. Bensaude, working with what was then known as *Coprinus fimetarius*, and Kniep, working with *S. commune*, were unaware of each other's discoveries because this was the time of the First World War and there was no scientific communication. Each showed that monosporous cultures were sterile and had a distinctly different morphology from the fertile mycelium with its wonderful clamp connections and that the mycelium required mating of different cultures in order to develop.

Kniep went on to show that four mating types segregated from a single fruiting body, consistent with there being two unlinked mating-type (MAT) loci, and desig-nated these *A* and *B* (55), referring to them as the *A* and *B* factors. These terms were retained until very recently, before the molecular structures of the loci were better understood. It was Kniep who recognized that the mating-type genes were multiallelic and many different mating types existed in the population (55). Tetrapolarity and multiallelism were confirmed later for *C. cinerea* by Brunswick in 1926 and Hanna in 1925 (see reference 87). This aspect of the *C. cinerea* mating system was probably not pursued by Bensaude, because, as she claimed in conversation with one of the authors in Lisbon in 1968, she had lost all but two of her strains in a cold winter working at the Sorbonne in Paris and was lucky that these had been compatible! John Raper in his book (87) gives an excellent account of these historical discoveries.

Kniep's remarkably thorough studies, involving analyzing progeny from many generations of crosses, led him to note that some 1 to 2% of the progeny of sexual crosses in *S. commune* had new mating types, and he naturally attributed this to mutation. He also showed by careful analysis of subsequent generations involving these new alleles that they reverted to the progenitor

Lorna A. Casselton, Department of Plant Sciences, University of Oxford, Oxford, OX1 3RB, United Kingdom. **Ursula Kües,** Institut für Forstbotanik, Georg-August-Universität Göttingen, Büsgenweg 2, Göttingen D-37077, Germany.

types at a similar frequency (55). Mutation, as shown much later, is very rare in the mating-type genes of these fungi, and a more likely explanation was provided by the studies of Haig Papazian (82), who showed that the *A* locus of *S. commune* was subdivided into two linked but distinct subloci, which were termed *Aα* and *Aβ*. Similarly, the *B* locus of *S. commune* (89) and the *A* locus of *C. cinerea* (21) were also shown to contain genes that are separable by recombination. Several estimates as to the actual numbers of *A* and *B* specificities in nature have been made, and these were based on the occurrence of different *A* and *B* alleles (specificities) in collections from the wild. For *C. cinerea*, these estimates vary from 164 *As* and 79 *Bs* (23) to more than 240 of each (54). In *S. commune*, there are estimates of 339 *As* and 64 *Bs* (89).

By showing that *A* mating specificity of *S. commune* derives from genes at two subloci, Papazian provided the first insight into how large numbers of mating types are generated in these fungi; the genes at both *Aα* and *Aβ* are multiallelic and functionally redundant, and it thus requires as few as 10 alleles at each to generate 100 different *A* or *B* mating specificities.

LIFE CYCLE

An understanding of the typical hymenomycete life cycle is important if we are to understand the roles that the *A* and *B* genes play in sexual development, and this is exemplified with *C. cinereus* in Fig. 17.1. This has been reviewed in detail by Kües (60). As we describe later in this chapter, all stages in this life cycle are accessible to mutational and molecular analysis by appropriate manipulation of the mating-type genes.

Haploid basidiospores germinate to give a self-sterile mycelium that is generally called a monokaryon because it has predominantly uninucleate cells. This mycelium constitutively produces abundant uninucleate asexual spores called oidia, a feature that facilitates the selection of mutants and has led to highly efficient DNA-mediated transformation systems. No special cells are required for mating, hyphal fusion is sufficient, and because this is mating type independent, a compatible mating is sensed only after cells have fused. If monokaryons have different *A* and *B* genes, this leads to the establishment of the fertile dikaryotic mycelium with two haploid nuclei, one from each mate, in each cell. Given the right environmental cues, the dikaryon produces the highly differentiated fruiting bodies that contain the basidia, the cells within which the dikaryotic nuclear pair fuses. The diploid nucleus immediately undergoes meiosis, and the resulting four haploid nuclei migrate into the basidiospores. The life cycle of *S. commune* is very similar to that of *C. cinerea*, the unmated mycelium is generally referred to as a homokaryon, and unlike that of *C. cinerea*, it lacks oidia (87).

The remarkable feature of the hymenomycete life cycle is the long delay between cell fusion and nuclear fusion, the so-called dikaryophase that predominates in nature. We see a similar dikaryophase following mating cell fusion in ascomycete fungi, but this is of limited extent and confined within the fruiting structure. What has fascinated hymenomycete geneticists is that the mating-type genes can be seen to play critical roles in the development and maintenance of the dikaryon. Following cell fusion, there is exchange of nuclei, and the donor nucleus in each case migrates through the established cells of the recipient monokaryon, a process that requires the dissolution of the complex dolipore septa that separate the cells of the hyphae (34). Once the migrating nucleus reaches the tip cells, nuclear pairing occurs and all subsequent cell divisions involve the formation of the characteristic clamp connection that is typical of the dikaryon of both *C. cinerea* and *S. commune* (Fig. 17.1). The clamp cell forms on the side of the tip cell, the nuclei take up positions adjacent to each other, one in the main hyphal cell and one in the clamp cell, and they then divide in synchrony. New cell walls are laid down generating three cells, a binucleate tip cell and uninucleate clamp and subterminal cells. The clamp cell grows backwards to fuse with a developing peg on the subterminal cell (4, 14, 43), and its nucleus is released into this cell to restore the binucleate state (14).

If crosses are made between strains with similar *A* or similar *B* alleles, cell fusion gives rise to heterokaryons (common *A* and common *B*) with characteristic phenotypes that show that the *A* and *B* genes regulate different developmental stages in dikaryon morphogenesis. Swiezynski and Day (100) used these incompatible matings to show that the *A* genes regulate nuclear pairing, clamp cell formation, and synchronized nuclear division and that the *B* genes regulate nuclear migration and clamp cell fusion. Similar studies with *S. commune* established the same roles for *A* and *B* (87). In *S. commune*, matings in which only the *B*-regulated pathway is active give rise to heterokaryons with what Papazian termed a "flat" phenotype, in which there is active nuclear migration, little aerial mycelium, extensive hyphal branching, and abnormal mitochondrial function (40). This characteristic phenotype is illustrated in chapter 18 of this volume by Fowler and Vaillancourt. The corresponding heterokaryon of *C. cinerea* has a less extreme phenotype, but electron microscopy revealed abnormalities in the mitochondria (L. A. Casselton, unpublished observations).

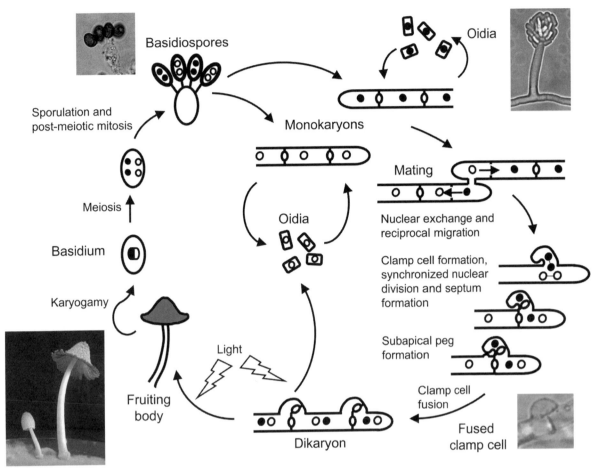

Figure 17.1 Life cycle of *Coprinopsis cinerea*. Photographs were taken by J. D. Granado, E. Polak, P. Srivilai, and W. Chaisaena.

THE MATING-TYPE GENES

The detailed knowledge we have of mating in the budding yeast *Saccharomyces cerevisiae* has provided a superb model for understanding sexual development in other fungi, and in particular basidiomycetes, because the genes that regulate mating in yeast (see reference 39) have direct counterparts in the *MAT* loci of these developmentally more complex fungi. In *S. cerevisiae*, mating-cell recognition is dependent on pheromone signaling between cells and on the generation, upon cell fusion, of a new diploid cell-specific transcription factor, formed by heterodimerization between two proteins encoded by genes in the *MATa* and *MATα* mating-type loci. The proteins, MATa1 and MATα2, are both members of the homeodomain protein family but have distinctly dissimilar homeodomain sequences. MATα2 has an atypical homeodomain (26), which we designated the HD1 type, whereas that of MATa1 is classed as typical and we termed it the HD2 type (61).

Genes encoding the homologues of these two homeodomain proteins are found in the *MAT* loci of all basidiomycete fungi. In *Cryptococcus neoformans*, they are *SXI1α* in *MATα* and *SXI2a* in *MATa* (41). In species with multiple mating types, the genes encoding both classes of proteins are found together in all versions of the *MAT* locus because each mate must be able to contribute one of each class of proteins into a compatible association. In *Ustilago maydis*, the genes are present as a divergently transcribed pair (*bE* and *bW*) in the *b* mating-type locus (35). This gene pair is the evolutionary unit found also in *C. cinerea* and *S. commune* (63, 98). As in *S. cerevisiae*, HD1 and HD2 proteins in basidiomycetes heterodimerize to form a new transcription factor following mating cell fusion (8, 48, 74). In the case of *C. cinerea* and *S. commune*, this is a dikaryon-specific regulator of the clamp cell pathway (33, 78). The mating pheromones and receptors may be encoded at the same locus as the homeodomain proteins, as in the bipolar *C. neoformans* (70) and *Ustilago hordei* (6, 7), or reside at the second *MAT* locus of tetrapolar forms such as *U. maydis* (11), *C. cinerea* (81), and *S. commune* (102, 104). Pheromone signaling is necessary for nuclear migration and clamp cell fusion in hymenomycetes. The remarkable conservation of function in the proteins that control mating in yeasts and mushrooms has helped in the elucidation of many aspects of basidiomycete mating pathways. There are of course differences in the way the genes are regulated, even in the structures of the proteins, but these relate to the different lifestyles of the fungi.

Several *A* and *B* loci of *C. cinerea* have been characterized, and in this species, the *A* and *B* mating specificities are determined by three closely linked subloci, each containing similar multiallelic and functionally redundant genes (37, 64, 66, 81, 84, 92). The *A* locus of *C. cinerea* extends over some 25 kb, and each sublocus contains representatives of three paralogous *HD1-HD2* gene pairs, whereas the *B* locus extends over 20 kb and contains three subloci with groups of paralogous genes encoding receptor and pheromone genes. Gene duplications and deletions often obscure this basic functional organization, and it is helpful to look at a theoretical archetypal locus that makes the compatible gene interactions easy to understand (Fig. 17.2A and B). The genes within each sublocus are maintained as a functional unit by having a unique DNA sequence that is very dissimilar from other allelic and paralogous versions. This lack of DNA homology extends into the flanking sequences that embed the genes, acts to prevent recombination within the sublocus, and keeps together sets of genes that are unable to activate sexual development in the absence of mating. None of the genes present in a single haploid genome can activate sexual development; for two mycelia to be compatible, it requires only that they differ in the alleles of genes in one sublocus. The many different *A* and *B* specificities found in nature have been generated by recombining the allelic versions of the groups of genes at each sublocus into all possible combinations (75, 84, 92). At the *A* locus, a detectable level of recombination still occurs between the first two subloci via a 7-kb sequence that is homologous in all backgrounds and separates the genes into the *A*α and *A*β loci identified by Day (21). In *S. commune*, the *A*α and *A*β loci are several map units apart (87), and only the *A*α locus has been characterized at the molecular level (98), but this is identical in organization to the *A*α locus of *C. cinerea*.

When one compares sequenced loci of *C. cinerea*, one can trace some of the evolutionary events that give rise to these complex loci (Fig. 17.3) (data from *Coprinus* sequencing project, Broad Institute of MIT and Harvard [http://www.broad.mit.edu], and references 63 and 92). Comparing *A42* and *A43* (Fig. 17.3A), it can be seen that the *A42* *A*α sublocus lacks an *HD1* gene but has an extra inactive *HD2* gene adjacent to the conserved *mip* gene that defines the border of the locus. This pseudogene is very similar in sequence to the adjacent gene and seems to have been the result of a recent duplication (U. Kües, unpublished data). An extra *HD1* gene is present between the two groups of genes in the *A*β complex. It was originally thought that this gene represented a fourth gene pair (hence the designations *a*, *b*, *c*, and *d* for gene pairs [62]), but in all backgrounds tested, this gene is nonfunctional. Interestingly, *A43* contains no extra genes but there is a footprint of this pseudogene. The *A43*α sublocus contains a complete gene pair, but transformation tests showed that the *HD2* gene is inactive (66). Nonetheless, with only four allelic versions of this sublocus, two with a functional gene pair, all combinations can generate at least one active HD1-HD2 heterodimer in mating. A population analysis identified 10 allelic versions of the *b* gene pair and 3 allelic versions of the *d* gene pair, together with the 4 *a* gene pairs, sufficient to generate 120 of the predicted 160 *A* mating specificities (75).

The three subloci at the *B* locus of *C. cinerea* appear to be contiguous (Fig. 17.3B), but recombination between the different alleles of the genes has clearly occurred to generate the many *B* mating specificities in nature (92). The genes encode lipopeptide mating pheromones and their cognate seven-transmembrane domain (7-TM) receptors. The number of genes within each group or sublocus is variable, a single receptor gene is associated with two to four pheromone genes, and the order and orientation of the genes vary. Thirteen loci conferring different *B* mating specificities have been characterized, and the allelic versions of the three groups of genes are sufficient to generate 70 unique combinations, close to the estimate of 79 different *B* specificities predicted by Day in 1961 (92).

Transformation analyses with many of the different pheromone genes identified were consistent with the three sets of genes being functionally independent and with the fact that most pheromones within each group activate all nonself receptors within the same group. Some intergroup activation could not be ruled out in this analysis, but conserved features of the mature pheromone sequences suggested that this is unlikely (Fig. 17.4). It is interesting therefore to compare this *B* locus with that of *S. commune*, described by Fowler and Vaillancourt in chapter 18 of this volume. In *S. commune* there are just two subloci, each containing a receptor gene and up to eight pheromone genes. Classical studies identified 9 *B*α and 9 *B*β alleles, but interestingly, not all

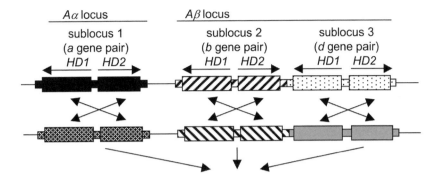

A The archetypal *A* mating type locus of *C. cinerea*

encode homeodomain proteins that heterodimerize in compatible matings to give a dikaryon-specific transcription factor that activates the clamp cell pathway

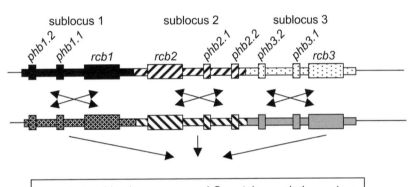

B The archetypal *B* mating type locus of *C. cinerea*

encode peptide pheromones and G-protein-coupled receptors that activate nuclear migration and clamp cell fusion in compatible matings

Figure 17.2 Archetypal organization of the *A* and *B* mating-type loci of *C. cinerea*. (A) *A* locus composed of three subloci each containing two divergently transcribed genes encoding dissimilar homeodomain proteins. The genes are distinguished as *HD1* and *HD2* based on the conserved but different protein homeodomains, and the three paralogous pairs of genes are designated the *a*, *b*, and *d* pairs. Horizontal arrows indicate the direction of transcription, and fill motifs are used to differentiate allelic and paralogous versions of the genes. Crossed arrows indicate the compatible gene pairs that encode subunits of an active transcription factor complex formed after cell fusion. (B) *B* locus composed of three subloci each containing a receptor gene and two pheromone genes. The three paralogous receptor genes are designated *rcb1*, *rcb2*, and *rcb3*. The corresponding pheromone genes are designated *phb1*, *phb2*, and *phb3* with different genes given the additional numbers (i.e., *phb 1.1* and *phb1.2*) to represent order with respect to *rcb*. Horizontal arrows indicate the direction of transcription; fill motifs are used to indicate different allelic and paralogous versions of the genes. Crossed arrows indicate receptor and pheromone combinations that can activate development upon mating.

combinations of these could recombine (57) and the reason for this became apparent from the analyses that Fowler and Vaillancourt describe. The activation specificity of 17 different pheromones was tested in a *B*-null background. The pheromones activated only certain subsets of receptors but were sufficient to activate all 18 of those predicted, but one of the *Bα* pheromones activated a *Bβ* receptor and five of the *Bβ* pheromones activated *Bα* receptors (30). The functional separation of the *Bα* and *Bβ* genes is thus not complete in *S. commune*,

(A)   Comparison of the sequenced *A42* and *A43* loci

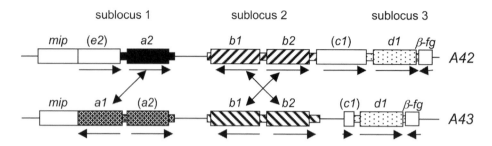

(B)   Comparison of sequenced *B42* and *B43* loci

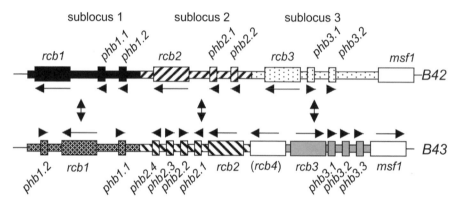

Figure 17.3   Comparison of sequenced mating-type loci of *C. cinerea*. (A) *A42* and *A43*. *a1*, *b1*, *c1*, and *d1* are *HD1* genes. *a2*, *b2*, and *e2* are *HD2* genes. Fill motifs denote allelic and paralogous versions of the genes. *A42* and *A43* have replicate alleles of the *HD1* gene, *d1*, in sublocus 3 but have different alleles of genes in sublocus 1 and 2. There are three compatible gene combinations indicated by diagonal arrows. (B) *B42* and *B43*. *rcb1*, *rcb2*, *rcb3*, and *rcb4* are receptor genes, and pheromone genes are designated *phb* followed by the sublocus and position numbers. *rcb4*, shown in parentheses, is predicted to be a pseudogene. Alleles of genes at all three subloci are different, as indicated by different fill motifs, and all generate compatible receptor-pheromone combinations (shown by vertical arrows). Horizontal arrows indicate the direction of gene transcription. *mip* and *β-fg* are non-mating-type genes that flank the *A* locus.

and lack of recombination between some *Bα* and *Bβ* complexes ensures that compatible combinations of receptors and pheromones are not generated in the same genome.

When comparing the *B42* and *B43* loci of *C. cinerea* (Fig. 17.3B), we see that, as at the *A* locus, extra genes can be found; one is seen in *B43* between the second and third subloci (deduced from the published *C. cinerea* sequence; T. Y. James, personal communication). This odd receptor gene has no associated pheromones, and it is most likely a pseudogene, but this has yet to be tested by transformation. Not shown is a footprint of another related receptor gene just downstream of *rcb1*, which was found in the unpublished sequence

of sublocus 1 from *B6*, an allele that is shared with *B42* (reference 37, L. A. Casselton and U. Kües, unpublished observation).

THE A AND B PROTEINS AND ALLELIC SPECIFICITY

The A proteins of the hymenomycetes, while clearly belonging to the same family as the MATa1 and MATα2 proteins of *S. cerevisiae*, differ in two significant ways: they have long C-terminal domains not present in the yeast proteins, and they have the ability to discriminate between large numbers of potential dimerization partners. Dimerization specificity resides in the N-terminal

(A) *C. cinerea*

Phero-mone	Precursor Sequences	Pheromone Activates
Phb3.1[5]	MSDTFTSLDIVLYGAAPRRDSDALDSAALFVNSQSVESSTIVPQLSSISVDEINDLPVDF**ERRTQGGNGLTFW**CVIA	Non-self sublocus 3 receptors
Phb3.2[3]	MDKEPLAQIRPARRIHLYRARLPSSTLGARSKRRPDSACRSNAGLVIRIGLIRDVPY**ERRTQGGSGPTWF**CIIQ	
Phb3.1[43]	MSDLFASLDLFLSSTEDNGCVCFDTNLSATTESQGCEILSKQASISTQELDGVLADF**ERRSGVGASWF**CTIA	
Phb3.2[42]	MSDAFTTLDTVDLFIEENEQEVVEVPSCPPPRRPSFSSADAESIFLTVEEVNDLPVDY**ERRTQGGGGLTWF**CVIA	
Phb2.3[43]	MSFSSLDAFVVDEELLQLAIDIPQPIPGQDQQPPINE**ERPGAGTMGAFC**IIN	Non-self sublocus 2 receptors
Phb2.2[42]	MDNFTVDLATLFEEFPELQEIQATASEHCSQDQYGSCEGPPINQ**ERPGSGVNRAF**CVIA	
Phb2.2[1]	MDTFTAFDDLNLECEVFEFLPEMSCTADASGFDQPPIDQ**ERPGTGSLGAF**CVIS	
Phb2.1[6]	MDTYSTFDPSLLEELGLTADILIVSSKPTPSLSTEPVDEVPRDE**ERAGPGDTPGGF**CVIA	

(B) *S. commune*

Phero-mone	Precursor Sequences	Pheromone Activates
Bap3(1)	MDAFQSILDVLSAALDEPVDAPLTAVAQHPDADAVFDTPTDF**ERVGTGGTATAF**CVVA	Bar2,3,5,6
Bap3(3)	MDSFATLPALEDTLLQALLDACAVPEDDALDAMLSSSRPSSDAVVSDA**ERHGSGNMTYF**CVVA	Bar4,7,Bbr8
Bbp2(4)	MDDFITLDFLEDTTPVFDFAPPTPNELTPEGYDEFMRMVANS**DSPDGYGGY**CVVA	Bar8, Bbr1
Bbp2(3)	MDTFTYVDLAAVAAAAVADEVPRDF**EDQITDYQSY**CIIC	Bar9, Bbr5
Bbp2(2)	..AGTSRPAEASARNPLGSSSASSSSASLAASTSDLLSASPSSAPTSPDDVIMSILADA**EHGYGGSNVHGW**CVVA	Bbr4,6,7

Figure 17.4 Pheromone precursor sequences from *C. cinerea* (A) and *S. commune* (B). The amino acid sequences are highly variable in length and sequence. The sequence of the mature pheromone is given in larger type; the predicted N-terminal recognition site, a charged doublet that is generally ER in *C. cinerea*, and the subterminal doublet are presented in bold. The CaaX motif is underlined.

domains. Switching the 5′ ends of sublocus 1 and sublocus 3 *HD1* genes of *C. cinerea* was sufficient to change the specificity of paralogous gene interactions (64) and switching sequences between the 5′ ends of an allelic pair of *HD1* genes of *S. commune* altered allele specificity (105). Studies with *C. cinerea* confirmed that the N-terminal domains enable highly specific dimerization in vitro (8). These domains are predicted to contain coiled-coil α-helices that mediate protein dimerization in other transcription factors. Interestingly, the predicted positions of these coiled-coil domains differ in the three paralogous versions of the HD1 proteins of *C. cinerea*, suggesting how paralogous proteins may be distinguished. Discriminating between different allelic versions of the proteins is more finely tuned, but studies with *U. maydis* show that certain types of amino acid substitutions in the critical N-terminal domains of the corresponding bE and bW proteins permitted normally incompatible proteins to dimerize (48).

Heterodimerization is an elegant way of regulating transcription factor function by bringing together different protein functional domains, thus ensuring that mating-dependent developmental pathways are activated only after compatible cell fusion. Studies with the

MATa1 and MATα2 proteins of *S. cerevisiae* revealed the essential role that dimerization plays in determining the affinity of homeodomains to bind DNA. The MATa1 (HD2) homeodomain, which normally has little affinity for DNA binding, is sufficient to bind with high affinity if fused to the short C-terminal tail of the MATα2 protein (46, 51). The protein-protein interaction thus appears to be more critical than having the MATα2 (HD1) homeodomain. Similarly for the *C. cinerea* and *S. commune* proteins, the HD2 homeodomain is critical to heterodimer function but the HD1 homeodomain can be deleted (3, 73). *C. cinerea* HD1 proteins have a negatively charged sequence in the long C-terminal domain of the HD1 protein, and this can activate transcription in an *S. cerevisiae* β-galactosidase reporter gene assay (3). The *S. cerevisiae* heterodimer, which lacks C-terminal domains, is a negative regulator that recruits other proteins for repressor function. Consistent with the prediction of an activation domain in the C-terminal domain of one of the basidiomycete proteins is the fact that the *U. maydis* B protein heterodimer has been shown to be a transcriptional activator (12). There is good genetic evidence to suggest that the A protein heterodimer of *C. cinerea* is also an activator (42). In a heterologous onion epidermal cell assay, two potential nuclear localization signals in the *C. cinerea* HD1 protein C-terminal domain were shown to be sufficient to localize a reporter protein to the nucleus. HD2 proteins lack nuclear localization signal sequences and could not localize to the nucleus in this assay (96). HD2 proteins must, therefore, be confined to the cytoplasm in unmated cells and can only localize to the nucleus after dimerization to a compatible HD1 partner. Dimerization via the N-terminal recognition domains occurs independently of DNA (8); thus, it can occur in the cell cytoplasm, enabling a compatible protein pair brought together by mating to localize to the nucleus.

In yeasts and filamentous ascomycetes there are two families of mating pheromones and receptors, but it is not known what biological advantage this may confer (69, 107). All predicted basidiomycete pheromones belong to the same family, typified by *S. cerevisiae* **a**-factor, and these activate corresponding G-protein coupled receptors belonging to the yeast (Ste3p) **a**-factor receptor family. Pheromone genes in *C. cinerea* and *S. commune* encode precursor molecules of 48 to 85 amino acids that have a C-terminal CaaX motif, a signal in corresponding *S. cerevisiae* **a**-factor precursor for C-terminal truncation and modification of the terminal cysteine residue by carboxymethylation and farnesylation (15, 19). The precursors of the *C. cinerea* pheromones are remarkably dissimilar in sequence, but alignment identified a conserved glutamic acid/arginine (ER) or aspartic acid/arginine (DR) motif 12 to 15 amino acids upstream of the CaaX motif (16). Processing at this conserved site predicts mature peptides of a size comparable to that of *S. cerevisiae* **a**-factor, which has 12 amino acids. The charged pair of predicted N-terminal amino acids is more variable in the *S. commune* proteins. The sequences of several pheromone precursors from both species are compared in Fig. 17.4.

In *C. cinerea*, as in *S. commune*, pheromones and receptors have been expressed heterologously in *S. cerevisiae* (29, 79, 80). Strains of yeast have been developed in which the pheromone signaling pathway is engineered to link receptor activation to reporter gene expression. Both qualitative and quantitative assays of receptor activation are possible using as reporter gene *HIS3*, which when activated permits histidine-independent growth of a *his3* mutant, or *lacZ*, which leads to induction of β-galactosidase (25). Despite the lack of sequence similarity between the hymenomycete pheromone precursors and **a**-factor, the processing pathway seems to be highly conserved because yeast can modify these proteins to active pheromone species that are secreted and activate appropriate *C. cinerea* or *S. commune* receptors heterologously expressed on the surface of its cells (28, 79). Significantly, processing of the *C. cinerea* precursors occurred only in *MATa* yeast cells, indicating that it was dependent on **a**-factor processing machinery not present in *MATα* cells that produce the alternative pheromone, α-factor (79).

Mushrooms appear to be unique among fungi in having evolved multiple versions of their pheromones and receptors, and this has been driven by the need to increase the numbers of mating types. Three alleles of the pheromones and receptors have been identified in the smut *Sporisorium reilianum* (94), but in the related *Ustilago* species and in *C. neoformans*, as in ascomycete fungi, there are just two alleles. These fungi secrete pheromones as chemoattractants to enable compatible cells to detect each other over a distance and orientate growth towards each other. Attempts to detect extracellular pheromones secreted by unmated cells of both *S. commune* (C. Raper, personal communication) and *C. cinerea* (79) have been unsuccessful. Hymenomycete fungi appear to sense compatibility only after cell fusion, and it is only then that pheromone-induced cell changes occur. The thousands of mating types generated by having multiallelic *A* and *B* genes ensure that nearly every chance fusion is compatible and pheromones are not required for mate attraction. The final step in pheromone secretion in *MATa* yeast cells is effected by the transporter protein Ste6 (19). A predicted homologue of *STE6* is found in the *C. cinerea* genome sequence (Casselton, unpublished), suggesting that pheromones are transported across membranes, even if this does not lead to secretion into the surrounding environment. The fusion of the clamp cell with the subterminal cell during tip cell division in the dikaryon is pheromone dependent, like mating-cell fusion in yeasts, and one would assume that pheromones have to activate receptors in separate cells (13). Fowler and Vaillancourt (chapter 18 of this volume) point out that unlike in other fungi, pheromone signaling is not involved in directing the growth of the cells that will fuse and the backward growth of the clamp cell to touch the subterminal cell is *A* gene regulated. They suggest that pheromone signaling may promote initial wall fusion by up-regulating agglutinins and other wall-bound proteins. As also pointed out by Fowler and Vaillancourt, an important goal of hymenomycete biology is to determine where the receptors and pheromones are localized. It is interesting that in *U. maydis* pheromone signaling arrests cells in G_2 of the cell cycle, rather than G_1 as occurs in *S. cerevisiae* (32). This may relate to the fact that nuclei are not prepared for fusion, but for synchronized coexistence in the dikaryotic cell. It would seem to be advantageous for nuclei to be able to enter directly into mitosis rather than having to replicate their DNA first.

A question that continues to intrigue is how pheromone and receptor specificity is determined. These signaling molecules are truly remarkable in that a given pheromone can activate several different receptors and a single receptor can be activated by several different pheromones. Having the sequences of so many receptors and pheromones and knowing the compatibility relationships should surely lead to insights into receptor-ligand binding. Because they are not secreted, the hymenomycete pheromones cannot be purified and sequenced, as they have been for another basidiomycete, *U. maydis* (95); their sequences can only be predicted. Based on the predicted ER processing site, Olesnicky et al. (79) designed a series of synthetic peptides of differing length, with and without farnesylation. They used the quantitative assay of β-galactosidase activity in *S. cerevisiae* to test the ability of synthetic peptides to activate an appropriate *C. cinerea* receptor. The most active peptide, and the one most likely to represent the native pheromone, was carboxymethylated and farnesylated and had the predicted processing ER signal at its N terminus. Based on the results of this experiment, it has been possible to concentrate attention on likely mature pheromone sequences and to look for amino acids that may act as specificity determinants in pheromone-receptor recognition (30, 80, 92).

Olesnicky et al. (80) showed by in vitro mutagenesis that for two pheromones at sublocus 3, the two amino acids adjacent to the C-terminal cysteine residue (the subterminal doublet) were critical for receptor specificity and that it was possible to reverse the receptor specificity by simply changing the order of these (FW to WF). Riquelme et al. (92) examined 29 predicted *C. cinerea* pheromone sequences and found that these could be grouped according to the subterminal doublet into the groups already defined by sublocus position within the *B* locus, good circumstantial evidence that the pheromones and receptors encoded by these three groups of genes are functionally independent (Fig. 17.4A). Fowler et al. (30) made groupings of pheromones based on receptor activation, and significantly, these groupings also correlate with the subterminal doublet (Fig. 17.4B). Since this doublet is shared between several pheromones with different allelic specificities, it cannot be the only specificity determinant. Multiple positions within the mature pheromone are likely to dictate specificity; in some cases this may appear to devolve onto a single amino acid, but in other cases, it is by no means clear, because pheromones with apparently identical activation spectra are entirely different in sequence (30, 92). However, several amino acid positions have all

been implicated in specific pheromone-receptor recognition (30, 80, 92) and it has to be assumed that the receptor contains a variable binding site that is able to differentiate various amino acids at different positions of the pheromone (92).

EVOLUTION OF COMPLEX MATING-TYPE LOCI

The genome sequence of *C. cinerea* reveals several clusters of genes of related function, suggesting that gene amplification has contributed to many aspects of the biology of this fungus. There are, for example, eight clusters of hydrophobin genes (103) and six clusters of laccase genes (53). Gene amplification has given rise to the complex mating-type loci that we find in hymenomycetes, but unlike the genes in other clusters, there appears to be some restriction on the numbers of functional genes at the *MAT* subloci and these genes have evolved to have very diverse allelic DNA sequences. Where extra genes are found, either from recent duplication events or by recombination, as described earlier (Fig. 17.3), selection has rendered these inactive. Sequence diversity is critical to maintain the integrity of the gene order and to restrict inappropriate recombination within the many different versions of the loci. Although we talk about allelic variants of the genes at each sublocus, these are not necessarily alleles in the strict sense of the term. A phylogeny analysis based on 13 different receptor protein sequences from *C. cinerea* suggests that the genes at the three subloci have a complex origin (Fig. 17.5) (92). The genes can be separated into two major clusters, indicative of early duplication of an ancestral gene and subsequent sequence diversification. Significantly, members of all three groups can be found in each of these two major clusters, indicating that the three groups have not evolved independently and that recombination has moved genes of different lineages between them. Sequence identity between receptor proteins ranges from 18 to 81%, and this is unlinked to pheromone specificity. Similar events seem to have occurred in the evolution of the *S. commune Bα* and *Bβ* families (44, 92). Pheromone specificity appears to have evolved faster than sequence diversification of the whole receptor, consistent with only a few residues in the receptor forming the pocket of interaction with the pheromone (92).

In contrast to genes at the *B* locus, a study of one of the *HD1* genes at the *A* locus (*b1*) by May et al. (76) shows that allelic versions of the same gene are very similar, with 67 to 78% identity in DNA sequence, an indication that these do have a common lineage. Sequence analysis moreover identifies regions where recombination has played a role in generating new alleles. In contrast to the sequence

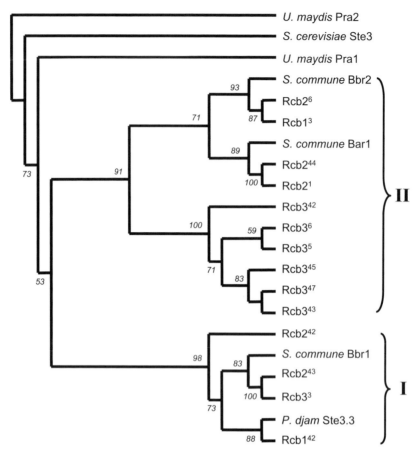

Figure 17.5 PAUP* maximum parsimony analysis of *C. cinerea* receptor proteins. A strict consensus tree is shown with bootstrap support values (percentage of 1,000 replicates) in italics. The nominated outgroup was *S. cerevisiae* Ste3p. Rcb alleles are distinguished by superscripts that denote the wild-type *B* specificity from which the genes were sequenced. Analysis by M. P. Challen.

diversity seen between different alleles, selection operates to maintain near identity of replicate alleles, as little as 1% variation in DNA sequence (5). At the detailed level, Fowler et al. (29) and Kothe et al. (59) have proposed useful models to show how new pheromone and receptor specificities could arise by gene duplication and recombination. However, it requires few allelic versions of genes at each of the subloci to generate large numbers of mating specificities and the driving force in evolution at hymenomycete *MAT* loci has not been to generate ever-increasing numbers of alleles, but to maintain a relatively small number of alleles at each sublocus and to recombine these into all possible allelic combinations (5, 76). Sublocus duplication has obviously been important. The *B* sublocus 1 in *C. cinerea*, with just two alleles, appears to be the result of a relatively recent duplication event, but these two alleles are sufficient to double the number of *B* mating specificities determined by the other subloci.

Analysis of the recently sequenced *C. cinerea* genome has unexpectedly identified additional members of the Ste3-type pheromone receptor family outside the *B* mating-type locus. There is no precedent for pheromone signaling other than in mating in fungi, so these genes are intriguing. They may simply be pseudogenes recombined out of the *MAT* locus, but as pointed out by T. James in chapter 19 of this volume, they may unexpectedly identify non-mating-type pheromone signaling pathways unique to hymenomycetes. He also points out that no corresponding *A*-type genes are found outside the *A* locus of the sequenced genome. It is perhaps relevant that we find very different recombination frequencies associated with the chromosome regions adjacent to the *MAT* loci of *C. cinerea*. For genes close to *A*, 1 centimorgan (cM) corresponds to 100 kb (63, 78), whereas for the region containing *B*, there is a >15-fold-higher recombination frequency with 1 cM corresponding to 6 kb (97) (compared with 1 cM equivalent to 28 kb on the chromosome that contains the *trp1* gene [31]). Indeed, the *B* locus resides on one of the smallest chromosomes, as detected by contour-clamped homogeneous electric field gel-electrophoresis (81), but the linkage map, based on recombination frequencies, is similar in length to that of the largest chromosome that contains the *A* locus. Low recombination frequency would act to maintain the organization around the *A* locus and hence the strong synteny of genes seen surrounding this locus in other hymenomycetes. Conversely, high recombination frequency would account for the much lesser synteny that is found around the *B* locus (see chapter 19 of this volume by James; T. Y. James and U. Kües, unpublished observation).

MUTATIONS IN THE *A* AND *B* GENES AND EVOLUTION OF HOMOTHALLISM

The majority of hymenomycetes are heterothallic, and it seems likely that as in ascomycetes, homothallic species were derived from heterothallic species (106). Mutations that confer self-compatibility potentially offer clues as to how homothallism may have arisen. They also identify critical functional domains of the proteins that might not be apparent from sequence analysis of wild-type genes.

When Dan Lewis originally chose *C. cinerea* as a new genetic model in the 1950s, he was interested in mutation rates and had hoped that it would be possible to induce mutations causing altered specificity in oidia (D. Lewis, personal communication). In the event, it required a much stronger selection technique to obtain mating-type mutants. During their study of heterokaryosis in *C. cinerea*, Swiezynski and Day (100) observed occasional fruiting of normally sterile common *A* heterokaryons and went on to show that these were the result of mutation in one of the *A* mating-type genes (22). Similar fruiting bodies on common *B* heterokaryons yielded *B* gene mutations in *S. commune* (83). Many more such mutations were obtained in both the *A* and *B* genes of *S. commune* (56, 58, 90, 91) and of *C. cinerea* (2, 22, 38, 99). With our current knowledge of just how polymorphic the sequences of the *A* and *B* genes are, it is not surprising that mutations obtained using this selection procedure never generated a new allele; mutation always resulted in self-compatibility. Mutations were generally dominant, and concomitant with a self-compatible phenotype, constitutive expression of the morphogenetic pathway was regulated by the mutant gene. *A* mutants produce unfused clamp cells, and *B* mutants, at least in *S. commune*, exhibit the typical flat morphology.

Molecular analysis of two independently derived *C. cinerea* mutations in the *A6* locus revealed a major intralocus deletion leaving just a single chimeric gene generated by in-frame fusion of an *HD2* gene to the 3′ end of an *HD1* gene (2, 65). These fusion genes were sufficient to activate *A*-regulated development in the mutant background and any other background including the wild-type *A6*. A third mutation in *A43* (99) was found to involve a similar gene fusion (84). As described earlier, heterodimerization between HD1 and HD2 proteins brings together different functional domains of the active transcription factor; the mutant chimeric *A* genes are predicted to encode a minimal heterodimer that has the essential HD2 DNA-binding domain and sufficient HD1 protein to provide the nuclear targeting sequences and potential activation domain. The critical dimerization domains in the N-terminal regions of the proteins, which normally distinguish between compatible and incompatible proteins, are no longer required once the proteins are fused. The chimeric protein activates *A*-regulated development without either N-terminal domain (8), confirming that this domain is not required for DNA binding, simply for association of the proteins prior to entry into the nucleus. Characterization of the *A6* fusion gene confirmed that dimerization between the homeodomain proteins was critical for mate recognition long before the actual protein interaction was demonstrated in vitro (61).

Fusions between mating-type genes in other fungi could similarly indicate that a direct interaction between the proteins is crucial to activating sexual development. In several self-compatible ascomycetes, gene fusions have clearly been instrumental in bringing about homothallism. This is reviewed by Debuchy and Turgeon (24). Unfortunately we know little, as yet, about the functions of these mating-type proteins, but it would be satisfying to think that protein dimerization is a common theme in mating.

Molecular characterization of *B* gene mutations conferring self-compatibility has shown that these can arise in both receptor and pheromone genes (29, 79, 80). Olesnicky et al. (79, 80) characterized amino acid replacements in two mutant receptor genes of *C. cinerea* and used the heterologous activation assay in *S. cerevisiae* to determine the functional defect of the proteins, something impossible to determine in the complex natural environment of the *C. cinerea* cell. One of the mutations, Q229P, a single substitution towards the extracellular end of the sixth transmembrane domain, resulted in a constitutively active receptor that no longer required activation by a compatible pheromone. The second mutation, R96H, a substitution at the intracellular end of the third transmembrane domain, resulted in a receptor that had altered specificity and could be activated by a normally incompatible pheromone encoded in the same sublocus (80). The *Bβ2* mutation obtained by Raper and Raper (86) is an example of a pheromone gene mutation, and self-compatibility is the result of a single amino acid substitution that enables the pheromone to activate its self-receptor (29).

Several attempts to create new wild-type *B* alleles in *S. commune* were made by taking a constitutive mutant (primary mutation) and selecting for secondary mutations that restored a fluffy wild-type phenotype (86, 88, 90). No new wild-type was ever generated, only loss of function mutants. T. Fowler and L. Vaillancourt (chapter 18 of this volume) describe the elegant studies of C. and J. Raper (86), who isolated a series of secondary mutants generated from a self-compatible *Bβ2* mutant and showed that these could be arranged in order of loss of function towards the other eight versions of the *Bβ* locus, culminating in complete loss of function. A full explanation for these varied phenotypes can now be obtained from the molecular perspective (29), an analysis that reveals a corresponding series of receptor and pheromone gene mutations and deletions spanning part or all of the two subloci.

It seems likely that constitutively active or compatible combinations of genes play a role in the evolution of bipolar and homothallic species of hymenomycetes. Bipolar species are found in all phylogenies, indicating that they have arisen independently many times (45). In bipolar *C. neoformans* and *U. hordei*, where there are only two versions of the *MAT* locus, this contains both homeobox genes and mating pheromone and receptor genes (6, 41, 70). In bipolar hymenomycetes, which have multiple mating types, a different evolutionary route is seen. Molecular analysis of bipolar *Coprinellus disseminatus* (*Coprinus disseminatus*) shows that the *MAT* locus contains only the paired genes that encode the *A* transcription factor. A putative *B* locus with genes encoding pheromones and receptors is found on another chromosome, but there are no allelic variants and the genes can no longer play a role in mate recognition (45). Nonetheless, *C. disseminatus* forms a typical dikaryon with clamp connections and one assumes that pheromone signaling still plays a critical role in maintaining the dikaryophase. The authors suggested that pheromone signaling is constitutive in bipolar species (45). In view of the sick phenotype that constitutive pheromone signaling induces in both *S. commune* (40) and, to a lesser extent, *C. cinerea* (67), it seems more likely that the activity of the pheromone and receptor genes is induced only after compatible cell fusion, possibly by the A protein heterodimer. A short step from bipolarity to homothallism would be to acquire a compatible pair of *A* genes, or a fusion gene at the *A* locus. Preliminary analysis of the homothallic *Agaricus subfloccosus* confirmed that the *A* genes are present (S. Burrow, L. A. Casselton, and M. P. Challen, unpublished data), but the putative mating-type locus has yet to be fully characterized.

AMBM (Amut Bmut) STRAINS AND DOWNSTREAM DEVELOPMENTAL PATHWAYS

Little work has focused on the direct downstream targets of the *A* and *B* genes in either *C. cinerea* or *S. commune*. In *S. cerevisiae*, pheromone signaling activates a mitogen-activated protein kinase cascade (9). Detailed studies of *U. maydis*, reviewed recently by Feldbrügge et al. (27) and reported in chapter 22 of this volume by Kahmann and Schirawski, have identified the corresponding basidiomycete components of this pathway and shown how this interacts with the cyclic AMP signaling pathway to regulate differential transcription of the *a* and *b* mating-type genes and how cross talk between these pathways regulates the growth and pathogenicity of the dikaryon (47, 77). Several targets of the bE/bW heterodimer homologue of the hymenomycete A proteins have also been identified (12). Genes encoding these *U. maydis* proteins have homologues in the *C. cinerea* genome sequence, and microarray analyses will soon make it possible to look at their transcription under different developmental regimes. For the present, a more attractive focus in hymenomycete research has been later stages in sexual development, the formation of the fruiting body, meiosis, and sporulation. This developmental program represents a dramatic switch from an undifferentiated filamentous mycelial phase to a multihyphal structure composed of many cell types (reviewed in reference 60). Several environmental cues, namely, light, temperature, and gravity, are essential to coordinate sequential stages in development. Exploitation of mutations in the *A* and *B* mating-type genes makes all these later stages in development accessible to genetic and molecular analysis.

Identifying mutations affecting fruiting would normally be difficult, because they occur when two genetically different nuclei are present. However, the combination of *A* and *B* mutations in the same haploid genome generates a homothallic mycelium that in most respects resembles a true dikaryon with binucleate cells and fused clamp connections (99). As shown originally by Raper et al. (90), *AmBm* dikaryons are fertile. A double-mutant strain of *C. cinerea*, generated by Swamy et al. (99), has been used extensively in screens for sexual developmental defects. This strain produces abundant oidia (85), unlike normal dikaryons (52), and these can be treated with mutagens or transformed to give insertional mutations (restriction enzyme-mediated integration [REMI]) (20, 36). A single oidium develops into a dikaryotic mycelium with identical haploid nuclei and produces fruiting bodies in which meiosis and sporulation occur. Any mutation arising in the *AmBm* (Amut Bmut) background will be expressed, hence the ability to detect recessive mutations affecting all stages in fruiting (Fig. 17.6).

Among the important genes identified using the *AmBm* background is *clp1* (42). The Clp1 predicted protein has no recognizable motifs that identify its likely function; however, it is essential for activating the *A* clamp cell pathway. Transcription of *clp1* is normally induced only by a compatible *A* gene interaction, but this requirement is bypassed if *clp1* is expressed from a constitutive promoter. Significantly, analysis of the *clp1* promoter identified an essential motif that resembles the HD1/HD2 heterodimer binding sites of *S. cerevisiae* MATa1/MATα2 and *U. maydis* bE/bW. Excitingly, the *U. maydis* homologue of this gene has recently been described and shown to have a similar role in clamp cell development, and it is confirmed as a direct target for the bE/bW heterodimer (93).

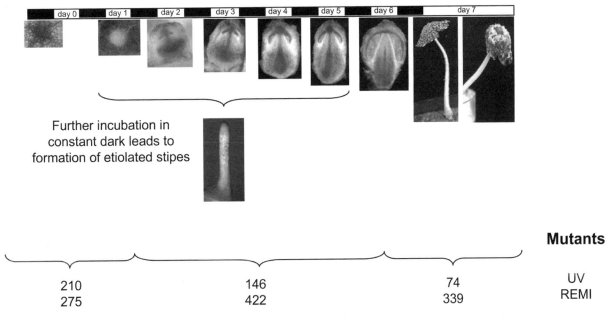

Figure 17.6 The self-fertile homokaryon AmutBmut of *C. cinerea* (*A43mut, B43mut,* and *pab1*) (99) forms fruiting bodies under suitable environmental conditions (day-night rhythm, 25 to 28°C, >80% humidity). Localized hyphal branching gives rise to loose aggregates (primary hyphal knots). A light signal is needed (day 0) for appearance of compact secondary hyphal knots (day 1 of development). Over the next 5 days, cap and stipe tissues differentiate. Tissue differentiation is light controlled: when light is lacking, structures known as etiolated stipes or dark stipes are formed with elongated stipe bases terminating in rudimentary stipes and caps. The light signal at day 5 gives rise to karyogamy, meiosis, and basidiospore formation at day 6 of development. During meiosis and basidiospore formation, the mushroom cap expands and the stipe elongates to give shortly after midnight a fully matured fruiting body. Fruiting bodies are short-lived and undergo autolysis on day 7 of development in order to disperse the black basidiospores. UV and REMI mutagenesis of oidia from this strain has yielded over 9,000 different clones that have been individually tested for fruiting behavior. Several hundreds of mutants were obtained and grouped into categories according to the developmental stage affected (U. Kües, J. D. Granado, and M. Aebi, unpublished data). Mutants in the Amut Bmut background mentioned in the text are characterized by lack of secondary hyphal knot formation (*cfs1*), formation of etiolated stipes in the light (*dst1*), formation of mushrooms with a short stipe that is unable to elongate (*eln3*), and white caps that lack basidiospores (*spo11*). The time course of fruiting-body development of homokaryon AmutBmut was kindly supplied by M. Navarro-González, and the photograph of the etiolated stipe was kindly supplied by W. Chaisaena.

Many mutants with defects in fruiting-body development in the *AmBm* background have been isolated (20, 36, 42, 68), providing an important resource for future studies on the whole pathway. So far, a gene required for fruiting-body initiation, *cfs1*, a gene required for light perception, *dst1*, and a gene required for stipe elongation, *eln3*, have been cloned and characterized. The Csf1 protein has homology to bacterial cyclopropane fatty acid synthases, enzymes shown to convert membrane-bound unsaturated fatty acids into cyclopropane fatty acids (71). Dst1 is a putative blue-light receptor similar to the WC-1 blue-light receptor protein of *Neurospora crassa* (101). The product of the *eln3* gene has a glycosyltransferase domain and is believed to localize to membranes (1). Many mutants blocked in sporulation and meiosis have been obtained in *AmBm* strains by standard or REMI mutagenesis techniques (17, 49, 50). Such mutants are readily detected because they produce white fruiting bodies lacking the characteristic black basidiospores (72). As yet there is only one published report on molecular characterization of a mutant (18), but the large numbers of complementation groups involved are indicative of just how many steps are likely to be involved in this developmental pathway.

CONCLUDING REMARKS

The classical genetical studies of John Raper and his colleagues, working with *S. commune*, and of Peter Day, working with *C. cinerea*, laid the foundations for the remarkable molecular revelations that we have described here. Their patient mapping experiments defined the map positions of the genes and foretold the complexity that we would find at the loci themselves. Like the molecular biologists, they were fascinated by the specificity of such large numbers of allelic variants of the genes. The protein families that are involved in mating in fungi, the homeodomain proteins and the pheromones and receptors, are ubiquitous in all eukaryotes, but it is the mushrooms that have exploited the properties of these proteins to generate thousands of mating types and shown us just how specific protein dimerization and ligand-receptor binding can be.

By characterizing the mating-type genes of these model species, we have made possible the cloning of mating-type genes from many other species, and we are beginning to unravel the basis of the different breeding systems that we find in the hymenomycetes and to elucidate how they may have evolved. It is a fitting tribute to John Raper that the sex life of fungi continues to excite so many researchers. With the molecular and genomic tools we now have, exploring deeper into the study of the *MAT* loci enhances our understanding of the importance of sex and its involvement in genome evolution.

We are particularly grateful to Mike Challen for the phylogeny analysis presented in Fig. 17.4 and to the coworkers of U.K. for the contribution of photographs of developmental structures of C. cinerea. L.A.C. gratefully acknowledges the support of the Leverhulme Trust with the award of an Emeritus Fellowship, and U.K. is supported by the Deutsche Bundesstiftung Umwelt (DBU).

References

1. **Arima, T., M. Yamamoto, A. Hirata, S. Kawano, and T. Kamada.** 2004. The *eln3* gene involved in fruiting body morphogenesis of *Coprinus cinereus* encodes a putative membrane protein with a general glycosyltransferase domain. *Fungal Genet. Biol.* **41**:805–812.

2. **Asante-Owusu, R. N.** 1994. Manipulation of the *A* mating type genes of *Coprinus cinereus*. Ph.D. thesis. University of Oxford, Oxford, United Kingdom.

3. **Asante-Owusu, R. N., A. H. Banham, H. U. Böhnert, E. J. C. Mellor, and L. A. Casselton.** 1996. Heterodimerization between two classes of homeodomain proteins in the mushroom *Coprinus cinereus* brings together potential DNA-binding and activation domains. *Gene* **172**:25–31.

4. **Badalyan, S. M., E. Polak, R. Hermann, M. Aebi, and U. Kües.** 2004. Role of peg formation in clamp cell fusion of homeobasidiomycete fungi. *J. Basic Microbiol.* **44**:167–177.

5. **Badrane, H., and G. May.** 1999. The divergence-homogenization duality in the evolution of the *b1* mating type of *Coprinus cinereus*. *Mol. Biol. Evol.* **16**:975–986.

6. **Bakkeren, G., and J. W. Kronstad.** 1994. Linkage of mating-type loci distinguishes bipolar from tetrapolar mating in basidiomycetous smut fungi. *Proc. Natl. Acad. Sci. USA* **91**:7085–7089.

7. **Bakkeren, G., G. Jiang, R. L. Warren, Y. Butterfield, H. Shin, R. Chiu, R. Linning, J. Schein, N. Lee, G. Hu, D. M. Kupfer, Y. Tang, B. A. Roe, S. Jones, M. Marra, and J. W. Kronstad.** 2006. Mating factor linkage and genome evolution in basidiomycetous pathogens of cereals. *Fungal Genet. Biol.* doi:10.1016/j.fgb.2006.04.002.

8. **Banham, A. H., R. N. Asante-Owusu, B. Göttgens, S. A. J. Thompson, C. S. Kingsnorth, E. J. C. Mellor, and L. A. Casselton.** 1995. An N-terminal dimerization domain permits homeodomain proteins to choose compatible partners and initiate sexual development in the mushroom *Coprinus cinereus*. *Plant Cell* **7**:773–783.

9. **Bardwell, L.** 2004. A walk-through of the yeast mating pheromone response pathway. *Peptides* **25**:1465–1476.

10. **Bensaude, M.** 1918. Recherches sur le cycle évolutif et la sexualité chez les Basidiomycètes. Ph.D. thesis. Faculté des Sciences de Paris, Imprimerie Nemourienne, Henri Bouloy, Nemours, France.

11. **Bölker, M., M. Urban, and R. Kahmann.** 1992. The *a* mating type locus of *U. maydis* specifies cell signalling components. *Cell* **68**:441–450.

12. Brachmann, A., G. Weinzierl, J. Kämper, and R. Kahmann. 2001. Identification of genes in the bW/bE regulatory cascade in *Ustilago maydis*. *Mol. Microbiol.* 42:1047–1063.

13. Brown, A. J., and L. A. Casselton. 2001. Mating in mushrooms: increasing the chances but prolonging the affair. *Trends Genet.* 17:393–400.

14. Buller, H. R. 1931. *Researches on Fungi. IV. Further Observations on the* Coprini *Together with Some Investigations on Social Organisation and Sex in the Hymenomycetes.* Hafner Publishing Co., New York, NY.

15. Caldwell, G. A., F. Naider, and J. M. Becker. 1995. Fungal lipopetide mating pheromones: a model system for the study of protein prenylation. *Microbiol. Rev.* 59:406–422.

16. Casselton, L. A., and N. S. Olesnicky. 1998. Molecular genetics of mating recognition in basidiomycete fungi. *Microbiol. Mol. Biol. Rev.* 62:55–70.

17. Casselton, L. A., and M. Zolan. 2002. The art and design of genetic screens: filamentous fungi. *Nat. Rev. Genet.* 3:683–697.

18. Celerin, M., S. T. Merino, J. F. Stone, A. M. Menzie, and M. F. Zolan. 2000. Multiple roles of Spo11 in meiotic chromosome behavior. *EMBO J.* 19:2739–2750.

19. Chen, P., S. K. Sapperstein, J. D. Choi, and S. Michaelis. 1997. Biogenesis of the *Saccharomyces cerevisiae* mating pheromone a-factor. *J. Cell Biol.* 136:251–269.

20. Cummings, W. J., M. Celerin, J. Crodian, L. K. Brunick, and M. E. Zolan. 1999. Insertional mutagenesis in *Coprinus cinereus*: use of a dominant selection marker to generate tagged sporulation defective mutants. *Curr. Genet.* 36:371–382.

21. Day, P. 1961. The structure of the *A* mating-type locus of *Coprinus lagopus*. *Genetics* 45:641–650.

22. Day, P. R. 1963. Mutations affecting the *A* mating-type locus in *Coprinus lagopus*. *Genet. Res.* 4:55–65.

23. Day, P. R. 1963. The structure of the *A* mating type factor in *Coprinus lagopus*: wild alleles. *Genet. Res.* 4:323–325.

24. Debuchy, R., and B. G. Turgeon. 2006. Mating-type structure, evolution, and function in Euascomycetes, p. 293–323. *In* U. Kües and R. Fischer (ed.), *The Mycota. 1. Growth, Differentiation and Sexuality*, 2nd ed. Springer-Verlag, Berlin, Germany.

25. Dowell, S. J., and A. J. Brown. 2002. Yeast assays for G-protein-coupled receptors. *Recept. Channels* 8:343–352.

26. Duboule, D. 1994. Guidebook to the Homeobox Genes. Oxford University Press, Oxford, England.

27. Feldbrügge, M., M. Bölker, G. Steinberg, J. Kämper, and R. Kahmann. 2006. Regulatory and structural networks orchestrating mating, dimorphism, cell shape and pathogenesis in *Ustilago maydis*, p. 375–391. *In* U. Kües and R. Fischer (ed.), *The Mycota. 1. Growth, Differentiation and Sexuality*, 2nd ed. Springer-Verlag, Berlin, Germany.

28. Fowler, T. J., S. M. DeSimone, M. F. Mitton, J. Kurjan, and C. A. Raper. 1999. Multiple sex pheromones and receptors of a mushroom-producing fungus elicit mating in yeast. *Mol. Biol. Cell* 10:2559–2572.

29. Fowler, T. J., M. J. Mitton, L. J. Vaillancourt, and C. A. Raper. 2001. Changes in mate recognition through alterations of pheromones and receptors in the multisexual mushroom fungus *Schizophyllum commune*. *Genetics* 158:1491–1503.

30. Fowler, T. J., M. F. Mitton, E. I. Rees, and C. A. Raper. 2004. Crossing the boundary between the *Bα* and *Bβ* mating-type loci in *Schizophyllum commune*. *Fungal Genet. Biol.* 41:89–101.

31. Freedman, T., and P. J. Pukkila. 1997. A physical assay for meiotic recombination in *Coprinus cinereus*. *Mol. Gen. Genet.* 254:372–378.

32. Garcia-Muse, T., G. Steinberg, and J. Perez-Martin. 2003. Pheromone-induced G(2) arrest in the phytopathogenic fungus *Ustilago maydis*. *Eukaryot. Cell* 2:494–500.

33. Giasson, L., C. A. Specht, C. Milgrim, C. P. Novotny, and R. C. Ullrich. 1989. Cloning and comparison of *Aα* mating-type alleles of the basidiomycete *Schizophyllum commune*. *Mol. Gen. Genet.* 218:72–77.

34. Giesy, R. M., and P. R. Day. 1965. The septal pores of *Coprinus lagopus* (Fr.) sensu Buller in relation to nuclear migration. *Am. J. Bot.* 52:287–293.

35. Gillissen, B., J. Bergemann, C. Sandmann, B. Schroeer, M. Bölker, and R. Kahmann. 1992. A two-component regulatory system for self/non-self recognition in *Ustilago maydis*. *Cell* 68:647–657.

36. Granado, J. D., K. Kertesz-Chaloupková, M. Aebi, and U. Kües. 1997. Restriction enzyme-mediated DNA integration in *Coprinus cinereus*. *Mol. Gen. Genet.* 256:28–36.

37. Halsall, J. R., M. J. Milner, and L. A. Casselton. 2000. Three subfamilies of pheromone and receptor genes generate multiple *B* mating specificities in the mushroom *Coprinus cinereus*. *Genetics* 154:1115–1123.

38. Haylock, R. W., A. Economou, and L. A. Casselton. 1980. Dikaryon formation in *Coprinus cinereus*: selection and identification of *B* factor mutants. *J. Gen. Microbiol.* 121:17–26.

39. Herskowitz, I. 1988. Life cycle of the budding yeast *Saccharomyces cerevisiae*. *Microbiol. Rev.* 52:536–553.

40. Hoffman, R. M., and J. R. Raper. 1974. Genetic impairment of energy conservation in development of *Schizophyllum*. Efficient mitochondria in energy-starved cells. *J. Gen. Microbiol.* 82:67–75.

41. Hull, C. M., M. J. Boily, and J. Heitman. 2005. Sex-specific homeodomain proteins Sxi1α and Sxi2a coordinately regulate sexual development in *Cryptococcus neoformans*. *Eukaryot. Cell* 4:526–535.

42. Inada, K., Y. Morimoto, T. Arima, Y. Murata, and T. Kamada. 2001. The *clp1* gene of the mushroom *Coprinus cinereus* is essential for *A*-regulated sexual development. *Genetics* 157:133–140.

43. Iwasa, M., S. Tanabe, and T. Kamada. 1998. The two nuclei in the dikaryon of the homobasidiomycete *Coprinus cinereus* change position after each conjugate division. *Fungal Genet. Biol.* 23:110–116.

44. James, T. Y., S. R. Liou, and R. Vilgalys. 2004. The genetic structure and diversity of the *A* and *B* mating-type genes from the tropical oyster mushroom, *Pleurotus djamor*. *Fungal Genet. Biol.* 41:813–825.

45. James, T. Y., P. Srivilai, U. Kües, and R. Vilgalys. 2006. Evolution of the bipolar mating system of the mushroom

Coprinellus disseminatus from its tetrapolar ancestors involves loss of mating-type specific pheromone receptor function. *Genetics* 172:1877–1891.

46. Johnson, A. D. 1995. Molecular mechanisms of cell-type determination in budding yeast. *Curr. Opin. Genet. Dev.* 5:552–558.

47. Kaffarnik, F., P. Müller, M. Leibundgut, R. Kahmann, and M. Feldbrügge. 2003. PKA and MAPK phosphorylation of Prf1 allows promoter discrimination in *Ustilago maydis*. *EMBO J.* 22:5817–5826.

48. Kämper, J., M. Reichmann, T. Romeis, M. Bölker, and R. Kahmann. 1995. Multiallelic recognition: nonself-dependent dimerization of the bE and bW homeodomain proteins in *Ustilago maydis*. *Cell* 81:73–83.

49. Kanda, T., A. Goto, K. Sawa, H. Arakawa, Y. Yasuda, and T. Takemaru. 1989. Isolation and characterization of recessive sporeless mutants in the basidiomycete *Coprinus cinereus*. *Mol. Gen. Genet.* 216:526–529.

50. Kanda, T., H. Arakawa, Y. Yasuda, and T. Takemaru. 1990. Basidiospore formation in a mutant of incompatibility factors and in mutants that arrest at metaphase I in *Coprinus cinereus*. *Exp. Mycol.* 14:218–226.

51. Ke, A., and C. Wolberger. 2003. Insights into binding cooperativity of MATa1/MATα2 from the crystal structure of a MATa1 homeodomain-maltose binding protein chimera. *Protein Sci.* 12:306–312.

52. Kertesz-Chaloupková, K., P. J. Walser, J. D. Granado, M. Aebi, and U. Kües. 1998. Blue light overrides repression of asexual sporulation by mating type genes in the basidiomycete *Coprinus cinereus*. *Fungal Genet. Biol.* 23:95–109.

53. Kilaru, S., P. J. Hoegger, and U. Kües. 2006. The laccase multi-gene family in *Coprinopsis cinerea* has seventeen different members that divide into two distinct subfamilies. *Curr. Genet.* 50:45–60.

54. Kimura, K. 1952. Studies on the sex of *Coprinus macrorhizus* Rea f. *microsporus* Hongo. I. Introductory experiments. *Biol. J. Okayama Univ.* 1:72–79.

55. Kniep, H. 1928. *Die Sexualität der niederen Pflanzen*. Fischer, Jena, Germany.

56. Koltin, Y. 1968. The genetic structure of the incompatibility factors of *Schizophyllum commune*. Comparative studies of primary mutations in the *B* factor. *Mol. Gen. Genet.* 102:196–203.

57. Koltin, Y., and J. R. Raper. 1967. The genetic structure of incompatibility factors of *Schizophyllum commune*: three functionally distinct classes of *B* factors. *Proc. Natl. Acad. Sci. USA* 58:1220–1226.

58. Koltin, Y., J. Stamberg, N. Bawnick, R. Tamarkin, and R. Werczberger. 1979. Mutational analysis of natural alleles in and affecting the *B* incompatibility factor of *Schizophyllum*. *Genetics* 93:383–391.

59. Kothe, E., S. Gola, and J. Wendland. 2003. Evolution of multispecific mating-type alleles for pheromone perception in the homobasidiomycete fungi. *Curr. Genet.* 42: 268–275.

60. Kües, U. 2000. Life history and developmental processes in the basidiomycete *Coprinus cinereus*. *Microbiol. Mol. Biol. Rev.* 64:316–353.

61. Kües, U., and L. A. Casselton. 1992. Homeodomains and regulation of sexual development in basidiomycetes. *Trends Genet.* 8:154–155.

62. Kües, U., and L. A. Casselton. 1993. The origin of multiple mating types in mushrooms. *J. Cell Sci.* 104:227–230.

63. Kües, U., W. V. J. Richardson, A. M. Tymon, E. S. Mutasa, B. Göttgens, S. Gaubatz, A. Gregoriades, and L. A. Casselton. 1992. The combination of dissimilar alleles of the *Aα* and *Aβ* gene complexes, whose proteins contain homeodomain motifs, determines sexual development in the mushroom *Coprinus cinereus*. *Genes Dev.* 4:568–577.

64. Kües, U., R. N. Asante-Owusu, E. S. Mutasa, A. M. Tymon, E. H. Pardo, S. F. O'Shea, and L. A. Casselton. 1994. Two classes of homeodomain proteins specify the multiple *A* mating types of the mushroom *Coprinus cinereus*. *Plant Cell* 6:1467–1475.

65. Kües, U., B. Göttgens, R. Stratmann, W. V. J. Richardson, S. F. O'Shea, and L. A. Casselton. 1994. A chimeric homeodomain protein causes self-compatibility and constitutive sexual development in the mushroom *Coprinus cinereus*. *EMBO J.* 13:4054–4059.

66. Kües, U., A. M. Tymon, W. V. J. Richardson, G. May, P. T. Geiser, and L. A. Casselton. 1994. A mating-type factors of *Coprinus cinereus* have variable numbers of specificity genes encoding two classes of homeodomain proteins. *Mol. Gen. Genet.* 245:45–52.

67. Kües, U., P. J. Walser, M. J. Klaus, and M. Aebi. 2002. Influence of activated *A* and *B* mating type pathways on developmental processes in the basidiomycete *Coprinus cinereus*. *Mol. Genet. Genomics* 268:262–271.

68. Kües, U., M. Navarro-González, P. Srivilai, W. Chaisaena, and R. Velagapudi. 2006. Mushroom biology and genetics. *In* U. Kües (ed.), *Wood Production, Wood Technology and Biotechnological Impacts*. Universitätsverlag Göttingen, Göttingen, Germany.

69. Kurjan, J. 1993. The pheromone response pathway in *Saccharomyces cerevisiae*. *Annu. Rev. Genet.* 27:147–179.

70. Lengeler, K. B., D. S. Fox, J. A. Fraser, A. Allen, K. Forrester, F. S. Dietrich, and J. Heitman. 2002. Mating-type locus of *Cryptococcus neoformans*: a step in the evolution of sex chromosomes. *Eukaryot. Cell* 1:704–718.

71. Liu, Y., P. Srivilai, S. Loos, M. Aebi, and U. Kües. 2006. An essential gene for fruiting initiation in the basidiomycete *Coprinopsis cinerea* is homologous to bacterial cyclopropane fatty acid synthase genes. *Genetics* 172: 873–884.

72. Lu, B. C., N. Gallo, and U. Kües. 2003. White-cap mutants and meiotic apoptosis in the fungus *Coprinus cinereus*. *Fungal Genet. Biol.* 39:82–93.

73. Luo, Y. H., R. C. Ullrich, and C. P. Novotny. 1994. Only one of the paired *Schizophyllum commune Aα* mating-type putative homeobox genes encodes a homeodomain essential for *Aα* regulated development. *Mol. Gen. Genet.* 244:318–324.

74. Magae, Y., C. Novotny, and R. Ullrich. 1995. Interaction of the *A* alpha Y mating-type and Z mating-type homeodomain proteins of *Schizophyllum commune* detected by the two-hybrid system. *Biochem. Biophys. Res. Commun.* 211:1071–1076.

75. May, G., and E. Matzke. 1995. Recombination and variation at the *A* mating-type locus of *Coprinus cinereus*. *Mol. Biol. Evol.* **12**:794–802.

76. May, G., F. Shaw, H. Badrane, and X. Vekemans. 1999. The signature of balancing selection: fungal mating compatibility gene evolution. *Proc. Natl. Acad. Sci. USA* **96**:172–177.

77. Müller, P., G. Weinzierl, A. Brachmann, M. Feldbrügge, and R. Kahmann. 2003. Mating and pathogenic development of the smut fungus *Ustilago maydis* are regulated by one mitogen-activated protein kinase cascade. *Eukaryot. Cell* **2**:1187–1199.

78. Mutasa, E. S., A. M. Tymon, B. Göttgens, F. M. Mellon, P. F. R. Little, and L. A. Casselton. 1989. Molecular organization of an *A*-mating type factor of the basidiomycete fungus *Coprinus cinereus*. *Curr. Genet.* **18**:233–229.

79. Olesnicky, N. S., A. J. Brown, S. J. Dowell, and L. A. Casselton. 1999. A constitutively active G-protein-coupled receptor causes mating self-incompatibility in the mushroom *Coprinus*. *EMBO J.* **18**:2756–2763.

80. Olesnicky, N. S., A. J. Brown, Y. Honda, S. L. Dyas, S. J. Dowell, and L. A. Casselton. 2000. Self-compatible *B* mutants in *Coprinus* with altered pheromone-receptor specificities. *Genetics* **156**:1025–1033.

81. O'Shea, S. F., P. T. Chaure, J. R. Halsall, N. S. Olesnicky, A. Leibrandt, I. F. Connerton, and L. A. Casselton. 1998. A large pheromone and receptor gene complex determines multiple *B* mating type specificities in *Coprinus cinereus*. *Genetics* **148**:1081–1090.

82. Papazian, H. 1954. Exchange of incompatibility factors between the nuclei of a dikaryon. *Science* **119**:691–693.

83. Parag, Y. 1962. Mutations in the *B* incompatibility factor of *Schizophyllum commune*. *Proc. Natl. Acad. Sci. USA* **48**:743–750.

84. Pardo, E. H., S. F. O'Shea, and L. A. Casselton. 1996. Multiple versions of the *A* mating type locus of *Coprinus cinereus* are generated by three paralogous pairs of multiallelic homeobox genes. *Genetics* **148**:87–94.

85. Polak, E., R. Hermann, U. Kües, and M. Aebi. 1997. Asexual development in *Coprinus cinereus*: structure and development of oidiophores and oidia in an *Amut Bmut* homokaryon. *Fungal Genet. Biol.* **22**:112–126.

86. Raper, C. A., and J. R. Raper. 1973. Mutational analysis of a regulatory gene for morphogenesis in *Schizophyllum*. *Proc. Natl. Acad. Sci. USA* **70**:1427–1431.

87. Raper, J. R. 1966. *Genetics of Sexuality in Higher Fungi*. Ronald Press, New York, NY.

88. Raper, J. R., and M. Raudaskoski. 1968. Secondary mutations at the *Bβ* incompatibility locus of *Schizophyllum*. *Heredity* **23**:109–117.

89. Raper, J. R., M. G. Baxter, and R. B. Middleton. 1958. The genetic structure of the incompatibility factors in *Schizophyllum commune*. *Proc. Natl. Acad. Sci. USA* **44**:887–900.

90. Raper, J. R., D. H. Boyd, and C. A. Raper. 1965. Primary and secondary mutations at the incompatibility loci in *Schizophyllum*. *Proc. Natl. Acad. Sci. USA* **53**:1324–1332.

91. Raudaskoski, M., J. Stamberg, N. Bawnik, and Y. Koltin. 1976. Mutational analysis of natural alleles at the *B* incompatibility factor of *Schizophyllum commune*: α2 and α6. *Genetics* **83**:507–516.

92. Riquelme, M., M. P. Challen, L. A. Casselton, and A. J. Brown. 2005. The origin of multiple *B* mating specificities in *Coprinus cinereus*. *Genetics* **170**:1105–1119.

93. Scherer, M., K. Heimel, V. Starke, and J. Kämper. 2006. The Clp1 protein is required for clamp formation and pathogenic development of *Ustilago maydis*. *Plant Cell* **18**:2388–2401.

94. Schirawski, J., B. Heinze, M. Wagenknecht, and R. Kahmann. 2005. Mating-type loci of *Sporisorium reilianum*: novel pattern with three *a* and multiple *b* specificities. *Eukaryot. Cell* **4**:1317–1327.

95. Spellig, T., M. Bölker, F. Lottspeich, R. W. Frank, and R. Kahmann. 1994. Pheromones trigger filamentous growth in *Ustilago maydis*. *EMBO J.* **13**:1620–1627.

96. Spit, A., R. H. Hyland, E. J. C. Mellor, and L. A. Casselton. 1998. A role for heterodimerization in nuclear localization of a homeodomain protein. *Proc. Natl. Acad. Sci. USA* **95**:6228–6233.

97. Srivilai, P. 2006. *Molecular Analysis of Genes Acting in Fruiting Body Development in Basidiomycetes*. Ph.D. thesis. Georg-August-University Göttingen, Göttingen, Germany.

98. Stankis, M. M., C. A. Specht, H. L. Yang, L. Giasson, R. C. Ullrich, and C. P. Novotny. 1992. The *Aα* mating type locus of *Schizophyllum commune* encodes two dissimilar multiallelic homeodomain proteins. *Proc. Natl. Acad. Sci. USA* **89**:7160–7173.

99. Swamy, S., I. Uno, and T. Ishikawa. 1984. Morphogenetic effects of mutations at the *A* and *B* incompatibility factors of *Coprinus cinereus*. *J. Gen. Microbiol.* **130**:3219–3224.

100. Swiezynski, K. M., and P. R. Day. 1960. Heterokaryon formation in *Coprinus lagopus*. *Genet. Res.* **1**:114–128.

101. Terashima, K., K. Yuki, H. Maraguchi, M. Akiyama, and T. Kamada. 2005. The *dst1* gene involved in mushroom photomorphogenesis of *Coprinus cinereus* encodes a putative photoreceptor for blue light. *Genetics* **171**:101–108.

102. Vaillancourt, L. J., M. Raudaskoski, C. A. Specht, and C. A. Raper. 1997. Multiple genes encoding pheromones and a pheromone receptor define the *Bβ1* mating-type specificity in *Schizophyllum commune*. *Genetics* **146**:541–551.

103. Velagapudi, R. 2006. *Extracellular Matrix Proteins in Growth and Fruiting Body Development of Straw and Wood Degrading Basidiomycetes*. Ph.D. thesis. Georg-August-University Göttingen, Göttingen, Germany.

104. Wendland, J., L. J. Vaillancourt, J. Hegner, K. B. Lengeler, K. J. Laddison, C. A. Specht, C. A. Raper, and E. Kothe. 1995. The mating-type locus of *Bα1* of *Schizophyllum commune* contains a pheromone receptor gene and putative pheromone genes. *EMBO J.* **14**:5271–5278.

105. Yue, C. L., M. Osier, C. P. Novotny, and R. C. Ullrich. 1997. The specificity determinant of the Y mating-type

proteins of *Schizophyllum commune* is also essential for Y-Z protein binding. *Genetics* **145**:253–260.

106. **Yun, S. H., T. Arie, I. Kaneko, O. C. Yoder, and B. G. Turgeon.** 2000. Molecular organization of mating type loci in heterothallic, homothallic, and asexual *Gibberella/Fusarium* species. *Fungal Genet. Biol.* **31:** 7–20.

107. **Zhang, L., R. A. Baasiri, and N. K. Van Alfen.** 1998. Viral regulation of fungal pheromone precursor gene expression. *Mol. Biol. Cell* **18**:953–959.

*Sex in Fungi: Molecular Determination
and Evolutionary Implications*
Edited by Joseph Heitman et al.
© 2007 ASM Press, Washington, D.C.

Thomas J. Fowler
Lisa J. Vaillancourt

18

Pheromones and Pheromone Receptors in *Schizophyllum commune* Mate Recognition:
Retrospective of a Half-Century of Progress and a Look Ahead

The small homobasidiomycete mushroom *Schizophyllum commune* has more than 15,000 different "sexes." The enormous number of mating specificities in this species results, in part, from a large set of pheromone and pheromone receptor genes that reside at one of the two mating-type loci. Before molecular investigations in the 1990s finally revealed the identity of the proteins that confer mating specificity, the effects of these molecules had been extensively explored for decades with classical genetics. With molecular tools in hand, many of these classical studies can be revisited, and their findings can be reinterpreted with the benefit of hindsight. This chapter describes the progress that has been achieved by combining recent molecular studies with older, classical investigations for understanding *S. commune* pheromone signaling. This chapter is dedicated to Carlene (Cardy) Raper, whose enthusiasm for *S. commune* as a model for basidiomycete mating type in the molecular age has inspired so many in this exciting and productive field of study.

MATING IN YEAST: A FAMILIAR TOUCHSTONE

The paradigm for fungal pheromone signaling in mating is the ascomycete yeast *Saccharomyces cerevisiae*. During *S. cerevisiae* mate recognition, the a-factor receptor and the α-factor receptor specifically bind to the small peptidyl pheromones a-factor and α-factor, respectively, and these interactions activate a signaling cascade that leads ultimately to mating competence, chemoattraction, and cell fusion. The two pheromone receptors belong to the seven-transmembrane-domain G-protein-coupled receptor (GPCR) superfamily and are prototypical members of a subset of GPCRs that includes all other known fungal pheromone receptors (24; www.gpcr.org). The yeast pheromones are small peptides, 12 (a-factor) and 13 (α-factor) amino acid residues in length, which are derived from larger precursor peptides. The details of processing and secretion differ for the two *S. cerevisiae* pheromones and have been explored in depth (6, 23, 33, 34, 68). The two pheromone precursors undergo vastly different posttranslational modifications, including protease

Thomas J. Fowler, Department of Biological Sciences, Southern Illinois University-Edwardsville, Edwardsville, IL 62026-1651. **Lisa J. Vaillancourt,** Department of Plant Pathology, University of Kentucky, Lexington, KY 40546-0312.

cleavages. The mature a-factor has a carboxymethylated and farnesylated C terminus, while the α-factor is
an unmodified peptide. Signal transduction by either activated pheromone receptor is effected through a heterotrimeric G protein and mitogen-activated protein kinase cascade. The molecular signals that emanate from
initial pheromone-receptor activation in *S. cerevisiae*
have been dissected in great detail (reviewed in references 8 and 11).

There are approximately 75 to 100 pheromones and
at least 18 pheromone receptors involved in the *S. commune* mating system (16), complicating initial pheromone recognition in *S. commune* in comparison with
S. cerevisiae. However, where the data are complete
enough for comparisons to be made, pheromone response pathways in the fungi are variations on a common theme (3, 5, 36, 70). Information learned from
S. cerevisiae and other fungi, such as the heterobasidiomycete corn smut pathogen *Ustilago maydis* and the
inky cap mushroom *Coprinus cinereus*, can be applied
to *S. commune*. *S. cerevisiae* in particular is a very useful tool for the study of certain molecular aspects of *S.
commune* mate recognition (12, 19).

CLASSICAL GENETIC STUDIES OF THE MATING SYSTEM OF *S. COMMUNE*

Haig Papazian, who at the time was in John Raper's
laboratory, revived the study of mating type in *S. commune* that had been initiated by Hans Kniep (see references 41 and 51 for a more complete history). Papazian
was intrigued by the very large number of mating specificities ("sexes") that were conferred by only two unlinked mating-type loci, *A* and *B*. Paired haploid strains
of *S. commune* must have different versions of both loci
in order to be fully compatible for mating. *S. commune*
has a tetrapolar incompatibility system, such that any
cross of compatible individuals, for example, *A1B1* ×
A2B2, gives rise to four combinations of the two loci in
the tetrad of progeny resulting from meiosis (*A1B1*,
A2B2, *A1B2*, and *A2B1*). If an individual progeny is
crossed to each of its three siblings, only one of the three
crosses will be fully compatible.

When mates differ at both *A* and *B*, their union initiates two different but complementary developmental
pathways and ultimately yields a dikaryon (Fig. 18.1).
The process of dikaryotization has been broken down
into its most obvious cellular features. After initial anastomosis of the individuals, compatible mates will reciprocally exchange nuclei. These nuclei migrate to hyphal
tip cells where an invading nucleus pairs with the resident nucleus of the cell. At the next division of the api-

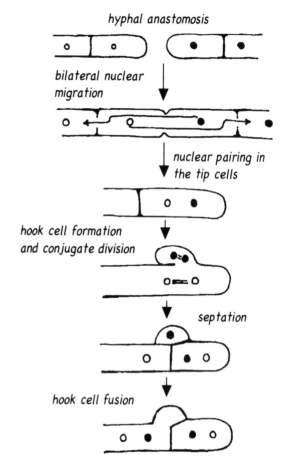

Figure 18.1 The mating process in *S. commune*. Hyphal anastomosis
is not regulated by the mating-type genes. Nuclear migration and
hook-cell fusion are controlled by the *B* locus, and all the other steps
in the process of dikaryon formation are controlled by *A*.

cal cell, the two nuclei undergo a synchronous mitosis.
In preparation for this nuclear division, the apical cell
initiates a structure, called a hook cell, from its cell wall
at a position adjacent to the paired nuclei. The hook
cell superficially resembles a branch initial but extends
in the direction opposite from the hyphal tip. One nucleus moves into the hook cell, and then both nuclei divide in synchrony. New cell walls are laid down across
the mitotic spindles to generate three cells: the hook cell
with one nucleus, the subterminal cell with a nucleus of
the opposite type, and the new apical cell with two nuclei, one of each nuclear type. The hook cell fuses to the
subterminal cell to make a "clamp connection" and releases the hook cell nucleus into the subterminal cell.
Each resulting daughter cell contains two nuclei, one derived from each of the original mates.

If mates differ at only one of the two mating-type
loci, only one of the developmental pathways is initiated,

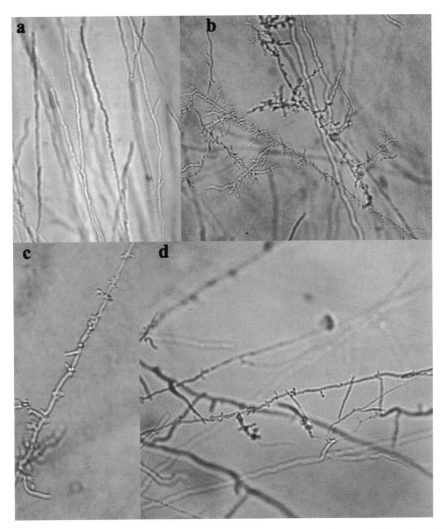

Figure 18.2 The flat phenotype. (a) Hyphae of an unmated homokaryon. (b through d) Hyphae of the common-*A* heterokaryon exhibiting a strong flat phenotype. Abundant short branches and hyphal wall distortions are typical hyphal features. Micrographs by T. Fowler.

and the result is a hemicompatible mating. Each hemicompatible mating produces a heterokaryon that is developmentally distinct from the dikaryon that results from a fully compatible mating. When *B* differs between mates but *A* does not (a "common-*A*" mating), the phenotype of the resulting heterokaryon is termed "flat" (41). Flat mycelia develop with few aerial hyphae and instead tend to submerge into the agar on solid media. The hyphae tend to be highly decorated with short lateral pegs, and the cell walls are often misshapen with bulges and bumps (Fig. 18.2). Nuclear exchange between mates occurs, but in the absence of *A*-regulated development the migrating nuclei are not subject to controlled pairing or synchronous division with resident nuclei, and hook cells are not initiated. While the flat

phenotype is diagnostic and readily identifiable in *S. commune*, its severity depends on the strains involved in the cross. It should also be noted that the flat phenotype is not a feature of all homobasidiomycetes (51). When mates have different versions of the *A* loci but the same version of *B* (the "common-*B*" mating), a different phenotype results (49). Exchanged nuclei do not migrate in these mates beyond the initial fusion cell because of a lack of *B*-regulated development. Hyphae with unfused hook cells (pseudoclamps) and binucleate tip cells form at the interface between the two mates. Detailed heterokaryon development resulting from comparison of fully compatible versus hemicompatible matings has provided a general understanding of the roles of *A* and *B*. Thus, *A*-regulated development leads to nuclear pairing, synchronous (conjugate) mitoses of the paired nuclei, hook cell formation, and septation, while B-regulated development initiates nuclear migration from the point of contact to the apical cells of the hyphae and also regulates hook cell fusion with the subterminal cell.

THE GENETIC RESOLUTION OF THE B LOCUS

John Raper and his colleagues genetically defined the B mating-type locus as consisting of two linked, recombinable subloci, which they termed Bα and Bβ (53). Nine interfertile haplotypes of each sublocus were predicted from a worldwide collection of B mating types (29, 66). The two subloci were shown to be functionally equivalent, such that either Bα or Bβ can direct B-regulated development in the formation of a dikaryon or a hemicompatible flat heterokaryon. A mate need only differ from a prospective partner at either Bα or Bβ in order to be fully compatible for B. With a few exceptions, each haplotype of Bα can be found in association with each haplotype of Bβ, and vice versa (28). The A locus was found to have a similar recombinable bipartite structure (42, 52, 54). These classical genetic approaches provided a basic explanation for the thousands of possible mating specificities within this species.

More than a decade of mutational analyses of the B subloci led to additional refinement of our understanding of the roles that B plays in mating and sexual development (25, 27, 30, 43, 48, 50). These roles were classified into *variable* and *constant* functions (46). The variable function is the unique identity conferred by each haplotype of Bα or Bβ. The constant functions are initiated whenever at least one B sublocus differs from that of its mating partner. These constant functions are the donation and acceptance of migrating nuclei and fusion of the hook cell to the subterminal cell. The variable and constant functions can be altered separately by mutagenesis (25, 30, 43, 55, 56, 61).

MOLECULAR ISOLATION OF GENES CONFERRING B MATING TYPE

Efficient protoplast transformation (65), along with other molecular tools, allowed the isolation of DNA harboring portions of Bα and Bβ (64). In a common-B heterokaryotic background, introduction of DNA from a different B haplotype initiated B-regulated development and resulted in the formation of a dikaryon. By using this selection strategy, portions of the Bα1 and Bβ1 haplotypes were isolated. These DNA fragments were used to initiate chromosome walks that encompassed nearly the entire Bα1 and Bβ1 loci (71, 72). Isolation of additional Bα and Bβ haplotypes was achieved through cross-hybridization

with DNA from previously isolated B haplotypes, or with DNA from regions flanking B (15, 16, 19, 72).

THE B LOCUS GENE PRODUCTS TURN OUT TO BE PHEROMONES AND RECEPTORS

DNA fragments from Bα and Bβ were tested for function by DNA-mediated transformation of *S. commune* and also characterized at the DNA sequence level (64, 71, 72). The unexpected discovery was that the Bα1 and Bβ1 loci contained genes predicted to encode lipopeptide pheromones and GPCRs. The predicted products were structurally similar to the *S. cerevisiae* **a**-factor pheromone and its GPCR, STE3p. The variable function of B that confers a unique identity to each haplotype arises from the particular combination of encoded pheromones and receptors. Finding that the B mating-type genes encoded pheromones and receptors was surprising because, unlike yeasts and dimorphic heterobasidiomycetes that use pheromones to sense and attract compatible mates, there is little evidence that hyphal anastomoses in *S. commune* are the result of a diffusible attractant linked to B (1). A second surprise was that each haplotype encoded multiple distinct pheromones and a single pheromone receptor (71, 72). Thus, B-regulated communication requires that each receptor recognize at least one pheromone encoded by each of the other eight haplotypes of the same sublocus, while at the same time being unresponsive to any of its own pheromones. Furthermore, since Bα and Bβ are recombinable, a lack of self-activation in the recombinants implies that each receptor must also be unresponsive to all of the pheromones produced by haplotypes of the other sublocus.

A convention for naming the pheromone and receptor genes of *S. commune* has been developed (72). For example, the receptor gene of the Bα1 haplotype is called *bar1*, which stands for B alpha receptor of haplotype 1; the receptor gene from the Bβ1 haplotype is *bbr1*; and so on. Because each B haplotype contains multiple pheromone genes, these have more complicated names. Thus, the first B alpha pheromone gene from haplotype α1 is *bap1(1)*, the second B beta pheromone gene from haplotype β1 is *bbp1(2)*, and so on. The protein products of these genes are designated Bar1, Bbr1, Bap1(1), and Bbp1(2), respectively.

THE ROLES OF PHEROMONES AND RECEPTORS IN NUCLEAR MIGRATION

Activation of B-regulated development after transformation with a pheromone or receptor gene results only if the introduced gene encodes a pheromone that is compatible with the transformant's endogenous receptor, or vice versa. When a strain is transformed with a compatible receptor or pheromone, the activation of *B*-regulated development in the absence of *A*-regulated development results in the production of the flat phenotype. At the time of this writing, seven pheromone receptor genes and more than 20 pheromone genes, representing seven different haplotypes (*Bα1*, *Bα2*, *Bα3*, *Bα8*, *Bβ1*, *Bβ2*, and *Bβ3*), have been identified by their ability to activate *B*-regulated development (15, 16, 18, 19, 71, 72). In addition, the entire sequence of one linked pair of subloci, *Bα3-Bβ2*, has been determined (Fig. 18.3) (15, 16). *Bα3-Bβ2* encompasses about 35 kb of DNA and includes 11 pheromone genes and two receptor genes that have also been functionally confirmed (15).

Transformation experiments with receptor versus pheromone genes led to the discovery of distinct roles for each during nuclear exchange (72). When a flat transformant expressing a nonself pheromone gene was mated with its untransformed homokaryotic progenitor strain, conversion of the progenitor strain to the flat phenotype resulted. This experiment demonstrated that pheromones extend their influence beyond the colony initially expressing the pheromone gene, i.e., pheromones are mobile. In contrast, when a flat transformant expressing a nonself receptor gene was mated with its progenitor strain, no apparent change in the phenotype of the progenitor resulted. This experiment demonstrated that receptors, unlike pheromones, are fixed. These observations are consistent with a model of secreted pheromones and membrane-bound receptors. Additional support for this interpretation came from experiments in which single pheromone or receptor genes were introduced into a sterile mutant strain with no *B* activity of its own. The mutant is the result of an X-ray-induced deletion of the entire *Bα3-Bβ2* locus, including the region between the subloci (50). The *B*-null mutant can express introduced pheromone or receptor genes (14, 16). Matings of *B*-null transformants, each carrying a single pheromone or receptor gene, with compatible testers result in unilateral mating reactions, in which dikaryosis is established in the mycelium of only one of the mating partners. When the *B*-null mutant expresses a pheromone transgene, the mate becomes dikaryotic, but when the *B*-null mutant carries a pheromone receptor transgene, it is the transformant that becomes dikaryotic. Thus, pheromones promote nuclear invasion of a mate expressing a compatible receptor. The receptor, on the other hand, is a "gatekeeper," permitting nuclei from a potential mate to invade only if the mate codes a compatible pheromone "password."

The complexity of the *B* loci, with each containing two receptors and multiple pheromones, suggested the possibility that multiple pheromone-receptor pairs might be necessary for mating. However, a *B*-null transformant expressing a single receptor gene can be dikaryotized by a second *B*-null strain with a compatible *A* haplotype that is expressing a single compatible pheromone gene (T. Fowler, unpublished data). Thus, a single pheromone/receptor pair is sufficient for the initiation of unilateral nuclear migration and dikaryosis, and additional pheromones and receptors are not necessary. The experiment also indicated that there are no additional genes located within the *B* locus itself that are necessary for *B*-regulated development.

CHARACTERIZATION OF *S. COMMUNE* PHEROMONES

All known basidiomycete mating pheromones, including those of *S. commune*, appear to be structurally similar to a-factor, the lipopeptide pheromone of *S. cerevisiae*. Key predicted features of each *S. commune* pheromone are (i) processing of pheromone precursor by protease(s) to produce a new amino terminus on the mature pheromone and (ii) modification and protease processing of a carboxy-terminal CaaX-box (Fig. 18.4) (16, 72). The CaaX motif (cysteine-aliphatic-aliphatic, one of several amino acid residues [7]) in these pheromone precursors is thought to signal farnesylation of the CaaX-box cysteine residue, cleavage of the C-terminal aaX amino acid residues, and methylation of the newly exposed carboxyl terminus of the cysteine. Removal of the CaaX motif by recombinant DNA methods resulted in loss of pheromone activity (71). *S. commune* is predicted to encode between 75 and 100 different small mating pheromones (16), and so far more than 20 pheromone genes have been cloned and tested for function in transformation experiments (Fig. 18.5) (15).

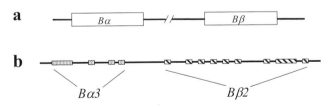

Figure 18.3 Arrangement of the *B* locus in *S. commune*. (a) The generalized arrangement of the two subloci *Bα* and *Bβ*, separated by a region of variable length, is shown. There are nine different haplotypes of *Bα* and nine different haplotypes of *Bβ* that can associate in most pairwise combinations. (b) The *Bα3-Bβ2* combination has been completely sequenced and tested in biological activity assays (15). Squares represent pheromone genes, and rectangles represent receptor genes.

```
Bbp2(7)  MSTGLSDGAGECIAYRTRPAQQTSPSCSSLSRRNHRARPRGRGFHRSR
Bbp2(8)     MDSFTTLSLLDGTMPTFEDDMPVSFLDAALSLSGDCFSSQSSASS

Bbp2(7)  AYGPKDKLTNPLGSNDRPSTKPADADVRRALASG**DKPCGYGGGY**CVVG
Bbp2(8)  SRSTPFASPSPSSSPSPNLLSSAKAATDPHLVVNA**DSPCGFGGGY**CVVA
                                           ▲
```

Figure 18.4 Precursors of two related *S. commune* pheromones. The pheromones Bbp2(7) and Bbp2(8) activate the same two receptors, Bbr3 and Bbr9, and have very similar mature sequences (predicted), shown in boldface type. These two pheromone precursors are typical of the *S. commune* pheromones. Amino acid pairs (DK and DS) are thought to mark the sites of protease processing for the N termini (arrow) of the mature pheromone and can be identified readily in most of the known pheromone sequences. The amino acid sequences of the N termini of these two precursors are very different although their C termini (the presumed mature pheromones) are not, also suggesting that the conserved, functional portions (bold) and the nonconserved, nonfunctional portions (italics) are separated at the predicted protease recognition sites. Underlined at the C termini is the CaaX-box motif that is processed to remove the final three amino acids. Following aaX cleavage, the new terminal cysteine is likely carboxymethylated and farnesylated.

S. cerevisiae and *S. commune* both use lipopeptide pheromones and GPCRs for mating compatibility signaling, and *S. cerevisiae* can function as a heterologous host for expression of *S. commune* pheromones and receptors (12, 19). The yeast bioassays are constructed so that **a**-mating-type cells express an *S. commune* pheromone, while α-mating-type cells express a compatible receptor. Mutations in the **a**-factor and **a**-factor receptor genes in the host yeast cells prevent the natural reciprocal response to the endogenous α-pheromone and α-pheromone receptor signaling. But yeast cells transformed with the genes for *S. commune* pheromones and receptors underwent cell cycle arrest and formed diploids (12). This response was specific to the heterologous genes; when either was omitted from the assay, the response was eliminated. Although pheromones have not been purified directly from *S. commune* tissues or culture media, expression of the pheromone genes in *S. cerevisiae* provides indirect evidence that the structural predictions for pheromones are accurate (12, 16). The *S. commune* pheromone activity secreted by the transformed yeast cells can be partially purified from spent culture medium with a hydrophobic resin (Fowler, unpublished), suggesting that the CaaX-box motif is directing the predicted hydrophobic C-terminal modifications.

Predictions about the sizes of mature *S. commune* pheromones are based on the observation that a pair of charged amino acids is often found between 10 and 20 amino acids away from the C terminus of the precursor peptides in both *C. cinereus* and *S. commune* (5). Synthetic *C. cinereus* pheromones had maximal activity when the pair of charged amino acid residues was incorporated as the last two residues on the amino terminus of the synthetic peptide. Pheromone activity was reduced if the amino terminus did not include these amino acids or if it was extended beyond this predicted protease cleavage point (38). When *S. commune* pheromone genes encoding predicted mature peptides with a high degree of similarity to one another are compared, a sharp transition can often be observed at the predicted cleavage site, beyond which there is relatively little amino acid similarity (Fig. 18.4). This is consistent with the idea that the charged pair of amino acids marks the shift from the functional portion of the pheromone near the C terminus to the nonfunctional, and therefore nonconserved, portion of the precursor.

The currently characterized collection of *S. commune* pheromones can be divided into five groups (I through V), based on the amino acid sequence similarity of their predicted mature pheromone peptides, and their receptor-activation spectra in transgene experiments (Fig. 18.5) (15). The pheromones and receptors from one group do not activate those from other groups. In the only exception to this rule, the *Bα4* receptor can be activated by one pheromone from group I [Bap3(1)], in addition to the group II pheromones (Fig. 18.5). An aromatic amino acid residue occupies the carboxy-subterminal position in all of the *S. commune* pheromones identified so far. Bap3(1) has a phenylalanine residue in this position, whereas all the other group I pheromones have a tryptophan residue. In contrast, all the group II pheromones have a phenylalanine in the carboxy-subterminal position, in common with Bap3(1), suggesting that this position may have an important discriminatory function in group I versus II pheromone/receptor interactions. Aromatic amino acids near the C terminus have been shown to be important specificity determinants in a *C. cinereus* pheromone (39). It seems likely that there is an additional purpose for the aromatic amino acids in the carboxy-terminal position in *S. commune* pheromones, because there are only three aromatic residues, and these would not be sufficient to distinguish among all of the pheromones. They could be involved in general (nonspecific) pheromone/receptor binding that precedes specific recognition and activation. In pheromone-binding models proposed for *Ustilago maydis*, aromatic residues are important for the initial insertion of pheromones into the receptors' binding pockets (67).

The predicted amino acid sequences of mature *S. commune* pheromones belonging to the same group can be aligned (Fig. 18.5) (15). These comparisons provide clues to the identities of the amino acid residues most likely to contribute to specificity in pheromone/receptor recognition. For example, the four group III phero-

Group	Pheromone		Bα Receptors									Bβ Receptors								
			1	2	3	4	5	6	7	8	9	1	2	3	4	5	6	7	8	9
I	Bap3(1)	ERVGTGGTATAFC	−	+	−	+	+	−	−	−	−	−	−	−	−	−	−	−	−	−
	Bap1(1)	EREGGSDCTAWC	−	+	+	−	+	+	−	−	−	−	−	−	−	−	−	−	−	−
	Bap1(3)	ERPGGSNCTAWC	−	+	+	−	+	+	−	−	−	−	−	−	−	−	−	−	−	−
	Bap2(3)	EKPGGSLTYAWC	+	−	+	−	+	+	−	−	−	−	−	−	−	−	−	−	−	−
	Bap3(2)	EKGGTSMAHAWC	+	+	−	−	−	+	−	−	−	−	−	−	−	−	−	−	−	−
II	Bbp2(6)	EREGDGNMTYFC	−	−	−	+	−	−	+	−	−	−	−	−	−	−	−	−	+	−
	Bap3(3)	ERHGSGNMTYFC	−	−	−	+	−	−	+	−	−	−	−	−	−	−	−	−	+	−
III	Bbp2(4)	DSPDGYFGGYC	−	−	−	−	−	−	−	+	−	+	−	−	−	−	−	−	−	−
	Bbp2(5)	DSPDGYFAGYC	−	−	−	−	−	−	−	+	−	+	−	−	−	−	−	−	−	−
	Bbp2(8)	DSPCGFGGGYC	−	−	−	−	−	−	−	−	−	−	−	+	−	−	−	−	−	+
	Bbp2(7)	DKPCGYGGGYC	−	−	−	−	−	−	−	−	−	−	−	+	−	−	−	−	−	+
IV	Bbp2(3)	DQITDYQSYC	−	−	−	−	−	−	−	−	+	−	−	−	−	+	−	−	−	−
	Bbp1(2)	DQIADYGSYC	−	−	−	−	−	−	−	−	+	−	−	−	−	+	−	−	−	−
V	Bbp1(1)	EHWRGGNTTAHGWC	−	−	−	−	−	−	−	−	−	−	+	−	+	−	+	+	−	−
	Bbp1(3)	EHTEAGEETTARGWC	−	−	−	−	−	−	−	−	−	−	+	−	+	−	+	+	−	−
	Bbp2(1)	EHGYGGSNVHGWC	−	−	−	−	−	−	−	−	−	−	−	−	+	−	+	+	−	−
	Bbp2(2)	EHEEDTDSNVHGWC	−	−	−	−	−	−	−	−	−	−	−	−	+	−	+	+	−	−

Figure 18.5 Some *S. commune* predicted mature pheromones and their biological activities. Predicted pheromones are grouped by sequence similarity (15), and the receptors activated by each pheromone are indicated by a plus sign (+). Only the peptide portion is shown; each pheromone is presumed to be carboxymethylated and farnesylated on the C-terminal cysteine. The asterisks above two group III amino acid positions were tested in site-directed mutagenesis studies (see the text for details).

mones are all predicted to have a mature length of 11 amino acids. Two of the pheromones activate receptor Bbr1 and the similar receptor Bar8, while the other two activate Bbr3 and Bbr9 (Fig. 18.5). Among the four group III pheromones, polymorphisms exist at 5 of the 11 positions. Two of the five are correlated with the receptor activation spectra, suggesting that these are involved in receptor specificity (Fig. 18.5, asterisks). To test this hypothesis, the genes encoding Bbp2(4) and Bbp2(7) were subjected to site-directed mutagenesis. Bbp2(4) normally activates Bbr1 but not Bbr3 and has a cysteine in the first polymorphic position and a glycine in the second. Bbp2(7) normally activates Bbr3 but not Bbr1 and has an aspartate in the first polymorphic position and a phenylalanine in the second. Swapping the cysteine at the first polymorphic position with an aspartate, or vice versa, produced no difference in receptor recognition of mutant compared to wild-type pheromones. However, swapping the residues at the second polymorphic position (glycine versus phenylalanine) did

have an effect. When this amino acid position was occupied by a glycine, Bbr3 was activated, and when the residue was a phenylalanine, Bbr1 was activated. Thus, the polymorphic amino acid at this position determines receptor specificity of group III pheromones. This is just one of several similar examples (15, 16).

DETERMINANTS OF SPECIFIC RECOGNITION OF PHEROMONES BY RECEPTORS

As a group, GPCR cell surface receptors in eukaryotes interact with ligands of a wide variety of chemical types and sizes (4). Understanding how these receptors recognize both agonists and antagonists is a major focus of academic and commercial interest. Fungal pheromone receptors belong to class D of the GPCR superfamily (20; www.GPCR.org). *S. commune* and other homobasidiomycetes with large numbers of pheromone receptors are sources of related but nonidentical GPCRs that can provide new insights into the molecular aspects of ligand recognition and receptor evolution. To date, seven pheromone receptor genes from *S. commune* have been identified, their activity in mating has been verified by transformation experiments, and their amino acid sequences and conformations have been predicted (15, 16, 71, 72). In general, the *S. commune* pheromone receptors have short extracellular N termini and long cytoplasmic C-terminal tails. In some cases, the cytoplasmic region is approximately one-half of the entire length of the receptor protein. Like all members of the GPCR superfamily, the receptors are predicted to have seven membrane-spanning α-helical regions, along with three extracellular and three intracellular loops connecting these transmembrane helices. In three dimensions, the seven-transmembrane domains of these receptors are thought to look like partially crushed and twisted barrels, with the seven transmembrane region "staves" somewhat out of parallel (40). The extracellular loops and the transmembrane regions are presumed to contain the determinants for the specific recognition of extracellular pheromones while the intracellular portion of the receptor interacts with a heterotrimeric G protein. Genes for two Gα subunits have been isolated from *S. commune*, but initial experiments do not confirm a direct role for either of these in pheromone signaling (74).

Each haplotype of the two *B* subloci contains a single pheromone receptor gene (as shown in Fig. 18.3, for example). The seven receptors that have been characterized so far cluster into small groups based on predicted amino acid sequence similarity. The receptors also cluster by function into the same groups: Bbr1, Bbr3, and Bar8 are activated by group III pheromones; Bar1, Bar2, and Bar3 are activated by group I pheromones; and Bbr2 is activated by group V pheromones (Fig. 18.5). The receptors from different groups are very dissimilar. For example, Bbr1 and Bbr2 are only 37% identical and 59% similar in their N-terminal halves, which contain the seven-transmembrane domain (16). The sequences of the C-terminal halves are not similar enough to be aligned at all. In contrast, within their groups the receptors are much more similar. For example, Bar8 shares 88.4% identity in the N-terminal half, and 67% identity overall, with Bbr1 (15).

Recombinant receptor chimeras and site-directed mutagenesis have been used to understand how *S. commune* pheromone receptors specifically recognize pheromones. Erika Kothe's group in Jena, Germany, has done the most extensive studies (17, 18). The *bar1* and *bar2* genes encode the *Bα* pheromone receptors Bar1 and Bar2, respectively. The two receptors are 72% identical in the N-terminal seven-transmembrane domain, which

is predicted to encompass the first 291 amino acids (18). The number of amino acid differences in this region is 81, of which 43 are nonconservative substitutions. The *bar1* and *bar2* genes were each subdivided into three parts by using conserved restriction sites, with each part including one of the three encoded extracellular loop regions. A mix-and-match approach was used to design six different receptor-gene chimeras containing combinations of the three portions of each gene (Fig. 18.6) (17). These six receptor chimeras and the two wild-type receptors were expressed individually in the *B*-null strain of *S. commune*, and then the transformants were mated with wild-type strains and with *B*-null transformants expressing various pheromone genes. These experiments were intended to identify the region of each receptor that determined its unique pheromone recognition spectrum. The region containing the third extracellular loop of the Bar1 receptor was associated with exclusion of activation by *Bα1* pheromones. This suggests that factors responsible for the prevention of self-activation in the *Bα1* haplotype occupy this part of the receptor. In contrast, no single region of Bar2 provided complete exclusion of *Bα2* pheromone activation. Furthermore, no single region from either receptor precisely conferred the pheromone specificity of the alternate receptor. However, some of the chimeric receptors did have altered activation spectra (Fig. 18.6) (17). One was constitutive and did not require activation by any pheromone to induce *B*-regulated development. Another was "promiscuous" and responded to at least one pheromone from every wild-type mate, including the "self" *Bα1* and *Bα2* haplotypes. The promiscuous chimeric receptor retained the ability to discriminate at some level, since it was not activated by absolutely every pheromone it encountered (18). A third type of chimera was responsive to a narrower range of pheromones than its wild-type progenitors. These analyses suggest that receptor/pheromone interactions leading to specific pheromone recognition involve regions of the protein that are distant from each other in the primary sequence, though they may be physically associated in three dimensions. It is of interest that all three extracellular loops and at least four of the transmembrane helices of the α-factor pheromone receptor of *S. cerevisiae* are implicated in interactions with α-factor (35, 37).

Site-directed mutagenesis has been used to test specific amino acid residues for their contribution to receptor function (18). One experiment involved a receptor chimera that was composed of the first third of Bar2 and the last two thirds of Bar1 (Fig. 18.6). In mating tests, this chimera was indistinguishable from the wild-type Bar1 receptor in that it responded to *Bα2* pheromones

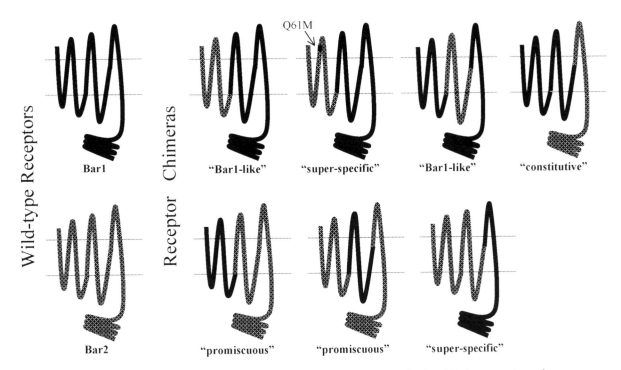

Figure 18.6 Bar1 and Bar2 pheromone receptor chimeras. *S. commune* Bar1 and Bar2 were engineered to produce hybrid receptors that exhibited a variety of phenotypes when exposed to wild-type *S. commune* pheromones in matings (redrawn based on references 17 and 18). Some receptor chimeras displayed the selectivity of a wild-type receptor, while other receptor chimeras had the ability to be activated by pheromones from all wild-type *B*α haplotypes ("promiscuous"). Still others, including a site-directed mutant, were "super-specific" and excluded more pheromones than the wild-type receptors. One receptor chimera was constitutively activated and required no pheromone for signaling. In these drawings, the dotted lines represent the boundaries of the plasma membrane in which the receptors are anchored. Above the upper dotted line is the extracellular space, and below the lower dotted line is the cytoplasm. The N termini of these receptors are extracellular, and the C termini are intracellular. The intracellular C-terminal portion of each receptor is not drawn representatively but is shortened for convenience.

but not to *B*α*1* pheromones. A glutamine located in the first extracellular loop of Bar2 was changed to the polymorphic Bar1 residue, a methionine, for both the wild-type Bar2 receptor and the chimera. The mutation had no noticeable effect on Bar2, but it caused the chimera to lose its ability to respond to *B*α*2* pheromones. Thus, and counterintuitively, changing the first extracellular loop to be less like Bar2 results in nonresponsiveness to *B*α*2* pheromones, a characteristic of Bar2 (Fig. 18.6).

Site-directed mutagenesis has also been used to investigate the role of individual amino acids in a *B*β receptor (16). Bbr2-2 is an X-ray-generated mutant of Bbr2. It is missing two amino acid residues in the third transmembrane helix, and it does not respond to several pheromones that activate the wild-type receptor. Site-directed mutagenesis was used to replace each of the missing amino acid residues and also to create a deletion of a different amino acid pair in the same helix (16). This experiment established that the integrity of the transmembrane helix, rather than the presence of specific amino acids, was important for maintaining the wild-type spectrum of pheromone response. The studies that use site-directed mutagenesis agree with the receptor chimera experiments in suggesting that the interactions of pheromones and receptors take place in three dimensions and are influenced by the overall conformation of the receptor in ways that are not yet predictable.

YEAST AS A TOOL FOR FUTURE INVESTIGATIONS OF PHEROMONE/RECEPTOR RECOGNITION

Functional assays of heterologous GPCRs expressed in *S. cerevisiae* have been in use for several years (9, 45). Several groups have recognized the potential for heterologous expression of basidiomycete pheromone receptors in *S. cerevisiae*, although each group has taken a different approach to developing the assays (12, 19, 38). In one such study, sterile yeast strains, with the endogenous **a**-factor or **a**-factor receptor gene deleted, were transformed with *S. commune* pheromone or receptor genes (12). Pheromone transformants were paired with other transformed strains that were expressing compatible or incompatible receptors and then assayed for diploid formation. This experiment confirmed that *S. commune* pheromone receptors and pheromones function in yeast, each expressing its expected specificity and inducing diploidization in compatible pairings (12, 16). Because the yeast assay is of relatively high throughput, it seemed that it would be a useful tool for identification of pheromone and receptor variants with altered specificities. Screens of mutant receptor alleles identified several that were constitutively active in the yeast assay. However, only one of these mutant receptors induced the flat phenotype when it was introduced into the *B*-null strain of *S. commune* (Fowler, unpublished). Similarly, several mutant pheromones that were active in the yeast heterologous assay had no effect in *S. commune* (Fowler, unpublished). This suggests that *S. commune* has additional factor(s), lacking in yeast, which impact the activities of the pheromones and receptors. Results of yeast heterologous assays, though potentially very useful, should be interpreted with caution.

REVISITING INDUCED MUTATIONS THAT AFFECTED THE *B* LOCI

Great effort was expended for many years to try to produce a new *B* specificity through random mutagenesis of wild-type loci and mutant derivatives (25, 27, 30, 43, 48, 50, 56, 57). A mutant with a new specificity would be self-sterile, but bilaterally fertile, with all other known *B* specificities, including the progenitor from which it was derived. A variety of chemical and radiation mutagenesis techniques were employed in these studies, and although several mutants that had altered *B* mating behavior were obtained, a novel *B* specificity was never found. It is interesting to reconsider the results of these experiments with the benefit of our current understanding of the molecular nature of the *B* loci and of pheromone and receptor function.

Several mutations in the *B* loci that were termed primary *B*-on were obtained. These were constitutively active for *B*-regulated functions and produced a flat phenotype in unmated homokaryons. One possibility that could produce this phenotype would be a mutant pheromone able to signal through the receptor of its own haplotype. If this mutant were crossed with its progenitor, we would expect induction of the *B*-on phenotype in the progenitor strain, because pheromones are mobile. The other possibility would be a mutant pheromone receptor that can be activated by a pheromone of its own haplotype, or one that does not require any pheromone ligand for signaling. If this type of mutant were crossed with its progenitor, we would expect it to be unable to induce the *B*-on phenotype in the progenitor strain, because receptors are not mobile. The majority of the reported primary *B*-on mutants can induce the flat phenotype in their progenitor strains (25, 30, 43, 61) and thus are predicted to be changes in pheromones. Two reports describe constitutive *B*-on strains that appear to be receptor mutants (30, 61). A primary *B*-on mutant was recently subjected to molecular analysis (16). The *Bβ2(1)* mutant, originally described by Parag (43), can induce the flat phenotype in its *Bβ2* progenitor and was shown to be the result of a missense mutation in the Bbp2(1) pheromone gene. A valine-to-alanine switch resulted in the mutant pheromone Bbp2(1-1), which can activate the Bbr2 receptor produced from the same haplotype.

John and Cardy Raper published an article in 1973 detailing the mating behavior of a series of secondary mutants that were generated from a *Bα3-Bβ2(1)* primary mutant (50). Secondary mutations were generated in this strain with X-irradiation and were initially selected for reversion from flat to normal hyphal growth. Mating tests with the secondary mutants revealed that they had mating peculiarities in addition to silencing of *B*-regulated development in the homokaryon and did not represent a true genetic reversion of the primary mutation. Mutants exhibited a gradient of mating activity, ranging from nearly normal interactions with all potential mating partners to complete infertility with strains of any other *B* mating type. The ability to donate fertilizing nuclei could be separated from the ability to accept fertilizing nuclei, which we now understand can be explained by the different functions of pheromones versus receptors. Studies of these mutants demonstrated that the *B* loci were divided into a larger number of independent units, with separate roles in nuclear migration, hook cell fusion, and mate recognition. This provided an important clue to the multipartite nature of the *B* loci, which we now see is a result of their containing multiple pheromone and receptor genes.

These primary and secondary mutants were revisited using a molecular approach, starting with characterization of *bbp2(1-1)* and its resulting missense mutation in the Bbp2(1) pheromone from the primary *B*-on mutant (see above). The wild-type *Bα3-Bβ2* is the only *B* complex for which the genes required for all known functions of the loci have been identified (15). The *Bα3* locus contains three pheromone genes and one receptor gene (Fig. 18.3). The *Bβ2* locus has eight pheromone genes and one receptor gene. In order to determine the nature and extent of the changes that had occurred in the secondary mutants, the wild-type *Bα3-Bβ2* was compared to the primary and secondary mutants with DNA sequence and Southern blot hybridization analyses. The secondary mutants had originally been selected for reversion of the flat hyphal phenotype (48). Molecular analysis revealed that this phenotypic reversion usually resulted from some type of damage to the *Bbr2* receptor gene, with an associated partial or complete loss of Bbr2 function (16). In one mutant, *Bβ2(1-3)*, a two-amino-acid deletion occurred in the receptor, which rendered it unresponsive to the mutant pheromone Bbp2(1-1), although it can still be activated by other pheromones. This is the mutant that was used for a site-directed mutagenesis experiment, as described above. In several of the other mutants, the receptor gene has been mostly or entirely deleted. In some cases, the mutant pheromone gene Bbp2(1-1) has been deleted as well. Although hyphal growth was restored to normal, mating behavior of most of the secondary mutants was significantly altered by these large deletions. The largest deletion extended across the whole *Bα3-Bβ2* complex, removing all of the *B* genes. The mutant, *Bα3(3)-Bβ2(1-8)*, is completely infertile in crosses with any wild-type strain. This is the *B*-null strain that has been such a useful recipient for transformation with individual pheromone and receptor genes, as described above.

One of the most interesting discoveries to come out of the reanalysis of this collection of secondary mutants was a transposable element named *scooter*, which was found inserted into the *bbr2* gene in one of the mutants (13). In the original study by the Rapers in 1973, some of the results with one mutant (called *def* or *f*) suggested the occurrence of infrequent recombination events near the *B* locus that influenced mating behavior (50). Recombination was not really a satisfying explanation, though, because the predicted reciprocal recombinants could not be identified with any haplotype except *Bβ2*. Excision of the transposon, *scooter*, from the *bbr2* gene is a consistent and much more plausible explanation for their results, but unfortunately this cannot be confirmed because the "recombinant" *Bβ2* strains no longer exist.

THE ROLE OF PHEROMONE SIGNALING IN *B*-REGULATED MATING PROCESSES

Two mating processes are known to depend on activation of the *B*-regulated developmental pathway: nuclear migration and hook cell fusion. Hook cell fusion is most closely analogous to fusion of individual haploid cells of opposite mating types in ascomycetes, including *S. cerevisiae* and *Schizosaccharomyces pombe*. But whereas in the ascomycetes the presence of opposite mating types is a prerequisite for fusion events at the start of mating, in the homobasidiomycetes, fusion of monokaryotic hyphae occurs regardless of mating type. It is only hook cell fusion that requires an interaction of compatible pheromones and receptors. Hook cells form normally in the absence of *B*-regulated development, so their initiation and directionality of growth apparently do not depend on pheromone or receptor interactions. This appears to be quite different from the case in ascomycetes, where shmoos and trichogynes grow along pheromone gradients. However, fusion of the hook cell does not occur in the absence of *B*, and therefore this must depend on the interaction of unlike pheromone with receptor. We do not understand much about this process, but it may be similar to what occurs in yeasts, where initial wall fusion is mediated by wall agglutinins and other wall-bound proteins that are up-regulated by pheromone signaling (44, 69). One big difference from the ascomycete systems is that the mating genes in *S. commune* must not only activate mating-dependent fusion in hook cells but also suppress the mating-independent fusion that normally occurs in homokaryotic mycelium. Since hook cells do not fuse in a common-*B* hemicompatible mycelium, we might assume that suppression of fusion is a function of *A* and not *B*. However, it was reported that the number of vegetative hyphal fusions was higher in common-*B* matings than in common-*A* matings, suggesting that *B*, and not *A*, suppresses fusion of vegetative hyphae (1, 59). This question deserves further investigation, now that the identities of the genes within *A* and *B* are known.

Nuclear migration, controlled by the *B* locus, depends on recognition of two nuclear types in a common cytoplasm, so that only the invading nucleus will be moved through the mycelium to the hyphal tip. In *S. commune* and other fungi, the determinants of nuclear identity are not clearly understood. Given the apparent lack of a role for pheromone receptors in initial anastomosis of *S. commune* matings, an early hypothesis held that receptors were retained in the nuclear membrane and so functioned to identify the different nuclear types through intracellular interaction with pheromones. However, preliminary immunolocalization results placed Bar1 (31) and Bbr1 (S. Horton and G. Palmer, personal communication) in the plasmalemma, not the nuclear membrane. In one study, recognition among nuclei of compatible mating types seemed to depend on the distance of the nuclei from one another (63). If the nuclei were far apart, then the mycelium containing them behaved as an *A*-off/*B*-off homokaryon, but if the nuclei were close together, then the mycelium behaved as a dikaryon. These authors suggested that pheromones are released into the space between the plasmalemma and the cell wall and do not travel very far. Thus, pheromones and receptors can interact only when nuclei of opposite mates are very close. This model is difficult to reconcile with transformation experiments (see above) that clearly show a difference between receptors (which behave as if they are stationary) and pheromones (which exert an influence at a distance). Recipient cells must be prepared for invasion of nuclei before they arrive, a process that includes dissolution of the septal pore and construction of a cytoskeletal "track" that will move the invading nucleus through the cell (59). It is logical to suppose that the role of a mobile pheromone is to carry the signal to cells that are ahead of the advancing nuclear front and to induce them to begin preparing for invasion, as suggested by Wendland et al. (72). Although it may be assumed that nuclear movement is identical to nuclear sorting, until we tag nuclei of each mate with a visible marker, we will not know if the nuclei that we see moving in flat strains are all invaders or if the activation of *B* in the absence of *A*-regulated development enables generalized nuclear movement of resident as well as invading nuclei. Separating the activities of *A* and *B* genes is a highly artificial, and perhaps misleading, exercise: in reality, there is likely to be considerable cross talk between these two pathways. For now, the question of where exactly the pheromones and receptors of *S. commune* interact awaits further research.

In the *S. commune* dikaryon, most cells have a 1:1 distribution of the two original parental nuclear types. However, studies showed that in *S. commune*, and some other fungi, an expectation of recovery of equal proportions of the two nuclei from dedikaryotized mycelia was seldom met (see reference 47 and references therein). Raper (47) studied recovery of nuclei from various dikaryons to determine whether the mating-type loci might be involved in this disparity. She found not only that *B* mating type was closely tied to asymmetric nuclear survival but also that the different versions of the *B* complex actually exhibited a hierarchy for nuclear survival. She proposed that, as in *S. cerevisiae* and other ascomycetes, pheromone signaling induces mitotic arrest of compatible nuclei in *S. commune*. In support of this hypothesis, nuclear division in nuclei derived from dikaryons was delayed in comparison to homokaryons of the same *B* mating types that had not been derived from a dikaryotic hypha. The rate of recovery of nuclei from dikaryons was faster for nuclei that carried dele-

tions of *B* than for nuclei carrying wild-type *B* genes. If we assume that signaling via *B*, and not just its physical presence, is responsible for the differences, this implies that an impaired ability to signal promotes nuclear recovery and suggests that weaker signals lead to a faster recovery while stronger signals delay recovery. These data suggest the existence of quantitative differences in signaling ability among the multitude of possible pairings of pheromone and receptor in *S. commune*, although all combinations exceed the threshold necessary for activation of *B*-regulated development. The precise nature of these differences is unknown but may reflect variation in the strength and stability of individual protein-protein interactions or in turnover rates of different pheromones and receptors. Results of the yeast heterologous experiments, in which pheromone/receptor combinations sufficient for activation of yeast sometimes did not induce development in *S. commune*, may also be relevant to this question.

Although the targets of the *B*-activated signaling pathway are still unknown, the characteristic flat phenotype of common-*A* heterokaryons in *S. commune* provides clues to the identities of some of these. Flat mycelia produce very few aerial hyphae. Hydrophobins have been shown to be critical for the production of aerial hyphae, and the production of certain hydrophobins is suppressed in common-*A* heterokaryons and in mutants in which *B* is constitutively activated (2). This suggests that one target of *B*-regulated signaling is a transcriptional repressor of the hydrophobin genes and further suggests that the lack of aerial hyphae production in the flat phenotype is a direct result of altered hydrophobin gene expression. The proliferation of short lateral pegs and the knotted appearance of hyphae in the flat mycelia are suggested to result from the production of lytic enzymes (Fig. 18.2) (59). One activity of *B* is destruction of the complex dolipore septa that usually separate the compartments of monokaryotic hyphae in order to facilitate nuclear migration (26, 58, 59, 60, 73). Constitutive activation of *B* presumably results in a situation in which all cells, not just those closest to the invading nuclei, are responding to pheromone signal. Since the pheromones activate their own transcription (71), the result might be a "runaway" reaction and overproduction of lytic enzymes, leading to wall weakening and production of numerous side branches. Chitin synthase mutants in *Aspergillus nidulans*, which also have defects in wall strength, display a highly branched phenotype that is rather reminiscent of the flat phenotype (21). The variation in the ability of particular combinations of pheromone and receptor to produce the flat phenotype in transformation recipients could relate to the signal strength of each combination, with those that produce weaker signals producing a less distinct phenotype. Thus, another target of *B*-regulated signaling surely includes a transcriptional activator that upregulates the expression of genes encoding chitinases and β-glucanases.

CONCLUDING REMARKS

It is amazing to consider the evolutionary process that has resulted in each *S. commune B* haplotype amassing genes for several pheromones, with at least one pheromone to activate the receptor from each of the other haplotypes of the same sublocus, but never its own receptor or any of the receptors of the other sublocus. At the same time, each receptor is responsive to at least one pheromone of each nonself haplotype of the same sublocus, but not to any of self pheromones and not to any of the pheromones encoded by the other sublocus. For such a thing to evolve there must have been significant selection pressure against self-activation and for the maximization of outbreeding. The evolutionary progression that produced multiple haplotypes of each sublocus that follows this pattern is not known, but models have been proposed that attempt to reconcile the current molecular understanding of the *B* locus with the expected negative consequences of self-activation or partially compatible haplotypes of evolutionary intermediates (16, 32). Phylogenetic relationships of pheromone receptors of several mushroom species, including *S. commune*, indicate that the subtypes of pheromone receptors that we can recognize today in mushroom fungi diverged in common ancestors of the modern homobasidiomycetes (22, 62).

Although the accumulated work of several generations of scientists has unraveled many of the mysteries of *S. commune* mating, the story remains incomplete, with some of the most intriguing questions still to be explained. The finer details of the pheromone and receptor interaction will be a challenge to resolve because the membrane-bound receptors and hydrophobic pheromones require the use of special biochemical techniques for lipophilic molecules. However, the opportunity to study such a multitude of individual pheromone/receptor interactions from the extremely important GPCR superfamily provides a strong impetus to overcome the technical obstacles. The location of the pheromones and receptors has yet to be determined during the active mating process. Fluorescent tagging of nuclei, along with the pheromones and receptors themselves, would allow these important questions to be addressed. Knowing where these molecules interact could provide clues to their still poorly understood roles in nuclear migra-

tion, nuclear identity, and hook cell fusion. There are still many questions about factors affecting the relative strength of pheromone signaling and about possible synergism or antagonism among molecules in a system that has so many players. The signaling cascade activated by the response to pheromone in *S. commune* is still uncharted. Classical mutational studies identified many modifier mutations that disrupt nuclear migration and map outside the *B* locus (10). Some of these are surely components of the signaling pathway downstream of the pheromones and receptors; others are likely to identify factors involved in the physical movement of nuclei. Identification of these components will be aided by the sequenced genome of *S. commune*, a project that is currently under way. Finally, the developmental pathway controlled by *B* is intimately intertwined with development that is controlled by *A*. Understanding the intricate details of the cross talk between the two mating specificities will require the attention of future generations of scientists using a variety of experimental approaches to the problem. Like us, they can draw inspiration from the insightful and elegant work of the Rapers and their colleagues.

References

1. **Ahmad, S. S., and P. G. Miles.** 1970. Hyphal fusions in the wood-rotting fungus *Schizophyllum commune. Genet. Res.* **15:**19–28.
2. **Ásgeirsdóttir, S. A., M. A. van Wetter, and J. G. H. Wessels.** 1995. Differential expression of genes under control of the mating-type genes in the secondary metabolism of *Schizophyllum commune. Microbiology* **141:**1281–1288.
3. **Banuett, F.** 1998. Signalling in the yeasts: an informational cascade with links to the filamentous fungi. *Microbiol. Mol. Biol. Rev.* **62:**249–274.
4. **Bockaert, J., and J. P. Pin.** 1999. Molecular tinkering of G protein-coupled receptors: an evolutionary success. *EMBO J.* **18:**1723–1729.
5. **Casselton, L. A., and N. S. Olesnicky.** 1998. Molecular genetics of mating recognition in basidiomycete fungi. *Microbiol. Mol. Biol. Rev.* **62:**55–70.
6. **Chen, P., S. K. Sapperstein, J. D. Choi, and S. Michaelis.** 1997. Biogenesis of the *Saccharomyces cerevisiae* mating pheromone a-factor. *J. Cell Biol.* **136:**251–269.
7. **Clarke, S.** 1992. Protein isoprenylation and methylation at carboxyl-terminal cysteine residues. *Annu. Rev. Biochem.* **61:**355–386.
8. **Dohlman, H. G.** 2002. G proteins and pheromone signaling. *Annu. Rev. Physiol.* **64:**129–152.
9. **Dowell, S. J., and A. J. Brown.** 2002. Yeast assays for G-protein-coupled receptors. *Recept. Channels* **8:**343–352.
10. **Dubovoy, C.** 1975. A class of genes affecting *B* factor-regulated development in *Schizophyllum commune. Genetics* **82:**423–428.

11. Elion, E. A. 2000. Pheromone response, mating and cell biology. *Curr. Opin. Microbiol.* 3:573–581.

12. Fowler, T. J., S. M. DeSimone, M. F. Mitton, J. Kurjan, and C. A. Raper. 1999. Multiple sex pheromones and receptors of a mushroom-producing fungus elicit mating in yeast. *Mol. Biol. Cell* 10:2559–2572.

13. Fowler, T. J., and M. F. Mitton. 2000. *Scooter*, a new active transposon in *Schizophyllum commune*, has disrupted two genes regulating signal transduction. *Genetics* 156:1585–1594.

14. Fowler, T. J., M. F. Mitton, and C. A. Raper. 1998. Gene mutations affecting specificity of pheromone/receptor mating interactions in *Schizophyllum commune*, p. 130–134. *In* L. J. L. D. Van Griensven and J. Visser (ed.), *Proceedings of the Fourth Meeting on the Genetics and Cellular Biology of Basidiomycetes*. Mushroom Experimental Station, Horst, The Netherlands.

15. Fowler, T. J., M. F. Mitton, E. I. Rees, and C. A. Raper. 2004. Crossing the boundary between the Bα and Bβ mating-type loci in *Schizophyllum commune*. *Fungal Genet. Biol.* 41:89–101.

16. Fowler, T. J., M. F. Mitton, L. J. Vaillancourt, and C. A. Raper. 2001. Changes in mate recognition through alterations of pheromones and receptors in the multisexual mushroom fungus *Schizophyllum commune*. *Genetics* 158:1491–1503.

17. Gola, S., J. Hegner, and E. Kothe. 2000. Chimeric pheromone receptors in the basidiomycete *Schizophyllum commune*. *Fungal Genet. Biol.* 30:191–196. (Erratum, 36:255, 2002.)

18. Gola, S., and E. Kothe. 2003. The little difference: in vivo analysis of pheromone discrimination in *Schizophyllum commune*. *Curr. Genet.* 42:276–283.

19. Hegner, J., C. Siebert-Bartholmei, and E. Kothe. 1999. Ligand recognition in multiallelic pheromone receptors from basidiomycete *Schizophyllum commune* studied in yeast. *Fungal Genet. Biol.* 26:190–197.

20. Horn, F. E., E. Bettler, L. Oliveira, F. Campagne, F. E. Cohen, and G. Vriend. 2003. GPCRDB information system for G protein-coupled receptors. *Nucleic Acids Res.* 31:294–297.

21. Ichinomiya, M., T. Motoyama, M. Fujiwara, M. Takagi, H. Horiuchi, and A. Ohta. 2002. Repression of chsB expression reveals the functional importance of class IV chitin synthase gene chsD in hyphal growth and conidiation of *Aspergillus nidulans*. *Microbiology* 148:1335–1347.

22. James, T. Y., S.-R. Liou, and R. Vilgalys. 2004. The genetic structure and diversity of the *A* and *B* mating-type genes from the tropical oyster mushroom, *Pleurotus djamor*. *Fungal Genet. Biol.* 41:813–825.

23. Julius, D., R. Schekman, and J. Thorner. 1984. Glycosylation and processing of prepro-alpha-factor through the yeast secretory pathway. *Cell* 36:309–318.

24. Kolakowski, L. F., Jr. 1994. GCRDb: a G-protein-coupled receptor database. *Recept. Channels* 2:1–7.

25. Koltin, Y. 1968. The genetic structure of incompatibility factors of *Schizophyllum commune*: comparative studies of primary mutations in the *B* factor. *Mol. Gen. Genet.* 102:196–203.

26. Koltin, Y., and A. S. Flexer. 1969. Alteration of nuclear distribution in *B*-mutant strains of *Schizophyllum commune*. *J. Cell Sci.* 4:739–749.

27. Koltin, Y., and J. R. Raper. 1966. *Schizophyllum commune*: new mutations in the *B* incompatibility factor. *Science* 154:510–511.

28. Koltin, Y., and J. R. Raper. 1967. The genetic structure of incompatibility factors of *Schizophyllum commune*: three functionally distinct classes of *B* factors. *Proc. Natl. Acad. Sci. USA* 58:1220–1226.

29. Koltin, Y., J. R. Raper, and G. Simchen. 1967. Genetic structure of incompatibility factors of *Schizophyllum commune*: the *B* factor. *Proc. Natl. Acad. Sci. USA* 57:55–63.

30. Koltin, Y., J. Stamberg, N. Bawnik, A. Tamarkin, and R. Werczberger. 1979. Mutational analysis of natural alleles in and affecting the *B* incompatibility factor of *Schizophyllum*. *Genetics* 93:383–391.

31. Kothe, E. 1996. Tetrapolar fungal mating types: sexes by the thousands. *FEMS Microbiol. Rev.* 18:65–87.

32. Kothe, E., S. Gola, and J. Wendland. 2003. Evolution of multispecific mating-type alleles for pheromone perception in the basidiomycete fungi. *Curr. Genet.* 42:268–275.

33. Kurjan, J. 1991. Cell-cell interactions involved in yeast mating, p. 113–144. *In* M. Dworkin (ed.), *Microbial Cell-Cell Interactions*. ASM Press, Washington, DC.

34. Kurjan, J., and I. Herskowitz. 1982. Structure of a yeast pheromone gene (*MFα*): a putative α-factor precursor contains four tandem copies of mature α-factor. *Cell* 30:933–943.

35. Lee, B.-K., S. Khare, F. Naider, and J. M. Becker. 2001. Identification of residues of the *Saccharomyces cerevisiae* G protein-coupled receptor contributing to α-factor binding. *J. Biol. Chem.* 276:37950–37961.

36. Lengeler, K. B., R. C. Davidson, C. D'Souza, T. Harashima, W.-C. Shen, P. Wang, X. Pan, M. Waugh, and J. Heitman. 2000. Signal transduction cascades regulating fungal development and virulence. *Microbiol. Mol. Biol. Rev.* 64:746–785.

37. Naider, F., and J. M Becker. 2004. The α-factor mating pheromone of *Saccharomyces cerevisiae*: a model for studying the interaction of peptide hormones and G protein-coupled receptors. *Peptides* 25:1441–1463.

38. Olesnicky, N. S., A. J. Brown, S. J. Dowell, and L. A. Casselton. 1999. A constitutively active G-protein-coupled receptor causes mating self-compatibility in the mushroom *Coprinus*. *EMBO J.* 18:2756–2763.

39. Olesnicky, N. S., A. J. Brown, Y. Honda, S. L. Dyos, S. J. Dowell, and L. A. Casselton. 2000. Self-compatible B mutants in *Coprinus* with altered pheromone-receptor specificities. *Genetics* 156:1025–1033.

40. Palczewski, K., T. Kumasaka, T. Hori, C. A. Behnke, H. Motoshima, B. A. Fox, I. L. Trong, D. C. Teller, T. Okada, R. E. Stenkemp, M. Yamamoto, and M. Miyano. 2000. Crystal structure of rhodopsin: a G protein-coupled receptor. *Science* 289:739–745.

41. Papazian, H. P. 1950. Physiology of the incompatibility factors in *Schizophyllum commune*. *Bot. Gaz.* 112:143–163.

42. Papazian, H. P. 1950. The incompatibility factors and a related gene in *Schizophyllum commune. Genetics* 36: 441–459.

43. Parag, Y. 1962. Mutations in the B incompatibility factor of *Schizophyllum commune. Proc. Natl. Acad. Sci. USA* 48:743–750.

44. Petersen, J., D. Weilguny, R. Egel, and O. Nielsen. 1995. Characterization of fus1 of *Schizosaccharomyces pombe*: a developmentally controlled function needed for conjugation. *Mol. Cell. Biol.* 15:3697–3707.

45. Price, L. A., E. M. Kajkowski, J. R. Hadcock, B. A. Ozenberger, and M. H. Pausch. 1995. Functional coupling of a mammalian somatostatin receptor to the yeast pheromone response pathway. *Mol. Cell. Biol.* 15:6188–6195.

46. Raper, C. A. 1983. Controls for development and differentiation of the dikaryon in basidiomycetes, p. 195–238, *In* J. W. Bennett and A. Ciegler (ed.), *Secondary Metabolism and Differentiation in Fungi.* Marcel Dekker, New York, NY.

47. Raper, C. A. 1985. B-mating-type genes influence survival of nuclei separated from heterokaryons of *Schizophyllum. Exp. Mycol.* 9:149–160.

48. Raper, C. A., D. H. Boyd, and J. R. Raper. 1965. Primary and secondary mutations at the incompatibility loci in *Schizophyllum. Proc. Natl. Acad. Sci. USA* 53:1324–1332.

49. Raper, C. A., and J. R. Raper. 1968. Genetic regulation of sexual morphogenesis in *Schizophyllum commune. J. Elisha Mitchell Sci. Soc.* 84:267–273.

50. Raper, C. A., and J. R. Raper. 1973. Mutational analysis of a regulatory gene for morphogenesis in *Schizophyllum. Proc. Natl. Acad. Sci. USA* 70:1427–1431.

51. Raper, J. R. 1966. *Genetics of Sexuality in Higher Fungi.* Ronald Press, New York, NY.

52. Raper, J. R., M. G. Baxter, and A. H. Ellingboe. 1960. The genetic structure of the incompatibility factors of *Schizophyllum commune*: the A factor. *Proc. Natl. Acad. Sci. USA* 46:833–842.

53. Raper, J. R., M. G. Baxter, and R. B. Middleton. 1958. The genetic structure of the incompatibility factors in *Schizophyllum commune. Proc. Natl. Acad. Sci. USA* 44: 889–900.

54. Raper, J. R., G. S. Krongelb, and M. G. Baxter. 1958. The number and distribution of incompatibility factors in *Schizophyllum. Am. Nat.* 92:221–232.

55. Raper, J. R., and M. Raudaskoski. 1968. Secondary mutations at the Bβ incompatibility locus of *Schizophyllum. Heredity* 23:109–117.

56. Raudaskoski, M. 1970. A new secondary Bβ mutation in *Schizophyllum* revealing functional differences in wild Bβ alleles. *Hereditas* 64:259–266.

57. Raudaskoski, M. 1972. Secondary mutations at the Bβ incompatibility locus and nuclear migration in the basidiomycete *Schizophyllum. Hereditas* 72:175–182.

58. Raudaskoski, M. 1984. Unusual structure of nuclei in a B-mutant strain of *Schizophyllum commune* with intercellular nuclear migration. *Nord. J. Bot.* 4:217–223.

59. Raudaskoski, M. 1998. The relationship between B-mating-type genes and nuclear migration in *Schizophyllum commune. Fungal Genet. Biol.* 24:207–227.

60. Raudaskoski, M., and Y. Koltin. 1973. Ultrastructural aspects of a mutant of *Schizophyllum commune* with continuous nuclear migration. *J. Bacteriol.* 116:981–988.

61. Raudaskoski, M., J. Stamberg, N. Bawnik, and Y. Koltin. 1976. Mutational analysis of natural alleles at the *B* incompatibility factor of *Schizophyllum commune: a2* and *a6. Genetics* 83:507–516.

62. Riquelme, M., M. P. Challen, L. A. Casselton, and A. J. Brown. 2005. The origin of multiple *B* mating specificities in *Coprinus cinereus. Genetics* 170:1105–1119.

63. Schuurs, T. A., H. J. P. Dalstra, J. M. J. Scheer, and J. G. H. Wessels. 1998. Positioning of nuclei in the secondary mycelium of *Schizophyllum commune* in relation to differential gene expression. *Fungal Genet. Biol.* 23:150–161.

64. Specht, C. A. 1996. Isolation of the Bα and Bβ mating-type loci of *Schizophyllum commune. Curr. Genet.* 28: 374–379.

65. Specht, C. A., A. Muñoz-Rivas, C. P. Novotny, and R. C. Ullrich. 1988. Transformation of *Schizophyllum commune*: parameters for improving transformation frequencies. *Exp. Mycol.* 12:357–366.

66. Stamberg, J., and Y. Koltin. 1972. The organization of the incompatibility factors in higher fungi: the effects of structure and symmetry on breeding. *Heredity* 30:15–26.

67. Szabó, Z., M. Tönnis, H. Kessler, and M. Feldbrügge. 2002. Structure-function analysis of lipopeptide pheromones from the plant pathogen *Ustilago maydis. Mol. Genet. Genomics* 268:362–370.

68. Tam, A., F. J. Nouvet, K. Fujimura-Kamada, H. Slunt, S. S. Sisodia, and S. Michaelis. 1998. Dual roles for Ste24p in yeast **a**-factor maturation: NH₂-terminal proteolysis and COOH-terminal CAAX processing. *J. Cell Biol.* 142:635–649.

69. Trueheart, J. A., J. D. Boeke, and G. R. Fink. 1987. Two genes required for cell fusion during yeast conjugation: evidence for a pheromone-induced surface protein. *Mol. Cell. Biol.* 7:2316–2328.

70. Vaillancourt, L. J., and C. A. Raper. 1996. Pheromones and pheromone receptors as mating-type determinants in basidiomycetes, p. 219–247. *In* J. K. Setlow (ed.), *Genetic Engineering*, vol. 18. Plenum Press, New York, NY.

71. Vaillancourt, L. J., M. Raudaskoski, C. A. Specht, and C. A. Raper. 1997. Multiple genes encoding pheromones and a pheromone receptor define the Bβ1 mating-type specificity in *Schizophyllum commune. Genetics* 146: 541–551.

72. Wendland, J., L. J. Vaillancourt, J. Hegner, K. B. Lengeler, K. J. Laddison, C. A. Specht, C. A. Raper, and E. Kothe. 1995. The mating-type locus Bα1 of *Schizophyllum commune* contains a pheromone receptor gene and putative pheromone genes. *EMBO J.* 14:5271–5278.

73. Wessels, J. G. H., and D. J. Niederpruem. 1967. Role of a cell-wall glucan-degrading enzyme in mating of *Schizophyllum commune. J. Bacteriol.* 94:1594–1602.

74. Yamagishi, K., T. Kimura, M. Suzuki, and H. Shinmoto. 2002. Suppression of fruit-body formation by constitutively active G-protein α-subunits ScGP-A and ScGP-C in the homobasidiomycete *Schizophyllum commune. Microbiology* 148:2797–2809.

*Sex in Fungi: Molecular Determination
and Evolutionary Implications*
Edited by Joseph Heitman et al.
© 2007 ASM Press, Washington, D.C.

Timothy Y. James

19

Analysis of Mating-Type Locus Organization and Synteny in Mushroom Fungi:
Beyond Model Species

The ability of a fungal individual to mate or outcross with another individual is dependent on its mating type. The mating type of an individual is determined by the phenotypic expression of its mating-type locus (*MAT*) or loci, and only individuals with different mating types are compatible. Species that have two *MAT* loci are termed tetrapolar, and those that have only a single *MAT* locus are bipolar. *MAT* loci correspond with regions of the genome that may be as small as a single gene (e.g., *Cochliobolus heterostrophus* [81]) or as large as 0.5 Mbp (e.g., *Ustilago hordei* [50]). As discussed in detail in this volume, the genes that function in determining mating type in both Ascomycota and Basidiomycota encode either transcription factors or pheromone receptors and their pheromone ligands (12, 32). *MAT* loci encode both genes whose expression directly determines mating-type specificity as well as those that do not determine the mating type of a cell but are nonetheless mating-type specific. For example, the gene *mtA-2* in *Neurospora crassa* is specific to the *A* mating-type allele of the *MAT* locus, functions in ascospore development, but is not utilized in mating-type determination (21). Likewise, the *a2* allele of the *Ustilago maydis* *MAT-a* locus harbors two genes (*lga2* and *rga2*) that are

specific to a single mating type but are involved in mitochondrial fusion rather than mating-type control (8, 83). In this review, genes that are in or near the *MAT* locus but do not function in determining mating-type specificity are referred to as non-mating-type *MAT*-linked genes.

The Basidiomycota appear to be divided into three major lineages: rusts (Urediniomycetes), smuts (Ustilaginomycetes), and mushroom-like fungi (Hymenomycetes, including homobasidiomycetes and jelly fungi; Fig. 19.1). Most of the described basidiomycete species are homobasidiomycetes (42), and these species are common components of soil ecosystems (33, 61). The basidiomycetes are unique among fungi in having species with tetrapolar mating systems. Mushroom fungi are further distinct from other basidiomycete species because there may be numerous (up to hundreds) mating-type alleles at both of the *MAT* loci. Most of the proliferation of mating types is due to the manner in which the mushroom *MAT* locus is composed of multiple, redundant subloci that can display recombination distances as high as 16 centimorgans (cM) (66). There is a great body of literature on the mating systems and mating-type number and distribution in mushrooms, in part because of the

Timothy Y. James, Department of Biology, Duke University, Durham, NC 27708.

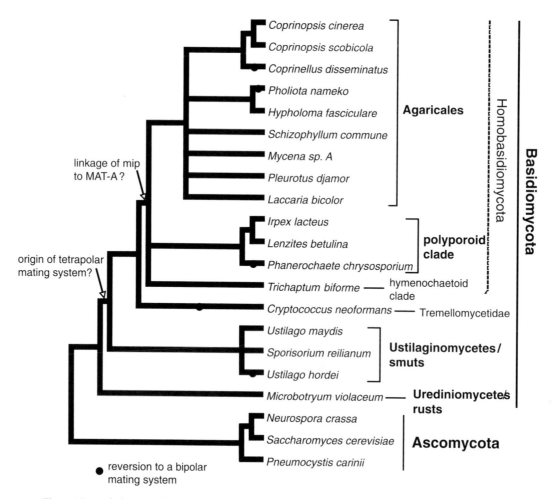

Figure 19.1   Phylogeny of the Basidiomycota. The cladogram depicts the current knowledge of the relationships among the fungi discussed in the text. Data are derived from references 30, 59, and 79, the Tree of Life Web project (http://tolweb.org/tree/phylogeny.html), and the *mor* Web project (http://mor .clarku.edu/).

utility of the mating-type locus to serve as a highly polymorphic marker (60, 82). Among homobasidiomycetes, an estimated 10% are homothallic (non-outcrossing), 25 to 35% are bipolar, and 55 to 65% are tetrapolar (68, 88). From an evolutionary perspective, mating-system switches in fungi are fascinating and none more so than the former genus *Coprinus*, in which homothallic, bipolar, and tetrapolar species interdigitate along the species phylogeny (7, 34). The molecular genetic bases for these mating-system switches have been less tractable than that observed in Ascomycetes (91), in part due to the complexity of the homobasidiomycete mating genes (41).

Coprinopsis cinerea (= *Coprinus cinereus*) and *Schizophyllum commune* are both model systems for studying mating genetics in mushrooms (10, 66). *C. cinerea* and *S. commune* are excellent model systems because they mate readily and can fruit directly on a petri dish of sim-

ple nutrient agar. *C. cinerea* has the added advantages of producing asexual propagules (oidia) and meiotic tetrads that can be isolated (44). Both species have been advanced as model systems by the development of auxotrophic mutants, *MAT* mutants, and genetic transformation (85).

During the 1990s the *MAT-A* and *MAT-B* loci of *C. cinerea* and *S. commune* were cloned. Cloning of the *MAT* loci was accomplished by various means (see chapters 16 and 17 by Stankis and Specht and by Casselton and Kües), but the determination that the cloned region actually carried *MAT* genes relied on transformation of the genes into a suitable strain followed by screening for a mating-compatible phenotype (27, 64, 76). The mushroom *MAT-A* locus encodes two types of dissimilar homeodomain proteins (HD1 and HD2 [48]). Dimerization between heteroallelic HD1 and HD2 proteins forms a transcription factor that activates a *MAT-A*-specific developmental pathway. The mushroom *MAT-B* locus encodes pheromone receptors and their lipopeptide pheromone ligands (64, 87). As with *MAT-A*, heteroallelic combinations of pheromones and pheromone receptors in mated cells cause the activation of the *MAT-B*-specific developmental pathway via a putative mitogen-activated protein kinase signaling cascade triggered by the G-protein-coupled pheromone receptors (10).

Cloning of these *MAT* loci revealed not only that they encode a large number of alleles but also that the alleles were highly dissimilar in sequence. The alleles were so different that they generally failed to cross-hybridize in Southern hybridizations (64, 77). This also holds at the amino acid level; for example, the *MAT-A* homeodomain proteins of *S. commune* are only 42 to 54% identical between alleles (78).

When trying to study the *MAT* genes of mushroom fungi other than *C. cinerea* and *S. commune*, there are three major obstacles. First, the loci are likely to be complex and may comprise several genes. Second, the loci display high sequence variation among *MAT* alleles and cannot be cloned by heterologous hybridization using probes from *C. cinerea* or *S. commune*. Lastly, the absence of a transformation system in which to test the cloned fragments prevents definitive proof that an isolated DNA region carries *MAT*. One way of getting around these obstacles is to target not the mating-type genes themselves but the more slowly evolving genes that they are tightly linked to, provided that synteny (or conserved gene order) has been maintained among the species of interest. This approach allows the investigator to isolate *MAT* loci and to determine homology of the isolated genes using syntenic arguments rather than by genetic transformation. This review focuses on how *MAT* genes may be cloned in nonmodel mushroom species.

SYNTENY AND *MAT-A* IN MUSHROOMS

In order that non-mating-type *MAT*-linked genes can be targeted as a proxy for *MAT* genes, it is essential that the linkage of non-mating-type *MAT*-linked genes to *MAT* be conserved among species. Because of the high amounts of genome rearrangements among closely related species (80), there was no reason, a priori, to assume that conserved gene order would hold near the *MAT* loci. Classical genetic studies, however, indicated early on that some markers might display highly conserved linkage to the *MAT-A* locus (66). Specifically, the earliest mapping studies of the *MAT-A* chromosomal regions in *C. cinerea* and *S. commune* indicated that they

were syntenic and very tightly linked to loci conferring para-aminobenzoic acid (*pab*) and adenine (*ade*) auxotrophy in mutants (66). The *pab* locus mapped to <1 cM from *MAT-A* in both species, and the *MAT-A* genes in *S. commune* were first cloned by a chromosomal walk from *pab*, ultimately found to be ~50 kbp from *MAT-A* (27). After *MAT-A* loci from *C. cinerea* and *S. commune* had been cloned, it was further apparent that they shared the presence of a metalloendopeptidase encoded immediately adjacent to *MAT-A* (13, 78). The metalloendopeptidase gene (*mip*) was not part of *MAT-A* as delimited by transformation assays (27). Mip is a mitochondrial matrix-localized enzyme that functions in the cleavage of the leader peptides of precursor proteins targeted to the mitochondrial matrix or inner membrane (37). The first use of non-mating-type *MAT*-linked genes in positional cloning of *MAT* genes was accomplished when Kües et al. (47) used a heterologous *mip* probe from *C. cinerea* to isolate the *MAT-A* locus from the related *Coprinopsis scobicola* (= *Coprinus bilanatus*). After recovering cosmid clones containing the *C. scobicola mip* gene, Kües et al. were able to delimit the *MAT-A* locus to an ~15-kbp region adjacent to *mip* through the use of genetic transformation in *C. cinerea* and *C. scobicola* host strains.

In order to develop *mip* as a marker for *MAT-A* in mushrooms, conservation of the genetic linkage between the two loci was explored. Because the *mip* gene of *C. cinerea* did not appear to hybridize to genomic DNA of other mushroom genera (U. Kües, personal communication), an approach based on PCR amplification of *mip* was attempted (39). The *mip* gene was successfully amplified from over 30 species of mushrooms throughout the diversity of homobasidiomycetes. The linkage relationships in several species were analyzed by studying cosegregation of *mip* and *MAT-A* in progeny arrays of single spore isolates obtained from single fruiting bodies (Table 19.1). The result of the cosegregation studies, as well as information derived from the sequencing of complete genomes, is that linkage of *mip* to *MAT-A* is completely conserved throughout homobasidiomycetes. Furthermore, in all known cases, the *mip* gene is directly adjacent and less than 1 kbp from the 3′ end of one of the mating-type genes.

How deep into evolutionary history was linkage between *MAT-A* and *mip* established? More data on heterobasidiomycetes, including rust and smut fungi, would be useful to address this question. The *b* mating-type locus (*MAT-b*) of *U. maydis*, which is homologous to *MAT-A* in mushrooms, displays no linkage to *mip*, nor does the single *MAT* locus of the pathogenic yeast *Cryptococcus neoformans* (25, 51) (Table 19.1). Synteny between *mip*

Table 19.1 Linkage of *mip* and *cla4* to *MAT-A* and *MAT-B* in basidiomycetes[a]

Genus and species	Clade[b]	Mating system	Ecology	*mip* linked to *MAT-A*?	*cla4* linked to *MAT-B*?	Method of inference	Reference(s)
Coprinellus disseminatus	Homobasidiomycetes: euagarics clade	Bipolar	Wood decay	+	?	Cosegregation/partial genome sequence	39, 41
Coprinopsis cinerea	Homobasidiomycetes: euagarics clade	Tetrapolar	Coprophilic	+	+	Genome sequence	Broad Institute website[c]
Coprinopsis scobicola	Homobasidiomycetes: euagarics clade	Secondarily homothallic	Coprophilic	+	?	Partial genome sequence	47
Cryptococcus neoformans	Heterobasidiomycetes: Tremellales	Bipolar	Saprophyte/pathogen	−	+	Genome sequence	52, 86; TIGR website[d]
Hypholoma fasciculare	Homobasidiomycetes: euagarics clade	Tetrapolar	Wood decay	+	?	Cosegregation	39
Irpex lacteus	Homobasidiomycetes: polyporoid clade	Tetrapolar	Wood decay	+	−	Cosegregation	39
Laccaria bicolor	Homobasidiomycetes: euagarics clade	Tetrapolar	Ectomycorrhizal	+	?	Genome sequence	JGI database[e]
Lenzites betulina	Homobasidiomycetes: polyporoid clade	Tetrapolar	Wood decay	+	?	Cosegregation	39
Mycena sp. A	Homobasidiomycetes: euagarics clade	Tetrapolar	Litter decay?	+	?	Cosegregation	39
Phanerochaete chrysosporium	Homobasidiomycetes: polyporoid clade	Bipolar	Wood decay	+	+	Genome sequence	JGI database[f]
Pleurotus djamor	Homobasidiomycetes: euagarics clade	Tetrapolar	Wood decay	+	+	Cosegregation	39, 40
Schizophyllum commune	Homobasidiomycetes: euagarics clade	Tetrapolar	Wood decay	+	+	Cosegregation/partial genome sequence	78 and Fig. 19.3
Trichaptum biforme	Homobasidiomycetes: hymenochaetoid clade	Tetrapolar	Wood decay	+	?	Cosegregation	39
Ustilago maydis	Ustilaginomycetes: Ustilaginales	Tetrapolar	Plant pathogen	−	+	Genome sequence	Broad Institute website[g]

[a]In this table, *MAT-A* refers to the homeodomain-encoding *MAT* region, and *MAT-B* refers to the pheromone/receptor gene locus. In smut fungi (e.g., *Ustilago maydis*) *MAT-A* (encoding the homeodomain genes) is actually referred to as the *b* mating-type locus and *MAT-B* is referred to as the *a* mating-type locus. +, linkage conserved; −, genes unlinked; ?, linkage relationship uncertain.
[b]Clade follows nomenclature of Hibbett and Binder (30) or GenBank.
[c]http://www.broad.mit.edu/annotation/genome/coprinus_cinereus/Home.html (Broad Institute).
[d]http://www.tigr.org/tdb/e2k1/cna1/ (The Institute for Genomic Research).
[e]http://genome.jgi-psf.org/Lacbi1/Lacbi1.home.html (Joint Genome Institute).
[f]http://genome.jgi-psf.org/Phchr1/Phchr1.home.html (Joint Genome Institute).
[g]http://www.broad.mit.edu/annotation/genome/ustilago_maydis/Home.html (Broad Institute).

and *MAT* is generally absent in Ascomycota but exists in the genome of *N. crassa*. Linkage in this species is over 100 kbp and quite possibly a statistical artifact. Further studies of Tremellales, Auriculariales, and other heterobasidiomycete genomes will be necessary to determine when the linkage relationship was established. Interestingly, the genome sequence of *U. maydis* shows that *mip* is syntenic (at a distance of ~150 kbp) with the *MAT-a* locus that encodes the pheromone and pheromone receptors (homologous to *MAT-B* of mushroom fungi).

The tight linkage between *mip* and *MAT-A* was also used to study the *MAT* loci of *Coprinellus disseminatus* and *Pleurotus djamor* (40, 41). The entire *MAT-A* loci and surrounding DNA regions were sequenced, thus allowing an assessment of conservation of gene order for the genomic region (Fig. 19.2). These gene order comparisons reveal synteny to be widespread at the *MAT-A* locus. Besides the *mip* gene, 11 additional genes have conserved linkage to *MAT-A* in *C. cinerea*, *C. disseminatus*, *Phanerochaete chrysosporium*, and *P. djamor*. The functions of 8 of the 11 additional conserved genes are uncertain (*chp1-6* [conserved hypothetical proteins], *ypl109*, and *β-fg* [Aβ-flanking gene]), but the function of three of them (*sec61*, *glgen*, and *glydh*) can be speculated

upon based on homology to genes in other organisms. *sec61* encodes the gamma subunit of a translocase involved in moving proteins from the cytoplasm to endoplasmic reticulum. *glgen* encodes an enzyme putatively functioning in lipopolysaccharide and glycogen synthesis. *glydh* encodes a putative glycine decarboxylating enzyme.

The *MAT-A* loci of *C. cinerea* and *S. commune* have both been divided into two subloci (Aα and Aβ) by fine mapping studies. A major difference between these two species was that the Aα and Aβ subloci are very close in *C. cinerea* (~7 kbp) whereas in *S. commune* the Aα and Aβ subloci are far apart. There are no non-mating-type *MAT*-linked genes harbored in the intervening region between the Aα and Aβ subunits of *C. cinerea*—hence the previous name of "homologous hole" for this region. In *C. cinerea*, a gene termed the *β-fg* was identified as a gene of unknown function on the other border of the *MAT-A* locus (48). This gene is among the 12 genes that show conserved linkage to *MAT-A* among the four homobasidiomycetes studied to date (Fig. 19.2).

In contrast to all other mushroom species for which data are available, in *S. commune* *MAT*-linked genes are housed between Aα and Aβ subloci, including *mip* and *pab1* (66). This arrangement of Aα and Aβ subloci could be explained by two large inversions moving the

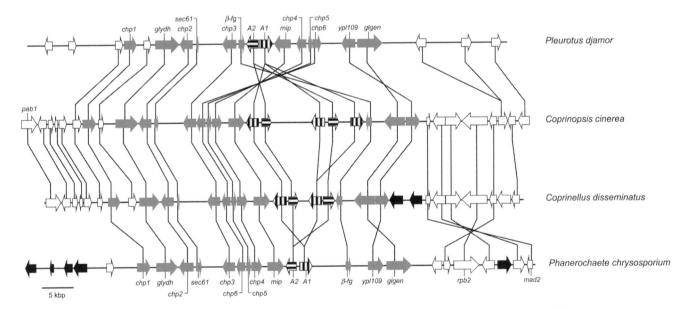

Figure 19.2 Schematic showing conserved gene order of the chromosomal region surrounding *MAT-A* in mushroom fungi. Arrows indicate genes and their direction of transcription. Vertical lines connect homologous genes between species. Genes with vertical stripes or horizontal stripes represent *MAT-A* homeodomain-type-1 (*HD1*) genes and homeodomain-type-2 (*HD2*) genes, respectively (14). Genes in gray are found in all four species, and genes in black are restricted to a single species. This map includes a number of additional genes for *P. djamor* that were missed in the previous annotation of the *MAT-A* region (40). *chp*, gene encoding a conserved hypothetical protein.

genes normally flanking the subloci into the region between the two subloci. The general rarity of non-mating-type *MAT*-linked genes between *MAT* subloci of mushrooms may relate to the theoretical prediction that increases in recombination between *MAT* subloci (thus generating recombinant *MAT* alleles at a higher frequency) will increase the chances of mating occurring among siblings and thereby reduce a species' outbreeding potential (74). In *S. commune*, a gene termed *X* was identified as the other *MAT-Aα* flanking gene (56). Disruption of *X* caused no obvious phenotype in development (56). Gene *X* may be specific to *S. commune*, as there are no clear homologues of *X* detectable in other mushroom genomes.

MAT-B ALSO DISPLAYS CONSERVED GENE ORDER

The *MAT-B* locus of mushroom fungi encodes small lipopeptide pheromones and G-protein-coupled pheromone receptors that have seven transmembrane-spanning domains (homologues of yeast *STE3* **a**-factor receptor). In *C. cinerea*, the *MAT-B* locus comprises three subgroups with redundant function (29). Each subgroup encodes 1 to 3 pheromone genes and a homoallelic receptor; the whole *MAT-B* locus of *C. cinerea* spans ~25 kbp, and no non-mating-type *MAT*-linked genes are encoded in the regions between the subgroups (29).

The sequencing of the first genome of a homobasidiomycete fungus (*Phanerochaete chrysosporium*) revealed homologues of the *MAT* proteins of mushrooms (57) and allowed an exploration of synteny at the *MAT-B* locus. Five genes encoding pheromone receptors (homologues of *STE3*) were detected in the *P. chrysosporium* genome. Three of these were found clustered into an ~12-kbp region (57), similar to the organization of *MAT-B* in *C. cinerea*. Further investigation of the *MAT-B*-like genomic region in *P. chrysosporium* demonstrated very close linkage to the gene *ste20* (38). Linkage to *ste20* was striking, as it is also linked to the *STE3* receptor homologues of both *Cryptococcus neoformans* (51) and *Pneumocystis carinii* (75). Thus, a conserved linkage between *MAT-B* and *ste20* was demonstrated for a wide diversity of organisms including both Ascomycota and Basidiomycota. Ste20 is a p21-activated kinase (PAK) required for mating in budding yeast (65). The closely related yeast protein Cla4 is also a PAK and is involved in budding and cytokinesis in yeast (15). Cla4 differs from Ste20 in possessing a pleckstrin homology domain, a region possibly involved in protein-protein interactions. The PAK genes linked to *STE3* in *Cryptococcus neoformans* and *Pneumocystis carinii* en-

code proteins with a pleckstrin homology domain and are phylogenetically more closely related to *cla4* of *Saccharomyces cerevisiae* and *U. maydis* (38, 53). Therefore, for the remainder of the chapter I refer to the *MAT-B* linked PAK as *cla4*.

The data from *Cryptococcus* and *Pneumocystis MAT* loci (or *STE3*-encoding regions) provided compelling evidence that these species have a cluster of genes all functioning in the process of mating (17, 75). Both *Cryptococcus* and *Pneumocystis MAT* loci also encode *ste12*, a key transcriptional activator of genes in the pheromone response pathway in *Saccharomyces* (3). The observation of pheromone receptors linked to genes that may be involved in the same developmental pathways suggested a possible coregulated cluster of genes that might be expected to show evolutionarily conserved gene order. Thus, the potential for cloning *MAT-B* from mushrooms by exploiting synteny with *cla4* was investigated. The linkage between *MAT-B* and *cla4* in additional mushroom species was explored by analyzing the cosegregation of the two loci among progeny of a fruited dikaryon. These data revealed linkage between the two loci in both of the agarics *Pleurotus djamor* (40) and *Schizophyllum commune* (Fig. 19.3), but the loci were unlinked in the polyporoid *Irpex lacteus* (Table 19.1).

By use of a positional cloning approach, the *MAT-B* locus of *Pleurotus djamor* was partially cloned by identifying cosmids from a genomic library that contained the *cla4* gene (40). *cla4* was determined to be ~29 kbp distant from a *STE3*-like pheromone receptor of the *P. djamor MAT-B* locus, as opposed to the ~6-kbp distance observed in the *Phanerochaete chrysosporium* genome. A search of the *Coprinopsis cinerea* genome demonstrates that *cla4* is ~95 kbp from *MAT-B*. In the heterobasidiomycetes *Cryptococcus neoformans* and *U. maydis*, *cla4* is ~5 and ~65 kbp, respectively, from the

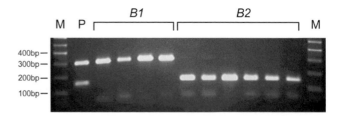

Figure 19.3 Cosegregation of *cla4* and *MAT-B* in *Schizophyllum commune*. Shown is an agarose gel (2%) of *cla4* amplicons digested with MseI. In lane P is the parental dikaryon (Guy.21.2) that is heterozygous at the *cla4* locus. The other lanes show the progeny of Guy.21.2. Lanes *B1* are of the *B1* mating type and possess a *cla4* allele that lacks an MseI cut site. Lanes *B2* are of the *B2* mating type and possess the MseI cut site that digests the 297-bp amplicon into two fragments of 148 and 149 bp. Lanes marked M contain DNA marker.

STE3 homologue in these species (53). These data demonstrate conserved synteny of *cla4* and *MAT-B* in the basidiomycetes, though this linkage is clearly looser than that of *mip* and *MAT-A* and there is at least one example in which synteny has been disrupted (*Irpex lacteus*). In ascomycetes other than *Pneumocystis carinii*, this tight linkage is not observed (based on a scan of several available genomes), though the genes are on the same chromosome in the ascomycetes *Magnaporthe grisea* and *Neurospora crassa* (at nearly 1 Mbp distance).

Comparisons of the genomic regions surrounding the *MAT-B* loci of the homobasidiomycetes (same species as that shown in Fig. 19.2) revealed no additional genes with conserved synteny to *MAT-B* shared by all four species. However, the *MAT-B* sequences available for *Pleurotus djamor* and *Coprinellus disseminatus* are much shorter than the available *MAT-A* sequences. Halsall and colleagues (29) identified a gene encoding a putative transporter of the major facilitator family (*mfs1*) directly flanking the *MAT-B* locus of *Coprinopsis cinerea*. Interestingly, alleles of *mfs1* differed by as much as 40% at the DNA level. A homologue of *C. cinerea mfs* is syntenic with the *MAT-B* locus of *Phanerochaete chrysosporium*, but the intervening distance is nearly 500 kbp. In general, comparative genomics of the *MAT-B* region suggests that rearrangements happen at a much greater frequency than those of the *MAT-A* region (data not shown).

APPROACHES TO CLONING *MAT* USING CONSERVED GENE ORDER

As opposed to the manner in which *MAT* genes were cloned in the model species, cloning *MAT* genes in nonmodel species lacks the advantage of directly testing the function of the cloned genes by transformation of the DNA into a mating-compatible host strain. This distinction means that care has to be given when assigning a cloned gene to the *MAT* locus. One strategy for cloning *MAT* using linked genes begins with amplifying the flanking genes (*mip* and *cla4*), using PCR with degenerate primers (39, 40). PCR amplification of these flanking genes is expected to be much easier than direct amplification of the *MAT* genes themselves. Primers for the *mip* gene have amplified a diversity of homobasidiomycetes, but this gene displays more sequence variation than *cla4*. Primers used successfully to amplify *cla4* are STE20-1F (5′-GTNATGGARTWYATGGARGG-3′) and STE20-2R (5′-ACNACTTCAGGNGCCATCCA-3′). These primers work readily on homobasidiomycetes but often amplify both PAKs (*cla4* and *ste20*).

After sequences of *mip* or *cla4* are obtained in the organism of interest, the fragment can be used to probe a genomic library. If genomic libraries are constructed using vectors capable of replicating a large DNA fragment (e.g., cosmids, bacterial artificial chromosomes, or phages), clones hybridizing to *mip* are very likely to contain all or part of the *MAT-A* locus. Alternatively, PCR-based approaches such as inverse PCR (62) or thermal asymmetric interlaced PCR (54) may be used to amplify and sequence the DNA regions immediately adjacent to *mip*. In the case of *cla4*, a strategy such as inverse PCR is unlikely to be fruitful given the larger physical distances typically observed between the gene and *MAT-B*.

Given that sequences homologous to *MAT* genes can be obtained for the species of interest, further effort may be required to demonstrate that the sequenced region actually harbors *MAT*. Ideally, it is desirable to show that the *MAT* homologues in nonmodel species cosegregate with *MAT* as determined by interstrain matings. For some species, such as ectomycorrhizal taxa for which single spore isolates cannot be obtained due to lack of spore germination (e.g., *Russula* and *Boletus* spp.), this may not be possible. Since the *MAT* genes are highly polymorphic in DNA sequence in all species for which they have been investigated (58, 69, 78), demonstration that the putative *MAT* genes are highly polymorphic in the species of interest can also provide evidence that they are indeed *MAT* genes. Finally, heterologous expression of the cloned *MAT* homologues, when transformed into *C. cinerea* or *S. commune*, may be used to confirm their function (41, 47). Although *MAT* genes of most mushroom species are unlikely to interact with those of the model host species, by transformation of two alleles into a single host strain or through mating of strains transformed with different alleles, the interactions between the gene products may be tested. This assay relies on common *MAT-A* and *MAT-B* downstream targets between the host and nonmodel species, which may not be valid for all mushroom species.

CLONING *MAT* USING DIRECT PCR AMPLIFICATION

Although most of this review has focused on how conserved gene order can be used to clone *MAT* from nonmodel species, a more direct option is through PCR amplification of the *MAT* homologues themselves. The *STE3 MAT-B* homologues display some relatively conserved amino acids among the seven transmembrane alpha-helices (amino terminus) that can be targeted by degenerate primers (1, 40, 41). These primers have successfully amplified *STE3* homologues from a wide range of mushrooms, from the wood ear fungus *Auricularia poly-tricha* (GenBank accession number AY226009) to the ash bolete *Gyrodon merulioides* (GenBank accession number AY226018). Phylogenetic analyses of the *MAT-B* receptors suggest two divergent groups of *STE3*-like pheromone receptors in homobasidiomycetes (1, 40, 69), and the PCR-amplified receptors group among the ones that have been demonstrated to have true *MAT*-determining function (data not shown). However, data suggest that *STE3*-like genes are plentiful in mushroom fungi, and non-mating-type-specific pheromone receptor-encoding genes have been detected in *Coprinellus disseminatus* (41), *Phanerochaete chrysosporium* (57), and even *Coprinopsis cinerea* (T. Y. James, unpublished observations). For example, at least four *STE3* homologues have been detected in *C. disseminatus*, mapping to at least two separate locations distinct from the *MAT* locus. Similarly, in *P. chrysosporium* five *STE3* homologues are present and found in three separate regions of the genome (57). Understanding the function of these non-mating-type-specific pheromone receptors is an exciting prospect that may reveal developmental pathways specific to mushroom fungi.

PCR amplification of the homeodomain transcription factors of *MAT-A* has been heroically accomplished with the mushroom *Pholiota nameko* (1). Here the authors targeted the few conserved amino acids of the homeodomain region (HD1) by degenerate PCR and isolated the entire HD1 gene from this species by genome walking using cassette-mediated PCR. A similar approach was also used to amplify and sequence a pheromone receptor homologue of the *MAT-B* locus. Aimi et al. (1) went further and amplified homologues of both *pab1* and *ade5* in *P. nameko* and used cosegregation analyses to show that the homeodomain-encoding gene (*hox1*) was linked to both *pab1* and *ade5* and, moreover, segregated 1:1 with the *MAT* locus in this bipolar mushroom species. The isolated pheromone receptor (*rcb1*), however, was not linked to *MAT*. Attempts to amplify homeodomain-encoding genes in other mushrooms have been unsuccessful (James, unpublished).

THE ORGANIZATION OF *MAT* IN HOMOBASIDIOMYCETES

The ancestor of the homobasidiomycetes is inferred to be tetrapolar (31). In tetrapolar species, the homeodomain proteins encoded by the *MAT-A* locus in a monokaryon are unable to heterodimerize to form an active transcription regulator. In compatible matings bringing together nuclei with different *A* mating types into a single dikaryotic cell, the protein products of the

two heteroallelic *MAT-A* loci are able to form the active heterodimer. It is unclear how *MAT* loci are organized in homothallic species, but the structure of *MAT* in heterothallic species suggests that recombination events between different *MAT* alleles could bring together compatible gene products into a single *MAT* allele capable of turning on dikaryotic development without the need for a mating partner. Recombination within a single *MAT-A* haplotype has been demonstrated to create a self-compatible *MAT* allele by the fusion of *HD1* and *HD2* genes from different subunits of the *Coprinopsis cinerea MAT-A* locus (46). This manner by which one of the two *MAT* loci can mutate to a self-compatible allele led Raper (66) to predict that this process led to the formation of bipolar species from tetrapolar species. He also suggested that homothallic species should originate from bipolar ancestors more readily than tetrapolar ancestors (i.e., bipolarity is a transition state from tetrapolar to homothallic). Data on a few species (*Coprinellus disseminatus* [41], *Pholiota nameko* [1], and *Phanerochaete chrysosporium* [38]) now suggest that bipolar mushroom species originate through the loss of mating-type-determining function of the pheromones or receptors of *MAT-B* (Fig. 19.1). These data agree with the facts that all previous attempts to mutate the *MAT* loci of mushrooms resulted only in self-compatible *MAT* alleles, rather than novel alleles (66), and that the self-compatible mutations observed at the *MAT-B* loci of *C. cinerea* and *Schizophyllum commune* occurred through single point mutations (24, 63).

Classical genetic studies determined that the *MAT* loci are composed of multiple, tightly linked, and redundant subloci (18, 67) and that recombination between these subloci, in part, generates the huge number of mating types. Molecular genetic studies have shown that the different subloci harbor the same genes; therefore, the diversity of mushroom mating types can also be explained by the hypervariability of the genes, particularly in their specificity-determining regions (2, 58). The number of subloci thus far observed at *MAT* loci ranges from one (e.g., *MAT-A* of *P. chrysosporium*) to three (e.g., *MAT-B* of *C. cinerea*); it is likely that species having four or more subloci will be found (45). These subloci can be close together (Fig. 19.2) or further apart (1 to 16 cM for *S. commune MAT-A*), though the close arrangement seems to be more common. Although previous observations suggested that the bipolar *MAT* locus may be indivisible into subloci (66), it is now apparent that this is not universally the case (1, 41). Conversely, tetrapolar species such as *Pleurotus djamor* may have only a single *MAT-A* sublocus. The *S. commune MAT-A* locus appears to be different from *MAT-A* loci of other mushroom species because genes are encoded between the *MAT-Aα* and *MAT-Aβ* subloci, presumably due to large chromosomal inversions. There is good evidence in *S. commune* that the physical distance between subloci at *MAT-B* may be much closer than the mapping distance suggests and that the products of the subloci may actually be able to activate each other (23). Future research will be needed to explain the observation that different *MAT* alleles of *S. commune* generate recombinant mating types at different frequencies.

WHY HAS LINKAGE BEEN CONSERVED BETWEEN *MAT* AND OTHER GENES?

Conserved gene order is generally observed to disappear or decay as organisms diverge (80). Nonetheless, some regions of the genome, such as the X chromosome of vertebrates, display extensive synteny, even among puffer fish and humans, which diverged over 400 million years ago (28). Other regions of the genome have undergone rapid and extensive rearrangements, in eukaryotes often as small inversions (20). Why have some genes, but not others, remained syntenic for such an extended period of time?

There are at least five hypotheses for the observation of conserved gene order near the *MAT* loci of basidiomycetes. The first and most obvious hypothesis is that the gene linkages are just due to chance or historical accident. This postulates that the observed gene order is due to a historical genome rearrangement that was unrelated to the functional coding ability of the genes. Since then, the maintenance of the gene order between species is a probabilistic function related to time of species divergence and distance between genes. There are a few arguments against this simple explanation, however. One is that the degree of synteny or at the least frequency of genome rearrangement at *MAT-A* differs from that at *MAT-B*, suggesting that the process is not random. A counterargument to this point is that different regions of the genome experience different levels of gene rearrangement, i.e., telomeric and centromeric portions of a chromosome appear to be more dynamic than other portions (20). A second argument against the "historical accident" hypothesis is that the linkage between *mip* and *MAT-A* has been maintained despite a small inversion near the *MAT* locus that switched whether the *HD1* or *HD2 MAT* gene was proximate to *mip* (Fig. 19.2). Lastly, the observed synteny between *cla4* and *MAT-B* has been conserved for a very long stretch of time (~400-million-year estimated divergence between *Pneumocystis* and *Phanerochaete* [5]), despite the rapid rate at which synteny has been observed to decline in fungi (22).

Another possible hypothesis for the conserved synteny between *MAT* and other genes is that the genes form a coregulated cluster. Groups of bacterial and eukaryotic genes that display conserved gene order have been demonstrated to physically interact or be coexpressed (16, 36). The clustering of genes that interact with each other suggests that their transcription factors may be distributed heterogenously over the genome/nucleus or that the genomic region may be coregulated by general mechanisms such as methylation or histone modification (36). Genes such as *mip* and the homeodomain *MAT-A* genes may be so tightly linked that they share regulatory regions. Severing these gene linkages by inversions or translocations could disrupt proper expression of one or both genes.

A third hypothesis is that the flanking non-mating-type *MAT*-linked genes are undergoing coevolution with the *MAT* genes. In order for coevolution to occur, the genome region would have to experience recombination suppression such that alleles at one gene could be correlated with the alleles at the linked genes. In this scenario, cophylogeny of alleles at the non-mating-type *MAT*-linked genes and the *MAT* genes is expected; in other words, the gene tree for the two genes should show an identical branching order among alleles. This is observed at sex-determining loci where recombination is suppressed over large regions. For example, at the self-incompatibility locus of the angiosperm *Brassica*, the dispensable gene *SLG* shows coevolution, cophylogeny, and even gene conversion with the self-incompatibility specificity-determining gene *SRK* (71). Similarly, in the large *MAT* locus of *Cryptococcus neoformans*, cophylogeny is observed between the specificity-determining genes (e.g., *STE3*) and the genes more recently recruited to the *MAT* locus (e.g., *ZNF1* [25]). In contrast, in homobasidiomycetes, recombination appears to occur very frequently outside the actual specificity-determining genes of the *MAT* locus (41, 55). Recombination between *mip* and *MAT-Aα* in *Schizophyllum commune* is also observed such that the two loci do not show strict cophylogeny (James, unpublished). This recombination disrupts the linkage disequilibrium needed to create coevolved complexes of alleles at physically linked genes.

A fourth hypothesis is that recombination is suppressed at the *MAT* region and this has a depressing effect on the frequency of genome rearrangements as well. This does not seem to be the case, however, as recombination at the *MAT* loci appears to be normal relative to other regions of the genome (23, 49, 55). In addition, the high number of small inversions, gene duplications, and deletions observed (Fig. 19.2) (51) suggests that rearrangements are frequent near the *MAT* loci.

A final hypothesis for linkage between *MAT* and other genes is related to how the *MAT* loci originate. Hurst and Hamilton (35) hypothesized that *MAT* loci originate as a mechanism to minimize cytoplasmic gene warfare between fusing gametes. They envisioned a scenario in which the fusion of isogamous gametes creates a conflict between mates for the control of mitochondrial genome inheritance. In the first step of this three-step model, a mutant mitochondrial gene arises that can destroy the mitochondrion of a mating partner. These destroyer mitochondrial genotypes can reduce the fitness of zygotes, particularly when two destroyer genotypes fuse. But, nonetheless, the destroyer phenotype is very likely to reach fixation. In the second step, a nuclear gene arises that suppresses the destroyer phenotype of its mitochondrion, thus partially alleviating the fitness loss due to destroyer mitochondria. This "suppressor" gene can be shown to result in a stable polymorphism of suppressor and nonsuppressor alleles. In the final step, a "choosy" gene arises which allows the cell to preferentially mate with a cell of the opposite suppressor phenotype. These suppressor/nonsuppressor matings have the highest fitness if the destroyer mitochondrial genotype is at fixation. The choosy gene may show a preference for mating with a suppressor or nonsuppressor genotype, and this scenario is mechanistically the easiest to imagine. Selection will then favor tight linkage between the choosy gene and the suppressor gene and lead to the formation of a *MAT* locus and uniparental mitochondrial inheritance.

Hurst and Hamilton (35) envisioned this origin of sexes to occur when both parental nuclear and cytoplasmic genotypes were combined in a zygote, such as the fusion of mating yeast cells. The case of the *MAT-a* locus of the basidiomycete yeast *Ustilago maydis* (homologous to *MAT-B* in mushrooms and encoding pheromones and pheromone receptor genes) provides an interesting situation in which the Hurst and Hamilton model may be tested. In this species, genes that appear to be involved in mitochondrial morphology and fusion (*lga2* and *rga2*) are found in only one of the two *MAT* alleles (*a2*) of the *MAT-a* locus (8). When overexpressed, *lga2* causes both mitochondrial fragmentation and mitochondrial DNA degradation (8). In this scenario, the *lga2* and *rga2* genes are suppressors of selfish mitochondria, and the pheromone/receptor genes are the choosy genes that help cells choose the proper partner by signaling with lipopeptide pheromones. This observation provides a compelling case for how a *MAT* locus can maintain genes that do not control mating-type specificity but actually control other processes such as mitochondrial fusion and inheritance. In this example,

selection to mediate nuclear/cytoplasmic conflict could have been the reason the pheromone/receptors became a *MAT* locus or could have merely tightened the linkage between the pheromone/receptors controlling mate preference and the *lga2*/*rga2* genes controlling mitochondrial morphology.

Additional support for an interaction between *MAT* and mitochondrial inheritance in basidiomycetous yeasts comes from the data on *Cryptococcus neoformans*. In this species with two mating types (**a** and α), laboratory crosses demonstrated that progeny almost exclusively inherit mitochondria from the *MAT***a** parent (90). Furthermore, Yan et al. (89) demonstrated that the homeodomain protein SXI1α encoded by *Cryptococcus MAT* controls the process of mitochondrial inheritance by demonstrating biparental inheritance of mitochondria in *SXI1α* deletion mutants. Since mitochondria from the *MAT*α parent are rarely inherited, the *MAT*α locus may encode a "suppressor" allele or gene that is missing from *MAT***a**.

In mushrooms, the non-mating-type *MAT*-linked gene most consistently linked to *MAT* is *mip*, a gene that also functions in the mitochondrion. In fact, a large number of genes that function in the mitochondrion are also linked to one or the other *MAT* loci in basidiomycetes (Table 19.2). One hypothesis for why this could occur is that the control of the sexual cycle (by *MAT*) is coregulated with the control of mitochondrial inheritance. In *Saccharomyces cerevisiae*, deletion of *mip* causes loss of functional mitochondrial genomes as well as severe defects in respiration (9). These data lead to the speculation that *mip* could be a suppressor locus of selfish mitochondrial genomes (70) and could control inheritance in heteroplasmic cells that result from mating. Hurst and Hamilton have argued that their theory does not apply to basidiomycetes since these fungi exchange nuclei but not cytoplasm following cell fusion. It now appears that mitochondria in mushroom species may, on occasion, be biparentally inherited, and recombinant mitochondria have now been detected (4, 72). Furthermore, reconciliation of their theory and the observation of mitochondrion-targeted genes linked to *MAT* loci in mushrooms is straightforward since the tetrapolar *MAT* loci of mushrooms and basidiomycetous yeasts are homologous and likely derived from a common origin.

Conservation of linkage between *MAT* and non-mating-type *MAT*-linked genes has also been observed in Ascomycetes. The two genes *sla2* (encoding an actin-binding protein involved in cytoskeleton assembly) and *apn2* (encoding a DNA lyase) flank the *MAT* locus in a large number of species including both hemiascomycetes and euascomycetes (11, 26, 84). Remarkably, the gene *sla2* also shows tight linkage to the *U. maydis*

Table 19.2 Non-mating-type *MAT*-linked genes in basidiomycetes that encode proteins targeted to mitochondria[a]

Gene	Taxon	Linked to *MAT-A* or *MAT-B*?	Function	References(s)
ETF1	*Cryptococcus neoformans*	A + B[b]	Mitochondrial electron transport	51
Glycine dehydrogenase (*glydh*)	Homobasidiomycetes	A	Amino acid transport and metabolism	—[c]
lga2	*Ustilago maydis*; *Sporisorium reilianum*	B	Mitochondrial morphology and fusion	8, 73
Mitochondrial intermediate peptidase (*mip*)	Homobasidiomycetes	A	N-terminal processing of nuclear encoded proteins targeted to the mitochondrial matrix or inner membrane	37
Mitochondrial ribosome small subunit component (*rps19*)	*Pleurotus djamor*	B	Protein translation	—[c]
rga2	*Ustilago maydis*; *Sporisorium reilianum*	B	Mitochondrial morphology and fusion	8, 73
RP041	*Cryptococcus neoformans*	A + B	Mitochondrial RNA polymerase	51
ypl109	Homobasidiomycetes	A	Possible role in ubiquinone biosynthesis	—[c]

[a]In this table, *MAT-A* refers to the homeodomain protein encoding *MAT* region, and *MAT-B* refers to the pheromone/receptor gene locus.
[b]The *MAT* locus in this bipolar species contains homologues of both *MAT-A* and *MAT-B* genes of homobasidiomycetes.
[c]—based on subcellular predictions using the software WoLF PSORT (http://wolfpsort.seq.cbrc.jp/).

MAT-b locus encoding homeodomain proteins. A functional link between *sla2* and mating is suggested by these data but yet to be established. It is worth noting that *sla2* and *cla4* have been demonstrated to interact in yeast based on yeast two-hybrid interactions (19).

CONCLUSIONS

Genomics tries to find higher-order meaning in the organization of genes within genomes. In this review, I discussed how mating-type genes can be cloned from nonmodel species and attempted to synthesize what is known about the organization of the *MAT* loci and neighboring genomic regions in mushroom fungi. Synteny appears to be more conserved at *MAT-A* than *MAT-B*. *MAT* genes in mushrooms can be cloned by both direct PCR amplification and positional cloning using non-mating-type *MAT*-linked genes by probing a large insert genomic library. Proper determination of whether a cloned gene is a mating-type gene is nontrivial but can utilize syntenic arguments and population genetics, as well as transformation into heterologous hosts. *MAT* loci show high gene order conservation, but the significance of this observation is unclear. Among the leading candidates for the conserved gene order are coregulation and remediation of nuclear/cytoplasmic genome conflict.

Model species are indispensable for understanding *MAT* at the molecular level and provide all of the information on how *MAT* genes function. However, in order to fully unravel the marvelous mysteries of mushroom mating we will need to explore *MAT* across the phylogenetic diversity of mushrooms. Observations, such as mushroom species having whorled (multiple) clamp connections per septum in both homokaryotic and heterokaryotic mycelium (e.g., *Stereum hirsutum*) and heterothallic species lacking clamp connections altogether (e.g., *Phanerochaete chrysosporium*), give an indication that a simplified mushroom mating system with *MAT-A/B*-controlled developmental pathways may not hold for the whole group. Further, the organization of *MAT* loci in homothallic species of mushrooms is unknown. As only a fraction of the edible and medicinal mushroom species can be readily fruited, access to *MAT* loci may provide information useful for breeding, production, or genetic engineering (43). Despite the previous notion that *Schizophyllum commune* was a member of the Aphyllophorales, molecular systematics suggests that it is related to fleshy mushrooms (Agaricales), the order to which *Coprinus* sensu lato also belongs (6). Thus, the molecular knowledge of *MAT* in mushrooms is primarily limited to one order. As we move deeper towards the base of the mushroom phylogeny, it is expected that the same genes will be used in *MAT* determination; however, novel gene arrangements and functions are likely to be uncovered.

I am indebted to Ursula Kües for her countless discussions on mating type in mushrooms. I thank her and Greg Bonito for comments on the manuscript.

References

1. Aimi, T., R. Yoshida, M. Ishikawa, D. P. Bao, and Y. Kitamoto. 2005. Identification and linkage mapping of the genes for the putative homeodomain protein (*hox1*) and the putative pheromone receptor protein homologue (*rcb1*) in a bipolar basidiomycete, *Pholiota nameko*. *Curr. Genet.* 48:184–194.

2. Badrane, H., and G. May. 1999. The divergence-homogenization duality in the evolution of the *b1* mating type gene of *Coprinus cinereus*. *Mol. Biol. Evol.* 16:975–986.

3. Banuett, F. 1998. Signalling in the yeasts: an informational cascade with links to the filamentous fungi. *Microbiol. Mol. Biol. Rev.* 62:249–274.

4. Baptistaferreira, J. L. C., A. Economou, and L. A. Casselton. 1983. Mitochondrial genetics of *Coprinus*—recombination of mitochondrial genomes. *Curr. Genet.* 7:405–407.

5. Berbee, M. L., and J. W. Taylor. 1993. Dating the evolutionary radiations of the true fungi. *Can. J. Bot.* 71:1114–1127.

6. Bodensteiner, P., M. Binder, J. M. Moncalvo, R. Agerer, and D. S. Hibbett. 2004. Phylogenetic relationships of cyphelloid homobasidiomycetes. *Mol. Phylogenet. Evol.* 33:501–515.

7. Boidin, J. 1971. Nuclear behavior in the mycelium and the evolution of the Basidiomycetes, p. 129–148. *In* R. H. Petersen (ed.), *Evolution in the Higher Basidiomycetes.* The University of Tennessee Press, Knoxville.

8. Bortfeld, M., K. Auffarth, R. Kahmann, and C. W. Basse. 2004. The *Ustilago maydis a2* mating-type locus genes *lga2* and *rga2* compromise pathogenicity in the absence of the mitochondrial p32 family protein Mrb1. *Plant Cell* 16:2233–2248.

9. Branda, S. S., and G. Isaya. 1995. Prediction and identification of new natural substrates of the yeast mitochondrial intermediate peptidase. *J. Biol. Chem.* 270:27366–27373.

10. Brown, A. J., and L. A. Casselton. 2001. Mating in mushrooms: increasing the chances but prolonging the affair. *Trends Genet.* 17:393–400.

11. Butler, G., C. Kenny, A. Fagan, C. Kurischko, C. Gaillardin, and K. H. Wolfe. 2004. Evolution of the *MAT* locus and its Ho endonuclease in yeast species. *Proc. Natl. Acad. Sci. USA* 101:1632–1637.

12. Casselton, L. A. 2002. Mate recognition in fungi. *Heredity* 88:142–147.

13. Casselton, L. A., R. N. Asante-Owusu, A. H. Banham, C. S. Kingsnorth, U. Kües, S. F. O'Shea, and E. H. Pardo. 1995. Mating type of sexual development in *Coprinus cinereus*. *Can. J. Bot.* 73:S266–S272.

14. Casselton, L. A., and N. S. Olesnicky. 1998. Molecular genetics of mating recognition in basidiomycete fungi. *Microbiol. Mol. Biol. Rev.* 62:55–70.

15. Cvrčková, F., C. De Virgilio, D. Manser, J. R. Pringle, and K. Nasmyth. 1995. Ste20-like protein kinases are required for normal localization of cell growth and for cytokinesis in budding yeast. *Genes Dev.* 9:1817–1830.

16. Dandekar, T., B. Snel, M. Huynen, and P. Bork. 1998. Conservation of gene order: a fingerprint of proteins that physically interact. *Trends Biochem. Sci.* 23:324–328.

17. Davidson, R. C., C. B. Nicholls, G. M. Cox, J. R. Perfect, and J. Heitman. 2003. A MAP kinase cascade composed of cell type specific and non-specific elements controls mating and differentiation of the fungal pathogen *Cryptococcus neoformans*. *Mol. Microbiol.* 49:469–485.

18. Day, P. R. 1960. The structure of the *A* mating-type locus in *Coprinus lagopus*. *Genetics* 45:641–651.

19. Drees, B. L., B. Sundin, E. Brazeau, J. P. Caviston, G. C. Chen, W. Guo, K. G. Kozminski, M. W. Lau, J. J. Moskow, A. Tong, L. R. Schenkman, A. McKenzie, P. Brennwald, M. Longtine, E. Bi, C. Chan, P. Novick, C. Boone, J. R. Pringle, T. N. Davis, S. Fields, and D. G. Drubin. 2001. A protein interaction map for cell polarity development. *J. Cell Biol.* 154:549–571.

20. Eichler, E. E., and D. Sankoff. 2003. Structural dynamics of eukaryotic chromosome evolution. *Science* 301:793–797.

21. Ferreira, A. V. B., S. Saupe, and N. L. Glass. 1996. Transcriptional analysis of the mtA idiomorph of *Neurospora crassa* identifies two genes in addition to mtA-1. *Mol. Gen. Genet.* 250:767–774.

22. Fischer, G., E. P. C. Rocha, F. Brunet, M. Vergassola, and B. Dujon. 2006. Highly variable rates of genome rearrangements between hemiascomycetous yeast lineages. *PLoS Genet.* 2:e32.

23. Fowler, T. J., M. F. Mitton, E. I. Rees, and C. A. Raper. 2004. Crossing the boundary between the *Bα* and *Bβ* mating-type loci in *Schizophyllum commune*. *Fungal Genet. Biol.* 41:89–101.

24. Fowler, T. J., M. F. Mitton, L. J. Vaillancourt, and C. A. Raper. 2001. Changes in mate recognition through alterations of pheromones and receptors in the multisexual mushroom fungus *Schizophyllum commune*. *Genetics* 158:1491–1503.

25. Fraser, J. A., S. Diezmann, R. L. Subaran, A. Allen, K. B. Lengeler, F. S. Dietrich, and J. Heitman. 2004. Convergent evolution of chromosomal sex-determining regions in the animal and fungal kingdoms. *PLoS Biol.* 2:2243–2255.

26. Fraser, J. A., and J. Heitman. 2006. Sex, *MAT*, and the evolution of fungal virulence, p. 13–33. *In* J. Heitman, S. G. Filler, J. E. Edwards, Jr., and A. P. Mitchell (ed.), *Molecular Principles of Fungal Pathogenesis.* ASM Press, Washington, DC.

27. Giasson, L., C. A. Specht, C. Milgrim, C. P. Novotny, and R. C. Ullrich. 1989. Cloning and comparison of *Aα* mating-type alleles of the basidiomycete *Schizophyllum commune*. *Mol. Gen. Genet.* 218:72–77.

28. Grützner, F., H. R. Crollius, G. Lütjens, O. Jaillon, J. Weissenbach, H. H. Ropers, and T. Haaf. 2002. Four-

hundred million years of conserved synteny of human Xp and Xq genes on three *Tetraodon* chromosomes. *Genome Res.* **12:**1316–1322.

29. **Halsall, J. R., M. J. Milner, and L. A. Casselton.** 2000. Three subfamilies of pheromone and receptor genes generate multiple *B* mating specificities in the mushroom *Coprinus cinereus. Genetics* **154:**1115–1123.

30. **Hibbett, D. S., and M. Binder.** 2002. Evolution of complex fruiting-body morphologies in homobasidiomycetes. *Proc. R. Soc. Lond. B* **269:**1963–1969.

31. **Hibbett, D. S., and M. J. Donoghue.** 2001. Analysis of character correlations among wood decay mechanisms, mating systems, and substrate ranges in homobasidiomycetes. *Syst. Biol.* **50:**215–242.

32. **Hiscock, S. J., and U. Kües.** 1999. Cellular and molecular mechanisms of sexual incompatibility in plants and fungi. *Int. Rev. Cytol.* **193:**165–295.

33. **Hogberg, P., A. Nordgren, N. Buchmann, A. F. S. Taylor, A. Ekblad, M. N. Hogberg, G. Nyberg, M. Ottosson-Lofvenius, and D. J. Read.** 2001. Large-scale forest girdling shows that current photosynthesis drives soil respiration. *Nature* **411:**789–792.

34. **Hopple, J. S., Jr., and R. Vilgalys.** 1999. Phylogenetic relationships in the mushroom genus *Coprinus* and dark-spored allies based on sequence data from the nuclear gene coding for the large ribosomal subunit RNA: divergent domains, outgroups, and monophyly. *Mol. Phylogenet. Evol.* **13:**1–19.

35. **Hurst, L. D., and W. D. Hamilton.** 1992. Cytoplasmic fusion and the nature of sexes. *Proc. R. Soc. Lond. B* **247:** 189–194.

36. **Hurst, L. D., C. Pal, and M. J. Lercher.** 2004. The evolutionary dynamics of eukaryotic gene order. *Nat. Rev. Genet.* **5:**299–310.

37. **Isaya, G., W. R. Sakati, R. A. Rollins, G. P. Shen, L. C. Hanson, R. C. Ullrich, and C. P. Novotny.** 1995. Mammalian mitochondrial intermediate peptidase—structure-function analysis of a new homolog from *Schizophyllum commune* and relationship to thimet oligopeptidases. *Genomics* **28:**450–461.

38. **James, T. Y.** 2003. *The Evolution of Mating-Type Genes in the Mushroom Fungi (Homobasidiomycetes).* Duke University, Durham, NC.

39. **James, T. Y., U. Kües, S. A. Rehner, and R. Vilgalys.** 2004. Evolution of the gene encoding mitochondrial intermediate peptidase and its cosegregation with the *A* mating-type locus of mushroom fungi. *Fungal Genet. Biol.* **41:**381–390.

40. **James, T. Y., S. R. Liou, and R. Vilgalys.** 2004. The genetic structure and diversity of the *A* and *B* mating-type genes from the tropical oyster mushroom, *Pleurotus djamor. Fungal Genet. Biol.* **41:**813–825.

41. **James, T. Y., P. Srivilai, U. Kües, and R. Vilgalys.** 2006. Evolution of the bipolar mating system of the mushroom *Coprinellus disseminatus* from its tetrapolar ancestors involves loss of mating-type-specific pheromone receptor function. *Genetics* **172:**1877–1891.

42. **Kirk, P. M., P. F. Cannon, J. C. David, and J. A. Stalpers (ed.).** 2001. *Ainsworth & Bisby's Dictionary of the Fungi,* 9th ed. CAB International, Wallingford, United Kingdom.

43. **Kothe, E.** 2001. Mating-type genes for basidiomycete strain improvement in mushroom farming. *Appl. Microbiol. Biotechnol.* **56:**602–612.

44. **Kües, U.** 2000. Life history and developmental processes in the basidiomycete *Coprinus cinereus. Microbiol. Mol. Biol. Rev.* **64:**316–353.

45. **Kües, U., and L. A. Casselton.** 1993. The origin of multiple mating types in mushrooms. *J. Cell Sci.* **104:**227–230.

46. **Kües, U., B. Göttgens, R. Stratmann, W. V. J. Richardson, S. F. O'Shea, and L. A. Casselton.** 1994. A chimeric homeodomain protein causes self-compatibility and constitutive sexual development in the mushroom *Coprinus cinereus. EMBO J.* **13:**4054–4059.

47. **Kües, U., T. Y. James, R. Vilgalys, and M. P. Challen.** 2001. The chromosomal region containing *pab-1, mip,* and the *A* mating type locus of the secondarily homothallic homobasidiomycete *Coprinus bilanatus. Curr. Genet.* **39:**16–24.

48. **Kües, U., A. M. Tymon, W. V. J. Richardson, G. May, P. T. Gieser, and L. A. Casselton.** 1994. *A* mating-type factors of *Coprinus cinereus* have variable numbers of specificity genes encoding two classes of homeodomain proteins. *Mol. Gen. Genet.* **245:**45–52.

49. **Larraya, L. M., G. Perez, E. Ritter, A. G. Pisabarro, and L. Ramirez.** 2000. Genetic linkage map of the edible basidiomycete *Pleurotus ostreatus. Appl. Environ. Microbiol.* **66:**5290–5300.

50. **Lee, N., G. Bakkeren, K. Wong, J. E. Sherwood, and J. W. Kronstad.** 1999. The mating-type and pathogenicity locus of the fungus *Ustilago hordei* spans a 500-kb region. *Proc. Natl. Acad. Sci. USA* **96:**15026–15031.

51. **Lengeler, K. B., D. S. Fox, J. A. Fraser, A. Allen, K. Forrester, F. Dietrich, and J. Heitman.** 2002. The mating type locus of *Cryptococcus neoformans:* a step in the evolution of sex chromosomes. *Eukaryot. Cell* **1:**704–718.

52. **Lengeler, K. B., P. Wang, G. M. Cox, J. R. Perfect, and J. Heitman.** 2000. Identification of the MATa mating-type locus of *Cryptococcus neoformans* reveals a serotype A MATa strain thought to have been extinct. *Proc. Natl. Acad. Sci. USA* **97:**14455–14460.

53. **Leveleki, L., M. Mahlert, B. Sandrock, and M. Bölker.** 2004. The PAK family kinase Cla4 is required for budding and morphogenesis in *Ustilago maydis. Mol. Microbiol.* **54:**396–406.

54. **Liu, Y.-G., and R. F. Whittier.** 1995. Thermal asymmetric interlaced PCR: automatable amplification and sequencing of insert end fragments from P1 and YAC clones for chromosome walking. *Genomics* **25:**674–681.

55. **Lukens, L., H. Yicun, and G. May.** 1996. Correlation of genetic and physical maps at the *A* mating-type locus of *Coprinus cinereus. Genetics* **144:**1471–1477.

56. **Marion, A. L., K. A. Bartholomew, J. Wu, H. L. Yang, C. P. Novotny, and R. C. Ullrich.** 1996. The Aα mating-type locus of *Schizophyllum commune:* structure and function of gene X. *Curr. Genet.* **29:**143–149.

57. **Martinez, D., L. F. Larrondo, N. Putnam, M. D. S. Gelpke, K. Huang, J. Chapman, K. G. Helfenbein, P. Ramaiya, J. C.**

Detter, F. Larimer, P. M. Coutinho, B. Henrissat, R. Berka, D. Cullen, and D. Rokhsar. 2004. Genome sequence of the lignocellulose degrading fungus *Phanerochaete chrysosporium* strain RP78. *Nat. Biotechnol.* **22**:695–700.

58. May, G., F. Shaw, H. Badrane, and X. Vekemans. 1999. The signature of balancing selection: fungal mating compatibility gene evolution. *Proc. Natl. Acad. Sci. USA* **96**:9172–9177.

59. Moncalvo, J.-M., R. Vilgalys, S. A. Redhead, J. E. Johnson, T. Y. James, M. C. Aime, V. Hofstetter, S. Verduin, E. Larsen, T. J. Baroni, R. G. Thorn, S. Jacobsson, H. Clémençon, and O. K. Miller, Jr. 2002. One hundred and seventeen clades of euagarics. *Mol. Phylogenet. Evol.* **23**:357–400.

60. Murphy, J. F., and O. K. Miller, Jr. 1997. Diversity and local distribution of mating alleles in *Marasmiellus praeacutus* and *Collybia subnuda* (Basidiomycetes, Agaricales). *Can. J. Bot.* **75**:8–17.

61. O'Brien, H. E., J. L. Parrent, J. A. Jackson, J. M. Moncalvo, and R. Vilgalys. 2005. Fungal community analysis by large-scale sequencing of environmental samples. *Appl. Environ. Microb.* **71**:5544–5550.

62. Ochman, H., A. S. Gerber, and D. L. Hartl. 1988. Genetic applications of an inverse polymerase chain reaction. *Genetics* **120**:621–623.

63. Olesnicky, N. S., A. J. Brown, Y. Honda, S. L. Dyos, S. J. Dowell, and L. A. Casselton. 2000. Self-compatible *B* mutants in *Coprinus* with altered pheromone-receptor specificities. *Genetics* **156**:1025–1033.

64. O'Shea, S. F., P. T. Chaure, J. R. Halsall, N. S. Olesnicky, A. Leibbrandt, I. F. Connerton, and L. A. Casselton. 1998. A large pheromone and receptor gene complex determines multiple *B* mating type specificities in *Coprinus cinereus. Genetics* **148**:1081–1090.

65. Ramer, S. W., and R. W. Davis. 1993. A dominant truncation allele identifies a gene, *STE20*, that encodes a putative protein kinase necessary for mating in *Saccharomyces cerevisiae. Proc. Natl. Acad. Sci. USA* **90**:452–456.

66. Raper, J. R. 1966. *Genetics of Sexuality in Higher Fungi.* Ronald Press, New York, NY.

67. Raper, J. R., M. G. Baxter, and A. H. Ellingboe. 1960. The genetic structure of the incompatibility factors of *Schizophyllum commune*: the *A* factor. *Proc. Natl. Acad. Sci. USA* **46**:833–842.

68. Raper, J. R., and A. S. Flexer. 1971. Mating systems and evolution of the Basidiomycetes, p. 149–167. *In* R. H. Petersen (ed.), *Evolution in the Higher Basidiomycetes.* University of Tennessee Press, Knoxville.

69. Riquelme, M., M. P. Challen, L. A. Casselton, and A. J. Brown. 2005. The origin of multiple *B* mating specificities in *Coprinus cinereus. Genetics* **170**:1105–1119.

70. Röhr, H., U. Kües, and U. Stahl. 1998. Organelle DNA of plants and fungi: inheritance and recombination. *Prog. Bot.* **60**:39–87.

71. Sato, K., T. Nishio, R. Kimura, M. Kusaba, T. Suzuki, K. Hatakeyama, D. J. Ockendon, and Y. Satta. 2002. Coevolution of the *S*-locus genes *SRK*, *SLG* and *SP11/SCR* in *Brassica oleracea* and *B. rapa. Genetics* **162**:931–940.

72. Saville, B. J., Y. Kohli, and J. B. Anderson. 1998. mtDNA recombination in a natural population. *Proc. Natl. Acad. Sci. USA* **95**:1331–1335.

73. Schirawski, J., B. Heinze, M. Wagenknecht, and R. Kahmann. 2005. Mating type loci of *Sporisorium reilianum*: novel pattern with three *a* and multiple *b* specificities. *Eukaryot. Cell* **4**:1317–1327.

74. Simchen, G. 1967. Genetic control of recombination and the incompatibility system in *Schizophyllum commune. Genet. Res.* **9**:195–210.

75. Smulian, A. G., T. Sesterhenn, R. Tanaka, and M. T. Cushion. 2001. The *ste3* pheromone receptor gene of *Pneumocystis carinii* is surrounded by a cluster of signal transduction genes. *Genetics* **157**:991–1002.

76. Specht, C. A. 1996. Isolation of the *Bα* and *Bβ* mating-type loci of *Schizophyllum commune. Curr. Genet.* **28**:374–379.

77. Specht, C. A., M. M. Stankis, C. P. Novotny, and R. C. Ullrich. 1994. Mapping the heterogeneous DNA region that determines the nine *Aα* mating-type specificities of *Schizophyllum commune. Genetics* **137**:709–714.

78. Stankis, M. M., C. A. Specht, H. Yang, L. Giasson, R. C. Ullrich, and C. P. Novotny. 1992. The *Aα* mating locus of *Schizophyllum commune* encodes two dissimilar multiallelic homeodomain proteins. *Proc. Natl. Acad. Sci. USA* **89**:7169–7173.

79. Stoll, M., D. Begerow, and F. Oberwinkler. 2005. Molecular phylogeny of *Ustilago*, *Sporisorium*, and related taxa based on combined analyses of rDNA sequences. *Mycol. Res.* **109**:342–356.

80. Suyama, M., and P. Bork. 2001. Evolution of prokaryotic gene order: genome rearrangements in closely related species. *Trends Genet.* **17**:10–13.

81. Turgeon, B. G., H. Bohlmann, L. M. Ciuffetti, S. K. Christiansen, G. Yang, W. Schafer, and O. C. Yoder. 1993. Cloning and analysis of the mating-type genes from *Cochliobolus heterostrophus. Mol. Gen. Genet.* **238**:270–284.

82. Ullrich, R. C., and J. R. Raper. 1974. Number and distribution of bipolar incompatibility factors in *Sistotrema brinkmannii. Am. Nat.* **108**:506–518.

83. Urban, M., R. Kahmann, and M. Bölker. 1996. The biallelic *a* mating type locus of *Ustilago maydis*: remnants of an additional pheromone gene indicate evolution from a multiallelic ancestor. *Mol. Gen. Genet.* **250**:414–420.

84. Waalwijk, C., T. van der Lee, I. de Vries, T. Hesselink, J. Arts, and G. H. J. Kema. 2004. Synteny in toxigenic *Fusarium* species: the fumonisin gene cluster and the mating type region as examples. *Eur. J. Plant Pathol.* **110**:533–544.

85. Walser, P. J., M. Hollenstein, M. J. Klaus, and U. Kües. 2001. Genetic analysis of basidiomycete fungi, p. 59–90. *In* N. J. Talbot (ed.), *Molecular and Cell Biology of Filamentous Fungi: a Practical Approach.* Oxford University Press, Oxford, United Kingdom.

86. Wang, P., C. B. Nichols, M. B. Lengeler, M. E. Cardenas, G. M. Cox, J. R. Perfect, and J. Heitman. 2002. Mating-type-specific and nonspecific PAK kinases play shared and divergent roles in *Cryptococcus neoformans. Eukaryot. Cell* **1**:257–272.

87. Wendland, J., L. J. Vaillancourt, J. Hegner, K. B. Lengeler, K. J. Laddison, C. A. Specht, C. A. Raper, and E. Kothe. 1995. The mating-type locus *Bα1* of *Schizophyllum commune* contains a pheromone receptor gene and putative pheromone genes. *EMBO J.* **14:**5271–5278.

88. Whitehouse, H. L. K. 1949. Multiple allelomorph heterothallism in the fungi. *New Phytol.* **48:**212–244.

89. Yan, Z., C. M. Hull, J. Heitman, S. Sun, and J. P. Xu. 2004. *SXI1α* controls uniparental mitochondrial inheritance in *Cryptococcus neoformans*. *Curr. Biol.* **14:**R743–R744.

90. Yan, Z., and J. P. Xu. 2003. Mitochondria are inherited from the MATa parent in crosses of the basidiomycete fungus *Cryptococcus neoformans*. *Genetics* **163:**1315–1325.

91. Yun, S.-H., M. L. Berbee, O. C. Yoder, and B. G. Turgeon. 1999. Evolution of the fungal self-fertile reproductive life style from self-sterile ancestors. *Proc. Natl. Acad. Sci. USA* **96:**5592–5597.

Sex in Fungi: Molecular Determination
and Evolutionary Implications
Edited by Joseph Heitman et al.
© 2007 ASM Press, Washington, D.C.

James B. Anderson
Linda M. Kohn

20

Dikaryons, Diploids, and Evolution

The regular association of unfused, haploid, gametic-type nuclei within the dikaryon is a striking outcome of evolution that is unique to the fungi. Many fungi with normally dikaryotic mycelia can also exist as diploids (12, 32, 44, 55, 59). What are the genetic and evolutionary implications of dikaryosis and diploidy? Remarkably, the majority of the genetic inferences about dikaryons, diploids, and their interactions reviewed by John Raper in *Genetics of Sexuality in Higher Fungi* in 1966 (66) remain essentially unchanged to the present. While the genetics of dikaryons and diploids have been established for more than 40 years, the evolutionary implications are only now becoming accessible through experiments. The purpose of this review is to examine the immediate and long-term consequences of dikaryosis, with comparison to diploidy. In keeping with the theme of Raper's book, this review focuses on fungi with prolonged dikaryotic phases, mainly the Hymenomycetes (nomenclature of higher taxa follows the Tree of Life Project, http://tolweb.org/Fungi). The four interrelated conclusions of this review are that (i) the known dynamics of nuclear migration and dikaryon formation suggest that mating in nature is asymmetric for male

and female function, (ii) the transmissions of nuclear and mitochondrial genomes follow different rules, (iii) dikaryons produce recombinant genotypes without fruiting, and (iv) dikaryons and diploids carry different expectations for evolution.

THE DIKARYON: MALES, FEMALES, NUCLEAR MIGRATION, AND ASYMMETRY IN MATING

The Dikaryon in the Ascomycota and Basidiomycota

In most eukaryotes, including plants and animals, plasmogamy and karyogamy occur in rapid succession to produce a diploid zygote nucleus. In the Ascomycota and Basidiomycota, plasmogamy and karyogamy are separated in time and the dikaryon occupies the extended period after plasmogamy and before karyogamy. Within the cells of the dikaryotic mycelium, a pair of haploid nuclei of the original, gametic genotypes physically associate with one another and divide synchronously through an indeterminate number of cell cycles. In most species of Basidiomycota, the dikaryon constitutes

James B. Anderson and Linda M. Kohn, Department of Ecology and Evolutionary Biology, University of Toronto, Mississauga, Ontario L5L 1C6, Canada.

the extended vegetative phase, while in the Ascomycota, the dikaryon is usually restricted to the ascogenous system within fruiting bodies. In the Basidiomycota, the dikaryon may or may not be associated with clamp connections, and in the Ascomycota, the dikaryon in the ascogenous system of the fruiting body may or may not be accompanied by croziers, which are likely homologous to clamp connections. While the dikaryons of the Ascomycota and Basidiomycota appear to be fundamentally similar in their coordinated nuclear division, the dikaryon of the Basidiomycota has been studied in more genetic detail because it is free-living and more amenable to experimentation.

In the Hymenomycetes, there is considerable variation on the theme of dikaryosis. For example, in *Schizophyllum commune* and *Coprinus cinereus*, the cells of dikaryotic mycelia are regularly binucleate with clamp connections. In *Heterobasidion annosum*, the cells are highly multinucleate and clamp connections are present only on some of the hyphae (13, 47), while in *Agaricus bisporus* the cells are also highly multinucleate, but with no clamp connections (65). Regardless of these morphological variations, the intricate association of two haploid nuclei of gametic genotypes remains constant.

Within the dikaryon there is molecular communication between the paired nuclei. The distance between nuclei has major consequences for gene expression (73, 87), especially for the production of hydrophobins, secreted proteins that have a strong effect on whether hyphae grow into the air or remain immersed in a hydrated substrate or host. These consequences are most likely mediated by the local deployment of *B* mating-type pheromone receptors on the cell surface nearest the nucleus within which they are encoded, as well as by the extent of binding with pheromones of the compatible *B* mating type produced by the other nucleus of the pair (17, 73, 87). There is coordination between the *A* and *B* mating-type genes, but many of the details of this cross talk and the regulation of downstream genes have yet to be clarified (9).

Mating and Development

Monokaryons stand ready to interact with a potential mate. In tetrapolar species, a dikaryon results whenever both the *A* and *B* mating types are compatible and their respective developmental pathways are both activated. Hyphal anastomosis within and between confronted mycelia of the same species is constitutive and not regulated by mating type. Recognition of sexual compatibility occurs only after hyphal fusion. In monokaryons, G-protein-coupled pheromone receptors encoded by the *B* mating-type genes are activated by binding with a

peptide pheromone from a compatible *B* mating type (9). G-protein-coupled pheromone receptors are of central importance in the early stages of mating in all higher fungi examined to date, and proteins in this family have a variety of important signaling functions in all eukaryotes (48).

When the *B* mating types are compatible, a monokaryon receives an incoming nucleus that divides and migrates through the existing hyphae. The long-range migration of nuclei is guided by microtubule tracks (68). The motive force for nuclear movement is most likely by dynein motor proteins, which are associated with the spindle pole body so that nuclei are pulled toward the minus ends of microtubules. This migration, along with dissolution of septa as migrating nuclei pass through, is controlled by the *B* mating types, but the exact nature of that control is unknown.

After migrating nuclei reach the hyphal tips, nuclear pairing and hook cell formation are controlled by the *A* mating-type genes, which encode pairs of homeodomain-proteins that activate transcription within the dikaryon when the proteins of a heterodimeric pair are derived from different mating types. Interestingly, fusion of the hook cell with the subapical peg that develops on the subapical cell is again controlled by the *B* mating-type genes; this is the cell fusion in the Basidiomycota that most resembles cell fusion in *Saccharomyces* (6), which is similarly guided by the interaction between mating-type pheromones and receptors. Just before fusion, the hook cell and the subapical cell are uninucleate and of opposite mating types. In the dikaryotic hypha of *C. cinereus* the positions of the two gametic-type nuclei switch regularly in the apical cell (37). This phenomenon, which may be a general feature of dikaryons, has also been observed in a dikaryon from a diploid-by-haploid mating of *Cryptococcus neoformans* (52).

Male and Female Functions

In both the Basidiomycota and the Ascomycota, a mycelium may act simultaneously as male and as female. In the Ascomycota, female gametangia are well differentiated within fruiting bodies and the males may be conidia, microconidia, or even a vegetative hypha. In the Hymenomycetes, there are also distinct male and female roles in fertilization, but there are no morphologically differentiated gametes or gametangia. The acceptance of fertilizing nuclei that migrate through existing mycelium can be regarded as a female role. Only the cytoplasm of those portions of monokaryons acting as females is transmitted to the new dikaryon. In contrast, the male contributes little or no cytoplasm to the dikaryon. On the male side, the source of the fertilizing nucleus in the

Hymenomycetes can be a germinating spore, another haploid monokaryon, a dikaryon, or even a diploid of the same species. The details of how exactly an "extra" nucleus of a dikaryon becomes available for fertilizing a monokaryon are not known.

How Mating Actually Happens; Asymmetry in Nature

The avidity of single nuclei in monokaryons for associations with nuclei of compatible mating types is convincingly illustrated by "spore trapping." To sample fertilizing nuclear genotypes, a petri dish with a haploid monokaryon is opened and placed outdoors for one or more days (2). In many species, such as *Pleurotus* species (86) and *S. commune* (39), the resident mycelium is almost invariably dikaryotized. Although the actual fertilization event in spore trapping, or anywhere else in nature, has never been witnessed directly, the simplest inference is that spores readily settle out from the air column, germinate, and then fuse with the resident monokaryon. Even monokaryotic mycelia placed outside the normal geographic range of the species can be fertilized, illustrating the ability of the vast numbers of fungal spores to disperse and potentially find an appropriate mate (33). A reasonable hypothesis is that most mating in nature is also asymmetric; spores germinate to form small monokaryotic colonies, which are soon fertilized by spores. If this hypothesis holds true, then the effective sizes of the two mates in nature are usually unequal, in contrast to the majority of experimental matings where the sizes of the two mates are often made equal.

TRANSMISSION OF NUCLEAR AND MITOCHONDRIAL GENOMES FOLLOWS DIFFERENT RULES

Nuclei Move Rapidly in Matings; Mitochondria Do Not

Where two equally sized monokaryotic mycelia are allowed to mate, each mycelium usually acts simultaneously as male and female, both donating and accepting incoming nuclei that subsequently migrate and proliferate (Fig. 20.1). Mitochondria do not accompany the rapid and long-range movement of nuclei (7, 34, 73, 78). Symmetric pairings of haploid monokaryons typical of laboratory matings therefore result in a mycelium of uniform dikaryotic genotype that is mosaic for mitochondrial type. In the zone where the two mycelia meet, there is limited cytoplasmic contact. Here, the often tubular mitochondria may physically mix and anastomose with one another (8). The mitochondrial genomes

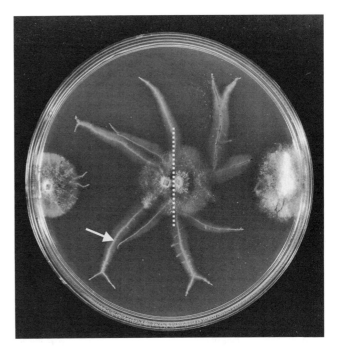

Figure 20.1 Mating of haploid strains of *Armillaria gallica*. Two mates were inoculated alone on either side; the pairing appears in the middle. The dotted line indicates the region of initial contact between mycelia, where fusion of hyphae from the different mates occurs. Nuclear migration proceeded bidirectionally, with each mate functioning both as donor (male) and recipient of nuclei (female). While the nuclei migrate rapidly (i.e., on the order of 10-fold faster than the mycelial growth rate), the mitochondria do not and the final mated colony is mosaic for parental mtDNA types. Heteroplasmy for mtDNA is restricted to those cells resulting from fusion of hyphae of the two mates near the center (dashed line). The rhizomorphs (arrow) carry the mtDNA type from the area of the mated colony from which they arose. In *A. gallica*, the initial dikaryon is established after nuclear migration becomes diploid, but the dynamics of nuclear migration and mitochondrial inheritance are otherwise typical of Homobasidiomycetes. Note that the colony morphology of the diploid is different from that of the mates; it has less aerial mycelium.

recombine and produce nonparental genotypes when mitochondrial DNAs (mtDNAs) of different descent are in close proximity (70). After cytoplasmic mixing, the different mtDNA genotypes rapidly sort to pure types in the growing dikaryon. The sorting of mtDNAs when heteroplasmic cells proliferate is not always random. In *C. neoformans*, mtDNA transmission from newly mated cells is uniparental and from the *MATa* parent (90).

In nature, the population structure of at least one species, *Armillaria gallica*, is highly recombined for mtDNA (70). Although the actual rate of recombination per nucleotide interval per generation may be very low, this apparently has been enough over time to erase most linkage disequilibrium in the mtDNA genome that may have existed in the past. For nuclear genomes, the population structure of basidiomycetes within large geographical areas approximates panmixia (38, 71), reflecting the high levels of outbreeding expected from sexual incompatibility systems and from the known dispersal ability of basidiospores (71). Since the extent of cytoplasmic mixing in matings is small, most of the resulting dikaryon has only one or the other parental mtDNA type from the beginning. Because matings in nature carry a small probability that the region of cytoplasmic mixing will be the one to proliferate and become fixed in the mycelium, a minority of resulting vegetative individuals will have a recombinant mtDNA. In addition to the homologous recombination in mtDNA, small selfish elements of the genome, such as the omega element in yeast (28, 30), may also be spread unilaterally; these elements have been described in the Ascomycota and may well also exist in the Basidiomycota.

The fundamental nature of mitochondrial genetics in fungi is more akin to population genetics of bacteria and viruses than conventional Mendelian genetics. mtDNAs exist as moderately sized populations, on the order of magnitude of 10^2 molecules per cell. Multiple rounds of exchange in heteroplasmic cells may occur, and the individual exchange events may be reciprocal or nonreciprocal. Unfortunately, there is as yet no means for isolating the immediate products of any mtDNA exchange event in any fungus, especially in filamentous fungi.

Are Individuals mtDNA Mosaics?

Although laboratory matings of monokaryotic mycelia of different mtDNA types establish a dikaryon or diploid that is mosaic for mtDNA type, more than one mtDNA type has not been detected within naturally existing individuals (71), even where mitochondrial mosaics of the same species have been demonstrated in the laboratory (Fig. 20.2). If the sizes of the mating individuals are strongly asymmetric in nature, as when a small monokaryotic colony is fertilized by a spore, then one or the other parental mtDNA type is more likely to predominate and the other type is either lost or not detected. Only the mtDNA of the monokaryon that acted as a female is transmitted.

Slow Movement of Mitochondria in Matings

Although mitochondria do not migrate rapidly (i.e., faster than the mycelial growth rate) along with nuclei in compatible matings of monokaryons, mitochondria may well move more slowly and gradually displace resident mitochondrial types (25). These outcomes are analogous to suppressive mtDNA types in *Saccharomyces cerevisiae*, where one mtDNA type may dis-

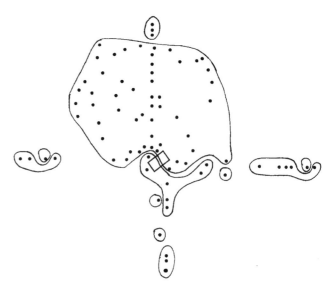

Figure 20.2 Unpublished spatial map of genetic individuals of *Armillaria gallica* in a mixed hardwood forest of Michigan. The rectangle (130 m in length) in the middle is a clear-cut site that was replanted with red pine and was the subject of intensive sampling of *Armillaria* in individuals (75–77). The dots represent collection points, and the lines encircle collections with identical multilocus genotypes. Each individual has a unique mtDNA type, such that the samples of each individual have only a single mtDNA. No mosaicism for mtDNA that may have been present in the initial mating has been detected in these or any other individuals of *A. gallica* from which multiple collections were made (67, 68).

place another over several rounds of budding due to a replicative advantage (18, 24).

What Is Good for a Mitochondrial Genome Can Be Bad for the Nuclear Genome

That nuclear and mitochondrial inheritance is uncoupled and that mated mycelia may be mosaic for mtDNA type raise the possibility of genomic conflict; namely, that the two original mtDNA types are in direct competition at the level of the entire dikaryon from the mating (1). With genomic conflict, enhanced transmission of one mtDNA type is always at the expense of the other. Further, enhanced transmission of one mtDNA type is at the expense of the fitness of the whole dikaryon; the individual mtDNA benefits, but the dikaryon suffers. The reason for the conflict is that the transmission of a mitochondrial genome is tied only to the female function, while the transmission of nuclear genes is tied to both the male and the female functions. If, for example, a mtDNA carries total male sterility, then the reproductive output of the entire mated colony may be greatly reduced because the unfertilized part of the colony will not make any contribution to the gene pool.

Such genomic conflict has been proposed by Aanen et al. (1) as an explanation for unilateral migration seen in many pairings of monokaryons from the wild. The model of Aanen et al. includes male-sterile and male-fertile mtDNA types, plus another set of nuclear determinants specifying either resistance or susceptibility to the cytoplasmic male sterility. In a resistant monokaryon, mtDNA-based male sterility is nullified and nuclear migration occurs regardless of whether the opposing mate is male sterile or male fertile. In a susceptible monokaryon, nuclear migration occurs only from male-fertile, and not from male-sterile, mates. This model explains to a remarkable extent the patterns of bilateral migration, unilateral migration, and no migration among mated monokaryons in the *Hebeloma crustuliniforme* complex (1).

Possible Mechanisms of Male Sterility

A possible mechanism for the male sterility is that mtDNA mutations somehow block either the mating pheromone receptors of the opposing mate or the production of the pheromones of the male-sterile strain. Alteration of the timing or efficiency of fruiting could also alter the proportions of spores with the two mtDNA types. Regardless of the mechanism of male sterility, an important part of the model is strong selection on the nuclear genome for resistance mutations that nullify male sterility (1).

At first glance, this model of genomic conflict leading to male sterility would not seem to be applicable to fungi like *S. commune*, in which the distribution of strains defective in nuclear migration appears to be different from that in the *H. crustuliniforme* complex. In *S. commune* many monokaryons from the wild are able to donate, but not to receive, nuclei regardless of their partner (66). Therefore, these strains are highly male fertile, but completely female sterile. This is not surprising, as the female function is in general more complex and presents a larger mutational target than the male function. The female function is therefore more easily lost through deleterious mutation than the male function (51). (For an excellent treatment of loss of female function in Ascomycota, see reference 53.) Before the cytoplasmic male sterility model is discounted, however, a more detailed analysis of the quantitative aspects of male function is needed even for fungi such as *S. commune*. Whether or not male sterility appears may depend on population history and whether selection has had enough time to act so that male sterility and the corresponding nuclear resistance type reach detectable frequencies.

One factor working against the spread of male sterility would be if all mating in nature were highly asymmetric to begin with; this would be the case whenever established monokaryons acting as the female were fertilized by spores acting as the male. Here male sterility would block all gene transmission, a dead end for that spore. Alternatively, if mating in nature occurred predominantly via the Buller phenomenon, then male sterility would block the ability of the dikaryon to fertilize monokaryons. Here there would be no benefit to the mtDNA, but there would be a cost to the nuclear genome of the dikaryon. Cytoplasmic male sterility would not evolve under these conditions.

DIKARYONS PRODUCE RECOMBINANT GENOTYPES WITHOUT FRUTING

Escape from Dikaryons and Reassociation of Nuclear Types

The association between the paired nuclei in the dikaryon is not absolute. Single nuclei may escape this association either through the production of uninucleate mitospores (oidia) as in *Coprinus cinereus* or through sectors of mycelium that contain only one genotype of haploid nucleus. In *Heterobasidion annosum* the association of paired nuclei is not universal in the mycelium, and monokaryotic sectors are common. Even in species where the dikaryotic association covers every cell, single nuclear genotypes might escape as monokaryons, especially when the nuclei are far apart. This raises the possibility that nuclei from different dikaryons can form new associations (42, 84). *H. annosum* commonly occurs on cut stump surfaces where many dikaryons may encounter one another. Here, new dikaryons are commonly formed by reassociation of nuclei formerly in different dikaryons (40). There are limits on this process, as each existing dikaryon remains highly resistant to modification by invading nuclei. How the multitude of dikaryotic genotypes on a substrate sorts out may depend on chance or on the competitive fitness of the individual dikaryons. Once a substrate is colonized, the prevailing dikaryons are thought to function as physiologically distinct individuals due to somatic incompatibility (89), rather than as a genetic mosaic functioning as a cooperative network (69).

The Buller Phenomenon

Dikaryons are well known for their ability to contribute a fertilizing nucleus to a monokaryon. This behavior was first described by A. H. R. Buller (10, 11), coined in the literature by Quintanilha (62) as the "Buller phenomenon," and comprehensively reviewed by Raper (66). In essence, all matings are a manifestation of the

Buller phenomenon. When nuclear migration is in progress in a mating of monokaryons, part of the colony is dikaryotic and part of the colony remains purely monokaryotic, until it too is eventually dikaryotized after nuclear migration is complete. In this transitional stage, the colony actually represents a dikaryon-monokaryon (di-mon) mating.

In tetrapolar species, di-mon matings can be classified according to the mating types of the dikaryon and monokaryon as compatible or incompatible depending on whether the dikaryon has at least one resident nuclear type that is fully compatible with the monokaryon (Table 20.1). In any di-mon interaction, there are three possibilities. First, the two nuclei of the dikaryon may simply replace the resident nucleus of the monokaryon (66). This can happen in either compatible or incompatible di-mon interactions. Second, one of the two nuclei of the dikaryon may dikaryotize the monokaryon; this is observed only in compatible di-mon interactions. Lastly, a recombinant nucleus may dikaryotize the monokaryon. This is the most common outcome in incompatible di-mon interactions, but it can also occur in compatible interactions. Even when both nuclei of the dikaryon are compatible with that of the monokaryon (16, 42, 56), somatic recombination for mating types may still occur; because these recombinants, also compatible with the monokaryon, are not necessary for fertilization, Papazian aptly described this as an ". . . apparently useless occurrence . . ." (57).

Nuclear Selection in Di-Mon Matings

In compatible di-mon interactions in which both nuclei of the dikaryon are compatible with that of the monokaryon, one nucleus of the dikaryon may dikaryotize the monokaryon more frequently than the other (15, 20, 22, 41, 42). These observations of nuclear selection include more than one species, and within species, they include strains that were either genetically heterogeneous or inbred to be isogenic for all but mating-type genes. Not surprisingly, these studies come to dif-

ferent conclusions as to the basis for selection. For compatible, highly isogenic di-mon matings of *S. commune*, nuclear selection is associated with the *B* mating types, with a subsidiary influence by the *A* mating types (22). The basis for this selection is not at the level of nuclei entering or migrating within the monokaryon. Both nuclear types of the dikaryon enter the monokaryon and migrate with approximately equal efficiency, but one dikaryotizes more frequently than the other. This internuclear selection based on the *B* mating types may share a common mechanism with another phenomenon described more recently by Carlene Raper (64); when certain dikaryons of *S. commune* are dedikaryotized, one nuclear type is recovered more frequently than the others and there is a linear hierarchy of biased recovery based on the *B* mating types.

In nonisogenic strains, especially those from the wild, other factors may be involved in nuclear selection in compatible di-mon matings. For example, among genetically heterogeneous strains of *Coprinus*, simple relatedness by descent is the main determining factor, with the less-related nuclear genotypes favored by selection over the more-related genotypes (41, 42). Mating types are not implicated in nuclear selection here, presumably because of modifiers of mating-type activity that might be expected in a highly heterogeneous genetic background. While nuclear selection is well documented in the laboratory, there is no evidence for or against nuclear selection in di-mon matings in nature.

Analysis of Somatic Recombinants in Di-Mon Matings

Much of the investigation of somatic recombination has focused on incompatible di-mon matings because recombinants with mating types compatible with that of the monokaryon can be readily selected (Table 20.1). Thus, even extremely rare recombination events embedded within a dikaryotic mycelium can be recovered. After the recombinant type is captured in the new dikaryon, any other markers can be assayed.

Early studies of somatic recombination were hampered by the difficulty of genotyping the fertilizing nucleus in di-mon matings from the resulting dikaryon. If the new dikaryon from an incompatible di-mon mating is merely fruited, the mating types of the spores are the same whether somatic recombination occurred or not (Table 20.2). Quintanilha's method (63) addressed this problem by pairing the dikaryotic product of incompatible di-mon matings with tester monokaryons of compatible mating types designed to separate and distinguish the mating types of the nuclei in this dikaryotic product (Table 20.2). The new dikaryons from these

Table 20.1 Classification of di-mon matings

Di-mon interaction	Dikaryon mating types	Monokaryon mating type
Compatible (legitimate[a])	*A1B1* + *A2B2*	*A1B1*
Compatible (legitimate[a])	*A3B3* + *A2B2*	*A1B1*
Incompatible (illegitimate[a])	*A1B2* + *A2B1*	*A1B1*

[a]Terminology of A. H. R. Buller.

Table 20.2 Genotyping nuclei in incompatible di-mon matings of the form $(A1B1 + A2B2)$ × $A1B2$; Quintinilha's test $(63)^a$

New dikaryon from di-mon mating	X tester	Dikaryon from di-mon mating with tester	B mating types expected in offspring
Expectation for nuclear replacement $(A1B1 + A2B2)^a$	X A1B3	A2B2 + A1B3	**B2**, B3
	X A2B3	A1B1 + A2B3	**B1**, B3
Expectation for somatic recombination $(A1B2 + A2B1)^a$	X A1B3	A2B1 + A1B3	**B1**, B3
	X A2B3	A1B2 + A2B3	**B2**, B3

^{a}Both dikaryons, either from nuclear replacement or somatic recombination, fruit to produce the same mating types among the offspring: A1B1, A2B2, A1B2, and A2B1. For each di-mon cross with a tester, the diagnostic B mating type is in bold type. (Table adapted from reference 66.)

tester di-mon matings were then fruited. Scoring the segregating mating types from these fruiting bodies allowed the mating type of the original fertilizing nucleus in the first di-mon mating to be deduced. This was an effective, but very laborious, process allowing analysis of only a limited number of recombination events in di-mon matings.

In Papazian's method (56) the dikaryons from fertilized monokaryons of di-mon matings were paired with tester monokaryons, and the resulting patterns of full compatibility versus incompatibility are morphologically distinguishable (Table 20.3). Further, the testers used in Papazian's method can carry auxotrophies to deduce whether or not the nucleus being assayed also contains a noncomplementing auxotrophy (19, 21).

Although not yet used widely in studies of somatic recombination, it is now easy to separate and genotype

Table 20.3 Genotyping nuclei in incompatible di-mon matings of the form $(A1B1 + A2B2)$ × $A1B2$; Papazian's test (56)

New dikaryon from di-mon mating	Compatibility^a with tester monokaryon:			
	A1B1	A1B2	A2B1	A2B2
Nuclear replacement $(A1B1 + A2B2)$	C	I	I	C
Somatic recombination $(A2B1 + A1B2)$	I	C	C	I

^{a}C, compatible; the homokaryon is quickly and uniformly dikaryotized. This can be readily seen in S. commune as a change in colony morphology. I, incompatible; the monokaryon is not uniformly dikaryotized but shows a "flat" reaction consistent with migration of the nucleus with a different B mating-type gene but with the same A mating type. Sectors of dikaryotic growth are commonly seen—somatic recombination generates a nucleus fully compatible with the monokaryon; this happens only after a delay. (Table adapted from reference 66.)

the component haploid nuclei of any dikaryon by protoplast formation and regeneration (3, 88); the direct recovery of haploids avoids the complication of the dominance or recessiveness of the markers and simplifies the genetic analysis. Of all three methods of analyzing di-mon matings, however, Papazian's method requires the least labor and therefore may permit genotyping of the largest numbers of potential recombinants in di-mon matings.

Patterns of Somatic Recombination in Di-Mon Matings

What is the nature of recombination in di-mon matings? Does it resemble parasexuality, meiotic recombination, or isolated transfer of specific elements between genomes? The surprising answer is all of the above, depending on the kind of interaction and the species. In *Coprinus*, somatic recombination appears to be essentially parasexual with reduction via intermediate stages of aneuploidy and little recombination among genes within chromosomes (61, 82, 83). In *S. commune* recombination in incompatible di-mon matings is entirely different, with two classes of recombinants that include intrachromosomal recombinants, i.e., those that appear to be derived from a meiotic-like process (Fig. 20.3) and those that appear to involve only the specific transfer of mating types between nuclei, and not other genes (19, 23). The meiotic-like recombinants show reassortment of genes on different chromosomes and numerous recombinants within chromosomes (Fig. 20.3). Here the recombinant mating types fertilizing the monokaryon have a variety of alleles from either of the two nuclear types of the original dikaryon. All of these recombinants in di-mon matings arise in the absence of fruiting bodies and basidia.

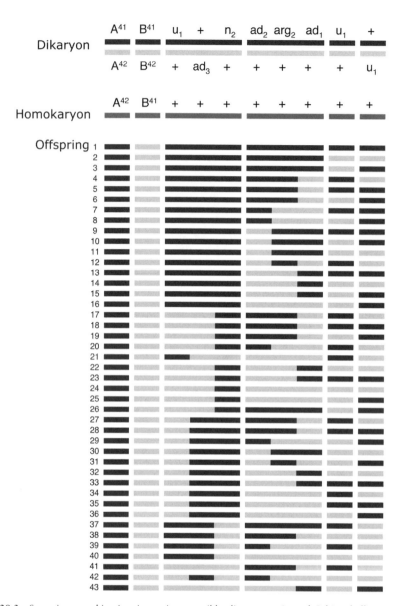

Figure 20.3 Somatic recombination in an incompatible di-mon mating of *Schizophyllum commune*, shown as a graphic representation of the data of Ellingboe (21). The dikaryon genotype is at the top. The homokaryon is below the dikaryon. The two haploid genomes of the dikaryon are marked as light/dark gray. Linkage between markers is indicated by a solid line; breaks in lines indicate that markers are not linked. The nuclei fertilizing the monokaryon were genotyped by the method of Papazian. Recombinants include numerous examples of unlinked and linked loci, as would be expected in meiosis. Interestingly, specific-factor transfer was not observed in this di-mon mating; recombination was exclusively meiotic-like.

In contrast to the meiotic-like recombinants, specific-factor transfer involves only the mating-type genes of the dikaryon; all of the other nonselected nuclear markers, including even the *pab* gene located between the *Aα* and *Aβ* loci, are from the monokaryon of the di-mon mating (19). Specific-factor transfer must therefore involve all three types of nuclei in the di-mon mating (19, 23). This would be difficult to explain by conventional crossing-over, because several nuclear fusions would be required, along with double crossovers bracketing each mating-type region transferred. Although the actual mechanism of specific-factor transfer remains unknown, the process may be general when the selection on certain genes is strong. For example, a very similar situation occurs with *Cryphonectria parasitica*, in which some heterokaryotic strains from the wild are capable of selfing because they are heteroallelic for mating types, but not for any other loci (54). These mating-type heterokaryons could have arisen from somatic exchange between strains with different multilocus genotypes between which only a mating type is somehow transferred into a new nuclear background. Like the specific-factor transfer described by Ellingboe (19), the mating-type heterokaryons in *C. parasitica* cannot easily be explained by conventional crossing-over. Loci responsible for the unusual hyperplastic growth phenotype in *S. commune* termed "mound" appear to be subject to a process similar to specific-factor transfer, but not involving mating types (49, 50).

Somatic Recombination in Lone Dikaryons

Although studied mostly in di-mon matings, somatic recombination is not limited to di-mon matings but also occurs in isolated dikaryons from which the nuclear components are recovered after a short period of growth, either with or without selection for recombinant types. There are indications that the somatic recombination in isolated dikaryons of *S. commune* is also meiotic-like (27, 58).

Somatic Recombination in Dikaryons: State of the Field

Despite the many early reports of somatic recombination, no unified explanation of somatic recombination in dikaryons has since emerged. There are several impediments to a complete understanding. First, somatic recombination is not associated with a specific developmental stage. In contrast, conventional meiosis occurs in well-defined basidia from which basidiospores containing the immediate meiotic products can be isolated. This makes it possible to calculate recombination frequencies and, with analysis of whole tetrads, to determine whether any individual recombination event is reciprocal or nonreciprocal. In somatic recombination, the exchange events are buried within the dikaryotic mycelium, and the fate of any individual recombinant, extinction or proliferation, is uncertain. Whether any or all of the cellular machinery associated with meiosis participates in this process is an open question. Further, after a recombination event, multiple rounds of division may occur and the recombinant genotypes may proliferate to different degrees, so that their frequency of recovery may not reflect the frequency of recombination. Yet another complication is whether the monokaryon merely selects for a nucleus of compatible mating type after somatic recombination has occurred, or actually induces somatic recombination.

Several conditions, duly noted by Raper (66), are needed for accurately characterizing somatic recombi-

nation further both in di-mon matings and in isolated dikaryons. First, larger numbers of recombinants from independent events are needed, and second, a densely marked genetic map is essential for a full evaluation of somatic recombination. Does somatic recombination fit the same mapping function relating recombination frequency to true map distance as in meiosis? Are the somatic recombination events reciprocal? The background genotype is also likely to be a major factor (66). Isogenic strains cannot be expected to show exactly the same correspondence between recombination frequency and physical distance as outbred strains. Modifiers of recombination frequency are well known in *S. commune* (45, 72, 79–81).

For specific-factor transfer, the nucleotide sequences surrounding mating types may provide clues about the process. For example, specific recombination hot spots flanking the mating-type locus in *Cryptococcus neoformans* have been identified (36). These hot spots show negative crossover inference; more double crossovers occur than are expected on the basis of the single-crossover frequencies, exactly what would be expected for the transfer of mating types, and not other genes, to new haploid backgrounds. In *S. commune*, the distribution of short nucleotide repeats and identity islands appears to play a role in determining the rate of recombination in the region between $B\alpha$ and $B\beta$ (26); whether or not sequence elements on the other sides of the B mating-type genes could promote mating-type transfer in conjunction with the elements between $B\alpha$ and $B\beta$ is an open question. In further characterizing specific-factor transfer, large numbers of di-mon pairings should be followed because this process is detected in some pairings, but not in others (21).

Evolutionary Significance of Somatic Recombination; Is There Any?

The diversity of somatic recombination processes in dikaryons, parasexual-like, meiotic-like, and specific-factor transfer, although interesting in their own right, are of unknown evolutionary significance. Although it is tempting to hypothesize that somatic recombination systems might somehow be adaptive at the levels of individuals or populations, it is also possible that somatic recombination has little or no effect on fitness and has a similarly negligible effect on the evolution of populations. The vast majority of somatic recombinants may simply be lost from mycelia, just as most mutant alleles are lost from most populations soon after they arise. In short-lived dikaryotic individuals, it is difficult to envision how somatic recombinants could rise to a sufficiently high frequency to confer an adaptive benefit or deficit. The survivability and ability of somatic recombinants to spread within a dikaryotic mycelium have not been studied.

Long-Term Changes in Dikaryons

If dikaryons have the capacity for internal genetic exchange, how do they behave as a population of cells over a long time? Dikaryons offer the advantage in evolutionary studies that the individual nuclear types can be recovered intact, most easily by protoplast formation and regeneration. This is not possible with diploids because of the genetic shuffling that occurs with any kind of reduction division, meiotic or nonmeiotic. In experimentally evolved dikaryotic lineages, the fitness of the paired nuclear types can also be measured to examine how dikaryons change over time with vegetative growth. From evolved dikaryons, the nuclear components can be paired with nuclei of other histories to measure their fitness at the level of the dikaryon or they can be measured alone as monokaryons. The time at which a mutation arose can be approximated after the fact by sampling the evolving lineages retained in a culture archive.

Only one such study examining long-term evolution in dikaryons has appeared. Clark and Anderson (14) evolved dikaryons in replicated populations all from a common ancestor with a uniform cytoplasm over 13,000 generations, defined as the time required for a hyphal tip cell to divide (ca. 90 min for *S. commune* on minimal medium). Selection was for high growth rate, which was considered a measure of fitness. (In nature fitness is more complicated than merely capturing resources through vegetative growth for eventual fruiting. Fruiting timing and efficiency, basidiospore production and germination, and mating efficiency are also important.) In addition to the replicate dikaryons, the original haploid components with the common cytoplasm were also evolved with replication.

Among the haploids, there was no overall change in growth rate, which was about twice that of dikaryons in the beginning of the experiment. Among the dikaryons, there were sharp increases in growth rates. Several dominant mutations for higher growth rate were identified. These mutations did not increase the frequency of cell division but rather increased the length of the cell compartments. The distance between nuclei also increased, and the colony margin changed from irregularly lobate to smooth, as in monokaryons. Along with the change in colony morphology, the production of a self-inhibitor (43) responsible for the slow growth and lobate colony margin was lost. In the wild-type dikaryons, the self-inhibitor is expressed when the mycelium is grown in light, but not in darkness. In the dark, even the wild type is fast growing with a smooth colony margin, exactly like the mutant dikaryons in the light or the dark. At the end of the evolution experiment, the growth rate of the mutant dikaryons nearly matched that of the monokaryons. Although gene expression has not been monitored in these evolved dikaryons growing under light, we speculate that it will be more "monokaryon-like" than dikaryon-like.

What might explain the relative lack of change in the growth rate of the monokaryons? In filamentous fungi, the effective population sizes are undoubtedly much smaller than those of planktonic unicells. This is because the different growing points of a mycelium are all related to a recent common ancestor cell by a short path of descent. In contrast, different cells drawn from a planktonic cell culture are on average much more distantly related by descent and the overall mutational diversity is higher, at least until a selective sweep homogenizes the populations. The dominant mutations for increased growth rate in the dikaryons would have to occur at a very high frequency for a response to selection in the dikaryotic lineages, which must be of small effective population size.

Detrimental mutations also accumulated over the course of the evolution experiment. Two different recessive lethal mutations were detected as the inability to recover one of the two nuclear types from dikaryons beyond a specific time in the experiment. Interestingly, one of these also exerted a dominant deleterious effect on dikaryotic growth in pairings of nuclei from all of the histories except the one with which it evolved. A compensatory mutation restoring growth occurred in the other nucleus coevolving in the dikaryon; the nuclear types that evolved together in this dikaryon were fitter when paired together than with any other nuclear types. The original recessive lethal mutation and the corresponding compensatory mutation were traced to particular times in the experimental lineages. This kind of compensation could be the basis for a coadaptive process between the haploid genomes of an asexually evolving dikaryon in which changes in one nucleus set the selective environment for change in the other nucleus in a continuing reciprocal process. The compensatory event here represented but one observation. More observations are needed before the generality of coadaptation in dikaryons can be tested.

Finally, the evolved dikaryons were tested for somatic recombination by separating and genotyping each nuclear type. Of 25 single nucleotide polymorphisms there were eight events of reciprocal transfer and two events of nonreciprocal transfer between the nuclei of

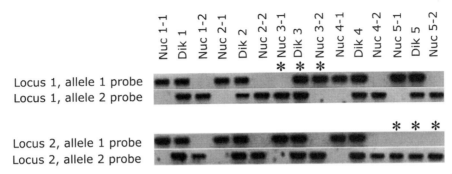

Figure 20.4 Reciprocal and nonreciprocal genetic exchange between nuclei during long-term growth. Two of 25 SNP markers were detected by Southern hybridization of amplified DNA of marker loci with allele-specific oligonucleotide probes (14). The original paired haploid nuclei carried different alleles for each of the 25 marker loci. Asterisks show reciprocal genetic exchange; circles show a nonreciprocal exchange.

6 of 12 dikaryons, with 8 of the 25 loci affected (Fig. 20.4). The recombination process was decidedly not meiotic-like, as a majority of the 25 markers were not reshuffled and most events were separated in time and occurred in different populations. Further, no recombinant mating types were detected during the evolution experiment, but none would necessarily be expected as the experiment was not like an incompatible di-mon mating where selection for recombinant mating types is strong. The process most resembles specific-factor transfer except that no selection on any of the 25 loci is expected. It is possible that many segregants in dikaryons cultured for long periods of time, as those seen over the short term, may not proliferate and persist.

One limitation of the study by Clark and Anderson (14) is that dikaryons of *S. commune* may not grow for 13,000 generations in the wild and are likely more short-lived. But many dikaryons of other species in the wild do persist for long periods of time such as fairy rings, for example, those of *Marasmius oreades*. These represent naturally occurring evolution experiments in that the mycelium grows over a long period of time from a common ancestor in a physically obvious growth pattern.

EXPECTATIONS FOR DIPLOIDS AND DIKARYONS IN EVOLUTION

By the time of publication of Raper's book in 1966, many of the fundamental genetic properties of dikaryons, including their exquisite control of nuclear migration, nuclear pairing, and formation of clamp connections by the mating-type genes, their ability to contribute a fertilizing nucleus to a haploid monokaryon on contact, and their capacity for somatic recombination, had been well worked out. About the evolutionary origin and maintenance of dikaryosis, one central question remains even now. Is the dikaryon maintained because of some selective advantage, or is the dikaryon merely an evolutionary holdover?

There are clues about the relative merits of dikaryosis and diploidy. What follows is a mixture of established fact and pure speculation. Many normally dikaryotic basidiomycetes have the capacity for diploidy (12, 32, 44, 55, 59). Although the total genetic complement of a dikaryon may be identical to that of its corresponding diploid state, the patterns of gene expression (5) and the phenotypic expression are different: for example, while dikaryons may have clamp connections, diploids of *S. commune*, *C. cinereus*, and *Armillaria* species do not. (Note that diploids of *Cryptococcus neoformans* have hook cells; see reference 74.) Diploids can be selected from compatible or incompatible confrontations of monokaryons as epigenetic states that are stable enough to persist for at least some time (12, 55, 59). Outside the Hymenomycetes, diploids of the distantly related *Microbotryum violaceum* and *Ustilago maydis* can easily be selected and maintained in culture. In the Hymenomycetes, the best example of a genetically based diploid in a normally dikaryotic fungus is the dominant mutation *dik−* in *S. commune* that causes dikaryons to become stably diploid after a short period of growth (29, 44, 67). Essentially the *dik−* mutation confers a life cycle typical of many *Armillaria* species on *S. commune*.

Life Cycle Variation for Diploidy and Dikaryosis

Species that produce dikaryons also vary in their propensity for forming diploids (Fig. 20.5). In most species of *Armillaria*, dikaryons are produced in matings, but these

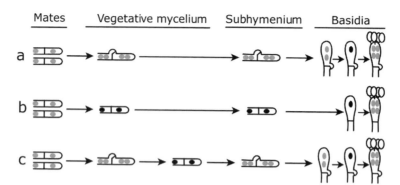

Figure 20.5 Generalized basidiomycete life cycles. Diploid nuclei are in black; haploid nuclei are in gray. (a) Typical life cycle of most homobasidiomycetes with dikaryotic vegetative phases. (b) Typical life cycle of *Armillaria mellea*. No dikaryotic stage has been observed, and matings produce only diploids, which carry through the vegetative phase and into the basidia. (c) Typical life cycle of most *Armillaria* species including *A. ostoyae*, *A. gallica*, and *A. tabescens*. A dikaryon forms in matings, but the duration of the dikaryon phase is variable. The nuclei fuse, leading to a persistent diploid. During fruiting there is a premeiotic reduction with dikaryons appearing in the prebasidial cells. Finally, the nuclei of the dikaryon fuse to form a diploid that immediately undergoes meiosis. *dik−* strains of *S. commune* form dikaryons that become diploid as in panel c; the events during fruiting and up to meiosis have not been characterized for *dik−* strains. Modified with the permission of Kari Korhonen, who created the original version (see reference 31).

quickly become stably diploid through nuclear fusion (32, 46, 85). Also, in most species of *Armillaria*, the subbasidial cells and basidia are again dikaryotic, a baffling observation given the uniformly diploid vegetative condition. In *Armillaria tabescens* (31) and *Armillaria gallica* (60) there is strong evidence for a premeiotic reduction mechanism in fruiting body tissues; this is yet another potential recombination system outside conventional meiosis awaiting full characterization with large numbers of recombinants evaluated in the context of a densely marked genetic map.

In *Armillaria* there is variation in the proportion of time spent as diploids and dikaryons. In *A. tabescens*, the dikaryotic phase is longer than in *A. ostoyae* or *A. gallica*. In *A. mellea* there is no dikaryon either after mating or in the subbasidial cells of fruiting bodies; diploids form during mating and persist through the production of basidia in fruiting bodies.

Evolutionary Merits of Diploidy and Dikaryosis

With their capacity for diploidy, why has dikaryosis predominated in the basidiomycetes? One possibility is that different conditions favor different ploidy states. A possible advantage of diploidy for extremely long-lived individuals inhabiting stable environments might be a lower mutation rate, enhanced genetic stability, and reduced need for the phenotypic flexibility afforded by dikaryosis (35). Within a diploid, an intact DNA template is always available for repair, but in the haploid nuclei of a dikaryon an intact template is available only after DNA replication in the G_2 phase of the cell cycle. Another possibility in species with long-lived diploids is that the mutation rate may have been driven low by natural selection. In two long-lived diploid individuals of *A. gallica*, no mutations have been detected despite a sampling and sequencing regimen that would have detected point mutations occurring at less than 10^9 per generation.

Given the potential for somatic recombination, diploids would have yet another bias toward stability. Genetic exchange in diploids does not create new combinations of alleles within nuclei, but in dikaryons new combinations of alleles arise with exchange (Fig. 20.6). This could be important among genes whose interactive control relationships extend only within nuclei and not for genes whose products can interact or complement at the cellular level (such as auxotrophies).

Another difference between diploids and dikaryons is that in tetrapolar species dikaryons mate readily with only two of the four possible sibling mating types, whereas diploids can mate readily with all four sibling mating types (4). The mating types in diploids are effectively codominant. Whether or not this is of evolutionary significance depends on the frequency with which the Buller phenomenon and its diploid counterpart occur in nature. But even if di-mon mating predominates in natural populations of Hymenomycetes, very little

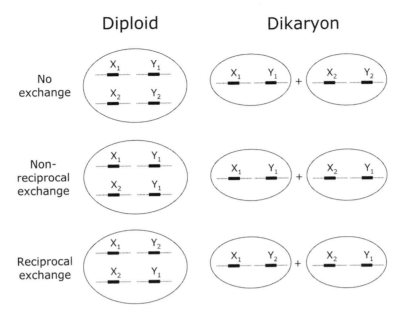

Figure 20.6 Genetic exchange in diploids and dikaryons affecting two loci of unknown linkage relationship. Both reciprocal and nonreciprocal exchanges produce new combinations of alleles within the nuclei of the dikaryon, but not in the diploid.

mating advantage would accrue to diploids because there are so many compatible mating types available; both dikaryons and diploids are capable of mating efficiently in populations. At the postzygotic level, however, the reproductive output from dikaryon-monokaryon matings may well be higher than that of diploid-monokaryon matings, in which the potential for irregular patterns of genetic segregation and lower meiospore viability is high. Here, the evolutionary advantage may well go to the dikaryon.

In addition to the genetic differences between diploids and dikaryons their different phenotypic responses may confer advantages or disadvantages depending on conditions. Nuclear spacing and associated variation in gene expression are inherent to dikaryons, but not to diploids. Dikaryons are therefore expected to be capable of supporting a greater range of phenotypes in response to environmental variation than diploids. With an enhanced range of phenotypes, dikaryons might be more adept than diploids in coping with heterogeneous environments.

CONCLUSION

While all of the above explanations for the advantages and disadvantages of dikaryosis and diploidy are plausible, none have been definitively tested. Neither have the evolutionary roles of nuclear mitochondrial genomic conflict or somatic recombination been clarified. In the next phase, the strongest inferences about dikaryons and evolution will come from a combination of molecular biology, genomics, and evolutionary analysis, both retrospective and experimental. It may now be possible to make fair comparisons of dikaryotic cell populations with and without opportunities for nuclear-mitochondrial genomic conflict, somatic recombination, and diploidy and to compare the evolutionary outcomes.

GLOSSARY

Buller phenomenon The process in which a monokaryon is fertilized by a dikaryon in a "di-mon" mating.
Cell The hyphal compartments bounded by septa. Since in most cases the cytoplasm is continuous between hyphal compartments, the hypha is sometimes considered to be acellular because the cytoplasm is continuous throughout.
Dikaryon A mycelium within which each cell contains paired, synchronously dividing nuclei, one of each given by the original gametic genotypes. See also the definition given by Papazian (57). (**Heterokaryon** is the term for a mycelium with more than one genotype of nucleus, regardless of the numbers of nuclei per cell.)
Female The capacity of a monokaryon to accept a fertilizing nucleus of compatible mating type, which migrates throughout its resident mycelium, resulting in a

dikaryon with the cytoplasm of the resident monokaryon.

Hook cell A developing clamp connection, arising from the base of the apical cell, but not fused with the subapical cell.

Male The capacity of a monokaryon to donate a fertilizing nucleus in a mating; the cytoplasm of the nuclear donor is not transferred to the opposing monokaryon.

Meiotic-like recombination Both chromosomal reassortment and crossing-over are common. Alleles of loci on different chromosomes are shuffled, as are alleles of loci on the same chromosomes.

Monokaryon A haploid mycelium derived from a single basidiospore. Many monokaryons have mainly uninucleate cells, but some also have cells with variable numbers of nuclei. See also the definition given by Papazian (57). Diploids may also have predominantly uninucleate cells and are also technically monokaryons. (**Homokaryon** is the term for a mycelium with only a single genotype of nucleus, regardless of the numbers of nuclei per cell.)

Parasexual recombination Chromosomal reassortment is common, but crossing-over is rare. Alleles of loci on different chromosomes are shuffled; alleles of loci on the same chromosomes remain parental with rare exceptions.

Somatic recombination Creation of nonparental combinations of alleles in a mycelium without fruiting-body formation.

Specific-factor transfer One or more genes under strong selection move between nuclei and into another genetic background without carryover of any additional genes.

Tetrapolar Two factors specify mating type such that the meiotic offspring of a dikaryon include four mating types compatible in two pairs: A1B1 and A2B2, and A1B2 and A2B1; synonym **bifactorial**. (Species with only one such factor are **bipolar**; synonym **unifactorial**.)

References

1. **Aanen, D. K., T. W. Kuyper, A. J. Debets, and R. F. Hoekstra.** 2004. The evolution of non-reciprocal nuclear exchange in mushrooms as a consequence of genomic conflict. *Proc. R. Soc. Lond. B* **271:**1235–1241.

2. **Adams, T. J. H., E. N. D. Williams, N. K. Todd, and A. D. M. Rayner.** 1984. A species-specific method of analyzing populations of basidiospores. *Trans. Br. Mycol. Soc.* **82:**359–361.

3. **Anderson, J. B., and R. Cenedese.** 1984. Extranuclear chloramphenicol resistance mutations in the basidiomycete *Sistotrema brinkmannii. Exp. Mycol.* **8:**256–260.

4. **Anderson, J. B., and R. C. Ullrich.** 1982. Diploids of *Armillaria mellea*—synthesis, stability, and mating behavior. *Can. J. Bot.* **60:**432–439.

5. **Babu, M. R., K. Choffe, and B. J. Saville.** 2005. Differential gene expression in filamentous cells of *Ustilago maydis. Curr. Genet.* **47:**316–333.

6. **Badalyan, S. M., E. Polak, R. Hermann, M. Aebi, and U. Kues.** 2004. Role of peg formation in clamp cell fusion of homobasidiomycete fungi. *J. Basic Microbiol.* **44:**167–177.

7. **Barroso, G., and J. Labarere.** 1997. Genetic evidence for nonrandom sorting of mitochondria in the basidiomycete *Agrocybe aegerita. Appl. Environ. Microbiol.* **63:**4686–4691.

8. **Birky, C. W.** 2001. The inheritance of genes in mitochondria and chloroplasts: laws, mechanisms, and models. *Annu. Rev. Genet.* **35:**125–148.

9. **Brown, A. J., and L. A. Casselton.** 2001. Mating in mushrooms: increasing the chances but prolonging the affair. *Trends Genet.* **17:**393–400.

10. **Buller, A. H. R.** 1930. The biological significance of conjugate nuclei in *Coprinus lagopus* and other hymenomycetes. *Nature* **126:**686–689.

11. **Buller, A. H. R.** 1931. *Researches on Fungi.* vol. IV. Longmans, Green, and Co., London, United Kingdom.

12. **Casselton, L. A.** 1965. Production and behaviour of diploids of *Coprinus lagopus. Genet. Res.* **6:**190–208.

13. **Chase, T. E., and R. C. Ullrich.** 1983. Sexuality, distribution, and dispersal of *Heterobasidion annosum* in pine plantations of Vermont. *Mycologia* **75:**825–831.

14. **Clark, T. A., and J. B. Anderson.** 2004. Dikaryons of the basidiomycete fungus *Schizophyllum commune*: evolution in long-term culture. *Genetics* **167:**1663–1675.

15. **Crowe, L. K.** 1963. Competition between compatible nuclei in the establishment of a dikaryon in *Schizophyllum commune. Heredity* **18:**525-533.

16. **Crowe, L. K.** 1960. The exchange of genes between nuclei of a dikaryon. *Heredity* **15:**397–405.

17. **Debuchy, R.** 1999. Internuclear recognition: a possible connection between euascomycetes and homobasidiomycetes. *Fungal Genet. Biol.* **27:**218–223.

18. **Dujon, B., P. P. Slonimski, and L. Weill.** 1974. Mitochondrial genetics IX: a model for recombination and segregation of mitochondrial genomes in *Saccharomyces cerevisiae. Genetics* **78:**415–437.

19. **Ellingboe, A. H.** 1963. Illegitimacy and specific factor transfer in *Schizophyllum commune. Proc. Natl. Acad. Sci. USA* **49:**286–292.

20. **Ellingboe, A. H.** 1964. Nuclear migration in dikaryotic-homokaryotic matings in *Schizophyllum commune. Am. J. Bot.* **51:**133–139.

21. **Ellingboe, A. H.** 1964. Somatic recombination in dikaryon K of *Schizophyllum commune. Genetics* **49:**247–251.

22. **Ellingboe, A. H., and J. R. Raper.** 1962. The Buller phenomenon in *Schizophyllum commune*: nuclear selection in fully compatible dikaryotic-homokaryotic matings. *Am. J. Bot.* **49:**454–459.

23. **Ellingboe, A. H., and J. R. Raper.** 1962. Somatic recombination in *Schizophyllum commune. Genetics* **47:**85–98.

24. **Fincham, J. R. S., P. R. Day, and A. Radford.** 1979. *Fungal Genetics,* 4th ed. University of California Press, Berkeley.

25. Fischer, M., and H. Wolfrath. 1997. Mitochondrial DNA in mon-mon and di-mon pairings of *Pleurotus ostreatus*. *Bot. Acta* **110**:172–176.

26. Fowler, T. J., M. F. Mitton, E. I. Rees, and C. A. Raper. 2004. Crossing the boundary between Ba and Bb mating-type loci in *Schizophyllum commune*. *Fungal Genet. Biol.* **41**:89–101.

27. Frankel, C. 1979. Meiotic-like recombination in vegetative dikaryons of *Schizophyllum commune*. *Genetics* **92**:1121–1126.

28. Gibb, E. A., and G. Hausner. 2005. Optional mitochondrial introns and evidence for a homing-endonuclease gene in the mtDNA rnl gene in *Ophiostoma ulmi* s. lat. *Mycol. Res.* **109**:1112–1126.

29. Gladstone, P. 1972. *Genetic Studies on Heritable Diploidy in* Schizophyllum. Ph.D. thesis. Harvard University, Cambridge, MA.

30. Goddard, M. R., and A. Burt. 1999. Recurrent invasion and extinction of a selfish gene. *Proc. Natl. Acad. Sci. USA* **96**:13880–13885.

31. Grillo, R., K. Korhonen, J. Hantula, and A. M. Hietala. 2000. Genetic evidence for somatic haploidization in developing fruit bodies of *Armillaria tabescens*. *Fungal Genet. Biol.* **30**:135–145.

32. Guillauimin, J. J., J. B. Anderson, and K. Korhonen. 1991. Life cycle, interfertility, and biological species. *In* C. G. Shaw III and G. A. Kile (ed.), *Armillaria Root Disease*. Forest Service, United States Department of Agriculture, Washington, DC.

33. Hallenberg, N., and N. Kuffer. 2001. Long-distance spore dispersal in wood-inhabiting Basidiomycetes. *Nord. J. Bot.* **21**:431–436.

34. Hintz, W. E. A., J. B. Anderson, and P. A. Horgen. 1988. Nuclear migration and mitochondrial inheritance in the mushroom *Agaricus bitorquis*. *Genetics* **119**:35–41.

35. Hodnett, B., and J. B. Anderson. 2000. Genomic stability of two individuals of *Armillaria gallica*. *Mycologia* **92**:894–899.

36. Hsueh, Y.-P., A. Idnurm, and J. Heitman. 2006. Recombination hotspots flank the Cryptococcus mating-type locus: implications for the evolution of a fungal sex chromosome. *PLoS Genet.* **2**:e184.

37. Iwasa, M., S. Tanabe, and T. Kamada. 1998. The two nuclei in the dikaryon of the homobasidiomycete *Coprinus cinereus* change position after each conjugate division. *Fungal Genet. Biol.* **23**:110–116.

38. James, T. Y., D. Porter, J. L. Hamrick, and R. Vilgalys. 1999. Evidence for limited intercontinental gene flow in the cosmopolitan mushroom, *Schizophyllum commune*. *Evolution* **53**:1665–1677.

39. James, T. Y., and R. Vilgalys. 2001. Abundance and diversity of *Schizophyllum commune* spore clouds in the Caribbean detected by selective sampling. *Mol. Ecol.* **10**:471–479.

40. Johannesson, H., and J. Stenlid. 2004. Nuclear reassortment between vegetative mycelia in natural populations of the basidiomycete *Heterobasidion annosum*. *Fungal Genet. Biol.* **41**:563–570.

41. Kimura, K. 1954. Diploidization in the Hymenomycetes. I. Preliminary experiments. *Biol. J. Okayama Univ.* **1**:226–233.

42. Kimura, K. 1958. Diploidization in the Hymenomycetes II. Nuclear behavior in the Buller Phenomenon. *Biol. J. Okayama Univ.* **4**:1–59.

43. Klein, K. K., J. Landry, T. Friesen, and T. Larimer. 1997. Kinetics of asymmetric mycelial growth and control by dikaryosis and light in *Schizophyllum commune*. *Mycologia* **89**:916–923.

44. Koltin, Y., and J. R. Raper. 1968. Dikaryosis: genetic determination in *Schizophyllum*. *Science* **160**:85–86.

45. Koltin, Y., and J. Stamberg. 1973. Genetic control of recombination in *Schizophyllum commune*—location of a gene controlling B-factor recombination. *Genetics* **74**:55–62.

46. Korhonen, K. 1978. Interfertility and clonal size in *Armillariella mellea*. *Karstenia* **18**:31–42.

47. Korhonen, K. 1978. Intersterility groups of *Heterobasidion annosum*. *Commun. Inst. For. Fenn.* **94**:1–25.

48. Lengeler, K. B., R. C. Davidson, C. D'Souza, T. Harashima, W. C. Shen, P. Wang, X. Pan, M. Waugh, and J. Heitman. 2000. Signal transduction cascades regulating fungal development and virulence. *Microbiol. Mol. Biol. Rev.* **64**:746–785.

49. Leonard, T. J., S. Dick, and R. F. Gaber. 1978. Internuclear genetic transfer in vegetative dikaryons of *Schizophyllum commune*. I. Di-mon mating analysis. *Genetics* **88**:13–26.

50. Leonard, T. J., R. F. Gaber, and S. Dick. 1978. Internuclear genetic transfer in dikaryons of *Schizophyllum commune*. II. Direct recovery and analysis of recombinant nuclei. *Genetics* **89**:685–693.

51. Leslie, J. F., and K. K. Klein. 1996. Female fertility and mating type effects on effective population size and evolution in filamentous fungi. *Genetics* **144**:557–567.

52. Lin, X., C. M. Hull, and J. Heitman. 2005. Sexual reproduction between partners of the same mating type in *Cryptococcus neoformans*. *Nature* **434**:1017–1021.

53. Makino, R., and T. Kamada. 2004. Isolation and characterization of mutations that affect nuclear migration for dikaryosis in *Coprinus cinereus*. *Curr. Genet.* **45**:149–156.

54. McGuire, I. C., R. E. Marra, and M. G. Milgroom. 2004. Mating-type heterokaryosis and selfing in *Cryphonectria parasitica*. *Fungal Genet. Biol.* **41**:521–533.

55. Mills, D. I., and A. H. Ellingboe. 1969. A common-AB diploid of *Schizophyllum commune*. *Genetics* **62**:271–279.

56. Papazian, H. P. 1954. Exchange of incompatibility factors between the nuclei of a dikaryon. *Science* **119**:691–693.

57. Papazian, H. P. 1950. Physiology of the incompatibility factors in *Schizophyllum commune*. *Bot. Gaz.* **112**:143–163.

58. Parag, Y. 1962. Studies on somatic recombination in dikaryons of *Schizophyllum commune*. *Heredity* **17**:305–318.

59. **Parag, Y., and B. Nachman.** 1966. Diploidy in tetrapolar heterothallic basidiomycete *Schizophyllum commune*. *Heredity* **21**:151–159.

60. **Peabody, R. B., D. C. Peabody, and K. M. Sicard.** 2000. A genetic mosaic in the fruiting stage of *Armillaria gallica*. *Fungal Genet. Biol.* **29**:72–80.

61. **Prudhomme, N.** 1961. Recombinaisons chromosomiques extra-basidiales chez un basidiomycete *Coprinus radiatus*. *Ann. Genet.* **4**:63–66.

62. **Quintanilha, A.** 1937. Contribution à l'étude génétique du phénomène de Buller. *C. R. Acad. Sci. Paris* **205**:745–747.

63. **Quintanilha, A.** 1939. Etude génétique du phénomène de Buller. *Bol. Soc. Broter.* **13**:425–486.

64. **Raper, C. A.** 1985. B-mating-type genes influence survival of nuclei separated from heterokaryons of *Schizophyllum*. *Exp. Mycol.* **9**:149–160.

65. **Raper, C. A., J. R. Raper, and R. E. Miller.** 1972. Genetic analysis of life cycle of *Agaricus bisporus*. *Mycologia* **64**:1088–1117.

66. **Raper, J. R.** 1966. *Genetics of Sexuality in Higher Fungi.* Ronald Press, New York, NY.

67. **Raper, J. R., and R. M. Hoffman.** 1974. *Schizophyllum commune*, p. 597–626. *In* R. C. King (ed.), *Handbook of Genetics*, vol. I. Plenum Press, New York, NY.

68. **Raudaskoski, M.** 1998. The relationship between B-mating-type genes and nuclear migration in *Schizophyllum commune*. *Fungal Genet. Biol.* **24**:207–227.

69. **Rayner, A. D. M.** 1991. The challenge of the individualistic mycelium. *Mycologia* **83**:48–71.

70. **Saville, B. J., Y. Kohli, and J. B. Anderson.** 1998. mtDNA recombination in a natural population. *Proc. Natl. Acad. Sci. USA* **95**:1331–1335.

71. **Saville, B. J., H. Yoell, and J. B. Anderson.** 1996. Genetic exchange and recombination in populations of the root-infecting fungus *Armillaria gallica*. *Mol. Ecol.* **5**:485–497.

72. **Schaap, T., and G. Simchen.** 1971. Genetic control of recombination affecting mating factors in a population of *Schizophyllum*, and its relation to inbreeding. *Genetics* **68**:67–75.

73. **Schuurs, T. A., H. J. Dalstra, J. M. Scheer, and J. G. Wessels.** 1998. Positioning of nuclei in the secondary mycelium of *Schizophyllum commune* in relation to differential gene expression. *Fungal Genet. Biol.* **23**:150–161.

74. **Sia, R. A., K. B. Lengeler, and J. Heitman.** 2000. Diploid strains of the pathogenic basidiomycete *Cryptococcus neoformans* are thermally dimorphic. *Fungal Genet. Biol.* **29**:153–163.

75. **Smith, M. L., J. N. Bruhn, and J. B. Anderson.** 1992. The fungus *Armillaria bulbosa* is among the largest and oldest living organisms. *Nature* **356**:428–431.

76. **Smith, M. L., J. N. Bruhn, and J. B. Anderson.** 1994. Relatedness and spatial distribution of *Armillaria* genets infecting red pine seedlings. *Phytopathology* **84**:822–829.

77. **Smith, M. L., L. C. Duchesne, J. N. Bruhn, and J. B. Anderson.** 1990. Mitochondrial genetics in a natural population of the plant pathogen *Armillaria*. *Genetics* **126**:575–582.

78. **Specht, C. A., C. P. Novotny, and R. C. Ullrich.** 1992. Mitochondrial-DNA of *Schizophyllum commune*—restriction map, genetic-map, and mode of inheritance. *Curr. Genet.* **22**:129–134.

79. **Stamberg, J.** 1968. Two independent gene systems controlling recombination in *Schizophyllum commune*. *Mol. Gen. Genet.* **102**:221–228.

80. **Stamberg, J.** 1969. Genetic control of recombination in *Schizophyllum commune*—occurrence and significance of natural variation. *Heredity* **24**:361–368.

81. **Stamberg, J.** 1969. Genetic control of recombination in *Schizophyllum commune*—separation of controlled and controlling loci. *Heredity* **24**:306–309.

82. **Swiezynski, K. M.** 1962. Analysis of an incompatible dimon mating in *Coprinus lagopus*. *Acta Soc. Bot. Pol.* **31**:169–184.

83. **Swiezynski, K. M.** 1963. Somatic recombination in two linkage groups in *Coprinus lagopus*. *Genet. Pol.* **4**:21–36.

84. **Swiezynski, K. M.** 1961. Exchange of nuclei between dikaryons in *Coprinus lagopus*. *Acta Soc. Bot. Pol.* **30**:535–552.

85. **Ullrich, R. C., and J. B. Anderson.** 1978. Sex and diploidy in *Armillaria mellea*. *Exp. Mycol.* **2**:119–129.

86. **Vilgalys, R., and B. L. Sun.** 1994. Assessment of species distributions in *Pleurotus* based on trapping of airborne basidiospores. *Mycologia* **86**:270–274.

83. **Wessels, J. G. H.** 1999. Fungi in their own right. *Fungal Genet. Biol.* **27**:134–145.

88. **Wessels, J. G. H., H. L. Hoeksema, and D. Stemerding.** 1976. Reversion of protoplasts from dikaryotic mycelium of *Schizophyllum commune*. *Protoplasma* **89**:317–321.

89. **Worrall, J. J.** 1997. Somatic incompatibility in basidiomycetes. *Mycologia* **89**:24–36.

90. **Yan, Z., and J. Xu.** 2003. Mitochondria are inherited from the *MAT*a parent in crosses of the basidiomycete fungus *Cryptococcus neoformans*. *Genetics* **163**:1315–1325.

Basidiomycetes: Plant and Animal Pathogenic Yeasts

Sex in Fungi: Molecular Determination and Evolutionary Implications
Edited by Joseph Heitman et al.
© 2007 ASM Press, Washington, D.C.

Flora Banuett

21

History of the Mating Types in *Ustilago maydis*

INTRODUCTION

General Features

Pioneering studies of *Ustilago maydis* and other smuts were carried out in Germany by O. Brefeld in the 1880s (24). These studies characterized the germination of teliospores and subsequent events, including the growth of the fungus in the plant, but the existence of heterothallism or "sex factors" was not documented. It is the remarkable work of J. J. Christensen, E. C. Stakmann, W. F. Hanna, J. B. Rowell, J. E. DeVay, and colleagues at University Farm, University of Minnesota, and H. O. Sleumer and R. Bauch in Germany that laid the foundations for our understanding of the mating types and their functional roles. This work was followed by that of R. Holliday, J. E. Puhalla, and P. R. Day, whose contributions set the stage for the molecular genetic analysis of the mating types by S. A. Leong, J. W. Kronstad, R. Kahmann, I. Herskowitz, F. Banuett, and their colleagues. Here I summarize their findings and try to capture and transmit the flavor of their work and their conclusions. I have chosen to arbitrarily divide the history of the *U. maydis* mating types into two major phases: (i) the discovery of heterothallism and the existence of "sex factors," and their functional roles, and (ii) the cloning and molecular analysis of the mating-type loci and the search for their targets. Both phases are intertwined with work on mating types in other smut fungi as well as in the mushrooms, and brief reference to that work will be made. The reader is referred to the appropriate chapters in this book for an in-depth analysis of mating types in other fungi. Before I begin this history, I present some general features of *U. maydis* and a synopsis of its life cycle as a framework to guide the reader in the work to be described.

Ustilago maydis (DeCandole) Corda, also known in the earlier literature as *Ustilago zeae* (Beckm.) Ungern and by other synonyms (see reference 26), is a basidiomycete fungus belonging to the class Ustilaginales (4). All members of the Ustilaginales are plant pathogens and cause severe diseases (smut diseases), particularly in cereal grains, resulting in great economic loss (2). Great interest has existed since the 1800s or earlier to understand smut diseases and their cause. *U. maydis* is the cause of corn smut disease. The disease is characterized by tumors that occur on all aerial plant parts. Tumors are not characteristic of other smut diseases. There are only two known hosts: maize (*Zea mays* L.) and teosinte

Flora Banuett, Department of Biological Sciences, California State University, 1250 Bellflower Bvd., Long Beach, CA 90840.

(*Zea mays* subsp. *parviglumis* and subsp. *mexicana*), the progenitor of maize (reviewed in references 8, 10, and 26). *U. maydis* is related to the rusts (class Uredinales) and to *Cryptococcus neoformans*, an opportunist pathogen of immunocompromised patients, and more distantly related to the jelly fungi (the Tremellales) and to the homobasidiomycetes *Schizophyllum commune* and *Coprinus cinereus*, both of which have been the subject of intense studies in the area of mating-type specificity.

Several features have contributed to making *U. maydis* amenable to studies of mating types, cell biology, and plant-pathogen interactions: the existence of a haploid unicellular phase that forms compact colonies on defined media and is amenable to manipulation by standard microbiological techniques, and the existence of meiosis, which allows segregation analysis (see references 8 and 10 for an extensive discussion of other features; see also references 36, 48, and 52). The development of DNA-mediated transformation and gene replacement by homologous recombination (37, 58, 93) ushered in the era of molecular genetics of the mating-type loci.

Life Cycle

The life cycle of *U. maydis* is characterized by three cell types (Fig. 21.1): a cigar-shaped haploid, unicellular form that divides by budding and is nonpathogenic; a dikaryotic, filamentous form that grows at the tip cell and is pathogenic; and a diploid spore, the teliospore, formed within the tumors induced by the fungus, that is not capable of vegetative growth but germinates and undergoes meiosis to produce the haploid yeast-like form (Fig. 21.1) (see references 8, 10, and 26). These morphological transitions entail several processes: conjugation, karyogamy, and meiosis, which are accompanied by changes in ploidy (haploid, dikaryotic, and diploid), growth habit (saprophytic, parasitic, or no vegetative growth), and pathogenicity (Fig. 21.1) (8, 17, 26). Understanding the life cycle, therefore, entails understanding the mechanisms that regulate cell type specialization and the processes that govern their transitions. Two unlinked mating-type loci, *a* and *b*, regulate the life cycle and cell type specialization. The *a* locus has two alleles, and the *b* locus is multiallelic (reviewed in references 8, 10, 36, and 51). The infectious cycle begins with fusion of haploid cells to form a dikaryon that exhibits filamentous growth (Fig. 21.1). Proliferation of this cell type in the plant induces the formation of tumors, within which the fungus undergoes a discrete developmental program that results in formation of teliospores (diploid spores) (Fig. 21.1) (reviewed in reference 8).

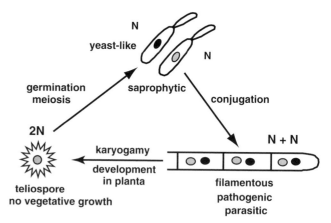

Figure 21.1 Morphological transitions in the life cycle of *U. maydis*. *U. maydis* exhibits three forms (cell types) during its life cycle: a haploid unicellular form; a filamentous dikaryon; and a diploid spore, the teliospore (reviewed in references 8, 10, and 26). The transition from one form to the other entails conjugation, karyogamy, and meiosis. These processes are accompanied by changes in ploidy, growth habit, and pathogenicity (see references 8, 10, and 26 for details). These morphological transitions are governed by two unlinked mating-type loci, *a* and *b*, by environment, by nutrients, and possibly by plant signals (reviewed in references 8, 10, 36, 51, and 52). (From reference 8, with permission.)

erence 8). Completion of the life cycle requires that the haploid cells that fuse carry different alleles at both *a* and *b* (reviewed in references 8, 10, and 51). I describe in later sections how the role of these loci was elucidated.

Although we have learned a great deal about how the mating-type loci control cell type specialization, much remains to be elucidated about the targets of these loci and how the regulation of these targets ultimately leads to the morphological transitions that characterize the life cycle and to a successful interaction with its host, which endows the fungus with the ability to undergo meiosis and to propagate itself successfully.

In other smut fungi, the *a* and *b* loci form a tightly linked complex known as the mating-type locus with two alleles (5, 6; see chapter 23 by G. Bakkeren and J. Kronstad). In *C. cinereus* and *S. commune* there are also two unlinked mating-type loci, but in contrast to *U. maydis*, both are multiallelic (see chapter 17 by L. A. Casselton and U. Kües and chapter 16 by M. M. Stankis and C. A. Specht). In *C. neoformans*, another basidiomycete, the mating-type locus with two alleles is a complex locus with genes for diverse functions (see reference 65 and references therein). A mating system determined by two unlinked loci is referred to as "tetrapolar," and one determined by one locus is known as "bipolar."

THE DISCOVERY OF HETEROTHALLISM AND SEX FACTORS IN *U. MAYDIS*

General Comments about Segregation Analysis and Nomenclature

Segregation Analysis

Because analysis of single meiotic segregants played a key role in earlier work, I shall describe briefly how this analysis is performed in *U. maydis*. *U. maydis* is a basidiomycete, and thus, the products of meiosis are not encased in specialized structures as in the Ascomycetes but are borne externally (basidiospores). The teliospore is a diploid spore that germinates to produce a short filament, the promycelium, into which the diploid nucleus migrates and undergoes the two meiotic divisions to form a septate filament consisting of four haploid cells, the primary meiotic segregants (68) (Fig. 21.2A). Each of these cells is capable of giving rise by mitosis (budding) to yeast-like cells (Fig. 21.2A). These are clones of the primary segregants. These yeast-like cells can be isolated individually by micromanipulation (Fig. 21.2B), the method of choice used by many earlier workers. Alternatively, each teliospore is allowed to undergo meiosis and form a microcolony of approximately 16 cells, which is then resuspended in water and spread on agar (Fig. 21.2C) followed by analysis of the colonies (50).

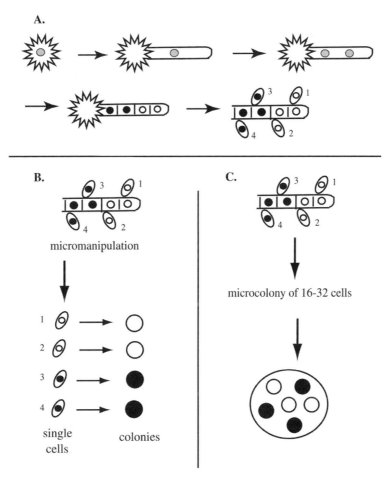

Figure 21.2 Teliospore germination and meiosis. The teliospore is a diploid spore, resistant to adverse environmental conditions. It is formed within the tumors following a discrete developmental program during which changes in morphology take place and karyogamy occurs (13). Upon germination, a short filament (the promycelium) emerges, into which the diploid nucleus migrates and undergoes meiosis to form a septate promycelium consisting of four haploid cells (68) (A). Clones of these primary meiotic products arise by budding to produce yeast-like cells (panel A), which can be removed with the aid of a micromanipulator for single tetrad analysis (panel B). Alternatively, the teliospore is allowed to germinate and form a microcolony, which is then resuspended in water and spread on an agar surface (panel C).

Lastly, teliospores can be germinated, and the products of different spores can be analyzed—random spore analysis, used commonly today.

Nomenclature

In earlier work "matings" meant inoculations into corn plants that result in formation of tumors (formerly galls) and production of teliospores (formerly chlamydospores). Earlier workers assumed that "matings" between haploid cells occurred sometime during passage through the plant, though this view was not universally accepted. Some had also inferred correctly that the teliospore is diploid, a view not widely accepted, and that haploid progeny arose by "reductional division" (meiosis). With the development of plate assays, "mating" became equated with the formation of aerial dikaryotic (or diploid) filaments (see below), but as we shall see later, there are instances where cells fuse and do not form filaments. The term "sporidia" was used to designate the haploid yeast-like cells that form on the promycelium and give rise to yeast-like cells by budding. This term continues to be used today by some researchers. I use "yeast-like cell" to refer to the "sporidium" because the term sporidium has the connotation of spore, which the yeast-like cells are not; they are clones of the primary meiotic segregants. "Sex factor" is used to indicate locus in some cases and allele in others. The meaning is clearly discerned from the context.

Heterothallism and Sex Factors

Stakman and Christensen (87) demonstrated for the first time that *U. maydis* is heterothallic. Pure lines (strains M2, M3, M4, and M7) collected in Minnesota when inoculated singly in corn plants did not induce disease symptoms but in certain combinations led to tumor formation and teliospore production. For example, coinoculation of M2 + M3, M2 + M7, M3 + M4, and M3 + M7 resulted in disease, whereas coinoculations of M3 + M4 or M3 + M7 did not. Stakman and Christensen (87) thus concluded that strains M3, M4, and M7 are all of one sex group and strain M2 is of a different sex group and therefore that *U. maydis* is heterothallic. Heterothallism was confirmed by coinoculation of mixtures of M2 with strains collected in different parts of the United States and Canada. In some crosses, progeny were obtained that induced disease when inoculated as pure cultures and were designated solopathogens. The yeast-like progeny resulting from inoculation with these strains did not cause disease when inoculated singly; only certain combinations resulted in disease. They inferred, correctly, that the

teliospore nucleus is diploid and gives rise to haploid progeny by reductional division. Diploid strains arise spontaneously with frequencies as high as 10% in certain crosses (27, 28, 49, 50).

Heterothallism had been demonstrated in other smut fungi. Kniep (55) was the first to report that the cells on promycelia of *Microbotryum violaceum* (formerly *Ustilago violacea*) belonged to two sexual groups. In *Ustilago longissima*, fusion between haploid cells occurs only in certain combinations, suggesting that this smut fungus is heterothallic (20).

U. maydis Has Two Unlinked Pairs of "Sex Factors" Unlike Other Smuts

Hanna (43) in a seminal paper demonstrated that *U. maydis* exhibits segregation of four sex types among the progeny of a single meiosis instead of two as found in most smut fungi. He germinated teliospores on a thin layer of agar on a microscope slide and isolated the individual yeast-like cells with the aid of a micromanipulator ("spore isolator"), carefully documenting the position of each cell on the promycelial cells, with no. 1 being the cell at the tip, and no. 4 the most basal cell (Fig. 21.2). A total of eight teliospores isolated at University Farm, Minnesota, or in Italy were analyzed. The four products of each teliospore were paired in all possible combinations with each other and injected into corn plants. Cultures of single strains were also inoculated. From this and additional analyses, he arrived at the following conclusions: (i) two pairs of factors—*Aa* and *Bb*—determine sex in *U. maydis*; (ii) the sex factors are on different chromosomes; (iii) the nucleus of the teliospore must be "genotypically" *Aa Bb*; (iv) the nucleus of each yeast-like cell receives one factor from each pair (for example, *AB, Ab, aB,* and *ab*) upon the divisions of the teliospore nucleus (reductional division); (v) first-division segregation of each factor results in a promycelium with two cells of one sex type and two of another (2 *AB* 2 *ab* or 2 *Ab* 2 *aB*; now referred to as parental ditype tetrad and nonparental ditype tetrad); (vi) first-division segregation for one of the factors and second-division segregation for the other factor will result in a promycelium with four cells of four different sex types (*AB, ab, Ab,* and *aB*; now referred to as a tetratype tetrad). We now know that second-division segregation will occur if there is recombination between the locus and the centromere. He also observed teliospores that produce three cells belonging to one sexual group and one to another. No explanation was provided, but these most likely represent gene conversion events. Hanna's work documents for the first time the existence of at least two unlinked "sex factors" in

U. maydis, a situation that differed from that described in other smut fungi.

It is likely that Hanna's work with *Coprinus lagopus* played an important role in the interpretation of the data found for *U. maydis*. In *C. lagopus*, a single basidium (the diploid cell that undergoes meiosis) can give rise to four basidiospores of four sexual groups (mating types) (44). Furthermore, as in *U. maydis*, of these four sex groups, only two result in compatible interactions. Pairings between mycelia derived from wild-type fruiting bodies of *C. lagopus* from different locations resulted in complete fertility between all combinations. He thus concluded that there were 24 distinct sexual groups represented in these *C. lagopus* populations. He did not extend similar studies to *U. maydis*, but Christensen (28) analyzed extensive pairings of *U. maydis* strains derived from teliospores obtained from several different localities and concluded that there were 19 sexual groups in *U. maydis*, though the evidence is not compelling.

The studies of Hanna (43) also demonstrated that *Sporisorium reilianum* (formerly *Sphacelotheca reilianum*), which causes anther smut of maize, is heterothallic and that the progeny of a single teliospore belong to four different sexual groups, a situation reminiscent of that in *U. maydis*. Recent analysis of the mating types of this smut fungus revealed some unexpected findings (80) (see below).

U. maydis has a Biallelic "Sex Factor" and a Multiallelic "Sterility Factor"

Even though disease development is a sensitive assay for strain compatibility and mating-type determination, it is time-consuming, and faster methods for analysis of strains were needed. Sleumer (83), in an attempt to correlate reactions between different strains under the microscope with those in the plant, analyzed the formation of mating pairs between different strains microscopically using different media. Three types of reactions were observed: (i) formation of aerial mycelium ("Suchfadenreaktion"); (ii) a weak reaction ("Wirrfadenreaktion"); and (iii) no reaction at all. Inoculation of corn plants with combinations of strains that produce each of these reactions showed that only those that produce aerial mycelium are capable of inducing tumors and producing teliospores. Those that produce a weak reaction are not capable of inducing disease. From this analysis he concluded that there are six sexual groups in *U. maydis*. Bauch (19) was the first one to develop a macroscopic plate assay using malt agar to record the reactions between haploid strains, a great improvement over the microscopic method. Extensive pairwise combinations of strains using the plate assay and inocula-

tions into corn plants led Bauch (19) to confirm the three types of reactions described by Sleumer (83) and to conclude that *U. maydis* has "two sex factors at one sex locus" and "6 factors at a sterility locus." For the first time, evidence was presented for the existence of multiple alleles at one of the mating-type loci, anticipating the work of Rowell and DeVay (78). Bauch (19) also conducted similar studies with *U. scorzonerae* and concluded that this smut fungus is bipolar.

The Genetic Basis of Mating Incompatibility: the *a* and *b* Loci

Rowell and DeVay (78) confirmed Hanna's conclusions that there were two sex factors (e.g., two loci) and conclusively demonstrated that one of these loci has two alleles and the other is multiallelic. Analysis of the progeny of a single teliospore indicated that there were at least four sex groups among which only two combinations are compatible. They designated the sex factors as the *a* and *b* mating-type loci. The analysis of a single teliospore (*a1/a2 b3/b5*, for example) resulted in four possible combinations of the *a* and *b* alleles in some tetrads: *a1 b3*, *a1 b5*, *a2 b3*, and *a2 b5*, which can combine only in two possible ways: *a1 b3 + a2 b5* and *a1 b5 + a2 b3*. Combinations of representative strains of the four groups from 11 different teliospores in all possible combinations indicated that there were only two forms (alleles) of the *a* factor, *a1* and *a2*, and that there were two major lines within which there were no compatible matings: "those having *a1* in common or those having *a2* in common." Matings between those having *a1* with those having *a2* divided the strains into many groups, in which those with identical *b*s were incompatible (meaning there was no disease), but those with different *b*s resulted in compatible combinations (meaning there was disease) (Table 21.1), hence the designation of incompatibility loci for the mating-type loci. Fifteen different *b* factors (alleles) were identified among these strains. Additional analysis included thousands of pairings between different strains. This work was a true genetic classic that conclusively demonstrated the genetic basis of sex determination in *U. maydis*.

In one particular cross between strains 10A (*a1 b1*) × 17D4 (*a2 b4*), seven different groups were obtained among the haploid isolates from different teliospores. Four corresponded to the expected haploid recombinants: *a2 b4*, *a2 b1*, *a1 b1*, and *a1 b4*. One group was solopathogenic, that is, it caused infection when injected as a pure culture into corn plants, and was thus inferred to be a diploid heterozygous at *a* and *b* (*a1/a2 b1/b4*). The other two strains behaved differently from all presumed haploid segregants: they mated with all *a1* and *a2* haploid strains that carried either *b1* or *b4*, respectively

Table 21.1 The b locus is multiallelic[a]

| a1 strain mating type (source) | Compatibility[b] with a2 strain mating type (source): | | | | | |
	a2 b2 (NyBb)	a2 b4 (NyBg)	a2 b3 (InC2)	a2 b5 (InC4)	a2 b7 (BrA4)	a2 b8 (BrA3)
a1 b2 (NyAd)	−	+	+	+	+	+
a1 b4 (NyCe)	+	−	+	+	+	+
a1 b3 (InB4)	+	+	−	+	+	+
a1 b5 (InC4)	+	+	+	−	+	+
a1 b7 (BrA1)	+	+	+	+	−	+
a1 b8 (BrA2)	+	+	+	+	+	−

[a]Haploid segregants derived from different teliospores collected in different localities were paired in different combinations (78). Source, locality where teliospores were collected: Ny, New York; In, Indiana; Br, Brazil. (From reference 78 with permission.)

[b]−, no disease reaction (incompatibility); +, disease reaction (compatibility).

(Table 21.2). The strains were inferred to be diploids homozygous for b1 or b4 and heterozygous for a (a1/a2 b1/b1 or a1/a2 b4/b4), and their genotype was conclusively demonstrated using crosses and alpha irradiation (77). Because these diploids were not pathogenic when injected singly into corn plants, Rowell and DeVay (78) concluded that the a locus is not involved in pathogenicity, a contention supported by later studies (16, 71, 76). These nonpathogenic diploid strains were the first to be documented in U. maydis. Diploids homozygous at a or b have played a key role in cloning the mating-type loci and in analysis of mutants (see below).

If a is not involved in pathogenicity, what then is its role?, was the question addressed by Rowell (76). Several investigators had claimed that cell fusions of U.

maydis haploid strains occurred in certain combinations (23, 83), but Rowell (76) was unable to repeat these observations with their assay conditions. Using a "maize-coleoptide agar assay" on which mixtures of strains are observed microscopically, Rowell (76) determined that strains that differ at both loci form a straight, fast-growing hypha (Sleumer's Suchfäden) after 3 to 4 h; if they differ only at a, slow-growing, sinuous, unstable hyphae develop (Sleumer's Wirrfäden); if they are identical at a, they remain yeast-like. Based on these observations, he concluded that "the a locus governs sporidial fusion and the b locus governs the vigor and viability of the dikaryon and apparently the compatibility of the paired nuclei" (76). For the first time, distinct functions were assigned to the a and b loci.

Table 21.2 Compatibility reactions among progeny from cross 10A4×17D4 and from solopathogen 410qq[a]

| Mating types | No. of isolates | | Compatibility[b] with tester strains: | | | | |
	10A4 x 17D4	410qq	a1 b1	a1 b4	a2 b4	a2 b1	Solopathogen
a2 b4	1	6	+	−	−	−	−
a2 b1	2	8	−	+	−	−	−
a1 b1	9	6	−	−	+	−	−
a1 b4	1	10	−	−	−	+	−
a1/a2 b4/b4	2	1	+	−	−	+	−
a1/a2 b1/b1	12	1	−	+	+	−	−
a1/a2 b1/b4	26	1	+	+	+	+	+

[a]Progeny from the cross between strains 10A4 (a1 b1) × 17D4 (a2 b4) and also derived from the solopathogen 410qq (a1/a2 b1/b4) were inoculated into corn plants with the indicated tester strains to determine mating type (78). Haploid progeny were of four classes (a2 b4, a2 b1, a1 b1, and a1 b4; four top rows), as expected. Diploids were also obtained and were of two classes: homozygous for b (a1/a2 b4/b4 and a1/a2 b1/b1; rows 5 and 6) and heterozygous at both loci (a1/a2 b1/b4; bottom row). The b homozygous diploids were shown to be nonpathogenic when inoculated as pure cultures but were able to induce disease in combination with tester strains carrying a different b allele, regardless of the a allele present (78). (From reference 78, with permission.)

[b]−, no disease reaction (incompatibility), +, indicates disease reaction (compatibility).

To determine if there is a correlation between the type of hypha formed on maize coleoptile agar and ability to induce tumors and produce teliospores, six representative strains paired in all possible combinations on coleoptile agar and by inoculation into corn plants established unequivocally that combinations of strains that have different *a* alleles fell into two groups: those that fuse to produce sinuous slow-growing hyphae are not pathogenic, and those that fuse to produce straight fast-growing hyphae are pathogenic. To further support this contention, 1,645 pairings between 669 strains were tested using both the maize coleoptile assay and induction of disease symptoms. In 97.3% of the cases, there was complete agreement between the two tests. In an additional 5,002 matings between 441 strains, there was disagreement in only 1.9% of the tests. The exceptions were usually strains that normally exhibit a mycelial phenotype as haploids or that have been subcultured for >10 years, as was the case for some of the strains used at Minnesota.

The seminal study of Rowell and DeVay (78) and Rowell (76, 77) was the culmination of decades of work by J. J. Christensen, E. C. Stakman, W. F. Hanna, and their colleagues and of H. O. Sleumer and R. Bauch and set the stage for molecular genetic studies of the mating-type loci.

New Developments That Rekindled Interest in the Study of Mating Types

Interest in the study of *U. maydis* resurfaced when Robin Holliday set out to study recombination in this fungus. Some factors which contributed to the advance of the genetic analysis of *U. maydis* were the use of defined media, the application of microbiological techniques such as replica plating, which had been recently developed by Lederberg and Lederberg (64), the isolation of mutants with clearly defined phenotypes that are due to alteration of single genes (69), the generation, for the first time, of diploid strains in the laboratory (49, 50), and the improvement of plate assays for determination of mating types and for mutational analysis of the mating types (31, 32, 71).

Generation of Diploids in the Laboratory

Holliday (49, 50), who was interested in using somatic segregation as a technique in construction of linkage maps in *U. maydis*, was the first one to generate solopathogenic and nonpathogenic diploid strains in the laboratory. The solopathogenic diploids were conclusively shown to be heterozygous at *a* and *b* after inoculation into corn plants and analysis of the haploid progeny which fell into the expected genotypic ratios (49,

50). Diploid solopathogenic strains as well as nonpathogenic diploids arise spontaneously with a frequency that can be as high as 10% in certain crosses, most likely due to failure of meiosis (27, 28, 50, 78). The reason for this failure in certain crosses and not others is not known. The *a1/a2 b1/b2* solopathogenic diploid was treated with UV irradiation to induce mitotic recombination (50). Two types of strains that exhibited distinct properties with respect to the original diploid and to haploid strains were obtained (49, 50). One type of strain was not pathogenic but induced disease in combination with *a1 b1* or *a2 b1* haploid strains and was assumed to be a diploid homozygous for *b2* and heterozygous for *a*. Rowell and DeVay (78) had conclusively demonstrated that nonpathogenic diploids were homozygous for *b*. The other strain was homozygous for *pan-1*, which had been shown to be tightly linked to the *a* locus (49). This strain was weakly pathogenic and was assumed to be a diploid homozygous for *a1*, the allele in coupling with the *pan-1* marker, and heterozygous for *b*. This claim, however, was not substantiated by analysis of the progeny. Conclusive demonstration that diploids homozygous at *a* and heterozygous at *b* are pathogenic was provided later (16, 71).

The Plate Mating Assay on Charcoal Agar

The development of an improved plate mating assay facilitated analysis of the mating type loci (71). This assay is reminiscent of that described by Bauch (19) (see above), except that the medium used is different. The plate assay developed by Puhalla (71) was later modified by the addition of activated charcoal (48). The inclusion of charcoal was due to a serendipitous observation of Day and Anagnostakis (31) that activated charcoal favors growth of the filamentous dikaryon for unknown reasons. It is possible that charcoal removes an inhibitor of cell fusion or filamentous growth produced by haploid strains. The use of charcoal did not become popular until much later (see, for example, references 16, 17, 18, 38, 82, and 96).

The plate mating assay requires close proximity of the strains to be tested (71). When haploid strains that carry different *a* and *b* alleles are costreaked or cospotted on charcoal agar, a white fuzziness develops (the fuzz reaction) due to formation of dikaryotic filaments; a Fuz$^+$ phenotype (Fig. 21.3) (16). If the strains carry identical *a* or identical *b* alleles, no fuzziness develops, a Fuz$^-$ phenotype (Fig. 21.3). Formation of dikaryotic filaments involves establishment of the dikaryon (cell fusion and migration of nuclei to establish the dikaryon) (see below), and maintenance of the dikaryon (filamentous growth and close apposition of the nuclei). Mutations that block

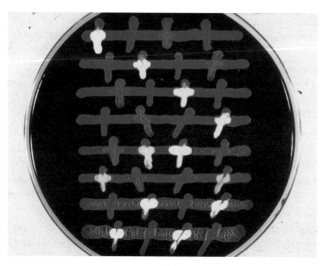

Figure 21.3 The fuzz reaction of haploid and Fuz− diploid strains on charcoal agar. Haploid and diploid strains carrying different *a* and *b* alleles produce a white fuzziness due to the formation of filaments in combination with haploid tester strains (16). Saturated cultures of tester strains (vertical lines) were cross-streaked versus haploid (top four horizontal lines) and Fuz− diploids (bottom four horizontal lines). The testers are, from left to right, *a1 b1* (FB1), *a2 b2* (FB2), *a1 b2* (FB6b), and *a2 b1* (FB6a). The horizontal lines represent, from top to bottom, strains *a2 b2*, *a1 b1*, *a2 b1*, *a1 b2*, and *a1/a2 b1/b1* (class I; FB-D12-3); *a1/a2 b2/b2* (class II; FB-D11-21); *a1/a1 b1/b2* (class III; FB-D11-7); and *a2/a2 b1/b2* (class IV; FB-D12-11) (see reference 16 for further details). (From reference 16, with permission.)

the first step also block the second step, but mutations that block the second step may not affect the first step. Formation of dikaryotic filaments cannot be equated with cell fusion because haploid strains that carry different *a* alleles and identical *b* alleles are able to fuse but do not develop filaments (see below) (15, 85), and their reaction cannot be distinguished from that of strains with identical *a* alleles. This assay, therefore, measures ability to form filaments, which entails cell fusion and growth of the dikaryon as a filament. Assays that specifically measure some of the early steps in cell fusion are described later.

On charcoal agar or on Puhalla's medium, diploids heterozygous at both *a* and *b* form mycelial colonies whereas *b* homozygotes do not (16, 21, 71, 86). Diploids homozygous at *a* and heterozygous at *b* were claimed to form mycelial colonies, but other studies indicated the contrary (16, 21, 70, 86). The charcoal agar assay has been very useful in isolation of mutants and cloning of the mating-type loci (see, for example, references 12, 18, 21, 38, 40, 56, and 82).

The plate assay allowed Puhalla (70) to determine that 18 different *b* alleles were represented in 62 different isolates from various locations in the United States and Canada. Thus far, this and the analysis of Rowell

and DeVay (78) have been the most comprehensive studies on the distribution of *b* alleles in the population. Whether the *b* alleles identified by Puhalla (70) are the same as those identified by Rowell and DeVay (78) or Holliday (49) is not known. To my knowledge, there are no studies on the distribution of *U. maydis b* alleles in Mexico, the center of origin of maize. It would be most interesting to know if *b* allele distribution is greater in association with greater diversity of its host, maize, and with the presence of its progenitor, teosinte.

The multiallelic mating-type loci of *C. lagopus* (34) and *S. commune* (72) were known to consist of two independent subunits (*Aα* and *Aβ*). Because recombination between them generated new incompatibility factors (33, 72), Puhalla (70) set out to determine if recombination occurred at the *b* locus of *U. maydis*. He analyzed large numbers of progeny from particular crosses using the plate mating assay. If recombination occurs, he reasoned, then the recombinant progeny would be of two classes: dual maters, that is, able to mate with both parental strains, or nonmaters, unable to mate with the parental strains. In approximately 5,000 progeny from one cross, a 1:1 segregation ratio for the *b* alleles was obtained. A few exceptional progeny displayed a dual-mater phenotype, as expected for a recombinant, but these strains were solopathogenic. Analysis of the progeny from these unusual strains showed a 1:1 segregation of the *b* alleles but no segregation for *a*, indicating that the strains were diploids homozygous for *a* and heterozygous for *b*. The lack of detectable recombination at the *b* locus may be due to extensive nucleotide polymorphisms among the *b* alleles (see below).

bmut Alleles and Possible Role of the *b* Locus in Meiosis

In order to gain a better understanding of the *b* locus, Day et al. (32) isolated mutant derivatives from diploids homozygous for different *b* alleles (*bD*, *bG*, and *bI*), which form yeast-like colonies on plate mating medium and are nonpathogenic (71). Six mutants designated *bmut* (three *bGmut*, two *bDmut*, and one *bImut*) were obtained, all of which induced tumors and produced teliospores when inoculated with their progenitors, with themselves and with each other, and thus behave as "universal" alleles. Prototrophic haploid *bmut* strains were solopathogenic. In crosses between *bmut* strains with wild-type *b* strains, the progeny segregate 1:1 (*b:bmut*). Tetrads had four, three, two, and one meiotic products, but some teliospores contained more than four meiotic products. In crosses between strains with the same *bmut* allele, survival of meiotic products was reduced, and only tetrads with one or two products were recovered, which appear to be diploids that arise due to failure of meiosis (65% of the time). Crosses between different *bmut* alleles produce teliospores with the expected number of meiotic products, for the most part, but also some teliospores with more than four meiotic products. Because meiosis as well as survival of meiotic products is altered, Day et al. (32) concluded that the *b* locus is necessary not only for tumor induction but also for proper meiosis and survival of meiotic products. These studies provide preliminary evidence for a role of the *b* locus in meiosis and spore survival.

Brief Personal Account of My Introduction to the Study of the Mating Types

After a visit to the United Kingdom where Ira Herskowitz had the opportunity to learn about the *U. maydis* mating-type loci from Robin Holliday, Ira and I became interested in pursuing work on the mating types of *U. maydis* and were encouraged to do so by Robin Holliday, Sally Leong, and David Perkins. We obtained from S. Leong strains 521 (*a1 b1*) and 518 (*a2 b2*), developed by Holliday (48), and crossed them to obtain from a single teliospore exhibiting a tetratype tetrad the four sister segregants designated FB1 (*a1 b1*), FB2 (*a2 b2*), FB6a (*a2 b1*), and FB6b (*a1 b2*) (Fig. 21.3). These are the standard tester strains used in my work and that of others. Auxotrophic derivatives of these strains were used to obtain diploid strains heterozygous at *a* and *b* by selection for prototrophy (see references 14 and 16).

As indicated above, diploids heterozygous at *a* and *b* (for example, *a1/a2 b1/b2*) exhibit a Fuz⁺ phenotype on charcoal agar. Nonfilamentous (Fuz⁻) derivatives of this diploid were obtained by treatment with a low dose of UV, which induces mitotic recombination (16). These Fuz⁻ diploids exhibited a dual-mater phenotype on charcoal agar and belonged to two distinct classes (Fig. 21.3): those that mated with haploid strains carrying a different *b* allele regardless of the *a* allele present and those that mated with strains that carried a different *a* allele regardless of the *b* allele present. The former were shown to be homozygous for *b* (*a1/a2 b1/b1* or *a1/a2 b2/b2*), and the latter were homozygous for *a* (*a1/a1 b1/b2* or *a2/a2 b1/b2*). The *b* homozygotes were nonpathogenic, as had been previously shown by Rowell and DeVay (78), Holliday (49, 50), Puhalla (71), and Day et al. (32). The *a* homozygotes were pathogenic. Meiotic analysis showed that only *a1 b1* and *a1 b2* progeny were obtained from this diploid. The studies by Banuett and Herskowitz (16) conclusively demonstrated with a set of congenic strains that different *a* alleles are necessary for maintenance of filamentous growth on charcoal agar but are not necessary for pathogenicity or meiosis. Heterozygosity at the *a* locus is not necessary for filamentous growth in planta (13). Because the *a* locus governs cell fusion, as concluded by Rowell (76) and conclusively demonstrated by others (see below), the dual-mater phenotype of the Fuz⁻ diploids suggests that heterozygosity at *a* or *b* does not inhibit mating as measured by filament formation (a qualitative assay). However, Laity et al. (63) presented evidence indicating that heterozygosity at *b* inhibits mating (see below).

I next describe cloning and molecular genetic analysis of the mating-type loci.

CLONING AND MOLECULAR GENETIC ANALYSIS OF THE MATING-TYPE LOCI

The *b* Locus Codes for a Combinatorial Homeodomain Protein That Is the Hallmark of the *U. maydis* Filamentous Pathogenic Dikaryon

I describe in this section the developments that have led to our current understanding of the molecular organization and mode of action of the *b* locus. The *b* locus of *U. maydis* was the first multiallelic mating-type locus cloned and understood at the molecular level and provided a framework in which to understand the more complex *A* multiallelic mating-type locus of other basidiomycetes.

Molecular Cloning of the *b* Alleles

Given that there are at least 25 naturally occurring *b* alleles and that any combination of different *b* alleles is sufficient to trigger pathogenicity (32, 70, 71, 78), the challenge was to determine how *U. maydis* discriminates self from nonself. Is recognition a cell surface phenomenon? Does it occur intracellularly? Does it involve nuclear-limited products, either RNA or protein? These were some of the challenges faced by those studying the multiallelic mating-type loci in the Basidiomycetes (see, for example, references 17, 62, and 89).

Cloning of bE

The *b1* and *b2* alleles were cloned by functional complementation of the Fuz⁻ phenotype of a diploid homozygous for the *b* locus, but no sequence information was provided (57). Schulz et al. (82) cloned and sequenced four alleles (*b1*, *b2*, *b3*, and *b4*) and showed that each allele contains an open reading frame (ORF) with coding capacity for a polypeptide of 473 amino acids that contains a homeodomain-related motif (Fig. 21.4). The

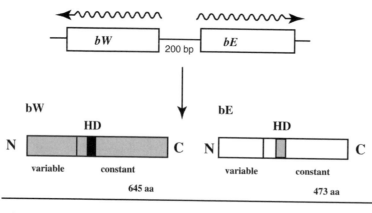

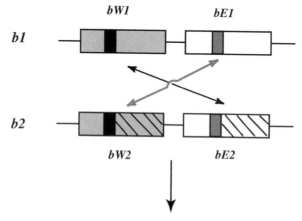

2 active heterodimers: bW1-bE2 and bW2-bE1

Figure 21.4   The molecular structure and organization of the *b* locus. The *b* locus is the major determinant of filamentous growth and pathogencity (reviewed in references 8, 10, 26, and 51) and is estimated to have 25 naturally occurring alleles (70, 78), each of which consists of two divergently transcribed genes, *bW* and *bE*, separated by an intergenic region of approximately 200 bp (40, 82). Each gene codes for a polypeptide containing a homeodomain-related motif (40, 82). The bW and bE polypeptides have the same structural organization, a variable region at the amino terminus and a constant region for the remainder of the polypeptide, but show no similarity except for the homeodomain region (40, 56, 82). The active b protein is found only in the dikaryon and consists of a bW and a bE subunit encoded by different *b* alleles, for example, bW1-bE2 or bW2-bE1 (40, 54). Each dikaryon has two different active heterodimers, but only one is necessary for activity (40). The variable region is the specificity determinant and controls dimerization of these polypeptides (54, 96, 97). The homeodomain of both bE and bW is required for activity (81). (From reference 8 with permission.)

ORF is interrupted by an intron of 74 nucleotides (nt) at position 1192. The presence of a homeodomain motif suggested that the *b* locus codes for a regulatory protein and that self-nonself recognition occurs intracellularly at the level of polypeptides. The bioinformatics tools available at the time did not detect the homeodomain in the b polypeptides. Ira and I uncovered its presence by "eye-balling" amino acid sequence alignments of the b polypeptides with the regulatory proteins encoded by the *MAT* locus of *Saccharomyces cerevisiae, Schizosac-*

charomyces pombe mat, and *Drosophila* Antennapedia and engrailed genes (82).

Molecular Organization of the bE Polypeptide
Alignment of the amino acid sequences of the b polypeptides revealed a molecular organization consisting of a variable region (63% identity) at the amino terminus and a constant region (93% identity) for the remainder of the protein (Fig. 21.5). The variable region encompasses approximately the first 110 amino acid

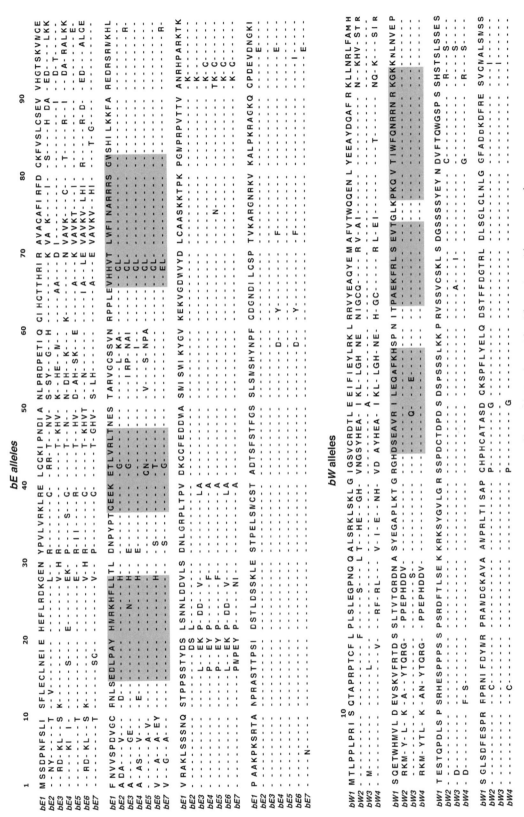

Figure 21.5 Amino acid sequence comparison of bE and bW polypeptides. The amino acid sequences of seven bE and four bW polypeptides are shown (40, 56, 82). The variable region of bE and bW encompasses the first 110 and 140 amino acids, respectively. The remainder of the polypeptide is the constant region (not shown in its entirety) and contains the homeodomain. The three helices of the homeodomain are shaded. The homeodomain of bE belongs to the atypical class of homeodomains, whereas that of bW belongs to the typical class (see reference 25).

residues (Fig. 21.5). The homeodomain extends from approximately amino acid residues 115 to 182. Kronstad and Leong (56) provided amino acid sequence information for additional alleles (b3=bH, b5=bJ, b6=bK, and b7=bL), which lent support to the proposed molecular structure and organization of the b locus.

Small deletions or insertions of the b2 ORF abolished b2 specificity, as measured by filamentous growth on charcoal agar and tumor induction in corn plants, when the altered b2 region was introduced into a diploid strain homozygous for b1 (82). A deletion encompassing part of the b1 ORF and extending 397 bp upstream of the ORF also abolished b activity (56). These observations indicated that b acts as a positive regulator of filamentous growth and pathogenicity. Introduction of an intact b2 fragment restored filamentous growth and pathogenicity to the a1/a2 b1/b1 Fuz⁻ diploid, supporting previous work that b is necessary for filamentous growth.

Models for Mode of Action of b

Two models were proposed to explain how b polypeptides might interact (82). In the first model, the variable region determines the association of polypeptides and activity of the protein. In one scenario of this model, only the variable regions of identical polypeptides interact, predicting that the active b protein is a homodimer and that it would be present in both the haploids and the dikaryon. In a second scenario, only the variable regions of different b polypeptides are able to interact, thus predicting that the active protein is a heteromer and most likely a positive regulator. In the second model, the constant region determines association of polypeptides. All polypeptides, therefore, can associate but activity is determined by the variable region which can do so in a variety of ways: "by determining the spacing of the recognition helix, by providing a site for interaction with another protein, or by creating a DNA recognition domain" (82). It was proposed that alignment of the corresponding positions between two polypeptide chains with each other could be used to monitor self from nonself. For example, identical polypeptides could have residues of identical charge facing each other, which would be thermodynamically unfavorable (82). Kronstad and Leong (56) proposed that the active species was a heterodimer that acted as a positive regulator, based on the analysis of a b-null mutation. They proposed that activity could be regulated at the level of localization: both homo- and heterodimers would be formed, but only heterodimers would enter the nucleus to activate target genes. Differences in localization could be due to interaction with a protein that

facilitates anchoring in the cytoplasm or differences in conformation of the variable regions. Recent models for self-nonself recognition are refinements of one of the models proposed by Schulz et al. (82) (see, for example, references 11, 40, 53, and 54).

Identification of bW

The molecular organization of the b locus turned out to be more complex than anticipated (Fig. 21.4). Gillissen et al. (40) deleted most of the b2 ORF, and the in vitro-generated null mutation was used to replace the b locus in a1 b1 and a2 b2 strains by homologous recombination. The null mutants were viable and exhibited no detectable phenotype. Remarkably, the a2 Δb2 strain retained b2 specificity when paired with a1 b1 strains on charcoal medium. This unexpected result contrasted with the null phenotype (loss of b specificity) described by Kronstad and Leong (56) and suggested the existence of a functionally redundant gene that conferred b2 specificity. Indeed a divergently transcribed gene was located upstream of the original b2 gene, and its location was conserved in the b1, b3, and b4 alleles (Fig. 21.4). A 200-bp intergenic region separates the two genes (40). The b-null deletion generated by Kronstad and Leong (56) removed the intergenic region and part of the 5′ end of this gene.

Molecular Organization of bW

The new gene contains an ORF with coding capacity for a polypeptide of 645 amino acids (40; see also reference 54) that has a homeodomain motif (Fig. 21.4). An intron of 270 nt interrupts the homeobox. Amino acid sequence comparisons of these polypeptides showed a molecular organization similar to that of the previously identified gene: a variable region extending for the first 130 amino acids (46% identity) and a constant region (96% identity) for the remainder of the polypeptide (Fig. 21.4 and 21.5). The homeodomain is located near the beginning of the constant region and extends from amino acid residues 143 to 193. This new gene was designated bW (for b West), and the previously identified one was designated bE (for b East) (Fig. 21.4). Sequence analysis of additional b alleles confirmed that the b locus consists of two divergently transcribed genes, bE and bW (R. Kahmann, personal communication). Each b allele thus has two separate genes: b1 contains bW1 and bE1, b18 contains bW18, and bE18, etc. These two genes appear to be inseparable by recombination and are thus inherited as a single genetic unit. The intergenic region and the variable regions of each gene exhibit nucleotide sequence polymorphisms, which may block homologous recombination between bW and bE. Although the molecular organization of the bW polypeptides is similar to that of the bE polypeptides, the two proteins share no similarity except for some invariant amino acid residues in the homeodomain region. The homeodomain of *bW* belongs to the typical class, whereas that of *bE* belongs to the atypical class (25) (Fig. 21.5). The homeodomains of the paired homeodomain proteins coded by the multiallelic *A* locus in *S. commune* and *C. cinereus* also fall into two classes (61, 88) and were designated HD2 and HD1, for the typical and atypical class, respectively (59). The homeodomains of *S. cerevisiae* α2 and **a**1 belong to the HD1 and HD2 class, respectively.

bE-bW from Different *b* Alleles is the Active b Protein

To determine whether the presence of different *bE*, *bW*, or different *bE* and *bW* results in the active b protein, Gillissen et al. (40) generated a set of *a1* and *a2* strains containing deletions of *bE*, *bW*, or both (Fig. 21.6). The strains were assayed for filament formation on charcoal agar and for tumor induction in corn plants to assess the presence of an active b species (Fig. 21.6). This elegant analysis demonstrated that only *bW* and *bE* from different alleles, for example, *bW1* and *bE2*, when present in the same cell trigger filament formation and tumor induction. All other combinations are inactive (Fig. 21.6). The active b protein is, therefore, a multimer of *bW* and *bE* contributed by different *b* alleles. Because there are two different *bW* and *bE* genes in a dikaryon, two possible active heterodimers can be formed, for example, bW1-bE2 and bW2-bE1, but only one of them is sufficient for *b* activity (40) (Fig. 21.6). The functional heteromultimer appears to act as a positive regulator of genes for filamentous growth and tumor induction, a conclusion drawn from the observation that both gene products are necessary to form filaments and to induce tumors. Because it is estimated that there are 25 *b* alleles and any combination of two different *b* alleles triggers pathogenic development, and given that there are two active heteromultimers in the dikaryon, then there are at least 1,200 possible active heteromers and 50 inactive ones that can be formed by combinations of these 25 *b* alleles. The rules by which a functional protein is generated remain to be elucidated.

A haploid strain engineered to carry *bE* and *bW* from different *b* alleles (for example, *a2 bW1 bE2* [RK1659] or *a1 bW1 bE2* [RK1645]) induces tumors, albeit with reduced efficiency, when a pure culture is inoculated into corn plants (54). This observation supports the above contentions. These haploids are weakly filamentous on charcoal agar but develop a strong reaction when paired with *a1 Δb* or *a2 Δb* strains, indicating that the *a* locus contributes to a full filamentous response as previously shown (16; see also reference 86). The weak-pathogenicity phenotype of this haploid may indicate that cell fusion itself activates genes that enhance pathogenicity.

The complex structure of the *b* locus in *U. maydis* and the proposed polypeptide interactions between genes at the *b* locus are also observed in the complex multiallelic mating-type loci of two other basidiomycetes, *S. commune* and *C. cinereus*. In these fungi, the *A* mating-type locus consists of two subunits (loci), Aα and Aβ. The Aα locus of *S. commune* was shown to contain two divergently transcribed genes, *Y* and *Z*, that code for dissimilar homeodomain proteins (88). These proteins interact to form an active heteromultimer (73). The allelic Z polypeptides show 42% identity to each other and 20 to 25% identity to the bE polypeptides of *U. maydis;* the allelic Y polypeptides show 49 to 54% identity to each other and 20 to 25% identity to the bW polypeptides of *U. maydis* (88). In *C. cinereus*, the Aα and Aβ loci contain at least 2 pairs of dissimilar homeodomain proteins that interact if they are from different *A* alleles and form active heteromultimers (59, 61; see chapters 16 and 17 for further details).

The Variable Region Mediates Protein-Protein Interactions between bW and bE

bE and bW Interact in the Yeast Two-Hybrid System

The genetic analysis showed that *bW* and *bE* from different *b* alleles are necessary for b activity (40). Yeast two-hybrid interactions and biochemical analysis demonstrated that the variable region mediates the interaction of bW and bE polypeptides (54). First, the entire ORF or functional carboxy-terminal truncations of *bW1*, *bW2*, *bE1*, and *bE2* were fused to the yeast Gal4 DNA binding domain or the Gal4 activation domain. Introduction of full-length bW1, bW2, bE1, or bE2 fused to Gal4 DNA binding domain or Gal4 activation domain alone into a yeast strain containing a *lacZ* reporter driven by the *gal4* binding site resulted in no detectable β-galactosidase activity. Introduction of both *bW1* and *bE1* did not activate the reporter, indicating that proteins from the same allele do not interact. In contrast, a combination of *bE1* and *bW2* or *bE2* and *bW1* activated expression of the reporter gene, demonstrating that only bW and bE polypeptides derived from different alleles are able to interact to form a stable complex. Next, combinations of b polypeptides containing various carboxy-terminal deletions were used to

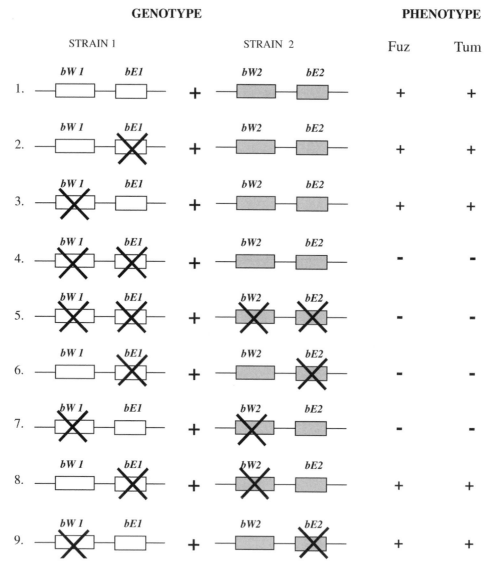

Figure 21.6 Deletion analysis of the *b* locus. The *bE* or *bW* or both genes were deleted in haploid strains (40), and the resulting mutants were tested for b activity (filament formation [Fuz phenotype] and tumor induction [Tum phenotype]) by cospotting on charcoal agar and inoculation into corn plants, respectively. The strains listed under strain 1 are derivatives of *a1 b1*, and those under strain 2 are derivatives of *a2 b2*. Mixtures of wild-type *a1 b1* and *a2 b2* strains result in a Fuz⁺ Tum⁺ phenotype (row 1). Deletion of one *bE* gene or one *bW* gene in one of the partners does not affect b activity (rows 2 and 3, respectively). On the other hand, b activity is abolished by deletion of both *bE* and *bW* in one of the partners (row 4), both *bE* and *bW* genes in both partners (row 5), *bE* in both partners (row 6), or *bW* in both partners (row 7). In contrast, deletion of *bE1* in one partner and *bW2* in the other (row 8) or of *bW1* in one partner and *bE2* in the other (row 9) does not abolish b activity and demonstrates that *bW* and *bE* from different *b* alleles are necessary for *b* activity (40). The dikaryon contains two active heterodimers, but only one is necessary for function (rows 8 and 9).

determine the region responsible for the interaction. Combinations of polypeptides derived from different alleles and containing the variable region only or portions thereof were still active in the yeast two-hybrid system, indicating that the variable region is sufficient for inter-

action of nonallelic bW and bE polypeptides and that the homeodomains are not necessary for this interaction. Biochemical experiments demonstrated that physical interactions occur only between bW and bE polypeptides derived from different *b* alleles (54). No interactions were detected between bW and bE derived from the same allele.

bmut *Alleles Identify the* b *Specificity Region*

The variable regions are 63 and 46% identical for the bE and bW polypeptides, respectively, suggesting that specific residues that differ among the different b polypeptides must play a key role in mediating the interaction (Fig. 21.5). The location of these residues may vary depending on the allele in question. To identify these residues, Kämper et al. (54) mutagenized the variable region of *bE2* and introduced the mutagenized plasmid into an *a1/a2 b2/b2* diploid strain, which is Fuz⁻, and screened for filament formation (Fuz⁺ phenotype) on charcoal agar. Filament formation was thus used as a measure that mutant *bE2* recognized *bW2* as nonself. Four mutants contained a single amino acid substitution (H87L, T94I, K90I, or A90V), one had two changes (A90V H121R), three had three changes (P32L A90V P135L; K74E K100R N148S; or K74E E92G T151S), and one had four changes (V13L F23V H60Q K100E). Inoculation of the *a1/a2 b2/b2* strain containing these mutations into corn plants led to tumor induction. These alleles, therefore, recognize *bW2* as nonself and were designated *bE2mut*. A subset of the mutations was introduced into an *a1 Δb* strain, and the resulting strains were cospotted on charcoal agar with *a2* strains carrying different *b* alleles. All pairwise combinations led to filamentous growth, indicating that the *bmut* alleles recognize other *b* alleles as nonself. In the yeast two-hybrid system, all *bE2mut* alleles in combination with *bW1* activated expression of the *lacZ* reporter gene. In combination with *bW2* alleles, they activated the *lacZ* reporter albeit at a much reduced level. Since filament formation on charcoal agar is not a quantitative assay, it is not known if differences in the extent of reporter expression translate into quantitative differences in filament formation. Could it be that the differences reflect a mutant allele whose specificity is in transition?

Using a reciprocal approach, Kämper et al. (54) isolated *bE2mut* alleles in yeast by screening for activation of the *lacZ* reporter gene in the presence of *bW2*. The mutants obtained were biologically active in *U. maydis*, indicating that the ability to interact in the yeast two-hybrid system correlates strongly with biological activity in *U. maydis*. The amino acids altered in the *bE2mut* alleles isolated in these studies (54) encompass the specificity region identified by Yee and Kronstad (96, 97; see below). In addition, some of the mutations fall within the homeodomain.

None of the *bE2mut* alleles isolated in these studies recognize themselves as self. This is not surprising given that they were screened for recognition of *bE2* and *bW2* as nonself. A new *b* allele specificity would have to recognize itself as self and all others as nonself. Some *b* chimeras described below recognize themselves as self and different *b* specificities, but not all, as nonself. It would be interesting to compare the amino acid sequence of, for example, all *b* alleles from different localities that exhibit *b1* specificity. Such a comparison may provide insights about residues that can be altered without changing specificity and, importantly, those that cannot be altered without a change in specificity.

A complementary approach to determine the *b* specificity region using *b* chimeric alleles is described next.

b1/b2 *Chimeras Delimit the* b *Specificity Region*

The extensive and careful analysis of Yee and Kronstad (96, 97) using chimeras between *bE1* and *bE2* and between *bW1* and *bW2* led to some important insights. The chimeras were generated in vitro and introduced at the *b2* locus by homologous recombination. The *bE1/bE2* chimeras carried an inactive *bW* gene, and the *bW1/bW2* chimeras carried an inactive *bE* gene; thus, only the *bE* or the *bW* component, respectively, is being assayed. The *bE1/bE2* chimeras fell into three classes (Fig. 21.7). Class I alleles have the same specificity as *b2*. Class II alleles are distinct from *b1* and *b2*; they are dual maters, that is, they exhibit filament formation on charcoal agar when cospotted with *b1* and *b2* strains. Class III alleles have switched specificity from *b2* to *b1* (Fig. 21.7). As clearly stated by the authors, the assumption is that differences in filament formation reflect differences in dimerization of b polypeptides. Class II strains in combination with some *b* alleles, for example, *bD*, *bG*, *bJ*, and *bL*, were compatible, whereas in combination with others, for example, *bI* and *bM*, they were incompatible (97). These chimeras, therefore, do not behave as "universal" alleles, as the *bmut* alleles described above do. The class II *bE1/bE2* chimeras define a region of approximately 40 amino acids in the variable region as the specificity region for *bE1* and *bE2* (97). The amino-terminal border of this region is between amino acids 31 and 39, and the carboxy-terminal border is between amino acids 79 and 87. The position of the specificity region may vary in different allelic combinations. The *bE1/bE2* chimeras encompassed "17 of the 36 potential positions in the variable region between codons 1 and 107" of *bE1* and *bE2* (97).

The *bW1/bW2* chimeras also fell into three classes (Fig. 21.7) and similarly to the *bE1/bE2* chimeras, class I had *b2* specificity; class II had *b1* and *b2* specificity; and class III had switched specificity from *b2* to *b1* (96). The specificity region was defined as a region between

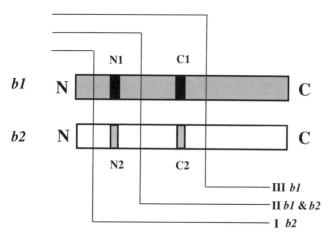

Figure 21.7 *b1/b2* chimeric alleles delimit the specificity region. Chimeras between *bE1* and *bE2* or between *bW1* and *bW2* were generated by homologous recombination at the *b2* locus (96, 97). Three classes of chimeras were generated by the crossover events indicated by the lines. Class I chimeras exhibit *b2* specificity, class II chimeras exhibit *b1* and *b2* specificity, and class III chimeras have switched from *b2* to *b1* specificity (96, 97). N2 and C2 refer to the borders of the *b2* specificity region; N1 and C1 refer to the borders of the *b1* specificity region. The class II chimeras have N1 C2 borders, that is, they share one border with *b1* and the other with *b2*. Because class II chimeras have both *b1* and *b2* specificity, it has been proposed that differences at only one border are sufficient for dimerization and generation of the active *b* heterodimer (96). The *b* specificity region for *bE1/bE2* encompasses a region of approximately 40 amino acids, and that of *bW1/bW2* encompasses a region of approximately 70 amino acids (96, 97). Similar conclusions were reached by Kämper et al. (54) from yeast two-hybrid interactions and analysis of *bE2mut* alleles.

codons 6 and 83, with an amino-terminal border between codons 6 and 9 and a less well-defined carboxy-terminal border, that is, class II and III specificities alternate between codons 76 and 83. For example, chimeras with endpoints at 76 and 77 are class III, at 79 are class II, at 80 and 81 are class III, and at 82 are class II. The class II *bW1/bW2* chimeras define a region of approximately 70 amino acids as the specificity region of *bW1/bW2*. The *bW1/bW2* chimeras encompassed "24 of the 43 potential positions between codons 1 and 109 of *bW1* and *bW2*" (96).

Class II *bE1/bE2* alleles were incompatible with class II *bW1/bW2* chimeric alleles, which led Yee and Kronstad (96) to propose that the borders that define class II alleles contain important residues for dimerization between bE and bW polypeptides. The borders for *bE1* and *bW1* were designated N1 C1; for *bE2* and *bW2*, they were designated N2 C2; and for class II *bE1/bE2* and *bW1/bW2* chimeric alleles, they were designated N1 C2 (Fig. 21.7) (96). According to this proposal, class II alleles have one border in common with *b1*, one

in common with *b2*, and both borders in common with each other. Given that class II alleles recognize *b1* and *b2* as nonself, it would thus appear that differences at only one border are sufficient to promote recognition as nonself (96). Some class III *bE1/bE2* chimeric alleles, which have switched from *b2* to *b1* specificity, differ from *b1* when crossed with some class III *bW1/bW2* chimeras (96). These observations point to important residues that may be responsible for the observed differences in *b* specificity between *b1* and these particular *b1/b2* chimeras with "*b1*" specificity.

Amino acid sequence comparisons of chimeric alleles that exhibit a different allele specificity when crossed with wild-type alleles led to the identification of pairs of chimeric alleles that differ at a single amino acid. This single amino acid difference appears to be sufficient to confer a different *b* allele specificity.

The extensive and carefully executed work of Yee and Kronstad (96) led to the conclusion that there are 6 positions for *bE1/bE2* (residues 31, 45, 79, 87, 89, and 90) and 10 positions for *bW1/bW2* (residues 2, 6, 9, 74, 76, 77, 79, 80, 81, and 82) that influence allele specificity. All of these amino acids, except no. 45, reside in the border regions of the specificity region and are mostly charged or polar. In contrast, the analysis of Kämper et al. (54) identified amino acid residues that are mostly hydrophobic. The reason for this is not clear but may reflect the type of mutant allele under study.

In *S. commune* and *C. cinereus*, pairs of homeodomain polypeptides coded by the *A* multiallelic locus interact to form an active species that controls growth of the dikaryon and sexual development (see chapters 16 and 17). The specificity region for these polypeptides in both fungi resides in the amino-terminal region of the protein (7, 98). In *S. commune*, analysis of chimeric alleles identified a region of approximately 40 amino acids as the specificity region (95) (see chapter 16 for further details).

The Variable Region Is Dispensable in bW-bE Fusion Proteins

A head-to-tail fusion of the bE1 and bW2 polypeptides (bE1-k-bW2) with a small linker region (k) separating the two polypeptides is active in promoting filamentous growth and pathogenicity when introduced into an *a2* Δ*b* strain (75). Remarkably, a fusion protein between bW1 and bE1 (bW1-k-bE1) is also active. Deletion of the variable region in bE1, bW2, or both in the bE1-k-bW2 fusion protein did not affect the biological activity of the fusion protein. The variable region, therefore, appears to be necessary only for mediating interactions between the bE and bW polypeptides from different *b* alleles. Precedent existed for a naturally occurring fusion protein between two homeodomain proteins at the *A* locus in *C. cinereus* (60). Because the *U. maydis* bE and bW genes are arranged divergently from each other, occurrence of a fusion protein in nature would necessitate a rearrangement of one of the genes.

The Homeodomains of bE and bW Are Necessary for b Protein Activity

Deletion of 10 amino acids from helix 3 in bE1, bW2, or both in the fusion protein (bE1-k-bW2) abolished activity, indicating that the homeodomain of both polypeptides is necessary for activity of this fusion protein. Additional mutational analysis of the native proteins confirmed this conclusion (81). An Ala substitution of residues Phe49, Trp48, or Arg53 in the putative helix 3 of the *bW2* homeodomain (Fig. 21.5) abolished function, indicating that the homeodomain of *bW* is essential for function of the bW-bE heterodimer. Ala substitution of residues Phe49 in *bE* did not abolish activity, whereas Ala substitution of Trp48 and Arg53 did. Additional Ala substitutions and small deletions in the homeodomain of bE abolished activity in most cases, indicating that the homeodomain of bE is necessary for activity of the bE-bW heterodimer (81). This observation contrasts with that in *S. cerevisiae*, *S. commune*, and *C. cinereus*, where only the HD2 homeodomain in one of the pair of polypeptides that form an active heteromer is necessary for activity (66, 92). The HD1 homeodomain is dispensable in these other fungi. Future X-ray crystallographic analysis of the *U. maydis* b heterodimer bound to its site will likely shed light on the reasons for the requirement of both homeodomains in *U. maydis*.

Genes in the *a* Locus Contain Binding Sites That Confer *b*-Dependent Regulation

I describe briefly work leading to identification of a *b* binding site. Use of new molecular biology tools has led to the identification of additional *b* targets (see chapter 22 by R. Kahmann and J. Schirawski). The fact that a fusion protein containing only the homeodomains of bW1 and bE2 is functional (75) provided an opportunity to identify possible targets of the bW-bE heterodimer. This fusion protein was insoluble in *Escherichia coli*, but deletion of amino acids in the carboxy-terminal part of the bW component rendered the protein soluble and still functional; it was designated Kon8 (74). Because the *lga2* gene, which locates in the *a2* allele, was shown to be upregulated in the presence of an active bE-bW heterodimer (see below) (90), the *lga2* upstream region was thus likely to contain a binding site for the *b* heterodimer. Purified Kon8 protein and a fragment (−241 to −1) upstream of the ATG codon in the *lga2* gene were used for DNA binding studies. Indeed, Kon8 binds to the *lga2* upstream region as demonstrated using electrophoretic mobility shift assays (74). Further analysis using DNase protection assays showed that a 29-bp region between positions −150 and −178 is protected on both strands. This result was confirmed by methylation interference analysis (74). Mutation of residues that are protected abolished DNA binding, indicating the importance of these nucleotides for interaction with the Kon8 protein. The region bound by Kon8 was designated *bbs1* (for b binding site 1) (74) and contains direct repeats of the sequence motif AC/GTGTG. In addition, a motif (GATG and ACA with a spacing of 9 nt) that characterizes the operator consensus site recognized by a1-α2 in yeast (see chapter 5 by A. E. Tsong, B. B. Tuch, and A. D. Johnson) is contained within *bbs1*. In the consensus operator site bound by a1- α2 in *S. cerevisiae*, the sequence GATG is recognized by the a1 component of the a1-α2 heterodimer and ACA by the α2 component (35, 41; see chapter 5). These motifs are each part of two similar half-sites of the operator site bound by a1-α2 and are arranged with dyad symmetry and separated by a fixed number of nucleotides (see chapter 5 and references therein). The motif found within *bbs1* in *U. maydis* may contribute to binding by the bW-bE heterodimer but is unlikely to be the sole motif contributing to binding (74). Additional regulatory elements located in a 456-bp promoter region may contribute to *b*-dependent regulation because deletion of *bbs1* still results in *b*-dependent induction (74). *mfa2*, a gene located in the *a2* allele (see below), contains a *bbs1* site, but in vitro binding by Kon8 could not be detected (74). Because *mfa2* is downregulated by bW-bE, it is possible that other proteins act together with bW-bE to regulate *mfa2* differently than *lga2*. The identification of other genes directly regulated by the *b* heterodimer will permit the generation of a consensus sequence and the application of new molecular genetic tools to identify all potential sites for the *b* heterodimer present in the genome (see chapter 22 for more details on targets of *b*).

The *a* Locus Codes for Components of a Pheromone Response Pathway

Two functions that were ascribed to the *a* locus were control of cell fusion of haploid strains—based on limited experimental data (78)—and maintenance of filamentous growth, a role shared with the *b* locus (16). Before the *a* locus was cloned, Banuett and Herskowitz (17) proposed that the *a* locus might code for cell-type-specific regulatory proteins that govern expression of *a1*-specific or *a2*-specific genes, in analogy with the regulatory proteins coded by the *MAT* alleles of *S. cerevisiae* (reviewed in references 17 and 46). In this view, filamentous growth was proposed to be regulated by a new combinatorial protein whose subunits would be coded by the *a1* and *a2* alleles, present in the dikaryon, similar to how the *S. cerevisiae* MATa and MATα alleles code for the **a**1 and α2 polypeptides, respectively, that form the **a**1-α2 combinatorial activity present only in the *a*/α diploid cell type (42; see also chapter 5). Other possibilities that were entertained were that the *a1* allele, for example, might code for a protein kinase that phosphorylates a set of inactive *a1*-specific proteins or for a specific protease that degrades *a2*-specific proteins in *a1* cells. As we shall see, the findings from cloning and sequencing the *a* locus were unexpected and unprecedented in the fungi at that time. We now know that the multiallelic *B* locus of *S. commune* and *C. cinereus* contains genes similar to those present in the *a* locus of *U. maydis* but their organization is more complex (see chapters 17 and 18).

Identification of the Pheromone Precursor Genes

Linkage of the *a* locus to the *pan* locus (49) was exploited by Froeliger and Leong (38) to clone the *a2* allele. Functional complementation of a *pan-1* mutation

followed by screening of the transformants containing the complementing fragments for ability to confer *a2* mating properties to an *a1 b1* strain identified a region that contained the *a2* allele. Further analysis indicated that the *a1* and *a2* regions were characterized by unique sequences of 4.5 to 5.0 kb and 7.5 to 8.0 kb for *a1* and *a2*, respectively, flanked by homologous sequences. Information about the gene(s) located in the nonhomologous regions was not provided. Bölker et al. (21) cloned both *a* alleles and confirmed observations of Froeliger and Leong (38). The *a1* allele was cloned by DNA-mediated transformation of an *a2 b2* strain and subsequent screening of the transformants for a dual-mater phenotype. A small fragment that confers *a1*-mating specificity was located near the left border of the nonhomologous region in the *a1* allele (Fig. 21.8). A similar fragment conferring *a2*-mating specificity was located near the right border of the nonhomologous region of the *a2* allele (Fig. 21.8). The nucleotide sequence of these regions did not reveal any long ORFs. cDNA cloning indicated that the spliced *a1* mRNA is 500 nt long and contains an ORF with coding capacity for a polypeptide of 40 amino acids. An intron of 90 nt is located downstream of the ORF. The spliced *a2* mRNA contains an ORF with coding capacity for a polypeptide of 38 amino acids. An intron of 101 nt is located downstream of the ORF in a position similar to that in *a1*. Both genes are preceded by a repeat of a 9-bp motif, ACAAAGGGA, later shown to confer pheromone-dependent regulation (90; see below). These short polypeptides contain the CAAX sequence at the carboxy terminus, a signature sequence for posttranslational

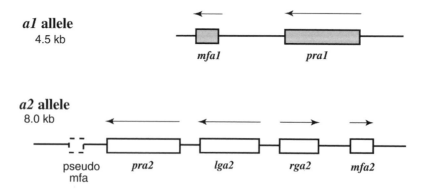

Figure 21.8 Molecular structure and organization of the *a* locus. The *a* locus has two alleles, *a1* and *a2* (78), and codes for components of a pheromone response pathway (21). Both *a1* and *a2* contain a pheromone precursor and a receptor gene. The pheromone precursor genes code for polypeptides of 40 and 38 amino acids for *a1* and *a2*, respectively, and the pheromone precursors contain a CAAX box at the C terminus (21). This sequence is a substrate for prenylation and carboxymethylation, and the modification is necessary for full activity of the mature pheromones (86). The modified pheromone precursors are processed further to yield lipopeptides of 13 and 9 amino acids for *a1* and *a2*, respectively. The *a2* allele contains additional genes: *lga2* and *rga2*, and a pseudopheromone gene, *mfa* (91), not present in the *a1* allele. (From reference 8 with permission.)

modification by carboxymethylation and prenylation (farnesylation or geranylgeranylation) (reviewed in reference 79). This posttranslational modification was first shown to occur in the tremerogens (pheromones) of the jelly fungus *Tremella mesenterica* and rhodotorucine A (a pheromone) produced by *Rhodosporidium toruloides* (see reference 3 and references therein). We now know that the CAAX sequence is present not only in other fungal pheromones but also in a diversity of eukaryotic proteins, including Ras and Cdc42, and that the modification is essential for targeting these proteins to the membrane and for their function (reviewed in reference 99). The presence of the CAAX sequence in the *U. maydis* a1 and a2 polypeptides suggested that these polypeptides are pheromone precursors for lipopeptides; thus, the genes were designated *mfa1* and *mfa2*, for mating factor a1 and mating factor a2, respectively (21). The a1 and a2 pheromone precursors are processed to shorter peptides of 13 (Gly-Arg-Asp-Asn-Gly-Ser-Pro-Ile-Gly-Tyr-Ser-Ser-X) and 9 (Asn-Arg-Gly-Gln-Pro-Gly-Tyr-Tyr-X) amino acids for *a1* and *a2*, respectively (86). If the modification of the Cys residue in the CAAX box of the pheromone precursors is blocked, pheromone activity is reduced 1,000-fold compared to the modified pheromone (86).

The generation of these short peptides from the larger pheromone precursors is likely to involve proteolytic processing. The genes involved in processing of a-factor in *S. cerevisiae* are present in the *U. maydis* genome, but their role remains to be elucidated (F. Banuett, unpublished observations).

Identification of the Receptor Genes
Insertional mutagenesis with *E. coli* transposon Tn5 was used to inactivate the pheromone precursor genes and to probe for additional genes in the *a* mating specificity region (21). The mutants were cospotted on charcoal agar with the strains indicated below, and filament formation was assayed as a measure of mating and thus a functional gene. A Tn5 insertion in the *mfa1* gene abolished filament formation when an *mfa1*::Tn5 *b2* strain was cospotted with *a2 b1*, indicating that the gene is necessary for *a* locus activity. In contrast, a positive reaction (Fuz$^+$ phenotype) was observed when the *mfa1*::Tn5 strain was cospotted with an *a1/a2 b1/b1* Fuz$^-$ diploid strain (see below), indicating that the mating defect of the *mfa1*::Tn5 *b2* strain was suppressed by the Mfa1 pheromone produced by the diploid strain, conferring mating competence to the mutant, which can now mate with the diploid strain. Another Tn5 insertion (*pra1*::Tn5) identified a gene upstream of *mfa1* (Fig. 21.8). This new gene contains an ORF with coding capacity for a protein of 357 amino acids. The ORF is interrupted by three introns of 79, 70, and 85 nucleotides. Likewise, the *a2* allele contains an ORF with coding capacity for a protein of 346 amino acids, interrupted by two introns of 68 and 72 nucleotides. The proteins exhibit some similarity to *S. cerevisiae* Ste3 (~20% identity), the receptor for a factor, and to *S. pombe* Map3 (~18%), the receptor for M pheromone. They also are 24% identical to each other (21). Because of their sequence similarity to the pheromone receptor proteins in the yeasts, the genes in *U. maydis* were designated *pra1* and *pra2* (for pheromone receptor a1 and pheromone receptor a2) and postulated to code for receptors for the a2 (Mfa2) and a1 (Mfa1) pheromones, respectively (21). The putative Pra1 and Pra2 receptors have seven potential transmembrane domains characteristic of G-protein-coupled receptors, a family of proteins that bind diverse ligands, including pheromones (reviewed in reference 9). Inactivation of *pra1* (*mfa1 pra1*::Tn5 *b2*) abolishes filament formation when this strain is combined with *a2 b1* or *a1/a2 b1/b1* on charcoal agar, indicating that the putative receptor genes are essential for *a* locus activity and, furthermore, that the presence of two receptor genes in the diploid partner does not suppress the mating defect of the haploid strain, in contrast to the behavior of the *mfa1*::Tn5 *pra1* strain (21). In other words, each mating partner needs to have a functional receptor gene.

The a2 Allele Contains Additional Genes
The complete nucleotide sequence of the *a1* and *a2* alleles revealed the presence of two additional ORFs located between *mfa2* and *pra2*, which are divergently transcribed (Fig. 21.8). The gene closest to *pra2* was designated *lga2* and contains an ORF with coding capacity for a polypeptide of 215 amino acids and is interrupted by four introns. The *rga2* gene is closest to *mfa2* and has an ORF with coding capacity for a polypeptide of 215 amino acids and is interrupted by one intron (91). These genes are absent from the *a1* allele. Lga2 and Rga2 exhibit no similarity to other proteins in the databases and localize to the mitochondrial matrix (22, 91). Inactivation of *lga2*, *rga2*, or both does not affect filament formation on charcoal agar. However, *lga2* and *rga2* alter the pathogenicity of strains lacking the mitochondrial protein Mrb1. Additional experiments led to the proposal that Lga2 and Rga2 interfere with mitochondrial fusion (22).

Downstream of *pra2* there is a pseudopheromone precursor gene with similarity to *mfa1* (Fig. 21.8). Urban et al. (91) propose that this gene might have been functional and coded for a pheromone distinct from *mfa2*. There are no *mfa* remnants in the *a1* allele.

Variations on a Theme: *Sporisorium reilianum*

S. reilianum is the cause of anther smut of maize and is closely related to *U. maydis*. As described earlier, Hanna presented preliminary evidence for the segregation of two mating types (43). Recent analysis has shown that mating type is determined by two unlinked loci, *Sra* and *Srb*. The *Sra* locus consists of at least three naturally occurring alleles, a situation that contrasts with that in *U. maydis*, and the *Srb* locus consists of at least five naturally occurring alleles (80). The molecular organization of these loci is similar to that of the *a* and *b* loci in *U. maydis* (80).

The Srb Locus

Each of the *Srb* alleles consists of two divergently transcribed genes, *bW* and *bE*, both of which code for homeodomain proteins that contain an amino-terminal variable region (~59% identity) and a constant region (~100% identity) for the remainder of the polypeptide. The constant region of *S. reilianum* bW1 and bE1 is 41.5 and 56.6% identical to that of *U. maydis* bW1 and bE1, respectively. The variable region of *S. reilianum* bW1 and bE1 is 23.4 and 24.5% identical to that of *U. maydis* bW1 and bE1, respectively. The *S. reilianum* bW and bE polypeptides are functional in *U. maydis*. For example, a *U. maydis a1 Δb* strain containing the *S. reilianum b1* allele is capable of filament formation and tumor induction in combination with strain *a2 b2* (80).

The Sra Locus

Each of the *Sra* alleles contains two pheromone precursor genes: *a1* contains *mfa1.2* and *mf1.3*; *a2* contains *mfa2.1* and *mfa2.3*; and *a3* contains *mfa3.1* and *mfa3.2*. Each allele also contains a single receptor gene (*pra1*, *pra2*, and *pra3*, respectively). *S. reilianum* Pra1 and Pra2 are most closely related to *U. maydis* Pra1 and Pra2 (69.7 and 63.4% amino acid identity, respectively). Pra3 is 24.6 and 24.8% identical to Pra1 and Pra2 of *U. maydis*. The amino acid sequence of Mf1.2 is identical to that of Mfa3.2 and most closely related to Mfa1 of *U. maydis* (58.5% identity); that of Mfa2.1 is almost identical to Mfa3.1 and most closely related to Mfa2 of *U. maydis* (41.0% identity); and that of Mfa1.3 is identical to Mfa2.3 and shows only weak similarity to the pheromone precursor genes of *U. maydis*, *U. hordei*, and *S. pombe* (80). It appears that each pheromone can activate only one mating partner (see reference 80 for details). The presence of two pheromone genes in each allele contrasts with the situation observed in *U. maydis*, where a single pheromone gene is present. The *S. reilianum a2* allele contains *lga2* and *rga2* genes, not present in the *a1* or *a3* alleles, and their protein products exhibit 26.9 and 30.0% identity to *U. maydis* Lga2 and Rga2, respectively.

Functional Assays for the Pheromone and Receptor Genes in *U. maydis*

Use of Haploid and Diploid Strains

Functionality of the *mfa* and *pra* genes is readily tested by introduction of these genes into haploid or diploid strains. For example, introduction of the *mfa1* gene into an *a2 b2* strain allows this strain to form filaments when combined with an *a2 b1* strain, even though both strains contain the same receptor (*pra2*), presumably because the Mfa1 pheromone is able to activate Pra2 present in both partners. The same is true if the *mfa2* gene is introduced into an *a1 b1* strain, which can now form filaments in combination with an *a1 b2* strain. In contrast, introduction of the *pra2* gene into an *a1 b1* strain does not result in filamentous growth when cospotted with an *a1 b2* strain, unless the latter also contains the *pra2* receptor gene. These results indicate that the pheromone receptor needs to be activated in both cells (21).

Diploids homozygous at *a* and heterozygous at *b* form filaments when combined with strains carrying a different *a* allele, for example, *a1 b1* induces filament formation of the diploid *a2/a2 b1/b2* (16). The formation of filaments in this combination is apparently due to activation of Pra2 by Mfa1 produced by the *a1 b1* strain and not due to cell fusion (86). Introduction of the *mfa2* gene into diploid strain *a1/a1 b1/b2* restores filamentous growth. Likewise, introduction of *mfa1* into *a2/a2 b1/b2* also restores filamentous growth, demonstrating that activation of a single receptor in each of these diploid strains is sufficient to trigger filamentous growth. Similar results are obtained upon introduction of *pra2* or *pra1* in these diploids. These observations were exploited in the purification of the pheromones (86). The activity of fractions derived from *a1 b1* or *a2 b2* strains was assayed by their ability to induce a Fuz$^+$ phenotype in strains *a2/a2 b1/b2* or *a1/a1 b1/b2*, respectively, on charcoal agar. Fractions derived from *a1 b1* elicited a response only in *a2/a2 b1/b2* strains, whereas those from *a2 b2* elicited a response in *a1/a1 b1/b2* only, indicating that the response is cell type specific (86). The purified a1 pheromone also elicits conjugation tube formation (see below) in *a2* strains only and purified a2 pheromone in *a1* strains only, providing additional evidence for a cell-type-specific response (86).

The finding that the *a* mating-type locus contains pheromone precursor genes and receptors led to the hypothesis that a pheromone response pathway involving a conserved mitogen-activated protein kinase (MAPK) cascade mediates the pheromone response, similar to that regulating the pheromone response in the yeasts. Indeed, many of the components of the pathway have now been identified (reviewed in references 9 and 36). Their role in cell fusion, filamentous growth, and pathogenicity is described in chapter 22.

The Pheromone Response Element (PRE) and the Regulator of Pheromone-Induced Genes, Prf1

The repeat sequence ACAAAGGGA is found either upstream, downstream, or within the ORF of the genes located in the *a* and *b* loci (90). In uninduced haploid cells of different mating types, the *mfa* and *pra* genes are expressed at a low level. The RNA level for these genes is induced dramatically in mixtures of cells of opposite mating types, presumably in response to activation of the pheromone response pathway (90). The *lga2* and *rga2* RNAs are hardly detectable in the uninduced state but are upregulated in mixtures of cells of opposite mating types. Interestingly, the RNA for the *bE* and *bW* genes, which is hardly detectable in the uninduced state, is also induced in mixtures of cells of opposite mating types. This pattern of expression is similar to that of *lga2* and *rga2*. It is possible that upregulation of the *b* genes by the pheromones in haploid cells prior to cell fusion leads to accumulation of high levels of the b polypeptides to ensure that upon cell fusion a high level of the bE-bW heterodimer is present in the dikaryon. This speculation on my part is based on the analysis of RNA levels; to my knowledge, b protein levels before and after cell fusion have not been determined.

In diploid strains heterozygous for both *a* and *b*, *mfa* and *pra* genes are downregulated to levels below those of the uninduced state, whereas *lga2* is induced further (90). Given that different *a* alleles are necessary for maintenance of filamentous growth (16, 21, 86), the low levels of *mfa* and *pra* gene expression must be sufficient for maintenance of filamentous growth. This downregulation is likely to be responsible for the attenuated mating observed in *b* heterozygotes (63). Conversely, unregulated expression in the absence of the *b* heterodimer may cause increased mating.

The repeat sequence ACAAAGGGA was shown to be necessary and sufficient to confer pheromone-dependent regulation to a GUS reporter gene and was designated PRE, for pheromone response element (90). The protein responsible for regulation of genes containing PREs is Prf1 (45), a transcription factor of the HMG class of regulatory proteins. Prf1 binds directly to the PRE sequence (45) (see chapter 22 for further details).

Conjugation Tube Formation and Cell Fusion Are *a*-Dependent Processes

Cell differentiation of *U. maydis* cells prior to cell fusion was reported by Sleumer (83) and subsequently by Bowman (23). According to Sleumer (83), early stages of cell fusion involved the formation of a connection between mating partners, followed by the emergence of a hyphal cell from the connection and transfer of cytoplasm and nuclei from the two mating partners to the "fusion hypha." This process left behind the two mating partners devoid of cytoplasm and nuclei. The resulting dikaryotic cell exhibits fast growth and consists of a tip cell containing all the cytoplasm, followed by small empty compartments (83). The findings of Bowman (23) agreed with most of Sleumer's observations, except that he indicated that the hyphae consist of compartments all of which contain cytoplasm and two nuclei, and in some cases three or four nuclei. The role of the sex factors in the morphological differentiation was not discussed. Rowell (76), in his classic study using a maize coleoptile assay, first proposed that the *a* locus governs cell fusion and the *b* locus governs growth of the dikaryotic hypha (see discussion above), but direct observation of cell fusion was not reported. Morphological differentiation of cells prior to cell fusion had been documented in the yeasts (*S. cerevisiae* and *S. pombe*), other smuts (*Ustilago hordei* and *M. violaceum*), and other basidiomycetes (*T. mesenterica* and *R. toruloides*) (1, 29, 30, 47, 67, 94).

Two independently developed assays—low-nitrogen, charcoal liquid medium (14) and water droplets or water agar (84, 85)—established that *a* controls conjugation tube formation and cell fusion independently of the *b* locus in *U. maydis*. Mixtures of cells carrying different *a* alleles respond by formation of a conjugation tube, a sinuous projection which arises from one of the cell poles, approximately 3 h after the cells are mixed. The sinuosity of the conjugation tube suggests that the axis of growth changes constantly, most likely in response to a changing gradient of pheromone produced by cells of opposite mating type. If the cells carry identical *a* alleles, no conjugation tubes develop. Because only mixtures of cells that differ at *a*, regardless of the *b* allele present, form conjugation tubes, it was concluded that the *a* locus is necessary for conjugation tube formation and that different *b* alleles are dispensable for this process. a2 cells respond faster than a1 cells (85). Similar observations were reported for *R. toruloides*, where cells of mating type A respond faster than cells of mating type *a* when placed in close proximity of each other (1). In cells with conjugation tubes, the nucleus locates near the point of emergence of the conjugation tube (14, 15) and appears to be arrested in the G$_2$ phase of the cell cycle (39). Little is known of the mechanisms that govern nuclear migration during cell fusion. The conjugation tubes fuse at their tips in a process that is independent of the *b* locus (14, 15, 85). If the cells that fuse carry different *b* alleles, then straight, fast-growing hyphae with a dikaryotic tip cell develop. If the cells that fuse have identical *b* alleles, the hyphae that develop are twisted and slow growing. These observations confirmed and expanded those of Rowell (76).

Strains with an inactive gene for the pheromone precursor can respond to the pheromone produced by cells of opposite mating type but they cannot induce a response in cells of opposite mating type, and strains with an inactive receptor gene can induce a response in cells of opposite mating type but cannot respond to cells of opposite mating type (85). These observations support the findings that strains with a defect in a MAPKK/MEK can elicit a response but are not able to respond to wild-type strains of opposite mating type (14).

The response in the low-nitrogen charcoal medium is rapid, beginning 2.5 to 3.0 h after mixing of cells; it is uniform, nearly 100% of the cells respond; and it initiates and progresses synchronously and can thus be used quantitatively (14, 15). The water agar assay is qualitative; it is not synchronous and not uniform but allows direct observation of single cells and their fusion (84, 85).

Diploid strains also undergo morphological transitions when grown in these media. Diploids heterozygous at *a* and *b* grow as yeast-like cells in rich medium and switch to filamentous growth in low-N charcoal medium. The transition involves several steps that include emergence of a projection from the cell apex, narrowing of the cell diameter, movement of the cytoplasm and nucleus from the yeast-like cell to the emerging projection, and changes in nuclear shape (15). The movement of the cytoplasm leaves behind a yeast-like cell devoid of cytoplasm separated from the emerging hypha by a septum (15). Diploids heterozygous at *a* form appendages from one or both cell ends that are similar in the early stages of development to those formed by strains heterozygous at both loci, but as they elongate, they become sinuous, have curled tips, and sometimes branch but never develop into straight hyphae as observed in the strain heterozygous at both loci. Diploids homozygous at *a* remain arrested as unbudded yeast-like cells (14, 15). These studies provided additional evidence that both *a* and *b* are required for filamentous growth and suggests a possible role for *a* early in this process.

Formation of conjugation tubes involves activation of a MAPK signaling cascade. Inactivation of a MAPKK/MEK abolishes the response, lending support to the view that a MAPK cascade regulates this **a**-dependent process (14; reviewed in reference 9; see chapter 22 and references therein). The MAPKK/MEK was also shown to be required for maintenance of filamentous growth, another *a*-locus dependent process, and for tumor induction, a process independent of the presence of different *a* alleles. These studies demonstrated for the first time that a MAPKK/MEK and thus a MAPK is required for filamentous growth and pathogenicity in a fungus other than the yeasts (see chapter 22 for a more exhaustive treatment of cell-cell signaling in *U. maydis*).

CONCLUDING REMARKS

In writing this chapter, I have tried to convey to the reader a historical perspective of the study of the mating types in *U. maydis*. In doing so, I have tried to describe key experiments that led to the discovery of heterothallism and the mating types and to the conclusion that *U. maydis* differs from most smuts, in possessing two mating-type loci: one locus has two alleles and governs cell fusion and maintenance of filamentous growth, and the other has a multiallelic locus that is the key determinant of filamentous growth, pathogenicity, and meiosis. Studies of the molecular organization of the mating-type loci in *U. maydis* have provided the framework in which to understand the more complex multiallelic mating-type loci of the mushrooms. I have also tried to convey to the reader the wealth of information and interesting observations that are hidden in the older literature, and which continue to provide inspiration for reexamination of old questions with modern molecular biological techniques and new and powerful imaging techniques. Due to space limitations, it has not been possible to cover the myriad of interesting observations on the subject. Despite the impressive progress made in our understanding of the molecular organization and function of the mating-type loci in *U. maydis*, much still remains to be learned. In particular, what are the rules that govern the formation of active b heterodimers? How does the *b* locus modulate the infectious process? What is the precise role of *b* during meiosis? Does the *a* locus regulate nuclear movement during cell fusion? Does *b* regulate karyogamy, a process that normally occurs during growth in the plant? Future studies will continue to provide new insights about the intricacies of the interplay between the *a* and *b* loci in regulation of the life cycle of this delightful blight.

I am grateful to Ira Herskowitz for his tremendous support, enthusiasm, confidence, and stimulating discussions that allowed me to develop and pursue without constraints my studies on different aspects of the life cycle of Ustilago maydis, *and for introducing me to bacteriophage* λ *and* Saccharomyces cerevisiae, *which provided the conceptual framework in which to think about* Ustilago maydis. *I also thank Robin Holliday, Sally Leong, David Perkins, and Robert Ullrich for strains, encouragement, and discussion when I initiated my studies on* U. maydis; *James DeVay for providing seeds of the famous B164 corn line that he used in his classic genetic studies of* U. maydis; *Carlene Raper and Lorna Casselton for basic lessons on the Basidiomycetes; and members of the Herskowitz lab, in particular, Tomoko Ogawa, Joe Gray, Fred Chang, Paul Sternberg, Sylvia Sanders, Aaron Mitchell, Matthias Peter, Lorraine Marsh, Brenda Andrews, Shai Shaham, Sandy Johnson, Megan Grether, and Marc Shuman, for stimulating discussions.*

References

1. **Abe, K., I. Kusaka, and S. Fukui.** 1975. Morphological change in the early stages of the mating process of *Rhodosporidium toruloides*. *J. Bacteriol.* **122:**710–718.

2. **Agrios, G. N.** 1994. *Plant Pathology*, 4th ed. Academic Press, New York, NY.

3. **Akada, R., K. Minomi, J. Kai, I. Yamashita, T. Miyakawa, and S. Fukui.** 1989. Multiple genes coding for precursors of rhodotorucine A, a farnesyl peptide mating pheromone of the basidiomycetous yeast *Rhodosporidium toruloides*. *Mol. Cell. Biol.* **9:**3491–3498.

4. **Alexopoulos, C. J., C. W. Mims, and M. Blackwell.** 1996. *Introductory Mycology*, 4th ed. Wiley, Hoboken, NJ.

5. **Bakkeren, G., G. Jiang, R. L. Warren, Y. Butterfield, H. Shin, R. Chiu, R. Linning, J. Schein, N. Lee, G. Hu, D. M. Kupfer, Y. Tang, B. A. Roe, S. Jones, M. Marra, and J. W. Kronstad.** 2006. Mating factor linkage and genome evolution in basidiomycetous pathogens of cereals. *Fungal Genet. Biol.* **43:**655–666.

6. **Bakkeren, G., and J. W. Kronstad.** 1994. Linkage of mating-type loci distinguishes bipolar from tetrapolar mating in basidiomycetous smut fungi. *Proc. Natl. Acad. Sci. USA* **91:**7085–7089.

7. **Banham, A. H., R. N. Asante-Owusu, B. Göttgens, S. Thompson, C. S. Kingsnorth, E. Mellor, and L. A. Casselton.** 1995. An N-terminal dimerization domain permits homeodomain proteins to choose compatible partners and initiate sexual development in the mushroom *Coprinus cinereus*. *Plant Cell* **7:**773–783.

8. **Banuett, F.** 2002. Pathogenic development in *Ustilago maydis*: a progression of morphological transitions that results in tumor formation and teliospore production, p. 349–398. *In* H. D. Osiewacz (ed.), *Molecular Biology of Fungal Development*. Marcel Dekker, New York, NY.

9. **Banuett, F.** 1998. Signalling in the yeasts: an informational cascade with links to the filamentous fungi. *Microbiol. Mol. Biol. Rev.* **62:**249–274.

10. **Banuett, F.** 1995. Genetics of *Ustilago maydis*, a fungal pathogen that induces tumors in maize. *Annu. Rev. Genet.* **29:**179–208.

11. **Banuett, F.** 1992. *Ustilago maydis*, the delightful blight. *Trends Genet.* **8:**174–180.

12. **Banuett, F.** 1991. Identification of genes governing filamentous growth and tumor induction by the plant pathogen *Ustilago maydis*. *Proc. Natl. Acad. Sci. USA* **88:**3922–3926.

13. **Banuett, F., and I. Herskowitz.** 1996. Discrete developmental stages during teliospore formation in the corn smut fungus, *Ustilago maydis*. *Development* **122:**2965–2976.

14. **Banuett, F., and I. Herskowitz.** 1994. Identification of Fuz7, a *Ustilago maydis* MEK/MAPKK homolog required for a-locus-dependent and -independent steps in the fungal life cycle. *Genes Dev.* **8:**1367–1378.

15. **Banuett, F., and I. Herskowitz.** 1994. Morphological transitions in the life cycle of *Ustilago maydis* and their genetic control by the *a* and *b* loci. *Exp. Mycol.* **18:**247–266.

16. **Banuett, F., and I. Herskowitz.** 1989. Different *a* alleles of *Ustilago maydis* are necessary for maintenance of filamentous growth but not for meiosis. *Proc. Natl. Acad. Sci. USA* **86:**5878–5882.

17. **Banuett, F., and I. Herskowitz.** 1988. *Ustilago maydis*, smut of maize, p. 427–455. *In* G. Sidhu (ed.), *Genetics of Plant Pathogenic Fungi. Advances in Plant Pathology*, vol. 6. Academic Press, London, United Kingdom.

18. **Barrett, K. J., S. E. Gold, and J. W. Kronstad.** 1993. Identification and complementation of a mutation to constitutive filamentous growth in *Ustilago maydis*. *Mol. Plant-Microbe Interact.* **6:**274–283.

19. **Bauch, R.** 1932. Die Sexualität von *Ustilago scorzonerae* und *Ustilago zeae*. *Phytopathol. Zeit.* **5:**315–321.

20. **Bauch, R.** 1923. Uber *Ustilago longissima* und ihre varietät *macrospora*. *Zeit. Botan.* **15:**241–279

21. **Bölker, M., M. Urban, and R. Kahmann.** 1992. The *a* mating type locus of *U. maydis* specifies cell signaling components. *Cell* **68:**441–450.

22. **Bortfeld, M., K. Auffarth, R. Kahmann, and C. W. Basse.** 2004. The *Ustilago maydis a2* mating-type locus genes *lga2* and *rga2* compromise pathogenicity in the absence of the mitochondrial p32 family protein Mrb1. *Plant Cell* **16:**2233–2248.

23. **Bowman, D. H.** 1946. Sporidial fusion in *Ustilago maydis*. *J. Agric. Res.* **72:**233–243.

24. **Brefeld, O.** 1883. Botanische Untersuchungen über Hefenpilze. Fortsetzung der Schimmelpilze, p. 1–208. Untersuchungen aus dem Gesammtgebiete der Mykologie. Heft V. *Die Brandpilze I (Ustilagineen)*. Verlag von Arthur Felix, Leipzig, Germany.

25. **Bürglin, T. R.** 1994. A comprehensive classification of homeobox genes, p. 27–71. *In* D. Deboule (ed.), *Guidebook to the Homeobox Genes*. Oxford University Press, Oxford, United Kingdom.

26. **Christensen, J. J.** 1963. Corn smut caused by *Ustilago maydis*. *Am. Phytopathol. Soc. Monogr.* No. 2.

27. **Christensen, J. J.** 1931. Studies of the genetics of *Ustilago zeae*. *Phytopathol. Zeit.* **4:**129–188.

28. **Christensen, J. J.** 1929. Mutation and hybridization in *Ustilago zeae*. Part II. Hybridization. *Minn. Agric. Exp. Stn. Tech. Bull.* **65:**89–108.

29. **Davey, J.** 1992. Mating pheromones of the fission yeast *Schizosaccharomyces pombe*: purification and structural characterization of M-factor and isolation and analysis of two genes encoding the pheromone. *EMBO J.* **11**:951–960.

30. **Day, A. W., and J. K. Jones.** 1968. The production and characteristics of diploids in *Ustilago violacea*. *Genet. Res.* **11**:63–81

31. **Day, P. R., and S. L. Anagnostakis.** 1971. Corn smut dikaryon in culture. *Nat. New Biol.* **231**:19–20.

32. **Day, P. R., S. L. Anagnostakis, and J. E. Puhalla.** 1971. Pathogenicity resulting from mutation at the *b* locus of *Ustilago maydis*. *Proc. Natl. Acad. Sci. USA* **68**:533–535.

33. **Day, P. R.** 1963. Mutations of the *A* mating type factor in *Coprinus lagopus*. *Genet. Res. Camb.* **4**:655–664.

34. **Day, P. R.** 1960. The structure of the *A* mating type locus in *Coprinus lagopus*. *Genetics* **45**:641–650.

35. **Dranginis, A. M.** 1990. Binding of yeast *a1* and *alpha2* as a heterodimer to the operator DNA of a haploid-specific gene. *Nature* **347**:682–685.

36. **Feldbrügge, M., J. Kämper, G. Steinberg, and R. Kahmann.** 2004. Regulation of mating and pathogenic development in *Ustilago maydis*. *Curr. Opin. Microbiol.* **7**:666–672.

37. **Fotheringham, S., and W. K. Holloman.** 1989. Cloning and disruption of *Ustilago maydis* genes. *Mol. Cell. Biol.* **9**:4052–4055.

38. **Froeliger, E. H., and S. A. Leong.** 1991. The *a* mating-type alleles of *Ustilago maydis* are idiomorphs. *Gene* **100**:113–122.

39. **Garcia-Muse, T., G. Steinberg, and J. Perez-Martin.** 2003. Pheromone-induced G2 arrest in the phytopathogenic fungus *Ustilago maydis*. *Eukaryot. Cell* **2**:494–500.

40. **Gillissen, B., J. Bergemann, C. Sandmann, B. Schroeer, M. Bölker, and R. Kahmann.** 1992. A two-component regulatory system for self/non-self recognition in *Ustilago maydis*. *Cell* **68**:647–657.

41. **Goutte, C., and A. D. Johnson.** 1994. Recognition of a DNA operator by a dimer composed of two different homeodomain proteins. *EMBO J.* **13**:1434–1442.

42. **Goutte, C., and A. D. Johnson.** 1993. Yeast a1 and alpha2 homeodomain proteins form a DNA-binding activity with properties distinct from those of either protein. *J. Mol. Biol.* **233**:359–371.

43. **Hanna, W. F.** 1929. Studies in the physiology and cytology of *Ustilago zeae* and *Sorosporium reilianum*. *Phytopathology* **19**:415–441.

44. **Hanna, W. F.** 1925. The problem of sex in *Coprinus lagopus*. *Ann. Bot.* **39**:431–457.

45. **Hartmann, H. A., R. Kahmann, and M. Bölker.** 1996. The pheromone response factor coordinates filamentous growth and pathogenicity in *Ustilago maydis*. *EMBO J.* **15**:1632–1641.

46. **Herskowitz, I.** 1989. A regulatory hierarchy for cell specialization in yeast. *Nature* **342**:749–757.

47. **Herskowitz, I.** 1988. Life cycle of the budding yeast *Saccharomyces cerevisiae*. *Microbiol. Rev.* **52**:536–553.

48. **Holliday, R.** 1974. *Ustilago maydis*, p. 575–595. *In* R. C. King (ed.), *Handbook of Genetics*, vol 1. Plenum, New York, NY.

49. **Holliday, R.** 1961. The genetics of *Ustilago maydis*. *Genet. Res.* **2**:204–230.

50. **Holliday, R.** 1961. Induced mitotic crossing-over in *Ustilago maydis*. *Genet. Res.* **2**:231–248.

51. **Kahmann, R., T. Romeis, M. Bölker, and J. Kämper.** 1995. Control of mating and development in *Ustilago maydis*. *Curr. Opin. Genet. Dev.* **5**:559–564.

52. **Kahmann, R., and C. Basse.** 2001. Fungal gene expression during pathogenesis-related development and host plant colonization. *Curr. Opin. Microbiol.* **4**:374–380.

53. **Kahmann, R., and M. Bölker.** 1996. Self/nonself recognition in fungi: old mysteries and simple solutions. *Cell* **85**:145–148.

54. **Kämper, J., M. Reichmann, T. Romeis, M. Bölker, and R. Kahmann.** 1995. Multiallelic recognition: nonself-dependent dimerization of the bE and bW homeodomain proteins in *Ustilago maydis*. *Cell* **81**:73–83.

55. **Kniep, H.** 1919. Untersuchungen über den Antherenbrand (*Ustilago violacea* Pers.). Ein Beitrag zum Sexualitätsproblem. *Zeit. Bot.* **11**:257–284.

56. **Kronstad, J. W., and S. A. Leong.** 1990. The *b* mating-type locus of *Ustilago maydis* contains variable and constant regions. *Genes Dev.* **4**:1384–1395.

57. **Kronstad, J. W., and S. A. Leong.** 1989. Isolation of two alleles of the *b* locus of *Ustilago maydis*. *Proc. Natl. Acad. Sci. USA* **86**:978–982.

58. **Kronstad, J. W., J. Wang, S. F. Covert, D. W. Holden, G. S. McKnight, and S. A. Leong.** 1989. Isolation of metabolic genes and demonstration of gene disruption in the phytopathogenic fungus *Ustilago maydis*. *Gene* **79**:97–106.

59. **Kües, U., R. N. Asante-Owusu, E. S. Mutasa, A. M. Tymon, E. H. Pardo, S. F. O'Shea, B. Göttgens, and L. A. Casselton.** 1994. Two classes of homeodomain proteins specify multiple *A* mating types of the mushroom *Coprinus cinereus*. *Plant Cell* **6**:1467–1475.

60. **Kües, U., B. Göttgens, R. Stratmann, W. V. Richardson, S. F. O'Shea, and L. A. Casselton.** 1994. A chimeric homeodomain protein causes self-compatibility and constitutive sexual development in the mushroom *Coprinus cinereus*. *EMBO J.* **13**:4054–4059.

61. **Kües, U., W. V. Richardson, A. M. Tymon, E. S. Mutasa, B. Göttgens, S. Gaubatz, A. Gregoriades, and L. A. Casselton.** 1992. The combination of dissimilar alleles of the *A alpha* and *A beta* gene complexes, whose proteins contain homeodomain motifs determines sexual development in the mushroom *Coprinus cinereus*. *Genes Dev.* **6**:568–577.

62. **Kuhn, J., and Y. Parag.** 1972. Protein-subunit aggregation model for self-incompatibility in higher fungi. *J. Theor. Biol.* **35**:77–91.

63. **Laity, C., L. Giasson, R. Campbell, and J. Kronstad.** 1995. Heterozygosity at the *b* mating-type locus attenuates fusion in *Ustilago maydis*. *Curr. Genet.* **27**:451–459.

64. **Lederberg, J., and W. M. Lederberg.** 1952. Replica plating and indirect selection of bacterial mutants. *J. Bacteriol.* **63**:399–406.

65. **Lengeler, K. B., D. S. Fox, J. A. Fraser, A. Allen, K. Forrester, F. S. Dietrich, and J. Heitman.** 2002. Mating-type

locus of *Cryptococcus neoformans*: a step in the evolution of sex chromosomes. *Eukaryot. Cell* 5:704–718.

66. **Luo, Y., R. C. Ullrich, and C. P. Novotny.** 1994. Only one of the paired *Schizophyllum commune A alpha* mating type, putative homeobox genes encodes a homeodomain essential for *A alpha*-regulated development. *Mol. Gen. Genet.* 244:318–324.

67. **Martínez-Espinoza, A. D., S. A. Gerhardt, and J. W. Sherwood.** 1993. Morphological and mutational analysis of mating in *Ustilago hordei. Exp. Mycol.* 17:200–214.

68. **O'Donnell, K. L, and K. J. McLaughlin.** 1984. Postmeiotic mitosis, basidiospore development and septation in *Ustilago maydis. Mycologia* 76:486–502.

69. **Perkins, D. D.** 1949. Biochemical mutants in the smut fungus *Ustilago maydis. Genetics* 34:607–626.

70. **Puhalla, J. E.** 1970. Genetic studies of the *b* incompatibility locus of *Ustilago maydis. Genet. Res.* 16:229–232.

71. **Puhalla, J. E.** 1968. Compatibility reactions on solid medium and interstrain inhibition in *Ustilago maydis. Genetics* 60:461–474.

72. **Raper, J. R., M. G. Baxter, and A. H. Ellingboe.** 1960. The genetic structure of the incompatibility factors of *Schizophyllum commune*: the *A* factor. *Proc. Natl. Acad. Sci. USA* 46:833–842.

73. **Robertson, C. I., K. A. McMahon, K. Toenjes, C. P. Novotny, and R. C. Ullrich.** 2002. Evidence for interaction of *Schizophyllum commune* Y mating-type proteins *in vivo. Genetics* 160:1461–1467.

74. **Romeis, T., A. Brachmann, R. Kahmann, and J. Kämper.** 2000. Identification of a target gene for the bE-bW homeodomain protein complex in *Ustilago maydis. Mol. Microbiol.* 37:54–66.

75. **Romeis, T., J. Kämper, and R. Kahmann.** 1997. Single-chain fusions of two unrelated homeodomain proteins trigger pathogenicity in *Ustilago maydis. Proc. Natl. Acad. Sci. USA* 94:1230–1234.

76. **Rowell, J. B.** 1955. Functional role of compatibility factors: an *in vitro* test for sexual compatibility with haploid lines of *Ustilago zeae. Phytopathology* 45:370–374.

77. **Rowell, J. B.** 1955. Segregation of sex factors in a diploid line of *Ustilago zeae* induced by alpha radiation. *Science* 121:304–306.

78. **Rowell, J. B., and J. E. DeVay.** 1954. Genetics of *Ustilago zeae* in relation to basic problems of its pathogenicity. *Phytopathology* 44:356–362.

79. **Schafer, W. R., and J. Rine.** 1992. Protein prenylation: genes, enzymes, targets, and functions. *Annu. Rev. Genet.* 26:209–237.

80. **Schirawski, J., B. Heinze, M. Wagenknecht, and R. Kahmann.** 2005. Mating type loci of *Sporisorium reilianum*: novel pattern with three *a* and multiple *b* specificities. *Eukaryot. Cell* 4:1317–1327.

81. **Schlesinger, R., R. Kahmann, and J. Kämper.** 1997. The homeodomains of the heterodimeric bE and bW proteins of *Ustilago maydis* are both critical for function. *Mol. Gen. Genet.* 254:514–519.

82. **Schulz, B., F. Banuett, M. Dahl, R. Schlesinger, W. Schäfer, T. Martin, I. Herskowitz, and R. Kahmann.** 1990. The *b* alleles of *U. maydis*, whose combinations program patho-

genic development, code for polypeptides containing a homeodomain-related motif. *Cell* 60:295–306.

83. **Sleumer, H. O.** 1932. Uber Sexualität und Zytologie von *Ustilago zeae* (Beckm.) Unger. *Zeit. Bot.* 25:209–263.

84. **Snetselaar, K. M.** 1993. Microscopic observation of *Ustilago maydis mating* interactions. *Exp. Mycol.* 17:345–355.

85. **Snetselaar, K. M., M. Bölker, and R. Kahmann.** 1996. *Ustilago maydis* mating hyphae orient their growth toward pheromone sources. *Fungal Genet. Biol.* 20:299–312.

86. **Spellig, T., M. Bölker, F. Lottspeich, R. W. Frank, and R. Kahmann.** 1994. Pheromones trigger filamentous growth in *Ustilago maydis. EMBO J.* 13:1620–1627.

87. **Stakman, E. C., and J. J. Christensen.** 1927. Heterothallism in *Ustilago zeae. Phytopathology* 17:827–834.

88. **Stankis, M. M., C. A. Specht, H. Yang, L. Giasson, R. C. Ullrich, and C. P. Novotny.** 1992. The *A alpha* mating locus of *Schizophyllum commune* encodes two dissimilar multiallelic homeodomain proteins. *Proc. Natl. Acad. Sci. USA* 89:7169–7173.

89. **Ullrich, R. C.** 1978. On the regulation of gene expression: incompatibility in *Schizophyllum. Genetics* 88:709–722.

90. **Urban, M., R. Kahmann, and M. Bölker.** 1996. Identification of the pheromone response element in *Ustilago maydis. Mol. Gen. Genet.* 251:31–37.

91. **Urban, M., R. Kahmann, and M. Bölker.** 1996. The biallelic *a* mating type locus of *Ustilago maydis*:remnants of an additional pheromone gene indicate evolution from a multiallelic ancestor. *Mol. Gen. Genet.* 250:414–420.

92. **Vershon, A. K., Y. S. Yin, and A. D. Johnson.** 1995. A homeodomain protein lacking specific side chains of helix 3 can still bind DNA and direct transcriptional repression. *Genes Dev.* 9:182–192.

93. **Wang, J., D. W. Holden, and S. A. Leong.** 1988. Gene transfer system for the phytopathogenic fungus *Ustilago maydis. Proc. Natl. Acad. Sci. USA* 85:865–869.

94. **Wong, G. J., and K. Wells.** 1985. Modified bifactorial incompatibility in *Tremella mesenterica. Trans. Br. Mycol. Soc.* 84:834–838.

95. **Wu, J., R. C. Ullrich, and C. P. Novotny.** 1996. Regions in the Z5 mating type gene of *Schizophyllum commune* involved in Y-Z binding recognition. *Mol. Gen. Genet.* 252:739–745.

96. **Yee, A. R., and J. W. Kronstad.** 1998. Dual sets of chimeric alleles identify specificity sequences for the bE and bW mating and pathogenicity genes of *Ustilago maydis. Mol. Cell. Biol.* 18:221–232.

97. **Yee, A. R., and J. W. Kronstad.** 1993. Construction of chimeric alleles with altered specificity at the *b* incompatibility locus of *Ustilago maydis. Proc. Natl. Acad. Sci. USA* 90:664–668.

98. **Yue, C., M. Osier, C. P. Novotny, and R. C. Ullrich.** 1997. The specificity determinant of the Y mating-type proteins of *Schizophyllum commune* is also essential for Y-Z protein binding. *Genetics* 145:253–260.

99. **Zhang, F. L., and P. J. Casey.** 1996. Protein prenylation: molecular mechanisms and functional consequences. *Annu. Rev. Biochem.* 65:241–269.

Sex in Fungi: Molecular Determination and Evolutionary Implications
Edited by Joseph Heitman et al.
© 2007 ASM Press, Washington, D.C.

Regine Kahmann
Jan Schirawski

22

Mating in the Smut Fungi:
From *a* to *b* to the Downstream Cascades

Smut fungi induce disease only in their dikaryotic stage, which is generated by mating. Mating is regulated by two loci, which harbor conserved genes. In the *a* locus these genes specify pheromones and receptors, while in the *b* locus two transcription factors are encoded. Despite this similarity in gene function, in the three smuts that have been molecularly analyzed (*Ustilago maydis*, *Ustilago hordei*, and *Sporisorium reilianum*) the mating-type loci differ substantially in locus structure. These differences are reviewed here in an evolutionary perspective. The processes controlled by the mating-type loci have been analyzed in some detail only in *U. maydis*. We focus here on the signaling cascades, which coordinate cyclic AMP (cAMP) and mitogen-activated protein kinase (MAPK) signaling in response to pheromone, as well as the transcriptional cascade triggered by the products of the *b* locus. Emphasis is also given to the developmental processes, which are initiated after mating and involve cell cycle arrest, mitotic divisions of the dikaryon, and pathogenesis.

Smut diseases of grasses are caused by Basidiomycetes of the order Ustilaginales. Because most smuts develop in kernel tissue, they cause considerable yield reductions. Smut fungi can initiate an infection only after formation of the dikaryotic stage by mating of compatible haploid cells. Smuts differ in host range as well as the site of symptom development, with symptoms in most cases being restricted to the inflorescences (1). One notable exception to this is *U. maydis*, a pathogen infecting maize and causing tumors on all aerial parts of the plant. Of the estimated 1,200 smut species only the corn smuts *U. maydis* and *S. reilianum*, as well as the barley smut *U. hordei*, have entered the molecular era with the establishment of efficient techniques for gene replacement as well as cytology and live-cell imaging. The close phylogenetic relationship of these three species is also reflected by the ease with which molecular tools can be transferred from one species to the other, i.e., promoters and autonomously replicating sequences from *U. maydis* that function in the other hosts. The 20.5-Mb genome of *U. maydis* has been sequenced and is publicly available through the Broad Institute (http://www.broad.mit.edu/annotation/fungi/ustilago_maydis/) and in manually annotated form through MIPS (http://MIPS.gsf.de/genre/proj/ustilago/) (32).

In their haploid form smuts can be propagated in the laboratory. As haploid cells are nonpathogenic and need to fuse with a compatible partner to generate the infective form, much effort has been directed to an understanding of the fusion process and the steps that control subsequent development of the dikaryon (18, 29, 37).

Regine Kahmann and Jan Schirawski, Max Planck Institute for Terrestrial Microbiology, 35043 Marburg, Germany.

Cell recognition and fusion are regulated by a pheromone/receptor system that is encoded at the *a* mating-type locus. After fusion the dikaryon is maintained and cells switch to filamentous growth if they are heterozygous for genes at the *b* mating-type locus. The *b* mating-type locus encodes two homeodomain proteins that function as transcriptional regulators after dimerization. The ability for complex formation is restricted to homeodomain proteins encoded by different alleles of the *b* locus. Thereby, it is guaranteed that active heterodimers are formed only in the dikaryon and do not arise in haploid cells.

In nature, mating occurs on the plant surface. The resulting dikaryon switches to filamentous growth and forms appressoria, which allow direct penetration of the plant cuticle in a process presumably aided by lytic enzymes (47). During this stage hyphae show tip growth but do not enlarge their cytoplasm and do not undergo nuclear division. In *U. maydis*, the dikaryon is arrested in the G$_2$ phase of the cell cycle (20) and this block is released only after penetration. During filamentous growth on the leaf surface, and during early stages after penetration, hyphal growth is accommodated by the formation of highly vacuolated sections in the rear part of the hyphae. These appear empty and are partitioned by regularly spaced septa (55). Between 1 and 2 days after penetration the growth mode changes, mitotic divisions occur, and branching is observed. Around 5 to 7 days after infection by *U. maydis* the plant reacts by forming tumor-like structures, which are likely induced by the fungus through interference with phytohormone signaling. In these structures the dikaryotic hyphae differentiate, the nuclei fuse, and hyphal fragmentation occurs followed by the formation of darkly pigmented spores (8, 50, 51). Tumor formation is unique to the *U. maydis*/maize system. In related smuts, the initial infection is asymptomatic and disease symptoms, i.e., the formation of spores, are restricted to sori, which develop in the flowers (21).

In the first part of our chapter we highlight variations in the organization of the mating-type loci in *U. maydis*, *S. reilianum*, and *U. hordei*, their consequences for mating, and finally evolutionary implications. In the following sections we review the signaling pathway underlying pheromone perception as well as the regulatory cascade triggered by the homeodomain heterodimer.

THE *a* AND *b* MATING-TYPE LOCI IN SMUT FUNGI: ORGANIZATION AND FUNCTIONAL/EVOLUTIONARY IMPLICATIONS

The *a* mating-type loci in smut fungi contain the genes responsible for cell-cell recognition, i.e., the genes encoding pheromones and pheromone receptors. The de-

tailed structure of these loci has been determined for *U. maydis*, for *S. reilianum*, and for the *MAT-1* allele in *U. hordei* (Fig. 22.1). For the *MAT-2* allele in *U. hordei* only partial information is available (Fig. 22.1). *U. maydis* and *U. hordei* have two alleles of the *a* mating system with one pheromone receptor and one active pheromone gene per locus. The pheromone and receptor genes in the *a2* locus differ in sequence from those in the *a1* locus (10). Sequences surrounding the pheromone and receptor genes are unrelated between the two *a* mating types. The *a1* locus of *U. maydis* encompasses 4 kb, while the *a2* locus spans an 8-kb region (Fig. 22.1). The *a1* and *a2* mating regions of *U. hordei* also contain one receptor and one pheromone gene each with nonidentical sequence (2, 6) (Fig. 22.1). Interestingly, *S. reilianum* has three different *a* mating types and this species has solved the problem of recognizing two mating partners by the presence of two pheromones in each *a* mating-type locus, with one each serving for recognition of one of the two mating partners (Fig 22.1) (46). The *a1* and *a2* loci of *S. reilianum* show synteny to the *a1* and *a2* loci of *U. maydis*, respectively. The *a2* loci of both organisms contain two additional genes, *lga2* and *rga2* (Fig. 22.1), which do not seem to be involved in cell-cell recognition but may play a role in mitochondrial inheritance (11). These two genes are conserved in order, orientation, and position in the *a2* loci of *U. maydis* and *S. reilianum* (Fig. 22.1). Whether or not they are present between the pheromone and pheromone receptor genes of the *MAT-2* locus of *U. hordei* remains to be determined.

The pheromone genes give rise to very small proteins of only 38 to 41 amino acids, which undergo several posttranslational modification reactions during maturation and secretion. They all have in common a CAAX motif (C, cysteine; A, aliphatic amino acid; X, any amino acid) at the C terminus. This motif is a signal for isoprenylation and carboxymethylation at the cysteine residue, which constitutes the C-terminal amino acid in the mature pheromones (15). The pheromone precursors are also processed at their N terminus, where a large part of the protein is cleaved off, leaving very small peptides of only 9 to 14 amino acids (33, 46, 53). The mature pheromones are secreted as small lipopeptide molecules, and removal of the lipid moiety drastically reduces their activity (53). One of the two pheromones produced by *S. reilianum a1* strains is identical in sequence to one of the pheromones produced by *S. reilianum a2* strains, while the other is identical to one of the pheromones of *S. reilianum a3* strains (46). This means that the six pheromone genes of *S. reilianum* give rise to only three different pheromones with one each for the recognition of each of the three *a* mating types.

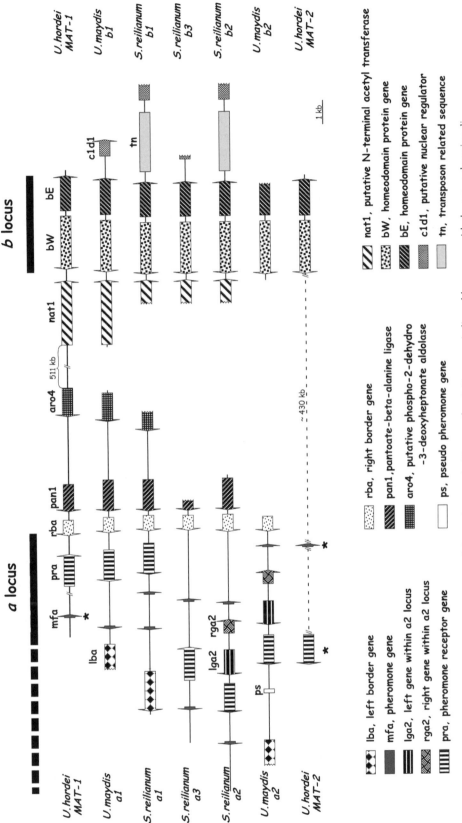

Figure 22.1 Genetic organization of the mating-type loci in smut fungi. Genes are indicated by arrows with the arrow denoting direction of transcription. Related genes are denoted by the same pattern, and respective gene functions are explained in the lower part of the figure. Asterisks (*) indicate that the relative order and orientation of these genes have not been determined. In *U. maydis* and *S. reilianum* the *a*- and *b*-specific sequences reside on different chromosomes, while they are linked in *U. hordei* by spacer regions (which are not drawn to scale and whose length is indicated). The black bars on top of the figure indicate the region of the *a* locus (that expands to different lengths in the different loci, indicated by a broken line) from the *lba* gene and the region of the *b* locus, which covers the two homeodomain protein genes, *bE* and *bW*. Sequence information was obtained from the following accession numbers: AF043940, AM118080, AACP01000083, AACP01000013, AJ884583, AJ884590, AJ884585, AJ884589, AJ884584, U37796, M84182, AF184070, AF184069, and Z18531.

The three different pheromones of *S. reilianum* show only very little sequence identity to each other. However, the pheromones involved in recognition by the *a2* receptors of *U. hordei*, *U. maydis*, and *S. reilianum* are related in primary sequence, as are the three pheromones recognized by the *a1* receptors in the three species (46). This indicates that these pheromones share a common ancestor. The lack of sequence relatedness of the *S. reilianum* pheromone involved in recognition of the third *a* mating type might indicate that this pheromone was acquired by recombination after mating with an as yet unidentified species. In line with this proposition is the fact that while the *a1* and *a2* mating-type loci of *U. maydis* are unrelated, the three *a* loci of *S. reilianum* show several stretches of high sequence conservation between different *a* alleles. This sequence conservation is localized around the pheromone genes, which indicates that they were exchanged between loci by recombination. Interestingly, the *a2* locus of *U. maydis* contains a pheromone pseudogene in a location similar to that of one of the two pheromone genes in the *S. reilianum a2* locus (57). This could indicate that the ancestral *a* mating-type locus had more than one active pheromone gene. During speciation these genes might have been recombined or lost, giving rise to the simple present loci as we find them, for example, in *U. maydis*. We consider it an attractive possibility that this process is still ongoing, and thus, the analysis of the *a* mating-type loci from other smut fungi might reveal more details on the events that have occurred during evolution.

The *b* mating-type loci encode two subunits of a homeodomain transcription factor. For *U. maydis* the *b* mating-type proteins bE and bW constitute a heterodimeric transcription factor that is active only if the two subunits are derived from different alleles (24, 31), a situation which naturally occurs after mating of two compatible partners. The bE and bW proteins are not related to each other in primary sequence; however, they share a common domain organization. Both proteins contain a homeodomain motif involved in DNA binding, which separates the conserved part of these proteins from the variable region (24, 34, 48). The current model proposes that dimerization is achieved via a limited number of hydrophobic and polar interactions between the variable regions of the bE and bW proteins (31). All three smut fungi analyzed have in common the divergently transcribed gene pair encoding homeodomain proteins (Fig. 22.1), and heterodimerization of bE and bW proteins to form active transcription factors is the general principle for interaction between the b proteins of smut fungi. Functional dimer formation has even been observed between b proteins from different species, which

may permit interspecies hybridization (6). The *a* and *b* mating-type genes are located in the same genomic context with conservation of neighboring genes (Fig. 22.1). While from *U. hordei* two different mating types are known (4), five different *b* alleles have been described in *S. reilianum* (45) and at least 19 exist in *U. maydis* (R. Kahmann and J. Kämper, unpublished data). While the genetic organization of the mating-type loci is highly conserved among the three smut fungi analyzed (Fig. 22.1), in *U. maydis* and in *S. reilianum* the *a* and *b* mating-type loci reside on different chromosomes and segregate independently during meiosis, giving rise to spores with four different mating types (tetrapolar system). In contrast, the regions encoding the *a* and *b* mating-type proteins in *U. hordei* are on the same chromosome, albeit some 500 kb apart (7, 38) (Fig. 22.1). In spite of this large distance, the mating-type determinants of *U. hordei* do not segregate independently during meiosis. Recombination appears to be suppressed over the entire *MAT* region, giving rise to only two mating types in the progeny. This provides a molecular explanation for the observed bipolar mating system in *U. hordei* (5) (see Bakkeren and Kronstad, chapter 23 this volume).

Comparison of the *MAT-1* sequence of *U. hordei* and the genome of *U. maydis* revealed that more than 90% of the genes harbored on the *U. hordei MAT-1* region have orthologues in the vicinity of either the *U. maydis a* mating-type genes or the *U. maydis b* mating-type genes. In addition, the *MAT-1* locus of *U. hordei* shows an accumulation of repetitive elements, which are not found in the syntenic regions in *U. maydis* (7). This suggests that the ancestral mating-type loci in smuts were tetrapolar (7). The mating-type loci of *U. hordei* could then have evolved by a translocation event between the chromosomes carrying the *a* and *b* loci. An accumulation of repetitive elements within the region separating the *a* and *b* mating-type genes could then have led to sufficiently different sequences to suppress recombination between *MAT-1* and *MAT-2*. In line with this reasoning is the observation that the *MAT-1* sequence is composed of about 50% repetitive DNA interspersed with small islands of genes and that the most abundant repetitive element found in the *U. hordei MAT-1* locus seems to be specific for the *MAT-1* region (7). The *MAT-2* region of *U. hordei* spans a 430-kb region, the sequence of which has not yet been determined (38). It is tempting to speculate that the *MAT-2* region could contain an accumulation of *MAT-2*-specific repeats. Alternatively, the mating-type regions might be recombination suppressed because they contain a centromere. One of the repetitive elements identified in the *MAT-1* region (Tuh3) (7) displays similarity to the *U. maydis* retrotransposon HobS. For *U.*

maydis it has been shown that mobile elements of the HobS-type cluster around the presumed centromeres in one location on each chromosome (32), and it is these regions that allow autonomous replication of plasmids in functional assays (32). This could indicate that the *MAT-1* region of *U. hordei* has evolved to become a centromeric region after the presumed initial translocation event. It would be interesting to functionally determine whether this region (or parts of it) provides activity for autonomous replication.

COMPONENTS OF THE PHEROMONE-CONTROLLED SIGNALING CASCADE

Pheromone perception in smut fungi elicits the formation of conjugation hyphae. These structures usually form at one pole of the cell, grow towards each other, and fuse (33, 46, 52, 53). *U. maydis* is the only smut fungus for which the signaling cascades that are activated after binding of the pheromones to their cognate receptors have been elucidated in quite some detail. In *U. maydis*, pheromone perception by the Pra1 and Pra2 receptors leads to elevated cAMP as well as MAPK signaling. Components of the cAMP pathway are the alpha subunit of a heterotrimeric G protein, Gpa3, the adenylyl cyclase Uac1, and the regulatory and catalytic protein kinase A (PKA) subunits Ubc1 and Adr1, respectively (Fig. 22.2). MAPK signaling is mediated by a cascade consisting of the MAPKKK Kpp4/Ubc4, the MAPKK Fuz7/Ubc5, and the MAPK Kpp2/Ubc3 (Fig. 22.2). Mutations in all of these components strongly affect the ability of strains to mate. cAMP and MAPK signaling pathways are strongly interconnected. This was revealed by showing that PKA signaling elevates pheromone gene expression and that the disruption of individual MAPK signaling components (Ubc4, Ubc5, and Ubc3) as well as the deletion of *ubc2* antagonizes filamentation of strains in which *uac1* is deleted (3, 25, 39, 40, 49). Ubc2 is related to Ste50p from *Saccharomyces cerevisiae* and is proposed to act as a scaffold for components of the MAPK module (49). The interaction between Ubc2 and Kpp4/Ubc4 could be mediated by the SAM domain of Kpp4. A potential interaction partner of Ubc2 and/or Kpp4 is Ras2 (Fig. 22.2). The gene for this small G protein is epistatic to genes in the MAPK module, and expression of a constitutively active version elicits increased pheromone gene expression (36, 41). A potential activator of Ras2 is the CDC25-like guanyl nucleotide exchange factor Sql2 (41). *U. maydis* possesses a second Ras protein, Ras1, which promotes cAMP-regulated gene expression. It was suggested that this small G protein is connected to PKA signaling by

interacting with Uac1 (41) (Fig. 22.2). Despite the fact that a large number of components involved in transmitting the pheromone signal have been identified in recent years, there are still significant gaps in understanding of the mode of signal transmission from the receptor to the downstream cascades as well as MAPK specificity. The availability of the fully annotated genome sequence is likely to allow closure of these gaps in the near future.

GENES AND PROCESSES CONTROLLED BY PHEROMONE SIGNALING

After pheromone stimulation cells react with a transcriptional as well as with a morphological response, and both require cAMP signaling as well as the MAPK module. One point of convergence between the two pathways is the pheromone response factor Prf1. Prf1 is an HMG-box transcription factor that binds to pheromone response elements of the consensus sequence 5'-ACAAAGGGA-3' (27). A large number of genes are regulated through Prf1, and many of these have putative Prf1 binding sites in their promoters (H. Eichhorn and R. Kahmann, unpublished data). Among the genes that require Prf1 for their basal as well as induced transcription after pheromone stimulation are the pheromone and receptor genes as well as the *b* genes. Consequently, *prf1* mutants are sterile. When *prf1* is deleted in a haploid pathogenic strain, the resulting mutant is unable to infect plants. Interestingly, this defect can be complemented by expressing an active homeodomain protein heterodimer. This illustrates that the crucial target for Prf1 after cell fusion is exclusively the *b* mating-type locus (27).

For full expression of the pheromone and receptor genes Prf1 needs to be phosphorylated by the PKA Adr1. To trigger the expression of the *b* locus genes, on the other hand, Prf1 needs to be phosphorylated by both Adr1 and the MAPK Kpp2 (Fig. 22.2). It is likely that the dual phosphorylation of Prf1 changes its regulatory potential, presumably through allowing interactions with additional transcription factors. However, as of yet, such factors have not been identified. In addition, *prf1* is subject to a very complex transcriptional regulation that adds on to the posttranscriptional regulation mentioned above. After pheromone signaling has been initiated, *prf1* is transcriptionally upregulated. This is likely to involve an autoregulatory system via two Prf1 binding sites in the *prf1* promoter. In addition, the HMG-box protein Rop1, which binds to three sites in the *prf1* gene promoter, is required for mating under laboratory conditions. Interestingly, *rop1* is dispensable

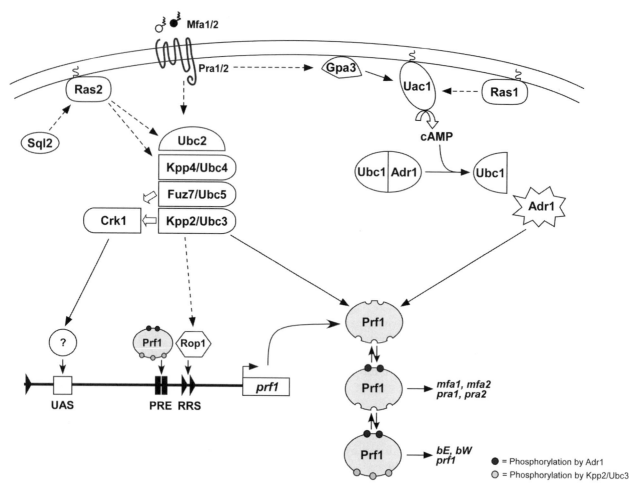

Figure 22.2 Pheromone-triggered signaling cascades in *U. maydis*. Lipopeptide pheromones Mfa1/2 bind to their cognate seven-transmembrane receptors present on either *a1* (Pra1) or *a2* (Pra2) cells. The left section displays the MAPK module consisting of Kpp4/Ubc4, Fuz7/Ubc5, and Kpp2/Ubc3 and includes Ras2 and Ubc2 as likely upstream components funneling into the MAPK module. The right-hand side depicts the known components of the cAMP signaling cascade. PKA signaling and MAPK signaling converge on Prf1. Phosphorylation of Prf1 by Adr1 and Kpp2/Ubc3 is indicated by black and gray circles, respectively, and the genes activated by the different forms are indicated. The transcriptional regulation of *prf1* is shown in the lower left section, with triangles denoting binding sites for Rop1 (RRS), black bars denoting binding sites for Prf1 (PRE), and an open bar denoting the upstream activating sequence (UAS). Broken lines depict hypothetical interactions. Details are described in the text.

for mating on the plant surface, suggesting the existence of alternative regulators that are activated under these conditions (14).

A potential candidate for such an alternative regulator is the MAPK Crk1, which was shown to activate *prf1* transcription via an upstream activating sequence (UAS) located about 1,600 bp upstream of the gene (23) (Fig. 22.2). Crk1 mutants are defective in mating, and this defect is complemented by constitutive expression of *prf1*, demonstrating that the mating defect is due solely to insufficient *prf1* gene expression. Crk1 is acti-

vated through Fuz7 and is thus likely to function in parallel to the MAPK Kpp2 (23) (Fig. 22.2) in activating an as yet unknown transcription factor.

Crk1 is also a prime candidate for mediating PKA and MAPK cross talk since *crk1* transcription is regulated antagonistically by PKA and MAPK signaling (22). Through the UAS the *prf1* gene is also subject to carbon source regulation, and glucose as well as fructose act as inducers (26). However, whether this is of relevance for mating on the plant surface needs to be investigated. How *rop1* and *crk1* are activated also remains a mystery.

As Crk1-related proteins in other organisms are involved in cell cycle regulation (22, 23), the idea has been put forward that the involvement of Crk1 could be somehow connected to the G_2 cell cycle arrest observed after pheromone stimulation. This G_2 arrest was shown to require integrity of the pheromone MAPK module. Since pheromone signaling was shown to downregulate expression of the B-type cyclin *clb1*, it has been speculated that the Clb1/Cdk1 complex could be the target of the cell cycle arrest (see reference 42). In addition, it has been shown that high levels of the mitotic cyclin Cln1 downregulate pheromone gene expression presumably through downregulating *prf1*, while reducing the levels of Cln1 leads to an induction of the pheromone gene even under repressive medium conditions (16). It is presently not known via which factors cyclin/Cdk1 complexes affect transcription of the *prf1* gene. The cell cycle machinery has also been shown to be an important target of the cAMP pathway. Cells in which the *cru1* gene, encoding a member of the Fizzy-related family of APC activators, is deleted showed reduced expression of the *mfa* gene and were consequently less efficient in cell fusion. The cAMP pathway was shown to have a positive role in the expression of *cru1*. In addition, a G_2 arrest was induced when *cru1* was overproduced (17). It has been speculated that Cru1 is needed to maintain the accurate length of the G_1 phase under changing environmental conditions and that this may be crucial for selecting the appropriate time point for entering into the sexual mode of development.

The connection between mating and cell cycle in *U. maydis* may also provide an explanation for the observation that the pheromone-induced formation of conjugation tubes is *prf1* independent. It has been speculated that this could indicate the existence of alternative transcription factors required for this developmental program. However, based on the observation that an induction of cell cycle arrest by various means leads to filamentation, this morphological program has also been interpreted as a default response to arrested cell division (19, 20, 54). Future experiments are needed to resolve these questions.

TARGETS OF THE bE/bW HETERODIMER AND THEIR ROLE DURING SEXUAL/PATHOGENIC DEVELOPMENT

The central role of the bE/bW transcription factor in triggering sexual as well as pathogenic development in *U. maydis* has stimulated the search for target genes, as these were assumed to play important roles during pathogenesis. The first targets of the bE/bW heterodimer were identified using a variety of differential techniques (9, 12, 44), and this was followed by microarray analysis. The Affymetrix DNA arrays for *U. maydis* represent about 93% of the predicted 6,902 genes (32; http://www.broad.mit.edu/annotation/fungi/ustilago_maydis/; Kämper and Kahmann, unpublished). For these genome-wide studies strains were constructed such that bE/bW gene expression could be induced via promoters that could be regulated through nitrogen or carbon source (12). These strains allowed following bE/bW-regulated development in a time-resolved fashion without interference from pheromone-induced responses. A set of 347 *b*-regulated genes was defined, in which 212 genes were upregulated while 135 genes were repressed after induction of the bE/bW heterodimer (M. Scherer and J. Kämper, unpublished data). Through DNA binding studies combined with bioinformatic predictions a potential *b* binding site, which has been proposed to include the motif TGA-N_9-TGA (12, 43, 45), was identified in the promoters of several of the bE/bW-regulated genes (Table 22.1). However, the majority of genes regulated by the bE/bW heterodimer lacked such a site. This led to the conclusion that the bE/bW heterodimer triggers a transcriptional cascade consisting of only a small number of direct targets. Among these targets then must be regulatory genes that subsequently control subsets of the complete set of bE/bW-regulated genes, the indirect targets (Table 22.1).

Among the first direct targets identified was *lga2* (43, 57). This gene resides in the *a2* locus and is divergently transcribed with the *rga2* gene (57). Orthologues of these two genes are also found in the *a2* locus of *S. reilianum* (46) (Fig. 22.1). *lga2* and *rga2* are upregulated through the pheromone cascade, and additional strong upregulation of *lga2* is observed through the bE/bW heterodimer (57). The Lga2 and Rga2 proteins localize to mitochondria (11). The deletion of either *lga2* alone or of *lga2* and *rga2* in combination did not affect mating (57). However, overexpression of *lga2* (as is expected to occur in the dikaryon) interfered with cell growth, induced mitochondrial fragmentation, and lowered mitochondrial respiratory activity. In wild-type cells these detrimental effects are counteracted by the Mrb1 protein, a member of the mitochondrial p32 protein family (11). It has been proposed that Lga2 interferes with mitochondrial fusion, and this may be prerequisite for the uniparental inheritance of mitochondria that is observed in *U. maydis* (11, 56) (see James, chapter 19). The other genes in the *a* locus, i.e., the pheromone and receptor genes, on the other hand, are downregulated by the bE/bW heterodimer. This effect is likely

Table 22.1 Targets of regulation by the bE/bW heterodimer

Target genes	Regulation by the bE/bW heterodimer[a]		Proposed function	Reference(s)
	Direct	Indirect		
biz1		X ↑	*b*-Induced zinc finger, involved in plant penetration	19
clp1	X ↑		Involved in clamp formation	45
kpp6		X ↑	MAPK	13
ant1/frb172		X ↑	Hypothetical K+/H+ antiporter	12
atr1/rb34		X ↓	Hypothetical acyltransferase	12
cap1/frb136		X ↓	Hypothetical capsule-associated protein	12
exc1/frb133		X ↑	Hypothetical exochitinase	12
frb110		X ↓	Hypothetical protein	12
frb124		X ↓	Hypothetical protein	12
frb63		X ↑	Hypothetical protein	12
pdi1/frb23		X ↑	Hypothetical disulfide isomerase	12
pma1/frb323		X ↑	Hypothetical plasma membrane ATPase	12
polX/frb52	X ↑		Hypothetical DNA polymerase X	12
pra1/2		X ↓	Pheromone receptor	58
rep1		X ↑	Repellent	59
lga2	X ↑		Interferes with mitochondrial fusion	11, 57
mfa1/2		X ↓	Pheromone precursor	58
egl1		X ↑	Endoglucanase	44
dik1		X ↑	Hypothetical protein	9
dik6	X ↑		Hypothetical protein	9
hum1		X ↑	Hydrophobin	9

[a]Regulation is either direct or indirect, as indicated by "X" in the appropriate column, and either positive (↑, upregulation) or negative (↓, downregulation).

to be indirect, as no potential binding sites could be detected in the respective promoters. Downregulation of the *a* mating-type genes provides an explanation for the observed inhibition of mating in strains expressing an active bE/bW heterodimer (35).

The functional classification of the other genes regulated by bE/bW revealed that cellular processes including restructuring of the cell wall and alterations in lipid metabolism are affected as well as genes involved in cell cycle control, mitosis, and DNA replication (Table 22.1) (Kämper and Scherer, unpublished). The latter is in line with the observation that the dikaryotic filaments are arrested in the G$_2$ phase of the cell cycle (20). Related to this, it was recently shown that the zinc-finger protein Biz1 triggers this arrest through downregulation of the cyclin *clb1*. Biz1 is induced through the bE/bW heterodimer, and the absence of potential bE/bW binding sites in its promoter suggests that it is an indirect target (19). The deletion of *biz1* had no effect in haploid cells (and mating was unaffected). However, in the filaments produced, more than two nuclei were often found, suggesting a defect in arresting mitosis. *biz1* mutants were unable to cause disease symptoms as a conse-

quence of a 10-fold reduction in appressorium formation as well as arrested growth immediately after penetration (19). Based on these and other studies it was suggested that cell cycle arrest is important though probably not sufficient for appressorium formation.

Among the bE/bW-regulated genes an unusual MAPK, Kpp6, was identified that shares partially redundant functions with Kpp2. *kpp6* expression is induced by pheromone as well as through an active bE/bW heterodimer, and the latter regulation appears to be indirect. Kpp6 is required for the penetration step after appressoria are formed (13).

Another bE/bW-regulated gene identified through the array analysis is *clp1* (45). This gene is related to *clp1* in *Coprinopsis cinerea*. In this organism it has been demonstrated that the expression of *clp1* triggers clamp cell formation in a process that ensures proper nuclear division in the dikaryotic filament (28). As *U. maydis* was not known to form clamps, the finding of a *clp1* orthologue was surprising. Intriguingly, it could subsequently be demonstrated that *clp1* is essential for proliferation of *U. maydis* after host plant infection. *clp1* mutants have no discernible phenotype during axenic culture and mate like wild-type cells. However, after plant penetration they arrest growth prior to the first mitotic division, fail to develop branch-like structures, and are nonpathogenic. The branch-like structures were shown to develop in the most apical cell of the dikaryotic hyphae at a point where new septa are later formed. One of the four mitotic nuclei became trapped in these structures after septum formation and moved from there to the subapical cell presumably through a septal pore (45). Thus, these branch-like structures in *U. maydis* carry out a clamp-related function in nuclear distribution. In contrast to the clamps seen, for example, in *C. cinereus* and *Schizophyllum commune*, however, the clamp-like structures in *U. maydis* do not need to fuse with the subapical cell to release the nucleus. This explains why the nuclear distribution process in *U. maydis* is independent of pheromone signaling while it requires pheromone-mediated recognition in the basidiomycetes with true clamps (see reference 30).

Based on the presence of putative binding sites of the bE/bW heterodimer in the *clp1* promoter region, *clp1* is likely to be a direct target of the homeodomain protein complex. Clp1 overexpression in a haploid strain has no effect, and thus, it is unlikely that Clp1 is a transcription factor. However, when *clp1* is overexpressed in a strain expressing bE and bW, filamentation is drastically reduced. This indicates that the cell cycle block is released. In fact, a number of genes that are upregulated through the bE/bW heterodimer are repressed upon *clp1* overexpression. The currently favored hypothesis is that Clp1 acts posttranscriptionally and affects the regulatory activity of the bE/bW heterodimer (45). During filamentation on the leaf surface, appressorium formation, and subsequent mitotic growth inside the infected tissue, these counteracting effects of the bE/bW heterodimer and Clp1 are likely to require a fair amount of fine-tuning with possible additional signaling inputs.

PERSPECTIVE

The analysis of the pheromone-controlled signaling cascade, as well as the regulatory cascade that is triggered by the homeodomain protein complex, is beginning to permit unprecedented insights into the processes which are controlled by the *a* and *b* mating-type loci. These processes are amazingly diverse, are connected on various levels, and are all likely to require additional fine-tuning. At present, only a few of these processes have been analyzed in some detail while others await discovery. What is clear, however, is that the mating-type loci control central developmental decisions, and the tools are now in place to elucidate this at a molecular level. One of the driving forces for many of the studies described in this chapter has been the assumption that finding the complete set of targets for the bE/bW heterodimer will explain sexual development and its connection to pathogenesis. It is now evident that this proposition is correct. Another area of great promise is comparative genomics of genes regulated by the mating-type loci, as this is likely to make it possible to distinguish between species-, family-, and genus-specific genes and processes.

We thank the DFG for support through SFB395 and the BMBF for funding through project no. 0312738.

References

1. Agrios, G. N. 2005. The smuts, p. 582–593. *In* G. N. Agrios (ed.), *Plant Pathology*, 5th ed. Elsevier Academic Press, Burlington, MA.

2. **Anderson, C. M., D. A. Willits, P. J. Kosted, E. J. Ford, A. D. Martinez-Espinoza, and J. E. Sherwood.** 1999. Molecular analysis of the pheromone and pheromone receptor genes of *Ustilago hordei. Gene* **240:**89–97.

3. **Andrews, D. L., J. D. Egan, M. E. Mayorga, and S. E. Gold.** 2000. The *Ustilago maydis ubc4* and *ubc5* genes encode members of a MAP kinase cascade required for filamentous growth. *Mol. Plant-Microbe Interact.* **13:**781–786.

4. **Bakkeren, G., and J. W. Kronstad.** 1993. Conservation of the *b* mating-type gene complex among bipolar and tetrapolar smut fungi. *Plant Cell* **5:**123–136.

5. **Bakkeren, G., and J. W. Kronstad.** 1994. Linkage of mating-type loci distinguishes bipolar from tetrapolar mating in basidiomycetous smut fungi. *Proc. Natl. Acad. Sci. USA* **91:**7085–7089.

6. **Bakkeren, G., and J. W. Kronstad.** 1996. The pheromone cell signaling components of the *Ustilago a* mating-type loci determine intercompatibility between species. *Genetics* **143:**1601–1613.

7. **Bakkeren, G., G. Jiang, R. L. Warren, Y. Butterfield, H. Shin, R. Chiu, R. Linning, J. Schein, N. Lee, G. Hu, D. M. Kupfer, Y. Tang, B. A. Roe, S. Jones, M. Marra, and J. W. Kronstad.** 2006. Mating factor linkage and genome evolution in basidiomycetous pathogens of cereals. *Fungal Genet. Biol.* **43:**655–666.

8. **Banuett, F., and I. Herskowitz.** 1996. Discrete developmental stages during teliospore formation in the corn smut fungus, *Ustilago maydis. Development* **122:**2965–2976.

9. **Bohlmann, R., F. Schauwecker, C. Basse, and R. Kahmann.** 1994. Genetic regulation of mating and dimorphism in *Ustilago maydis*, p. 239–245. *In* M. J. Daniels (ed.), *Advances in Molecular Genetics of Plant–Microbe Interactions*, vol. 3. Kluwer, Dordrecht, The Netherlands.

10. **Bölker, M., M. Urban, and R. Kahmann.** 1992. The *a* mating type locus of *U. maydis* specifies cell signaling components. *Cell* **68:**441–450.

11. Bortfeld, M., K. Auffarth, R. Kahmann, and C. W. Basse. 2004. The *Ustilago maydis a2* mating-type locus genes *lga2* and *rga2* compromise pathogenicity in the absence of the mitochondrial p32 family protein Mrb1. *Plant Cell* **16:**2233–2248.

12. Brachmann, A., G. Weinzierl, J. Kämper, and R. Kahmann. 2001. Identification of genes in the *bW/bE* regulatory cascade in *Ustilago maydis. Mol. Microbiol.* **42:**1047–1063.

13. Brachmann, A., J. Schirawski, P. Müller, and R. Kahmann. 2003. An unusual MAP kinase is required for efficient penetration of the plant surface by *Ustilago maydis. EMBO J.* **22:**2199–2210.

14. Brefort, T., P. Müller, and R. Kahmann. 2005. The high-mobility-group domain transcription factor Rop1 is a direct regulator of *prf1* in *Ustilago maydis. Eukaryot. Cell* **4:**379–391.

15. Caldwell, G. A., F. Naider, and J. M. Becker. 1995. Fungal lipopeptide mating pheromones: a model system for the study of protein prenylation. *Microbiol. Rev.* **59:**406–422.

16. Castillo-Lluva, S., and J. Pérez-Martín. 2005. The induction of the mating program in the phytopathogen *Ustilago maydis* is controlled by a G1 cyclin. *Plant Cell* **17:**3544–3560.

17. Castillo-Lluva, S., T. Garcia-Muse, and J. Pérez-Martín. 2004. A member of the Fizzy-related family of APC activators is regulated by cAMP and is required at different stages of plant infection by *Ustilago maydis. J. Cell Sci.* **117:**4143–4156.

18. Feldbrügge, M., J. Kämper, G. Steinberg, and R. Kahmann. 2004. Regulation of mating and pathogenic development in *Ustilago maydis. Curr. Opin. Microbiol.* **7:**666–672.

19. Flor-Parra, I., M. Vranes, J. Kämper, and J. Pérez-Martín. 11 August 2006. Biz1, a zinc finger protein required for plant invasion by *Ustilago maydis*, regulates the levels of a mitotic cyclin. *Plant Cell* **18:**2369–2387. [Epub ahead of print.]

20. Garcia-Muse, T., G. Steinberg, and J. Pérez-Martín. 2003. Pheromone-induced G2 arrest in the phytopathogenic fungus *Ustilago maydis. Eukaryot. Cell* **2:**494–500.

21. Garcia-Pedrajas, M. D., and S. E. Gold. 2004. Kernel knowledge: smut of corn. *Adv. Appl. Microbiol.* **56:**263–290.

22. Garrido, E., and J. Pérez-Martín. 2003. The *crk1* gene encodes an Ime2-related protein that is required for morphogenesis in the plant pathogen *Ustilago maydis. Mol. Microbiol.* **47:**729–743.

23. Garrido, E., U. Voss, P. Müller, S. Castillo-Lluva, R. Kahmann, and J. Pérez-Martín. 2004. The induction of sexual development and virulence in the smut fungus *Ustilago maydis* depends on Crk1, a novel MAPK protein. *Genes Dev.* **18:**3117–3130.

24. Gillissen, B., J. Bergemann, C. Sandmann, B. Schroeer, M. Bölker, and R. Kahmann. 1992. A two-component regulatory system for self/non-self recognition in *Ustilago maydis. Cell* **68:**647–657.

25. Gold, S., G. Duncan, K. Barrett, and J. Kronstad. 1994. cAMP regulates morphogenesis in the fungal pathogen *Ustilago maydis. Genes Dev.* **8:**2805–2816.

26. Hartmann, H. A., J. Krüger, F. Lottspeich, and R. Kahmann. 1999. Environmental signals controlling sexual development of the corn smut fungus *Ustilago maydis* through the transcriptional regulator Prf1. *Plant Cell* **11:**1293–1306.

27. Hartmann, H. A., R. Kahmann, and M. Bölker. 1996. The pheromone response factor coordinates filamentous growth and pathogenicity in *Ustilago maydis. EMBO J.* **15:**1632–1641.

28. Inada, K., Y. Morimoto, T. Arima, Y. Murata, and T. Kamada. 2001. The clp1 gene of the mushroom *Coprinus cinereus* is essential for A-regulated sexual development. *Genetics* **157:**133–140.

29. Kahmann, R., and J. Kämper. 2004. *Ustilago maydis*: how its biology relates to pathogenic development. *New Phytol.* **164:**31–42.

30. Kamada, T. 2002. Molecular genetics of sexual development in the mushroom *Coprinus cinereus. Bioessays* **24:**449–459.

31. Kämper, J., M. Reichmann, T. Romeis, M. Bölker, and R. Kahmann. 1995. Multiallelic recognition: nonself-dependent dimerization of the bE and bW homeodomain proteins in *Ustilago maydis. Cell* **81:**73–83.

32. Kämper, J., R. Kahmann, M. Bölker, L.-J. Ma, T. Brefort, B. J. Saville, F. Banuett, J. W. Kronstad, S. E. Gold, O. Müller, M. H. Perlin, H. A. B. Wösten, R. deVries, J. Ruiz-Herrera, C. G. Reynaga-Peña, K. Snetselaar, M. McCann, J. Pérez-Martín, M. Feldbrügge, C. W. Basse, G. Steinberg, J. I. Ibeas, W. Holloman, P. Guzman, M. Farman, J. E. Stajich, R. Sentandreu, J. M. González-Prietro, J. C. Kennell, L. Molina, J. Schirawski, A. Mendoza-Mendoza, D. Greilinger, K. Münch, N. Rössel, M. Scherer, M. Vranes, O. Ladendorf, V. Vincon, U. Fuchs, B. Sandrock, S. Meng, E. C. H. Ho, M. J. Cahill, K. J. Boyce, J. Klose, S. J. Klosterman, H. J. Deelstra, L. Ortiz-Castellanos, W. Li, P. Sanchez-Alonso, P. H. Schreier, I. Häuser-Hahn, M. Vaupel, E. Koopmann, G. Friedrich, H. Voss, T. Schlüter, D. Platt, C. Swimmer, A. Gnirke, F. Chen, V. Vysotskaia, G. Mannhaupt, U. Güldener, M. Münsterkötter, D. Haase, M. Oesterheld, H.-W. Mewes, E. W. Mauceli, D. DeCaprio, C. M. Wade, J. Butler, S. Young, D. D. Jaffe, S. Calvo, C. Nusbaum, J. Galagan, and B. Birren. 2006. Living in pretend harmony: insights from the genome of the biotrophic fungal plant pathogen *Ustilago maydis. Nature* **444:**97–101.

33. Kosted, P. J., S. A. Gerhardt, C. M. Anderson, A. Stierle, and J. E. Sherwood. 2000. Structural requirements for activity of the pheromones of *Ustilago hordei. Fungal Genet. Biol.* **29:**107–117.

34. Kronstad, J. W., and S. A. Leong. 1990. The *b* mating-type locus of *Ustilago maydis* contains variable and constant regions. *Genes Dev.* **4:**1384–1395.

35. Laity, C., L. Giasson, R. Campbell, and J. Kronstad. 1995. Heterozygosity at the *b* mating-type locus attenuates fusion in *Ustilago maydis. Curr. Genet.* **27:**451–459.

36. Lee, N., and J. W. Kronstad. 2002. *ras2* controls morphogenesis, pheromone response, and pathogenicity in the fungal pathogen *Ustilago maydis. Eukaryot. Cell* **1:**954–966.

37. Lee, N., C. A. D'Souza, and J. W. Kronstad. 2003. Of smuts, blasts, mildews, and blights: cAMP signaling in

phytopathogenic fungi. *Annu. Rev. Phytopathol.* **41:**399–427.

38. Lee, N., G. Bakkeren, K. Wong, J. E. Sherwood, and J. W. Kronstad. 1999. The mating-type and pathogenicity locus of the fungus *Ustilago hordei* spans a 500-kb region. *Proc. Natl. Acad. Sci. USA.* **96:**15026–15031.

39. Mayorga, M. E., and S. E. Gold. 1999. A MAP kinase encoded by the *ubc3* gene of *Ustilago maydis* is required for filamentous growth and full virulence. *Mol. Microbiol.* **34:**485–497.

40. Mayorga, M. E., and S. E. Gold. 2001. The *ubc2* gene of *Ustilago maydis* encodes a putative novel adaptor protein required for filamentous growth, pheromone response and virulence. *Mol. Microbiol.* **41:**1365–1379.

41. Müller, P., J. D. Katzenberger, G. Loubradou, and R. Kahmann. 2003. Guanyl nucleotide exchange factor Sql2 and Ras2 regulate filamentous growth in *Ustilago maydis.* *Eukaryot. Cell* **2:**609–617.

42. Pérez-Martín, J., S. Castillo-Lluva, C. Sgarlata, I. Flor-Parra, N. Mielnichuk, J. Torreblanca, and N. Carbo. 2006. Pathocycles: *Ustilago maydis* as a model to study the relationships between cell cycle and virulence in pathogenic fungi. *Mol. Genet. Genomics* **276:**211–229.

43. Romeis, T., A. Brachmann, R. Kahmann, and J. Kämper. 2000. Identification of a target gene for the bE-bW homeodomain protein complex in *Ustilago maydis. Mol. Microbiol.* **37:**54–66.

44. Schauwecker, F., G. Wanner, and R. Kahmann. 1995. Filament-specific expression of a cellulase gene in the dimorphic fungus *Ustilago maydis. Biol. Chem. Hoppe-Seyler* **376:**617–625.

45. Scherer, M., K. Heimel, V. Starke, and J. Kämper. 2006. The Clp1 protein is required for clamp formation and pathogenic development of *Ustilago maydis. Plant Cell* **18:**2388–2401.

46. Schirawski, J., B. Heinze, M. Wagenknecht, and R. Kahmann. 2005. Mating type loci of *Sporisorium reilianum*: novel pattern with three *a* and multiple *b* specificities. *Eukaryot. Cell* **4:**1317–1327.

47. Schirawski, J., H. U. Böhnert, G. Steinberg, K. Snetselaar, L. Adamikowa, and R. Kahmann. 2005. Endoplasmic reticulum glucosidase II is required for pathogenicity of *Ustilago maydis. Plant Cell* **17:**3532–3543.

48. Schulz, B., F. Banuett, M. Dahl, R. Schlesinger, W. Schäfer, T. Martín, I. Herskowitz, and R. Kahmann.

1990. The *b* alleles of *U. maydis*, whose combinations program pathogenic development, code for polypeptides containing a homeodomain-related motif. *Cell* **60:**295–306.

49. Smith, D. G., M. D. Garcia-Pedrajas, W. Hong, Z. Yu, S. E. Gold, and M. H. Perlin. 2004. An *ste20* homologue in *Ustilago maydis* plays a role in mating and pathogenicity. *Eukaryot. Cell* **3:**180–189.

50. Snetselaar, K. M. 1993. Microscopic observation of *Ustilago maydis* mating interactions. *Exp. Mycol.* **17:**345–355.

51. Snetselaar, K. M., and C. W. Mims. 1994. Light and electron microscopy of *Ustilago maydis* hyphae in maize. *Mycol. Res.* **98:**347–355.

52. Snetselaar, K. M., M. Bölker, and R. Kahmann. 1996. *Ustilago maydis* mating hyphae orient their growth toward pheromone sources. *Fungal Genet. Biol.* **20:**299–312.

53. Spellig, T., M. Bölker, F. Lottspeich, R. W. Frank, and R. Kahmann. 1994. Pheromones trigger filamentous growth in *Ustilago maydis. EMBO J.* **13:**1620–1627.

54. Steinberg, G., R. Wedlich-Söldner, M. Brill, and I. Schulz. 2001. Microtubules in the fungal pathogen *Ustilago maydis* are highly dynamic and determine cell polarity. *J. Cell Sci.* **114:**609–622.

55. Steinberg, G., M. Schliwa, C. Lehmler, M. Bölker, R. Kahmann, and J. R. McIntosh. 1998. Kinesin from the plant pathogenic fungus *Ustilago maydis* is involved in vacuole formation and cytoplasmic migration. *J. Cell Sci.* **111:**2235–2246.

56. Trueheart, J., and I. Herskowitz. 1992. The *a* locus governs cytoduction in *Ustilago maydis. J. Bacteriol.* **174:**7831–7833.

57. Urban, M., R. Kahmann, and M. Bölker. 1996. The biallelic *a* mating type locus of *Ustilago maydis*: remnants of an additional pheromone gene indicate evolution from a multiallelic ancestor. *Mol. Gen. Genet.* **250:**414–420.

58. Urban, M., R. Kahmann, and M. Bölker. 1996. Identification of the pheromone response element in *Ustilago maydis. Mol. Gen. Genet.* **251:**31–37.

59. Wösten, H. A., R. Bohlmann, C. Eckerskorn, F. Lottspeich, M. Bölker, and R. Kahmann. 1996. A novel class of small amphipathic peptides affect aerial hyphal growth and surface hydrophobicity in *Ustilago maydis. EMBO J.* **15:**4274–4281.

Sex in Fungi: Molecular Determination
and Evolutionary Implications
Edited by Joseph Heitman et al.
© 2007 ASM Press, Washington, D.C.

Guus Bakkeren
James W. Kronstad

23

Bipolar and Tetrapolar Mating Systems in the Ustilaginales

INTRODUCTION

Mating, Sexual Development, and Phytopathogenesis

The smut fungi are attractive experimental models to investigate basidiomycete mating systems and to explore the role of mating-type functions in pathogenic development. A fascinating aspect of these fungi is that their ability to cause disease on host plants is dependent on mating interactions between haploid cells leading to formation of an infectious dikaryon. Sex and pathogenesis are thus intimately intertwined because the infectious dikaryon requires a host for proliferation and for the eventual formation of sexual spores (teliospores). *Ustilago maydis*, the corn pathogen, has emerged as the primary model for studying smut fungi and is discussed in other chapters in this book. Here we discuss other species, *Ustilago hordei* in particular, that have provided useful comparative information leading to insights into the genetic basis of bipolar versus tetrapolar mating systems in the smut fungi as a group. We first discuss the importance of smut fungi and the interactions of these pathogens with host plants to provide

context for appreciating the role of mating in disease. We then focus on the details of the mating system in *U. hordei*, including the structure and function of the mating-type loci, the genomic organization of these elements, and the sequence of the 527-kb *MAT-1* locus. Comparisons between the tetrapolar mating system in *U. maydis* and the bipolar system of *U. hordei* allowed the development of a detailed view of the genomic basis of mating-system organization. This work sets the stage of a broader examination of the interconnections between genomic organization, mating systems, and pathogenesis in these fungi. In particular, comparisons suggest an evolutionary drive towards larger but genetically less complex mating loci: this results in the genesis of sex chromosomes which promote inbreeding within the species and presumably bestow a selective advantage (chapter 2). The latter might be particularly suited to species occupying specialized niches such as pathogens. However, in several homobasidiomycete mushrooms, such as hymenomycete lineages in the genus *Coprinus*, bipolar mating systems are frequently found and seem to be derived from tetrapolar organizations by losing pheromones and receptors as mating-type determinants;

Guus Bakkeren, Pacific Agri-Food Research Centre, Agriculture and Agri-Food Canada, Summerland, BC, V0H 1Z0, Canada. **James W. Kronstad,** The Michael Smith Laboratories, Department of Microbiology and Immunology, University of British Columbia, Vancouver, BC, V6T 1Z4, Canada.

in *Coprinellus* (*Coprinus*) *disseminatus* these are not linked to *MAT* (chapter 19). In this context it is rather striking that fusion of *MAT* loci in some phytopathogenic ascomycete lineages, which confers a change from hetero- to homothallism, is persistent over time and might even have been transfered laterally (chapter 6).

Overview of the Smut Fungi

Smut fungi, or smuts for short, have attracted interest for centuries because they form conspicuous fruiting structures on many kinds of plants, and these structures contain black masses of teliospores that give the infected tissue a "sooty" or "smutted" appearance (touched by fire or "Brand" in German and Dutch). In the order Ustilaginales, smuts (family of Ustilaginaceae) are distinguished from related bunt ("bu[r]nt") fungi (family of Tilletiaceae) based on distinct features of the promycelium: septate with lateral and terminal basidiospores (sporidia) in the smuts and a continuous promycelium with terminal basidiospores in the bunts. However, these characteristics proved imprecise and later research proposed only one family for the smuts, the Ustilaginaceae (19). Fischer and Holton (19) consolidated the many described species into 1,162 species belonging to 33 genera, whereas recently Vanky (85) recognized approximately 1,200 species in more than 50 genera. New species are being described continuously (11, 60, 86).

Smuts infect over 4,000 species of angiosperms, including both monocots and dicots in over 75 families (19). However, each smut species generally has a rather narrow host range. Interestingly, a disproportionate share of the host species are in the family Gramineae, and because these plants include the world's most important agricultural crops (cereals and forage grasses), smuts infecting these hosts have been studied extensively for over a century. Some of the earliest official reports of (corn) smut came from France in the mid-18th century (14). Among the best known are *U. maydis* (infecting corn), which is the best characterized at the molecular level and whose complete genome has been sequenced (discussed in chapter 22), *Sporosorium reiliana* (infecting sorghum and corn), and *Ustilago scitaminea* (infecting sugarcane). A group of small grain-infecting smuts have also been studied extensively. These include *U. hordei* (barley and oats), *U. avenae* (oats), *U. kolleri* (oats), *U. nuda* (barley), and *U. tritici* (wheat). In contrast to most species in this group, the last two have a distinctive mode of germination and initiate disease by infecting embryos via flowers rather than young seedlings. *U. hordei* has emerged as the representative for the small-grain-infecting group. Several smuts are also pathogenic on forage grasses (e.g., *Ustilago bullata*), and it is sometimes possible to find grasses that serve as common hosts for several smut species (82). These common hosts have potential utility for testing mating interactions between different smut species. Other fungal pathogens of cereal crops such as the rusts and bunts share similar life cycle features with the smut fungi. The bunts are most similar and include well-known pathogens such as *Tilletia foetida* (syn. *T. laevis*) and *T. caries* (syn. *T. tritici*, causing common bunt or stinking smut on wheat) and *T. controversa* and *T. indica* (causing dwarf and Karnal or partial bunt of wheat, respectively).

As mentioned, the smut and bunt fungi are particularly interesting because of the role of mating in formation of the infectious dikaryon that invades host tissue. The mating systems in these fungi include bipolar and tetrapolar classifications based on genetic experiments, and these fungi provide an opportunity to explore mechanisms of mating in the context of infection and host range. A fascinating additional layer of complexity exists in that infection by these fungi is often governed by so-called avirulence genes in the pathogens and corresponding resistance genes in the host plants. Fungal isolates within a species can carry different complements of avirulence genes constituting many specific so-called "races." Similarly, a host species may harbor different combinations of resistance genes that recognize specific avirulence genes to block infection; this gives rise to many "cultivars" within that species. Genetically, a specific avirulence gene interacts with one specific resistance gene; dominance triggers defense and blocks disease, but a recessive allele in either or both partners is not recognized. This "gene-for-gene" concept actually emerged from research on the rust *Melampsora lini* on flax (20) and the smut *U. tritici* on wheat (57). Over the years, many pathosystems have been shown to follow this concept, including many species in the Ustilaginales with their respective hosts (19). *U. hordei* has emerged as the model for the genetic analysis of avirulence genes, and one avirulence gene has been isolated (25, 50, 70, 79, 80, 81). The molecular basis of recognition and subsequent defense is being worked out and reflects the intimate coevolution of these pathogens with their host plants (13). Finally, some smut and bunt fungi are interesting because they provoke profound physiological and/or morphological changes in their hosts. For example, *U. maydis* incites conspicuous tumors in which massive sporulation occurs, *Tilletia buchloeana* induces a sex change such that pistils develop in an otherwise staminate floret of male buffalo grass (38), and *Microbotryum violaceum* transforms female flowers of *Silene alba* into male ones, eventually replacing pollen with spores (84).

Interactions with Host Plants and the Mode of Infection of *U. hordei*

Infection by smut fungi such as *U. hordei* is initiated when seeds are contaminated by dispersed teliospores. Seed and teliospore germination occur together, and the spore produces a basidium with subsequent meiosis to produce four haploid basidiospores (sporidia). In many but not all smuts, the basidiospores grow by budding and can be cultured for molecular genetic manipulations. These cells are not infectious in the absence of compatible mating partners. Most smuts infect their hosts systemically and target meristematic tissues; interestingly, the fungal dikaryon initially grows passively with the plant with subsequent proliferation and sporulation in certain tissues in response to plant developmental changes (e.g., flowering). Teliospores form in sori on stems, leaves, and anthers or in ovaries to replace seeds (particularly in the *Gramineae* [19, 71]). The complete path of infection of *U. hordei* was recently described in a detailed microscopic study (35, 36). The initial mating interaction of haploid cells on the plant surface is a key event, and an interesting observation was that the mating process on the host surface appeared to be more profuse than is seen in vitro. Specifically, many more conjugation tubes (mating hyphae) emerged from basidiospores (often more than two per cell) on the plant surface, and it is possible that stimulatory chemical or physical factors from the plant are sensed to maximize mating partner detection (36). After mating, the dikaryotic filament grows over the leaf surface and can cover considerable distances (on average more than 30 μm) over many epidermal cells without apparent differentiation. The hyphal tip forms a characteristic crook when it grows over a juncture between the long anticlinal walls of epidermal cells. The fungal tip becomes slightly swollen immediately adjacent to this bend, and an appressorium-like structure indicates the site of direct penetration. After penetration, both intra- and intercellular mycelium can be found and the intracellular hyphae do not compromise host cell plasma membranes. Necrotic reactions or other abnormal cell morphologies are not observed, suggesting that *U. hordei* evades and/or actively suppresses host defense mechanisms to establish a biotrophic relationship. Intercellular hyphae can also send several branches into plant cells, some of which seem to terminate within cells. These invading hyphae become encased in an electron-dense interfacial matrix that separates the hyphae and the host plasma membrane. These cells may function as haustoria to acquire nutrients, possibly through induced transporters as occurs with rust fungi (87). Fungal hyphae eventually reach the apical meristem and nodes of the coleoptile, but scant fungal biomass is present, making extensive nutrient uptake and elaborate haustorial structures as seen in rusts unnecessary. However, fungal proliferation occurs when the meristematic region differentiates into a floral meristem and sporulation is initiated. Teliospores are formed inside the spikelets to replace the seeds. Economically, infection results in complete yield loss and massive spore contamination of harvested grain. In contrast to infection of small-grain cereals (as described for *U. hordei*), infection by *U. maydis* occurs on any aboveground part of corn plants and local symptoms (tumors) are induced. The paths of infection for *U. maydis* in corn and for *Sphacelotheca reilianum* in sorghum have been thoroughly studied (72–74, 90). Bunt fungi can also infect locally or systemically (10).

Mating Systems in Basidiomycete Fungi and Smuts

The majority of smut fungi are heterothallic, and many possess a clearly defined sexual stage. Some species, however, produce an ephemeral haploid stage with no separate sporidial cells, making it difficult to establish compatibility relationships and blurring the distinctions between homo- and heterothallism (30). In the species that have been used for extensive experimentation (*U. maydis* and *U. hordei*), the basidiospores produced upon teliospore germination can be isolated and tested for mating specificity on culture medium (28, 29, 67). Successful mating results in the initiation of thin conjugation tubes in response to the exchange of pheromones (encoded by the *a* mating-type gene complex) and can be observed microscopically. These tubes fuse, and a thick, straight-growing, dikaryotic filament is produced if the mating partners have compatible specificity at the *b* mating-type gene complex (encoding a heterodimeric homeodomain transcription factor). When compatible haploid cells are mixed on rich medium supplemented with activated charcoal, a positive mating reaction can be scored as a white, "fuzzy" reaction (Fig. 23.1). Pairings of basidiospores dissected from single basidia (tetrad analysis), or randomly isolated from nature, yield information on the complexity of compatible mating types present in populations and can define whether a bipolar or tetrapolar system is represented (Fig. 23.1).

Among the homobasidiomycetes, approximately 65% of the species regulate sexual compatibility via a

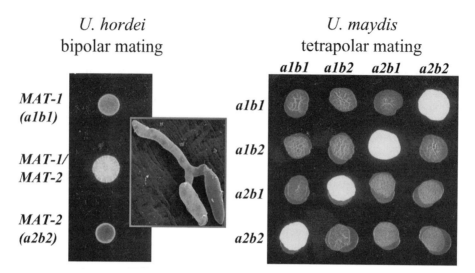

Figure 23.1 Examples of mating interactions on charcoal medium. Aliquots of cultures of haploid basidiospores having the mating genotype indicated are cospotted on complete medium plates supplemented with activated charcoal (28). When both the *a* and the *b* loci are of different allelic specificity, a straight-growing, dikaryotic hypha is produced (inset) and the ensuing colony will have a white "fuzzy" appearance. Note that for *U. hordei* there are only two mating-type alleles (*MAT-1* and *MAT-2*) in nature but that for *U. maydis a* and *b* specificities can assort in all combinations among progeny from genetic crosses.

tetrapolar (bifactorial) mating system. In these fungi, two genetic loci such as the *a* and *b* loci in smut fungi can each possess two or more allelic specificities. For some mushroom fungi such as *Schizophyllum commune* and *Coprinus cinereus* (*Coprinopsis cinerea*), this can result in thousands of mating specificities (chapters 17 and 18). Another 10% of the species are considered to be homothallic (39, 63). Finally, 25% of the species have a bipolar (unifactorial) mating system in which compatibility is governed by a single genetic locus designated *MAT* with alleles *MAT-1* and *MAT-2* (also sometimes called *a* and *A*). This locus can have two or multiple alleles, the latter being common in the homobasidiomycetes and the nonparasitic heterobasidiomycetes. The terms *MAT-1* and *MAT-2* were adopted for *U. hordei* (4). Among the parasitic heterobasidiomycetes, such as the phytopathogenic rust, bunt, and smut fungi, the bipolar mating system is predominant with certain exceptions (19). These exceptions include a multiallelic bipolar system for *T. controversa* (27) and the occurrence of a tetrapolar mating system in several species including *U. maydis*, *Ustilago longissima* (19, 62), and *S. reilianum* (68). Other chapters in this book describe the best-studied tetrapolar system found in *U. maydis*, where the *a* mating-type gene complex has two known specificities and *b* has at least 25 different specificities. Interestingly, recent detailed mating-type analysis of *S. reilianum* revealed at least three specificities for the *a* locus and at least five for *b* (68).

THE MATING-TYPE LOCI OF BIPOLAR SMUT FUNGI

Identification of the *a* and *b* Mating-Type Loci in *U. hordei*

The isolation of the *a* and *b* mating-type gene complexes from *U. maydis* presented an opportunity to search for related sequences in other *Ustilago* species (23, 42). A DNA hybridization survey of a variety of basidiomycete and ascomycete fungi for the presence of sequences with similarity to the *U. maydis a* and *b* genes revealed related sequences in the *Ustilago* species that infect small-grain cereal crops, as well as in *U. longissima* and *S. reilianum* (3). Strongly cross-hybridizing sequences were not found in *Cryptococcus neoformans*, *M. violaceum*, *Puccinia graminis*, *S. commune*, *T. controversa*, *T. caries*, or *Neurospora crassa* (3). Of course, subsequent studies revealed that genes with functions comparable to those of the *a* and *b* gene complexes do exist in other basidiomycetes including *C. neoformans*, *S. commune*, and *C. cinereus*, but these sequences are too divergent to be detected by hybridization (15, 44, 49, 53, 58, 64, 75, 88). One would expect a similar situation for the *Tilletia* and rust species, but these loci remain to be characterized. The hybridization results for the group of *Ustilago* species with bipolar mating systems were interesting because of the similarities observed in the pattern of hybridizing DNA fragments.

Specifically, hybridization with the *bE* gene of *U. maydis* detected a restriction fragment length polymorphism (RFLP) (1.5- or 2.8-kb BamHI fragments) among the strains of *U. hordei*, *U. kolleri*, *U. avenae*, *U. aegilopsidis*, and *U. nigra*, indicating the presence of similar gene complexes in this group. Surprisingly, the RFLP band of 2.8 kb was associated with a positive hybridization result and the 1.5-kb band was associated with a negative result with the *U. maydis* a1 sequences. In particular, the two different mating-type strains of *U. hordei* that were tested showed this association (3). These results led to speculation that the *a* and *b* sequences might be genetically linked in *U. hordei* (and the other bipolar *Ustilago* species). As described below, subsequent work revealed that this was the case and provided an explanation for the differences between bipolar and tetrapolar mating systems in *Ustilago*.

FUNCTIONAL ANALYSIS OF THE MATING-TYPE LOCI IN BIPOLAR SMUTS (*U. HORDEI*)

The cloning and analysis of the *b* genes from *U. hordei* revealed that they are very similar in structure to those in *U. maydis*, consisting of a gene complex with divergently transcribed *bE* and *bW* genes (4). Just as for *U. maydis*, the carboxy-terminal ends of the predicted UhbE and UhbW allelic proteins were more conserved (92% identity) than the "variable" amino termini (51% identity); however, comparisons to the *U. maydis* counterparts revealed 64 and 43% similar domains, respectively. The homeodomain, WFxRxR, which is important for function (26), was conserved in all of the proteins. Subsequent work showed that *U. hordei* b genes were able to induce filamentous growth in haploid *U. maydis* strains and these transformants were weakly virulent when inoculated on corn (4). These experiments confirmed cross-species functionality of the *b* mating-type genes after fusion.

The presence of orthologs of *U. maydis* a locus sequences in the bipolar smuts and the demonstration of the involvement of diffusible small-molecular-weight factors (pheromones) in *U. hordei* mating (similar to the process in *U. maydis* [52]) led to attempts to isolate these regions. To isolate the *a* locus, the *pan1* gene from *U. maydis* was used as a probe because it was known to be linked to *a* in *U. maydis* (23). This probe identified a cosmid paMAT-1 in *U. hordei* that was tested for mating-type activity by transformation into a *MAT-2* strain. This clone induced erratic behavior including the formation of long, meandering mating hyphae and cell aggregates in the absence of cells of the opposite mating type (a so-called dual-mater phenotype [5]). Sequence

analysis identified the pheromone receptor gene *Uhpra1* on an 8.5-kb SphI fragment which coded for a predicted protein that was 64% identical and 82% similar to Pra1 of *U. maydis*. Similarly, Uhpra2 was found to be 60% identical and 79% similar to Umpra2 and when transformed in a *MAT-1* strain induced the same dual-mater phenotype (1). The *U. hordei* pheromone genes, *Uhmfa1* and *Uhmfa2*, were also cloned and analyzed, and again, their functions were analogous to those of their homologs in *U. maydis* (1, 40). As for *U. maydis*, *Uhmfa1* and *Uhmfa2* are expressed at basal levels but are up-regulated when cells of opposite mating type are encountered. Interestingly, mating inhibition factors (MIFs) have been found in *U. hordei* that are truncated and/or "undecorated" forms of the pheromones, probably oxidative degradation products. Normally, translated Uhmfa is a preprotein which is processed by clipping and becomes farnesylated and carboxy-methylated (41, 69). It was suggested that a gradient of MIFs during teliospore germination might prevent immediate fusion of siblings within tetrads, thereby allowing time for nonsibling interactions (outbreeding). *U. hordei* MIFs also reportedly have an inhibitory effect on the germination of teliospores of several *Tilletia* species, which might confer a competitive advantage during coincident infection of the same host (41).

Smut pheromones are produced and secreted in low quantities by basidiospores, thereby creating a gradient thought to alert and guide potential mates of opposite sexual persuasion. It has been shown that the *U. hordei* *a* locus genes, *Uhmfa1* and *Uhpra1*, when introduced by transformation, are necessary and sufficient to make *U. maydis* intercompatible with *U. hordei* MAT-2, but not *MAT-1*, strains. In addition, *U. hordei* strains transformed with the *U. maydis* a1 locus also became intercompatible with *U. maydis* a2, but not a1, strains. The interspecies hybrids produced dikaryotic hyphae but were not fully virulent on either corn or barley (6). This shows that within these smuts, the machinery and pathways to transmit signals from the receptor once pheromone is bound are conserved. Although the Mfa-Pra interaction is thought to be species specific, many examples of natural intercompatible combinations exist and some are interfertile and infectious on common hosts. The interspecies mating capabilities within the small-grain-infecting smuts were mentioned earlier, but some hybrids do not produce viable or pathogenic progeny (19, 37, 82). Likewise, *U. maydis* and *S. reilianum* can also interbreed and produce teliospores on corn, although no pathogenic progeny have been reported (66). Partial mating capabilities have been observed between *U. scitaminea* and both *U. hordei* and *U. maydis*, indicating that there may be a continuum of pheromone recognition specificities among the smut fungi (6). A distinction has to be made between fusion of basidiospores, due to sometimes promiscuous responses of pheromone receptors to pheromones, the differentiation to the filamentous cell type brought about by the productive interaction between different *b* alleles from the participating species ("Fuz" reaction), and the subsequent invasion of (common) host tissues resulting ultimately in the production of teliospores that can produce viable, pathogenic offspring. The latter steps point to true hybridization and fertility/fecundity. These issues are intertwined with difficulties in taxonomic placement, which is often based on host range, the species concept based on productive hybridization, and the "biological species" concept which defines a species as organisms that share "morphological" or "taxonomic" characteristics but are, or have become, (partially) intersterile due to geographic isolation.

PHYSICAL ASSOCIATION OF THE *a* AND *b* GENE COMPLEXES AT THE *MAT-1* LOCUS OF *U. HORDEI*

The detection of an RFLP at the *b* gene complex in the *Ustilago* species that infect small-grain cereals and that was potentially genetically linked to the *a* locus provided the impetus for further examination of the genetic and physical linkage in *U. hordei*. Initially, a hybridization probe for *UhbE1* was found to detect an RFLP that cosegregated with mating type (*MAT-1* or *MAT-2*) in 86 meiotic progeny from two teliospore populations. No recombinant progeny were obtained, thus indicating tight linkage. The cloning and characterization of the *a* gene complexes from *U. hordei* (described above) provided the specific hybridization probes to also analyze linkage from the perspective of the *a* locus. Both the *a* and *b* probes hybridized to RFLPs that showed 100% cosegregation with mating type in the 86 progeny. Additionally, a 2.1-kb fragment from the paMAT-1 cosmid hybridized specifically to *MAT-1*, indicating that regions of nonhomology or divergence existed between *MAT-1* and *MAT-2*. The probes were also used to demonstrate physical linkage between the *a* and *b* sequences. Electrophoretically separated chromosomes were blotted, and the *a* and *b* probes were found to hybridize to the same chromosome of ~3 Mbp. In contrast, *a* and *b* probes from *U. maydis* detected the *a* locus on a 1.5 Mbp chromosome (later identified as ~1-Mbp chromosome V) and the *b* locus on a 2-Mbp chromosome (later identified as ~2.5-Mbp chromosome I) (Fig. 23.2) (5). Evidence that recombination might be suppressed at *MAT* came from the analysis of 2,182 random progeny from three collections of teliospores. A

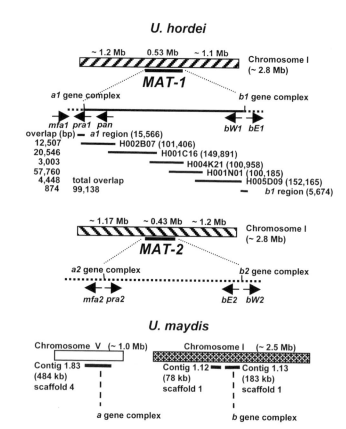

Figure 23.2 Genomic organization of the mating-type loci in *U. hordei* and *U. maydis*. In *U. hordei* the MAT locus is defined as the region delimited by the known *a* and *b* mating-type gene complexes, a distance of 526,707 bp for *MAT-1*. The five BAC clones, yielding a total of 625,845 bp of sequence, are indicated (8). In both *MAT-1* and *MAT-2* (430 kb) strains, the locus sits roughly in the middle of the largest chromosome (chromosome I). Note that the *MAT-2* region has not been sequenced but that its length and the orientation of the complexes have been determined (47). In *U. maydis*, the gene complexes are found on two different chromosomes: the *a* locus on contig 1.83 (scaffold 4; Broad Institute http://www.broad.mit.edu/annotation/fungi/ustilago_maydis/) harbored by chromosome V, and the *b* locus on contig 1.13 (scaffold 1) on chromosome I; contig 1.12 is likely linked to contig 1.13 based on the synteny found with *U. hordei* (represented by the solid black lines). Gene names are as follows: *mfa*, mating pheromone gene; *pra*, a1 pheromone receptor gene; *pan*, pantoate b-alanine ligase gene; *bW1*, bWest1 gene; *bE1*, bEast1 gene (4, 6, 16). Drawing is not to scale. Reprinted from *Fungal Genetics & Biology* (reference 8, Fig. 1) © 2006, with permission from Elsevier.

screen of the progeny for altered mating specificity, e.g., showing a failure to mate with either parental type (possibly due to genotypes *a1b2* and *a2b1*), was unsuccessful, suggesting that the combinations of the mating types *a1b1* and *a2b2* were maintained through meiosis.

Unlike the situation in smut fungi with tetrapolar mating systems (such as *U. maydis*), the physical linkage of *a* and *b* in *U. hordei* would ensure that every fusion event mediated by the *a* genes at *MAT-1* and *MAT-2* during mating would bring together *b* gene complexes of opposite specificity. This result indicates that bipolar smut fungi would require only two *b* specificities, one linked to each *a* specificity. This hypothesis was confirmed by DNA sequence analysis of alleles coding for the variable N-terminal 121 amino acids of bE and 171 amino acids of bW. This analysis revealed only two classes having identical protein sequences among many *U. hordei* isolates from a worldwide collection, each belonging to either *MAT-1* or *MAT-2* (5). That is, only two classes of *b* gene complexes were found among 18 isolates and there was very little variability within each class. Five additional *Ustilago* species similarly had only two classes. No amino acid changes were found when comparing *U. hordei* with *U. nigra* (likely synonymous to *U. hordei*) and *U. aegilopsidis*, while *U. avenae* and *U. kolleri* (82) had the same one conservative base-pair change. *U. bullata*, a more distantly related forage grass pathogen (7), had 3 amino acid changes out of 171 for bW1 and 4 out of 121 for bE1. These results led to the conclusion that most bipolar smut fungi have only two allelic specificities (*MAT-1* with *a1* and *b1* and *MAT-2* with *a2* and *b2*). Moreover, most can probably functionally interact since tests on charcoal mating plates indicated positive interactions in the predicted combinations and several of these have been productively hybridized on a common grass host, *Agropyron tsukushiense* (82).

The size of the *MAT* locus in *U. hordei* and the extent of recombination suppression between the *a* and *b* regions were explored in more detail by tagging each gene complex with a different gene encoding a selectable marker and the site for the rare-cutting restriction enzyme I-SceI (Fig. 23.3) (47). Specifically, an *a* region replacement DNA cassette was prepared that contained the gene for phleomycin (phleo) resistance linked to the I-SceI site and a *b* region replacement cassette was prepared with the gene for hygromycin B (hyg) resistance linked to the same I-SceI site. A set of single- and double-tagged strains was then constructed starting with *MAT-1* (*a1b1*) and *MAT-2* (*a2b2*) parents and containing a single phleo cassette at *a1*, a single hyg cassette at *b2*, the cassettes at the *a1* and *b1* gene complexes, and the cassettes at the *a2* and *b2* gene complexes (Fig. 23.3). Chromosome-sized DNA was then prepared from these strains and digested with I-SceI to either cleave the chromosome at *a* or *b* or release the segment of DNA between *a* and *b* at both *MAT-1* and *MAT-2*. The digested DNA was separated by pulsed-field gel electrophoresis, and the DNA segments were identified by Southern hybridization using specific *a* and *b* region clones. These experiments revealed that the distance be-

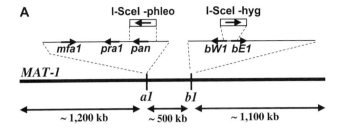

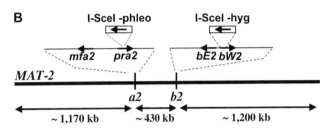

Figure 23.3 Strategy to prove linkage and recombination suppression, and to measure the physical distance between the mating-type gene complexes in *U. hordei*. The two thick lines represent the *MAT* chromosomes with the respective locations of the *a* and *b* mating-type gene complexes for *MAT-1* (A) and *MAT-2* (B). The organization of the mating-type gene complexes is enlarged (compare with Fig. 23.2) and the positions of the integrated constructs used to tag them are indicated (boxes). I-SceI represents the 18-bp recognition sequence for the rare-cutting, intron-homing enzyme from *Saccharomyces cerevisiae* which was linked to the selectable markers for phleomycin (phleo) to tag the *a* complex, and for hygromycin B (hyg) to tag the *b* complex. Digestion with the enzyme I-SceI led to the estimate of the respective distances as indicated by the double-headed arrows (see the text for details) (47). See the legend to Fig. 23.2 for explanation of gene names.

tween *a* and *b* at *MAT-1* is approximately 500 kb and the distance is approximately 430 kb at *MAT-2*. Additionally, the use of the single-tagged strains revealed that the *MAT* locus is close to the center of an ~2.8-kb chromosome. The use of hybridization probes from the *a* or *b* gene complexes and flanking the I-SceI sites also provided evidence that the *MAT-1* and *MAT-2* loci had undergone rearrangements relative to each other. Specifically, the orientation of the *bE-bW* complex relative to the *a* complex differed between *MAT-1* and *MAT-2*. Similarly, the hybridization experiments showed that the organization of the *pra* and *mfa* genes differed between *a1* and *a2* (Fig. 23.3) and revealed that regions of nonhomology existed between these gene complexes. Combined with the size difference between *MAT-1* and *MAT-2*, these experiments indicated that substantial rearrangements existed between *MAT-1* and *MAT-2*.

The single-tagged strains also allowed a direct examination of the frequency of recombination in the interval between *a* and *b*. Specifically, two strains tagged with phleo at *a1* were crossed with each of two strains tagged with hyg at *b2* by mixing the strains together on barley seedlings. The strategy was to estimate the frequency of recombination by collecting teliospores from the cross, germinating the spores to obtain meiotic progeny, and selecting double-resistant recombinant progeny on medium containing both hygromycin B and phleomycin. The prediction was that only progeny with the genotype *a1b2* (phleo[r], hyg[r]) would be able to form colonies. A sample of 1×10^4 to 2×10^4 random progeny yielded 34 double-resistant progeny (affectionately called drp or "drips"), suggesting that recombination was indeed greatly suppressed in the 400- to 500-kb interval. However, none of the drp strains displayed clear *a1b2* or *a2b1* mating specificity; the strains either failed to give a positive mating reaction with any tester strain or showed mating with more than one specificity. Subsequent hybridization experiments revealed that 13 of the 34 strains had both the *b1* and *b2* mating-type sequences and 32 contained both *a1* and *a2*. These results thus indicated that the drp did not arise from simple recombination events between *a* and *b* but rather from rearrangements at these loci or retention of part or all of the *MAT* chromosome homologs to establish aneuploid or diploid strains. Strains of the expected *a1b2* and *a2b1* mating types have been constructed artificially, so it is known that these combinations are viable and functional (6).

There are many examples of suppression of recombination between orthologous areas in sex-determining regions including the mammalian (human) X and Y chromosomes (46), the mating-type loci in the unicellular green alga *Chlamydomonas reinhardtii* (17), and fungal mating-type loci from species such as the model ascomycete *N. crassa* (24) (chapter 1) and the ascomycetous chestnut blight fungus *Cryphonectria parasitica* (43). In the last species, for example, suppression of recombination occurred in an otherwise highly polymorphic region near the *MAT* locus and it was suggested that this was caused by an ancient inversion, although in this case the *MAT* locus did not seem part of a sex chromosome because recombination between this region and *MAT* was observed.

GENOMIC ANALYSIS OF BIPOLAR AND TETRAPOLAR MATING SYSTEMS IN SMUT FUNGI

The large size of the *U. hordei MAT* region and the suppression of recombination in this area prompted us to propose that the region might function to maintain a set of genes that function together in sexual development

and, potentially, in pathogenesis (47). That is, the ~500-kb region might function as a pathogenicity island. We therefore characterized the *MAT* region in more detail with the goal of specifically identifying the genes present at the locus and testing the prediction (8). The genome sequence has not been determined for *U. hordei*, so a physical map was initially constructed for a *MAT-1* strain by using 2,304 large insert bacterial artificial chromosome (BAC) clones (average insert size, 113 kb) to allow characterization of the genome and to isolate the ~500-kb *MAT-1* region. A map containing 38 contigs was constructed by HindIII restriction enzyme fingerprinting of all clones, and an overlapping set of five BAC clones spanning the *MAT-1* locus was identified on a 1.34-Mb contig (Fig. 23.2). These five clones were sequenced, and the assembled sequence was combined with existing sequence from the *a1* and *b1* gene complexes to identify a genomic region of 526,707 bp (GenBank accession no. AM118080) that extended from the *Uhpra1* pheromone receptor gene at the left side of the *a1* locus to the *UhbE1* gene on the right end.

The examination and annotation of the 527-kb *MAT-1* sequence revealed the presence of 47 candidate protein-coding genes, of which 20 were designated hypothetical proteins. The remaining 27 genes encoded proteins with similarities to proteins of known function. Contrary to our prediction (47) and unlike the findings with the *MAT* locus of *C. neoformans* (49), the majority of the functions were not obviously related to mating or pathogenicity, at least at the level of sequence inspection. These included functions for metabolism (trehalose phosphatase, a-mannosidase, and ferric reductase), gene expression (ribosomal proteins and TATA-binding protein), and signaling (GTPases and glycogen synthase kinase). The region of the *a1* gene complex contained the known *Uhpra1* and *Uhmfa1* genes along with an ortholog of the *rba2* gene found at the *a* locus in *U. maydis*. Genes adjacent to the *a1* locus included the *pan1* gene for pantothenic acid biosynthesis (known to be linked to the *a* locus in *U. hordei* and *U. maydis*), as well as genes predicted to encode a ribosomal protein, a DAHP synthase, and an oligopeptide transporter. In addition to the mating functions, there are hints of functions that are potentially conserved in their association with the *a* locus. For example, a gene related to pantothenic acid biosynthesis is also associated with the *MAT* locus in *C. neoformans* (21) and genes for oligopeptide transporters are found near *MAT* in *U. hordei* and *U. maydis* or are mating regulated in *S. commune* (48). The *b1* gene complex of *UhbW1* and *UhbE1* also has an associated gene predicted to encode an N-terminal acetyltransferase.

The characterization of the *MAT-1* sequence and the annotation of the genes in the region revealed a remarkable accumulation of repeated DNA elements such that ~50% of the total sequence was repetitive (Color Plate 2). This remarkable feature of the locus resulted in a pattern of islands of one to four genes separated by extensive stretches of repeated sequences. The *a1* region had the largest collection of contiguous gene sequences with seven genes, and the largest stretch of repeats spanned approximately 64 kb. The repetitive sequences fell into four classes: (i) gypsy-type retrotransposons, (ii) copia-type retrotransposons, (iii) partial and intact copies of long terminal repeats (LTRs) with and without associated retrotransposons, and (iv) putative DNA transposons. In total, there were ~100 copies of retrotransposons in the region, although only 12 of these appeared to be intact, and the predicted coding regions of the gag-pol proteins in these contained numerous stop codons. Therefore, it was not clear whether any of these elements had retained function. There was sequence evidence that the repetitive elements had been mobile in the past. For example, LTR elements were present in the coding regions of two genes encoding a predicted isocitrate dehydrogenase and a vacuolar protein-sorting function. Additionally, there were numerous examples of LTRs interrupting transposons, retrotransposons, or other LTR elements. One particularly striking example involved a transposon designated Tho1 of 4,132 bp in length that was apparently interrupted five times by LTR1 (the most commonly found LTR), once by LTR10, once by a retroelement called Tuh5, and once by a protein-coding gene with similarity to a gene in *U. maydis* (Color Plate 2) (see reference 8 for more details).

Given the repetitive nature of *MAT-1*, it was of interest to determine whether the rest of the *U. hordei* genome showed a similar repetitive character. The BAC clones and physical map provided an opportunity to attempt to answer this question. First, the ends of the BAC clones from the tiling set of the contig carrying *MAT-1* were sequenced and tested for similarity to the repeats in *MAT-1*. The tiling path included 3 clones from the *b1* gene complex side of *MAT-1*, the 5 clones within the region (Fig. 23.2), and 14 clones on the *a1* locus side. This analysis revealed that the LTR, transposon, and retrotransposon sequences were distributed across the approximately 1.3-Mb contig, although it was not possible to determine their density. To examine the whole genome, seven types of repetitive sequences from *MAT-1* were hybridized to filters containing all 2,304 BAC clones from the map construction. Three patterns of hybridization were detected: (i) widespread distribution such that the majority of the BAC clones

hybridized, (ii) intermediate distribution (e.g., one probe detected 249 BAC clones), and (iii) hybridization at a higher frequency to BAC clones that contained the *MAT-1* sequence (8). Overall, these results indicated that the *U. hordei* genome is likely to be highly repetitive but that some types of repeats have accumulated preferentially at *MAT-1*. The highly repetitive nature of the *MAT-1* sequence was consistent with the difficulties in completing and assembling the sequence; in this regard, the physical map and the end sequences of the mapped BAC clones proved invaluable in determining the orientation and positions of sequence contigs. However, these findings also indicate that it might be challenging to complete the sequence of the entire *U. hordei* genome. Overall, the repetitive nature of the *MAT-1* sequence might provide clues with regard to the lack of recombination in the region. That is, a scrambling of repetitive elements between *MAT-1* and *MAT-2* may interfere with pairing during meiosis and reduce opportunities for recombination. It is also possible that large inversions may be present given the orientation differences we observed for the mating-type gene complexes (47).

The availability of the *MAT-1* sequence and the regions around the *a1* and *b1* complexes allowed a comparison to be made with the corresponding regions in the recently sequenced genome of *U. maydis*. The opportunity for this analysis was suggested during the process of annotating the *U. hordei* genes when it became apparent that for many, the closest orthologs were *U. maydis* genes, and that the genes in the same island in *MAT-1* often matched consecutive genes in the *U. maydis* genome (Color Plate 2). Initially, the *U. maydis* sequence contigs carrying the *a* and *b* gene complexes were identified by BLAST: the *a* complex was present on contig 1.83 on chromosome V, and the *b* complex was on contig 1.13 on chromosome I. Synteny seemed to be maintained in the unique regions of the *MAT-1* locus when contig 1.13 was linked to contig 1.12 (Color Plate 2). The alignments of the *MAT-1* sequence with contigs 1.83, 1.12, and 1.13 of *U. maydis* dramatically illustrated the tremendous accumulation of repetitive elements at the *MAT-1* locus and the substantial rearrangements (deletions, insertions, inversions, and translocations) that have occurred since the two species diverged from a common ancestor (Color Plate 2) (8). For example, the distribution of the single-copy regions across the 527-kb sequence of *MAT-1* corresponds to an ~80-kb region around the *a* genes and an ~150-kb region around the *b* genes in *U. maydis*. Remarkably, the order of genes around the *a* gene complex in *U. maydis* was generally conserved in the *MAT-1* sequence of *U. hordei*. Similarly, there is evidence of conservation of synteny for the region containing the *b* genes between *U. maydis* and *U. hordei*. However, a pattern of interdigitation of *a*-associated and *b*-associated genes exists such that the *MAT-1* locus contains interspersed genes that have locations on different chromosomes in *U. maydis*. One interpretation of this pattern is that a tetrapolar arrangement of *a* and *b* gene complexes was present in the common ancestor of these fungi and that an interchromosomal rearrangement fused the regions together to establish a progenitor *MAT* locus for the bipolar pattern. Subsequently, the *MAT* locus accumulated repetitive elements and these contributed to a large number of inversion and transposition events to generate the current arrangement of genes at *MAT-1*. There was also evidence of rearrangements that might have involved other regions of the *U. maydis* genome. That is, 4 of the 47 genes identified in *MAT-1* were not found to be associated with the contigs carrying the *a* and *b* gene complexes in *U. maydis* (8).

One question concerns the position of the original fusion event that joined the *a* and *b* gene complexes together. From Color Plate 2 it is apparent that *a*-associated genes from *U. maydis* are distributed in the *MAT-1* locus from positions 1 to ~332 kb. Thus, one could use the 332-kb coordinate as the right boundary of the putative fusion event. The left boundary is more difficult to establish because of the interdigitation of *a*- and *b*-associated genes described earlier that extends from ~90 to 332 kb. Inspection of this interval did not reveal obvious features that may have participated in the rearrangements, although among the five types of retroelements and four types of transposons, the eight sequences related to the Tho4 transposon are found only in this interval and are roughly distributed with the gene islands (G. Bakkeren and J. Kronstad, unpublished observations). In addition, if one focuses on the clusters of genes around the 258- and 332-kb positions in *MAT-1*, it appears that these genes (in an inverted orientation) represent the ends of the *U. maydis* 1.12 and 1.83 contigs that appear at *MAT-1* (Color Plate 2). Interestingly, these regions are flanked by exceptionally large clusters (10 to 20 kb) of the LTR1 element (Bakkeren and Kronstad, unpublished). These interpretations are therefore suggestive of the action of transposable elements in the translocation events that might have taken place in a common ancestor. It is also possible that the comparison has revealed rearrangement events that occurred in the *U. maydis* sequences after separation. Finally, given the highly repetitive nature of the *MAT-1* sequence, it is possible that some of the observed differences in gene order might be due to assembly errors.

An analysis of the sequence contigs carrying the mating-type regions of *U. maydis* was also performed, and it was found that these regions contained very few repetitive elements. Initially, the sequences of each of the different types of repetitive elements from *MAT-1* (LTRs, retroelements, and transposons) were used to search the *U. maydis* contigs and the complete genome sequence. Short sequences with weak similarity were found that were related to 2 of the 12 LTR elements from *MAT-1* and one of the transposon sequences. In addition, sequences related to each of the five retrotransposons from *MAT-1* were also present in the *U. maydis* genome. Interestingly, none of the sequences related to the elements from *MAT-1* were present on contigs 1.12, 1.13, and 1.83, which are associated with the *a* and *b* loci in *U. maydis*. A closer examination of these contigs identified very few repetitive sequences, and these were short (i.e., <100 bp in length). Taken together, these results indicate that the mating-type regions of *U. hordei* and *U. maydis* are dramatically different with a paucity of repetitive elements in *U. maydis* and approximately 50% representation of these sequences in the *MAT-1* region of *U. hordei*.

IMPLICATIONS OF BIPOLAR AND TETRAPOLAR MATING FOR PATHOGENIC FUNGI

The analysis of the genomic organization of the *a* and *b* gene complexes provides a simple explanation for the observed differences in the mating systems of smut fungi (5). However, this analysis also revealed the highly repetitive nature of the *MAT-1* sequence with repeated sequences spread across the whole genome of *U. hordei* (although the overall proportion is not known [8]). The accumulation of repetitive sequences in sex-determining regions of a variety of organisms is well documented, and there are clear examples in the fungi *Microbotryum violaceum* and *Cryptococcus neoformans*, which exhibit bipolar mating. Indeed, Hood (31, 33) has shown that the chromosomes carrying the mating-type locus in *M. violaceum* are dimorphic and rich in repetitive sequences. Specifically, sequence analysis of random genomic fragments revealed that DNA from the sex chromosomes was twice as likely to contain transposable elements than DNA from autosomal chromosomes. Overall, the genome of *M. violaceum* appears to contain 15% or more repetitive DNA (33). A similar situation exists with the bipolar *MAT* loci in *Cryptococcus* species, but the level of repetitive sequences appears to be lower at 13.2% for *MATa* and 17.3% for *MATα* in *C. neoformans* var. *grubii* (21). This level can be compared with a

5% overall genome content of transposons, although this estimate comes from the *C. neoformans* var. *neoformans* genome (51). The shared features raise the possibility that a bipolar mating system might have contributed to the accumulation of repetitive elements at the mating-type locus and throughout the genome in these fungi. It has been postulated that the lack of "purifying recombination" in sex-determining regions leads to the accumulation of transposable elements and repeats (12), offering an explanation for the increased abundance of such elements compared to the rest of the genome.

The examination of the mating-type regions for the tetrapolar system of *U. maydis* revealed a striking departure from the paradigm of repeat accumulation (8). If *U. maydis* and *U. hordei* are truly representative, the results suggest that the *MAT* loci (sex chromosomes) in these species have evolved by quite different paths. Several possible contributing factors come to mind. For example, it is possible that the extant features are the result of differences in the potential for inbreeding versus outbreeding. For inbreeding species, a lack of different parental stretches of DNA for "purifying recombination" might sustain transposable element loads. Inbreeding may have an influence similar to that of asexuality with regard to the accumulation of transposable elements for *U. hordei*. Arkhipova and Meselson (2) hypothesized that, relative to asexual organisms, sexual activity may limit the proliferation of transposable elements within a genome even though new elements may be introduced through sex. Tetrapolar mating is generally thought to promote outbreeding because teliospore germination generates progeny that only have a one-in-four chance of being compatible. In *U. hordei*, one-half of the progeny from a teliospore would be compatible, thus potentially favoring inbreeding. Other possibilities for the observed differences between *U. hordei* and *U. maydis* could involve the interconnections between mating and the need for host infection to complete sexual development. For example, the length of the saprobic phase on the host surface before mating and initiation of infection might influence the window of opportunity for different smuts to find a compatible nonsibling partner. Thus, outbreeding potential could be influenced by inoculation density and the extent of mixing of gametes on the plant surface. This is further complicated by whether the smut species cause local or systemic disease after infection of seedlings, flowers, or older plants and whether infection occurs underground or aboveground. Thus, the differences in the pathogenic lifestyles of these fungi could have a major influence on mate detection.

Additional features of importance include the role of the host genetic background in allowing completion of the sexual cycle after mating. Different smuts that infect small-grain cereals and grasses have the potential for interactions with compatible isolates of different species (e.g., *U. hordei* and *U. bullata*) on plants that serve as common hosts (18). However, even if successful cell fusion and dikaryon formation occur, the contributions of avirulence genes in the backgrounds of the mating partners may preclude successful sexual development because of a host defense response. A clear gene-for-gene system has not been described for *U. maydis*, while *U. hordei* and other smuts that infect small-grain cereals are genetically well characterized with regard to their avirulence genes. Finally, another consideration is that the different smut species may be more or less likely to bypass mating altogether because it is known that teliospore germination can directly lead to an infectious dikaryon (e.g., due to "partial or delayed [meiotic] reduction as described for *U. maydis* [14]).

The evidence to date with *U. maydis* suggests the hypothesis that species with tetrapolar mating have less repetitive DNA. An interesting analysis would be to trace repeat content in the genomes and at the mating-type loci in a set of smut species with different levels of evolutionary separation. Recent molecular phylogeny studies using ribosomal sequences and whole-genome scans for DNA length polymorphism profiles have started to resolve issues regarding taxonomic placement within the smuts (7, 54, 61, 76, 77). For example, *M. violaceum* resided previously in the genus *Ustilago* (*U. violacea*), *U. maydis* has been placed in the genus *Sporisorium*, and *U. hordei* is considered a representative of the "true" *Ustilago* species. These relationships and the emerging molecular view of smut phylogeny will be a valuable guide to choose species for further sequence analysis.

We have attempted to assess the evolutionary distance between *U. hordei* and *U. maydis*. A rough estimate comes from calculating mutation rates between several genes present on the recently sequenced *U. hordei MAT-1* locus (8) and homologs in the *U. maydis* database (MIPS: Munich Information Center for Protein Sequences, http://mips.gsf.de/genre/proj/ustilago). No precise clock exists for the smuts, but substitution rates at fourfold degenerate sites were estimated at 2.2×10^{-6} mutations per year per kb for mammals (45) and at 11×10^{-6} mutations per year per kb for *Drosophila* (78). In general, no good estimate exists for fungi, but some researchers have used 1×10^{-6} mutations per year per kb for the internal transcribed spacer region, which might not be as precise (9). Assuming a "clock" of between 2.2×10^{-6} and 11×10^{-6} mutations per year per kb, our limited set of data would suggest the divergence between *U. maydis* and *U. hordei* to be between 21 and 27 million years (Bakkeren and Kronstad, unpublished).

As mentioned, there may be special evolutionary considerations for fungi that spend their lives intimately associated with their plant hosts and whose evolution was coincident with the evolution, domestication, and widespread monoculture cultivation of cereal crops. Interestingly, the timing of divergence of bipolar and tetrapolar mating systems might be coincident with the divergence of host plants. The cereal plants such as barley, corn, rice, and wheat are thought to have diverged from a common ancestor 50 to 70 million years ago. In thinking about parallel evolution of smuts and their hosts, one curious feature of the *MAT-1* locus is the discovery of gene islands in a sea of repetitive sequences. This organization is remarkably similar to the pattern found in the highly repetitive genomes of cereal crops (59, 65, 89). Overall, these observations reinforce the unique aspects of mating-type evolution in smut fungi in the context of interactions with host plants.

UNANSWERED QUESTIONS AND FUTURE WORK

Is *MAT* Carried on Sex Chromosomes in *U. hordei*?

It is clear that recombination is suppressed within the *U. hordei MAT* locus, and an open question is whether this suppression extends past *MAT* to include part or all of the chromosome arms. That is, the situation may be similar to that of *M. violaceum*, where a length polymorphism for the chromosomes harboring the two opposite mating-type loci, that is, dimorphic sex chromosomes, has been described (31). The absence of recombination has also been reported for almost the entire chromosome carrying the mating-type locus in *Neurospora tetrasperma* (24, 55).

Does the Genome of *U. hordei* Contain Fewer Genes than That of *U. maydis*?

The abundance of repetitive sequences identified at the *MAT-1* locus potentially extends throughout the genome of *U. hordei* (8). The genome sizes for *U. hordei* (~19.6 Mb) and *U. maydis* (~20.5 Mb) are similar (at least by the inaccurate method of electrophoretic karyotyping), so it would be interesting to compare the overall level of repetitive DNA, particularly in the context of

potential deleterious effects of transposable elements. There is clear evidence for gene disruption by elements within *MAT-1* (8), and it would be interesting to determine whether *U. hordei* has fewer genes overall or fewer active genes. The prediction from the analysis of the *U. hordei* genome is that other small-grain smut fungi will also have repetitive genomes.

What Is the Evolutionary History and Level of Activity of the Transposable Elements at the *MAT-1* Locus?

The abundance of repeats at the *MAT-1* locus provides a rich opportunity to examine the biology of transposable elements in fungi. In particular, there is an opportunity to use the elements to trace patterns of activity. For example, some elements may be preferentially inserted into others in *MAT-1* and such a pattern has been described for transposable elements in *Magnaporthe grisea* where Maggy and MGL (*M. grisea* LINE retrotransposon) are found inserted into *Pot2*, but the reciprocal arrangement is not observed (83). More detailed analysis of the variation in the *MAT-1* LTR sequences, both solo and associated with retrotransposons, may provide insight into the evolutionary history of the elements and of different parts of the *MAT* locus. In particular, the examination of intraelement LTR sequences may shed light on whether certain elements may have transposed more recently than others. Finally, Hood et al. (34) presented evidence for ripping in *M. violaceum*. This process could contribute to the inactivation and control of transposable elements, and the occurrence of ripping at *MAT-1* needs to be examined in *U. hordei*.

How Do the *MAT-1* and *MAT-2* Regions Differ in *U. hordei*?

The *MAT-2* locus is estimated to be 430 kb in length, and this region remains to be sequenced for comparison with *MAT-1*. This type of comparative analysis has been quite informative for the *MATa* and *MATα* loci of *Cryptococcus neoformans* (22). Hood et al. (32) also presented evidence that the A1 and A2 sex chromosomes in *M. violaceum* contain different densities of functional genes. One could imagine that *MAT-2* contains the same genes as *MAT-1* or that rearrangements have resulted in the complement of genes identified at *MAT-1* being distributed elsewhere in the genome with some retained at *MAT-2*. Related questions include how many of the genes at *MAT-1* and *MAT-2* are essential, how many are actually involved in mating or virulence, and how many are pseudogenes? The evidence for insertion of LTRs and transposable elements into some of the genes in *MAT-1* raises the question of whether functional copies exist elsewhere in the genome or whether the strains have lost specific metabolic functions. The local movement of transposable elements might explain the linkage between proline auxotrophy and mating type in some smut fungi such as *U. nuda* (56).

The reasons for suppression of recombination for the *MAT-1* and *MAT-2* intervening sequences also need to be explored. One initial experiment would be to use a *MAT-1* strain and swap *a2* for *a1* and *b2* for *b1* to create a strain that would potentially mate with a *MAT-1* (*a1 b1*) strain but that would have the same intervening region. The frequency of recombination could then be tested, and one could assess whether heterozygosity for genes contained with *MAT-1* and *MAT-2* is important for sexual development and virulence.

What Is the Evolutionary History of Bipolar and Tetrapolar Mating Systems in Smut Fungi?

The sequence comparisons for *U. hordei* and *U. maydis* suggest that the tetrapolar system is ancestral and gave rise to the bipolar system in the Ustilaginales. One wonders then whether the bipolar system arose once or several times and whether the history of changes in mating systems can be traced in the phylogeny of the smuts. The approach to answer some of these questions would be to link a detailed phylogenetic analysis (e.g., by multilocus sequence typing) with the selective analysis and sequencing of mating-type regions. In particular, the phylogeny of the unique genes at *MAT* would be interesting to compare among different smuts.

References

1. Anderson, C. M., D. A. Willits, P. J. Kosted, E. J. Ford, A. D. Martinez-Espinoza, and J. E. Sherwood. 1999. Molecular analysis of the pheromone and pheromone receptor genes of *Ustilago hordei*. *Gene* **240**:89–97.
2. Arkhipova, I., and M. Meselson. 2005. Deleterious transposable elements and the extinction of asexuals. *Bioessays* **27**:76–85.
3. Bakkeren, G., B. Gibbard, A. Yee, E. Froeliger, S. Leong, and J. Kronstad. 1992. The *a* and *b* loci of *Ustilago maydis* hybridize with DNA sequences from other smut fungi. *Mol. Plant-Microbe Interact.* **5**:347–355.
4. Bakkeren, G., and J. W. Kronstad. 1993. Conservation of the *b* mating-type gene complex among bipolar and tetrapolar smut fungi. *Plant Cell* **5**:123–136.
5. Bakkeren, G., and J. W. Kronstad. 1994. Linkage of mating-type loci distinguishes bipolar from tetrapolar mating in basidiomycetous smut fungi. *Proc. Natl. Acad. Sci. USA* **91**:7085–7089.
6. Bakkeren, G., and J. W. Kronstad. 1996. The pheromone cell signaling components of the *Ustilago a* mating-type
7. loci determine intercompatibility between species. *Genetics* **143**:1601–1613.
7. Bakkeren, G., J. W. Kronstad, and C. A. Lévesque. 2000. Comparison of AFLP fingerprints and ITS sequences as phylogenetic markers in Ustilaginomycetes. *Mycologia* **92**:510–521.
8. Bakkeren, G., G. Jiang, R. Warren, Y. Butterfield, H. Shin, R. Chiu, R. Linning, J. Schein, N. Lee, G. Hu, D. M. Kupfer, Y. Tang, B. A. Roe, S. Jones, M. Marra, and J. W. Kronstad. 2006. Mating factor linkage and genome evolution in basidiomycetous pathogens of cereals. *Fungal Genet. Biol.* **43**:655–666.
9. Berbee, M. L., B. P. Payne, G. Zhang, R. G. Roberts, and B. G. Turgeon. 2003. Shared ITS DNA substitutions in isolates of opposite mating type reveal a recombining history for three presumed asexual species in the filamentous ascomycete genus *Alternaria*. *Mycol. Res.* **107**:169–182.
10. Carris, L. M., L. A. Castlebury, and B. J. Goates. 2006. Nonsystemic bunt fungi—*Tilletia indica* and *T. horrida*: a review of history, systematics, and biology. *Annu. Rev. Phytopathol.* **44**:113–133.
11. Castlebury, L. A., and L. M. Carris. 1999. *Tilletia walkeri*, a new species on *Lolium multiflorum* and *L. perenne*. *Mycologia* **91**:121–131.
12. Charlesworth, B., and C. H. Langley. 1989. The population genetics of *Drosophila* transposable elements. *Annu. Rev. Genet.* **23**:251–287.
13. Chisholm, S. T., G. Coaker, B. Day, and B. J. Staskawicz. 2006. Host-microbe interactions: shaping the evolution of the plant immune response. *Cell* **124**:803–814.
14. Christensen, J. J. 1963. *Corn Smut Caused by* Ustilago maydis. Monograph no. 2. American Phytopathological Society, St. Paul, MN.
15. Chung, S., M. Karos, Y. C. Chang, J. Lukszo, B. L. Wickes, and K. J. Kwon-Chung. 2002. Molecular analysis of *CPRalpha*, a *MATalpha*-specific pheromone receptor gene of *Cryptococcus neoformans*. *Eukaryot. Cell* **1**:432–439.
16. Feldbrugge, M., J. Kamper, G. Steinberg, and R. Kahmann. 2004. Regulation of mating and pathogenic development in *Ustilago maydis*. *Curr. Opin. Microbiol.* **7**:666–672.
17. Ferris, P. J., and U. W. Goodenough. 1994. The mating-type locus of *Chlamydomonas reinhardtii* contains highly rearranged DNA sequences. *Cell* **76**:1135–1145.
18. Fischer, G. W. 1951. Induced hybridization in graminicolous smut fungi. I. *Ustilago hordei* X *U. bullata*. *Phytopathology* **41**:839–853.
19. Fischer, G. W., and C. S. Holton. 1957. *Biology and Control of the Smut Fungi*. Ronald Press, New York, NY.
20. Flor, H. H. 1942. Inheritance of pathogenicity in *Melampsora lini*. *Phytopathology* **32**:653–669.
21. Fraser, J. A., S. Diezmann, R. L. Subaran, A. Allen, K. B. Lengeler, F. S. Dietrich, and J. Heitman. 2004. Convergent evolution of chromosomal sex-determining regions in the animal and fungal kingdoms. *PLoS Biol.* **2**:e384.
22. Fraser, J. A., and J. Heitman. 2005. Chromosomal sex-determining regions in animals, plants and fungi. *Curr. Opin. Genet. Dev.* **15**:645–651.

23. Froeliger, E. H., and S. A. Leong. 1991. The *a* mating-type alleles of *Ustilago maydis* are idiomorphs. *Gene* **100:** 113–122.

24. Gallegos, A., D. J. Jacobson, N. B. Raju, M. P. Skupski, and D. O. Natvig. 2000. Suppressed recombination and a pairing anomaly on the mating-type chromosome of *Neurospora tetrasperma*. *Genetics* **154:**623–633.

25. Gaudet, D. A., and R. L. Kiesling. 1991. Variation in aggressiveness among and within races of *Ustilago hordei* on barley. *Phytopathology* **81:**1385–1390.

26. Gillissen, B., J. Bergemann, C. Sandmann, B. Schroeer, M. Boelker, and R. Kahmann. 1992. A two-component regulatory system for self/non-self recognition in *Ustilago maydis*. *Cell* **68:**647–657.

27. Hoffmann, J. A., and E. L. Kendrick. 1968. *Phytopathology* **59:**79–83.

28. Holliday, R. 1961. The genetics of *Ustilago maydis*. *Genet. Res.* **2:**204–230.

29. Holliday, R. 1965. Induced mitotic crossing-over in relation to genetic replication in synchronously dividing cells of *Ustilago Maydis*. *Genet Res.* **10:**104–120.

30. Holton, C. S., J. A. Hoffmann, and R. Duran. 1968. Variation in the smut fungi. *Annu. Rev. Phytopathol.* **6:**213–242.

31. Hood, M. E. 2002. Dimorphic mating-type chromosomes in the fungus *Microbotryum violaceum*. *Genetics* **160:** 457–461.

32. Hood, M. E., J. Antonovics, and B. Koskella. 2004. Shared forces of sex chromosome evolution in haploid-mating and diploid-mating organisms: *Microbotryum violaceum* and other model organisms. *Genetics* **168:**141–146.

33. Hood, M. E. 2005. Repetitive DNA in the automictic fungus *Microbotryum violaceum*. *Genetica* **124:**1–10.

34. Hood, M. E., M. Katawczik, and T. Giraud. 2005. Repeat-induced point mutation and the population structure of transposable elements in *Microbotryum violaceum*. *Genetics* **170:**1081–1089.

35. Hu, G.-G., R. Linning, and G. Bakkeren. 2003. Ultrastructural comparison of a compatible and incompatible interaction triggered by the presence of an avirulence gene during early infection of the smut fungus, *Ustilago hordei*, in barley. *Physiol. Mol. Plant Pathol.* **62:**155–166.

36. Hu, G. G., R. Linning, and G. Bakkeren. 2002. Sporidial mating and infection process of the smut fungus, *Ustilago hordei*, in susceptible barley. *Can. J. Bot.* **80:**1103–1114.

37. Huang, H. Q., and J. Nielsen. 1984. Hybridization of the seedling-infecting *Ustilago* spp. pathogenic on barley and oats, and a study of the genotypes conditioning the morphology of their spore walls. *Can. J. Bot.* **62:**603–608.

38. Huff, D. R., D. Zagory, and L. Wu. 1987. Report of buffalograss bunt (*Tilletia buchloeana*) in Oklahoma. *Plant Dis.* **71:**651.

39. Koltin, Y., J. Stamberg, and P. A. Lemke. 1972. Genetic structure and evolution of the incompatibility factors in higher fungi. *Bacteriol. Rev.* **36:**156–171.

40. Kosted, P. J., S. A. Gerhardt, C. M. Anderson, A. Stierle, and J. E. Sherwood. 2000. Structural requirements for activity of the pheromones of *Ustilago hordei*. *Fungal Genet. Biol.* **29:**107–117.

41. Kosted, P. J., S. A. Gerhardt, and J. E. Sherwood. 2002. Pheromone-related inhibitors of *Ustilago hordei* mating and *Tilletia tritici* teliospore germination. *Phytopathology* **92:**210–216.

42. Kronstad, J. W., and S. A. Leong. 1989. Isolation of two alleles of the *b* locus of *Ustilago maydis*. *Proc. Natl. Acad. Sci. USA* **86:**978–982.

43. Kubisiak, T. L., and M. G. Milgroom. 2006. Markers linked to vegetative incompatibility (vic) genes and a region of high heterogeneity and reduced recombination near the mating type locus (*MAT*) in *Cryphonectria parasitica*. *Fungal Genet. Biol.* **43:**453–463.

44. Kues, U., and L. A. Casselton. 1992. Molecular and functional analysis of the *A* mating type genes of *Coprinus cinereus*. *Genet. Eng.* (NY) **14:**251–268.

45. Kumar, S., and S. Subramanian. 2002. Mutation rates in mammalian genomes. *Proc. Natl. Acad. Sci. USA* **99:** 803–808.

46. Lahn, B. T., and D. C. Page. 1999. Four evolutionary strata on the human X chromosome. *Science* **286:**964–967.

47. Lee, N., G. Bakkeren, K. Wong, J. E. Sherwood, and J. W. Kronstad. 1999. The mating-type and pathogenicity locus of the fungus *Ustilago hordei* spans a 500-kb region. *Proc. Natl. Acad. Sci. USA* **96:**15026–15031.

48. Lengeler, K. B., and E. Kothe. 1999. Mated: a putative peptide transporter of *Schizophyllum commune* expressed in dikaryons. *Curr. Genet.* **36:**159–164.

49. Lengeler, K. B., D. S. Fox, J. A. Fraser, A. Allen, K. Forrester, F. S. Dietrich, and J. Heitman. 2002. Mating-type locus of *Cryptococcus neoformans*: a step in the evolution of sex chromosomes. *Eukaryot. Cell* **1:**704–718.

50. Linning, R., D. Lin, N. Lee, M. Abdennadher, D. Gaudet, P. Thomas, D. Mills, J. W. Kronstad, and G. Bakkeren. 2004. Marker-based cloning of the region containing the *UhAvr1* avirulence gene from the basidiomycete barley pathogen *Ustilago hordei*. *Genetics* **166:**99–111.

51. Loftus, B. J., E. Fung, P. Roncaglia, D. Rowley, P. Amedeo, D. Bruno, J. Vamathevan, M. Miranda, I. J. Anderson, J. A. Fraser, J. E. Allen, I. E. Bosdet, M. R. Brent, R. Chiu, T. L. Doering, M. J. Donlin, C. A. D'Souza, D. S. Fox, V. Grinberg, J. Fu, M. Fukushima, B. J. Haas, J. C. Huang, G. Janbon, S. J. Jones, H. L. Koo, M. I. Krzywinski, J. K. Kwon-Chung, K. B. Lengeler, R. Maiti, M. A. Marra, R. E. Marra, C. A. Mathewson, T. G. Mitchell, M. Pertea, F. R. Riggs, S. L. Salzberg, J. E. Schein, A. Shvartsbeyn, H. Shin, M. Shumway, C. A. Specht, B. B. Suh, A. Tenney, T. R. Utterback, B. L. Wickes, J. R. Wortman, N. H. Wye, J. W. Kronstad, J. K. Lodge, J. Heitman, R. W. Davis, C. M. Fraser, and R. W. Hyman. 2005. The genome of the basidiomycetous yeast and human pathogen *Cryptococcus neoformans*. *Science* **307:**1321–1324.

52. Martinez, E. A. D., S. A. Gerhardt, and J. E. Sherwood. 1993. Morphological and mutational analysis of mating in *Ustilago hordei*. *Exp. Mycol.* **17:**200–214.

53. McClelland, C. M., J. Fu, G. L. Woodlee, T. S. Seymour, and B. L. Wickes. 2002. Isolation and characterization of the *Cryptococcus neoformans* MATa pheromone gene. *Genetics* **160:**935–947.

54. Menzies, J.-G., G. Bakkeren, F. Matheson, J.-D. Procunier, and S. Woods. 2003. Use of inter-simple sequence repeats and amplified fragment length polymorphisms to analyze genetic relationships among small grain-infecting species of *Ustilago*. *Phytopathology* **93**:167–175.

55. Merino, S. T., M. A. Nelson, D. J. Jacobson, and D. O. Natvig. 1996. Pseudohomothallism and evolution of the mating-type chromosome in *Neurospora tetrasperma*. *Genetics* **143**:789–799.

56. Nielsen, J. 1968. Isolation and culture of monokaryotic haplonts of *Ustilago nuda*, the role of proline in their metabolism, and the inoculation of barley with resynthesized dikaryons. *Can. J. Bot.* **46**:1193–1200.

57. Oort, A. J. P. 1944. Onderzoekingen over stuifbrand II. Overgevoeligheid van tarwe voor stuifbrand, *Ustilago tritici*. [Hypersensitiveness of wheat to loose smut.] *Tijdschrift over plantenziekten* **50**:73–106.

58. Pardo, E. H., S. F. O'Shea, and L. A. Casselton. 1996. Multiple versions of the *A* mating-type locus of *Coprinus cinereus* are generated by three paralogous pairs of multiallelic homeobox genes. *Genetics* **144**:87–94.

59. Paterson, A. H., J. E. Bowers, D. G. Peterson, J. C. Estill, and B. A. Chapman. 2003. Structure and evolution of cereal genomes. *Curr. Opin. Genet. Dev.* **13**:644–650.

60. Piepenbring, M., and R. Bauer. 1997. Erratomyces, new genus of Tilletiales with species on Leguminosae. *Mycologia* **89**:924–936.

61. Piepenbring, M., M. Stoll, and F. Oberwinkler. 2002. The generic position of *Ustilago maydis*, *Ustilago scitaminea*, and *Ustilago esculenta* (Ustilaginales). *Mycol. Prog.* **1**:71–80.

62. Puhalla, J. E. 1970. Genetic studies of the *b* incompatibility locus of *Ustilago maydis*. *Genet. Res.* **16**:229–232.

63. Raper, J. R., and A. S. Flexer. 1971. Mating systems and evolution of the basidiomycetes, p. 149–167 *In* R. H. Petersen (ed.), *Evolution in the Higher Basidiomycetes; an International Symposium*. University of Tennessee Press, Knoxville.

64. Riquelme, M., M. P. Challen, L. A. Casselton, and A. J. Brown. 2005. The origin of multiple B mating specificities in *Coprinus cinereus*. *Genetics* **170**:1105–1119.

65. Rostoks, N., Y. J. Park, W. Ramakrishna, J. Ma, A. Druka, B. A. Shiloff, P. J. SanMiguel, Z. Jiang, R. Brueggeman, D. Sandhu, K. Gill, J. L. Bennetzen, and A. Kleinhofs. 2002. Genomic sequencing reveals gene content, genomic organization, and recombination relationships in barley. *Funct. Integr. Genomics* **2**:51–59.

66. Rowell, J. B., and J. E. DeVay. 1954. Genetics of *Ustilago zeae* in relation to basic problems of its pathogenicity. *Phytopathology* **44**:356–362.

67. Rowell, J. B. 1955. Functional role of compatibility factors and an in vitro test for sexual compatibility with haploid lines of *Ustilago zeae*. *Phytopathology* **45**:370–374.

68. Schirawski, J., B. Heinze, M. Wagenknecht, and R. Kahmann. 2005. Mating-type loci of *Sporisorium reilianum*: novel pattern with three *a* and multiple *b* specificities. *Eukaryot. Cell* **4**:1317–1327.

69. Sherwood, J. E., P. J. Kosted, C. M. Anderson, and S. A. Gerhardt. 1998. Production of a mating inhibitor by *Ustilago hordei*. *Phytopathology* **88**:456–464.

70. Sidhu, G., and C. Person. 1972. Genetic control of virulence in *Ustilago hordei*. II. Identification of genes for host resistance and demonstration of gene-for-gene relations. *Can. J. Genet. Cytol.* **14**:209–213.

71. Smith, I. M. 1988. Basidiomycetes I, Ustilaginales, p. 462–472. *In* I. M. Smith, J. Dunez, D. H. Phillips, R. A. Lelliott, and S. A. Archer (ed.), *European Handbook of Plant Diseases*. Blackwell Scientific Publications, Oxford, United Kingdom.

72. Snetselaar, K. M., and C. W. Mims. 1992. Sporidial fusion and infection of maize seedlings by the smut fungus *Ustilago maydis*. *Mycologia* **84**:193–203.

73. Snetselaar, K. M., and C. W. Mims. 1993. Infection of maize stigmas by *Ustilago maydis*: light and electron microscopy. *Phytopathology* **83**:843–850.

74. Snetselaar, K. M., and C. W. Mims. 1994. Light and electron microscopy of *Ustilago maydis* hyphae in maize. *Mycol. Res.* **98**:347–355.

75. Stankis, M. M., C. A. Specht, H. Yang, L. Giasson, R. C. Ullrich, and C. P. Novotny. 1992. The A alpha mating locus of *Schizophyllum commune* encodes two dissimilar multiallelic homeodomain proteins. *Proc. Natl. Acad. Sci. USA* **89**:7169–7173.

76. Stoll, M., M. Piepenbring, D. Begerow, and F. Oberwinkler. 2003. Molecular phytogeny of *Ustilago* and *Sporisorium* species (Basidiomycota, Ustilaginales) based on internal transcribed spacer (ITS) sequences. *Can. J. Bot.* **81**:976–984.

77. Stoll, M., D. Begerow, and F. Oberwinkler. 2005. Molecular phylogeny of *Ustilago*, *Sporisorium*, and related taxa based on combined analyses of rDNA sequences. *Mycol. Res.* **109**:342–356.

78. Tamura, K., S. Subramanian, and S. Kumar. 2004. Temporal patterns of fruit fly (Drosophila) evolution revealed by mutation clocks. *Mol. Biol. Evol.* **21**:36–44.

79. Tapke, V. F. 1937. Physiologic races of *Ustilago hordei*. *J. Agric. Res.* **55**:683–692.

80. Tapke, V. F. 1945. New physiologic races of *Ustilago hordei*. *Phytopathology* **35**:970–976.

81. Thomas, P. L. 1976. Interaction of virulence genes in *Ustilago hordei*. *Can. J. Genet. Cytol.* **18**:141–149.

82. Thomas, P. L., and H. Q. Huang. 1985. Inheritance of virulence on barley in the hybrids *Ustilago aegilopsidis* times *Ustilago hordei* and *Ustilago aegilopsidis* times *Ustilago nigra*. *Can. J. Genet. Cytol.* **27**:312–317.

83. Thon, M. R., S. L. Martin, S. Goff, R. A. Wing, and R. A. Dean. 2004. BAC end-sequences and a physical map reveal transposable element content and clustering patterns in the genome of *Magnaporthe grisea*. *Fungal Genet. Biol.* **41**:657–666.

84. Thrall, P. H., A. Biere, and J. Antonovics. 1993. Plant life-history and disease susceptibility; the occurrence of *Ustilago violacea* on different species within the Caryophyllaceae. *J. Ecol.* **81**:489–498.

85. Vanky, K. 1987. *Illustrated Genera of Smut Fungi*. Cryptogamic Studies, vol. 1. Gustav Fisher Verlag, New York, NY.

86. Vanky, K. 2003. Taxonomical studies on Ustilaginales. XXIII. *Mycotaxon* **85**:1–65.

87. Voegele, R. T., and K. Mendgen. 2003. Rust haustoria: nutrient uptake and beyond. *New Phytol.* **159:**93–100.

88. Wendland, J., L. J. Vaillancourt, J. Hegner, K. B. Lengeler, K. J. Laddison, C. A. Specht, C. A. Raper, and E. Kothe. 1995. The mating-type locus *B alpha 1* of *Schizophyllum commune* contains a pheromone receptor gene and putative pheromone genes. *EMBO J.* **14:**5271–5278.

89. Whitelaw, C. A., W. B. Barbazuk, G. Pertea, A. P. Chan, F. Cheung, Y. Lee, L. Zheng, S. van Heeringen, S. Karamycheva, J. L. Bennetzen, P. SanMiguel, N. Lakey, J. Bedell, Y. Yuan, M. A. Budiman, A. Resnick, S. Van Aken, T. Utterback, S. Riedmuller, M. Williams, T. Feldblyum, K. Schubert, R. Beachy, C. M. Fraser, and J. Quackenbush. 2003. Enrichment of gene-coding sequences in maize by genome filtration. *Science* **302:**2118–2120.

90. Wilson, J. M., and R. A. Frederiksen. 1970. Histopathology of the interaction of *Sorghum bicolor* and *Sphacelotheca reilianum. Phytopathology* **60:**828–832.

Zygomycetes, Chytridiomycetes, and Oomycetes: the Frontiers of Knowledge

Sex in Fungi: Molecular Determination
and Evolutionary Implications
Edited by Joseph Heitman et al.
© 2007 ASM Press, Washington, D.C.

Alexander Idnurm
Timothy Y. James
Rytas Vilgalys

24

Sex in the Rest:
Mysterious Mating in the Chytridiomycota and Zygomycota

THE IMPORTANCE OF THE "OTHER" FUNGI

Of the estimated 1.5 million species comprising the fungal kingdom (30), the sexual reproduction of a small subset has been defined at a molecular level. All of these species are members of two phyla, the Ascomycota and Basidiomycota, together known as the Dikarya because many members have a binucleate phase in their life cycle (31). A comparison of the mating systems in the Dikarya reveals that the mechanisms used to control mating are largely conserved in both groups. Although there are various systems of outcrossing or self-fertilization and variation in the number of loci and alleles at each locus, the group has conserved the use of G-protein-coupled peptide pheromone receptors and DNA-binding transcription factors (the homeodomain, α-box, and HMG-class proteins). At present, little is known about the molecular mechanisms regulating mating of fungi outside the Dikarya.

Although the Dikarya comprise the majority of all known species, the greatest evolutionary diversity in the kingdom lies among the so-called basal fungal lineages that first branched from the earliest common ancestor of the Fungi (Fig. 24.1) (31). These basal lineages of "other fungi" form the earliest lineages in the Fungal Tree of Life (31, 42) and include several phyla including Chytridiomycota, Glomeromycota, Microsporidia, and Zygomycota. These "other" fungal species encompass an enormous diversity, and at least two of these phyla are paraphyletic (Chytridiomycota and Zygomycota) and thus will eventually need to be further divided in order to establish stable monophyletic phyla (Fig. 24.1).

Are mechanisms that control mating, such as mating-type loci, conserved throughout the Fungi? If so, what was the original sex determination mechanism like and how has it diversified throughout the various fungal lineages? If not, how is mating controlled in these basal lineages and when did Dikarya-like mating-type loci evolve? This chapter discusses mating biology of basal fungi with emphasis on chytridiomycetes and zygomycetes (mating in Microsporidia, Glomeromycota, and Zygomycota is also discussed in other chapters in this volume).

The chytrids and zygomycetes are ecologically diverse. Many of them are parasites, on hosts such as plants, algae, invertebrates, or other fungi. These nutritional requirements have precluded their isolation into

---

Alexander Idnurm, Department of Molecular Genetics and Microbiology, Duke University Medical Center, Durham, NC 27710. **Timothy Y. James,** Department of Biology, Duke University, Durham, NC 27708. **Rytas Vilgalys,** Department of Biology, Duke University, Durham, NC 27708.

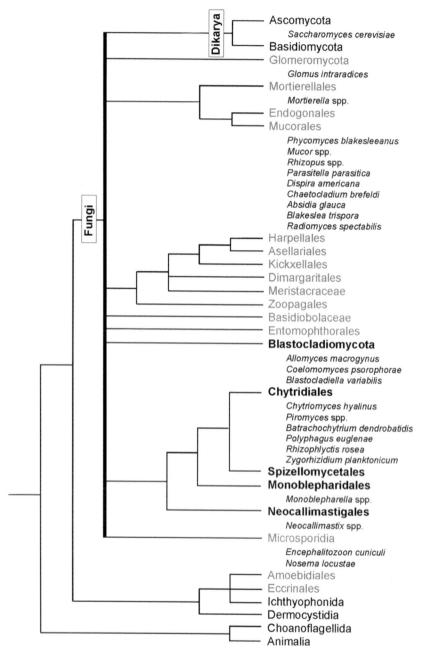

Figure 24.1 A phylogenetic tree of the fungal kingdom, illustrating both diversity and ambiguity of relationships, with species discussed in the text listed. While some groups such as the Dikarya are well supported by phylogenetic studies, the relationships at the base of the tree remain ambiguous (represented by the thick line). Previously, taxa in gray have been classified as members of the Zygomycota and taxa in bold have been classified as members of the Chytridiomycota. Tree based on references 31 and 84.

axenic culture. The primary reason we know little about these fungi is that they have been largely overlooked by mycologists since the dawn of the molecular era because few have been identified as model systems to investigate general aspects of biology. Further, at present the absence

of a system for genetic transformation, lack of mating in some lineages (e.g., Chytridiales), lack of dioecious species (e.g., Blastocladiales), or zygospore dormancy (e.g., Mucorales) has reduced their attractiveness for mating studies at the molecular level. Nonetheless, in addition to providing an understanding on the evolution of sex in the kingdom, these groups warrant study since representatives can cause disease, can be used in biotechnological and food industries (91), and were used in the earliest studies of mating in fungi.

Both chytrids and zygomycetes are capable of causing diseases in humans and other animals. A number of zygomycete fungi cause human diseases that are particularly difficult to treat with current antifungal agents (14, 69). The decline of amphibians worldwide has been attributed to various factors, including anthropogenic disturbance and a disease caused by a chytrid, *Batrachochytrium dendrobatidis* (66, 85). One zygomycete, *Mucor amphibiorum*, causes disease in both endemic Australian species like the platypus and introduced species like the cane toad. Curiously, there is increasing evidence for a skew between virulence and distribution of + mating type of *M. amphibiorum* that may also be seen in plant-pathogenic *Mucor piriformis* (48, 79). Such mating-type allele skews are also observed in human pathogenic ascomycetes and basidiomycetes (24).

The earliest research on and advances in understanding of the mating properties of fungi were conducted on Zygomycota and Chytridiomycota members. The terms "heterothallism" and "homothallism," originally coined for the Mucoralean zygomycete fungi such as *Rhizopus* (5), were quickly adopted to describe fungal mating systems across the kingdom. The first demonstration of a diffusible pheromone involved in mating was also in a zygomycete, *Mucor mucedo* (6). The first purified, characterized, and artificially synthesized pheromone was sirenin from the chytrid *Allomyces macrogynus* (44–46). The zygomycetes have some remarkable biological features that are linked to mating. For example, only two naturally occurring systems of genetic transformation have been demonstrated: from *Agrobacterium tumefaciens* and related bacteria to host plants, and from three mycoparasitic zygomycetes *Chaetocladium brefeldi*, *Dispira americana*, and *Parasitella parasitica* to other species of Mucorales. For these fungi, the DNA transfer is a modified form of mating, such that only other Mucorales species of the opposite mating type are susceptible to the fungus (75, 91). Both nuclear encoded genes and autonomously replicating plasmids enter the host, probably at least into the nuclei, but the transgenes are unstable with passage of the fungus (36).

Finally, there is an anthropocentric reason to study these fungi. Although the ascomycetes and basidiomycetes comprise an estimated 96 to 97% of the species diversity in the Fungi (37), they are likely to represent only a fraction of the phylogenetic or genetic diversity (10). Given the vital role fungi have played in human nutrition and health (from bread to penicillin), what pharmaceutical resources remain untapped in this diversity? A drug discovery made in these little-studied species would be greatly enhanced by a better understanding of their basic biology. For instance, the commonly used anticancer agent camptothecin, worth over a billion dollars a year, is purified from seeds and bark of two tree species, making the drug extremely expensive to produce (41). Last year a zygomycetous fungus associated with the bark of one such tree was isolated and shown to produce this metabolite, thereby providing a new ecologically sound source for fermentation production (67).

WHAT IS KNOWN ABOUT SEX IN THE BASAL FUNGI?

Most of the knowledge of sex in basal fungi comes from studies of behaviors of axenic cultures or careful observations made using light microscopy on natural substrates for these fungi. Almost nothing is known about the molecular genetics of many of these fungi, and even less about the molecular mechanisms governing mating. However, these groups have benefited from extensive biochemical analyses that elucidated the nature of the pheromones in *A. macrogynus* (a chytrid) and the Mucorales fungi. The advances of whole-genome sequences for these organisms promise to enable the molecular genetic basis of sex determination to catch up with the biochemical discoveries, many of which predate by decades what is now well-established knowledge on mating systems in ascomycetes and basidiomycetes. The following sections discuss historical landmarks in understanding sexuality of these groups and what is known about the sexuality of the chytrids and zygomycetes.

Chytridiomycota

Chytridiomycetes have long been recognized to be divisible into several major groups based on life cycles and sexual mechanisms (78). The most distinctive group of Chytridiomycetes is the Blastocladiales that exhibit an alternation of generations between haploid and diploid thalli and may also exhibit dimorphic gametangia and anisogamy, whereby a larger female motile spore fuses with a smaller male motile spore. Filamentous Monoblepharidales (e.g., Monoblepharella) undergo oogamy in which a motile male gamete fuses with a nonmotile female gamete. In the large, heterogeneous group of the Chytridiales, plasmogamy can occur by fusion of zoospores, by fusion of differentiated gametangia, or by fusion of somatic structures. Sexuality in the two most recently established orders, Spizellomycetales and Neocallimastigales, has not been described. Despite some knowledge of the mating behavior in chytrids, sexuality in most species and genera is largely unknown, and very little research has been accomplished since Sparrow's monograph was published in 1960 (78).

A major issue with understanding the sexuality and life cycle of Chytridiomycetes is the ploidy level of various life history stages and the determination of where in the life cycle meiosis occurs. Chytridiomycetes, other than Blastocladiales, are described in textbooks as being haploid for the predominant part of their life cycle, with diploidy exclusive to the zygote. There is recent molecular evidence that some chytrids (e.g., *Batrachochytrium dendrobatidis*) may be vegetatively diploid and possibly asexual (40, 52). Knowledge of the ploidy of the species under investigation is a prerequisite for mating studies, yet for most chytrids ploidy is unknown. However, in at least the Chytridialean fungus, *Zygorhizidium planktonicum*, Doggett and Porter demonstrate zygotic meiosis (18), suggesting a largely haploid life history. Further clarification of the ploidy of most chytridiomycetes awaits a thorough and systematic approach using molecular techniques.

Chytridiomycetes are irregularly septate or coenocytic; however, they do undergo controlled cell divisions during spore production. Chytridiomycetes reproduce by two types of sporangia, thin-walled zoosporangia that cleave into asexually produced zoospores (Fig. 24.2d) and thick-walled resting sporangia (or resting spores) that may be asexually or sexually produced (Fig. 24.2f). Sexual processes (plasmogamy) in the Chytridiales are always known to result in the formation of a resting spore (Fig. 24.2e). After a period of dormancy, the resting spore germinates and produces a germ sporangium from an outgrowth or more rarely may be directly converted to a sporangium. Karyogamy may occur in the resting spore (*Chytriomyces hyalinus*) or in the germ sporangium (*Polyphagus euglenae*) (78). Meiosis is believed to occur within the resting spore (82) or to be coincident with germination of the resting spore (18). In species that undergo fusion of motile gametes, zygotes are presumably of a single diploid genotype, as each contributing zoospore is uninucleate. In at least some species of Chytridiales that undergo conjugation of vegetative structures, only a single nucleus from each contributing thallus migrates into the resting spores to create the diploid fusion nucleus (18, 50). In contrast, asexually produced resting spores of the rumen chytrid *Neocallimastix* sp. were shown to contain numerous, apparently diploid, nuclei (92).

Few data exist on whether the majority of chytridiomycetes are homothallic or heterothallic. By demonstration that single-spore-derived cultures are able to undergo plasmogamy and form sexual structures, *Allomyces* spp., *Chytriomyces hyalinus*, and *Zygorhizidium planktonicum* (18, 19, 51) all have the ability to undergo homothallic reproduction. In the case of at least *Allomyces* spp., it has been clearly demonstrated that isolates are

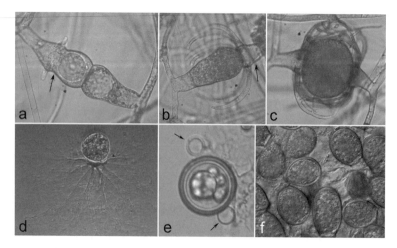

Figure 24.2 Sexual and vegetative structures of zygomycetes and chytridiomycetes. (a through c) Mating and zygospore formation in the homothallic species *Radiomyces spectabilis*: fusion of undifferentiated gametangia; the arrow indicates septum delimiting one gametangium (a); young zygospore produced between two suspensors, only one of which is appendaged (arrow) (b); mature zygospore (c). (d) Asexual zoosporangium of *Chytriomyces hyalinus* subtended by filamentous rhizoids. (e) Resting spore of *C. hyalinus*. Arrows indicate the two empty zoospore cysts which presumably donated nuclei to the resting spore. (f) Thick-walled resting spores of *Allomyces arbusculus*.

also able to outcross as even interspecific hybrids may be formed (20). Obligate heterothallism is suggested in some species such as *Rhizophlyctis rosea* (15), *Dictyomorpha dioica* (53), and *Coelomomyces psorophorae* (87), in which isolates cultured separately failed to produce resting spores, but certain crosses in which pairs of isolates were cocultured led to the production of resting spores. Filamentous species such as *Allomyces* and *Monoblepharella* are monoecious and produce both female and male gametangia on the same mycelium. There are a few documented cases of dioecism, such as in *Blastocladiella variabilis* and *Zygorhizidium planktonicum*, in which the male and female gametophytes are of different color or different size. *B. variabilis* is obligately outcrossing (28), whereas *Z. planktonicum* can self-fertilize (18). *Z. planktonicum*, in which both male and female gametangia can be produced in the same single-sporangium-derived, and presumably haploid, culture, is particularly interesting as sex determination is phenotypic rather than genotypic. In *C. psorophorae*, Whisler et al. (87) used the "+ and −" mating-type terminology, but there has been no clear demonstration that a mating-type locus exists in any chytrid fungus by investigation of the segregation of genetic factors at meiosis. In fact, the pattern of sexuality in some chytrids appears to be distinct from that of the Dikarya. Namely, thalli that are genetically identical may develop into male or female gametangia that produce gametes capable of undergoing fusion. In the absence of mating-type switching, it is hard to envision the control of such a process by mating-type loci, unless the strains are diploid and heterozygous for the factors controlling sex determination.

In order for planogametic (motile) copulation to occur in water molds, gametes must be able to use a chemoattractant to locate and swim towards a mate. Pheromones involved in gamete attraction have been characterized in *Allomyces macrogynus*. The female pheromone, which acts only on male gametes and at very low concentrations (10^{-10} M), was identified over 50 years ago and named sirenin after the Sirens of Greek mythology whose attractive song drew sailors to their death (9, 43). The active molecule was purified from large volume cultures (45), its sesquiterpene structure was subsequently elucidated (44, 56), and the efficiency of chemoattraction for various synthesized analogs was determined (65). It is worth remembering that these discoveries were made decades before the peptide pheromones used in regulating mating in the Ascomycota and Basidiomycota were identified. A hormone (parasin) is also produced by male gametes, but its structure is unknown (63). The mechanism of action of sirenin to control chemotaxis has not been fully eluci-

dated, although exposure to sirenin causes a 2.5-fold-increased uptake in Ca^{2+} ions into male gametes, suggesting that calcium signaling may be involved (64). In addition to the analysis of sirenin as a pheromone, the mating properties of *Allomyces* species have also been extensively studied through light and electron microscopy (19, 38, 62, 76). Fusion of gametes occurs at the posterior (flagellar) end to create a dinucleate cell, with some ambiguity as to how soon the nuclei fuse, that may reflect differences in the species used in these studies (29, 61). Thus, while there is excellent knowledge on pheromones and cell morphology, at present little is known at the molecular level in *A. macrogynus* as to how these pheromone molecules are synthesized and detected and how the signal is transduced to affect gamete tropism.

Blastocladiales are exceptional among Fungi by having a life cycle with alternation of generations in which a diploid sporophyte can undergo extensive vegetative growth including asexual reproduction. Mating structures have been observed in a number of species, including fossilized specimens morphologically similar to extant Blastocladiales species (68). From an evolutionary perspective, the Blastocladiales are very distinctive from other chytrid groups, and a new phylum has recently been proposed for these fungi (32, 77). By analogy to other fungi, the other orders of chytridiomycetes were presumed to be vegetatively haploid, with the diploid phase restricted only to the zygote. Evidence regarding the ploidy level of the vegetative phases of most chytrids is lacking; however, one study on the frog pathogen *Batrachochytrium dendrobatidis* suggested that this fungus reproduces solely clonally and is vegetatively diploid (52). This inference was based on the observation of fixed heterozygous multilocus sequence markers among a global collection of strains. These data argue for a rethinking of the evolution of ploidy in basal fungi and beg for further investigations on ploidy level and population genetics of chytrids.

Zygomycota

The Zygomycota comprise nine orders (4), eight of which include species that undergo homothallic or heterothallic sex. Among the Zygomycota, the Mucorales are the best studied: it is generally inferred that similar patterns of mating occur in the other zygomycetes, although this may not be the case. Mating in heterothallic species (Fig. 24.2 and 24.3) is characterized by the conjugation of two hyphae that are usually morphologically indistinguishable (Fig. 24.2a). In homothallic species, two hyphae originating from a single mycelium undergo fusion (Fig. 24.2b). Mating results in the formation of a

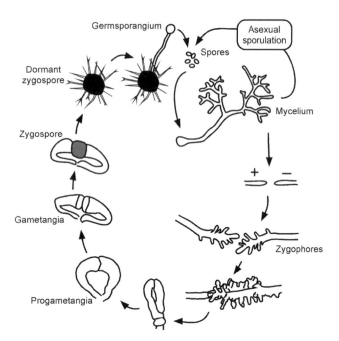

Figure 24.3 Diagram of the sexual cycle of the heterothallic fungus *Phycomyces blakesleeanus* (Mucorales, Zygomycota). Based on reference 12.

pigmented zygospore that is often decorated with spines on its surface or surrounded by decorative structures derived from the parental hyphal cells (Fig. 24.2c). The zygospores usually remain dormant for months to over a year. This aspect of zygomycete biology slowed research on mating properties until dormancy was understood and could be taken into consideration in experimental systems. Nevertheless, what happens within the zygospore still remains a black box (or more literally a black sphere). It appears that at least in *Phycomyces blakesleeanus*, of the many nuclei that first form the zygospore, only one pair of nuclei eventually undergo fusion to yield a zygote. The zygospore germinates into a single sporangium, similar in morphology to the sporangia formed during asexual reproduction from the mycelium. The germsporangiophore contains multiple germspores derived from mitotic events after the reduction of a single diploid nucleus, although exceptions do occur to give progeny from a single zygospore that must be derived from multiple diploid nuclei (13, 47).

In the Mucorales, species within a single genus can exhibit both homothallism and heterothallism. The +/− nomenclature was defined to describe arbitrarily two parental strains that cannot be distinguished based on morphology (5, 72). Because there is cross-reactivity in mating reactions between species, + or − designations are made based on pairings with +/− tester strains

of *Rhizopus nigricans*. Presumably most zygomycetes use the same diffusible hormones, discussed below, for gametangial attraction.

The Mucorales and many Zygomycota are coenocytic and thus contain more than one nucleus type from each parent. The coenocytic condition may facilitate development of homothallism from heterothallic species. Indeed, this has been tested experimentally by fusing protoplasts of + and − mating-type individuals in attempts to create such homothallic strains. Such strains have been produced in two species of *Mucor* and in *Absidia glauca* (58, 90). However, other factors can limit sexual development, as similar experiments fusing + and − strains in *P. blakesleeanus* result in the production of immature mating structures known as pseudophores (7). As yet, the +/− locus has not been cloned from any species of Mucorales, and its identification will be a key step towards extending understanding of mating in the Fungi.

The possible location of the +/− locus in the zygomycete genome has been investigated through resolution of chromosomal DNA and genetic crosses. The first resolution of chromosomal DNA by pulsed-field gel electrophoresis in a zygomycete, *Absidia glauca*, suggested that chromosome length differences may exist between mating types (34). An analysis of 10 strains of *Mucor circinelloides* further supported a correlation in chromosomal DNA sizes and mating type (17). This information would suggest the existence of possible sex chromosomes in mucoralean fungi. The situation was made further complex by the discovery of a + specific protein, PSSP$_{15}$, that is encoded on a small 1.5-kb circular extrachromosomal DNA fragment (27). The DNA fragment hybridized to three + strains of *A. glauca* but not to three − strains. The + mating type and production of PSSP$_{15}$ are also correlated in the offspring of hybrids created by protoplast fusion (89). There are still no other homologs of this gene in the public databases (including relatives *P. blakesleeanus* and *Rhizopus oryzae*); however, its secreted nature and the presence of eight cysteine residues are reminiscent of hydrophobin organization in other fungi. The final piece of chromosomal evidence for the location of the +/− locus comes from genetic crosses. Mendelian segregation of + and − alleles has been demonstrated in *P. blakesleeanus* (13, 21, 22) and *Rhizopus stolonifer* (25). In *P. blakesleeanus* the locus is linked to the *madE* mutation (conferring reduced phototropism), a lysine auxotrophic marker, and a colonial morphology mutation (60). This discovery suggested that the +/− locus may comprise a defined region of the chromosome, equivalent to the mating-type loci seen in the Dikarya.

The genetic map for the zygomycete *P. blakesleeanus* suggests that the mating process in this fungus is similar to that of the Dikarya, involving fusion of haploid nuclei and meiosis. Due to the inherent restrictions in studying the zygospore, this observation remains to be characterized cytologically. One piece of evidence for meiosis is that levels of recombination during nuclear reduction are higher than would be expected from mitotic recombination alone, at least for the one region in the genome where physical and genetic distance can be compared. In this region, where the *carRA* (encoding a bifunctional enzyme) and *carB* genes required for carotene biosynthesis are adjacent (2), mutations affecting the three enzyme functions appear to be closely linked in genetic segregation analysis (70, 86). However, recent cloning of these genes and characterization of mutations in the parent strains (2, 71) used in segregation analysis suggest that recombination at the 1% level occurs over much shorter physical distances in *P. blakesleeanus* (0.3 to 0.5 kb/centimorgan) than those estimated for meiotic recombination in other fungi (ca. 1 to 10 kb/centimorgan). Further genetic mapping, especially aided by the upcoming genome sequence, should further resolve rates of recombination in this zygomycete.

Unlike the basidiomycetes and ascomycetes that use small-peptide pheromones, the Mucorales use a trisporic acid-based set of pheromones. This property is likely conserved in other zygomycetes as it has also been reported in *Mortierella* species (Mortierellales) (73). The trisporic acids are derived from β-carotene, with the link between carotene and mating having been studied for a century (3, 5, 23, 39). The trisporic acid pheromones are described in chapter 26 (by Johannes Wöstemeyer and Christine Schimek). However, as this is the only area with an understanding at the genetic level of control of mating in the basal fungi, the pertinent points are emphasized again here. Although all strains can initiate synthesis of trisporic acid, completion of the synthesis cannot be achieved by either + or − mating-type strains, alone. Late in the pathway, acids that are secreted by one partner can be modified by the other one and the modified compounds are then recognized as pheromones by the original partner in a chemical courtship. One gene encoding 4-dihydromethyltrisporate dehydrogenase, involved in trisporic acid synthesis, has been characterized (16, 74). The protein was purified based on its activity from a − strain of *Mucor mucedo*, and fragments of the peptide were sequenced (71). This purification enabled the design of primers for amplification of the genomic DNA of the gene that was used to obtain cDNA and genomic DNA clones. Southern blot hybridization studies demonstrated that the gene *TSP1* is also present in other Mucorales species, including *Absidia glauca*, *Blakeslea trispora*, and *Parasitella parasitica*. *TSP1* is present in both + and − strains of *Mucor*

mucedo, clearly indicating that it is not a component of the +/− locus (16). However, other enzymes regulate trisporic acid metabolism, and one hypothesis is that the +/− locus comprises a regulatory and/or enzymatic element for these pheromones. One of the most curious features of the protein encoded by *TSP1* is that it is transcribed and translated at equal levels in + and − strains, yet activity is present only in the − mating type isolate (74). Thus, there is evidence for posttranslational regulation of sexual differentiation in *M. mucedo*.

To summarize what is known about the molecular determinants of mating in the Zygomycota, at present a single gene has been identified that is involved in the trisporic acid pheromone signaling pathway. All information on the mating properties comes from a single order of these fungi (the Mucorales), and thus, one can at present only extrapolate that similar mechanisms may exist in the rest of the phylum. These assumptions must be considered tenuous at best, as the other orders appear to be predominantly if not exclusively homothallic (4) and the phylum Zygomycota, as currently defined, is not monophyletic (88). Nevertheless, we have an excellent understanding of the chemical nature of the trisporic acid-derived pheromones, and the completion of two Mucorales genome sequences promises exciting discoveries into the nature of the other enzymes involved and genes regulated during mating.

Sex in Microsporidia

Microsporidia are obligate intracellular parasites of animals (often insects) and humans, characterized by the absence or remnants of mitochondria and the presence of a specialized structure, the polar tube, with which they infect cells. The microsporidia have highly reduced genomes, and the morphological and genetic divergence of these organisms has meant that placement of this group among other eukaryotes is ambiguous (33, 35). At this point it is still unresolved as to which other fungal phyla are most closely related to microsporidia or if microsporidia are basal to all other fungi (31). Mating in microsporidia is unknown. However, many genera are characterized by a diplokaryotic nuclear stage comprising two similar nuclei that divide synchronously. In some species, the diplokaryotic phase alternates with a monokaryotic phase, suggestive of nuclear fusion and reduction. Furthermore, there is a report of a mating-type locus-like structure comprising two adjacent homeodomain proteins in the genome of the microsporidium *Encephalitozoon cuniculi* (8). Thus, at present a sexual cycle for microsporidia remains a mystery and will be difficult to further analyze due to the obligate pathogenic lifestyle of these organisms.

SEXUAL DIFFERENCES BETWEEN THE DIKARYA AND BASAL FUNGI

Several features differ between mating in the Dikarya and mating in the chytridiomycetes and zygomycetes. The last two groups use nonpeptide pheromones, lack a vegetative phase with dikaryotic cells, and produce thick-walled zygospores or resting spores following plasmogamy. The origin of peptide pheromones may have been an adaptation for terrestrial growth in Dikarya and may reflect the metabolic cost of a carbohydrate versus a peptide signal in limited aqueous environments. It is apparent that the production of pheromones is a complex process requiring multiple biochemical modifications of either trisporic acids or a small peptide (26).

Another major difference between the groups is the general lack of macroscopic fruiting structures. Massive sexual reproduction through fructification is a characteristic of higher fungi represented by mushrooms and toadstools (Basidiomycota) or morels and truffles (Ascomycota). This in part may be attributed to structural support gained from the dikaryotic phase of their life cycle. The dikaryotic state, comprising two coordinated dividing nuclei, is prolonged in some basidiomyctes and short in most ascomyctes. One advantage of dikaryons is the ability to fuse the two nuclei and undergo meiosis: thus, one single cell fusion event between two strains can give rise to progeny deriving from multiple meiotic events, while in the basal lineages one cell fusion event leads to a single meiotic event and a maximum of four recombinant progeny. A distinguishing feature of many basidiomyctes is the presence of clamp connections to maintain the dikaryotic state. Curiously, during zygospore production in a number of homothallic Mucorales species apposed suspensors structures are produced that are reminiscent of clamps, including the production of a septum (which is a rare event in this order). It is tempting to speculate that this structure is the precursor of the basidiomycete clamp connection.

Morphological dioecism exists in some chytrid lineages but is not observed in the zygomycetes, though it is present in some ascomycetes.

No structure analogous to a resting spore is found in higher fungi. They have asci and basidia, but not resistant sporangia encoding dormancy that enforces temporal dissemination, only resistant haploid spores that germinate under favorable conditions. The zygospore is similar to the resting spore of chytrids because it is converted into a sporangium upon germination. Zygospores and resting spores also share a number of functional similarities: they are thick-walled and desication tolerant, they undergo a period of dormancy, and they are associated with limiting nutrients.

As mentioned above, ploidy level in basal fungal species is key to understanding their life cycle yet poorly understood. Very few crossing or mapping studies have been conducted. The observation that at least one chytrid is vegetatively diploid (*B. dendrobatidis*) begs for further investigations into ploidy and demands a reappraisal of where and when meiosis occurs in the proposed life cycles of chytrids. Sexuality in basal fungi is inherently more difficult to study than in most Dikarya because there is no prolonged diploid or dikaryotic life phase. An exception is the Blastocladiales with alternating diploid and haploid thalli (e.g., *Allomyces*), and thus, these species should serve as excellent model systems for studies on mating in basal fungi, with the caveat that these species are homothallic, thereby precluding genetic crosses.

WHY IS LITTLE KNOWN ABOUT THE MOLECULAR CONTROLS OF MATING?

Our lack of knowledge about mating in the basal Fungi stems from several causes. First, there are far too few researchers to cover so many species. More people study mating in *Saccharomyces cerevisiae* than all those studying mating in all chytrids and zygomycetes. While the early mating research was performed with the Chytridiomycota and Zygomycota, starting in the mid-1970s the genetic tractability of model Ascomycota and Basidiomycota and the ease of molecular studies with these fungi enabled them to surpass rapidly the basal fungi. Second, the basal Fungi comprise an enormous diversity, such that one cannot guess necessarily what will happen in one fungal group based on another, as is often possible in the ascomycetes and basidiomycetes. Furthermore, the relationships among these fungi are not well resolved (Fig. 24.1) (83). There is still much debate about whether some taxa should even be considered fungi (e.g., the Microsporidia and Trichomycetes). Many species have been observed by very few researchers, and the taxonomic expertise of these early mycologists is waning. In addition, there is very little information about the population biology and life history of the basal fungi. This information can often be used to provide clues about the mechanisms of reproduction (sexual versus asexual) in the wild. Finally, and perhaps more difficult to address, is whether mating or mating types even exist in these fungi or if the systems represent forms of vegetative incompatibility as seen in the Dikarya or self-incompatibility observed in the plant kingdom. How does one define sex? Sex could be considered the fusion of genetic material from two parents (resulting in a change in chromosome copy number) and the resolution of the fusion product, usually mediated via meiotic reduction, to the initial DNA content. This process becomes challenging to address experimentally in systems such as all chytrids, for which it is still unknown whether the normal vegetative state is haploid or diploid. Even for the best-studied basal fungus (*P. blakesleeanus*) for mating, a leading geneticist, Enrique Cerdá-Olmedo, pointed out that we do not even know if this is a form of meiosis or a form of parasexual mating coupled with mitotic recombination (11, 47).

APPROACHES TOWARDS A MOLECULAR UNDERSTANDING OF SEXUAL DEVELOPMENT

When will more information emerge about chytrid or zygomycete mating? A number of steps need to be taken to increase the probability of success by addressing the issues above. There is a need for increased studies on the basic biology including systematics. Phylogenetic resolution may come through the role of the Fungal Tree of Life (AFTOL) project, which aims to understand the relationships among fungi (31, 42). Whole-genome resources are also emerging for many basal fungal lineages. The first basal fungal genome was sequenced from the microsporidium *Encephalitozoon cuniculi* (33). The zygomycete *Rhizopus oryzae* genome is nearing completion (by the Broad Institute), and the genome sequences of the related Mucoralean species *Phycomyces blakesleeanus* is in progress (Department of Energy Joint Genome Initiative [JGI]). Genome sequencing of the mycorrhizal fungus *Glomus intraradices* is also in progress (JGI), and two chytrids, a *Piromyces* species and the amphibian pathogen *B. dendrobatidis*, are scheduled for sequencing (Broad Institute and JGI), as is a second microsporidium species, *Nosema locustae* (Marine Biological Laboratory, Woods Hole, MA). It is likely that cloning of key genes for mating, at least in the Mucorales fungi, will emerge through map-based identification of the +/− locus in a segregating population. A preliminary genetic map based on mutant markers was generated for *P. blakesleeanus*, including a linkage group with the +/− locus (1, 60). Mapping should be aided by the genome sequence project through more facile development of molecular markers. An alternative approach could be random mutagenesis and cloning by complementation or from DNA sequences flanking insertion mutations. Mutant strains affected in mating were isolated in the zygomycetes *P. blakesleeanus* (80, 81) and *Mucor mucedo* (93, 94) and in the chytrid *Allomyces macrogynus* (55, 59), some over 30 years ago, yet the genes responsible have never been identified. An insertional mutagenesis approach is a challenge given ploidy issues (some chytrids) or nuclear content per spore or hypha (zygomycetes) and because gene transformation in many of these species has never been attempted. Furthermore, transformation has been notoriously difficult in those species in which it has been attempted (49, 57). An RNA interference approach to knock down gene expression with a cDNA library transformed into the fungus, as is possible in *Mucor circinelloides*, may enable another forward genetic approach (54). Thus, while we currently know little about sex determination in the basal fungal lineages, resources and tools are becoming increasingly available to elucidate these mechanisms.

For researchers interested in the issues of mating and sex in the Fungi, the next frontier of discovery is mating systems and sex determination in the basal zygomycetes and chytrids. This is an ideal time to become involved in these discoveries as more genomic resources become available and because the field is wide open, requiring individuals from diverse fields such as biochemistry, functional genomics, bioinformatics, cytology, taxonomy, and descriptive biology. Ideally, we would like to reconstruct the entire molecular genetic mechanism that occurs, including gametangia formation, control of plasmogamy, and regulation of gene expression differences in haploids versus diploids. Hopefully we may soon learn whether a mating-type locus exists, what it encodes, and what genes the locus regulates. Further studies of basal fungal lineages promise to yield important clues about the evolution of sex-determining mechanisms in the Fungi and other Opisthokont lineages.

A.I. thanks his postdoctoral mentor Joseph Heitman for encouraging zygomycete research, and NIAID R01 grant AI063443 (to J. Heitman) for financial support. Research in R.V.'s laboratory is partially supported by NSF's Assembling the Tree of Life initiative (DEB-0228668).

References

1. **Alvarez, M. I., M. I. Peláez, and A. P. Eslava.** 1980. Recombination between ten markers in *Phycomyces. Mol. Gen. Genet.* **179:**447–452.

2. **Arrach, N., R. Fernández-Martín, E. Cerdá-Olmedo, and J. Avalos.** 2001. A single gene for lycopene cyclase, phtoene synthase, and regulation of carotene biosynthesis in *Phycomyces. Proc. Natl. Acad. Sci. USA* **98:**1687–1692.

3. **Barnett, H. L., V. G. Lilly, and R. F. Krause.** 1956. Increased production of carotene by mixed + and − cultures of Choanephora cucurbitarum. *Science* **123:**141.

4. **Benny, G. L., R. A. Humber, and J. B. Morton.** 2001. Zygomycota: Zygomycetes, p. 113–146. *In* D. J. McLaughlin, E. G. McLaughlin, and P. A. Lemke (ed.), *The Mycota VIIA, Systematics and Evolution.* Springer-Verlag, Berlin, Germany.

5. **Blakeslee, A. F.** 1904. Sexual reproduction in the Mucorineae. *Proc. Am. Acad. Arts Sci.* **40:**205–319.

6. **Burgeff, H.** 1924. Untersuchungen über Sexualität under Parasititismus bei Mucorineen. I. *Bot. Abh. Heft* **4:** 5–135.

7. **Burgeff, H.** 1914. Untersuchungen über Variabilität, Sexualität und Erblichkeit bei Phycomyces nitens Kuntze. *Flora* **107:**259–316.

8. **Bürglin, T. R.** 2003. The homeobox genes of *Encephalitozoon cuniculi* (Microsporidia) reveal a putative mating-type locus. *Dev. Genes Evol.* **213:**50–52.

9. **Carlile, M. J., and L. Machlis.** 1965. The response of male gametes of Allomyces to the sexual hormone sirenin. *Am. J. Bot.* **52:**478–483.

10. **Cavalier-Smith, T.** 2001. What are fungi?, p. 3–37. *In* P. A. Lemke (ed.), *The Mycota*, vol. 7A. Springer-Verlag, New York, NY.

11. **Cerdá-Olmedo, E.** 1974. *Phycomyces*, p. 343–357. *In* R. C. King (ed.), *Handbook of Genetics*. Plenum Press, New York, NY.

12. **Cerdá-Olmedo, E.** 2001. *Phycomyces* and the biology of light and color. *FEMS Microbiol. Rev.* **25:**503–512.

13. **Cerdá-Olmedo, E.** 1975. The genetics of *Phycomyces blakesleeanus*. *Genet. Res.* **25:**285–296.

14. **Chayakulkeeree, M., M. A. Ghannoum, and J. R. Perfect.** 2006. Zygomycosis: the re-emerging fungal infection. *Eur. J. Clin. Microbiol. Infect. Dis.* **25:**215–229.

15. **Couch, J. N.** 1939. Heterothallism in the Chytridiales. *J. Elisha Mitchell Sci. Soc.* **55:**409–414.

16. **Czempinski, K., V. Kruft, J. Wöstemeyer, and A. Burmester.** 1996. 4-Dihydromethyltrisporate dehydrogenase from *Mucor mucedo*, an enzyme of the sexual hormone pathway: purification, and cloning of the corresponding gene. *Microbiology* **142:**2647–2654.

17. **Díaz-Mínguez, J. M., M. A. López-Matas, and A. P. Eslava.** 1999. Complementary mating types of *Mucor circinelloides* show electrophoretic karyotype heterogeneity. *Curr. Genet.* **36:**383–389.

18. **Doggett, M. S., and D. Porter.** 1996. Sexual reproduction in the fungal parasite, *Zygorhizidium planktonicum*. *Mycologia* **88:**720–732.

19. **Emerson, R.** 1941. An experimental study on the life cycles and taxonomy of *Allomyces*. *Lloydia* **4:**77–144.

20. **Emerson, R., and C. M. Wilson.** 1954. Interspecific hybrids and the cytogenetics and cytotaxonomy of Euallomyces. *Mycologia* **46:**393–434.

21. **Eslava, A. P., M. I. Alvarez, P. V. Burke, and M. Delbrück.** 1975. Genetic recombination in sexual crosses of Phycomyces. *Genetics* **80:**445–462.

22. **Eslava, A. P., M. I. Alvarez, and M. Delbrück.** 1975. Meiosis in Phycomyces. *Proc. Natl. Acad. Sci. USA* **72:** 4076–4080.

23. **Feofilova, E. P.** 2006. Heterothallism of Mucoraceous fungi: a review of biological implications and uses in biotechnology. *Appl. Biochem. Microbiol.* **42:**439–454.

24. **Fraser, J. A., and J. Heitman.** 2006. Sex, *MAT*, and the evolution of fungal virulence, p. 13–33. *In* J. Heitman, S. G. Filler, J. E. Edwards, Jr., and A. P. Mitchell (ed.), *Molecular Principles of Fungal Pathogenesis*. ASM Press, Washington, DC.

25. **Gauger, W. L.** 1977. Meiotic gene segregation in *Rhizopus stolonifer*. *J. Gen. Microbiol.* **101:**211–217.

26. **Gooday, G. W., and D. J. Adams.** 1993. Sex hormones and fungi. *Adv. Microb. Physiol.* **34:**69–145.

27. **Hänfler, J., H. Teepe, C. Weigel, V. Kruft, R. Lurz, and J. Wöstemeyer.** 1992. Circular extrachromosomal DNA codes for a surface protein in the (+) mating type of the zygomycete *Absidia glauca*. *Curr. Genet.* **22:**319–325.

28. **Harder, R., and G. Sörgel.** 1938. Über einen neuen plano-isogamen Phycomyceten mit Generationswechsel und seine phylogenetische Bedeutung. *Nachrichten Gesell. Wiss. Göttingen, Mat.-Physik Kl., Fachgruppe VI (Biol.) (N.F.)* **3.**

29. **Hatch, W. R.** 1938. Conjugation and zygote germination in *Allomyces arbuscula*. *Ann. Bot.* **2:**583–614.

30. **Hawksworth, D. L.** 2001. The magnitude of fungal diversity: the 1.5 million species estimate revisited. *Mycol. Res.* **105:**1422–1432.

31. **James, T. Y., F. Kauff, C. Schoch, P. B. Matheny, V. Hofstetter, C. Cox, G. Celio, C. Gueidan, E. Fraker, J. Miadlikowska, H. T. Lumbsch, A. Rauhut, V. Reeb, A. E. Arnold, A. Amtoft, J. E. Stajich, K. Hosaka, G.-H. Sung, D. Johnson, B. O'Rourke, M. Crockett, M. Binder, J. M. Curtis, J. C. Slot, Z. Wang, A. W. Wilson, A. Schüßler, J. E. Longcore, K. O'Donnell, S. Mozley-Standridge, D. Porter, P. M. Letcher, M. J. Powell, J. W. Taylor, M. M. White, G. W. Griffith, D. R. Davies, R. A. Humber, J. B. Morton, J. Sugiyama, A. Y. Rossman, J. D. Rogers, D. H. Pfister, D. Hewitt, K. Hansen, S. Hambleton, R. A. Shoemaker, J. Kohlmeyer, B. Volkmann-Kohlmeyer, R. A. Spotts, M. Serdani, P. W. Crous, K. W. Hughes, K. Matsuura, E. Langer, G. Langer, W. A. Untereiner, R. Lücking, B. Büdel, D. M. Geiser, A. Aptroot, P. Diederich, I. Schmitt, M. Schultz, R. Yahr, D. Hibbett, F. Lutzoni, D. McLaughlin, J. Spatafora, and R. Vilgalys.** 2006. Reconstructing the early evolution of Fungi using a six-gene phylogeny. *Nature* **443:**818–822.

32. **James, T. Y., P. M. Letcher, J. E. Longcore, S. E. Mozley-Standridge, D. Porter, M. J. Powell, G. W. Griffith, and R. Vilgalys.** 2006. A molecular phylogeny of the flagellated Fungi (Chytridiomycota) and description of a new phylum (Blastocladiomycota). *Mycologia* **98:**860–871.

33. **Katinka, M. D., S. Duprat, E. Cornillot, G. Méténier, F. Thomarat, G. Prensier, V. Barbe, E. Peyretaillade, P. Brottier, P. Wincker, F. Delbac, H. El Alaoui, P. Peyret, W. Saurin, M. Gouy, J. Weissenbach, and C. P. Vivarès.** 2001. Genome sequence and gene compaction of the eukaryote parasite *Encephalitozoon cuniculi*. *Nature* **414:** 450–453.

34. **Kayser, T., and J. Wöstemeyer.** 1991. Electrophoretic karyotype of the zygomycete *Absidia glauca*: evidence for differences between mating types. *Curr. Genet.* **19:**279–284.

35. **Keeling, P. J.** 2003. Congruent evidence from α-tubulin and β-tubulin gene phylogenies for a zygomycete origin of microsporidia. *Fungal Genet. Biol.* **38:**298–309.

36. **Kellner, M., A. Bumester, A. Wöstemeyer, and J. Wöstemeyer.** 1993. Transfer of genetic information from the mycoparasite *Parasitella parasitica* to its host *Absidia glauca*. *Curr. Genet.* **23:**334–337.

37. Kirk, P. M., P. F. Cannon, J. C. David, and J. A. Staplers (ed.). 2001. *Ainsworth & Bisby's Dictionary of the Fungi*, 9th ed. CAB International, Wallingford, United Kingdom.

38. Kneip, H. 1929. *Allomyces jaranicus* n. sp. ein anisogamer Phycomycet mit Planogameten. *Ber. Dtsch. Bot. Ges.* 47:199–212.

39. Kuzina, V., and E. Cerdá-Olmedo. 2006. Modification of sexual development and carotene production by acetate and other small carboxylic acids in *Blakeslea trispora* and *Phycomyces blakesleeanus*. *Appl. Environ. Microbiol.* 72:4917–4922.

40. Liu, Y., M. Hodson, and B. Hall. 2003. Heterozygosity in chytrids. *Fungal Genet. Newsl.* 50(Suppl.):459.

41. Lorence, A., and C. L. Nessler. 2004. Camptothecin, over four decades of surprising findings. *Phytochemistry* 65:2735–2749.

42. Lutzoni, F., F. Kauff, C. J. Cox, D. McLaughlin, G. Celio, B. Dentinger, M. Padamsee, D. Hibbett, T. Y. James, E. Baloch, M. Grube, V. Reeb, V. Hofstetter, C. Schoch, A. E. Arnold, J. Miadlikowska, J. Spatafora, D. Johnson, S. Hambleton, M. Crockett, R. Shoemaker, S. Hambleton, M. Crockett, R. Shoemaker, G.-H. Sung, R. Lücking, T. Lumbsch, K. O'Donnell, M. Binder, P. Diederich, D. Ertz, C. Gueidan, K. Hansen, R. C. Harris, K. Hosaka, Y.-W. Lim, B. Matheny, H. Nishida, D. Pfister, J. Rogers, A. Rossman, I. Schmitt, H. Sipman, J. Stone, J. Sugiyama, R. Yahr, and R. Vilgalys. 2004. Assembling the fungal tree of life: progress, classification, and evolution of subcellular traits. *Am. J. Bot.* 91:1446–1480.

43. Machlis, L. 1958. Evidence for a sexual hormone in *Allomyces*. *Physiol. Plantarum* 11:181–192.

44. Machlis, L., W. H. Nutting, and H. Rapoport. 1968. The structure of sirenin. *J. Am. Chem. Soc.* 90:1674–1676.

45. Machlis, L., W. H. Nutting, M. W. Williams, and H. Rapoport. 1966. Production, isolation, and characterization of sirenin. *Biochemistry* 5:2147–2152.

46. Mandai, T., K. Hara, M. Kawada, and J. Nokami. 1983. A new total synthesis of DL-sirenin. *Tetrahedron Lett.* 24:1517–1518.

47. Mehta, B. J., and E. Cerdá-Olmedo. 2001. Intersexual partial diploids of Phycomyces. *Genetics* 158:635–641.

48. Michailides, T. J., and R. A. Spotts. 1986. Mating types of *Mucor piriformis* isolated from soil and pear fruit in Oregon orchards. *Mycologia* 78:766–770.

49. Michielse, C. B., K. Salim, P. Ragas, A. F. J. Ram, B. Kudla, B. Jarry, P. J. Punt, and C. A. M. J. J. van den Hondel. 2004. Development of a system for integrative and stable transformation of the zygomycete *Rhizopus oryzae* by *Agrobacterium*-mediated DNA transfer. *Mol. Genet. Genomics* 271:499–510.

50. Miller, C. E., and D. P. Dylewski. 1981. Syngamy and resting body development in *Chytriomyces hyalinus* (Chytridiales). *Am. J. Bot.* 68:342–349.

51. Moore, E. D., and C. E. Miller. 1973. Resting body formation by rhizoidal fusion in *Chytriomyces hyalinus*. *Mycologia* 65:145–154.

52. Morehouse, E. A., T. Y. James, A. R. Ganley, R. Vilgalys, L. Berger, P. J. Murphy, and J. E. Longcore. 2003. Multilocus sequence typing suggests the chytrid pathogen of amphibians is a recently emerged clone. *Mol. Ecol.* 12:395–403.

53. Mullins, J. T. 1961. The life cycle and development of *Dicytomorpha gen. nov.* (formerly *Pringsheimiella*), a genus of the aquatic fungi. *Am. J. Bot.* 48:377–387.

54. Nicolás, F. E., S. Torres-Martínez, and R. M. Ruiz-Vázquez. 2003. Two classes of small antisense RNAs in fungal RNA silencing triggered by non-integrative transgenes. *EMBO J.* 22:3983–3991.

55. Nielsen, T. A. B., and L. W. Olson. 1982. Nuclear control of sexual differentiation in Allomyces macrogynus. *Mycologia* 74:303–312.

56. Nutting, W. H., H. Rapoport, and L. Machlis. 1968. The structure of sirenin. *J. Am. Chem. Soc.* 90:6434–6438.

57. Obraztsova, I. N., N. Prados, K. Holzmann, J. Avalos, and E. Cerdá-Olmedo. 2004. Genetic damage following introduction of DNA in *Phycomyces*. *Fungal Genet. Biol.* 41:168–180.

58. Ohnuki, T., Y. Etoh, and T. Beppu. 1982. Intraspecific and interspecific hybridization of *Mucor pusillus* and *Mucor miehei* by protoplast fusion. *Agric. Biol. Chem.* 46:451–458.

59. Olson, L. W., T. A. B. Nielsen, H. P. Heldt-Hansen, and N. G. Grant. 1982. Maleness, its inheritance and control in the aquatic Phycomycete Allomyces macrogynus. *Trans. Br. Mycol. Soc.* 78:331–336.

60. Orejas, M., M. I. Peláez, M. I. Alvarez, and A. P. Eslava. 1987. A genetic map of *Phycomyces blakesleeanus*. *Mol. Gen. Genet.* 210:69–76.

61. Pommerville, J. 1982. Morphology and physiology of gamete mating and gamete fusion in the fungus *Allomyces*. *J. Cell Sci.* 53:193–209.

62. Pommerville, J., and M. S. Fuller. 1976. The cytology of the gamete and fertilization of *Allomyces macrogynus*. *Arch. Microbiol.* 109:21–30.

63. Pommerville, J. C., and L. W. Olson. 1987. Evidence for a male-produced pheromone in *Allomyces macrogynus*. *Exp. Mycol.* 11:245–248.

64. Pommerville, J. C., J. B. Strickland, and K. E. Harding. 1990. Pheromone interactions and ionic communication in gametes of aquatic fungus *Allomyces macrogynus*. *J. Chem. Ecol.* 16:121–131.

65. Pommerville, J. C., J. B. Strickland, D. Romo, and K. E. Harding. 1988. Effects of analogs of the fungal sexual pheromone sirenin on male gamete motility in *Allomyces macrogynus*. *Plant Physiol.* 88:139–142.

66. Pounds, J. A., M. R. Bustamante, L. A. Coloma, J. A. Consuegra, M. P. L. Fogden, P. N. Foster, E. La Marca, K. L. Masters, A. Merino-Viteri, R. Puschendorf, S. R. Ron, G. A. Sánchez-Azofeifa, C. J. Still, and B. E. Young. 2006. Widespread amphibian extinctions from epidemic disease driven by global warming. *Nature* 439:161–167.

67. Puri, S. C., V. Verma, T. Amna, G. N. Qazi, and M. Spiteller. 2005. An endophytic fungus from *Nothapodytes foetida* that produces camptothecin. *J. Nat. Prod.* 68:1717–1719.

68. Remy, W., T. N. Taylor, and H. Hass. 1994. Early Devonian fungi: Blastocladalean fungus with sexual reproduction. *Am. J. Bot.* 81:690–702.

69. Ribes, J. A., C. L. Vanover-Sams, and D. J. Baker. 2000. Zygomycetes in human disease. *Clin. Microbiol. Rev.* **13:**236–301.

70. Roncero, M. I. G., and E. Cerdá-Olmedo. 1982. Genetics of carotene biosynthesis in *Phycomyces*. *Curr. Genet.* **5:**5–8.

71. Sanz, C., M. I. Alvarez, M. Orejas, A. Velayos, A. P. Eslava, and E. P. Benito. 2002. Interallelic complementation provides genetic evidence for the multimeric organization of the *Phycomyces blakesleeanus* phytoene dehydrogenase. *Eur. J. Biochem.* **269:**902–908.

72. Satina, S., and A. F. Blakeslee. 1929. Criteria of male and female in bread moulds (Mucors). *Proc. Natl. Acad. Sci. USA* **15:**735–740.

73. Schimek, C., K. Kleppe, A.-R. Saleem, K. Voigt, A. Burmester, and J. Wöstemeyer. 2003. Sexual reactions in Mortierellales are mediated by the trisporic acid system. *Mycol. Res.* **107:**736–747.

74. Schimek, C., A. Petzold, K. Schultze, J. Wetzel, F. Wolschendorf, A. Burmester, and J. Wöstemeyer. 2005. 4-Dihydromethyltrisporate dehydrogenase, an enzyme of the sex hormone pathway in *Mucor mucedo*, is constitutively transcribed but its activity is differently regulated in (+) and (−) mating types. *Fungal Genet. Biol.* **42:**804–812.

75. Schultze, K., C. Schimek, J. Wöstemeyer, and A. Burmester. 2005. Sexuality and parasitism share common regulatory pathways in the fungus Parasitella parasitica. *Gene* **348:**33–44.

76. Seale, T. W., and R. B. Runyan. 1979. Scanning electron microscopy of surface ultrastructure changes during meiosporangium maturation and meiospore liberation in the aquatic fungus *Allomyces arbuscula*. *J. Bacteriol.* **140:**276–284.

77. Seif, E., J. Leight, Y. Liu, I. Roewer, L. Forget, and B. F. Lang. 2005. Comparative mitochondrial genomics in zygomycetes: bacteria-like RNase P RNAs, mobile elements and a close source of the group I intron invasion in angiosperms. *Nucleic Acids Res.* **33:**734–744.

78. Sparrow, F. K. 1960. *Aquatic Phycomycetes*, 2nd ed. University of Michigan Press, Ann Arbor.

79. Stewart, N. J., and B. L. Munday. 2005. Possible differences in pathogenicity between cane toad-, frog- and platypus-derived isolates of *Mucor amphibiorum*, and a platypus-derived isolate of *Mucor circinelloides*. *Med. Mycol.* **43:**127–132.

80. Sutter, R. P. 1975. Mutations affecting sexual development in *Phycomyces blakesleeanus*. *Proc. Natl. Acad. Sci. USA* **72:**127–130.

81. Sutter, R. P., A. B. Grandin, B. D. Dye, and W. R. Moore. 1996. (−) mating type-specific mutants of *Phycomyces* defective in sex pheromone biosynthesis. *Fungal Genet. Biol.* **20:**268–279.

82. Sykes, E. E., and D. Porter. 1981. Meiosis in the aquatic fungus *Catenaria allomycis*. *Protoplasma* **105:**307–320.

83. Tanabe, Y., M. Saikawa, M. M. Watanabe, and J. Sugiyama. 2004. Molecular phylogeny of Zygomycota based on EF-1a and RPB1 sequences: limitations and utility of alternative markers to rDNA. *Mol. Phylogenet. Evol.* **30:**438–449.

84. Tanabe, Y., M. M. Watanabe, and J. Sugiyama. 2005. Evolutionary relationships among basal fungi (Chytridiomycota and Zygomycota): insights from molecular phylogenetics. *J. Gen. Appl. Microbiol.* **51:**267–276.

85. Tehler, A., J. S. Farris, D. L. Lipscomb, and M. Kallersjo. 2000. Phylogenetic analyses of the fungi based on large rDNA data sets. *Mycologia* **92:**459–474.

86. Torres-Martínez, S., F. J. Murillo, and E. Cerdá-Olmedo. 1980. Genetics of lycopene cyclization and substrate transfer in beta-carotene biosynthesis in *Phycomyces*. *Genet. Res.* **36:**299–309.

87. Whisler, H. C., S. L. Zebold, and J. A. Shemanchuk. 1975. Life history of *Coelomomyces psorophorae*. *Proc. Natl. Acad. Sci. USA* **72:**693–696.

88. White, M. M., T. Y. James, K. O'Donnell, M. Cafaro, Y. Tanabe, and J. Sugiyama. 2006. Phylogeny of the Zygomycota based on nuclear ribosomal sequence data. *Mycologia* **98:**872–884.

89. Wöstemeyer, A., H. Teepe, and J. Wöstemeyer. 1990. Genetic interactions in somatic inter-mating type hybrids of the zygomycete *Absidia glauca*. *Curr. Genet.* **17:**163–168.

90. Wöstemeyer, J., and E. Brockhausen-Rohdemann. 1987. Inter-mating type protoplast fusion in the zygomycete *Absidia glauca*. *Curr. Genet.* **12:**435–441.

91. Wöstemeyer, J., A. Wöstemeyer, A. Burmester, and K. Czempinski. 1995. Relationships between sexual processes and parasitic interactions in the host-pathogen system *Absidia glauca-Parasitella parasitica*. *Can. J. Bot.* **73**(Suppl. 1):S243–S250.

92. Wubah, D. A., M. S. Fuller, and D. E. Akin. 1991. Resistant body formation in *Neocallimastix* sp., an anaerobic fungus from the rumen of a cow. *Mycologia* **83:**40–47.

93. Wurtz, T., and H. Jockusch. 1978. Morphogenesis in *Mucor mucedo*: mutations affecting gamone response and organ differentiation. *Mol. Gen. Genet.* **159:**249–257.

94. Wurtz, T., and H. Jockusch. 1975. Sexual differentiation in *Mucor*: trisporic acid response mutants and mutants blocked in zygospore development. *Dev. Biol.* **43:**213–220.

Sex in Fungi: Molecular Determination
and Evolutionary Implications
Edited by Joseph Heitman et al.
© 2007 ASM Press, Washington, D.C.

Teresa E. Pawlowska

25

How the Genome Is Organized in the Glomeromycota

INTRODUCTION

Arbuscular mycorrhizal (AM) fungi (phylum Glomeromycota) are one of the most important yet least understood groups of soil microorganisms. Recently, these obligate symbionts of plants have been brought from relative obscurity into the evolutionary spotlight. It was discovered that the oldest known fossils of fungi dating back to the Ordovician closely resemble modern asexual spores of AM fungi (73). Based on this morphological similarity and the lack of compelling evidence of sexual reproduction in extant AM fungi, Glomeromycota are hypothesized to represent an assemblage of ancient clonal lineages (63). This hypothesis challenges the predictions of the evolutionary theory, which asserts that clonal lineages are destined for rapid extinction owing to accumulation of deleterious mutations (60). Starting in the late 1990s, a series of studies has been initiated with the aim of understanding the evolutionary, population, and transmission genetics of Glomeromycota (8, 43, 49, 79, 88, 98). These studies triggered an ongoing explosion of exciting and thought-provoking forays into the evolutionary history and biology of this group of fungi (9, 45, 54, 67, 68, 72, 82, 85, 89). A number of hypotheses on genome organization of AM fungi have emerged offering various explanations for the long-term evolutionary survival of Glo-

meromycota despite their putative asexuality (8, 9, 45, 54, 66–68). In this chapter, I discuss recent findings concerning the genome structure in Glomeromycota in the context of the evolutionary theory predictions.

Obligate biotrophy, horizontal transmission between hosts, and the apparent absence of sexual reproduction are the life history traits that make AM fungi vulnerable to the vicissitudes of selective forces. Despite these limitations, arbuscular mycorrhiza, the mutualistic symbiosis that Glomeromycota form with plant roots, has been a staggering evolutionary and ecological success that dates back to the early Devonian (74) and involves members of 95% of the extant plant families (84). In exchange for plant-assimilated carbon, AM fungi facilitate plant uptake of mineral nutrients from the soil. They also moderate the negative effects of plant pathogens (12), drought (4), and toxic metal stress (39) and may contribute up to 50% of the total soil microbial biomass (64).

Spores as Surrogate Individuals in Glomeromycota

For practical reasons, which include the paucity of morphologically distinct structures (58) and the recalcitrance to in vitro cultivation (32), spore morphology

Teresa E. Pawlowska, Department of Plant Pathology, Cornell University, 334 Plant Science Building, Ithaca, NY 14853-5904.

and development were historically the basis for taxonomy and systematics in Glomeromycota (58). For the same reasons, spores are treated as surrogate individuals in studies devoted to population genetics of these fungi (68, 79, 85, 89).

In the absence of morphological evidence of sexual reproduction in Glomeromycota, spores are believed to be the means of asexual propagation. Spores in AM fungi are uncommonly large and harbor hundreds of nuclei (Fig. 25.1a). In addition, they are characterized by an unusual intraindividual polymorphism of the nuclear ribosomal RNA coding genes. The ribosomal DNA (rDNA) polymorphism has been detected in all AM fungal morphospecies examined to date (for examples, see references 54, 68, and 76). This polymorphism is clearly a product of the genetic processes that have historically shaped the genomes of Glomeromycota and continue to operate in their modern populations. Consequently, genome organization in AM fungi cannot be considered in isolation from the evolutionary and population genetic framework.

Organization of Intrasporal Genetic Variation

Two contrasting patterns of organization of the intraindividual genetic variation have been inferred from the studies of AM fungi (Fig. 25.1b and c): (i) containment of the intraindividual variation in every nucleus resulting in the overall genetic homogeneity of nuclei, referred to as homokaryosis (67, 68); and (ii) distribution of the intraindividual variation among different nuclei, also known as nuclear heterogeneity, heterokaryosis, or multigenomic structure (8, 45, 54). (i) Homokaryosis, while not excluding the possibility of incidental genetic mosaicism due to mutations, would represent a genetic

makeup analogous to that of other multicellular eukaryotic organisms. This genetic makeup is concordant with a clonal as well as a recombinant reproductive mode potentially existing in the extant populations of Glomeromycota. (ii) Heterokaryosis, on the other hand, is interpreted as a product of a novel non-Mendelian genetic process (8, 9, 45). In this process, fusions of genetically nonidentical hyphae are proposed to lead to the formation of a chimeric mycelium that retains and reassorts parental nuclei and distributes them into spores.

THE HYPOTHESIS OF NUCLEAR HOMOGENEITY

Variation Patterns in Protein-Coding Gene Sequences.

The intraindividual genetic polymorphism in Glomeromycota is not limited to rDNA. Other DNA sequences, including some protein-coding genes, exhibit intraindividual variation (54, 67, 68). By analogy to multicellular eukaryotes (36), one can speculate that in the absence of other processes, including sexual or parasexual recombination, mutations alone can generate nuclear mosaicism in the individuals of AM fungi. According to the evolutionary theory, the coexistence of multiple sequence variants generated by neutral mutations in an initially single-copy DNA sequence is expected to depend on the rate at which these mutations occur in every generation and on the size of the nuclear populations (52). The gain of new alleles generated by mutation is likely to be moderated by genetic drift, i.e., random sampling of nuclei from generation to generation. Bottlenecks, periods of reduced population size experienced by the nuclei during spore formation and/or

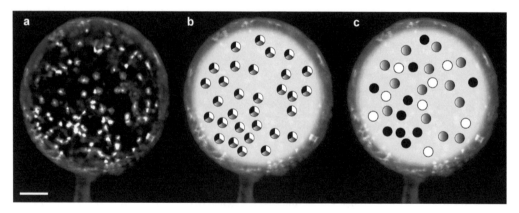

Figure 25.1 Spores of Glomeromycota harbor hundreds of nuclei. (a) Optical section through a spore of *Glomus intraradices* stained with DAPI (4′, 6′-diamidino-2-phenylindole) to visualize nuclei, scale bar 10 μm; (b) model of homokaryosis; (c) model of heterokaryosis.

germination, are also expected to exert negative effects on the extent of the intraindividual allelic variation (94). These principles together with life history data available for a population of *Glomus etunicatum* in a maize field in Berkeley, CA, and data on the extent of polymorphism of genetic markers variable within its individual spores were used by Pawlowska and Taylor to develop a simple simulation model describing variation patterns of genetic markers that would be expected in a heterokaryotic fungus (68). According to the predictions of this model, the extent of the intraindividual polymorphism in the DNA sequence of a genetic marker whose variants reside in different nuclei will diminish over generations owing to genetic drift and bottlenecks suffered by nuclei during their progression through the life cycle. These predictions were then applied to explain the DNA sequence variation exhibited in *G. etunicatum* by a genetic marker referred to as PLS (for *POL1*-like sequence), because of its sequence similarity to a *Saccharomyces cerevisiae* gene encoding the catalytic subunit of DNA polymerase alpha, *POL1*. In contrast to *POL1*, which is a single-locus gene in yeast and other organisms, PLS is present in individual spores of *G. etunicatum* in multiple copies that are phylogenetically differentiated into two distinct clusters, PLS1 and PLS2. Pawlowska and Taylor (68) analyzed patterns of PLS1 variation in multiple individual spores per isolate, as well as among several different isolates of *G. etunicatum*, including isolates from geographically distant locations in California and Minnesota (Fig. 25.2). Spores representing the same isolate contained identical sets of 13 PLS1 variants. While the number of variants remained constant among geographically distinct isolates, their DNA sequences displayed signs of divergence (Fig. 25.2). No independent assortment of any of the variants among the spores was detected. Consequently, the patterns of PLS1 variation were interpreted as consistent with homokaryosis, i.e., containment of the entire intrasporal variation in each nucleus.

An alternative interpretation of the PLS1 data generated by Pawlowska and Taylor (68) was offered by Bever and Wang (9) and by Hijri and Sanders (45). Under this alternative scenario, AM fungal individuals are proposed to exhibit a multigenomic structure with separate nuclei carrying different PLS variants. Such genetically distinct nuclei would be brought together by fusions of hyphae among genetically differentiated individuals. This, in effect, would lead to a complete intermixing of different types of nuclei within a population, and each individual would possess a set of 13 different types of nuclei. The interpretation by Bever and Wang (9) was based on a simulation model of het-

erokaryosis that, in addition to parameters considered by Pawlowska and Taylor (68), included also occurrence of hyphal fusions among genetically differentiated individuals. However, while hyphal fusions between genetically distinct individuals of Glomeromycota are not entirely impossible, no data are currently available that would support their occurrence (37, 38; see also discussion later in the chapter). Consequently, hyphal fusions were not included in the original model of heterokaryosis proposed by Pawlowska and Taylor (67, 68). In contrast to the modeling approach of Bever and Wang (9), Hijri and Sanders (45) offered empirical data on the number of PLS variants per nucleus that support heterokaryosis in *G. etunicatum*. These authors used flow cytometry to obtain an estimate of *G. etunicatum* nuclear DNA content (37.45 Mb) and then utilized this information to infer the number of PLS copies per nucleus by quantitative PCR (45). Using this approach, a maximum of two PLS variants per nucleus has been inferred. This conclusion is in a striking disagreement with the phylogenetic evidence of Pawlowska and Taylor (68), who found that a constant number of 13 PLS1 variants were present in individual spores of *G. etunicatum* isolates from California and Minnesota (Fig. 25.2). If the number of the PLS copies was indeed limited to only one or two per nucleus, spores from geographically distant locations would not be expected to contain exactly 13 PLS1 variants.

While the distribution of the PLS1 variants among the nuclei of *G. etunicatum* remains a point of contention, it should be mentioned that the intraindividual variability of the PLS1 marker is not unique. Another gene, *BiP*, exists in multiple sequence variants per individual in *Glomus intraradices* (54) and in *G. etunicatum* (H. den Bakker and T.E. Pawlowska, unpublished data). The extent of the intraindividual polymorphism of the PLS1 and rRNA genes combined with the uncommonly large genome size estimates available at the time (10, 48) led Pawlowska and Taylor (68) to speculate that genomes of Glomeromycota may be polyploid. However, genomic data emerging from phylogenetic and molecular biology studies of AM fungi, including a model species, *G. intraradices*, suggest that the genes present in multiple variants (24, 25, 54) are rather an exception and most genes are represented by single unique sequences (40, 56). If this is the case, then such genome structure would not be unlike genome structures of other eukaryotes, which in addition to single-copy genes harbor duplicate genes and multicopy gene families.

A solution to the dilemma of how multiple gene copies are distributed among the nuclei of Glomeromycota is expected to emerge from the complete genome sequence of *G. intraradices*, which is anticipated to be

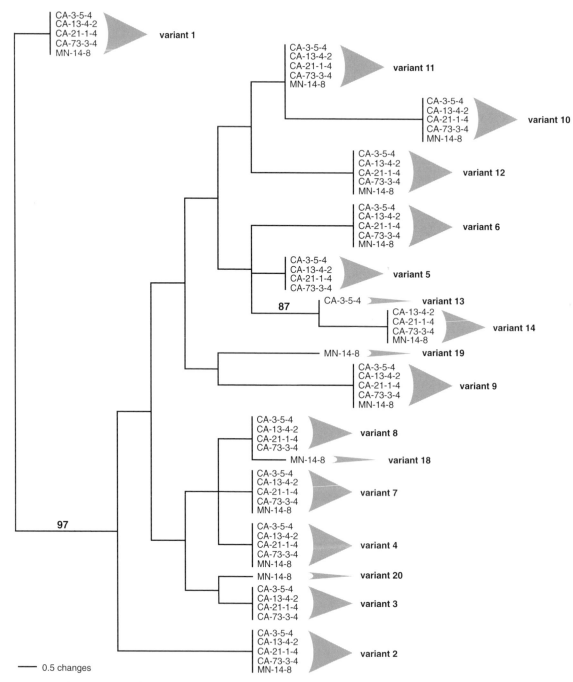

Figure 25.2 A phylogram representing the single most parsimonious tree of the PLS1 variants detected in isolates of *G. etunicatum* from California (CA-3-5-4, CA-13-4-2, CA-21-1-4, and CA-73-3-4) and Minnesota (MN-14-8). The numbering of variants corresponds to the numbers assigned by Pawlowska and Taylor (68). Each isolate contains exactly 13 PLS1 variants (variants 1, 2, 3, 4, 5, 6, 7, 8, 9, 10, 11, 12, and 14 are present in CA-13-4-2, CA-21-1-4, and CA-73-3-4; variants 1, 2, 3, 4, 5, 6, 7, 8, 9, 10, 11, 12, and 13 are present in CA-3-5-4; and variants 1, 2, 4, 6, 7, 9, 10, 11, 12, 14, 18, 19, and 20 are present in MN-14-8). Numbers above branches indicate bootstrap support values (1,000 replicates).

available in 2007. This genome sequence and future projects to sequence genomes of other AM fungi are also likely to shed light on the source of the huge diversity of genome sizes in various species of Glomeromy-

cota that was uncovered recently. In Glomeraceae (83), the genome size estimates range from 14.1 Mb in *G. intraradices* (44) through 37.45 Mb in *G. etunicatum* (45) and 177.6 Mb in *Glomus geosporum* (48) to 375.0 Mb in *Glomus caledonium* (48). In Gigasporaceae, the range extends from 128.3 Mb in *Scutellospora pellucida* (48) through 795.0 Mb in *Scutellospora castanea* (45) to 1,065.9 Mb in *Scutellospora gregaria* (48). Some of this variation is most likely contributed by repetitive DNA sequences, e.g., in *S. castanea*, the reassociation kinetics analysis revealed that over one-half of the genome consists of repetitive DNA (45).

Concerted Evolution of rDNA and Its Vagaries

In typical multicopy genes, such as the rRNA genes, the fate of mutations in individual gene copies is confounded by the homogenizing action of concerted evolution. To better understand the patterns of the rDNA variation in Glomeromycota, I first explore general mechanisms behind the process of concerted evolution and possible causes of deviations from the molecular pattern of concerted evolution.

Concerted evolution is a recombination-based process responsible for rDNA sequence uniformity within an individual and among individuals of a recombining population (2, 30). It results in a molecular pattern of similarity of the repeated rDNA units that is greater within species than among species. The basic eukaryotic rRNA gene repeat unit usually consists of the 18S, 5.8S, and 26/28S rRNA genes separated by two internal transcribed spacers and flanked by external spacers (91). The tandem repeats of rRNA gene units are arrayed in nucleolar organizer regions (NORs), which may be present in one chromosomal location or distributed among multiple chromosomes. In interphase nuclei, all NORs converge into the nucleolus, which is the site of rRNA transcription and ribosome assembly. Shared nucleolar localization brings together NORs from homologous as well as nonhomologous chromosomes and facilitates sequence homogenization (3). The underlying mechanisms of concerted evolution involve recombination processes collectively referred to as mitotic or somatic recombination. These processes include unequal crossing-over (2, 86) and gene conversion (47) that are responsible for DNA repair to correct errors occurring during replication (100) and transcription (53, 92). Meiosis also offers opportunity for sequence homogenization, although in many species, including budding yeast (97) and *Drosophila* (81), meiotic recombination within the rRNA gene array is suppressed.

The paradigm of concerted evolution allows for treating hundreds of tandemly repeated rRNA genes as one locus. For this reason, rRNA genes have been widely used in phylogeny reconstructions (46). In the process of inferring rDNA-based phylogenies, deviations from the expected molecular pattern of the intraspecific and intraindividual homogeneity of rDNA sequences have been detected in many taxa. Interspecific hybridization is a common source of rDNA sequence heterogeneity (50, 57, 59). Hybridization events are usually followed by rapid homogenization of the parental rDNA sequences (95) even in asexual taxa (47). In some cases, however, sequence heterogeneity is maintained over extended periods of time. Hybrid genomes may retain vestiges of parental nuclear dynamics and organize multiple nucleoli (31). Usually, however, only some of the parental NORs retain transcriptional activity while the remaining rRNA genes are silenced (70) and, as a consequence, may degenerate into pseudogenes and follow their own evolutionary trajectories (13, 57). Impediments to meiotic recombination between NORs on homologous chromosomes that are sequestered for most of the life cycle in haploid nuclei may be another source of rDNA divergence (51). Such sequestration is typical for dikaryotic Basidiomycota, which maintain two genetically distinct nuclei in their hyphal compartments. The result may be a preferential sequence homogenization within individual chromosomal lineages (51).

Divergent rRNA gene clusters may also arise as a consequence of duplication events (18). A striking example of such a pattern comes from the apicomplexans *Cryptosporidium* and *Plasmodium* (77). Instead of hundreds of tandemly repeated rRNA gene copies arrayed in one or more NORs typical for many other eukaryotes, their genomes contain several single 18S-5.8S-28S rRNA units distributed among different chromosomes (1, 34, 96). Individual units are structurally divergent and expressed differentially during distinct developmental stages of the parasitic life cycle (93). Based on the phylogenetic analysis of rRNA genes in *Cryptosporidium parvum* and its relatives, Rooney (77) concluded that the patterns of rRNA gene variation in these organisms are consistent with the model of gene birth-and-death (62) rather than the model of concerted evolution. In the birth-and-death model, new genes are created by duplication events. Some duplicated genes are maintained in the genome for a long time, whereas others are lost or become pseudogenes due to the accumulation of mutations (62). Sequence similarity among the genes that remain functional is maintained by a strong purifying selection rather than by the homogenizing processes of concerted evolution. In many eukaryotes, one of the ribosome components, 5S rRNA, is encoded by genes that are located in the genome separately from the tandemly arrayed 18S-5.8S-26/28S rRNA units and form a disperse gene family. Such dispersed gene organization makes the 5S rRNA genes a likely subject for the birth-and-death evolutionary process, which, in fact, has recently been confirmed in Pezizomycotina (78).

Departures from rDNA Concerted Evolution in Glomeromycota

Intraindividual polymorphisms of rRNA gene sequences have been detected in all AM fungal species that have been examined so far. These polymorphisms were traditionally taken as evidence of nuclear heterogeneity. A systematic study of rRNA gene sequence polymorphisms performed by Pawlowska and Taylor in several isolates of *G. etunicatum* revealed, however, that individuals within each examined isolate as well as different isolates from the same geographic location carried the same set of rDNA sequence variants (68). This outcome would not be expected in a heterokaryotic fungus with distinct rRNA gene variants residing in different nuclei. Pawlowska and Taylor extended their analysis of rDNA polymorphism to individual nuclei that were microdissected from spores (68). This approach confirmed that all rRNA gene variants detectable in a spore are also present in each microdissected nucleus. Based on these observations, Pawlowska and Taylor concluded that the pattern of rDNA variation in *G. etunicatum* suggests a departure from the concerted evolution (68). Importantly, the departure from concerted evolution in *G. etunicatum* is not limited to variation at the intraindividual level. It also extends to the variation within and among taxa closely related to *G. etunicatum*, which is apparent from the phylogeny reconstruction of the *G. etunicatum* lineage based on the rDNA sequences by Rodriguez et al. (76). Distinctly paralogous variants of rRNA genes share less sequence similarity within a taxon than they share with their orthologues in sister taxa. The origin of the paralogous rRNA gene copies is unclear. It may be explained either by a duplication event followed by sequence divergence that predates emergence of the distinct taxa of the *G. etunicatum* lineage or by a hybridization event also predating speciation.

Despite obvious departures from the concerted evolution model, firm evidence is available that recombination processes normally responsible for concerted evolution of rRNA genes are not completely abolished in *G. etunicatum* (68) and other AM fungi (33, 54). However, it is unclear whether the signatures of recombination in rDNA sequences are attributable exclusively to the somatic recombination processes that typically underlie concerted evolution, or whether they can be traced to genetic exchanges among individuals in a population. Without better understanding of the origin of the molecular pattern of the departure from concerted evolution of rDNA, it is unclear how this departure relates to the reproductive mode of Glomeromycota. Based on the data from other organisms exhibiting similar patterns of molecular evolution of rRNA genes, it seems that rDNA polymorphism could be maintained regardless of whether AM fungi are clonal or engage in genetic exchanges. In fact, both clonal (79, 85, 89) and recombinant (89) population structures have been inferred in AM fungi, and the occurrence of recombination has been proposed to be a response to unfavorable environmental conditions (89). The absence of genetic exchanges has been reported from populations of *Glomus caledonium*, *G. mosseae* (79, 85), and *G. geosporum* (85). In another study, which examined several different populations of *G. claroideum* and *Glomus* sp., recombination was detected exclusively in populations experiencing the stress of elevated soil heavy-metal concentrations (89). Unfortunately, in this case the use of anonymous dominant genetic markers precluded determination of the specific nature of recombination processes that contributed to the observed pattern of the lack of association among alleles of the markers (89).

THE HYPOTHESIS OF MULTIGENOMIC STRUCTURE

Are Glomeromycota Chimeric Organisms?

A very different image of genome organization and genetic and evolutionary processes is presented by studies that argue for a multigenomic structure of AM fungal individuals. The hypothesis of nuclear heterogeneity in AM fungi relies on several lines of reasoning, which in addition to the inference of a low number of PLS copies per nucleus in *G. etunicatum* (45) discussed earlier, also include (i) the interpretation of patterns of heritability of spore size and shape in *Scutellospora pellucida* (8) and (ii) the differential distribution of DNA-DNA fluorescent in situ hybridization (FISH) signals among spore nuclei in *S. castanea* (54). (i) Spores of *S. pellucida* exhibit a striking morphological variation of shapes that range from globose to highly elongated and cylindrical (8). Based on significant interisolate differences of shapes and sizes among spores representing independently established single-spore isolates, this morphological variation was interpreted as heritable (8). To explain the observed patterns of morphological variation, the authors of the study proposed that each isolate was derived from a founder spore that carried a particular ratio of two distinct types of nuclei, i.e., "nuclei with genes for round spores" and "nuclei with genes for oblong spores" (8). In other words, founder spores with a large proportion of nuclei containing genetic determinants of the round spore shape would be expected to predominantly produce round spores, whereas spores dominated by nuclei with genetic determinants of the oblong spore shape would mostly give rise to oblong spores. It should be noted, however, that like all other species of AM fungi, the species of *S. pellucida* is defined by its phenotypic characteristics. Consequently, the true phylogenetic history and relatedness of distinct *S. pellucida* isolates remain uncertain. Because of this uncertainty, it cannot be excluded that the observed interisolate morphological variation may be a product of low relatedness and divergent characteristics of the studied isolates rather than of different ratios of two distinct types of nuclei present in the founder spores. (ii) Evidence for nuclear heterogeneity includes also hybridization patterns of two FISH probes (T2 and T4) targeted towards a polymorphic region of the internal transcribed spacer 2 of the rRNA genes in *S. castanea* (54). In this experiment, approximately 40% of nuclei were reported to contain only the T2 sequence, between 6 and 9% of nuclei contained only the T4 sequence, 8 to 9% of nuclei were identified as carrying both types of sequences, and the remaining nuclei were proposed to contain rDNA sequence(s) different from both T2 and T4.

The ongoing maintenance of the intraindividual nuclear heterogeneity has been attributed to a novel genetic process whereby individual mycelia fuse to form a chimeric mycelium harboring genetically distinct populations of nuclei that are subsequently reassorted during spore formation (8, 9, 45). Together with mutations (54), this process could be a potential source of genetic variability in AM fungi needed for coping with the vagaries of the environment. It could also buffer the effects of deleterious mutations and allow for purging them. In the absence of typical sexual reproduction, it might be responsible for the long-term evolutionary survival of Glomeromycota. In the following paragraphs, I summarize principles of chimera formation in other groups of fungi and focus on phenomena related to chimerism in Glomeromycota.

General Principles of Chimera Formation

Virtually all eukaryotes, including protists (20), animals (75), plants (16), and fungi (16), engage in interactions that may lead to formation of chimeric individuals during (i) sexual reproduction and (ii) vegetative growth. (i) In sexual reproduction, successful mating is possible if the interacting partners are sexually compatible, which often means that they are distinct genetically. Chimera formation is universally a component of the sexual cycle. In most cases, the coexistence of genetically nonidentical parental nuclei is limited to a brief period between the plasmogamy and karyogamy that leads to the zygote formation. In fungi, however, the period of coexistence of parental nuclei in the common cytoplasm

(heterokaryosis) may be significantly extended. In the case of Basidiomycota, it may last almost indefinitely, thus almost blurring the distinction between the sexual and the vegetative phases of the life cycle. (ii) During vegetative growth of fungi, fusions of hyphae within an expanding colony are important for mitigation of physical damage and predation of the mycelial structures (17). They are also critical for the maintenance of intramycelial communication and homeostasis, which involve translocation of nutrients and water (14, 15, 71). Fusions between individual mycelia and formation of chimeric colonies allow for an almost instantaneous doubling of each partner's size that leads to the significant improvement of their survivorship and fecundity (15). While hyphal fusions within the same mycelium occur between entities that are genetically identical, formation of chimeric colonies requires partners that are genetically compatible but not necessarily identical, although in fact, they often are. In fungi that are not known to reproduce sexually, interactions between mycelia that are vegetatively compatible and genetically nonidentical may be a prelude to a parasexual process (69), where a chimeric individual may benefit from functional diploidy and mitotic genetic exchanges.

Significance of Self versus Nonself Discrimination

The sophistication of genetic controls involved in sexual and vegetative interindividual interactions suggests that such interactions may entail severe fitness costs if one of the partners exhibits undesirable characteristics. For example, mating-type genes exert control over the elaborate cytological mechanisms of clamp connection and crozier cell formation that ensure fair partitioning of resources between the nuclear lineages contributing to the mated dikaryotic mycelia (16, 19, 23). Otherwise, parasitic nuclei may monopolize the access to the sexual reproductive and dispersal structures (16, 26, 99). Similarly, an indiscriminate partner choice during vegetative encounters carries a risk of acquiring infective cytoplasmic elements (61), including viruses (11, 21, 90), or mitochondria containing senescence plasmids (27). Other possible consequences of fusions with undesirable partners may involve disruptive intraindividual competition for resources (16, 42).

Mechanisms of Fair Resource Partitioning in Sexual Chimeras

In mated filamentous Basidiomycota and Ascomycota, nuclei contributed by the interacting sexual partners coexist in the common cytoplasm of highly structured dikaryotic hyphae. In Basidiomycota, the dikaryotic phase involves large expanses of regularly septate mycelium with clamp connections that provide a mechanism for unbiased distribution of the interacting nuclear partners, allowing the simultaneously dividing pairs of nuclei to maintain their dikaryotic status in all hyphal compartments, including each probasidium (19). While fusion of haploid basidiomycete mycelia may be independent of the mating type, the expression of genes at mating-type loci is required for the growth of the dikaryon (19). In Ascomycota, the dikaryotic state is limited to the ascogenous hyphae that are often multinucleate, and yet a strict 1:1 ratio is maintained between both parental nuclear types (28). Mating-type gene involvement (99) and physical cytoskeletal microtubule linkage between the nuclei of opposing mating types (87) have been shown to mediate the recognition between these two types of nuclei. Croziers formed by ascogenous hyphae play a role similar to clamp connections in ensuring fair access of the interacting partners to the ascus initial (99).

The existence of the dikaryotic phase in the life cycles of Basidiomycota and Ascomycota unifies them in a group known as the Dikaryomycota that is considered to be a sister lineage of Glomeromycota (55). In contrast to Dikaryomycota, however, the mycelium of Glomeromycota seems to be free of any cytological evidence of nuclear partitioning mechanisms. The mycelium of Glomeromycota is a quintessential coenocyte with septae forming only in response to injury (29, 35) or senescence (5). While some nuclei assume stationary positions at regular intervals along the hyphae, others seem to be able to migrate freely throughout the mycelium (5, 6).

Vegetative Chimerism

Given the potential negative effects of vegetative chimerism mentioned earlier, one would imagine that Glomeromycota have mechanisms regulating its occurrence. Experimental evidence shows that indeed different lineages of Glomeromycota exhibit distinct patterns of interindividual interactions. In Glomeraceae, hyphal fusions occur readily within an individual mycelium (29) and among individuals of the same isolate (37). However, interactions between genetically distinct individuals are avoided, and avoidance measures are deployed even before the interacting hyphae come into physical contact (38). For example, vegetative incompatibility responses observed between genetically differentiated isolates of *Glomus mosseae* included retraction of protoplasm from the tips of interacting hyphae and septum formation (38). In Gigasporaceae, including species of *Scutellospora* (one of the taxa in which the existence of heterokaryosis was proposed), occurrence of hyphal fusions within an individual mycelium is lim-

ited to situations when a hypha is injured and a hyphal bypass is formed to restore cytoplasmic flow over the site of injury (29). No fusions between different hyphae emerging from the same spore or from spores representing the same isolate have been recorded in Gigasporaceae despite frequently observed physical contacts among such hyphae (37). These observations are in apparent conflict with the proposed role of hyphal fusions in the maintenance of the multigenomic status by Glomeromycota.

Chimerism in the Parasexual Cycle

In vegetatively compatible but otherwise genetically divergent asexual Ascomycota, hyphal fusions may be a prelude to the parasexual cycle (69). Traditionally, parasexual exchanges have been invoked as a possible source of genetic variation in Glomeromycota (89), but so far no evidence of the parasexual process in this group of fungi has been obtained. The parasexual cycle involves temporary heterokaryosis followed by karyogamy. Once diploid nuclei are formed, they proliferate mitotically, which enables somatic recombination events between nonsister chromatids of homologous chromosomes. Sister chromatid nondisjunctions, also occurring during nuclear divisions, lead initially to aneuploidy, and eventually to the formation of haploid recombinant nuclei. New genotypes generated during the parasexual process are then dispersed as uninucleate asexual spores, e.g., conidia. While the parasexual cycle is routinely exploited in laboratory manipulations of asexual fungi, its occurrence in nature seems to be severely limited by vegetative incompatibility barriers (22).

SIGNIFICANCE OF SPORE MULTINUCLEATE STRUCTURE

The understanding of the developmental processes involved in the formation of the presumably asexual spores may hold the key to some of the mysteries enveloping evolutionary genetic processes in Glomeromycota. While mature spores are unicellular and multinucleate, it is unclear whether their multinucleate status is achieved by clonal propagation of one nucleus that migrates into a spore primordium or by an influx of nuclei randomly sampled from the supporting mycelium. This distinction is crucial for the understanding of the evolutionary forces that shaped the modern AM fungi. Evolutionary theory asserts that unicellular and, by extension, uninucleate propagules reduce the intraorganismal conflicts that arise as a consequence of somatic mutations (genetic mosaicism) (7). Specifically, they are expected to be more effective than multicellular/multinucleate propagules in eliminating selfish mutations, i.e., mutations that benefit a mutant cell/nucleus at the expense of the organism (80). This prediction underscores the role of selfish mutations in the evolution of predominantly unicellular/uninucleate structures as propagules of multicellular organisms despite their obvious vulnerability to environmental challenges and the cost of re-creating complex multicellular individuals (41). In contrast, the evolution of multicellular/multinucleate propagules is expected to be favored under conditions when uniformly deleterious mutations, i.e., mutations deleterious at the cell/nucleus level as well as at the organism level, are eliminated more effectively at the cell/nucleus level than at the organism level (65, 80). Such multicellular/multinucleate propagules are predicted to substantially reduce the mutation load of a population and might be the means of averting the extinction expected in an asexual lineage. The applicability of these predictions to Glomeromycota is unclear at this time because of the paucity of data on their life history and nuclear behavior. It is clear, however, that understanding the processes involved in spore formation will shed light on the evolutionary biology and genetics of this group of fungi.

CONCLUSION

The conflicting inferences about the genetic makeup and genome structure of Glomeromycota are clearly a reflection of the scarcity of experimental data (66). Against the backdrop of evolutionary paradigms on the ephemeral nature of clonal lineages (60), the universal character of rDNA concerted evolution (46), tight genetic regulation of chimera formation (16), and the selective advantages of uninucleate propagules (7), the enigma of genetic processes in Glomeromycota stands out as deserving further exploration and deeper understanding. Some of the outstanding questions will be resolved by the examination of genome sequence data of representative AM fungi when these become available in the near future. Others require transmission and population genetics as well as evolutionary biology studies to demystify this extraordinary group of organisms.

I am grateful to Henk den Bakker and Jean-Luc Jany for discussions during preparation of the manuscript, to Lorna Casselton, whose suggestions improved clarity of the manuscript, and to Rima Shamieh for editing. This work was supported by NSF grant MCB0538363.

References

1. Abrahamsen, M. S., T. J. Templeton, S. Enomoto, J. E. Abrahante, G. Zhu, C. A. Lancto, M. Q. Deng, C. Liu, G. Widmer, S. Tzipori, G. A. Buck, P. Xu, A. T. Bankier, P. H. Dear, B. A. Konfortov, H. F. Spriggs, L. Iyer, V. Anantharaman, L. Aravind, and V. Kapur. 2004. Complete genome sequence of the apicomplexan, *Cryptosporidium parvum. Science* 304:441–445.

2. Arnheim, N., M. Krystal, R. Schmickel, G. Wilson, O. Ryder, and E. Zimmer. 1980. Molecular evidence for genetic exchanges among ribosomal genes on nonhomologous chromosomes in man and apes. *Proc. Natl. Acad. Sci. USA* 77:7323–7327.

3. Arnheim, N., D. Treco, B. Taylor, and E. M. Eicher. 1982. Distribution of ribosomal gene length variants among mouse chromosomes. *Proc. Natl. Acad. Sci. USA* 79:4677–4680.

4. Augé, R. M. 2004. Arbuscular mycorrhizae and soil/plant water relations. *Can. J. Soil Sci.* 84:373–381.

5. Bago, B., W. Zipfel, R. M. Williams, H. Chamberland, J. G. Lafontaine, W. W. Webb, and Y. Piché. 1998. In vivo studies on the nuclear behavior of the arbuscular mycorrhizal fungus *Gigaspora rosea* grown under axenic conditions. *Protoplasma* 203:1–15.

6. Bago, B., W. Zipfel, R. M. Williams, and Y. Piché. 1999. Nuclei of symbiotic arbuscular mycorrhizal fungi as revealed by in vivo two-photon microscopy. *Protoplasma* 209:77–89.

7. Bell, G., and V. Koufopanou. 1991. The architecture of the life-cycle in small organisms. *Philos. Trans. R. Soc. London B* 332:81–89.

8. Bever, J. D., and J. Morton. 1999. Heritable variation and mechanisms of inheritance of spore shape within a population of *Scutellospora pellucida*, an arbuscular mycorrhizal fungus. *Am. J. Bot.* 86:1209–1216.

9. Bever, J. D., and M. Wang. 2005. Arbuscular mycorrhizal fungi: hyphal fusion and multigenomic structure. *Nature* 433:E3–E4.

10. Bianciotto, V., and P. Bonfante. 1992. Quantification of the nuclear DNA content of two arbuscular mycorrhizal fungi. *Mycol. Res.* 96:1071–1076.

11. Biella, S., M. L. Smith, J. R. Aist, P. Cortesi, and M. G. Milgroom. 2002. Programmed cell death correlates with virus transmission in a filamentous fungus. *Proc. R. Soc. Ser. B Biol.* 269:2269–2276.

12. Borowicz, V. A. 2001. Do arbuscular mycorrhizal fungi alter plant-pathogen relations? *Ecology* 82:3057–3068.

13. Buckler, E. S., A. Ippolito, and T. P. Holtsford. 1997. The evolution of ribosomal DNA: divergent paralogues and phylogenetic implications. *Genetics* 145:821–832.

14. Buller, A. H. R. 1958. *Researches on Fungi*, vol. V. Hafner Publishing Co., New York, NY.

15. Buller, A. H. R. 1958. *Researches on Fungi*, vol. IV. Hafner Publishing Co., New York, NY.

16. Buss, L. W. 1982. Somatic cell parasitism and the evolution of somatic tissue compatibility. *Proc. Natl. Acad. Sci. USA* 79:5337–5341.

17. Carlile, M. J. 1994. The success of the hypha and mycelium, p. 3–19. *In* N. A. R. Gow and G. M. Gadd (ed.), *The Growing Fungus*. Chapman & Hall, London, United Kingdom.

18. Carranza, S., J. Baguna, and M. Riutort. 1999. Origin and evolution of paralogous rRNA gene clusters within the flatworm family Dugesiidae (Platyhelminthes, Tricladida). *J. Mol. Evol.* 49:250–259.

19. Casselton, L. A., and N. S. Olesnicky. 1998. Molecular genetics of mating recognition in basidiomycete fungi. *Microbiol. Mol. Biol. Rev.* **62**:55–70.

20. Castillo, D. I., G. T. Switz, K. R. Foster, D. C. Queller, and J. E. Strassmann. 2005. A cost to chimerism in *Dictyostelium discoideum* on natural substrates. *Evol. Ecol. Res.* **7**:263–271.

21. Caten, C. E. 1972. Vegetative incompatibility and cytoplasmic infection in fungi. *J. Gen. Microbiol.* **72**:221–229.

22. Clutterbuck, A. J. 1996. Parasexual recombination in fungi. *J. Genet.* **75**:281–286.

23. Coppin, E., R. Debuchy, S. Arnaise, and M. Picard. 1997. Mating types and sexual development in filamentous ascomycetes. *Microbiol. Mol. Biol. Rev.* **61**:411–428.

24. Corradi, N., G. Kuhn, and I. R. Sanders. 2004. Monophyly of beta-tubulin and H+-ATPase gene variants in *Glomus intraradices*: consequences for molecular evolutionary studies of AM fungal genes. *Fungal Genet. Biol.* **41**:262–273.

25. Corradi, N., and I. R. Sanders. 2006. Evolution of the P-type II ATPase gene family in the fungi and presence of structural genomic changes among isolates of *Glomus intraradices*. *BMC Evol. Biol.* **6**:21.

26. Debets, A. J. M., and A. J. F. Griffiths. 1998. Polymorphism of *het*-genes prevents resource plundering in *Neurospora crassa*. *Mycol. Res.* **102**:1343–1349.

27. Debets, F., X. Yang, and A. J. F. Griffiths. 1994. Vegetative incompatibility in *Neurospora*: its effect on horizontal transfer of mitochondrial plasmids and senescence in natural populations. *Curr. Genet.* **26**:113–119.

28. Debuchy, R. 1999. Internuclear recognition: a possible connection between euascomycetes and homobasidiomycetes. *Fungal Genet. Biol.* **27**:218–223.

29. de la Providencia, I. E., F. A. de Souza, F. Fernández, N. S. Delmas, and S. Declerck. 2005. Arbuscular mycorrhizal fungi reveal distinct patterns of anastomosis formation and hyphal healing mechanisms between different phylogenic groups. *New Phytol.* **165**:261–271.

30. Dover, G. 1982. Molecular drive: a cohesive mode of species evolution. *Nature* **299**:111–117.

31. Fankhauser, G., and R. R. Humphrey. 1943. The relation between number of nucleoli and number of chromosome sets in animal cells. *Proc. Natl. Acad. Sci. USA* **29**:344–350.

32. Fortin, J. A., G. Bécard, S. Declerck, Y. Dalpé, M. St-Arnaud, A. P. Coughlan, and Y. Piché. 2002. Arbuscular mycorrhiza on root-organ cultures. *Can. J. Bot.* **80**:1–20.

33. Gandolfi, A., I. R. Sanders, V. Rossi, and P. Menozzi. 2003. Evidence of recombination in putative ancient asexuals. *Mol. Biol. Evol.* **20**:754–761.

34. Gardner, M. J., N. Hall, E. Fung, O. White, M. Berriman, R. W. Hyman, J. M. Carlton, A. Pain, K. E. Nelson, S. Bowman, I. T. Paulsen, K. James, J. A. Eisen, K. Rutherford, S. L. Salzberg, A. Craig, S. Kyes, M. S. Chan, V. Nene, S. J. Shallom, B. Suh, J. Peterson, S. Angiuoli, M. Pertea, J. Allen, J. Selengut, D. Haft, M. W. Mather, A. B. Vaidya, D. M. A. Martin, A. H. Fairlamb, M. J. Fraunholz, D. S. Roos, S. A. Ralph, G. I. McFadden, L. M.

Cummings, G. M. Subramanian, C. Mungall, J. C. Venter, D. J. Carucci, S. L. Hoffman, C. Newbold, R. W. Davis, C. M. Fraser, and B. Barrell. 2002. Genome sequence of the human malaria parasite *Plasmodium falciparum*. *Nature* **419**:498–511.

35. Gerdemann, J. W. 1955. Wound healing of hyphae in a phycomycetous mycorrhizal fungus. *Mycologia* **47**:916–918.

36. Gill, D. E., L. Chao, S. L. Perkins, and J. B. Wolf. 1995. Genetic mosaicism in plants and clonal animals. *Annu. Rev. Ecol. Syst.* **26**:423–444.

37. Giovannetti, M., D. Azzolini, and A. S. Citernesi. 1999. Anastomosis formation and nuclear and protoplasmic exchange in arbuscular mycorrhizal fungi. *Appl. Environ. Microbiol.* **65**:5571–5575.

38. Giovannetti, M., C. Sbrana, P. Strani, M. Agnolucci, V. Rinaudo, and L. Avio. 2003. Genetic diversity of isolates of *Glomus mosseae* from different geographic areas detected by vegetative compatibility testing and biochemical and molecular analysis. *Appl. Environ. Microbiol.* **69**:616–624.

39. Göhre, V., and U. Paszkowski. 2006. Contribution of the arbuscular mycorrhizal symbiosis to heavy metal phytoremediation. *Planta* **223**:1115–1122.

40. Govindarajulu, M., P. E. Pfeffer, H. R. Jin, J. Abubaker, D. D. Douds, J. W. Allen, H. Bucking, P. J. Lammers, and Y. Shachar-Hill. 2005. Nitrogen transfer in the arbuscular mycorrhizal symbiosis. *Nature* **435**:819–823.

41. Grosberg, R. K., and R. R. Strathmann. 1998. One cell, two cell, red cell, blue cell: the persistence of a unicellular stage in multicellular life histories. *Trends Ecol. Evol.* **13**:112–116.

42. Hartl, D. L., E. R. Dempster, and S. W. Brown. 1975. Adaptive significance of vegetative incompatibility in *Neurospora crassa*. *Genetics* **81**:553–569.

43. Hijri, M., M. Hosny, D. van Tuinen, and H. Dulieu. 1999. Intraspecific ITS polymorphism in *Scutellospora castanea* (Glomales, Zygomycota) is structured within multinucleate spores. *Fungal Genet. Biol.* **26**:141–151.

44. Hijri, M., and I. R. Sanders. 2004. The arbuscular mycorrhizal fungus *Glomus intraradices* is haploid and has a small genome size in the lower limit of eukaryotes. *Fungal Genet. Biol.* **41**:253–261.

45. Hijri, M., and I. R. Sanders. 2005. Low gene copy number shows that arbuscular mycorrhizal fungi inherit genetically different nuclei. *Nature* **433**:160–163.

46. Hillis, D. M., and M. T. Dixon. 1991. Ribosomal DNA: molecular evolution and phylogenetic inference. *Q. Rev. Biol.* **66**:410–453.

47. Hillis, D. M., C. Moritz, C. A. Porter, and R. J. Baker. 1991. Evidence for biased gene conversion in concerted evolution of ribosomal DNA. *Science* **251**:308–310.

48. Hosny, M., V. Gianinazzi-Pearson, and H. Dulieu. 1998. Nuclear DNA content of 11 fungal species in Glomales. *Genome* **41**:422–428.

49. Hosny, M., M. Hijri, E. Passerieux, and H. Dulieu. 1999. rDNA units are highly polymorphic in *Scutellospora castanea* (Glomales, Zygomycetes). *Gene* **226**:61–71.

50. Hugall, A., J. Stanton, and C. Moritz. 1999. Reticulate evolution and the origins of ribosomal internal tran-

scribed spacer diversity in apomictic *Meloidogyne*. *Mol. Biol. Evol.* **16**:157–164.

51. James, T. Y., J. M. Moncalvo, S. Li, and R. Vilgalys. 2001. Polymorphism at the ribosomal DNA spacers and its relation to breeding structure of the widespread mushroom *Schizophyllum commune*. *Genetics* **157**:149–161.

52. Kimura, M., and J. F. Crow. 1964. Number of alleles that can be maintained in finite population. *Genetics* **49**:725–738.

53. Kobayashi, T., and A. R. D. Ganley. 2005. Recombination regulation by transcription-induced cohesin dissociation in rDNA repeats. *Science* **309**:1581–1584.

54. Kuhn, G., M. Hijri, and I. R. Sanders. 2001. Evidence for the evolution of multiple genomes in arbuscular mycorrhizal fungi. *Nature* **414**:745–748.

55. Lutzoni, F., F. Kauff, C. J. Cox, D. McLaughlin, G. Celio, B. Dentinger, M. Padamsee, D. Hibbett, T. Y. James, E. Baloch, M. Grube, V. Reeb, V. Hofstetter, C. Schoch, A. E. Arnold, J. Miadlikowska, J. Spatafora, D. Johnson, S. Hambleton, M. Crockett, R. Shoemaker, S. Hambleton, M. Crockett, R. Shoemaker, G. H. Sung, R. Lucking, T. Lumbsch, K. O'Donnell, M. Binder, P. Diederich, D. Ertz, C. Gueidan, K. Hansen, R. C. Harris, K. Hosaka, Y. W. Lim, B. Matheny, H. Nishida, D. Pfister, J. Rogers, A. Rossman, I. Schmitt, H. Sipman, J. Stone, J. Sugiyama, R. Yahr, and R. Vilgalys. 2004. Assembling the fungal tree of life: progress, classification and evolution of subcellular traits. *Am. J. Bot.* **91**:1446–1480.

56. Maldonado-Mendoza, I. E., G. R. Dewbre, and M. J. Harrison. 2001. A phosphate transporter gene from the extra-radical mycelium of an arbuscular mycorrhizal fungus *Glomus intraradices* is regulated in response to phosphate in the environment. *Mol. Plant-Microbe Interact.* **14**:1140–1148.

57. Marquez, L. M., D. J. Miller, J. B. MacKenzie, and M. J. H. van Oppen. 2003. Pseudogenes contribute to the extreme diversity of nuclear ribosomal DNA in the hard coral *Acropora*. *Mol. Biol. Evol.* **20**:1077–1086.

58. Morton, J. B. 1990. Evolutionary relationships among arbuscular mycorrhizal fungi in the Endogonaceae. *Mycologia* **82**:192–207.

59. Muir, G., C. C. Fleming, and C. Schlotterer. 2001. Three divergent rDNA clusters predate the species divergence in *Quercus petraea* (Matt.) Liebl. and *Quercus robur* L. *Mol. Biol. Evol.* **18**:112–119.

60. Muller, H. J. 1964. The relation of recombination to mutational advance. *Mutat. Res.* **1**:2–9.

61. Nauta, M. J., and R. F. Hoekstra. 1994. Evolution of vegetative incompatibility in filamentous Ascomycetes. I. Deterministic models. *Evolution* **48**:979–995.

62. Nei, M., and A. P. Rooney. 2005. Concerted and birth-and-death evolution of multigene families. *Annu. Rev. Genet.* **39**:121–152.

63. Normark, B. B., O. P. Judson, and N. A. Moran. 2003. Genomic signatures of ancient asexual lineages. *Biol. J. Linn. Soc.* **79**:69–84.

64. Olsson, P. A., I. Thingstrup, I. Jakobsen, and F. Bååth. 1999. Estimation of the biomass of arbuscular mycorrhizal fungi in a linseed field. *Soil Biol. Biochem.* **31**:1879–1887.

65. Otto, S. P., and M. E. Orive. 1995. Evolutionary consequences of mutation and selection within an individual. *Genetics* **141**:1173–1187.

66. Pawlowska, T. E. 2005. Genetic processes in arbuscular mycorrhizal fungi. *FEMS Microbiol. Lett.* **251**:185–192.

67. Pawlowska, T. E., and J. W. Taylor. 2005. Arbuscular mycorrhizal fungi: hyphal fusion and multigenomic structure. Reply. *Nature* **433**:E4.

68. Pawlowska, T. E., and J. W. Taylor. 2004. Organization of genetic variation in individuals of arbuscular mycorrhizal fungi. *Nature* **427**:733–737.

69. Pontecorvo, G. 1956. Parasexual cycle in fungi. *Annu. Rev. Microbiol.* **10**:393–400.

70. Pontes, O., R. J. Lawrence, N. Neves, M. Silva, J. H. Lee, Z. J. Chen, W. Viegas, and C. S. Pikaard. 2003. Natural variation in nucleolar dominance reveals the relationship between nucleolus organizer chromatin topology and rRNA gene transcription in Arabidopsis. *Proc. Natl. Acad. Sci. USA* **100**:11418–11423.

71. Rayner, A. D. M., G. S. Griffith, and A. M. Ainsworth. 1994. Mycelial interconnectedness, p. 21–40. *In* N. A. R. Gow and G. M. Gadd (ed.), *The Growing Fungus*. Chapman & Hall, London, United Kingdom.

72. Redecker, D., M. Hijri, H. Dulieu, and I. R. Sanders. 1999. Phylogenetic analysis of a dataset of fungal 5.8S rDNA sequences shows that highly divergent copies of internal transcribed spacers reported from *Scutellospora castanea* are of ascomycete origin. *Fungal Genet. Biol.* **28**:238–244.

73. Redecker, D., R. Kodner, and L. E. Graham. 2000. Glomalean fungi from the Ordovician. *Science* **289**:1920–1921.

74. Remy, W., T. N. Taylor, H. Hass, and H. Kerp. 1994. Four hundred-million-year-old vesicular arbuscular mycorrhizae. *Proc. Natl. Acad. Sci. USA* **91**:11841–11843.

75. Rinkevich, B. 2005. Natural chimerism in colonial urochordates. *J. Exp. Mar. Biol. Ecol.* **322**:93–109.

76. Rodriguez, A., J. P. Clapp, L. Robinson, and J. C. Dodd. 2005. Studies on the diversity of the distinct phylogenetic lineage encompassing *Glomus claroideum* and *Glomus etunicatum*. *Mycorrhiza* **15**:33–46.

77. Rooney, A. P. 2004. Mechanisms underlying the evolution and maintenance of functionally heterogeneous 18S rRNA genes in apicomplexans. *Mol. Biol. Evol.* **21**:1704–1711.

78. Rooney, A. P., and T. J. Ward. 2005. Evolution of a large ribosomal RNA multigene family in filamentous fungi: birth and death of a concerted evolution paradigm. *Proc. Natl. Acad. Sci. USA* **102**:5084–5089.

79. Rosendahl, S., and J. W. Taylor. 1997. Development of multiple genetic markers for studies of genetic variation in arbuscular mycorrhizal fungi using AFLP(TM). *Mol. Ecol.* **6**:821–829.

80. Roze, D., and R. E. Michod. 2001. Mutation, multilevel selection, and the evolution of propagule size during the origin of multicellularity. *Am. Nat.* **158**:638–654.

81. Schlötterer, C., and D. Tautz. 1994. Chromosomal homogeneity of *Drosophila* ribosomal DNA arrays suggests intrachromosomal exchanges drive concerted evolution. *Curr. Biol.* **4**:777–783.

82. Schüßler, A. 1999. Glomales SSU rRNA gene diversity. *New Phytol.* **144**:205–207.

83. Schüßler, A., D. Schwarzott, and C. Walker. 2001. A new fungal phylum, the Glomeromycota: phylogeny and evolution. *Mycol. Res.* **105**:1413–1421.

84. Smith, S. E., and D. J. Read. 1997. *Mycorrhizal Symbiosis*, 2nd ed. Academic Press, San Diego, CA.

85. Stukenbrock, E. H., and S. Rosendahl. 2005. Clonal diversity and population genetic structure of arbuscular mycorrhizal fungi (*Glomus* spp.) studied by multilocus genotyping of single spores. *Mol. Ecol.* **14**:743–752.

86. Szostak, J. W., and R. Wu. 1980. Unequal crossingover in the ribosomal DNA of *Saccharomyces cerevisiae*. *Nature* **284**:426–430.

87. Thompson-Coffe, C., and D. Zickler. 1994. How the cytoskeleton recognizes and sorts nuclei of opposite mating-type during the sexual cycle in filamentous ascomycetes. *Dev. Biol.* **165**:257–271.

88. Trouvelot, S., D. van Tuinen, M. Hijri, and V. Gianinazzi-Pearson. 1999. Visualization of ribosomal DNA loci in spore interphasic nuclei of glomalean fungi by fluorescence *in situ* hybridization. *Mycorrhiza* **8**:203–206.

89. Vandenkoornhuyse, P., C. Leyval, and I. Bonnin. 2001. High genetic diversity in arbuscular mycorrhizal fungi: evidence for recombination events. *Heredity* **87**:243–253.

90. van Diepeningen, A. D., A. J. M. Debets, and R. F. Hoekstra. 1997. Heterokaryon incompatibility blocks virus transfer among natural isolates of black *Aspergilli*. *Curr. Genet.* **32**:209–217.

91. Venema, J., and D. Tollervey. 1999. Ribosome synthesis in *Saccharomyces cerevisiae*. *Annu. Rev. Genet.* **33**:261–311.

92. Voelkel-Meiman, K., R. L. Keil, and G. S. Roeder. 1987. Recombination-stimulating sequences in yeast ribosomal DNA correspond to sequences regulating transcription by RNA polymerase I. *Cell* **48**:1071–1079.

93. Waters, A. P. 1994. The ribosomal RNA genes of *Plasmodium*. *Adv. Parasitol.* **34**:33–79.

94. Watterson, G. A. 1984. Allele frequencies after a bottleneck. *Theor. Popul. Biol.* **26**:387–407.

95. Wendel, J. F., A. Schnabel, and T. Seelanan. 1995. Bidirectional interlocus concerted evolution following allopolyploid speciation in cotton (*Gossypium*). *Proc. Natl. Acad. Sci. USA* **92**:280–284.

96. Xu, P., G. Widmer, Y. P. Wang, L. S. Ozaki, J. M. Alves, M. G. Serrano, D. Puiu, P. Manque, D. Akiyoshi, A. J. Mackey, W. R. Pearson, P. H. Dear, A. T. Bankier, D. L. Peterson, M. S. Abrahamsen, V. Kapur, S. Tzipori, and G. A. Buck. 2004. The genome of *Cryptosporidium hominis*. *Nature* **431**:1107–1112.

97. Zamb, T. J., and T. D. Petes. 1982. Analysis of the junction between ribosomal RNA genes and single-copy chromosomal sequences in the yeast *Saccharomyces cerevisiae*. *Cell* **28**:355–364.

98. Zézé, A., E. Sulistyowati, K. Ophelkeller, S. Barker, and S. Smith. 1997. Intersporal genetic variation of *Gigaspora margarita*, a vesicular arbuscular mycorrhizal fungus, revealed by M13 minisatellite-primed PCR. *Appl. Environ. Microbiol.* **63**:676–678.

99. Zickler, D., S. Arnaise, E. Coppin, R. Debuchy, and M. Picard. 1995. Altered mating-type identity in the fungus *Podospora anserina* leads to selfish nuclei, uniparental progeny, and haploid meiosis. *Genetics* **140**:493–503.

100. Zou, H., and R. Rothstein. 1997. Holliday junctions accumulate in replication mutants via a RecA homolog-independent mechanism. *Cell* **90**:87–96.

Sex in Fungi: Molecular Determination
and Evolutionary Implications
Edited by Joseph Heitman et al.
© 2007 ASM Press, Washington, D.C.

Johannes Wöstemeyer
Christine Schimek

Trisporic Acid and Mating in Zygomycetes

26

SEXUAL MORPHOGENESIS

Although the sexual morphogenesis and behavior of the zygomycetes were already known and the first attempts at studying the molecular basis of the mating reaction had been conducted (47, 63, 65), this group was not discussed in Raper's original account, as at that time the then-recognized taxon "Phycomycetes" was not considered to belong within the "higher fungi." Within the Eumycota, the Zygomycota form one of the basal groups. Typically they are largely unseptate, and their cell walls partially consist of chitosan, the product of chitin deacetylation. Many members of the group are saprotrophs living as primary colonizers on carbon-rich substrates, whereas others are obligate entomopathogenic endoparasites. Apart from the obvious damage caused by saprotroph zygomycetes as causative agents of postharvest fruit rot, zygomycetes have comparatively little impact on human affairs, as only a few of the thermophilic species are opportunistic mammalian parasites.

The Zygomycota received their name by analogy to the other fungal groups based on the overt morphology of their sexual apparatus. Zygote formation in this group occurs within a structure arranged between two suspensor hyphae, and the whole apparatus typically resembles an oxen yoke (from the Greek ζυγοζ [zygos]) (Fig. 26.1). Sexuality is based on the interaction of two thalli of different mating types in the heterothallic species (Fig. 26.1a and c) and on the interaction of two hyphae or hyphal branches in the same mycelium in homothallic species (Fig. 26.1b) (73). Zygospore formation starts with the detection and recognition of a suitable mating partner, and localized hyphal swellings are subsequently formed. These progametangia arise between parental hyphae either positioned opposed to each other (Fig. 26.1a and c) or growing apposedly alongside each other for some time, and the mode of contact is a species-specific trait. In species with opposed progametangia, the first contact is usually established between aerial hyphae that are often indistinguishable from neighboring hyphae (Fig. 26.1c). The first contact may be accidental, as distinct search movements or tropisms are unknown. It is hypothesized that the parental hyphae are pushed apart by the pressure of the growing progametangia.

In others, e.g., the well-known species *Mucor mucedo*, distinct and specialized aerial hyphae are

Johannes Wöstemeyer and Christine Schimek, Chair of General Microbiology and Microbial Genetics, Friedrich-Schiller-Universität Jena, 07743 Jena, Germany.

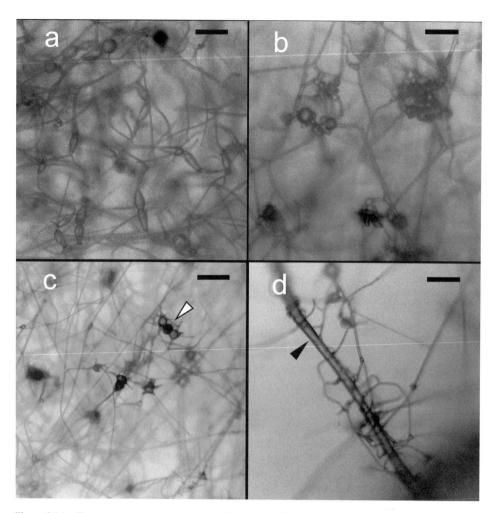

Figure 26.1 Zygomycete mating structures in 5-day-old cultures grown on solid medium in continuous darkness. (a) *Mucor mucedo* progametangial and gametangial stages formed between + and − zygophores; bar, 250 μm. (b) *Mycotypha africana*, homothallic zygospores and earlier developmental stages formed between hyphae of the same individual; bar, 150 μm. (c) *Parasitella parasitica* zygospore formation in stages ranging from progametangia to mature zygospore (white arrowhead) between + and − hyphae; bar, 150 μm. (d) *Parasitella parasitica* early stages of sikyospore formation with a young sporangiophore of *Mucor mucedo* (black arrowhead); bar, 150 μm.

formed as a reaction to signals received from a putative partner (Fig. 26.1a). Within a distance of 2 mm, these zygophores of the different mating types perform a directed growth reaction termed zygotropism, which guides them to contact each other at or near the tip (39). Zygophores resemble young asexual sporangiophores in shape but are much smaller and not responsive to light or gravitational stimuli (39). At the contact site, the hyphal tips swell and thus form the progametangia. When the hyphal tips touch each other, a firm connection between the two walls is rapidly established, and in the subsequent stages a homogeneous cell wall is apparent by electron microscopy (28). Apposed progametangia are a feature restricted to species where the first contact occurs in the substrate or the surfacing substrate mycelium (Fig. 26.2). Hyphal tips meet and coil around each other, or irregular lobed hyphal tips (which are zygophores in *Phycomyces blakesleeanus*) clasp each other. From the resulting hyphal knot the progametangia emerge and continue to grow alongside each other (Fig. 26.2a and b) (23). Only in the later stages the suspensor pair may separate again to form the typical yoke between the tightly connected hyphal tips (Fig. 26.2c) (22, 34), or the zygospore develops at the tip of the aligned progametangia (2, 8, 25). Neither progametangia nor zygophores are necessarily separated from the

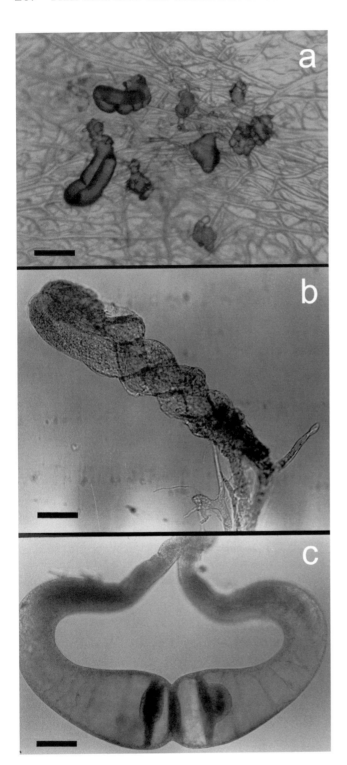

Figure 26.2 Early sexual morphogenesis in *Phycomyces blakesleeanus*. (a) Zygophore knots and young apposed progametangia; bar, 150 μm. (b) Apposed progametangia coiling around each other during elongation; bar, 100 μm. (c) Gametangia are delimited by septa at the tips of the now open yoke structure; bar, 100 μm.

vegetative mycelium. Only gametangia are delimited by a septum (Fig. 26.2c) but remain connected to the progametangial residues, then called suspensors, by plasmodesmata (45). In the next stage, the fusion wall dissolves, and the contents of the former gametangia mix within the arising fusion area. Large amounts of cytoplasm and numerous nuclei are transported into the gametangia and, later, the fusion zone. A new zygospore wall is deposited from inside, while the outer wall of the zygosporangium is continuous with the walls of the respective suspensors (59). During maturation, a marked differentiation into diverse wall layers occurs (28), which leaves the remaining outer wall covered in characteristic relief ornamentation. It becomes impregnated with sporopollenin (43) and can be removed in mature structures. Later events, i.e., reduction of nuclei within the zygospore, nuclear fusion, and meiosis, are for that reason hidden to microscopic analysis and have consequently never been studied in great detail (54). Whether the emerging structure represents a zygospore proper or a zygosporangium including a single zygospore depends on the respective author's interpretation of the available ultrastructural data.

Sexual reactions have been described for members of all groups within the Zygomycetes, one of the two classes within the Zygomycota. Nevertheless, zygospore formation is a comparatively rare event under laboratory conditions, and sexual stages are unknown in many isolates. Zygomycetes appear to rely on the sexual cycle mainly for the production of dormant or resting spores, probably with the effect to survive a period of environmental limitation. Some Entomophthorales commonly form azygospores as resting spores (62). These are morphologically similar to zygospores but are produced without a genetic partner by a single mating type. If a partner is present at all, it is only physiologically involved. The usual and by far more effective asexual reproductive cycle culminates in the formation of one-spored to oligosporic sporangioles or multispored sporangia borne at the tip of exclusively aerial sporangiophores. Chlamydospores and arthrospores may also be formed to various degrees. In the multinucleate haploid vegetative mycelium that may well proceed within the substratum layer, septa are only rarely formed, usually to delimit specialized hyphal compartments, e.g., gametangia, sporangiophores, or sporangia. Prior to germination, an obligatory period of dormancy is required, lasting at least 53 days in *Phycomyces blakesleeanus* (30) and around 60 days in *Choanephora cucurbitarum* (103), and germination under laboratory conditions is an infrequent event.

MATING-TYPE SYSTEM

Based on his original observation on *Rhizopus nigricans* that zygospores developed only at the line where two different isolates met, Blakeslee (9) postulated that zygomycetes belong to one of only two mating types, (+) and (−), a view which he was able to verify subsequently by more extensive studies (10, 11). None of the occasionally occurring features such as preference for a certain location for progametangial fusion or the actual size of the progametangia and gametangia could be unequivocally attributed to one of the mating types in any species (66). Therefore, as no distinguishing morphological mark exists to define mating type, classification of newly isolated strains still requires mating reactions with defined tester strains. Despite intensive studies, a switch of mating type has never been unambiguously proven, and the phenomenon of "relative sexuality," where one strain reacts as (+) with one possible partner but as (−) towards another strain of known (+) behavior, was positively discounted (10). Homothallism has been attributed to differential distribution of nuclei within the mycelium, but following a comprehensive study on homothallic *Mucor* and *Zygorhynchus* species, Satina and Blakeslee (66) reached the conclusion that the abilities to react as (+) and (−) are equally present in all hyphae. These findings indicate that mating-type definition in zygomycetes may not rely on the expression of one single gene or mating-type locus in the way known from Ascomycota or Basidiomycota. Genetical analysis is compatible with the view of a single mating-type locus containing the alleles (+) and (−), but considering the absolute lack of supportive molecular data for such a mating-type locus, the mating behavior of any strain seems to be defined solely by the regulation of gene expression and the physiological reaction to the partner present in the actual mating situation. The repeated observations of strains being "neutral" in mating, i.e., not being inducible to zygospore formation, may be explained either as the result of mutation or the possibility that the culture conditions permissive to zygophore formation have not yet been elucidated. It may also be interpreted as the effect of some hitherto unknown incompatibility mechanism.

Within the Zygomycetes, interspecific reactions are quite common. These usually abortive events lead to the formation of zygophores, progametangia, and in certain cases also to the formation of gametangia by one or both partners, but mature zygospores are never formed. In several strains, the production of azygospores was observed as the result of an interspecific mating reaction. In these reactions the progametangium or suspensor of the unsuitable mating partner becomes detached from the mating apparatus in the later stages of development (55, 60, 72).

A further reaction reminiscent of zygospore formation is a specific parasitic process. Several zygomycetes act as facultative biotrophic fusion parasites exclusively on members of the same group. As a result of the interaction, a structure named the sikyospore is formed between *Parasitella parasitica* (Fig. 26.1d) or *Chaetocladium jonesii* and their respective hosts. As the susceptibility to parasitic interaction with *Parasitella parasitica* can be strictly mating type specific, at least with the host *Absidia glauca*, this process also underlies mating communication and regulation (98). Although the whole process resembles azygospore formation in the presence of a putative partner, sikyospores differ from azygospores in three aspects. Sikyospores are (i) never formed without a partner/host being present, (ii) they involve participation of both partners without severance during the later stages, and (iii) they are structurally different from the zygospores formed by either the parasite or the host (17).

TRISPOROIDS: A FAMILY OF SIGNAL COMPOUNDS FOR SEXUAL COMMUNICATION

Neither the recognition of a possible mating partner nor the initiation of the complex processes leading to the formation of zygospore or sikyospore is conceivable without the involvement of one or more metabolic signals. Moreover, from the observed interspecific mating reactions it became obvious quite early that the signal system responsible for the initial steps must necessarily be the same within the Mucorales, the group studied in greatest detail by Blakeslee, Burgeff, and other researchers at the beginning of the 20th century. The molecular basis of this signaling became apparent with the identification of trisporic acid in culture extracts from *Mucor mucedo* (5) and with the proof of its inductive activity in the initiation of zygophore formation (18, 37, 38, 63, 87, 88). Trisporic acid was then already known to be the substance responsible for the increase in β-carotene production in mated cultures of *Blakeslea trispora* (20). In this species, both carotene and trisporic acid production is much higher than in all other species investigated so far; this fact is immensely helpful to research, as usually *Blakeslea trispora* products are used for analysis and testing of other strains. A short overview of the early history of the detection of trisporoids as the sexual signals in Mucorales has been published by Gooday and Carlile (42).

Chemical analysis revealed trisporic acid to be a C_{18} degradation product of β-carotene (4, 12). Further metabolic and structural analyses also included *Phycomyces blakesleeanus*, contributed by Sutter and his group, and in one study *Zygorhynchus heterogamus* and *Zygorhynchus moelleri* (95). Besides trisporic acid, the respective culture extracts were found to contain a number of related compounds, which were identified as metabolic precursors of trisporic acid (13, 14, 16, 57, 58, 64, 79, 81, 82). Despite the small number of species analyzed, from the high similarity of the mating reaction throughout this group it was nevertheless concluded that trisporic acid-based signaling would be the general principle of sexual communication in Mucorales. This opinion is supported by the finding that presentation with culture extracts or direct confrontation with several other species induced zygophore formation in *Mucor mucedo*, which is used as the tester strain because of its easily recognizable zygophores (for an example, see reference 78). With positive reactions also found with culture extracts from several *Mortierella* species, belonging to the order Mortierellales, the hypothesis was expanded to include the whole class Zygomycetes (71). In fact, trisporic acid induces zygophore formation in both mating types of *Mucor mucedo* (91) whereas the precursors affect specifically one mating type (12). It was therefore easily imaginable that these precursors constitute the mating-type-specific signal pheromones involved in partner recognition. This view was corroborated by the trisporic acid biosynthesis scheme that was soon thereafter elucidated (12, 13). This scheme is based on the results of feeding experiments using ^{14}C-labeled substrates and was formulated basically for *Blakeslea trispora*, with a later addition also including *Phycomyces blakesleeanus* (84) (Fig. 26.3). Trisporoid biosynthesis commences with the cleavage of β-carotene. It is still not quite clear whether this cleavage is performed asymmetrically or symmetrically. Both routes may be used alternatively depending on the oxidative stress status of the mycelium (36). Feeding of the C_{20} cleavage products retinol and retinal establishes that the products of such a reaction at least enter into the trisporoid synthesis pathway in *Blakeslea trispora* (16). The same applies to a subsequent postulated intermediate, β-C_{18}-ketone (13, 16). This ketone has never been isolated from culture extracts, and it is hypothesized to be immediately processed and not freely occurring within the cytoplasm in any significant amount. By oxidation at C4 and reduction of the double bond at C11–C12, β-C_{18}-ketone in turn is converted into 4-dihydrotrisporin, the first compound exhibiting the typical trisporoid structure with a C_{14} backbone (Fig. 26.3), a stretch of three conjugated double bonds between C5–C6, C7–C8, and C9–C10, and an oxygen moiety at C4.

All subsequent intermediates in trisporic acid synthesis are derived from 4-dihydrotrisporin by mating-type-specific modifications of the substituents at C4, C1, C2, C3, and C13. In the (+) mating type, 4-dihydrotrisporin becomes converted into 4-dihydromethyl trisporate, and it is converted into trisporin in the (−) mating type (58). The next synthetic step then may lead to methyl trisporate (+) or trisporol (−) (84). With either possible product, trisporoid synthesis comes to an end within the respective mating type. Trisporic acid itself can be formed only from the precursor of the complementary mating type. The necessary exchange of intermediates has been described as cooperative or complementary biosynthesis (12, 77, 89). The final step involves demethylation in the (+) type. In the (−) mating type, trisporic acid synthesis is achieved by oxidation of the hydroxyl group at the C1 residue of trisporol (Fig. 26.3). Feeding experiments helped to identify only two mating-type-specific enzymatic reactions in the entire process: the oxidation mentioned just above occurs only in the (+) mating type, whereas the oxidation at C4, converting 4-dihydromethyl trisporate into methyl trisporate, is a (−)-type-specific reaction (12). The chemistry of trisporic acid synthesis has been reviewed in more detail elsewhere (41, 69). Only one of the enzymes has been isolated and characterized thus far: the dehydrogenase catalyzing the oxidation of 4-dihydromethyl trisporate. The protein with M_r of 36,000 (24) with its activity maximum at pH 8.0 requires NADP as a cosubstrate (94). In the presence of the authentic substrate, enzyme activity can be assessed either directly as production of NADPH from NADP (70) or via coupling of the reduction reaction to nitroblue tetrazolium chloride as the final electron acceptor, assayed either in situ (71, 94) or in protein preparations separated by native polyacrylamide gel electrophoresis (70). Enzyme activity was detected only in the (−) mating type in *Mucor mucedo* and was also confined to the (−)-acting suspensor in the homothallic species *Zygorhynchus moelleri* (94).

CHEMICAL AND REGULATORY DIVERSITY IN TRISPOROID SIGNALING

As stated above, trisporids act throughout the entire class of zygomycetes and interspecific attempts at mating reactions are frequently observed. Nevertheless, completion of zygospore formation takes place only between active and compatible partners of the same species, ensuring species specificity of the mating reaction. One way to establish such specificity would lie in modifications of the signaling system itself. And indeed,

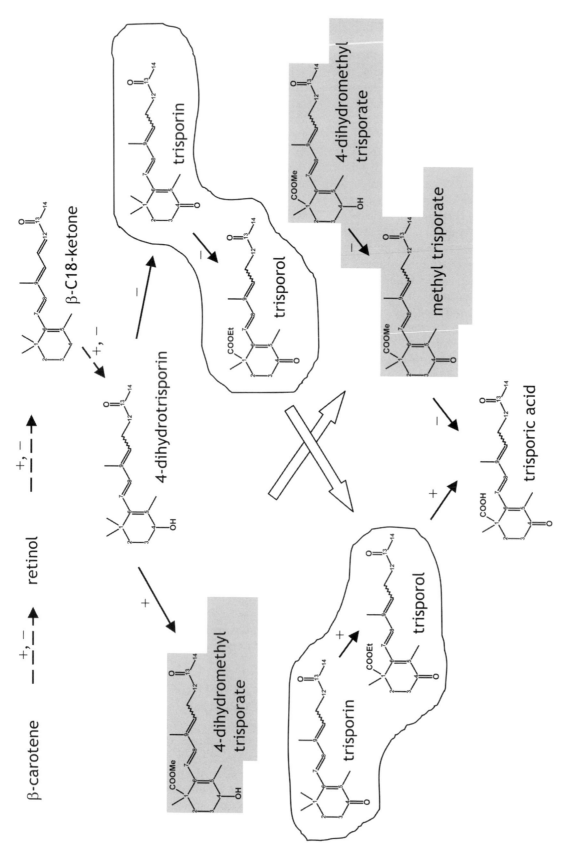

Figure 26.3   Simplified schematic representation of the trisporic acid biosynthesis pathway in *Blakeslea trispora*. Shown are the structures of the B-derivative series of trisporoids. In the A derivatives, the residues at C13, R₁₃, are H, H; in the C derivatives, R₁₃ = H, OH; in the D derivatives, R₂ = H, OH, R₁₃ = O; and in the E derivatives, R₃ = H, OH, R₁₃ = H, OH. The respective mating types where the reactions occur are indicated by a plus sign (+) or a minus sign (−) set beside the reaction arrow. The open arrows indicate an exchange of precursors between the mating types. The determination of the exact order of reactions between β-carotene and 4-dihydrotrisporin is under way (dashed arrows).

trisporoids occur in a number of natural derivatives that might be applied to such an end. These derivatives are characterized by the oxygen substitutions carried at C2 and C3 of the ring moiety (84) and at C13 in the long isoprenoid side chain. Five well-characterized derivatives have been described, but still more active combinations might occur as biosynthetic intermediates (67; D. Schachtschabel, unpublished data) or possible degradation products. To date, the synthesis steps introducing this diversity have not yet been unequivocally identified, but it seems probable that the derivatives might be converted from one to another during biosynthesis (Schachtschabel, unpublished). In every species analyzed, at least two of the derivatives, B and C (5), or B, C, and D or E (84), were found to be produced. Based on these observations, several scenarios for the realization of specificity using derivative patterns have been formulated: either the respective mating partner recognizes several of the derivatives and specificity is therefore communicated via the presented ratio between these derivatives (84), or only one of the derivatives is active as a signaling molecule and is specifically recognized. It may be considered supportive to this idea that in all instances, the B derivative of trisporic acid and of its pheromone precursors proved to cause stronger reactions than the C derivative (12, 14, 79, 91). As a third possibility, the diverse derivatives may be synthesized successively so that only one of the derivatives is present at the time of the partner's utmost susceptibility. Finally, the action of signals other than trisporoids in partner recognition is certainly possible. Especially the interactions at the hyphal surface during contact and fusion might well be supported by specialized receptors. Hitherto, no evidence for any kind of interindividual peptide or steroid signal has been detected.

At another organizational level, species specificity of the zygomycete mating reaction might also be influenced by the regulatory mechanisms involved in trisporoid biosynthesis. This has been studied to some detail for the expression and activity of the (−) mating-type-specific enzyme 4-dihydromethyltrisporate dehydrogenase. The corresponding gene, *tsp1* (24 [therein called *tdh*]), was found to be transcribed in *Mucor mucedo* (70), *Parasitella parasitica* (75), and *Absidia glauca* (unpublished), irrespective of the mating type or the actual mating situation. Nevertheless, the translation product could be detected only in the (−) mating type in *Parasitella parasitica*, whereas in *Mucor mucedo* and *Absidia glauca*, translation takes place. In *Mucor mucedo*, activity itself is restricted to the sexually stimulated (−) type, so here a regulation must occur at the posttranslational level (70). In *Absidia glauca*, finally, enzyme activity was detected in both mating types (J. Wetzel, un-

published data). This and similar regulatory differences in the regulation of other crucial biosynthesis steps would serve in establishing species specificity, as a necessary transcript or enzyme might not be present to react to a signal appearing from a partner belonging to another species.

PHYSIOLOGICAL ASPECTS OF TRISPOROID SYNTHESIS AND ACTION

Trisporoid synthesis has hitherto been studied mainly in cultures near the end of their developmental cycle. As the focus lay on elaborating the chemical nature and conversions, these experiments were strongly oriented towards the product side of the process, and a lot of care was therefore invested into finding optimal conditions for trisporoid production. To date, few data exist to clarify the role of particular trisporoids in the sexual process, and trisporoid biosynthesis in general is also not yet fully understood at the physiological level. Production conditions in laboratory environments may not necessarily mirror those encountered in a natural environment, and research is appreciably hampered by the shortage of data on the natural life cycle of many species. Nevertheless, from the collected observations a number of conclusions can be drawn concerning the factors influencing sexual morphogenesis.

For trisporoid production in *Blakeslea trispora* in the standard protocol, the strains are first grown individually to obtain mycelial mass and are then transferred to a production medium. Under natural conditions, a similar process is also conceivable, and it has been found that a certain age and a certain level of competence need to be reached before the mycelium may enter into sexual reactions, including trisporic acid production (37, 100). Under favorable conditions, mycelia show the first signs of sexual interaction at an age of 40 to 48 h and the reaction is completed within the next 2 days (46). This feature is almost certainly related to the developmental and nutritional status of the mycelium, as the formation of the sexually committed hyphae and structures requires a large amount of resources lost to further vegetative growth or asexual reproduction. In fact, a marked influence of the commitment for either developmental route is known from several species. In the homothallic species *Zygorhynchus heterogamus*, zygospore formation occurs in areas of reduced sporangiophore formation (28), and in *Mucor mucedo* (37) and *Choanephora cucurbitarum* (6), zygosporogenesis was also reported to occur at the expense of sporangiophore production.

As to the nutrient conditions favoring sexual development, the zygomycetes seem to divide into two types:

in *Syzigites megalocarpus* (93), *Mucor piriformis* and *Gilbertella persicaria* (56), and *Thamnidium elegans* (48) rich nutrient supply promotes zygospore formation, whereas in *Choanephora cucurbitarum* (6) and *Phycomyces blakesleeanus* (27, 52) nutrient limitation, especially of nitrogen, was found to be supportive. Many species also show marked preferences for some carbon sources over others. In *Phycomyces blakesleeanus* and *Blakeslea trispora*, zygospore production increases in the presence of millimolar concentrations of acetate (53). The differences in nutrient requirements are thought to reflect the physiological adaptation of the diverse species to the climate and nutritional supply of their respective habitats. Similar diversity was observed for temperature stimuli, with a drop to 6 to 7°C being mandatory for the initiation of zygospores in *Thamnidium elegans* (48) and to 10°C being optimal for *Mucor piriformis* (56). In *Mucor hiemalis*, *Zygorhynchus macrocarpus*, *Z. moelleri*, *Gilbertella persicaria*, and *Rhizopus sexualis*, but not in *Phycomyces blakesleeanus*, low temperatures are inhibitory to the early stages of sexual reactions (46, 56). The temperature-sensitive step has been proposed by Hawker and coworkers (46) to be the synthesis of a volatile signal, presumably the mating-type-specific pheromone, at a temperature at which vegetative growth still continues. In *Phycomyces blakesleeanus*, where zygospore development does not continue above 25°C, the formation of early stages actually increased at that temperature, which was discussed as a possible temperature-related increase in pheromone production (101). Concurrently with the observations on sexual morphogenesis, trisporic acid synthesis is also influenced by the carbon source and the quantity and quality of the nitrogen source (for an example, see reference 76).

Another environmental factor influencing sexual reactions is the light regime. Inhibitory effects of light, especially of continuous light, have been observed. In addition to many of the species mentioned in the previous paragraph (6, 29, 50, 56), light also inhibits zygospore formation in *Dicranophora fulva* (26) and in a number of *Mucor* species, but not in *Mucor hiemalis* (49). These results are not actually surprising, as trisporoids are exceptionally sensitive to UV irradiation, and any reaction involving the passage of such compounds between the mating partners should preferentially occur predominantly in the dark. In other species, e.g., *Absidia glauca*, no light inhibition of zygospore production is apparent. Here, zygospores are formed between strong aerial hyphae which meet each other accidentally, and thus a degradation of the pheromones useful for partner recognition would not be expected to be essential for the success of a mating reaction. For *Phycomyces*, the action

spectrum for the inhibition of zygospore production demonstrated several different photoinhibitory events during sexual development. The photoreceptors involved are neither carotene nor flavins (102) as in the other photoreactions of this fungus (33, 51). Unfortunately, the spectrum was not continued into the region of maximal absorbance of all trisporoids beyond 350 nm, so it remains probable but presently unsolved whether or not trisporic acid or its precursors are directly affected. Although trisporic acid accumulates in mixed cultures of *Blakeslea trispora*, its degradation by UV irradiation is conceivably the mode of inactivation, which is considered necessary to withdraw any regulatory compound from circulation. Actual metabolic inactivation products of trisporic acid have not been identified. As the precursors are physiologically inactive in the mating type they are produced in, inactivation of these compounds would seem to be redundant. Nevertheless in *Phycomyces blakesleeanus* and *Blakeslea trispora*, certain compounds with a shortened or circularized side chain were identified (80, 83). These apotrisporoids have been interpreted as inactivation products of trisporin. Inactivation of trisporin is supposed to counteract accumulation of this compound and thus prevent the possible production of trisporic acid, which may occur to a minimal degree in single growing *Blakeslea trispora* (+) (81; Schachtschabel, unpublished).

If produced in surplus amounts, trisporic acids are secreted into the surrounding medium. This especially applies to species where zygospore formation (*Blakeslea trispora*), or at least zygophore formation and contact (*Phycomyces blakesleeanus*), occurs in the substrate. It is quite obvious that a species with substrate zygophores would never undergo a mating reaction with a species requiring contact between aerial hyphae. This is corroborated by the mode of host contact established by *Parasitella parasitica*, whose sexual reactions occur in aerial mycelium without apparent zygophores. All pseudosexual/parasitic interactions occur likewise with aerial hyphae of the host mycelia. Probably the location of the contact organs also governs whether the volatile precursor 4-dihydromethyl trisporate or the less volatile methyl trisporate serves as the pheromone for partner finding. Nonvolatile compounds might diffuse through the medium and be taken up in higher concentrations, whereas for contact between aerial hyphae, a volatile compound would be necessary that is supposedly present and sufficient in smaller amounts. In *Absidia glauca* with exclusively aerial contact, trisporic acid is produced only in minimal amounts, and likewise in *Mucor mucedo*, with its specialized aerial zygophores, trisporoids are best isolated from the mycelium, not the surrounding substrate (38).

A related factor in mating is the location where sexual morphogenesis takes place. In *Phycomyces blakesleeanus*, growth arrest was described as a necessary prelude to the formation of zygospores (27). As growth arrest was also observed when strains of the same mating type encountered each other, this fact may not be directly related to mating reactions or trisporoid synthesis. On the other hand, in *Phycomyces blakesleeanus* the zygophores are formed from vegetative hyphal tips at the mycelial front where the chance to intercept a partner hypha is highest, and the rearrangement of the tip growth mechanism will require some time. Possibly growth arrest provides the necessary time for zygophore differentiation, which then proceeds in the presence of the partner's pheromone. In *Mucor mucedo*, the zygophores grow at exactly the position where the stimulus first reached the vegetative hypha, but no obvious growth arrest has been described before the mycelial fronts of complementary mating types intermingle. Instead, the stimulus apparently induces hyphal branching, stimulating the zygophore to enter into the aerial space. In stimulation experiments with trisporoids under laboratory conditions, the formation of mature zygophores requires about 8 to 12 h and the zygophores develop locally near the site of stimulation and shortly behind the mycelial front. As the mycelia of the mating partners continue growth, the (+) and (−) zygophores, with their length of up to 2 mm, are then within easy reach of each other.

Besides the induction of zygophore formation in both mating types, trisporic acid activities encompass mainly metabolic activation. Trisporic acid (20, 32, 85, 86, 88) and methyl trisporate B and C (44) stimulate the synthesis of β-carotene. Moreover, trisporic acid enhances ergosterol production (40, 90) and generally leads to an increase in terpenoid content (15, 85). As trisporic acid production is analyzed mostly in liquid cultures whereas zygophores and zygospores are formed only on a solid substrate, the kinetics of the two reactions cannot be easily compared. The rise in β-carotene synthesis is caused by direct stimulation of phytoene synthase, lycopene cyclase, and phytoene desaturase transcription (74). As the respective transcription levels were determined in mated cultures of *Blakeslea trispora*, it cannot be excluded that one of the precursors necessarily present in the culture mix caused the stimulatory effect. The metabolic effects of trisporic acid may depend on a direct signaling function of the molecule by activating a trisporic acid binding protein that in turn acts as transcription factor. Similar regulatory functions have been shown for the chemically related retinoid compounds involved in multiple developmental regulation processes in animals (for an example, see reference 7). Trisporic acid furthermore promotes the production of its own precursors. This effect is driven not only by the increase in β-carotene synthesis, but also by a transcriptional upregulation of a putative β-carotene dioxygenase as a reaction to trisporic acid B that has been observed in *Blakeslea trispora* (M. Richter, unpublished data). The metabolic activities of the various precursors have never been studied in any detail. They all induce zygophore formation in the complementary mating partner, and some of them also positively affect β-carotene synthesis. These reactions may be caused by their conversion to trisporic acid in the mating partner, thus suppressing similar reactions in the original hypha. On the other hand, the respective precursors might regulate responses in the receptive hypha. Whatever applies, the expressed mating-type specificity will depend mainly on the existence of specific receptors or precursor binding proteins. For the physiological activity of trisporids, the overall dimensions of the various compounds were found to be critical, which reflects the specificity of the presumed binding proteins. Compounds with slightly longer or shorter isoprenoid side chains are inactive in zygophore induction (12, 67, 80). The polarity of the functional groups at C4 and C13 and the number and position of double bonds are also important, whereas an oxygen substitution at the ring moiety proved not to be essential for function (67).

GENES INVOLVED IN SEXUAL DIFFERENTIATION

Functional studies of sexual differentiation and development in zygomycetes are severely compromised by deficits in the toolkit of genetic analysis. Classical genetic characterization has been attempted, especially for *Phycomyces blakesleeanus* but, due to poor germination frequencies of zygospores, has never reached the same level of refinement as for the ascomycetous model systems *Neurospora crassa* and *Aspergillus nidulans*. An additional, more fundamental problem lies in the intrinsically complicated karyotic situation in fused gametangia with hundreds of nuclei. Although it has been shown that approximately 80% of the offspring of each germ sporangium derive from a single meiotic event (31), the origin of a single progeny spore from a meiotic germ sporangium is not certain.

In addition to these biological peculiarities of the meiotic system, the techniques for genetic manipulation and analysis in vitro are far from satisfactory. Several mucoralean zygomycetes can be transformed with satisfactory rates. Many different vectors have been constructed, based on the dominantly selectable, Tn5-derived Neo[R] marker (97) or on complementation of *leuA* mutants by the wild-type allele (3, 92). It was also possible to construct plasmids carrying the versatile reporter gene coding for the green fluorescent protein under the control of the strong and essentially constitutive promoters for the translational elongation factor EF1α or for actin (68). While holding a great potential for studying promoter activity in appropriate transformants, it is not yet possible to visualize reporter activity under the control of developmentally regulated promoters. Unfortunately, all *Mucor*-like fungi tend to propagate plasmid DNA extrachromosomally. Integrative events can hardly be predicted and planned and have been obtained only in rare cases in *Absidia glauca* (19) and, somewhat more reliably, in *Mucor circinelloides* (3).

Due to these general experimental constraints, genetic analysis of sexual differentiation in vivo is performed by protoplast fusion and, at the molecular level, by analytical rather than manipulatory approaches.

Protoplast fusion between strains of complementary mating types is suitable for elucidating if both sets of genetic information are compatible in a common individual. In *Phycomyces blakesleeanus*, somatic intermating-type hybridization does not result in zygospores, but rather in so-called pseudophores that are interpreted as abortive structures of sexual conjugation (21). Different mating types in a single individual are more pronounced in *Mucor pusillus*, where normal zygospores are formed following somatic fusion (61).

In *Absidia glauca*, the situation has been studied in more detail (99). Although it is a strictly heterothallic species, somatic fusions between auxotrophically labeled mating types develop zygospores of normal size and appearance in large numbers. Both mating types are thus fully compatible in a single organism. These somatic fusions are stable only under selective conditions. Comparing the segregation behavior of several genetic markers indicated that nuclei belonging to complementary mating types can undergo fusion. Under nonselective conditions for the auxotrophic traits of the fusion partners, a strong bias for one of the genetic markers is observed, whereas the distribution of the distinguishable ribosomal DNA clusters ranges around 50% under all culture conditions. This highly biased segregation behavior can best be explained by assuming nuclear fusion followed by segregation of the diploid chromosome set towards aneuploidy. Nuclear fusions, revealed by different types of segregating recombinants, have been observed for different fusions with a variety of auxotrophies, including events that involve a change in mating type with respect to the initial auxotrophic markers of the parental strains (96). Thus, the loci determining mating type and controlling sexual development are interchangeable between mating types and do not depend on the mating-type-specific genetic background.

Due to the problems in performing genetic crosses and segregation analyses, there is little information on the genes determining mating types. Genetic data are available for *Rhizopus stolonifer* (35), where Mendelian segregation was obtained for two loci, mating type and sulfur utilization. Arguably the best genetic analysis has been performed with *Phycomyces blakesleeanus* (1). A mutant collection with 10 different markers, including mating type, was used for crosses. According to recombination frequencies, it was possible to ascribe these markers to six linkage groups. All genetic data are compatible with the classical assumption of a single mating-type locus with two alleles, specifying either the (+) or the (−) behavior in crosses. There are, however, no clues to the nature of this locus and no indications of mating-type control mediated by a master regulator or general transcription factor. There are also no mutants known with the expected pleiotropic phenotypes that could correspond to a defect in a putative mating-type locus. In zygomycetes, different models can be outlined for explaining the observed mating behavior. The obligatory complementarity between the mating types in synthesizing the sexual pheromone trisporic acid would be sufficient to account for the strict dependency of developing sexual structures on the presence of both mating types. In this sense, there is no mandatory need for an additional mating-type locus. Instead, the genes themselves coding for the trisporoid pathway could represent the mating-type locus. Unfortunately, the degree of genetic analysis is not sufficiently refined to test this hypothesis strictly. Only a single gene from the pathway, *tsp1* for 4-dihydromethyltrisporate dehydrogenase, has been cloned, sequenced, and analyzed for regulatory features (24, 70, 75). Although *tsp1* has been examined in several species, *Mucor mucedo*, *Parasitella parasitica*, and *Absidia glauca*, there is no indication for clustering with other trisporic acid pathway genes. The partial genomic sequence of *Rhizopus oryzae* also provides no evidence for clustering of trisporic acid genes at a single chromosomal location. There is, however, some clustering around the *tsp1* gene. In all three species mentioned, a putative gene for an acylthioesterase has been found in the immediate vicinity and also several unidentified short open reading frames are remarkably conserved (75). It is conceivable that knowing the *tsp1* sequence alone is not sufficient for identifying additional trisporoid biosynthesis genes. Cloning of a second gene, *tsp2* for 4-dihydrotrisporin dehydrogenase from *Mucor mucedo*, is under way. Possibly, predicting additional trisporoid-associated genes based on sequence data alone will be feasible afterwards and lead us one step further towards the goal of understanding the elusive nature of mating-type determination in the zygomycetes.

References

1. Alvarez, M. I., M. I. Pelaez, and A. P. Eslava. 1980. Recombination between 10 markers in *Phycomyces*. *Mol. Gen. Genet.* **179**:447–452.

2. Ansell, P. J., and T. W. K. Young. 1983. Light and electron microscopy of *Mortierella indohii* zygospores. *Mycologia* **75**:64–69.

3. Arnau, J., L. P. Jepsen, and P. Strømann. 1991. Integrative transformation by homologous recombination in the zygomycete *Mucor circinelloides*. *Mol. Gen. Genet.* **225**:193–198.

4. Austin, D. J., J. D. Bu'Lock, and D. Drake. 1970. The biosynthesis of trisporic acids from β-carotene via retinal and trisporol. *Experientia* **26**:348–349.

5. Austin, D. J., J. D. Bu'Lock, and G. W. Gooday. 1969. Trisporic acids: sexual hormones from *Mucor mucedo* and *Blakeslea trispora*. *Nature* **223**:1178–1179.

6. Barnett, H. L., and V. G. Lilly. 1956. Factors affecting the production of zygospores by *Choanephora cucurbitarum*. *Mycologia* **48**:617–627.

7. Bastien, J., and C. Rochette-Egly. 2004. Nuclear retinoid receptors and the transcription of retinoid target genes. *Gene* **17**:1–16.

8. Benjamin, R. K. 1958. Sexuality in the Kickxellaceae. *Aliso* **4**:149–169.

9. Blakeslee, A. F. 1904. Sexual reproduction in the Mucorineae. *Proc. Am. Acad. Arts Sci.* **40**:205–319.

10. Blakeslee, A. F., and J. L. Cartledge. 1927. Sexual dimorphism in Mucorales. II. Interspecific reactions. *Bot. Gaz.* **84**:51–57.

11. Blakeslee, A. F., J. L. Cartledge, D. S. Welch, and A. D. Bergner. 1927. Sexual dimorphism in Mucorales. I. Intraspecific reactions. *Bot. Gaz.* **84**:27–50.

12. Bu'Lock, J. D., B. E. Jones, and N. Winskill. 1976. The apocarotenoid system of sex hormones and prohormones in Mucorales. *Pure Appl. Chem.* **47**:191–202.

13. Bu'Lock, J. D., B. E. Jones, S. A. Quarrie, and N. Winskill. 1973. The biochemical basis of sexuality in Mucorales. *Naturwissenschaften* **60**:550–551.

14. Bu'Lock, J. D., D. Drake, and D. J. Winstanley. 1972. Specificity and transformations of the trisporic acid series of fungal sex hormones. *Phytochemistry* **11**:2011–2018.

15. Bu'Lock, J. D., and A. U. Osagie. 1973. Prenols and ubiquinones in single-strain and mated cultures of *Blakeslea trispora*. *J. Gen. Microbiol.* **76**:77–83.

16. Bu'Lock, J. D., B. E. Jones, D. Taylor, N. Winskill, and S. A. Quarrie. 1974. Sex hormones in Mucorales. The incorporation of C_{20} and C_{18} precursors into trisporic acids. *J. Gen. Microbiol.* **80**:301–306.

17. Burgeff, H. 1924. Untersuchungen über Sexualität und Parasitismus bei Mucorineen. *Bot. Abhandlungen* **4**:1–135.

18. Burgeff, H., and M. Plempel. 1956. Zur Kenntnis der Sexualstoffe bei Mucorineen. *Naturwissenschaften* **43**:473–474.

19. Burmester, A., A. Wöstemeyer, and J. Wöstemeyer. 1990. Integrative transformation of a zygomycete, *Absidia glauca*, with vectors containing repetitive DNA. *Curr. Genet.* **17**:155–161.

20. Caglioti, L., G. Cainelli, B. Camerino, R. Mondelli, A. Prieto, A. Quilico, T. Salvatori, and A. Selva. 1966. The structure of trisporic-C acid. *Tetrahedron Suppl.* **7**:175–187.

21. Cerdá-Olmedo, E. 1974. *Phycomyces*, p. 343–357. *In* R. C. King (ed.), *Handbook of Genetics* 1. Plenum, New York, NY.

22. Cerdá-Olmedo, E., and E. D. Lipson. 1987. A biography of Phycomyces, p. 7–26. *In* E. Cerdá-Olmedo and E. D. Lipson (ed.), Phycomyces. Cold Spring Harbor Laboratory, Cold Spring Harbor, NY.

23. Chang, C. W., H. C. Yang, and L. S. Leu. 1984. Zygospore formation of *Choanephora cucurbitarum*. *Trans. Mycol. Soc. Jpn.* **25**:67–74.

24. Czempinski, K., V. Kruft, J. Wöstemeyer, and A. Burmester. 1996. 4-Dihydromethyltrisporate dehydrogenase from *Mucor mucedo*, an enzyme of the sexual hormone pathway: purification, and cloning of the corresponding gene. *Microbiology* **142**:2647–2654.

25. Degawa, Y., and S. Tokumasu. 1998. Two new homothallic species of *Mortierella*, *M. cogitans*, and *M. microzygospora*, and their zygospore formation. *Mycologia* **90**:1040–1046.

26. Dobbs, C. G. 1938. The life history and morphology of *Dicranophora fulva*. *Trans. Br. Mycol. Soc.* **21**:167–192.

27. Drinkard, L. C., G. E. Nelson, and R. P. Sutter. 1982. Growth arrest: a prerequisite for sexual development in *Phycomyces blakesleeanus*. *Exp. Mycol.* **6**:52–59.

28. Edelmann, R. E., and K. L. Klomparens. 1995. Zygosporogenesis in *Zygorhynchus heterogamus*, with a proposal for standardization of structural nomenclature. *Mycologia* **87**:304–318.

29. Edelmann, R. E., and K. L. Klomparens. 1995. Low temperature scanning electron microscopy of the ultrastructural development of zygospores and sporangiospores in *Mycotypha africana*, and the effects of cultural conditions on sexual versus asexual reproduction. *Mycol. Res.* **99**:539–548.

30. Eslava, A. P., and M. I. Alvarez. 1987. Crosses, p. 361–365. *In* E. Cerdá-Olmedo and E. Lipson (ed.), Phycomyces. Cold Spring Harbor Press, Cold Spring Harbor, NY.

31. Eslava, A. P., M. I. Alvarez, and M. Delbrück. 1975. Meiosis in *Phycomyces*. *Proc. Natl. Acad. Sci. USA* **72**:4076–4080.

32. Feofila, E. P., T. V. Fateeva, and V. A. Arbuzov. 1976. Mechanism of the action of trisporic acids on carotene-synthesizing enzymes of a (−) strain of *Blakeslea trispora*. *Microbiologiya* **45**:153–155.

33. Galland, P., and N. Tölle. 2003. Light-induced fluorescence changes in *Phycomyces*: evidence for blue light-receptor associated flavo-semiquinones. *Planta* **217**:971–982.

34. **Gams, W., and S. T. Williams.** 1963. Heterothallism in *Mortierella parvispora* Linnemann. *Nova Hedwigia* **5**:347–357.

35. **Gauger, W. L.** 1977. Meiotic gene segregation in *Rhizopus stolonifer*. *J. Gen. Microbiol.* **101**:211–217.

36. **Gessler, N. N., A. V. Sokolov, and T. A. Belozerskaya.** 2002. Initial stages of trisporic acid synthesis in *Blakeslea trispora*. *Appl. Biochem. Microbiol.* **38**:536–543.

37. **Gooday, G. W.** 1968. Hormonal control of sexual reproduction in *Mucor mucedo*. *New Phytol.* **67**:815–821.

38. **Gooday, G. W.** 1968. The extraction of a sexual hormone from the mycelium of *Mucor mucedo*. *Phytochemistry* **7**:2103–2105.

39. **Gooday, G. W.** 1973. Differentiation in the Mucorales. *Symp. Soc. Gen. Microbiol.* **23**:269–293.

40. **Gooday, G. W.** 1978. Functions of trisporic acid. *Philos. Trans. R. Soc. Lond. B* **284**:509–520.

41. **Gooday, G. W.** 1983. Hormones and sexuality in fungi, p. 239–266. *In* J. W. Bennett and A. Ciegler (ed.), *Secondary Metabolism and Differentiation in Fungi*. Dekker, New York, NY.

42. **Gooday, G. W., and M. J. Carlile.** 1997. The discovery of fungal sex hormones. III. Trisporic acid and its precursors. *Mycologist* **11**:1263–130.

43. **Gooday, G. W., P. Fawcett, D. Green, and G. Shaw.** 1973. The formation of fungal sporopollenin in the zygospore wall of *Mucor mucedo*: a role for the sexual carotenogenesis in the Mucorales. *J. Gen. Microbiol.* **74**:233–239.

44. **Govind, N. S., and E. Cerdá-Olmedo.** 1986. Sexual activation of carotenogenesis in *Phycomyces blakesleeanus*. *J. Gen. Microbiol.* **132**:2775–2780.

45. **Hawker, L. E., M. A. Gooday, and C. E. Bracker.** 1966. Plasmodesmata in fungal cell walls. *Nature* **212**:635.

46. **Hawker, L. E., P. M. Hepden, and S. M. Perkins.** 1957. The inhibitory effects of low temperature on early stages of zygospore production in *Rhizopus sexualis*. *J. Gen. Microbiol.* **17**:758–767.

47. **Hepden, P. M., and L. E. Hawker.** 1961. A volatile substance controlling early stages of zygospore formation in *Rhizopus sexualis*. *J. Gen. Microbiol.* **24**:155–164.

48. **Hesseltine, C. W., and P. Anderson.** 1956. The genus *Thamnidium* and a study of the formation of its zygospores. *Am. J. Bot.* **43**:696–703.

49. **Hesseltine, C. W., and R. Rogers.** 1987. Dark-period induction of zygospores in *Mucor*. *Mycologia* **79**:289–297.

50. **Hocking, D.** 1967. Zygospore initiation, development and germination in *Phycomyces blakesleeanus*. *Trans. Br. Mycol. Soc.* **50**:207–220.

51. **Idnurm, A., J. Rodriguez-Romero, L. M. Corrochano, C. Sanz, E. A. Iturriaga, A. P. Eslava, and J. Heitman.** 2006. The *Phycomyces* madA gene encodes a blue-light photoreceptor for phototropism and other light responses. *Proc. Natl. Acad. Sci. USA* **103**:4546–4551.

52. **Komarova, G. V., A. N. Kozlova, G. I. El-Registan, S. A. Egorova, and N. A. Krasilnikov.** 1972. Cultivation of *Phycomyces blakesleeanus* and study of sexual reproduction. *Mikrobiologiya* **41**:93–98.

53. **Kuzina, V., and E. Cerdá-Olmedo.** 2006. Modification of sexual development and carotene production by acetate and other small carboxylic acids in *Blakeslea trispora* and

Phycomyces blakesleeanus. *Appl. Environ. Microbiol.* **72**:4917–4922.

54. **Laane, M. M.** 1974. Nuclear behaviour during vegetative stage and zygospore formation in *Absidia glauca*. *Norw. J. Bot.* **21**:125–135.

55. **Mehrotra, B. S., and V. S. Mehrotra.** 1990. Imperfect mating reaction in *Rhizomucor pusillus* on crossing with *Mucor hiemalis*. *Natl. Acad. Sci. Lett.* **13**:113–114.

56. **Michailides, T. J., L.-Y. Guo, and D. P. Morgan.** 1997. Factors affecting zygosporogenesis in *Mucor piriformis* and *Gilbertella persicaria*. *Mycologia* **89**:603–609.

57. **Miller, M. L., and R. P. Sutter.** 1984. Methyl trisporate E. A sex pheromone in *Phycomyces blakesleeanus*. *J. Biol. Chem.* **259**:6420–6422.

58. **Nieuwenhuis, M., and H. van den Ende.** 1975. Sex specificity of hormone synthesis in *Mucor mucedo*. *Arch. Microbiol.* **102**:167–169.

59. **O'Donnell, K. L., J. J. Ellis, C. W. Hesseltine, and G. R. Hooper.** 1977. Zygosporogenesis in *Gilbertella persicaria*. *Can. J. Bot.* **55**:662–675.

60. **O'Donnell, K. L., J. J. Ellis, C. W. Hesseltine, and G. R. Hooper.** 1977. Morphogenesis of azygospores induced in *Gilbertella persicaria* (+) by imperfect hybridization with *Rhizopus stolonifer* (−). *Can. J. Bot.* **55**:2721–2727.

61. **Ohnuki, T., Y. Etoh, and T. Beppu.** 1982. Intraspecific and interspecific hybridization of *Mucor pusillus* and *Mucor miehei* by protoplast fusion. *Agric. Biol. Chem.* **46**:451–458.

62. **Perry, D. F., and J.-P. Latge.** 1982. Dormancy and germination of *Conidiobolus obscurus* azygospores. *Trans. Br. Mycol. Soc.* **78**:221–225.

63. **Plempel, M.** 1957. Die Sexualstoffe der Mucoraceae. Ihre Abtrennung und die Erkaerung ihrer Funktion. *Arch. Mikrobiol.* **26**:151–174.

64. **Plempel, M.** 1962. Die zygotropische Reaktion bei Mucorineen. *Planta* **55**:254–258.

65. **Plempel, M.** 1963. Die chemischen Grundlagen der Sexualreaktionen bei Zygomyceten. *Planta* **59**:492–508.

66. **Satina, S., and A. F. Blakeslee.** 1930. Imperfect sexual reactions in homothallic and heterothallic Mucors. *Bot. Gaz.* **90**:299–311.

67. **Schachtschabel, D., C. Schimek, J. Wöstemeyer, and W. Boland.** 2005. Biological activity of trisporoids and trisporoid analogues in *Mucor mucedo* (−). *Phytochemistry* **66**:1358–1365.

68. **Schilde, C., J. Wöstemeyer, and A. Burmester.** 2001. Green fluorescent protein as a reporter for gene expression in the mucoralean fungus *Absidia glauca*. *Arch. Microbiol.* **175**:1–7.

69. **Schimek, C., and J. Wöstemeyer.** 2006. Pheromone action in the fungal groups *Chytridiomycota*, and *Zygomycota*, and in the *Oomycota*, p. 215–231. *In* U. Kües and R. Fischer (ed.), *The Mycota I*, 2nd ed. Springer-Verlag, Berlin, Germany.

70. **Schimek, C., A. Petzold, K. Schultze, J. Wetzel, F. Wolschendorf, A. Burmester, and J. Wöstemeyer.** 2005. 4-Dihydromethyltrisporate dehydrogenase, an enzyme of the sex hormone pathway in *Mucor mucedo*, is constitutively transcribed but its activity is differently regulated in (+) and (−) mating types. *Fungal Genet. Biol.* **42**:804–812.

71. Schimek, C., K. Kleppe, A.-R. Saleem, K. Voigt, A. Burmester, and J. Wöstemeyer. 2003. Sexual reactions in Mortierellales are mediated by the trisporic acid system. *Mycol. Res.* **107:**736–747.

72. Schipper, M. A. A. 1976. Induced azygospore formation in *Mucor* (*Rhizomucor*) *pusillus* by *Absidia corymbifera*. *Antonie Leeuwenhoek* **42:**141–144.

73. Schipper, M. A. A., and J. A. Stalpers. 1980. Various aspects of the mating system in Mucorales. *Persoonia* **11:** 53–63.

74. Schmidt, A. D., T. Heinekamp, M. Matuschek, B. Liebmann, C. Bollschweiler, and A. A. Brakhage. 2005. Analysis of mating-dependent transcription of *Blakeslea trispora* carotenoid biosynthesis genes *carB* and *carRA* by quantitative real-time PCR. *Appl. Microbiol. Biotechnol.* **67:**549–555.

75. Schultze, K., C. Schimek, J. Wöstemeyer, and A. Burmester. 2005. Sexuality and parasitism share common regulatory pathways in the fungus *Parasitella parasitica*. *Gene* **348:** 33–44.

76. Sutter, R. P. 1975. Mutations affecting sexual development in *Phycomyces blakesleeanus*. *Proc. Natl. Acad. Sci. USA* **72:**127–130.

77. Sutter, R. P. 1977. Regulation of the first stage of sexual development in *Phycomyces blakesleeanus* and in other mucoraceous fungi, p. 251–272. *In* D. H. O'Day and P. A. Horgen (ed.), *Eukaryotic Microbes as Model Development Systems*. Dekker, New York, NY.

78. Sutter, R. P., and J. P. Whitaker. 1981. Sex pheromone metabolism in *Blakeslea trispora*. *Naturwissenschaften* **68:**147–148.

79. Sutter, R. P., and J. P. Whitaker. 1981. Zygophore-stimulating precursors (pheromones) of trisporic acids active in (−)-*Phycomyces blakesleeanus*. *J. Biol. Chem.* **256:**2334–2341.

80. Sutter, R. P., and P. D. Zawodny. 1984. Apotrisporin: a major metabolite of *Blakeslea trispora*. *Exp. Mycol.* **8:**89–92.

81. Sutter, R. P., D. A. Capage, T. L. Harrison, and W. A. Keen. 1973. Trisporic acid biosynthesis in separate plus and minus cultures of *Blakeslea trispora*: identification of two mating-type-specific components. *J. Bacteriol.* **114:** 1074–1082.

82. Sutter, R. P., T. L. Harrison, and G. Galasko. 1974. Trisporic acid biosynthesis in *Blakeslea trispora* via mating type-specific precursors. *J. Biol. Chem.* **249:**2282–2284.

83. Sutter, R. P. 1986. Apotrisporin-E: a new sesquiterpenoid isolated from *Phycomyces blakesleeanus* and *Blakeslea trispora*. *Exp. Mycol.* **10:**256–258.

84. Sutter, R. P., J. Dadok, A. A. Bothner-By, R. R. Smith, and P. K. Mishra. 1989. Cultures of separated mating types of *Blakeslea trispora* make D and E forms of trisporic acids. *Biochemistry* **28:**4060–4066.

85. Thomas, D. M., and T. W. Goodwin. 1967. Studies on carotenogenesis in *Blakeslea trispora*. I. General observations on synthesis in mated and unmated strains. *Phytochemistry* **6:**355–360.

86. Vail, W. J., C. Morris, and V. G. Lilly. 1967. Hormone-like substances which increase carotenogenesis in + and − sexes of *Choanephora cucurbitarum*. *Mycologia* **59:**1069–1074.

87. van den Ende, H. 1967. Sexual factors of the Mucorales. *Nature* **215:**211–212.

88. van den Ende, H. 1968. Relationship between sexuality and carotene synthesis in *Blakeslea trispora*. *J. Bacteriol.* **96:**1298–1303.

89. van den Ende, H. 1976. *Sexual Interactions in Plants*. Academic Press, London, United Kingdom.

90. van den Ende, H. 1978. Sexual morphogenesis in the Phycomycetes, p. 257–274. *In* J. E. Smith and D. R. Berry (ed.), *The Filamentous Fungi*. Arnold, London, United Kingdom.

91. van den Ende, H., A. H. Wiechmann, D. J. Reyngoud, and T. Hendriks. 1970. Hormonal interactions in *Mucor mucedo* and *Blakeslea trispora*. *J. Bacteriol.* **101:** 423–428.

92. Wada, M., T. Beppu, and S. Horinouchi. 1996. Integrative transformation of the zygomycete *Rhizomucor pusillus* by homologous recombination. *Appl. Microbiol. Biotechnol.* **45:**652–657.

93. Wenger, C. J., and V. G. Lilly. 1966. The effects of light on carotenogenesis, growth, and sporulation of *Syzigites megalocarpus*. *Mycologia* **58:**671–680.

94. Werkman, B. A. 1976. Localization and partial characterization of a sex-specific enzyme in homothallic and heterothallic Mucorales. *Arch. Microbiol.* **109:** 209–213.

95. Werkman, B. A., and H. van den Ende. 1974. Trisporic acid synthesis in homothallic and heterothallic Mucorales. *J. Gen. Microbiol.* **82:**273–278.

96. Wöstemeyer, A., H. Teepe, and J. Wöstemeyer. 1990. Genetic interactions in somatic inter-mating type hybrids of the zygomycete *Absidia glauca*. *Curr. Genet.* **17:**163–168.

97. Wöstemeyer, J., A. Burmester, and C. Weigel. 1987. Neomycin resistance as dominantly selectable marker for transformation of the zygomycete *Absidia glauca*. *Curr. Genet.* **12:**625–627.

98. Wöstemeyer, J., A. Wöstemeyer, A. Burmester, and K. Czempinski. 1995. Relationships between sexual processes and parasitic interactions in the host-pathogen system *Absidia glauca–Parasitella parasitica*. *Can. J. Bot.* **73:**S243–S250.

99. Wöstemeyer, J., and E. Brockhausen-Rohdemann. 1987. Inter-mating type protoplast fusion in the zygomycete *Absidia glauca*. *Curr. Genet.* **12:**435–441.

100. Wurtz, T., and H. Jockusch. 1978. Morphogenesis in *Mucor mucedo*: mutations affecting gamone response and organ differentiation. *Mol. Gen. Genet.* **159:**249–257.

101. Yamazaki, Y., and T. Ootaki. 1996. Roles of extracellular fibrils connecting progametangia in mating of *Phycomyces blakesleeanus*. *Mycol. Res.* **100:**984–988.

102. Yamazaki, Y., H. Kataoka, A. Miyazaki, M. Watanabe, and T. Ootaki. 1996. Action spectra for photoinhibition of sexual development in *Phycomyces blakesleeanus*. *Photochem. Photobiol.* **64:**387–392.

103. Yu, M. Q., and W. H. Ko. 1997. Factors affecting germination and the mode of germination of zygospores of *Choanephora cucurbitarum*. *J. Phytopathol.* **145:**357–361.

Sex in Fungi: Molecular Determination
and Evolutionary Implications
Edited by Joseph Heitman et al.
© 2007 ASM Press, Washington, D.C.

Howard S. Judelson

27

Sexual Reproduction in Plant Pathogenic Oomycetes:
Biology and Impact on Disease

The sexual cycle is a significant component of the life histories of oomycetes, a group of fungus-like eukaryotes that include many devastating plant pathogens. Sexual reproduction is important not only as a source of genetic variation, but also because it creates thick-walled sexual spores called oospores. These can be important for surviving harsh environments such as freezing or dry conditions and for resisting microbial attack. Oospores are the most durable of all oomycete propagules and are particularly valuable to the pathogenic species which generally persist poorly as saprophytes in soil or plant debris. The relative importance of the sexual cycle in each pathosystem can vary depending on factors that include climate, host, cropping practices, patterns of pathogen migration, and the inherent biology of each species.

This chapter describes the biology of sexual reproduction in phytopathogenic oomycetes and its role in disease. Examples are provided across a spectrum of pathosystems in which the sexual cycle appears to play essential, minor, or intermediate roles. How sexuality confers long-term evolutionary advantages to oomycetes is also discussed.

IMPORTANCE AND TAXONOMY OF OOMYCETES

Oomycetes are a diverse group, encompassing both saprophytes and species pathogenic on plants or animals. Of the approximately 500 recognized species, those with the greatest impact on humans are the plant pathogens. These infect both mono- and dicotyledonous crops, ornamentals, and native plants, causing foliar blights or root, crown, or fruit rots. Annual losses from these diseases exceed tens of billions of dollars, and greater losses are avoided only through intensive applications of crop protection chemicals and resistance breeding efforts (2).

Most oomycetes causing significant diseases on plants are grouped within the orders Peronosporales and Saprolegniales. Belonging to the former is the best-studied oomycete, the potato late blight agent *Phytophthora infestans*. This species is notorious for its role in causing the Irish potato famine in the 1840s (88) and is still responsible for major losses worldwide (99). More than 60 other destructive *Phytophthora* species also exist. These include *P. capsici*, which is a limiting factor in the production of peppers, tomatoes, and cucurbits; *P. sojae*, which causes soybean root rot; *P. palmivora* and *P. megakarya*, which

Howard S. Judelson, Department of Plant Pathology and Microbiology, University of California, Riverside, CA 92521.

are responsible for black pod of cacao; and the newly discovered species *P. ramorum*, which is infecting deciduous trees and shrubs in North America and Europe (17, 20, 40). Related to *Phytophthora* is the genus *Pythium*, which includes more than 101 species that cause root and seed rots as well as foliage diseases (68). *Pythium aphanidermatum*, for example, is a major pathogen of a broad range of plants including tomato, cotton, and grasses. Other important members of the Peronosporales include *Bremia*, *Hyaloperonospora*, *Peronospora*, *Peronosclerospora*, *Plasmopara*, *Pseudoperonospora*, and *Sclerospora*, which cause downy mildew on crops and ornamentals (73), and *Albugo*, which causes white rust on crucifers. Within the Saprolegniales, the genus *Aphanomyces* includes significant damping-off pathogens such as *A. cochlioides*, which is a major problem on sugar beet (111).

Oomycetes were once grouped in taxonomic schemes with true fungi (i.e., ascomycetes and basidiomycetes) due to their shared patterns of filamentous growth and the ability of many to feed on decaying matter. However, contemporary phylogenies assign oomycetes to another branch of the eukaryotic tree, forming the kingdom Stramenopila along with brown algae (kelps) and diatoms (6, 100). Notable features distinguishing oomycetes from true fungi include the absence of significant amounts of chitin in the oomycete cell walls, the use of the β-1,3-glucan mycolaminarin as the major storage carbohydrate, and the absence of a free haploid stage, as oomycete hyphae are diploid (98).

The diploidy of oomycetes (and the absence of a free haploid stage) underlies just one reason why the sexual cycles of oomycetes must operate distinctly from those of true fungi, as is described in the next section. An important technical consequence of diploidy is that mutagenesis strategies as used traditionally in studies of fungi are not easily applied to oomycetes. This helps explain why our understanding of the molecular genetics of mating in oomycetes is rudimentary. For example, unlike most species described in this volume, mating-type loci have not yet been cloned from any oomycete. Another limiting factor is that many oomycetes, namely the downy mildews and white rusts, are obligate pathogens not culturable apart from their hosts. Others such as *Phytophthora*, however, can be propagated on artificial media and represent therefore the best models for laboratory studies. Most knowledge of sexual processes consequently originates from such species.

BIOLOGY AND GENETICS OF SEXUAL REPRODUCTION

Oomycetes display either heterothallism or homothallism. Homothallic species are capable of sexual reproduction in single culture. In contrast, heterothallics generally require the interaction of two sexual compatibility types (mating types). These have been named A1 and A2 in *Phytophthora*, B1 and B2 in *Bremia*, and P1 and P2 in *Plasmopara* (30, 74, 112). Sexual reproduction in such species therefore usually involves outcrossing. The sexual cycle of a typical oomycete is illustrated in Fig. 27.1.

Cytology of Mating

Sexual reproduction starts in both homothallic and heterothallic species with the development and pairing of male and female gametangia, which appear initially as swollen hyphal tips with no obvious morphological or cytological differentiation (21, 41). The oogonium provides most of the cytoplasm and nutrient reserves for the oospore and is considered the female gamete, while the male is the antheridium. Both organs, but especially the oogonium, expand after contact. Numerous vesicles are observed on the antheridial side of the contact region, suggesting that an adhesive may be secreted. Nuclei within the oogonial and antheridial initials are diploid as in vegetative hyphae but then undergo meiosis. After the antheridial initial penetrates the developing oogonium, a haploid postmeiotic nucleus moves from antheridium to oogonium, forming a single diploid zygote; the remaining unfertilized nuclei within the oogonium then degrade (41). Little if any cytoplasm enters the oogonium from the antheridium, and a plug forms to separate the oogonium from the rest of the thallus. After fertilization the oogonial cytoplasm becomes rich in lipid bodies, proteins, β-linked glucose polymers, and vacuoles. Ribosomes and cytochromes disappear during maturation, indicating that the metabolic activity of the oospore will be at a very low level (41, 63). A thick multilayered wall develops which is important for protection against environmental stresses and microbial degradation; the presence of the wall, surrounding a cytoplasmic matrix rich in lipid reserves, is considered to make the oospore extremely well adapted as a resting structure. The precise structure of the wall varies in different taxa, but in *Phytophthora*, *Pythium*, and *Plasmopara* it ranges in thickness from 200 to 800 nm, containing up to 12 uneven and often confluent dense layers which culminate in three layers in the mature oospore (41, 90, 107). This thickness compares to a diameter of average oospores of about 30 μm. The wall usually pulls away from the exterior boundary of the oogonium, giving the appearance of an endospore (Fig. 27.2).

The gametangia of homothallic species are often monoclinous, with male and female organs developing from the same thallus. This situation favors self-fertilization. However, diclinous antheridia (not directly connected to the oogonia) can also form; consequently, a homothallic species can sometimes outcross with another strain (Fig.

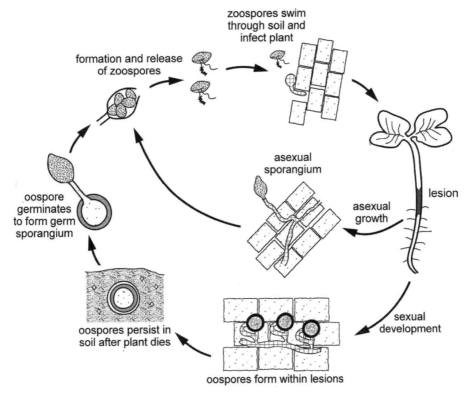

Figure 27.1 Simplified life cycle of homothallic oomycete plant pathogen.

27.3) (26, 108). No physical or genetic barrier to outcrossing therefore appears to exist, at least in *Phytophthora* and *Pythium*.

While heterothallic strains of many saprophytic oomycetes are dioecious, the pathogenic species are typically bisexual. However, a system of relative sexuality exists in which some strains may act more male or female (29). Sexual choice results from an apparently complex interaction between strains, as the predominance of male or female differentiation by one strain

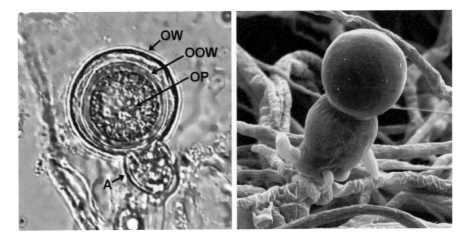

Figure 27.2 Light and scanning electron microscopy of *P. infestans* oospores. Note in the left image the thick inner oospore wall, which has retracted from the external layer of the oosphere, giving the appearance of an endospore. Shown are the antheridium (A), outer oospore wall (OOW), oogonial wall (OW), and ooplast (OP). Reprinted from *Fungal Genetics and Biology* (47) and *Nature Reviews Microbiology* (48) with permission of the publishers.

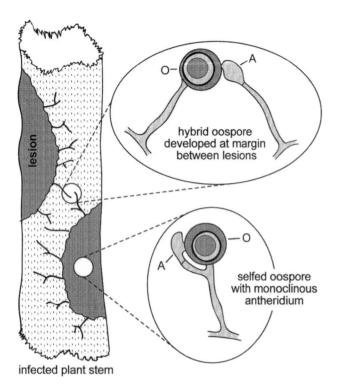

Figure 27.3 Selfing and outcrossing by homothallic oomycetes. Oospores produced within a lesion caused by a single strain are necessarily selfs, but hybrids can form when plants are coinfected by two strains. The structures shown are representative of *Pythium*, where both monoclinous and diclinous antheridia can form. Note that the oospores illustrated are "paragynous," in which the antheridium (A) contacts the side of the oogonium (O). This contrasts with the "amphigynous" oospores shown in Fig. 27.2 for *P. infestans*, in which the oogonium grows through the antheridium.

varies depending on its particular mating partner. This is unrelated to mating type and appears to be a polygenic trait (46). An interesting consequence of bisexuality is that both selfed and outcrossed progeny can develop in cultures containing both mating types, once cross-stimulation is enabled by the mating hormones which are discussed below.

Oospores usually form readily in artificial media in the laboratory, but in nature multiple factors influence oosporogenesis. Plant stress is reported to stimulate oospore generation in several species including *Peronospora farinosa* (27), which may be related to the sensing of cellular carbon/nitrogen ratios (62). Different cultivars of potato support distinct levels of oospore production by *P. infestans*, possibly due to variation in sterol content (61); these compounds cannot be synthesized by most oomycetes and are required for sexual reproduction (54). Oosporogenesis by *P. infestans* in tomato and potato was found to increase with falling

temperatures, but this is not a universal oomycete phenomenon as dry, hot conditions are believed to be stimulatory in some graminaceous downy mildews (10, 14). An interesting negative correlation between sexual and asexual sporulation is reported in several oomycetes. In *P. infestans*, asexual sporulation is largely absent within mating zones (21), and with several downy mildews oosporogenesis appears to be stimulated by climatic conditions unfavorable for asexual reproduction (80). However, in the *Arabidopsis* downy mildew pathogen *Hyaloperonospora parasitica*, asexual and sexual sporulation can occur in parallel (20, 42, 56).

Germination of Oospores

Various factors are reported to stimulate germination including plant extracts, light, carbon dioxide, and alternating temperature and wetness regimes, but the requirements do not seem to have universal effects on different oomycetes (68, 78, 86, 95, 104). Also suggested to enhance germination in some species is partial degradation of the oospore wall, which would normally occur in soil or plant debris due to microbial action (39). This can be mimicked in the laboratory by cell wall-degrading enzymes or by passage through the digestive tract of snails (96, 97).

Prior to germination, it is typical for oospores to experience an extended stage of constitutive (endogenous) dormancy. This describes a reversible hypometabolic state resulting from the presence of self-inhibitors, barriers to nutrient penetration, or metabolic blocks (101); consequently many oospores fail to germinate even under favorable conditions. The dormant phase also apparently aids the proper maturation of the oospore. For example, in *Phytophthora cactorum* 4-day-old oospores were reported to show no germination while 67% germinated after 25 days, and similar results were described for *Pythium aphanidermatum* (5, 97). Such delays in germination are presumed to enhance ecological fitness by preventing the simultaneous germination of all propagules, which would be deleterious if occurring at a time unsuited for continued growth, such as when young plants are absent.

Germination is characterized by the restoration of the cytoplasm of the oospore to an "active" configuration and the utilization of lipid reserves (86, 90). In most but not all species, conversion to germinability is accompanied by a reduction in wall thickness (68). Either single or multiple germ tubes emerge from the oospore, which can either infect plants directly or form what is known as a germ sporangium. These are structurally similar to asexual conidia or sporangia and can germinate and form plant infection structures such as appressoria (Fig. 27.1).

Genetic Basis of Mating Behavior

The molecular determinants of heterothallism or homothallism, including the basis of mating type in the case of the heterothallics, have not yet been established for any oomycete. Overall, the majority of oomycetes are homothallic. Also, homothallic genera such as *Albugo* tend to occupy the basal clades of molecular phylogenies of the Peronosporales (87). It can therefore be proposed that homothallism is the ancestral state within the order. Nevertheless, both homothallics and heterothallics are found within many groups such as the downy mildews, *Pythium*, and *Phytophthora* (as illustrated in Fig. 27.4). This suggests that heterothallism has evolved multiple times and/or that heterothallism has frequently reverted to self-fertility. Each of these two types of mating behaviors may also be determined by more than one genetic or molecular mechanism.

In the heterothallic species *P. infestans* and *P. parasitica*, a single mating-type locus was defined genetically, and A1 and A2 types were shown to behave as heterozygote and homozygote, respectively (22, 47, 50). The *P. infestans* locus was placed on a contig of genomic clones, and genes within the region were discovered (84). It was encouraging to note that many genes identified from that *P. infestans* contig also showed linkage to mating type in *P. parasitica*. However, subsequent attempts to precisely define the mating-type determinant have been hindered by the modest tools currently available for manipulating genes in *Phytophthora* and by genetic aberrations that have confounded fine-scale mapping studies. In *P. infestans*, such abnormalities include balanced lethal loci, skewed segregation ratios, and translocations, although not all isolates show each defect (47, 50, 84, 105). In both *Phytophthora* and *Bremia*, cytogenetic analyses suggested that the mating-type locus is associated with a reciprocal translocation (74, 92). Until mating-type loci are cloned, the role of such aberrations will remain unclear, but they may reflect mechanisms to restrict recombination

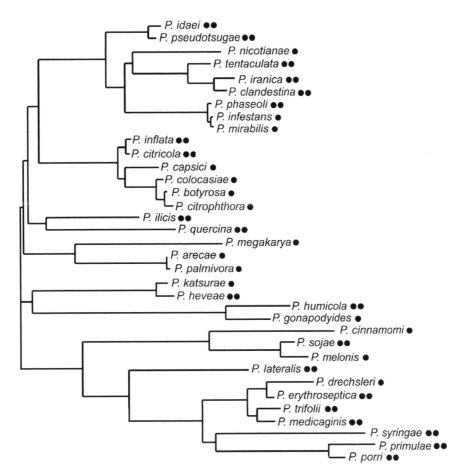

Figure 27.4 Phylogram of representative *Phytophthora* species, indicating whether mating behavior is heterothallic (one filled circle) or homothallic (two circles). Tree generated using data of Cooke et al. (15).

within a complex locus along the lines of an incipient sex chromosome.

At least within *Phytophthora*, where a mating-type locus has been genetically defined, one may speculate that the development of homothallism is due to novel combinations or alterations of those loci, as seen in ascomycetes (113). Consistent with this is the observation that homothallic species of *Phytophthora* synthesize both mating hormones (55). Certain strains of normally heterothallic species can also exhibit homothallic behavior. In the case of *Phytophthora drechsleri*, *P. infestans*, and *Bremia lactucae* this appears to be associated with rearrangements or trisomy at the mating-type locus (28, 74, 76). Environmental factors may also induce sexual reproduction (or at least apomixis) as observed in *Phytophthora* hyphae exposed to fungicides (37), volatiles from the oomycete-degrading fungus *Trichoderma* (11), or physical damage (85). It therefore appears that the regulation of sexual development is leaky. This may confer an evolutionary advantage by enabling single heterothallic strains to occasionally form oospores for resisting harsh environments, in addition to the genetic benefits conferred by recombination.

Mating Hormones

The mating-type locus is thought to regulate the synthesis and/or response to mating hormones. Within the plant pathogenic oomycetes, these compounds have been studied mainly in *Phytophthora* (55). The hormones appear to form the sole basis of heterothallism, as there is no evidence for the type of self-incompatibility seen in true fungi. The hormone produced by A1 strains of *Phytophthora nicotianae* (α1) was recently purified and shown to be a 16-carbon polar diterpene (83). The A2 hormone (α2) is not yet purified but is also believed to be a small, slightly polar compound. These are structurally and biologically distinct from the peptide mating hormones of true fungi and the steroid sex hormones of the saprophytic oomycete *Achlya* (72).

In *Phytophthora* the mating hormones appear to be conserved throughout the genus. Therefore, an A1 of one species can induce oospores in A2 strains of another species, and vice versa (55). Ecological consequences include the formation of apomictic or selfed oospores in an otherwise asexual population and the generation of species hybrids. For example, interspecific crosses are possible between *P. infestans* and *Phytophthora mirabilis*, although the resulting hybrids were shown to be limited in host range (34).

Gene Expression during Mating

A glimpse into molecular aspects of oosporogenesis has been provided by expression-profiling and subtraction cloning studies with *P. infestans* (21, 82). Approximately 101 genes up-regulated more than 10-fold during matings between A1 and A2 isolates were identified, of which approximately one-quarter appear to be expressed exclusively during sexual development. Many are also expressed at high levels during conditions conducive to oosporogenesis in a homothallic species, *Phytophthora phaseoli*, and in a homothallic variant of *P. infestans* in which self-fertility was apparently conferred by tertiary trisomy at the mating-type locus (82). Approximately two-thirds of the genes yielded informative matches in BLASTX searches, suggesting roles in regulation, cell biology, metabolism, structure, and meiosis. A disproportionate number matched RNA-regulating proteins. These included a *Puf* RNA binding protein, an RNase, a DOM3Z RNase regulator, an mRNA cap-removing enzyme, and a putative KH-domain RNA-binding protein. This suggests that posttranscriptional regulation may play key roles in oosporogenesis. For example, *Puf* proteins are known to regulate differentiation by repressing the translation of specific RNAs (110). The RNases and decapping enzymes may help commit *P. infestans* to sexual development by degrading specific substrates. Alternatively, such enzymes may simply remove mRNAs as ribosomes disassemble during late oosporogenesis. However, not induced during mating were exosome proteins, which comprise the normal eukaryotic RNA-degrading machinery (13).

SIGNIFICANCE OF OOSPORES IN DISEASE

Oospores play important roles in pathosystems as infectious propagules, as structures for surviving harsh environments, and as part of a mechanism for maintaining genetic fitness in the long term. Some of these functions are shared with various types of asexual spores, which depending on species typically include sporangia or conidia and sometimes chlamydospores (20). The relative importance of sexual and asexual spores varies for each pathosystem, depending on the biology of both host and pathogen, agronomic practices, climate, and other issues.

Oospores as Survival Structures

Compared to asexual spores, oospores are much more robust and can usually persist between growing seasons of the host plant. The durability of the different sexual and asexual forms can be compared using data assembled for *P. cactorum* by Malajczuk (66). In relatively mild storage environments, survival times of 14 days were reported for vegetative hyphae, up to 35 days for asexual sporangia, >106 days for chlamydospores (another form of asexual spore), and at least 1 year for oospores. The value of oospores as survival structures becomes even more apparent considering that under sunny field conditions, asexual spores would likely survive less than a few days; unlike conidia of true fungi, oomycete asexual spores are fully hydrated and metabolically active. Survival of oospores for more than 5 years has been reported for other species of *Phytophthora* (40, 103), with the apparent oomycete record of 25 years belonging to the onion downy mildew pathogen *Peronospora destructor* (71). Therefore, in many pathosystems oospores commence disease outbreaks at the start of a new growing season, while asexual spores cause most subsequent secondary infections. However, as detailed later, there are exceptions to this pattern.

Role of Oospores in Disease Transmission

While asexual spores typically form on plant surfaces, oospores develop within the host and then become incorporated into soil or surface litter along with plant debris. This generates a reservoir of inoculum that can become activated when new plants appear, leading to primary infections of below-ground parts of the plant such as roots, crowns, or tubers. Oospores can sometimes also be transported by fungus gnats (44), within infected seeds (45), and in the case of some graminaceous downy mildews, by wind (8).

Oospores in soil are recalcitrant to most control efforts, being relatively insensitive to fumigation. Heating soil can reduce oospore viability, but achieving adequate temperatures at sufficient depths is generally impractical (19). Germination rates can decline over time, but not rapidly enough in most cases to make crop rotation programs effective (59). Germination is stimulated by water and plant exudates, but since this occurs asynchronously a low but continuously infective population is maintained over years (23). Therefore, once oospores are present, growers are usually forced to accept that disease will occur each season.

Pathosystems that rely exclusively on oospores as inoculum are usually monocyclic. This means that only one round of sexual sporulation and infection occurs per season. This is not due to the length of the sexual cycle itself, which typically requires only about 1 week. Instead, this mostly reflects the constitutive dormancy and low germination rates of oospores. In contrast, diseases in which asexual spores play major roles are typically polycyclic, with the time required for a cycle of asexual sporulation being similar to that of oosporogenesis. Specific approaches for retarding the rate of oospore formation in planta do not exist, although systemic fungicides may reduce their number or viability (24).

EXAMPLES OF SPECIFIC PATHOSYSTEMS

In many diseases the importance of oospores as inoculum is assumed; however, in only a few cases have detailed assessments actually been made. An absence of good data is particularly a problem with many of the obligate pathogens, where a lack of ex planta culture methods has hindered efforts to identify molecular markers useful for epidemiological and population studies. Also, some plants are colonized by multiple oomycetes which are hard to distinguish morphologically. For example, many vegetables can host multiple *Pythium* spp., and some graminaceous plants can be infected by related downy mildews. Descriptions of a few pathosystems in which the importance of the sexual cycle has been assessed are presented below.

Late Blight of Potato

Potatoes, as well as tomatoes and several other *Solanum* spp., are hosts of the heterothallic species *P. infestans*. This is an interesting system in which to examine the role of sexual reproduction: not only is this the most-studied oomycete disease, but also the geographical distribution of A1 and A2 mating types has changed over time, which allows the impact of the sexual cycle to be assessed.

The center of genetic diversity of *P. infestans* (although not necessarily of its origin) is thought to be the central highlands of Mexico. For example, in the Toluca valley A1 and A2 isolates are present in an approximately 1:1 ratio (33, 38). Oospores are consequently ubiquitous and at the end of the season become incorporated into the soil along with foliar debris (38). Oospores may also persist below ground within potato tubers, many of which remain unharvested at season's end. Survival of oospores through the winter is facilitated by the cool and dry climate of the region. The absence of significant rainfall helps protect oospores from microbial degradation and germination before a host is present; the latter is important since like many oomycetes (except some *Pythium* spp.) *P. infestans* survives poorly as a saprophyte. A new cycle of potato cultivation begins just before the rainy season, which stimulates oospore germination. New potato plants as well as wild *Solanum* spp. can then become infected by oospores (23). Asexual sporangia emerging from these plants then serve as secondary inoculum for the disease, spreading blight through the rest of the season.

Although in the Toluca valley oospores are considered to be an important source of primary inoculum, they are probably not the only primary inoculum. While most asexual sporangia do not overwinter, viable *P. infestans* hyphae may persist within tubers, which could lead to infected foliage on volunteer potato plants or wild *Solanum* spp. early in the season. Nevertheless, a major role for oospores is supported by population studies using molecular markers that indicate a moderate rate of heterozygosity within field isolates, consistent with outcrossing (33).

In contrast to the situation in Mexico, elsewhere *P. infestans* existed asexually until just a few decades ago. Analyses of herbarium specimens indicate that several genotypes escaped from Mexico to North America and the Old World in the 19th century, but only the clonal lineage (a group of near-identical asexually derived strains) known as US-1 persisted (69, 99). Sexual reproduction was therefore undescribed outside Mexico, since all members of that lineage are A1. Asexual propagules of *P. infestans*, such as those emanating from potato cull piles, volunteer plants, or seed potatoes, therefore represented the primary inoculum for outbreaks of late blight. Importantly, this inoculum would typically need to travel from their point of origin to distant potato fields before significant amounts of disease could occur, slowing epidemic progression.

The situation changed dramatically in the 1990s as new strains of *P. infestans* were introduced from Mexico, apparently on shipments of potato and tomato (99). In Europe, new genotypes which included A2 strains became established, profoundly altering the ecology and epidemiology of *P. infestans*. Not only were the new isolates more aggressive in terms of infection frequency, rate of asexual sporulation, and latent period (time between infection and sporulation), but also they were often resistant to phenylamides, which is a class of fungicides previously effective against *P. infestans* (36). In addition, for the first time oospores began to be generated in significant numbers within Europe. To date, oospore production has been confirmed in commercial fields in Denmark, Finland, Germany, The Netherlands, Norway, Poland, Russia, Sweden, Switzerland, and several other countries (cited by Flier et al. [24]). In such regions widespread infections are therefore more likely to occur earlier in the season, although measurements of genotypic diversity indicate that sexual reproduction is still less common in Europe than in Mexico (114). Also, in some parts of Europe the A1 mating type still predominates (16, 53).

Curiously, although A2 isolates were also introduced into North America, the sexual cycle does not appear to be as rampant there as in Europe or Mexico (99). Nevertheless, there is evidence that sexual reproduction has occurred based on the appearance of recombinant genotypes (35). The particularly aggressive clonal lineage US-11, for example, is believed to have resulted from

sexual recombination (31). Fields contaminated extensively by oospores are not yet reported, but control of late blight is more difficult due to the abundance of phenylamide-insensitive and more aggressive genotypes (99). This demonstrates that even rare occurrences of the sexual cycle can have a dramatic effect on disease, if more fit isolates are generated.

An interesting point is whether the new genotypes are actually more aggressive than the strains that first escaped from Mexico in the 19th century. A tenet of evolutionary genetic theory is that sexual reproduction helps maintain the integrity of a genome (75). It is possible that the asexually propagated A1 populations prevalent until the end of the 20th century had gradually lost fitness. That genetic defects accumulated is suggested by the frequent detection of strains that are polyploid, exhibit low fertility, or yield many offspring reduced in pathogenicity (3, 70, 102).

In addition to generating more aggressive strains of *P. infestans*, the occurrence of sexual recombination dictates resistance breeding strategies. For example, major gene resistance (i.e., determined by race-specific R genes that recognize pathogenic *Avr* alleles) would not be durable, as recombination would yield virulent *avr/avr* strains, for example from crosses between *Avr/avr* isolates (89). Indeed, during the 1950s and 1960s many resistant lines of potato developed in other countries were found to become blighted during field tests in Mexico. It should be noted, however, that variation in avirulence phenotypes can also arise within asexual populations through single base changes or deletion (67, 106).

Phytophthora capsici Blights

P. capsici colonizes several solanaceous species such as pepper and tomato, as well as cucurbits. Its story provides an interesting comparison to that of *P. infestans*. *P. capsici* is believed to be the only heterothallic *Phytophthora* that regularly completes the sexual cycle in the United States, where its mating types occur at approximately equal frequencies (40). One study of alleles within populations measured Wright's fixation index (*F*), in which a value of 0 indicates no heterozygosity and 1 indicates complete heterozygosity. For *P. capsici*, a value of 0.92 was calculated, compared to 0.20 for *P. infestans* in central Mexico. It was suggested that populations of soil-inhabiting species such as *P. capsici* may be more prone to inbreeding, because its aerial dispersal options may be more limited (33). Unlike *P. infestans*, which infects mostly foliage, *P. capsici* more commonly colonizes the root or crown of its hosts.

In a study of infected squash fields in Michigan, the percentage of unique genotypes detected at the beginning of the growing season, which are presumably generated by sexual recombination, started at 100% (58). Later in the season only 30% were unique. Therefore, the model proposed for *P. capsici* in that region is that most primary infections result from oospores, with asexual sporangia serving later as secondary inoculum. Along the lines of the clonal lineages described for *P. infestans*, certain recombinant genotypes of *P. capsici* predominated within fields, presumably spread within each season by asexual sporangia (40). However, unlike in potato late blight, for which some genotypes persist year to year and spread over wide areas, the *P. capsici* lineages were confined in space to single fields or in time to single years (60). This implies that overwintering through oospores is much more critical for *P. capsici*.

Outcrossing is also proposed to have contributed to an increase in levels of insensitivity to phenylamide fungicides in *P. capsici*. Resistance is due to the combined effects of a major semidominant locus (*Mex*) plus minor genes, as in a quantitative trait (49). Once spontaneous mutations within *Mex* generated partially resistant *Mex/mex* heterozygotes, field applications of phenylamides are proposed to have selected for highly resistant *Mex/Mex* strains within sexually recombining populations (60). A similar phenomenon may have occurred in *P. infestans*, although mitotic gene conversion also may explain how highly resistant strains develop.

Phytophthora sojae Root and Stem Rot

This homothallic species causes root and stem rot on soybean. Lesions can develop from either oospores or asexual sporangia. However, secondary infections by asexual spores are relatively limited since plants become more resistant with age, so the disease is effectively monocyclic (93). Therefore, the sexual cycle plays a crucial role in the soybean disease.

Sexual reproduction, as well as asexual events, has contributed to the genetic diversity of *P. sojae* in many soybean-growing regions. *P. sojae* is believed to have originated in North America as a pathogen of wild lupines, although this is not proven. The first strains isolated from soybean belonged to a single pathotype based on reactions against cultivars expressing as many as 13 different *Rps* resistance genes. However, numerous pathotypes are now described, which is of major importance since *Rps* genes are used widely to control the disease (64). Many new pathotypes appear to have developed as a result of spontaneous mutations in *Avr* genes, but several resulted from outcrossing based on DNA polymorphism analysis (25). When two *P. sojae* isolates are mixed in culture, up to 10% of oospores can be genetic hybrids, similar to that observed in other ho-

mothallic oomycetes such as *Pythium ultimum* (26, 108). The rate of outcrossing in nature is probably much lower since two isolates would need to infect the same plant. However, there would be strong selection for virulent hybrids in the presence of the cognate *Rps* gene. A recent survey of *P. sojae* in Ohio reported 202 pathotypes, which was a dramatic increase from previous surveys there and elsewhere (18, 64, 94). Multiple pathotypes were detected in the same field, indicating that opportunities for outcrossing are high.

While outcrossing is possible, many hybrids exhibit genetic abnormalities as evidenced by distorted segregation of markers in F2 populations (65, 109). A likely explanation is that chromosome structure had diverged within each of the inbred parental lines. This would be consistent with the operation of Muller's ratchet through the accumulation of harmful or deleterious mutations in geographically or genetically isolated lines of *P. sojae*, as a step towards speciation (77).

Downy Mildew of Grape

Plasmopara viticola is a major impediment to grape production in many parts of the world. Like *P. infestans*, it is another pathogen native to the Americas that was exported to Europe. A role of oospores in the survival and epidemiology of *P. viticola* has been proposed for many decades, with oospores in fallen leaves and berries representing an overwintering reservoir of inoculum. Despite this knowledge, only in 2001 was it discovered that *P. viticola* is heterothallic, not homothallic (112). More recently, it was learned that the role of the sexual cycle has been greatly underestimated. It was assumed previously that oospores played a significant role only in initiating disease at the start of the vegetative growth season of the grapevine (7). The progress of disease throughout the season was largely attributed to infections by asexual spores, similar to the situation described above for *P. infestans* and *P. capsici*. However, recent studies of *P. viticola* populations within Europe and the northeastern United States indicated that oospores are a predominant source of inoculum throughout the season. Based on analyses of microsatellite markers, as many as 85% of isolates collected within single vineyards had unique genotypes, suggesting they arose from independent sexual recombination events (32, 51). Only in limited regions did secondary infections play the major role, as evidenced by the dominance of clonal (asexual) populations (91). Continued oospore germination throughout the year was also observed. The high frequency of sexual reproduction in grape-growing regions was further suggested by a survey that detected the two mating types, P1 and P2, in a 1:1 ratio (112).

This discovery will have a major effect on disease-forecasting models, which are used to time fungicide applications and play a major role in the economics of grape production (7). Models will need to be biased more towards conditions conducive for oospore germination and less to environments favorable for zoospore release from asexual spores. However, these must be tailored to each growing region since oospores may not be the primary inoculum throughout all of the world. For example, sexual reproduction appears to be rare in Western Australia (52). This is likely because the disease has been there less than one decade, and both mating types may not be evenly distributed due to founder effects. This is similar to the situation with *P. infestans* in North America and underscores the point that factors other than the inherent biology of each species influence the frequency of sexual reproduction.

Sorghum Downy Mildew

Peronosclerospora sorghi causes downy mildew on sorghum and maize and is responsible for frequent epidemics in many tropical and subtropical parts of the world, especially Africa (45). Detailed information on the population structure of the pathogen is somewhat limited, due both to difficulties in working with this obligate pathogen and to the fact that it is easily confused with related downy mildews such as *Peronosclerospora maydis* and *Peronosclerospora sacchari*. Nevertheless, an interesting picture has emerged in which host, pathogen genotype, and environment interact to determine the importance of sexual reproduction.

As with most of the pathogens described above, *P. sorghi* is theoretically capable of producing both long-lived oospores and ephemeral asexual spores. However, asexual sporulation occurs under a very narrow range of conditions, with specific temperature and humidity requirements (10). Asexual sporulation consequently occurs rarely in some regions of the world, and in such locations oospores are the main inoculum. This is the situation in arid regions of the United States and Africa, for example, where infections commonly result from oospores that reside in the soil or are introduced through contaminated seed or wind (8, 45). AFLP marker analysis of isolates from the United States revealed genotypes consistent with a population reproducing at least in part through the sexual cycle (81). In contrast, in humid regions such as the southern part of Nigeria, asexual conidia appear to be the sole source of infection as oospores are found only rarely (4).

Overlain on the effects of climate on the extent of sexual reproduction are differences in the abilities of different *P. sorghi* genotypes to produce oospores. One

genotype found mainly on maize is reported to rarely produce oospores (1), while a strain capable of infecting sorghum and maize produced oospores in abundance (8). Several strains able to produce oospores on sorghum failed to so do on maize (9). Possibly, host metabolites may influence the rate of oospore production, as was suggested for *P. infestans* on different potato cultivars (61). It has been proposed that the strains impaired in oospore production evolved in geographic regions in which host plants, including weeds, were present year-round (45). Advantages of asexual propagation in such environments may have selected for defects in genes for oosporogenesis.

CONCLUSIONS AND PROSPECTS

The sexual cycle significantly impacts diseases caused by oomycetes, generating both durable oospores and strains altered in fitness, host specificity, and levels of fungicide sensitivity. The role of sexual processes is most obvious in diseases caused by homothallics since their oospores tend to be the primary inoculum, with the situation being more complex with heterothallics where the distribution of mating types must be considered. In the case of species such as *P. viticola* both mating types have enjoyed worldwide distribution throughout known history, but as seen for *P. infestans* this is not always the case. Many heterothallic oomycetes currently exhibit uneven distributions of mating type, such as *P. palmivora*, where A1 strains are rare; *P. megakarya*, where the A2 type is rare; and *P. ramorum*, where A1 and A2 types are largely restricted to Europe and the United States (although this situation may be changing [20, 43]). In such cases, a redistribution of the mating types is likely to generate sexually active populations that are harder to control.

The simultaneous presence of both mating types is no guarantee that sexual populations will develop. The movement of pathogens from their centers of origin frequently involves a genetic bottleneck, in which only limited genotypes enter previously disease-free areas. Not only will the introduced population be prone to lower fitness due to its reduced adaptive potential, but also if only a single mating type was introduced, then deleterious mutations may accumulate that reduce fertility, according to Muller's ratchet (77). Even if the missing mating type is introduced and fertility is maintained, recombination will be unlikely if strains of each mating type are adapted to different hosts. This is believed to have occurred with *P. infestans*-like species in Ecuador specialized to colonize either cultivated potato or wild *Solanum* spp. (79). Nevertheless, in the long term even a rare case of successful recombination could dramatically change the ecology or epidemiology of a pathosystem.

Reproductive isolation is not limited to heterothallics passed through genetic bottlenecks. A similar process could occur with homothallics as mutation followed by inbreeding would cause previously clonal populations to diverge. In the extreme this would lead to speciation (57). Consistent with this concept is the observation that homothallic species predominate in both *Pythium*, *Phytophthora*, and the downy mildews (20, 68, 73). Since mating hormones appear to be conserved within a genus, at least in *Phytophthora*, it is notable that speciation would not necessarily block sexual fusions. This might explain how new species of *Phytophthora* are generated through hybridization processes (12).

While sexual reproduction challenges efforts to control oomycete pathogens, better knowledge of the mechanisms of oospore formation and germination could lead to new management strategies. For example, mating hormone analogues might trick isolates into switching from vegetative growth to sexual development; this would likely lead to abortive selfed oospores in heterothallics (46) and also divert cellular energies from asexual sporulation (21). Cultivars less likely to support the generation of overwintering inoculum might also result from understanding the role of plant metabolites in oosporogenesis (61). Soil populations of oospores might also be reduced by applying compounds stimulating oospore germination near the end of a growing season.

Work in my laboratory has been supported by the United States Department of Agriculture and National Science Foundation.

References

1. Adenle, V. O., and K. F. Cardwell. 2000. Seed transmission of maize downy mildew (*Peronosclerospora sorghi*) in Nigeria. *Plant Pathol.* 49:628–634.

2. Agrios, G. N. 2004. *Plant Pathology*, 5th ed. Academic Press, San Diego, CA.

3. Al-Kherb, S. M., C. Fininsa, R. C. Shattock, and D. S. Shaw. 1995. The inheritance of virulence of *Phytophthora infestans* to potato. *Plant Pathol.* 44:552–562.

4. Anaso, A. B. 1989. Survival of downy mildew pathogen of maize in Nigerian guinea savanna. *Appl. Agric. Res.* 4:258–263.

5. Ayers, W. A., and R. D. Lumsden. 1975. Factors affecting production and germination of oospores of 3 *Pythium* spp. *Phytopathology* 65:1094–1100.

6. Baldauf, S. L., A. J. Roger, I. Wenk-Siefert, and W. F. Doolittle. 2000. A kingdom-level phylogeny of eukaryotes based on combined protein data. *Science* 290:972–977.

7. Blaise, P., R. Dietrich, and C. Gessler. 1999. Vinemild: an application oriented model of *Plasmopara viticola* epidemics on *Vitis vinifera*. *Acta Hortic.* 499:187–192.

8. Bock, C. H., M. J. Jeger, B. D. L. Fitt, and J. Sherington. 1997. Effect of wind on the dispersal of oospores of *Peronosclerospora sorghi* from sorghum. *Plant Pathol.* 46:439–449.

9. Bock, C. H., M. J. Jeger, L. K. Mughogho, K. F. Cardwell, E. Mtisi, G. Kaula, and D. Mukansabimana. 2000. Variability of *Peronosclerospora sorghi* isolates from different geographic locations and hosts in Africa. *Mycol. Res.* 104:61–68.

10. Bonde, M. R. 1982. Epidemiology of downy mildew diseases of maize sorghum and pearl millet. *Trop. Pest Manag.* 28:49–60.

11. Brasier, C. M. 1971. Induction of sexual reproduction in single A2 isolates of *Phytophthora* species by *Trichoderma viride*. *Nat. New Biol.* 231:283.

12. Brasier, C. M., S. A. Kirk, J. Delcan, D. E. Cooke, T. Jung, and W. A. Man in't Veld. 2004. *Phytophthora alni* sp. nov. and its variants: designation of emerging heteroploid hybrid pathogens spreading on Alnus trees. *Mycol. Res.* 108:1172–1184.

13. Butler, J. S. 2002. The yin and yang of the exosome. *Trends Cell Biol.* 12:90–96.

14. Cohen, Y., S. Farkash, Z. Reshit, and A. Baider. 1997. Oospore production of *Phytophthora infestans* in potato and tomato leaves. *Phytopathology* 87:191–196.

15. Cooke, D. E. L., A. Drenth, J. M. Duncan, G. Wagels, and C. M. Brasier. 2000. A molecular phylogeny of *Phytophthora* and related oomycetes. *Fungal Genet. Biol.* 30:17–32.

16. Cooke, D. E. L., V. Young, P. R. J. Birch, R. Toth, F. Gourlay, J. P. Day, S. F. Carnegie, and J. M. Duncan. 2003. Phenotypic and genotypic diversity of *Phytophthora infestans* populations in Scotland (1995-97). *Plant Pathol.* 52:181–192.

17. Davidson, J. M., S. Werres, M. Garbelotto, E. M. Hansen, and D. M. Rizzo. 2003. Sudden Oak Death and associated diseases caused by *Phytophthora ramorum*. *Plant Health Prog.* 1:1–21.

18. Dorrance, A. E., S. A. McClure, and A. DeSilva. 2003. Pathogenic diversity of *Phytophthora sojae* in Ohio soybean fields. *Plant Dis.* 87:139–146.

19. Drenth, A., E. M. Janssen, and F. Govers. 1995. Formation and survival of oospores of *Phytophthora infestans* under natural conditions. *Plant Pathol.* 44:86–94.

20. Erwin, D. C., and O. K. Ribeiro. 1996. *Phytophthora diseases worldwide*. APS Press, St. Paul, MN.

21. Fabritius, A.-L., C. Cvitanich, and H. S. Judelson. 2002. Stage-specific gene expression during sexual development in *Phytophthora infestans*. *Mol. Microbiol.* 45:1057–1066.

22. Fabritius, A.-L., and H. S. Judelson. 1997. Mating-type loci segregate aberrantly in *Phytophthora infestans* but normally in *Phytophthora parasitica*: implications for models of mating-type determination. *Curr. Genet.* 32:60–65.

23. Fernandez-Pavia, S. P., N. J. Grunwald, M. Diaz-Valasis, M. Cadena-Hinojosa, and W. E. Fry. 2004. Soilborne oospores of *Phytophthora infestans* in central Mexico survive winter fallow and infect potato plants in the field. *Plant Dis.* 88:29–33.

24. Flier, W. G., G. J. T. Kessel, and H. T. A. M. Schepers. 2004. The impact of oospores of *Phytophthora infestans* on late blight epidemics. *Plant Breed. Seed Sci.* **50**:5–13.

25. Förster, H., B. M. Tyler, and M. D. Coffey. 1994. *Phytophthora sojae* races have arisen by clonal evolution and by rare outcrosses. *Mol. Plant-Microbe Interact.* **7**:780–791.

26. Francis, D. M., and D. A. St. Clair. 1993. Outcrossing in the homothallic oomycete, *Pythium ultimum*, detected with molecular markers. *Curr. Genet.* **24**:100–106.

27. Frinking, H. D., J. L. Harrewijn, and C. F. Geerds. 1984. Factors governing oospore production by *Peronospora farinosa* f. sp. *spinaciae* in cotyledons of spinach [*Spinacia oleracea*]. *Neth. J. Plant Pathol.* **91**:215–224.

28. Fyfe, A. M., and D. S. Shaw. 1992. An analysis of self-fertility in field isolates of *Phytophthora infestans*. *Mycol. Res.* **96**:390–394.

29. Gallegly, M. E. 1968. Genetics of pathogenicity of *Phytophthora infestans*. *Annu. Rev. Plant Pathol.* **6**:375–396.

30. Gallegly, M. E. 1960. Genetics of *Phytophthora*. *Phytopathology* **60**:1135–1141.

31. Gavino, P. D., C. D. Smart, R. W. Sandrock, J. S. Miller, P. B. Hamm, T. Y. Lee, R. M. Davis, and W. E. Fry. 2000. Implications of sexual reproduction for Phytophthora infestans in the United States: generation of an aggressive lineage. *Plant Dis.* **84**:731–735.

32. Gobbin, D., M. Jermini, B. Loskill, I. Pertot, M. Raynal, and C. Gessler. 2005. Importance of secondary inoculum of *Plasmopara viticola* to epidemics of grapevine downy mildew. *Plant Pathol.* **54**:522–534.

33. Goodwin, S. 1997. The population genetics of *Phytophthora*. *Phytopathology* **87**:462–473.

34. Goodwin, S. B., and W. E. Fry. 1994. Genetic analyses of interspecific hybrids between *Phytophthora infestans* and *Phytophthora mirabilis*. *Exp. Mycol.* **18**:20–32.

35. Goodwin, S. B., C. D. Smart, R. W. Sandrock, K. L. Deahl, Z. K. Punja, and W. E. Fry. 1998. Genetic charge within populations of *Phytophthora infestans* in the United States and Canada during 1994 to 1996: role of migration and recombination. *Phytopathology* **88**:939–949.

36. Goodwin, S. B., L. S. Sujkowski, and W. E. Fry. 1994. Metalaxyl-resistant clonal genotypes of *Phytophthora infestans* in the United States and Canada were probably introduced from northwestern Mexico. *Phytopathology* **84**:1079.

37. Groves, C. T., and J. B. Ristaino. 2000. Commercial fungicide formulations induce in vitro oospore formation and phenotypic change in mating type in *Phytophthora infestans*. *Phytopathology* **90**:1201–1208.

38. Grunwald, N. J., and W. G. Flier. 2005. The biology of *Phytophthora infestans* at its center of origin. *Annu. Rev. Phytopathol.* **43**:171–190.

39. Hancock, J. G. 1981. Longevity of *Pythium ultimum* in moist soils. *Phytopathology* **71**:1033–1037.

40. Hausbeck, M. K., and K. H. Lamour. 2004. *Phytophthora capsici* on vegetable crops: research progress and management challenges. *Plant Dis.* **88**:1292–1303.

41. Hemmes, D. E. 1983. Cytology of *Phytophthora*, p. 9–40. *In* D. C. Erwin, S. Bartnicki-Garcia, and P. H. Tsao (ed.), Phytophthora, *Its Biology, Taxonomy, Ecology, and Pathology.* APS Press, St. Paul, MN.

42. Inaba, T., and T. Morinaka. 1983. The relationship between conidium and oospore production in soybean leaves infected with *Peronospora manshurica*. *Ann. Phytopathol. Soc. Jpn.* **49**:554–557.

43. Ivors, K., M. Garbelotto, I. D. E. Vries, C. Ruyter-Spira, B. T. Hekkert, N. Rosenzweig, and P. Bonants. 2006. Microsatellite markers identify three lineages of *Phytophthora ramorum* in US nurseries, yet single lineages in US forest and European nursery populations. *Mol. Ecol.* **15**: 1493–1505.

44. Jarvis, W. R., J. L. Shipp, and R. B. Gardiner. 1993. Transmission of *Pythium aphanidermatum* to greenhouse cucumber by the fungus gnat *Bradysia impatiens* (Diptera: Sciaridae). *Ann. Appl. Biol.* **122**:23–29.

45. Jeger, M. J., E. Gilijamse, C. H. Bock, and H. D. Frinking. 1998. The epidemiology, variability and control of the downy mildews of pearl millet and sorghum, with particular reference to Africa. *Plant Pathol.* **47**:544–569.

46. Judelson, H. S. 1997. Expression and inheritance of sexual preference and selfing potential in Phytophthora infestans. *Fungal Genet. Biol.* **21**:188–197.

47. Judelson, H. S. 1996. Genetic and physical variability at the mating type locus of the oomycete, *Phytophthora infestans*. *Genetics* **144**:1005–1013.

48. Judelson, H. S., and F. A. Blanco. 2005. The spores of *Phytophthora*: weapons of the plant destroyer. *Nat. Microbiol. Rev.* **3**:47–58.

49. Judelson, H. S., and S. Roberts. 1999. Multiple loci determining insensitivity to phenylamide fungicides in *Phytophthora infestans*. *Phytopathology* **89**:754–760.

50. Judelson, H. S., L. J. Spielman, and R. C. Shattock. 1995. Genetic mapping and non-Mendelian segregation of mating type loci in the oomycete, *Phytophthora infestans*. *Genetics* **141**:503–512.

51. Kennelly, M. M., C. Eugster, D. M. Gadoury, C. D. Smart, R. C. Seem, D. Gobbin, and C. Gessler. 2004. Contributions of oosporic inoculum to epidemics of grapevine downy mildew (*Plasmopara viticola*). *Phytopathology* **94**:S50.

52. Killigrew, B. X., D. Sivasithamparam, and E. S. Scott. 2005. Absence of oospores of downy mildew of grape caused by *Plasmopara viticola* as the source of primary inoculum in most western Australian vineyards. *Plant Dis.* **89**:777.

53. Knapova, G., and U. Gisi. 2002. Phenotypic and genotypic structure of *Phytophthora infestans* populations on potato and tomato in France and Switzerland. *Plant Pathol.* **51**:641–653.

54. Ko, W. H. 1998. Chemical stimulation of sexual reproduction in *Phytophthora* and *Pythium*. *Bot. Bull. Acad. Sinica.* **39**:81–86.

55. Ko, W. H. 1988. Hormonal heterothallism and homothallism in *Phytophthora*. *Annu. Rev. Phytopathol.* **26**:57–73.

56. Koch, E., and A. Slusarenko. 1990. *Arabidopsis* is susceptible to infection by a downy mildew fungus. *Plant Cell* **2**: 437–45.

57. Kohn, L. M. 2005. Mechanisms of fungal speciation. *Annu. Rev. Phytopathol.* **43**:279–308.

58. Lamour, K. H., and M. K. Hausbeck. 2001. The dynamics of mefenoxam insensitivity in a recombining population of *Phytophthora capsici* characterized with amplified fragment length polymorphism markers. *Phytopathology* 91:553–557.

59. Lamour, K. H., and M. K. Hausbeck. 2003. Effect of crop rotation on the survival of *Phytophthora capsici* in Michigan. *Plant Dis.* 87:841–845.

60. Lamour, K. H., and M. K. Hausbeck. 2002. The spatiotemporal genetic structure of *Phytophthora capsici* in Michigan and implications for disease management. *Phytopathology* 92:681–684.

61. Langcake, P. 1974. Sterols in potato leaves and their effects on growth and sporulation of *Phytophthora infestans*. *Trans. Br. Mycol. Soc.* 63:573–586.

62. Leal, J. A., M. E. Gallegly, and V. G. Lilly. 1967. The relation of the carbon-nitrogen ratio in the basal medium to sexual reproduction in species of *Phytophthora*. *Mycologia* 59:953–964.

63. Leary, J. V., J. R. Roheim, and G. A. Zentmyer. 1974. Ribosome content of various spore forms of *Phytophthora* spp. *Phytopathology* 64:404–408.

64. Leitz, R. A., G. L. Hartman, W. L. Pedersen, and C. D. Nickell. 2000. Races of *Phytophthora sojae* on soybean in Illinois. *Plant Dis.* 84:487.

65. MacGregor, T., M. Bhattacharyya, B. Tyler, R. Bhat, A. F. Schmitthenner, and M. Gijzen. 2002. Genetic and physical mapping of *Avr1a* in *Phytophthora sojae*. *Genetics* 160:949–959.

66. Malajczuk, N. 1983. Microbial antagonism to *Phytophthora*, p. 197–218. *In* D. C. Erwin, S. Bartnicki-Garcia, and P. H. Tsao (ed.), Phytophthora, *Its Biology, Taxonomy, Ecology, and Pathology*. APS Press, St. Paul, MN.

67. Malcolmson, J. F. 1969. Races of *Phytophthora infestans* occurring in Great Britain. *Trans. Br. Mycol. Soc.* 53:417–423.

68. Martin, F. N., and J. E. Loper. 1999. Soilborne plant diseases caused by *Pythium* spp: ecology, epidemiology, and prospects for biological control. *Crit. Rev. Plant Sci.* 18:111–181.

69. May, K. J., and J. B. Ristaino. 2004. Identity of the mitochondrial DNA haplotype of *Phytophthora infestans* in historic specimens of the Irish potato famine. *Mycol. Res.* 108:1–9.

70. Mayton, H., C. D. Smart, B. C. Moravec, E. S. G. Mizubuti, A. E. Muldoon, and W. E. Fry. 2000. Oospore survival and pathogenicity of single oospore recombinant progeny from a cross involving US-17 and US-8 genotypes of *Phytophthora infestans*. *Plant Dis.* 84:1190–1196.

71. McKay, R. 1957. The longevity of the oospores of onion downy mildew *Peronospora destructor* (Berk.) Casp. *Sci. Proc. Royal Dublin Soc. New Ser.* 27:295–307.

72. McMorris, T. C. 1978. Antheridiol and the oogoniols steroid hormones which control sexual reproduction in *Achlya*. *Philos. Trans. R. Soc. Lond. B* 284:459–470.

73. Michelmore, R. W., T. Ilott, S. H. Hulbert, and B. Farrara. 1988. The downy mildews. *Adv. Plant Pathol.* 6:53–79.

74. Michelmore, R. W., and E. R. Sansome. 1982. Cytological studies of heterothallism and secondary homothallism in *Bremia lactucae*. *Trans. Br. Mycol. Soc.* 79:291–298.

75. Mooney, S. M. 1995. H. J. Muller and R. A. Fisher on the evolutionary significance of sex. *J. Hist. Biol.* 28:133–149.

76. Mortimer, A. M., D. S. Shaw, and E. R. Sansome. 1977. Genetical studies of secondary homothallism in *Phytophthora dreschsleri*. *Arch. Microbiol.* 111:255–259.

77. Muller, H. J. 1964. The relation of recombination to mutational advance. *Mutat. Res.* 1:29.

78. Nelson, E. B. 1990. Exudate molecules initiating fungal responses to seeds and roots. *Plant Soil* 129:61–74.

79. Oliva, R. F., L. J. Erselius, N. E. Adler, and G. A. Forbes. 2002. Potential of sexual reproduction among host-adapted populations of *Phytophthora infestans sensu lato* in Ecuador. *Plant Pathol.* 51:710–719.

80. Pegg, G. F., and M. J. Mence. 1970. The biology of *Peronospora viciae* on pea: laboratory experiments on the effects of temperature, relative humidity and light on the production germination and infectivity of sporangia. *Ann. Appl. Biol.* 66:417–428.

81. Perumal, R., T. Isakeit, M. Menz, S. Katile, E. G. No, and C. W. Magill. 2006. Characterization and genetic distance analysis of isolates of *Peronosclerospora sorghi* using AFLP fingerprinting. *Mycol. Res.* 110:471–478.

82. Prakob, W., and H. Judelson. 8 January 2007. Gene expression during oosporogenesis in heterothallic and homothallic *Phytophthora*. *Fungal Genet. Biol.* doi: 10.1016/j.fgb.206.11.011. [Epub ahead of print.]

83. Qi, J., T. Asano, M. Jinno, K. Matsui, Y. Atsumi, Y. Sakagami, and M. Ojika. 2005. Characterization of a *Phytophthora* mating hormone. *Science* 309:1828.

84. Randall, T. A., A. Ah Fong, and H. Judelson. 2003. Chromosomal heteromorphism and an apparent translocation detected using a BAC contig spanning the mating type locus of *Phytophthora infestans*. *Fungal Genet. Biol.* 38:75–84.

85. Reeves, R. J., and R. M. Jackson. 1974. Stimulation of sexual reproduction in *Phytophthora cinnamomi* by damage. *J. Gen. Microbiol.* 84:303–310.

86. Ribeiro, O. K. 1983. Physiology of asexual sporulation and spore germination in *Phytophthora*, p. 55–70. *In* D. C. Erwin, S. Bartnicki-Garcia, and P. H. Tsao (ed.), Phytophthora, *Its Biology, Taxonomy, Ecology, and Pathology*. APS Press, St. Paul, MN.

87. Riethmuller, A., H. Voglmayr, M. Göker, M. Weiss, and F. Oberwinkler. 2002. Phylogenetic relationships of the downy mildews (Peronosporales) and related groups based on nuclear large subunit ribosomal DNA sequences. *Mycologia* 94:834–849.

88. Ristaino, J. B. 2002. Tracking historic migrations of the Irish potato famine pathogen, *Phytophthora infestans*. *Microbes Infect.* 4:1369–1377.

89. Romero, S., and D. C. Erwin. 1969. Variation in pathogenicity among single oospore cultures of *Phytophthora infestans*. *Phytopathology* 59:1310–1317.

90. Ruben, D. M., and M. E. Stanghellini. 1978. Ultrastructure of oospore germination in *Pythium aphanidermatum*. *Am. J. Bot.* 65:491–501.

91. **Rumbou, A., and C. Gessler.** 2004. Genetic dissection of *Plasmopara viticola* population from a Greek vineyard in two consecutive years. *Eur. J. Plant Pathol.* **110:**379–392.

92. **Sansome, E.** 1980. Reciprocal translocation heterozygosity in heterothallic species of *Phytophthora* and its significance. *Nature* **241:**344–345.

93. **Schmitthenner, A. F.** 1999. Phytophthora rot of soybean, p. 39–42. *In* G. L. Hartman, J. B. Sinclair, and J. C. Rupe (ed.), *Compendium of Soybean Diseases*, 4th ed. APS Press, St. Paul, MN.

94. **Schmitthenner, A. F., M. Hobe, and R. G. Bhat.** 1994. *Phytophthora sojae* races in Ohio over a 10-year interval. *Plant Dis.* **78:**269–276.

95. **Shang, H., C. R. Grau, and R. D. Peters.** 2000. Oospore germination of *Aphanomyces euteiches* in root exudates and on the rhizoplanes of crop plants. *Plant Dis.* **84:**994–998.

96. **Shattock, R. C., P. W. Tooley, and W. E. Fry.** 1986. Genetics of *Phytophthora infestans*: determination of recombination, segregation, and selfing by isozyme analysis. *Phytopathology* **76:**410–413.

97. **Shaw, D. S.** 1967. A method of obtaining single-oospore cultures of *Phytophthora cactorum* using live water snails. *Phytopathology* **57:**454.

98. **Shaw, D. S., and I. A. Khaki.** 1971. Genetical evidence for diploidy in *Phytophthora*. *Genet. Res.* **17:**165–167.

99. **Smart, C. D., and W. E. Fry.** 2001. Invasions by the late blight pathogen: renewed sex and enhanced fitness. *Biol. Invasions* **3:**235–243.

100. **Sogin, M. L., and J. D. Silberman.** 1998. Evolution of the protists and protistan parasites from the perspective of molecular systematics. *Int. J. Parasitol.* **28:**11–20.

101. **Sussman, A. S., and H. A. Douthit.** 1973. Dormancy in microbial spores. *Annu. Rev. Plant Physiol.* **24:**311–352.

102. **Tooley, P. W., and C. D. Therrien.** 1987. Cytophotometric determination of the nuclear DNA content of 23 Mexican and 18 non-Mexican isolates of *Phytophthora infestans*. *Exp. Mycol.* **11:**19–26.

103. **Turkensteen, L. J., W. G. Flier, R. Wanningen, and A. Mulder.** 2000. Production, survival and infectivity of oospores of *Phytophthora infestans*. *Plant Pathol.* **49:**688–696.

104. **Van Der Gaag, D., and H. D. Frinking.** 1997. Factors affecting germination of oospores of *Peronospora viciae* f. sp. *pisi* in vitro. *Eur. J. Plant Pathol.* **103:**573–580.

105. **van der Lee, T., A. Testa, A. Robold, J. W. van 't Klooster, and F. Govers.** 2004. High density genetic linkage maps of *Phytophthora infestans* reveal trisomic progeny and chromosomal rearrangements. *Genetics* **157:** 949–956.

106. **Van der Lee, T., A. Testa, J. van 't Klooster, G. van den Berg-Velthuis, and F. Govers.** 2001. Chromosomal deletion in isolates of Phytophthora infestans correlates with virulence on R3, R10, and R11 potato lines. *Mol. Plant-Microbe Interact.* **14:**1444–1452.

107. **Vercesi, A., R. Tornaghi, S. Sant, S. Burruano, and F. Faoro.** 1999. A cytological and ultrastructural study on the maturation and germination of oospores of *Plasmopara viticola* from overwintering vine leaves. *Mycol. Res.* **103:**193–202.

108. **Whisson, S. C., A. Drenth, D. J. Maclean, and J. A. Irwin.** 1994. Evidence for outcrossing in *Phytophthora sojae* and linkage of a DNA marker to two avirulence genes. *Curr. Genet.* **27:**77–82.

109. **Whisson, S. C., A. Drenth, D. J. MacLean, and J. A. G. Irwin.** 1995. *Phytophthora sojae* avirulence genes, RAPD, and RFLP markers used to construct a detailed genetic linkage map. *Mol. Plant-Microbe Interact.* **8:** 988–995.

110. **Wickens, M., D. S. Bernstein, J. Kimble, and R. Parker.** 2002. A PUF family portrait: 3′UTR regulation as a way of life. *Trends Genet.* **18:**150–157.

111. **Windels, C. E.** 2000. Aphanomyces root rot on sugar beet. *Plant Health Prog.* doi:10.1094/PHP-2000-0720-01-DG.

112. **Wong, F. P., H. N. Burr, and W. F. Wilcox.** 2001. Heterothallism in *Plasmopara viticola*. *Plant Pathol.* **50:** 427–432.

113. **Yun, S.-H., M. L. Berbee, O. C. Yoder, and B. G. Turgeon.** 1999. Evolution of the fungal self-fertile reproductive life style from self-sterile ancestors. *Proc. Natl. Acad. Sci. USA* **96:**5592–5597.

114. **Zwankhuizen, M. J., F. Govers, and J. C. Zadoks.** 2000. Inoculum sources and genotypic diversity of *Phytophthora infestans* in Southern Flevoland, the Netherlands. *Eur. J. Plant Pathol.* **106:**667–680.

The Implications of Sex

Sex in Fungi: Molecular Determination
and Evolutionary Implications
Edited by Joseph Heitman et al.
© 2007 ASM Press, Washington, D.C.

Jianping Xu

28

Origin, Evolution, and Extinction of Asexual Fungi:
Experimental Tests Using *Cryptococcus neoformans*

The evolution of sex and sexual reproduction is one of the most enduring issues in evolutionary biology. Like zoologists and botanists, mycologists have long recognized the importance of sexual reproduction in their study organisms. In the opening statement of his 1966 book *Genetics of Sexuality in Higher Fungi*, John R. Raper remarked that "sexuality in fungi has long been recognized as one of the more perplexing yet intriguing facets of the biology of this large and varied group of microorganisms" (39). Over the last 40 years, while mycologists, molecular biologists, geneticists, and population and evolutionary biologists have made significant discoveries in almost all aspects of fungal sexuality (see other chapters in this book), Raper's statement rings truer than ever.

The 1966 book covered almost all aspects of research on sexuality in basidiomycetes from the beginning of the 20th century to the 1960s, including extensive results on many areas from Raper's own research group. In the book, based on experimental observations and logical reasoning, Raper proposed many hypotheses that have been experimentally tested and confirmed or that are in the processes of being tested. His original

and outstanding contributions to our understanding of fungal sexuality are the more remarkable considering how few molecular genetic tools were available during that time—tools that we have come to take for granted today. His insights on fungal sexuality and his landmark publication in 1966 not only summarized over half a century's research progresses on the subject but also impacted significantly on subsequent generations of mycologists and biologists from other areas such as genetics, molecular biology, and population and evolutionary biology.

In this chapter, I discuss one area that Raper discussed repeatedly in his book, the transitions among reproductive systems in fungi. Several authors in this book discuss the transitions among various forms of sexual reproduction in fungi. Different from these reviews, my focus is on the transition between sexual and asexual reproduction. Specifically, I review and discuss recent experimental studies using the model basidiomycetous yeast *Cryptococcus neoformans* to test the effects of spontaneous mutations and biological interactions that may have contributed to the distribution of asexual fungal strains and species in nature. To begin

Jianping Xu, Department of Biology, McMaster University, 1280 Main Street West, Hamilton, Ontario, Canada L8S 4K1.

with, I first introduce some background information on fungal sexuality and spontaneous mutations.

FUNGAL SEXUALITY: DEFINITIONS AND VARIETY

The study of fungal sexuality could be traced back to over 100 years ago when Blakeslee discovered obligatory cross-fertilization in the Mucorales (1). Since then, a wide variety of sexual reproductive structures and mating systems have been identified in fungi. For example, in basidiomycetes, reproductive systems ranged from seemingly completely asexual, to homothallism, secondary homothallism (or pseudohomothallism), bipolar heterothallism with two mating-type alleles, bipolar heterothallism with multiple mating-type alleles, tetrapolar heterothallism with two alleles for one locus and multiple alleles for the other, and tetrapolar heterothallism with multiple alleles for both loci (39).

In fungi, the term homothallism refers to the sexual reproductive system in which a haploid sexual spore is capable of completing the sexual reproductive cycle by itself. In contrast, heterothallism refers to sexual reproductive systems where each haploid sexual spore is incapable of completing the sexual reproductive cycle by itself and fusion between genetically different mating partners is required before such cycles can be completed. Among the different forms of heterothallism, bipolar and tetrapolar refer to the number of segregating loci that control mating compatibility, with bipolar and tetrapolar corresponding to one and two loci, respectively. However, each segregating locus may be highly complex and contain multiple tightly linked genes. Secondary homothallism may be considered an intermediate form between homothallism and heterothallism, that is, the sexual spore may be able to complete the sexual cycle by itself or can mate with a partner to make outbred progeny. In secondary homothallic species, most sexual spores are diploid or dikaryotic with the majority of these spores capable of each completing the sexual reproductive life cycle by itself, without the need of mating. However, for the minority haploid sexual spores in secondary homothallic strains, mating between genetically compatible partners is required before the sexual reproductive life cycle can be completed. In addition, among heterothallic species, the number of alleles at an individual mating-type locus may vary from two to hundreds of alleles. For example, there are two functional alleles at the mating-type locus in the basidiomycete yeast *C. neoformans* (5, 15). In contrast, there are over 400 alleles at one mating-type locus and 65 alleles at another in the filamentous basidiomycete *Schizophyllum commune* (39).

VARIATION IN SEXUALITY AND ASEXUALITY

Among the currently named 100,000 or so fungal species, over 17,000 distributed among 1,680 or so genera have no known sexual states (16). However, phylogenetic analyses over the last 2 decades suggest that most of these apparently "asexual fungi" have close relatives capable of sexual reproduction (for examples, see references 15, 25, 30, 34, and 42). In addition, aside from variation in reproductive life cycles among species, there is also extensive variation among strains and populations within a species in their ability to mate and produce sexual progeny. A good example of a basidiomycete with such intraspecific variation in life cycle is the commercial button mushroom *Agaricus bisporus*. There is extensive variation among strains and populations of *A. bisporus* in their reproductive life cycles, with some strains having the homothallic life cycle and some having a bipolar heterothallic life cycle (4). Closely related species such as those in the genus *Agaricus* may also show very different sexual reproductive systems (46). In bipolar, secondarily homothallic strains of *A. bisporus*, homokaryotic progeny isolated from a single cross show extensive variation in their ability to successfully backcross to the two parental strains (45, 56). Among homokaryons isolated from wild strains of *A. bisporus*, the ability to mate and establish heterokaryons also differs among each other, with significant genotype-environment interactions in determining the outcome of such crosses (46, 55, 56). Similar observations have also been found in other species, including the bipolar heterothallic basidiomycete yeast *C. neoformans* (for an example, see reference 58).

At the population level, evidence for clonality and asexual reproduction has been found in many groups of microorganisms, including both sexual and asexual fungi (20, 23, 27, 38, 43, 53). Signatures of clonality include overrepresentations of certain genotypes, significant linkage disequilibria, Hardy-Weinberg disequilibria (excess heterozygosity or homozygosity in diploid or dikaryotic species), phylogenetic compatibility, and congruent phylogenies among genes (51). Evidence for clonality in fungal species is not surprising because most fungi are capable of asexual reproduction through the generation of asexual spores and/or the fragmentation and dispersion of vegetative hyphae/thalli. However, despite such abilities, molecular population genetic studies over the last decade have indicated that populations of almost all analyzed "asexual fungi" as well as sexual fungi contained signatures of recombination (3, 20, 38, 42, 52, 54). In contrast to the signatures of asexuality, those of sexuality in nature include linkage equilibrium, Hardy-Weinberg equilibrium, phylogenetic incompatibility, and incongruent phylogenies (51). Indeed, current research results indicated little definitive evidence for ancient asexuality in any fungi (for a carefully characterized asexual fungus, see chapter 12).

HYPOTHESES ABOUT FUNGAL ASEXUALITY

Taken together, the above observations suggest three experimentally testable hypotheses on the evolution of fungal asexuality. I would like to emphasize from the outset that these hypotheses are not new. Indeed, as noted above, at least two of these (hypotheses I and III) have received broad support from many phylogenetic, population genetic, and genomic studies (for an example, see reference 42). However, direct experimental tests of these hypotheses are scarce. Below is a brief description of the three hypotheses.

The first hypothesis is that fungi can lose sex easily. The second is that the reason for the loss of sex might be due to some kind of cost associated with sex and sexual reproduction in fungi. For example, maintaining the genes and physiological pathways for sexual reproduction under certain environmental conditions might be a costly burden and reduce the organism's overall fitness in such environments. Other potential costs in fungi may include segregation and recombination load and the cost of interacting with sexual partners, same as those discussed for plants and animals (33). The third hypothesis is that once the ability to undergo sexual reproduction is completely lost, the asexual fungi are prone to the accumulation of mutations. Since most mutations are either neutral or slightly deleterious, mutations accumulated in asexual lineages may drive these asexual fungi into extinction. There is abundant experimental evidence for inevitable loss of fitness in severely bottlenecked populations of bacteria, fungi, plants, and animals (21, 22, 40, 61).

Before discussing recent experimental tests of these hypotheses, I first briefly introduce how mutations can be accumulated and used to test the above hypotheses.

SPONTANEOUS MUTATIONS

Spontaneous mutation is the ultimate source of all heritable variations in all organisms. It can occur in both replicating and nonreplicating genetic materials in cells or viral particles. To cope with constant generations of spontaneous mutations, cells have devised extensive mechanisms to prevent, recognize, correct, and/or modify the damages and mistakes in their genetic materials. In biological literature, the importance of spontaneous mutations is probably best illustrated by examining the theoretical models used to explain some of the fundamental biological phenomena. These phenomena include the wide distribution of diploidy in higher plants and animals; the diverse arrays of mating systems and mate choices in eukaryotes; the existence and evolution of aging and senescence; the long-term survival of small and fragmented populations; and the origin and maintenance of sex and recombination. All widely recognized models explaining the origin and evolution of these phenomena have genome-wide rate and effect of spontaneous mutations as their key parameters (for examples, see references 7, 10, 11, 23, 28, and 36).

The genome-wide mutation rate and the average effect per mutation are most commonly estimated using mutation accumulation (MA) experiments (12, 35, 61). In these experiments, spontaneous mutations are allowed to accumulate in replicate lines in the absence of selection for the trait under investigation (Fig. 28.1). There are two typical underlying assumptions in these

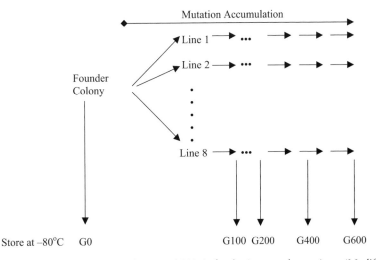

Figure 28.1 Schematic representation of a typical MA in facultative sexual organisms. (Modified from reference 47, reprinted with permission from *Genetics*.)

MA experiments. The first is that in the absence of selection, deleterious mutations will continuously accumulate in the population, leading to a lower fitness value for the trait under consideration. The second assumption is that divergence among replicate lines in trait values should increase over time. The relative rates of decrease and divergence among lines in trait values allow estimation of various mutational parameters, including the genomic mutation rate, the average effect per mutation, and the mutational heritability (17, 31). Like bacteria, fungi are ideal for mutation accumulation experiments as they can be clonally propagated, manipulated, and stored efficiently in the laboratory.

THE QUANTITATIVE NATURE OF FUNGAL SEXUALITY

Sexuality in fungi is typically considered a qualitative trait. Two strains are considered either capable or not capable of mating with each other to produce meiotic progeny. However, similar to those in plants and animals, sexuality in fungi is increasingly recognized as a quantitative trait (for examples, see references 18, 37, 45, 46, and 56). Commonly recognized sexual fitness components include mating ability, the number of sexual progeny, and the fitness of sexual progeny. In most organisms, these traits are all quantifiable. In addition, in organisms such as many eukaryotic microbes where sexual zygotes can be separated from the parents and the progeny, the fitness of these zygotes may also be estimated and quantified (45, 48). The comparison in sexual fitness components between MA lines and the original clones allows estimates on the rate and effects of spontaneous mutations affecting these individual traits.

C. NEOFORMANS AS A MODEL EXPERIMENTAL SYSTEM TO STUDY FUNGAL SEXUALITY

C. neoformans is an encapsulated basidiomycetous yeast. It is ubiquitous in the environment such as in soil, bird droppings, and trees and can cause significant morbidity and mortality, especially in immunocompromised hosts (5). It has a bipolar heterothallic life cycle controlled by one locus with two alternative alleles, a and α (26). Under suitable conditions, unicellular yeasts carrying opposite mating types can mate to form dikaryotic or diploid cells (26, 41, 60). These mated cells can grow in filamentous form (i.e., hyphae) with meiosis occurring at the terminal cell (basidium) of the hyphae. It should be noted that filamentous form is necessary but not sufficient for proper meiosis and basidiospore formation in C. neoformans. On media conducive for mating and filamentation, the filamentous phase may be maintained indefinitely and sexual basidiospores may be produced continuously.

Several features make C. neoformans an ideal organism for molecular genetic studies (18, 44; see also other chapters on C. neoformans in this book). First, it has a small haploid genome and can be grown easily in the laboratory on a variety of media and growth conditions. Second, it has a well-characterized mating system and sexual cycle. Third, strains can be relatively easily transformed with a variety of methods. And fourth, many isogenic strains with a variety of auxotrophic and drug-resistant markers are available in both mating-type backgrounds (18, 44). These same features also make C. neoformans an ideal system for experimental investigations of the evolution of fungal sexuality. For example, the existence of abundant auxotrophic markers makes quantifying mated cells and parental cells very convenient. In this review, the relative number of mated cells in a cell mixture is used as a measure of mating ability. Similarly, the proportion of mated cells capable of forming hyphae provides a quantitative estimate for filamentation.

In the following three sections, I summarize and discuss three recent studies that tested the three hypotheses on fungal asexuality: the loss of sex (47), the cost of sex (50), and the fitness consequences of fungal asexual clones (49) in experimental populations of C. neoformans.

TESTING HYPOTHESIS ABOUT THE LOSS OF SEX

To investigate the effects of spontaneous mutations on the potential loss of sex, we compared the sexual fitness of MA lines and the starting clones for two critical steps during sexual reproduction in C. neoformans: mating and filamentation. In C. neoformans, mating refers to the fusion of cells of opposite sexes (this step is a universal first step in eukaryotic sexual reproduction). In contrast, filamentation is a necessary step preceding meiosis and sexual spore development in C. neoformans and a few other fungal species but is not shared by the majority of sexual fungi or other sexual eukaryotes (44, 47). For example, in filamentous fungi, filamentation is a constitutive process throughout much of their life cycle (except spores) and not a unique component reserved for sexual reproduction. In most unicellular fungi (such as ascomycete yeasts), filamentation is typically not a required process preceding meiosis and sexual spore formation.

Mutation Accumulation

The basic protocol for accumulating spontaneous mutations was established in the 1960s by Mukai and colleagues for Drosophila melanogaster (35). That protocol was subsequently adapted for organisms with different reproductive modes. Our protocol here followed that established for the model bacterium *Escherichia coli* (22). Here, two haploid isogenic strains were used to accumulate spontaneous mutations: JEC50 (*MATα* and *ade2*) and MCC3 (*MATa*, *cna*, and *ura5*) (41, 47). MA lines were grown and maintained on the rich medium YEPD agar (1% yeast extract, 2% Bacto-peptone, 2% dextrose, and 2% Bacto-agar in distilled water) at 25°C. A single colony (the founder colony, generation 0) was picked and streaked onto YEPD agar to establish eight independent lines for each of the two strains (Fig. 28.1). The 16 MA lines were maintained by repeating the picking-streaking-incubating procedure of a single colony for each line every 3 days. By the end of the third day, each colony increased from 1 cell to about 10^6 cells, representing approximately 20 mitotic divisions (± 1 cell division). These 16 MA lines were grown under identical conditions for a total of 30 repeated transfers, equivalent to a total of approximately 600 mitotic divisions. Cells from generations 0, 100, 200, 400, and 600 (abbreviated G0, G100, G200, G400, and G600, respectively) were stored in glycerol (18%) in a −80°C freezer. This protocol intensified genetic drift by forcing each line through a bottleneck of one random cell in each growth cycle. Because the abilities for mating and filamentation were not selected in this nutrient-rich environment, mutations affecting these two traits were expected to accumulate, with independently maintained MA lines accumulating different numbers and types of mutation.

Sexual Fitness of MA Clones and Data Analysis

The abilities of mating and filamentation were assayed and calculated for the two starting clones and the 64 evolved clones by using quantitative measures of the two phenotypes as discussed above and described in detail by Xu (47). These data were then used to estimate the genome-wide mutation rate ($\hat{U}$) related to these phenotypes, the average effect per mutation ($\hat{a}$), and the mutational heritability (h_m^2). Since the two strains of *C. neoformans* examined here are haploid, estimation procedures follow those for haploids as described by Lynch and Walsh (31). Briefly, for haploid organisms, the minimum mutation rate ($\hat{U}_{min}$) and the maximum mutational effects ($\hat{a}_{max}$) were calculated as follows:

$$\hat{U}_{min} = (\Delta M)^2/[\Delta V(1 + C_{\Delta M})(1 + C_{\Delta V})]$$
$$\hat{a}_{max} = \Delta V/[\Delta M(1 + C_{\Delta M})]$$

where ΔM and ΔV denote estimates of the rates of change of mean and variance and $C_{\Delta M}$ and $C_{\Delta V}$ are squared coefficients of sampling variance (ratios of sampling variance to squared estimates) of ΔM and ΔV, respectively. The

mutational heritability (h_m^2) was obtained as mutational variance scaled by environmental variance (17, 47).

MA Clones Had Reduced Sexual Fitness

After 30 asexual transfers on the rich medium YEPD, all 16 MA lines showed reduced abilities for mating and filamentation, with the mean abilities decreased significantly, by over 67 and 24%, respectively (47). Furthermore, these two sexual fitness traits were positively correlated among the derived MA clones. The summary results are presented in Table 28.1 and Fig. 28.2 and 28.3 and briefly described below.

Estimates for the three mutational parameters for mating ability differed between the two mating-type backgrounds. The $\hat{U}_{min}$ for mating was 0.0172 as estimated from the eight MA lines derived from JEC50 and 0.0772 for those derived from MCC3, a difference over fivefold. Consequently, the mutational effects differed over fivefold, but in reverse direction (Table 28.1). The mutational heritabilities for mating were 0.009 for MCC3 and 0.0293 for JEC50, respectively.

Different from the patterns of mating among MA clones, the rates of reduction in filamentation and the among-line divergence were more uniform across the 600 mitotic generations than mating ability (47). Similar to those for mating ability, estimates for the three mutational parameters for filamentation differed between the two strain backgrounds. The $\hat{U}_{min}$ for filamentation was 0.0036 as estimated from the eight MA lines derived from JEC50 and 0.0232 for those from MCC3, a >6-fold difference. The mutational effects had a <2-fold difference, in reverse direction (Table 28.1). The mutational heritability for filamentation was 0.0108 for MCC3 and 0.0343 for JEC50 (47).

These estimates of mutation rates ($\hat{U}_{min}$) of 0.0172 to 0.0772 for mating and 0.0036 to 0.0232 for filamentation were similar to those determined for a variety of life history and reproductive traits in plants and animals (for examples, see references 12, 17, 31, and 61). Typically studied traits in plants and animals are the number and the fitness of sexual progeny. However, because of differences in sexual reproductive processes between fungi and complex plants and animals and the difficulty in obtaining quantitative data with regard to individual processes during sexual production in plants and animals, data comparable to those discussed here for *C. neoformans* are lacking for plants and animals.

THE COST OF SEX

Theoretical and empirical investigations of sexually dimorphic eukaryotes such as insects, birds, and mammals suggest that sex and sexual reproduction can be

Table 28.1 Decreases in mating and filamentation abilities in mutation accumulation lines of *C. neoformans* over 600 asexual generations[a]

Trait and measures/parameters	JEC50 (*MATα*)	MCC3 (*MATa*)
Mating		
Mean decrease (range)	0.6761 (0.0751–0.8636)	0.7879 (0.6062–0.8764)
$\hat{U}_{min}$	0.0172	0.0772
$\hat{a}_{max}$	0.0604	0.0110
h_m^2	0.0293	0.0090
Filamentation		
Mean loss (range)	0.2488 (0.0456–0.5431)	0.4045 (0.2674–0.5795)
$\hat{U}_{min}$	0.0036	0.0232
$\hat{a}_{max}$	0.0948	0.0519
h_m^2	0.0343	0.0108

[a]Modified from reference 47; reprinted with permission of *Genetics*.

costly (33). The most discussed costs of sex in the literature are the twofold cost of producing males and the cost associated with producing exaggerated mating behaviors and morphological modifications. These costs have been little discussed in facultative sexual microbes such as the majority of fungi. Indeed, in eukaryotic sexual microbes, sexual gamete dimorphism similar to those in plants and animals can be found. For example, the protoperitheciating partner in *Neurospora* and many other filamentous ascomycetes invests more in the production of ascospores than that in the production of conidia from opposite mating partners, but with both contributing equal nuclear genetic materials to the next generation. The differential investments in parental resources in mating suggest potential conflicts and potentially different costs in sexual reproduction among mating partners in these fungi. In addition, the costs associated with segregation and recombination are universal among all sexual eukaryotes (6, 10).

Unlike obligate sexual organisms such as the majority of higher plants and animals where sex (gamete fusion or mating, and meiosis) and reproduction (the maintenance or increase in population size) are intrinsically linked, these two processes can be separated in facultative sexual microbes (48). The ability to separate these two processes in facultative sexual microbes makes these organisms excellent models to investigate the origin and evolution of individual processes. In many microbial eukaryotes, mating between compatible partners occurs only under selective conditions. In these conditions, both mating (i.e., the formation of zygotes) and vegetative growth of individual mating partners may occur. Using appropriately marked strains, all three cell types (i.e., the two parental types and the zygotes) in such a mating mixture can be directly counted using se-

lective media. From these cell counts, the mating success rate and the relative vegetative fitness of the parental strains can be estimated. The vegetative fitness of individual parental strains in the presence of an active mating partner may change in comparison to that in the absence of such a partner. If the vegetative fitness were lower in the presence of an active mating partner, the reduction would constitute one type of cost of sex, i.e., the cost of interacting with a mating partner.

Mating Interaction in *C. neoformans*

C. neoformans is a good model organism in which to study the cost of sex because when parental cells of opposite mating types of the facultative sexual yeast *C. neoformans* are mixed on a nitrogen-limiting medium and incubated at a temperature around 22 to 25°C, each parental yeast cell may take one of the following two reproductive paths, one leading to sexual reproduction and the other to asexual reproduction.

In the first path, the parental yeast cells of opposite mating types may fuse to form zygotes. In *C. neoformans*, both the *MATa* and the *MATα* cells can secrete sex pheromones and release them into the surrounding medium (19). In response to pheromones from the opposite mating types, the *MATα* cells may make conjugation tubes and the *MATa* cells may expand in size (19, 32). These morphologically differentiated *MATa* and *MATα* cells then may fuse and form diploid zygotes. The zygotes can be detected after about 12 h of incubation and are abundant at about 20 to 24 h (47, 59, 60). At this time (i.e., after 20 to 24 h of coincubation), the zygotes may begin producing hyphal filaments that gradually intertwine among each other to form microscopic mycelial mats. After a few more days of incubation, the mycelial mats will be visible with the naked eye

A.

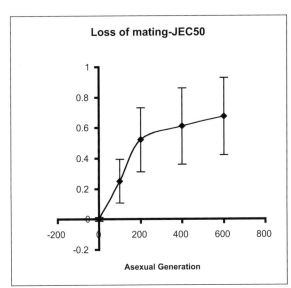

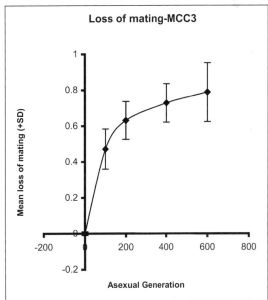

B.

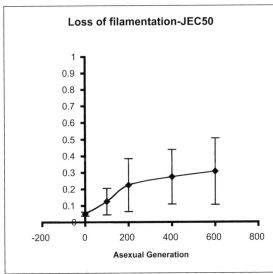

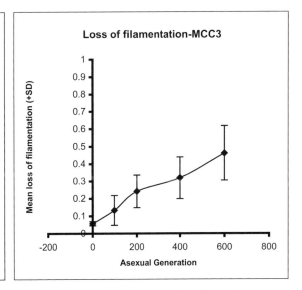

Figure 28.2 The mean loss of sex and the among-line divergence over 600 mitotic generations for each of the two parental strains JEC50 and MCC3: (A) mating ability and (B) filamentation ability. In all four graphs, the *x* axis shows the number of asexual generations, while the *y* axis represents the mean and standard deviation of losses of mating or filamentation. For each data point, the mean and standard deviation were obtained from eight MA lines (from reference 47, reprinted with permission from *Genetics*).

and the ends of hyphae may enlarge and form basidia within which nuclear fusion and meiosis may occur. Four chains of basidiospores, the sexual spores of *C. neoformans*, are then produced from each basidium. Each basidiospore contains a recombinant haploid nu-

cleus, the product of meiosis. After about 20 to 24 h the zygotes form hyphal filaments that can intertwine and complicate counting; therefore, an accurate estimate of zygote number is often difficult to obtain beyond this time point. In addition, extended incubation will result

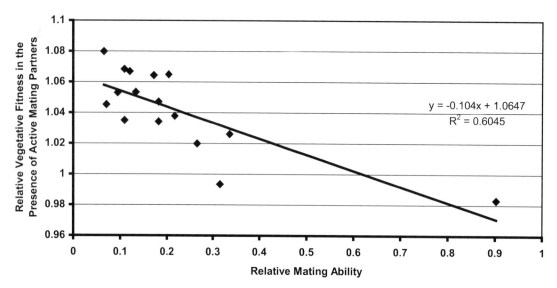

Figure 28.3 A negative correlation between mating ability (*x* axis) and the relative vegetative fitness in the presence of active mating partners (*y* axis, the inverse of cost of sex) for the 16 MA clones at G600 ($r = -0.777$; $P < 0.0001$). (From reference 50, reprinted with permission from *Genetics*.)

in the production of chlamydospores as well as basidiospores and associated mitospores, all of which can complicate the estimates of parental and mated cells (47, 50, 59, 60). Therefore, mating success rate is typically estimated within 24 h of incubation (18, 47, 50).

In the second reproductive path, the parental cells do not fuse with those of opposite mating types but instead reproduce asexually through mitosis and budding. However, because these cells are exposed to their mating partners, their rates of asexual reproduction may change and a reduced rate of asexual reproduction would constitute a cost of sex or of attempted sex.

There Is a Cost of Interacting with Mating Partners

To test for the existence of a potential cost of interacting with mating partners in *C. neoformans*, the same two strains used above, JEC50 (*MATα ade2*) and MCC3 (*MATa cna ura5*), were used as parental strains and tested in three medium conditions: one nitrogen-limiting and two nitrogen-rich media (50). The nitrogen-limiting medium was SSΔN agar (0.17% Difco yeast nitrogen base without amino acids or ammonium sulfate, 2% sucrose, 2% Bacto-agar, and 0.02 g per liter of uracil and adenine). The two nitrogen-rich media were SS (0.17% Difco yeast nitrogen base with amino acids and ammonium sulfate, 2% sucrose, 2% Bacto-agar, and 0.02 g per liter of uracil and adenine) and the common laboratory yeast growth medium YEPD (2% yeast extract, 1% peptone, 2% dextrose, and 2% agar).

In mating mixtures on the nitrogen-limiting medium, SSΔN agar, the mean generation times of strains JEC50 and MCC3 increased by over 10.1 and 8.9%, respectively, relative to those in the absence of compatible mating partners (Table 28.2) (50). While both reductions were statistically significant ($P < 0.01$), there was no difference in the amount of reduction between the two strains with different mating types ($P > 0.9$). This ~10% reduction in vegetative fitness of the two original clones in the presence of compatible mating partners was medium specific. On nitrogen-rich agar media such as SS and YEPD, no mating was observed and there was no difference for either parent in vegetative growth rates between those with and those without compatible mating partners (50).

Relationship between the Cost of Sex and Mating Ability

The above results suggested that the cost of sex (i.e., the reduction of vegetative fitness in mating mixtures) was due to the presence of active mating partners. Therefore, we hypothesized that this cost of sex might be reduced for strains with reduced mating abilities. To test for this hypothesis, the 16 MA lines of JEC50 and MCC3 (eight for each strain) obtained previously and discussed above (47) were examined. These MA lines included 64 MA clones that had shown variable degrees of reduced mating abilities compared to the original clones (47).

Table 28.2 A cost of interacting with active mating partners in *C. neoformans*[a]

Trait	Value[b] for strain:	
	JEC50	MCC3
Mating ability	6.111% ± 0.508%	6.111% ± 0.508%
Vegetative fitness on SSΔN (generation time, in hours)		
Without active mating partners	4.112 ± 0.092	6.834 ± 0.124
With active mating partners	4.527 ± 0.108	7.442 ± 0.126
Cost of sex[c]	0.101 ± 0.006	0.089 ± 0.011

[a]Modified from reference 50; reprinted with permission from *Genetics*.
[b]All values represent the means ± standard deviations of 8 repeats.
[c]The cost of sex is calculated as (Gp − Ga)/Ga, where Gp is the generation time in the presence of active mating partners and Ga is the generation time in the absence of active mating partners.

Indeed, when the MA clones with reduced mating abilities were mixed with the original clones of opposite mating types on SSΔN agar, their cost of sex was reduced relative to those of the original clones (50). A summary result of the 16 MA clones from the 30th transfer (i.e., the ones after 600 mitotic generations of mutation accumulation) is presented in Table 28.2. As expected, among these MA clones, there was a negative correlation between mating success rate and the relative vegetative fitness in the presence of active mating partners among the evolved clones (Pearson correlation coefficient $r = -0.777$; $P < 0.0001$; Fig. 28.3). It should be noted that due to accumulated genetic differences among MA lines in vegetative fitness, the relative vegetative fitness for each clone used here was standardized to that of the corresponding MA clone in the absence of an active mating partner, not to that of the original clone. This result supports the hypothesis that the cost of sex is greater for strains with higher mating ability

(50). Negative correlations between sexual fitness and vegetative fitness in other species have also been observed (for examples, see references 11, 21, and 24). The general pattern is also consistent with results from molecular genetic studies of fungi and other microbial eukaryotes in that mating pheromones and the initiation of sexual reproduction often lead to the arrest of cell cycles (14).

In addition, our study was able to deduce the reduction in two components of the cost of sex among the MA clones. One component was the cost of producing mating signals that would exert its effect on its mating partners. The second was associated with responding to mating signals from its mating partners. Compared to the original clones, the MA clones from G600 imposed an average 74.5% (standard deviation, ±17.2%, $n = 16$) of the cost of their ancestral clones imposed on their mating partner (Table 28.3). For the second component, the MA clones from G600 had an average 42%

Table 28.3 Reductions in the cost of sex in asexually evolved MA lines after 30 transfers in experimental populations of *C. neoformans*[a]

Traits	Value[b] for MA lines of progenitor strain:	
	JEC50	MCC3
Vegetative fitness on SSΔN		
Without mating partners	4.251 ± 0.179*	7.009 ± 0.181*
With mating partners	4.422 ± 0.253**	7.285 ± 0.194**
Cost of sex	0.040 ± 0.022***	0.039 ± 0.023***
Relative cost of sex on SSΔN compared to parental clones[c]		
Cost in response to mating partners	0.398 ± 0.219***	0.442 ± 0.257***
Cost exerted on mating partners	0.715 ± 0.179***	0.775 ± 0.148***

[a]Modified from reference 50; reprinted with permission from *Genetics*.
[b]*, **, and *** denote significant differences between the MA clones and their respective progenitor clones at $P < 0.05$, $P < 0.01$, and $P < 0.001$, respectively.
[c]The costs of sex in the founder clones are scaled to 1 for each of the two strains.

Table 28.4 Estimates of mutational parameters on vegetative fitness in *C. neoformans* using two MA conditions and four testing environments[a]

MA condition	Mutation parameter	Value under test condition:			
		T25/SD	T25/YEPD	T37/SD	T37/YEPD
T25/YEPD	$\hat{U}_{min}$ (10^{-3})	1.815	1.127	5.662	1.982
	$\hat{a}_{max}$	0.253	0.007	0.147	0.085
	h_m^2 (10^{-3})	2.124	0.213	1.185	1.962
T37/YEPD	$\hat{U}_{min}$ (10^{-3})	5.334	1.253	0.702	1.363
	$\hat{a}_{max}$	0.057	0.029	0.156	0.014
	h_m^2 (10^{-3})	0.326	0.445	0.646	0.091

[a]Modified from reference 49; reprinted with permission from *Genetics*.

($\pm23.2\%$) of the cost of their ancestral clones (Table 28.3). There was no significant difference between the two mating types in the reductions of the two components of the cost of sex. Overall, the rate of reduction was significantly greater for responding to active mating partners than that for effects exerted on mating partners (Student's *t* test, $t = 4.606$, $P < 0.001$). These two reductions of costs were not correlated ($r = 0.276$, $P = 0.301$) (50).

DECREASES IN VEGETATIVE FITNESS IN ASEXUAL CLONES OF *C. NEOFORMANS*

As described in the introduction and in "Hypotheses about Fungal Asexuality," because little evidence is available for ancient asexuality in fungi, it has long been hypothesized that asexual fungi may be typically short-lived on an evolutionary timescale. At present, most experimental investigations of fitness consequences of asexuality in fungi have come from the baker's yeast, *Saccharomyces cerevisiae*. In these investigations, irreversible losses of vegetative fitness have been seen in all MA experiments (61). Comparable experimental investigations of asexual (nonrecombining) genomes in microbes or fragments of asexual genomes in animals such as *D. melanogaster* have shown that, indeed, such genomes or genomic fragments can accumulate significant spontaneous mutations, most of which are deleterious to fitness (31). At present, little is known about the fitness effects of asexual reproduction in the majority of fungi.

To examine the effects of asexual reproduction on vegetative fitness in the basidiomycete *C. neoformans*, an MA experiment similar to that described above in the "loss of sex" section was carried out (49). Indeed, as expected, after 30 rounds of severe population bottlenecks over the 600 mitotic generations, all MA lines

showed significant reductions in vegetative fitness compared to their original clones (Table 28.4). Interestingly, a variety of environment-specific effects on both the accumulation of spontaneous mutations and the nature of these mutations were also observed. These effects and their potential utility in inferring microbial ecology in nature are briefly discussed below.

Instead of accumulating spontaneous mutations in only one environment and testing the effects of such mutations in the same or a different environment as most MA studies did, two different MA conditions and four testing environments were investigated in the study on *C. neoformans* (49). Specifically, eight MA lines were established from a single clone of the model lab strain JEC21 on the nutrient-rich medium YEPD (to allow maximum accumulation of mutations) for each of two temperatures, 25 and 37°C. Cells from generations 100, 200, 400, and 600 for each of the 16 MA lines were stored and assayed for vegetative fitness on each of four conditions: (i) 25°C on SD (a minimal medium); (ii) 25°C on YEPD; (iii) 37°C on SD; and (iv) 37°C on YEPD.

The analyses identified that both MA conditions and assay environments had a significant influence on the estimates of genomic mutation rates, average effect per mutation, and mutational heritability (Table 28.4). Significant genotype-environment interactions were detected among the newly accumulated spontaneous mutations (Fig. 28.4; Table 28.5). Overall, clones from MA lines maintained at 37°C showed lower decline in vegetative fitness than those maintained at 25°C, especially in the 37°C testing environments (Fig. 28.4). Our results indicated that the relative fitness of individual MA clones was highly dependent on environmental conditions and no clone showed the highest or lowest fitness in all testing environments. As expected, the SD minimal medium testing conditions revealed more deleterious mutations than the rich YEPD media. The results

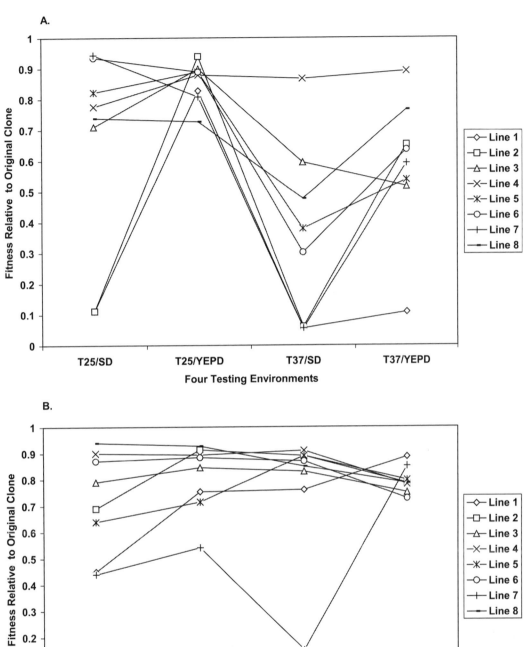

Figure 28.4 Relative mean fitness of MA lines grown in four different conditions, on the *x* axis from left to right: (i) 25°C on SD medium; (ii) 25°C on YEPD medium; (iii) 37°C on SD medium; (iv) 37°C on YEPD medium. Only clones from G600 are shown here. Panel A shows MA lines maintained at 25°C on YEPD medium; panel B shows MA lines maintained at 37°C on YEPD medium (from reference 49, reprinted with permission from *Genetics*).

Table 28.5 Three-way analysis of variance and evidence for significant genotype-environment interactions of spontaneous mutations on vegetative fitness in experimental populations of *C. neoformans*[a]

Source of variation	df	Value under MA condition:			
		T25/YEPD		T37/YEPD	
		Mean square	F	Mean square	F
L	7	0.1241	29.188***	0.0618	15.096***
T	1	0.6362	149.70***	0.0032	0.7805(ns)
M	1	0.4084	96.104***	0.0293	7.1543**
L × T	7	0.0396	9.3194***	0.0139	3.4151**
L × M	7	0.0488	11.483***	0.0274	6.6967***
T × M	1	0.0012	0.2858(ns)	0.0090	2.2015(ns)
L × T × M	7	0.0451	10.615***	0.0197	4.8172***
Within treatment (random variation)	288	0.0043		0.0041	

[a] Only data from G600 were used in these analyses. (From reference 49, reprinted with permission from *Genetics*.) Abbreviations and symbols: L, MA lines; T, temperature; M, medium; ns, not significant; **, $P < 0.01$; ***, $P < 0.001$.

suggested that a high temperature environment might be very important for maintaining the ability to grow at a high temperature.

Results from comparisons between clinical and environmental samples of *C. neoformans* were also consistent with laboratory experimental population analyses (Tables 28.6 and 28.7). Specifically, clinical strains were better able to grow at 37°C than environmental strains (Table 28.6), and there was a significant interaction between strain source and incubation temperature or vegetative growth rates between the two samples of *C. neoformans* (Table 28.7). Taking together the results from MA experiments and the natural population comparisons, it was suggested that frequent exposures to a high-temperature environment (i.e., 37°C) might be important for *C. neoformans* to maintain the ability to grow at 37°C.

Since the ability to grow vigorously at 37°C is required for any pathogen to cause disease in humans, the results described above thus call into question several long-standing views about *C. neoformans* ecology. These long-standing views include the following: (i) *C. neoformans* cells infecting humans and other warm-blooded mammals do not contribute to the pathogenic populations in nature, (ii) humans and other warm-blooded animals do not serve any reservoir function for *C. neoformans*, and (iii) humans and other warm-blooded animals were only unfortunate/occasional hosts of this fungal pathogen.

The hypothesis that human and other warm-blooded animals might play greater than expected roles in the ecology and epidemiology of pathogenic strains of *C. neoformans* was also supported by data from population genetic studies. Epidemiological surveys have identified that a few clones or clonal lineages dominate the global clinical strains of *C. neoformans* and that clonal dispersals are common over wide geo-

Table 28.6 Summary of vegetative fitness comparisons between clinical and environmental samples of *C. neoformans* at 25 and 37°C[a]

Population sample	No. of strains	Mean colony size ± SD (mm)	
		Testing environment	
		T25/YEPD	T37/YEPD
Environmental	45	0.251 ± 0.091	0.516 ± 0.181
Clinical	35	0.192 ± 0.057	0.637 ± 0.096

[a] From reference 49; reprinted with permission from *Genetics*.

Table 28.7 Two-way analysis of variance table showing a significant interaction between strain source and incubation temperature in vegetative growth of strains of *C. neoformans*[a]

Source of variation	df	Mean square	F[b]
A: Sample source (clinical versus environmental)	1	0.0003	0.0205 (ns)
B: Incubation temp (25 versus 37°C)	1	1.1662	79.877***
C: A × B interaction	1	4.1382	284.02***
D: Within subgroup (error term)	156	0.0146	

[a]From reference 49; reprinted with permission from *Genetics*.
[b]ns, not significant; ***, $P < 0.001$.

graphic areas (for examples, see references 2, 23, and 57). If pathogenic strains of *C. neoformans* spent most of their time in soil, trees, and bird droppings, if warm-blooded animals were only accidental hosts, and if warm-blooded animals did not contribute to environmental populations, it would be expected that geographic populations should diverge from each other rather rapidly. Such divergences should lead to geographic structuring and the lack of clonal dispersal over long distances. While endemic elements of *C. neoformans* may exist in certain regions (e.g., MATa strains in Africa [29] and South America [14]), current molecular population genetic studies identified abundant evidence for long-distance clonal dispersal, consistent with the hypothesis that warm-blooded animals such as humans might act as important reservoirs and selective agents for pathogenic strains (2, 23, 57). Such a reservoir and the selective pressure for high-temperature growth could purge less pathogenic or nonpathogenic strains from human populations, leading to limited genetic diversity, extensive clonality, and long-distance clonal dispersal (also by humans) of these pathogenic strains that we see today (for a more detailed discussion on the relationship between reservoir, transmission, and the patterns of population genetic variation in *C. neoformans*, please see reference 49).

CONCLUSIONS AND PERSPECTIVES

This review summarized our recent investigations using mutation accumulation experiments to examine the origin, evolution, and extinction of fungal asexuality. These experiments provided some of the first quantitative estimates in fungi with regard to the rate of loss of sex, the cost of sex, and the fitness consequences of asexuality under controlled laboratory conditions.

It should be emphasized that there are limitations in applying these laboratory results and techniques to natural fungal populations. In their natural habitats, most microorganisms, including *C. neoformans*, likely experi-

ence fluctuations in temperature, water availability, pH, nutrient levels, and interacting biotic components, etc. All these external factors can significantly influence the rate of sexual and asexual reproduction and the size and density of individual populations. While some of these parameters (e.g., generation time, mating ability, the potential cost of sex, and other fitness traits) may be individually estimated in natural environments, there are currently no effective methods to estimate all parameters to allow accurate comparisons between laboratory research results and those from natural environments.

While the potential usefulness of these results in explaining natural fungal population patterns remains to be fully evaluated, the approaches described here can be powerful tools in addressing a variety of evolutionary genetic questions such as the origin and maintenance of diploidy, dikaryons, and fungal senescence. Furthermore, with increasing accessibility to genomic information and high-throughput technology such as microarrays in *C. neoformans*, it may be possible to identify the specific spontaneous mutations influencing the loss of sex, the cost of sex, and the fitness of asexual clones. Indeed, microarray technologies have been used to identify beneficial mutations selected using strong selective pressures in several microbial species, including *E. coli* (8) and *Candida albicans* (9). It should be noted that in these studies, directional selection using either glucose limitation (for *E. coli* [8]) or the antifungal drug fluconazole (for *C. albicans* [9]) often resulted in a limited number of mutations. In contrast, the mutational targets for traits discussed in this review might be much greater and more complex. Further molecular analyses of the specific mutations accumulated here should help elucidate the quantitative genetic loci and potential environmental factors influencing sexual and asexual reproduction in *C. neoformans* and other fungi.

My research is supported by grants from the Natural Sciences and Engineering Research Council (NSERC) of Canada, the Premier's Research Excellence Award, and Genome Canada.

References

1. Blakeslee, A. F. 1904. Sexual reproduction in the Mucorineae. *Proc. Am. Acad. Sci.* 40:206–219.

2. Brandt, M. E., L. C. Hutwagner, L. A. Klug, W. S. Baughman, D. Rimland, E. A. Graviss, R. J. Hamill, C. Thomas, P. G. Pappas, A. L. Reingold, R. W. Pinner, and the CDC Cryptococcal Disease Active Surveillance Group. 1996. Molecular subtype distribution of *Cryptococcus neoformans* in four areas of the United States. *J. Clin. Microbiol.* 34:912–917.

3. Burt, A., D. A. Carter, G. L. Koenig, T. J. White, and J. W. Taylor. 1996. Molecular markers reveal cryptic sex in the human pathogen *Coccidioides immitis* (Ascomycota). *Proc. Natl. Acad. Sci. USA* 93:770–773.

4. Callac, P., S. Hocquart, M. Imbernon, C. Desmerger, and J. M. Olivier. 1998. Bsn-t alleles from French field strains of *Agaricus bisporus. Appl. Environ. Microbiol.* 64: 2105–2110.

5. Casadevall, A., and J. R. Perfect. 1998. Cryptococcus neoformans. ASM Press, Washington, DC.

6. Charlesworth, B., and N. H. Barton. 1996. Recombination load associated with selection for increased recombination. *Genet. Res.* 67:27–41

7. Charlesworth, B., and D. Charlesworth. 1998. Some evolutionary consequences of deleterious mutations. *Genetica* 103:3–19.

8. Cooper, T. F., D. E. Rozen, and R. E. Lenski. 2003. Parallel changes in gene expression after 20,000 generations of evolution in *Escherichia coli. Proc. Natl. Acad. Sci. USA* 100:1072–1077.

9. Cowen, L. E., A. Nantel, M. S. Whiteway, D. Y. Thomas, D. C. Tessier, L. M. Kohn, and J. B. Anderson. 2002. Population genomics of drug resistance in *Candida albicans. Proc. Natl. Acad. Sci. USA* 99:9284–9289.

10. Crow, J. F. 1992. Mutation, mean fitness, and genetic load. *Oxf. Surv. Evol. Biol.* 9:3–42.

11. Da Silva, J., and G. Bell. 1992. The ecology and genetics of fitness in *Chlamydomonas* VI. Antagonism between natural selection and sexual selection. *Proc. R. Soc. Lond. B.* 249:227–233.

12. Drake, J. W., B. Charlesworth, D. Charlesworth, and J. F. Crow. 1998. Rates of spontaneous mutation. *Genetics* 148:1667–1686.

13. Escandon, P., A. Sanchez, M. Martinez, W. Meyer, and E. Castaneda. 2006. Molecular epidemiology of clinical and environmental isolates of the *Cryptococcus neoformans* species complex reveals a high genetic diversity and the presence of the molecular type VGII mating type a in Colombia. *FEMS Yeast Res.* 6:625–635.

14. Fields, S. 1990. Pheromone response in yeast. *Trends Biochem. Sci.* 15:270–273.

15. Geiser, D. M., W. E. Timberlake, and M. L. Arnold. 1996. Loss of meiosis in *Aspergillus. Mol. Biol. Evol.* 13: 809–817.

16. Hawkesworth, D. L., P. M. Kirk, B. C. Sutton, and D. N. Pegler. 1995. *Ainsworth and Bisby's Dictionary of the Fungi*, 8th ed. International Mycological Institute, Surrey, England.

17. Houle, D., B. Morikawa, and M. Lynch. 1996. Comparing mutational variabilities. *Genetics* 143:1467–1483.

18. Hull, C. M., M.-J. Boily, and J. Heitman. 2005. Sex-specific homeodomain proteins Sxi1alpha and Sxi2a coordinately regulate sexual development in *Cryptococcus neoformans. Eukaryot. Cell* 4:526–535.

19. Hull, C. M., and J. Heitman. 2002. Genetics of *Cryptococcus neoformans. Annu. Rev. Genet.* 36:557–615.

20. James, T. J. 2005. The population genetics of phycomycetes, p. 117–148. *In* J. Xu (ed.), *Evolutionary Genetics of Fungi.* Horizon Bioscience, Norfolk, United Kingdom.

21. Keightley, P. D., and A. Caballero. 1997. Genomic mutation rates for lifetime reproductive output and lifespan in *Caenorhabditis elegans. Proc. Natl. Acad. Sci. USA* 94: 3823–3827.

22. Kibota, T. T., and M. Lynch. 1996. Estimate of the genomic mutation rate deleterious to overall fitness in *Escherichia coli. Nature* 381:694–696.

23. Kidd, S. E., H. Guo, K. H. Bartlett, J. Xu, and J. W. Kronstad. 2005. Comparative gene genealogies indicate that two clonal lineages of *Cryptococcus gattii* in British Columbia resemble strains from other geographical areas. *Eukaryot. Cell* 4:1629–1638.

24. Kondrashov, A. S. 1997. Evolutionary genetics of life cycles. *Annu. Rev. Ecol. Syst.* 28:391–435.

25. Kurtzman, C. P., and C. J. Robnett. 1998. Identification and phylogeny of ascomycetous yeasts from analysis of nuclear large subunit (26S) ribosomal DNA partial sequences. *Antonie Leeuwenhoek* 73:331–371.

26. Kwon-Chung, K. J. 1976. Morphogenesis of *Filobasidiella neoformans*, the sexual state of *Cryptococcus neoformans. Mycologia* 67:821–833.

27. Lan, L., and J. Xu. 2006. Multiple gene genealogical analyses suggest divergence and recent clonal dispersal in the opportunistic human pathogen *Candida guilliermondii. Microbiology* 152:1539–1549.

28. Lande, R. 1995. Mutation and conservation. *Conserv. Biol.* 9:782–791.

29. Litvintseva, A. P., R. Thakur, R. Vilgalys, and T. G. Mitchell. 2006. Multilocus sequence typing reveals three genetic subpopulations of *Cryptococcus neoformans* var. *grubii* (serotype A), including a unique population in Botswana. *Genetics* 172:2223–2238.

30. LoBuglio, K. F., J. I. Pitt, and J. W. Taylor. 1993. Phylogenetic analysis of two ribosomal DNA regions indicates multiple independent losses of a sexual Talaromyces state among asexual *Penicillium* species in subgenus *Biverticillium. Mycologia* 85:592–604.

31. Lynch, M., and B. Walsh. 1998. *Genetics and Analysis of Quantitative Traits.* Sinauer Associates, Sunderland, MA.

32. McClelland, C. M., Y. C. Chang, A. Varma, and K. J. Kwon-Chung. 2004. Uniqueness of the mating system in *Cryptococcus neoformans. Trends Microbiol.* 12:208–212.

33. Michod, R. E., and B. R. Levin. 1988. *The Evolution of Sex: An Examination of Current Ideas.* Sinauer Associates Inc., Sunderland, MA.

34. **Moncalvo, J.-M.** 2005. Molecular Systematics: major fungal phylogenetic groups and fungal species concepts, p. 1–34. *In* J. Xu (ed.), *Evolutionary Genetics of Fungi.* Horizon Bioscience, Norfolk, United Kingdom.

35. **Mukai, T.** 1964. The genetic structure of natural populations of *Drosophila melanogaster.* I. Spontaneous mutation rate of polygenes controlling viability. *Genetics* **50:**1–19.

36. **Partridge, L., and N. H. Barton.** 1993. Optimality, mutation and the evolution of aging. *Nature* **362:**305–311.

37. **Pringle, A., and J. W. Taylor.** 2002. The fitness of filamentous fungi. *Trends Microbiol.* **10:**474–481.

38. **Pujol, C., A. Dodgson, and D. R. Soll.** 2005. Population genetics of ascomycetes pathogenic to humans and animals, p. 149–188. *In* J. Xu (ed.), *Evolutionary Genetics of Fungi.* Horizon Bioscience, Norfolk, United Kingdom.

39. **Raper, J. R.** 1966. *Genetics of Sexuality in Higher Fungi.* The Ronald Press Company, New York, NY.

40. **Schultz, S. T., M. Lynch, and J. H. Willis.** 1999. Spontaneous deleterious mutations in *Arabidopsis thaliana.* *Proc. Natl. Acad. Sci. USA* **96:**11393–11398.

41. **Sia, R. A., K. B. Lengeler, and J. Heitman.** 2000. Diploid strains of the pathogenic basidiomycete *Cryptococcus neoformans* are thermally dimorphic. *Fungal Genet. Biol.* **29:**153–163.

42. **Taylor, J. T., D. J. Jacobson, and M. C. Fisher.** 1999. The evolution of asexual fungi: reproduction, speciation and classification. *Annu. Rev. Phytopathol.* **37:**197–246.

43. **Tibayrenc, M.** 1999. Toward an integrated genetic epidemiology of parasitic protozoa and other pathogens. *Annu. Rev. Genet.* **33:**449–477.

44. **Wang, P., and D. S. Fox.** 2005. *Cryptococcus neoformans* evolves as a model of choice for studying signal transduction, p. 321–338. *In* J. Xu (ed.), *Evolutionary Genetics of Fungi.* Horizon Bioscience, Norfolk, United Kingdom.

45. **Xu, J.** 1995. Analysis of inbreeding depression in *Agaricus bisporus.* *Genetics* **141:**137–145.

46. **Xu, J.** 1996. *Mating and Population Genetic Analyses of the Basidiomycete Fungus,* Agaricus bisporus. Ph.D. thesis. University of Toronto, Toronto, Canada.

47. **Xu, J.** 2002. Estimating the spontaneous mutation rate of loss of sex in the human pathogenic fungus *Cryptococcus neoformans.* *Genetics* **162:**1157–1167.

48. **Xu, J.** 2004. The prevalence and evolution of sex in microorganisms. *Genome* **47:**775–780.

49. **Xu, J.** 2004. Genotype-environment interactions of spontaneous mutations affecting vegetative fitness in the human pathogenic fungus *Cryptococcus neoformans.* *Genetics* **168:**1177–1188.

50. **Xu, J.** 2005. Cost of interacting with sexual partners in a facultative sexual microbe. *Genetics* **171:**1597–1604.

51. **Xu, J.** 2005. Fundamentals of fungal molecular population genetic analyses, p. 87–116. *In* J. Xu (ed.), *Evolutionary Genetics of Fungi.* Horizon Bioscience, Norfolk, United Kingdom.

52. **Xu, J., M. Cheng, Q. Tan, and Y. Pan.** 2005. Molecular population genetics of basidiomycete fungi, p. 189–220. *In* J. Xu (ed.), *Evolutionary Genetics of Fungi.* Horizon Bioscience, Norfolk, Unitd Kingdom.

53. **Xu, J., and T. G. Mitchell.** 2002. Strain variation and clonality in *Candida* spp. and *Cryptococcus neoformans,* p. 739–750. *In* R. A. Calderone and R. L. Cihlar (ed.), *Fungal Pathogenesis: Principles and Clinical Applications.* Marcel Dekker, Inc. New York, NY.

54. **Xu, J., and T. G. Mitchell.** 2003. Comparative gene genealogical analyses of strains of serotype AD identify recombination in populations of serotypes A and D in the human pathogenic yeast *Cryptococcus neoformans.* *Microbiology* **149:**2147–2154.

55. **Xu, J., P. A. Horgen, and J. B. Anderson.** 1993. Media and temperature effects on mating interactions of *Agaricus bisporus.* *Cultivated Mushroom Res. Newsl.* **1:**25–32.

56. **Xu, J., P. A. Horgen, and J. B. Anderson.** 1996. Variation in mating interactions in *Agaricus bisporus.* *Cultivated Mushroom Res. Newsl.* **3:**23–30.

57. **Xu, J., R. J. Vilgalys, and T. G. Mitchell.** 2000. Multiple gene genealogies reveal recent dispersion and hybridization in the human pathogenic fungus *Cryptococcus neoformans.* *Mol. Ecol.* **9:**1471–1481.

58. **Yan, Z., X. Li, and J. Xu.** 2002. Geographic distribution of mating type alleles of *Cryptococcus neoformans* in four areas of the United States. *J. Clin. Microbiol.* **40:**965–972.

59. **Yan, Z., C. M. Hull, J. Heitman, S. Sun, and J. Xu.** 2004. *SXI1α* controls uniparental mitochondrial inheritance in *Cryptococcus neoformans.* *Curr. Biol.* **14:**R743–R744.

60. **Yan, Z., and J. Xu.** 2003. Mitochondria are inherited from the MATa parent in crosses of the basidiomycete fungus *Cryptococcus neoformans.* *Genetics* **163:**1315–1325.

61. **Zeyl, C.** 2005. Rates and effects of spontaneous mutations in fungi, p. 289–320. *In* J. Xu (ed.), *Evolutionary Genetics of Fungi.* Horizon Bioscience, Norfolk, United Kingdom.

*Sex in Fungi: Molecular Determination
and Evolutionary Implications*
Edited by Joseph Heitman et al.
© 2007 ASM Press, Washington, D.C.

Dee Carter, Nathan Saul,
Leona Campbell, Tien Bui,
and Mark Krockenberger

29

Sex in Natural Populations of *Cryptococcus gattii*

SEX AND FUNGAL PATHOGENS

Fungal pathogens of humans, once considered to be an interesting sideline, have emerged as a serious threat to human health. With this emergence has come an increased interest in the natural life histories of these organisms, many of which normally exist as environmental saprophytes. Fungal sex is of importance to both the pathogen and the medical mycologist in two ways: sex, by recombination, increases the rate of adaptive change, potentially allowing the fungus to more quickly evolve new virulence, host range, and drug resistance phenotypes. Sex also is frequently the means by which resistant, airborne propagules are produced, allowing the fungus to persist, to expand its range, and to contact the human host. In this chapter we review what is known about sexual recombination in naturally occurring populations of *Cryptococcus gattii*. As it is readily isolated from its environmental source, *C. gattii* allows a unique insight into how ecology and infection interface in a primary fungal pathogen.

CRYPTOCOCCUS GATTII AND CRYPTOCOCCOSIS

C. gattii is one of only a handful of fungal species, and the only member of the Basidiomycotina, that are capable of causing disseminated disease in immunocompe-

tent people and can be considered primary pathogens. This encapsulated yeast, together with the closely related *Cryptococcus neoformans*, causes cryptococcosis, a normally benign illness that can develop into potentially life-threatening meningitis and meningoencephalitis. Cryptococcosis is difficult to treat and even with treatment can result in permanent neurological impairment, including decreased mental capacity, memory loss, blindness, hearing loss, and hydrocephalus (14). Unlike *C. neoformans*, which is an extremely important AIDS pathogen, *C. gattii* rarely causes disease in AIDS patients, and most cases of *C. gattii* cryptococcosis occur in otherwise normal, healthy hosts.

A wide range of animals can be infected with *C. gattii*, including humans, cats, dogs, sheep, koalas, birds, dolphins, lizards, caterpillars, and nematodes. However, *C. gattii* is a fundamentally environmental organism, and disease is thought to be an accidental outcome of adaptations to environmental pressures and predators (11). *C. gattii* occurs in the environment associated with decaying wood of certain tree species and was first isolated from a *Eucalyptus camaldulensis* tree in South Australia (16). While it has since been isolated from numerous additional tree species (33, 34, 40), *E. camaldulensis* remains the most reliable source of this fungus (29). *E. camaldulensis* is native to Australia and is found throughout most of the Australian continent. It is

Dee Carter, Leona Campbell, and Tien Bui, School of Molecular and Microbial Biosciences, University of Sydney, NSW 2006, Australia. Nathan Saul, School of Molecular and Microbial Biosciences and Faculty of Veterinary Science, University of Sydney, NSW 2006, Australia. Mark Krockenberger, Faculty of Veterinary Science, University of Sydney, NSW 2006, Australia.

therefore thought that *C. gattii* originated in Australia and has been exported to other parts of the world with its host tree, which is a very important plantation timber and reforestation species (24).

Most *C. gattii* cryptococcosis has occurred in the warmer parts of Australia, Africa, Asia, and the Americas, and until recently *C. gattii* was thought to be restricted by geoclimatic factors to the tropical and subtropical parts of the world (30). Since 2000, however, there has been a large, ongoing outbreak of *C. gattii* cryptococcosis on Vancouver Island, Canada, which has caused hundreds of cases of animal disease and more than 60 human cases and has claimed four human lives (37). *Eucalyptus* species are rare on Vancouver Island, and all examined have failed to yield viable *C. gattii* propagules, which have instead been isolated from a range of native temperate tree species as well as from the soil and air (2). The outbreak seems to be predominantly due to a particularly virulent clone of *C. gattii*, and it appears that the fungus can expand into new ecological zones, with serious consequences for human and animal health (18, 26).

SEXUAL REPRODUCTION AND BASIDIOSPORE FORMATION

C. gattii is heterothallic and obligately outcrossing, with the two mating types designated MATa and MATα. Mating type is determined by an unusually large locus of over 100 kb encompassing more than 20 genes, many of which have no known role in mating (17). The overall control of mating identity and sexual development is coordinated through the sex-specific homeodomain proteins Sxi1α and Sxi2a (23).

C. gattii propagates asexually as a budding yeast. Mating is stimulated by starvation and low water availability and occurs when fertile cells of opposite mating types are sufficiently close to one another. MATa and MATα cells signal and respond via a pheromone-receptor system and grow toward one another until they contact and fuse to form a diploid. Septate dikaryotic hyphae with fused clamp connections then form within 1 to 3 days, and eventually the hyphal tips differentiate to form basidia. Karyogamy and meiosis occur in the basidia, and recombinant nuclei are released within basidiospores, which form four chains from the basidium tip. Depending on the strain, spores can be abundant or sparse (30, 42). In laboratory crosses many *C. gattii* isolates appear to be unable to mate, even when specially developed hyperfertile mating partners are used (9, 19).

In *C. neoformans*, basidiospores can also be produced asexually from a single parent in a process termed monokaryotic or haploid fruiting (43). Physically this process resembles sexual reproduction, except that the clamp connections remain unfused and each cell contains a single nucleus. The resulting basidiospores are genetically identical to the parent strain and are all of a single mating type. Initial studies indicated that this process might be restricted to MATα cells, which helped explain an overwhelming bias of cells of α mating type in most clinical and environmental collections. Later studies, however, have found that this can also occur in *C. neoformans* MATa cells (39). Basidiospore production via monokaryotic fruiting has never been observed in *C. gattii* (43; J. Heitman, personal communication), and the role of asexual basidiospores in the life cycle of this species remains unknown.

Cryptococcosis in humans usually begins as a primary pulmonary disease resulting from the inhalation of an infectious propagule (10). Encapsulated yeast cells are too large to penetrate into the lower reaches of the lung alveoli, and when desiccated to a suitably small size, they rapidly lose viability (42). Basidiospores, whether sexually or asexually produced, are resistant to environmental stresses, readily aerosolized, and small enough to penetrate the alveoli. It is therefore likely that the basidiospore is the infectious propagule that initiates cryptococcosis (38, 42).

METHODS FOR DETECTING RECOMBINATION IN *C. GATTII* POPULATIONS

C. gattii mating structures are microscopic, and it has not been possible to observe these in environmental samples. Fertility tests can assess whether individual isolates are mating competent but cannot determine whether sex is occurring in a given population. Evidence for sexual recombination is therefore best obtained by indirect measures that assess the extent to which alleles at unlinked loci are shared among members of a population. The distribution of alleles in recombining and in asexual populations is distinctly different (Fig. 29.1A). In an asexual population, not all possible combinations of the multilocus genotype will be present, and as the genome is inherited as an intact unit, unlinked loci will appear to be associated. In a fully recombining population alleles will be randomly assorted among the population members so that all genotype combinations are theoretically possible. Note also that a compact phylogenetic tree can be fitted to the data from an asexual population whereas in sexually reproducing populations, allele swapping means that there is no clear pathway relating isolates to one another and no resulting

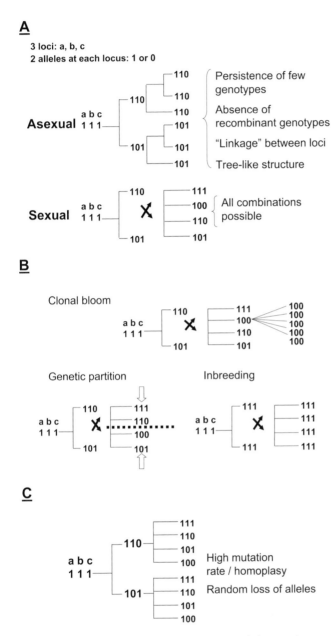

Figure 29.1 (A) Diagrammatic representation of the genetic structure of asexually and sexually reproducing populations. Here, genotypes consist of three loci, a, b, and c, with two alleles, 1 or 0, at each locus. In each population, the ancestral genotype (111) mutates over time to give new genotypes 110 and 101. In the asexual population these genotypes are passed intact to the offspring to give two clonal lineages, and recombinant genotypes are absent. The sexual population undergoes genetic reassortment to produce recombinant genotypes 111 and 100 in addition to the parental genotypes. The clonal population can be fitted to a phylogenetic tree with clear branches linking each genotype, whereas in the recombining population all isolates are equally related in a brush-like structure. (B) Complications that can cause a sexual population to appear asexual: clonal expansion of one genotype can overwhelm genetic evidence of recombination. Genetic partitioning can result in the fixation of alleles in each subgroup—here the populations above and below the partition (repre-

phylogenetic structure. Of course, many fungi including *C. gattii* can reproduce both sexually and asexually, and statistical methods must be used to determine if the population is significantly different from one in which completely panmictic recombination is occurring (5). There are a number of different methods that have been used to assess recombination in fungal populations (see reference 44 for a comprehensive review). We have recently reviewed the phylogenetic and population genetic approaches that have been used to assess recombination in *C. gattii* (7).

The definition of a population, in ecological terms, is a collection of organisms belonging to a single species, present in one place at one time (6). In practice, "place" may be determined by the biological question under study and could range from pancontinental to a single, confined environmental niche. As well as being geographically determined, populations of microbial pathogens can be defined according to whether they occur in human or animal hosts or on colonized environmental substrates. Choosing a population that maximizes the chances of sex is clearly essential when looking for evidence of sexual recombination in an organism that can reproduce both sexually and asexually; sampled too widely, individual isolates within a population may never have the chance to meet and mate and may even belong to genetically isolated groups (see below), and if the sample is too restrictive there may be oversampling of very closely related asexually derived clone mates (although these can be pruned before testing for recombination) (Fig. 29.1B). In both cases, it may be erroneously concluded that the population is not sexually reproducing. In practice, finding a recombining population is a stepwise process of refining the sampling procedure while taking into account the biotic and abiotic factors that may influence how the organism moves in the environment.

Genetic partitioning within a population can also lead to problems in assessing recombination. This occurs when a population consists of two or more groups, which for biological, geophysical, or temporal reasons are unable to exchange genetic information. Founder effects or local extinction events can result in some loci

sented by a dashed line) are fixed at locus b for alleles 1 and 0, respectively. Two or more alleles that become fixed in this way will appear to be linked and will skew the analysis in favor of clonality. Inbreeding of highly related genotypes will also give the appearance of a clonally derived population. (C) Asexual populations can appear to be sexual if there is a particularly high mutation rate, or if experimental artifact results in the random loss of alleles. (Modified from reference 7 with permission from the publisher.)

becoming fixed for a single allele in each subpopulation, and if the population is analyzed as a whole, these loci will appear to be genetically linked (Fig. 29.1B). *C. gattii* isolates have been found to partition into four distinct molecular genotypes, designated VGI-VGIV, or AFLP4-AFLP7 (3, 35, 36). These cluster to some degree according to the country of origin (15, 25). In Australia VGI predominates in clinical samples, and this is the only genotype to have been isolated from *E. camaldulensis*. VGII occurs in Australia to a lesser extent and has never been isolated from *E. camaldulensis* but is relatively frequent among clinical isolates from the Northern Territory (NT), where an exceptionally high level of *C. gattii* cryptococcosis occurs among the indigenous Australian population (12, 13). Interestingly, the genotype responsible for the outbreak on Vancouver Island is also VGII (27). There is no documented evidence of gene flow among the different genotypes, and each genotype effectively behaves as a separate cryptic species (18). It is therefore important to use the molecular type of isolates as one of the defining features of a population before proceeding to assess recombination in *C. gattii*.

Excessive inbreeding can also complicate analysis. Prolonged inbreeding results in homogenized genotypes, which will appear to be clonally related (Fig. 29.1B). Finally, there are some circumstances in which asexual populations can appear to be sexual. Particularly high mutation rates resulting in homoplasy, or a random loss of alleles due to amplification or scoring artifacts, can cause alleles to be randomly distributed throughout the population members (Fig. 29.1C). The latter is thought to account for conflicting data on recombination and clonality in *Candida albicans* (1). The use of a number of loci and care in their choice and analysis can generally prevent this from occurring, and it is much more likely to erroneously conclude that recombination is absent from a population than that it is present.

SEX IN NATURAL POPULATIONS OF *C. GATTII*

What constitutes a "natural" population of *C. gattii*? We set out with the assumption that the most natural population for a given organism would be found in the environmental niche in which it is able to complete its life cycle, which for *C. gattii* VGI would appear to be woody detritus associated with some *Eucalyptus* species. The structure of such a population would be expected to contrast with one that had undergone rapid recent expansion on a favored environmental substrate, as might be the case in the *C. gattii* population causing the outbreak on Vancouver Island. Populations obtained from infected animals living in close proximity to an environmental source might also be considered natural, with the advantage that the infecting isolates would be derived from an infectious propagule that might be sexually produced; the animal host would effectively be serving as a biological sampling device (R. Malik, M. Krockenberger, and P. Canfield, "The contribution of veterinary medicine to understanding the global epidemiology of cryptococcosis," presented at the 6th International Conference on *Cryptococcus* and Cryptococcosis, Boston, MA, 24 to 28 June 2005). Human-derived populations are complicated by the propensity for humans to travel and the potential for some fungal species to establish latent infection that might become apparent only many years after exposure. However, clinical isolates derived from human communities in which travel is restricted due to cultural or socioeconomic factors might also be expected to reflect the natural population in the environment. In *C. gattii* VGII, only four isolates have been obtained from local native trees in Arnhemland, NT, despite extensive sampling (D. Ellis, unpublished data). Our analysis of VGII has therefore been restricted to populations consisting of veterinary and clinical isolates.

SEX IN AN ENVIRONMENTAL *C. GATTII* POPULATION COMPRISING BOTH MATING TYPES

As in *C. neoformans*, the α mating type predominates in most populations of *C. gattii*. To date we have found two environmental populations of *C. gattii* VGI where this is not the case: a population collected from a highly colonized tree in Balranald, New South Wales, which is predominantly MAT**a**, and a population derived from *E. camaldulensis* trees growing along a riverbank in Renmark, South Australia, which consists of MAT**a** and MATα cells in approximately equal ratios (20). As equal numbers of MATα and MAT**a** cells are expected to result from sexual recombination, we targeted the latter in the first attempt to find a sexually recombining *C. gattii* population.

Renmark is located near the junction of the states of South Australia, New South Wales, and Victoria. This is a semiarid region, and *E. camaldulensis* trees are common along watercourses that were originally surrounded by areas of native mallee bushlands but that are now largely deforested. Trees along approximately 5 km of a watercourse were sampled, with the rationale that this sample range would allow genetic exchange but would not result in oversampling clone mates. Nearly 50% of the sampled trees yielded *C. gattii* colonies, and most positive samples came from tree hollows, with some hollows containing both MAT**a** and MATα strains. A total of 27 isolates, 13 MAT**a** and 14 MATα, from 13 different trees, were included in the analysis.

Preliminary analysis of the genetic diversity of the isolates indicated that all isolates had the VGI molecular genotype and diversity was very low. AFLP analysis, which is highly discriminatory, was therefore used to develop polymorphic loci and produce a multilocus genotype for each isolate. Although each strain was distinguished by a unique multilocus genotype and there was no clear partitioning of the population according to mating type, both phylogenic and population genetic analysis found the population to be significantly asexual. Interestingly, however, when canonical variant analysis was used to map genotypes according to their tree of origin, it was clear that each tree was colonized by its own distinct subpopulation. There was no evidence for dispersal among trees, and those located less than 10 m apart were just as genetically distant as those separated by several kilometers. From this analysis we concluded that the population had been sampled too widely and sexual recombination among isolates in the population under study had been precluded by the inability of propagules to disperse among the trees in this sampling site (21).

Recently we have performed extensive sampling of hollows present on two trees that harbored both mating types. Population and phylogenetic analysis again indicated that the populations found in each hollow were not recombining. Multidimensional scaling was then used to map isolate genotypes in isometric space and to assess differentiation between isolates grouped according to tree of origin and mating type (Fig. 29.2). As found above, there was a significant differentiation of the populations associated with the two trees ($R = 0.665$, $P = 0.01$). However, there was no genetic difference between the MAT**a** and the MATα groups found on each tree ($R = -0.055$, $P = 0.77$ for tree no. 13; $R = 0.065$, $P = 0.16$ for tree no. 15). If the populations were reproducing strictly asexually, the two mating types would be expected to be functioning as one or more separate clonal lineages and would be genetically isolated. The current data suggest that recombination is occurring, but among isolates that are so genetically similar that their loci remain in linkage disequilibrium (Fig. 29.1B).

The above result is supported by data published by Fraser et al. (18), who analyzed a subset of the original Renmark strains by multilocus sequence typing (MLST) (Table 29.1). There are two remarkable features in their data: first, genetic diversity is extremely restricted, such

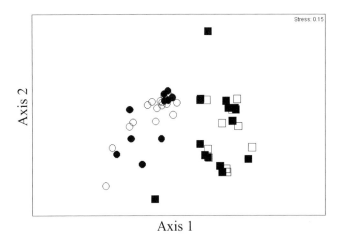

Figure 29.2 Nonmetric multidimensional scaling showing the genetic relationships among MAT**a** and MATα isolates derived from two *E. camaldulensis* tree hollows. ●, tree no. 13, MAT**a**; ○, tree no. 13, MATα; ■, Tree #15, MAT**a**; □, Tree #15, MATα. While isolates from the two trees are distinct, there is no clear separation of the MATα and MAT**a** isolates obtained from either tree.

that over the 5,740 and 6,925 bp sequenced from the MATα and MAT**a** strains, respectively, there is only one polymorphic locus. Second, GAPDH alleles 5 and 14 occur in both MATα and MAT**a** cells. These alleles are distinguished by a single nucleotide polymorphism, and homoplasy at this site is possible, but it is also plausible that the polymorphic site has been exchanged between the two mating types by recombination. Recombination hot spots flanking the *Cryptococcus MAT* locus have recently been identified, and it was proposed that these may allow highly inbred populations to maintain both mating types, and with them the ability to produce spores (22).

Taken together, the above results suggest that recombination is occurring within colonized *E. camaldulensis* hollows; however, this recombination is between strains that are so genetically similar that there is insufficient polymorphism to detect it. Unanswered questions remain as to how sexual exchange occurs, however, as all analyses conducted with specially developed "supermating" tester strains have failed to find any evidence of fertility among the Renmark isolates (N. Saul, unpublished data). In addition, if basidiospores are produced, these seem to be unable to disperse and colonize hollows in adjacent trees. The physical enclosure of the hollows may prevent spore escape, and it will be interesting to analyze isolates associated with external debris to determine if these are more genetically connected. The current data suggest that infective *C. gattii* propagules may not readily be dispersed from colonized tree hollows.

Table 29.1 MLST genotypes for 20 *C. gattii* VGI isolates from Renmark[a]

Isolate[b]	Allele (size in bp)								
	SXI1α (1,354)	SXI2a (2,529)	IGS (740)	TEF1 (700 bp)	GPD1 (547)	LAC (554)	CAP10 (568)	PLB1 (600)	MPD1 (677)
E566		1	3	4	5	5	4	5	3
E567		1	3	4	5	5	4	5	3
E572		1	3	4	5	5	4	5	3
E555		1	3	4	5	5	4	5	3
E312		1	3	4	5	5	4	5	3
E287		1	3	4	5	5	4	5	3
E283		1	3	4	5	5	4	5	3
E276		1	3	4	5	5	4	5	3
E554[R]		1	3	4	14	5	4	5	3
E316[R]	7		3	4	5	5	4	5	3
E275[R]	7		3	4	5	5	4	5	3
E286	7		3	4	14	5	4	5	3
E280	7		3	4	14	5	4	5	3
E307	7		3	4	14	5	4	5	3
E569	7		3	4	14	5	4	5	3
E310	7		3	4	14	5	4	5	3
E306	7		3	4	14	5	4	5	3
E549	7		3	4	14	5	4	5	3
E278	7		3	4	14	5	4	5	3
E296	7		3	4	14	5	4	5	3

[a]Adapted from Fraser et al. (18).
[b]Superscript R denotes putative recombinant strains.

SEX IN CLINICAL AND VETERINARY POPULATIONS OF *C. GATTII*

In the Australasian region, isolates of *C. gattii* obtained from the environment to date have been almost entirely VGI, whereas clinical and veterinary isolates, although predominantly VGI, can also be VGII. There are two areas in this region in which *C. gattii* cryptococcosis is particularly high: Arnhemland, in the north of Australia's NT, and Papua New Guinea (PNG). Indigenous populations living in these areas have a 10- to 20-fold-greater risk of contracting cryptococcosis due to *C. gattii* than persons living elsewhere in Australia (12). What is also interesting about these two areas is that *E. camaldulensis* is not commonly found. Attempts to isolate *C. gattii* from the PNG environment have been completely unsuccessful (31), and only four environmental isolates have been obtained from NT, all from local *Eucalyptus* species (Ellis, unpublished). In addition, these four isolates and a significant proportion of the clinical isolates from Arnhemland are of the less common VGII molecular type (13). The indigenous communities living in these areas have limited access to long-distance travel, and it is likely that infections have been acquired locally. We therefore targeted clinical *C. gattii* isolates from these regions to assess recombina-

tion in VGII populations and in VGI populations in the absence of the *E. camaldulensis* host.

Veterinary isolates from the Sydney region were also included in the study. Animals are unlikely to travel long distance, making this another geographically confined population. We had also previously established that an unexpectedly high number of the isolates in this population were molecular genotype VGII (C. Halliday, unpublished data), and the relationship of these Sydney isolates with those from the NT was of interest.

AFLP analysis identified 55 polymorphic loci, and each strain had a unique multilocus genotype. Phylogenetic analysis of the AFLP data found isolates to group according to molecular type (VGI, VGII, or VGIII), although there were some outliers that might represent one or more hitherto unknown molecular genotypes or might be hybrids between the VG types (Fig. 29.3). PNG isolates in particular had very limited genetic diversity, and their appearance on the phylogram suggested a clonal bloom, although the presence of two MATa strains could indicate inbreeding in this population, which, in the absence of polymorphic markers, would be difficult to detect. Recombination analysis revealed that each geographic population had a strongly clonal genetic structure (8).

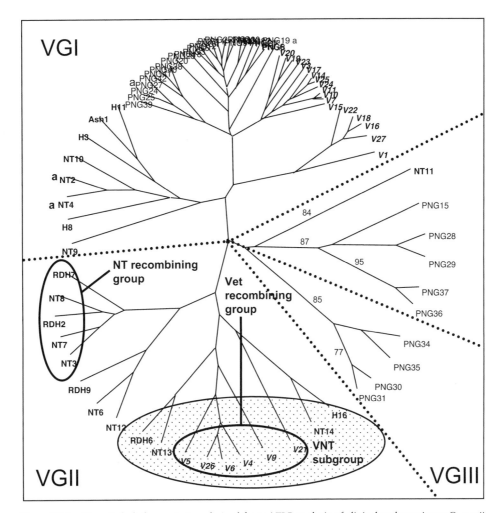

Figure 29.3 Unrooted phylogenetic tree derived from AFLP analysis of clinical and veterinary *C. gattii* isolates. VGI isolates group together on the top of the phylogram, whereas VGII and VGIII isolates form separate clusters in the bottom left and right corners, respectively. A group of isolates without a clear genotype affiliation lies between the VGI and VGIII clusters. PNG isolates are indicated by plain font, NT isolates are indicated in boldface, and veterinary isolates are italicized. **a** indicates MAT**a** isolates. Limited genetic diversity, shown by the tight clustering of isolates on short branches, is particularly evident in the PNG and veterinary VGI populations. The recombining VGII populations are circled. Stippling depicts the VNT-gene flow subgroup, which contains NT and veterinary isolates that did not appear to be genetically differentiated. (Modified from reference 7 with permission from the publisher.)

VGII isolates were considerably more diverse, and the veterinary isolates grouped within a cluster that also contained NT isolates. When analyzed as a whole, the NT isolates were significantly clonal; however, when the analysis was restricted to a group of five isolates that appeared closely related on the phylogram there was strong evidence of recombination. The group of six veterinary VGII isolates likewise had a recombining genetic structure. Recombination within VGII populations was supported by fertility tests, which found VGII isolates to be predominantly fertile when paired with appropriate mater strains, with many producing basidia and abundant basidiospores. In contrast, all VGI veterinary and NT isolates tested to date have been infertile or weakly fertile, with only limited filamentation and no apparent basidiospore production (9).

SEX, GENETIC DIFFERENTIATION, AND LONG-DISTANCE MIGRATION

Each of the VGI populations (PNG-clinical, NT-clinical, and Sydney-veterinary) formed one or more distinct clades on the phylogram, indicating that each of these consisted of genetically distinct isolates. Theta analysis, which is an estimate of Wright's Fst test for genetic differentiation (41), confirmed that each geographic population was genetically isolated from the others ($P <$ 0.01 for each pairwise comparison). In contrast, within the VGII isolates the veterinary isolates did not form a distinct group on the phylogram but were found within a clade that also contained NT isolates (Fig. 29.3). An overall test of genetic differentiation between the VGII isolates from Sydney and the NT returned a significant result ($P < 0.01$). However, when the analysis was confined to the veterinary and NT (VNT) isolates that branched together on the phylogram, the significance of theta became borderline ($P = 0.039$; VNT subgroup [Fig. 29.3]), suggesting that the two populations are not strongly differentiated. The similarity of the isolates in the VNT-gene flow subgroup was confirmed by MLST analysis performed by Fraser et al. (18) (Table 29.2), where only one of the nine loci tested was polymorphic and there was no partitioning of alleles according to geographic origin. Two of the environmental VGII isolates obtained from the NT also had the shared Vet/NT genotype, and the same MLST profile was seen in a clinical isolate from the Caribbean Islands and in two isolates from Vancouver Island that belong to the smaller of two genotypes responsible for the outbreak, known as the "minor," or VGIIb genotype (18, 26). In contrast, there was a completely different profile in the recombining NT group (Table 29.2), which had private alleles at seven of the nine loci.

The level of resolution of MLST markers is considerably less than that for AFLPs, and the isolates that share an MLST genotype may still be quite different. Nonetheless, the level of genetic similarity across isolates separated by a vast geographic distance remains striking. Fraser and colleagues (18) proposed that *C. gattii* of the minor genotype may have been introduced to the Pacific Northwest of North America via imported *Eucalyptus* trees, where it underwent mating with a local strain to produce the virulent "major"/VGIIa genotype that has caused the majority of infections on Vancouver Island. We have recent preliminary evidence that the Vet/NT genotype may also be present in Western Australia, suggesting a pancontinental distribution that could also be due to movement on colonized plant substrate. Alternatively, with the finding of sexual recombination and fertility in the VGII isolates it is tempting to speculate that basidiospores might be allowing long-range spore transfer. Long-distance dispersal, thought to be due to the direct movement of spores across Australia and between Australia and other continents, has been reported in some fungal plant pathogens (4, 28). The ability of *C. gattii* basidiospores to survive extended periods in the environment is unknown and would be an interesting area for further study. Interestingly, when the entire VNT-gene flow subgroup was analyzed for evidence of recombination, this was strongly rejected ($P < 0.001$). Thus, while there may be some movement of isolates between regions, and sexual exchange is occurring in the Sydney population, there is no evidence of sex across the much greater distance separating Sydney and the NT.

SEX AND CRYPTOCOCCOSIS

What are the implications of sex for *C. gattii* and cryptococcosis? The VGI genotype presents an enigmatic situation whereby, despite the presence of both mating types, Australian environmental and clinical populations appear to be limited in diversity, genetically isolated, clonally derived, and infertile in laboratory crosses (Table 29.3). Sex, if it occurs, seems to be limited to inbreeding between very genetically similar parents and may therefore not result in an increased level of diversity or adaptive potential. The nature of the infectious propagule remains unknown, and there is no evidence among the clinical VGI populations that this has been produced from sexually recombining isolates. However, VGI is the genotype responsible for most *C. gattii* infections in Australia and causes almost all of the cryptococcal infections in PNG, a region with particularly high levels of infection (31, 32). We aim to target additional natural populations from Australia to provide a fine-scale map of diversity and dispersal of this widespread genotype.

In contrast, among the VGII populations we have evidence of greater genetic diversity, genetic connectivity, fertility, and recombination. We have clear evidence that at least some veterinary and clinical infections have resulted from sexually produced propagules, and it is likely that these are basidiospores. Interestingly, the four VGII isolates obtained from environmental sources to date are significantly less fertile than the majority of the clinical and environmental VGII isolates, suggesting that sexual fecundity is linked with infectivity in this genotype (9). There is evidence that sex between VGII isolates has resulted in a novel genotype with elevated virulence and an extended geographic range, which is responsible for the ongoing outbreak on Vancouver Island (18, 19). Fraser et al. (18) suggested that this might be the product of a MATα-MATα cross, rather than the usual MATa-MATα mating. Although most VGII isolates mate successfully with MATa testers, we are yet to find VGII MATa strains among Australian environmental, clinical, and veterinary isolates, and α-α mating re-

Table 29.2 Division of VGII isolates into Vet/NT and NT-only subgroups

Genotype designation	Isolate designation	Allele (size in bp)									Recombining group[c]	VNT gene flow subgroup[d]	Source/location
		SXI1α (1,354)[a,b]	IGS (740)[b]	TEF1 (700)[b]	GPD1 (547 bp)[b]	LAC (554)[b]	CAP10 (568)[b]	PLB1 (600)[b]	MPD1 (677)[b]	URA5 (779)[a]			
VGII-NT only	RDH-2	C/20	6	6	2	7	5	1	5	D	NT		Clinical/NT
	NT-8	Biii/24	6	6	2	7	5	1	5	D	NT		Clinical/NT
	NT-3	Biii/24	6	6	2	7	5	1	5	D	NT		Clinical/NT
	RDH-7	Biii/24	6	6	2	7	5	1	5	D	NT		Clinical/NT
	NT-7[e]	Biv/21	6								D	NT		Clinical/NT
	RDH-9	Biii/24	6	6	2	7	5	1	5	D	NT		Clinical/NT
VGII-Vet/NT	V5	Bi/19	10	5	6	4	1	2	5	C	Vet	x	Veterinary/Sydney
	V26	Bi/19	10	5	6	4	1	2	5	C	Vet	x	Veterinary/Sydney
	V21	Bi/19	10	5	6	4	1	2	5	C	Vet	x	Veterinary/Sydney
	V9	Bi/19	10	5	6	4	1	2	5	C	Vet	x	Veterinary / Sydney
	V4	Bi/19	10	5	6	4	1	2	5	C	Vet	x	Veterinary/Sydney
	V20	Bi/19	10	5	6	4	1	2	5	C		x	Veterinary/ Sydney
	RDH-6	Bi/19	10	5	6	4	1	2	5	C		x	Clinical/NT
	NT-6	Bi/19	10	5	6	4	1	2	5	C		x	Clinical/NT
	NT-13	Bi/19	10	5	6	4	1	2	5	C		x	Clinical/NT
	NT-12	20	10	5	6	4	1	2	5	C		x	Clinical/NT
	Ram5	Bi/19	10	5	6	4	1	2	5	C			Environmental/NT
	Ram15	Bi/19	10	5	6	4	1	2	5	C			Environmental/NT
	99/473	19	10	5	6	4	1	2	5	ND			Clinical/Caribbean Islands
	R272	19	10	5	6	4	1	2	5	ND			Clinical/Vancouver Island
	RB31	19	10	5	6	4	1	2	5	ND			Environmental/ Vancouver Island

[a]Data from Campbell et al. (9).
[b]Data from Fraser et al. (18).
[c]Data from Campbell et al. (8).
[d]Vet and clinical isolates in this group were not genetically differentiated (Campbell et al. [9]).
[e]Isolate NT7 belongs in the recombining NT group, but based on the data from Fraser et al. (18), it shared MLST alleles with the Vet/NT group. Resequencing of IGS and URA5 found alleles 6 and D, respectively, which are characteristic of the recombining NT group and are consistent with AFLP data.

Table 29.3 Summary of sexual characteristics in natural *C. gattii* populations studied

Characteristic	VGI		VGII	
	Environmental	Clinical	NT only	Vet/NT
MATα + MATa	Yes	Few MATa	No MATa found	No MATa found
Genetic diversity	Low	Low	Moderate	Moderate
Recombining population structure	No	No	Yes	Yes, Vet isolates only
Inbreeding?	Possible	Possible	No	No
Fertility/basidiospore production	Very limited/no	Limited/no	Moderate–high/yes	Moderate–high/yes
Genetic connectivity with other populations	Restricted to individual trees	Restricted to geographic regions	Restricted to NT region	NT and Sydney populations connected; similar genotypes seen in Caribbean and Canada

mains a possibility among the Australian isolates of this genotype.

As a fungal species that can reproduce both sexually and asexually, the population dynamics driving the evolution of *C. gattii* represent an ongoing process of clonal expansion and cryptic speciation, balanced with recombination and possible interspecific hybridization (45). MLST and AFLP analyses allow the structure of natural populations of *C. gattii* to be investigated at broad and finer scales. Ongoing and future studies will allow us to assess how *C. gattii* moves in the environment and its potential to expand into new regions and to contact new environmental, human, and animal hosts. Such expansions may also allow it to come into contact with sexually compatible fungal genotypes, which, as proposed in the Vancouver Island outbreak, can have important consequences in the emergence of new pathogenic strains.

The work presented in this chapter was supported by a grant from the Howard Hughes Medical Institute under the International Scholars Program (No. 55000640) to D.C. and an Australian Research Council Linkage grant (No. LP0560572) to M.K.

References

1. Anderson, J. B., C. Wickens, M. Khan, L. E. Cowen, N. Federspiel, T. Jones, and L. M. Kohn. 2001. Infrequent genetic exchange and recombination in the mitochondrial genome of *Candida albicans*. *J. Bacteriol.* **183**:865–872.

2. Bartlett, K., M. W. Fyfe, and L. A. MacDougall. 2003. Environmental *Cryptococcus neoformans* var. *gattii* in British Columbia, Canada. *Am. J. Respir. Crit. Care Med.* **167**:A499.

3. Boekhout, T., B. Theelen, M. Diaz, J. W. Fell, W. C. Hop, E. C. Abeln, F. Dromer, and W. Meyer. 2001. Hybrid genotypes in the pathogenic yeast *Cryptococcus neoformans*. *Microbiology* **147**:891–907.

4. Brown, J. K. M., and M. S. Hovmøller. 2002. Aerial dispersal of pathogens on the global and continental scales and its impact on plant disease. *Science* **297**:537–541.

5. Burt, A. C., D. A. Carter, G. L. Koenig, T. J. White, and J. W. Taylor. 1996. Molecular markers reveal cryptic sex in the human pathogen *Coccidioides immitis*. *Proc. Natl. Acad. Sci. USA* **93**:770–773.

6. Callow, P. 1998. *The Encyclopedia of Ecology and Environmental Management.* Blackwell Science Ltd., Oxford, England.

7. Campbell, L. T., and D. A. Carter. 2006. Looking for sex in the fungal pathogens *Cryptococcus neoformans* and *Cryptococcus gattii*. *FEMS Yeast Res.* **6**:588–598.

8. Campbell, L. T., B. Currie, M. Krockenberger, R. Malik, W. Meyer, J. Heitman, and D. A. Carter. 2005. Clonality and recombination in genetically differentiated subgroups of *Cryptococcus gattii*. *Eukaryot. Cell* **4**:1403–1409.

9. Campbell, L. T., J. A. Fraser, C. B. Nichols, D. A. Carter, and J. Heitman. 2005. Identification of Australian clinical and environmental isolates of *Cryptococcus gattii* that retain sexual fecundity. *Eukaryot. Cell* **4**:1410–1419.

10. Casadevall, A., and J. R. Perfect. 1999. Cryptococcus neoformans. ASM Press, Washington, DC.

11. Casadevall, A., J. N. Steenbergen, and J. D. Nosanchuk. 2003. 'Ready made' virulence and 'dual use' virulence factors in pathogenic environmental fungi—the *Cryptococcus neoformans* paradigm. *Curr. Opin. Microbiol.* **6**:332–337.

12. Chen, S., T. Sorrell, G. Nimmo, B. Speed, B. Currie, D. Ellis, D. Marriott, T. Pfeiffer, D. Parr, and K. Byth. 2000. Epidemiology and host- and variety-dependent characteristics of infection due to *Cryptococcus neoformans* in Australia and New Zealand. *Clin. Infect. Dis.* **31**:499–508.

13. Chen, S. C., B. J. Currie, H. M. Campbell, D. A. Fisher, T. J. Pfeiffer, D. H. Ellis, and T. C. Sorrell. 1997. *Cryptococcus neoformans* var. *gattii* infection in northern

Australia: existence of an environmental source other than known host eucalypts. *Trans. R. Soc. Trop. Med. Hyg.* **91:**547–550.

14. Diamond, R. D., and J. E. Bennett. 1974. Prognostic factors in cryptococcal meningitis. A study in 111 cases. *Ann. Intern. Med.* **80:**176–181.

15. Ellis, D., D. Marriott, R. A. Hajjeh, D. Warnock, W. Meyer, and R. Barton. 2000. Epidemiology: surveillance of fungal infections. *Med. Mycol.* **38:**173–182.

16. Ellis, D. H., and T. J. Pfeiffer. 1990. Natural habitat of *Cryptococcus neoformans* var. *gattii. J. Clin. Microbiol.* **28:**1642–1644.

17. Fraser, J. A., S. Diezmann, R. L. Subaran, A. Allen, K. B. Lengeler, F. S. Dietrich, and J. Heitman. 2004. Convergent evolution of chromosomal sex-determining regions in the animal and fungal kingdoms. *PLoS Biol.* **2:**1–13.

18. Fraser, J. A., S. S. Giles, E. C. Wenink, S. G. Geunes-Boyer, J. R. Wright, S. Diezmann, A. Allen, J. E. Stajich, F. S. Dietrich, J. R. Perfect, and J. Heitman. 2005. Same-sex mating and the origin of the Vancouver Island *Cryptococcus gattii* outbreak. *Nature* **437:**1360–1364.

19. Fraser, J. A., R. L. Subaran, C. B. Nichols, and J. Heitman. 2003. Recapitulation of the sexual cycle of the primary fungal pathogen *Cryptococcus neoformans* var. *gattii:* implications for an outbreak on Vancouver Island, Canada. *Eukaryot. Cell* **2:**1036–1045.

20. Halliday, C. L., T. Bui, M. Krockenberger, R. Malik, D. H. Ellis, and D. A. Carter. 1999. Presence of alpha and a mating types in environmental and clinical collections of *Cryptococcus neoformans* var. *gattii* strains from Australia. *J. Clin. Microbiol.* **37:**2920–2926.

21. Halliday, C. L., and D. A. Carter. 2003. Clonal reproduction and limited dispersal in an environmental population of *Cryptococcus neoformans* var. *gattii* isolates from Australia. *J. Clin. Microbiol.* **41:**703–711.

22. Hsueh, Y.-P., A. Idnurn, and J. Heitman. 2006. Recombination hotspots flank the *Cryptococcus* mating-type locus: implications for the evolution of a fungal sex chromosome. *PLoS Genet.* **2:**e184.

23. Hull, C. M., M. J. Boily, and J. Heitman. 2005. Sex-specific homeodomain proteins Sxi1alpha and Sxi2a coordinately regulate sexual development in *Cryptococcus neoformans. Eukaryot. Cell* **4:**526–535.

24. James, S. A., and D. T. Bell. 1995. Morphology and anatomy of leaves of *Eucalpytus camaldulensis* clones: variation between geographically separated locations. *Aust. J. Bot.* **43:**415–433.

25. Kidd, S. E. 2003. *Molecular Epidemiology and Characterisation of Genetic Structure to Assess Speciation within the* Cryptococcus neoformans *Complex.* Ph.D. thesis. University of Sydney, Sydney, Australia.

26. Kidd, S. E., H. Guo, K. H. Bartlett, J. Xu, and J. W. Kronstad. 2005. Comparative gene genealogies indicate that two clonal lineages of *Cryptococcus gattii* in British Columbia resemble strains from other geographical areas. *Eukaryot. Cell* **4:**1629–1638.

27. Kidd, S. E., F. Hagen, R. L. Tscharke, M. Huynh, K. H. Bartlett, M. Fyfe, L. MacDougall, T. Boekhout, K. J. Kwon-Chung, and W. Meyer. 2004. A rare genotype of *Cryptococcus gattii* caused the cryptococcosis outbreak on Vancouver Island (British Columbia, Canada). *Proc. Natl. Acad. Sci. USA* **101:**17258–17263.

28. Kolmer, J. A. 2005. Tracking wheat rust on a continental scale. *Curr. Opin. Plant Biol.* **8:**441–449.

29. Krockenberger, M. B., P. J. Canfield, and R. Malik. 2002. *Cryptococcus neoformans* in the koala (*Phascolarctos cinereus*): colonization by *C. n.* var. *gattii* and investigation of environmental sources. *Med. Mycol.* **40:**263–272.

30. Kwon-Chung, K. J., and J. E. Bennett. 1992. *Medical Mycology.* Lea & Febiger, Malvern, PA.

31. Laurenson, I. F., D. G. Lalloo, S. Naraqi, R. A. Seaton, A. J. Trevett, A. Matuka, and I. H. Kevau. 1997. *Cryptococcus neoformans* in Papua New Guinea: a common pathogen but an elusive source. *J. Med. Vet. Mycol.* **35:**437–440.

32. Laurenson, I. F., A. J. Trevett, D. G. Lalloo, N. Nwokolo, S. Naraqi, J. Black, N. Tefurani, A. Saweri, B. Mavo, J. Igo, and D. A. Warrell. 1996. Meningitis caused by *Cryptococcus neoformans* var. *gattii* and var. *neoformans* in Papua New Guinea. *Trans. R. Soc. Trop. Med. Hyg.* **90:**57–60.

33. Lazéra, M. S., M. A. Cavalcanti, L. Trilles, M. M. Nishikawa, and B. Wanke. 1998. *Cryptococcus neoformans* var. *gattii*—evidence for a natural habitat related to decaying wood in a pottery tree hollow. *Med. Mycol.* **36:**119–122.

34. Lazéra, M. S., B. Wanke, and M. M. Nishikawa. 1993. Isolation of both varieties of *Cryptococcus neoformans* from saprophytic sources in the city of Rio de Janeiro, Brazil. *J. Med. Vet. Mycol.* **31:**449–454.

35. Meyer, W., A. Castaneda, S. Jackson, M. Huynh, E. Castaneda, and The IberoAmerican Cryptococcal Study Group. 2003. Molecular typing of IberoAmerican *Cryptococcus neoformans* isolates. *Emerg. Infect. Dis.* **9:**189–195.

36. Meyer, W., K. Marszewska, M. Amirmostofian, R. P. Igreja, C. Hardtke, K. Methling, M. A. Viviani, A. Chindamporn, S. Sukroongreung, M. A. John, D. H. Ellis, and T. C. Sorrell. 1999. Molecular typing of global isolates of *Cryptococcus neoformans* var. *neoformans* by polymerase chain reaction fingerprinting and randomly amplified polymorphic DNA—a pilot study to standardize techniques on which to base a detailed epidemiological survey. *Electrophoresis* **20:**1790–1799.

37. Stephen, C., S. J. Lester, W. C. Black, M. Fyfe, and S. Raverty. 2002. Multispecies outbreak of cryptococcosis on southern Vancouver Island, British Columbia. *Can. Vet. J.* **43:**792–794.

38. Sukroongreung, S., K. Kitiniyom, C. Nilakul, and S. Tantimavanich. 1998. Pathogenicity of basidiospores of *Filobasidiella neoformans* var. *neoformans. Med. Mycol.* **36:**419–424.

39. Tscharke, R. L., M. Lazera, Y. C. Chang, B. L. Wickes, and K. J. Kwon-Chung. 2003. Haploid fruiting in *Cryptococcus neoformans* is not mating type alpha-specific. *Fungal Genet. Biol.* **39:**230–237.

40. Vilcins, I., M. Krockenberger, H. Agus, and D. Carter. 2002. Environmental sampling for *Cryptococcus neofor-*

mans var. *gattii* from the Blue Mountains National Park, Sydney, Australia. *Med. Mycol.* **40:**53–60.

41. **Weir, B. S.** 1996. *Genetic Data Analysis II.* Sinauer Associates, Sunderland, MA.

42. **Wickes, B. L.** 2002. The role of mating type and morphology in *Cryptococcus neoformans* pathogenesis. *Int. J. Med. Microbiol.* **292:**313–329.

43. **Wickes, B. L., M. E. Mayorga, U. Edman, and J. C. Edman.** 1996. Dimorphism and haploid fruiting in *Crypto-*

coccus neoformans: association with the α-mating type. *Proc. Natl. Acad. Sci. USA* **93:**7327–7331.

44. **Xu, J.** 2005. Fundamentals of fungal molecular population genetic analyses, p. 87–116. *In* J. Xu (ed.), *Evolutionary Genetics of Fungi.* Horizon Bioscience, Norfolk, England.

46. **Xu, J. P., R. Vilgalys, and T. G. Mitchell.** 2000. Multiple gene genealogies reveal recent dispersion and hybridization in the human pathogenic fungus *Cryptococcus neoformans*. *Mol. Ecol.* **9:**1471–1481.

Sex in Fungi: Molecular Determination
and Evolutionary Implications
Edited by Joseph Heitman et al.
© 2007 ASM Press, Washington, D.C.

Matthew R. Goddard

30

Why Bother with Sex?
Answers from Experiments with Yeast and Other Organisms

Of a truth, he is past helping who does not regard with wonder and admiration the adaptations which have been worked out in this connection in the course of evolution!

August Weismann, 1904, *The Evolution Theory*,
volume 2, p. 228

THE PROBLEM

My interpretation of the modern synthesis of evolution by Natural Selection leads me to believe that an organism's traits are, in general, adaptive. This may not mean that every trait is adaptive; it means that Natural Selection hones harshly those traits it sees clearly. While it is obvious to understand why the traits conferring a more efficient reproduction are adaptive, for other traits the adaptive function is somewhat cryptic. The function of sex falls into the latter class because, on the face of it, reproducing sexually is less efficient than reproducing asexually. To engage in a meiotic division, which comprises genetic recombination and the formation of haploid gametes, which must then fuse with other haploid gametes to reconstitute the diploid again, takes time and

energy. Mitosis achieves this in one simpler step. Moreover, sex incurs further costs. Organisms that survive to reproduction have a successful collection of alleles by definition, but sex and recombination ensure that these genotypes are destroyed. Barring de novo mutation, asexuals guarantee that successful genotypes are replicated faithfully. There are further hindrances. There appears to be a twofold cost for sex in organisms where one parent contributes more reserves to the zygote than the other parent (anisogamy): a female wastes one-half of her output on nonreproductive males (52). Of course, this twofold cost may be lessened if the male helps with external resources to care for the female and her young (55). In higher organisms, where it must be remembered that sex is mostly obligate, yet further costs and dangers are imposed by mate finding and courtship.

There appear to be many costs to sex, but it is plain that sex is not the property of a single obscure taxon—its ubiquity is no accident (4). Most eukaryotes engage in sex to some degree or other (52). The popularity of sex implies that it arose very early and has been retained by the vast majority of subsequent organisms (58). To suggest otherwise, one would have to infer an independent gain of sex by clades a great many times during the

---

Matthew R. Goddard, School of Biological Sciences, University of Auckland, Private Bag 92019, Auckland 1142, New Zealand.

course of evolution. The very early move from asexuality to sexuality may be considered one of the major transitions in evolution: it was a move to a more organized state (53). At this point I think it is worth defining exactly what sex is. At its heart, sex revolves around genetic recombination (3), but I think that it is more sophisticated than just recombination. Sex, in my view, comprises recombination, random assortment, and syngamy. These three cornerstones ensure that genetic material is mixed and redistributed among the members of an interbreeding population. How each of these traits came to be established is a hard question to answer, and it has been addressed in many other places (see reference 44). The maintenance of sex, however, is a question that we can address: why does sex persist in the vast majority of extant eukaryotic taxa? I believe that sex is maintained because it is adaptive and must in general allow sexual organisms to fare better than their asexual competitors.

Suggestions concerning the function of sex are many, and a huge array of different models have been constructed which attempt to find conditions under which sexuality flourishes. To delve into the intricacies of all these ideas is a huge task and well beyond the scope of this chapter. These ideas have been considered many times (for excellent reviews see references 3, 4, 41, 52, 56, 59, and 70). However, I think that to appreciate experiments in this area, a brief look at the elegant and classic early texts is justified, since these constitute the core of our understanding about the maintenance of sex. Most modern theories are finely tuned sophisticated versions of these original ideas. The following refers to some of the landmark papers that instigated the field and is intended to give the reader a general background.

POSSIBLE SOLUTIONS

Weismann

In 1904 August Weismann published a series of lectures on his ideas concerning evolution by Natural Selection (68). One of the topics he addressed was the origin and persistence of sexual reproduction, or amphimixis (the union of two germplasms [gametes]). In reading the relevant chapters (p. 192–287), it is clear that Weismann had a powerful and deep insight into the intricate workings of Natural Selection. He was one of the first to note that sex is not essential for life and to ask why it exists. Even before genes and the mutational process were identified, Weismann had a strong feeling for Mendelian inheritance (which was only acknowledged later) rather than a blending idea; he also correctly saw that "Amphimixis,

that is the union of two germ plasms, does not itself cause variation of the determinants, it only arranges the ids [genes] in ever new combinations" (p. 195). Weismann had a clear understanding that, as we now know well, mutation provides raw genetic variation and that sex mixes these variants in the same way that a deck of cards may be shuffled. Weismann also correctly saw that the traits (genes) responsible for adaptation to any environment must be many for sex to be of advantage. He used several examples to highlight the extent of biological variation. Indeed, his view was that subtle degrees of biological variation and adaptation are possible only because of sex: "I still regard amphimixis as the means by which a continual new combination of variations is effected, a process without which the evolution of this world of organisms so endlessly diverse in form and so inconceivably complex, could not have taken place" (p. 194). Weismann further suggested that sex increases variance: "Thus amphimixis, together with the preparatory reduction of the ids, secures the constant recurrence of individual peculiarities through the ceaseless new combinations of individual characters already existing in the species" (p. 194). This increased genetic variance is the primary advantage of sex since it allows Natural Selection to proceed more rapidly: "In my opinion this indirect effect of amphimixis, that is, the increasing of the possibilities of adaptation by new combinations of individual variational tendencies, is the main one" (p. 227). In situations where many beneficial alleles are available Weismann suggested that sex allows "harmonious re-adjustments" to be made: sex speeds adaptation since it allows many alleles to increase in frequency simultaneously whereas, because of competition between lineages with different adaptive mutations, asexuality does not allow this. Weismann considered not only beneficial mutations but also detrimental ones, or "rebellious determinants" as he termed them. He says that sex makes possible "adaptations which are continually required, but that it also leads, by continual crossing of individuals, simultaneously, with the elimination of the less fit, to a gradually increasing constancy of the species" (p. 230). In this respect Weismann considered both sides of the coin, believing that sex increases variation, which therefore allows Natural Selection to proceed more efficiently. This view is in line with others (1, 4, 25, 33, 36, 41), and Burt has recently brought Weismann's ideas back into focus and provides an excellent overview of, and extension to, the original ideas (9).

Fisher and Muller

In 1930 Fisher succinctly discussed the topic of sex in his Genetical Theory of Natural Selection (21). Fisher's standpoint was that the very existence of sex implies its advantage. He suggested that this advantage lay in the fact that sex is able to combine together different independent beneficial alleles, and this allows a greater rate of adaptation than in asexuals. Asexuals have only the stepwise mutation process to draw upon and are forced to fix alleles one at a time. Fisher said that for this idea to hold, there must be genetic variance, more than one locus responsible for adaptation, and directional selection. Even though many of Weismann's thoughts were echoed, Fisher did not explicitly link Weismann's ideas concerning sex to his own.

Muller published a brilliant paper in 1932 entitled "Some genetic aspects of sex" (58). He begins the first section saying that he views sex as a "luxury" and that it is not vital for life. Muller postulated that sex arose very early in the eukaryotes and persisted as it was competitively superior; he noted that the core of the tree of life seems sexual, but that ephemeral asexuals are thrown off. He identified that recombination is the essence of sexuality, but that it is an "accessory to the primary process of gene mutation." Muller explained that in the long run sex allows "better adjustments to be made." I interpret this to mean that sexuals are able to sample the adaptive landscape more thoroughly (i.e., have a greater genetic variance). Muller listed two possible benefits of sex. First, and less importantly in his opinion, if the adaptive landscape is undulating, with many shifting peaks, then sex allows the storage of mutations adapted to alternative peaks. Second, and more importantly, sex combines adaptive alleles, allowing multiple beneficial alleles to increase in frequency simultaneously. Asex suffers from clonal interference: the same suite of adaptive mutations arises in an asexual population but here creates competition between different lineages. Competition between these asexual lineages, within which the different beneficial alleles are locked, results in the elimination of all but one (or a few) of these beneficial alleles. In line with Fisher, Muller suggested that in order to progress, an asexual population must wait for a second adaptive mutation to arise in the genetic background of the first: a much slower process. Muller fully acknowledged Weismann for the idea of increased variance being the adaptive aspect of sex, but only after S. Holmes pointed him to Weismann's work.

These concepts form the core of the Fisher-Muller theory, which says that asex must fix mutations one at a time while sex allows multiple beneficial alleles to permeate through populations. In turn, this allows adaptation to proceed more rapidly. These ideas seem intuitively correct, but there are some aspects that need consideration. First, Fisher and Muller discuss only beneficial mutations and seem to discount the importance of detrimental mutations (at least at this stage). Second, these ideas require simple directional selection: the advantage of sex evaporates in the absence of abundant beneficial alleles. Last, there needs to be negative linkage disequilibrium between adaptive mutations: sex must serve to bring together adaptive alleles rather than break them apart (52).

Crow, Kimura, and John Maynard Smith

In 1965 Crow and Kimura quantified Muller's argument (14). They constructed a model and considered a number of situations, including epistatic interactions and linkage of mutations. They concluded that sex is of most benefit when evolution proceeds by many small incremental mutations. If mutations are few and far between, and mostly of large effect, then sex becomes less advantageous. Crow and Kimura also investigated the impact of population size on sex. Smaller populations have lower effective mutation rates, and the supply of beneficial mutations becomes a limiting factor. Sex is irrelevant if most beneficial mutations are fixed before the next arises. The smaller the population, the less the benefit of sex.

John Maynard Smith entered the field in 1968 with his incisive intellect and argued that, since Crow and Kimura's model treated mutation as a unique rather than recurrent event, their conclusions were wrong (51). His model, which included recurrent mutations of the same type, suggested that in large asexual populations at equilibrium, individuals with two beneficial mutations will exist. Sex is not needed to bring the individual mutations together, and it may be that sex instead serves to destroy these genotypes. Crow and Kimura responded to Maynard Smith's criticism and agreed that large asexual populations with a relatively high mutation rate could fix two adaptive mutations at the same speed as a sexual one. However, Crow and Kimura maintained that the Fisher-Muller model was pertinent, as in reality beneficial mutations are relatively rare and population sizes are not infinite. Further, they thought it likely that adaptation would proceed due to many mutations. Indeed, the higher the number of beneficial alleles, the greater the postulated advantage of sex. If there are many beneficial mutations, the probability of a multiple mutant arising in an asexual situation, even in a very large population, is vanishingly rare (that is, the mutation rate raised to the number of loci).

Detrimental Mutations

In 1964 Muller published an additional theory for the maintenance of sex: Muller's ratchet (57). This initiated a branch of models which suggest that deleterious mutations, rather than beneficial ones, promote the maintenance of sex. In short, Muller's ratchet proposes that sexual populations may more easily rid themselves of deleterious mutations than asexual populations. The ratchet is significant only in small populations and operates simply because there is a reasonable probability that the small class of genotypes with the fewest deleterious mutations will be lost stochastically. Once lost, apart from very rare back mutations, asexuals have no way of reconstituting genomes with fewer mutations. After a time the next least mutationally loaded class will be lost. Sex allows genomes with fewer mutations to be created quickly. For this reason the ratchet turns and sees the inexorable decline in fitness of asexual but not sexual populations. A similar process may operate in larger populations if the decrease in fertility of mutationally loaded individuals is also accounted for (23). This "mutational meltdown" process proposes that population size will decline as fecundity decreases due to the accumulation of deleterious mutations: as the population size decreases, the effects of the classic Muller's ratchet are enhanced.

Detrimental mutation hypotheses are attractive since they are free of the restrictions which hinder models considering only beneficial mutations. The notion of a beneficial mutation implies some degree of suboptimal adaptation to a particular environment. Sex's benefit, as typified by the Fisher-Muller hypothesis, revolves around the repeated incorporation of adaptive mutations, and this means constant directional selection and a reasonable pace of evolution. As Kondrashov (41) points out, the assumption that all sexual populations undergo fast evolution or live in environments sharply fluctuating in space or time seems unreasonable. On the other hand, mutation is an inherent feature of DNA replication and models based around this process do not require environmental restrictions. Kondrashov championed mutational theories and extended Muller's ratchet to infinite populations with the construction of a "mutational deterministic" hypothesis (41). This hypothesis finds an advantage to sex in clearing deleterious mutations, but it does so only when the mutation rate is more than about one per genome per generation and when deleterious mutations interact with positive epistasis (the greater the number of mutations, the larger the drop in fitness). Unfortunately we have very few estimates of mutation rate, and the ones that we do have fluctuate, from well below one for yeast (*Saccharomyces cervisiae*), *Drosophila*, and *Caenorhabditis elegans*, to above one for hominids (see Table 1 in refer-

ence 37). Neither do we have a very good feeling for how mutations interact, although studies with *Escherichia coli* suggest that, on average, deleterious mutations do not act in the manner required for the deterministic hypothesis to work (17, 24). A recent indirect study with *Arabidopsis* suggests that both positive and negative epistatic interactions between loci may be important in adaptation (49), but exactly how this relates to detrimental mutations is unclear.

One idea incorporates the effects of both detrimental and beneficial mutations and suggests that sex is an advantage since it frees beneficial mutations from deleterious backgrounds (43). Asexuals are unable to extract beneficial alleles from the "rubbish" background of deleterious mutations. One formulation of this is the "ruby in the rubbish" model put forth by Peck (60), which shows that with a reasonable deleterious mutation load, the likelihood of a beneficial mutation becoming established is greater in sexual than asexual populations. Again, it seems that in general sex is beneficial because of its ability to promote the "constant recurrence of individual peculiarities through the ceaseless new combinations of individual characters already existing in the species" (reference 68, p. 194).

Parasites

No tour through theories concerning the evolution of sex, however brief, would be complete without a mention of the Red Queen hypothesis. This is another intuitive and attractive idea which suggests that sex is of benefit when considering the effects of frequency-dependent selection, mainly from parasites (4, 31). If the ability of an organism to parasitize another is genetically determined, then the production of rare host genotypes may be advantageous. Rare host genotypes may afford a degree of resistance to parasites which Natural Selection has ensured are adapted to common host genotypes. The frequency of these initially resistant genotypes will increase due to their greater resistance, but the likelihood of parasitic invasion through adaptation also increases as these genotypes become more common (47). This is out-of-phase frequency-dependent selection. When very common, this genotype will be no better off than any other common genotype before it, hence the term "Red Queen" (who was running to stay in the same place). Asexuality suffers because offspring are exposed to the same degree of parasitic load as their parent. Sexuals, however, produce continually variable offspring, and this ensures that parasites are kept guessing. Again then we see Weismann's concept in this theory—sex is beneficial because it produces a greater genetic variance, and this allows selection to proceed more efficiently. In this case the selection is not strictly directional, but circular, and it is the greater variance (higher likelihood of rare combinations being produced) that allows sexuals to escape parasitic invasion more effectively. There are some excellent and compelling studies that show correlations between the degree of parasitism and mode of reproduction (45, 46), but there are few data to suggest whether this is because of frequency-dependent selection or not.

The Group Selection Problem

One potential stumbling block for Weismann's concept was highlighted by Williams, who pointed out that theories based on this principle invoke group selection since sex is not necessarily of any benefit to the individual (69, 70). I agree with Williams that adaptation should be attributed to no higher level than is necessary, but sex is a rare case where it may be appropriate to consider that the advantage accrues at the level of the group (9). Indeed, this is precisely what we are considering: competition between groups of sexuals and asexuals. In sexual populations individuals are merely transient aggregations of genes sourced from a common pool. That is not to say that there may not be advantages at other levels. One method that has allowed movement beyond arguments based on group selection is consideration of the fate of loci that modify the rate of recombination (3). Even if there is no clear immediate individual advantage, there is an advantage at the level of the gene (the modifier of recombination). An allele that promotes recombination is more likely to find itself in genotypes that have higher than average fitness, because of the variance it creates (59). The more tightly such modifiers are linked to beneficial alleles, the greater their increase in frequency. There are clearly situations where modifiers may be selected for, but the advantage at the level of the gene is probably only around the order of 0.1 to 1% (9). The cumulative advantage of sex in increasing the mean fitness of a group is much larger than this.

The reasoning behind the Weismann concept, thus far presented verbally, can also be demonstrated by way of a simulation. Figure 30.1 shows the results of a 700-generation experiment conducted in silico. The only difference between these two populations was the presence or absence of recombination, random assortment, and syngamy. One can see that the variance in fitness (represented by the standard error of the mean) of the sexual population is greater, and this translates to a greater rate of adaptation than in the asexual population. The details of the simulation are explained in the figure legend. This extremely simple simulation demonstrates that sex can lead to a considerable advantage.

Back Full Circle to Weismann

There are myriad other models that attempt to explain the maintenance of sex, some of which are not based on increasing variance. However, those based implicitly or explicitly on Weismann's principle seem to be at the core of most common hypotheses (9). Solutions from the field of population genetics, which rely on the interaction between variation and selection, apply to any sexual population. The latest models show that sex is of benefit when there is a nonrandom association between loci, and this advantage depends on the interaction between genes that contribute to fitness (3, 52, 59). In a population that is statically fully randomized with respect to all relevant loci, sex does nothing. However, if there are correlations between loci, sex will destroy these. Sex is beneficial when there are negative associations between relevant loci (3, 9, 52). Sex randomizes, and it breaks down these negative associations and increases genetic variance. But why should there be negative associations between loci? One reason is that selection favors this situation. If mutations interact with negative epistasis, then selection will tend to disperse beneficial mutations. In this situation additional beneficial mutations increase fitness by an ever-diminishing amount, and eventually fitness increases are negligible. This means that selection will act more strongly on novel beneficial alleles in genomes with few rather than many preexisting beneficial alleles. The few data we have suggest that this situation is met only rarely in nature (3). An alternative reason for the negative association of alleles is due to the effects of finite population size. In this situation it is much more likely that mutations will occur in different individuals, rather than the same individual (14). Negative fitness correlations may also occur because of random drift (the Hill-Robertson effect [32]). Overall then, in finite populations, sex can mean that detrimental mutations are concentrated and weeded out by selection; beneficial mutations are separated from backgrounds containing detrimental mutations; and beneficial mutations are concentrated and flourish rapidly. The increased additive genetic variance (the proportion of genetic variation that is the summation of the effect of all individual genes influencing a trait) that sex conveys allows Natural Selection to proceed more rapidly and adaptation to occur faster than in asexuals (9). It thus seems to me that many of the aforementioned ideas can be encompassed by the Weismann concept. Indeed, this concept may be the most

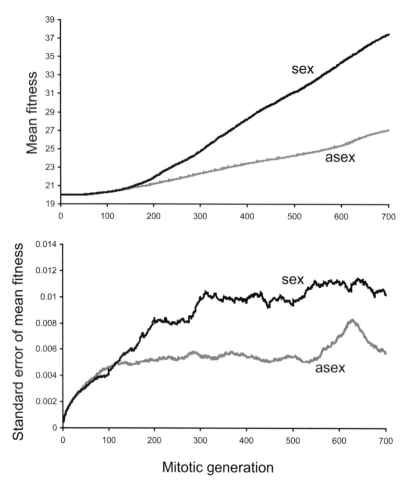

Figure 30.1 Each individual in this simulation contains two homologous chromosomes, each with 10 loci with an initial value of one. Each population comprises 10,000 individuals. An individual's fitness is simply the sum of values at each locus, and so there are no dominant/recessive or epistatic effects. All individuals undergo mitotic divisions with random mutation. Every generation, each allele has a 1% chance of mutating—if an allele is chosen for mutation, the value mutated to is drawn randomly from a normal distribution with a mean of 0.01 and variance of 0.1. Thus, most mutations are detrimental but beneficial ones do occur, with larger beneficial mutations being increasingly rare. Selection, which is simply reproduction weighted by fitness, occurs after mutation. The sexual populations have the option of also reproducing via a meiotic division with recombination, which produces two haploid gametes. Recombination occurs at five random points along the chromosomes. The haploid gametes then mate at random to reconstitute diploids once again. There is no timing cost to reproducing sexually. Both sexual and asexual populations were propagated for 700 mitotic generations, and sex was imposed every 25 mitotic generations in the sexual line. The mean and standard error of fitness were recorded for both populations and are plotted (black, sexual; gray, asexual). Replicate runs produced very similar results.

general statement we can make about the advantage to sex, with each case differing in its exact composition of balance between mechanisms.

ANSWERS FROM EXPERIMENTS
As always, the possibilities presented through verbal and mathematical reasoning need to be assessed empirically to determine which of these ideas are relevant

within the natural world. Many traits appear correlated with sex (parasitism rates of sexual and asexual New Zealand snails and Mexican fish being excellent examples [45, 46]), but unfortunately these relationships are unable to explain the function of sex as they are correlations only. Empirical data which would allow us to directly support or reject ideas concerning sex are scant and contradictory (3, 4, 9). At first glance, this may seem bizarre given the importance and long history of the subject. Closer consideration reveals that experiments which rigorously test the advantage of sex and recombination are hard to conduct. If we wish to examine the effects of sex, we are faced immediately with a problem: the choice of experimental organism. The vast majority of higher organisms have to have outcrossed sex to reproduce and therefore afford no asexual control. There are hermaphrodites of various forms, but these are still largely engaged in most aspects of meiotic division, and most undergo recombination and random assortment. In general, one cannot force the same genotype to reproduce both hermaphroditically and nonhermaphroditically. Therefore, one has to compare two closely related species. As with any experiment, one must endeavor to hold as many factors as possible constant so as to observe the effects of the one manipulated factor. But with two closely related species, the experiment is complicated by the existence of many other differences between them, above and beyond their mode of reproduction. Many plants, however, can reproduce either asexually (vegetatively) or sexually (seed production), and so here there exist possibilities to control for different genetic backgrounds.

Most of the important experiments that I am aware of, that have *directly* tested ideas about the advantage to sex, are described in Table 30.1. There are not many. This book is concerned with fungi, and so I concentrate on those experiments that employ *S. cerevisiae*. I briefly mention the others, but I refer the interested reader to the original studies.

Experiments with Organisms Other than Fungi

Malmberg conducted one of the first experiments concerned with recombination in 1977 (48). He used the T4 bacteriophage to test the Fisher-Muller theory and to test the idea that asexuality promotes epistatic interactions between loci due to coevolution. This was an elegant study which manipulated the probability of multiple-phage infection and therefore the opportunity for interphage recombination, by varying the phage:bacteria ratio. Malmberg measured various phage fitness parameters (infectivity, offspring number, and life cycle time) over a number of generations and showed that fitness increased at a significantly greater rate in the lines with higher recombination. However, the advantage took some time to manifest and seemed relatively short-lived, disappearing after 15 growth cycles. Even so, these data support the Fisher-Muller model, and more generally the Weismannian concept. Further, as predicted by theory, Malmberg found evidence for synergistic interactions between alleles in the asexual lines. Another study with a phage, this time φ6, suggested that increased recombination (again manipulated by multiple infection rates) allowed the more efficient removal of deleterious mutations (11). Unfortunately, in

Table 30.1 Relevant studies that have directly tested hypotheses concerning the maintenance of sex[a]

Organism (reference)	Hypothesis tested	Generations (or other)	Sex of advantage?
Bacteriophage T4 (48)	F-M[a] and epistasis	Recombination, then 15–20	Yes
Bacteriophage φ 6 (11)	Muller's ratchet	40 bottlenecks	Yes
Drosophila melanogaster (22)	Recombination modifier	330 days	Yes
Drosophila melanogaster (42)	Recombination modifier	49	Yes
Drosophila melanogaster (61)	F-M	11	Yes
Anthoxanthum odoratum (38)	Twofold cost	2	Yes
Anthoxanthum odoratum (1)	Frequency dependent	3 years	Yes
Chlamydomonas reinhardtii (15)	Mutational and Weismann's	1 sexual, then 500–600 mitotic	No
Chlamydomonas reinhardtii (13)	F-M	1 sexual, then ~50 mitotic	Yes in the long run
Chlamydomonas reinhardtii (36)	Weismann's	3 sexual, over ~120 mitotic	Yes
Chlamydomonas reinhardtii (12)	F-M and population size	150 mitotic, 1 sexual, 50 mitotic	Yes in large populations
Saccharomyces cerevisiae (71)	Environmental fluctuations	32 sexual (inbred) over 800 mitotic	N/A
Saccharomyces cerevisiae (5)	F-M	1 sexual (inbred), then 100–300 mitotic	Possibly
Saccharomyces cerevisiae (72)	F-M and mutation clearance	8 sexual over 400–600 mitotic	F-M, no; mutation clearance, yes
Saccharomyces cerevisiae (27)	F-M	1 sexual (inbred), then 500 mitotic	Yes in the long run
Saccharomyces cerevisiae (29)	Heterogeneous environments	4–6 sexual over 200–300 mitotic	Uncertain
Saccharomyces cerevisiae (25)	Weismann's	12 sexual over 300 mitotic	Yes

[a]F-M, Fisher-Muller theory.

both these studies, increasing the probability of multiple infection also increased intrahost competition and confounded the effects of recombination to some degree (36).

In 1982 Flexon and Rodell conducted experiments with dichlorodiphenyltrichloroethane (DDT) resistance using *Drosophila* and were the first to demonstrate a correlation between recombination rate and selection (22). They gradually increased the concentration of DDT and compared the change in resistance of experimental lines to control lines unexposed to DDT. They determined the recombination rate of the lines by crossing to a marked reference strain, and they also tested chromosomes individually for the presence of resistance alleles. They found that the recombination rate of the selected lines was significantly higher than that of the control lines. They reported that the change in DDT resistance brought about by directional selection was accompanied by a significant increase in the intrachromosomal recombination rate. They then examined the chromosomes and discovered that the greater the chromosomal response to DDT (number of beneficial mutations) the greater the chromosomal recombination rate. These data support the idea that genes which increase the rate of recombination are maintained as they produce new beneficial combinations; those linked to such beneficial combinations will hitchhike to higher frequencies. Other experiments with *Drosophila* show that adaptation to daily temperature fluctuations (73) and two-way directional selection for geotaxis (42) also resulted in increased recombination. These studies provide evidence for the modifier of recombination ideas touched on in the previous section and show that genes which increase the rate of recombination can increase in frequency in response to selection. These data are completely congruent with Weismann's hypothesis—the production of variance by modifiers of recombination is adaptive, and these will be swept along with the loci directly selected for. Rice and Chippindale recently conducted an experiment with sophisticated manipulations, again with *Drosophila*, looking at the interaction between recombination and the efficacy of (artificial) selection (61). They followed the fate of an artificially introduced beneficial mutation (w^+) in many replicate populations (17 with and without recombination). The beneficial allele increased more rapidly in the lines with recombination over the 11 generations of the experiment. In agreement with Weismann, the authors suggest that this is because recombination allows selection to work more efficiently to promote beneficial mutations because of an increase in variance.

Antonovics and Ellstrand conducted experiments with clones of *Anthoxanthum odoratum* (sweet vernal

grass) in 1983 and first set out to test the frequency-dependent hypothesis for the advantage to sex (1). The number of samples is initially impressive (1,920 plants), but the noise and lack of conformity in the data meant the study could only suggest that rare clones were at a fitness advantage, and the authors do not state whether this advantage was due to evading parasitic invasion. A few years later Antonovics published another coauthored study examining the short-term effects of sex with the same experimental system (38). This study concluded that sexually generated progeny were on average 1.43 times more productive than their asexually generated siblings, although the nature of the advantage is not clear. Both these experiments are excellent in the respect that they examine individuals in natural situations, but they suffer from a lack of control over breeding system (outcrossing rate), too small a sample size, and very short periods of experimentation (only two to three generations).

Eukaryotic microbes of the same genotype can also reproduce sexually or asexually and seem to be attractive experimental organisms since large populations may be propagated through many generations in a reasonable amount of time. In addition, microbes are easy to store in suspended animation. To this end, the alga *Chlamydomonas reinhardtii* has been employed to test ideas about sex.

Work with *Chlamydomonas* suggested that deleterious mutations show positive epistasis but did not directly demonstrate whether or not sex was advantageous (16). Another study demonstrated that an immediate effect of sex for long-standing clones was to reduce mean fitness (15). Colegrave et al. (13) showed that when a genetically heterogeneous mix of strains was subjected to novel carbon sources, a single episode of sex increased the variance in fitness (measured as division time). Sex caused an immediate drop in mean fitness but subsequently allowed a higher fitness to be achieved in comparison to the asexual lines. After 50 generations of vegetative growth the actions of selection caused the two treatments to converge in fitness and variance. In contrast to theoretical predictions, the effects of sex were not seen to be greater in more complex environments (where more loci are presumably involved in adaptation). Kaltz and Bell conducted a similar experiment, again starting with heterogeneous strains, but this time passed the populations through three rounds of sex (36). Their data showed that sexual populations were able to adapt more rapidly to novel environments, when compared with asexuals, and this was associated with a greater variance in fitness. They also showed that the advantage of sex may be compounded over more than one sexual episode. In agreement with the Colegrave et al. study, they showed that sex causes a brief drop in fitness, but in contrast, this study showed that sex conferred a greater advantage in environments with more novel factors. Last, an excellent further study by Colegrave (12) examined the effects of sex on adaptation with differing beneficial mutation supply rates. He achieved this by manipulating population size: the smaller the population, the less the supply of mutations. As had been clearly predicted by Weismann and Fisher, this study showed that sex is of more advantage in large populations where multiple beneficial mutations are likely to exist and where clonal interference is a problem for asexuals. However, these populations are not so large as to mean that multiple mutants are readily accessible to asexuals. In general the ideas we have about the advantage of sex are congruent with these experiments. They demonstrate that although there may be short-term disadvantages, sex allows a greater rate of adaptation for a period after it occurs.

Experiments with Yeast

S. cerevisiae also has the attractive microbial experimental properties mentioned earlier and further is extremely well genetically characterized, and may be genetically manipulated and dissected. These facts mean that *S. cerevisiae* is often the organism of choice for many researchers interested in population genetics and evolution.

Wolf, Wohrmann, and Tomiuk were the first to use *S. cerevisiae* to test ideas concerning the maintenance of sex, choosing to look at heterogeneous environments (71). Their study propagated diploids for 32 rounds of batch culture and sporulation. Sporulation was induced by placing the cultures on media with a nonfermentable carbon source which was also lacking nitrogen (Fig. 30.2). Three different environmental factors were manipulated: temperature (28 or 36°C), aeration level (low or high), and carbon source (glucose or glycerol). Combinations of these factors allowed the authors to construct six different environments to which they subjected

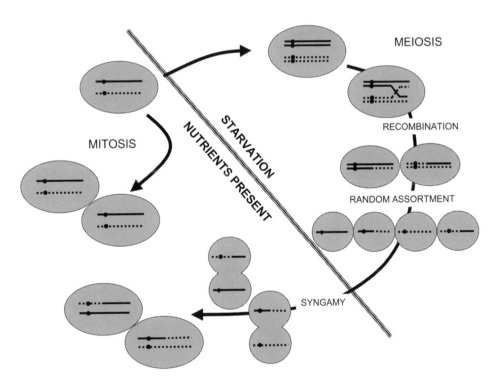

Figure 30.2 A generalized life cycle of *S. cerevisiae*. A diploid cell will divide mitotically when provided with sufficient nutrients and produce cells that, barring de novo mutation, are genetically identical to the mother (top left). If starved, a diploid that is heterozygous at the mating-type locus (*MAT*) will begin a single meiotic division. Recombination occurs after DNA duplication. The chromosomes are then randomly assorted among four haploid spores. The spores form a distinctive tetrad structure and are encapsulated within an ascus. Each spore is one of two mating types (a or α), which is dictated by the idiomorph at the *MAT* locus. These spores germinate when nutrients are encountered again, and will mate with cells of the opposite mating type to reconstitute a diploid (bottom left).

the replicate populations. In order to manipulate the degree of environmental heterogeneity, the number and periodicity of transfers between successive different environments was also altered, from a constant environment to up to six different environments with changes daily. The sporulation rate of these populations was monitored during the course of the experiment by microscopic examination and determination of the ratio of sporulated and nonsporulated cells, with the frequency changes at four heterozygous loci also being tracked. The authors used sporulation rate as a measure of recombination intensity or the amount of sex. The sporulation rate appeared to change in an inconsistent way between the different treatments (which emulated different degrees of temporal heterogeneity). The study concludes that the overall positive correlation between sporulation rate and genetic heterogeneity (as measured by the Shannon's information measure, H) is evidence for sex increasing variance and accelerating adaptation. Unfortunately it is not clear to me that this necessarily follows. To begin with, no data on rates of adaptation are presented. We are therefore uncertain as to the adaptive trajectories of any of the populations, and it is possible that an increase in sporulation rate could be correlated with a decrease in fitness. Having said that, if genetic heterogeneity can be substituted for fitness variance, then selection should be more effective in the more variable populations. Moreover, I fear that sporulation rates are under different selective considerations than sex. We have no knowledge of the outcrossing rates during the experiment. Yeasts form four ascospores contained within an ascus, and if left undisturbed, as they were in this experiment, yeasts will naturally inbreed (8). If this is the case, then variance will barely increase. It seems that one area of intense selection will be the emergence and subsequent mating of spores, and intra-ascus mating will be more efficient than outcrossing in this respect. In summary, this experiment unfortunately sheds little light on ideas concerning the maintenance of sex.

In 1996 Birdsell and Wills constructed strains of yeast that were either completely homozygous at all loci or heterozygous to an unknown degree (5). They examined the effects of recombination by making these strains either homozygous or heterozygous at the mating-type loci (MAT). Haploid yeast may be one of two mating types, either a or α, which is dictated by the idiomorph (these are not strictly allelic) found at the MAT locus. Haploids of the opposite mating type are attracted to a pheromone excreted by the other, and two haploids then fuse to reconstitute a diploid (Fig. 30.2). Birdsell and Wills engineered diploids and made them homozygous at MAT. Diploids, whether homo- or heterozygous at MAT, will grow mitotically when supplied with sufficient nutrients. Starvation induces strains heterozygous at MAT to go through one meiotic division and produce four haploid spores encapsulated within an ascus. Strains homozygous at MAT do not go through meiosis when starved. The authors termed cells heterozygous at MAT sexual and those homozygous asexual. The sexual strains were sporulated (having an average of around 35% sporulation efficiency) and then allowed to mate within the ascus, meaning that these populations were extremely inbred. After mating, the diploid sexual progeny were mixed with the asexuals and their frequencies were measured using a mating pheromone halo test. The study included treatments where no meiosis took place for control purposes. After sporulation, all sexuals and asexuals, with both homozygous and heterozygous genetic backgrounds, were competed in batch culture for between 100 and 300 generations. The paper does not state the medium that was used or the conditions that this was conducted under. First, and most importantly, the study found that sexual strains (those heterozygous at MAT) were competitively superior regardless of whether or not the competition was preceded by meiosis. The inclusion of meiosis (with inbreeding) before competition allowed the sexual lines greater superiority at some point during the competition. Meiosis provided an initial advantage of around 10% for heterozygous backgrounds, but it was not clear that sex was of such immediate advantage in homozygous backgrounds. This study used the presence or absence of sporulation to control for sex. It examined the effect of a single inbred round of sex in different genetic backgrounds in four duplicated lines. Overall the data suggest that heterozygosity at MAT in itself carries an advantage, and this unfortunately confounds subsequent inferences to some degree. It may be that this advantage of MAT heterozygosity can be further enhanced by the actions of recombination, especially if there is genetic variation as found in heterozygous backgrounds. Although these data are somewhat hard to interpret (due to the confounding effects of MAT status), they are not in conflict with Weismann's hypothesis.

A year later Zeyl and Bell published a study that removed the confounding effects of the differences at MAT and examined the effects of sex on the mean fitness in two environments (72). The aim of this experiment was to assess the importance of sex in a situation where beneficial alleles are important (i.e., under directional selection) and where detrimental alleles are important (i.e., under purifying selection). Haploid and diploid isogenic lines were propagated with either glucose or galactose as carbon sources, to which the strains were well and poorly adapted, respectively. Sex was induced by starvation. The asexual treatments were not sporulated but held at 4°C where little growth occurs, while the sexuals sporulated. The populations passed through approximately 400 and 600 generations on glucose and galactose media, respectively, and the sexual treatments were sporulated eight times during the course of the experiment. After each round of sporulation the cultures were subjected to conditions designed to destroy unsporulated cells and promote the disruption of acsi and therefore encourage random mating. No data are presented on the efficiency of the treatment to force the populations to deviate from extreme inbreeding. The mean fitness of each population was measured by comparing the growth rates of the derived strains to that of the ancestral strains either by direct competition or, more commonly, by measuring colony size. All the populations propagated on galactose significantly adapted to this environment; however, the presence of sex had no effect on the change in fitness of these populations. While there was no change in mean fitness of the asexual populations propagated on glucose, the sexual populations demonstrated an increase in fitness when later tested on galactose. It seems that under these conditions sex is of no benefit in speeding adaptation, but these results, along with additional data on the effect of inbreeding depression, suggest that sex may help purge deleterious mutations in a constant environment. Sex would only serve to increase the rate of adaptation if the supply of beneficial mutations was plentiful, but Zeyl and Bell show that adaptation to galactose likely occurred through a single mutation of large effect. With only one locus under consideration, one would not expect sex to be of any significant advantage. However, this study provides further empirical support for ideas suggesting that sex is advantageous when eliminating deleterious mutations.

Greig et al. employed a slightly different approach in order to investigate the effects of sex (27). In this study diploids were made constitutively asexual by deleting the *IME1* gene (a positive regulator of meiosis) and replacing it with the *KanMX4* resistance marker to distinguish the asexuals from the sexuals. Strains lacking *IME1* are unable to go through meiosis when placed on sporulation media. Like Birdsell and Wills (5), Greig et al. were interested in the effects of sex in different types of genetic backgrounds, and this study involved three homozygous and three heterozygous backgrounds. The sexual and asexual strains (with differing genetic backgrounds) were mixed in roughly equal proportions and

then exposed to sporulation media where the sexuals went through one round of meiosis. No treatment to prevent inbreeding was included. After sporulation the cultures were transferred to standard laboratory media and incubated at a stressful 37°C: yeasts routinely grow at 30°C. The authors tracked the change in proportion of the sexuals and asexuals during batch culture over 500 generations and looked at the change in fitness by direct competition with the ancestor. An immediate cost to sex was seen in all treatments. This is presumably in part due to physical processes—the asexuals may still grow on the sporulation media but the sexuals do not, and the sexuals are further penalized since they must also allow time for the spores to germinate and mate once nutrient-rich conditions are encountered again. Further, the cost to sex was higher in heterozygous backgrounds, which suggests that recombinational load (the destruction of coadapted gene complexes) also plays a part in the cost to sex. However, the initial costs that sex imposed were ultimately overcome in 10 of the 12 treatments with heterozygous backgrounds. There seemed to be no long-term cost or benefit to sex in homozygous backgrounds. In the heterozygous backgrounds, the lines that demonstrated the biggest initial cost to sex enjoyed the clearest long-term advantage. This is consistent with the idea of sex increasing variance: the destruction of coadapted alleles carries an initial cost but has the advantage of producing rare combinations conferring greater adaptation, which eventually dominate the population.

In general these studies with yeasts (and other organisms) appear to support the broader Weismannian concept. At the very least, they do not refute it. However, these studies do not unconditionally demonstrate that sex speeds adaptation or that sex provides an advantage in constant environments. Although they are excellent experiments, and let us not forget that they are hard to conduct, there are certain aspects that leave one pondering whether or not their results reflect the true effect of sex. In contrast to most natural populations, many of these studies employed only a few episodes of meiosis, which were often extremely inbred. Inbreeding severely restricts the randomizing process of sex. Experiments with such low incidences of meiosis and high levels of inbreeding may mask any effects of sex. One of the more concerning aspects of these experiments, to my mind, is the difference in treatment between sexual and asexual lines. In studies with these microbial model systems, sex is unavoidably correlated with starvation and sporulation. Asexuals are either not exposed to the sporulation environment or they have engineered mutations which prevent sporulation from occurring. There is strong evidence that starvation in yeast and *Chlamydomonas* increases the mutation rate (26, 50). Additionally, the process of sporulation (meiosis) means the expression of a whole host of genes not expressed by the nonsporulating asexuals. Last, there may be differences in treatment after sporulation: if the spores are to be randomized, and unsporulated cells killed, then the sexual populations will also be subjected to a barrage of enzymes, chemicals, and conditions that the asexuals will not. Such discrepancies in treatment, which in some cases are known to elevate mutation rates, are likely to increase genetic variance and therefore have significant effects on the course of evolution that are independent of sexual reproduction. Ideally one needs to hold all factors constant apart from the three cornerstones of sex: recombination, random assortment, and syngamy.

A Hint from Nature

At the turn of the last century Guilliermond described a strain of *Saccharomyces pastorianus* in which the nucleus divided only once before spore formation and conjugation between spores did not occur (30). In 1972 Grenwal and Miller described 3 strains (of 140) of *S. cerevisiae* that stably produced only two spored asci (dyads) where the nucleus divided only once (28). They proved that each of the two spores in a dyad contained the full complement of DNA (i.e., it was diploid) and that each of these spores could germinate and sporulate independently without conjugation. Grenwal and Miller concluded that this was an apomictic (without mixing) strain, and during sporulation it appeared to go through a mitotic-like, as opposed to a meiotic, division. Here we have a hint that one may be able to use naturally described phenotypes to control for at least one of the confounding factors associated with experimental investigations into the function of sex. It seems that these strains experience only a single meiotic division, and two genes, *SPO12* and *SPO13*, are responsible for this apomictic phenotype (39). It appears that *SPO12* has a mitotic role, and this consequently seems an unattractive locus to artificially manipulate (34). However, *SPO13* is meiosis specific (it is not expressed during mitosis) (7) and acts to stabilize sister chromosome cohesion (63, 67). Strains deficient in *SPO13* go through only one meiotic division. This means that the idiomorphs at *MAT* can potentially go to opposite poles and produce diploid spores that are heterozygous at *MAT* and thus able to sporulate (35). However, this happens in only approximately one-half of the spores: the remainder will segregate reductionally and will be homozygous at *MAT* and thus not sporulate (64). In addition, strains deficient for *SPO13* still undergo normal

levels of meiotic recombination. These traits mean that a *spo13* mutant is not optimal for use in experimentation. However, strains that are also deficient for *SPO11* go through a 100% nonreductional division, and this means that all spores are heterozygous for *MAT* (64). This is extraordinarily fortuitous since *SPO11* controls meiotic recombination (40). *SPO11* encodes an endonuclease that causes double-stranded chromosome breaks which initiate recombination during meiosis. *SPO11* is also meiosis specific and does not affect mitotic recombination (40). Strains deficient for only *SPO11* sporulate, but the absence of chiasmata means that chromosomes are not stabilized and that aberrant segregation occurs and results in less than 1% of the spores being viable (6, 40). However, a *spo13* mutant rescues this phenotype and so the *spo11 spo13* double mutant produces 100% viable spores which have not experienced any meiotic recombination or random assortment—these spores are exact copies of the mother cell (64). Since these spores are all diploid and heterozygous at *MAT*, syngamy is not necessary either. This double mutant may be exposed to exactly the same conditions as a wild-type strain and sporulates just like the wild type, but it produces spores in a strictly asexual manner, and these do not mate upon germination.

The *spo11 spo13* system

The *spo11 spo13* system was developed in the *S. cerevisiae* Y55 background (25). Y55 was selected because of its high sporulation rates and relative "wild-type" nature, i.e., a lack of autotrophies. One can engineer a *spo11 spo13* double mutant and compare this to the wild type. However, it is possible that any difference in adaptive trajectories observed between sexual and asexual lines are not due to the presence of sex but to some hangover from engineering. In order to minimize this difference, a protocol was derived to produce strains with identical histories of engineering (25). Figure 30.3 shows an overview of the strain construction procedure.

Starting with a prototrophic haploid which was *ho* (unable to switch mating type), the *URA3* gene was first precisely deleted using PCR-based gene targeting (66). To achieve this, the 5' and 3' intergenic spacer regions on each side of *URA3* were independently amplified with high-fidelity PCR and ligated: products of the correct 5' and 3' orientation were then selected using high-fidelity PCR. This *ura3Δ* PCR product was then transformed into the haploid, and those cells which involved this in a homologous recombination event at the correct location produced 5-fluoroorotic acid-resistant (5-FOA) colonies, which were picked (8). The mating type of this

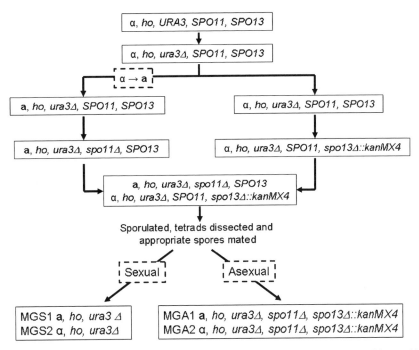

Figure 30.3 An overview of the construction protocol for the *spo11 spo13* system used by Goddard et al. (25). Solid boxes represent the genotypes of strains constructed at each step. Genes in uppercase denote wild-type alleles, while those in lowercase represent mutants, with the Greek delta symbol indicating a precise deletion of the gene. The status of only the mating type (either **a** or α), mating type switching (*HO*), a component of uracil anabolism (*URA3*), *SPO11*, and *SPO13* loci are shown, and these strains are assumed to be of the wild type at all other loci. A detailed explanation may be found in the text.

strain was then switched using the pSC11 plasmid as described by Burke et al. (8), and this resulted in two strains of opposite mating types which were identical at all loci except *MAT*. Next, *SPO11* was disrupted in the **a** strain with a PCR product containing *URA3* (successful transformants were prototrophic). A *spo11*Δ product was constructed, by the same 5' and 3' intergenic ligation and PCR method described for the *ura3*Δ construction, and successful *spo11*Δ transformants were again selected on 5-FOA. *SPO13* was precisely replaced with the *kanMX4* module in the α strain as described in Wach et al. (66). All engineered junctions were confirmed by independent PCRs. The *a ho ura3*Δ *spo11*Δ haploid was mated to the α *ho ura3*Δ *spo13*Δ::*kanMX4* haploid and produced a strain that was heterozygous at only the *SPO11*, *SPO13*, and *MAT* loci. Since this diploid contained one wild-type copy of *SPO11* and *SPO13*, it sporulated normally and produced four recombined haploid spores. Tetrad dissection and analyses were then employed to identify those haploid progeny which contained either the wild-type or engineered *spo11* and *spo13* loci. One *a ho ura3*Δ *spo11*Δ *spo13*Δ::*kanMX4* and one α *ho ura3*Δ genotype were found from the same tetrad and were named MGA1 and MGS1, respectively; one α *ho ura3*Δ *spo11*Δ

*spo13*Δ::*kanMX4* and one *a ho ura3*Δ genotype were found from another tetrad, and these were named MGA2 and MGS2, respectively. The MGA1 and MGA2 and the MGS1 and MGS2 haploids were mated to yield strains that were identical at all but the *SPO11* and *SPO13* loci and shared exactly the same engineering history.

One cannot assume that the removal of two genes, even if they are expressed only during meiosis, will have no effect on mitotic fitness. Competition experiments were thus conducted, under a variety of conditions, to discern the relative fitness of the wild-type and asexual sporulator. First, six populations with equal frequencies of sexuals and asexuals were propagated in YPD (standard rich yeast media) in batch culture for 60 generations. The frequency of the asexuals was assayed every 10 generations by scoring the proportion of the populations resistant to the G418 toxin: the *kanMX4* module carried by the asexuals confers resistance to this toxin. The change in frequency allows one to estimate the relative competitive ability of one strain compared with that of the other. The relative fitness of the asexual strain, calculated from six replicates, was effectively the same as that of the sexual strain (the asexuals were 0.02% ± 0.3% less fit; the standard error of the mean

[SEM] encompassed zero). The relative fitness of the asexual strain was also assayed under continuous culture with carbon limitation, both with and without elevated temperature (30 or 37°C) and salt concentrations (0.2 M NaCl). The fitness advantage of the sexual strain over the asexual strain in both environments was similar and very near zero (30°C, no NaCl: 0.3% ± 0.16% [mean ± SEM]; 37°C, plus NaCl: 0.3% ± 0.23%; n = 8) (25). The deletion of two genes involved in meiosis had not affected sporulation efficiency: the proportion of cells that produced spores was very similar in both the sexual and engineered asexual strains (sexual, 76% ± 4%, mean ± SEM; asexual, 73% ± 4%; t test, P = 0.62) (25).

Overall the *spo11 spo13* system seems ideally suited to applications concerning the maintenance of sex. This system has no effect on mitotic fitness or sporulation rates (at least in the Y55 genetic background) and can hold all factors constant apart from recombination, random assortment, and syngamy.

EXPERIMENTS USING *SPO11 SPO13* STRAINS

Grimberg and Zeyl conducted an excellent study examining the interplay between sex and adaptation to heterogeneous environments (29). Although they used strains with mutations in *SPO11* and *SPO13*, they did not use the system described above. They also examined the adaptive trajectories of mutator strains, but I will not discuss this here. Grimberg and Zeyl rightly point out that many experiments concerning sex have been conducted in very simplistic laboratory environments but that several ideas concerning sex require more complicated spatial or temporal heterogeneity (e.g., Red Queen and sibcompetition). For this reason they decided to propagate a pathogenic strain of *S. cerevisiae* in laboratory test tubes (simple environment) and in mouse brains (heterogeneous environment). This study used the genetic background of YJM145 (isolated from humans), and replaced *SPO11* and *SPO13* with hygromycin and G418 markers, respectively. Their engineering protocol differed from that described in the previous section. The sporulation efficiency of these sexuals and asexuals is slightly lower than that of the Y55 background, at around 60%, but this is still a reasonable sporulation rate. In line with data from Goddard et al. (25), the mitotic fitness of the ancestral sexual and asexual strains in the test tube environment was equivalent. No data are presented for the relative fitness of the ancestral strains in mouse brains. An equal mix of these strains was propagated under serial batch culture conditions in defined, glucose-limited media (five replicates) and was also injected into mice (four replicates) and

serially transferred to the next mouse by harvesting the cultures from the brain. Each environment had two treatments: either sporulation was imposed roughly every 50 generations, or it was not. For the sporulation treatment the cultures were harvested and sporulation was preceded by growing in YPD overnight; the cultures were then placed in sporulation media for 5 days. After this, unsporulated cells were killed, and asci were softened and disrupted with zymolyase, β-mercaptoethanol, and a detergent. The appropriate serial transfer regime was resumed after mating overnight in YPD. No data on the efficiency of this procedure for eliminating unsporulated cells, or for promoting random mating, are presented, apart from the observation that only 1.4% of the sexual populations in mice appeared to have mated. The experiments were continued until either the sexual or asexual strains became fixed. In the test tube environment with no sex the wild-type (sex capable) strains climbed steadily in frequency in all five replicates and became fixed within 250 generations. This implies that the asexual strains are at a ~3% disadvantage, either because of the presence of the hygromycin and/or G418 resistance markers, or because of some pleiotropic effect of deleting *SPO11* and *SPO13*. However, this effect vanished when sex was induced once every 50 generations. Here, the frequencies fluctuated, and in three of the five replicates the asexuals went to fixation. These dynamics were reversed in the mouse brain. The frequency fluctuation was greater, but the asexual strains became fixed in all four populations within 200 generations in the absence of sporulation and mating. The inclusion of sex meant that the sexuals became fixed in half the populations (two of four): the induction of sex had opposite effects in test tubes and in mice. Interestingly, when sporulation and mating were induced in the mice treatment, the two populations where the sexual strains won had a much lower fitness than the two where the asexuals won: in stark contrast to theory, sex resulted in a smaller increase in fitness than asex. This was reversed in test tubes. In the populations where the sexuals became fixed (only two of five), the sexuals achieved a final higher fitness than the asexuals when sporulation and mating were induced. The contrast in fitness dynamics of the sexual and asexual strains without sporulation and mating in each environment is concerning, as is the uncertainty of the postsporulation treatment to eliminate unsporulated cells and encourage random mating. Overall these data provide no strong support for an advantage to sex but do suggest that sex is of more benefit in heterogeneous environments. Sex sometimes won in the simpler test tube environment, and when it did it conferred a higher mean fitness than when the asexuals won.

The same year, Goddard, Godfray, and Burt published a study that used the *spo11 spo13* system described earlier (25). The aim of this experiment was to test the core of Weismann's concept with as few confounding effects as possible: does sex really allow a greater rate of adaptation? To do this, we propagated sexual and asexual strains in glucose-limited chemostats in two environments. Continuous culture has some advantages over serial batch transfer. To begin with, one can precisely define the environment and hold this constant; in batch transfer the environment changes as the population expands (nutrients are depleted and waste accumulates). Also, continuous culture allows one to hold population size relatively constant and to manipulate generation time; again these are not controlled in batch culture. We termed one environment benign since it comprised standard defined yeast media at 30°C. The other we termed harsh as this included an osmotic and thermal stress: 0.2 M NaCl was added to the medium and this was incubated at 37°C. The harsh environment proved to be stressful by the significant decrease in growth rate that it imposed. Eight separate populations of sexuals and asexuals, constructed as described in the earlier section, were exposed to each environment, and sporulation was imposed every 4 days (50 and 25 generations in the benign and harsh environments, respectively). In all populations sporulation was followed by a treatment designed to destroy unsporulated cells and to break apart asci to encourage random mating. This involved using lyticase, β-glucuronidase, dithiothreitol, and sodium dodecyl sulfate (a surfactant), followed by sonication. The effectiveness of the treatment was assayed, and the results showed that on average 87% ± 3% (SEM) of the diploid progeny were produced from random mating. No haploid or unsporulated cells survived this treatment. The change in relative fitness of each population was estimated by direct competition with the ancestor across the 300 mitotic generations interspersed with either 12 or 6 episodes of sporulation and mating. The fitness dynamics of the sexual and asexual populations are shown in Fig. 30.4. There was no significant change in fitness of either the sexual or asexual populations in the benign environment. This strongly suggests that the production of beneficial mutations was rare: it seems that the strains were near optimally adapted to these conditions. Even though the continuous culture conditions may be novel, this lack of adaptation is not too surprising given that these are the standard media and temperature conditions under which laboratory strains are routinely cultured. In contrast, both the sexuals and asexuals significantly adapted to the harsh environment. The asexual strain increased in fitness by 80%, but the sexuals increased by 94%. Not only did the sexuals reach a higher final fitness, but they also got there significantly faster ($P = 0.0004$; Fig. 30.4). This, to my mind, provides very strong evidence that the net effect of sex is to confer a greater rate of adaptation and fully supports Weismann's original contention.

The mechanisms that underlie the advantage of sex are not completely discernible in this experiment, but some may be ruled out. It is unlikely that environmental variation was the parameter behind the advantage since this was controlled precisely. It is also unlikely that the clearance of detrimental mutations was driving the advantage of sex. There was no detectable difference in fitness between sexual and asexual populations in the environment where beneficial mutations were of no importance (benign). If detrimental mutations were significant enough to decrease fitness, and sex served to ameliorate this to some degree, then one would expect to have seen a difference between sexuals and asexuals in the benign environment. These populations were too large to suffer Muller's ratchet, and estimates of the yeast mutation rate are far below the one per genome per generation required for the mutation deterministic hypothesis to apply. The rapid and consistent increase in fitness in the harsh environment points toward the more efficient incorporation of beneficial mutations as the most likely candidate for sex's benefit. Sex, in principle, was advantageous as it broke down negative associations between alleles, although it cannot be determined whether these associations were due to negative epistasis between beneficial mutations or due to the effects of a finite population. The detail behind the genetic mechanisms needs further investigation.

FUTURE DIRECTIONS

The body of experimental evidence is such that the general form of Weismann's original concept is supported. There can be little doubt that sex is of benefit in the longer term in allowing populations to adapt more rapidly than asexual populations. However, the underlying mechanisms of the advantage of sex are far from being answered empirically. Much work is needed in this area if we are to understand the extent to which theoretical models apply in natural situations.

Along with elucidating the underlying genetics of adaptation, ideas concerned with the types of selection that promote sex may be addressed. The effect of sex under environmental heterogeneity is an area ripe for experimentation. These ideas are pertinent for many reasons. Selection due to parasites is an attractive concept

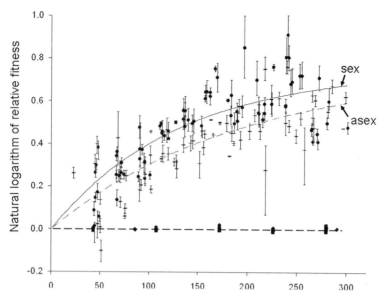

Figure 30.4 The change in natural logarithm of fitness of asexual and sexual populations of yeast in benign and harsh environments. Points represent fitness measurements for individual populations with twice log-likelihood error bars (these approximate 95% confidence limits); the error bars for the benign treatment are plotted but are mostly too small to be discriminated. The fitted model for the harsh environment is plotted for asexual (—) and sexual (●) treatments (parameters: $a_1 = 0.761$, a_2(asexual) = -5.287, a_2(sexual) = -4.901). Asexual strains in the benign environment are represented by a square (■); sexual strains in a benign environment are represented by a diamond (♦). Reproduced from reference 25 with slight modification.

and can be tested with an appropriate model system. The effect of sex under conditions of spatial and temporal environmental heterogeneity also needs to be considered carefully. Some models propose that sex is of advantage in such heterogeneous environments in that it reduces competition (4), while others suggest that sex serves to retard adaptation (20). The experiments that support Weismann's hypothesis have, to date, examined the effect of sex mostly in homogeneous environments. In these simple one-dimensional environments, sexual populations adapt more rapidly and outcompete asexuals. Yet in reality, the majority of populations inhabit complex environments. Even in the simplest of conceivable structured environments, an environment that comprises just two niches, models show that sex retards adaptation (20). This prediction remains in the latest highly complex models (18, 19, 54, 65). These models assume that alleles which afford a degree of advantage in one niche do not afford an advantage in another. Individuals carrying alleles adapted to any one niche will flourish there. However, sexual reproduction permits unions between individuals carrying alleles adapted to different niches, and such matings result in maladaptive hybrids (20). Models suggest that sex retards adapta-

tion unless there is reproductive isolation between niches (10, 18, 62, 65). With some means of isolation, individuals carrying alleles adapted to one environment are more likely to mate with similarly adapted individuals. A mating barrier serves to isolate and concentrate alleles adapted to the same niche and deters the formation of maladapted hybrids. If mating were restricted for some reason, then two populations would form and subsequently adaptively diverge into each niche (62).

The natural world contains a plethora of sexual species which inhabit structured (heterogeneous) environments. If the theories are correct, this must mean that reproductive isolation of some form occurs frequently. Much work has gone into attempting to understand how restricted mating occurs (65): Butlin has published a review of the connection between sex and speciation and points out that recombination lies at the heart of the two (10). It seems that unless there is some sort of reproductive isolation of sexuals, asexuals (who are not hindered by recombination) will adapt more rapidly and effectively to the differing niches available in heterogeneous environments (2). In sympatric situations it is hard to understand how reproductive isolation can arise fast enough to allow sex to be maintained (65). Questions concerning the effects of sex in heterogeneous environments not only are important in helping understand why and how sex is maintained in the real world, but also have a very large bearing on ideas concerning the origin of species. As yet we have very few empirical data in these areas.

I thank Basil for help with manuscript preparation and Mary for her tireless and precise proofreading.

References

1. Antonovics, J., and C. Ellstrand. 1984. Experimental studies of the evolutionary significance of sexual reproduction. I. A test of the frequency-dependent selection hypothesis. *Evolution* 38:103–115.

2. Barraclough, T. G., C. W. Birky, and A. Burt. 2003. Diversification in sexual and asexual organisms. *Evolution* 57:2166–2172.

3. Barton, N. H., and B. Charlesworth. 1998. Why sex and recombination. *Science* 281:1986–1990.

4. Bell, G. 1982. *The Masterpiece of Nature*. University of California Press, Berkeley.

5. Birdsell, J., and C. Wills. 1996. Significant competitive advantage conferred by meiosis and syngamy in the yeast *Saccharomyces cerevisiae*. *Proc. Natl. Acad. Sci. USA* 93:908–912.

6. Bruschi, C. V., and R. E. Esposito. 1982. Recombination process in a sporulation-defective mutant of *S. cerevisiae*: role of Holliday structure resolution. *Recent Adv. Yeast Mol. Biol.* 1:254–268.

7. Buckingham, L. E., H.-T. Wang, R. Elder, R. McCarroll, M. Slater, and R. E. Esposito. 1990. Nucleotide sequence and promoter analysis of *SPO13*, a meiosis-specific gene of *Saccharomyces cerevisiae*. *Proc. Natl. Acad. Sci. USA* 87:9406–9410.

8. Burke, D., D. Dawson, and T. Stearns. 2000. *Methods in Yeast Genetics. A Cold Spring Harbor Laboratory Course Manual*. Cold Spring Harbor Press, Cold Spring Harbor, NY.

9. Burt, A. 2000. Sex, recombination, and the efficacy of selection—was Weismann right? *Evolution* 54:337–351.

10. Butlin, R. K. 2005. Recombination and speciation. *Mol. Ecol.* 14:2621–2635.

11. Chao, L. 1990. Fitness of RNA virus decreased by Muller's ratchet. *Nature* 348:454–455.

12. Colegrave, N. 2002. Sex releases the speed limit on evolution. *Nature* 420:664–666.

13. Colegrave, N., O. Kaltz, and G. Bell. 2002. The ecology and genetics of fitness in *Chlamydomonas*. VIII. The dynamics of adaptation to novel environments after a single episode of sex. *Evolution* 56:14–21.

14. Crow, J. F., and M. Kimura. 1965. Evolution in sexual and asexual populations. *Am. Nat.* 99:439–450.

15. Da Silver, J., and G. Bell. 1996. The ecology and genetics of fitness in *Chlamydomonas*. VII. The effect of sex on the variance in fitness and mean fitness. *Evolution* 50:1705–1713.

16. de Visser, J. A. G. M., R. F. Hoekstra, and H. Van den Ende. 1996. The effect of sex and deleterious mutations on fitness in *Chlamydomonas*. *Proc. R. Soc. Lond. Ser. B* 267:123–129.

17. de Visser, J. A. G. M., C. Zeyl, P. J. Gerrish, J. L. Blanchard, and R. E. Lenski. 1999. Diminishing returns from mutation supply rate in asexual populations. *Science* 283:404–406.

18. Dieckmann, U., and M. Doebeli. 2004. Adaptive dynamics of speciation: sexual populations. *In* U. Dieckmann, M. Doebeli, J. A. J. Metz, and D. Tautz (ed.), *Adaptive Speciation*. Cambridge University Press, Cambridge, United Kingdom.

19. Doebeli, M., and U. Dieckmann. 2003. Speciation along environmental gradients. *Nature* 421:259–264.

20. Felsenstein, J. 1981. Skepticism toward santa rosa, or why are there so few kinds of animals? *Evolution* 35:124–128.

21. Fisher, R. A. 1930. *The Genetical Theory of Natural Selection*. Oxford University Press, Oxford, United Kingdom.

22. Flexon, P. B., and C. F. Rodell. 1982. Genetic recombination and directional selection for DDT resistance in *Drosophila melanogaster*. *Nature* 298:672–674.

23. Gabriel, W., M. Lynch, and R. Burger. 1993. Muller's Ratchet and mutational meltdowns. *Evolution* 47:1744–1757.

24. Gerrish, P. J., and R. E. Lenski. 1998. The fate of competing beneficial mutations in an asexual population. *Genetica* 102:127–144.

25. Goddard, M. R., H. J. C. Godfray, and A. Burt. 2005. Sex increases the efficacy of natural selection in experimental yeast populations. *Nature* 434:636–640.

26. Goho, S., and G. Bell. 2000. Mild environmental stress elicits mutations affecting fitness in *Chlamydomonas*. *Proc. R. Soc. Lond. Ser. B* 267:123–129.

27. Greig, D., R. H. Borts, and E. J. Louis. 1998. The effect of sex on adaptation to high temperature in heterozygous and homozygous yeast. *Proc. R. Soc. Lond. Ser. B* 265:1017–1023.

28. Grewal, N. S., and J. J. Miller. 1972. Formation of asci with two diploid spores by diploid cells of *Saccharomyces*. *Can. J. Microbiol.* 18:1897–1905.

29. Grimberg, B., and C. Zeyl. 2005. The effects of sex and mutation rate on adaptation in test tubes and to mouse hosts by *Saccharomyces cerevisiae*. *Evolution* 59:431–438.

30. Guilliermond, M. A. 1905. Recherches sur la germination des spores et la conjugaison chez les levures. *Rev. Gen. Bot.* 17:337–376.

31. Hamilton, W. D., R. Axelrod, and R. Tanese. 1990. Sexual reproduction as an adaptation to resist parasites (a review). *Proc. Natl. Acad. Sci. USA* 87:3566–3573.

32. Hill, W. G., and A. Robertson. 1966. Effect of linkage on limits to artificial selection. *Genet. Res.* 8:269.

33. Hoekstra, R. F. 2005. Why sex is good. *Nature* 434:571–573.

34. Hoyton, J. H., and L. H. Johnston. 1993. *Spo12* is a limiting factor that interacts with the cell cycle protein kinase Dbf2 and Dbf20, which are involved in mitotic chromatid disjunction. *Genetics* 135:963–971.

35. Hugerat, Y., and G. Simchen. 1993. Mixed segregation and recombination of chromosomes and YACs during single-division meiosis in *spo13* strains of *Saccharomyces cerevisiae*. *Genetics* **135**:297–308.

36. Kaltz, O., and G. Bell. 2002. The ecology and genetics of fitness in *Chlamydomonas*. XII. Repeated sexual episodes increase the rates of adaptation to novel environments. *Evolution* **56**:1743–1753.

37. Keightley, P. D., and A. Eyre-Walker. 2000. Deleterious mutations and the evolution of sex. *Science* **290**:331–333.

38. Kelly, S. E., J. Antonovics, and J. Schmitt. 1988. A test of the short-term advantage of sexual reproduction. *Nature* **331**:714–716.

39. Klapholz, S., and R. E. Esposito. 1980. Isolation of *SPO12-1* and *SPO13-1* from a natural variant of yeast that undergoes a single meiotic division. *Genetics* **96**:567–588.

40. Klapholz, S., C. S. Waddell, and R. E. Esposito. 1985. The role of the *SPO11* gene in meiotic recombination in yeast. *Genetics* **110**:187–216.

41. Kondrashov, A. S. 1988. Deleterious mutations and the evolution of sexual reproduction. *Nature* **336**:435–440.

42. Korol, A. B., and K. G. Iliada. 1994. Increased recombination frequencies resulting from directional selection for geotaxis in *Drosophila*. *Heredity* **72**:64–68.

43. Lenski, R. E. 2001. Genetics and evolution. Come fly, and leave the baggage behind. *Science* **294**:533–534.

44. Lenski, R. E. 1999. A distinction between the origin and maintenance of sex. *J. Evol. Biol.* **12**:1034–1035.

45. Lively, C. 1987. Evidence from a New Zealand snail for the maintenance of sex by parasitism. *Nature* **328**:519–521.

46. Lively, C., C. Craddock, and R. C. Vrijenhoek. 1990. Red Queen hypothesis supported by parasitism in sexual and clonal fish. *Nature* **344**:864–866.

47. Lively, C. M. 1996. Host-parasite coevolution and sex. *BioScience* **46**:107–114.

48. Malmberg, R. L. 1977. The evolution of epistasis and the advantage of recombination in populations of bacteriophage T4. *Genetics* **86**:607–621.

49. Malmberg, R. L., S. Held, A. Waits, and R. Mauricio. 2005. Epistasis for fitness-related quantitative traits in Arabidopsis thaliana grown in the field and in the greenhouse. *Genetics* **171**:2013–2027.

50. Marini, A., N. Matmati, and G. Morpurgo. 1999. Starvation in yeast increases non-adaptive mutation. *Curr. Genet.* **35**:77–81.

51. Maynard Smith, J. 1968. Evolution in sexual and asexual populations. *Am. Nat.* **102**:469–473.

52. Maynard Smith, J. 1978. *The Evolution of Sex*. Cambridge University Press, Cambridge, United Kingdom.

53. Maynard Smith, J., and E. Szathmary. 1995. *The Major Transitions in Evolution*. Oxford University Press, Oxford, United Kingdom.

54. Maynard-Smith, J. 1966. Sympatric speciation. *Am. Nat.* **100**:637–650.

55. Michiels, N. K., L. W. Beukeboom, J. M. Greeff, and A. J. Pemberton. 1999. Individual control over reproduction: an underestimated element in the maintenance of sex? *J. Evol. Biol.* **12**:1036–1039.

56. Michod, R. E., and B. R. Levin. 1988. *The Evolution of Sex: an Examination of Current Ideas*. Sinauer, Sunderland, MA.

57. Muller, H. J. 1964. The relation of recombination to mutational advance. *Mutat. Res.* **1**:2–9.

58. Muller, H. J. 1932. Some genetic aspects of sex. *Am. Nat.* **66**:118–138.

59. Otto, S. P., and T. Lenormand. 2002. Resolving the paradox of sex and recombination. *Nat. Rev. Genet.* **3**:252–261.

60. Peck, J. R. 1994. A ruby in the rubbish: beneficial mutations, deleterious mutations and the evolution of sex. *Genetics* **137**:597–606.

61. Rice, W. R., and A. K. Chippindale. 2001. Sexual recombination and the power of natural selection. *Science* **294**:555–559.

62. Schluter, D. 2000. *The Ecology of Adaptive Radiation*. Oxford University Press, Oxford, United Kingdom.

63. Shonn, M. A., R. McCarroll, and A. W. Murray. 2002. Spo13 protects meiotic cohesion at centromeres in meiosis I. *Genes Dev.* **16**:1659–1671.

64. Steele, D. F., M. E. Morris, and S. Jinks-Robertson. 1991. Allelic and ectopic interactions in recombination-defective yeast strains. *Genetics* **127**:53–60.

65. Turelli, M., N. H. Barton, and J. A. Coyne. 2001. Theory and speciation. *Trends Ecol. Evol.* **16**:330–343.

66. Wach, A., A. Brachat, C. Rebischung, S. Steiner, K. Pokorni, S. Heesen, and P. Philippsen. 1998. PCR-based gene targeting in *Saccharomyces cerevisiae*. *In* A. J. P. Brown and M. Tuite (ed.), *Methods in Microbiology*, *Yeast Gene Analysis*, vol. 26. Academic Press, London, United Kingdom.

67. Wang, H., S. Frackman, J. Kowalisyn, R. E. Esposito, and R. Elder. 1987. Developmental regulation of *SPO13*, a gene required for separation of homologous chromosomes at meiosis I. *Mol. Cell. Biol.* **7**:1425–1435.

68. Weismann, A. 1904. *The Evolution Theory*, vol. 2. Edward Arnold, London, United Kingdom.

69. Williams, G. C. 1966. *Adaptation and Natural Selection*. Princeton University Press, Princeton, NJ.

70. Williams, G. C. 1975. *Sex and Evolution*. Princeton University Press, Princeton, NJ.

71. Wolf, H. G., K. Wohrmann, and J. Tomiuk. 1987. Experimental evidence for the adaptive value of sexual reproduction. *Genetica* **72**:151–159.

72. Zeyl, C., and G. Bell. 1997. The advantage of sex in evolving yeast populations. *Nature* **388**:465–468.

73. Zhuchenko, A., A. B. Korol, and L. P. Kovtyukh. 1985. Change in cross-over frequency in *Drosophila* during selection for resistance to temperature fluctuations. *Genetica* **67**:73–78.

*Sex in Fungi: Molecular Determination
and Evolutionary Implications*
Edited by Joseph Heitman et al.
© 2007 ASM Press, Washington, D.C.

Clifford Zeyl

31

Ploidy and the Sexual Yeast Genome in Theory, Nature, and Experiment

Saccharomyces cerevisiae is a supermodel organism in the exploding field of genomics. We try to understand the molecular biology by which a genome is converted into phenotypes—the roles played by each gene, mechanisms of gene regulation, responses to stress, the developmental processes of gamete formation, mating, and sporulation, and so on. Behind this research often lies the implicit assumption that the genome has evolved to optimize yeast phenotypes. Evolutionary geneticists consider this perspective inaccurate for sexual organisms, in very interesting ways (19). Sex and the relative frequencies of outcrossing and inbreeding have pervasive effects on genome evolution, extending well beyond their direct fitness effects. Ploidy is another aspect of an organism's genome with potentially important but poorly understood effects. In an attempt to follow up on the previous chapter, I consider the effects of sex and ploidy on yeast genetics and genomics and emphasize experimental studies. Published studies of the natural history, population genetics, and ecology of yeast are few but valuable, so I also refer to such research on both *S. cerevisiae* and its sister species *S. paradoxus*.

THE YEAST MATING SYSTEM

The previous chapter (chapter 30, by M. R. Goddard) considered sex as an adaptive trait and examined experimental tests of hypotheses for how it could be adaptive.

Once established, sex can have numerous long-term effects on the genetics and genomics of a population, driven by its life cycle. Much of this chapter is focused on how the life cycle of *S. cerevisiae* has shaped its genome.

The mating system of *Saccharomyces cerevisiae* establishes the potential for extensive inbreeding. A haploid yeast individual is one of two mating types, *MATa* or *MATα* that mate spontaneously on contact with each other. When starved of nitrogen and restricted to carbon sources that must be respired (in the lab, typically acetate), diploid cells undergo meiosis and sporulation to produce tetrads of haploid spores, two of each mating type, enclosed in an ascus.

Because in many microbes the production of resistant or dormant stages is associated with sex, one hypothesis has been that sex evolved or is maintained as the mechanism for producing spores. From an evolutionary point of view this is an unsatisfying explanation, because more efficient competitors that can sporulate without sex should be only a regulatory mutation or two away. As with other traits, however, the ecological and demographic roles of sporulation in nature are unknown.

When the return of favorable conditions permits the germination of ascospores, the spores from a single meiosis can mate, leading to very high rates of inbreeding. Further inbreeding can result from the fact that most yeasts are homothallic, which means that mitosis

Clifford Zeyl, Department of Biology, Wake Forest University, P.O. Box 7325, Winston-Salem, NC, 27109.

and budding by a haploid cell yield daughter cells of the opposite mating type. The ensuing mother-daughter mating results in more inbreeding. Homothallism is inconvenient in the lab, because homothallic genotypes cannot be maintained as haploids, so in lab strains the *HO* gene responsible for mating-type switching is typically deleted (the molecular mechanism of this switching is itself of interest in connection with genome evolution, and is discussed below). But it is intriguing that most wild isolates are homothallic (for an example, see reference 59, but with the exception of possibly domesticated winery strains [85]) and therefore seem to be giving up most of the benefits that sex is hypothesized to offer. An interesting exception is strain S288c, one of the most widely used lab strains and the subject of the inaugural genome sequencing project (33). It was derived from a heterothallic strain isolated from a now-famous rotting fig in Mercedes, CA (72). This strain is also a notoriously poor sporulator, so it may not be typical of natural genotypes.

Until recently the conclusion that yeasts are highly inbred was based on the behavior of lab strains. Our knowledge of the population genetics of *S. cerevisiae* in nature has lagged far behind the intricate detail of its molecular genetics and genomics and the variety and precision of techniques for studying them. The study of wild *S. cerevisiae* strains has been impeded in part by uncertainty about whether or not they are from truly wild populations. *S. cerevisiae* has usually been found in vineyards or orchards, raising the suspicion that they are more or less domesticated and that their population structure is an artifact of human activity (89). *S. paradoxus*, a close but genetically distinct relative of *S. cerevisiae*, has emerged as an excellent substitute because it is also a good lab pet but is found worldwide, typically in association with red oak trees, in populations that appear untainted by domestication (105). The opportunity to apply the full range of genetic and genomic techniques to wild yeast is now luring evolutionary and population geneticists, and those with a more ecological interest, to track down and study natural populations (72).

SEX IN NATURAL YEAST POPULATIONS: INBREEDING AND OUTCROSSING

As outlined above, one would predict high levels of inbreeding from the yeast life cycle, but it is unknown how accurately lab observations characterize the natural populations in which the life cycle and sexual systems actually undergo selection. In a pioneering study that begins to answer this question, Johnson et al. (59) estimated that outcrossing occurs only 0.011 times as frequently as inbreeding among *S. paradoxus* genotypes in Silwood Park, England. Another recent study of natural isolates (1) also suggests frequent mother-daughter matings in homothallic populations, with levels of homozygosity and linkage disequilibrium that indicate infrequent recombination with outcrossing. The effect of outcrossing depends on the nucleotide diversity available for recombination, since the less sequence variation there is, the fewer new genotypes can be generated by recombination. Johnson et al. (59) found a fairly typical nucleotide diversity at synonymous sites of 0.3%. They concluded that mitotic reproduction (cloning), inbreeding, and outcrossing all have significant effects on population structure.

The signature of recombination can be seen in the evolutionary ancestries of different segments of the genome. In an asexual population, the evolutionary histories of all the genes are the same, and phylogenetic trees drawn from different genes will usually be the same. But recombination combines alleles from different lineages, producing chromosomes that have different ancestries in different regions. Even the rare recombination inferred by Johnson et al. (59) scrambled the phylogenies of their 28 isolates at different loci.

Ruderfer et al. (104) used this genome-shuffling effect of recombination over time to estimate the number of outcrossing recombination events among three strains of *S. cerevisiae*. They used genealogies inferred for over 25,000 single-nucleotide polymorphisms among the three complete genome sequences (using an *S. paradoxus* genome as an outgroup to identify ancestral alleles). The four genomes showed an estimated 314 crossovers since common ancestry, and a median size of 2 kb of recombined fragments. Using the rough and speculative approximation of one to eight generations per day, this amounts to one outcrossing event per 17 to 137 years, or roughly once per 50,000 cell divisions. Available evidence from both the laboratory and the woods, then, indicates that *S. cerevisiae* is highly inbred.

As the following discussion will show, a balance of frequent inbreeding with occasional outcrossing may actually be an efficient mechanism for adaptation. Explaining the origin and maintenance of sex for *S. cerevisiae* in particular does not face the challenge that it does for most eukaryotes, because yeasts are isogamous. This means that they do not pay the familiar twofold cost of sex, because the mating types do not produce specialized gametes. In species with large, costly female gametes and numerous, small male gametes, producing female clones should yield twice as many offspring on average, if female gametes and clones are equally viable and fecund. *S. cerevisiae* does not pay the cost of producing males; the mating types are ecologically and genetically equivalent.

The mechanisms of adaptation in some evolution experiments suggest an additional benefit to regular inbreeding with occasional outcrossing. Adaptation to organic phosphate limitation in two independent chemostat experiments (2, 51) occurred by chromosomal rearrangements. Later experiments with glucose-limited chemostats also consistently turned up chromosomal rearrangements, which were mapped using evolved and ancestral genomes competing for hybridization to a microarray of open reading frames (ORFs) (array-based comparative genomic hybridization [35]). The chromosomal mutations had caused a variety of changes to hexose transport and metabolism, and some rearrangements had evolved independently in replicate populations. The experimental populations were asexual, but outcrossing would probably have been an impediment to adaptation by chromosomal rearrangement, because in their first few generations such mutations would be heterozygous and would likely cause deleterious changes in gene dosage. Generations of inbreeding could allow an adaptive chromosomal mutation to reach high enough frequencies that outcrossing would then produce homozygotes. Chromosomal rearrangements have also played a prominent role in the evolution of wine-making strains of *S. cerevisiae* (100).

SEX AND EPISTASIS

Epistasis, while inconvenient, is increasingly acknowledged as an unavoidable factor in population and quantitative genetics. Here, I am using epistasis to refer to specific interactions between particular alleles (or nonlinear phenotypic effects of segregating alleles, also termed "statistical epistasis," which may be different from the physiological epistasis that describes molecular interactions between invariant gene products [16]). Synergistic epistasis among deleterious mutations can hypothetically set up an advantage for sex (see chapter 30). But epistasis between adaptive mutations can also play a role in determining the fitness effects of recombination. A potentially significant cost of sex is the breakup of epistatic, coadapted combinations of alleles. Inbreeding avoids this cost, but exclusive inbreeding would prevent the assembly of optimal allele combinations in the first place. Periodic outcrossing may allow such optimal combinations to be generated and then tested by generations of mitosis and inbreeding. If they increase in frequency due to higher fitness, they are less likely to be broken down by subsequent outcrossing.

On the other hand, if epistasis is weak and negative, combinations of adaptive alleles will become less frequent than predicted from random association (negative linkage disequilibrium), and recombination will be favored because it will increase the frequency of high-fitness genotypes combining adaptive alleles (94).

An ecological view of the positive epistasis described above would be specialization, whereby different lineages assemble distinct combinations of adaptive mutations in the course of adaptation to different niches within an environment. Sex and recombination may therefore impede specialization. To my knowledge this possible impedance has not been tested experimentally, although there has been theoretical work on similar ideas in the context of phenotypic diversification (4) and of sympatric speciation through the evolution of specialization and assortative mating (30). In asexual populations evolving in the artificial simplicity of experimental environments, the traditional expectation would be a series of selective sweeps, in which adaptive mutations would occur and sweep to fixation in the population, eliminating all genetic variation. Instead, genetic and ecological polymorphism has evolved in several microbial evolution experiments (for examples, see references 53, 101, and 103). These examples arose in bacterial populations lacking mechanisms for DNA exchange, but there is every reason to think that they also evolve in long-term asexual yeast experiments. Results from unpublished yeast experiments in my lab include negative frequency-dependent fitness and colony size polymorphisms in at least three populations (S. Campbell, unpublished data; M. Baliga, unpublished data). An interesting possibility is that these polymorphisms result from specialization of different lineages for different aspects of the culture environment, and that recombination between these lineages would disrupt synergistically (positively) epistatic allele combinations and produce genotypes that would not be well adapted to either niche (see chapter 30). Sex could prove maladaptive; alternatively, specialization in sexual experimental yeast populations may provide an opportunity to observe the evolution of assortative mating.

Epistasis is usually studied as an interaction among nuclear loci. But nuclear loci also interact epistatically with organelle loci, and yeasts are particularly well suited to the study of these interactions because their ability to survive without respiration (although rather debilitated) makes it possible to combine a variety of nuclear and mitochondrial genotypes. Mitochondrial inheritance is biparental in the yeast sexual cycle. Parental mitochondrial chromosomes recombine following mating, and the recombined chromosomes then segregate during mitosis (43). Outcrossing therefore breaks up nuclear-mitochondrial associations, impeding the evolution of epistasis. In asexual populations, on the other hand, lineages of nuclear and mitochondrial genotypes remain paired, permitting epistatic allele combinations to evolve. This prediction can be tested with fitness comparisons of the four possible pairwise combinations of ancestral and evolved nuclear and mitochondrial genomes from an asexual yeast evolution experiment (119). To construct mispairings (for example, an ancestral nuclear genome with an evolved mitochondrial genome) the *MIP1* gene is deleted from the nuclear donor. This nuclear gene encodes the DNA polymerase that replicates mtDNA, so its deletion produces a genotype lacking mtDNA. Using the appropriate crosses and fitness assays, significant positive epistasis was found in an experimentally evolved genotype. The evolved mitochondrial genotype made no fitness contribution when paired with the ancestral nuclear genome but significantly increased fitness in combination with the nuclear genome with which it evolved (Fig. 31.1). Disrupting coevolved organelle pairings between two unrelated yeast strains had a similar result. Inbreeding would have the same effect as mitotic cloning on mitochondrial inheritance and the opportunity for nuclear-mitochondrial coadaptation, so this form of epistasis may be common in natural yeast populations given the evidence discussed above for frequent inbreeding.

SEX AND PARASITIC DNA

One of the most interesting consequences of sex is the opportunity it creates for parasitic genetic elements to spread through a population. These are also referred to as selfish DNA. All DNA is selfish in the sense that it is selected for its ability to maximize its rate of transmission to the next generation, but parasitic DNA is different in that its evolutionary success results from non-Mendelian transmission and despite the absence of any fitness benefit, or even a fitness cost. Invasions by several types of parasitic DNA have played major roles in shaping the yeast genome.

Genetic parasitism is made possible by outcrossing, as illustrated in Fig. 31.2. In an asexual population, any fitness cost causes the extinction of parasitic DNA (disregarding genetic drift), due to selection against any contaminated lineage. Sex would not change that outcome for an element that behaved in the usual Mendelian way, with one-half of the offspring of heterozygotes carrying the element. But if a DNA sequence somehow subverts meiosis so that all offspring of a carrier inherit it, then the infection of new lineages through random mating can override selection against infected

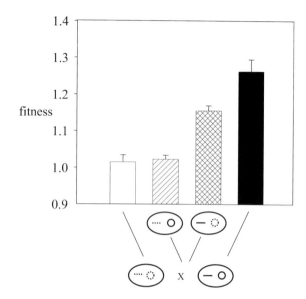

Figure 31.1 Nuclear-mitochondrial epistasis for fitness. Dashed lines and circles represent ancestral nuclear and mitochondrial genomes, respectively; solid lines and circles represent evolved genomes.

genotypes, and the DNA sequence can spread as a genetic parasite.

In the classic demonstration of this process, the yeast 2μm plasmid played the role of genetic parasite (40). The 2μm plasmid is a circular 6-kb element that is widespread among lab and industrial yeast strains despite reducing fitness by about 1% and having no other detectable phenotype (40, 110). Its mechanism of cheating in meiosis is that it replicates independently of the chromosomes, maintaining a copy number of about 50 per cell. This results in the transmission of the plasmid to all four meiotic progeny after a cross between haploids with and without it. In a control experiment, asexual populations were founded by isogenic genotypes with and without the plasmid at equal frequencies, and as expected the frequency of cells with the plasmid declined to about 0.35 over 140 mitotic generations (40). But when introduced at a frequency of 0.1, the plasmid spread to about 40% of the cells after four sexual cycles. A parasitic element would not be expected to invade an inbreeding sexual population, since new lineages could not be infected, and this prediction was supported by the failure of plasmids introduced at low frequencies to spread in inbreeding populations.

The 1% fitness cost of the 2μm plasmid is small, but not irrelevant, being probably similar to that of many of the slightly deleterious spontaneous mutations that continually arise. But what is the greatest fitness cost that parasitic DNA could impose on its host and still spread?

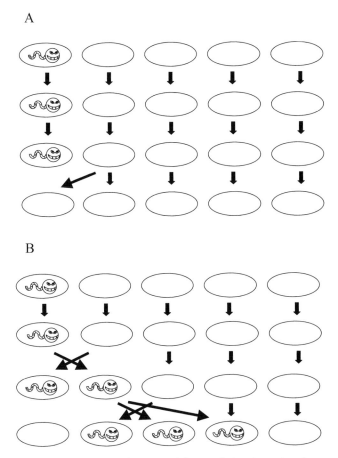

Figure 31.2 Schematic depiction of the population dynamics of parasitic DNA. (A) In asexual populations, selection opposes the increase in frequency of parasitic DNA. (B) In sexual populations, cycles of mating and meiosis enable parasitic DNA to spread despite selection, by infecting new lineages.

This depends on its efficiency of transmission through a sexual cycle. The condition for a genetic parasite to increase in frequency from an initially very low frequency is $(1 - s)(\mathbf{a} + \alpha) > 1$, where s is the fitness cost and $\mathbf{a}$ and α are the rates of transmission by infected individuals of the two mating types (fathers and mothers, in the more familiar world of animals). In the extreme case, a parasite that is transmitted to all offspring by every infected parent can reduce fitness by up to 50% and still spread (Fig. 31.3). The permissible cost increases as the element spreads (55) because uninfected mates become increasingly rare. For all its hypothesized benefits and its ubiquity, outcrossing sex therefore carries a risk of substantial declines in population mean fitness by allowing genetic parasites to spread.

Because mitochondria are uniparentally inherited in animals, the sum of transmission rates by males and fe-males is 1, preventing invasion by genetic parasites (Fig. 31.3). Animal mitochondrial genomes are concise and, as a rule, free of parasitic sequences. Holding off invasion by parasitic elements has been hypothesized to be the selective advantage responsible for the evolution of anisogamy and uniparental organelle inheritance from isogamous systems like that of yeast (18, 19). Supporting this hypothesis, the biparentally inherited mitochondrial chromosomes of yeast have been repeatedly infected. They are much larger than mammalian mitochondrial genomes (80 kb as opposed to about 17 kb in mammals), and much of this extra DNA appears parasitic.

Before conventional introns were found in eukaryotic nuclear genes, the yeast mitochondrial ω element was discovered as a mobile intron. It has no detectable phenotypic effect, probably because it is excised from transcripts by a mechanism that is more autonomous than the passive, spliceosome-mediated excision of normal introns (22). Mobile introns have subsequently been found in many genomes, particularly those of algal and fungal organelles, which in many cases are biparentally inherited. Mobile introns are classified as group I or group II introns. The yeast mitochondrial genome contains 11 group I introns and 3 group II introns. Paradoxically, these introns contain their own ORFs. Group I ORFs encode endonucleases, which provide a mechanism for invading biparentally inherited organelle genomes (32). The endonucleases very specifically cut alleles lacking the intron at the DNA sequences homologous to those

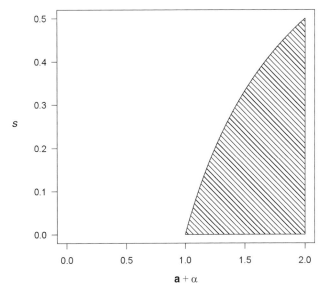

Figure 31.3 Tolerable fitness costs for genetic parasites (s) as a function of their rates of transmission by both mating types or genders ($\mathbf{a} + \alpha$)

where the parent copy resides. DNA repair synthesis then copies the mobile intron into the cut site, so that in crosses between genotypes with and without the mobile intron, all the mitochondrial chromosomes segregating from the cross carry the intron. Alleles already carrying the intron are spared further endonuclease activity by the presence of the intron, which interrupts the recognition sequence. With mechanisms for both invasion and self-censoring, mobile introns are therefore well adapted to taking advantage of the biparental inheritance of their host chromosomes while minimizing if not eliminating selection against them.

Group II introns are also self-splicing, although they do so more efficiently with the help of nuclear proteins. The mechanism of group II intron splicing is so similar to that of nuclear introns that they are hypothesized to be the ancestors of the latter (21). They contain ORFs that encode reverse transcriptases, enzymes that synthesize DNA by using RNA as a template. These reverse transcriptases specifically copy intron RNA, which can then be integrated into target sites in the mitochondrial chromosome (64, 84). At least in some cases, the reverse transcriptases are less site specific than the homing endonucleases of group I introns, and occasionally a group II intron will transpose into a new site where permitted by sequence similarity to its usual location (86). This transposition can result in a cost to the host: where multiple copies of an intron are present in the same genome, recombination between copies can cause deletions. Group II introns therefore use a different mechanism from group I introns, but with a similar result: unoccupied sites transmitted by one parent of a cross are invaded by an intron from the other parent, and a greater than Mendelian fraction of offspring inherit the intron.

To pursue the analogy between organismal and DNA parasites, ω can be seen as highly specialized, apparently able to infect only one target sequence in yeast. Retrotransposons, on the other hand, are more generalist. These nuclear elements can copy and insert themselves into numerous chromosomal sites, with a general preference for the upstream regions and 5′ ends of genes, particularly tRNA genes (23, 65, 66). Five families of retrotransposons have been recognized, Ty1 through Ty5 (reviewed in references 13, 14, and 65), which, along with their degenerating remains, comprise about 3% of the yeast genome (33, 65). Retrotransposons vary in copy number and location among genotypes but are ubiquitous among both lab and wild strains (for examples, see references 37 and 98). They are homologous to animal retroviruses and have the same life cycle except that unlike retroviruses they are usually unable to leave their host cells and infect new ones by horizontal transfer. There is evidence that Ty elements originally infected yeast by infrequent horizontal transfer: Ty1 and Ty2 are more closely related to one class of plant and animal retrotransposons than they are to Ty3 and Ty4 (14, 65). Ty elements are bounded at each end by long terminal repeats (LTRs), sequences of a few hundred base pairs that are essential structures in the retrotransposition process. In between the LTRs are several kilobases that encode proteins responsible for reverse transcription of Ty RNAs and their insertion into new chromosomal locations. Different transcript molecules are used for translation of Ty proteins and as templates for reverse transcription, and Ty transcripts are strikingly stable and therefore abundant among cellular RNAs.

Retrotransposition is the mechanism by which Ty elements achieve their biased transmission and infective spread (there is another type of nuclear mobile element, transposons, which use a different, DNA-mediated mechanism, but they are not found in yeast genomes). A solitary Ty element will be inherited by the Mendelian 50% of progeny in a cross with an uninfected genome. By multiplying its copy number, retrotransposition increases the transmission rates of a Ty element, again compensating for its fitness costs. Those fitness costs arise from the mutagenic effect of its activity. Ty insertion into promoters or ORFs can alter gene expression or protein structure, respectively. These effects are further discussed below.

The view of retrotransposons as parasitic DNA was supported by an experiment analogous to the 2μm plasmid experiment (117). Ty3 elements were introduced at a frequency of 1% into populations of a genotype lacking the element (10). The Ty3 elements were inducible by galactose (52, 67), permitting experimental induction or repression of transposition using media containing galactose or glucose, respectively. The retrotransposon was introduced into sexual and asexual populations grown in both glucose and galactose. Repressed Ty3 elements were never seen again in asexual populations, although they did become abundant in sexual populations, which is one of a few unexplained complications of the results. The main result was that Ty3 spread to frequencies over 0.8 within eight sexual cycles in most sexual populations where it was induced. This spread was probably artificially fast, because the galactose-driven retrotransposition rate would have been unnaturally high, but it demonstrates the principle within a practical time frame.

Although induced Ty3 elements remained at undetectably low frequencies in most asexual populations, there were two striking exceptions in which particular insertions showed major increases in frequency (117).

In both cases, genotypes carrying those insertions had higher competitive fitness than Ty3-free genotypes from the same population. Although the possibility that the fitness increases were unrelated to Ty3 activity cannot be ruled out, those insertions may have caused adaptive mutations.

The possibility that Ty3 element insertion can increase fitness introduces the second effect of retrotransposition: mutagenesis. The hypothesis that transposable elements are constructive agents of adaptive change has sometimes been seen as an alternative to thinking of them as parasites (38, 106). There is no shortage of examples in which transposable elements have caused adaptive mutations (although I am unaware of any such examples for the mobile introns of mitochondrial genomes). The genomic analyses of experimental evolution in glucose-limited chemostats (35), described above in the context of inbreeding and chromosomal rearrangements, showed numerous changes in gene expression. Regulatory mutations had clearly played an important role in adaptation (36). In at least some cases, the regulatory mutations were associated with the chromosomal rearrangements, and the breakpoints of all these rearrangements were at Ty sequences or solo copies of their LTRs (36). When multiple copies of a Ty element are present at nonhomologous loci, recombination between them can cause chromosomal mutations such as those found by Dunham et al. (35). (Recombination also occurs within copies between their LTRs, littering the genome with solo copies of the LTRs). Insertions that inactivate particular genes can also be adaptive, although this may be less typical in natural settings than in experimental lab conditions where many yeast genes are likely to be unnecessary or even unwanted. Blanc and Adams (12) tracked down two adaptive mutations where new copies of a single genomic Ty1 element had been inserted in an earlier evolution experiment. One insertion was upstream of FAR3, which is involved in the cell cycle arrest that occurs in response to mating pheromones. The insertion silenced FAR3, so this gene apparently is a liability in asexual chemostat culture. The effect of the second insertion was obscure, with no apparent effects on gene expression.

As with sex itself, it is important to distinguish the origin and initial spread of a mobile genetic element from its subsequent evolutionary effects. It is unlikely that Ty elements are widespread due to their mutagenic effects, because most of the mutations they cause reduce fitness. Wilke et al. (113) found that Ty1 insertions reduced mean fitness but that some insertion mutants increased greatly in frequency, indicating that Ty1 insertions also increased variance in fitness and produced a few adaptive mutations. Reducing the average and increasing the variation in fitness are typical of mutagens, although the distributions of fitness effects of different mutagens probably vary. If only a small fraction of Ty-induced mutations are adaptive, then the element would have to be well established in a population before it would begin generating adaptive variation.

COEVOLUTION AND DOMESTICATION OF PARASITIC DNA

There is actually some reason to expect that the spectrum of mutations induced by Ty activity may include a greater frequency of beneficial mutations than that caused by mutagens that are not genetically transmitted, such as X rays or chemical mutagens. Parasitic DNA coevolves with its host genome, and some of its behavior can be predicted and understood by analogy with organismal parasites. Whatever the mechanism by which a parasitic element spreads, it can be more successful by reducing the fitness cost it imposes.

Ty elements show some signs of having adapted to the yeast genome, which is a challenging environment for mobile elements because it has relatively little noncoding DNA. The strong preference of Ty elements for regions upstream of coding sequences and for tRNA genes (which are typically present in multiple, partly redundant copies) may have evolved as a mechanism for avoiding ORFs in areas where insertions are more likely to be maladaptive to the host (19). Insertions in upstream regions may also be more likely to have an adaptive effect on gene expression. Here, however, there is a complication that besets all studies of mutations and their fitness effects: the genotypes and fitness effects that we observe have been filtered from the complete set of mutations by selection. It may be that Ty insertion is random, but selection eliminates insertions into ORFs. No cases of dominant mutations caused by Ty1 insertion are known. Working with the assumption that all insertions have recessive effects, Ji et al. (58), confirmed that the preferential insertion of Ty1 copies near tRNA genes (and LTRs of existing Ty insertions) precedes selection.

Mobile elements whose evolutionary success depends on sexual transmission would be expected to restrict their activity as much as possible to host sexuality. The classic example of this is the P element of *Drosophila melanogaster*, which is active only in the germline of its host. This strategy is unavailable to Ty elements in their unicellular host, and transposition during cycles of mitotic reproduction could impede their spread by fruitlessly reducing host fitness. Instead, Ty3 elements are induced to transpose in haploid cells by mating pheromones (66), thereby restricting their activity to times when mating and the opportunity to spread to more chromosomes are imminent.

However, if there are indications that the host may be in trouble, then increased transposition rates may be advantageous regardless of the immediate prospects for sex, in the hope that the host may mate before expiring. Some organismal parasites do this, monitoring host condition and diverting more resources to producing dispersal stages when host condition is deteriorating (for examples, see references 17, 61, and 108). Similarly, some agents of DNA damage, such as UV irradiation, increase transposition rates of Ty elements (by as much as 17 times [15]).

The point of this section (perhaps digression) on parasitic DNA is that, given the opportunity by occasional horizontal transmission, parasitic DNA can invade sexual yeast populations, and then become an important agent of variation and change. But the dependence of parasitic DNA on sexual transmission and its coevolution with the genome of its host extend all the way back to the evolution of sexuality itself. There is both theoretical and molecular genetic support for the idea that sex originated as a strategy encoded by parasitic DNA for its own spread.

From a theoretical point of view, their reliance on sex suggests that when possible, parasitic elements could benefit by encoding it themselves, converting asexual hosts to sexual ones (56, 60, 102). The yeast mating system appears to have evolved by the domestication of genetic parasites. To review briefly Goddard's description of the genetics of mating in *S. cerevisiae* (chapter 30), mating compatibility is determined by two alternative sequences, *MATa* and *MATα*, each encoding proteins that regulate the genes whose activities are specific to one or the other mating type. Inactive copies, called *HMLα* and *HMRa*, flank a chromosomal site where one of the alleles is present and expressed. In homothallic strains, mating-type switching between parent and daughter cells is caused by a form of transposition by the alternate *MAT* sequence into the active site. This transposition is induced by an endonuclease encoded by the *HO* gene, located on a separate chromosome. The endonuclease is derived from another fascinating group of parasitic sequences, the inteins. Inteins are protein analogues of mobile introns: they are translated into protein and then excised from the protein to prevent any ill effects. This permits them to insert themselves into ORFs as long as they are in the correct reading frame. Phylogenetic analysis of intein and HO sequences indicates that *HO* entered the yeast genome (presumably by horizontal transfer), as an intein, approximately 135 million years ago (62). HO no longer inserts itself into host genes nor excises it-

self from their translated proteins, but instead induces mating-type switching in homothallic yeast strains. The endonuclease it encodes cuts within the active site where mating type is determined, resulting in random transposition of one of the *MAT* sequences into that active site (54). In this domesticated state, HO is advantageous to its host if homothallism is an advantage, but many of the hypothesized advantages of sex are discarded in a homothallic system because of the resulting inbreeding. On the other hand, mating-type switching increases the pool of potential mates because in a heterothallic population one mating type may become rare or extinct, particularly if local populations are repeatedly founded by a small number of cells. Homothallism may therefore have increased the efficiency with which the ancestral HO was sexually transmitted (62). The homothallic system of *S. cerevisiae* appears to have been derived from an ancestral heterothallic system (20), perhaps with the invasion and subsequent evolution of HO.

The *MATa* and *MATα* sequences themselves may also be the descendants of horizontally transmitted parasitic DNA. Their sequences are highly divergent, to the point where any homology is completely obscured, and they are termed idiomorphs instead of alleles (83). This divergence could be maintained by selection against recombination between *MAT* alleles, which would disrupt their regulatory roles and presumably cause sterility. Less straightforward is the fact that homologues to both *MATa* and *MATα* are found in other fungi, some fairly distantly related, so that the *MATa* allele of *S. cerevisiae* is more similar to *MATa* alleles in other fungi than to the *MATα* allele of *S. cerevisiae* (20). This pattern suggests horizontal transfer, the repeated movement of each allele from one species to another. Horizontal transmission may seem less plausible than the conventional evolution of one *MAT* allele from the other, perhaps following a duplication (39). But from an evolutionary perspective, it is even more unlikely that homologous DNA sequences would repeatedly diverge from each other within a genome while repeatedly and independently converging on similar pairs of sequences in different lineages. Incoming alleles would somehow have to be integrated into the regulatory circuitry responsible for mating type, but these circuits can change even as their phenotypic output remains the same. As an example, the same two mating types are seen in *Candida albicans* as in *S. cerevisiae*, similarly determined by *MATa* and *MATα* sequences, but by different mechanisms. In *S. cerevisiae*, **a** is the default mating type, and one of the two proteins encoded by *MATα* represses the **a** genes and induces α genes (109). In *C. albicans*, a MATa2 protein, absent from *S. cerevisiae*, induces **a** gene expression while

MATα1 induces α genes. I suggest that similar regulatory changes could have accommodated horizontally received mating-type alleles.

Even in their current domesticated state, the *MAT* alleles act like mobile introns. When a double-stranded cut is made in the active site by the Ho endonuclease, *MATa* or *MATα* is copied into that site by gene conversion. In the past they may have helped to orchestrate mating as a strategy for sexual transmission (6).

Horizontal transmission, once thought to be a rare and unimportant event in evolutionary genetics, seems to have played an important role in the evolution of the yeast mating system by the domestication of initially parasitic sequences. For mobile introns, horizontal transmission has even been more frequent than transposition (45). Is there anything about parasitic DNA that would make it particularly likely to be horizontally transmitted? The mechanism of horizontal transmission is unknown, but the one current intein in the yeast genome, *VDE*, offers a possible explanation. A phylogenetic analysis of the distributions of *VDE* and its recognition sequences among 24 yeast species shows that horizontal transmission has been frequent. Its recognition sequence is strikingly well conserved compared to other yeast sequences, which has been interpreted as an adaptation for integration into homologous sites in the genomes of new hosts (71).

As implied by this recurrent theme, horizontal transmission may be essential to the long-term evolutionary viability of parasitic DNA. Once established, the traits that enabled a parasitic element to spread through outcrossing become irrelevant, as they simply reinfect lineages already carrying it. If the element makes no contribution to host fitness (as with, for example, the mitochondrial ω element), then disabling mutations are neutral and accumulate over time. In the case of ω, mutations would eventually inactivate the endonuclease required by the element to insert itself into unoccupied alleles. Eukaryote genomes are riddled with such decaying elements, and unless they can find a new population to invade, they seem doomed to extinction. Goddard and Burt (45) further explored the third possibility of horizontal transmission to another species. Their phylogenetic analysis of the occurrence of the ω element among saccharomycete yeasts showed significant differences between host and element phylogenies and indicated that the element has been transferred between closely related Saccharomycetes approximately every 2 million years. Goddard and Burt inferred a cycle of infection by horizontal transfer, decay due to mutation, and complete loss of the element, with the element staving off extinction only through horizontal transfer.

ANTAGONISM BETWEEN SEXUAL AND NATURAL SELECTION

The phenotypic outcome of tension between natural, or vegetative selection and sexual selection is familiar in the forms of extravagant plumage in birds and noisy violence in sea lion rookeries. This tension takes a more subtle form in microbes like *S. cerevisiae*, for which sex is not reproduction and in fact interrupts reproduction. Although the complete sexual cycle produces four haploid cells from two, continued mitotic reproduction for the time taken to complete this cycle would yield more than four offspring. Reproduction occurs by mitosis and budding, and alleles that improve those functions are favored by vegetative selection. Sex occurs in two stages: mating when compatible haploids meet, and meiosis when conditions prohibit rapid reproduction by diploids. Up until now I have implicitly applied the term "fitness" to the entire life cycle, but this section requires a distinction between vegetative selection/fitness, which affect only true reproduction, and sexual selection/fitness. For yeast, sexual fitness involves mating and its components, such as pheromone production, reception, and response, and the coupled processes of meiosis and sporulation.

In birds, plumage is shaped by vegetative selection for shorter tails and by sexual selection through a female preference for longer tails. In facultatively sexual microbes, this conflict can take one of two more generic forms. One is the selection of pleiotropic alleles, which enhance one life history component while impairing the other, resulting in a negative correlation between them. A second is the accumulation of mutations that have no effect on one component and degrade the other. In the case of yeast, mitosis and reproduction are necessary reproductive stages of the life cycle, but the genes encoding sex could be eroded by mutation if there were many asexual generations between sexual cycles. This erosion is different from the more general mutation accumulation predicted by some hypotheses of sex, in that it involves only mutations that do *not* reduce vegetative fitness. In facultatively sexual organisms, sexual fitness can decline, regardless of population size, as a result of mutations that impose no vegetative cost.

An illustration of the erosion of sex by mutation accumulation was given for the fungus *Cryptococcus neoformans* by Xu (116; chapter 28) who estimated the per-generation, per-genome rates of mutations that degrade mating and meiotic proficiency as ≥ 0.0172 and ≥ 0.0036, respectively. These estimates came from a type of study known as a mutation accumulation experiment, in which the aim is to obtain as unbiased a sample of random mutations as possible by minimizing selection, in replicate experimental populations. Selection is minimized by minimizing effective population size. In microbes, this minimization can be accomplished by successive transfers of single colonies, each representing a population bottlenecked to a single cell. In Xu's experiment, over approximately 600 mitotic generations, average mating ability decreased by two-thirds relative to the ancestor, implying that natural populations must experience some consistent selective pressure to maintain their mating proficiency. A comparable result was obtained in a mutation-accumulation experiment in which 50 populations were propagated by serial transfer of single colonies, also for about 600 generations (118). Mutations with detectable effects on vegetative fitness were very scarce, but only 12 of the 50 lines retained the ability to sporulate.

A more direct and interesting conflict between vegetative and sexual selection can arise through the selection of pleiotropic alleles, those with opposite effects on vegetative and sexual fitness. This conflict is a specific case of the more general observation that life history components are negatively correlated in populations that are well adapted to their environments and are thus more or less uniformly proficient at acquiring its resources (8). Alleles that increase both vegetative and sexual fitness will be fixed, alleles that reduce both will be eliminated, and the remaining genetic variation will increase one at the expense of the other, producing a negative correlation. Assuming that pleiotropic alleles exist, selection acting on genetic variation in facultatively sexual microbes should change the two traits in opposite directions, and this has been observed in *Chlamydomonas reinhardtii*. In this unicellular alga, the environmental cues for the sexual cycle are the inverse of those for yeast: mating occurs only upon nitrogen starvation and produces dormant, resistant diploid spores which undergo meiosis and return to mitotic reproduction when favorable conditions return. Mating can be prevented by a continual supply of nitrogen, and chloroform can be used to select for spores and kill unmated, haploid cells. Mating ability almost disappeared from long-term asexually selected populations, in which continual mitotic reproduction was enforced (27), and under selection for sexually produced spores the mating system itself evolved from heterothallism to a libertine homothallism (9). Analogous declines in sexual fitness occurred in yeast populations subjected to 2,000 generations of asexual selection, and in a separate experiment to a previously pathogenic strain propagated asexually in minimal medium and in mouse brains (49). It was unclear, however, whether this was due to antagonistic pleiotropy. After 10 cycles of sexual selection on populations derived from a genotype that had already been propagated asexually for 2,000 generations, sexual fit-

ness had increased and vegetative fitness had declined (120). On the other hand, high vegetative fitness and low sexual fitness did not cosegregate in crosses between evolved asexual and ancestral genotypes, as would be expected if the same alleles encoded both traits.

Quantitative analysis of those crosses indicated that alleles reducing mating ability had been selected at very few loci, perhaps two, during the 2,000 mitotic generations. However, genomic studies implicate hundreds of genes in the process of meiosis and sporulation. The yeast genome-sequencing project was followed by another large-scale undertaking, the systematic construction of a set of isogenic mutants, each with one of the approximately 5,900 ORFs surgically deleted (114; see http://www-sequence.stanford.edu/group/yeast_deletion _project/deletions3.html). Deletions were performed by transformations using PCR products with ~45-bp 5' and 3' ends, homologous to each end of the target ORF, flanking the Kan MX4 gene conferring resistance to the antibiotic kanamycin (G418). Transformants in which this sequence has replaced the target ORF by homologous recombination were selected using kanamycin. The entire collection, or a subset, can now be pooled and screened for some deficiency (or possibly proficiency) of interest, and all the genes affecting that trait can be identified by their molecular barcodes. The barcodes are 20-nucleotide sequences that do not exist anywhere else in the yeast genome, and the performance of a mutant carrying a barcode can be quantified relative to those of the other mutants.

Using this approach, Deutschbauer et al. (29) estimated that 361 genes were involved in sporulation; 102 deletions actually increased sporulation ability, while 261 deletions reduced it. An additional 158 mutants were impaired in growth immediately following spore germination, independent of vegetative growth in general. But the pleiotropy documented in these genes was positive and thus boring: 92 of the 261 sporulation-deficient mutants also paid fitness costs ≥5% and would be quickly selected out of any yeast population regardless of its degree of sexuality. In a 14-gene subset of the data (http://www.pnas.org/cgi/content/full/202604399/ DC1), sporulation ability and growth rate are positively correlated ($r = 0.56$, $P = 0.037$). The effects on both vegetative and sexual fitness result from the involvement of those genes in general processes such as carbon use and responses to starvation. Deletions with negative pleiotropy seem not to have been pursued.

The interaction between sexual and vegetative selection has an interesting twist in yeast. In the yeast mutation accumulation experiment referred to above (118), most of the lines that were incapable of sporulation also formed the small colonies typical of the *petite* phenotype, which is caused by an inability to respire and almost always results from large deletions of the mitochondrial chromosome. Sporulation requires respiration, and when this requirement is relaxed, *petites* arise at the relatively high frequency of about 1%. So in this case, sexual selection imposes strong selection for mitochondria that can respire. When sexual selection is absent and vegetative selection is weak, *petite* mitochondria actually increase in frequency in populations because within cells they have a replication or transmission advantage over wild types (107). Vegetative selection usually overrides this within-cell selection to maintain the ability to respire.

Sporulation proficiency varies widely among yeast isolates. At least a few lab strains are unenthusiastic about sporulating (my personal observations), but pathogenic strains isolated from immunocompromised human patients (82) and more typical saprophytic strains of both *S. cerevisiae* and *S. paradoxus* are efficient sporulators (59, 105). This observation and the ease with which sexual functions deteriorate in the lab suggest that natural yeast populations are under strong selection to remain capable of sex. Based on the information discussed here, it is unclear whether such sexual selection typically constrains vegetative adaptation.

To summarize this chapter so far: apart from the twofold cost of meiosis, which does not apply to yeast, there are substantial costs of sex. These include the breakup of positively epistatic combinations of adaptive alleles, an impediment to adaptive chromosomal rearrangements, the spread of parasitic DNA, and possible antagonism between sexual and vegetative selection. Although the machinery and resources required for sex may seem wasted when they are so often used for inbreeding, the yeast combination of frequent inbreeding and rare outcrossing may be a good balance between these costs of sex and its benefits. It has been suggested that, genetically, occasional sex may provide most of the benefits of obligate sex, with lower costs (48). High rates of inbreeding reduce the susceptibility of *S. cerevisiae* to new genetic parasites and permit adaptation by chromosome rearrangements. The resulting homozygosity also may allow the recessive harmful mutations that continually threaten to erode adaptive traits to be more efficiently removed by selection, by logic that is similar to a hypothesis discussed below regarding ploidy.

THE EVOLUTION OF PLOIDY

Ploidy is one of the most fundamental features of genome structure, but as with sex, there is still no widely accepted explanation for the observed variation in ploidy. Since sex is an alternation between haploid and diploid states, sex and ploidy are often intertwined. Studies of yeasts can be illuminating here too, because yeasts can be propagated with or without sex and as haploids or as diploids. This is especially valuable when attempting to explain the occurrence and relative prominence of haploidy and diploidy among eukaryotes. Diplophase predominates in the life cycles of plants and animals, but the variety of life cycles among microbes is a reminder that a satisfying theory of ploidy must provide not just an advantage for diploidy in plants and animals but also an explanation of the taxa in which haploidy is more prominent, as Bell (7) has pointed out. His review indicated that haploidy is primitive among eukaryotes and that diploidy is associated with larger size and greater complexity.

For both sex and ploidy, intuitively appealing explanations have turned out to be questionable both theoretically and experimentally, leaving this chapter short on firm conclusions. Recent and prominent contenders are discussed below. There are hypotheses that propose immediate physiological and ecological benefits to one ploidy or the other, advantages to heterozygosity rather than diploidy per se, and longer-term effects of ploidy on rates of adaptation.

PHYSIOLOGICAL EFFECTS OF PLOIDY

As with meiosis and sporulation, there are physiological as well as genetic effects of ploidy. Nongenetic hypotheses are more compelling for ploidy than for meiosis because there is no doubt that ploidy has widespread effects on cell structure and physiology that are unlikely to be matched by a few regulatory mutations. In addition to having one-half as much DNA to replicate, haploid cells are smaller and have larger ratios of surface area to volume when nutrients are scarce (110), potentially enabling more efficient nutrient uptake from the environment (74). This difference implies that haploids will be at an advantage when nutrients are scarce, but not when they are in rich media.

The actual evidence is not straightforward, and results depend on the particular yeast strains or growth conditions used. In one set of experiments, diploids outcompeted haploids in rich medium, while haploids were superior in minimal medium, as expected. However, when different strains were used, diploids were always superior (44, 88; discussed by Mable [78]). Adams and Hansche (3) pitted haploids against diploids in a variety of media in chemostats and observed that nutrient scarcity per se did not confer any consistent haploid advantage, although organic phosphate limitation did. In rich media, there was no detectable effect of ploidy. A more recent attempt to resolve the question appears to have finished off the physiological hypothesis: contrary to its predictions, haploids had higher growth rates and stationary phase densities than diploids in rich medium (although no competitive fitness difference was resolved), while haploid and diploid growth rates and competitive abilities were equal in minimal medium (78).

There are other physiological differences between haploid and diploid yeasts that, as with sporulation and meiosis, are interesting but probably not the focus of selection on ploidy. Both haploids and diploids can adopt alternatives to the usual yeast form of growth, in which budded cells detach from their parent cell. Nitrogen starvation can induce diploid strains to form pseudohyphae, in which cells remain attached to each other to form filaments. Haploids starved of glucose can grow invasively, penetrating and growing into agar media. Pseudohyphal growth may represent a strategy for exploration in search of nitrogen. Invasive and pseudohyphal growth are thought to be important components of virulence both in pathogenic yeasts such as *Candida albicans* (for an example, see reference 76) and in strains of *S. cerevisiae* that were isolated from immunocompromised human patients and may be emerging pathogens (46). However, pseudohyphal and invasive growth are under the control of specific regulatory pathways (26, 42, 73) and so could presumably be selected upon independently of ploidy.

HETEROZYGOSITY

The *MAT* alleles, perhaps descended from horizontally transmitted genetic parasites, now determine mating type. They also regulate many genes, some involved in functions specific to one mating type, specific to haploids, or specific to diploids (reviewed by Herskowitz et al. [54] and by Nelson [90]). For example, haploid *MATa* and *MATα* cells produce only the pheromone and pheromone receptor appropriate to their mating type, as directed by their *MAT* genotype. *MATa/MATα* heterozygosity is required for the ability to sporulate and undergo meiosis, serving to some extent as a proxy for diploidy since natural diploids would be produced only by mating between compatible haploids. Artificially constructed *MATa/MATa* and *MATα/MATα* diploids behave as haploids, capable of mating but not meiosis.

MAT heterozygosity is certainly more than just an indicator of diploidy. Birdsell and Wills (11) looked for an advantage of meiosis and mating by placing diploids that were heterozygous and homozygous at the *MAT* locus in direct competition. Competitions between heterozygotes and homozygotes in which no meiosis actually occurred were used as controls. An advantage of *MAT* heterozygosity was observed in *both* treatments

(although not in all replicates, in the case of the controls). Whatever the effect of meiosis on relative fitness in this experiment, *MAT* heterozygosity itself was an advantage.

This is a specific and apparently isolated example of heterosis, or overdominance, a fitness interaction between alleles at a locus in which heterozygotes have higher fitness than either homozygote. When surprisingly high levels of within-population genetic variation were encountered in the 1960s and there ensued a vigorous debate on whether this variation was maintained by neutral or selective processes, overdominance was a favorite mechanism proposed for selection (75). But despite its appeal, few examples of overdominance have been convincingly demonstrated.

An indirect consequence of heterozygosity has also been hypothesized as a benefit: it may store genetic variation in a population, facilitating adaptation to environmental change. The support for this idea is more intuitive than theoretical or empirical. Reproductive success is determined by the current fitness of a genotype, not its potential future use to a population as a vessel of variation. As with sex, it is possible that selection can act on entire populations: those lacking heterozygosity may go extinct with greater frequency or likelihood than heterozygous populations. But the link between heterozygosity and probability of extinction has not been demonstrated, and despite the tacit assumption in much of conservation biology that a more heterozygous population is a fitter or more resilient one, the correlation between fitness and heterozygosity is weak and inconsistent (28). Overdominance hypotheses also suffer from the defect of proposing universal advantages for diploidy, leaving unexplained the many life histories in which a haploid phase is more prominent. Finally, they are a poor fit for populations such as those of yeasts, at least some of which are diploid but highly inbred and thus have very low rates of heterozygosity (59), except at the mating-type and tightly linked loci.

MASKING AND PURGING DELETERIOUS MUTATIONS

Both sex and ploidy affect the frequencies and distributions within populations of deleterious mutations. Hypotheses about sex and ploidy that are built on these effects have the attraction that harmful mutations are a universal fact of life, giving such hypotheses at least the potential to be widely applicable. Two long-standing hypotheses regarding ploidy are based on the effects of selection on deleterious mutations.

The first is that harmful mutations are at least partly masked in diploids, which therefore do not pay their full fitness cost. Barring rare processes such as gene conversion, mutations in diploid genomes arise in a heterozygous state. Deleterious mutations are kept at low frequencies, so they remain limited almost entirely to heterozygotes. Diploidy could therefore protect an individual from the fitness cost of the mutations that it carries, if they are recessive enough (24, 87).

Just how recessive the mutations have to be has been a topic of some theoretical interest. Dominance and recessiveness are quantified using the dominance coefficient, h. For a new mutation, $h = 0$ would mean complete recessiveness, $h = 1$ represents complete dominance, and most mutations would lie somewhere in between. The masking hypothesis requires that a new mutation on average has h values of <0.5, meaning that the fitness of heterozygotes is greater than the average fitness of the two homozygotes. Unbiased and reliable estimates of h are hard to obtain, but those that are available support the masking hypothesis. Deleterious mutations are mostly recessive in *Drosophila* (25) and in other organisms (77). Korona (70) used a mutation accumulation experiment to see whether the same is true of yeast. He started with a haploid mutant that lacked the *MSH2* gene and thus had no DNA mismatch repair system, accelerating the accumulation of mutations 100- to 200-fold. He then used crosses to construct diploids that were homozygous and heterozygous for the accumulated mutations and obtained an estimated h value of 0.08. This result implies strong support for the masking hypothesis, with the caveat that *MSH2* deletion causes mostly substitutions and deletions of single nucleotides (80), and it is not known whether the distribution of h is the same for these mutations as it is for the random mutations that occur in wild types. In addition to low dominance coefficients, high mutation rates also favor diploidy and masking (95).

Mable and Otto (79) tested for masking by exposing haploid and diploid cells to the mutagen ethyl methanesulfonate, at a concentration that would be predicted to damage four to eight bases per haploid genome. This mutagenesis reduced growth rates, but by less in diploids than in haploids, an immediate advantage predicted by the masking hypothesis.

The organization of the yeast genome has been interpreted as a masking device (68). Essential genes have a greater than random linkage to centromeres and to each other. The resulting linkage disequilibrium could result in some masking by maintaining greater heterozygosity in these genes than would otherwise result from high rates of inbreeding.

There is therefore some evidence in support of the masking hypothesis. However, it has been understood for some time that diploids pay for the short-term benefit of masking mutations by accumulating a greater load of mutations than haploids in the longer term (50). Not only are mutations less efficiently removed by selection, but also they presumably arise at twice the frequency. From these observations came the counterpart to the masking hypothesis, the purging hypothesis, which holds that haploidy is advantageous because it allows selection to remove mutations more efficiently.

Experimental tests of the purging hypothesis have been surprisingly few, given its simplicity. In the same study (described above) that demonstrated an immediate diploid advantage of masking, Mable and Otto (79) found that haploids did not recover higher growth rates any faster than diploids, in contrast to the prediction of purging. A complication of this result was that the ploidy of experimental populations itself changed during the experiment.

Theoretical studies have been more prominent. The premise of the purging hypothesis, that diploids end up with a greater mutation load and lower fitness, is solid enough, so the emphasis has been on rescuing diploids from a theoretical demise. One mechanism is strong synergistic epistasis among accumulating deleterious mutations (69). This mechanism works by analogy to Kondrashov's deterministic mutation hypothesis for the evolution of sex (see chapter 30): mutation accumulation in diploids is curtailed by increasingly efficient selection against mutants as the number of mutations and their negative epistatic interactions increase.

Other studies have shown that even in the absence of epistasis, diploidy can be favored by the right life cycles and mating systems. These studies depict one locus that determines the fraction of the life cycle spent in each ploidy, and other loci at which deleterious mutations occur. Haploidy is favored when recombination between the ploidy and fitness loci is reduced. With a greater fraction of the life cycle spent in haplophase, selection against deleterious mutations is more efficient. The genomes carrying the alleles that confer this longer haplophase therefore come to have fewer mutations. With low recombination rates, these alleles become tightly linked to fitter genomes (57, 96).

S. cerevisiae defies these predictions. With its high rates of inbreeding and potentially some asexual reproduction as well, available evidence indicates only rare outcrossing, which should favor haploidy. But under the usual assumption that homothallism is the natural state of most wild yeast strains, haploidy is a brief stage in their life cycle. The restriction of a resistant ascospore stage to the diploid phase complicates the effects of ploidy. But as noted above, it is unclear how this constraint would have evolved if selection were acting on ploidy, if changes in gene regulation could permit haploids to make resistant spores. Of course *S. cerevisiae* alone is not a definitive test, and the purging hypothesis is wide open for empirical testing, either experimentally or by comparative work with a clade such as the Chlorophyte algae in which there is a great deal of variation in life cycles (96).

RATES OF ADAPTATION

Although deleterious mutations are much more frequent than beneficial ones, it may be the rates of haploids and diploids to adapt, rather than their abilities to resist decay, that determine their prevalence. Since diploids have twice as much DNA and double the copy number of each gene, adaptive mutations should occur twice as often in diploids as in haploids. On the other hand, unless h is well above 0.5 the masking effect that shields diploids from the cost of harmful mutations can also deprive them of the full benefit of adaptive ones, greatly impeding selection.

Faced with these conflicting effects of ploidy, the decisive parameter is population size, at least for asexuals (92). In small populations, the rate of adaptation is limited by the supply of beneficial mutations, and diploids adapt faster. In large populations, beneficial mutations arise frequently enough that it is their selection rather than their occurrence that limits the rate of adaptation, and therefore haploids adapt faster.

But before reviewing a test of this prediction, there is an earlier and less straightforward experiment to consider. Paquin and Adams (97) compared adaptation in haploid and diploid chemostat populations by counting rapid increases in the frequencies of genetic markers, inferring a selective sweep by an adaptive mutation for each such increase. They concluded that sweeps were 60% more common in diploids than in haploids. Adjusting for a difference in population size (4.9×10^9 for haploids and 4.5×10^9 for diploids) leaves a difference of 14% (93). But a further analysis of the results (93) does not support a simple diploid advantage. The fitness gains conferred by adaptive mutations were not measured directly but inferred from competitions between the genotypes of successive sweeps. The estimated fitness increments (which averaged 10%) are inconsistent with the durations of the inferred selective sweeps, which should have taken much longer. One possible explanation for this discrepancy is that, instead of replacing each other by sweeps, genotypes carrying adaptive mutations coexisted.

The predicted effect of population size has been tested directly in a serial transfer experiment (121). The results supported Orr and Otto's (92) prediction: among large populations (effective population size, 1.3×10^7), fitness

increases were 45% greater in haploids than diploids, but among small populations (1.4×10^4) no difference was detected. Because of the masking effect, the mutations selected in a diploid population might be expected to be more dominant than those selected in a haploid population. The estimated dominance coefficients of adaptive mutations in this experiment are consistent with this intuitive prediction: for haploids $h = 0.2$, and for diploids $h = 0.75$. However, distributions of h would also be affected by the covariance of h with the selective advantage of adaptive mutations (for example, more adaptive mutations may tend to be more dominant), which is unknown.

GENOME DUPLICATION

Nearly all book chapters conclude with some sort of prediction, whether this is prudent or not. For almost any organism of genetic inquiry, but especially for a genetic and genomic supermodel, genomics are an inevitable part of both the present and future. The following section is not so much a prediction (since it has already occurred) as a preview of genomic research complementing experimental studies to provide remarkable insight into evolutionary history.

Hypotheses of the evolutionary effects of ploidy, like the discussion above, usually compare haploidy with diploidy. Polyploidy also makes periodic appearances in evolutionary history. These were often dismissed in the past as unimportant in evolution, but Ohno (91) proposed that genome duplication, followed by functional divergence of the duplicated gene pairs, is an important source of evolutionary novelty. There are increasingly numerous examples of new lineages founded by initially polyploid ancestors, and a search for duplicated genes in the yeast genome convinced Wolfe and Shields (115) that *S. cerevisiae* was one of them. The fraction of the yeast genome actually duplicated was rather low: 55 duplicate regions averaging 55 kb long, containing 376 gene pairs in total. They interpreted this as the effect of extensive gene deletion following the original duplication, viewing *S. cerevisiae* as a "degenerate tetraploid." Skeptics saw independent duplications of individual chromosomal segments as a better explanation for the observed patterns (reviewed by Piskur [99]). The genome sequences of *Kluyveromyces waltii* and *Ashbya gossypii* have since confirmed that a duplication of the entire genome occurred in a common ancestor that they share with *S. cerevisiae* dating about 100 million years ago (31, 34, 63). The members of some duplicated gene pairs (such as *CYC1-CYC7* and *COX5A-COX5B* and some sugar transporters) are regulated differently under aerobic and anaerobic conditions. This led Wolfe and Shields (115) to suggest that genome duplication may even have played a role in the evolution of vigorous anaerobic fermentation by yeast, noting further that the estimated time of genome duplication corresponded roughly with the radiation of the angiosperms and their sugary fruit.

The initial stages of evolution by genome duplication are unclear. Artificially constructed tetraploid versions of modern *S. cerevisiae* do not fare well, suffering genome instability (81) and lower fitness than either haploids or diploids in both nutrient-rich and poor media (78). The fact that of the original duplication only 8 to 10% of the current yeast genome remains duplicated (63) suggests that much pruning of gene copies has occurred since genome duplication. Perhaps genome duplication is, over evolutionary timescales, a recurrent event with few but important successes. In the context of the largely experimental focus of this chapter, the genomic research summarized here illustrates the payoffs of working with an organism that is a model for both genomic and evolutionary experimental inquiry.

A POSSIBLE FUTURE

The opportunities for experimental evolution in *S. cerevisiae* are probably evident from this chapter—a great deal of basic theory remains untested. This empirical gap is being remedied, with several tests of hypotheses on sex now published and no doubt a few to follow shortly. Fewer experiments have tested hypotheses about ploidy—for example, a comparison of genetic loads of haploid and diploid populations following a long-term experiment would provide insight into the masking and purging hypotheses. It might also be interesting to see whether, by analogy to the trade-off between sexual and vegetative fitness discussed above, there are alleles with antagonistic effects in haploids and diploids.

A major area for explanation is the part of the living world that exists outside laboratories. Several theories of sex rely on particular descriptions of the environment—variable in space or time, or on the whole constant in their selective pressures (5). Reasonable constancy can be achieved in the lab, but the nature of environmental variation that is experienced by yeasts or that selects on their life histories and mating systems is unknown. The ability to regulate and replicate an experimental environment must at some point be traded off for realism. At our stage of understanding the evolution of sex and ploidy, lab experiments can still make valuable contributions, but ultimately ecology must be incorporated into the study of yeast.

Much of the research on *S. cerevisiae*, both the abundant molecular genetic work and the experimental evolu-

tionary research described here and by Goddard (chapter 30) has been driven by its value as a model eukaryote, and it has been a spectacularly successful model. But it would be another biological triumph to see how complete a picture of this particular species we could get by meeting the accumulating wealth of its genetics, genomics, molecular biology and proteomics with evolutionary and ecological understanding. The past couple of decades have astonished us with how little we knew, and it will be all the more exciting if that turns out to be true once again.

References

1. Aa, E., J. P. Townsend, R. I. Adams, K. M. Nielsen, and J. W. Taylor. 2006. Population structure and gene evolution in *Saccharomyces cerevisiae*. *FEMS Yeast Res.* 6:702–715.
2. Adams, J., S. Puskas-Rosza, J. Simlar, and C. M. Wilke. 1992. Adaptation and major chromosomal changes in populations of *Saccharomyces cerevisiae*. *Curr. Genet.* 22:13–19.
3. Adams, J., and P. E. Hansche. 1974. Population studies in microorganisms I. Evolution of diploidy in *Saccharomyces cerevisiae*. *Genetics* 76:327–338.
4. Barraclough, T. G., C. W. Birky, Jr., and A. Burt. 2003. Diversification in sexual and asexual organisms. *Evolution* 57:2166–2172.
5. Bell, G. 1982. *The Masterpiece of Nature: the Evolution and Genetics of Sexuality*. University of California, Berkeley.
6. Bell, G. 1993. The sexual nature of the eukaryote genome. *J. Hered.* 84:351–359.
7. Bell, G. 1994. The comparative biology of the alternation of generations. *Lect. Math. Life Sci.* 25:1–25.
8. Bell, G. 1997. *Selection: the Mechanism of Evolution*. Chapman & Hall, New York, NY.
9. Bell, G. 2005. Experimental sexual selection in *Chlamydomonas*. *J. Evol. Biol.* 18:722–734.
10. Bilanchone, E. W., J. A. Claypool, P. T. Kinsey, and S. B. Sandmeyer. 1993. Positive and negative regulatory elements control expression of the yeast retrotransposon Ty3. *Genetics* 134:685–700.
11. Birdsell, J., and C. Wills. 1996. Significant competitive advantage conferred by meiosis and syngamy in the yeast *Saccharomyces cerevisiae*. *Proc. Natl. Acad. Sci. USA* 93:908–912.
12. Blanc, V. M., and J. S. Adams. 2003. Identification of mutations increasing fitness in laboratory populations. *Genetics* 165:975–983.
13. Boeke, J. D. 1989. Transposable elements in *Saccharomyces cerevisiae*, p. 335–374. *In* D. E. Berg and M. M. Howe (ed.), *Mobile DNA*. American Society for Microbiology, Washington, DC.
14. Boeke, J. D., and S. B. Sandmeyer. 1991. Yeast transposable elements, p. 193-261. *In The Molecular and Cellular Biology of the Yeast Saccharomyces: Genome Dynamics, Protein Synthesis, and Energetics*, vol. I. Cold Spring Harbor, Plainview, NY.

15. Bradshaw, V. A., and K. McEntee. 1989. DNA damage activates transcription and transposition of yeast Ty retrotransposons. *Mol. Gen. Genet.* **218**:465–474.

16. Brodie, E. D. I. 2000. Why evolutionary genetics does not always add up, p. 3–19. *In* J. B. Wolf, E. D. Brodie III, and M. J. Wade (ed.), *Epistasis and the Evolutionary Process*. Oxford University Press, New York, NY.

17. Brown, M. J. F., R. Loosli, and P. Schmid-Hempel. 2000. Condition-dependent expression of virulence in trypanosomes infecting bumblebees. *Oikos* **91**:421–427.

18. Bulmer, M. 1994. *Theoretical Evolutionary Ecology*. Sinauer, Sunderland, MA.

19. Burt, A., and R. Trivers. 2006. *Genes in Conflict*. Harvard University Press, Cambridge, MA.

20. Butler, G., C. Kenny, A. Fagan, C. Kurischko, C. Gaillardin, and K. H. Wolfe. 2004. Evolution of the MAT locus and its Ho endonuclease in yeast species. *Proc. Natl. Acad. Sci. USA* **101**:1632–1637.

21. Cech, T. R. 1986. The generality of self-splicing RNA: relationship to nuclear mRNA splicing. *Cell* **44**:207–210.

22. Cech, T. R. 1987. The chemistry of self-splicing RNA and RNA enzymes. *Science* **236**:1532–1539.

23. Chalker, D. L., and S. B. Sandmeyer. 1990. Transfer RNA genes are genomic targets for *de novo* transposition of the yeast retrotransposon Ty3. *Genetics* **126**:837–850.

24. Crow, J., and M. Kimura. 1965. Evolution in sexual and asexual populations. *Am. Nat.* **99**:439–450.

25. Crow, J. F., and M. J. Simmons. 1983. The mutation load in Drosophila, p. 1–35. *In* M. Ashburner, H. L. Carson, and J. N. Thompson (ed.), *The Genetics and Biology of Drosophila*, vol. 3. Academic Press, London, United Kingdom.

26. Cullen, P. J., and G. F. Sprague, Jr. 2000. Glucose depletion causes haploid invasive growth in yeast. *Proc. Natl. Acad. Sci. USA* **97**:13619–13624.

27. Da Silva, J., and G. Bell. 1992. The ecology and genetics of fitness in Chlamydomonas VI. Antagonism between natural selection and sexual selection. *Proc. R. Soc. Lond. B* **249**:227–283.

28. David, P. 1998. Heterozygosity-fitness correlations: new perspectives on old problems. *Heredity* **80**:531–537.

29. Deutschbauer, A. M., R. M. Williams, A. M. Chu, and R. W. Davis. 2002. Parallel phenotypic analysis of sporulation and postgermination growth in Saccharomyces cerevisiae. *Proc. Natl. Acad. Sci. USA* **99**:15530–15535.

30. Dieckmann, U., and M. Doebeli. 1999. On the origin of species by sympatric speciation. *Nature* **400**:354–357.

31. Dietrich, F. S., S. Voegeli, S. Brachat, K. G. A. Lerch, S. Steiner, C. Mohr, P. L. R. Pohlmann, S. Choi, R. A. Wing, A. Flavier, T. D. Gaffney, and P. Philippsen. 2004. The *Ashbya gossypii* genome as a tool for mapping the ancient *Saccharomyces cerevisiae* genome. *Science* **304**:304–307.

32. Dujon, B. 1989. Group I introns as mobile genetic elements: facts and mechanistic speculations—a review. *Gene* **82**:91–114.

33. Dujon, B. 1996. The yeast genome project: what did we learn? *Trends Genet.* **12**:263–270.

34. Dujon, B., D. Sherman, G. Fischer, P. Durrens, S. Casaregola, I. Lafontaine, J. de Montigny, C. N. C. Marck, E.

Talla, N. Goffard, L. Frangeul, M. Aigle, V. Anthouard, V. B. A. Babour, S. Barnay, S. Blanchin, J.-M. Beckerich, E. Beyne, C. Bleykasten, J. B. A. Boisrame, L. Cattolico, F. Confanioleri, A. de Daruvar, L. Despons, E. Fabre, H. F.-D. C. Fairhead, A. Groppi, F. Hantraye, C. Hennequin, N. Jauniaux, P. Joyet, A. K. R. Kachouri, R. Koszul, M. Lemaire, I. Lesur, L. Ma, H. Muller, J.-M. Nicaud, S. O. M. Nikolski, O. Ozier-Kalogeropoulos, S. Pellenz, S. Potier, G.-F. Richard, M.-L. Straub, D. S. A. Suleau, F. Tekaia, M. Wesolowski-Louvel, E. Westhof, B. Wirth, I. Z. M. Zeniou-Meyer, M. Bolotin-Fukuhara, A. Thierry, C. Bouchier, B. Caudron, and C. G. C. Scarpelli, J. Weissenbach, P. Wincker, and J.-L. Souciet. 2004. Genome evolution in yeasts. *Nature* **430**:35–44.

35. Dunham, M. J., H. Badrane, T. Ferea, J. Adams, P. O. Brown, F. Rosenzweig, and D. Botstein. 2002. Characteristic genome rearrangements in experimental evolution of *Saccharomyces cerevisiae*. *Proc. Natl. Acad. Sci. USA* **99**:16144–16149.

36. Ferea, T. L., D. Botstein, P. O. Brown, and R. F. Rosenzwieg. 1999. Systematic changes in gene expression patterns following adaptive evolution in yeast. *Proc. Natl. Acad. Sci. USA* **96**:9721-9726.

37. Fingerman, E. G., P. G. Dombrowski, C. A. Francis, and P. D. Sniegowski. 2003. Distribution and sequence analysis of a novel Ty3-like element in natural *Saccharomyces paradoxus* isolates. *Yeast* **20**:761–770.

38. Finnegan, D. J. 1989. Eukaryotic transposable elements and genome evolution. *Trends Genet.* **5**:103–107.

39. Fraser, J. A., and J. Heitman. 2004. Evolution of fungal sex chromosomes. *Mol. Microbiol.* **51**:299–306.

40. Futcher, B., E. Reid, and D. A. Hickey. 1988. Maintenance of the 2 micron circle plasmid of *Saccharomyces cerevisiae* by sexual transmission: an example of a selfish DNA. *Genetics* **118**:411–415.

41. Galitski, T., A. J. Saldanha, C. A. Styles, E. S. Lander, and G. R. Fink. 1999. Ploidy regulation of gene expression. *Science* **285**:251–254.

42. Gancedo, J. M. 2001. Control of pseudohyphae formation in *Saccharomyces cerevisiae*. *FEMS Microbiol. Rev.* **25**:107–123.

43. Gingold, E. B. 1988. The replication and segregation of yeast mitochondrial DNA, p. 149-170. *In* S. A. Boffey and D. Lloyd (ed.), *The Division and Segregation of Organelles*. Cambridge Univ. Press, New York, NY.

44. Glazunov, A. V., A. V. Boreiko, and A. Esser. 1989. Relative competitiveness of haploid and diploid yeast cells growing in a mixed population. *Mikrobiologiia* **68**:769–777.

45. Goddard, M. R., and A. Burt. 1999. Recurrent invasion and extinction of a selfish gene. *Proc. Natl. Acad. Sci. USA* **96**:13880–13885.

46. Goldstein, A. L., and J. H. McCusker. 2001. Development of *Saccharomyces cerevisiae* as a model pathogen: a system for the genetic identification of gene products required for survival in the mammalian host environment. *Genetics* **159**:499–513.

47. Goodnight, C. J. 1988. Epistasis and the effect of founder events on the additive genetic variance. *Evolution* **42**:441–454.

48. Green, R. F., and D. L. G. Noakes. 1995. Is a little bit of sex as good as a lot? *J. Theor. Biol.* **174:**87–96.

49. Grimberg, B., and C. Zeyl. 2005. The effects of sex and mutation rate on adaptation in test tubes and to mouse hosts by *Saccharomyces cerevisiae*. *Evolution* **59:**431–438.

50. Haldane, J. B. S. 1924. A mathematical theory of natural and artificial selection. I. *Trans. Camb. Phil. Soc.* **23:**19–41.

51. Hansche, P. E., V. Beres, and P. Lange. 1978. Gene duplication in *Saccharomyces cerevisiae*. *Genetics* **88:**673–687.

52. Hansen, L. J., D. L. Chalker, and S. B. Sandmeyer. 1988. Ty3, a yeast retrotransposon associated with tRNA genes, has homology to animal retroviruses. *Mol. Cell. Biol.* **8:**5245–5246.

53. Helling, R. B., C. N. Vargas, and J. Adams. 1987. Evolution of *Escherichia coli* during growth in a constant environment. *Genetics* **116:**349–358.

54. Herskowitz, I., J. Rine, and J. Strathern. 1992. Mating-type determination and mating-type interconversion in *Saccharomyces cerevisiae*, p. 583–656. *In* E. W. Jones, J. R. Pringle, and J. R. Broach (ed.), *The Molecular and Cellular Biology of the Yeast Saccharomyces*, vol. II. *Gene Expression*. Cold Spring Harbor Laboratory Press, Plainview, NY.

55. Hickey, D. A. 1982. Selfish DNA: a sexually transmitted nuclear parasite. *Genetics* **101:**519–531.

56. Hickey, D. A., and M. R. Rose. 1988. The role of gene transfer in the evolution of eukaryotic sex, p. 161-175. *In* R. E. Michod and B. R. Levin (ed.), *The Evolution of Sex*. Sinauer, Sunderland, MA.

57. Jenkins, C. D., and M. Kirkpatrick. 1995. Deleterious mutation and the evolution of genetic life cycles. *Evolution* **49:**512–520.

58. Ji, H., D. P. Moore, M. A. Blomberg, L. T. Braiterman, D. F. Voytas, G. Natsoulis, and J. D. Boeke. 1993. Hotspots for unselected Ty1 transposition events on yeast chromosome III are near tRNA genes and LTR sequences. *Cell* **73:**1007–1018.

59. Johnson, L. J., V. Koufopanou, M. R. Goddard, R. Hetherington, S. M. Schafer, and A. Burt. 2004. Population genetics of the wild yeast *Saccharomyces paradoxus*. *Genetics* **166:**43–52.

60. Joshi, N. V. 1990. Transposable-mediated evolution of sex: a population genetic model. *J. Genet.* **69:**127–139.

61. Kalz, O., and C. Koella. 2003. Host growth conditions regulate the plasticity of horizontal and vertical transmission in *Holospora undulata*, a bacterial parasite of the protozoan *Paramecium caudatum*. *Evolution* **57:**1535–1542.

62. Keeling, P. J., and A. J. Roger. 1995. The selfish pursuit of sex. *Nature* **375:**283.

63. Kellis, M., B. W. Birren, and E. S. Lander. 2004. Proof and evolutionary analysis of ancient genome duplication in the yeast *Saccharomyces cerevisiae*. *Nature* **428:**617–624.

64. Kennell, J. C., J. V. Moran, P. S. Perlman, R. A. Butow, and A. C. Lambowitz. 1993. Reverse transcriptase activity associated with maturase-encoding group II introns in yeast mitochondria. *Cell* **73:**133–146.

65. Kim, J. M., S. Vanguri, J. D. Boeke, A. Gabriel, and D. F. Voytas. 1998. Transposable elements and genome organization: a comprehensive survey of retrotransposons revealed by the complete *Saccharomyces cerevisiae* genome sequence. *Genome Res.* **8:**464–478.

66. Kinsey, P. T., and S. B. Sandmeyer. 1995. Ty3 transposes in mating populations of yeast: a novel transposition assay for Ty3. *Genetics* **139:**81–94.

67. Kirchner, J., S. B. Sandmeyer, and D. Forrest. 1992. Transposition of a Ty3 GAGS-POL3 fusion mutant is limited by availability of capsid protein. *J. Virol.* **66:**6081–6092.

68. Knop, M. 2006. Evolution of the hemiascomycete yeasts: on life styles and the importance of inbreeding. *Bioessays* **28:**696–708.

69. Kondrashov, A. S., and J. F. Crow. 1991. Haploidy or diploidy: which is better ? *Nature* **351:**314–315.

70. Korona, R. 1999. Unpredictable fitness transitions between haploid and diploid strains of genetically loaded yeast *Saccharomyces cerevisiae*. *Genetics* **151:**77–85.

71. Koufopanou, V., M. R. Goddard, and A. Burt. 2002. Adaptation for horizontal transfer in a homing endonuclease. *Mol. Biol. Evol.* **19:**239–246.

72. Landry, C. R., J. P. Townsend, D. L. Hartl, and D. Cavalieri. 2006. Ecological and evolutionary genomics of *Saccharomyces cerevisiae*. *Mol. Ecol.* **15:**575–591.

73. Lengeler, K. B., R. C. Davidson, C. D'Souza, T. Harashima, W. C. Shen, et al. 2000. Signal transduction cascades regulating fungal development and virulence. *Microbiol. Mol. Biol. Rev.* **64:**746–785.

74. Lewis, W. M. 1985. Nutrient scarcity as an evolutionary cause of haploidy. *Am. Nat.* **125:**692–701.

75. Lewontin, R. C. 1974. *The Genetic Basis of Evolutionary Change*. Columbia University Press, New York, NY.

76. Lo, H. J., J. R. Kohler, B. DiDomenico, D. Loebenberg, A. Cacciapuoti, et al. 1997. Nonfilamentous *C. albicans* mutants are avirulent. *Cell* **90:**939–949.

77. Lynch, M., J. Blanchard, D. Houle, T. Kibota, S. Schultz, L. Vassilieva, and J. Willis. 1999. Perspective: spontaneous deleterious mutation. *Evolution* **53:**645–663.

78. Mable, B. K. 2001. Ploidy evolution in the yeast *Saccharomyces cerevisiae*: a test of the nutrient limitation hypothesis. *J. Evol. Biol.* **14:**157–170.

79. Mable, B. K., and S. P. Otto. 2001. Masking and purging mutations following EMS treatment in haploid, diploid and tetraploid yeast (*Saccharomyces cerevisiae*). *Genet. Res.* **77:**9–26.

80. Marsischky, G. T., N. Filosi, M. F. Kane, and R. Kolodner. 1996. Redundancy of *Saccharomyces cerevisiae MSH3* and *MSH6* in *MSH2*-dependent mismatch repair. *Genes Dev.* **10:**407–420.

81. Mayer, V. W., and A. Aguilera. 1990. High levels of chromosome instability in polyploids of *Saccharomyces cerevisiae*. *Mutat. Res.* **231:**177–186.

82. McCusker, J. H., K. V. Clemons, D. A. Stevens, and R. W. Davis. 1994. Genetic characterization of pathogenic *Saccharomyces cerevisiae* isolates. *Genetics* **136:**1261–1269.

83. Metzenberg, R. L., and N. L. Glass. 1990. Mating type and mating strategies in *Neurospora*. *Bioessays* **12:**53–59.

84. Moran, J. V., S. Zimmerly, R. Eskes, J. C. Kennell, A. M. Lambowitz, R. A. Butow, and P. S. Perlman. 1995. Mobile

group II introns of yeast mitochondrial DNA are novel site-specific retroelements. *Mol. Cell. Biol.* **15**:2828–2838.

85. **Mortimer, R. K.** 2000. Evolution and variation of the yeast (Saccharomyces) genome. *Genome Res.* **10**:403–409.

86. **Mueller, M. W., M. Allmaler, R. Eskes, and R. J. Schweyen.** 1993. Transposition of group II intron al1 in yeast and invasion of mitochondrial genes at new locations. *Nature* **366**:174–178.

87. **Muller, H. J.** 1932. Some genetic aspects of sex. *Am. Nat.* **66**:118–138.

88. **Naidkhardt, K., and A. V. Glasunov.** 1991. Competition of isogenic haploid and diploid cells of the yeast *Saccharomyces cerevisiae* and *Pichia pinus* growing in mixed populations. *Mikrobiologiia* **60**:686–692. (In Russian.)

89. **Naumov, G. I., E. S. Naumova, and P. Sniegowski.** 1998. *Saccharomyces paradoxus* and *Saccharomyces cerevisiae* are associated with exudates of North American oaks. *Can. J. Microbiol.* **44**:1045–1050.

90. **Nelson, M. A.** 1996. Mating systems in ascomycetes: a romp in the sac. *Trends Genet.* **12**:69–74.

91. **Ohno, S.** 1970. *Evolution by Gene Duplication.* George Allen and Unwin, London, United Kingdom.

92. **Orr, H. A., and S. P. Otto.** 1994. Does diploidy increase the rate of adaptation? *Genetics* **136**:1475–1480.

93. **Otto, S. P.** 1994. The role of deleterious and beneficial mutations in the evolution of ploidy levels, p. 69–96. *In* M. Kirkpatrick (ed.), *Theories for the Evolution of Haploid-Diploid Life Cycles.* American Mathematical Society, Providence, RI.

94. **Otto, S. P., and T. Lenormand.** 2002. Resolving the paradox of sex and recombination. *Nat. Rev. Genet.* **3**:252–261.

95. **Otto, S. P., and D. B. Goldstein.** 1992. Recombination and the evolution of diploidy. *Genetics* **131**:745–751.

96. **Otto, S. P., and J. C. Marks.** 1996. Mating systems and the evolutionary transition between haploidy and diploidy. *Biol. J. Linn. Soc.* **57**:197–218.

97. **Paquin, C., and J. Adams.** 1983. Frequency of fixation of adaptive mutations is higher in evolving diploid than haploid yeast populations. *Nature* **302**:495–500.

98. **Philippsen, P. H., J. Eibel, J. Gafner, and A. Stotz.** 1983. Ty elements and the stability of the yeast genome. *In* M. Korhola and E. Vaisanen (ed.), *Gene Expression in Yeast.* Foundation for Biotechnical and Industrial Fermentation Research, Helsinki, Finland.

99. **Piskur, J.** 2001. Origin of the duplicated regions in the yeast genomes. *Trends Genet.* **17**:302–303.

100. **Querol, A., M. T. Fernández-Espinar, M. del Olmo, and E. Barrio.** 2003. Adaptive evolution of wine yeast. *Int. J. Food Microbiol.* **86**:3–10.

101. **Rainey, P. B., and M. Travisano.** 1998. Adaptive radiation in a heterogeneous environment. *Nature* **394**:69–72.

102. **Rose, M. R.** 1983. The contagion mechanism for the evolution of sex. *J. Theo. Biol.* **101**:137–146.

103. **Rozen, D. E., and R. E. Lenski.** 2000. Long-term experimental evolution in *Escherichia coli*. VIII. Dynamics of a balanced polymorphism. *Am. Nat.* **155**:24–35.

104. **Ruderfer, D. M., S. C. Pratt, H. S. Seidel, and L. Kruglyak.** 2006. Population genomic analysis of outcrossing and recombination in yeast. *Nat. Genet.* **38**:1077–1081.

105. **Sniegowski, P., P. G. Dombrowski, and E. Fingerman.** 2002. *Saccharomyces cerevisiae* and *Saccharomyces paradoxus* coexist in a natural woodland site in North America and display different levels of reproductive isolation from European conspecifics. *FEMS Yeast Res.* **1**:299–306.

106. **Syvanen, M.** 1984. The evolutionary implications of mobile genetic elements. *Trends Genet.* **18**:271–293.

107. **Taylor, D., C. Zeyl, and E. Cooke.** 2002. Conflicting levels of selection in the accumulation of mitochondrial defects in *Saccharomyces cerevisiae*. *Proc. Natl. Acad. Sci. USA* **99**:3690–3694.

108. **Thomas, F., S. P. Brown, M. Sukhedo, and F. Renaud.** 2002. Understanding parasite strategies: a state-dependent approach? *Trends Parasitol.* **18**:387–390.

109. **Tsong, A., B. T. Tuch, H. Li, and A. D. Johnson.** 2006. Evolution of alternative transcriptional circuits with identical logic. *Nature* **443**:415–420.

110. **Van der Sand, S. T., W. Greenhalf, D. C. Gardner, and S. G. Oliver.** 1995. The maintenance of self-replicating plasmids in *Saccharomyces cerevisiae*: mathematical modeling, computer simulations, and experimental tests. *Yeast* **11**:641–658.

111. **Weiss, R. L., J. R. Kukora, and J. Adams.** 1975. The relationship between enzyme activity, cell geometry, and fitness in *Saccharomyces cerevisiae*. *Proc. Natl. Acad. Sci. USA* **72**:794–798.

112. **Wilke, C., and J. Adams.** 1992. Fitness effects of Ty transposition in *Saccharomyces cerevisiae*. *Genetics* **131**:31–42.

113. **Wilke, C. M., E. Maimer, and J. Adams.** 1992. The population biology and evolutionary significance of Ty elements in *Saccharomyces cerevisiae*. *Genetics* **86**:155–173.

114. **Winzeler, E. A., D. D. Shoemaker, A. Astromoff, H. Liang, K. Anderson, B. Andre, R. Bangham, R. Benito, J. D. Boeke, H. Bussey, A. M. Chu, C. Connelly, K. Davis, F. Dietrich, S. W. Dow, M. El Bakkoury, F. Foury, S. H. Friend, E. Gentalen, G. Giaever, J. H. Hegemann, T. Jones, M. Laub, H. Liao, N. Liebundguth, D. J. Lockhart, A. Lucau-Danila, M. Lussier, N. M'Rabet, P. Menard, M. Mittmann, C. Pai, C. Rebischung, J. L. Revuelta, L. Riles, C. J. Roberts, P. Ross-MacDonald, B. Scherens, M. Snyder, S. Sookhai-Mahadeo, R. K. Storms, S. Veronneau, M. Voet, G. Volckaert, T. R. Ward, R. Wysocki, G. S. Yen, K. Yu, K. Zimmermann, P. Philippsen, M. Johnston, and R. W. Davis.** 1999. Functional characterization of the *S. cerevisiae* genome by gene deletion and parallel analysis. *Science* **285**:901–906.

115. **Wolfe, K. H., and D. C. Shields.** 1997. Molecular evidence for an ancient duplication of the entire yeast genome. *Nature* **387**:708–713.

116. **Xu, J.** 2002. Estimating the spontaneous mutation rate of loss of sex in the human pathogenic fungus *Cryptococcus neoformans*. *Genetics* **162**:1157–1167.

117. Zeyl, C., G. Bell, and D. M. Green. 1996. Sex and the spread of retrotransposon Ty3 in experimental populations of *Saccharomyces cerevisiae*. *Genetics* **143**:1567–1577.

118. Zeyl, C., and J. A. G. M. DeVisser. 2001. Estimates of the rate and distribution of fitness effects of spontaneous mutation in *Saccharomyces cerevisiae*. *Genetics* **157**:53–61.

119. Zeyl, C., B. Andreson, and E. Weninck. 2005. Nuclear-mitochondrial epistasis for fitness in *Saccharomyces cerevisiae*. *Evolution* **59**:910–914.

120. Zeyl, C., C. Curtin, K. Karnap, and E. Beauchamp. 2005. Tradeoffs between sexual and vegetative fitness in *Saccharomyces cerevisiae*. *Evolution* **59**:2109–2115.

121. Zeyl, C., T. Vanderford, and M. Carter. 2003. An evolutionary advantage of haploidy in large yeast populations. *Science* **299**:555–558.

*Sex in Fungi: Molecular Determination
and Evolutionary Implications*
Edited by Joseph Heitman et al.
© 2007 ASM Press, Washington, D.C.

Duur K. Aanen
Rolf F. Hoekstra

32

Why Sex Is Good:
On Fungi and Beyond

A SHORT HISTORY OF IDEAS ON THE SIGNIFICANCE OF SEXUAL REPRODUCTION

Sexual reproduction is less efficient than asexual reproduction. It requires more time and energy, and in most species individuals need to find a mating partner—to mention but only three of its most conspicuous disadvantages. On the other hand, sexual reproduction is prevalent among eukaryotes, and much of organismal design has evolved to facilitate and promote sexual reproduction. Sex has prompted the evolution of the male-female distinction and of specialized structures and behavior. All of which suggests that sexual reproduction must confer important benefits, despite its apparent disadvantages. Ever since Weismann (47, 48), the genetic consequences of sex have been regarded as the basis for its functional explanation. The genetic essence of sex is the production of genetically variable offspring due to the meiotic recombination involved in the formation of gametes. Weismann argued that the increased genetic variation among the offspring would allow more efficient natural selection and in this way accelerate evolution. For quite some time this explanation was accepted. Later it was recognized that the increased

genetic variability among offspring also allows a more efficient elimination of deleterious mutations (35). These two aspects are still prominent in the discussions about the significance of sex. However, in the 1970s the explanations of sex came under scrutiny because it was realized that on top of other disadvantages there is also a direct twofold cost of sex since asexual females can potentially produce twice as many daughters as sexual females (which, in a randomly mating population, "waste" on average 50% of their resources to the production of males), so that an asexual mutant in a sexual population would initially double in frequency each generation (32, 33; see also below). Therefore, the challenge to theoreticians was to come up with quantitative models that could predict a >2-fold advantage to sex as compared to asex. A proliferation of models ensued, characterized by increasing sophistication and specificity, and this theoretical literature has been reviewed extensively (for examples, see references 2, 5, 19a, 28, 34, 36, and 45). A general conclusion emerging from the theory is that a prerequisite for the validity of explanations for sex, i.e., that increased genetic variation among offspring favors both natural selection and elimination

Duur K. Aanen and Rolf F. Hoekstra, Laboratory of Genetics, Wageningen University and Research Center, Arboretumlaan 4, 6703 BD Wageningen, The Netherlands.

of deleterious mutations, is that the genetic variation for fitness in the population is lower than would result from a random distribution of the relevant alleles. In other words, deleterious and advantageous alleles at different loci should be associated (negative linkage disequilibria). Sex is then advantageous because it breaks down these associations by recombination.

One reason why it is so difficult to identify a general evolutionary explanation of sex is that various possible explanations do not exclude each other: they may be valid at the same time. This point of view has been stressed by West et al. (49), who argue that the different theories have been viewed as competing, while advantage may be gained from studying to what extent different mechanisms may complement or reinforce each other. After all, it is not uncommon in evolution that traits, once arisen, have been put to different uses in different lineages. Examples are gill arches and mammalian forelimbs. In a similar way, sex may be maintained in different groups of organisms for reasons that differ between these groups and also from the reasons that have led to its evolution. Although perhaps somewhat unattractive from a scientific point of view—because it implies that a central and conspicuous aspect of biology like sexual reproduction would not have a simple general explanation—this so-called pluralist approach may well be relevant to the understanding of sex in fungi, as is detailed below.

What Use Is Sex in Fungi?

The distribution of sexual reproduction in fungi provides a nice illustration of the difficulties to understand sexuality in functional evolutionary terms. Since sex is essentially the fusion of two haploid nuclei into a diploid nucleus with subsequent meiotic recombination, most basidiomycota are obligately sexual, while in other groups, such as the arbuscular-mycorrhiza-forming Glomeromycota and in "Deuteromycota" (related to Ascomycota), sexuality has never been observed, implying that sex is either absent, very rare, or difficult to observe in nature and/or to induce in the laboratory (because almost every "asexual" fungus that has been subjected to a population genetic study shows evidence of recombination). Most other fungi are somewhere between these extremes, having both the options of sexual and asexual spore production. The general explanations of sex mentioned above, namely, that under certain conditions meiotic recombination can result in speeding up adaptive evolution and in more efficient elimination of deleterious mutations, are of course also applicable to fungi. The problem is whether these advantages of sex are sufficient to counterbalance its disadvantages, and this balance may differ between species. Ideally, a satisfactory functional explanation of sex in fungi should explain

not only why sex is good but also why it is so rare in some groups and very common in others. We are still far from that level of understanding.

In the remainder of this chapter we discuss aspects of fungal biology that are relevant for an understanding of the occurrence of sex in fungi. First, we ask if the genetic cost of sex is the same in fungi as in animals and plants. Second, we discuss "sexual hang-ups" in fungi: traits or processes that are linked to the sexual cycle, without being part of the sexual cycle itself. An example is the production of resting spores or survival structures, which is often coupled to the sexual cycle. Once such a trait is strictly coupled to sex, selection to retain the trait automatically selects for the maintenance of sex. Finally we discuss the suitability of fungi as experimental model systems to test theories about the function of sex.

THE COST OF SEX

Maynard Smith (32) argued that in a population consisting of an equal number of females and males, a rare mutation causing females to produce only daughters by asexual reproduction would initially double in frequency. He called this "the twofold cost of sex," which was defined as the initial rate of increase in frequency of an asexual mutant introduced into a sexually reproducing population (17). Later, the cost of sex was also theoretically analyzed for various other mating systems, such as hermaphroditism, and for varying degrees of selfing (17, 25). Much of the literature on the cost of sex, including that referred to above, is primarily based on the situation in animals and plants. Sex in these groups has several characteristics that are not shared by other groups of organisms. First, in most animals and plants sex is always connected to reproduction and most reproduction is sexual. This situation means that the cost of sex identified for these groups is "the cost of frequent sex." In many other taxa, however, only a fraction of the reproduction is sexual, so that the actual cost of sex is lower. This is especially so if sexual reproduction is limited to circumstances where the opportunity costs of sex are lowest (13, 31; see below). Second, many organisms are isogamic in contrast to plants and animals. In isogamic organisms, both gametes contribute limiting resources to the zygote. Everything else being equal, there is no cost of fusion between equally sized cells, followed by recombination and division. Therefore, there is no "cost of males" (or male function) for such organisms.

The Costs of Sex in Fungi

Fungi exhibit a huge variation in some determinants for the cost of sex. First, some fungi are isogamic whereas others are anisogamic. Fusion between same-sized single cells is found among yeast-forming ascomycetes and basidiomycetes and in some chytridiomycetes. Another type of isogamy is when specialized hyphal structures fuse to form specialized sexual structures, which is found in zygomycetes. A unique type of isogamy is found in the hymenomycetes, which will be discussed below. Here, sex starts with the fusion of haploid mycelia followed by reciprocal nuclear exchange. A second major factor influencing the cost of sex is the mating system. Many fungi are homothallic, i.e., capable of self-fertilization, but often cross-fertilization is also possible (11). Other fungi are obligatorily heterothallic, and mating is regulated by mating-type factors. Still other fungi have secondarily developed mechanisms that allow self-fertilization, for example, by producing spores that contain both mating types. The general effect of selfing in hermaphroditic organisms is a reduction of the cost of sex (17).

Most fungi have asexual reproduction as an alternative to sexual reproduction. Asexual reproduction is usually much faster than sexual reproduction. For unicellular organisms like yeasts, it has been estimated that under optimal growth conditions, in the time taken for two cells to conjugate and reorganize their nuclei (with no increase in cell number), a cell could divide asexually up to eight times, which would result in up to 256 cells, which translates to a 256-fold cost of sex (13)! The difference in speed between asexual and sexual reproduction has been ascribed to the mechanical components of sexual reproduction (meiosis, gametic union, and nuclear fusion [31]). This "cellular-mechanical cost of sex" (31) can be considered an opportunity cost, relative to the yield of the alternative option, asexual reproduction. This cost is therefore variable, because it depends on the growth conditions: the opportunity costs of sex are high when the potential for asexual growth is good, and they are low when the growth conditions become worse, because the alternative then becomes less advantageous. Strikingly, many fungi do indeed have sex when growth conditions become worse, for example, at the end of the growth season, or when the local density becomes high (see below).

These considerations show that the actual cost of sex in fungi (accounting for its specific characteristics, timing, and frequency) is much lower than the cost of sex for animals and plants that is most prominently discussed in the literature.

The 1.5-Fold Cost of Outcrossing in Homobasidiomycetes: the Cost of Female Fertility

Sex in mushroom-forming basidiomycetes is unique. First, in contrast to most other fungi, sexual reproduction is the only possible reproduction mode, at least re-

production associated with specialized dispersal structures (as opposed to asexual mycelial growth, where sometimes also asexual spores are formed by fragmentation of single cells). Therefore, the cost of sex can be directly compared with the alternative of asexual reproduction. Second, sex in basidiomycetes starts with hyphal fusions between two monokaryons (haploid mycelia arising after spore germination). Subsequently, haploid nuclei are exchanged and reciprocally migrate throughout the existing mycelia to give rise to a dikaryon (14, 22). The dikaryon can reproduce sexually via the production of fruiting bodies (basidiomata or mushrooms) bearing basidia. In these basidia, nuclear fusion occurs, followed by meiosis and the formation of basidiospores. Although the two monokaryons in a mating at first sight do not have specialized male and female organs, monokaryons exhibit two clearly distinct behaviors, the acceptance of a nucleus, and the donation of a nucleus. The acceptance of a nucleus (with a contribution of the cytoplasm of that monokaryon to the newly established dikaryon) can be considered a female role, while the donation of a nucleus (without a contribution of the cytoplasm of that monokaryon to the newly established dikaryon) can be considered a male role (Fig. 32.1a and b [1]), and each monokaryon is a hermaphrodite. Therefore, mating between monokaryons is essentially anisogamous. The dikaryon retains a male potential, as it can donate one of its nuclei after fusing with a compatible monokaryon to form a new dikaryon (10) by the "Buller" phenomenon (38).

To calculate the cost of heterothallic sex in a basidiomycete, we consider the relative advantage of a dominant gene causing asexuality while producing an equal number of spores in a population of sexuals. All else being equal, such a gene would initially double its frequency via the "female" route (it would monopolize the spores) but have zero fitness via the "male" route. The net difference between heterothallic sexual and purely asexual reproduction and hence, by definition, the cost of sex, would therefore be zero. However, a mutation causing asexual or (secondarily) homothallic reproduction while maintaining male fertility would initially have a 1.5-fold advantage, because such a gene would double in frequency via the 'female' route, while maintaining male fertility (Fig. 32.1c). The 1.5-fold cost calculated in this way is the cost of female outcrossing. (The same calculation has been made for hermaphrodite plants [17].)

With secondary homothallism, the basidiospores contain two compatible nuclei. The dikaryons resulting from such spores retain their male fertility via Buller matings. Transitions from heterothallism to secondary homothallism have occurred repeatedly in many basidiomycete

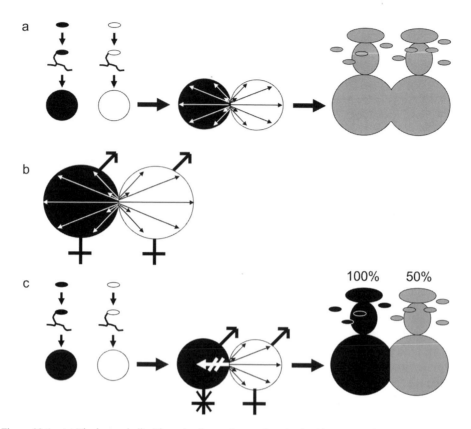

Figure 32.1 (a) The heterothallic life cycle of a mushroom-forming basidiomycete. (b) The monokaryons exhibit two clearly distinct behaviors in mating: accepting a nucleus and donating a nucleus, which can be considered female and male roles, respectively. (c) An asexual (or [secondarily] homothallic) mutant that still can donate a nucleus has a 1.5-fold advantage, because it doubles its fitness via the female route while maintaining its fitness via the male route. The percentages indicate the representation of the hypothetical asexuality allele in the spores resulting from the female route (100%) and the male route (50%).

groups (29, 30, 37). The initial 50% fitness gain of a homothallic mutant in a heterothallic population might ultimately explain such transitions. However, there are several reasons why the fitness difference between a heterothallic and a secondarily homothallic in reality is less than one-half. First, the fitness difference of 50% is only the *initial* difference and decreases as the fraction of homothallics increases, reaching zero in a population consisting of homothallics only. Second, many species are "amphithallic" instead of strictly secondarily homothallic, meaning that only a *fraction* of the spores of a single fruiting body has compatible nuclei, whereas the remaining fraction has incompatible nuclei and requires outbreeding to reproduce sexually (37). This implies that the female fitness is less than double for such amphithallic mutants. Third and last, but not least, in some instances the transition to secondary homothallism is accompanied by a twofold reduction in the number of spores. This means that there is an initial *disadvantage*

of secondary homothallism, with a relative fitness of three-quarters only. (This calculation may be an underestimation of the actual fitness of a two-spored homothallic, however. Although the spore *number per basidium* is halved, the spores of two-spored homothallics are generally bigger [37] and may therefore have a higher fitness. Second, if resources are saved because of the lower number *per basidium*, these resources are still available for the formation of additional spores via subsequent meioses in other basidia.) The observation of several species in which heterokaryotic hyphae of secondarily homothallic strains can fuse with and donate nuclei to homokaryotic outcrossing strains shows that the above-mentioned advantages are not just hypothetical (e.g., *Stereum hirsutum* [3] and *Mycocalia denudata* [12]). Obviously, other factors may also have contributed to the evolution of (secondary) homothallism, one advantage of homothallism being the absence of the need to find a compatible mate (11).

SEXUAL HANG-UPS IN FUNGI

Sex and the Production of Survival Structures

As a rule, sexually and asexually produced spores are ecologically distinct. Commonly, sexual spores have the capacity to survive adverse conditions, while asexual spores are destined to germinate quickly or act as fertilizing agents (spermatia). In fact, the specialization of sexual propagules into dormant or survival structures is seen not only in fungi but also in many plants and animals that have both sexual and asexual reproductive options (32). Related to this ecological specialization is the tendency of the sexual cycle to be induced when food is scarce and/or population density is high (6). Thus, a possible explanation of the coupling between sex and the production of survival structures may be that it is best to produce genetically variable offspring when conditions deteriorate, maximizing success in later (and possibly changed), more favorable conditions ("tangled bank hypothesis" [6]).

Here we want to suggest a somewhat different explanation of the association between sex and the production of dormant structures. Consider an initial situation of a species with both a sexual and an asexual reproductive cycle. Assuming that sex is for some reason advantageous, natural selection is expected to adjust the timing of sex such that its opportunity cost is minimal, namely, when conditions are unfavorable for somatic growth and asexual spore production. These are precisely the conditions when the production of survival structures is important. Therefore, the optimal times for both sex and the production of dormant propagules coincide. A next step then would be selection to regulate both processes in such a way that they are induced by the same signal(s). In this way both become developmentally linked. This coupling, once arisen and fixed within the species, is difficult to break. Presumably multiple genes are involved in (the fine-tuning of) this developmental regulation. Therefore, a single mutation that would cleanly unlink the two processes is very unlikely and would face a selective disadvantage because either sex or the formation of survival structures would be lost or suboptimally timed. This scenario implies that natural selection favoring a lower rate of sexual recombination would not be effective, being opposed by selection to retain the production of resistant spores. The coupling between sexual reproduction and the formation of survival structures might also be one of the contributing factors to the evolution of homothallism in fungi (11): whereas a reversal to asexuality is not possible once the coupling has evolved, a transition from heterothallism to homothallism still is.

The Sexual Cycle as a Selection Arena

Recent studies with *Aspergillus nidulans* have revealed more stringent intraorganismal selection against deleterious mutations in sexual reproduction than in asexual reproduction. Bruggeman et al. (8) showed that many auxotrophic mutations confer sexual self-sterility as a pleiotropic effect under conditions where asexual spore production is normal. This result points to the sexual cycle as an arena of selection: *A. nidulans* overproduces dikaryotic fruiting initials and prevents those carrying deleterious mutations from developing. Thus linked to the sexual reproduction cycle is an internal mechanism of progeny choice, or selection arena (45). Although the exact mechanism of this process is still to be elucidated, it seems likely that the maternal mycelium only invests in further development of young fruiting bodies after they have proven to be able of independent (initial) development. In a mutation accumulation experiment, again with *A. nidulans*, slightly deleterious mutations accumulated at a lower rate in the sexual than in the asexual pathway (9). In this experiment the sexual spores were formed exclusively within a single colony and were genetically identical (apart from mutations) to the asexually produced spores. Therefore, whatever fitness difference exists between sexually and asexually produced offspring, it cannot be a consequence of genetic recombination, which is thought to be the key factor in the evolutionary success of sex. Thus, the result of this experiment also points to more stringent early intraorganismal selection of potential offspring in the sexual than in the asexual cycle.

Other observations in mycelial fungi suggest that selection arena mechanisms linked to the sexual cycle may be common in fungal biology. Natural isolates of *Podospora anserina* show senescence (39, 46), which is correlated with increasing instability of the mitochondrial genome during vegetative growth (24). Sexual progeny from senescing cultures of *P. anserina* are juvenile with normal mitochondria. This rejuvenation of mitochondria is probably a consequence of selection against deleterious mitochondria so that only dikaryons with relatively healthy mitochondria are able to complete fruiting-body development. Similarly, matings between mitochondrial mutant strains of *P. anserina* yield progeny with only wild-type mitochondria (44). The selection arena hypothesis may also account for observations that viruses are readily transmitted via asexual spores, but rarely via sexual spores (41), for example, in *Ophiostoma ulmi* (40), in *Cryphonectria parasitica* (4), and in *A. nidulans* (18). This would be an example of a selection arena in the sexual cycle if virus-infected dikaryons display lower developmental vigor than virus-free dikaryons and will not be sustained by the maternal mycelium. However, the few studies on virus transmission in basidiomycetes do not confirm this. They have generally shown high rates of virus transmission via the sexual spores (e.g., *Rhizoctonia solani* [15], *Agaricus bisporus* [42], and *Ustilago maydis* [19]). In *Heterobasidion annosum* transmission of virus via asexual spores and transmission via basidiospores have been compared, and it was found that transmission was highest via the basidiospores (26, 27).

FUNGAL MODEL SYSTEMS FOR TESTING THEORIES OF THE EVOLUTION OF SEX

The scientific literature on the evolutionary function of sex is dominated by theory, with relatively few reports of experimental work. This is not surprising, since many of the key ideas involve genomic characteristics that are hard to measure, such as the amount and type of epistatic interactions among mutations, the amount of linkage disequilibrium, and distributions of fitness effects of deleterious and beneficial mutations. Only recently, genomic information has become available in sufficient detail to allow measurement of the crucial parameters. This is most easily done in small genomes, and it is therefore not surprising that viruses (16, 43), bacteria (21), and also fungi (20, 50) are prominent among the model systems used. For competition experiments involving directly a comparison between sex and asex, an even smaller number of good experimental systems is available. Ideally, such experiments should contrast sexual and asexual reproduction, **all else being equal**. This ceteris paribus condition excludes in practice animals and higher plants. However, fungi can come very close to the requirements. Many fungal species have both a sexual and an asexual reproductive cycle, while the switch between them can often easily be manipulated under controlled laboratory conditions. In particular, *Saccharomyces cerevisiae*, for which very sophisticated molecular and genomic tools are available, allows precise experimental comparisons between sex and asex. Elegant experiments by Birdsell and Wills (7) and Zeyl and Bell (50) compared isogenic sexual and asexual yeast strains in rate of adaptation. However, these experiments were still open to the criticism that the ceteris paribus criterion had not been met completely, either because of genetic differences at the *MAT* locus (7) or because of a difference in treatment necessary for the induction of meiosis in the sexual strain (50). In a more recent experiment Goddard et al. (23; see also chapter 30 of this volume) succeeded in engineering a strain by deleting *SPO11* and *SPO13* to respond to the usual starvation

trigger for meiosis and sporulation by sporulating only, producing two diploid spores without having undergone meiosis. They compared rates of adaptation over 300 generations in sexual and asexual chemostat cultures. In a standard glucose-limited medium no difference between sexuals and asexuals was observed, while in a harsh medium with high osmolarity and a stressful temperature of 37°C the sexuals adapted more quickly than the asexuals to the novel environment. Thus, this experiment provides evidence that sex can accelerate adaptive evolution.

SUMMARIZING CONCLUSIONS

The main advantages of sexual reproduction compared to asexuality are likely to be acceleration of adaptive evolution and more efficient elimination of deleterious mutations, while major disadvantages are the costs in terms of time and energy expenditure and of the production of male structures. There are several reasons why the actual cost of sex in fungi is lower than the cost of sex for animals and plants, which has been discussed most prominently in the literature. First, sexual reproduction in most fungi is facultative rather than obligate, so that its timing can be adjusted to when the opportunity costs are lowest. Second, many fungi are (secondarily) homothallic, which reduces the cost of sex and avoids the problem of mate finding. Third, many fungi are isogamous. The unique life cycle of mushroom-forming basidiomycetes, with an almost universally observed obligate coupling between sex and reproduction and distinct male and female roles of the monokaryon, results in a ≤1.5-fold cost of sex. Strikingly, in many fungi sex occurs at the end of the growing season, when the conditions for somatic growth become adverse, and furthermore, sex is often linked to the formation of specialized survival structures. In this chapter, we have proposed that this link may be causally related to the optimal timing of both processes, viz., at the end of the growing season. This is because at that moment, both the opportunity costs of sex are lowest and the formation of resistant spores is most relevant. Once such a link between sex and an essential process such as the formation of resistant spores has evolved, natural selection favoring a lower rate of sexual recombination is not effective, being opposed by selection to retain the production of resistant spores. Another possible example of a link between sex and an essential process, discussed in this chapter, is the presence of an efficient selection arena in the sexual, but not in the asexual, cycle. A major challenge is to explain existing patterns of the distribution and frequency of sex among different species on the basis of the balance between these advantages and disadvantages. Fungal model systems such as yeast, *Aspergillus*, and *Schizophyllum* are eminently suitable for experimentally testing hypotheses on the evolutionary significance of sex.

We thank Fons Debets, Arjan de Visser, and Thom Kuyper for stimulating discussion and critically reading this chapter.

References

1. Aanen, D. K., T. W. Kuyper, A. J. M. Debets, and R. F. Hoekstra. 2004. The evolution of non-reciprocal nuclear exchange in mushrooms as a consequence of genomic conflict. *Proc. R. Soc. London B* 271:1235–1241.

2. Agrawal, A. F. 2006. Evolution of sex: why do organisms shuffle their genotypes? *Curr. Biol.* 16:R696–R704.

3. Ainsworth, A. M., A. D. M. Rayner, S. J. Broxholme, and J. R. Beeching. 1990. Occurrence of unilateral genetic transfer and genomic replacement between strains of *Stereum hirsutum* from nonoutcrossing and outcrossing populations. *New Phytol.* 115:119–128.

4. Anagnostakis, S. L. 1988. Cryphonectria parasitica, cause of chestnut blight, p. 123–136. *In* D. S. Ingram, P. H. Williams, and G. S. Sidhu (ed.), *Advances in Plant Pathology*, vol. 6. *Genetics of Plant Pathogenic Fungi*. Academic Press, New York, NY.

5. Barton, N. H., and B. Charlesworth. 1998. Why sex and recombination? *Science* 281:1986–1990.

6. Bell, G. 1982. *The Masterpiece of Nature: the Evolution and Genetics of Sexuality*. CroomHelm, London, United Kingdom.

7. Birdsell, J., and C. Wills. 1996. Significant competitive advantage conferred by meiosis and syngamy in the yeast Saccharomyces cerevisiae. *Proc. Natl. Acad. Sci. USA* 93:908–912.

8. Bruggeman, J., A. J. M. Debets, and R. F. Hoekstra. 2004. Selection arena in Aspergillus nidulans. *Fungal. Genet. Biol.* 41:181–188.

9. Bruggeman, J., A. J. M. Debets, P. J. Wijngaarden, J. deVisser, and R. F. Hoekstra. 2003. Sex slows down the accumulation of deleterious mutations in the homothallic fungus Aspergillus nidulans. *Genetics* 164:479–485.

10. Buller, A. H. R. 1931. *Researches on Fungi*, vol. IV. Longmans, Green and Co., London, United Kingdom.

11. Burnett, J. 2003. *Fungal populations and species*. Oxford University Press, Oxford, England.

12. Burnett, J. H., and M. E. Boulter. 1963. The mating systems of fungi. II. Mating systems of the Gasteromycetes Mycocalia denudata and M. duriaeana. *New Phytol.* 62:217–236.

13. Burt, A. 2000. Perspective: sex, recombination, and the efficacy of selection—was Weismann right? *Evolution* 54:337–351.

14. Casselton, L. A., and A. Economou. 1985. Dikaryon formation, p. 213–229. *In* D. Moore, L. A. Casselton, D. A. Wood, and J. C. Frankland (ed.), *Developmental Biology of Higher Fungi*. Cambridge University Press, Cambridge, England.

15. Castanho, B., E. E. Butler, and R. J. Shepherd. 1978. Association of double-stranded-RNA with Rhizoctonia decline. *Phytopathology* 68:1515–1519.

16. Chao, L., T. T. Tran, and T. T. Tran. 1997. The advantage of sex in the RNA virus phi6. *Genetics* 147:953–959.

17. Charlesworth, B. 1980. The cost of sex in relation to mating system. *J. Theor. Biol.* 84:655–671.

18. Coenen, A., F. Kevei, and R. F. Hoekstra. 1997. Factors affecting the spread of double-stranded RNA viruses in Aspergillus nidulans. *Genet. Res.* 69:1–10.

19. Day, P. R., and J. A. Dodds. 1979. Viruses of plant pathogenic fungi, p. 201–238. *In* P. A. Lemke (ed.), *Viruses and Plasmids in Fungi*. Marcel Dekker, New York, NY.

19a. de Visser, J. A. G. M., and S. F. Elena. 2007. The evolution of sex: empirical insights into the roles of epistasis and drift. *Nat. Rev. Genet.* 8:139–149.

20. de Visser, J. A. G. M., R. F. Hoekstra, and H. vanden-Ende. 1997. Test of interaction between genetic markers that affect fitness in Aspergillus niger. *Evolution* 51:1499–1505.

21. Elena, S. F., and R. E. Lenski. 1997. Test of synergistic interactions among deleterious mutations in bacteria. *Nature* 390:395–398.

22. Elliott, C. G. 1994. *Reproduction in Fungi: Genetical and Physiological Aspects*. Chapman & Hall, London, England.

23. Goddard, M. R., H. Charles, J. Godfray, and A. Burt. 2005. Sex increases the efficacy of natural selection in experimental yeast populations. *Nature* 434:636–640.

24. Griffiths, A. J. F. 1992. Fungal senescence. *Annu. Rev. Genet.* 26:351–372.

25. Hoekstra, R. F., and E. N. Vanloo. 1986. The cost of sex in hermaphrodite populations with variation in functional sex. *J. Theor. Biol.* 122:441–452.

26. Ihrmark, K., H. Johannesson, E. Strenstrom, and J. Stenlid. 2002. Transmission of double-stranded RNA in Heterobasidion annosum. *Fungal. Genet. Biol.* 36:147–154.

27. Ihrmark, K., E. Stenstrom, and J. Stenlid. 2004. Double-stranded RNA transmission through basidiospores of Heterobasidion annosum. *Mycol. Res.* 108:149–153.

28. Kondrashov, A. S. 1993. Classification of hypotheses on the advantage of amphimixis. *J. Hered.* 84:372–387.

29. Kühner, R. 1977. Variation of nuclear behavior in Homobasidiomycetes. *Trans. Br. Mycol. Soc.* 68:1–16.

30. Lamoure, D. 1989. Indexes of useful informations for intercompatibility tests in Basidiomycetes V-Agaricales sensu-lato. *Cryptogam. Mycol.* 10:41–80.

31. Lewis, W. M. 1983. Interruption of synthesis as a cost of sex in small organisms. *Am. Nat.* 121:825–834.

32. Maynard Smith, J. 1978. *The Evolution of Sex*. Cambridge University Press, Cambridge, England.

33. Maynard Smith, J. 1971. What use is sex? *J. Theor. Biol.* 30:319–335.

34. Michod, R., and B. R. Levin (ed.). 1987. *The Evolution of Sex: a Critical Review of Current Ideas*. Sinauer Associates, Sunderland, MA.

35. Muller, H. J. 1964. The relation of recombination to mutational advance. *Mutat. Res.* 1:2–9.

36. Otto, S. P., and T. Lenormand. 2002. Resolving the paradox of sex and recombination. *Nat. Rev. Genet.* **3:**252–261.

37. Petersen, R. H. 1995. There's more to a mushroom than meets the eye—mating studies in the Agaricales. *Mycologia* **87:**1–17.

38. Quintanilha, A. 1937. Contribution à l'étude génétique du phénomène de Buller. *C. R. Acad. Sci. Paris* **205:**747.

39. Rizet, G. 1953. Sur l'impossibilite d'obtenir la multiplication vegetative ininterrompue et illimitee de l'ascomycete *Podospora anserina. C. R. Hebd. Seances Acad. Sci.* **237:**838–840.

40. Rogers, H. J., K. W. Buck, and C. M. Brasier. 1986. Transmission of double-stranded-RNA and a disease factor in *Ophiostoma ulmi. Plant Pathol.* **35:**277–287.

41. Rogers, H. J., K. W. Buck, and C. M. Brasier. 1988. *dsRNA and disease factors of the aggressive subgroup of* Ophiostoma ulmi. Dekker, New York, NY.

42. Romaine, C. P., P. Ulhrich, and B. Schlagnhaufer. 1993. Transmission of La France isometric virus during basidiosporogenesis in *Agaricus bisporus. Mycologia* **85:**175–179.

43. Sanjuan, R., A. Moya, and S. F. Elena. 2004. The contribution of epistasis to the architecture of fitness in an RNA virus. *Proc. Natl. Acad. Sci. USA* **101:**15376–15379.

44. Silliker, M. E., M. R. Liotta, and D. J. Cummings. 1996. Elimination of mitochondrial mutations by sexual reproduction: two Podospora anserina mitochondrial mutants yield only wild-type progeny when mated. *Curr. Genet.* **30:**318–324.

45. Stearns, S. C. (ed.). 1987. *The Evolution of Sex and Its Consequences.* Birkhauser Verlag, Basel, Switzerland.

46. van der Gaag, M., A. J. M. Debets, J. Oosterhof, M. Slakhorst, J. Thijssen, and R. F. Hoekstra. 2000. Sporekilling meiotic drive factors in a natural population of the fungus Podospora anserina. *Genetics* **156:**593–605.

47. Weismann, A. 1886. Zur Frage nach der Vererbung erworbener Eigenschaften. *Biol. Zentbl.* **6:**33–48.

48. Weismann, A. 1904. *The Evolution Theory.* Edward Arnold, London, United Kingdom.

49. West, S. A., C. M. Lively, and A. F. Read. 1999. A pluralist approach to sex and recombination. *J. Evol. Biol.* **12:**1003–1012.

50. Zeyl, C., and G. Bell. 1997. The advantage of sex in evolving yeast populations. *Nature* **388:**465–468.

Index

A

Absidia glauca, 413, 434, 437, 438, 440
Adaptation, rates of, 520
Agaricus bisporus var. *bisporus*, 46–47
Agaricus bisporus var. *eurotetrasporus*, 47
Agrocybe aegerita, 46
Allomyces macrogynus, 409, 411, 415
Allomyces sp., 410–411
Alternaria alternata, *MAT* locus of, 102–103
AMBM (Amut Bmut) strains, of *Coprinellus cinerea*, 294–296
Amphimixis, 490
Arbuscular mycorrhizal (AM) fungi, *see* Glomeromycota
Armillaria
   A. gallica, 335–336
   diploidy in, 343–344
Ascochyta lentis, 98
Ascomycetes, *see also specific species*
   cell identity control, 60–66
   dikaryon in, 333–335
   evolution of *MAT*
      Candida species, 8–13
      filamentous ascomycetes (Pezizomycotina), 13–15
      HMG domain protein function, 10, 13
      HO endonuclease gene, 7–8
      Saccharomyces lineage, 5–8
   male and female functions, 334–335
   MAT and genome comparisons in hemiascomycetous yeasts, 247–261
   MAT locus location, 3–5
   phylogenetic relationships, 4
   sexual cycles, 39–46
      heterothallism
         diagram of, 36

heterothallic fungi with homothallic life cycles, 45–46
homothallism
   homothallic fungi with heterothallic life cycles, 45
   primary, 41–45
   pseudohomothallism, 39, 40
   primary homothallism, 41–45
      MAT absence, 44–45
      one *MAT* idiomorph in genome, 43–44
      two *MAT* idiomorphs in same genome, 41–43
Ascospore formation in *Emericella nidulans*, genetics of, 134
Asexual reproduction
   in *Aspergillus* spp., 135–138
   in *Candida albicans*, 216
   in *Cryptococcus gattii*, 479
   in Dothidomycetes, 101–103
   evolution of, 201–210
   in Glomeromycota, 419–420
   hypotheses about, 463
   long-term costs of, 202
   in *Penicillium marneffei*, 207–209
   speed of, 529
   variation in, 462–463
   vegetative fitness decrease in *Cryptococcus neoformans*, 470–473
Ashbya gossypii
   genome duplication, 520
   homothallism, 44
   MAT locus, 5, 6, 7, 8, 193
   Sir proteins in, 192
Ashbya waltii, 250–256, 258
Aspergillus

asexuality in, 135–138
evolution of homothallism and heterothallism in, 138–139
HMG-domain genes, 15
MAT genes
   expression, 132
   presence and distribution, 136–137
   structure, 15, 131–132
   sexual development, genetics of, 126–132
   COP9 signal, 131
   cpcA, 127
   cpcB, 127
   DopA protein, 130
   environmental signal perception, 126–127
   FphA, 126
   G-protein subunits, 128–129
   lsdA, 127
   MAP kinase cascade, 129
   MAT1 and *MAT2*, 131–132
   MedA protein, 130
   NosA protein, 130–131
   NsdD protein, 130
   overview, 134–135
   pheromone precursors (*PpgA/PpgB*), 127–128
   pheromone receptors (*PreA/GprB* and *PreB/GprA*), 128
   phoA, 127
   RosA protein, 130–131
   signal transduction pathways, 127–130
   StuA protein, 130
   transcription factors/regulatory proteins, 130–132
   VeA, 126–127

Aspergillus (Continued)
  sexual development, morphology of,
    124–125
  sexual development, physiology of,
    133–134
    ascospore production, 134
    carbohydrate metabolism, 133
    lipid metabolism, 133
    overview, 134–135
    oxidation state, 133–134
    psi hormonal signaling, 133
  taxonomy, 123–124
  teleomorphs, 124
Aspergillus flavus, 135–137
Aspergillus fumigatus
  asexuality and, 135–136
  genome sequence data, 136
  MAT genes, 135–139
  transgenic *MAT* strain, 41
Aspergillus nidulans
  GprD receptor, 40
  homothallism, 42, 43
  outcrossing, 45
  sexual cycle as a selection arena, 531
Aspergillus niger, 135–138
Aspergillus oryzae, 135–139
AXL1, in *Saccharomyces cerevisiae*, 76

B

Basal fungi, 407–415
  Chytridiomycota, 409–411
  Dikarya compared, 414
  importance of, 407–409
  Microsporidia, 413
  Zygomycota, 411–413, 431–441
Basidiomycetes, *see also specific species*
  cell identity control, 66–70
  cost of sex, 529–530
  dikaryon in, 333–335
  life cycle, 20
  male and female functions, 334–335
  mating-type loci
    cloning, 318–319, 323–324
    organization and synteny, 317–318
  MAT (mating-type locus), evolution of,
    19–32
    Coprinellus disseminatus, 26
    Coprinopsis cinerea, 24–25
    Cryptococcus neoformans, 27–30
    homeodomain locus, 22
    Microbotryum violaceum, 27
    multiallelic bipolar basidiomycetes,
      26–27
    pheromone/receptor locus, 21
    Pholiata nameko, 26
    recombination, 29–30, 31
    Schizophyllum commune, 25–26
    Sporisorium reilianum, 23–24, 25
    tetrapolar mating system, 20, 22–23
    transitions between heterothallic and
      homothallic cycles, 30–32
    Ustilago hordei, 26–27
    Ustilago maydis, 22–24
  phylogeny, 318–319, 323–324
  regulation of mating in, 20–21
  sexual cycles
    heterothallic fungi with homothallic life
      cycles, 47–50
    heterothallism, diagram of, 36

primary homothallism, 47
  secondary homothallism, 46–47
Batrachochytrium dendrobatidis, 409, 410
Bensaude, Mathilde, 283
BGL1 β-glucosidase gene, 96
Biofilm formation, by *Candida albicans*,
  223–224, 225
Bipolaris sacchari, *MAT* locus of, 101–103
Birth-and-death model, 423
Blakeslea trispora, 413, 434–439
Blastocladiella variabilis, 411
Botryotinia fuckeliana, 108
Botrytis, *MAT* locus, 14
Bottlenecks, in Glomeromycota, 420–421
Bremia, 446, 449, 450
Buller phenomenon, 337–338, 345, 529

C

Camptothecin, 409
Candida
  CUG codon reassignment, 237
  evolution of *MAT* locus, 8–13, 235–242
  MTL locus structure in, 240–242
  phylogeny, 235–237
  teleomorphs, 237
Candida albicans
  Candida dubliniensis, in vitro mating
    with, 229
  cell identity control, 62–64
  centromere structure, 193–194
  mating-type regulation, 79–87
    asg regulon evolution, 85–87
    regulatory circuit, 81–85
    Saccharomyces cerevisiae compared,
      81–87
    white-opaque switching and, 80–81,
      82–83, 84–85
  MAT locus, 79–80
  MTL locus
    discovery of, 62, 216
    evolution and structure of, 240–241
    heterozygosity and virulence, 224–228
    white-opaque switching and, 216–217,
      222, 223
  overview of mating in, 79–80
  recombination, 215, 216
  Saccharomyces cerevisiae compared, 216,
    217, 220–222
  Sir proteins in, 192
  white-opaque switching, 63–64, 80–81,
    82–83, 84–85, 217–224
    biofilm formation, 223–224, 225
    features of, 218–219
    reasons for, 222–223
    temperature and, 219–220
Candida dubliniensis
  Candida albicans, in vitro mating with,
    229
  switching and mating in, 228–229
Candida glabrata
  MAT and genome comparison in
    hemiascomycetes, 250–257
  mating in, 230–231
  MAT locus, 5, 6, 7, 8
  Saccharomyces cerevisiae compared,
    230–231, 240
  silencing of mating-type loci, 194
  Sir proteins in, 192, 194

Candida guilliermondii (teleomorph *Pichia
    guilliermondii*), 239–240, 241, 242
Candida krusei (teleomorph *Issatchenkia
    orientalis/Pichia kudriavzevii*), 240
Candida lusitaniae (teleomorph *Clavispora
    lusitaniae*), 237–239, 241, 242
Candida parapsilosis, 215, 229–230, 241,
    242, 258
Carbohydrate metabolism, in *Emericella
    nidulans*, 133
β-Carotene, 434–436, 439
Cell identity, *MAT* locus and, 59–70
  ascomycetes, 60–66
    Candida albicans, 62–64
    Cochliobolus heterostrophus, 64–65
    Neurospora crassa, 65–66
    Podospora anseria, 65
    Saccharomyces cerevisiae, 60–61
    Schizosaccharomyces pombe, 61–62
  asidiomycetes
    Coprino cinerea, 66–68
    Cryptococcus neoformans, 69–70
    Ustilago maydis, 66, 67
  basidiomycetes, 66–70
  overview, 59–60
cenH, in *Schizosaccharomyces pombe*, 190
Ceratocystis coerulescens, 108
Chaetocladium brefeldi, 409
Chaetocladium jonesii, 434
Chimerism, in Glomeromycota
  formation, general principles of, 425
  overview, 424–425
  in parasexual cycle, 426
  resource partitioning mechanisms, 425–426
  vegetative, 426
Chlamydomonas reinhardtii, 496, 516
Choanephora cucurbitarum, 433, 437–438
Chromocrea spinulosa, 107–108
Chromosomal rearrangements, 509
Chytridiomycetes
  diseases caused by, 409
  sexual mechanisms, 409–411
Chytriomyces hyalinus, 410
cla4 gene, 320, 322–323, 325, 327
Clr4, in *Schizosaccharomyces pombe*,
    190–191
Coccidioides, *MAT* locus, 14
Cochliobolus
  mating-type structure and function,
    95–107
    conserved motifs in MAT1-1-1 and
      MAT1-2-1, 107
    conversion self-compatible to self-
      incompatible, 105–107
    conversion self-incompatible to self-
      compatible, 105
    phylogenetic and structural analyses,
      104
    self-compatible species, 99–101
    self-incompatible species, 95–98
  MAT locus
    evolution, 15
    structure, 14, 15
  primary homothallism, 42
  reproductive biology of, 96
  Sordaria compared, 174
  transgenic *MAT* strains, 41
Cochliobolus carbonum, 97–98
Cochliobolus cymbopogonis, 99–101, 102

Cochliobolus ellisii, 99–101
Cochliobolus heterostrophus
    cell identity control, 64–65
    MAT locus, structure/organization of,
        95–97, 99–101
    transgenic studies with *Cochliobolus
        luttrellii*, 105–107
Cochliobolus homomorphus, 99–101, 102,
    104
Cochliobolus kusanoi, 99–100, 102
Cochliobolus luttrellii, 99–101, 102, 104,
    105–107
Cochliobolus victoriae, 97–98
Coelomomyces psorophorae, 411
Concerted evolution of rDNA, 423–424
Coniochaeta tetraspora, 107–108
Conjugation tube formation, *Ustilago may-
    dis*, 371–372
con mutation, in *Schizophyllum commune*,
    271
Coprinellus disseminatus, 26, 320, 321,
    323–324
Coprinopsis cinerea
    cell identity control, 66–68
    homeodomain locus, 22, 24–25
    life cycle, 284–285
    mating-type genes, 285–296
        AMBM (Amut Bmut) strains, 294–296
        cloning of, 318–319
        comparison of sequenced, 288
        evolution of, 291–293
        mutations, 293–296, 324
        organization of, 286, 287, 319,
            321–324
        overview, 285–288
        pheromones-receptors, 290–292
        origin of multiple mating types in,
            283–296
    pheromone/receptor locus, 21
    segmental duplication and evolution of
        mating types in, 24–25
Coprinopsis scobicola, 319, 320
Coprinus cinereus
    homothallism in, 49–50
    somatic recombination, 339
COP9 signalosome, in aspergilli, 131
Corn smut disease, 351
Cost of sex, 528–530
cpcA, in aspergilli, 127
cpcB, in aspergilli, 127
Crivellia
    C. *papaveracea*, 98
    MAT locus
        phylogenetic and structural analyses,
            105
        structure, 101, 102
Crow, J. F., 491
Croziers, 95, 172, 173
Cryphonectria parasitica
    heterokaryons, 340–341
    self-fertility, 108
Cryptococcus gattii, 477–486
    cryptococcosis, 477–478, 484, 486
    MLST (multilocus sequence typing), 481,
        482, 484
    phylogeny, 482–484
    recombination, methods for detecting, 478
    sexual reproduction
        basidiospore formation, 478

clinical and veterinary populations,
    482–483
environmental population with both
    mating types, 480
natural populations, 480, 486
Cryptococcus neoformans
    cell identity control, 69–70
    fitness in asexual clones, 470–473
    mating interaction, 466–470
    mating-type loci
        organization and synteny, 322, 325
        overview, 27–30
    mating-type switching, 36, 47–49
    mitochondrial inheritance, 326
    as model system, 464
    mtDNA transmission, 335
    mutation accumulation, 464–466,
        515–516
    recombination hot spots, 341
    sexual cycle of, 48
CsnD, in aspergilli, 131
CsnE, in aspergilli, 131
CUG codon reassignment, in *Candida*,
    237

D

Debaryomyces hansenii
    characteristics of, 250
    MAT locus
        evolution, 9
        structure, 10, 11, 193, 258, 260
        phylogeny of, 8, 250
        Sir proteins in, 192
Dicranophora fulva, 438
Didymella rabiei, 98
Didymella zeae-maydis, 101
4-Dihydromethyltrisporate dehydrogenase,
    437, 440
4-Dihydrotrisporin dehydrogenase, 440
Dikarya
    basal fungi compared to, 407, 414
    Glomeromycota compared, 426
Dikaryon
    in Ascomycota, 333–335
    in Basidiomycota, 333–335
    Buller phenomenon, 337–338
    in *Coprinellus cinerea*, 284
    Di-Mon (dikaryon-monokaryon) matings
        analysis of somatic recombinants,
            338–339, 340
        nuclear selection in, 338
        patterns of somatic recombination,
            339–341
    evolution and, 343–345
    life cycle variation, 343–344
    long-term changes in, 342–343
    male potential, 529
    mitochondrial DNA and, 335–337
    nuclear escape from, 337
    nuclear reassociation, 337
    overview of, 333–334
    in *Schizophyllum commune*, 268–270,
        284, 302–303, 312
    somatic recombination in, 341–342
    in *Ustilago maydis*, 359
Diploids
    adaptation rate, 520
    dikaryosis and, 343–344
    evolution and, 343–345, 517

masking of deleterious mutations,
    518–520
of oomycetes, 446
physiological effects of ploidy, 517–518
Ustilago maydis, 357, 370–372
Dispira americana, 409
DopA, in aspergilli, 130
Dothideomycetes, *see also Cochliobolus het-
    erostrophus*
    evolution of *MAT*, 12, 103–105
    mating-type (*MAT*) locus structure, 12, 14
        asexual species, 101–103
        self-compatible species, 99–101
        self-incompatible species, 98–99
Double-stranded break repair, in *Saccha-
    romyces cerevisiae MAT* switching,
        161–163
Downy mildew
    of grape, 453–454
    of sorghum, 454
Dynein motor proteins, 334

E

Effective population size
    estimating in fungi, 202–204
    Penicillium marneffei, 206, 210
    in Wright-Fisher model of evolution, 202
Emericella nidulans, *see also Aspergillus
    nidulans*
    MAT locus, 131–132
    sexual development
        genetics of, 126–132
        morphology of, 124–125
        overview of, 133–134
        physiology of, 133–134
Emericella species, morphology of sexual
    development, 124
Encephalitozoon cuniculi, 415
Epistasis, 509–510
Eremothecium coryli, homothallism in, 44
Euascomycetes, postfertilization functions of
    mating-type genes in, 116
Eurotiomycetes, *MAT* locus in, 4, 12, 14
Eurotium species, morphology of sexual
    development, 124
Evolution
    asexual reproduction, 201–210
    asg regulon, 85–87
    concerted, 423–424
    Glomeromycota, 419–424, 426–427
    homothallism, mutations in *A* and *B* genes
        and, 293–294
    homothallism and heterothallism in
        aspergilli, 138–139
    horizontal transmission, role of, 515
    maintenance of sex, 489–505
    mating type in the Dothideomycetes,
        103–105
    mating-type locus (*MAT*)
        ascomycetes, 3–15
        basidiomycetes, 19–32
        Candida species complex, 235–242
        complex, 291–293
        silencing in Hemiascomycetes, 189–197
    mating-type switching, 36, 38–39
    merits of diploidy and dikaryosis,
        344–345
    of ploidy, 517
Schizosaccharomyces pombe, 152–154

Evolution *(Continued)*
  of sex, fungal models and, 532
  of silencing at the mating-type loci in
    Hemiascomycetes, 189–197
Evolutionary distance, between *Ustilago
  hordei* and *Ustilago maydis*, 399–400

F

FadA, in aspergilli, 128
Filamentous ascomycetes, cell identity con-
  trol, 64–66
Filobasidiella depauperata, 49
Fisher, R. A., 490–491
Fisher-Muller theory, 491, 492, 495
Fission yeast, *see Schizosaccharomyces
    pombe*
Fkh1, in *Saccharomyces cerevisiae*, 165–166,
  167
FMR1, in *Podospora anserina*, 109–115
FphA, in aspergilli, 126
FPR1, in *Podospora anserina*, 109–115
Frequency-dependent selection, 492–493
Fungal Tree of Life (AFTOL) project, 415

G

Genetic drift, role in evolution of asexuality,
  202
Genetics of Sexuality in Higher Fungi
  (Raper), 267, 268, 333, 461
Genome duplication, 520–521
Gibberella species
  homothallism, 42
  MAT locus, 15
Gilbertella persicaria, 438
Git3, in *Schizosaccharomyces pombe*, 147
Glomerella species, *MAT* locus in, 14
Glomeromycota, 419–427
  chimerism
    formation, general principles of, 425
    overview, 424–425
    in parasexual cycle, 426
    resource partitioning mechanisms,
      425–426
    vegetative, 426
  evolution and, 419–424, 426–427
  multigenomic structure, 424–426
  multinucleate spore structure, 426–427
  phylogeny of PLSI variants, 422
  polymorphism, intraindividual, 420–421,
    424
  ribosomal DNA (rDNA), 420, 423–425
  spores as surrogate individuals in studies
    of, 419–420
Glomus caledonium, 423, 424
Glomus claroideum, 424
Glomus etunicatum, 421–422, 424
Glomus geosporum, 422, 424
Glomus intraradices, 415, 421–422
Glomus mosseae, 424, 426
β-Glucosidase gene, 99
Gpa1, in *Schizosaccharomyces pombe*, 147
GpgA, in aspergilli, 128
gprA, in aspergilli, 128
gprB, in aspergilli, 128
gprC, D, E, G, K genes, in aspergilli, 40,
  129–130
Grape, downy mildew of, 453–454

H

Haploids, adaptation rate and, 520
Hebeloma crustuliniforme, 337
HEGs (homing endonuclease genes), 8
Hemiascomycetes
  evolution of silencing at mating-type loci,
    189–197
  MAT and genome comparisons, 247–261
    conservation of synteny, 256
    gene identification, 249–251
    HO endonuclease, 257–258
    MAT and silenced cassettes, 251–257
    multiple sequence alignments, 252–253
    overview, 258–261
    species with multiple *MAT* loci,
      249–258
    species with single *MAT* locus, 258
  phylogeny of, 191
Heterobasidion annosum, 337
Heterochromatin, silencing of storage cassettes in *Schizosaccharomyces pombe*,
  151–152
Heterokaryosis
  chimerism and, 425
  in Glomeromycota, 420–421
  in parasexual cycle, 426
Heterothallic-homothallic transitions, *see*
  Mating-type switching
Heterothallism
  defined, 35, 462
  homothallism transition, 30–32, 35–53
  in Mucorales, 411–412
  oomycetes, 446, 447
  recognition and karyogamy in
    ascomycetes, 94
  in *Schizosaccharomyces pombe*, 145
  Ustilago maydis, discovery in, 354
Heterozygosity, of *MAT* alleles, 518
HMG-domain proteins, 10, 13, 15
Ho endonuclease
  acquisition of gene, 7–8, 38
  in hemiascomycetous yeasts, 257–258
  in *Saccharomyces cerevisiae*, 37–38,
    161–163, 231, 514
  silencing of *HML* and *HMR* cassettes and,
    37–38
Homeodomain locus, in basidiomycetes,
  22
Homing endonuclease genes (HEGs), 8
Homobasidiomycetes
  cost of sex, 529–530
  internuclear recognition in, 114
  MAT loci organization, 324–325
Homokaryosis, in Glomeromycota, 420–421
Homothallism
  defined, 35, 462
  heterothallism transition, 30–32, 35–51
  mechanisms of, 35
  in Mucorales, 411–412
  mutations in *A* and *B* genes and evolution
    of, 293–294
  oomycetes, 446–448
  outcrossing, 45
  primary
    ascomycetes, 41–45
      MAT absence, 44–45
      one *MAT* idiomorph in genome,
        43–44

two *MAT* idiomorphs in same
    genome, 41–43
  in basidiomycetes, 47
  pseudohomothallism, 39, 40
  recognition and karyogamy in
    ascomycetes, 94
  Saccharomyces cerevisiae, 37–38, 514
  secondary, 529–530
    in ascomycetes, 39
    in basidiomycetes, 46–47
    overview, 39
  Sordaria macrospora, 171–184
  in Zygomycetes, 434
Hook cell formation, 334
Hst1p, in *Saccharomyces cerevisiae*, 196
Hydrophobins, 334

I

Inbreeding, in natural yeast populations,
  508–509
Internuclear recognition model, 112–114
Introns, 511–512, 515
Irpex lacteus, 320, 322–323
Isogamy, 528–529

K

KAR4 gene, in *Saccharomyces cerevisiae*,
  164–165
Kimura, M., 491
Kluyveromyces lactis
  asg (a-specific gene) regulation, 85
  MAT and genome comparison in
    hemiascomycetes, 250–257
  mating-type switching, 38, 195
  MAT locus structure, 5, 6, 7, 8
  silencing of mating-type loci, 194–195
  Sir proteins in, 192, 193, 195
Kluyveromyces thermotolerans, 249–256,
  258
Kluyveromyces waltii
  characteristics of, 250
  genome duplication, 520
  HO gene absence, 258
  mating-type locus, 193, 250–256
  phylogeny of, 250
  sequence identity boxes, 255
Kniep, Hans, 283–284

L

Leotiomycetes, *MAT* locus in, 14
Leptosphaeria biglobosa, 98
Leptosphaeria maculans, 98
Linkage, of mating-type loci in mushroom
  fungi, 317–328
Linkage disequilibria, 528
Lipid metabolism, in *Emericella nidulans*,
  133
Lodderomyces elongisporus, 45
Loss-of function mutations in mating-type
  loci, 202
lsdA, in aspergilli, 127

M

Magnaporthe grisea, *MAT* locus in, 5, 14
Male sterility, 336–337
MAPK, *see* Mitogen-activated protein kinase
  (MAPK)

Mating inhibition factors (MIFs), in *Ustilago hordei*, 393
Mating-type-like (*MTL*) locus, in *Candida albicans*, 62–64, 216–217, 222–228, 240–241
Mating-type locus (*MAT*) locus
  Alternaria alternata, 102–103
  in aspergilli, 131–132, 136–139
  basdiomycetes
    cloning *MAT* genes, 318–319, 323–324
    organization and synteny, 317–318
  in *Candida* species complex, 8–13, 235–242
  cell identity control, 59–70
    ascomycetes, 60–66
    basidiomycetes, 66–70
  in hemiascomycetous yeast genomes, 247–261
  Kluyveromyces lactis, 5, 6, 7, 8, 194–195
  Kluyveromyces waltii, 193, 250–256
  in *Saccharomyces cerevisiae*, 37–38, 76–78
  in sexual *Aspergillus* spp., 136–139
  in *Ustilago hordei*, 392–401
Mating-type locus (*MAT*) locus, evolution of
  ascomycetes
    Candida species, 8–13
    filamentous ascomycetes (Pezizomycotina), 13–15
    HMG domain protein function, 10, 13
    HO endonuclease gene, 7–8
    Saccharomyces lineage, 5–8
  basidiomycetes, 19–32
    Coprinellus disseminatus, 26
    Coprinopsis cinerea, 24–25
    Cryptococcus neoformans, 27–30
    homeodomain locus, 22
    Microbotryum violaceum, 27
    multiallelic bipolar basidiomycetes, 26–27
    pheromone/receptor locus, 21
    Pholiata nameko, 26
    recombination, 29–30, 31
    Schizophyllum commune, 25–26
    Sporisorium reilianum, 23–24, 25
    tetrapolar mating system, 20, 22–23
    transitions between heterothallic and homothallic cycles, 30–32
    Ustilago hordei, 26–27
    Ustilago maydis, 22–24
Mating-type regulation
  Candida albicans, 79–87
  evolution of, 75–76
  Saccharomyces cerevisiae, 75–78
Mating-type switching
  in ascomycetes, 39–46
  in basidiomycetes, 46–50
  in *Candida albicans*, 63–64
  cell identity control and, 61, 63–64
  in *Cryptococcus neoformans*, 36, 47–49
  in *Kluyveromyces lactis*, 195
  in *Saccharomyces cerevisiae*, 36, 37–38, 78, 159–168
    donor preference, genetic control of, 163–167
    donor preference, microscopic analysis of, 167–168
    gene conversion, 161–162

microscopic analysis of, 162–163
    overview, 159–161
    RE (recombination enhancer), 163–167
  in *Saccharomyces* lineage, 6–7
  in *Schizosaccharomyces pombe*, 61, 145, 147–151
  secondary homothallism, 39, 46–47
  in *Schizosaccharomyces pombe*, 38–39
Maynard Smith, John, 491
Mcm1
  in *Candida albicans*, 85
  in *Saccharomyces cerevisiae*, 77–78, 165
  in *Sordaria macrospora* homologue, 182–183
MedA, in aspergilli, 130
Mei2, in *Schizosaccharomyces pombe*, 147, 153
Mei3, in *Schizosaccharomyces pombe*, 147
Meiosis
  meiotic silencing by unpaired DNA (MSUD), 118
  Ustilago maydis, 358–359
mib (metalloendopeptidase) gene, 319–321, 323, 325–327
Microbotryum violaceum, 27, 47, 390–391, 398, 400
Microsporidia, sex in, 413
MIFs (mating inhibition factors), in *Ustilago hordei*, 393
Mitochondria
  basdiomycetes mating-type loci and, 326–327326
  cytoplasmic mixing, 335–336
  epistasis, 509–510
  male sterility and, 336–337
  mosaics, 336
  recombination, 335–336
  rejuvenation of, 531
  slow movement in matings, 336
Mitogen-activated protein kinase (MAPK)
  in aspergilli, 129
  in *Schizosaccharomyces pombe*, 146–147
  in *Ustilago maydis*, 371–372, 377, 381–383
MLMT (multilocus microsatellite typing), of *Penicillium marneffei*, 205
MLST (multilocus sequence typing)
  of *Cryptococcus gattii*, 481, 482
  of *Penicillium marneffei*, 204–205
Mortierella species, 413, 435
Motor proteins, dynein, 334
Mucorales
  mating in, 411–413
  mycoparasitism, 409
  trisporoid, 434–439
  zygospore dormancy, lack of, 408
Mucor amphibiorum, 409
Mucor circinelloides, 415, 440
Mucor hiemalis, 438
Mucor mucedo, 409, 413, 415, 431–432, 434–435, 437–440
Mucor piriformis, 409, 438
Mucor pusillus, 440
Muller's ratchet, 491–492
Multilocus genotyping, of *Penicillium marneffei*, 204–205
Multilocus microsatellite typing (MLMT), of *Penicillium marneffei*, 205

Multilocus sequence typing (MLST)
  of *Cryptococcus gattii*, 481, 482
  of *Penicillium marneffei*, 204–205
Mushroom fungi; *see also* Basidiomycetes; Homobasidiomycetes; *specific species*
  mating-type loci, organization and synteny of, 317–328
Mutation accumulation, 464–466, 515–516
Mutations
  evolution of asexuality and, 202
  in Glomeromycota, 420, 426–427
  masking and purging deleterious, 518–520
  Muller's ratchet, 491–492
Mutualistic symbiosis, 419
Mycosphaerella graminicola, 98
Mycosphaerella zeae-maydis, 101
Mycotypha africana, 432

N

Natural selection, role in evolution of asexuality, 202
NEJ1, in *Saccharomyces cerevisiae*, 76
Nematospora coryli, homothallism in, 44
Neocallimastix sp., 410
Neosartorya species
  morphology of sexual development, 124
  N. fischeri, 42, 43
Neurospora africana, 40–41, 44
Neurospora crassa
  cell identity control, 65–66
  immunity to MSUD (meiotic silencing by unpaired DNA), 118
  mating, 39–41
  nucleus-restricted expression, 113
  postfertilization functions of mating-type genes, 116
  Sordaria macrospora compared, 171–177, 179–184
Neurospora terricola, 44
Neurospora tetrasperma, 171
NosA, in aspergilli, 130–131
Nosema locustae, 415
noxA gene, in *Emericella nidulans*, 133–134
NsdD, in aspergilli, 130
Nuclear migration
  into ascogenous hyphae, 112
    internuclear recognition model, 112–114
    random segregation model, 114–115
  Buller phenomenon, 337–338
  male sterility and, 336–337
  motive force for, 334
  Schizophyllum commune, 268, 269, 302, 304–305
Nucleolar organizer regions (NORs), 423

O

Oomycetes, 445–455
  cytology of mating, 446–448
  gene expression during mating, 450
  genetic basis of mating behavior, 449–450
  importance of, 445–446
  mating hormones, 450
  oospores
    disease transmission, role in, 451
    germination of, 448
    as survival structures, 450–451

Oomycetes, *(Continued)*
  pathology
    downy mildew of grape, 453–454
    late blight of potato, 451–452
    Phytophthora capsici blights, 452–453
    Phytophthora sojae root and stem rot, 453
    sorghum downy mildew, 454
    taxonomy, 446
Oospores
  disease transmission, role in, 451
  germination of, 448
  as survival structures, 450
Orc1p, 195–196
Outcrossing, in natural yeast populations, 508–509
Oxidation state and hyphal differentiation, 133–134
Oxysterol binding protein, in *Candida albicans*, 62, 63

P

pab (*para*-aminobenzoic acid) gene, 319, 321, 324, 340
Papazian, Haig, 284
Parasexual cycle, 426
Parasitella parasitica, 409, 413, 432, 434, 437, 440
Parasitic DNA
  coevolution and domestication of, 513–515
  sex and, 510–513
Pat1 protein, in *Schizosaccharomyces pombe*, 148, 153
PCR, cloning *MAT* using direct amplification, 323–324
Penicillium marneffei
  biology of, 204
  effective population size, 206, 210
  evolutionary trajectory, 209–210
  multilocus genotyping, 204–205
  reproduction, 207–209
  spatial components of genetic diversity, 205–206
Peronosclerospora sorghi, 454
Peronospora destructor, 451
Peronospora farinosa, 448
Petromyces species, morphology of sexual development in, 124
Pezizomycotina, mating-type locus in, 13–15
Phaeosphaeria nodorum, 98
Phanerochaete chrysosporium, 320, 321–324, 327
Pheromone/receptor systems
  in aspergilli, 127–128
  in basidiomycetes, 21
  Candida albicans, 84, 220–223
  Chytridiomycetes, 411
  Coprinellus cinerea, 290–292
  Coprinopsis cinerea, 66–68
  Cryptococcus neoformans, 69–70
  Dikarya and basal fungi compared, 414
  Mucorales, 413
  Saccharomyces cerevisiae, 76, 301–302
  Schizophyllum commune
    characterization of pheromones, 305–307
    determinants of recognition by receptors, 307–309
    discovery of, 304

gene nomenclature, 304
    nuclear migration, role in, 304–305
    number of, 286, 302
    pheromone precursor sequences, 289, 290
    role in *B*-regulated mating processes, 311–313
    yeast studies, 309–310
  Schizosaccharomyces pombe, 61–62, 146–148
  signaling, in smut fungi, 381–383
  Sordaria macrospora, 179–180
  trisporic acid, 434–439
  Ustilago maydis, 66, 67, 367–372
phoA, in aspergilli, 127
Pholiota nameko, 26, 324
Phosphatidyl inositol-3 kinase, in *Candida albicans*, 62, 63
Phycomyces blakesleeanus, 412–413, 415, 432–433, 435, 438–440
Phylogenetic tree of the fungal kingdom, 408
Phytophthora cactorum, 448
Phytophthora capsici, 452–453
Phytophthora infestans, 445, 447–452, 454
Phytophthora parasitica, 449
Phytophthora phaseoli, 450
Phytophthora sojae, 453
Phytophthora species, 445–447, 449–451
Pichia angusta, 250, 258, 260
Piromyces species, 415
Plasmid DNA, 510
Plasmopara, 446
Plasmopara viticola, 453
Pleurotus djamor, 320, 321–324
Ploidy, physiological effects of, 517–518
Pneumocystis carinii
  homothallism, 44
  mating-type loci organization and synteny, 322
Podospora anserina
  cell identity control, 65
  mating-type structure and function, 109–117
    control of fertilization, 111
    internuclear recognition model, 112–114
    mating-type structure, 109
    mutations, phenotype of, 110
    nucleus migration into hyphae, 112
    postfertilization developmental steps during fruiting-body formation, 109–111
    postfertilization functions of mating-type genes, 116–117
    random segregation model, 114–115
  MAT locus
    Cochliobolus heterostrophus compared, 41
    gene mutations, 109–111
    structure/organization, 97, 109
  pseudohomothallism, 171
  senescence in, 531
  Sordaria macrospora mating-type genes in, 176–178
Poly(A) polymerase, in *Candida albicans*, 62, 63
Polymorphism, intraindividual in Glomeromycota, 420–421, 424
Polyphagus euglenae, 410

Population size, effective
  estimating in fungi, 202–204
  Penicillium marneffei, 206, 210
  in Wright-Fisher model of evolution, 202
Potato, late blight of, 451–452
ppgA, in aspergilli, 127–128
ppgB, in aspergilli, 128
ppg1/ppg2, in *Sordaria macrospora*, 179–180, 184
ppoA gene, 133
preB, in aspergilli, 128
pre1/pre2, in *Sordaria macrospora*, 179–180, 184
Promoters, in *Schizophyllum commune*, 276
Pseudohomothallism, 39, 40, 171
Psi hormonal signaling, in *Emericella nidulans*, 133
Pyrenopeziza, 14
Pyrenopeziza brassicae, 98
Pyrenophora teres, 98
Pythium, 446–449
Pythium aphanidermatum, 448

R

Radiomyces spectabilis, 410
Random drift, 493
Random segregation model, 114–115
Raper, Carlene (Cardy), 267, 272, 301, 310, 338
Raper, John, 267–268, 270–273, 276–278, 280, 283, 304, 310, 333, 461
Rap1p, in *Saccharomyces cerevisiae*, 189–190
Ras1, in *Schizosaccharomyces pombe*, 147
rDNA, concerted evolution of, 423–424
Receptors, *see* Pheromone/receptor systems
Recombination
  in basidiomycetes, 29–30, 31
  evolution of *MAT* and, 29–30
  mitochondrial DNA, 336
  outcrossing, 508
Recombination enhancer (RE), in *Saccharomyces cerevisiae*, 163–167
Red Queen hypothesis, 492–493
Regulation of mating-type
  Candida albicans, 79–87
  evolution of, 75–76
  Saccharomyces cerevisiae, 75–78
Retrotransposition, 512
Rhizopus oryzae, 415, 434, 440
Rhizopus sexualis, 438
Rhizopus stolonifer, 440
RITS (RNA-induced transcriptional silencing), 190
RME1, in *Saccharomyces cerevisiae*, 76
RosA, in aspergilli, 130–131
Rum1, in *Schizosaccharomyces pombe*, 146

S

Saccharomyces cerevisiae
  Candida albicans compared, 216, 217, 220–222
  Candida glabrata compared, 230–231, 240
  cell identity control, 60–61
  cell types, 76
  centromere structure, 193
  chromosomal rearrangements, 509
  genome duplication, 520–521

Ho endonuclease, 514
life cycle, 497
maintenance of sex experiments, 497
mating-type regulation, 75–78
    asg expression, 77–78
    Candida albicans compared, 81–87
    hsg expression, 78
    mating-type switching, 78
    regulatory circuit, 81, 82
    αsg expression, 78
    transcriptional switching, 78
mating-type switching
    donor preference, genetic control of,
        163–167
    donor preference, microscopic analysis
        of, 167–168
    gene conversion, 161–162
    microscopic analysis of, 162–163
    overview, 36, 37–38, 78, 159–161, 514
    RE (recombination enhancer), 163–167
MAT loci
    evolution of, 514–515
    MAT and genome comparison in
        hemiascomycetes, 250–257
    mitochondria and, 326
    pheromone-receptor system, 301–302
    phylogeny, 191
    regulation of mating type in, 247, 249
    sexual cycle of an *ho* mutant of, 248
    silencing in, 78, 189–190
    spo11 spo13 system, 500–503
    whole genome duplication, 191
Saccharomyces kluyveri, 250, 258
Saccharomyces lineage, evolution of mating-
    type cassettes in, 5–8
Saccharomyces pastorianus, 500
*sak*A, in *Emericella nidulans*, 130
Schizophyllum commune
    A genes
        classical genetics studies, 302–304
        cloning of *A*α, 277–278
        flanking genes and boundaries, 277
        isolation of, 271–274
        mutations, 271, 293
        number of alleles, 270–271
        overview, 270, 274
        promoter regions, 276
        Y and Z gene products, 274–276
    B genes
        arrangement of, 305
        classical genetics studies, 302–304
        evolution of, 291
        gene products, 304
        isolation of, 278–280, 304
        mutations, 271, 293, 294, 310–311,
            324
        number of alleles, 270–271, 286
        overview, 270
        pheromone signaling and, 311–313
        recombination, 287–288
    classical genetics studies, 302–304
    flat phenotype, 284, 303
    life cycle, 268–270, 284
    mating process, 302
    mating-type loci
        A genes, 270–278
        B genes, 270–271, 278–280
        cloning of, 277–278, 318–319
        discovery of, 283–284

historical perspective, 267–280
mutations, 271, 293, 294, 310–311,
    324
organization and synteny of, 319,
    321–322, 324–325
promoter regions, 276
origin of multiple mating types in,
    283–296
overview, 25–26
pheromones/receptors
    characterization of pheromones,
        305–307
    determinants of recognition by
        receptors, 307–309
    discovery of, 304
    gene nomenclature, 304
    nuclear migration, role in, 304–305
    number of, 286, 302
    pheromone precursor sequences, 289,
        290
    role in *B*-regulated mating processes,
        311–313
    yeast studies, 309–310
somatic recombination, 339–341
Schizosaccharomyces kambucha, 152
Schizosaccharomyces pombe
    cell identity control, 61–62
    centromere structure, 193
    evolution, 152–154
    heterothallic strains, 145
    homothallic switch, 147–148
    life cycle, overview of, 143–144
    mating-type switching, 145, 147–151
        bias of donor cassette choice,
            150–151
        copy transposition from *mat2/3* to
            mat1, 149–151
        imprinting *mat1* DNA, 148–149
        overview, 38–39
        silencing of storage cassettes *mat2* and
            mat3, 151–152
    mat region, organization of, 144–145
    pheromone communication, 146–148
    silencing in, 190–191
    Sir proteins in, 192
    transcriptional regulation, 145–146
Sclerotinia trifolium, 108
Scutellospora castanea, 423, 424–425
Scutellospora gregaria, 423
Scutellospora pellucida, 423, 424–425
SDSA (synthesis-dependent strand anneal-
    ing), 161
Selection
    antagonism between sexual and natural
        selection, 515–517
    frequency-dependent, 492–493
    natural selection role in evolution of
        asexuality, 202
Septoria passerinii, 98, 103
Sex Inducer genes, of *Cryptococcus neofor-
    mans*, 69–70
Sex (sexual reproduction)
    cost of, 201–202, 465–470, 528–530
    epistasis and, 509–510
    evolutionary implications of, 527–528,
        532
    loss of, 464
    maintenance of
        Crow and Kimura and, 491

experiments on, 494–505
    Fisher and Muller, 490–491, 495
    Maynard Smith and, 491
    Muller's ratchet, 491–492
    Red Queen hypothesis, 492–493
    Weismann and, 490, 493
in natural populations in *Cryptococcus
    gattii*, 477–486
in natural yeast populations, 508–509
oomycetes, 445–455
parasitic DNA and, 510
quantitative nature of, 464
survival structure production, 531
Sexual fitness, 465
Sexual selection, antagonism with natural
    (vegetative) selection, 515–517
SfaD, in aspergilli, 128
Sikyospore, 434
Silencing of mating-type loci, 189–197
    Candida glabrata, 194
    comparative genomics of, 190–191,
        192–194
    evolution of silenced chromatin,
        196–197
    Kluyveromyces lactis, 194–195
    overview, 189
    Saccharomyces cerevisiae, 189–190
    Schizosaccharomyces pombe, 190–191
    Sir proteins, role of, 192–196
    whole-genome duplication, impact of,
        195–196
Silent mating cassettes
    defects in, 61
    in *Saccharomyces cerevisiae*, 36, 37–38,
        61, 78
    in *Saccharomyces* lineage, 6–7
Sirenin, 411
Sir proteins
    functions of, 192–194, 195–196
    hemiascomycetes phylogeny and, 191
    in *Saccharomyces cerevisiae*, 37, 189–190
    where they act, 193–194
Sistotrema brinkmannii, homothallism and,
    47
SMIP gene, in *Schizophyllum commune*, 277
SMR1, in *Podospora anserina*, 109, 111,
    113–114, 116–117
SMR2, in *Podospora anserina*, 109–115
Smut fungi
    bipolar and tetrapolar mating systems,
        389–401
    mating-type loci
        evolutionary history, 400–401
        functional analysis of, 393–394
        future studies and questions, 400–401
        genomic analysis, 396–398
        identification of, 392–393
        organization of, 394–396
    mating-type loci organization, 378–381
    overview, 377–378, 390–391
    pheromone-controlled signaling cascade,
        381–385
    phytopathogenesis, 389–390, 391,
        398–400
Somatic recombination
    in dikaryons, 341–342
    Di-Mon matings, 338–341
Sordaria brevicollis, 45–46, 175
Sordaria fimicola, outcrossing of, 45